Grundlehren der mathematischen Wissenschaften

A Series of Comprehensive Studies in Mathematics

Volume 362

Editors-in-Chief

Alain Chenciner, Observatoire de Paris, Paris, France
S. R. S. Varadhan, New York University, New York, NY, USA

Series Editors

Henri Darmon, McGill University, Montréal, Canada
Pierre de la Harpe, University of Geneva, Geneva, Switzerland
Frank den Hollander, Leiden University, Leiden, The Netherlands
Nigel J. Hitchin, University of Oxford, Oxford, UK
Nalini Joshi, University of Sydney, Sydney, Australia
Antti Kupiainen, University of Helsinki, Helsinki, Finland
Gilles Lebeau, Côte d'Azur University, Nice, France
Jean-François Le Gall, Paris-Saclay University, Orsay, France
Fang-Hua Lin, New York University, New York, NY, USA
Shigefumi Mori, Kyoto University, Kyoto, Japan
Bào Châu Ngô, University of Chicago, Chicago, IL, USA
Denis Serre, École Normale Supérieure de Lyon, Lyon, France
Michel Waldschmidt, Sorbonne University, Paris, France

Grundlehren der mathematischen Wissenschaften (subtitled Comprehensive Studies in Mathematics), Springer's first series in higher mathematics, was founded by Richard Courant in 1920. It was conceived as a series of modern textbooks. A number of significant changes appear after World War II. Outwardly, the change was in language: whereas most of the first 100 volumes were published in German, the following volumes are almost all in English. A more important change concerns the contents of the books. The original objective of the Grundlehren had been to lead readers to the principal results and to recent research questions in a single relatively elementary and accessible book. Good examples are van der Waerden's 2-volume Introduction to Algebra or the two famous volumes of Courant and Hilbert on Methods of Mathematical Physics.

Today, it is seldom possible to start at the basics and, in one volume or even two, reach the frontiers of current research. Thus many later volumes are both more specialized and more advanced. Nevertheless, most volumes of the series are meant to be textbooks of a kind, with occasional reference works or pure research monographs. Each book should lead up to current research, without over-emphasizing the author's own interests. Proofs of the major statements should be enunciated, however the presentation should remain expository. Examples of books that fit this description are Maclane's Homology, Siegel & Moser on Celestial Mechanics, Gilbarg & Trudinger on Elliptic PDE of Second Order, Dafermos's Hyperbolic Conservation Laws in Continuum Physics ... Longevity is an important criterion: a GL volume should continue to have an impact over many years. Topics should be of current mathematical relevance, and not too narrow.

The tastes of the editors play a pivotal role in the selection of topics.

Authors are encouraged to follow their individual style, but keep the interests of the reader in mind when presenting their subject. The inclusion of exercises and historical background is encouraged.

The GL series does not strive for systematic coverage of all of mathematics. There are both overlaps between books and gaps. However, a systematic effort is made to cover important areas of current interest in a GL volume when they become ripe for GL-type treatment.

As far as the development of mathematics permits, the direction of GL remains true to the original spirit of Courant. Many of the oldest volumes are popular to this day and some have not been superseded. One should perhaps never advertise a contemporary book as a classic but many recent volumes and many forthcoming volumes will surely earn this attribute through their use by generations of mathematicians.

Wolfgang Lück • Tibor Macko

Surgery Theory

Foundations

With Contributions by Diarmuid Crowley

Wolfgang Lück
Mathematical Institute
University of Bonn
Bonn, Germany

Tibor Macko
Faculty of Mathematics, Physics and Informatics,
Comenius University, and Institute of Mathematics,
Slovak Academy of Sciences
Bratislava, Slovakia

ISSN 0072-7830 ISSN 2196-9701 (electronic)
Grundlehren der mathematischen Wissenschaften
ISBN 978-3-031-56333-1 ISBN 978-3-031-56334-8 (eBook)
https://doi.org/10.1007/978-3-031-56334-8

Mathematics Subject Classification (2020): 57-02

This work was supported by Deutsche Forschungsgemeinschaft (Gottfried Wilhelm Leibniz Prize, GZ 2047/1, Projekt-ID 390685813), European Research Council (KL2MG-interactions (no. 662400)), Agentúra Ministerstva Školstva, Vedy, Výskumu a Športu SR (VEGA 1/0101/17, VEGA 1/0596/21), Agentúra na Podporu Výskumu a Vývoja (APVV-16-0053) and Návraty: Topológia vysokorozmerných variet (VEGA 1/0101/17, VEGA 1/0596/21).

© The Editor(s) (if applicable) and The Author(s), under exclusive license to Springer Nature Switzerland AG 2024

This work is subject to copyright. All rights are solely and exclusively licensed by the Publisher, whether the whole or part of the material is concerned, specifically the rights of translation, reprinting, reuse of illustrations, recitation, broadcasting, reproduction on microfilms or in any other physical way, and transmission or information storage and retrieval, electronic adaptation, computer software, or by similar or dissimilar methodology now known or hereafter developed.

The use of general descriptive names, registered names, trademarks, service marks, etc. in this publication does not imply, even in the absence of a specific statement, that such names are exempt from the relevant protective laws and regulations and therefore free for general use.

The publisher, the authors, and the editors are safe to assume that the advice and information in this book are believed to be true and accurate at the date of publication. Neither the publisher nor the authors or the editors give a warranty, expressed or implied, with respect to the material contained herein or for any errors or omissions that may have been made. The publisher remains neutral with regard to jurisdictional claims in published maps and institutional affiliations.

This Springer imprint is published by the registered company Springer Nature Switzerland AG
The registered company address is: Gewerbestrasse 11, 6330 Cham, Switzerland

If disposing of this product, please recycle the paper.

Preface

Surgery theory was created in the sixties to solve classification problems for manifolds and since then has led to an enormous number of striking results. It has many interactions with other areas of mathematics, such as algebra, differential geometry, geometric group theory, algebraic K-theory, number theory, and the theory of operator algebras. Surgery theory also promises to be a major tool in geometry and topology in the future.

Surgery theory was initiated by Kervaire and Milnor in their paper [216] on the classification of homotopy spheres. Surgery theory for simply connected closed manifolds was developed systematically in Browder's book [55]. The book of Wall [414] established surgery theory for arbitrary fundamental groups. It also contains numerous new results on the classification of closed manifolds. The main tool in surgery theory is the surgery exact sequence due to Browder, Novikov, Sullivan, and Wall. It combines the homotopy theory of manifolds with the L-theory of quadratic forms over the group rings of their fundamental groups in order to obtain classification results about manifolds. The work of Kirby and Siebenmann [219] made it possible to do surgery also in the topological category. Quinn [337] gave a description of the surgery exact sequence as the long exact sequence of homotopy groups of a fibration and identified one of its maps as the so-called assembly map. Ranicki [344] developed a chain complex version of algebraic L-theory, answering a request by Wall [414, Chapter 17G], and later provided an algebraic description of the assembly map [348]. The Farrell–Jones Conjecture [150] about the assembly maps and its proofs for a large class of groups allow computations of L-groups of infinite groups and open the door to many applications of surgery theory for compact manifolds with infinite fundamental groups.

The goal of this book is to present an accurate, comprehensible, complete, and detailed introduction to surgery theory, which is useful for various groups of readers, such as experts in surgery theory, experienced mathematicians, who may not be experts in surgery, but just want to learn or use it, and also, of course, advanced undergraduate and graduate students. This is quite a challenge since surgery theory is sophisticated and complicated, requiring a large amount of previous knowledge, and since a lot of material has been accumulated, but not all details are well documented. We tried to find a reasonable compromise between the intention to present many details in full generality, to fix some bugs in the literature, to motivate the constructions, theorems, and proofs, and the desire to allow the reader to just browse through the book to get a first impression, or find a solution to a specific problem or an answer to a specific question, without necessarily going through all of the text.

Throughout the book we rely on the basics of the surgery theory developed in recent decades. None of the main theorems or concepts are new, but there are places where our approach to certain details is novel.

Bonn, February, 2024
Wolfgang Lück
Tibor Macko

Contents

1 Introduction 1
   1.1 Some Classical Problems that Can Be Attacked by Surgery Theory   1
   1.2 Overview of the Contents of this Book 3
   1.3 Outlook 8
   1.4 How to Use this Book 9
   1.5 Prerequisites 11
   1.6 Acknowledgement 12

2 The s-Cobordism Theorem 15
   2.1 Introduction 15
   2.2 Handlebody Decompositions 19
   2.3 Handlebody Decompositions and CW-Structures 25
   2.4 Reducing the Handlebody Decomposition 29
   2.5 Handlebody Decompositions and Whitehead Groups 34
   2.6 Notes 37

3 Whitehead Torsion 39
   3.1 Introduction 39
   3.2 Whitehead Groups 41
   3.3 Algebraic Approach to Whitehead Torsion 44
   3.4 Geometric Approach to Whitehead Torsion 52
   3.5 Reidemeister Torsion and (Generalised) Lens Spaces 57
       3.5.1 Lens Spaces 57
       3.5.2 Generalised Lens Spaces 58
       3.5.3 Homotopy Classification of Generalised Lens Spaces 59
       3.5.4 Reidemeister Torsion 63
       3.5.5 Simple Homotopy Classification of Generalised Lens
            Spaces 65
       3.5.6 A Variant of Reidemeister Torsion 66
       3.5.7 The Classification of Lens Spaces 69

		3.5.8	The Homeomorphism Classification of Generalised Lens Spaces	72
	3.6		The Spherical Space Form Problem	72
	3.7		Notes	73

4 The Surgery Step and ξ-Bordism 75
   4.1 Introduction 75
   4.2 The CW-Version of the Surgery Step 76
   4.3 Motivation for the Surgery Step 79
   4.4 Motivation for the Bundle Data 86
   4.5 Immersions and Embeddings 88
   4.6 The Surgery Step 94
   4.7 The Pontrjagin–Thom Construction 98
   4.8 Notes 103

5 Poincaré Duality 105
   5.1 Introduction 105
   5.2 Local Coefficients 106
   5.3 Passage to Group Rings 108
   5.4 The Determinant Line Bundle 111
   5.5 The Intrinsic Fundamental Class 114
   5.6 Poincaré Duality 120
      5.6.1 Rings with Involutions 120
      5.6.2 Poincaré Complexes 121
      5.6.3 The Signature 131
      5.6.4 The Degree of a Map 138
   5.7 Notes 139

6 The Spivak Normal Structure 141
   6.1 Introduction 141
   6.2 Spherical Fibrations 142
   6.3 The Thom Isomorphism 153
   6.4 The Normal Fibration of a Connected Finite CW-Complex 163
   6.5 The Spivak Normal Structure and Its Main Properties 166
   6.6 Existence of Poincaré Complexes without Vector Bundle Reduction 170
   6.7 Characterisation of Poincaré Duality in Terms of the Normal Fibration 175
   6.8 The Existence of the Spivak Normal Fibration 177
   6.9 Spanier–Whitehead Duality 178
   6.10 The Uniqueness of the Spivak Normal Fibration 181
   6.11 Notes 191

7 Normal Maps and the Surgery Problem 193
   7.1 Introduction 193
   7.2 An Informal Preview 194
   7.3 Rank k Normal Invariants 196

	7.4	Normal Maps 201
	7.5	Notes 217

8 The Even-Dimensional Surgery Obstruction 219
8.1 Introduction 219
8.1.1 The Nature of the Surgery Obstruction 219
8.1.2 The Intrinsic Versions 221
8.1.3 The Simply Connected Case 221
8.1.4 The Surgery Problem Relative Boundary and Wall's Realisation Theorem 222
8.2 Normal Γ-Maps 223
8.3 Intersection and Self-intersection Pairings 227
8.3.1 Intersections of Immersions 228
8.3.2 Self-intersections of Immersions 232
8.4 Surgery Kernels and Forms 238
8.4.1 Surgery Kernels 239
8.4.2 Symmetric Forms and Surgery Kernels 245
8.4.3 Quadratic Forms and Surgery Kernels 252
8.5 Even-Dimensional L-Groups 258
8.5.1 The Definition of $L_{2k}(R)$ via Forms 258
8.5.2 The L-Groups of $\mathbb{Z}$ in Dimensions $n = 4k$ 262
8.5.3 The L-Groups of $\mathbb{Z}$ in Dimensions $n = 4k+2$ 265
8.6 The Even-Dimensional Surgery Obstruction in the Universal Covering Case 270
8.6.1 Normal Maps of Pairs 270
8.6.2 Surgery Kernels for Maps of Pairs 271
8.6.3 Proof of Theorem 8.112 278
8.7 The Intrinsic L-Group and Surgery Obstruction in Even Dimensions 285
8.7.1 Conjugation Invariant Functors 286
8.7.2 Half-Conjugation Invariant Functors 288
8.7.3 The Definition of Intrinsic L-Groups in Even Dimensions . 292
8.7.4 The Intrinsic Even-Dimensional Surgery Obstruction 293
8.7.5 Bordism Invariance in Even Dimensions for a Not Necessarily Cylindrical Target 299
8.7.6 The Simply Connected Case in Even Dimensions 300
8.7.7 The Role of the Bundle Data in the Highly Connected Case 305
8.8 The Even-Dimensional Surgery Obstruction Relative Boundary ... 308
8.8.1 Normal Maps of Pairs 308
8.8.2 Normal Bordism for Normal Maps of Pairs 309
8.8.3 The Intrinsic Even-Dimensional Surgery Obstruction Relative Boundary 311
8.8.4 Kernels for Maps of Triads 315
8.9 Realisation of Even-Dimensional Surgery Obstructions 317
8.10 Notes 325

9 The Odd-Dimensional Surgery Obstruction ... 327
   9.1 Introduction ... 327
   9.2 Odd-Dimensional L-Groups ... 329
      9.2.1 The Definition of $L_{2k+1}(R)$ via Formations ... 330
      9.2.2 Presentations of Formations ... 334
      9.2.3 The Definition of $L_{2k+1}(R)$ via Automorphisms ... 339
      9.2.4 The Odd-Dimensional L-Groups of $\mathbb{Z}$ ... 344
   9.3 The Odd-Dimensional Surgery Obstruction in the Universal Covering Case ... 360
      9.3.1 The Kernel Formation ... 361
      9.3.2 The Bordism Formation ... 370
      9.3.3 Proof of Theorem 9.60 ... 379
   9.4 The Intrinsic Surgery Obstruction in Odd Dimensions ... 386
      9.4.1 The Definition of Intrinsic L-Group in Odd Dimensions .. 386
      9.4.2 The Intrinsic Odd-Dimensional Surgery Obstruction ... 386
      9.4.3 Bordism Invariance in Odd Dimensions for Not Necessarily Cylindrical Target ... 387
      9.4.4 The Simply Connected Case in Odd Dimensions ... 388
   9.5 The Odd-Dimensional Surgery Obstruction Relative Boundary ... 389
   9.6 Realisation of Odd-Dimensional Surgery Obstructions ... 389
   9.7 Notes ... 392

10 Decorations and the Simple Surgery Obstruction ... 393
   10.1 Introduction ... 393
   10.2 Decorated L-Groups ... 393
   10.3 Reidemeister U-Torsion ... 398
   10.4 The Simple Surgery Obstruction ... 402
      10.4.1 The Even-Dimensional Case ... 403
      10.4.2 The Odd-Dimensional Case ... 404
      10.4.3 Main Properties of the Simple Surgery Obstruction ... 404
   10.5 Notes ... 406

11 The Geometric Surgery Exact Sequence ... 407
   11.1 Introduction ... 407
   11.2 The Geometric Structure Set ... 407
   11.3 The Set of Normal Maps ... 410
   11.4 The Surgery Obstruction Groups ... 411
   11.5 The Geometric Surgery Exact Sequence ... 413
   11.6 The Piecewise Linear and Topological Categories ... 415
   11.7 Rigidity ... 419
   11.8 Group Structures in the Surgery Exact Sequences ... 420
   11.9 Some Information about G/O, G/PL and G/TOP ... 422
   11.10 Functorial Properties of the Geometric Surgery Exact Sequence ... 424
   11.11 Notes ... 427

12 Homotopy Spheres ... 431
12.1 Introduction ... 431
12.2 The Group of Homotopy Spheres ... 433
12.3 The Surgery Exact Sequence for Homotopy Spheres ... 435
12.4 The J-Homomorphism and Stably Framed Bordism ... 442
12.5 The Computation of bP_{n+1} ... 446
12.6 The Group Θ_n/bP_{n+1} ... 450
12.7 The Kervaire–Milnor Braid ... 450
12.8 Notes ... 453

13 The Geometric Surgery Obstruction Group and Surgery Obstruction 459
13.1 Introduction ... 459
13.2 Surgery on and Normal Bordisms for Pairs ... 460
  13.2.1 Short Review of Normal Bordism for Manifolds Relative Boundary ... 460
  13.2.2 Surgery on the Boundary ... 461
13.3 The π-π-Theorem ... 462
  13.3.1 Algebraic Proof of the π-π-Theorem 13.4 ... 464
  13.3.2 Preliminaries for the Proof of the π-π-Theorem 13.4 ... 464
  13.3.3 Proof of the π-π-Theorem 13.4 in the Even-Dimensional Case ... 466
  13.3.4 Proof of the π-π-Theorem 13.4 in the Odd-Dimensional Case ... 472
13.4 The Geometric Surgery Obstruction Group and Surgery Obstruction 481
  13.4.1 The Construction of the Geometric Surgery Groups ... 481
  13.4.2 Making the Reference Map a π_0- and π_1-Isomorphism ... 482
  13.4.3 The Geometric Surgery Obstruction ... 489
  13.4.4 Identifying the Geometric Surgery Obstruction Groups with the Algebraic L-Groups ... 490
13.5 The Geometric Rothenberg Sequence ... 491
13.6 The Geometric Shaneson Splitting ... 497
13.7 Notes ... 499

14 Chain Complexes ... 501
14.1 Introduction ... 501
14.2 Modules over Associative Rings ... 502
14.3 Modules over Rings with Involution ... 502
14.4 Some Basic Chain Complex Constructions ... 504
14.5 Further Chain Complex Constructions over Rings with Involution ... 507
14.6 Homotopy Theory of Chain Complexes ... 510
  14.6.1 Chain Homotopy ... 510
  14.6.2 Pushouts ... 512
  14.6.3 Mapping Cylinders, Mapping Cones and Suspensions ... 513
  14.6.4 Cofibrations and Homotopy Cofibrations ... 520
  14.6.5 Homotopy Pushouts ... 524

		14.6.6	Homotopy Cocartesian Squares 525
		14.6.7	Pullbacks 527
		14.6.8	Homotopy Fibres 528
		14.6.9	Relating the Homotopy Cofibre and the Homotopy Fibre . 531
		14.6.10	Homotopy Pullbacks 538
		14.6.11	Homotopy Colimits of Sequences 542
		14.6.12	Homotopy Limits of Inverse Systems of Maps 551
		14.6.13	Chain Complexes of Projective Modules..............551
	14.7	Notes	554

15 Algebraic Surgery 555
15.1 Introduction 555
15.2 Overview 557
15.2.1 Statement of the Main Results 558
15.2.2 Some Basic Notions and Explanations 559
15.2.3 Outline of the Proofs..................................565
15.3 Structured Chain Complexes..566
15.3.1 Basic Definitions 567
15.3.2 The Symmetric Construction 577
15.3.3 Products 581
15.3.4 Homotopy Invariance 583
15.3.5 The Suspension Map 585
15.3.6 The Quadratic Construction 595
15.3.7 Algebraic Poincaré Complexes 618
15.3.8 Equivariant S-Duality and Umkehr Maps...............619
15.3.9 Wu Classes 623
15.3.10 Homotopy $\mathbb{Z}[\mathbb{Z}/2]$-Chain Maps 626
15.4 Forms and Formations 632
15.4.1 Homotopy Theory of Highly Connected Structured Chain Complexes 633
15.4.2 Revision of Forms and Formations 636
15.4.3 Complexes versus Forms and Formations 638
15.5 L-Groups in Terms of Chain Complexes 649
15.5.1 Pairs 649
15.5.2 Cobordisms 666
15.5.3 The Algebraic Thom Construction 683
15.5.4 The Algebraic Boundary 693
15.5.5 Algebraic Surgery 708
15.6 The Identification of the L-Groups 724
15.7 The Identification of the Surgery Obstructions 740
15.7.1 Euler Classes 742
15.7.2 Self-Intersections 744
15.7.3 Putting it Together....................................745
15.8 Simple L-groups and Simple Surgery Obstruction in Terms of Chain Complexes 751

	15.9	Applications of Algebraic Surgery 753
		15.9.1 Product formulas 753
		15.9.2 The Algebraic Surgery Transfers for Fibrations 754
		15.9.3 The Algebraic Rothenberg Sequence 755
		15.9.4 The Algebraic Shaneson Splitting..................... 757
		15.9.5 The Algebraic Surgery Exact Sequence 758
		15.9.6 The Total Surgery Obstruction 763
	15.10	Notes 763

16 Brief Survey of Computations of L-Groups 765
16.1 Introduction 765
16.2 More Decorated L-Groups 766
16.3 Finite Groups 767
16.4 Torsionfree Groups 771
16.5 The Farrell–Jones Conjecture 773
    16.5.1 The K-theoretic Farrell–Jones Conjecture with Coefficients in Additive G-Categories 774
    16.5.2 The L-theoretic Farrell–Jones Conjecture with Coefficients in Additive G-Categories with Involution 776
    16.5.3 Reducing the Family 777
    16.5.4 The Full Farrell–Jones Conjecture 779
16.6 Notes 781

17 The Homotopy Type of G/TOP, G/PL and G/O 783
17.1 Introduction 783
17.2 Localisation 784
17.3 G/TOP 785
17.4 G/PL 789
17.5 G/O 791
17.6 H-Space Structures on G/TOP and G/PL 791
17.7 Splitting Invariants 795
17.8 Milnor Manifolds and Kervaire Manifolds 800
17.9 Notes 802

18 Computations of Topological Structure Sets of some Prominent Closed Manifolds 803
18.1 Introduction 803
18.2 Fake Spaces 804
18.3 Products of Spheres 807
18.4 Complex Projective Spaces 809
18.5 Quaternionic Projective Spaces 810
18.6 The Join and the Transfer 811
18.7 The ρ-Invariant 813
18.8 Real Projective Spaces 818
18.9 Lens Spaces 822

       18.10 Aspherical Manifolds 828
       18.11 Some Torus Bundles over Lens Spaces 829
       18.12 Notes 831

19   **Topological Rigidity**.. 833
       19.1 Introduction 833
       19.2 Topological Rigidity and the Surgery Exact Sequence 833
       19.3 Aspherical Closed Manifolds 838
             19.3.1 Homotopy Classification of Aspherical CW-Complexes .. 838
             19.3.2 Low Dimensions 839
             19.3.3 Non-Positive Curvature 839
             19.3.4 Torsionfree Discrete Subgroups of Almost Connected
                     Lie Groups 840
             19.3.5 Hyperbolisation 840
             19.3.6 Exotic Aspherical Closed Manifolds 841
             19.3.7 Non-Aspherical Closed Manifolds 843
       19.4 The Borel Conjecture 843
       19.5 Non-Aspherical Topologically Rigid Closed Manifolds 844
             19.5.1 Spheres 845
             19.5.2 3-Manifolds 845
             19.5.3 Products of Two Spheres 845
             19.5.4 Homology Spheres 846
             19.5.5 Connected Sums 846
             19.5.6 Homology Isomorphisms and Topological Rigidity 846
       19.6 Smooth Rigidity 847
       19.7 Notes 847

20   **Modified Surgery** 849
       20.1 Introduction 849
       20.2 Stable Diffeomorphisms 850
       20.3 Bordism Groups Associated to Spaces over BO 851
       20.4 Modified Surgery 857
       20.5 Normal k-Type 859
             20.5.1 Embedding Spaces and Gauss Maps up to Contractible
                     Choice 859
             20.5.2 The Moore–Postnikov Factorisation 860
             20.5.3 Normal k-Types 861
       20.6 Classification up to Stable Diffeomorphisms 865
       20.7 Tangential Approach 868
             20.7.1 Passing to the Tangent Bundle 868
             20.7.2 Connected Sum 869
       20.8 Applications 870
             20.8.1 Complete Intersections 870
             20.8.2 Homogeneous Spaces 871
             20.8.3 4-Manifolds 872

20.9 Notes 873

21 Solutions of the Exercises 875

References 921

Notation 939

Index 945

Chapter 1
Introduction

1.1 Some Classical Problems that Can Be Attacked by Surgery Theory

In this section we give a list of concrete, classical, and prominent problems that have been (partially) solved by surgery theory. The list shall illustrate the high potential of surgery theory and the (partial) solutions of these problems will constitute the contents of this book.

The following two problems represent the prototype of surgery problems, which, however, cannot be solved in full generality.

Problem 1.1 (Recognising manifolds) Let X be a connected finite CW-complex. Under which conditions is X homotopy equivalent to a closed manifold that is topological, PL (= piecewise linear), or smooth?

Problem 1.2 (Classifying manifolds) Let M and N be closed manifolds that are both topological, PL, or smooth. Under which conditions can one decide whether they are homeomorphic, PL homeomorphic, or diffeomorphic respectively. What are possible obstructions and under which conditions are they sufficient?

Exact formulations of both of these problems may vary. We sometimes consider modifications and use slightly different descriptions. Problem 1.1 may also be called an "existence problem" because we are asking whether there exists a manifold in the homotopy type of X. The variation of Problem 1.2 where we assume to begin with that M and N are homotopy equivalent may be called a "uniqueness problem" since we are asking how unique the manifolds in a given homotopy type are.

The next conjecture, a special case of Problem 1.2 in the topological category, is known to be true for all $n \geq 1$. Its proof uses surgery theoretic methods, except in dimension 3 where the proof relies on Ricci flow.

Conjecture 1.3 ((Generalised) Poincaré Conjecture) If M is a closed topological manifold homotopy equivalent to the standard n-sphere S^n, then M is homeomorphic to S^n.

The following problem, a special case of Problem 1.2 in the smooth category, triggered the development of surgery theory and will be discussed in Chapter 12.

Problem 1.4 (Homotopy spheres)   Classify all oriented homotopy spheres, that means closed oriented smooth manifolds homotopy equivalent to the standard n-sphere, up to orientation preserving diffeomorphism.

Another problem, which is completely solved, is the following.

Problem 1.5 (Fake complex projective spaces)   Classify all fake complex projective spaces, that means topological manifolds homotopy equivalent to the standard n-dimensional complex projective space, up to homeomorphism.

One can ask the corresponding question for other prominent manifolds, for example for fake real projective spaces or fake lens spaces. There are solutions in many interesting cases, as we will see in Chapter 18.

The next conjecture about aspherical manifolds, that means connected manifolds whose universal covering is contractible, will be treated in Chapter 19. It is known to be true if the fundamental group is contained in a large class of groups, which encompasses hyperbolic groups, CAT(0)-groups, solvable groups, and lattices in almost connected Lie groups, but is open in general. It is the topological version of Mostow rigidity.

Conjecture 1.6 (Borel Conjecture)   Let M and N be closed aspherical topological manifolds. Then:

(i) The fundamental groups $\pi_1(M)$ and $\pi_1(N)$ are isomorphic if and only if M and N are homeomorphic;
(ii) Any homotopy equivalence $f \colon M \to N$ is homotopic to a homeomorphism;
(iii) Any map $f \colon M \to N$ inducing an isomorphism between the fundamental groups is homotopic to a homeomorphism.

The next problem, which is essentially solved, triggered surgery theory for non-simply connected manifolds, see Section 3.6. It is a kind of generalisation of the Space Form Problem asking which finite groups occur as fundamental groups of closed Riemannian manifolds with constant positive sectional curvature.

Problem 1.7 (Spherical Space Form Problem)   Which finite groups can act freely and topologically or smoothly respectively on a standard sphere, or, equivalently, occur as fundamental groups of closed manifolds whose universal covering is homeomorphic or diffeomorphic respectively to a standard sphere.

1.2 Overview of the Contents of this Book

Chapters 2 to 11 contain the core of surgery theory that leads to a general method for solving Problems 1.1 and 1.2, while Chapter 12 illustrates the method on the most prominent example of homotopy spheres. The following Chapters 13 to 17 contain additional theoretical tools that are needed to effectively solve Problems 1.1 and 1.2 in other cases, in particular in the topological category. Chapters 18 and 19 illustrate how all this is applied to the concrete examples from the list in the previous section. Finally, Chapter 20 is about an alternative approach called modified surgery.

Although it may not be obvious at first glance, Problems 1.1 and 1.2 are closely linked. The general slogan is that Problem 1.2 is a relative version of Problem 1.1. The general approach to both of these problems is to split them into several steps and to treat the steps separately. Both of these ideas are explained with more details at the appropriate places in the following brief survey of the individual chapters. In the book itself, each chapter has its own more detailed introduction.

In the overview below we will often only treat the smooth category. Nearly all notions and statements carry over to the PL and the topological category.

Chapter 2: The s-Cobordism Theorem

We state and prove the s-Cobordism Theorem 2.1. Roughly speaking, it says that a compact smooth cobordism W from the closed smooth manifold M_0 to the closed smooth manifold M_1, for which the inclusion $M_i \to W$ is a simple homotopy equivalence for $i = 0, 1$ and $\dim(M_0) \geq 5$ holds, is diffeomorphic relative M_0 to the cylinder $M_0 \times [0, 1]$. This implies that M_0 and M_1 are diffeomorphic. Hence the s-Cobordism Theorem is highly relevant for the solution of Problem 1.2, namely, it is a cornerstone in the Surgery Program, see Remark 2.9, designed to solve Problem 1.2 by splitting it into three steps. Roughly, first find a (simple) homotopy equivalence, then construct a cobordism compatible with the (simple) homotopy equivalence, and finally improve this cobordism to an s-cobordism. If we get a positive answer in all three steps, then by the s-Cobordism Theorem we obtain a diffeomorphism.

The s-Cobordism Theorem 2.1 (in the topological category) implies the (Generalised) Poincaré Conjecture 1.3 in dimensions ≥ 5.

Chapter 3: Whitehead Torsion

We give a systematic treatment of Whitehead torsion, which is the obstruction for a homotopy equivalence of finite CW-complexes to be a simple homotopy equivalence. This is relevant since it appears in the s-Cobordism Theorem 2.1. We also explain the classification of lens spaces by their Reidemeister torsion in Section 3.5. This yields a solution of Problem 1.2 in a very specific case where it can exceptionally be achieved without surgery theory from later chapters.

Chapter 4: The Surgery Step and ξ-Bordism

This chapter contains the first step towards a solution of Problem 1.1. Namely, we solve the following Problem 4.2: Given a map $f : M \to X$ from a closed n-dimensional smooth manifold M to a CW-complex X of finite type, can we modify it, without changing the target, to a map $f' \colon M' \to X$ with a closed n-dimensional smooth manifold as source such that f' is k-connected where $n = 2k$ or $n = 2k + 1$? The basic idea is to modify the procedure of making a map of finite CW-complexes highly connected by attaching cells to a finite sequence of so-called surgery steps, so that the source still remains a closed smooth manifold, a procedure commonly called "surgery below middle dimension". We will see in the process that the presence of certain bundle data is desirable in order to make this work. These bundle data will be formalised in the technical notion of a normal map (also called a surgery problem) in Chapter 7.

Chapter 5: Poincaré Duality

We explain the notion of a finite Poincaré complex. This is relevant for Problem 1.1 since any finite CW-complex that is homotopy equivalent to a closed manifold is a finite Poincaré complex.

Chapter 6: The Spivak Normal Structure

Recall that a closed smooth manifold has a normal bundle, which is unique up to stable isomorphism. Its underlying sphere bundle is a spherical fibration unique up to stable fibre homotopy equivalence. In this chapter we show that any finite Poincaré complex possesses a Spivak normal structure, which is a spherical fibration coming with a certain collapse map and is unique up to stable fibre homotopy equivalence. Since this structure is homotopy invariant, we discover another obstruction for a finite Poincaré complex X to be homotopy equivalent to a closed manifold. Namely, its Spivak normal structure must have a vector bundle reduction, that means it must come from a vector bundle since the Spivak normal structure of a closed smooth manifold comes from its normal vector bundle by the Pontrjagin–Thom construction.

Chapter 7: Normal Maps and the Surgery Problem

We define the notion of a normal map of degree one motivated by the previous sections. Roughly speaking, a normal map of degree one is a map $f \colon M \to X$ from a closed smooth manifold M to a finite Poincaré complex X of degree one, which comes with bundle data, namely, a bundle map from the normal bundle of M to a vector bundle reduction ξ of the Spivak normal structure on X. The surgery problem, see Problem 7.40, now asks whether we can modify it by surgery to a normal map whose underlying map $f' \colon M' \to X$ is a homotopy equivalence. We also show that the set of smooth normal bordism classes of smooth normal maps with the target a fixed finite Poincaré complex X (also called the set of smooth normal invariants) can be identified with the set of homotopy classes of maps from X to a certain space G/O, see Theorem 7.34. Analogous statements hold in the PL category and the topological category, see Theorem 11.24.

Summarising the development of the chapters so far, we see that the solution of Problem 1.1 is split into three steps as formulated in the Surgery Program for recognising manifolds 7.47. Roughly speaking, the first step is to check the necessary homotopical and homological condition on X, that means it must be a finite Poincaré complex. The second step is to find a normal map of degree one from some closed smooth manifold M to X, which exists if the Spivak normal fibration has a vector bundle reduction. In the third step one tries to improve the map to a homotopy equivalence. Surgery below middle dimension from Chapter 4 yields first improvements towards the third step.

Now the slogan that Problem 1.2 is the relative version of Problem 1.1 becomes more apparent, see Remark 7.46. Moreover, the steps in Surgery Program 2.9 and in Surgery Program for recognising manifolds 7.47 correspond to each other as explained in the discussion after Remark 7.47.

Chapter 8: The Even-Dimensional Surgery Obstruction

Not every surgery problem can be solved; there are so-called surgery obstructions. In this chapter we construct the even-dimensional L-groups and the surgery obstructions taking values in them and show that in dimension ≥ 5 the vanishing of the surgery obstruction is equivalent to the existence of a solution of the surgery problem. The L-groups are defined in terms of quadratic forms over the group ring of the fundamental group of X. If the dimension n is divisible by four and X is simply connected, then the surgery obstruction group is $\mathbb{Z}$ and the surgery obstruction is the difference of the signatures of M and X. In this particular case the surgery obstruction is independent of the bundle data, but that is not true in general. If the dimension n is even but not divisible by four and X is simply connected, the surgery obstruction group is $\mathbb{Z}/2$ and the surgery obstruction is the so-called Arf invariant, which definitely depends on the bundle data. This chapter yields the final step in the solution of Problem 1.1 in the even-dimensional case.

Chapter 9: The Odd-Dimensional Surgery Obstruction

We construct the odd-dimensional L-groups and the surgery obstructions taking values in them and show that in dimension ≥ 5 the vanishing of the surgery obstruction is equivalent to the existence of a solution of the surgery problem. The L-groups are defined in terms of automorphisms of quadratic forms, or, equivalently, in terms of formations over the group ring of the fundamental group of X. If X is simply connected and odd-dimensional, the surgery obstruction groups vanish and there are no surgery obstructions. This chapter yields the final step in the solution of Problem 1.1 in the odd-dimensional case.

Chapter 10: Decorations and the Simple Surgery Obstruction

We develop the simple version of the surgery obstruction groups and the surgery obstructions. The difference to the previous constructions is that we want the underlying map $f \colon M \to X$ to be a simple homotopy equivalence, while before we were only aiming at a homotopy equivalence. This is relevant in view of the s-Cobordism Theorem 2.1. In the definition of the surgery obstruction groups we now take finitely generated free modules coming with a basis as the underlying modules of quadratic forms and then consider the Whitehead torsion of the various isomorphisms appearing in the previous constructions.

Chapter 11: The Geometric Surgery Exact Sequence

We introduce the surgery exact sequence in Theorems 11.22 and 11.25, see also Remark 11.23. It is the realisation of the Surgery Program 2.9, and thus yields a general method for solving Problem 1.2. The surgery exact sequence is the main theoretical tool in solving the classification problem of manifolds of dimensions greater than or equal to five.

Its simple topological version for an n-dimensional closed topological manifold N from Theorem 11.25 aims at the computation of the simple structure set $\mathcal{S}^{TOP,s}(N)$. Elements in $\mathcal{S}^{TOP,s}(N)$ are represented by simple homotopy equivalences $f\colon M \to N$ with a closed topological manifold M as source and N as target. Two such elements $f\colon M \to N$ and $f'\colon M' \to N$ represent the same element if there is a homeomorphism $h\colon M \to M'$ such that $f' \circ h$ and f are homotopic. The surgery exact sequence is an exact sequence of abelian groups of the shape

$$\mathcal{N}^{TOP}(N \times [0,1], N \times \{0,1\}) \to L_{n+1}^s(\mathbb{Z}\pi, w) \to \mathcal{S}^{TOP,s}(N)$$
$$\to \mathcal{N}^{TOP}(N) \to L_n^s(\mathbb{Z}\pi, w)$$

where $L_n^s(\mathbb{Z}\pi, w)$ is the simple algebraic L-group of the integral group ring of the fundamental group π of N with the orientation homomorphism w, the normal invariants $\mathcal{N}^{TOP}(N)$ are given by the surgery problems with target N, and the first and the fourth map are given by taking surgery obstructions. Here one needs to require either $n \geq 5$ or that $n = 4$ and the fundamental group is good in the sense of Freedman, see [157, 158], and Remark 8.30

Note that a closed topological manifold N has the property that any simple homotopy equivalence $M \to N$ from a closed topological manifold to N is homotopic to a homeomorphism if and only if the structure set $\mathcal{S}^{TOP,s}(N)$ consists of precisely one element, namely the one given by id_N.

The surgery exact sequence in the smooth category from Theorem 11.22 is in general not an exact sequence of abelian groups, only of pointed sets, see Section 11.8.

Chapter 12: Homotopy Spheres

This chapter is devoted to the classification of oriented homotopy spheres up to oriented diffeomorphism, where a homotopy sphere is a smooth closed manifold that is homotopy equivalent to S^n. This boils down to calculating the structure set $\mathcal{S}(S^n)$ in the smooth category. The input is the geometric surgery exact sequence in the smooth category from Chapter 11, calculations of the L-groups in the simply connected case from Chapters 8 and 9, and homotopy theoretic results about the so-called J-homomorphisms, which shed light on the normal invariants. The maps in the sequence are determined by studying the signature and the Arf invariant of surgery problems.

Information gained by these calculations yields results about various classifying spaces, which are organised in the so-called Kervaire–Milnor braid, see Section 12.7.

1.2 Overview of the Contents of this Book

Chapter 13: The Geometric Surgery Obstruction Group and Surgery Obstruction

This chapter is equivalent to the famous Chapter 9 in Wall's book [414]. We give a geometric approach to the L-groups and the surgery obstruction based on bordism theory. This is convenient in some situations where the necessary algebra is hard to analyse or not even available, such as controlled or equivariant surgery.

Chapter 14: Chain Complexes

This chapter has two goals. Firstly, we summarise the sign conventions that we use in the subsequent chapter about algebraic surgery and in fact throughout the book. These have been well thought through, and we hope that they will become standard. Unfortunately, in the literature many different sign conventions are used. The second goal of this chapter is to review some basic homotopy theory of chain complexes. Both of these topics provide background for the next chapter.

Chapter 15: Algebraic Surgery

We introduce a chain complex version of the L-groups and of the surgery obstructions and identify them with the L-groups and surgery obstructions from Chapters 8 and 9, see Theorem 15.3 and Theorem 15.4. So the chapter contains a presentation of Ranicki's theory of algebraic surgery where forms and formations are uniformly generalised to algebraic Poincaré chain complexes and their algebraic cobordism theory. One drawback of Ranicki's presentation in the original sources is that he very often describes certain constructions, such as algebraic surgery, only by writing down formulas without giving any structural insight. In our exposition we try to give certain general chain complex constructions that shall motivate the outcome and lead finally to explicit formulas. Moreover, we always use our sign conventions whereas Ranicki uses different sign conventions in different papers.

Chapter 16: Brief Survey of Computations of L-Groups

We give a brief survey of computations of L-group of group rings $\mathbb{Z}\pi$. For finite π the calculations were mostly done in the previous century and involve using representation theory and number theory. For infinite π nowadays the main tool is the Farrell–Jones Conjecture, which will be extensively treated in the book in preparation [261].

Chapter 17: The Homotopy Type of G/TOP, G/PL and G/O

We review how to determine the homotopy types of the classifying spaces G/PL and G/TOP, see Section 11.9 and Theorem 17.6. This leads to the computation of the set of normal invariants in the topological category in terms of singular cohomology after localising at 2 and in terms of KO-theory after inverting 2, see (11.41) and Theorem 11.24.

Chapter 18: Computations of Topological Structure Sets of some Prominent Closed Manifolds

We discuss how surgery theory and in particular the surgery exact sequence lead to computations of topological structure sets. We treat products of spheres, complex and real projective spaces, lens spaces, and tori. This leads to the classification of closed topological manifolds that are homotopy equivalent to these spaces up to homeomorphism. In particular we completely solve Problem 1.5. We have treated this problem for the standard sphere in the smooth category already in Chapter 12 about homotopy spheres.

Chapter 19: Topological Rigidity

A closed topological manifold N is called topologically rigid if any homotopy equivalence $M \to N$ with a closed topological manifold M as source and N as target is homotopic to a homeomorphism. In this chapter we want to study the question of which closed topological manifolds are topologically rigid. Section 19.4 is devoted to the Borel Conjecture predicting that any aspherical closed manifold is topologically rigid. Examples of non-aspherical closed manifolds that are topologically rigid are discussed in Section 19.5.

In Section 19.6 we briefly discuss the rarity of smooth rigidity in high dimensions.

Chapter 20: Modified Surgery

In this chapter we digress from the main line of the book, which treats the classification of manifolds with a given homotopy type via classical surgery, and discuss aspects of the use of surgery theory to classify manifolds with less homotopy theoretic input. Specifically we discuss variations of the Surgery Program, see Remark 2.9, which were pioneered by Kreck [225] and are often called modified surgery. Modified surgery might not have the general structural impact as surgery has on the classification of manifolds, on prominent conjectures such as those of Borel or Novikov, or on index theory and C^*-algebras, but leads in some special but very interesting cases to better and beautiful results, for instance for complete intersections, homogeneous spaces, and 4-manifolds.

1.3 Outlook

Here is a (not necessarily complete) list (in alphabetical order) of topics that we were not able to treat in this book in detail or at all, but which are very interesting. Some of them could be part of sequels to this book (not necessarily written by the authors of this book). For some items we include references where these topics have already been addressed and where further references can be found.

- Algebraic surgery in the setting of ∞-categories and the relation of algebraic surgery to hermitian K-theory, see [69, 70, 71, 271];
- ANR-homology manifolds and the Quinn obstruction, see [67, 338, 339];

- Applications of surgery theory to knot theory, see [245, 351];
- Applications to questions from differential geometry, in particular to the existence of Riemannian metrics with positive scalar curvature, see [362];
- Automorphism groups of closed manifolds, see [430];
- Computations of the L-groups of group rings of finite groups, see [177];
- Controlled surgery theory [328, 330];
- Equivariant surgery theory [76, 136, 137, 138, 139, 262, 263, 335];
- Finite H-spaces, see [6, 34];
- Full presentation of the proof of the Spherical Space Form Problem, see [120, 277];
- Mapping surgery to analysis, see [184, 185, 186, 440];
- Parametrised surgery, see [163, 162, 195];
- Poincaré surgery, Poincaré embeddings, and LS-groups, see [179],[221], [414, Chapter 11];
- Stratified surgery theory; see [424];
- Surgery in dimension 4, see [37, 159];
- Surgery in the topological category, see [219];
- The Novikov Conjecture, see [153, 154, 226];
- The total surgery obstruction, see [234, 343, 348];
- UNil-terms and splitting obstructions, see [77, 78, 79, 104].

1.4 How to Use this Book

As mentioned before, the potential readers may vary from established experts on surgery theory to advanced students without any previous knowledge about surgery theory. Obviously the various groups of readers have rather different expectations and needs. On the one hand we want to give correct and complete definitions, theorems and proofs, but we also want to allow the reader to browse through the text and get a first impression or a global picture. This leads of course to some tension that we tried to solve as explained below.

A typical example is the notion of a normal map and normal bordism. The definition is quite lengthy, see for instance Definition 7.13 and Definition 7.15. This is actually necessary, as none of the items occurring there can be dropped when one wants to set up the theory and give accurate proofs in full generality. But when one is working or thinking about a problem or wants to get a first impression, one should work with an extract of these definitions as explained for instance in Section 7.2. It can also be useful to make simplifying assumptions, for instance, that all manifolds are orientable, or, equivalently, that the orientation homomorphism $w \colon \pi \to \{\pm 1\}$ is trivial, or even that every manifold is simply connected. Then one can ignore the local coefficient systems and work with ordinary homology, and one does not have to deal with group rings but only with the ring $\mathbb{Z}$ of integers. In daily life one may get as far as to say that a normal map of degree one is a map $f \colon M \to X$ of degree one with connected orientable source and target covered by bundle data, without really memorising what the bundle data are.

Another example of how one can use the book in different ways concerns Chapter 2 on the s-Cobordism Theorem. The minimal approach is to go through the introduction and ignore the rest; that suffices completely to go on with the book. Or one may want to get the full proof and therefore go through all the material of Chapter 2 and at least parts of Chapter 3. This is explained in the Guide of Chapter 2.

The rather long Chapter 15 on algebraic surgery can also be read in rather different ways. One may ignore all the motivations and structural explanations, just concentrate on the formulas, completely leave out the proofs, and just browse through the definitions and main theorems. Or one may want to understand all the details and get an insight into why the definitions and proofs are set up as they stand, and therefore read everything. Again here a reader should first go through the introduction and the guide at its end (and then through the Overview given in Section 15.2), before she or he decides which parts of the chapter she or he wants to read in which reading modus.

One can apply surgery theory in the smooth, PL or topological category. In other words one may consider smooth compact manifolds and try to classify them up to diffeomorphism, compact PL manifolds and try to classify them up to PL homeomorphism, or compact topological manifolds and try to classify them up to homeomorphism. When we explain some technical constructions, such as the surgery step, the bundle data, and so on, we will work in the smooth category since there all the notions such as tangent bundle, normal bundle, transversality and so on are well documented. All this carries over to the PL category and the topological category. We will not go into the sophisticated details of how this can be done since it would go beyond the scope of this book. For the topological category the seminal work of Kirby–Siebenmann [219] is needed. A good reference for the PL category is Rourke–Sanderson [367]. So basic tools such as the surgery exact sequence do exist in all three categories.

The classification results do of course depend on whether we are working in the smooth, PL, or topological category. It turns out that the nicest results occur in the topological category. The reader will have to accept the fact that we develop surgery theory in detail only in the smooth category, but will also apply it to the topological category without further explanations.

Very often we will make the assumption that the dimensions of the manifolds under consideration are greater than or equal to 5. The problem is that the so-called Whitney trick applies only under this dimension assumption. The problem with the Whitney trick can be solved and hence surgery can also be carried out in dimension 4, provided that we work in the topological category and the fundamental group π is good in the sense of Freedman, see Remark 8.30. All of the results presented in this book with the dimension condition ≥ 5 extend to dimension 4 in the topological category if the fundamental group is good. The reader has to live with the fact that we do not explain what is behind these ideas of Freedman, but refer for instance to [37, 159].

The book contains a number of exercises. They come in two flavours. A few of them contain additional information or a computation that may be needed later. Most of them are not needed for the exposition of the book, but give some illuminating insight. Moreover, the reader may test whether she or he has understood the text, or improve her or his understanding by trying to solve the exercises. Note that hints to the solutions of the exercises are given in Chapter 21.

Readers wishing to find a specific topic are advised to first look at the Overview of the Contents of this Book in Section 1.2, in order to find the right chapter and then that chapter's introduction. Each introduction to a chapter concludes with a guide, which may help the reader to figure out how to access the contents of that chapter. The extensive index at the end of the book can also be used to find the right spot for a specific topic. The index contains an item Theorem, under which all theorems with their names appearing in the book are listed, and analogously there is an item Conjecture.

We have successfully used parts of this book for seminars, reading courses, and advanced courses for students.

The reader may also consult other monographs on surgery theory such as [55, 93, 252, 352, 414]. Further surveys article or more information can be found for instance in [72, 73, 145, 146, 153, 154, 219, 226, 348, 424, 425].

1.5 Prerequisites

We require that the reader is familiar with the basics of the following concepts and notions. Readers can learn these topics from the suggested references, but there are many more books and monographs available.

- *CW*-complexes, Cellular Approximation Theorem, Whitehead Theorem, see [45, 178, 399];
- Covering theory, universal coverings, see [45, 178, 399];
- Homology, cohomology, cup and cap-product, signature, characteristic classes [45, 178, 198, 308, 399];
- Homotopy groups, fibrations, cofibrations, Hurewicz Theorem, see [45, 178, 399];
- Topological and smooth manifolds, see [45, 49, 224, 237, 241];
- Vector bundles, normal and tangent bundle of a smooth manifold, see [45, 49, 189, 224, 308];
- Classifying spaces for groups and for vector bundles, fibre bundles, and fibrations, see [198, 279, 289, 308];
- Transversality, regular values, immersions, and submersions, see [45, 49, 189, 224];
- Group rings, modules and chain complexes over a non-commutative ring, see [59, 238, 327, 421];
- Bordism of manifolds, bordism ring, see [189, 308].

1.6 Acknowledgement

The origin of this book dates back to 2001 when the Summer school on Topology of high-dimensional manifolds was held at ICTP in Trieste. Wolfgang Lück gave a series of lectures and the resulting notes [252] were published as *A basic introduction to surgery theory* in the Proceedings of the school.

However, the idea that the notes could be expanded into a book that would provide a more comprehensive tour of the foundations of the subject persisted. In 2011 Wolfgang Lück approached Diarmuid Crowley and Tibor Macko with an offer to contribute to this endeavour. At the time all three were based in Bonn, and the two younger mathematicians joined the project.

After his moves to Aberdeen and later to Melbourne, Diarmuid Crowley left the team working on the book. The authors are grateful to him for his contributions.

Diarmuid Crowley was scientifically involved with the material covered in Chapters 4 to 9, and 11 and 12, and specifically with versions of these chapters prior to the introduction of the intrinsic fundamental class. His involvement was most significant in Chapters 6, 8, and 9. He also contributed material which persists in Chapters 4, 11 and 12. Specific contributions of his that we would like to mention are:

- Material on the construction of Poincaré complexes without vector bundle reduction in Section 6.6 and the proof of the uniqueness of the Spivak normal fibration covered in Sections 6.9 and 6.10;
- Material on the groups $L_{2k}(\mathbb{Z})$ in Sections 8.5.2 and 8.5.3, the role of the bundle data in the surgery obstruction for highly connected normal maps in Section 8.7.7 and the discussion of surgery kernels for maps of pairs and triads in Sections 8.6.2, 8.8.2, and 8.8.4;
- Chapter 9, in particular the proof that $L_{4k+1}(\mathbb{Z}) = 0$ in Section 9.2.4;
- Material on the spaces G/O, G/PL and G/TOP in Section 11.9 and on the algebraic properties of surgery exact sequences in Section 11.8.

His insights, through our many conversations over time, have contributed to our understanding of the subject and many of its subtleties.

The authors also want to thank the participants of the one year course *Introduction to Surgery Theory*, which took place during the Covid pandemic via ZOOM in 2020 and 2021. There were many very fruitful discussions during the lectures and the tutorials, which helped to improve the exposition a lot. Special thanks go to Frieder Jäckel, Dominik Kirstein, and Christian Kremer.

There are many more mathematicians who made very useful comments about the book, including Spiros Adams-Florou, Serhii Dylda, Ian Hambleton, Fabian Hebestreit, Samuel Kalužný, Daniel Kasprowski, Matthias Kreck, Markus Land, Mark Powell, Andrew Ranicki, Ajay Raj, Arunima Ray, Julia Semikina, Wolfgang Steimle, Peter Teichner, Simona Veselá, Shmuel Weinberger, Christoph Winges, and the (unknown) referees. We are grateful to Philipp Kühl for helping us with the pictures.

Finally the first author wants to thank in particular his wife Sibylle for her patience.

1.6 Acknowledgement

This book project was funded by the Leibniz-Award of the first author granted by the Deutsche Forschungsgemeinschaft (DFG, German Research Foundation), the ERC Advanced Grant "KL2MG-interactions" (no. 662400) of the first author granted by the European Research Council, and by the Deutsche Forschungsgemeinschaft (DFG, German Research Foundation) under Germany's Excellence Strategy – GZ 2047/1, Projekt-ID 390685813, Hausdorff Center for Mathematics at Bonn.

The second author was supported by the following grants: "Topology of high-dimensional manifolds" in the scheme "Returns", VEGA 1/0101/17 and VEGA 1/0596/21 of the Ministry of Education of Slovakia, and APVV-16-0053 of the Slovak Research and Development Agency.

Chapter 2
The s-Cobordism Theorem

2.1 Introduction

In this chapter we want to discuss and prove the following theorem (in the smooth category).

Theorem 2.1 (*s-Cobordism Theorem*) *Let M_0 be a closed connected smooth manifold with $\dim(M_0) \geq 5$ and fundamental group $\pi = \pi_1(M_0)$. Then:*

(i) *Let $(W; M_0, f_0, M_1, f_1)$ be a smooth h-cobordism over M_0. Then W is trivial over M_0 if and only if its Whitehead torsion $\tau(W, M_0) \in \text{Wh}(\pi)$ vanishes;*
(ii) *For any $x \in \text{Wh}(\pi)$ there is a smooth h-cobordism $(W; M_0, f_0, M_1, f_1)$ over M_0 with $\tau(W, M_0) = x \in \text{Wh}(\pi)$;*
(iii) *The function assigning to a smooth h-cobordism $(W; M_0, f_0, M_1, f_1)$ over M_0 its Whitehead torsion yields a bijection from the diffeomorphism classes relative M_0 of smooth h-cobordisms over M_0 to the Whitehead group $\text{Wh}(\pi)$.*

The analogous statements hold in the PL category and in the topological category.

Here are some explanations. In the sequel we work in the smooth category unless explicitly stated otherwise. An *n-dimensional cobordism* (sometimes also just called a bordism) $(W; M_0, f_0, M_1, f_1)$ consists of a compact n-dimensional manifold W, closed $(n-1)$-dimensional manifolds M_0 and M_1, a disjoint decomposition $\partial W = \partial_0 W \coprod \partial_1 W$ of the boundary ∂W of W, and diffeomorphisms $f_0 \colon M_0 \to \partial_0 W$ and $f_1 \colon M_1 \to \partial_1 W$. If we want to specify M_0, we say that W is a *cobordism over M_0*. If $\partial_0 W = M_0$, $\partial_1 W = M_1$ and f_0 and f_1 are given by the identity or if f_0 and f_1 are obvious from the context, we briefly write $(W; \partial_0 W, \partial_1 W)$. Note that the choices of the diffeomorphisms f_i do play a role, although they are often suppressed in the notation. Two cobordisms $(W; M_0, f_0, M_1, f_1)$ and $(W'; M_0, f_0', M_1', f_1')$ over M_0 are *diffeomorphic relative M_0* if there is a diffeomorphism $F \colon W \to W'$ with $F \circ f_0 = f_0'$. We call a cobordism $(W; M_0, f_0, M_1, f_1)$ an *h-cobordism* if the inclusions $\partial_i W \to W$ for $i = 0, 1$ are homotopy equivalences. We call an h-cobordism over M_0 *trivial* if it is diffeomorphic relative M_0 to the trivial h-cobordism $(M_0 \times [0,1]; M_0 \times \{0\}, M_0 \times \{1\})$. We will discuss the Whitehead group in Sections 2.5 and 3.2.

Exercise 2.2 Let G be a finitely presented group and $n \geq 6$. Show that $\text{Wh}(G)$ is trivial if every n-dimensional h-cobordism with G as fundamental group is trivial.

Exercise 2.3 Classify all two-dimensional connected h-cobordisms.

We will later see that the Whitehead group of the trivial group vanishes. Thus the s-Cobordism Theorem 2.1 implies the following theorem.

Theorem 2.4 (h-Cobordism Theorem) *Every h-cobordism over a simply connected closed smooth manifold M_0 with $\dim(M_0) \geq 5$ is trivial.*

The analogous statement holds in the PL *category and in the topological category.*

Theorem 2.5 ((Generalised) Poincaré Conjecture) *The (Generalised) Poincaré Conjecture holds for a closed topological manifold M with $\dim(M) \geq 5$, namely, if M is simply connected and its homology $H_p(M)$ is isomorphic to $H_p(S^n)$ for all $p \in \mathbb{Z}$, then M is homeomorphic to S^n.*

The (Generalised) Poincaré Conjecture also holds in the PL *category.*

Proof. We begin with $\dim(M) \geq 6$. As M is simply connected and $H_*(M) \cong H_*(S^n)$, one can conclude from the Hurewicz Theorem and the Whitehead Theorem, see [178, Theorem 4.32 on page 366 and Corollary 4.33 on page 367] or [431, Theorem IV.7.13 on page 181 and Theorem IV.7.17 on page 182], that there is a homotopy equivalence $f: M \to S^n$. Let $D_i^n \subset M$ for $i = 0, 1$ be two embedded disjoint disks. Put $W = M \setminus (\text{int}(D_0^n) \coprod \text{int}(D_1^n))$. Then W turns out to be a simply connected h-cobordism. By Theorem 2.4 there is a diffeomorphism $F: (\partial D_0^n \times [0, 1], \partial D_0^n \times \{0\}, \partial D_0^n \times \{1\}) \to (W, \partial D_0^n, \partial D_1^n)$ that is the identity on $\partial D_0^n = \partial D_0^n \times \{0\}$ and induces some (unknown) diffeomorphism $f_1: \partial D_0^n \times \{1\} \to \partial D_1^n$. By the *Alexander trick* one can extend $f_1: \partial D_0^n = \partial D_0^n \times \{1\} \to \partial D_1^n$ to a homeomorphism $\overline{f_1}: D_0^n \to D_1^n$. Namely, any homeomorphism $f: S^{n-1} \to S^{n-1}$ extends to a homeomorphism $\overline{f}: D^n \to D^n$ by sending $t \cdot x$ for $t \in [0, 1]$ and $x \in S^{n-1}$ to $t \cdot f(x)$. Now define a homeomorphism $h: D_0^n \times \{0\} \cup_{i_0} \partial D_0^n \times [0, 1] \cup_{i_1} D_0^n \times \{1\} \to M$ for the canonical inclusions $i_k: \partial D_0^n \times \{k\} \to \partial D_0^n \times [0, 1]$ for $k = 0, 1$ by $h|_{D_0^n \times \{0\}} = \text{id}$, $h|_{\partial D_0^n \times [0,1]} = F$ and $h|_{D_0^n \times \{1\}} = \overline{f_1}$. Since the source of h is obviously homeomorphic to S^n, Theorem 2.5 follows.

In the case $\dim(M) = 5$ one uses the fact that M is the boundary of a contractible 6-dimensional manifold W and applies the s-cobordism theorem to W with an embedded disk removed. See also [439]. □

Remark 2.6 (The Poincaré Conjecture does not hold in the smooth category) Note that the proof of the Poincaré Conjecture in Theorem 2.5 works only in the topological category and PL category, but not in the smooth category. In other words, we cannot conclude the existence of a diffeomorphism $h: S^n \to M$. The proof in the smooth case breaks down when we apply the Alexander trick. The construction of $\overline{f}$ given by coning f yields only a PL homeomorphism $\overline{f}$ and not a diffeomorphism if we start with a diffeomorphism f. The map $\overline{f}$ is smooth outside the origin of D^n but not necessarily at the origin. We will see that not every diffeomorphism $f: S^{n-1} \to S^{n-1}$ can be extended to a diffeomorphism $D^n \to D^n$ and that there exist

2.1 Introduction

so-called *exotic spheres*, i.e., closed smooth manifolds which are homeomorphic to S^n but not diffeomorphic to S^n for some n. The classification of these exotic spheres is one of the early very important achievements of surgery theory and one motivation for its further development, see Chapter 12 and in particular Remark 12.36.

Figure 2.7 (Poincaré Conjecture).

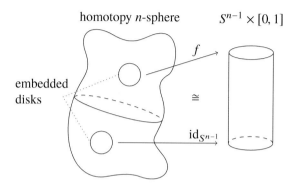

Exercise 2.8 Show that any diffeomorphism $f : S^1 \to S^1$ can be extended to a diffeomorphism $F : D^2 \to D^2$.

Remark 2.9 (Surgery Program) In some sense the s-Cobordism Theorem 2.1 is one of the first theorems where diffeomorphism classes of certain manifolds are determined by an algebraic invariant, namely the Whitehead torsion. Moreover, the Whitehead group $\mathrm{Wh}(\pi)$ depends only on the fundamental group $\pi = \pi_1(M_0)$ whereas the diffeomorphism classes of h-cobordisms over M_0 a priori depend on M_0 itself. The s-Cobordism Theorem 2.1 is one step in a program to decide whether two closed manifolds M and N are diffeomorphic, which is in general a very hard question. The idea is to construct an h-cobordism $(W; M, f, N, g)$ with vanishing Whitehead torsion. Then W is diffeomorphic to the trivial h-cobordism over M, which implies that M and N are diffeomorphic. So the *Surgery Program* is:

(i) Construct a homotopy equivalence $f : M \to N$;
(ii) Construct a cobordism $(W; M, N)$ and a map $(F, f, \mathrm{id}) : (W; M, N) \to (N \times [0, 1]; N \times \{0\}, N \times \{1\})$;
(iii) Modify W and F relative boundary by so-called *surgery* so that F becomes a homotopy equivalence and thus W becomes an h-cobordism. During these processes one should make certain that the Whitehead torsion of the resulting h-cobordism is trivial.

The advantage of this approach will be that it can be reduced to problems in homotopy theory and algebra, which can sometimes be handled by well-known techniques. In particular one will sometimes get computable obstructions for two homotopy equivalent manifolds to be diffeomorphic. Often surgery theory has proved to be very useful, when one wants to distinguish two closed manifolds that have very similar properties. The classification of homotopy spheres, see Chapter 12, is one example. Moreover, surgery techniques can be applied to problems that are of a different nature than diffeomorphism or homeomorphism classifications.

In this chapter we want to present the proof of the s-Cobordism Theorem and explain why the notion of Whitehead torsion comes in. We will encounter a typical situation in mathematics. We will consider an h-cobordism and try to prove that it is trivial. We will introduce modifications that are designed to reduce the number of handles and that we can apply to a handlebody decomposition without changing the diffeomorphism type. If we could get rid of all handles, the h-cobordism would be trivial. When attempting to cancel all handles, we run into an algebraic difficulty. A priori this difficulty could be a lack of a good idea or technique. But it will turn out to be a genuine obstruction and lead us to the definition of Whitehead torsion and the Whitehead group.

Figure 2.10 (Surgery Program).

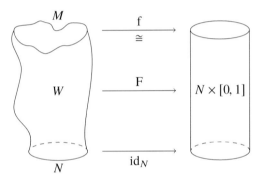

The rest of this chapter is devoted to the proof of the s-Cobordism Theorem 2.1 in the smooth category. Its proof is interesting and illuminating and it motivates the definition of Whitehead torsion. The definition of Whitehead torsion itself and the final step in the proof will appear in Chapter 3.

Guide 2.11 It is not necessary to go through the remainder of this chapter to comprehend the following chapters. It suffices to understand the statement of the s-Cobordism Theorem 2.1 and to read through Remark 2.9 about the Surgery Program.

It is even possible for the first reading to pretend that every s-cobordism is trivial and the Whitehead group $Wh(G)$ is trivial, which is known to be true in the simply connected case and for many torsionfree groups G.

2.2 Handlebody Decompositions

In this section we explain basic facts about handles and handlebody decompositions.

Definition 2.12 (Handlebody) The *n-dimensional handle of index q* or briefly *q-handle* is $D^q \times D^{n-q}$. Its *core* is $D^q \times \{0\}$. The *boundary of the core* is $S^{q-1} \times \{0\}$. Its *cocore* is $\{0\} \times D^{n-q}$ and its *transverse sphere* is $\{0\} \times S^{n-q-1}$.

Consider an n-dimensional manifold M with boundary ∂M. Given an embedding $\phi^q : S^{q-1} \times D^{n-q} \hookrightarrow \partial M$, we say that the manifold $M + (\phi^q)$ defined by $M \cup_{\phi^q} D^q \times D^{n-q}$ is obtained from M by attaching a handle of index q by ϕ^q.

Figure 2.13 (Handlebody).

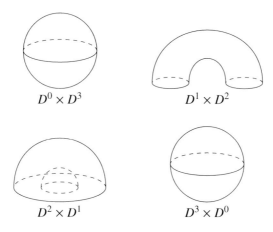

Obviously $M + (\phi^q)$ carries the structure of a topological manifold. To get a smooth structure, one has to use the technique of straightening the angle to get rid of the corners at the place where the handle is glued to M. The boundary $\partial(M + (\phi^q))$ can be described as follows. Delete from ∂M the interior of the image of ϕ^q. We obtain a manifold with boundary together with a diffeomorphism from $S^{q-1} \times S^{n-q-1}$ to its boundary induced by $\phi^q|_{S^{q-1} \times S^{n-q-1}}$. If we use this diffeomorphism to glue $D^q \times S^{n-q-1}$ to it, we obtain a closed manifold, namely, $\partial(M + (\phi^q))$.

Let W be a compact manifold whose boundary ∂W is the disjoint sum $\partial_0 W \coprod \partial_1 W$. Then we want to construct W from $\partial_0 W \times [0, 1]$ by attaching handles as follows. Note that the following construction will not change $\partial_0 W = \partial_0 W \times \{0\}$. If $\phi^q : S^{q-1} \times D^{n-q} \hookrightarrow \partial_0 W \times \{1\}$ is an embedding, we get by attaching a handle the compact manifold $W_1 = \partial_0 W \times [0, 1] + (\phi^q)$ that is given by $\partial_0 W \times [0, 1] \cup_{\phi^q} D^q \times D^{n-q}$. Its boundary is a disjoint sum $\partial_0 W_1 \coprod \partial_1 W_1$ where $\partial_0 W_1$ is the same as $\partial_0 W$. Now we can iterate this process where we attach a handle to W_1 at $\partial_1 W_1$. Thus we obtain a compact manifold with boundary

$$W = \partial_0 W \times [0, 1] + (\phi_1^{q_1}) + (\phi_2^{q_2}) + \cdots + (\phi_r^{q_r})$$

whose boundary is the disjoint union $\partial_0 W \coprod \partial_1 W$ where $\partial_0 W$ is $\partial_0 W \times \{0\}$. We call a description of W as above a *handlebody decomposition* of W relative $\partial_0 W$.

Figure 2.14 (Handlebody decomposition).

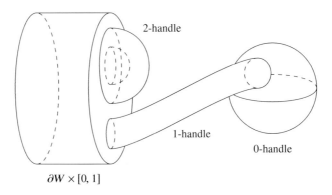

From Morse theory, see [189, Chapter 6], [298, part I], we obtain the following lemma.

Lemma 2.15 *Let W be a compact manifold whose boundary ∂W is the disjoint sum $\partial_0 W \coprod \partial_1 W$. Then W possesses a handlebody decomposition relative $\partial_0 W$, i.e., W is up to diffeomorphism relative $\partial_0 W = \partial_0 W \times \{0\}$ of the form*

$$W = \partial_0 W \times [0, 1] + (\phi_1^{q_1}) + (\phi_2^{q_2}) + \cdots + (\phi_r^{q_r}).$$

In order to show that W is diffeomorphic to $\partial_0 W \times [0, 1]$ relative $\partial_0 W = \partial_0 W \times \{0\}$, we must get rid of the handles. For this purpose we have to find modifications of the handlebody decomposition that reduce the number of handles without changing the diffeomorphism type of W relative $\partial_0 W$.

Lemma 2.16 (Isotopy Lemma) *Let W be an n-dimensional compact manifold whose boundary ∂W is the disjoint sum $\partial_0 W \coprod \partial_1 W$. Given isotopic embeddings $\phi^q, \psi^q \colon S^{q-1} \times D^{n-q} \hookrightarrow \partial_1 W$, there is a diffeomorphism $W + (\phi^q) \to W + (\psi^q)$ relative $\partial_0 W$.*

Proof. Let $i \colon S^{q-1} \times D^{n-q} \times [0, 1] \to \partial_1 W$ be an isotopy from ϕ^q to ψ^q. Then one can find a diffeotopy $H \colon W \times [0, 1] \to W$ with $H_0 = \mathrm{id}_W$ such that the composition of H with $\phi^q \times \mathrm{id}_{[0,1]}$ is i and H is stationary on $\partial_0 W$, see [189, Theorem 1.3 in

2.2 Handlebody Decompositions

Chapter 8 on page 184]. Thus $H_1 \colon W \to W$ is a diffeomorphism relative $\partial_0 W$ and satisfies $H_1 \circ \phi^q = \psi^q$. It induces a diffeomorphism $W + (\phi^q) \to W + (\psi^q)$ relative $\partial_0 W$. □

Lemma 2.17 (Diffeomorphism Lemma) *Let W resp. W' be a compact manifold whose boundary ∂W is the disjoint sum $\partial_0 W \coprod \partial_1 W$ resp. $\partial_0 W' \coprod \partial_1 W'$. Let $F \colon W \to W'$ be a diffeomorphism that induces a diffeomorphism $f_0 \colon \partial_0 W \to \partial_0 W'$. Let $\phi^q \colon S^{q-1} \times D^{n-q} \hookrightarrow \partial_1 W$ be an embedding.*

Then there is an embedding $\overline{\phi}^q \colon S^{q-1} \times D^{n-q} \hookrightarrow \partial_1 W'$ and a diffeomorphism $F' \colon W + (\phi^q) \to W' + (\overline{\phi}^q)$ that induces f_0 on $\partial_0 W$.

Proof. Put $\overline{\phi}^q = F \circ \phi^q$. □

Lemma 2.18 *Let W be an n-dimensional compact manifold whose boundary ∂W is the disjoint sum $\partial_0 W \coprod \partial_1 W$. Suppose that $V = W + (\psi^r) + (\phi^q)$ for $q \leq r$. Then V is diffeomorphic relative $\partial_0 W$ to $V' = W + (\overline{\phi}^q) + (\psi^r)$ for appropriate $\overline{\phi}^q$.*

Proof. By transversality and the assumption $(q-1) + (n-1-r) < n-1$, we can show that the embedding $\phi^q|_{S^{q-1} \times \{0\}} \colon S^{q-1} \times \{0\} \hookrightarrow \partial_1(W + (\psi^r))$ is isotopic to an embedding that does not meet the transverse sphere of the handle (ψ^r) attached by ψ^r [189, Theorem 2.3 in Chapter 3 on page 78]. This isotopy can be embedded in a diffeotopy on $\partial_1(W+(\psi^r))$. Thus the embedding $\phi^q \colon S^{q-1} \times D^{n-q} \hookrightarrow \partial_1(W+(\psi^r))$ is isotopic to an embedding whose restriction to $S^{q-1} \times \{0\}$ does not meet the transverse sphere of the handle (ψ^r). Since we can isotope an embedding $S^{q-1} \times D^{n-q} \hookrightarrow W + (\psi^r)$ so that its image becomes arbitrarily close to the image of $S^{q-1} \times \{0\}$, we can isotope $\phi^q \colon S^{q-1} \times D^{n-q} \hookrightarrow \partial_1(W + (\psi^r))$ to an embedding that does not meet a closed neighbourhood $U \subset \partial_1(W + (\psi^r))$ of the transverse sphere of the handle (ψ^r). There is an obvious diffeotopy on $\partial_1(W + (\psi^r))$ that is stationary on the transverse sphere of (ψ^r) and moves any point on $\partial_1(W + (\psi^r))$, which belongs to the handle (ψ^r) but not to U, to a point outside the handle (ψ^r). Thus we can find an isotopy of ϕ^q to an embedding $\overline{\phi}^p$ that does not meet the handle (ψ^r) at all. Obviously $W + (\psi^r) + (\overline{\phi}^q)$ and $W + (\overline{\phi}^q) + (\psi^r)$ agree. By the Isotopy Lemma 2.16 there is a diffeomorphism relative $\partial_0 W$ from $W + (\psi^r) + (\overline{\phi}^q)$ to $W + (\psi^r) + (\phi^q)$. □

Example 2.19 (Cancelling handles) Here is a standard situation where attaching first a q-handle and then a $(q+1)$-handle does not change the diffeomorphism type of an n-dimensional compact manifold W with the disjoint union $\partial_0 W \coprod \partial_1 W$ as boundary ∂W. Let $0 \leq q \leq n-1$. Consider an embedding

$$\mu \colon S^{q-1} \times D^{n-q} \cup_{S^{q-1} \times S^{n-1-q}_+} D^q \times S^{n-1-q}_+ \hookrightarrow \partial_1 W$$

where S^{n-1-q}_+ is the upper hemisphere in $S^{n-1-q} = \partial D^{n-q}$. Note that the source of μ is diffeomorphic to D^{n-1}. Let $\phi^q \colon S^{q-1} \times D^{n-q} \to \partial_1 W$ be its restriction to $S^{q-1} \times D^{n-q}$. Let $\phi^{q+1}_+ \colon S^q_+ \times S^{n-q-1}_+ \hookrightarrow \partial_1(W + (\phi^q))$ be the embedding given by

$$S^q_+ \times S^{n-q-1}_+ = D^q \times S^{n-q-1}_+ \subset D^q \times S^{n-q-1} = \partial(\phi^q) \subset \partial_1(W + (\phi^q)).$$

It does not meet the interior of W. Let $\phi_-^{q+1}: S^q \times S_+^{n-1-q} \hookrightarrow \partial_1(W + (\phi^q))$ be the embedding obtained from μ by restriction to $S_-^q \times S_+^{n-1-q} = D^q \times S_+^{n-1-q}$. Then ϕ_-^{q+1} and ϕ_+^{q+1} fit together to yield an embedding $\psi^{q+1}: S^q \times D^{n-q-1} = S_-^q \times S_+^{n-q-1} \cup_{S^{q-1} \times S_+^{n-q-1}} S_+^q \times S_+^{n-q-1} \hookrightarrow \partial_1(W+(\phi^q))$. Then it is not difficult to check that $W+(\phi^q)+(\psi^{q+1})$ is diffeomorphic relative $\partial_0 W$ to W since up to diffeomorphism $W+(\phi^q)+(\psi^{q+1})$ is obtained from W by taking the boundary connected sum of W and D^n along the embedding μ of $D^{n-1} = S_+^{n-1} = S^{q-1} \times D^{n-q} \cup_{S^{q-1} \times S_+^{n-1-q}} D^q \times S_+^{n-1-q}$ into $\partial_1 W$.

Figure 2.20 (Handle cancellation).

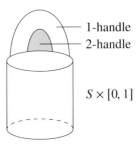

This cancellation of two handles of consecutive index can be generalised as follows.

Lemma 2.21 (Cancellation Lemma) *Let W be an n-dimensional compact manifold whose boundary ∂W is the disjoint sum $\partial_0 W \coprod \partial_1 W$. Let $\phi^q: S^{q-1} \times D^{n-q} \hookrightarrow \partial_1 W$ be an embedding. Let $\psi^{q+1}: S^q \times D^{n-q-1} \hookrightarrow \partial_1(W+(\phi^q))$ be an embedding. Suppose that $\psi^{q+1}(S^q \times \{0\})$ is transversal to the transverse sphere of the handle (ϕ^q) and meets the transverse sphere in exactly one point.*

Then there is a diffeomorphism relative $\partial_0 W$ from W to $W + (\phi^q) + (\psi^{q+1})$.

Proof. Given any neighbourhood $U \subset \partial(\phi^q)$ of the transverse sphere of (ϕ^q), there is an obvious diffeotopy on $\partial_1(W + (\phi^q))$ that is stationary on the transverse sphere of (ϕ^q) and moves any point on $\partial_1(W + (\phi^q))$, which belongs to the handle (ϕ^q) but not to U, to a point outside the handle (ϕ^q). Thus we can achieve that ψ^{q+1} maps the lower hemisphere $S_-^q \times \{0\}$ to points outside (ϕ^q) and on the upper hemisphere $S_+^q \times \{0\}$ it is given by the obvious inclusion $D^q \times \{x\} \hookrightarrow D^q \times D^{n-q} = (\phi^q)$ for some $x \in S^{n-q-1}$ and the obvious identification of $S_+^q \times \{0\}$ with $D^q \times \{x\}$. Now it is not hard to construct a diffeomorphism relative $\partial_0 W$ from $W + (\phi^q) + (\psi^{q+1})$ to W modelling the standard situation of Example 2.19. □

The Cancellation Lemma 2.21 will be our only tool to reduce the number of handles. Note that one can never get rid of one handle alone, there must be at least two handles involved simultaneously. The reason for this is that the Euler

2.2 Handlebody Decompositions

characteristic $\chi(W, \partial_0 W)$ is independent of the handle decomposition and can be computed by $\sum_{q \geq 0}(-1)^q \cdot p_q$, where p_q is the number of q-handles, see Section 2.3.

We call an embedding $S^q \times D^{n-q} \hookrightarrow M$ for $q < n$ into an n-dimensional manifold *trivial* if it can be written as the composition of an embedding $D^n \hookrightarrow M$ and a fixed standard embedding $S^q \times D^{n-q} \hookrightarrow D^n$. We call an embedding $S^q \hookrightarrow M$ for $q < n$ *trivial* if it can be extended to a trivial embedding $S^q \times D^{n-q} \hookrightarrow M$. We conclude from the Cancellation Lemma 2.21 the following result.

Lemma 2.22 *Let $\phi^q : S^{q-1} \times D^{n-q} \hookrightarrow \partial_1 W$ be a trivial embedding. Then there is an embedding $\phi^{q+1} : S^q \times D^{n-1-q} \hookrightarrow \partial_1(W + (\phi^q))$ such that W and $W + (\phi^q) + (\phi^{q+1})$ are diffeomorphic relative $\partial_0 W$.*

Consider a compact n-dimensional manifold W whose boundary is the disjoint union $\partial_0 W \coprod \partial_1 W$. In view of Lemma 2.15 and Lemma 2.18 we can write it as

$$W \cong \partial_0 W \times [0,1] + \sum_{i=1}^{p_0}(\phi_i^0) + \sum_{i=1}^{p_1}(\phi_i^1) + \cdots + \sum_{i=1}^{p_n}(\phi_i^n) \tag{2.23}$$

where $\cong$ means diffeomorphic relative $\partial_0 W$.

Notation 2.24 Put for $-1 \leq q \leq n$

$$W_q := \partial_0 W \times [0,1] + \sum_{i=1}^{p_0}(\phi_i^0) + \sum_{i=1}^{p_1}(\phi_i^1) + \cdots + \sum_{i=1}^{p_q}(\phi_i^q);$$

$$\partial_1 W_q := \partial W_q - \partial_0 W \times \{0\};$$

$$\partial_1^\circ W_q := \partial_1 W_q - \coprod_{i=1}^{p_{q+1}} \phi_i^{q+1}(S^q \times \mathrm{int}(D^{n-1-q})).$$

Note for the sequel that $\partial_1^\circ W_q \subset \partial_1 W_{q+1}$.

Lemma 2.25 (Elimination Lemma) *Fix an integer q with $1 \leq q \leq n-3$. Suppose that $p_j = 0$ for $j < q$, i.e., W looks like*

$$W = \partial_0 W \times [0,1] + \sum_{i=1}^{p_q}(\phi_i^q) + \sum_{i=1}^{p_{q+1}}(\phi_i^{q+1}) + \cdots + \sum_{i=1}^{p_n}(\phi_i^n).$$

Fix an integer i_0 with $1 \leq i_0 \leq p_q$. Suppose that there is an embedding $\psi^{q+1} : S^q \times D^{n-1-q} \hookrightarrow \partial_1^\circ W_q$ with the following properties:

(i) *The restriction $\psi^{q+1}|_{S^q \times \{0\}}$ is isotopic in $\partial_1 W_q$ to an embedding $\psi_1^{q+1} : S^q \times \{0\} \hookrightarrow \partial_1 W_q$ that meets the transverse sphere of the handle $(\phi_{i_0}^q)$ transversally and in exactly one point and is disjoint from the transverse sphere of ϕ_i^q for $i \neq i_0$;*

(ii) *The restriction $\psi^{q+1}|_{S^q \times \{0\}}$ is isotopic in $\partial_1 W_{q+1}$ to a trivial embedding $\psi_2^{q+1} : S^q \times \{0\} \hookrightarrow \partial_1^\circ W_{q+1}$.*

Then W is diffeomorphic relative $\partial_0 W$ to a manifold of the shape

$$\partial_0 W \times [0,1] + \sum_{\substack{i=1,2,\ldots,p_q,\\ i\neq i_0}} (\phi_i^q) + \sum_{i=1}^{p_{q+1}} (\overline{\phi}_i^{q+1}) + (\psi^{q+2}) + \sum_{i=1}^{p_{q+2}} (\overline{\phi}_i^{q+2}) + \cdots + \sum_{i=1}^{p_n} (\overline{\phi}_i^n).$$

Proof. Since $\psi^{q+1}|_{S^q \times \{0\}}$ is isotopic to ψ_1^{q+1} and ψ_2^{q+1} is trivial, we can extend ψ_1^{q+1} and ψ_2^{q+1} to embeddings denoted in the same way $\psi_1^{q+1} : S^q \times D^{n-q-1} \hookrightarrow \partial_1 W_q$ and $\psi_2^{q+1} : S^q \times D^{n-1-q} \hookrightarrow \partial_1^\circ W_{q+1}$ with the following properties: ψ^{q+1} is isotopic to ψ_1^{q+1} in $\partial_1 W_q$, ψ_1^{q+1} does not meet the transverse spheres of the handles (ϕ_i^q) for $i \neq i_0$, $\psi_1^{q+1}|_{S^q \times \{0\}}$ meets the transverse sphere of the handle $(\phi_{i_0}^q)$ transversally and in exactly one point, ψ^{q+1} is isotopic to ψ_2^{q+1} within $\partial_1 W_{q+1}$, and ψ_2^{q+1} is trivial, see [189, Theorem 1.5 in Chapter 8 on page 180]. Because of the Diffeomorphism Lemma 2.17 we can assume without loss of generality that there are no handles of index $\geq q+2$, i.e., $p_{q+2} = p_{q+3} = \cdots = p_n = 0$. It suffices to show for appropriate embeddings $\overline{\phi}_i^{q+1}$ and ψ^{q+2} that

$$\partial_0 W \times [0,1] + \sum_{i=1}^{p_q} (\phi_i^q) + \sum_{i=1}^{p_{q+1}} (\phi_i^{q+1})$$
$$\cong \partial_0 W \times [0,1] + \sum_{i=1,2,\ldots,p_q, i\neq i_0} (\phi_i^q) + \sum_{i=1}^{p_{q+1}} (\overline{\phi}_i^{q+1}) + (\psi^{q+2})$$

where $\cong$ means diffeomorphic relative $\partial_0 W$. Because of Lemma 2.22 there is an embedding ψ^{q+2} satisfying

$$\partial_0 W \times [0,1] + \sum_{i=1}^{p_q} (\phi_i^q) + \sum_{i=1}^{p_{q+1}} (\phi_i^{q+1})$$
$$\cong \partial_0 W \times [0,1] + \sum_{i=1}^{p_q} (\phi_i^q) + \sum_{i=1}^{p_{q+1}} (\phi_i^{q+1}) + (\psi_2^{q+1}) + (\psi^{q+2}).$$

We conclude from the Isotopy Lemma 2.16 and the Diffeomorphism Lemma 2.17 for appropriate embeddings ψ_k^{q+2} for $k = 1, 2$

$$\partial_0 W \times [0,1] + \sum_{i=1}^{p_q} (\phi_i^q) + \sum_{i=1}^{p_{q+1}} (\phi_i^{q+1}) + (\psi_2^{q+1}) + (\psi^{q+2})$$
$$\cong \partial_0 W \times [0,1] + \sum_{i=1}^{p_q} (\phi_i^q) + \sum_{i=1}^{p_{q+1}} (\phi_i^{q+1}) + (\psi_1^{q+1}) + (\psi_1^{q+2})$$
$$\cong \partial_0 W \times [0,1] + \sum_{i=1,2,\ldots,p_q, i\neq i_0} (\phi_i^q) + (\phi_{i_0}^q) + (\psi^{q+1}) + \sum_{i=1}^{p_{q+1}} (\phi_i^{q+1}) + (\psi_2^{q+2}).$$

We get from the Diffeomorphism Lemma 2.17 and the Cancellation Lemma 2.21 for appropriate embeddings $\overline{\phi}_i^{q+1}$ and ψ_3^{q+2}

$$\partial_0 W \times [0,1] + \sum_{i=1,2,\ldots,p_q, i\neq i_0}^{p_q} (\phi_i^q) + (\phi_{i_0}^q) + (\psi^{q+1}) + \sum_{i=1}^{p_{q+1}} (\phi_i^{q+1}) + (\psi_2^{q+2})$$

$$\cong \partial_0 W \times [0,1] + \sum_{i=1,2,\ldots,p_q, i\neq i_0}^{p_q} (\phi_i^q) + \sum_{i=1}^{p_{q+1}} (\overline{\phi}_i^{q+1}) + (\psi_3^{q+2}).$$

This finishes the proof of the Elimination Lemma 2.25. □

2.3 Handlebody Decompositions and CW-Structures

Next we explain how we can associate to a handlebody decomposition (2.23) a CW-pair $(X, \partial_0 W)$ such that there is a bijective correspondence between the q-handles of the handlebody decomposition and the q-cells of $(X, \partial_0 W)$. The key ingredient is that the projection $(D^q \times D^{n-q}, S^{q-1} \times D^{n-q}) \to (D^q, S^{q-1})$ is a homotopy equivalence and actually, as we will explain later, a simple homotopy equivalence.

Recall that a (relative) CW-complex (X, A) consists of a pair of topological spaces (X, A) together with a filtration

$$X_{-1} = A \subset X_0 \subset X_1 \subset \ldots \subset X_q \subset X_{q+1} \subset \ldots \subset \cup_{q \geq 0} X_q = X$$

such that for any $q \geq 0$ there exists a pushout of spaces

$$\begin{array}{ccc} \bigsqcup_{i \in I_q} S^{q-1} & \xrightarrow{\bigsqcup_{i \in I_q} \phi_i^q} & X_{q-1} \\ \downarrow & & \downarrow \\ \bigsqcup_{i \in I_q} D^q & \xrightarrow{\bigsqcup_{i \in I_q} \Phi_i^q} & X_q \end{array}$$

and X carries the colimit topology with respect to this filtration. The map ϕ_i^q is called the *attaching map* and the map (Φ_i^q, ϕ_i^q) is called the *characteristic map* of the q-cell belonging to $i \in I_q$. The pushouts above are not part of the structure, only their existence is required. Only the filtration $\{X_q \mid q \geq -1\}$ is part of the structure. The path components of $X_q - X_{q-1}$ are called the *open cells*. The open cells coincide with the sets $\Phi_i^q(D^q - S^{q-1})$. The closure of an open cell $\Phi_i^q(D^q - S^{q-1})$ is called a *closed cell* and turns out to be $\Phi_i^q(D^q)$.

Suppose that X is connected with fundamental group π. Let $p \colon \widetilde{X} \to X$ be the *universal covering* of X, i.e., a covering with simply connected total space. Put $\widetilde{X}_q = p^{-1}(X_q)$ and $\widetilde{A} = p^{-1}(A)$. Then $(\widetilde{X}, \widetilde{A})$ inherits a CW-structure from (X, A) by the filtration $\{\widetilde{X}_q \mid q \geq -1\}$. The *cellular $\mathbb{Z}\pi$-chain complex* $C_*(\widetilde{X}, \widetilde{A})$ has as q-th

$\mathbb{Z}\pi$-chain module the singular homology $H_q(\widetilde{X_q}, \widetilde{X_{q-1}})$ with $\mathbb{Z}$-coefficients and the π-action coming from the deck transformations. The q-th differential d_q is given by the composition

$$H_q(\widetilde{X_q}, \widetilde{X_{q-1}}) \xrightarrow{\partial_q} H_{q-1}(\widetilde{X_{q-1}}) \xrightarrow{i_q} H_{q-1}(\widetilde{X_{q-1}}, \widetilde{X_{q-2}})$$

where ∂_q is the boundary operator of the long exact sequence of the pair $(\widetilde{X_q}, \widetilde{X_{q-1}})$ and i_q is induced by the inclusion. If we choose for each $i \in I_q$ a lift $(\widetilde{\Phi_i^q}, \widetilde{\phi_i^q}) \colon (D^q, S^{q-1}) \to (\widetilde{X_q}, \widetilde{X_{q-1}})$ of the characteristic map (Φ_i^q, ϕ_i^q), we obtain a $\mathbb{Z}\pi$-basis $\{b_i \mid i \in I_q\}$ for $C_q(\widetilde{X}, \widetilde{A})$ if we define b_i as the image of a generator in $H_q(D^q, S^{q-1}) \cong \mathbb{Z}$ under the map $H_q(\widetilde{\Phi_i^q}, \widetilde{\phi_i^q}) \colon H_q(D^q, S^{q-1}) \to H_q(\widetilde{X_q}, \widetilde{X_{q-1}}) = C_q(\widetilde{X}, \widetilde{A})$. We call $\{b_i \mid i \in I_q\}$ *the cellular basis*. Note that we have made several choices in defining the cellular basis. We call two $\mathbb{Z}\pi$-bases $\{\alpha_j \mid j \in J\}$ and $\{\beta_k \mid k \in K\}$ for $C_q(\widetilde{X}, \widetilde{A})$ *equivalent* if there is a bijection $\phi \colon J \to K$ and elements $\epsilon_j \in \{\pm 1\}$ and $\gamma_j \in \pi$ for $j \in J$ such that $\epsilon_j \cdot \gamma_j \cdot \alpha_j = \beta_{\phi(j)}$. The equivalence class of the basis $\{b_i \mid i \in I_q\}$ constructed above only depends on the CW-structure on (X, A) and is independent of all further choices such as (Φ_i^q, ϕ_i^q), its lift $(\widetilde{\Phi_i^q}, \widetilde{\phi_i^q})$, and the generator of $H_q(D^q, S^{q-1})$.

Now suppose we are given a handlebody decomposition (2.23). We want to define a finite n-dimensional relative CW-complex $(X, \partial_0 W)$ and a homotopy equivalence

$$(f, \mathrm{id}) \colon (W, \partial_0 W) \xrightarrow{\simeq} (X, \partial_0 W). \tag{2.26}$$

For this purpose we construct by induction over $q = -1, 0, 1, \ldots, n$ a sequence of spaces $X_{-1} = \partial_0 W \subset X_0 \subset X_1 \subset X_2 \subset \ldots \subset X_n$ together with homotopy equivalences $f_q \colon W_q \to X_q$ such that $f_q|_{W_{q-1}} = f_{q-1}$ and $(X, \partial_0 W)$ is a CW-complex with respect to the filtration $\{X_q \mid q = -1, 0, 1, \ldots, n\}$. Then f will be f_n. The induction beginning with $f_1 \colon W_{-1} = \partial_0 W \times [0, 1] \to X_{-1} = \partial_0 W$ is given by the projection. The induction step from $(q-1)$ to q is done as follows. We attach for each handle (ϕ_i^q) for $i = 1, 2, \ldots, p_q$ a cell D^q to X_{q-1} by the attaching map $f_{q-1} \circ \phi_i^q|_{S^{q-1} \times \{0\}}$. In other words, we define X_q by the pushout

$$\begin{array}{ccc}
\coprod_{i=1}^{p_q} S^{q-1} & \xrightarrow{\coprod_{i=1}^{p_q} f_{q-1} \circ \phi_i^q|_{S^{q-1} \times \{0\}}} & X_{q-1} \\
\downarrow & & \downarrow \\
\coprod_{i=1}^{p_q} D^q & \longrightarrow & X_q.
\end{array}$$

2.3 Handlebody Decompositions and CW-Structures

Recall that W_q is the pushout

$$\begin{array}{ccc} \coprod_{i=1}^{p_q} S^{q-1} \times D^{n-q} & \xrightarrow{\coprod_{i=1}^{p_q} \phi_i^q} & W_{q-1} \\ \downarrow & & \downarrow \\ \coprod_{i=1}^{p_q} D^q \times D^{n-q} & \longrightarrow & W_q. \end{array}$$

Define a space Y_q by the pushout

$$\begin{array}{ccc} \coprod_{i=1}^{p_q} S^{q-1} & \xrightarrow{\coprod_{i=1}^{p_q} \phi_i^q|_{S^{q-1} \times \{0\}}} & W_{q-1} \\ \downarrow & & \downarrow \\ \coprod_{i=1}^{p_q} D^q & \longrightarrow & Y_q. \end{array}$$

Define $(g_q, f_{q-1}): (Y_q, W_{q-1}) \to (X_q, X_{q-1})$ by the pushout property applied to homotopy equivalences given by $f_{q-1}: W_{q-1} \to X_{q-1}$ and the identity maps on S^{q-1} and D^q. Define $(h_q, \mathrm{id}): (Y_q, W_{q-1}) \to (W_q, W_{q-1})$ by the pushout property applied to homotopy equivalences given by the obvious inclusions $S^{q-1} \to S^{q-1} \times D^{n-q}$ and $D^q \to D^q \times D^{n-q}$ and the identity on W_{q-1}. The resulting maps are homotopy equivalences of pairs since the left vertical arrows in the three pushouts above are cofibrations, see [60, page 249]. Choose a homotopy inverse $(h_q^{-1}, \mathrm{id}): (W_q, W_{q-1}) \to (Y_q, W_{q-1})$. Define f_q by the composite $g_q \circ h_q^{-1}$.

In particular we see that the inclusions $W_q \to W$ are q-connected since the inclusion of the q-skeleton $X_q \to X$ is q-connected for a CW-complex X.

Denote by $p: \widetilde{W} \to W$ the universal covering with $\pi = \pi_1(W)$ as group of deck transformations. Let $\widetilde{W}_q$ be the preimage of W_q under p. Note that this is the universal covering for $q \geq 2$ since each inclusion $W_q \to W$ induces an isomorphism on the fundamental groups. Define the *handlebody $\mathbb{Z}\pi$-chain complex* $C_*(\widetilde{W}, \widetilde{\partial_0 W})$ to be the $\mathbb{Z}\pi$-chain complex whose q-th chain group is $H_q(\widetilde{W}_q, \widetilde{W}_{q-1})$ and whose q-th differential is given by the composition

$$H_q(\widetilde{W}_q, \widetilde{W}_{q-1}) \xrightarrow{\partial_q} H_{q-1}(\widetilde{W}_{q-1}) \xrightarrow{i_q} H_{q-1}(\widetilde{W}_{q-1}, \widetilde{W}_{q-2})$$

where ∂_q is the boundary operator of the long homology sequence associated to the pair $(\widetilde{W}_q, \widetilde{W}_{q-1})$ and i_q is induced by the inclusion. The map $(f, \mathrm{id}): (W, \partial_0 W) \xrightarrow{\simeq} (X, \partial_0 W)$ of (2.26) induces an isomorphism of $\mathbb{Z}\pi$-chain complexes

$$C_*(\widetilde{f}, \mathrm{id}_{\widetilde{\partial_0 W}}): C_*(\widetilde{W}, \widetilde{\partial_0 W}) \xrightarrow{\cong} C_*(\widetilde{X}, \widetilde{\partial_0 W}). \tag{2.27}$$

Each handle (ϕ_i^q) determines an element

$$[\phi_i^q] \in C_q(\widetilde{W}, \widetilde{\partial_0 W}) \tag{2.28}$$

after choosing a lift $(\widetilde{\Phi}_i^q, \widetilde{\phi}_i^q) \colon (D^q \times D^{n-q}, S^{q-1} \times D^{n-q}) \to (\widetilde{W}_q, \widetilde{W}_{q-1})$ of its characteristic map $(\Phi_i^q, \phi_i^q) \colon (D^q \times D^{n-q}, S^{q-1} \times D^{n-q}) \to (W_q, W_{q-1})$, namely, the image under the map $H_q(\widetilde{\Phi}_i^q, \widetilde{\phi}_i^q)$ of the preferred generator in $H_q(D^q \times D^{n-q}, S^{q-1} \times D^{n-q}) \cong H_0(\{*\}) = \mathbb{Z}$. This element is only well defined up to multiplication by an element $\gamma \in \pi$. The elements $\{[\phi_i^q] \mid i = 1, 2, \ldots, p_q\}$ form a $\mathbb{Z}\pi$-basis for $C_q(\widetilde{W}, \widetilde{\partial_0 W})$. Its image under the isomorphism (2.27) is a cellular $\mathbb{Z}\pi$-basis.

If W has no handles of index ≤ 1, i.e., $p_0 = p_1 = 0$, one can express $C_*(\widetilde{W}, \widetilde{\partial_0 W})$ also in terms of homotopy groups as follows. Fix a base point $z \in \partial_0 W$ and a lift $\widetilde{z} \in \widetilde{\partial_0 W}$. All homotopy groups are taken with respect to these base points. Let $\pi_*(W_*, W_{*-1})$ be the $\mathbb{Z}\pi$-chain complex whose q-th $\mathbb{Z}\pi$-module is $\pi_q(W_q, W_{q-1})$ for $q \geq 2$ and zero for $q \leq 1$ and whose q-th differential is given by the composition

$$\pi_q(W_q, W_{q-1}) \xrightarrow{\partial_q} \pi_{q-1}(W_{q-1}) \xrightarrow{\pi_{q-1}(i)} \pi_{q-1}(W_{q-1}, W_{q-2}).$$

The $\mathbb{Z}\pi$-action comes from the canonical $\pi_1(A)$-action on the group $\pi_q(Y, A)$ and the $\pi_1(Y)$ action on $\pi_q(Y)$, see [178, Section 4.A] or [431, Theorem I.3.1 on page 164], and the identification $\pi_1(W_{q-1}) \to \pi_1(W) = \pi$ coming from the inclusions $W_{q-1} \to W$. Note that $\pi_q(Y, A)$ is abelian for any pair of spaces (Y, A) for $q \geq 3$ and is abelian also for $q = 2$ if A is simply connected or empty. For $q \geq 2$ the Hurewicz homomorphism is an isomorphism, see [178, Theorem 4.32 on page 366] or [431, Corollary IV.7.11 on page 181], $\pi_q(\widetilde{W}_q, \widetilde{W}_{q-1}) \to H_q(\widetilde{W}_q, \widetilde{W}_{q-1})$ and the projection $p \colon \widetilde{W} \to W$ induces isomorphisms $\pi_q(\widetilde{W}_q, \widetilde{W}_{q-1}) \to \pi_q(W_q, W_{q-1})$. Thus we obtain an isomorphism of $\mathbb{Z}\pi$-chain complexes

$$C_*(\widetilde{W}, \widetilde{\partial_0 W}) \xrightarrow{\cong} \pi_*(W_*, W_{*-1}). \tag{2.29}$$

Fix a path w_i in W from a point in the transverse sphere of (ϕ_i^q) to the base point z. Then the handle (ϕ_i^q) determines an element

$$[\phi_i^q] \in \pi_q(W_q, W_{q-1}). \tag{2.30}$$

It is represented by the obvious map $(D^q \times \{0\}, S^{q-1} \times \{0\}) \to (W_q, W_{q-1})$ together with w_i. It agrees with the element $[\phi_i^q] \in C_q(\widetilde{W}, \widetilde{\partial_0 W})$ defined in (2.28) under the isomorphism (2.29) if we use the lift of the characteristic map determined by the path w_i.

Exercise 2.31 Let M be a closed odd-dimensional manifold. Show that for any handlebody decomposition the number of handles of odd index is equal to the number of handles of even index.

Exercise 2.32 Let M be a closed connected manifold that possesses a handlebody decomposition without handles of index one. Show that M is simply connected.

Exercise 2.33 Show that a handlebody decomposition for $W = S^n \times S^n$ must have at least one 0-handle, two n-handles and one $2n$-handle. Describe one.

2.4 Reducing the Handlebody Decomposition

In the next step we want to get rid of the handles of index zero and one in the handlebody decomposition (2.23).

Lemma 2.34 *Let W be an n-dimensional manifold for $n \geq 6$ whose boundary is the disjoint union $\partial W = \partial_0 W \coprod \partial_1 W$. Then the following statements are equivalent:*

(i) *The inclusion $\partial_0 W \to W$ is 1-connected;*
(ii) *We can find a diffeomorphism relative $\partial_0 W$*

$$W \cong \partial_0 W \times [0,1] + \sum_{i=1}^{p_2}(\phi_i^2) + \sum_{i=1}^{p_3}(\overline{\phi}_i^3) + \cdots + \sum_{i=1}^{p_n}(\overline{\phi}_i^n).$$

Proof. (ii) $\Rightarrow$ (i) has already been proved in Section 2.3. It remains to conclude (ii) provided that (i) holds.

We first get rid of all 0-handles in the handlebody decomposition (2.23). It suffices to give a procedure to reduce the number of handles of index 0 by one. As the inclusion $\partial_0 W \to W$ is 1-connected, the inclusion $\partial_0 W \to W_1$ induces a bijection on the set of path components. Given any index i_0, there must be an index i_1 such that the core of the handle $\phi_{i_1}^1$ is a path connecting a point in $\partial_0 W \times \{1\}$ with a point in $(\phi_{i_0}^0)$. We conclude from the Diffeomorphism Lemma 2.17 and the Cancellation Lemma 2.21 that (ϕ_{i_0}) and $(\phi_{i_1}^1)$ cancel one another, i.e., we have

$$W \cong \partial_0 W \times [0,1] + \sum_{i=1,2,\ldots,p_0,\, i \neq i_0}(\phi_i^0) + \sum_{i=1,2,\ldots,p_1,\, i \neq i_1}(\phi_i^1) + \cdots + \sum_{i=1}^{p_n}(\phi_i^n).$$

Hence we can assume $p_0 = 0$ in (2.23).

Next we want to get rid of the 1-handles assuming that the inclusion $\partial_0 W \to W$ is 1-connected. It suffices to give a procedure to reduce the number of handles of index 1 by one. We want to do this by constructing an embedding $\psi^2 \colon S^1 \times D^{n-2} \hookrightarrow \partial_1^\circ W_1$ that satisfies the two conditions of the Elimination Lemma 2.25, and then applying the Elimination Lemma 2.25. Consider the embedding $\psi_+^2 \colon S_+^1 = D^1 = D^1 \times \{x\} \hookrightarrow D^1 \times D^{n-1} = (\phi_1^1)$ for some fixed $x \in S^{n-2} = \partial D^{n-1}$. The inclusion $\partial_1^\circ W_0 \to \partial_1 W_0 = \partial_0 W \times \{1\}$ induces an isomorphism on the fundamental group since $\partial_1^\circ W_0$ is obtained from $\partial_1 W_0 = \partial_0 W \times \{1\}$ by removing the interior of a finite number of embedded $(n-1)$-dimensional disks. Since by assumption the inclusion $\partial_0 W \to W$ is 1-connected, the inclusion $\partial_1^\circ W_0 \to W$ induces an epimorphism on the fundamental groups. Therefore we can find an embedding $\psi_-^2 \colon S_-^1 \hookrightarrow \partial_1^\circ W_0$ with $\psi_-^2|_{S^0} = \psi_+^2|_{S^0}$ such that the map $\psi_0^2 \colon S^1 = S_+^1 \cup_{S^0} S_-^1 \to \partial_1 W_1$ given by $\psi_+^2 \cup \psi_-^2$ is nullhomotopic in W. One can isotope the attaching maps $\phi_i^2 \colon S^1 \times D^{n-2} \to \partial_1 W_1$ of the 2-handles (ϕ_i^2) so that they do not meet the image of ψ_0^2 because the sum of the dimension of the source of ψ_0^2 and of $S^1 \times \{0\} \subset S^1 \times D^{n-2}$ is less than the dimension $(n-1)$ of $\partial_1 W_1$ and one can always shrink inside D^{n-2}. Thus we can assume without loss of generality by the Isotopy Lemma 2.16 and the Diffeomorphism Lemma 2.17 that the image of ψ_0^2 lies in $\partial_1^\circ W_1$. The inclusion $\partial_1 W_2 \to W$ is 2-connected. Hence ψ_0^2 is nullhomotopic in $\partial_1 W_2$. Let $h \colon D^2 \to \partial_1 W_2$ be a nullhomotopy for ψ_0^2. Since $2 \cdot \dim(D^2) < \dim(\partial_1 W_2)$, we can change h relative to S^1 into an embedding. (Here we need for the first time the assumption $n \geq 6$.) Since D^2 is contractible, the normal bundle of h and thus of $\psi_0^2 = \psi_+^2 \cup \psi_-^2$ are trivial. Therefore we can extend ψ_0^2 to an embedding $\psi^2 \colon S^1 \times D^{n-1} \hookrightarrow \partial_1^\circ W_1$ that is isotopic to a trivial embedding in $\partial_1 W_2$, meets the transverse sphere of the handle (ϕ_1^1) transversally and in exactly one point, and does not meet the transverse spheres of the handles (ϕ_i^1) for $2 \leq i \leq p_1$. Now Lemma 2.34 follows from the Elimination Lemma 2.25. □

Now consider an h-cobordism $(W; \partial_0 W, \partial_1 W)$. Because of Lemma 2.34 we can write it as

$$W \cong \partial_0 W \times [0,1] + \sum_{i=1}^{p_2} (\phi_i^2) + \sum_{i=1}^{p_3} (\phi_i^3) + \cdots .$$

Lemma 2.35 (Homology Lemma) *Suppose $n \geq 6$. Fix $2 \leq q \leq n-3$ and $i_0 \in \{1, 2, \ldots, p_q\}$. Let $f \colon S^q \hookrightarrow \partial_1 W_q$ be an embedding. Then the following statements are equivalent:*

(i) *The embedding f is isotopic to an embedding $g \colon S^q \hookrightarrow \partial_1 W_q$ such that g meets the transverse sphere of $(\phi_{i_0}^q)$ transversally and in exactly one point and is disjoint from transverse spheres of the handles (ϕ_i^q) for $i \neq i_0$;*

(ii) *Let $\widetilde{f} \colon S^q \to \widetilde{W}_q$ be a lift of f under $p|_{\widetilde{W}_q} \colon \widetilde{W}_q \to W_q$. Let $[\widetilde{f}]$ be the image of the class represented by $\widetilde{f}$ under the obvious composition*

$$\pi_q(\widetilde{W}_q) \to \pi_q(\widetilde{W}_q, \widetilde{W}_{q-1}) \to H_q(\widetilde{W}_q, \widetilde{W}_{q-1}) = C_q(\widetilde{W}).$$

Then there is a $\gamma \in \pi$ with

$$[\widetilde{f}] = \pm \gamma \cdot [\phi_{i_0}^q].$$

Proof. (i) $\Rightarrow$ (ii) We can isotope f so that $f|_{S_+^q} \colon S_+^q \to \partial_1 W_q$ looks like the canonical embedding $S_+^q = D^q \times \{x\} \hookrightarrow D^q \times S^{n-1-q} = \partial(\phi_{i_0}^q)$ for some $x \in S^{n-1-q}$ and $f(S_-^q)$ does not meet any of the handles (ϕ_i^q) for $i = 1, 2, \ldots, p_q$. One easily checks that then (ii) is true.

(ii) $\Rightarrow$ (i) We can isotope f so that it is transverse to the transverse spheres of the handles (ϕ_i^q) for $i = 1, 2, \ldots, p_q$. Since the sum of the dimension of the source of f and of the dimension of the transverse spheres is the dimension of $\partial_1 W_q$, the intersection of the image of f with the transverse sphere of the handle (ϕ_i^q) consists of finitely many points $x_{i,1}, x_{i,2}, \ldots, x_{i,r_i}$ for $i = 1, 2, \ldots, p_q$. Fix a base point $y \in S^q$.

2.4 Reducing the Handlebody Decomposition

It yields a base point $z = f(y) \in W$. Fix for each handle (ϕ_i^q) a path w_i in W from a point in its transverse sphere to z. Let $u_{i,j}$ be a path in S^q with the property that $u_{i,j}(0) = y$ and $f(u_{i,j}(1)) = x_{i,j}$ for $1 \le j \le r_i$ and $1 \le i \le p_q$. Let $v_{i,j}$ be any path in the transverse sphere of (ϕ_i^q) from $x_{i,j}$ to $w_i(0)$. Then the composite $f(u_{i,j}) * v_{i,j} * w_i$ is a loop in W with base point z and thus represents an element denoted by $\gamma_{i,j}$ in $\pi = \pi_1(W, z)$. It is independent of the choice of $u_{i,j}$ and $v_{i,j}$ since S^q and the transverse sphere of each handle (ϕ_i^q) are simply connected. The tangent space $T_{x_{i,j}} \partial_1 W_q$ is the tangent space of the handle (ϕ_i^q) at $x_{i,j}$ and is the direct sum of $T_{f^{-1}(x_{i,j})} S^q$ and the tangent space of the transverse sphere $\{0\} \times S^{n-1-q}$ of the handle (ϕ_i^q) at $x_{i,j}$. All these three tangent spaces come with preferred orientations. We define elements $\epsilon_{i,j} \in \{\pm 1\}$ by requiring that it is 1 if these orientations fit together and -1 otherwise. Now one easily checks that

$$[\widetilde{f}] = \sum_{i=1}^{p_q} \sum_{j=1}^{r_i} \epsilon_{i,j} \cdot \gamma_{i,j} \cdot [\phi_i^q]$$

where $[\phi_i^q]$ is the element associated to the handle (ϕ_i^q) after the choice of the path w_i, see (2.28) and (2.30). We have by assumption $[\widetilde{f}] = \pm \gamma \cdot [\phi_{i_0}^q]$ for some $\gamma \in \pi$. We want to isotope f so that f does not meet the transverse spheres of the handles (ϕ_i^q) for $i \ne i_0$ and does meet the transverse sphere of $(\phi_{i_0}^q)$ transversally and in exactly one point. Therefore it suffices to show in the case that the number $\sum_{j=1}^{p_q} r_i$ of all intersection points of f with the transverse spheres of the handles (ϕ_i^q) for $i = 1, 2, \ldots, p_i$ is bigger than one that we can change f by an isotopy so that this number becomes smaller. We have

$$\pm \gamma \cdot [\phi_{i_0}^q] = \sum_{i=1}^{p_q} \sum_{j=1}^{r_i} \epsilon_{i,j} \cdot \gamma_{i,j} \cdot [\phi_i^q].$$

Recall that the elements $[\phi_i^q]$ for $i = 1, 2, \ldots, p_q$ form a $\mathbb{Z}\pi$-basis. Hence we can find an index $i \in \{1, 2, \ldots, p_q\}$ and two different indices j_1, j_2 in $\{1, 2, \ldots, r_i\}$ such that the composite of the paths $f(u_{i,j_1}) * v_{i,j_1} * v_{i,j_2}^- * f(u_{i,j_2}^-)$ is nullhomotopic in W and hence in $\partial_1 W_q$ and the signs ϵ_{i,j_1} and ϵ_{i,j_2} are different. Now by the Whitney trick, see [301, Theorem 6.6 on page 71], [435], we can change f by an isotopy so that the two intersection points x_{i,j_1} and x_{i,j_2} disappear, the other intersection points of f with transverse spheres of the handles (ϕ_i^q) for $i \in \{1, 2, \ldots, p_q\}$ remain and no further intersection points are introduced. For the application of the Whitney trick we need the assumption $n - 1 \ge 5$ and when $q = 2$ or $q = n - 3$ an additional assumption on fundamental groups. One of the referees pointed out that we did not address the second assumption. In our situation it turns out that this requirement is always fulfilled, a subtlety explained in the book by Scorpan [373, pages 51-53] in dimensions $(n - 1) \ge 5$. (What happens for $(n - 1) = 4$ is discussed in [373, pages 57-58].) This finishes the proof of the Homology Lemma 2.35. □

Lemma 2.36 (Modification Lemma) *Let $f\colon S^q \hookrightarrow \partial_1^\circ W_q$ be an embedding and let $x_j \in \mathbb{Z}\pi$ be elements for $j = 1, 2 \ldots, p_{q+1}$. Then there is an embedding $g\colon S^q \hookrightarrow \partial_1^\circ W_q$ with the following properties:*

(i) *f and g are isotopic in $\partial_1 W_{q+1}$;*
(ii) *For a given lift $\widetilde{f}\colon S^q \to \widetilde{W}_q$ of f one can find a lift $\widetilde{g}\colon S^q \to \widetilde{W}_q$ of g such that we get in $C_q(\widetilde{W})$*

$$[\widetilde{g}] = [\widetilde{f}] + \sum_{j=1}^{p_{q+1}} x_j \cdot d_{q+1}[\phi_j^{q+1}]$$

where d_{q+1} is the $(q+1)$-th differential in $C_*(\widetilde{W}, \widetilde{\partial_0 W})$.

Proof. Any element in $\mathbb{Z}\pi$ can be written as a sum of elements of the form $\pm \gamma$ for $\gamma \in \pi$. Hence it suffices to prove for a fixed number $j \in \{1, 2 \ldots, p_q\}$, a fixed element $\gamma \in \pi$, and a fixed sign $\epsilon \in \{\pm 1\}$ that one can find an embedding $g\colon S^q \hookrightarrow \partial_1^\circ W_q$ that is isotopic to f in $\partial_1 W_{q+1}$ and satisfies for an appropriate lifting $\widetilde{g}$

$$[\widetilde{g}] = [\widetilde{f}] + \epsilon \cdot \gamma \cdot d_{q+1}[\phi_j^{q+1}].$$

Consider the embedding $t_j\colon S^q = S^q \times \{z\} \subset S^q \times S^{n-2-q} \hookrightarrow \partial(\phi_j^{q+1}) \subset \partial_1 W_q$ for some point $z \in S^{n-2-q} = \partial D^{n-1-q}$. It is in $\partial_1 W_{q+1}$ isotopic to a trivial embedding. Choose a path w in $\partial_1^\circ W_q$ connecting a point in the image of f with a point in the image of t_j. Without loss of generality we can arrange w to be an embedding. Moreover, we can thicken $w\colon [0,1] \to \partial_1^\circ W_q$ to an embedding $\overline{w}\colon [0,1] \times D^q \hookrightarrow \partial_1^\circ W_q$ such that $\overline{w}(\{0\} \times D^q)$ and $\overline{w}(\{1\} \times D^q)$ are embedded q-dimensional disks in the images of f and t_j and $\overline{w}((0,1) \times D^q)$ does not meet the images of f and t_j. Now one can form a new embedding, the connected sum $g := f \sharp_w t_j \colon S^q \to \partial_1^\circ W_q$. It is essentially given by restriction of f and t_j to the part of S^q which is not mapped under f and t_j to the interior of the disks $\overline{w}(\{0\} \times D^q)$, $\overline{w}(\{1\} \times D^q)$, and $\overline{w}|_{[0,1] \times S^{q-1}}$. Since t_j is isotopic to a trivial embedding in $\partial_1 W_{q+1}$, the embedding g is isotopic in $\partial_1 W_{q+1}$ to f. Recall that we have fixed a lifting $\widetilde{f}$ of f. This determines a unique lifting of $\widetilde{g}$, namely, we require that $\widetilde{f}$ and $\widetilde{g}$ coincide on those points where f and g already coincide. For an appropriate element $\gamma' \in \pi$ one gets $[\widetilde{g}] = [\widetilde{f}] + \gamma' \cdot d_{q+1}([\phi_j^{q+1}])$ since $t_j\colon S^q \to \partial_1 W_q \subset W_q$ is homotopic to $\phi_j^{q+1}|_{S^q \times \{0\}}\colon S^q \times \{0\} = S^q \to W_q$ in W_q. We can change the path w by composing it with a loop representing $\gamma \cdot (\gamma')^{-1} \in \pi$. Then we get for the new embedding g that

$$[\widetilde{g}] = [\widetilde{f}] + \gamma \cdot d_{q+1}([\phi_j^{q+1}]).$$

If we compose t_j with a diffeomorphism $S^q \to S^q$ of degree -1, we still get an embedding g that is isotopic to f in $\partial_1 W_{q+1}$ and satisfies

$$[\widetilde{g}] = [\widetilde{f}] - \gamma \cdot d_{q+1}([\phi_j^{q+1}]).$$

This finishes the proof of the Modification Lemma 2.36. □

2.4 Reducing the Handlebody Decomposition

Lemma 2.37 (Normal Form Lemma) *Let* $(W; \partial_0 W, \partial_1 W)$ *be a compact h-cobordism of dimension* $n \geq 6$. *Let* q *be an integer with* $2 \leq q \leq n-3$.

Then there is a handlebody decomposition that has only handles of index q *and* $(q+1)$, *i.e., there is a diffeomorphism relative* $\partial_0 W$

$$W \cong \partial_0 W \times [0,1] + \sum_{i=1}^{p_q} (\phi_i^q) + \sum_{i=1}^{p_{q+1}} (\phi_i^{q+1}).$$

Proof. In the first step we show that we can arrange $W_{-1} = W_{q-1}$, i.e., $p_r = 0$ for $r \leq q-1$. We do this by induction over q. The induction beginning with $q = 2$ has already been carried out in Lemma 2.34. In the induction step from q to $(q+1)$ we must explain how we can decrease the number of q-handles, provided that there are no handles of index $< q$. In order to get rid of the handle (ϕ_1^q), we want to attach a new $(q+1)$-handle and a new $(q+2)$-handle such that (ϕ_1^q) and the new $(q+1)$-handle cancel and the new $(q+1)$-handle and the new $(q+2)$-handle cancel each other. The effect will be that the number of q-handles is decreased by one at the cost of increasing the number of $(q+2)$-handles by one.

Fix a trivial embedding $\overline{\psi}^{q+1} \colon S^q \times D^{n-1-q} \hookrightarrow \partial_1^\circ W_q$. As the inclusion $\partial_0 W \to W$ is a homotopy equivalence, $H_p(\widetilde{W}, \widetilde{\partial_0 W}) = 0$ for all $p \geq 0$. Since the p-th homology of $C_*(\widetilde{W}, \widetilde{\partial_0 W})$ is $H_p(\widetilde{W}, \widetilde{\partial_0 W}) = 0$, the $\mathbb{Z}\pi$-chain complex $C_*(\widetilde{W}, \widetilde{\partial_0 W})$ is acyclic. Since $C_{q-1}(\widetilde{W}, \widetilde{\partial_0 W})$ is trivial, the q-th differential of $C_*(\widetilde{W}, \widetilde{\partial_0 W})$ is zero and hence the $(q+1)$-th differential d_{q+1} is surjective. We can choose elements $x_j \in \mathbb{Z}\pi$ such that

$$[\phi_1^q] = \sum_{i=1}^{p_{q+1}} x_j \cdot d_{q+1}([\phi_i^{q+1}]).$$

Since $\alpha := \overline{\psi}^{q+1}|_{S^q \times \{0\}} \to \partial_1^\circ W_q$ is nullhomotopic, $[\widetilde{\alpha}] = 0$ in $H_q(\widetilde{W_q}, \widetilde{W_{q-1}})$. We conclude from the Modification Lemma 2.36 that we can find an embedding $\psi^{q+1} \colon S^q \times D^{n-1-q} \hookrightarrow \partial_1^\circ W_q$ such that $\beta := \psi^{q+1}|_{S^q \times \{0\}}$ is isotopic in $\partial_1 W_{q+1}$ to α and we get

$$[\widetilde{\beta}] = [\widetilde{\alpha}] + \sum_{i=1}^{p_{q+1}} x_j \cdot d_{q+1}([\phi_i^{q+1}]) = [\phi_1^q].$$

Because of the Homology Lemma 2.35 the embedding $\beta = \psi^q|_{S^q \times \{0\}}$ is isotopic in $\partial_1 W_q$ to an embedding $\gamma \colon S^q \hookrightarrow \partial_1 W_q$ that meets the transverse sphere of (ϕ_1^q) transversally and in exactly one point and is disjoint from the transverse spheres of all other handles of index q. By construction ψ^{q+1} is isotopic in $\partial_1 W_{q+1}$ to the trivial embedding $\overline{\psi}^{q+1}$. Now we can apply the Elimination Lemma 2.25. This finishes the proof that we can arrange $W_{-1} = W_{q-1}$.

Next we explain the *dual handlebody decomposition*. Suppose that W is obtained from $\partial_0 W \times [0,1]$ by attaching one q-handle (ϕ^q), i.e., $W = \partial_0 W \times [0,1] + (\phi^q)$. Then we can interchange the role of $\partial_0 W$ and $\partial_1 W$ and try to build W from $\partial_1 W$ by handles. It turns out that W can be written as

$$W = \partial_1 W \times [0,1] + (\psi^{n-q}) \tag{2.38}$$

by the following argument.

Let M be the manifold with boundary $S^{q-1} \times S^{n-1-q}$ obtained from $\partial_0 W$ by removing the interior of $\phi^q(S^{q-1} \times D^{n-q})$. We get

$$W \cong \partial_0 W \times [0,1] \cup_{S^{q-1} \times D^{n-q}} D^q \times D^{n-q}$$
$$= M \times [0,1] \cup_{S^{q-1} \times S^{n-1-q} \times [0,1]}$$
$$\left(S^{q-1} \times D^{n-q} \times [0,1] \cup_{S^{q-1} \times D^{n-q} \times \{1\}} D^q \times D^{n-q} \right).$$

Inside $S^{q-1} \times D^{n-q} \times [0,1] \cup_{S^{q-1} \times D^{n-q} \times \{1\}} D^q \times D^{n-q}$ we have the following submanifolds

$$X := S^{q-1} \times 1/2 \cdot D^{n-q} \times [0,1] \cup_{S^{q-1} \times 1/2 \cdot D^{n-q} \times \{1\}} D^q \times 1/2 \cdot D^{n-q};$$
$$Y := S^{q-1} \times 1/2 \cdot S^{n-1-q} \times [0,1] \cup_{S^{q-1} \times 1/2 \cdot S^{n-1-q} \times \{1\}} D^q \times 1/2 \cdot S^{n-1-q}.$$

The pair (X, Y) is diffeomorphic to $(D^q \times D^{n-q}, D^q \times S^{n-1-q})$, i.e., it is a handle of index $(n-q)$. Let N be obtained from W by removing the interior of X. Then W is obtained from N by adding an $(n-q)$-handle, the so-called *dual handle*. One easily checks that N is diffeomorphic to $\partial_1 W \times [0,1]$ relative $\partial_1 W \times \{1\}$. Thus (2.38) follows.

Suppose that W is relatively $\partial_0 W$ of the shape

$$W \cong \partial_0 W \times [0,1] + \sum_{i=1}^{p_0} (\phi_i^0) + \sum_{i=1}^{p_1} (\phi_i^1) + \cdots + \sum_{i=1}^{p_n} (\phi_i^n).$$

Then we can conclude inductively using the Diffeomorphism Lemma 2.17 and (2.38) that W is diffeomorphic relative to $\partial_1 W$ to

$$W \cong \partial_1 W \times [0,1] + \sum_{i=1}^{p_n} (\overline{\phi}_i^0) + \sum_{i=1}^{p_{n-1}} (\overline{\phi}_i^1) + \cdots + \sum_{i=1}^{p_0} (\overline{\phi}_i^n). \tag{2.39}$$

This corresponds to replacing a Morse function f by $-f$. The effect is that the number of q-handles now becomes the number of $(n-q)$-handles.

Now applying the first step to the dual handlebody decomposition for q replaced by $(n-q-1)$ and then considering the dual handlebody decomposition of the result finishes the proof of the Normal Form Lemma 2.37. □

2.5 Handlebody Decompositions and Whitehead Groups

Let $(W, \partial_0 W, \partial_1 W)$ be an n-dimensional compact h-cobordism for $n \geq 6$. By the Normal Form Lemma 2.37 we can fix a handlebody decomposition for some fixed number $2 \leq q \leq n-3$

2.5 Handlebody Decompositions and Whitehead Groups

$$W \cong \partial_0 W \times [0,1] + \sum_{i=1}^{p_q} (\phi_i^q) + \sum_{i=1}^{p_{q+1}} (\phi_i^{q+1}).$$

Recall that the $\mathbb{Z}\pi$-chain complex $C_*(\widetilde{W}, \widetilde{\partial_0 W})$ is acyclic. Hence the only non-trivial differential $d_{q+1} \colon H_{q+1}(\widetilde{W_{q+1}}, \widetilde{W_q}) \to H_q(\widetilde{W_q}, \widetilde{W_{q-1}})$ is bijective. Recall that $\{[\phi_i^{q+1}] \mid i = 1, 2, \ldots, p_{q+1}\}$ is a $\mathbb{Z}\pi$-basis for $H_{q+1}(\widetilde{W_{q+1}}, \widetilde{W_q})$ and $\{[\phi_i^q] \mid i = 1, 2, \ldots, p_q\}$ is a $\mathbb{Z}\pi$-basis for $H_q(\widetilde{W_q}, \widetilde{W_{q-1}})$. In particular $p_q = p_{q+1}$. The matrix A, which describes the differential d_{q+1} with respect to these bases, is an invertible (p_q, p_q)-matrix over $\mathbb{Z}\pi$. Since we are working with left modules, d_{q+1} sends an element $x \in \mathbb{Z}G^n$ to $x \cdot A \in \mathbb{Z}G^n$, or equivalently, $d_{q+1}([\phi_i^{q+1}]) = \sum_{j=1}^n a_{i,j}[\phi_j^q]$.

Next we define an abelian group $\mathrm{Wh}(\pi)$ as follows. It is the set of equivalence classes of invertible matrices of arbitrary size with entries in $\mathbb{Z}\pi$ where we call an invertible (m,m)-matrix A and an invertible (n,n)-matrix B over $\mathbb{Z}\pi$ equivalent if we can pass from A to B by a sequence of the following operations:

(i) B is obtained from A by adding the k-th row multiplied by x from the left to the l-th row for $x \in \mathbb{Z}\pi$ and $k \neq l$;
(ii) B is obtained by taking the direct sum of A and the $(1,1)$-matrix $I_1 = (1)$, i.e., B looks like the block matrix $\begin{pmatrix} A & 0 \\ 0 & 1 \end{pmatrix}$;
(iii) A is the direct sum of B and I_1. This is the inverse operation to (ii);
(iv) B is obtained from A by multiplying the i-th row from the left by a trivial unit, i.e., by an element of the shape $\pm\gamma$ for $\gamma \in \pi$;
(v) B is obtained from A by interchanging two rows or two columns.

The group structure is given on representatives A and B as follows. By taking the direct sum $A \oplus I_m$ and $B \oplus I_n$ with the identity matrices I_m and I_n of size m and n for appropriate m and n one can arrange that $A \oplus I_m$ and $B \oplus I_n$ are invertible matrices of the same size and can be multiplied. Define $[A] \cdot [B]$ by $[(A \oplus I_m) \cdot (B \oplus I_n)]$. The zero element $0 \in \mathrm{Wh}(\pi)$ is represented by I_n for any positive integer n. The inverse of $[A]$ is given by $[A^{-1}]$. We will show later in Lemma 3.8 that the multiplication is well defined and yields an abelian group $\mathrm{Wh}(\pi)$.

Exercise 2.40 Show that the Whitehead group of the trivial group vanishes.

Exercise 2.41 Let $t \in \mathbb{Z}/5$ be a generator. Consider the $(1,1)$-matrix given as $(1 - t - t^{-1})$ over $\mathbb{Z}[\mathbb{Z}/5]$. Show that it represents a non-trivial element in $\mathrm{Wh}(\mathbb{Z}/5)$.

Lemma 2.42 (i) *Let $(W, \partial_0 W, \partial_1 W)$ be an n-dimensional compact h-cobordism for $n \geq 6$ and A be the matrix defined above. If $[A] = 0$ in $\mathrm{Wh}(\pi)$, then the h-cobordism W is trivial relative $\partial_0 W$.*
(ii) *Consider an element $u \in \mathrm{Wh}(\pi)$, a closed manifold M of dimension $n - 1 \geq 5$ with fundamental group π and an integer q with $2 \leq q \leq n - 3$. Then we can find an h-cobordism of the shape*

$$W = M \times [0,1] + \sum_{i=1}^{p_q}(\phi_i^q) + \sum_{i=1}^{p_{q+1}}(\phi_i^{q+1})$$

such that $[A] = u$.

Proof. (i) Let B be a matrix which is obtained from A by applying one of the operations (i), (ii), (iii), (iv), and (v). It suffices to show that we can modify the given handlebody decomposition in normal form of W with associated matrix A such that we get a new handlebody decomposition in normal form whose associated matrix is B.

We begin with (i). Consider $W' = \partial_0 W \times [0,1] + \sum_{i=1}^{p_q}(\phi_i^q) + \sum_{j=1, j \neq l}^{p_{q+1}}(\phi_j^{q+1})$. Note that we get from W' our h-cobordism W if we attach the handle (ϕ_l^{q+1}). By the Modification Lemma 2.36 we can find an embedding $\overline{\phi}_l^{q+1} : S^q \times D^{n-1-q} \hookrightarrow \partial_1 W'$ such that $\overline{\phi}_l^{q+1}$ is isotopic to ϕ_l^{q+1} and we get

$$\left[\widetilde{\overline{\phi}_l^{q+1}}|_{S^q \times \{0\}}\right] = \left[\widetilde{\phi_l^{q+1}}|_{S^q \times \{0\}}\right] + x \cdot d_{q+1}([\phi_k^{q+1}]).$$

If we attach to W' the handle $(\overline{\phi}_l^{q+1})$, the result is diffeomorphic to W relative $\partial_0 W$ by the Isotopy Lemma 2.16. One easily checks that the associated invertible matrix B is obtained from A by adding the k-th row multiplied by x from the left to the l-th row.

The claim for the operations (ii) and (iii) follows from the Cancellation Lemma 2.21 and the Homology Lemma 2.35. The claim for the operation (iv) follows from the observation that we can replace the attaching map of a handle $\phi^q : S^q \times D^{n-1-q} \hookrightarrow \partial_1 W_q$ by its composition with $f \times \mathrm{id}$ for some diffeomorphism $f : S^q \to S^q$ of degree -1 and that the base element $[\phi_i^q]$ can also be changed to $\gamma \cdot [\phi_i^q]$ by choosing a different lift along $\widetilde{W_q} \to W_q$. Operation (v) can be realised by interchanging the numeration of the q-handles and $(q+1)$-handles.

(ii) Fix an invertible matrix $A = (a_{i,j}) \in \mathrm{GL}(n, \mathbb{Z}\pi)$. Choose trivial pairwise disjoint embeddings $\phi_i^2 : S^1 \times D^{n-2} \hookrightarrow M_0 \times \{1\}$. Consider

$$W_2 = M_0 \times [0,1] + (\phi_1^2) + (\phi_2^2) + \cdots + (\phi_n^2).$$

As the embeddings ϕ_i^2 are trivial, there are embeddings $\phi_i^3 : S^2 \times D^{n-3} \hookrightarrow \partial_1 W_2$ and lifts $\widetilde{\phi}_i^3 : S^2 \times D^{n-3} \to \widetilde{\partial_1 W_2}$ such that in $\pi_2(\widetilde{W_2}, \widetilde{\partial_0 W})$

$$[\widetilde{\phi}_i^3|_{S^2 \times \{0\}}] = \sum_{j=1}^n a_{i,j} \cdot [\phi_j^2].$$

Put $W = W_2 + (\phi_1^3) + (\phi_2^3) + \cdots + (\phi_n^3)$. One easily checks that W is an h-cobordism over M_0 with a handlebody decomposition realising the matrix A. This finishes the proof Lemma 2.42. □

If π is trivial, then $\text{Wh}(\pi)$ is trivial. Hence Lemma 2.42 (i) implies already the h-Cobordism Theorem 2.4.

Remark 2.43 (Strategy to finish the proof of the s-Cobordism Theorem 2.1)
As soon as we have shown that $[A] \in \text{Wh}(\pi)$ agrees with the Whitehead torsion $\tau(W, M_0)$ of the h-cobordism W over M_0 and that this invariant depends only on the diffeomorphism type of W relative M_0, the s-Cobordism Theorem 2.1 (i) will follow.

Obviously Lemma 2.42 (ii) implies the s-Cobordism Theorem 2.1 (ii). We will later see that assertion (iii) of the s-Cobordism Theorem 2.1 follows from assertions (i) and (ii) if we have more information about the Whitehead torsion, namely the sum and the composition formulas. All this will be carried out in Section 3.3

2.6 Notes

The h-Cobordism Theorem 2.4 is due to Smale [379]. The s-Cobordism Theorem 2.1 is due to Barden, Mazur, and Stallings, see [20, 291]. In the PL category proofs can be found in [367, 6.19 on page 88]. Its topological version follows from Kirby and Siebenmann [219, Conclusion 7.4 on page 320]. More information about the s-Cobordism Theorem can be found for instance in [215], [301], [367, page 87-90]. The s-Cobordism Theorem is known to be false for $\dim(M_0) = 4$ in general, by the work of Donaldson [135], but it is true for $n = \dim(M_0) = 4$ for good fundamental groups in the topological category by results of Quinn and Freedman [37, 157, 158, 159]. Counterexamples in the case $\dim(M_0) = 3$ are constructed by Matsumoto and Siebenmann [288] and Cappell and Shaneson [82] where the relevant 4-dimensional s-cobordism is a topological manifold. It is not known whether one can choose the s-cobordism to be smooth in these counterexamples. It follows from Kwasik and Schultz [236] and Perelman's proof of the Thurston Geometrisation Conjecture, see [222, 311], that every h-cobordism between two orientable closed 3-manifolds is an s-cobordism. The Poincaré Conjecture, see Theorem 2.5, is known in all dimensions where dimension 3 is due to the work of Perelman, see [222, 310, 311, 331, 332, 333], and dimension 4 is due to Freedman, see [37, 157, 158, 159]. The first proof of the Poincaré Conjecture in the topological category in dimension ≥ 5 was given by Newman [319] using engulfing theory. The smooth version of the Poincaré Conjecture holds in dimensions ≤ 3, is open in dimension 4, and is discussed in dimensions ≥ 5 in Remark 12.36.

Chapter 3
Whitehead Torsion

3.1 Introduction

In this chapter we will assign to a homotopy equivalence $f \colon X \to Y$ of finite CW-complexes its Whitehead torsion $\tau(f)$ in the Whitehead group $\operatorname{Wh}(\Pi(Y))$ associated to Y. The main properties of this invariant are summarised in Theorem 3.1 below. For its formulation we need the following notion.

Suppose that the following diagram is a pushout

$$\begin{array}{ccc} A & \xrightarrow{f} & B \\ i \downarrow & & \downarrow j \\ X & \xrightarrow{g} & Y \end{array}$$

the map i is an inclusion of CW-complexes, and f is a *cellular* map of CW-complexes, i.e., respects the filtration given by the CW-structures. Then Y inherits a CW-structure by defining Y_n as the union of $j(B_n)$ and $g(X_n)$. If we equip Y with this CW-structure, we call the pushout above a *cellular pushout*.

Theorem 3.1 (Basic properties of Whitehead torsion)

(i) *Sum formula*

Let the following two diagrams

$$\begin{array}{ccc} X_0 & \xrightarrow{i_1} & X_1 \\ i_2 \downarrow & & \downarrow j_1 \\ X_2 & \xrightarrow{j_2} & X \end{array} \qquad \begin{array}{ccc} Y_0 & \xrightarrow{k_1} & Y_1 \\ k_2 \downarrow & & \downarrow l_1 \\ Y_2 & \xrightarrow{l_2} & Y \end{array}$$

be cellular pushouts of finite CW-complexes. Denote by l_0 the map $l_1 \circ k_1 = l_2 \circ k_2 \colon Y_0 \to Y$. Let $f_i \colon X_i \to Y_i$ be homotopy equivalences for $i = 0, 1, 2$

*satisfying $f_1 \circ i_1 = k_1 \circ f_0$ and $f_2 \circ i_2 = k_2 \circ f_0$. Denote by $f \colon X \to Y$ the map induced by f_0, f_1 and f_2 and the pushout property.
Then f is a homotopy equivalence and*

$$\tau(f) = (l_1)_* \tau(f_1) + (l_2)_* \tau(f_2) - (l_0)_* \tau(f_0);$$

(ii) *Homotopy invariance*

Let $f \simeq g \colon X \to Y$ be homotopic maps of finite CW-complexes. Then the homomorphisms $f_*, g_* \colon \mathrm{Wh}(\Pi(X)) \to \mathrm{Wh}(\Pi(Y))$ agree. If additionally f and g are homotopy equivalences, then

$$\tau(g) = \tau(f);$$

(iii) *Composition formula*

Let $f \colon X \to Y$ and $g \colon Y \to Z$ be homotopy equivalences of finite CW-complexes. Then

$$\tau(g \circ f) = g_* \tau(f) + \tau(g);$$

(iv) *Product formula*

Let $f \colon X' \to X$ and $g \colon Y' \to Y$ be homotopy equivalences of connected finite CW-complexes. Then

$$\tau(f \times g) = \chi(X) \cdot j_* \tau(g) + \chi(Y) \cdot i_* \tau(f)$$

where the integers $\chi(X)$ and $\chi(Y)$ denote the Euler characteristics, the homomorphism $j_* \colon \mathrm{Wh}(\Pi(Y)) \to \mathrm{Wh}(\Pi(X \times Y))$ is induced by the map $j \colon Y \to X \times Y$, $y \mapsto (x_0, y)$ for some base point $x_0 \in X$, and i_* is defined analogously;

(v) *Topological invariance*

Let $f \colon X \to Y$ be a homeomorphism of finite CW-complexes. Then

$$\tau(f) = 0.$$

Given an h-cobordism $(W; M_0, f_0, M_1, f_1)$ over M_0, we define its Whitehead torsion $\tau(W, M_0)$ by the Whitehead torsion of the inclusion $\partial_0 W \to W$, see Definition 3.24. This is the invariant appearing in the s-Cobordism Theorem 2.1.

We will give some information about the Whitehead group $\mathrm{Wh}(\Pi(Y))$ in Section 3.2. We will present the algebraic definition of Whitehead torsion and the proof of Theorem 3.1 in Section 3.3.

A geometric approach to the Whitehead torsion is summarised in Section 3.4. There we will introduce the notion of a simple homotopy equivalence, show that a homotopy equivalence of finite CW-complexes is simple if and only if its Whitehead torsion vanishes, and every element in the Whitehead group $\mathrm{Wh}(\Pi(Y))$ of a finite CW-complex occurs as the Whitehead torsion of a homotopy equivalence of finite CW-complexes with Y as target.

A similar invariant, the Reidemeister torsion, will be treated in Section 3.5. It will be used to classify lens spaces.

Exercise 3.2 Let $(W; M_0, f_0, M_1, f_1)$ be an h-cobordism. Show that the h-cobordism $(W \times S^n; M_0 \times S^n, f_0 \times \mathrm{id}_{S^n}, M_1 \times S^n, f_1 \times \mathrm{id}_{S^n})$ is trivial if n is odd and $n \geq 5$.

Guide 3.3 It is not necessary to go through the remainder of this chapter to understand the following chapters. The minimal requirement is to understand the statement of Theorem 3.1, its relevance for the s-Cobordism Theorem 2.1 and that one can assign to a $\mathbb{Z}G$-chain homotopy equivalence of finite based free $\mathbb{Z}G$-chain complexes its Whitehead torsion, which takes values in the Whitehead group $\mathrm{Wh}(G)$. It is even possible for the first reading to pretend that every homotopy equivalence is simple, every h-cobordism is trivial, and the Whitehead group $\mathrm{Wh}(G)$ is trivial. All this is known to be true in the simply connected case and for many torsionfree groups G.

3.2 Whitehead Groups

In this section we define $K_1(R)$ for an associative ring R with unit and the Whitehead group $\mathrm{Wh}(G)$ of a group G and relate the definitions of this section with the one of Section 2.5. Furthermore we give some basic information about its computation.

Let R be an associative ring with unit. Denote by $\mathrm{GL}(n, R)$ the group of invertible (n, n)-matrices with entries in R. Define the group $\mathrm{GL}(R)$ by the colimit of the system $\cdots \subset \mathrm{GL}(n, R) \subset \mathrm{GL}(n+1, R) \subset \cdots$ indexed by the natural numbers where the inclusion $\mathrm{GL}(n, R)$ to $\mathrm{GL}(n+1, R)$ is given by stabilisation

$$A \mapsto \begin{pmatrix} A & 0 \\ 0 & 1 \end{pmatrix}.$$

Define $K_1(R)$ to be the abelianisation $\mathrm{GL}(R)/[\mathrm{GL}(R), \mathrm{GL}(R)]$ of $\mathrm{GL}(R)$. Let $\widetilde{K}_1(R)$ be the cokernel of the map $K_1(\mathbb{Z}) \to K_1(R)$ induced by the canonical ring homomorphism $\mathbb{Z} \to R$. The map $\det \colon K_1(\mathbb{Z}) \to \{\pm 1\}$, $[A] \mapsto \det(A)$ is an isomorphism because $\mathbb{Z}$ is a ring with Euclidean algorithm. Hence $\widetilde{K}_1(R)$ is the same as the quotient of $K_1(R)$ by the cyclic subgroup of order at most two generated by the class of the $(1, 1)$-matrix (-1).

Definition 3.4 (Whitehead group) Define the *Whitehead group* $\mathrm{Wh}(G)$ of a group G to be the cokernel of the map $G \times \{\pm 1\} \to K_1(\mathbb{Z}G)$ which sends $(g, \pm 1)$ to the class of the invertible $(1, 1)$-matrix $(\pm g)$.

It is conjectured that the Whitehead group $\mathrm{Wh}(G)$ vanishes for any torsionfree group G. This is a special case of the K-theoretic Farrell–Jones Conjecture, which has been proved for a large class of groups. This class includes hyperbolic groups, CAT(0)-groups, solvable groups, lattices in almost connected Lie groups, arithmetic groups, and fundamental groups of (not necessarily compact) manifolds (possibly with boundary) of dimension ≤ 3, and mapping class groups. It is closed under taking subgroups, finite free products, finite direct products and directed colimits over directed systems (with arbitrary structure maps). Proofs can be found in [22, 23, 24, 27, 28, 30, 204, 261, 370, 420], see also Section 16.5.

The Whitehead group satisfies $\mathrm{Wh}(G * H) = \mathrm{Wh}(G) \oplus \mathrm{Wh}(H)$, see [385]. The Bass–Heller–Swan decomposition, see [361, Theorem 3.2.22 on page 149], yields for a group G an isomorphism of the shape

$$\mathrm{Wh}(G \times \mathbb{Z}) \cong \widetilde{K}_0(\mathbb{Z}G) \oplus \mathrm{Wh}(G) \oplus NK_1(\mathbb{Z}G) \oplus NK_1(\mathbb{Z}G). \tag{3.5}$$

If G is finite, then $\mathrm{Wh}(G)$ is very well understood, see [324]. Namely, $\mathrm{Wh}(G)$ is finitely generated, its rank as an abelian group is the number of conjugacy classes of unordered pairs $\{g, g^{-1}\}$ in G minus the number of conjugacy classes of cyclic subgroups, and its torsion subgroup is isomorphic to the kernel $SK_1(G)$ of the change of coefficients homomorphism $K_1(\mathbb{Z}G) \to K_1(\mathbb{Q}G)$. For a finite cyclic group G the Whitehead group $\mathrm{Wh}(G)$ is torsionfree. For instance the Whitehead group $\mathrm{Wh}(\mathbb{Z}/p)$ of a cyclic group of order p for an odd prime p is the free abelian group of rank $(p-3)/2$ and $\mathrm{Wh}(\mathbb{Z}/2) = 0$. The Whitehead group of the symmetric group S_n is trivial. The Whitehead group of $\mathbb{Z}/4 \times \mathbb{Z}$ is not finitely generated as an abelian group due to (3.5) and the fact that $NK_1(\mathbb{Z}[\mathbb{Z}/4])$ is not finitely generated, see [31, Theorem 10.6 (d) on page 695].

Next we want to relate the definitions above to the one of Section 2.5. Denote by $E_n(i, j)$ for $n \geq 1$ and $1 \leq i, j \leq n$ the (n, n)-matrix whose entry at (i, j) is one and is zero elsewhere. Denote by I_n the identity matrix of size n. An *elementary* (n, n)-matrix is a matrix of the form $I_n + r \cdot E_n(i, j)$ for $n \geq 1$, $1 \leq i, j \leq n$, $i \neq j$ and $r \in R$. Let A be an (n, n)-matrix. The matrix $B = A \cdot (I_n + r \cdot E_n(i, j))$ is obtained from A by adding the i-th column multiplied by r from the right to the j-th column. The matrix $C = (I_n + r \cdot E_n(i, j)) \cdot A$ is obtained from A by adding the j-th row multiplied by r from the left to the i-th row. Let $E(R) \subset GL(R)$ be the subgroup generated by all elements in $GL(R)$ that are represented by elementary matrices.

Lemma 3.6 *We have* $E(R) = [GL(R), GL(R)]$. *In particular,* $E(R) \subset GL(R)$ *is a normal subgroup and* $K_1(R) = GL(R)/E(R)$.

Proof. For $n \geq 3$, pairwise distinct numbers $1 \leq i, j, k \leq n$ and $r \in R$ we can write $I_n + r \cdot E_n(i, k)$ as a commutator in $GL(n, R)$, namely

$$I_n + r \cdot E_n(i, k) = (I_n + r \cdot E_n(i, j)) \cdot (I_n + E_n(j, k)) \cdot$$
$$(I_n + r \cdot E_n(i, j))^{-1} \cdot (I_n + E_n(j, k))^{-1}.$$

This implies $E(R) \subset [GL(R), GL(R)]$.

Let A and B be two elements in $GL(n, R)$. Let $[A]$ and $[B]$ be the elements in $GL(R)$ represented by A and B. Given two elements x and y in $GL(R)$, we write $x \sim y$ if there are elements e_1 and e_2 in $E(R)$ with $x = e_1 y e_2$, in other words, if the classes of x and y in $E(R) \backslash GL(R) / E(R)$ agree. One easily checks

$$[AB] \sim \left[\begin{pmatrix} AB & 0 \\ 0 & I_n \end{pmatrix}\right] \sim \left[\begin{pmatrix} AB & A \\ 0 & I_n \end{pmatrix}\right] \sim \left[\begin{pmatrix} 0 & A \\ -B & I_n \end{pmatrix}\right] \sim \left[\begin{pmatrix} 0 & A \\ -B & 0 \end{pmatrix}\right]$$

3.2 Whitehead Groups

since each step is given by multiplication from the right or left by a block matrix of the form $\begin{pmatrix} I_n & 0 \\ C & I_n \end{pmatrix}$ or $\begin{pmatrix} I_n & C \\ 0 & I_n \end{pmatrix}$ and such a block matrix is obviously obtained from I_{2n} by a sequence of column and row operations and hence its class in $\mathrm{GL}(R)$ belongs to $\mathrm{E}(R)$. Analogously we get

$$[BA] \sim \left[\begin{pmatrix} 0 & B \\ -A & 0 \end{pmatrix} \right].$$

Since the element in $\mathrm{GL}(R)$ represented by $\begin{pmatrix} 0 & -I_n \\ I_n & 0 \end{pmatrix}$ belongs to $\mathrm{E}(R)$, we conclude

$$\left[\begin{pmatrix} 0 & A \\ -B & 0 \end{pmatrix} \right] \sim \left[\begin{pmatrix} A & 0 \\ 0 & B \end{pmatrix} \right] \sim \left[\begin{pmatrix} 0 & B \\ -A & 0 \end{pmatrix} \right].$$

This shows

$$[AB] \sim [BA]. \tag{3.7}$$

For elements $x \in \mathrm{GL}(R)$ and $e \in \mathrm{E}(R)$, we get $xex^{-1} \sim ex^{-1}x = e$ and hence $xex^{-1} \in \mathrm{E}(R)$. Therefore $\mathrm{E}(R)$ is normal. Given a commutator $xyx^{-1}y^{-1}$ for elements $x, y \in \mathrm{GL}(R)$, we conclude for appropriate elements e_1, e_2, e_3 in $\mathrm{E}(R)$

$$xyx^{-1}y^{-1} = e_1 yxe_2 x^{-1} y^{-1} = e_1 yxx^{-1}y^{-1}(yx)e_2(yx)^{-1} = e_1 e_3 \in \mathrm{E}(R).$$

This finishes the proof of Lemma 3.6. □

Lemma 3.8 *The definition of* $\mathrm{Wh}(G)$ *from Section 2.5 makes sense and yields an abelian group, which can be identified with* $\mathrm{Wh}(G)$ *from Definition 3.4.*

Proof. Note that the operation (i) appearing in the definition of $\mathrm{Wh}(G)$ in Section 2.5 corresponds to multiplication by an elementary matrix from the left. Since $\mathrm{E}(R)$ is normal by Lemma 3.6, two invertible matrices A and B over $\mathbb{Z}G$ are equivalent under the equivalence relation appearing in the definition of $\mathrm{Wh}(G)$ as explained in Section 2.5 if and only if their classes $[A]$ and $[B]$ in $\mathrm{Wh}(G)$ as defined in this section agree. Now the claim follows from Lemma 3.6. □

Remark 3.9 ($K_1(R)$ **in terms of automorphisms**) Often $K_1(R)$ is defined in a more conceptual way in terms of automorphisms as follows. Namely, $K_1(R)$ is defined as the abelian group whose generators $[f]$ are conjugacy classes of automorphisms $f \colon P \to P$ of finitely generated projective R-modules P and which satisfies the following relations. For any commutative diagram of finitely generated projective R-modules with exact rows and automorphisms as vertical arrows

$$\begin{array}{ccccccccc}
0 & \longrightarrow & P_0 & \xrightarrow{i} & P_1 & \xrightarrow{p} & P_2 & \longrightarrow & 0 \\
& & {\scriptstyle f_0} \downarrow \cong & & {\scriptstyle f_1} \downarrow \cong & & {\scriptstyle f_2} \downarrow \cong & & \\
0 & \longrightarrow & P_0 & \xrightarrow{i} & P_1 & \xrightarrow{p} & P_2 & \longrightarrow & 0
\end{array}$$

we get the relation $[f_0] - [f_1] + [f_2] = 0$. If $f, g\colon P \to P$ are automorphisms of a finitely generated projective R-module P, then $[g \circ f] = [g] + [f]$. Using Lemma 3.6 one easily checks that sending the class $[A]$ of an invertible (n, n)-matrix A to the class of the automorphism $R_A\colon R^n \to R^n$, $x \mapsto xA$ defines an isomorphism from $\mathrm{GL}(R)/[\mathrm{GL}(R), \mathrm{GL}(R)]$ to the abelian group defined above.

Exercise 3.10 Using the ring homomorphism $f\colon \mathbb{Z}[\mathbb{Z}/5] \to \mathbb{C}$, which sends the generator of $\mathbb{Z}/5$ to $\exp(2\pi i/5)$, and the norm of a complex number, define a homomorphism of abelian groups

$$\phi\colon \mathrm{Wh}(\mathbb{Z}/5) \to \mathbb{R}^{>0}.$$

Show that $1 - t - t^{-1}$ is a unit in $\mathbb{Z}[\mathbb{Z}/5]$ whose class in $\mathrm{Wh}(\mathbb{Z}/5)$ is an element of infinite order. (Actually $\mathrm{Wh}(\mathbb{Z}/5)$ is an infinite cyclic group with this class as generator).

3.3 Algebraic Approach to Whitehead Torsion

In this section we give the definition and prove the basic properties of the Whitehead torsion using an algebraic approach via chain complexes. This will enable us to finish the proof of the s-Cobordism Theorem 2.1. The idea, which underlies the notion of Whitehead torsion, will become more transparent in Section 3.4 where we will develop a geometric approach to Whitehead torsion and link it to the strategy of proof of the s-Cobordism Theorem 2.1.

We begin with some input about chain complexes. A more thorough treatment of conventions and homotopy theory of chain complexes is contained in Chapter 14, here we only discuss the notions needed in this chapter. Let $f_*\colon C_* \to D_*$ be a chain map of R-chain complexes for some ring R. Define $\mathrm{cyl}_*(f_*)$ to be the chain complex with p-th differential

$$C_{p-1} \oplus C_p \oplus D_p \xrightarrow{\begin{pmatrix} -c_{p-1} & 0 & 0 \\ -\mathrm{id} & -c_p & 0 \\ f_{p-1} & 0 & d_p \end{pmatrix}} C_{p-2} \oplus C_{p-1} \oplus D_{p-1}. \qquad (3.11)$$

Define $\mathrm{cone}_*(f_*)$ to be the quotient of $\mathrm{cyl}_*(f_*)$ by the obvious copy of C_*. Hence the p-th differential of $\mathrm{cone}_*(f_*)$ is

$$C_{p-1} \oplus D_p \xrightarrow{\begin{pmatrix} -c_{p-1} & 0 \\ f_{p-1} & d_p \end{pmatrix}} C_{p-2} \oplus D_{p-1}. \qquad (3.12)$$

3.3 Algebraic Approach to Whitehead Torsion   45

Given a chain complex C_*, define ΣC_* to be the quotient of $\mathrm{cone}_*(\mathrm{id}_{C_*})$ by the obvious copy of C_*, i.e., the chain complex with p-th differential

$$C_{p-1} \xrightarrow{-c_{p-1}} C_{p-2}. \qquad (3.13)$$

Definition 3.14 (Mapping cylinder and mapping cone) Let $\mathrm{cyl}_*(f_*)$ be the *mapping cylinder*, $\mathrm{cone}_*(f_*)$ be the *mapping cone* of the chain map f_*, and ΣC_* be the *suspension* of the chain complex C_*.

These algebraic notions of mapping cylinder, mapping cone and suspension are modelled on their geometric counterparts. Namely, the cellular chain complex of a mapping cylinder of a cellular map of CW-complexes is the mapping cylinder of the chain map induced by f. From the geometry it is also clear why one obtains obvious exact sequences such as $0 \to C_* \to \mathrm{cyl}_*(f_*) \to \mathrm{cone}_*(f_*) \to 0$ and $0 \to D_* \to \mathrm{cone}_*(f_*) \to \Sigma C_* \to 0$.

A *chain contraction* γ_* for an R-chain complex C_* is a collection of R-homomorphisms $\gamma_p : C_p \to C_{p+1}$ for $p \in \mathbb{Z}$ such that $c_{p+1} \circ \gamma_p + \gamma_{p-1} \circ c_p = \mathrm{id}_{C_p}$ holds for all $p \in \mathbb{Z}$.

Exercise 3.15 Let $f_* : C_* \to D_*$ be a chain map. Show that f_* is a chain homotopy equivalence if and only if $\mathrm{cone}_*(f_*)$ is contractible.

We call an R-chain complex C_* *finite* if there is a number N with $C_p = 0$ for $|p| > N$ and each R-chain module C_p is a finitely generated R-module. We call an R-chain complex C_* *projective*, *free* or *based free* respectively if each R-chain module C_p is projective, free, or free with a preferred (not necessarily ordered) basis respectively. Suppose that C_* is a finite based free R-chain complex which is *contractible*, i.e., which possesses a chain contraction. Put $C_{\mathrm{odd}} = \bigoplus_{p \in \mathbb{Z}} C_{2p+1}$ and $C_{\mathrm{ev}} = \bigoplus_{p \in \mathbb{Z}} C_{2p}$. Let γ_* and δ_* be two chain contractions. Define R-homomorphisms

$$(c_* + \gamma_*)_{\mathrm{odd}} : C_{\mathrm{odd}} \to C_{\mathrm{ev}};$$
$$(c_* + \delta_*)_{\mathrm{ev}} : C_{\mathrm{ev}} \to C_{\mathrm{odd}}.$$

Let A be the matrix of $(c_* + \gamma_*)_{\mathrm{odd}}$ with respect to the given bases. Let B be the matrix of $(c_* + \delta_*)_{\mathrm{ev}}$ with respect to the given bases. Put $\mu_n = (\gamma_{n+1} - \delta_{n+1}) \circ \delta_n$ and $\nu_n = (\delta_{n+1} - \gamma_{n+1}) \circ \gamma_n$. One easily checks that $(\mathrm{id} + \mu_*)_{\mathrm{odd}}$, $(\mathrm{id} + \nu_*)_{\mathrm{ev}}$ and both compositions $(c_* + \gamma_*)_{\mathrm{odd}} \circ (\mathrm{id} + \mu_*)_{\mathrm{odd}} \circ (c_* + \delta_*)_{\mathrm{ev}}$ and $(c_* + \delta_*)_{\mathrm{ev}} \circ (\mathrm{id} + \nu_*)_{\mathrm{ev}} \circ (c_* + \gamma_*)_{\mathrm{odd}}$ are given by upper triangular matrices whose diagonal entries are identity maps. Hence A and B are invertible and their classes $[A], [B] \in \widetilde{K}_1(R)$ satisfy $[A] = -[B]$. Since $[B]$ is independent of the choice of γ_*, the same is true for $[A]$. Thus we can associate to a finite based free contractible R-chain complex C_* its *Reidemeister torsion*

$$\Delta(C_*) = [A] \in \widetilde{K}_1(R). \qquad (3.16)$$

Let $f_* \colon C_* \to D_*$ be a chain homotopy equivalence of finite based free R-chain complexes. Its mapping cone $\operatorname{cone}(f_*)$ is a contractible finite based free R-chain complex.

Definition 3.17 (Whitehead torsion of a chain homotopy equivalence) Define the *Whitehead torsion* of the chain homotopy equivalence of finite based free R-chain complexes $f_* \colon C_* \to D_*$ by

$$\tau(f_*) := \Delta(\operatorname{cone}_*(f_*)) \in \widetilde{K}_1(R).$$

Of course the Reidemeister torsion $\Delta(C_*)$ of a finite based free contractible R-chain complex in the sense of (3.16) is the Whitehead torsion of the chain homotopy equivalence $0_* \to C_*$ in the sense of Definition 3.17. We follow the convention that the torsion of a chain complex, which is contractible or may come with extra information about its homology, is attributed to Reidemeister whereas the torsion of a chain homotopy equivalence is attributed to Whitehead.

A sequence $0 \to C_* \xrightarrow{i_*} D_* \xrightarrow{q_*} E_* \to 0$ of finite based free R-chain complexes is called *based exact* if for any $p \in \mathbb{Z}$ the basis B for D_p can be written as a disjoint union $B' \coprod B''$ such that the image of the basis of C_p under i_p is B' and the image of B'' under q_p is the basis for E_p.

Lemma 3.18 (i) *Consider a commutative diagram of finite based free R-chain complexes*

$$\begin{array}{ccccccccc}
0 & \to & C'_* & \to & D'_* & \to & E'_* & \to & 0 \\
& & \downarrow f_* & & \downarrow g_* & & \downarrow h_* & & \\
0 & \to & C_* & \to & D_* & \to & E_* & \to & 0
\end{array}$$

whose rows are based exact. Suppose that two of the chain maps f_*, g_* and h_* are R-chain homotopy equivalences.
Then all three are R-chain homotopy equivalences and

$$\tau(f_*) - \tau(g_*) + \tau(h_*) = 0;$$

(ii) *Let $f_* \simeq g_* \colon C_* \to D_*$ be homotopic R-chain homotopy equivalences of finite based free R-chain complexes. Then*

$$\tau(f_*) = \tau(g_*);$$

(iii) *Let $f_* \colon C_* \to D_*$ and $g_* \colon D_* \to E_*$ be R-chain homotopy equivalences of finite based free R-chain complexes. Then*

$$\tau(g_* \circ f_*) = \tau(g_*) + \tau(f_*).$$

Proof. (i) The claim about the R-chain homotopy equivalences is proved in a more general setting later, see Proposition 14.55. Alternatively one may argue as follows. A chain map of finite projective chain complexes is a homotopy equivalence if and

3.3 Algebraic Approach to Whitehead Torsion

only if it induces an isomorphism on homology, (actually bounded below instead of finite suffices), see Proposition 14.52 (ii). The five lemma and the long homology sequence of a short exact sequence of chain complexes imply that all three chain maps f_*, h_* and g_* are chain homotopy equivalences if two of them are.

Let $u_* \colon F_* \to G_*$ be an isomorphism of contractible finite based free R-chain complexes. Since the choice of a chain contraction does not affect the values of the Whitehead torsion, we can compute $\tau(F_*)$ and $\tau(G_*)$ with respect to chain contractions which are compatible with u_*. Then one easily checks in $\widetilde{K}_1(R)$

$$\tau(G_*) - \tau(F_*) = \sum_{p \in \mathbb{Z}} (-1)^p \cdot [u_p], \tag{3.19}$$

where $[u_p]$ is the element represented by the matrix of u_p with respect to the given bases.

To prove the sum formula, it suffices to show for a based free exact sequence $0 \to C_* \xrightarrow{i_*} D_* \xrightarrow{q_*} E_* \to 0$ of contractible finite based free R-chain complexes

$$\Delta(C_*) - \Delta(D_*) + \Delta(E_*) = 0. \tag{3.20}$$

Let ϵ_* be a chain contraction for E_*. Choose for any $p \in \mathbb{Z}$ an R-homomorphism $\sigma_p \colon E_p \to D_p$ satisfying $q_p \circ \sigma_p = \mathrm{id}$. Define $s_p \colon E_p \to D_p$ by $d_{p+1} \circ \sigma_{p+1} \circ \epsilon_p + \sigma_p \circ \epsilon_{p-1} \circ e_p$. One easily checks that the collection of the s_p-s defines a chain map $s_* \colon E_* \to D_*$ with $q_* \circ s_* = \mathrm{id}$. Thus we obtain an isomorphism of contractible based free R-chain complexes

$$i_* \oplus s_* \colon C_* \oplus E_* \to D_*.$$

Since the matrix of $i_p \oplus s_p$ with respect to the given basis is a block matrix of the shape $\begin{pmatrix} I_m & * \\ 0 & I_n \end{pmatrix}$ we get $[i_p \oplus s_p] = 0$ in $\widetilde{K}_1(R)$. Now (3.19) implies $\Delta(C_* \oplus E_*) = \Delta(D_*)$. Since obviously $\Delta(C_* \oplus E_*) = \Delta(C_*) + \Delta(E_*)$, we obtain (3.20) and thus assertion (i) follows.

(ii) If $h_* \colon f_* \simeq g_*$ is a chain homotopy, we obtain an isomorphism of based free R-chain complexes

$$\begin{pmatrix} \mathrm{id} & 0 \\ h_{*-1} & \mathrm{id} \end{pmatrix} \colon \mathrm{cone}_*(f_*) = C_{*-1} \oplus D_* \to \mathrm{cone}_*(g_*) = C_{*-1} \oplus D_*.$$

We conclude from (3.19)

$$\tau(g_*) - \tau(f_*) = \sum_{p \in \mathbb{Z}} (-1)^p \cdot \left[\begin{pmatrix} \mathrm{id} & 0 \\ h_{*-1} & \mathrm{id} \end{pmatrix}\right] = 0.$$

(iii) Define a chain map $h_* \colon \Sigma^{-1} \operatorname{cone}_*(g_*) \to \operatorname{cone}_*(f_*)$ by

$$\begin{pmatrix} 0 & 0 \\ -\operatorname{id} & 0 \end{pmatrix} \colon D_p \oplus E_{p+1} \to C_{p-1} \oplus D_p.$$

There is an obvious based exact sequence of contractible finite based free R-chain complexes $0 \to \operatorname{cone}_*(f_*) \to \operatorname{cone}(h_*) \to \operatorname{cone}(g_*) \to 0$. There is also a based exact sequence of contractible finite based free R-chain complexes $0 \to \operatorname{cone}_*(g_* \circ f_*) \xrightarrow{i_*} \operatorname{cone}_*(h_*) \to \operatorname{cone}_*(\operatorname{id} \colon D_* \to D_*) \to 0$ where i_p is given by

$$\begin{pmatrix} f_{p-1} & 0 \\ 0 & \operatorname{id} \\ \operatorname{id} & 0 \\ 0 & 0 \end{pmatrix} \colon C_{p-1} \oplus E_p \to D_{p-1} \oplus E_p \oplus C_{p-1} \oplus D_p.$$

We conclude from assertion (i)

$$\tau(h_*) = \tau(f_*) + \tau(g_*);$$
$$\tau(h_*) = \tau(g_* \circ f_*) + \tau(\operatorname{id}_* \colon D_* \to D_*);$$
$$\tau(\operatorname{id}_* \colon D_* \to D_*) = 0.$$

This finishes the proof of Lemma 3.18. □

Now we can pass to CW-complexes. Let $f \colon X \to Y$ be a homotopy equivalence of connected finite CW-complexes. Let $p_X \colon \widetilde{X} \to X$ and $p_Y \colon \widetilde{Y} \to Y$ be the universal coverings. Fix base points $\widetilde{x} \in \widetilde{X}$ and $\widetilde{y} \in \widetilde{Y}$ such that f maps $x = p_X(\widetilde{x})$ to $y = p_Y(\widetilde{y})$. Let $\widetilde{f} \colon \widetilde{X} \to \widetilde{Y}$ be the unique lift of f satisfying $\widetilde{f}(\widetilde{x}) = \widetilde{y}$. We abbreviate $\pi = \pi_1(Y, y)$ and identify $\pi_1(X, x)$ in the sequel with π by $\pi_1(f, x)$. After the choice of the base points $\widetilde{x}$ and $\widetilde{y}$ we get unique operations of π on $\widetilde{X}$ and $\widetilde{Y}$. The lift $\widetilde{f}$ is π-equivariant. It induces a $\mathbb{Z}\pi$-chain homotopy equivalence $C_*(\widetilde{f}) \colon C_*(\widetilde{X}) \to C_*(\widetilde{Y})$. We can apply Definition 3.17 to it and thus obtain an element

$$\tau(f) \in \operatorname{Wh}(\pi_1(Y, y)). \tag{3.21}$$

So far this definition depends on the various choices of base points. We can get rid of these choices as follows. If y' is a second base point, we can choose a path w from y to y' in Y. Conjugation with w yields a homomorphism $c_w \colon \pi_1(Y, y) \to \pi_1(Y, y')$ which induces $(c_w)_* \colon \operatorname{Wh}(\pi_1(Y, y)) \to \operatorname{Wh}(\pi_1(Y, y'))$. If v is a different path from y to y', then c_w and c_v differ by an inner automorphism of $\pi_1(Y, y)$. Since an inner automorphism of $\pi_1(Y, y)$ induces the identity on $\operatorname{Wh}(\pi_1(Y, y))$, we conclude that $(c_w)_*$ and $(c_v)_*$ agree. Hence we get a unique isomorphism $t(y, y') \colon \operatorname{Wh}(\pi_1(Y, y)) \to \operatorname{Wh}(\pi_1(Y, y'))$ depending only on y and y'. Moreover $t(y, y) = \operatorname{id}$ and $t(y, y'') = t(y', y'') \circ t(y, y')$. Therefore we can define $\operatorname{Wh}(\Pi(Y))$ independently of a choice of a base point by $\coprod_{y \in Y} \operatorname{Wh}(\pi_1(Y, y))/\sim$ where $\sim$ is the obvious equivalence relation generated by $a \sim b \Leftrightarrow t(y, y')(a) = b$ for $a \in \operatorname{Wh}(\pi_1(Y, y))$ and $b \in \operatorname{Wh}(\pi_1(Y, y'))$. Define $\tau(f) \in \operatorname{Wh}(\Pi(Y))$ by the element represented by the element introduced

3.3 Algebraic Approach to Whitehead Torsion

in (3.21). Note that $\mathrm{Wh}(\Pi(Y))$ is isomorphic to $\mathrm{Wh}(\pi_1(Y, y))$ for any base point $y \in Y$. Using Lemma 3.18 it is not hard to check that $\tau(f)$ depends only on $f\colon X \to Y$ and not on the choice of the universal coverings and base points.

Finally we want to drop the assumption that Y is connected. Note that f induces a bijection $\pi_0(f)\colon \pi_0(X) \to \pi_0(Y)$.

Definition 3.22 (Whitehead group of a space and Whitehead torsion of a map)
Let $f\colon X \to Y$ be a homotopy equivalence of finite CW-complexes. Define the *Whitehead group* $\mathrm{Wh}(\Pi(Y))$ of Y and the *Whitehead torsion* $\tau(f) \in \mathrm{Wh}(\Pi(Y))$ by

$$\mathrm{Wh}(\Pi(Y)) = \bigoplus_{C \in \pi_0(Y)} \mathrm{Wh}(\Pi(C));$$

$$\tau(f) = \bigoplus_{C \in \pi_0(Y)} \tau\left(f|_{\pi_0(f)^{-1}(C)}\colon \pi_0(f)^{-1}(C) \to C\right).$$

In the notation $\mathrm{Wh}(\Pi(Y))$ one should think of $\Pi(Y)$ as the fundamental groupoid of Y. Note that a map $f\colon X \to Y$ induces a homomorphism $f_*\colon \mathrm{Wh}(\Pi(X)) \to \mathrm{Wh}(\Pi(Y))$ such that $\mathrm{id}_* = \mathrm{id}$, $(g \circ f)_* = g_* \circ f_*$ and $f \simeq g \Rightarrow f_* = g_*$.

A detailed discussion of base point issues will be given in Subsection 8.7.1 and for the more complicated case of L-theory in Subsection 8.7.2.

Next we give the proof of Theorem 3.1.

Proof of Theorem 3.1. (i), (ii), and (iii) follow from Lemma 3.18.

(iv) Because of assertion (iii) we have

$$\tau(f \times g) = \tau(f \times \mathrm{id}_Y) + (f \times \mathrm{id}_Y)_* \tau(\mathrm{id}_X \times g).$$

Hence it suffices to treat the case $g = \mathrm{id}_Y$. Now one proceeds by induction over the cells of Y using assertions (i), (ii), and (iii).

(v) This (in comparison with the other assertions much deeper result) is due to Chapman [94, 95]. This finishes the proof of Theorem 3.1. □

Exercise 3.23 Consider a square of connected finite CW-complexes such that the vertical arrows are coverings and the horizontal arrows are homotopy equivalences

$$\begin{array}{ccc} \overline{X} & \xrightarrow{\overline{f}}_{\simeq} & \overline{Y} \\ p_X \downarrow & & \downarrow p_Y \\ X & \xrightarrow{\simeq}_{f} & Y. \end{array}$$

Show that $\tau(\overline{f})$ vanishes if $\tau(f)$ vanishes. Is the converse true in general?

Definition 3.24 (Whitehead torsion of an *h*-cobordism) We define the Whitehead torsion of an *h*-cobordism $(W; M_0, f_0, M_1, f_1)$

$$\tau(W, M_0) \in \mathrm{Wh}(\Pi(M_0))$$

by $\tau(W, M_0) := (i_0 \circ f_0)_*^{-1}(\tau(i_0 \circ f_0 \colon M_0 \to W))$ where we equip W and M_0 with some *CW*-structure, for instance one coming from a smooth triangulation.

This is independent of the choice of *CW*-structure by Theorem 3.1 (v). Because of Theorem 3.1 (v) two *h*-cobordisms over M_0, which are diffeomorphic relative M_0, have the same Whitehead torsion.

Let R be a *ring with involution* $i \colon R \to R$, $r \mapsto \overline{r}$, i.e., a map satisfying $\overline{r+s} = \overline{r}+\overline{s}$, $\overline{r \cdot s} = \overline{s} \cdot \overline{r}$ and $\overline{1} = 1$. Given a (m, n)-matrix $A = (a_{i,j})$ define the (n, m)-matrix A^* by $(\overline{a_{j,i}})$. We obtain an involution

$$*\colon K_1(R) \to K_1(R), \quad [A] \mapsto [A^*]. \tag{3.25}$$

Let P be a left R-module. Define the *dual R-module* P^* to be the left R-module whose underlying abelian group is $P^* = \hom(P, R)$ and whose left R-module structure is given by $(rf)(x) := f(x)\overline{r}$ for $f \in P^*$ and $x \in P$. Then the induced involution on $K_1(R)$ corresponds to $[f \colon P \to P] \mapsto [f^* \colon P^* \to P^*]$ if one defines $K_1(R)$ as in Remark 3.9.

Our main example of a ring with involution is the group ring AG for some commutative associative ring A with unit and a group G together with a homomorphism $w \colon G \to \{\pm 1\}$. The so-called *w-twisted involution* is defined by

$$*\colon AG \to AG, \quad \sum_{g \in G} a_g \cdot g \mapsto \sum_{g \in G} w(g) \cdot a_g \cdot g^{-1}. \tag{3.26}$$

Now consider a connected manifold M. Then the first Stiefel–Whitney class yields a homomorphism $w_1(M) \colon \pi_1(M) \to \{\pm 1\}$. Thus we obtain an involution on $K_1(\mathbb{Z}[\pi_1(M)])$ coming from the $w_1(M)$-twisted involution on $\mathbb{Z}[\pi_1(M)]$. It induces an involution on the Whitehead group

$$*\colon \mathrm{Wh}(\Pi_1(M)) \to \mathrm{Wh}(\Pi_1(M)). \tag{3.27}$$

This extends in the obvious way to the non-connected case by considering each component of M separately.

Lemma 3.28 (i) *Let $(W; M_0, f_0, M_1, f_1)$ and $(W'; M_0', f_0', M_1', f_1')$ be h-cobordisms over M_0 and M_0' and let $g \colon M_1 \to M_0'$ be a diffeomorphism. Let $W \cup W'$ be the h-cobordism over M_0 obtained from W and W' by gluing with the diffeomorphism $f_0' \circ g \circ f_1^{-1} \colon \partial_1 W \to \partial_0 W'$. Denote by $u \colon \mathrm{Wh}(\Pi(M_0')) \to \mathrm{Wh}(\Pi(M_0))$ the isomorphism given by the composite $(f_0)_*^{-1} \circ (i_0)_*^{-1} \circ (i_1)_* \circ (f_1)_* \circ (g_*)^{-1}$ where $i_k \colon \partial_k W \to W$ is the inclusion for $k = 0, 1$. Then*

$$\tau(W \cup W', M_0) = \tau(W, M_0) + u\left(\tau(W', M_0')\right);$$

3.3 Algebraic Approach to Whitehead Torsion   51

(ii) *Let* $(W; M_0, f_0, M_1, f_1)$ *be an h-cobordism over* M_0 *and denote by* $v \colon \mathrm{Wh}(\Pi(M_1)) \to \mathrm{Wh}(\Pi(M_0))$ *the isomorphism given by the composite* $(f_0)_*^{-1} \circ (i_0)_*^{-1} \circ (i_1)_* \circ (f_1)_*$. *Then*

$$*(\tau(W, M_0)) = (-1)^{\dim(M_0)} \cdot v(\tau(W, M_1)).$$

Proof. (i) follows from Theorem 3.1.

(ii) Let $C_*(\widetilde{W}, \widetilde{\partial_k W})$ be the cellular $\mathbb{Z}\pi$-chain complex with respect to a triangulation of W of the universal covering $\widetilde{W} \to W$ for $\pi = \pi_1(W) = \pi_1(\partial_0 W) = \pi_1(\partial_1 W)$. Put $n = \dim(W)$. The dual $\mathbb{Z}\pi$-chain complex $C^{n-*}(\widetilde{W}, \widetilde{\partial_1 W})$ has as p-th chain module the dual module of $C_{n-p}(\widetilde{W}, \widetilde{\partial_1 W})$ and its p-th differential is the dual of $c_{n-p+1} \colon C_{n-p+1}(\widetilde{W}, \widetilde{\partial_1 W}) \to C_{n-p}(\widetilde{W}, \widetilde{\partial_1 W})$. Poincaré duality yields a $\mathbb{Z}\pi$-chain homotopy equivalence for $n = \dim(W)$

$$- \cap [W, \partial W] \colon C^{n-*}(\widetilde{W}, \widetilde{\partial_1 W}) \to C_*(\widetilde{W}, \widetilde{\partial_0 W}).$$

Inspecting the proof of Poincaré duality using dual cells [243] shows that this is a base preserving chain map if one passes to subdivisions. This implies that this $\mathbb{Z}\pi$-chain homotopy equivalence has trivial Whitehead torsion. We conclude $(f_k)_* \circ (i_k)_*(\tau(W, M_k)) = \tau(\widetilde{W}, \widetilde{\partial_k W})$ for $k = 0, 1$ from Lemma 3.18 and get by a direct inspection

$$\tau(C_*(\widetilde{W}, \widetilde{\partial_0 W})) = \tau(C^{n-*}(\widetilde{W}, \widetilde{\partial_1 W})) = (-1)^{n-1} \cdot *(\tau(C_*(\widetilde{W}, \widetilde{\partial_1 W}))).$$

This finishes the proof of Lemma 3.28   □

Next we can finish the proof of the *s*-Cobordism Theorem 2.1.

Proof of the s-Cobordism Theorem 2.1. We have to show that the class of the matrix $[A] \in \mathrm{Wh}(\pi)$ appearing in Lemma 2.42 (i) agrees with $(-1)^{pq} \cdot \tau(W, M_0)$ and that $\tau(W, M_0)$ depends only on the diffeomorphism type of W relative M_0, as explained in Remark 2.43, where $\tau(W, M_0)$ is the Whitehead torsion of the homotopy equivalence $M_0 \xrightarrow{f_0} \partial_0 W \xrightarrow{i_0} W$. Since f_0 is a diffeomorphism, $\tau(W, M_0)$ agrees with the Whitehead torsion $\tau(i_0)$ of $i_0 \colon \partial_0 W \to W$ by Theorem 3.1 (iii) and (v).

Actually, $\tau(W, M_0)$ depends only on the homeomorphism type of W relative M_0 by Theorem 3.1 (v) and by Lemma 3.18 (i).

Now we fix a handlebody decomposition, which is in normal form in the sense of Lemma 2.37. Using Theorem 3.1 (i) and (iv), one can show that the homotopy equivalence $f \colon W \to X$ of (2.26) satisfies $\tau(f) = 0$. We conclude $\tau(W, M_0) = \tau(i_0 \colon \partial_0 W \to W) = \tau(C_*(\widetilde{X}, \widetilde{\partial_0 W}))$ from Lemma 3.18 (i). We conclude $(-1)^{pq} \cdot [A] = \tau(C_*(\widetilde{X}, \widetilde{\partial_0 W}))$ from the existence of the base preserving isomorphism (2.27). Now assertions (i) and (ii) of Theorem 2.1 follow from Lemma 2.42.

In order to prove Theorem 2.1 (iii), we have to show for two *h*-cobordisms $(W; M_0, M_1, f_0, f_1)$ and $(W'; M_0, M_1', f_0', f_1')$ over M_0 satisfying $\tau(W, M_0) = \tau(W', M_0)$ that they are diffeomorphic relative M_0. Choose an *h*-cobordism $(W''; M_1, M_2'', f_1'', f_2'')$ over M_1 such that $\tau(W''; M_1, M_2'', f_1'', f_2'')$ is the image of

$-\tau(W, M_0)$ under the isomorphism $(i_1 \circ f_1)_*^{-1} \circ (i_0 \circ f_0)_* \colon \mathrm{Wh}(\Pi(M_0)) \to \mathrm{Wh}(\Pi(M_1))$ where $i_k \colon \partial_k W \to W$ for $k = 0, 1$ is the inclusion. We can glue W and W'' along M_1 to get an h-cobordism $W \cup_{M_1} W''$ over M_0. From Lemma 3.28 (i) we get $\tau(W \cup_{M_1} W'', M_0) = 0$. Hence there is a diffeomorphism $G \colon W \cup_{M_1} W'' \to M \times [0, 1]$, which induces the identity on $M_0 = M_0 \times \{0\}$ and a diffeomorphism $g_1 \colon M_2 \to M_0 \times \{1\} = M_0$. Now we can form the h-cobordism $W \cup_{M_1} W'' \cup_{g_1} W'$. Using G we can construct a diffeomorphism relative M_0 from $W \cup_{M_1} W'' \cup_{g_1} W'$ to W'. Similarly one can show that $W'' \cup_{g_1} W'$ is diffeomorphic relative M_1 to the trivial h-cobordism over M_1. Hence there is also a diffeomorphism relative M_0 from $W \cup_{M_1} W'' \cup_{g_1} W'$ to W. We conclude that W and W' are diffeomorphic relative M_0. This finishes the proof of the s-Cobordism Theorem 2.1. □

Exercise 3.29 Let $(W; M_0, f_0, M_1, f_1)$ be an h-cobordism. Suppose that the Whitehead torsion $\tau = \tau(W; M_0, f_0, M_1, f_1) \in \mathrm{Wh}(\Pi(M_0))$ is non-trivial and satisfies $\tau + (-1)^{\dim(W)} \cdot v(\tau) = 0$ for the isomorphism v defined in Lemma 3.28 ii. Show that W is non-trivial and that M_0 and M_1 are simple homotopy equivalent.

We call a cobordism $(W; M_0, f_0, M_1, f_1)$ an *s-cobordism* if the inclusions $\partial_i W \to W$ for $i = 0, 1$ are simple homotopy equivalences.

3.4 Geometric Approach to Whitehead Torsion

In this section we introduce the concept of a simple homotopy equivalence $f \colon X \to Y$ of finite CW-complexes geometrically. We will show that the obstruction for a homotopy equivalence $f \colon X \to Y$ of finite CW-complexes to be simple is the Whitehead torsion. This proof is a CW-version or homotopy version of the proof of the s-Cobordism Theorem 2.1.

We have the inclusion $S^{n-2} \subset S^{n-1}_+ \subset S^{n-1} \subset D^n$ where $S^{n-1}_+ \subset S^{n-1}$ is the upper hemisphere. The pair (D^n, S^{n-1}_+) carries an obvious relative CW-structure. Namely, attach an $(n-1)$-cell to S^{n-1}_+ by the attaching map $\mathrm{id} \colon S^{n-2} \to S^{n-2}$ to obtain S^{n-1}. Then we attach to S^{n-1} an n-cell by the attaching map $\mathrm{id} \colon S^{n-1} \to S^{n-1}$ to obtain D^n. Let X be a CW-complex. Let $q \colon S^{n-1}_+ \to X$ be a map satisfying $q(S^{n-2}) \subset X_{n-2}$ and $q(S^{n-1}_+) \subset X_{n-1}$. Let Y be the space $D^n \cup_q X$, i.e., the pushout

$$\begin{array}{ccc} S^{n-1}_+ & \xrightarrow{q} & X \\ {\scriptstyle i}\downarrow & & \downarrow{\scriptstyle j} \\ D^n & \xrightarrow{g} & Y \end{array}$$

where i is the inclusion. Then Y inherits a CW-structure by putting $Y_k = j(X_k)$ for $k \leq n-2$, $Y_{n-1} = j(X_{n-1}) \cup g(S^{n-1})$ and $Y_k = j(X_k) \cup g(D^n)$ for $k \geq n$. Note that Y is obtained from X by attaching one $(n-1)$-cell and one n-cell. Since the map $i \colon S^{n-1}_+ \to D^n$ is a homotopy equivalence and cofibration, the map $j \colon X \to Y$ is a homotopy equivalence and cofibration. We call j an *elementary expansion* and say

3.4 Geometric Approach to Whitehead Torsion

that Y is obtained from X by an elementary expansion. There is a map $r: Y \to X$ with $r \circ j = \mathrm{id}_X$. This map is unique up to homotopy relative $j(X)$. We call any such map an *elementary collapse* and say that X is obtained from Y by an elementary collapse.

Figure 3.30 (Elementary expansion).

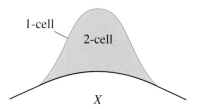

An elementary expansion is the CW-version or homotopy version of the construction in Example 2.19 where we have added a q-handle and a $(q + 1)$-handle to W without changing the diffeomorphism type of W. This corresponds to an elementary expansion for the CW-complex X, which we have assigned to W in Section 2.3.

Definition 3.31 (Simple homotopy equivalence) Let $f: X \to Y$ be a map of finite CW-complexes. We call it a *simple homotopy equivalence* if there is a sequence of maps
$$X = X[0] \xrightarrow{f_0} X[1] \xrightarrow{f_1} X[2] \xrightarrow{f_2} \ldots \xrightarrow{f_{n-1}} X[n] = Y$$
such that each f_i is an elementary expansion or elementary collapse and f is homotopic to the composition of the maps f_i.

Exercise 3.32 Consider the simplicial complex X with four vertices v_0, v_1, v_2 and v_3, the edges $\{v_0, v_1\}$, $\{v_1, v_2\}$, $\{v_0, v_2\}$ and $\{v_2, v_3\}$ and one 2-simplex $\{v_0, v_1, v_2\}$. Describe a sequence of elementary collapses and expansions transforming it to the one-point-space $\{\bullet\}$.

The idea of the definition of a simple homotopy equivalence is that such a map can be written as a composition of elementary maps, which are obviously homotopy equivalences. This is similar to the idea in knot theory that two knots are equivalent if one can pass from one knot to the other by a sequence of elementary moves, the so-called Reidemeister moves. A Reidemeister move obviously does not change the equivalence class of a knot and, indeed, it turns out that one can pass from one knot to a second knot by a sequence of Reidemeister moves if and only if the two knots are equivalent. The analogous statement is not true for homotopy equivalences $f: X \to Y$ of finite CW-complexes since there is an obstruction for f to be simple, namely, its Whitehead torsion.

Lemma 3.33 (i) *Let $f: X \to Y$ be a simple homotopy equivalence. Then its Whitehead torsion $\tau(f) \in \mathrm{Wh}(\Pi(Y))$ vanishes;*
(ii) *Let X be a finite CW-complex. Then for any element $x \in \mathrm{Wh}(\Pi(X))$ there is an inclusion $i: X \to Y$ of finite CW-complexes such that i is a homotopy equivalence and $i_*^{-1}(\tau(i)) = x$.*

Proof. (i) Because of Theorem 3.1 it suffices to prove for an elementary expansion $j \colon X \to Y$ that its Whitehead torsion $\tau(j) \in \mathrm{Wh}(\Pi(Y))$ vanishes. We can assume without loss of generality that Y is connected. In the sequel we write $\pi = \pi_1(Y)$ and identify $\pi = \pi_1(X)$ by $\pi_1(f)$. The following diagram of based free finite $\mathbb{Z}\pi$-chain complexes

$$
\begin{array}{ccccccccc}
0 & \longrightarrow & C_*(\widetilde{X}) & \xrightarrow{C_*(\widetilde{j})} & C_*(\widetilde{Y}) & \xrightarrow{\mathrm{pr}_*} & C_*(\widetilde{Y},\widetilde{X}) & \longrightarrow & 0 \\
& & {\scriptstyle \mathrm{id}_*}\big\uparrow & & {\scriptstyle C_*(\widetilde{j})}\big\uparrow & & {\scriptstyle 0_*}\big\uparrow & & \\
0 & \longrightarrow & C_*(\widetilde{X}) & \xrightarrow{\mathrm{id}_*} & C_*(\widetilde{X}) & \xrightarrow{\mathrm{pr}_*} & 0 & \longrightarrow & 0
\end{array}
$$

has based exact rows and $\mathbb{Z}\pi$-chain homotopy equivalences as vertical arrows. We conclude from Lemma 3.18 (i)

$$\tau\bigl(C_*(\widetilde{j})\bigr) = \tau\bigl(\mathrm{id}_* \colon C_*(\widetilde{X}) \to C_*(\widetilde{X})\bigr) + \tau\bigl(0_* \colon 0 \to C_*(\widetilde{Y},\widetilde{X})\bigr) = \tau\bigl(C_*(\widetilde{Y},\widetilde{X})\bigr).$$

The $\mathbb{Z}\pi$-chain complex $C_*(\widetilde{Y},\widetilde{X})$ is concentrated in two consecutive dimensions and its only non-trivial differential is $\mathrm{id} \colon \mathbb{Z}\pi \to \mathbb{Z}\pi$ if we identify the two non-trivial $\mathbb{Z}\pi$-chain modules with $\mathbb{Z}\pi$ using the cellular basis. This implies $\tau(C_*(\widetilde{Y},\widetilde{X})) = 0$ and hence $\tau(j) := \tau(C_*(\widetilde{j})) = 0$.

(ii) We can assume without loss of generality that X is connected. Put $\pi = \pi_1(X)$. Choose an element $A \in \mathrm{GL}(n,\mathbb{Z}\pi)$ representing $x \in \mathrm{Wh}(\pi)$. Choose $n \geq 2$. In the sequel we fix a zero-cell in X as base point. Put $X' = X \vee \bigvee_{j=1}^n S^n$. Let $b_j \in \pi_n(X')$ be the element represented by the inclusion of the j-th copy of S^n into X for $j = 1,2\ldots,n$. Recall that $\pi_n(X')$ is a $\mathbb{Z}\pi$-module. Choose for $i = 1,2\ldots,n$ a map $f_i \colon S^n \to X'$ such that $[f_i] = \sum_{j=1}^n a_{i,j} \cdot b_j$ holds in $\pi_n(X')$. Attach to X' for each $i \in \{1,2\ldots,n\}$ an $(n+1)$-cell by $f_i \colon S^n \to X'$. Let Y be the resulting CW-complex and $i \colon X \to Y$ be the inclusion. Then i is an inclusion of finite CW-complexes and induces an isomorphism on the fundamental groups. In the sequel we identify π and $\pi_1(Y)$ by $\pi_1(i)$. The cellular $\mathbb{Z}\pi$-chain complex $C_*(\widetilde{Y},\widetilde{X})$ is concentrated in dimensions n and $(n+1)$ and its $(n+1)$-differential is given by the matrix A with respect to the cellular basis. Hence $C_*(\widetilde{Y},\widetilde{X})$ is a contractible finite based free $\mathbb{Z}\pi$-chain complex with $\tau(C_*(\widetilde{Y},\widetilde{X})) = [A]$ in $\mathrm{Wh}(\pi)$. We conclude that $i \colon X \to Y$ is a homotopy equivalence with $i_*^{-1}(\tau(i)) = x$. This finishes the proof of Lemma 3.33. □

Note that Lemma 3.33 (ii) is the CW-analogue of Theorem 2.1 (ii).

Recall that the *mapping cylinder* $\mathrm{cyl}(f)$ of a map $f \colon X \to Y$ is defined by the pushout

$$
\begin{array}{ccc}
X \times \{0\} & \xrightarrow{f} & Y \\
\downarrow & & \downarrow \\
X \times [0,1] & \longrightarrow & \mathrm{cyl}(f).
\end{array}
\qquad (3.34)
$$

3.4 Geometric Approach to Whitehead Torsion

There are natural inclusions $i_X \colon X = X \times \{1\} \to \mathrm{cyl}(f)$ and $i_Y \colon Y \to \mathrm{cyl}(f)$ and a natural projection $p \colon \mathrm{cyl}(f) \to Y$. Note that i_X is a cofibration and $p \circ i_X = f$ and $p_Y \circ Y = \mathrm{id}_Y$. Define *the mapping cone* $\mathrm{cone}(f)$ by the quotient $\mathrm{cyl}(f)/i_X(X)$.

Figure 3.35 (Mapping cylinder and mapping cone).

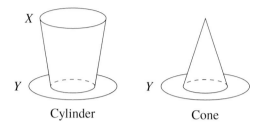

Lemma 3.36 *Let $f \colon X \to Y$ be a cellular map of finite CW-complexes and let $A \subset X$ be a CW-subcomplex. Then the inclusion $\mathrm{cyl}(f|_A) \to \mathrm{cyl}(f)$ and (in the case $A = \emptyset$) $i_Y \colon Y \to \mathrm{cyl}(f)$ is a composition of elementary expansions and hence a simple homotopy equivalence.*

Proof. It suffices to treat the case where X is obtained from A by attaching an n-cell by an attaching map $q \colon S^{n-1} \to X$. Then there is an obvious pushout

$$
\begin{array}{ccc}
S^{n-1} \times [0,1] \cup_{S^{n-1} \times \{0\}} D^n \times \{0\} & \longrightarrow & \mathrm{cyl}(f|_A) \\
\downarrow & & \downarrow \\
D^n \times [0,1] & \longrightarrow & \mathrm{cyl}(f)
\end{array}
$$

and an obvious homeomorphism

$$(D^n \times [0,1], S^{n-1} \times [0,1] \cup_{S^{n-1} \times \{0\}} D^n \times \{0\}) \to (D^{n+1}, S^n_+).$$

□

Lemma 3.37 *A map $f \colon X \to Y$ of finite CW-complexes is a simple homotopy equivalence if and only if $i_X \colon X \to \mathrm{cyl}(f)$ is a simple homotopy equivalence.*

Proof. This follows from Lemma 3.36 since a composition of simple homotopy equivalence and a homotopy inverse of a simple homotopy equivalence is again a simple homotopy equivalence. □

Fix a finite CW-complex X. Consider two pairs of finite CW-complexes (Y, X) and (Z, X) such that the inclusions of X into Y and Z are homotopy equivalences. We call them equivalent if there is a chain of pairs of finite CW-complexes

$$(Y, X) = (Y[0], X), (Y[1], X), (Y[2], X), \ldots, (Y[n], X) = (Z, X),$$

such that for each $k \in \{1, 2, \ldots, n\}$ either $Y[k]$ is obtained from $Y[k-1]$ by an elementary expansion or $Y[k-1]$ is obtained from $Y[k]$ by an elementary expansion.

Denote by $\mathrm{Wh}^{\mathrm{geo}}(X)$ the equivalence classes $[Y, X]$ of such pairs (Y, X). This becomes an abelian group under the addition $[Y, X] + [Z, X] := [Y \cup_X Z, X]$. The zero element is given by $[X, X]$. The inverse of $[Y, X]$ is constructed as follows. Choose a map $r\colon Y \to X$ with $r|_X = \mathrm{id}$. Denote by $p\colon X \times [0, 1] \to X$ the projection. Then $[(\mathrm{cyl}(r) \cup_p X) \cup_r X, X] + [Y, X] = 0$. A map $g\colon X \to X'$ induces a homomorphism $g_*\colon \mathrm{Wh}^{\mathrm{geo}}(X) \to \mathrm{Wh}^{\mathrm{geo}}(X')$ by sending $[Y, X]$ to $[Y \cup_g X', X']$. We have $\mathrm{id}_* = \mathrm{id}$ and $(g \circ h)_* = g_* \circ h_*$. In other words, we obtain a covariant functor on the category of finite CW-complexes with values in abelian groups. More information about this construction can be found for instance in [98, § 6 in Chapter II].

The next result may be viewed as the homotopy theoretic analogue of the s-Cobordism Theorem 2.1 (iii) where $\mathrm{Wh}^{\mathrm{geo}}(X)$ plays the role of the set of the diffeomorphism classes relative M_0 of h-cobordism over M_0

Theorem 3.38 (Comparing geometric and algebraic Whitehead torsion)

(i) *Let X be a finite CW-complex. The map*

$$\tau\colon \mathrm{Wh}^{\mathrm{geo}}(X) \to \mathrm{Wh}(\Pi(X))$$

sending $[Y, X]$ to $i_*^{-1}\tau(i)$ for the inclusion $i\colon X \to Y$ is a natural isomorphism;
(ii) *A homotopy equivalence $f\colon X \to Y$ is a simple homotopy equivalence if and only if $\tau(f) \in \mathrm{Wh}(\Pi(Y))$ vanishes.*

Proof. (i) The map τ is a well-defined homomorphism by Theorem 3.1 and Lemma 3.33 (i). It is surjective by Lemma 3.33 (ii).

We give only a sketch of the proof of injectivity which is similar but much easier than the proof of the s-Cobordism Theorem 2.1 (i). Consider an element $[Y, X]$ in $\mathrm{Wh}^{\mathrm{geo}}(X)$ with $i_*^{-1}\tau(i) = 0$ for the inclusion $i\colon X \to Y$. We want to show that $[Y, X] = [X, X]$ by reducing the number of cells, which must be attached to X to obtain Y, to zero without changing the class $[Y, X]$ in $\mathrm{Wh}^{\mathrm{geo}}(X)$. This corresponds in the proof of the s-Cobordism Theorem 2.1 (i) to reducing the number of handles in the handlebody decomposition to zero without changing the diffeomorphism type of the s-cobordism.

In the first step one arranges that Y is obtained from X by attaching only cells in two dimensions r and $(r+1)$ for some integer r. This is analogous to, but much easier to achieve than in the case of the s-Cobordism Theorem 2.1 (i), see the Normal Form Lemma 2.37. Details of this construction for CW-complexes can be found in [98, page 25-26].

Let $A \in \mathrm{GL}(n, \mathbb{Z}\pi)$ be the matrix describing the $(r+1)$-differential in the $\mathbb{Z}\pi$-chain complex $C_*(\widetilde{Y}, \widetilde{X})$. As in the proof of the s-Cobordism Theorem 2.1 (i), see also [98, Chapter II, Section §8], one shows that we can modify (Y, X) without changing its class $[Y, X] \in \mathrm{Wh}^{\mathrm{geo}}(X)$ so that the new matrix B is obtained from A by applying one of the operations (i), (ii), (iii), (iv), and (v) introduced in Section 2.5. Since one can reduce A by a sequence of these operations to the trivial matrix if and only if its class $[A] \in \mathrm{Wh}(\Pi(X))$ vanishes and this class $[A]$ is $i_*^{-1}\tau(i)$, the map τ is injective. Hence τ is a natural isomorphism of abelian groups.

(ii) This follows from Lemma 3.33 (i), Lemma 3.37 and the obvious fact that

3.5 Reidemeister Torsion and (Generalised) Lens Spaces 57

$i \colon X \to Y$ is a simple homotopy equivalence if $[Y, X] = 0$ in $\mathrm{Wh}^{\mathrm{geo}}(X)$. This finishes the proof of Theorem 3.38. □

3.5 Reidemeister Torsion and (Generalised) Lens Spaces

In this section we deal with Reidemeister torsion, which was defined earlier than Whitehead torsion and motivated the definition of Whitehead torsion. Reidemeister torsion was the first invariant in algebraic topology that could distinguish spaces that are homotopy equivalent but not homeomorphic. Namely, it can be used to classify lens spaces up to homeomorphism and diffeomorphism.

More generally, we will also consider generalised lens spaces since some of the facts and constructions about lens spaces directly carry over to generalised lens spaces. Moreover, generalised lens spaces can be classified up to homeomorphism using surgery, which we will explain in detail in Section 18.9.

3.5.1 Lens Spaces

We start by introducing the family of spaces which we want to classify completely using Reidemeister torsion. Let G be a non-trivial cyclic group of finite order $|G|$. Let V be a unitary finite-dimensional G-representation. Define its *unit sphere* SV and its *unit disk* DV to be the G-subspaces $SV = \{v \in V \mid ||v|| = 1\}$ and $DV = \{v \in V \mid ||v|| \le 1\}$ of V. Note that an m-dimensional complex vector space has a preferred orientation as a real vector space, namely the one given by the $\mathbb{R}$-basis $\{b_1, ib_1, b_2, ib_2, \ldots, b_m, ib_m\}$ for any $\mathbb{C}$-basis $\{b_1, b_2, \ldots, b_m\}$. Any $\mathbb{C}$-linear automorphism of a complex finite-dimensional vector space preserves this orientation. Thus SV and DV are compact oriented Riemannian manifolds with isometric orientation preserving G-action. We call a unitary G-representation V *free* if the induced G-action on its unit sphere $SV = \{v \in V \mid ||v|| = 1\}$ is free. Then $SV \to G\backslash SV$ is a covering and the quotient space $L(V) := G\backslash SV$ inherits from SV the structure of a closed oriented Riemannian manifold.

Definition 3.39 (Lens space) We call the closed oriented Riemannian manifold $L(V)$ the *lens space* associated to the free finite-dimensional unitary representation V of the finite cyclic group G.

One can specify these lens spaces also by numbers as follows.

Notation 3.40 Let $\mathbb{Z}/t$ be the cyclic group of order $t \ge 2$. The 1-dimensional unitary representation V_k for $k \in \mathbb{Z}/t$ has as underlying vector space $\mathbb{C}$ and $l \in \mathbb{Z}/t$ acts on it by multiplication with $\exp(2\pi ikl/t)$. Note that V_k is free if and only if $k \in \mathbb{Z}/t^\times = \{\bar{n} \in \mathbb{Z}/t \mid (n, t) = 1\}$, and is trivial if and only if $k = 0$ in $\mathbb{Z}/t$. Define the lens space $L(t; k_1, \ldots, k_b)$ for an integer $b \ge 1$ and elements $k_1, \ldots, k_b$ in $\mathbb{Z}/t^\times$ by $L(\bigoplus_{i=1}^{b} V_{k_i})$.

These lens spaces form a very interesting family of manifolds, which can be completely classified as we will explain below. Two lens spaces $L(V)$ and $L(W)$ of the same dimension $n \geq 3$ have the same homotopy groups, namely, their fundamental group is G and their p-th homotopy group is isomorphic to $\pi_p(S^n)$ for $p \geq 2$. They also have the same homology with integral coefficients, namely $H_p(L(V)) \cong \mathbb{Z}$ for $p = 0, n$, $H_p(L(V)) \cong G$ for p odd and $1 \leq p < n$ and $H_p(L(V)) = 0$ for all other values of p. Also their cohomology groups agree. Nevertheless, not all of them are homotopy equivalent. Moreover, there are homotopy equivalent lens spaces that are not homeomorphic (and hence also not diffeomorphic), see Example 3.72.

3.5.2 Generalised Lens Spaces

Next we introduce a more general class of spaces.

Definition 3.41 (Generalised lens space) Let G be a non-trivial finite cyclic group coming with a free topological action α on S^n. The sphere S^n together with the free topological G-action given by α is here and in the sequel denoted by S^n_α. Then the associated *generalised lens space* $L(\alpha)$ is defined to be $G \backslash S^n_\alpha$.

Suppose that $n \neq 4$. Since then $L(\alpha)$ is a topological manifold of dimension n, it has a CW-structure by [159, Section 9.2] or [219, Theorem 2.2 in III.2 on page 107] and hence S^n_α has a free G-CW-structure.

Any lens space $L(V)$ is diffeomorphic to a generalised lens space, namely, take $\alpha \colon G \to \mathrm{aut}(SV)$ to be the G-action induced on the unit sphere SV, which can be identified with S^n up to diffeomorphism.

If n is odd, then the free action of G on S^n is orientation preserving by the Lefschetz Fixed Point Theorem. Since S^n has a standard orientation, a generalised lens space comes with a preferred orientation given by a fundamental class $[L(\alpha)] \in H_n(L(\alpha))$.

Remark 3.42 (n even and $n = 1$) Suppose that n is even. We conclude from the equality for Euler characteristics $2 = \chi(S^n) = |G| \cdot \chi(L(\alpha))$ and the Lefschetz Fixed Point Theorem that $G \cong \mathbb{Z}/2$ and the action on S^n is orientation reversing. Then $L(\alpha)$ is homotopy equivalent to $\mathbb{RP}^n$, the group of homotopy classes of self-homotopy equivalences of $\mathbb{RP}^n$ is trivial if n is even, and $\mathrm{Wh}(\mathbb{Z}/2)$ is zero. We also note that if n is odd, then the group of homotopy classes of self-homotopy equivalences of $\mathbb{RP}^n$ is $\mathbb{Z}/2$.

If $n = 1$, then $L(\alpha)$ is diffeomorphic to S^1.

Therefore the homotopy classification of generalised lens paces is easy for even n and for $n = 1$ and we will concentrate in the sequel on the case where n is odd and $n \geq 3$.

3.5.3 Homotopy Classification of Generalised Lens Spaces

Next we give the homotopy classification of generalised lens spaces for $n \geq 3$ and n odd. We start by giving an explicit identification

$$\pi_1(L(\alpha), x) = G. \tag{3.43}$$

Given a point $x \in L(\alpha)$, we obtain an isomorphism $s(x): \pi_1(L(\alpha), x) \xrightarrow{\cong} G$ by sending the class of a loop w in $L(\alpha)$ with base point x to the element g in G, for which there is a lift $\widetilde{w}$ in SV of w with $\widetilde{w}(1) = g \cdot \widetilde{w}(0)$. One easily checks using elementary covering theory that this is a well defined isomorphism. If y is another base point, we obtain a homomorphism $t(x, y): \pi_1(L(\alpha), x) \to \pi_1(L(\alpha), y)$ by conjugation with any path v in $L(\alpha)$ from x to y. Since $\pi_1(L(\alpha), x)$ is abelian, $t(x, y)$ is independent of the choice of v. One easily checks $t(x, x) = \text{id}$, $t(y, z) \circ t(x, y) = t(x, z)$ and $s(y) \circ t(x, y) = s(x)$. Hence we can in the sequel identify $\pi_1(L(\alpha), x)$ with G and ignore the choice of the base point $x \in L(\alpha)$.

Let $p: EG \to BG$ be a model for the *universal principal G-bundle*. It has the property that for any principal G-bundle $q: E \to B$ there is a map $f: B \to BG$ called the *classifying map of q* which is up to homotopy uniquely determined by the property that the pullback of p with f is isomorphic over B to q. Equivalently p can be characterised by the properties that p is a principal G-bundle, BG is a CW-complex and EG is contractible. The space BG is called the *classifying space for G*.

Consider a generalised lens space $L(\alpha)$ of odd dimension n. Denote by $f(\alpha): L(\alpha) \to BG$ the classifying map of the principal G-bundle $S^n_\alpha \to L(\alpha)$. Define the element

$$c(L(\alpha)) \in H_n(BG) \tag{3.44}$$

by the image of the fundamental class $[L(\alpha)] \in H_n(L(\alpha))$ associated to the preferred orientation on $L(\alpha)$ under the map $H_n(f(\alpha)): H_n(L(\alpha)) \to H_n(BG)$ induced by $f(\alpha)$ on homology with integer coefficients. The map $f(\alpha): L(\alpha) \to BG$ is n-connected since it induces an isomorphism on π_1 and it has a lift $S^n_\alpha \to EG$ and this lift is n-connected. Hence $H_n(f(\alpha))$ is surjective. As $H_n(L(\alpha))$ is infinite cyclic with $[L(\alpha)]$ as generator, $c(L(\alpha))$ generates $H_n(BG)$. Note that n is odd and that $H_n(BG)$ is isomorphic to $\mathbb{Z}/|G|$ for a cyclic group G of finite order $|G|$.

A map $f: L(\alpha) \to L(\beta)$ of generalised lens spaces of the same dimension $n = \dim(L(\alpha)) = \dim(L(\beta))$ for two free topological G-actions α and β on S^n induces a homomorphism $\pi_1(f, x): \pi_1(L(\alpha)) \to \pi_1(L(\beta))$. Under the identification (3.43) this is an endomorphism $\pi_1(f)$ of G. Define

$$e(f) \in \mathbb{Z}/|G| \tag{3.45}$$

to be the element for which $\pi_1(f)$ sends $g \in G$ to $g^{e(f)}$. Note that $e(f)$ depends only on the homotopy class of f and satisfies $e(g \circ f) = e(g) \cdot e(f)$ and $e(\text{id}) = 1$. In particular $e(f) \in \mathbb{Z}/|G|^\times$ for a homotopy equivalence $f\colon L(\alpha) \to L(\beta)$. Define the *degree*

$$\deg(f) \in \mathbb{Z} \qquad (3.46)$$

of f to be the integer for which $H_n(f)$ sends $[L(\alpha)] \in H_n(L(\alpha))$ to $\deg(f) \cdot [L(\beta)] \in H_n(L(\beta))$.

Given two spaces X and Y, define their *join* $X * Y$ by the pushout

$$\begin{array}{ccc} X \times Y & \longrightarrow & X \times \mathrm{cone}(Y) \\ \downarrow & & \downarrow \\ \mathrm{cone}(X) \times Y & \longrightarrow & X * Y. \end{array}$$

If X and Y are G-spaces, $X * Y$ inherits a G-operation by the diagonal operation. Given two free finite-dimensional unitary G-representations V and W, there is a G-homeomorphism

$$S(V \oplus W) \cong SV * SW. \qquad (3.47)$$

Given a free topological G-action α on S^m and a free topological G-action β on S^n,

$$\alpha * \beta \colon G \to \mathrm{aut}(S^{m+n+1}) \qquad (3.48)$$

denotes the obvious free topological G-action on $S^m * S^n = S^{m+n+1}$.

Theorem 3.49 (Homotopy classification of (generalised) lens spaces) *Let $L(\alpha)$ and $L(\beta)$ be two generalised lens spaces of the same dimension $n \geq 3$ for odd n. Then*

(i) *The map*

$$e \times \deg \colon [L(\alpha), L(\beta)] \to \mathbb{Z}/|G| \times \mathbb{Z}, \quad [f] \mapsto (e(f), \deg(f))$$

is injective, where $[L(\alpha), L(\beta)]$ is the set of homotopy classes of maps from $L(\alpha)$ to $L(\beta)$;

(ii) *An element $(u, d) \in \mathbb{Z}/|G| \times \mathbb{Z}$ is in the image of $e \times \deg$ if and only if we get in $H_n(BG)$*

$$d \cdot c(L(\beta)) = u^{(n+1)/2} \cdot c(L(\alpha));$$

(iii) *The generalised lens spaces $L(\alpha)$ and $L(\beta)$ are homotopy equivalent if and only if there is an element $e \in \mathbb{Z}/|G|^\times$ satisfying in $H_n(BG)$*

$$\pm c(L(\beta)) = e^{(n+1)/2} \cdot c(L(\alpha)).$$

The generalised lens spaces $L(\alpha)$ and $L(\beta)$ are oriented homotopy equivalent if and only if there is an element $e \in \mathbb{Z}/|G|^\times$ satisfying in $H_n(BG)$

$$c(L(\beta)) = e^{(n+1)/2} \cdot c(L(\alpha));$$

3.5 Reidemeister Torsion and (Generalised) Lens Spaces

(iv) *We get in* $H_{2b-1}(B\mathbb{Z}/t)$ *for* $k_1, \ldots, k_b \in \mathbb{Z}/t^\times$

$$c\left(L\left(\bigoplus_{i=1}^{b} V_{k_i}\right)\right) = \prod_{i=1}^{b} k_i^{-1} \cdot c\left(L\left(\bigoplus_{i=1}^{b} V_1\right)\right);$$

(v) *The lens spaces* $L(t; k_1, \ldots, k_b)$ *and* $L(t; l_1, \ldots, l_b)$ *are homotopy equivalent if and only if there is an* $e \in \mathbb{Z}/t^\times$ *such that we get*

$$\prod_{i=1}^{b} k_i = \pm e^b \cdot \prod_{i=1}^{b} l_i \in \mathbb{Z}/t^\times.$$

The lens spaces $L(t; k_1, \ldots, k_b)$ *and* $L(t; l_1, \ldots, l_b)$ *are oriented homotopy equivalent if and only if there is an* $e \in \mathbb{Z}/t^\times$ *such that we get*

$$\prod_{i=1}^{b} k_i = e^b \cdot \prod_{i=1}^{b} l_i \in \mathbb{Z}/t^\times.$$

Proof. (i) Obviously $e(f)$ and $\deg(f)$ depend only on the homotopy class of f. Consider two maps $f_0, f_1: L(\alpha) \to L(\beta)$ with $e(f_0) = e(f_1)$ and $\deg(f_0) = \deg(f_1)$. Choose lifts $\widetilde{f_0}, \widetilde{f_1}: S_\alpha^n \to S_\beta^n$. Let $a: G \to G$ be the endomorphism sending g to $g^{e(f_0)} = g^{e(f_1)}$. Then both lifts $\widetilde{f_k}$ are a-equivariant, which means $\widetilde{f_k}(\alpha(g, x)) = \beta(a(g), \widetilde{f_k}(x))$. The projections induce maps $H_n(S^n) \to H_n(L(\alpha))$ and $H_n(S^n) \to H_n(L(\beta))$, which send the standard fundamental class $[S^n]$ to $|G| \cdot [L(\alpha)]$ and $|G| \cdot [L(\beta)]$. Hence $\deg(f_k) = \deg(\widetilde{f_k})$ for $k = 0, 1$. Thus $\deg(f_0) = \deg(f_1)$ implies $\deg(\widetilde{f_0}) = \deg(\widetilde{f_1})$. Let $a^* S_\beta^n$ be the G-space obtained from S_β^n by restricting the group action with a. Then $\widetilde{f_0}$ and $\widetilde{f_1}: S_\alpha^n \to a^* S_\beta^n$ are G-maps with $\deg(\widetilde{f_0}) = \deg(\widetilde{f_1})$. It suffices to show that they are G-homotopic.

We outline an elementary proof of this fact in the special case of lens spaces, i.e., $S_\alpha^n = SV$ and $S_\beta^n = SW$. There is a G-CW-structure on SV such that there is exactly one equivariant cell $G \times D^k$ in each dimension. It induces a G-CW-structure on $SV \times [0, 1]$ using the standard CW-structure on $[0, 1]$. We want to define inductively for $k = -1, 0, \ldots, n$ G-maps $h_k: SV \times \{0, 1\} \cup SV_k \times [0, 1] \to a^* SW$ such that h_k extends h_{k-1} and $h_{-1} = \widetilde{f_0} \amalg \widetilde{f_1}: SV \times \{0, 1\} \to a^* SW$. Then h_n will be the desired G-homotopy between $\widetilde{f_0}$ and $\widetilde{f_1}$. Note that $SV \times \{0, 1\} \cup SV_k \times [0, 1]$ is obtained from $SV \times \{0, 1\} \cup SV_{k-1} \times [0, 1]$ by attaching one equivariant cell $G \times D^{k+1}$ with some attaching G-map $q_k: G \times S^k \to SV$. In the induction step we must show that the composite $h_{k-1} \circ q_k: G \times S^k \to a^* SW$ can be extended to a G-map $G \times D^{k+1} \to a^* SW$. This is possible if and only if its restriction to $S^k = \{1\} \times S^k$ can be extended to a (non-equivariant) map $D^{k+1} \to SW$. Since SW is n-connected, this can be done for any map $S^k \to SW$ for $k < n$. In the final step we run into the obstruction that a map $S^n \to SW$ can be extended to a map $D^{n+1} \to SW$ if and only if its degree is zero. Now one has to check that the degree of the map $(h_{n-1} \circ q_n)|_{\{1\} \times S^n}: \{1\} \times S^n = S^n \to SW$ is exactly the difference $\deg(\widetilde{f_0}) - \deg(\widetilde{f_1})$ and hence zero.

In the general case one has to apply equivariant obstruction theory, see for instance [398, Theorem 4.11 in Chapter II on page 126].

(ii) Given a map $f: L(\alpha) \to L(\beta)$, let $\pi_1(f): G \to G$ be the map induced on the fundamental groups under the identification (3.43). Then the following diagram

$$\begin{array}{ccc} H_n(L(\alpha)) & \xrightarrow{H_n(f)} & H_n(L(\beta)) \\ {\scriptstyle H_n(f(\alpha))}\downarrow & & \downarrow{\scriptstyle H_n(f(\beta))} \\ H_n(BG) & \xrightarrow[H_n(B\pi_1(f))]{} & H_n(BG) \end{array}$$

commutes. This implies

$$\deg(f) \cdot c(L(\beta)) = H_n(B\pi_1(f))\big(c(L(\alpha))\big). \tag{3.50}$$

Given $s \in \mathbb{Z}$, let $m_s: \mathbb{Z}/t \to \mathbb{Z}/t$ be multiplication by s. Let $s, k_1, \ldots, k_b, l_1, \ldots, l_b$ be integers which are prime to t. Fix integers $k'_1, \ldots, k'_b$ such that we get in $\mathbb{Z}/t$ the equation $\overline{k_i} \cdot \overline{k'_i} = \overline{1}$. Define a m_s-equivariant map $d_i: V_{k_i} \to V_{l_i}$ by $z \mapsto z^{k'_i l_i s}$. It has degree $k'_i l_i s$. The m_s-equivariant map $*_{i=1}^b d_i: *_{i=1}^b SV_{k_i} \to *_{i=1}^b SV_{l_i}$ yields under the identification (3.47) an m_s-equivariant map $\widetilde{d}: S(\bigoplus_{i=1}^b V_{k_i}) \to S(\bigoplus_{i=1}^b V_{l_i})$ of degree $s^b \cdot \prod_{i=1}^b k'_i l_i$. Taking the quotient under the G-action yields a map $d: L(t; \overline{k_1}, \ldots, \overline{k_b}) \to L(t; \overline{l_1}, \ldots, \overline{l_b})$ of $n = (2b-1)$-dimensional lens spaces with $\deg(d) = s^b \cdot \prod_{i=1}^b k'_i l_i$ and $e(d) = s$. We conclude from (3.50) in the special case $k_i = l_i$ for $i = 1, \ldots, b$ that $H_n(Bm_s): H_n(B\mathbb{Z}/t) \to H_n(B\mathbb{Z}/t)$ is multiplication by s^b since $c\big(L\big(\bigoplus_{i=1}^b V_{k_i}\big)\big)$ is a generator of $H_n(B\mathbb{Z}/t)$. If we apply this in the case where $\pi_1(f): G \to G$ is given by $g \mapsto g^s$, then $H_n(B\pi_1(f)): H_n(BG) \to H_n(BG)$ is multiplication by s^b for $b = (n+1)/2$. Thus (3.50) becomes

$$\deg(f) \cdot c(L(\beta)) = e(f)^{(n+1)/2} \cdot c(L(\alpha)). \tag{3.51}$$

We conclude from (3.51) in the special case $s = 1$ that we get in $H_n(B(\mathbb{Z}/t))$ the equation

$$\prod_{i=1}^b l_i \cdot c\left(L\left(\bigoplus_{i=1}^b V_{l_i}\right)\right) = \prod_{i=1}^b k_i \cdot c\left(L\left(\bigoplus_{i=1}^b V_{k_i}\right)\right). \tag{3.52}$$

Next we show for a map $f: L(\alpha) \to L(\beta)$ and $m \in \mathbb{Z}$ that we can find another map $f': L(\alpha) \to L(\beta)$ with $e(f) = e(f')$ and $\deg(f') = \deg(f) + m \cdot |G|$. This follows from equivariant obstruction theory [398, Theorem 4.11 in Chapter II on page 126]. We outline an elementary proof of this fact in the special case of lens spaces, i.e., $S_\alpha^n = SV$ and $S_\beta^n = SW$. Fix some embedded disk $D^n \subset SV$ such that $g \cdot D^n \cap D^n \neq \emptyset$ implies $g = 1$. Let $\widetilde{f}: SV \to SW$ be a lift of f. Recall that f is m_e-equivariant for $m_e: G \to G$, $g \mapsto g^e$, with $e = e(f_0) = e(f_1)$. Define a new m_e-equivariant map $\widetilde{f}': SV \to SW$ as follows. Outside $G \cdot D^n$ the maps $\widetilde{f}$ and $\widetilde{f}'$

3.5 Reidemeister Torsion and (Generalised) Lens Spaces

agree. Let $\frac{1}{2}D^n$ be $\{x \in D^n \mid \|x\| \le 1/2\}$. Define $\widetilde{f}'$ on $G \cdot \left(D^n - \frac{1}{2}D^n\right)$ by sending $(g, t \cdot x)$ for $g \in G$, $t \in [1/2, 1]$ and $x \in S^{n-1} = \partial D^n$ to $g^e \cdot \widetilde{f}((2t-1)x)$. For $0 \in D^n$ the origin, let $c \colon (\frac{1}{2}D^n, \partial \frac{1}{2}D^n) \to (SW, \widetilde{f}(0))$ be a map such that the induced map $(\frac{1}{2}D^n)/\partial(\frac{1}{2}D^n) \to SW$ has degree m. Define $f'|_{G \cdot \frac{1}{2}D^n} \colon G \cdot \frac{1}{2}D^n \to SW$ by sending (g, x) to $g^e \cdot c(x)$. One easily checks that $\widetilde{f}'$ has degree $\deg(\widetilde{f}) + m \cdot |G|$. Then $G \backslash \widetilde{f}'$ is the desired map f'.

Note that there is at least one map $f \colon L(\alpha) \to L(\beta)$ with $e(f) = e$ for any given $e \in \mathbb{Z}/|G|$. It can be constructed by induction over the skeletons of S^n_α since G acts freely on S^n_α and S^n_β is $(n-1)$-connected. Now assertion (ii) follows.

(iii) Let $f \colon L(\alpha) \to L(\beta)$ be a map. Then f is a homotopy equivalence if and only if it induces isomorphisms on all homotopy groups. This is the case if and only if $\pi_1(f)$ is an automorphism and a lift $\widetilde{f} \colon S^n_\alpha \to S^n_\beta$ induces an isomorphism on all homology groups. This follows from the Hurewicz Theorem, see [178, Theorem 4.32 on page 366] or [431, Theorem IV.7.13 on page 181], and the fact that a covering induces an isomorphism on π_n for $n \ge 2$. Hence f is a homotopy equivalence if and only if $e(f) \in \mathbb{Z}/|G|^\times$ and $\deg(f) = \deg(\widetilde{f}) = \pm 1$. Recall that f is an oriented homotopy equivalence if and only if f is a homotopy equivalence and $\deg(f) = 1$. Now the claim follows from assertion (ii).

(iv) This follows follows from (3.52)

(v) This follows follows from (3.52) and assertion (iii). This finishes the proof of Theorem 3.49. □

Lemma 3.53 (i) *Every generalised lens space of odd dimension is oriented homotopy equivalent to some lens space;*
(ii) *Every closed manifold that is homotopy equivalent to a lens space is homeomorphic to a generalised lens spaces of odd dimension.*

Proof. (i) This follows from Theorem 3.49 (iii) since by Theorem 3.49 (iv) every element in $H_1(B\mathbb{Z}/t)$ can be realised by $c(L(t, k_1, \ldots, k_b))$.

(ii) This follows from the fact that the generalised Poincaré Conjecture holds in all dimensions, see Theorem 2.5. □

3.5.4 Reidemeister Torsion

Let G be a finite cyclic group. Denote by $\mathbb{Z}G$ and $\mathbb{Q}G$ its integral and rational group ring. Let $N = \sum_{g \in G} g \in \mathbb{Z}G$ be the *norm element*. Consider the diagram of commutative rings

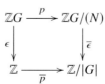

where $\mathbb{Z}G/(N)$ is obtained from $\mathbb{Z}G$ by dividing out the ideal generated by N, p and $\bar{p}$ are the obvious projections, ϵ is the augmentation map sending $\sum_{g \in G} \lambda_g \cdot g$ to $\sum_{g \in G} \lambda_g$, and $\bar{\epsilon}$ is uniquely determined by the property that the diagram commutes. One easily checks that it is a cartesian diagram of commutative rings. It induces an isomorphism

$$p \times \epsilon \colon \mathbb{Q}G \xrightarrow{\cong} \mathbb{Q}G/(N) \times \mathbb{Q} \tag{3.54}$$

where p denotes again the natural projection and ϵ again the augmentation homomorphism. As $\mathbb{Q}G$ is a commutative semisimple ring, it is a product of fields by Wedderburn's Theorem. Since K_1 commutes with finite products and the determinant induces for any field F an isomorphism $\det \colon K_1(F) \xrightarrow{\cong} F^\times$, we obtain isomorphisms

$$K_1(p) \times K_1(\epsilon) \colon K_1(\mathbb{Q}G) \xrightarrow{\cong} K_1(\mathbb{Q}G/(N)) \times K_1(\mathbb{Q});$$

$$\det \colon K_1(\mathbb{Q}G/(N)) \xrightarrow{\cong} (\mathbb{Q}G/(N))^\times;$$

$$\det \colon K_1(\mathbb{Q}) \xrightarrow{\cong} \mathbb{Q}^\times.$$

Next recall that for any commutative ring R the determinant induces a homomorphism $\det \colon K_1(R) \to R^\times$, which is split surjective since any unit in $R^\times$ yields a 1×1-matrix, which represents an element in $K_1(R)$, see also [361, Proposition 2.2.1]. We conclude from [324, Theorem 5.6 on page 142], see also [5] and [411], that the homomorphism $K_1(\mathbb{Z}G) \to K_1(\mathbb{Q}G)$ is injective and hence the map $\det \colon K_1(\mathbb{Z}G) \to \mathbb{Z}G^\times$ is bijective. From [324, Theorem 7.4 on page 180] we also obtain that $\mathrm{Wh}(G)$ is a finitely generated free abelian group. Put

$$D(G) := \mathbb{Q}G/(N)^\times / T(G), \tag{3.55}$$

where $T(G)$ is the image of the subgroup $\{\pm g \mid g \in G\}$ of trivial units under the composite $\mathbb{Z}G^\times \xrightarrow{i} \mathbb{Q}G^\times \xrightarrow{p} \mathbb{Q}G/(N)^\times$ where i is the obvious inclusion. This composite induces a map

$$\mu \colon \mathrm{Wh}(G) \to D(G). \tag{3.56}$$

The following diagram commutes

$$\begin{array}{ccccc}
K_1(\mathbb{Z}G) & \xrightarrow[\cong]{\det} & \mathbb{Z}G^\times & \xrightarrow{p \times \epsilon^\times} & \mathbb{Z}G/(N)^\times \times \mathbb{Z}^\times \\
\downarrow & & \downarrow & & \downarrow \\
K_1(\mathbb{Q}G) & \xrightarrow[\det]{\cong} & \mathbb{Q}G^\times & \xrightarrow[p \times \epsilon^\times]{\cong} & \mathbb{Q}G/(N)^\times \times \mathbb{Q}^\times
\end{array}$$

where the vertical arrows are induced by the inclusions. Now one easily checks

Lemma 3.57 *The map μ is injective.*

Let X be a free finite G-CWcomplex, or, equivalently, X is a compact free G-space such that $G \backslash X$ is a finite CW-complex. Suppose that G-acts trivially on its homology $H_n(X, \mathbb{Z})$. Let $C_*(X)$ be the finite free cellular $\mathbb{Z}G$-chain complex.

Then $\mathbb{Q}G/(N) \otimes_{\mathbb{Z}G} C_*(X)$ is a finite free $\mathbb{Q}G/(N)$-chain complex. Its homology $H_n(\mathbb{Q}G/(N) \otimes_{\mathbb{Z}G} C_*(X)) \cong \mathbb{Q}G/(N) \otimes_{\mathbb{Q}G} H_n(\mathbb{Q} \otimes_{\mathbb{Z}} C_*(X))$ is trivial, as G acts trivial on $H_n(\mathbb{Q} \otimes_{\mathbb{Z}} C_*(X))$ and $\mathbb{Q}G$ is semisimple. Hence it is contractible. Choose a cellular $\mathbb{Z}G$-basis for $C_*(X)$. It induces a basis on the finite free $\mathbb{Q}G/(N)$-chain complex $\mathbb{Q}G/(N) \otimes_{\mathbb{Z}G} C_*(X)$. Its Reidemeister torsion defines an element in $\widetilde{K}_1(\mathbb{Q}G/(N))$, see (3.16). The image of this class under the obvious surjection $\widetilde{K}_1(\mathbb{Q}G/(N)) \to D(G)$ given by the composite of the map $\det \colon K_1(\mathbb{Q}G/(N)) \to \mathbb{Q}G/(N)^\times$ with the obvious projection $\mathbb{Q}G/(N)^\times \to D(G)$ is denoted by

$$\Delta(X) \in D(G). \tag{3.58}$$

The passage to $D(G)$ ensures that the choice of the cellular $\mathbb{Z}G$-basis does not matter. Let $f \colon X \to Y$ be a G-homotopy equivalence of finite free G-CW-complexes such that G acts on $H_*(X; \mathbb{Z})$ trivially and hence also on $H_*(Y; \mathbb{Z})$. So $\Delta(X)$ and $\Delta(Y)$ in $D(G)$ are defined. Let $\tau(f) \in \mathrm{Wh}(\Pi_1(X))$ be the Whitehead torsion of $C_*(f) \colon C_*(X) \to C_*(Y)$ with respect to some cellular basis, see Definition 3.22. We can consider $\tau(f)$ as an element in $\mathrm{Wh}(G)$ by the identification (3.43). Lemma 3.18 implies

$$\mu(\tau(f)) = \Delta(Y) - \Delta(X). \tag{3.59}$$

We conclude from Lemma 3.57 and from (3.59) that $\tau(f)$ is trivial if and only if $\Delta(X) = \Delta(Y)$.

Definition 3.60 (Reidemeister torsion of generalised lens spaces) Define the *Reidemeister torsion* of an odd-dimensional generalised lens space $L(\alpha)$

$$\Delta(L(\alpha)) \in D(G)$$

by $\Delta(S_\alpha^n)$ in the sense of (3.58).

3.5.5 Simple Homotopy Classification of Generalised Lens Spaces

Let α and β be free topological actions of the finite cyclic group G on S^n. Let $f \colon L(\alpha) \to L(\beta)$ be a homotopy equivalence. For $e \in \mathbb{Z}/|G|^\times$ let $m_e \colon G \xrightarrow{\cong} G$ be the group automorphism sending g to g^e. Let $(m_e)_* \colon D(G) \to D(G)$ be the homomorphism induced by m_e. We have defined the element $e(f) \in \mathbb{Z}/|G|^\times$ in (3.45). Recall that $m_{e(f)}$ is the automorphism induced by $\pi_1(f)$ under the identification (3.43). Let $\tau(f) \in \mathrm{Wh}(\Pi(L(\beta)))$ be the Whitehead torsion of f, which we can view because of (3.43) to be an element in $\mathrm{Wh}(G)$.

Theorem 3.61 (Simple Homotopy classification of generalised lens spaces)

(i) *We get in $D(G)$*

$$\mu(\tau(f)) = \Delta(L(\beta)) - (m_{e(f)})_*(\Delta(L(\alpha)));$$

(ii) *The homotopy equivalence* $f: L(\alpha) \to L(\beta)$ *is simple if and only if* $\Delta(L(\beta)) = (m_{e(f)})_*(\Delta(L(\alpha)))$ *in* $D(G)$ *holds;*
(iii) *The generalised lens spaces* $L(\alpha)$ *and* $L(\beta)$ *are simple homotopy equivalent if and only if there is an element* $e \in \mathbb{Z}/|G|^\times$ *such that we get* $\pm c(L(\beta)) = e^{(n+1)/2} \cdot c(L(\alpha))$ *in* $H_n(BG)$ *and* $\Delta(L(\beta)) = (m_e)_*(\Delta(L(\alpha)))$ *in* $D(G)$.
The generalised lens spaces $L(\alpha)$ *and* $L(\beta)$ *are simple oriented homotopy equivalent if and only if there is an element* $e \in \mathbb{Z}/|G|^\times$ *such that we obtain* $c(L(\beta)) = e^{(n+1)/2} \cdot c(L(\alpha))$ *in* $H_n(BG)$ *and* $\Delta(L(\beta)) = (m_e)_*(\Delta(L(\alpha)))$ *in* $D(G)$.

Proof. (i) The homotopy equivalence $f: L(\alpha) \to L(\beta)$ induces an $m_{e(f)}$-equivariant homotopy equivalence $\widetilde{f}: S_\alpha^n \to S_\beta^n$, or equivalently a G-homotopy equivalence $\widetilde{f}: (m_{e(f)})_* S_\alpha^n \to S_\beta^n$ where $(m_{e(f)})_* S_\alpha^n$ is obtained from S_α^n by induction with respect to the automorphism $m_{e(f)}$. From (3.59) we obtain that

$$\mu(\tau(f)) = \Delta(S_\beta^n) - \Delta((m_{e(f)})_* S_\alpha^n).$$

The desired result follows from the definition $\Delta(L(\beta)) = \Delta(S_\beta^n)$ and from $\Delta((m_{e(f)})_* S_\alpha^n) = (m_{e(f)})_* \Delta(S_\alpha^n) = (m_{e(f)})_* \Delta(L(\alpha))$ where the first equation is obtained by observing the effect of the automorphism $(m_{e(f)})$ on the differentials in the underlying based contractible chain complex and the second equation is again a definition.
(ii) This follows from assertion (i) and Lemma 3.57.
(iii) This follows from Theorem 3.49 (i) and (iii) and assertion (ii). □

3.5.6 A Variant of Reidemeister Torsion

Let X be a finite CW-complex with finite fundamental group π. Let U be an orthogonal finite-dimensional π-representation. Denote by $H_p(X; U)$ the homology of X with coefficients in U, i.e., the homology of the $\mathbb{R}$-chain complex $U \otimes_{\mathbb{Z}\pi} C_*(\widetilde{X})$. Suppose that X is U-acyclic, i.e., $H_p(X; U) = 0$ for all $p \geq 0$. If we fix a cellular basis for $C_*(\widetilde{X})$ and some orthogonal $\mathbb{R}$-basis for U, then $U \otimes_{\mathbb{Z}\pi} C_*(\widetilde{X})$ is a contractible based free finite $\mathbb{R}$-chain complex and defines an element $\tau(U \otimes_{\mathbb{Z}\pi} C_*(\widetilde{X})) \in \widetilde{K}_1(\mathbb{R})$, see (3.16). Define the *Reidemeister torsion*

$$\Delta(X; U) \in \mathbb{R}^{>0} \tag{3.62}$$

to be the image of $\tau(U \otimes_{\mathbb{Z}\pi} C_*(\widetilde{X})) \in \widetilde{K}_1(\mathbb{R})$ under the homomorphism $\widetilde{K}_1(\mathbb{R}) \to \mathbb{R}^{>0}$ sending the class $[A]$ of $A \in \mathrm{GL}(n, \mathbb{R})$ to $\det(A)^2$. Note that for any trivial unit $\pm \gamma$ the automorphism of U given by multiplication by $\pm \gamma$ is orthogonal and that the square of the determinant of any orthogonal automorphism of U is 1. Therefore

3.5 Reidemeister Torsion and (Generalised) Lens Spaces

$\Delta(X; U) \in \mathbb{R}^{>0}$ is independent of the choice of cellular basis for $C_*(\widetilde{X})$ and the orthogonal basis for U and hence is an invariant of the CW-complex X and U.

Exercise 3.63 Show that the 3-dimensional real projective space $\mathbb{R}P^3$ is a lens space. Let $\mathbb{R}^-$ be the non-trivial orthogonal $\mathbb{Z}/2$-representation. Show that $\mathbb{R}P^3$ is $\mathbb{R}^-$-acyclic and compute the Reidemeister torsion $\Delta(\mathbb{R}P^3; \mathbb{R}^-)$.

Lemma 3.64 *Let $f: X \to Y$ be a homotopy equivalence of connected finite CW-complexes and let U be an orthogonal finite-dimensional $\pi = \pi_1(Y)$-representation. Suppose that Y is U-acyclic. Let f^*U be the orthogonal $\pi_1(X)$-representation obtained from U by restriction with the isomorphism $\pi_1(f)$. Let*

$$\det_U : \mathrm{Wh}(\Pi(Y)) \to \mathbb{R}^{>0}$$

be the map sending the class $[A]$ of $A \in \mathrm{GL}(n, \mathbb{Z}\pi_1(Y))$ to the real number $\det(\mathrm{id}_U \otimes_{\mathbb{Z}\pi} R_A : U \otimes_{\mathbb{Z}\pi} \mathbb{Z}\pi^n \to U \otimes_{\mathbb{Z}\pi} \mathbb{Z}\pi^n)^2$ where $R_A : \mathbb{Z}\pi^n \to \mathbb{Z}\pi^n$ is the $\mathbb{Z}\pi$-automorphism induced by A. Then

$$\frac{\Delta(Y, U)}{\Delta(X, f^*U)} = \det_U(\tau(f)).$$

Proof. This follows from Lemma 3.18 (i) applied to the based exact sequence of contractible based free finite $\mathbb{R}$-chain complexes $0 \to U \otimes_{\mathbb{Z}\pi} C_*(\widetilde{Y}) \to U \otimes_{\mathbb{Z}\pi} \mathrm{cone}_*(C_*(\widetilde{f})) \to \Sigma(U \otimes_{\mathbb{Z}\pi_1(f)} C_*(\widetilde{X})) \to 0$. □

Lemma 3.65 *Let G be a cyclic group of finite order $|G|$.*

(i) *Let V be a free unitary finite-dimensional G-representation. Let U be an orthogonal finite-dimensional G-representation with $U^G = 0$. Then $L(V)$ is U-acyclic and the Reidemeister torsion $\Delta(L(V); U) \in \mathbb{R}^{>0}$ is defined;*

(ii) *Let V, V_1, V_2 be free unitary finite-dimensional G-representations. Let U, U_1 and U_2 be orthogonal finite-dimensional G-representations with $U^G = U_1^G = U_2^G = 0$. Then*

$$\Delta(L(V_1 \oplus V_2); U) = \Delta(L(V_1); U) \cdot \Delta(L(V_2); U);$$
$$\Delta(L(V); U_1 \oplus U_2) = \Delta(L(V); U_1) \cdot \Delta(L(V); U_2).$$

Proof. (i) Let X be a finite CW-complex with fundamental group G. We show that X is U-acyclic if G acts trivial on $H_p(\widetilde{X})$ and U is an orthogonal finite-dimensional G-representation with $U^G = 0$. We have to show that $H_p(C_*(\widetilde{X}) \otimes_{\mathbb{Z}G} U) \cong H_p(\widetilde{X}) \otimes_{\mathbb{Z}} \mathbb{R} \otimes_{\mathbb{R}G} U$ vanishes. By assumption $H_p(\widetilde{X}) \otimes_{\mathbb{Z}} \mathbb{R}$ is a direct sum of copies of the trivial G-representation $\mathbb{R}$ and U is a direct sum of non-trivial irreducible representations. Since for any non-trivial irreducible G-representation W the tensor product $\mathbb{R} \otimes_{\mathbb{R}G} W$ is trivial, the claim follows.

(ii) The claim for the second equation is obvious, it remains to prove the first. Since G acts trivially on the homology of SV and SW and hence on the homology of $SV \times SW$, $SV \times DV$ and $DV \times SW$, we conclude from Theorem 3.1 (i), Lemma 3.18 (i), Lemma 3.64 and (3.47)

$$\Delta(L(V \oplus W); U)$$
$$= \Delta(G\backslash(DV \times SW); U) \cdot \Delta(G\backslash(SV \times DW); U) \cdot (\Delta(G\backslash(SV \times SW); U))^{-1}.$$

Hence it remains to show

$$\Delta(G\backslash(DV \times SW); U) = \Delta(L(W); U);$$
$$\Delta(G\backslash(SV \times DW); U) = \Delta(L(V); U);$$
$$\Delta(G\backslash(SV \times SW); U) = 1.$$

These equations will follow from the slightly more general formula (3.66) below. Let D_* be a finite $\mathbb{R}G$-chain complex such that G-acts trivially on its homology. Assume that D_* comes with an $\mathbb{R}$-basis. Then $C_*(SV) \otimes_\mathbb{Z} D_*$ with the diagonal G-action is a finite $\mathbb{R}G$-chain complex such that G acts trivially on its homology and there is a preferred $\mathbb{R}G$-basis. Let $\chi_\mathbb{R}(D_*)$ be the Euler characteristic of D_*, i.e.,

$$\chi_\mathbb{R}(D_*) := \sum_{p \in \mathbb{Z}} (-1)^p \cdot \dim_\mathbb{R}(D_p) = \sum_{p \in \mathbb{Z}} (-1)^p \cdot \dim_\mathbb{R}(H_p(D_*)).$$

Then

$$\Delta(C_*(SV) \otimes_\mathbb{Z} D_* \otimes_{\mathbb{R}G} U) = \Delta(L(V); U)^{\chi_\mathbb{R}(D_*)} \qquad (3.66)$$

where $\Delta(C_*(SV) \otimes_\mathbb{Z} D_* \otimes_{\mathbb{R}G} U) \in \mathbb{R}^{>0}$ is the Reidemeister torsion of the acyclic based free finite $\mathbb{R}$-chain complex $C_*(SV) \otimes_\mathbb{Z} D_* \otimes_{\mathbb{R}G} U$, which is defined as the image of the Whitehead torsion $\tau(C_*(SV) \otimes_\mathbb{Z} D_* \otimes_{\mathbb{R}G} U)$ in $\widetilde{K}_1(\mathbb{R})$, see (3.16), under the homomorphism $\widetilde{K}_1(\mathbb{R}) \to \mathbb{R}^{>0}$, $[A] \mapsto \det(A)^2$.

It remains to prove (3.66). Since $\mathbb{R}G$ is semisimple, there is an $\mathbb{R}G$-chain homotopy equivalence $p_* \colon D_* \to H_*(D_*)$ where we consider $H_*(D_*)$ as an $\mathbb{R}G$-chain complex with the trivial differential. Equip $H_*(D_*)$ with an $\mathbb{R}$-basis. Then we get in $\widetilde{K}_1(\mathbb{R}G)$ from Lemma 3.18 (i) and $\chi(L(V)) = 0$

$$\tau\left(\mathrm{id}_{C_*(SV)} \otimes p_* \colon C_*(SV) \otimes_\mathbb{Z} D_* \to C_*(SV) \otimes_\mathbb{Z} H_*(D_*)\right)$$
$$= \sum_{q \in \mathbb{Z}} (-1)^q \cdot \tau\left(\mathrm{id}_{C_q(SV)} \otimes p_* \colon C_q(SV) \otimes_\mathbb{Z} D_* \to C_q(SV) \otimes_\mathbb{Z} H_*(D_*)\right)$$
$$= \sum_{q \in \mathbb{Z}} (-1)^q \cdot \dim_{\mathbb{Z}G}(C_p(SV)) \cdot \tau\left(\mathrm{id}_{\mathbb{Z}G} \otimes p_* \colon \mathbb{Z}G \otimes_\mathbb{Z} D_* \to \mathbb{Z}G \otimes_\mathbb{Z} H_*(D_*)\right)$$
$$= \chi(L(V)) \cdot \tau\left(\mathrm{id}_{\mathbb{Z}G} \otimes p_* \colon \mathbb{Z}G \otimes_\mathbb{Z} D_* \to \mathbb{Z}G \otimes_\mathbb{Z} H_*(D_*)\right)$$
$$= 0 \cdot \tau\left(\mathrm{id}_{\mathbb{Z}G} \otimes p_* \colon \mathbb{Z}G \otimes_\mathbb{Z} D_* \to \mathbb{Z}G \otimes_\mathbb{Z} H_*(D_*)\right)$$
$$= 0.$$

The chain complex version of Lemma 3.64 shows

$$\Delta(C_*(SV) \otimes_\mathbb{Z} D_* \otimes_{\mathbb{R}G} U) = \Delta(C_*(SV) \otimes_\mathbb{Z} H_*(D_*) \otimes_{\mathbb{R}G} U).$$

3.5 Reidemeister Torsion and (Generalised) Lens Spaces

Note that the left side and hence also the right side of the equation above is independent of the choice of the $\mathbb{R}$-basis for $H_*(D_*)$ so that we can use any $\mathbb{R}$-basis on $H_*(D_*)$. (This is a consequence of $\chi(L(V)) = 0$.) For every p choose an isomorphism $\mathbb{R}^{d_p} \xrightarrow{\cong} H_p(D_*)$ and equip $H_p(D_*)$ with the image of the standard basis on $\mathbb{R}^{d_p}$. Now one easily checks

$$\Delta(C_*(SV) \otimes_\mathbb{Z} H_*(D_*) \otimes_{\mathbb{R}G} U) = \Delta\left(\bigoplus_{p \geq 0} \bigoplus_{i=1}^{d_p} \Sigma^p(C_*(SV) \otimes_{\mathbb{Z}G} U)\right)$$

$$= \prod_{p \geq 0} \left(\prod_{i=1}^{d_p} \Delta(C_*(SV) \otimes_{\mathbb{Z}G} U)\right)^{(-1)^p}$$

$$= \Delta(C_*(SV) \otimes_{\mathbb{Z}G} U)^{\chi_\mathbb{R}(D_*)}$$

$$= \Delta(L(V); U)^{\chi_\mathbb{R}(D_*)}.$$

This finishes the proof of (3.66) and hence of Lemma 3.65. □

Remark 3.67 Let V be a free unitary finite-dimensional G-representation. Let U be an orthogonal finite-dimensional G-representation with $U^G = 0$. Then we have defined $\Delta(L(V)) \in D(G)$ in (3.58) and $\Delta(L(V); U) \in \mathbb{R}$ in (3.62).

Define a homomorphism

$$\overline{\det}_U : D(G) \to \mathbb{R}$$

by sending $x \in D(G)^\times$ represented by $\tilde{x} = \sum_{g \in G} \lambda_g \cdot g \in \mathbb{Z}G$ to the real number $\det(m_{\tilde{x}} : U \xrightarrow{\cong} U)^2$ for $m_{\tilde{x}}(u) = \sum_{g \in G} \lambda_g \cdot gu$. Then the composite of $\overline{\det}_U$ with the map $\mu : \text{Wh}(G) \to D(G)$ of (3.56) is the homomorphism $\det_U : \text{Wh}(G) \to \mathbb{R}$ which comes from the one appearing in Lemma 3.64 and the identification (3.43). Now one easily checks

$$\overline{\det}_U(\Delta(L(V))) = \Delta(L(V); U).$$

3.5.7 The Classification of Lens Spaces

Theorem 3.68 (Diffeomorphism classification of lens spaces) *Let $L(V)$ and $L(W)$ be two lens spaces of the same dimension $n \geq 3$. Then the following statements are equivalent:*

(i) *There is an automorphism $a: G \to G$ such that V and a^*W are isomorphic as orthogonal G-representations;*
(ii) *There is an isometric diffeomorphism $L(V) \to L(W)$;*
(iii) *There is a diffeomorphism $L(V) \to L(W)$;*
(iv) *There is a homeomorphism $L(V) \to L(W)$;*
(v) *There is a simple homotopy equivalence $L(V) \to L(W)$;*

(vi) *There is an automorphism* $a\colon G \to G$ *such that for any orthogonal finite-dimensional representation* U *with* $U^G = 0$

$$\Delta(L(W), U) = \Delta(L(V), a^*U)$$

holds;

(vii) *There is an automorphism* $a\colon G \to G$ *such that for any non-trivial 1-dimensional unitary G-representation* U

$$\Delta(L(W), \operatorname{res} U) = \Delta(L(V), a^* \operatorname{res} U)$$

holds where the orthogonal representation res U *is obtained from* U *by restricting the scalar multiplication from* $\mathbb{C}$ *to* $\mathbb{R}$.

Proof. The implications (i) $\Rightarrow$ (ii) $\Rightarrow$ (iii) $\Rightarrow$ (iv) $\Rightarrow$ (v) $\Rightarrow$ (vi) $\Rightarrow$ (vii) are obvious or follow directly from Theorem 3.1 (v) and Lemma 3.64. Hence it remains to prove the implication (vii) $\Rightarrow$ (i).

Fix a generator $g \in G$, or equivalently, an identification $G = \mathbb{Z}/|G|$. Choose $e \in \mathbb{Z}/|G|^\times$ such that $a\colon G \to G$ sends g to g^e. Recall that for $k \in \mathbb{Z}/|G|$ we denote by V_k the 1-dimensional unitary G-representation, for which g acts by multiplication by $\exp(2\pi i k/|G|)$. In the sequel we consider $k \in \mathbb{Z}/|G|^\times$ and hence its inverse $k^{-1} \in \mathbb{Z}/|G|^\times$ is defined. Recall that then G acts freely on SV_k. The based $\mathbb{Z}G$-chain complex of SV_k is concentrated in dimension 0 and 1 and its first differential is $g^{k^{-1}} - 1\colon \mathbb{Z}G \to \mathbb{Z}G$. If g acts on U by multiplication by the $|G|$-th root of unity ζ, then we conclude

$$\Delta(L(V_k); U) = ||\zeta^{k^{-1}} - 1||^2;$$
$$\Delta(L(V_k); a^*U) = ||\zeta^{ek^{-1}} - 1||^2.$$

We can write for appropriate numbers $b \in \mathbb{Z}$, $b \geq 1$, $k_i, l_j \in \mathbb{Z}/G^\times$

$$V = \bigoplus_{i=1}^{b} V_{k_i};$$
$$W = \bigoplus_{j=1}^{b} V_{l_j}.$$

We conclude from Lemma 3.65

$$\Delta(L(V); U) = \prod_{i=1}^{b} (\zeta^{k_i^{-1}} - 1) \cdot (\zeta^{-k_i^{-1}} - 1);$$
$$\Delta(L(W); a^*U) = \prod_{i=1}^{b} (\zeta^{el_j^{-1}} - 1) \cdot (\zeta^{-el_j^{-1}} - 1).$$

3.5 Reidemeister Torsion and (Generalised) Lens Spaces

This implies that for any $|G|$-th root of unity ζ with $\zeta \neq 1$ the following equation holds

$$\prod_{i=1}^{b}(\zeta^{k_i^{-1}} - 1) \cdot (\zeta^{-k_i^{-1}} - 1) = \prod_{i=1}^{b}(\zeta^{el_j^{-1}} - 1) \cdot (\zeta^{-el_j^{-1}} - 1). \quad (3.69)$$

We will need the following number theoretic result due to Franz, whose proof can be found for instance in [128].

Theorem 3.70 (Franz' Independence Lemma) *Let $t \geq 2$ be an integer and $S = \{j \in \mathbb{Z} \mid 0 < j < t, (j,t) = 1\}$. Let $(a_j)_{j \in S}$ be a sequence of integers indexed by S such that $\sum_{j \in S} a_j = 0$, $a_j = a_{t-j}$ for $j \in S$ and $\prod_{j \in S}(\zeta^j - 1)^{a_j} = 1$ holds for every t-th root of unity $\zeta \neq 1$.*

Then $a_j = 0$ for $j \in S$.

Put $t = |G|$. For $j \in S = \{j \in \mathbb{Z} \mid 0 < j < |G|, (j, |G|) = 1\}$, let x_j be the number of elements in the sequence $k_1^{-1}, -k_1^{-1}, k_2^{-1}, -k_2^{-1}, \ldots, k_b^{-1}, -k_b^{-1}$, which agree with the class of j in $\mathbb{Z}/|G|^{\times}$. We conclude $\sum_{j \in S} x_j = 2b$. Obviously $x_j = x_{|G|-j}$ for $j \in S$. Define analogously a sequence $(y_j)_{j \in S}$ for the sequence $el_1^{-1}, -el_1^{-1}, el_2^{-1}, -el_2^{-1}, \ldots, el_b^{-1}, -el_b^{-1}$. Put $a_j = x_j - y_j$ for $j \in S$. Then $\sum_{j \in S} a_j = 0$, $a_j = a_{|G|-j}$ for $j \in S$ and we conclude from (3.69) that $\prod_{j \in S}(\zeta^j - 1)^{a_j} = 1$ for any $|G|$-th root of unity $\zeta \neq 1$. We conclude from Franz' Independence Lemma 3.70 that $a_j = 0$ and hence $x_j = y_j$ holds for $j \in S$. This implies that there is a permutation $\sigma \in \Sigma_b$ together with signs $\epsilon_i \in \{\pm 1\}$ for $i = 1, 2, \ldots, b$ such that $k_i^{-1} = \epsilon_i \cdot l_{\sigma(i)}^{-1}$ and hence $k_i = \epsilon_i \cdot l_{\sigma(i)}$ holds in $\mathbb{Z}/|G|^{\times}$ for $i = 1, 2, \ldots, b$. We conclude that V_i and $a^*W_{\sigma(i)}$ are isomorphic as orthogonal G-representations. Hence V and a^*W are isomorphic as orthogonal representations. This finishes the proof of Theorem 3.68. □

Corollary 3.71 *Two lens spaces $L(t; k_1, \ldots, k_b)$ and $L(t; l_1, \ldots, l_b)$ are homeomorphic if and only if there are $e \in \mathbb{Z}/t^{\times}$, signs $\epsilon_i \in \{\pm 1\}$, and a permutation $\sigma \in \Sigma_b$ such that $\overline{k_i} = \epsilon_i \cdot e \cdot \overline{l_{\sigma(i)}}$ holds in $\mathbb{Z}/t^{\times}$ for $i = 1, 2, \ldots, b$.*

Proof. Note that $\bigoplus_{i=1}^{b} V_{k_i}$ and $\bigoplus_{i=1}^{b} V_{l_i}$ are isomorphic as orthogonal representations if and only if there are signs $\epsilon_i \in \{\pm 1\}$ and a permutation $\sigma \in \Sigma_b$ such that $\overline{k_i} = \epsilon_i \cdot \overline{l_{\sigma(i)}}$ holds in $\mathbb{Z}/t^{\times}$ for $i = 1, 2, \ldots, b$. If $m_e \colon \mathbb{Z}/t \to \mathbb{Z}/t$ is multiplication by $e \in \mathbb{Z}/t^{\times}$, then the restriction $m_e^*\left(\bigoplus_{i=1}^{b} V_{l_i}\right)$ is $\bigoplus_{i=1}^{b} V_{e \cdot l_i}$. Now apply Theorem 3.68. □

Example 3.72 (Distinguished examples of lens spaces) We conclude from Theorem 3.49 and Corollary 3.71 the following facts:

(i) Any homotopy equivalence $L(7; k_1, k_2) \to L(7; k_1, k_2)$ has degree 1. Thus $L(7; k_1, k_2)$ possesses no orientation reversing self-diffeomorphism;
(ii) $L(5; 1, 1)$ and $L(5; 2, 1)$ have the same homotopy groups, homology groups and cohomology groups, but they are not homotopy equivalent;
(iii) $L(7; 1, 1)$ and $L(7; 2, 1)$ are oriented homotopy equivalent, but not homeomorphic.

Exercise 3.73 Prove the various claims appearing in Example 3.72.

3.5.8 The Homeomorphism Classification of Generalised Lens Spaces

Note that we have presented a full classification of lens spaces in Theorem 3.68 based on the notion of Reidemeister torsion without using surgery theory. Now one may ask whether one can classify generalised lens spaces as well. This is possible in the topological category, as we will explain next. The proof of the following claims will be explained in Section 18.9 and rely on surgery theory, mainly on the surgery exact sequence in the topological category and computations about the algebraic L-groups of $\mathbb{Z}G$.

In order to state the classification, we need another invariant, the so-called ρ-invariant, which will be explained in Definition 18.31

$$\rho_G(L(\alpha)) \in \overline{\mathbb{Q}R_\mathbb{C}(G)}^{(-1)^d}.$$

We will show in Theorem 18.54 following [414, Theorem 14E.7 on page 224] that, for a cyclic group of odd order G and two generalised lens spaces $L(\alpha)$ and $L(\beta)$ of the same dimension $n \geq 5$, there is an orientation preserving homeomorphism $L(\alpha) \to L(\beta)$ inducing the identity on G with respect to the identification (3.43) if and only if the equations $\Delta(L(\alpha)) = \Delta(L(\beta))$ and $\rho_G(L(\alpha)) = \rho_G(L(\beta))$ hold.

3.6 The Spherical Space Form Problem

Let p and q be two not necessarily distinct prime numbers. A finite group satisfies the *pq-condition* if every subgroup of order pq is cyclic.

Problem 3.74 (Space Form Problem) When does a finite group G admit an orthogonal free action on the standard sphere S^{n-1} for some natural number n?

Note that a finite group G admits an orthogonal free action on the standard sphere S^{n-1} if and only if there exists a closed $(n-1)$-dimensional Riemannian manifold with constant sectional curvature equal to 1 whose fundamental group is isomorphic to G. This follows from the facts that a simply connected closed $(n-1)$-dimensional Riemannian manifold with constant sectional curvature equal to 1 is isometrically diffeomorphic to the standard sphere S^{n-1} and any isometric self-diffeomorphism of S^{n-1} comes from an orthogonal automorphism of $\mathbb{R}^n$ by restriction. The proof of the next result and more information on the Space Form Problem 3.75 can be found in Wolf [437].

Theorem 3.75 (Space Form Problem) *A finite group G which admits an orthogonal free action on the standard sphere S^{n-1}, for some natural number n, satisfies the pq-condition for any two primes p and q.*

If the finite group G is solvable and satisfies the pq-condition for any two primes p and q, then G admits an orthogonal free action on the standard sphere S^{n-1} for some natural number n.

Note that finite cyclic groups are examples of groups appearing in Theorem 3.75 because of the existence of lens spaces.

The Space Form Problem 3.74 and its solution triggered the following problem.

Problem 3.76 (Spherical Space Form Problem) When does a finite group G admit a free topological action on the standard sphere S^{n-1} for some natural number n?

If G is finite cyclic, a spherical space form in the sense of Problem 3.76 is just a generalised lens space.

The solution of Problem 3.76 was finally achieved by Madsen–Thomas–Wall [282, 283].

Theorem 3.77 (Spherical Space Form Problem) *A finite group G admits a free topological action on the standard sphere S^{n-1} for some natural number n if and only if it satisfies the $2p$-condition and the p^2-condition for every prime p.*

Note that the action appearing in Theorem 3.77 can actually be taken to be smooth, see [283, Theorem 0.6].

There is also information about the possible dimension of the standard sphere S^{n-1} appearing in Theorem 3.77.

The Spherical Space Form Problem 3.76 and the search for a solution has been very influential and was one of the main motivations for Wall to investigate surgery problems in the non-simply connected case. Further key words connected to it are Swan complexes, periodic resolutions, finiteness obstructions, periodic group cohomology, Dress induction, and the computations of L-groups $L_n^\epsilon(\mathbb{Z}G)$ for finite groups G. Note that the solution of the Spherical Space Form Problem is a typical application of surgery to an existence problem.

Further information and references to the literature can be found for instance in the survey articles [120, 172, 276, 277], or [93, Chapter 6.6].

3.7 Notes

Reidemeister torsion can be used to show that for any finite group G two finite-dimensional (not necessarily free) orthogonal representations V and W have G-diffeomorphic unit spheres SV and SW if and only if they are isomorphic as orthogonal representations, see [127]. This generalises Theorem 3.68. The corresponding statement is false if one replaces G-diffeomorphic by G-homeomorphic, see [81, 83, 176].

Let $(W; L, L')$ be an h-cobordism of lens spaces, which is compatible with the orientations and the identifications of $\pi_1(L)$ and $\pi_1(L')$ with G. Then W is diffeomorphic relative L to $L\times[0, 1]$ and L and L' are diffeomorphic, see [302, Corollary 12.13 on page 410]. On the other hand, there are examples of lens spaces L and L' and of an h-cobordism $(W; L, M)$ such that L and L' are not h-cobordant, but there is a simple homotopy equivalence $M \to L'$ with the property that f^*TL' and TM are isomorphic vector bundles, see [302, Example 12.14 on page 410].

We refer to [98] and [302] for more information about Whitehead torsion and lens spaces.

Chapter 4
The Surgery Step and ξ-Bordism

4.1 Introduction

In this chapter manifolds are assumed to be smooth and we will take the first steps towards the solution of the following problem.

Problem 4.1 Let X be a finite CW-complex. When is X homotopy equivalent to a closed n-dimensional manifold?

To do this we will address the following related problem. Recall that a CW-complex is said to be of *finite type* if it has a finite number of cells in each dimension.

Problem 4.2 Let $f \colon M \to X$ be a map from a closed n-dimensional manifold M to a CW-complex X of finite type. Can we modify it without changing the target to a map $f' \colon M' \to X$ with a closed n-dimensional manifold as source such that f' is k-connected where $n = 2k$ or $n = 2k + 1$?

Since a closed manifold is homotopy equivalent to a finite CW-complex, it is reasonable to assume that X is finite in Problem 4.1 from the very beginning. We weaken this condition in Problem 4.2 to being of finite type since this case is useful in other situations as well. In Chapter 5 we will see that X needs to be a finite n-dimensional Poincaré complex and the degree of the map f needs to be ± 1 in order for us to have a chance of solving Problem 4.1.

In Section 4.2 we review the CW-analogue of Problem 4.2, namely, the technique of making a map between CW-complexes highly connected by attaching cells. However, this procedure destroys the manifold structure on the source. In Section 4.3 we explain how to modify the source so that it remains a closed manifold. Thus we replace the process of attaching a single cell by the process called a surgery step. We will see that a surgery step can be carried out if we require that our map f is covered by bundle data, i.e., we need a vector bundle ξ over X and a bundle map $TM \oplus \underline{\mathbb{R}^a} \to \xi$ covering f for some natural number a where $\underline{\mathbb{R}^a}$ denotes the a-dimensional trivial bundle, see Section 4.4. All this will be formalised and carried out in detail in Section 4.6. We will see that the outcome of a surgery step inherits the necessary bundle data and thus we can continue doing surgery. Moreover, the surgery step does not change the bordism class of f, taking the bundle data into account, and so Problem 4.2 can be solved, see Theorem 4.44.

In Chapters 8 and 9 the surgery obstructions will arise as obstructions which prevent the map f from being $(k+1)$-connected, after it has been made k-connected if $n = 2k$ or $n = 2k+1$. If f is $(k+1)$-connected, X is a finite n-dimensional Poincaré complex for $n = 2k$ or $n = 2k+1$, and f has degree ± 1, then Poincaré duality will imply that f is a homotopy equivalence.

Finally we present the Pontrjagin–Thom construction in Section 4.7, which expresses the normal bordism set of ξ-maps for a vector bundle ξ over a CW-complex in terms of the homotopy groups of the Thom space associated to ξ, see Theorem 4.55. This is a fundamental link between manifold theory and homotopy theory. We also show that a ξ-map can always be made highly connected within its ξ-bordism class by surgery below the middle dimension, see Theorem 4.59.

Guide 4.3 For a reader who is learning surgery theory for the first time we recommend to go through all of the material of this section except Section 4.5, which can be used as a black box, just take notice of Theorem 4.30.

The material of the preceding chapters is not needed to read this chapter.

4.2 The *CW*-Version of the Surgery Step

We first consider the CW-complex version of Problem 4.2.

Let $f \colon Y \to X$ be a map of CW-complexes. We want to find a procedure which leaves X fixed and changes Y and f into a map $f' \colon Y' \to X$ which is a homotopy equivalence. In addition, this procedure should have the potential to be carried over to the case where Y is a manifold and the resulting source Y' is also a manifold. The Whitehead Theorem, see [178, Theorem 4.5 on page 346] or [431, Theorem V.3.1 and Theorem V.3.5 on page 230], says that f is a homotopy equivalence if and only if it is k-connected for all $k \geq 0$. Recall that k-connected means that the map $\pi_j(Y, y) \to \pi_j(X, f(y))$ induced by f is bijective for $j < k$ and surjective for $j = k$ for all base points y in Y. Hence we would expect the procedure to produce maps where the homotopy groups $\pi_k(f)$ of f become closer and closer to being trivial for all k. It is reasonable to try to work out an inductive procedure where f is already k-connected and we would like to make it $(k+1)$-connected. Recall that there is a long exact homotopy sequence of a map $f \colon Y \to X$

$$\cdots \to \pi_{l+1}(Y) \to \pi_{l+1}(X) \to \pi_{l+1}(f) \to \pi_l(Y) \to \pi_l(X) \to \cdots$$

where $\pi_{l+1}(f) \cong \pi_{l+1}(\mathrm{cyl}(f), Y)$ consists of pointed homotopy classes of commutative squares of pointed spaces with j the inclusion

4.2 The CW-Version of the Surgery Step

$$\begin{array}{ccc} S^l & \xrightarrow{q} & Y \\ {\scriptstyle j}\downarrow & & \downarrow{\scriptstyle f} \\ D^{l+1} & \xrightarrow{Q} & X. \end{array}$$

The map f is k-connected if and only if $\pi_l(f) = 0$ for $l \leq k$. Suppose that f is k-connected. In order to achieve that f is $(k+1)$-connected, we must arrange that $\pi_{k+1}(f) = 0$ without changing $\pi_l(f)$ for $l \leq k$. Consider an element ω in $\pi_{k+1}(f)$ given by a square as above. Define Y' to be the pushout

$$\begin{array}{ccc} S^k & \xrightarrow{q} & Y \\ {\scriptstyle j}\downarrow & & \downarrow{\scriptstyle j'} \\ D^{k+1} & \xrightarrow{q'} & Y'. \end{array}$$

Then the universal property of the pushout gives a map $f': Y' \to X$. We will say that f' is obtained from f by attaching a $(k+1)$-cell.

One easily checks that $\pi_l(f') = \pi_l(f)$ for $l \leq k$ and that there is a natural map $\pi_{k+1}(f) \to \pi_{k+1}(f')$ which is surjective and whose kernel contains the class $\omega \in \pi_l(f)$. Recall that in general $\pi_l(f)$ is an abelian group for $l \geq 3$ and a group for $l = 2$, but carries no group structure for $l = 0, 1$. Moreover $\pi_1(Y)$ acts on $\pi_l(f)$. Hence $\pi_l(f)$ is a $\mathbb{Z}\pi_1(Y)$-module for $l \geq 3$. For $k \geq 3$ the kernel of the epimorphism $\pi_{k+1}(f) \to \pi_{k+1}(f')$ is the $\mathbb{Z}\pi_1(Y)$-submodule generated by ω, see [178, Proposition 4.13 on page 353] or [431, Section V.1]. We see that if $\pi_{l+1}(f)$ is finitely generated over $\mathbb{Z}\pi_1(Y)$, then we can achieve that $\pi_{l+1}(f)$ becomes zero by a finite sequence of cell attachments. Suppose that Y and X are CW-complexes satisfying certain finiteness conditions, for instance that they are homotopic to CW-complexes of finite type, then one can achieve that $\pi_{k+1}(f) = 0$ in a finite number of cell attachments by the following result.

Lemma 4.4 *Consider an integer $k \geq 0$. Let $f : Y \to X$ be a map of CW-complexes such that Y has the homotopy type of a CW-complex with finite $(k-1)$-skeleton and X has the homotopy type of a CW-complex with finite k-skeleton. Suppose that f is $(k-1)$-connected.*

(i) *Suppose that X is connected and $k \geq 2$. Moreover, assume that the map $\pi_1(f): \pi_1(Y) \to \pi_1(X)$ is bijective if $k = 2$.*
   Then $\pi_k(f)$ is a finitely generated $\mathbb{Z}\pi_1(Y)$-module;
(ii) *One can make f k-connected by attaching finitely many cells to Y.*

Proof. (i) Note that both Y and X are connected. Of course we can assume without loss of generality that Y itself has a finite $(k-1)$-skeleton and that X itself has a finite k-skeleton. We use $\pi_1(f)$ to identify $\pi := \pi_1(Y) = \pi_1(X)$. The map f lifts to π-map $\widetilde{f}: \widetilde{Y} \to \widetilde{X}$ between the universal coverings. The obvious map $\pi_k(\widetilde{f}) \to \pi_k(f)$ is bijective and compatible with the π-operations, which are given by the π-actions on the universal covering and the operation of the fundamental group on the homotopy groups. The Hurewicz homomorphism induces an isomorphism of $\mathbb{Z}\pi$-modules $\pi_k(\widetilde{f}) \xrightarrow{\cong} H_k(\widetilde{f}) := H_k(\mathrm{cyl}(\widetilde{f}), \widetilde{Y})$ since $\widetilde{Y}$ and $\widetilde{X}$ are simply connected, see [178, Theorem 4.32 on page 366] or [431, Corollary IV.7.10 on page 181]. In particular we see that $\pi_k(f)$ is indeed an abelian group, which is true in general only for $k \geq 3$. Since f and hence $\widetilde{f}$ is $(k-1)$-connected, the $\mathbb{Z}\pi$-chain complex D_*, which is defined to be the mapping cone $D_* := \mathrm{cone}_*(C_*(\widetilde{f}))$ of the $\mathbb{Z}\pi$-chain map $C_*(\widetilde{f}) \colon C_*(\widetilde{Y}) \to C_*(\widetilde{X})$, is $(k-1)$-connected. Thus $C_*(\widetilde{f})$ yields an exact sequence of $\mathbb{Z}\pi$-modules

$$0 \to \ker(d_k) \to D_k \xrightarrow{d_k} D_{k-1} \xrightarrow{d_{k-1}} \cdots \xrightarrow{d_1} D_0 \to 0$$

such that D_i finitely generated free for $0 \leq i \leq k$. Hence $\ker(d_k)$ is a finitely generated projective $\mathbb{Z}\pi$-module which is stably free, i.e., after adding a finitely generated free $\mathbb{Z}\pi$-module it becomes free. Since $H_k(\widetilde{f}) = H_k(D_*)$ is a quotient of $\ker(d_k)$, it is finitely generated. Hence $\pi_k(f)$ is finitely generated as a $\mathbb{Z}\pi$-module.

(ii) We begin with $k = 0$. Attaching zero-cells means taking the disjoint union of Y with finitely many points. In this way one can obviously achieve that $\pi_0(f) \colon \pi_0(Y) \to \pi_0(X)$ is surjective.

Next we treat the case $k = 1$. Since Y has a finite 0-skeleton, $\pi_0(Y)$ is finite. If two path components of Y are mapped to the same path component in X, one can attach a 1-cell in the obvious manner to connect these components. Thus we can achieve that $\pi_0(f)$ is bijective. Since X has finite 1-skeleton, $\pi_1(X)$ is finitely generated. By attaching 1-cells trivially to Y, which is the same as taking the one-point union of Y with S^1, we can achieve that $\pi_1(f, y)$ is an epimorphism for every base point $y \in Y$.

Next we treat the case $k = 2$ following [225, Section 3, Lemma 3]. From the case $k = 1$, we may assume that $f \colon Y \to X$ induces a bijection on π_0 and $\pi_1(f, y)$ is an epimorphism for every base point $y \in Y$. It now suffices to consider each connected component separately, so we assume that X and Y are connected. Since Y has a finite 1-skeleton and X has a finite 2-skeleton, the induced map $\pi_1(f) \colon \pi_1(Y) \to \pi_1(X)$ is an epimorphism where $\pi_1(Y)$ is finitely generated and $\pi_1(X)$ is finitely presented. We choose a finite set of generators for $\pi_1(Y)$ and a finite presentation for $\pi_1(X)$:

$$\pi_1(Y) = \langle y_1, \ldots, y_l \rangle;$$
$$\pi_1(X) = \langle x_1, \ldots, x_k \mid r_1, \ldots, r_m \rangle.$$

For each $i = 1, \ldots, k$ attach a 1-cell trivially to Y, call the resulting CW-complex Y', denote the element of $\pi_1(Y')$ represented by the i-th summand by z_i and extend the map f to a map $f' \colon Y' \to X$ so that $\pi_1(f')(z_i) = x_i$. Then $\pi_1(Y')$ is finitely generated by $\langle y_1, \ldots, y_l, z_1, \ldots, z_k \rangle$. With these choices write $w_i(x_1, \ldots, x_k) = \pi_1(f')(y_i)$ for each $i = 1, \ldots, l$. Consider the words $a_i = y_i^{-1} \cdot w_i(z_1, \ldots, z_k)$ for $i = 1, \ldots, l$ and $b_j = r_j(z_1, \ldots, z_k)$ for $j = 1, \ldots, n$, and for each word attach one 2-cell to Y' along a loop represented by that word. We call the resulting CW-complex Y'' and extend f' to $f'' \colon Y'' \to X$ using the fact that $\pi_1(f')$ maps a_i and b_j to 0. We claim that $\pi_1(f'') \colon \pi_1(Y'') \to \pi_1(X)$ is an isomorphism: Consider the subgroup $\pi \subseteq \pi_1(Y'')$ generated by the set $\{z_1, \ldots, z_k\}$. The relations a_i introduced to $\pi_1(Y'')$ ensure

4.3 Motivation for the Surgery Step

$\pi = \pi_1(Y'')$ whereas the relations b_j ensure that the homomorphism $\pi_1(f'')\colon \pi \to \pi_1(X)$ is an isomorphism. To complete the case $k = 2$ we must now make our map surjective on π_2. By assertion (i) the $\mathbb{Z}\pi$-module $\pi_2(f'')$ is finitely generated, so by attaching a finite number of trivial 2-cells, we arrive at a map $f'''\colon Y''' \to X$ which is 2-connected.

The cases $k \geq 3$ follow directly from assertion (i). This finishes the proof of Lemma 4.4. □

Exercise 4.5 Show that the map $f\colon \bigvee_{i \in I} S^1 \to \{\bullet\}$ can be made 2-connected by attaching finitely many cells if and only if the set I is finite.

In fact, in the category of *CW*-complexes one can achieve the desired homotopy equivalence $f'\colon Y' \to X$ directly by the following construction. Namely, consider the projection pr$\colon \operatorname{cyl}(f) \to X$ of the mapping cylinder of f to X. Obviously the mapping cylinder is in general not a manifold, even if $f\colon Y \to X$ is a smooth map of closed manifolds. Neither is there a chance that the space Y' obtained from Y by attaching a cell is a manifold, even if Y is a closed manifold and f smooth. But when Y is a manifold, the step of attaching a cell to the source of the map $f\colon Y \to X$ can be modified, and the modified procedure can be carried out in the category of manifolds as sources of maps. This will be explained in the next section.

4.3 Motivation for the Surgery Step

Suppose that M is a closed manifold of dimension n, X is a *CW*-complex, and $f\colon M \to X$ is a k-connected map. Consider $\omega \in \pi_{k+1}(f)$ represented by a diagram

$$\begin{array}{ccc} S^k & \xrightarrow{q} & M \\ {\scriptstyle j}\downarrow & & \downarrow{\scriptstyle f} \\ D^{k+1} & \xrightarrow{Q} & X. \end{array}$$

In general we cannot attach a single $(k+1)$-cell to M without destroying its manifold structure. But we can glue two manifolds together along a common boundary so that the resulting space is a manifold. Suppose that the map $q\colon S^k \to M$ extends to an embedding

$$q^{\text{th}}\colon S^k \times D^{n-k} \hookrightarrow M$$

(the superscript "th" means "thickened"). Let int(im(q^{th})) be the interior of the image of qth and note that the space $M - \operatorname{int}(\operatorname{im}(q^{\text{th}}))$ is a manifold with boundary im($q^{\text{th}}|_{S^k \times S^{n-k-1}}$). We can get rid of the boundary by attaching $D^{k+1} \times S^{n-k-1}$ along $q^{\text{th}}|_{S^k \times S^{n-k-1}}$. Denote the resulting manifold

$$M' := \left(D^{k+1} \times S^{n-k-1}\right) \cup_{q^{\text{th}}|_{S^k \times S^{n-k-1}}} \left(M - \operatorname{int}(\operatorname{im}(q^{\text{th}}))\right).$$

The manifold M' is said to be obtained from M by performing *surgery along* q^{th} and this process is called a *k-surgery* on M. Here and elsewhere we will apply without further comment the technique of straightening the angle in order to get a well-defined smooth structure, see [48, Definition 13.11 on page 145 and (13.12) on page 148] and [189, Chapter 8, Section 2].

There is also a canonically defined map $f' \colon M' \to X$, as we now explain. The inclusion
$$(S^k \times D^{n-k}) \cup (D^{k+1} \times \{0\}) \to D^{k+1} \times D^{n-k}$$
admits a retraction
$$r \colon D^{k+1} \times D^{n-k} \to (S^k \times D^{n-k}) \cup (D^{k+1} \times \{0\})$$
which is part of a deformation retraction. If we define
$$\overline{Q} := (f \circ q^{\text{th}} \cup Q) \circ r \colon D^{k+1} \times D^{n-k} \to X, \tag{4.6}$$
then the restriction of f to $M - \text{int}(\text{im}(q^{\text{th}}))$ extends to a map $f' \colon M' \to X$ using $\overline{Q}|_{D^{k+1} \times S^{n-k-1}}$, and we say that $f' \colon M' \to X$ is obtained from the map $f \colon M \to X$ by surgery along q^{th} and ω.

Exercise 4.7 Consider a map $f \colon M \to X$ from a closed n-dimensional manifold M to a finite CW-complex X. Suppose that by a finite sequence of surgery steps f can be converted to a homotopy equivalence $f' \colon M' \to X$. Show that in this case $\chi(M) - \chi(X) \equiv 0 \bmod 2$.

We next illustrate the surgery step by considering a canonical example of surgery on the 2-torus. Further examples of elementary surgery steps are given at the end of the section. An important example of a map $f \colon M \to X$, on which we may wish to do surgery, is the *Hopf collapse map*, which has $X = S^n$.

Example 4.8 (Hopf collapse map) Let M be a closed connected n-manifold and let $X = S^n$ be the n-sphere. Fix a base point $x_0 \in S^n$ and an embedded open n-disk $\text{int}(D^n) \subset M$. A map $f \colon M \to S^n$ is called a *Hopf collapse map* if it maps $\text{int}(D^n)$ homeomorphically onto $S^n - \{x_0\}$ and $M - \text{int}(D^n)$ to $x_0 \in S^n$. For $0 \leq k < n$, a Hopf collapse map is k-connected if and only if M is $(k-1)$-connected.

We illustrate the surgery procedure in the case where M is the 2-torus $T^2 = S^1 \times S^1$, X is the 2-sphere and $f \colon T^2 \to S^2$ is a Hopf collapse map. We fix $y_0 \in S^1$ so that $S^1 := S^1 \times \{y_0\} \subset T^2$ satisfies $f(S^1) = x_0$. We define $\omega \in \pi_2(f)$ by extending $f|_{S^1}$ to the constant map at x_0 on all of D^2. The following diagram illustrates the effect of surgery on the source.

4.3 Motivation for the Surgery Step

Figure 4.9 (Source of a surgery step for $M = T^2$).

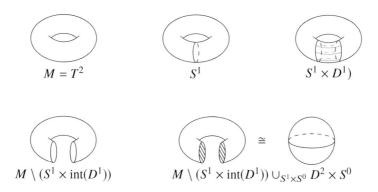

Exercise 4.10 Show that the map $f' \colon S^2 \to S^2$ obtained by carrying out the surgery step on the Hopf collapse map $f \colon T^2 \to S^2$ as described in Figure 4.9 and Example 4.8 above is a homotopy equivalence.

Remark 4.11 (**Homotopy theoretic effect of the surgery step below the middle dimension**) Note that the inclusion $M - \operatorname{int}(\operatorname{im}(q^{\mathrm{th}})) \to M$ is $(n-k-1)$-connected since $S^k \times S^{n-k-1} \to S^k \times D^{n-k}$ is $(n-k-1)$-connected. So the passage from M to $M - \operatorname{int}(\operatorname{im}(q^{\mathrm{th}}))$ will not affect $\pi_j(f)$ for $j < n-k-1$. All in all we see that $\pi_l(f) = \pi_l(f')$ for $l \leq k$ and that there is an epimorphism $\pi_{k+1}(f) \to \pi_{k+1}(f')$ whose kernel contains ω, provided that $2(k+1) \leq n$.

The condition $2(k+1) \leq n$ can be viewed as a consequence of Poincaré duality. Roughly speaking if we change something in a manifold in dimension l, Poincaré duality also forces a change in dimension $(n-l)$. This phenomenon is one reason why there are surgery obstructions to converting any map $f \colon M \to X$ into a homotopy equivalence in a finite number of surgery steps.

Remark 4.12 (**Trace of surgery**) It is important to note that the maps $f \colon M \to X$ and $f' \colon M' \to X$ are bordant as manifolds with reference map to X. The relevant bordism is given by

$$W = \left(D^{k+1} \times D^{n-k}\right) \cup_{q^{\mathrm{th}}} (M \times [0,1])$$

where we think of qth as an embedding $S^k \times D^{n-k} \hookrightarrow M \times \{1\}$. In other words, W is obtained from $M \times [0,1]$ by attaching a handle $D^{k+1} \times D^{n-k}$ to $M \times \{1\}$. Then M appears in W as $M \times \{0\}$ and M' as the other component of the boundary of W. The manifold W is called the *trace of surgery* along the embedding q^{th}.

If we define the map $F \colon W \to X$ by setting $F|_{D^{k+1} \times D^{n-k}} = \overline{Q}$ where $\overline{Q}$ is defined in (4.6), and by defining $F|_{M \times [0,1]}(x,t) = f(x)$ for $(x,t) \in M \times [0,1]$, then F restricted to M is f and F restricted to M' is f'. Hence $f \colon M \to X$ and $f' \colon M' \to X$ are bordant as manifolds with maps to X.

Remark 4.13 (Surgery and bordism) Recall from Lemma 2.15 that every bordism W with boundary $\partial W = \partial_0 W \coprod \partial_1 W$ can be obtained from $\partial_0 W \times [0, 1]$ by the addition of finitely many handles. Since the addition of a handle is precisely the trace of a single surgery, it follows that two manifolds M_0 and M_1 are bordant if and only if M_1 can be obtained from M_0 by finitely many surgeries.

Figure 4.14 immediately below displays the surgery step and its trace for the special of a 0-surgery on case $M = S^1 \sqcup S^1$ where we start from an embedding $S^0 \times D^1 \hookrightarrow S^1$.

Figure 4.14 (Surgery along $S^0 \times D^1 \hookrightarrow S^1 \sqcup S^1$).

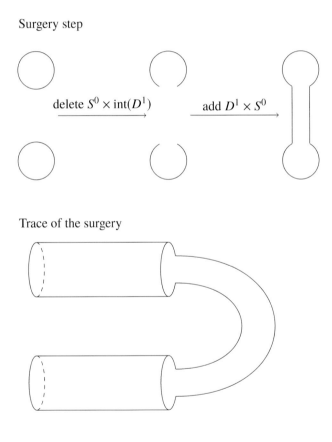

Figure 4.15 immediately below gives a schematic representation of the trace of a surgery. For obvious reasons, this fundamental image in surgery theory is often called the *surgeon's suitcase*.

4.3 Motivation for the Surgery Step

Figure 4.15 (Surgeon's suitcase).

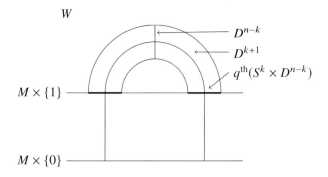

Exercise 4.16 Let $f\colon M \to N$ be a map of closed manifolds and suppose that we can convert f in finitely many surgery steps into a homotopy equivalence $f'\colon M' \to N$. Show that M and N have the same Stiefel–Whitney numbers.

Remark 4.17 (Homotopy theoretic model of the trace of a surgery) We note that the bordism W between the source M and the outcome M' of a surgery also provides us with an illuminating homotopy theoretic interpretation of the process of surgery. Namely we obtain homotopy equivalences

$$M \cup_q D^{k+1} \simeq W \simeq M' \cup_{q'} D^{n-k}$$

where $q\colon S^k \to M$ is the map we started with and $q'\colon S^{n-k-1} \to M'$ is the map given by the inclusion $\{0\} \times S^{n-k-1} \to M'$. Hence we can say that the outcome M' is obtained up to homotopy from M by attaching one $(k+1)$-dimensional cell and detaching one $(n-k)$-dimensional cell. This viewpoint is useful when studying the effect of surgery on homology, as we will do in Chapters 8 and 9.

As a consequence of the above, we see that the inclusions $M \to W$ and $M' \to W$ are respectively k-connected and $(n-k-1)$-connected. This leads to an alternative proof of the connectivity statements in the first paragraph of Remark 4.11.

We conclude this section by discussing some elementary examples of the surgery step, recalling that our goal is to make the map $f\colon M \to X$ as highly connected as possible. If X is path connected, then an essential first step is to make M path connected: so suppose that $M = M_{-1} \sqcup M_1$ is a disjoint union of two path connected n-manifolds. For $i \in \{\pm 1\}$, taking base points $x_i \in M_i$ and a path in X from $f(x_{-1})$ to $f(x_1)$ defines $\omega \in \pi_1(f)$. We then find an embedding

$$q^{\text{th}}\colon S^0 \times D^n \hookrightarrow M \qquad (4.18)$$

where $q^{\text{th}}(i,(0,\ldots,0)) = x_i$ and M', the manifold obtained from surgery along q^{th}, is path connected. This is illustrated in Figure 4.14 above when $M = S^1 \sqcup S^1$. More generally, if M consists of a finite number of path components then a finite number of 0-surgeries can be used to obtain a path connected manifold.

The operation of 0-surgery is the basis of the *connected sum* of manifolds. The connected sum of manifolds requires that we work with orientations on manifolds, which is a topic we will consider in detail in Chapter 5. Hence here we only give a brief review of orientation and refer to Chapter 5 for more details. Let $x \in M$ be a point in the interior of a connected n-manifold M with neighbourhood $U \cong \mathbb{R}^n$. By excision, the inclusion $(U, U \setminus \{x\}) \to (M, M \setminus \{x\})$ induces an isomorphism

$$H_n(U, U \setminus \{x\}; \mathbb{Z}) \to H_n(M, M \setminus \{x\}) \cong \mathbb{Z}$$

and choice of generator for $H_n(M, M \setminus \{x\})$ is called a *local orientation* of M at x. The manifold M is *orientable* if there is a class in $H_n(M, \partial M; \mathbb{Z})$ which maps to a generator of $H_n(M, M \setminus \{x\})$ under the canonical map $H_n(M, \partial M; \mathbb{Z}) \to H_n(M, M \setminus \{x\})$ for all $x \in M$. If M is orientable, then $H_n(M, \partial M; \mathbb{Z}) \cong \mathbb{Z}$ and an orientation of M is a choice of generator $[M] \in H_n(M, \partial M; \mathbb{Z})$. If M is oriented, then M^- denotes M with orientation $-[M]$. We fix once and for all the standard orientation $[D^n]$ on D^n, using the conventions of Remark 5.37.

If N is another oriented n-manifold, then an embedding $f \colon N \hookrightarrow M$ is *orientation preserving* if it carries the local orientations of N to those of M; otherwise it is called *orientation reversing*. A foundational result of Palais, and independently Cerf [326, 90], states that two embeddings $p_0, p_1 \colon D^n \hookrightarrow \mathrm{int}(M)$ into an oriented manifold M are ambient isotopic if and only if they are either both orientation preserving or orientation reversing. Here, an ambient isotopy of M is a diffeomorphism $F \colon M \times I \to M \times I$ such that $F|_{M \times \{t\}} \colon M \times \{t\} \to M \times \{t\}$ is a diffeomorphism for all $t \in I$ and $F_0 = \mathrm{id}_M$, and p_0, p_1 are ambient isotopic if and only if there is an ambient isotopy of M such that $F|_{M \times 1} \circ p_0 = p_1$. (If M is not orientable, then any two embeddings p_0, p_1 as above are ambient isotopic.) It follows that the outcome of the connected sum operation defined below is well defined, up to orientation preserving diffeomorphism, see [216, Lemma 2.1].

Definition 4.19 (Connected sum) Let M_{-1} and M_1 be connected oriented n-manifolds. Their *connected sum*, $M_{-1} \# M_1$, is any manifold obtained from 0-surgery on an orientation preserving embedding q^{th} as in (4.18), where $\{1\} \times D^n$ is oriented via $[D^n]$ and $\{-1\} \times D^n$ is oriented via $-[D^n]$.

More generally, let (M_{-1}, x_{-1}) and (M_1, x_1) be locally oriented n-manifolds. Their *connected sum*, $M_{-1} \# M_1$, is any manifold obtained from 0-surgery on an orientation preserving embedding q^{th} as in (4.18) where $\{1\} \times D^n$ is oriented using $[D^n]$ and $\{-1\} \times D^n$ is oriented using $-[D^n]$.

Remark 4.20 (Orientations and connected sum) It is not hard to show that if either of M_{-1} or M_1 is either non-orientable or admits an orientation reversing diffeomorphism, then the diffeomorphism type of $M_{-1} \natural M_1$ is independent of the embeddings used for the 0-surgery. However, there are examples where M_{-1} and M_1 are connected oriented closed manifolds and $M_{-1} \natural M_1$ is not diffeomorphic to $(M_{-1}^-) \natural M_1$, see Exercise 5.91.

4.3 Motivation for the Surgery Step

Remark 4.21 (Ambient isotopy and surgery) The connected sum is well defined because of the following more general fact. If M' is obtained from M via surgery along q^{th} and M'' is obtained from M via surgery along q'^{th} where q^{th} and q'^{th} are ambient isotopic, then M' and M'' are diffeomorphic; we leave the proof for the reader.

In fact, by focussing on embeddings for $2k \leq n$, the part of classical surgery theory covered in this book can be reduced to situations where the classification of embeddings $q^{\text{th}} \colon S^k \times D^{n-k} \hookrightarrow M$ is solved, or where enough is known for the purposes of surgery. The general classification of embeddings is a subject which lies beyond the scope of this book, even though it is also a subject closely related to the development of surgery and about which surgery has much to say. Here we remark only that studying the outcome of surgery M' is often a good method for defining and classifying embeddings up to isotopy. We refer the reader to [414, Chapter 11], [80] and [347] for an entry to the literature on the role of surgery theory in the classification of embeddings.

We have seen that 0-surgery on n-disks in different path components of M can be used to decrease the complexity of M, as measured by the number of path-components of M. This corresponds to the fact that the class ω in $\pi_1(X, M)$ maps to an element of $\pi_0(M)$ different from the element given by the base point, and this element is killed by the surgery. On the other hand, if we have a class $\omega \in \pi_{k+1}(X, M)$ which maps trivially to $\pi_k(M)$, then doing surgery on ω increases the complexity of M, as measured by the group π_{k+1}. Namely, ω lifts to a class in $\bar{\omega} \in \pi_{k+1}(X)$ which is not in the image of the induced map $\pi_{k+1}(M) \to \pi_{k+1}(X)$ and we use surgery to produce M' where $\pi_{k+1}(M') \to \pi_{k+1}(X)$ has $\bar{\omega}$ in its image. A general example of this sort of surgery is called *trivial surgery*, which occurs when the embedding q^{th} factors through an embedding of the n-disk in M. Hence we define the standard embeddings

$$q^{k,n}_{\text{std}} \colon S^k \hookrightarrow D^n, \quad (x_1, \ldots, x_{k+1}) \to \sqrt{\frac{1}{2}}(x_1, \ldots, x_{k+1}, 0, \ldots, 0);$$

$$q^{\text{th},k,1}_{\text{std}} \colon S^k \times D^1 \hookrightarrow D^{k+1}, \quad ((x_1, \ldots, x_{k+1}), y_1) \to \sqrt{\frac{2+y_1}{4}}(x_1, \ldots, x_{k+1}),$$

and

$$q^{\text{th},k,n}_{\text{std}} \colon S^k \times D^{n-k} \hookrightarrow D^n, \quad ((x_1, \ldots, x_{k+1}), (y_1, \ldots, y_{n-k})) \to$$

$$\phi\left(q^{\text{th},k,1}_{\text{std}}(x_1, \ldots, x_{k+1}, y_1), y_2, \ldots, y_{n-k}\right)$$

where $\phi \colon D^{k+1} \times D^{n-k-1} \to D^n$ is the diffeomorphism given by

$$\phi((x_1, \ldots, x_{k+1}), (y_2, \ldots, y_{n-k})) \mapsto \frac{(x_1, \ldots, x_{k+1}, y_2, \ldots, y_{n-k})}{\sqrt{x_1^2 + \ldots x_{k+1}^2 + y_2^2 + \ldots y_{n-k}^2}}.$$

Note that these standard embeddings are designed so that the image does not meet the boundary. This is the reason why we do not use the obvious inclusion $S^{n-1} \to D^n$ for $q_{std}^{n-1,n}$.

Definition 4.22 (Trivial surgery (with trivial framing)) If an embedding $q^{th} \colon S^k \times D^{n-k} \hookrightarrow M$ is such that $q = q^{th}|_{S^k \times \{0\}} \colon S^k \hookrightarrow M$ can be factored as $q = p \circ q_{std}^{k,n}$ for some embedding $p \colon D^n \hookrightarrow M$, then surgery along q^{th} is called a *trivial surgery*. If q^{th} itself can be factored as $q^{th} = p \circ q_{std}^{th,k,n}$, then surgery along q^{th} is called a *trivial surgery with trivial framing*.

Trivial surgery is an important first step in a variety of surgery procedures including surgery below the middle dimension, discussed in Section 4.6, and also the Wall realisation of odd-dimensional surgery obstructions, see Theorem 9.115. We leave the reader to verify that if $p \colon D^n \to S^n$ is the inclusion of the northern hemisphere, then surgery on $p \circ q_{std}^{th,k,n} \colon S^k \times D^{n-k} \to S^n$ is diffeomorphic to $S^{k+1} \times S^{n-k-1}$.

More generally, an application of the ambient tubular neighbourhood theorem [189, Theorem 1.8 in Chapter 8 on page 181] shows that any embedding $q^{th} \colon S^k \times D^{n-k} \hookrightarrow S^n$ with $q^{th}|_{S^k \times \{0\}} = q_{std}^{k,n}$ differs up to isotopy from $q_{std}^{th,n}$ by precomposition with a diffeomorphism

$$\phi_\alpha \colon S^k \times D^{n-k} \to S^k \times D^{n-k}, \quad (x,y) \mapsto (x, \alpha(x)(y)),$$

where $\alpha \colon S^k \to SO(n-k)$ is a smooth map. It follows that surgery on q^{th} is diffeomorphic to $S^{n-k-1} \tilde{\times}_\alpha S^{k+1}$, the total space of the $(n-k-1)$-sphere bundle over S^{k+1} with clutching function α. One easily checks

Lemma 4.23 (i) *Let M' be obtained from M by a trivial k-surgery. Then M' is diffeomorphic to $M \sharp (S^{n-k-1} \tilde{\times}_\alpha S^{k+1})$;*
(ii) *Let M' be obtained from M by a trivial k-surgery with trivial framing. Then M' is diffeomorphic to $M \sharp (S^{n-k-1} \times S^{k+1})$.*

4.4 Motivation for the Bundle Data

We keep the notation and assumptions from the previous section. In order to be able to carry out the surgery step to kill $\omega \in \pi_{k+1}(f)$, it remains to figure out whether we can arrange that q is an embedding and that it extends to an embedding q^{th}. Assume $2k \leq n-1$. Then, as we shall see in Theorem 4.29 in the next section, we can change q up to homotopy so that it becomes an embedding. The existence of the extension q^{th} is equivalent to the triviality of the normal bundle of the embedding $q \colon S^k \hookrightarrow M$. As the following exercise shows, there is no reason why this normal bundle should be trivial in general.

4.4 Motivation for the Bundle Data

Exercise 4.24 Let $M = \mathbb{CP}^2$, $X = S^4$ and $f\colon \mathbb{CP}^2 \to S^4$ be a Hopf collapse map. Show that $\pi_3(f) \cong \mathbb{Z}$ and that, for $\omega \in \pi_3(f)$ a generator, ω can be represented by a map $(Q, q)\colon (D^3, S^2) \to (S^4, \mathbb{CP}^2)$ where $q\colon S^2 \hookrightarrow \mathbb{CP}^4$ is an embedding but that the normal bundle of q is necessarily non-trivial.

One way to ensure the triviality of the normal bundle of an embedding is to assume that we have certain bundle data at our disposal. Namely, we assume that there is a bundle ξ over X and that the map f is covered by a bundle map $\bar{f}\colon TM \oplus \underline{\mathbb{R}}^a \to \xi$ for some natural number a. We shall often write such bundle maps as $(f, \bar{f})\colon TM \oplus \underline{\mathbb{R}}^a \to \xi$ and call them *tangential ξ-maps*, see Definition 4.42 for a precise formulation. Their use is justified by the observation that in the desired case when f is a homotopy equivalence, we can choose a homotopy inverse f^{-1} of f, take ξ to be the pullback $(f^{-1})^*TM$ and take $a = 0$, for then there is indeed a bundle map $\bar{f}\colon TM \to \xi$ covering f.

If we want to kill an element $\omega \in \pi_{k+1}(f)$ represented by a diagram

then $q^*TM \oplus \underline{\mathbb{R}}^a$ is isomorphic to $q^*f^*\xi = j^*Q^*\xi$ and hence is trivial since $Q^*\xi$ is a bundle over the contractible space D^{k+1}. As $\nu(q) \oplus TS^k$ is isomorphic to q^*TM, the bundle $\nu(q) \oplus \underline{\mathbb{R}}^b$ is trivial for some $b \geq 0$. Since we assume $2k \leq n - 1$, the natural map $BO(n-k) \to BO(n-k+a)$ is $(k+1)$-connected. Hence $\nu(q)$ itself is trivial. So the thickening q^{th} exists and we are able to carry out one surgery step.

We have to ensure that we can repeat this process. So we must arrange that bundle data are also available for the resulting map $f'\colon M' \to X$. To do this, we need to be careful in choosing the embedding q^{th}. For this we will need some information about embeddings and immersions and their relationship to the bundle data, which we will give in the next section. Our actual approach to the surgery step will be based on immersions, i.e., using the bundle data and $\omega \in \pi_{k+1}(f)$ we will extend q to an immersion $S^k \times D^{n-k} \looparrowright M$ and then deal with the problem of whether this immersion can also be arranged to be an embedding. This differs from the strategy above, which first made the map q into an embedding and then analysed the normal bundle. The advantage of working with immersions is that the preservation of the bundle data will be guaranteed automatically.

When $n = 2k$ we can in general only change q up to homotopy into an immersion and the surgery obstruction will be built from the further obstruction to change it up to regular homotopy into an embedding.

4.5 Immersions and Embeddings

Let M and N be manifolds, one or both possibly with boundary and with $\dim(M) \leq \dim(N)$. Given vector bundles ξ over M and η over N, we have so far only considered bundle maps $(q, \overline{q}) \colon \xi \to \eta$ which are fibrewise isomorphisms. Now we need to consider more generally bundle monomorphisms, i.e., we only will require that the map is fibrewise injective. Consider two bundle monomorphisms $(q_0, \overline{q}_0), (q_1, \overline{q}_1) \colon \xi \to \eta$. Let $\xi \times [0, 1]$ be the vector bundle $p_\xi \times \mathrm{id} \colon E \times [0, 1] \to M \times [0, 1]$, where $p_\xi \colon E \to M$ is the projection associated to ξ. A homotopy of bundle monomorphisms $(h, \overline{h})$ from $(q_0, \overline{q}_0)$ to $(q_1, \overline{q}_1)$ is a bundle monomorphism $(h, \overline{h}) \colon \xi \times [0, 1] \to \eta$ whose restriction to $M \times \{j\}$ is $(q_j, \overline{q}_j)$ for $j = 0, 1$. Homotopy evidently defines an equivalence relation on bundle monomorphisms. We denote by $\pi_0(\mathrm{Mono}(\xi, \eta))$ the set of homotopy classes of bundle monomorphisms.

An immersion $q \colon M \looparrowright N$ is a smooth map such that the differential $Tq \colon TM \to TN$ is a bundle monomorphism. Note that if $m \in \partial M$, then $T_m M$ and $Tq_m \colon T_m M \to T_{f(m)} N$ are still well defined. For example, every embedding $q \colon M \hookrightarrow N$ is an immersion and moreover, an immersion is locally an embedding. However, immersions are in general not embeddings and, as we explain below, they are not even regularly homotopic to embeddings in general. Like an embedding, an immersion $q \colon M \looparrowright N$ has a *normal bundle*, which is the quotient bundle $\nu(q)$ over M defined by

$$\nu(q) := q^*TN/TM. \qquad (4.25)$$

A *regular homotopy* $h \colon M \times [0, 1] \to N$ from an immersion $q_0 \colon M \looparrowright N$ to an immersion $q_1 \colon M \looparrowright N$ is a (continuous, but not necessarily smooth) homotopy $h \colon M \times [0, 1] \to N$ such that $h_0 = q_0$, $h_1 = q_1$, $h_t \colon M \looparrowright N$ is a (smooth) immersion for each $t \in [0, 1]$ and the derivatives $Th_t \colon TM \to TN$ of h_t fit together to define a (continuous) homotopy of bundle monomorphisms

$$TM \times [0, 1] \to TN, \quad (v, t) \mapsto Th_t(v)$$

between Tq_0 and Tq_1. Regular homotopy evidently defines an equivalence relation on the set of immersions from M to N. We denote the set of regular homotopy classes of such immersions by

$$\pi_0(\mathrm{Imm}(M, N)). \qquad (4.26)$$

Exercise 4.27 Show that if $f_0, f_1 \colon M \looparrowright N$ are regularly homotopic immersions of closed manifolds, then there is a smooth homotopy $H \colon M \times [0, 1] \to N$ which is a regular homotopy between f_0 and f_1.

Exercise 4.28 Let $f \colon \mathbb{R} \looparrowright \mathbb{R}^2$ be an immersion such that $f(x) = (x, 0)$ holds for $|x| \geq 1$ and $\|f(x)\| \leq 1$ for $|x| \leq 1$. Let $h \colon \mathbb{R} \times [0, 1] \to \mathbb{R}^2$ be the homotopy defined by $h(x, t) := tf(t^{-1}x)$ if $|x| \leq t$ and $t \neq 0$ and by $h(x, t) := (x, 0)$ otherwise. Show that h is continuous and h_t is an immersion for $t \in [0, 1]$. Prove that h is a regular homotopy if and only if $f(x) = (x, 0)$ holds for all $x \in \mathbb{R}$.

4.5 Immersions and Embeddings

The next result is due to Whitney [433, 435].

Theorem 4.29 (Whitney's Approximation Theorem) *Let M and N be closed manifolds of dimensions m and n. Then any map $p\colon M \to N$ is arbitrarily close to an immersion, provided that $2m \leq n$, and arbitrarily close to an embedding, provided that $2m \leq n - 1$.*

For a proof of the following result, we refer to Smale [377] for the case of immersions of spheres into Euclidean space and to Hirsch [188, Theorems 5.7 and 5.9] for the general statement. A modern treatment of Hirsch–Smale theory in the setting of the h-principle can be found in Eliashberg–Mishachev [142, 8.2.1].

Theorem 4.30 (Immersions and bundle monomorphisms) *Let M be an m-manifold and N an n-manifold.*

(i) *Suppose that $1 \leq m < n$. Then taking the differential of an immersion yields a bijection*
$$T\colon \pi_0(\mathrm{Imm}(M,N)) \xrightarrow{\cong} \pi_0(\mathrm{Mono}(TM, TN));$$

(ii) *Suppose that $1 \leq m \leq n$ and that M has a handlebody decomposition consisting of q-handles for $q \leq n - 2$. Then taking the differential of an immersion yields a bijection*
$$T\colon \pi_0(\mathrm{Imm}(M,N)) \xrightarrow{\cong} \mathrm{colim}_{a\to\infty}\, \pi_0(\mathrm{Mono}(TM \oplus \underline{\mathbb{R}^a}, TN \oplus \underline{\mathbb{R}^a}))$$

where the colimit is given by stabilisation with the identity.

Example 4.31 (Turning the sphere "inside out") Let $f_0\colon S^2 \hookrightarrow \mathbb{R}^3$ be the standard embedding. Then we obtain another embedding $f_1\colon S^2 \hookrightarrow \mathbb{R}^3$ by precomposing with the antipodal map $a\colon S^2 \to S^2$ sending x to $-x$. We claim that f_0 and f_1 are regularly homotopic as immersions. In view of Theorem 4.30 (i) it suffices to show that $\pi_0(\mathrm{Mono}(TS^2, T\mathbb{R}^3))$ consists of precisely one element, or, equivalently, that two bundle monomorphisms $TS^2 \to T\mathbb{R}^3$ can be connected by a homotopy through bundle monomorphisms.

Consider bundle monomorphisms $(q_i, \bar{q}_i)\colon TS^2 \to T\mathbb{R}^3$ for $i = 0, 1$. Equip the vector bundles TS^2 and $T\mathbb{R}^3$ with the standard orientations and Riemannian metrics. Since $T_x q_i(T_x S^2) \subset T_{q_i(x)} \mathbb{R}^3$ is a 2-dimensional vector subspace of the Euclidean vector space $T_{q_i(x)} \mathbb{R}^3$, there is precisely one vector $v_i(x) \in T_{q_i(x)} \mathbb{R}^3$ whose norm is one and for which the orientation on $T_x q_i(T_x S^2) \oplus \mathbb{R} v_i(x) = T_{q_i(x)} \mathbb{R}^3$, induced by the orientation of $T_x S^2$ and $v_i(x)$, and the orientation on $T_{q_i(x)} \mathbb{R}^3$ agree. Hence we can find a map of vector bundles $(q_i, \bar{g}_i)\colon TS^2 \oplus \mathbb{R} \to T\mathbb{R}^3$ that is fibrewise an orientation preserving isomorphism, covers q_i, and extends $\bar{q}_i$. Since q_i is homotopic to the constant map c with value 0, we can find a homotopy of bundle maps that are fibrewise orientation preserving isomorphisms from $(q_i, \bar{g}_i)$ to $(c, \bar{c}_i)$ for $i = 0, 1$. It suffices to show that $(c, \bar{c}_0)$ and $(c, \bar{c}_1)$ are strongly fibre homotopic as bundle maps that are fibrewise isomorphisms, because then by restriction we get a homotopy of bundle monomorphisms between $(q_0, \bar{q}_0)$ and $(q_1, \bar{q}_1)$. Now $(c, \bar{c}_0)$ and $(c, \bar{c}_1)$ differ by an orientation preserving bundle automorphism $(\mathrm{id}, \bar{u})\colon S^2 \times T_0 \mathbb{R}^3 \to S^2 \times T_0 \mathbb{R}^3$

covering the identity. This is the same as a map $\bar{u}\colon S^2 \to \mathrm{GL}(3,\mathbb{R})^+$ where we identify $T_0\mathbb{R}^3 = \mathbb{R}^3$ and $\mathrm{GL}(3,\mathbb{R})^+$ is the Lie group of orientation preserving linear automorphisms of $\mathbb{R}^3$. The inclusion $\mathrm{SO}(3) \to \mathrm{GL}(3,\mathbb{R})^+$ is a homotopy equivalence by the Gram–Schmidt process. Since $\pi_2(\mathrm{SO}(3)) \cong \pi_2(S^3)$ is known to be zero, there is a strong homotopy of bundle maps that are fibrewise isomorphisms and cover the identity from $(\mathrm{id},\bar{u})$ to $(\mathrm{id},\overline{\mathrm{id}})$. This proves that $\pi_0(\mathrm{Imm}(S^2,\mathbb{R}^3))$ consists of precisely one element.

Why does a regular homotopy between f_0 and f_1 justify the phrase that one can turn the sphere inside out? Note that any immersion $f\colon S^2 \hookrightarrow \mathbb{R}^3$ comes with a preferred orientation of its normal bundle, which corresponds to the collection of normal vectors $v(x)$ emanating from $f(x)$ for $x \in S^2$ coming from the construction above. If f happens to be an embedding, then $\mathbb{R}^3 \setminus \mathrm{im}(f)$ has precisely two path components. One of them is characterised by the property that the normal vector $v(x)$ at $f(x)$ for $x \in S^2$ points into it and this component is called the outside component. The other one is called the inside component. Thus we can associate to an embedding $s\colon S^2 \hookrightarrow \mathbb{R}^3$ an outer and an inner component of $\mathbb{R}^3 \setminus \mathrm{im}(f)$. Now one easily checks that $\mathrm{im}(f_0)$ and $\mathrm{im}(f_1)$ and hence $\mathbb{R}^3 \setminus \mathrm{im}(f_0)$ and $\mathbb{R}^3 \setminus \mathrm{im}(f_1)$ agree, but the outer component of f_0 is the inner component of f_1. If one has an immersion, one cannot talk about an outside and inside component anymore, but one still has the orientation of the normal bundles and hence the normal vectors $v(x)$. Moreover a regular homotopy between f_0 and f_1 will also respect these normal vectors.

There are movies that picture a regular homotopy h from f_0 and f_1. The blue and red colours indicate the position of these normal vectors emanating from $h_t(x)$ for $t \in [0,1]$ and $x \in S^2$ deciding whether one can see them emanating from the image of the immersion or not, provided that $f^{-1}(f(x))$ consist of precisely one point.

For surgery, we shall apply Theorem 4.30 to immersions $q\colon S^k \hookrightarrow M$ and $q^{\mathrm{th}}\colon S^k \times D^{n-k} \hookrightarrow M$. Put

$$\pi_0(\mathrm{Imm}_k(M)) := \pi_0(\mathrm{Imm}(S^k, M)). \tag{4.32}$$

Define

$$\pi_0(\mathrm{Imm}_k^{\mathrm{fr}}(M)) \tag{4.33}$$

as follows. Consider pairs (q,u) where $q\colon S^k \hookrightarrow M$ is an immersion and $u\colon \nu(q) \xrightarrow{\cong} \mathbb{R}^{n-k}$ is a trivialisation of the normal bundle of q. Let (q_m, u_m) for $m=0,1$ be two such pairs. We call them *equivalent* if there is a regular homotopy $h\colon S^k \times [0,1] \to M$ between q_0 and q_1 such that for the immersion $h \times \mathrm{pr}_{[0,1]}\colon S^k \times [0,1] \hookrightarrow M \times [0,1]$, where $\mathrm{pr}_{[0,1]}\colon S^k \times [0,1] \to [0,1]$ is the projection, there exists a trivialisation $U\colon \nu(h \times \mathrm{pr}_{[0,1]}) \xrightarrow{\cong} \mathbb{R}^{n-k}$ of vector bundles over $S^k \times [0,1]$ whose restriction to $S^k \times \{m\}$ agrees with u_m for $m=0,1$. We define $\pi_0(\mathrm{Imm}_k^{\mathrm{fr}}(M))$ to be the equivalence classes $[(q,u)]$ of such pairs (q,u).

4.5 Immersions and Embeddings

We have the forgetful map

$$\zeta \colon \pi_0(\mathrm{Imm}_k^{\mathrm{fr}}(M)) \to \pi_0(\mathrm{Imm}_k(M))$$

and the map

$$\rho \colon \pi_0(\mathrm{Imm}(S^k \times D^{n-k}, M)) \to \pi_0(\mathrm{Imm}_k^{\mathrm{fr}}(M)),$$

which sends the class of an immersion $Q \colon S^k \times D^{n-k} \hookrightarrow M$ to the pair $(Q \circ i, u)$ where $i \colon S^k \to S^k \times D^{n-k}$ sends x to $(x, 0)$ and u comes from the isomorphism $\nu(Q \circ i) \xrightarrow{\cong} \nu(i)$ induced by TQ and the standard framing $\nu(i) \cong \underline{\mathbb{R}^{n-k}}$.

The group structure on $O(n-k)$ induces a group structure on the space $\mathrm{map}(S^k, O(n-k))$ by post-composition and hence on $[S^k, O(n-k)]$. Next we define an $[S^k, O(n-k)]$-action on $\pi_0(\mathrm{Imm}_k^{\mathrm{fr}}(M))$. Let $a \colon S^k \to O(n-k)$ be a representative of the class $[a]$ in $[S^k, O(n-k)]$ and (q, u) a representative of the class $[(q,u)]$ in $\pi_0(\mathrm{Imm}_k^{\mathrm{fr}}(M))$. Then a yields a bundle automorphism $\widehat{a} \colon (x, v) \mapsto (x, a(x)v)$ of the trivial vector bundle $\underline{\mathbb{R}^{n-k}}$ over S^k, and we define $[a] \cdot [(q, u)] := [(q, \widehat{a} \circ u)]$. One easily checks that this definition is independent of the choices of the representatives and indeed defines an $[S^k, O(n-k)]$-action on $\pi_0(\mathrm{Imm}_k^{\mathrm{fr}}(M))$.

Let $q \colon M \hookrightarrow N$ be an immersion. A *thickening* of q is an immersion $q^{\mathrm{th}} \colon E_{\nu(q)} \hookrightarrow N$ for the total space $E_{\nu(q)}$ of the normal bundle $\nu(q)$ of q such that for the zero section $s \colon M \to E_{\nu(q)}$ we have $q^{\mathrm{th}} \circ s = q$ and the differential Tq^{th} induces the identity $\mathrm{id} \colon \nu(q) \to \nu(q)$, in other words, the following diagram commutes

$$\begin{array}{ccccc} TM \oplus \nu(q) & \xrightarrow{Tq \oplus j} & s^*TE_{\nu(q)} & \xrightarrow{s^*Tq^{\mathrm{th}}} & s^*(q^{\mathrm{th}})^*TN = q^*TN \\ \downarrow & & & & \downarrow \\ \nu(q) & & \xrightarrow{\mathrm{id}} & & \nu(q) \end{array}$$

where $Ts \oplus j \colon TM \oplus \nu(q) \xrightarrow{\cong} s^*TE_{\nu(q)}$ is the canonical bundle isomorphism and the vertical arrows are the canonical projections. If q is an embedding, this reduces to the well-known notion of a tubular neighbourhood. The proof of the existence of a tubular neighbourhood, which is based on the exponential map for N, extends directly to immersions, see [49, Theorem 12.11]. Two thickenings q_0^{th} and q_1^{th} of q are called *isotopic* if and only if there is a smooth map $h \colon E_{\nu(q)} \times [0, 1] \to M$ such that $h_t \colon E_{\nu(q)} \hookrightarrow M$ is a thickening of q for all $t \in [0, 1]$ and $h_m = q_m^{\mathrm{th}}$ for $m = 0, 1$. Two thickenings are always isotopic.

Lemma 4.34

(i) The map $\rho \colon \pi_0(\mathrm{Imm}(S^k \times D^{n-k}, M)) \to \pi_0(\mathrm{Imm}_k^{\mathrm{fr}}(M))$ is a bijection;

(ii) Let $Q_0 \colon S^k \times D^{n-k} \hookrightarrow M$ be an immersion. Denote by $i \colon S^k = S^k \times \{0\} \to S^k \times D^{n-k}$ the inclusion. Let $h \colon S^k \times [0, 1] \to M$ be a regular homotopy from $q_0 := Q_0 \circ i$ to the immersion $q_1 \colon S^k \hookrightarrow M$.
Then there is an immersion $Q_1 \colon S^k \times D^{n-k} \hookrightarrow M$ with $Q_1 \circ i = q_1$ such that Q_0 and Q_1 are regularly homotopic as immersions $S^k \times D^{n-k} \hookrightarrow M$.

Proof. (i) The inverse of ρ sends the class $[(q, u)]$ of the pair (q, u) to the composite

$$S^k \times D^{n-k} \xrightarrow{\text{id}_{S^k} \times j} S^k \times \mathbb{R}^{n-k} \xrightarrow{u^{-1}} E_{\nu(q)} \xrightarrow{q^{\text{th}}} M$$

where $j \colon D^{n-k} \to \mathbb{R}^{n-k}$ is the inclusion and q^{th} some thickening of q.

(ii) We get an immersion $\overline{h} := h \times \mathrm{pr}_{[0,1]} \colon S^k \times [0, 1] \hookrightarrow M \times [0, 1]$ from the regular homotopy h. Let $\mathrm{pr} \colon S^k \times [0, 1] \to S^k$ be the canonical projection. For $m = 0, 1$, let $k_m \colon S^k \to S^k \times [0, 1]$ be the map sending x to (x, m). Choose a bundle isomorphism $w \colon \nu(\overline{h}) \to (k_0 \circ \mathrm{pr})^* \nu(\overline{h})$ of vector bundles over $S^k \times [0, 1]$, see [198, Theorem 4.3 in Chapter 3 on page 28]. We can arrange that $k_0^* w \colon k_0^* \nu(\overline{h}) \to k_0^* (k_0 \circ \mathrm{pr})^* \nu(\overline{h}) = k_0^* \nu(\overline{h})$ is the identity, otherwise compose w with the inverse of $(k_0 \circ \mathrm{pr})^* w$. From the construction of the map ρ, we get a trivialisation $u \colon \nu(q_0) \xrightarrow{\cong} \mathbb{R}^{n-k}$ of vector bundles over S^k. The pullback construction yields an isomorphism $\mathrm{pr}^* u \colon \mathrm{pr}^* \nu(q_0) \xrightarrow{\cong} \mathrm{pr}^* \mathbb{R}^{n-k} = \mathbb{R}^{n-k}$ of vector bundles over $S^k \times [0, 1]$. Since we have the natural identification $\nu(q_0) \xrightarrow{\cong} k_0^* \nu(\overline{h})$, we regard the pullback $\mathrm{pr}^* u$ as a vector bundle isomorphism $(k_0 \circ \mathrm{pr})^* \nu(\overline{h}) \xrightarrow{\cong} \mathbb{R}^{n-k}$ over $S^k \times [0, 1]$. Consider the composite

$$a \colon \nu(\overline{h}) \xrightarrow{w} (k_0 \circ \mathrm{pr})^* \nu(\overline{h}) \xrightarrow{\mathrm{pr}^* u} \mathbb{R}^{n-k},$$

which is an isomorphism of vector bundles over $S^k \times [0, 1]$ such that the pullback bundle map $k_0^* a \colon k_0^* \nu(\overline{h}) = \nu(q_0) \to k_0^* \mathbb{R}^{n-k} = \mathbb{R}^{n-k}$ is u. Define

$$H \colon S^k \times [0, 1] \times D^{n-k} \xrightarrow{\text{id}_{S^k \times [0,1]} \times j} S^k \times [0, 1] \times \mathbb{R}^{n-k} \xrightarrow{a^{-1}} \nu(\overline{h}) \xrightarrow{\overline{h}^{\text{th}}} M \times [0, 1]$$

where $\overline{h}^{\text{th}}$ is some thickening of $\overline{h}$. For $\epsilon \in (0, 1]$ define the composite

$$H[\epsilon] \colon S^k \times D^{n-k} \times [0, 1] \xrightarrow{\cong} S^k \times [0, 1] \times D^{n-k} \xrightarrow{\text{id}_{S^k \times [0,1]} \times (\epsilon \cdot \text{id}_{D^{n-k}})}$$
$$S^k \times [0, 1] \times D^{n-k} \xrightarrow{H} M \times [0, 1] \xrightarrow{\mathrm{pr}_M} M$$

where the first map switches the second and the last factor and pr_M is the obvious projection. Choose $\epsilon > 0$ small enough so that $H[\epsilon]$ is a regular homotopy. Put $Q_1 := H[\epsilon]_1$. Then H is a regular homotopy of immersions $S^k \times D^{n-k} \hookrightarrow M$ from $H[\epsilon]_0$ to Q_1 and $Q_1 \circ i = q_1$. By construction the immersions $H[\epsilon]_0$ and Q_0 have the same image under $\rho \colon \pi_0(\mathrm{Imm}(S^k \times D^{n-k}, M)) \to \pi_0(\mathrm{Imm}_k^{\text{fr}}(M))$ and so assertion (i) implies that they are regularly homotopic. Hence Q_0 and Q_1 are regularly homotopic. □

Remark 4.35 Theorem 4.30 and Lemma 4.34 can be obtained from more general statements about spaces. Namely, there are spaces $\mathrm{Imm}(M, N)$, $\mathrm{Mono}(TM, TN)$, $\mathrm{Imm}_k^{\text{fr}}(M)$, and $\mathrm{Imm}_k(M)$ such that the sets described above are indeed just the sets of path components of these spaces. Then there are maps $\mathrm{Imm}(M, N) \to \mathrm{Mono}(TM, TN)$ and $\mathrm{Imm}(S^k \times D^{n-k}, N) \to \mathrm{Imm}_k^{\text{fr}}(M)$, which are weak homo-

4.5 Immersions and Embeddings

topy equivalences and hence induce bijections after applying π_0, Moreover, there is a map(S^k, O($n-k$))-principal bundle $p\colon \mathrm{Imm}_k^{\mathrm{fr}}(M) \to \mathrm{Imm}_k(M)$. The fact that paths can be lifted along p with prescribed initial point implies assertion (ii) of Lemma 4.34.

Example 4.36 (Immersions of S^k in D^n) Consider Theorem 4.30 (ii) when $M = S^k$, $N = D^n$, and $1 \le k \le n-2$. Assume that $k \ge 2$. Identify $TS^k \oplus \underline{\mathbb{R}} = \underline{\mathbb{R}^{k+1}}$ and $TD^n \oplus \underline{\mathbb{R}} = \underline{\mathbb{R}^{n+1}}$. For any immersion $q\colon S^k \looparrowright D^n$ define

$$ST(q) = \mathrm{pr} \circ (Tq \oplus \mathrm{id}_{\underline{\mathbb{R}}}) \colon S^k \times \mathbb{R}^{k+1} \to \mathbb{R}^{n+1}$$

where $\mathrm{pr}\colon \underline{\mathbb{R}^{n+1}} \to \mathbb{R}^{n+1}$ is the projection. The map $ST(q)$ associates to every point x in the sphere S^k a $(k+1)$-frame of linearly independent vectors $(v_1(x), \ldots, v_{k+1}(x))$ in $\mathbb{R}^{n+1}$. From Theorem 4.30 (ii) and Lemma 4.34 (i) we obtain bijections

$$\pi_0(\mathrm{Imm}_k(D^n)) \xrightarrow{\cong} [S^k, \mathrm{GL}(n+1,\mathbb{R})/\mathrm{GL}(n-k,\mathbb{R})]$$

and

$$\pi_0(\mathrm{Imm}_k^{\mathrm{fr}}(D^n)) \xrightarrow{\cong} [S^k, \mathrm{GL}(n+1,\mathbb{R})].$$

The Gram–Schmidt process ensures that the inclusion $\mathrm{O}(j) \to \mathrm{GL}(j,\mathbb{R})$ is a homotopy equivalence. We have $[S^k, \mathrm{O}(j)] \cong [S^k, \mathrm{O}]$ for all $j \ge k+2$. So we obtain the following commutative diagram

$$\begin{array}{ccc} \pi_0(\mathrm{Imm}_k^{\mathrm{fr}}(D^n)) & \xrightarrow{\cong} & [S^k, \mathrm{O}] \\ \downarrow{\zeta} & & \downarrow{\pi_*} \\ \pi_0(\mathrm{Imm}_k(D^n)) & \xrightarrow{\cong} & [S^k, \mathrm{O}/\mathrm{O}(n-k)] \end{array}$$

where $\pi\colon \mathrm{O} \to \mathrm{O}/\mathrm{O}(n-k)$ is the natural quotient map. Since $\mathrm{O}/\mathrm{O}(n-k)$ is simply connected, we have $[S^k, \mathrm{O}/\mathrm{O}(n-k)] \equiv \pi_k(\mathrm{O}/\mathrm{O}(n-k))$.

Also when $k = 1$, the above argument works without first stabilising Tq since TS^1 is trivial.

In the case where $n = 2k$, the groups $\pi_k(\mathrm{O}/\mathrm{O}(k))$ are known [214] and, provided that $1 \le k \le n-2$, we conclude from Theorem 4.30 (i) that there are isomorphisms

$$\pi_0(\mathrm{Imm}_k(D^{2k})) \xrightarrow{\cong} \pi_k(\mathrm{O}/\mathrm{O}(k)) \cong \begin{cases} \mathbb{Z} & k=1 \text{ or } k \text{ is even}; \\ \mathbb{Z}/2 & k \ge 3 \text{ is odd}. \end{cases}$$

Exercise 4.37 Let $f\colon S^1 \looparrowright D^2$ be an immersion. Its differential induces a map $Tf\colon S^1 \to \mathbb{R}^2 - \{0\}$. If we compose it with the map $\mathrm{pr}\colon \mathbb{R}^2 - \{0\} \to S^1$ sending x to $x/||x||$, we obtain a map $\mathrm{pr}\circ f\colon S^1 \to S^1$. Show that

$$\pi_0(\mathrm{Imm}_1(D^2)) \xrightarrow{\cong} \mathbb{Z}, \quad [f] \mapsto \deg(\mathrm{pr}\circ f)$$

is bijective.

Example 4.38 Suppose that $1 \leq k \leq n - 2$. Then under the bijection

$$\pi_0(\mathrm{Imm}_k(D^n)) \xrightarrow{\cong} \pi_k(O/O(n-k))$$

of Example 4.36, the boundary map

$$\partial: \pi_k(O/O(n-k)) \to \pi_{k-1}(O(n-k))$$

in the long exact sequence of the fibration $O(n-k) \to O \to O/O(n-k)$ corresponds to taking the isomorphism class of the normal bundle of a regular homotopy class of immersions $S^k \hookrightarrow D^n$.

4.6 The Surgery Step

Now we can carry out the surgery step (in the smooth category.) The key technical statement for this step is the following theorem.

Theorem 4.39 (Surgery step) *Let $f: M \to X$ be a map from a closed manifold M to a CW-complex X covered by a bundle map $\bar{f}: TM \oplus \mathbb{R}^a \to \xi$. Fix a base point $m \in M$ and a generator o of the infinite cyclic group $O_\xi(f(m))$. Let $s \in S^k$ be the base point. Consider $\omega \in \pi_{k+1}(f, m)$ for $k \leq n - 2$ and $n = \dim(M)$. Let $Tj^{\mathrm{th}} \oplus n_\nu: T(S^k \times D^{n-k}) \oplus \mathbb{R} \to T(D^{k+1} \times D^{n-k})$ be the bundle map covering the standard inclusion $j^{\mathrm{th}}: S^k \times D^{n-k} \to D^{k+1} \times D^{n-k}$ that is given by the differential Tj^{th} of jth and the outward normal field of the boundary of $D^{k+1} \times D^{n-k}$, see Definition 5.31.*

(i) *The element ω in $\pi_{k+1}(f, m)$ is given by a commutative diagram*

$$\begin{array}{ccc} S^k & \xrightarrow{q} & M \\ \downarrow & & \downarrow f \\ D^{k+1} & \xrightarrow{Q} & X \end{array}$$

together with a path w from m to $q(s)$.
We can find a commutative diagram of vector bundles

$$\begin{array}{ccc} T(S^k \times D^{n-k}) \oplus \mathbb{R}^{a+b} & \xrightarrow{\overline{q^{\mathrm{th}}}} & TM \oplus \mathbb{R}^{a+b} \\ {\scriptstyle Tj^{\mathrm{th}} \oplus n_\nu \oplus \mathrm{id}_{\mathbb{R}^{a+b-1}}} \downarrow & & \downarrow \bar{f} \\ T(D^{k+1} \times D^{n-k}) \oplus \mathbb{R}^{a+b-1} & \xrightarrow{\overline{Q^{\mathrm{th}}}} & \xi \oplus \mathbb{R}^b \end{array}$$

covering a commutative diagram

4.6 The Surgery Step

$$\begin{array}{ccc} S^k \times D^{n-k} & \xrightarrow{q^{\text{th}}} & M \\ {}^{j^{\text{th}}}\downarrow & & \downarrow f \\ D^{k+1} \times D^{n-k} & \xrightarrow{Q^{\text{th}}} & X \end{array}$$

such that the restriction of the last diagram to $S^k \times \{0\} \to D^{k+1} \times \{0\}$ represents ω, the bundle map $\overline{Q^{\text{th}}}$ restricted to $(s,0)$ sends the standard generator of $O_{T(D^{k+1} \times D^{n-k}) \oplus \mathbb{R}^{a+b-1}}(s,0)$ to the image of the generator $o \in O_\xi(f(m))$ under the isomorphism $O_\xi(f(m)) \xrightarrow{\cong} O_\xi(f \circ q(s)) \cong O_{\xi \oplus \mathbb{R}^b}(f \circ q(s))$ coming from $f \circ w$, and $q^{\text{th}} \colon S^k \times D^{n-k} \looparrowright M$ is an immersion;

(ii) The regular homotopy class of the immersion q^{th} appearing in assertion (i) is uniquely determined by the properties above and depends only on the element ω in $\pi_{k+1}(f,m)$ and the generator o of $O_\xi(f(m))$;

(iii) Suppose that the regular homotopy class of the immersion q^{th} appearing in (i) is such that the immersion $q := q^{\text{th}}|_{S^k \times \{0\}} \colon S^k \looparrowright M$ is regularly homotopic to an embedding. Then one can arrange q^{th} in assertion (i) to be an embedding. If $2k \leq n-1$, one can always find an embedding in the regular homotopy class of q;

(iv) Suppose that the map q^{th} appearing in assertion (i) is an embedding (this can always be arranged if $2k \leq n-1$) and let W be the trace of surgery along q^{th}; i.e., the manifold obtained from $M \times [0,1]$ by attaching a handle $D^{k+1} \times D^{n-k}$ along $q^{\text{th}} \colon S^k \times D^{n-k} \to M = M \times \{1\}$. Let $F \colon W \to X$ be the map induced by $M \times [0,1] \xrightarrow{\text{pr}} M \xrightarrow{f} X$ and $Q^{\text{th}} \colon D^k \times D^{n-k} \to X$. After possibly stabilising $\overline{f}$, the bundle maps $\overline{f}$ and $\overline{Q^{\text{th}}}$ induce a bundle map $\overline{F} \colon TW \oplus \mathbb{R}^{a+b} \to \xi \oplus \mathbb{R}^b$ covering $F \colon W \to X$. Thus we get a normal map $(F, \overline{F}) \colon TW \oplus \mathbb{R}^{a+b} \to \xi \oplus \mathbb{R}^b$, which extends the stabilised normal map $(f, \overline{f} \oplus (f \times \text{id}_{\mathbb{R}^b})) \colon TM \oplus \mathbb{R}^{a+b} \to \xi \oplus \mathbb{R}^b$.

Proof. (i) Choose a commutative diagram of smooth maps

$$\begin{array}{ccc} S^k & \xrightarrow{q} & M \\ {}^{j}\downarrow & & \downarrow f \\ D^{k+1} & \xrightarrow{Q} & X \end{array}$$

together with a path w from m to $q(s)$ representing ω. Of course it can be extended to a diagram

$$\begin{array}{ccc} S^k \times D^{n-k} & \xrightarrow{q^{\text{th}}} & M \\ {}^{j^{\text{th}}}\downarrow & & \downarrow f \\ D^{k+1} \times D^{n-k} & \xrightarrow{Q^{\text{th}}} & X. \end{array}$$

As $D^{k+1} \times D^{n-k}$ is contractible, there is a bundle map $\overline{Q^{\text{th}}}: T(D^{k+1} \times D^{n-k}) \oplus \underline{\mathbb{R}}^{a-1} \to \xi$ covering Q^{th} that is uniquely determined up to isotopy by requiring that $\overline{Q^{\text{th}}}$ restricted to $(s, 0)$ sends the standard generator of $O_{T(D^{k+1} \times D^{n-k}) \oplus \underline{\mathbb{R}}^{a+b-1}}(s, 0)$ to the image of the generator $o \in O_\xi(f(m))$ under the isomorphism $O_\xi(f(m)) \xrightarrow{\cong} O_\xi(f \circ q(s)) \cong O_{\xi \oplus \mathbb{R}^b}(f \circ q(s))$ coming from $f \circ w$. There is precisely one bundle map $\overline{q^{\text{th}}}$ covering q^{th} such that the following diagram commutes

$$\begin{array}{ccc}
T(S^k \times D^{n-k}) \oplus \underline{\mathbb{R}}^a & \xrightarrow{\overline{q^{\text{th}}}} & TM \oplus \underline{\mathbb{R}}^a \\
{\scriptstyle Tj^{\text{th}} \oplus n \oplus \mathrm{id}_{\underline{\mathbb{R}}^{a-1}}} \downarrow & & \downarrow {\scriptstyle \overline{f}} \\
T(D^{k+1} \times D^{n-k}) \oplus \underline{\mathbb{R}}^{a-1} & \xrightarrow[\overline{Q^{\text{th}}}]{} & \xi.
\end{array}$$

Now we use $k \leq n - 2$. From Theorem 4.30 (ii) we obtain an immersion $q_0^{\text{th}}: S^k \times D^{n-k} \looparrowright M$, such that $(q_0^{\text{th}}, Tq_0^{\text{th}}): T(S^k \times D^{n-k}) \to TM$ and $(q^{\text{th}}, \overline{q^{\text{th}}})$ define the same element in

$$\mathrm{colim}_{c \to \infty} \pi_0(\mathrm{Mono}(T(S^k \times D^{n-k}) \oplus \underline{\mathbb{R}}^c, TM \oplus \underline{\mathbb{R}}^c)).$$

After possibly stabilising $\overline{f}$ and thus enlarging a to $a + b$, we can achieve the desired diagram by a cofibration argument.

(ii) Let $q^{\text{th}}: S^k \times D^{n-k} \looparrowright M$ be an immersion from part (i). The stable homotopy class of the bundle monomorphism $(q^{\text{th}}, Tq^{\text{th}})$ is uniquely determined by the commutativity of the following diagram of vector bundles:

$$\begin{array}{ccc}
T(S^k \times D^{n-k}) \oplus \underline{\mathbb{R}}^{a+b} & \xrightarrow{T\overline{q^{\text{th}}} \oplus (q^{\text{th}} \times \mathrm{id}_{\underline{\mathbb{R}}^{a+b}})} & TM \oplus \underline{\mathbb{R}}^{a+b} \\
{\scriptstyle Tj^{\text{th}} \oplus n \oplus \mathrm{id}_{\underline{\mathbb{R}}^{a+b-1}}} \downarrow & & \downarrow {\scriptstyle \overline{f}} \\
T(D^{k+1} \times D^{n-k}) \oplus \underline{\mathbb{R}}^{a+b-1} & \xrightarrow[\overline{Q^{\text{th}}}]{} & \xi \oplus \underline{\mathbb{R}}^b.
\end{array}$$

This is because $\overline{Q^{\text{th}}}$ is uniquely determined up to isotopy. Now apply Theorem 4.30 (ii).

(iii) By Lemma 4.34 (ii) there is a regular homotopy of framed immersions from q^{th} to a framed embedding q_0^{th}. By a cofibration argument we can use this regular homotopy from q^{th} to q_0^{th} to change all the diagrams for Q^{th} up to homotopy to diagrams for Q_0^{th}.

If $2k < n$, we can find an embedding q_0 which is arbitrarily close to q by Theorem 4.29 and hence regularly homotopic to q since the condition of being an immersion is an open condition.

(iv) We leave it to the reader to check that the construction of $(F, \overline{F})$ makes sense. $\square$

4.6 The Surgery Step

Remark 4.40 There is a preferred choice for the generator $o \in O_\xi(f(m))$, when dealing with a normal map of degree one, which we will introduce in Definition 7.13. It comes from the choice of a fundamental class of the target.

Since Theorem 4.39 is central but also rather complicated, we give some explanation. Assertion (i) allows us to blow up a representative of an element $\omega \in \pi_{k+1}(f)$ to an immersion $\overline{q^{\text{th}}} \colon S^k \times D^{n+k} \looparrowright M$ with a preferred nullhomotopy $\overline{Q^{\text{th}}}$ and everything is covered with appropriate bundle data. Assertion (ii) tells us that up to regular homotopy $\overline{q^{\text{th}}}$ is unique, ensuring that we have made the right choice. So we are prepared to carry out the surgery step, provided that we can arrange $\overline{q^{\text{th}}}$ to be an embedding. Assertion (iii) tells us that this can be arranged if the immersion $q := q^{\text{th}}|_{S^k \times \{0\}} \colon S^k \looparrowright M$ is regularly homotopic to an embedding and that this can always be achieved in the case $2k \leq n - 1$. Assertion (iv) ensures that the bundle data carry over to the trace of the surgery and hence onto the result of the surgery.

Definition 4.41 (The surgery step, its outcome and its trace) Let $f \colon M \to X$ be a map from a closed n-dimensional manifold M to a CW-complex which is homotopy equivalent to a CW-complex with finite k-skeleton as target. Suppose that f is covered by a bundle map $(f, \overline{f}) \colon TM \oplus \mathbb{R}^a \to \xi$. Consider an element $\omega \in \pi_{k+1}(f)$ for $k \leq n - 2$.

(i) The bundle map $(f', \overline{f'}) \colon TM' \oplus \mathbb{R}^{a+b} \to \xi \oplus \mathbb{R}^b$ appearing in Theorem 4.39 (iv) is the *outcome of surgery on* $(f, \overline{f})$ *along* (ω, q^{th}), provided that it exists, i.e., that the map q^{th} appearing in assertion (i) of Theorem 4.39 is an embedding, which can always be arranged in the case $2k \leq n - 1$;
(ii) The bundle map $(F, \overline{F}) \colon TW \oplus \mathbb{R}^{a+b} \to \xi \oplus \mathbb{R}^b$ appearing in Theorem 4.39 (iv) is the *trace of surgery on* $(f, \overline{f})$ *along* (ω, q^{th});
(iii) The step taken in passing from $(f, \overline{f})$ to $(f', \overline{f'})$ is the *surgery step* along (ω, q^{th}).

Definition 4.42 (Tangential ξ-map) Let ξ be an l-dimensional vector bundle over a CW-complex X. A *tangential ξ-map* $(M, f, \overline{f})$ consists of the following data:

- A closed manifold M of dimension n;
- A map $f \colon M \to X$;
- A bundle map $(f, \overline{f}) \colon TM \oplus \mathbb{R}^a \to \xi$ for $a = l - n$.

Definition 4.43 (Bordism of tangential ξ-maps) Let ξ be an l-dimensional vector bundle over a CW-complex X. Let $(M_0, f_0, \overline{f}_0)$ and $(M_1, f_1, \overline{f}_1)$ be tangential ξ-maps with $n = \dim(M_0) = \dim(M_1) < l$. A *tangential bordism* of tangential ξ-maps $(W, F, \overline{F})$ from $(M_0, f_0, \overline{f}_0)$ to $(M_1, f_1, \overline{f}_1)$ consists of:

- A compact manifold W of dimension $n + 1$ whose boundary ∂W is the disjoint union $\partial_0 W \amalg \partial_1 W$;
- Diffeomorphisms $u_i \colon M_i \to \partial_i W$ for $i = 0, 1$;
- A map $F \colon W \to X$ satisfying $f_i = F \circ u_i$;
- A bundle map $(F, \overline{F}) \colon TW \oplus \mathbb{R}^{a-1} \to \xi$ such that

$$(F, \overline{F}) \circ \big((u_i, \overline{u_i}) \oplus (u_i \times \text{id}_{\mathbb{R}^{a_i-1}})\big) = (f_i, \overline{f_i})$$

holds where $(u_i, \overline{u_i}): TM_i \oplus \underline{\mathbb{R}} \to TW$ is the bundle map induced by Tu_i with respect to some inward normal vector field for $i = 0$ and for some outward normal vector field for $i = 1$.

From Lemma 4.4, Remark 4.11, and Theorem 4.39 we obtain the following theorem.

Theorem 4.44 (Making a tangential ξ-map highly connected) *Consider natural numbers k and $n \geq 3$ such that $k \leq m$ holds, where m is given by $n = 2m$ or $n = 2m + 1$. Let ξ be an l-dimensional vector bundle over the CW-complex X that is homotopy equivalent to a CW-complex with finite k-skeleton. Let $(f, \overline{f}): TM \oplus \underline{\mathbb{R}^{a+b}} \to \xi \oplus \underline{\mathbb{R}^b}$ be a tangential $\xi \oplus \underline{\mathbb{R}^b}$-map with a closed n-dimensional manifold M as source for some natural numbers a and b.*

Then we can carry out a finite sequence of surgery steps to obtain a tangential $\xi \oplus \underline{\mathbb{R}^{b'}}$-map $(f', \overline{f'}): TN \oplus \underline{\mathbb{R}^{a+b'}} \to \xi \oplus \underline{\mathbb{R}^{b'}}$ satisfying:

(i) *The map $f': N \to X$ is k-connected;*
(ii) *There are natural numbers c and c' with $b + c = b' + c'$ such that the bundle maps $(f, \overline{f}) \oplus (f \times \mathrm{id}_{\mathbb{R}^c}): TM \oplus \underline{\mathbb{R}^{a+b}} \oplus \underline{\mathbb{R}^c} \to \xi \oplus \underline{\mathbb{R}^b} \oplus \underline{\mathbb{R}^c}$ and $(f', \overline{f'}) \oplus (f' \times \mathrm{id}_{\mathbb{R}^{c'}}): TN \oplus \underline{\mathbb{R}^{a+b'}} \oplus \underline{\mathbb{R}^{c'}} \to \xi \oplus \underline{\mathbb{R}^{b'}} \oplus \underline{\mathbb{R}^{c'}}$ are bordant as tangential $\xi \oplus \underline{\mathbb{R}^{b'}} \oplus \underline{\mathbb{R}^{c'}}$-maps.*

Exercise 4.45 A smooth manifold M is called stably parallelisable if the vector bundle $TM \oplus \underline{\mathbb{R}^a}$ is trivial for some natural number a. Let M be a closed stably parallelisable manifold of dimension n. Define k by requiring $n = 2k$ or $n = 2k + 1$. Show that there is a $(k-1)$-connected stably parallelisable manifold N together with a stably parallelisable bordism between M and N.

4.7 The Pontrjagin–Thom Construction

Let (M, i) be an embedding $i: M^n \hookrightarrow \mathbb{R}^{n+k}$ of a closed n-dimensional manifold M into $\mathbb{R}^{n+k}$. Note that $T\mathbb{R}^{n+k}$ comes with an explicit trivialisation $\mathbb{R}^{n+k} \times \mathbb{R}^{n+k} \xrightarrow{\cong} T\mathbb{R}^{n+k}$ and the standard Euclidean inner product induces a Riemannian metric on $T\mathbb{R}^{n+k}$. Denote by $\nu(M) = \nu(i)$ the *normal bundle* of i, which is the orthogonal complement of TM in $i^*T\mathbb{R}^{n+k}$. For a vector bundle $\xi: E \to X$ with Riemannian metric define its *disk bundle* $p_{DE}: DE \to X$ by $DE = \{v \in E \mid ||v|| \leq 1\}$ and its *sphere bundle* $p_{SE}: SE \to X$ by $SE = \{v \in E \mid ||v|| = 1\}$, where p_{DE} and p_{SE} are the restrictions of p. Its *Thom space* $\mathrm{Th}(\xi)$ is defined by DE/SE. It has a preferred base point $\infty := SE/SE$. The Thom space can be defined without a choice of a Riemannian metric as follows. Put $\mathrm{Th}(\xi) = E \cup \{\infty\}$ for some extra point ∞. Equip $\mathrm{Th}(\xi)$ with the topology for which $E \subset \mathrm{Th}(E)$ is an open subset and a basis of open neighbourhoods for ∞ is given by the complements of closed subsets $A \subset E$ for which $A \cap E_x$ is compact for each fibre E_x. If X is compact, E is locally compact and $\mathrm{Th}(\xi)$ is the one-point-compactification of E. The advantage of this definition is that any bundle map $(f, \overline{f}): \xi_0 \to \xi_1$ of vector bundles $\xi_0: E_0 \to X_0$ and $\xi_1: E_1 \to X_1$

4.7 The Pontrjagin–Thom Construction

canonically induces a pointed map $\mathrm{Th}(\overline{f})\colon \mathrm{Th}(\xi_0) \to \mathrm{Th}(\xi_1)$. Note that we require that $\overline{f}$ induces a bijective map on each fibre. Denote by $\underline{\mathbb{R}^k}$ the trivial vector bundle with fibre $\mathbb{R}^k$. We mention that there are homeomorphisms

$$\mathrm{Th}(\xi \times \eta) \cong \mathrm{Th}(\xi) \wedge \mathrm{Th}(\eta); \tag{4.46}$$
$$\mathrm{Th}(\xi \oplus \underline{\mathbb{R}^k}) \cong \Sigma^k \mathrm{Th}(\xi), \tag{4.47}$$

where $\wedge$ stands for the *smash product* of pointed spaces

$$X \wedge Y := (X \times Y)/(X \times \{y\} \cup \{x\} \times Y) \tag{4.48}$$

and $\Sigma^k Y = S^k \wedge Y$ is the *(reduced) suspension*; a proof can be found in [393, Proposition 12.28]. Let $(N(M), \partial N(M))$ be a tubular neighbourhood of M. Recall that there is a diffeomorphism

$$u\colon (D\nu(M), S\nu(M)) \to (N(M), \partial N(M))$$

that is up to isotopy relative M uniquely determined by the property that its restriction to M is i and its differential at M is $\epsilon \cdot \mathrm{id}$ for small $\epsilon > 0$ under the canonical identification $T(D\nu(M))|_M = TM \oplus \nu(M) = i^*T\mathbb{R}^{n+k}$. The *Thom collapse map*

$$c\colon S^{n+k} = \mathbb{R}^{n+k} \amalg \{\infty\} \to \mathrm{Th}(\nu(M)) \tag{4.49}$$

is the pointed map that is given by the diffeomorphism u^{-1} on the interior of $N(M)$ and sends the complement of the interior of $N(M)$ to the preferred base point ∞.

Figure 4.50 (Pontrjagin–Thom construction).

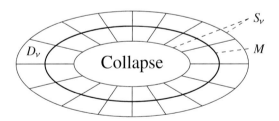

The homology group $H_{n+k}(\mathrm{Th}(\nu(M))) \cong H_{n+k}(N(M), \partial N(M))$ is infinite cyclic since $N(M)$ is a compact orientable $(n+k)$-dimensional manifold with boundary $\partial N(M)$. The Hurewicz homomorphism

$$h\colon \pi_{n+k}(\mathrm{Th}(\nu(M))) \to H_{n+k}(\mathrm{Th}(\nu(M)))$$

sends the class $[c]$ of c to a generator. This follows from the fact that any point in the interior of $N(M)$ is a regular value of c and has precisely one point in its preimage.

Next we apply this construction to bordism. Fix a space X together with a k-dimensional vector bundle ξ over X. Let us recall the definition of the bordism set

$$\Omega_n(\xi) \tag{4.51}$$

of normal ξ-bordism classes of normal ξ-maps.

Definition 4.52 (Normal ξ-map) A *normal ξ-map* $(M, i, f, \overline{f})$ is a quadruple consisting of

- A closed manifold M of dimension n;
- An embedding $i \colon M \hookrightarrow \mathbb{R}^{n+k}$;
- A map $f \colon M \to X$;
- A bundle map $(f, \overline{f}) \colon \nu(i) \to \xi$ covering f where $\nu(i)$ is the normal bundle of the embedding i.

Definition 4.53 (Bordism of normal ξ-maps) A *normal ξ-bordism* from the normal ξ-map $(M_0, i_0, f_0, \overline{f_0})$ to the normal ξ-map $(M_1, i_1, f_1, \overline{f_1})$ is a quadruple $(W, I, F, \overline{F})$ consisting of

- A compact manifold W of dimension $(n+1)$ whose boundary ∂W is the disjoint union $\partial_0 W \amalg \partial_1 W$;
- An embedding of manifolds with boundary $I \colon W \hookrightarrow \mathbb{R}^{n+k} \times [0, 1]$ sending $\partial_m W$ to $\mathbb{R}^{n+k} \times \{m\}$ for $m = 0, 1$;
- Diffeomorphisms $u_m \colon M_m \to \partial_m W$ and $U_m \colon \mathbb{R}^{n+k} \to \mathbb{R}^{n+k} \times \{m\}$ for $m = 0, 1$ satisfying $I \circ u_m = U_m \circ i_m$;
- A map $F \colon W \to X \times [0, 1]$ satisfying $j_m \circ f_m = F \circ u_m$ for $m = 0, 1$ where $j_m \colon X \to X \times [0, 1]$ sends x to (x, m);
- A bundle map $(F, \overline{F}) \colon \nu(I) \to \xi$ covering F such that $\overline{F} \circ \nu(u_m, U_m) = \overline{f}_m$ holds for $m = 0, 1$ where $(u_m, \nu(u_m, U_m)) \colon \nu(i_m) \to \nu(I)$ is the obvious bundle map induced by Tu_m and TU_m.

Remark 4.54 Note that in the definition above the following implicit identification

$$\nu(\partial W \subseteq \mathbb{R}^{n+k} \times \{0, 1\}) = \nu(W \subseteq \mathbb{R}^{n+1} \times [0, 1])|_{\partial W}$$

is used, which is based on the convention that at $\{0\}$ we take the inward normal field and at $\{1\}$ the outward normal vector field to get identifications

$$T\mathbb{R}^n \times [0, 1]|_{\mathbb{R}^n \times \{0,1\}} = T\mathbb{R}^n \times \{0, 1\} \oplus \mathbb{R};$$
$$TW|_{\partial W} = T\partial W \oplus \mathbb{R}.$$

This convention guarantees that we can stack two cobordisms together to prove transitivity of the bordism relation.

4.7 The Pontrjagin–Thom Construction

Theorem 4.55 (Pontrjagin–Thom Construction) *Let $\xi\colon E \to X$ be a k-dimensional vector bundle over a CW-complex X. Then the map*

$$P_n(\xi)\colon \Omega_n(\xi) \to \pi_{n+k}(\mathrm{Th}(\xi)),$$

which sends the bordism class of $(M, i, f, \overline{f})$ to the homotopy class of the composite $S^{n+k} \xrightarrow{c} \mathrm{Th}(\nu(M)) \xrightarrow{\mathrm{Th}(\overline{f})} \mathrm{Th}(\xi)$, is a well-defined bijection, which is natural in ξ.

Proof. The details can be found in [50, Satz 3.1 on page 28, Satz 4.9 on page 35] or [189, Section 7.2 on page 172]. The basic idea becomes clear after we have explained the construction of the inverse for a finite CW-complex X. Consider a pointed map $(S^{n+k}, \infty) \to (\mathrm{Th}(\xi), \infty)$. We can change f up to homotopy relative $\{\infty\}$ so that f becomes transverse to X. Note that transversality makes sense although X is not a manifold, one needs only the fact that X is the zero-section in a vector bundle. Put $M = f^{-1}(X)$. The transversality construction yields a bundle map $\overline{f}\colon \nu(M) \to \xi$ covering $f|_M$. Let $i\colon M \to \mathbb{R}^{n+k} = S^{n+k} - \{\infty\}$ be the inclusion. Then the inverse of $P_n(\xi)$ sends the class of f to the class of $(M, i, f|_M, \overline{f})$. □

Let $\Omega_n(X)$ be the bordism group of pairs (M, f) of oriented closed n-dimensional manifolds M together with reference maps $f\colon M \to X$. Let ξ_k be the universal oriented k-dimensional vector bundle over $\mathrm{BSO}(k)$. Recall that for a finite-dimensional vector space V we denote the trivial vector bundle with fibre V by $\underline{V}$. Let $\overline{j_k}\colon \xi_k \oplus \underline{\mathbb{R}} \to \xi_{k+1}$ be a bundle map covering a map $j_k\colon \mathrm{BSO}(k) \to \mathrm{BSO}(k+1)$. Up to homotopy of bundle maps this map is unique. Denote by γ_k the bundle $X \times E_k \to X \times \mathrm{BSO}(k)$ obtained by crossing the projection of the universal bundle ξ_k with id_X, and by $(i_k, \overline{i_k})\colon \gamma_k \oplus \underline{\mathbb{R}} \to \gamma_{k+1}$ the bundle map $\mathrm{id}_X \times (j_k, \overline{j_k})$. The bundle map $(i_k, \overline{i_k})$ is unique up to homotopy of bundle maps and hence induces a well defined map $\Omega_n(\overline{i_k})\colon \Omega_n(\gamma_k) \to \Omega_n(\gamma_{k+1})$ which sends the class of $(M, i, f, \overline{f})$ to the class of the quadruple coming from the embedding $j\colon M \xrightarrow{i} \mathbb{R}^{n+k} \hookrightarrow \mathbb{R}^{n+k+1}$ and the canonical isomorphism $\nu(i) \oplus \underline{\mathbb{R}} = \nu(j)$. Consider the homomorphism

$$V_k \colon \Omega_n(\gamma_k) \to \Omega_n(X)$$

which sends the class of $(M, i, f, \overline{f})$ to $(M, \mathrm{pr}_X \circ f)$, where pr_X is the projection $X \times \mathrm{BSO}(k) \to X$ and we equip M with the orientation determined by $\overline{f}$. Let $\mathrm{colim}_{k \to \infty} \Omega_n(\gamma_k)$ be the colimit of the directed system indexed by $k \geq 0$

$$\cdots \xrightarrow{\Omega_n(\overline{i_{k-1}})} \Omega_n(\gamma_k) \xrightarrow{\Omega_n(\overline{i_k})} \Omega_n(\gamma_{k+1}) \xrightarrow{\Omega_n(\overline{i_{k+1}})} \cdots.$$

Since $V_{k+1} \circ \Omega_n(\overline{i_k}) = V_k$ holds for all $k \geq 0$, we obtain a map

$$V\colon \mathrm{colim}_{k \to \infty} \Omega_n(\gamma_k) \xrightarrow{\cong} \Omega_n(X). \tag{4.56}$$

This map is bijective because of the classifying property of γ_k and the facts that for $k > n + 1$ any closed manifold M of dimension n can be embedded into $\mathbb{R}^{n+k}$ and two such embeddings are isotopic.

We see a sequence of spaces $\text{Th}(\gamma_k)$ together with maps

$$\text{Th}(\overline{i_k}): \Sigma\,\text{Th}(\gamma_k) = \text{Th}(\gamma_k \oplus \mathbb{R}) \to \text{Th}(\gamma_{k+1}).$$

They induce homomorphisms

$$s_k: \pi_{n+k}(\text{Th}(\gamma_k)) \to \pi_{n+k+1}(\Sigma\,\text{Th}(\gamma_k)) \xrightarrow{\pi_{n+k+1}(\text{Th}(\overline{i_k}))} \pi_{n+k+1}(\text{Th}(\gamma_{k+1}))$$

where the first map is the suspension homomorphism. We now define the group $\operatorname{colim}_{k\to\infty} \pi_{n+k}(\text{Th}(\gamma_k))$ to be the colimit of the directed system

$$\cdots \xrightarrow{s_{k-1}} \pi_{n+k}(\text{Th}(\gamma_k)) \xrightarrow{s_k} \pi_{n+k+1}(\text{Th}(\gamma_{k+1})) \xrightarrow{s_{k+1}} \cdots.$$

From Theorem 4.55 we obtain a bijection

$$P: \operatorname{colim}_{k\to\infty} \Omega_n(\gamma_k) \xrightarrow{\cong} \operatorname{colim}_{k\to\infty} \pi_{n+k}(\text{Th}(\gamma_k)).$$

This implies together with (4.56) the following theorem.

Theorem 4.57 (Pontrjagin–Thom construction and oriented bordism) *There is an isomorphism of abelian groups, natural in X,*

$$P: \Omega_n(X) \xrightarrow{\cong} \operatorname{colim}_{k\to\infty} \pi_{n+k}(\text{Th}(\gamma_k)).$$

Remark 4.58 (Spectra) Note that this is the beginning of the theory of spectra and stable homotopy theory. A *spectrum* **E** consists of a sequence of spaces $(E_k)_{k\in\mathbb{Z}}$ together with maps $\sigma_k: \Sigma E_k \to E_{k+1}$, which are called the structure maps of **E**. The n-th *stable homotopy group* $\pi_n(\mathbf{E})$ is then defined as the colimit $\operatorname{colim}_{k\to\infty} \pi_{n+k}(E_k)$ with respect to the directed system given by the composites

$$\pi_{n+k}(E_k) \to \pi_{n+k+1}(\Sigma E_k) \xrightarrow{\pi_{n+k+1}(\sigma_k)} \pi_{n+k+1}(E_{k+1})$$

where the first map is the suspension homomorphism. Theorem 4.57 is a kind of milestone in homotopy theory since it is the prototype of a result where the computation of geometrically defined objects are translated into a computation of (stable) homotopy groups. It applies to all other kinds of bordism groups where one puts additional structures on the manifolds, for instance a Spin-structure. The bijection is always of the same type, but the sequence of bundles ξ_k depends on the additional structure. If we want to deal with the unoriented bordism ring, we have to replace the bundle $\xi_k \to \text{BSO}(k)$ by the universal k-dimensional vector bundle over $\text{BO}(k)$.

Theorem 4.59 (Making a normal ξ-map highly connected) *Consider natural numbers k and $n \geq 3$ such that $k \leq m$ holds where m is given by $n = 2m$ or $n = 2m+1$. Let ξ be an l-dimensional vector bundle over a CW-complex X, which is homotopy equivalent to a CW-complex with finite k-skeleton. Consider an element $x \in \Omega_n(\xi)$. Then there exists a natural number c such that the image of x under* the stabilisation map $\Omega_n(\xi) \to \Omega_n(\xi \oplus \underline{\mathbb{R}^c})$ has a representative $(M, i, f, \overline{f})$ with the property that f is k-connected.

Proof. This is a modification of the proof of Theorem 4.44. One just has to ensure that in any of the finitely many surgery steps carried out in Theorem 4.44 one can use the normal bundle data to get the necessary information about the trivialisations of the tangent bundle data pulled back to the spheres S^k. But this follows from the fact that the normal bundle and the tangent bundle of M are stably inverse to one another. □

4.8 Notes

Our approach to the surgery step is based on immersions. We used Hirsch–Smale theory and the bundle data to determine a unique regular homotopy class of immersion with appropriate bundle data so that we could carry out the surgery step, provided that the immersion can be arranged to be an embedding. This approach is used for instance by Wall [414]. When X is simply connected, another approach can be found in Kervaire–Milnor [216] and Browder [55] where one first tries to find an appropriate framed embedding and then possibly changes the framing to be able to extend the bundle data to the bordism. For a comparison of these two strategies we refer to [352, Chapter 10].

The concept of surgery arises in the setting of Morse theory, as it describes the change in the topology of the preimage of the regular value of a smooth function as one passes over a single critical value of the function. Early explorations of the effect of surgery can be found in work of Wallace [416], Milnor [297] and Wall [403]. An important first application of surgery was the Kervaire–Milnor description of homotopy spheres [216], which we will review in Chapter 12.

Chapter 5
Poincaré Duality

5.1 Introduction

Recall that our goal is to find a solution to the following problem.

Problem 4.1. Let X be a finite CW-complex. When is X homotopy equivalent to a closed n-dimensional manifold?

In this chapter we deal with the necessary condition that X has to be a finite n-dimensional Poincaré complex. It is necessary since any closed n-dimensional manifold has this structure and the existence of this structure depends only on the homotopy type of a space.

In this chapter we will for simplicity consider only smooth manifolds unless explicitly stated otherwise.

We will recall the notion of a local coefficient system in Section 5.2. In Section 5.5 we will define the *intrinsic fundamental class* of a closed n-dimensional manifold M, $[[M]] \in H_n(M; \mathcal{O}_M)$ where $\mathcal{O}_M$ is a canonical infinite cyclic local coefficient system defined using the tangent bundle TM. It is intrinsic since it is invariant under any diffeomorphism and no orientability conditions are required, see Lemma 5.30.

We introduce the notion of a finite Poincaré complex in Section 5.6. Classical Poincaré duality for a closed oriented connected manifold relates its cohomology to its homology. We will need a more advanced notion of Poincaré duality for a finite CW-complex X, which is due to Wall [408] and is a chain complex version for the universal covering of X allowing arbitrary orientation homomorphisms $w \colon \pi_1(X) \to \{\pm 1\}$. As mentioned above, any closed n-dimensional manifold is a finite n-dimensional Poincaré complex, see Theorem 5.51. We will also treat finite Poincaré pairs, which correspond to compact manifolds with boundary, see Theorem 5.75. We will also introduce the signature, which was one of the first surgery obstructions, see Remark 7.43.

Guide 5.1 It is crucial to understand the notion of a Poincaré complex X, see Definition 5.43, and in particular that it is defined by a chain map of chain complexes over the group ring $\mathbb{Z}\pi_1(X)$ constructed using the universal covering of X. A reader who is not familiar with local coefficient systems and the associated homology groups may pass directly to Section 5.6 and from the very beginning make the choice (BP), see Notation 5.6, and take the point of view explained in Convention 5.49. Background material on local coefficients can be found in [178, Chapter 3.H]. One may also assume for simplicity that the orientation homomorphism $w \colon \pi \to \{\pm 1\}$ is trivial. In all events the reader should absorb the material on the signature, which is a fundamental concept in surgery theory and the study of manifolds. The reader may ignore all aspects concerning the notion of simple Poincaré complex and of Whitehead torsion until she or he will later encounter Chapter 10.

The material of the preceding chapters is not needed to read this chapter.

5.2 Local Coefficients

To avoid certain inconsistencies and bugs which appear in the literature and can lead to unpleasant difficulties and confusion, we will be very careful concerning base point issues. Often in algebraic topology these can be ignored since constructions are often independent of such choices or such choices can be made without harm. In surgery theory this is different, as we will see at various places.

It is obvious, at least in bordism theory, that one cannot just choose a base point. The spaces which occur may have several path components. Of course here one could say that one specifies for each path component one base point. But if one has two different connected manifolds, for each of which one has fixed a base point, and then considers a connected bordism between them, the question arises which base point one should take for the bordism.

If X is a path connected space and $x, y \in X$ are two points, the fundamental groups $\pi_1(X, x)$ and $\pi_1(X, y)$ are known to be isomorphic. For any homotopy class relative endpoints $[u]$ of paths u from x to y, one obtains an isomorphism $c_{[u]} \colon \pi_1(X, x) \to \pi_1(X, y)$ by sending the class $[v]$ of a loop at x to the class $[u^- * v * u]$ of the loop at y given by the concatenation of paths $u^- * v * u$ for u^- the inverse path of u. The problem is that $c_{[u]}$ depends on $[u]$. If $[u']$ is a second choice of a homotopy class relative endpoints of paths from x to y, then $u^- * u'$ defines an element $[u^- * u'] \in \pi_1(X, y)$. If $c_{[u^- * u']}$ is the inner automorphism of $\pi_1(X, y)$ given by $[u^- * u']$, then $c_{[u']} = c_{[u^- * u']} \circ c_{[u]}$. Since an inner automorphism of a group G induces the identity on $\mathrm{Wh}(G)$, we can assign to X the Whitehead group $\mathrm{Wh}(\pi)$ as explained after (3.21). Later, for a group G together with a homomorphism $w \colon G \to \{\pm 1\}$, we will introduce algebraic L-groups $L_n(\mathbb{Z}G, w)$. These groups have the property that an inner automorphism given by the element $g \in G$ induces $w(g) \cdot \mathrm{id}$, see Lemma 8.152 and Lemma 9.38. Hence we can no longer argue as we did for the Whitehead group. This problem will manifest itself at several other places, e.g., when we change the bundle data by bundle automorphisms of the target, see Exercise 7.30, when we change the fundamental classes by multiplication by -1 on one component and not on the other, see Remark 7.45, or when we deal with the functorial properties of the

5.2 Local Coefficients

L-groups, see Subsection 8.7.3. The discussion above explains why we must spend some time to establish a base point free approach to surgery theory.

Recall that the *fundamental groupoid* $\Pi(X)$ of a CW-complex X has as objects points in X and as morphisms from $x_0 \in X$ to $x_1 \in X$ the homotopy classes $[v]$ relative endpoints of paths v from x_0 to x_1. Composition of $[v_0]: x_0 \to x_1$ and $[v_1]: x_1 \to x_2$ is given by concatenation of paths, namely, $[v_1] \circ [v_0] = [v_0 * v_1]: x_0 \to x_2$. A *local coefficient system* is a covariant functor $O_X: \Pi(X) \to \mathbb{Z}$-MOD to the abelian category of $\mathbb{Z}$-modules. The *universal covering functor*

$$\widetilde{X}: \Pi(X) \to \text{SPACES} \tag{5.2}$$

is defined to be the contravariant functor sending an object x to the set $\widetilde{X}(x)$ of homotopy classes $[u]$ relative endpoints of paths $u: [0, 1] \to X$ with $u(0) = x$. The topology on $\widetilde{X}(x)$ is determined by the requirement that the map $p_X: \widetilde{X}(x) \to X$ sending $[u]$ to $u(1)$ is a covering. (It is actually a model for the universal covering.) A morphism $[v]: x \to y$ in $\Pi(X)$ is sent to

$$\widetilde{X}([v]): \widetilde{X}(y) \to \widetilde{X}(x), \quad [u] \mapsto [v * u].$$

Composition of $\widetilde{X}$ with the functor sending a CW-complex to its cellular chain complex yields a contravariant functor $C_*^{\mathbb{Z}\Pi(X)}(X)$ from $\Pi(X)$ to the category of $\mathbb{Z}$-chain complexes. The *tensor product over* $\Pi(X)$ yields a $\mathbb{Z}$-chain complex $C_*^{\mathbb{Z}\Pi(X)}(X) \otimes_{\mathbb{Z}\Pi(X)} O_X$. The n-th chain module is given as

$$\left(\bigoplus_{x \in X} C_n(\widetilde{X}(x)) \otimes_{\mathbb{Z}} O_X(x) \right) \Big/ A,$$

where A is the abelian subgroup generated by elements of the form $C_n(\widetilde{X}([u]))(a) \otimes b - a \otimes O_X([u])(b)$ for any morphism $[u]: x \to y$ in $\Pi(X)$, $a \in C_n(\widetilde{X}([y]))$ and $b \in O_X(x)$. More details of this notion can be found for instance in [250, 9.2 on page 166].

Define the *homology with coefficients in the coefficient system* O_X

$$H_n(X; O_X) := H_n\left(C_*^{\mathbb{Z}\Pi(X)}(X) \otimes_{\mathbb{Z}\Pi(X)} O_X \right) \tag{5.3}$$

to be the homology of the $\mathbb{Z}$-chain complex $C_*^{\mathbb{Z}\Pi(X)}(X) \otimes_{\mathbb{Z}\Pi(X)} O_X$.

Remark 5.4 (Induced maps) Given CW-complexes X and Y, coefficient systems $O_X: \Pi(X) \to \mathbb{Z}$-MOD and $O_Y: \Pi(Y) \to \mathbb{Z}$-MOD and a map $f: X \to Y$, one would like to define an induced homomorphism $H_n(X; O_X) \to H_n(Y; O_Y)$. But to do this, we need more data; the map f is not enough. Specifically, if f^*O_Y is defined to be $O_Y \circ \Pi(f)$ for the obvious functor $\Pi(f): \Pi(X) \to \Pi(Y)$, then one also needs a natural transformation $\overline{f}: O_X \to f^*O_Y$ of covariant functors $\Pi(X) \to \mathbb{Z}$-MOD and then one can define a homomorphism

$$H_n(f, \overline{f}): H_n(X; O_X) \to H_n(Y; O_Y).$$

An *infinite cyclic local coefficient system* O_X is a local coefficient system O_X such that $O_X(x)$ is an infinite cyclic group for all $x \in X$. It defines an element

$$w_1(O_X) \in H^1(X; \mathbb{F}_2) = \hom(H_1(X; \mathbb{Z}), \{\pm 1\}) \tag{5.5}$$

by sending $g \in H_1(X; \mathbb{Z})$ to ± 1 if, for one (and hence all) choice of $x \in X$ and loop u at x such that the image of $[u]$ under the Hurewicz homomorphism $\pi_1(X, x) \to H_1(X; \mathbb{Z})$ is g, the automorphism $O_X([u]) \colon O_X(x) \to O_X(x)$ is $\pm \operatorname{id}$. If O'_X is a second infinite cyclic local coefficient system, then O_X and O'_X are isomorphic as covariant functors $\Pi(X) \to \mathbb{Z}\text{-MOD}$ if and only if $w_1(O_X) = w_1(O'_X)$. The automorphism group of O_X is isomorphic to $H^0(X; \mathbb{F}_2) = \operatorname{map}(\pi_0(X), \{\pm 1\})$, namely, a function $\epsilon \colon \pi_0(X) \to \{\pm 1\}$ defines an automorphism of O_X by assigning to an element $x \in X$ the automorphism $\epsilon(C(x)) \cdot \operatorname{id} \colon O_X(x) \to O_X(x)$ for $C(x)$ the path component of X containing x.

5.3 Passage to Group Rings

We describe a choice of base points and related notation, choice (BP) below, which will enable us to pass from local coefficients systems to modules over group rings.

Let Γ be a discrete group and let X be a connected *CW*-complex. A *Γ-covering* $\widehat{p} \colon \widehat{X} \to X$ is a principal Γ-bundle. We note that $\widehat{X}$ need not be connected; e.g., if $\Gamma \neq \{e\}$ and $\widehat{p}$ is trivial. A Γ-covering $\widehat{p}$ defines after a choice of $\widehat{x} \in \widehat{X}$ with $\widehat{p}(\widehat{x}) = x$ a homomorphism

$$\phi(\widehat{p}, \widehat{x}) \colon \pi_1(X, x) \to \Gamma,$$

which is uniquely determined by the property that for any path $\widehat{u} \colon [0, 1] \to \widehat{X}$ with $\widehat{u}(0) = \widehat{x}$ and $\widehat{p} \circ \widehat{u}(1) = x$ the element $g \in \pi_1(X, x)$ represented by the loop $\widehat{p} \circ \widehat{u}$ satisfies the equation $\widehat{u}(1) = \phi(\widehat{p}, \widehat{x})(g) \cdot \widehat{u}(0)$.

Notation 5.6 (Choice (BP)) Let (X, O_X) be a connected *CW*-complex together with an infinite cyclic local coefficient system. For a group Γ and a Γ-covering $\widehat{p} \colon \widehat{X} \to X$ over X, a *choice (BP)* consists of:

- Base points $\widehat{x} \in \widehat{X}$ and $x \in X$ with $\widehat{p}(\widehat{x}) = x$;
- An isomorphism $o_x \colon O_X(x) \xrightarrow{\cong} \mathbb{Z}$;
- A group homomorphism $w \colon \Gamma \to \{\pm 1\}$ such that the composition of maps $\pi_1(X, x) \xrightarrow{\phi(\widehat{p}, \widehat{x})} \Gamma \xrightarrow{w} \{\pm 1\}$ is $w_1(O_X)$.

We note that it is not always possible to choose w as above, but we confine our attention to the case where such a w exists. Having made a choice (BP), we denote by $\mathbb{Z}^w$ the $\mathbb{Z}\Gamma$-module whose underlying abelian group is $\mathbb{Z}$ and on which $g \in \Gamma$ acts by $w(g) \cdot \operatorname{id}$. In the special case where $\widehat{p} \colon \widehat{X} \to X$ is a model for the universal covering $\widetilde{X} \to X$, $\Gamma = \pi_1(X, x)$, $\phi = \operatorname{id}$, and $w = w_1(O_X)$, we can always make a choice (BP) for the universal covering of X.

5.3 Passage to Group Rings

Suppose that we have made a choice (BP). We get an isomorphism of $\mathbb{Z}$-chain complexes

$$C_*(\widetilde{X}(x)) \otimes_{\mathbb{Z}[\mathrm{aut}_{\Pi(X)}(x)]} O_X(x) \xrightarrow{\cong} C_*^{\mathbb{Z}\Pi(X)}(X) \otimes_{\mathbb{Z}\Pi(X)} O_X \tag{5.7}$$

induced by the obvious map $C_n(\widetilde{X}(x)) \otimes_{\mathbb{Z}} O_X(x) \to C_*^{\mathbb{Z}\Pi(X)}(X) \otimes_{\mathbb{Z}\Pi(X)} O_X$ where $\widetilde{X} \colon \Pi(X) \to \mathrm{SPACES}$ is the universal covering functor defined in (5.2). We have the group isomorphism $\alpha \colon \pi_1(X, x) \xrightarrow{\cong} \mathrm{aut}_{\Pi(X)}(x)$ sending $[u]$ to $[u^-]$. Recall that Γ operates from the left on $\widehat{X}$ and $\mathrm{aut}_{\Pi(X)}(x)$ operates from the right on $\widetilde{X}(x)$ since $\widetilde{X}$ is contravariant. We can consider $\widetilde{X}(x)$ as a left $\pi_1(X, x)$-space and $C_*(\widetilde{X}(x))$ as a left $\mathbb{Z}[\pi_1(X, x)]$-chain complex using α. From $o_x \colon O_X(x) \xrightarrow{\cong} \mathbb{Z}$ we obtain an isomorphism of $\mathbb{Z}$-chain complexes

$$\mathbb{Z}^{w_1(O_X)} \otimes_{\mathbb{Z}[\pi_1(X,x)]} C_*(\widetilde{X}(x)) \xrightarrow{\cong} C_*(\widetilde{X}(x)) \otimes_{\mathbb{Z}[\mathrm{aut}_{\Pi(X)}(x)]} O_X(x). \tag{5.8}$$

From (5.7) and (5.8) we obtain an isomorphism of $\mathbb{Z}$-chain complexes

$$\mathbb{Z}^{w_1(O_X)} \otimes_{\mathbb{Z}[\pi_1(X,x)]} C_*(\widetilde{X}(x)) \xrightarrow{\cong} C_*^{\mathbb{Z}\Pi(X)}(X) \otimes_{\mathbb{Z}\Pi(X)} O_X. \tag{5.9}$$

Let $q \colon \widetilde{X}(x) \to \widehat{X}$ be the map that is uniquely determined by the properties $q(c_x) = \widehat{x}$ and $\widehat{p} \circ q = \widetilde{p}$ where $c_x \in \widetilde{X}(x)$ is the base point represented by the constant loop with value x and $\widetilde{p} \colon \widetilde{X}(x) \to X$ is the canonical projection. We obtain a Γ-homeomorphism of left Γ-spaces

$$\widehat{q} \colon \Gamma \times_\phi \widetilde{X}(x) \xrightarrow{\cong} \widehat{X}, \quad g \times y \mapsto g \cdot q(y).$$

It induces an isomorphism of $\mathbb{Z}\Gamma$-chain complexes

$$\mathbb{Z}\Gamma \otimes_{\mathbb{Z}[\pi_1(X,x)]} C_*(\widetilde{X}(x)) \xrightarrow{\cong} C_*(\widehat{X}) \tag{5.10}$$

and hence an isomorphism of $\mathbb{Z}$-chain complexes

$$\mathbb{Z}^{w \circ \phi} \otimes_{\mathbb{Z}[\pi_1(X,x)]} C_*(\widetilde{X}(x)) \xrightarrow{\cong} \mathbb{Z}^w \otimes_{\mathbb{Z}\Gamma} C_*(\widehat{X}). \tag{5.11}$$

Combining (5.9) and (5.10) yields a $\mathbb{Z}$-isomorphism of $\mathbb{Z}$-chain complexes

$$\mathbb{Z}^w \otimes_{\mathbb{Z}\Gamma} C_*(\widehat{X}) \xrightarrow{\cong} C_*^{\mathbb{Z}\Pi(X)}(X) \otimes_{\mathbb{Z}\Pi(X)} O_X, \tag{5.12}$$

which induces a preferred isomorphism

$$\beta_{\widehat{p},\widehat{x},o_x,w} \colon H_n^\Gamma(X; \mathbb{Z}^w) := H_n\bigl(\mathbb{Z}^w \otimes_{\mathbb{Z}\Gamma} C_*(\widehat{X})\bigr)$$
$$\xrightarrow{\cong} H_n(X; O_X) := H_n\bigl(C_*^{\mathbb{Z}\Pi(X)}(X) \otimes_{\mathbb{Z}\Pi(X)} O_X\bigr). \tag{5.13}$$

In the special case where the choice (BP) is for the universal covering of X, we get a preferred isomorphism

$$\beta_{\widetilde{p},\widetilde{x},o_x} : H_n(X; \mathbb{Z}^{w_1(O_X)}) := H_n\big(\mathbb{Z}^{w_1(O_X)} \otimes_{\mathbb{Z}\pi_1(X,x)} C_*(\widetilde{X})\big)$$
$$\xrightarrow{\cong} H_n(X; O_X) := H_n\big(C_*^{\mathbb{Z}\Pi(X)}(X) \otimes_{\mathbb{Z}\Pi(X)} O_X\big). \quad (5.14)$$

Let $\widehat{p}' \colon \widehat{X}' \to X$ be another Γ-covering together with a Γ-homeomorphism $\widehat{v} \colon \widehat{X} \to \widehat{X}'$ satisfying $\widehat{p}' \circ \widehat{v} = \widehat{p}$. Let $\widehat{x}' \in \widehat{X}'$, $x' \in X$ with $\widehat{p}'(\widehat{x}') = x'$, the isomorphism $o'_{x'} \colon O_X(x') \xrightarrow{\cong} \mathbb{Z}$ and the homomorphism $w' \colon \Gamma \to \{\pm 1\}$ be a choice (BP) for $\widehat{p}'$. Let $\widehat{u}$ be any path from $\widehat{v}(\widehat{x})$ to $\widehat{x}'$ in $\widehat{X}'$. Let u be the path from x to x' in X given by $\widehat{p}' \circ \widehat{u}$. Choose a homotopy $h \colon X \times [0,1] \to X$ such that $h_0 = \mathrm{id}$ and $h(x,t) = u(t)$ for every $t \in [0,1]$. Then $h_1 \colon X \to X$ maps x to x'. Let $\widehat{h}_1 \colon \widehat{X} \to \widehat{X}'$ be the map uniquely determined by $\widehat{h}_1(\widehat{x}) = \widehat{x}'$ and $\widehat{p}' \circ \widehat{h}_1 = h_1 \circ \widehat{p}$. Then $\widehat{h}_1$ is Γ-equivariant and we have $\phi' = \phi \circ c_{[u]}$, where $\phi \colon \pi_1(X,x) \to \Gamma$ and $\phi' \colon \pi_1(X,x') \to \Gamma$ are the homomorphisms associated to the choices (BP) and $c_{[u]} \colon \pi_1(X,x) \to \pi_1(X,x')$ is given by conjugation with u. Let

$$\gamma_{\widehat{p},\widehat{p}',\widehat{x},\widehat{x}'} \colon H_n\big(\mathbb{Z}^w \otimes_{\mathbb{Z}\Gamma} C_*(\widehat{X})\big) \xrightarrow{\cong} H_n\big(\mathbb{Z}^{w'} \otimes_{\mathbb{Z}\Gamma} C_*(\widehat{X}')\big)$$

be the isomorphism induced by the $\mathbb{Z}$-chain map

$$\mathrm{id} \otimes_{\mathbb{Z}\Gamma} C_*(\widehat{h}_1) \colon \mathbb{Z}^w \otimes_{\mathbb{Z}\Gamma} C_*(\widehat{X}) \to \mathbb{Z}^w \otimes_{\mathbb{Z}\Gamma} C_*(\widehat{X}') = \mathbb{Z}^{w'} \otimes_{\mathbb{Z}\Gamma} C_*(\widehat{X}')$$

where the identification $\mathbb{Z}^w \otimes_{\mathbb{Z}\Gamma} C_*(\widehat{X}') = \mathbb{Z}^{w'} \otimes_{\mathbb{Z}\Gamma} C_*(\widehat{X}')$ comes from the fact that $w' \circ \phi' \circ c_{[u]} = w \circ \phi$ holds and $\mathbb{Z}^w \otimes_{\mathbb{Z}\Gamma} C_*(\widehat{X}')$ and $\mathbb{Z}^{w'} \otimes_{\mathbb{Z}\Gamma} C_*(\widehat{X}')$ depend only on $w \circ \phi$ and $w' \circ \phi'$ respectively, see (5.11). The map $\gamma_{\widehat{p},\widehat{p}',\widehat{x},\widehat{x}'}$ is indeed independent of the choice of h and $\widehat{u}$ since for a loop $\widehat{v}$ at $\widehat{x}'$ in $\widehat{X}'$ the loop $\widehat{p}' \circ \widehat{v}$ at x' defines an element in $\pi_1(X',x')$ which lies in the kernel of $\phi' \colon \pi_1(X,x') \to \Gamma$. Let $\epsilon \in \{\pm 1\}$ be the element for which the composite $\mathbb{Z} \xrightarrow{o_x} O_X(x) \xrightarrow{O_X([u])} O_X(x') \xrightarrow{(o'_{x'})^{-1}} \mathbb{Z}$ is multiplication by ϵ. Then the following diagram

$$\begin{array}{c}
H_n\big(\mathbb{Z}^w \otimes_{\mathbb{Z}\Gamma} C_*(\widehat{X})\big) \\
\Big\downarrow {\scriptstyle \epsilon \cdot \gamma_{\widehat{p},\widehat{p}',\widehat{x},\widehat{x}'}} \qquad \xrightarrow{\beta_{\widehat{p},\widehat{x},o_x,w}} \\
\qquad\qquad H_n\big(C_*^{\mathbb{Z}\Pi(X)}(X) \otimes_{\mathbb{Z}\Pi(X)} O_X\big) \\
\qquad\qquad \xrightarrow{\beta_{\widehat{p}',\widehat{x}',o_{x'},w'}} \\
H_n\big(\mathbb{Z}^{w'} \otimes_{\mathbb{Z}\Gamma} C_*(\widehat{X}')\big)
\end{array} \quad (5.15)$$

commutes. In the special case where we consider a choice (BP) for two models for the universal covering of X, we obtain the commutative diagram

5.4 The Determinant Line Bundle

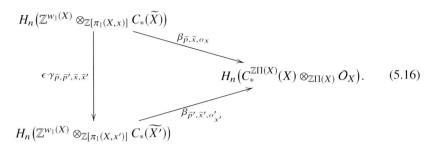
(5.16)

Exercise 5.17 If $\underline{A}\colon \Pi(X) \to \mathbb{Z}\text{-MOD}$ is the constant functor with the abelian group A as value, then $H_n(X;\underline{A})$ agrees with the homology $H_n(X;A)$ of X with coefficients in A.

Exercise 5.18 Let X be a connected CW-complex. Fix $x \in X$. Consider the local coefficient system $\widetilde{O}\colon \Pi(X) \to \mathbb{Z}\text{-MOD}$ that sends $y \in X$ to the free abelian group generated by the set of homotopy classes relative endpoints of paths from x to y. Show that $H_n(X;\widetilde{O})$ is $\mathbb{Z}$-isomorphic to $H_n(\widetilde{X};\mathbb{Z})$.

5.4 The Determinant Line Bundle

Definition 5.19 (Determinant line bundle) Let ξ be a d-dimensional vector bundle. Define its *determinant line bundle* $\det(\xi)$ to be its d-th exterior power $\Lambda^d \xi$.

For a real line bundle μ over a CW-complex X, which is given by a map $p_\mu\colon E_\mu \to X$, we define the $\mathbb{Z}/2$-principal bundle $\widehat{\mu}$,

$$p_{\widehat{\mu}}\colon E_{\widehat{\mu}} := (E_\mu - \{0\})/\sim \, \to X, \tag{5.20}$$

where the equivalence relation $\sim$ is given by identifying $e \in E_\mu$ with $r \cdot e$ for all $r > 0$ and the $\mathbb{Z}/2$-action comes from $e \mapsto -e$ for $e \in E_\mu$. The fibre transport yields a covariant functor

$$O_\mu\colon \Pi(X) \to \mathbb{Z}\text{-MOD} \tag{5.21}$$

by sending an object x to the infinite cyclic group $\mathbb{Z} \times_{\mathbb{Z}/2} p_{\widehat{\mu}}^{-1}(x)$ where the generator of $\mathbb{Z}/2$ acts on $\mathbb{Z}$ by $-\operatorname{id}$. For a vector bundle ξ we put

$$O_\xi := O_{\det(\xi)}. \tag{5.22}$$

In the sequel, for a finite-dimensional vector space V, we will denote by $\underline{V}$ the trivial bundle with fibre V.

Lemma 5.23

(i) Let ξ_m for $m = 0, 1, 2$ be vector bundles over X. Let $i\colon \xi_0 \to \xi_1$ and $p\colon \xi_1 \to \xi_2$ be maps of vector bundles over X where we do not require that the induced maps on the fibres are bijective. Suppose that for any $x \in X$ the sequence $0 \to (\xi_0)_x \xrightarrow{i_x} (\xi_1)_x \xrightarrow{p_x} (\xi_2)_x \to 0$ of linear maps is exact.
Then there is a preferred and natural isomorphism of one-dimensional vector bundles over X,

$$\kappa\colon \det(\xi_0) \otimes \det(\xi_2) \xrightarrow{\cong} \det(\xi_1),$$

which induces a natural isomorphism of infinite cyclic local coefficient systems

$$O_{\xi_0} \otimes O_{\xi_2} \xrightarrow{\cong} O_{\xi_1}.$$

In particular, we get for two vector bundles ξ and η over X preferred and natural isomorphisms

$$\det(\xi) \otimes \det(\eta) \xrightarrow{\cong} \det(\xi \oplus \eta);$$
$$O_\xi \otimes O_\eta \xrightarrow{\cong} O_{\xi \oplus \eta},$$

and for a vector bundle ξ over X and a natural number a preferred and natural isomorphisms

$$\det(\xi) \xrightarrow{\cong} \det(\xi \oplus \underline{\mathbb{R}}^a);$$
$$O_\xi \xrightarrow{\cong} O_{\xi \oplus \underline{\mathbb{R}}^a},$$

and

$$\det(\xi) \xrightarrow{\cong} \det(\underline{\mathbb{R}}^a \oplus \xi);$$
$$O_\xi \xrightarrow{\cong} O_{\underline{\mathbb{R}}^a \oplus \xi};$$

(ii) Let $\underline{u}, \underline{v}\colon \xi \xrightarrow{\cong} \eta$ be isomorphisms of vector bundles over X covering the identity that are isotopic. Then the induced isomorphisms of infinite cyclic coefficient systems $O_{\underline{u}}$ and $O_{\underline{v}}$ from O_ξ to O_η agree;

(iii) Let ξ be a vector bundle over X. Then there is a preferred and natural isomorphism of infinite cyclic local coefficient systems

$$O_{\xi^*} \xrightarrow{\cong} O_\xi;$$

(iv) Let M be an n-dimensional manifold (possibly with boundary) together with an embedding $i\colon M \to \mathbb{R}^{n+k}$. Then there is a preferred and natural isomorphism of infinite cyclic local coefficient systems

$$\kappa\colon O_{TM} \xrightarrow{\cong} O_{\nu(i)},$$

5.4 The Determinant Line Bundle

where $\nu(i)$ is the normal bundle of i;

(v) Let $(i, \partial i)\colon (M, \partial M) \to (N, \partial N)$ be an embedding of manifolds with boundary. Then there is a preferred and natural isomorphism of vector bundles over ∂M, where $j_M\colon \partial M \to M$ is the inclusion

$$\kappa\colon \nu(\partial i) \xrightarrow{\cong} (j_M)^* \nu(i),$$

which induces an isomorphism, denoted in the same way,

$$\kappa\colon \det(\nu(\partial i)) \xrightarrow{\cong} (j_M)^* \det(\nu(i)).$$

Proof. (i) Let $0 \to V_0 \xrightarrow{j} V_1 \xrightarrow{q} V_2 \to 0$ be an exact sequence of finite-dimensional vector spaces. Put $d_m := \dim(V_m)$ for $m = 0, 1, 2$. Choose a section $s\colon V_2 \to V_1$, i.e., a linear map s with $q \circ s = \mathrm{id}_{V_2}$. Then we obtain an isomorphism

$$\Lambda^{d_1}(j \oplus s)\colon \Lambda^{d_1}(V_0 \oplus V_2) \xrightarrow{\cong} \Lambda^{d_1}(V_1).$$

One easily checks that it is independent of the choice of s. The exterior product $\wedge$ induces an isomorphism

$$\Lambda^{d_0}(V_0) \otimes \Lambda^{d_2}(V_2) \xrightarrow{\cong} \Lambda^{d_1}(V_0 \oplus V_2).$$

The composite of the two isomorphisms above yields a preferred and natural isomorphism of 1-dimensional vector spaces

$$\Lambda^{d_0}(V_0) \otimes \Lambda^{d_2}(V_2) \xrightarrow{\cong} \Lambda^{d_1}(V_1).$$

Applying this construction fibrewise yields the desired isomorphism κ.

(ii) We can assume without loss of generality that X is connected. It suffices to show that $O_{\underline{v}^{-1} \circ \underline{u}} \colon O_\xi \to O_\xi$ is the identity. Hence we can assume without loss of generality that $\underline{v}$ is the identity. If $\underline{u}$ is isotopic to the identity on ξ, then $\det(\underline{u})$ is isotopic to the identity on $\det(\xi)$. Hence there is a continuous function $f\colon X \to \mathbb{R}^{>0}$ such that $\det(\underline{u})_x = f(x) \cdot \mathrm{id}_{\Lambda^d \xi_x}$, where d is the dimension of ξ. Now the claim follows directly from the definitions.

(iii) For a 1-dimensional vector bundle η there is in general no canonical isomorphism $\eta \xrightarrow{\cong} \eta^*$. However, there is a canonical isomorphism of infinite cyclic local coefficient systems $O_\eta \xrightarrow{\cong} O_{\eta^*}$ since for a 1-dimensional vector space V, $v \in V$ with $v \neq 0$, and $\lambda \in \mathbb{R}$ with $\lambda \neq 0$ we get $\lambda^{-1} \cdot v^* = (\lambda v)^*$ where $v^* \in V^*$ is defined by $v^*(w) \cdot v = w$ for all $w \in V$.

(iv) Assertion (i) applied to the exact sequence $0 \to TM \xrightarrow{Ti} i^* T\mathbb{R}^{n+k} \xrightarrow{p} \nu(i) \to 0$ and the standard isomorphism $\det(T\mathbb{R}^{n+k}) \xrightarrow{\cong} \mathbb{R}^{n+k}$ coming from the standard trivialisation $T\mathbb{R}^{n+k} \xrightarrow{\cong} \mathbb{R}^{n+k}$ yield an isomorphism of 1-dimensional vector bundles $\det(TM) \otimes \det(\nu(i)) \xrightarrow{\cong} \mathbb{R}$. This is the same as an isomorphism $\det(TM) \xrightarrow{\cong} \det(\nu(i))^*$, where $\det(\nu(i))^*$ is the dual vector bundle. It induces an isomorphism of infinite cyclic local coefficient systems $\mathcal{O}_M := \mathcal{O}_{\det(TM)} \xrightarrow{\cong} \mathcal{O}_{\det(\nu(i))^*}$. Now the desired isomorphism is obtained by the composite of this isomorphism with the isomorphism $\mathcal{O}_{\det(\nu(i))^*} \xrightarrow{\cong} \mathcal{O}_{\det(\nu(i))} =: \mathcal{O}_{\nu(i)}$ coming from assertion (iii).

(v) The desired isomorphism κ is induced by the following commutative diagram

$$\begin{array}{ccccc}
T\partial M & \xrightarrow{T\partial i} & (\partial i)^* T\partial N & \xrightarrow{p_0} & \nu(\partial i) \\
\downarrow{Tj_M} & & \downarrow{(\partial i)^* Tj_N} & & \downarrow{\kappa} \\
j_M^* TM & \xrightarrow{j_M^* Ti} & j_M^* i^* TN = (\partial i)^* j_N^* TN & \xrightarrow{(\partial i)^* p_1} & (j_M)^* \nu(i)
\end{array}$$

where p_m is the canonical projection and $j_N : \partial N \to N$ is the inclusion. □

5.5 The Intrinsic Fundamental Class

In this section we assign to any compact connected manifold M of dimension n an intrinsic fundamental class that is a preferred generator $[[M]]$ of the infinite cyclic group $H_n(M, \partial M; \mathcal{O}_M)$. Intrinsic means that it is a diffeomorphism invariant where no kind of compatibility conditions for orientations are required for the diffeomorphism.

We will often identify $\det(\xi)$ with $\det(\underline{\mathbb{R}}^a \oplus \xi)$ or with $\det(\xi \oplus \underline{\mathbb{R}}^a)$ using Lemma 5.23 (i).

Exercise 5.24 Show for a vector bundle ξ over a CW-complex X that $w_1(\xi) = w_1(\det(\xi))$ holds in $H^1(X; \mathbb{Z}/2)$.

Consider a compact n-dimensional manifold M with boundary ∂M. Define

$$\mathcal{O}_M : \Pi(M) \to \mathbb{Z}\text{-MOD} \qquad (5.25)$$

by $\mathcal{O}_{\det(TM)}$. We call it the *local orientation coefficient system of M*.

Exercise 5.26 Let M be a closed n-dimensional manifold. Consider the $\mathbb{Z}/2$-principal bundle $\widehat{p_M} : \widehat{M} \to M$ given by its orientation covering. Define an infinite cyclic local coefficient system using the fibre transport

$$\mathcal{O}_{\widehat{p_M}} : \Pi(M) \to \mathbb{Z}\text{-MOD}, \quad x \mapsto \mathbb{Z} \times_{\mathbb{Z}/2} \widehat{p_M}^{-1}(x).$$

Show that $\mathcal{O}_{\widehat{p_M}}$ and $\mathcal{O}_M$ are isomorphic.

Example 5.27 Let V be a finite-dimensional vector space. Next we explain how an orientation on V yields preferred generators of both infinite cyclic groups $H_n(V, V \setminus \{0\}; \mathbb{Z})$ and $\widehat{\det(V)} \times_{\mathbb{Z}/2} \mathbb{Z}$, where $\widehat{\det(V)}$ is the transitive free $\mathbb{Z}/2$-set $\Lambda^{\dim(V)} V / \sim$ for the equivalence relation $\sim$ given by identifying $x \in \Lambda^{\dim(V)} V$ with

5.5 The Intrinsic Fundamental Class

$r \cdot x$ for all $r > 0$, and the $\mathbb{Z}/2$-action comes from $x \mapsto -x$ for $x \in \Lambda^{\dim(V)}V$. Obviously the orientation of V yields a generator $[V]$ of $\widehat{\det(V)} \times_{\mathbb{Z}/2} \mathbb{Z}$. Equip $\mathbb{R}^n$ with the standard basis and thus with the standard orientation. Choose an orientation preserving isomorphism $u \colon \mathbb{R}^n \xrightarrow{\cong} V$. It induces an isomorphism $H_n(u) \colon H_n(\mathbb{R}^n, \mathbb{R}^n \setminus \{0\}) \xrightarrow{\cong} H_n(V, V \setminus \{0\})$ and we define $[V] \in H_n(V, V \setminus \{0\})$ to be the image of the standard generator of $H_n(\mathbb{R}^n, \mathbb{R}^n \setminus \{0\})$. This is independent of the choice of u.

The tensor product of these two generators defines a generator $[[V]]$ of the infinite cyclic group $H_n(V, V \setminus \{0\}; \mathbb{Z}) \otimes_{\mathbb{Z}} (\widehat{\det(V)} \times_{\mathbb{Z}/2} \mathbb{Z})$. The generator $[[V]]$ is independent of the choice of orientation on V since a change of the orientation on V changes both these generators simultaneously by multiplying by (-1) and $(-1) \cdot (-1) = 1$.

Lemma 5.28 *Consider a connected compact n-dimensional manifold M with boundary ∂M. Then the abelian group $H_n(M, \partial M; \mathcal{O}_M)$ is an infinite cyclic group and there is a preferred generator*

$$[[M, \partial M]] \in H_n(M, \partial M; \mathcal{O}_M).$$

Proof. Choose a point $x \in M \setminus \partial M$. The inclusion $j_x \colon M = (M, \partial M) \to (M, M \setminus \{x\})$ induces an isomorphism

$$(j_x)_* \colon H_n(M, \partial M; \mathcal{O}_M) \xrightarrow{\cong} H_n(M, M \setminus \{x\}; \mathcal{O}_M).$$

This claim is proved as follows. If M is orientable, or equivalently, if $\mathcal{O}_M$ is isomorphic to the constant coefficient theorem, then this is equivalent to the well-known statement that $(j_x)_* \colon H_n(M, \partial M; \mathbb{Z}) \xrightarrow{\cong} H_n(M, M \setminus \{x\}; \mathbb{Z})$ is an isomorphism.

Suppose that M is not orientable. Then its orientation covering $p \colon \overline{M} \to M$ has a connected orientable closed manifold as total space $\overline{M}$. Note that p is given by the $S\Lambda^n TM \to M$. Moreover, $p^*\mathcal{O}_M$ is isomorphic to $\mathcal{O}_{\overline{M}}$ and we obtain a commutative diagram where $\overline{x} \in \overline{M}$ is an element with $p(\overline{x}) = x$

$$\begin{array}{ccc} H_n(\overline{M}, \partial \overline{M}; \mathcal{O}_{\overline{M}}) & \xrightarrow{\cong} & H_n(\overline{M}, \overline{M} \setminus \{\overline{x}\}; \mathcal{O}_{\overline{M}}) \\ p_* \downarrow & & \downarrow p_* \cong \\ H_n(M, \partial M; \mathcal{O}_M) & \longrightarrow & H_n(M, M \setminus \{x\}; \mathcal{O}_M). \end{array}$$

The right vertical arrow is bijection by excision. We know already that the upper horizontal arrow is an isomorphism. Hence the lower horizontal arrow is an epimorphism onto an infinite cyclic group. The pair $(M, \partial M)$ is homotopy equivalent to an n-dimensional finite CW-pair (X, A) with one n-dimensional cell. This follows for instance from Morse theory. Hence $H_n(M, \partial M; \mathcal{O}_M)$ is a $\mathbb{Z}$-submodule of an infinite cyclic group and hence either trivial or infinite cyclic. Hence the lower horizontal arrow is an isomorphism.

Note that this implies that also the left vertical arrow is bijective. This can also been seen directly, see for instance [178, Example 3H.3 on page 329] in the closed case.

Let $\exp_x \colon T_x M \to M$ be an exponential map, i.e., an embedding such that $\exp_x(0) = x$ and $T_0 \exp_x \colon T_0 T_x M \to T_x M$ is the identity under the canonical identification $T_x M \xrightarrow{\cong} T_0 T_x M$. The induced map

$$(\exp_x)_* \colon H_n(T_x M, T_x M \setminus \{0\}; \exp_x^* \mathcal{O}_M) \xrightarrow{\cong} H_n(M, M \setminus \{x\}; \mathcal{O}_M)$$

is an isomorphism because of excision. Since $T_x M$ is simply connected and $T_0 \exp_x = \mathrm{id}_{T_x M}$, there is a canonical isomorphism between the constant functor $\Pi(T_x M) \to \mathbb{Z}\text{-MOD}$ with value $\mathcal{O}_M(x)$ and the covariant functor $(\exp_x)^* \mathcal{O}_M$. It induces an isomorphism

$$\alpha_x \colon H_n(T_x M, T_x M \setminus \{0\}; \mathcal{O}_M(x)) \xrightarrow{\cong} H_n(T_x M, T_x M \setminus \{0\}; \exp_x^* \mathcal{O}_M).$$

The canonical map

$$\beta_x \colon H_n(T_x M, T_x M \setminus \{0\}; \mathbb{Z}) \otimes_{\mathbb{Z}} \mathcal{O}_M(x) \xrightarrow{\cong} H_n(T_x M, T_x M \setminus \{0\}; \mathcal{O}_M(x))$$

is an isomorphism by the universal coefficient theorem. Define an isomorphism

$$\lambda_x \colon H_n(T_x M, T_x M \setminus \{0\}; \mathbb{Z}) \otimes \mathcal{O}_M(x) \xrightarrow{\cong} H_n(M, \partial M; \mathcal{O}_M)$$

by the composite of isomorphisms

$$H_n(T_x M, T_x M \setminus \{0\}; \mathbb{Z}) \otimes_{\mathbb{Z}} \mathcal{O}_M(x) \xrightarrow{\beta_x} H_n(T_x M, T_x M \setminus \{0\}; \mathcal{O}_M(x))$$
$$\xrightarrow{\alpha_x} H_n(T_x M, T_x M \setminus \{0\}; \exp_x^* \mathcal{O}_M)$$
$$\xrightarrow{(\exp_x)_*} H_n(M, M \setminus \{x\}; \mathcal{O}_M)$$
$$\xrightarrow{((j_x)_*)^{-1}} H_n(M, \partial M; \mathcal{O}_M).$$

The map λ_x is independent of the choice of the exponential map $\exp_x$ since two such exponential maps are isotopic relative $\{0\}$. Let u be a path in M from x to y. The fibre transport induces a linear isomorphism $t_{[u]} \colon T_x M \xrightarrow{\cong} T_y M$, unique up to isotopy of linear isomorphisms. Since u defines a morphism $[u] \colon x \to y$ in $\Pi(M)$, it induces an isomorphism $\mathcal{O}_M([u]) \colon \mathcal{O}_M(x) \to \mathcal{O}_M(y)$. Then the following diagram

5.5 The Intrinsic Fundamental Class

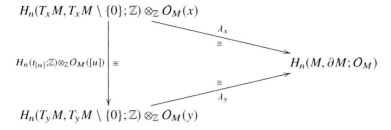

commutes. Moreover, the vertical arrow $H_n(t_{[u]}; \mathbb{Z}) \otimes_{\mathbb{Z}} O_M([u])$ is independent of $[u]$ since for any path v from y to y we have $O_M([v]) = w_1([v]) \cdot \mathrm{id}$ and $t_{[v]} = w_1([v]) \cdot \mathrm{id}$ and $w_1([v]) \cdot w_1([v]) = 1$ where $w_1 \colon \pi_1(M, y) \to \{\pm 1\}$ is given by the first Stiefel–Whitney class of TM.

Applying Example 5.27 to the special case $V = T_x M$ yields a generator $[[T_x M]]$ of the infinite cyclic group $H_n(T_x M, T_x M \setminus \{0\}; \mathbb{Z}) \otimes_{\mathbb{Z}} O_M(x)$. Define

$$[[M, \partial M]] = \lambda_x([[T_x M]]).$$

This is independent of the choice of $x \in X$ since the vertical isomorphism $H_n(t_{[u]}; \mathbb{Z}) \otimes O_M([u])$ above sends $[[T_x M]]$ to $[[T_y M]]$ and is independent of $[u]$. □

Now suppose that M is a compact n-dimensional manifold with boundary ∂M which is not necessarily connected. Then we get a natural isomorphism

$$\bigoplus_{C \in \Pi_0(C)} j_C \colon \bigoplus_{C \in \pi_0(M)} H_n(C, \partial C; O_C) \xrightarrow{\cong} H_n(M, \partial M; O_M)$$

where j_C is induced by the inclusion $C \to M$ and we use the identification $TC = TM|_C$. Note that we have a preferred generator $[[C, \partial C]]$ for each infinite cyclic group $H_n(C, \partial C; O_C)$.

Definition 5.29 (Intrinsic fundamental class) Define the *intrinsic fundamental class*

$$[[M, \partial M]] \in H_n(M, \partial M; O_M)$$

to be $\sum_{C \in \pi_0(C)} j_C([[C, \partial C]])$.

The name intrinsic fundamental class is justified by the next result.

Lemma 5.30 *Let $(f, \partial f) \colon (M, \partial M) \to (N, \partial N)$ be a diffeomorphism of compact n-dimensional manifolds. Its differential Tf induces an equivalence $\widehat{f} \colon O_M \to f^* O_N$ of covariant functors $\Pi(M) \to \mathbb{Z}\text{-MOD}$.*

Then the induced isomorphism

$$H_n(f; \widehat{f}) \colon H_n(M, \partial M; O_M) \xrightarrow{\cong} H_n(N, \partial N; O_N)$$

sends $[[M, \partial M]]$ to $[[N, \partial N]]$.

Proof. One easily checks that it suffices to treat the case where M and N are connected. In the sequel we use the notation of the proof of Lemma 5.28. If x is any point in $M \setminus \partial M$, then the following diagram commutes

$$\begin{array}{ccc}
H_n(T_xM, T_xM \setminus \{0\}; \mathbb{Z}) \otimes O_M(x) & \xrightarrow{\lambda_x}_{\cong} & H_n(M, \partial M; O_M) \\
{\scriptstyle H_n(T_xf; \mathbb{Z}) \otimes \widehat{f}(x)} \Big\downarrow \cong & & \cong \Big\downarrow {\scriptstyle H_n(f, \widehat{f})} \\
H_n(T_{f(x)}N, T_{f(x)}N \setminus \{0\}; \mathbb{Z}) \otimes O_N(f(x)) & \xrightarrow{\lambda_{f(x)}}_{\cong} & H_n(N, \partial N; O_N).
\end{array}$$

If we fix orientations for T_xM and $T_{f(x)}N$, then $H_n(T_xf; \mathbb{Z})$ respects the induced orientations if and only if $\widehat{f}(x)$ respects the induced orientations. This implies that $H_n(f, \widehat{f}) \colon H_n(M, \partial M; O_M) \to H_n(N, \partial N; O_N)$ sends $[[M, \partial M]]$ to $[[N, \partial N]]$. □

Definition 5.31 (Outward normal vector field) Let M be a manifold with boundary. An *outward normal vector field* is a smooth map $v \colon \partial M \to TM|_{\partial M}$ such that the composite of v with the projection $TM|_{\partial M} \to \partial M$ is the identity, for every $x \in \partial M$ the vector $v(x)$ does not belong to $T_x \partial M$ and there is a smooth curve $u \colon (-1, 0] \to M$ with $u(0) = x$ and with the differential $du_0(1) = v(x)$.

Let $n_v \colon \mathbb{R} \to TM$ be the fibrewise injective bundle map covering the inclusion $\partial M \to M$ which sends 1 in the fibre over $x \in \partial M$ to $v(x)$.

Figure 5.32 (Outward normal vector field).

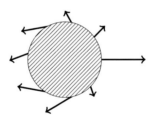

Exercise 5.33 Let M be a manifold with boundary ∂M. Let $i \colon \partial M \to M$ be the inclusion. Show:

(i) There exists an outward normal vector field v;
(ii) Let v be an outward normal vector field on M. Then the bundle map $n_v \oplus Ti \colon \mathbb{R} \oplus T\partial M \to TM$ covering $i \colon \partial M \to M$ is fibrewise an isomorphism;
(iii) Let v and v' be outward normal vector fields on M. Then there exists a preferred isotopy of monomorphisms of vector bundles covering $i \colon \partial M \to M$ between n_v and $n_{v'}$;
(iv) Let v and v' be outward normal vector fields on M. Then there exists a preferred isotopy of fibrewise bijective maps of vector bundles covering $i \colon \partial M \to M$ between $n_v \oplus Ti$ and $n_{v'} \oplus Ti$ that leaves Ti fixed;
(v) If TM comes with a Riemannian metric, then there is precisely one outward normal vector field v that satisfies $\langle v(x), v(x) \rangle_x = 1$ and $\langle v(x), T_x i(v) \rangle_x = 0$ for all $x \in \partial M$ and $v \in T_x \partial M$.

5.5 The Intrinsic Fundamental Class

Now one easily checks the following lemma.

Lemma 5.34 *Let M be a compact n-dimensional manifold with boundary ∂M. Choose an outward normal vector field v.*

Then we get a preferred isomorphism of infinite cyclic coefficient systems

$$\widehat{i}\colon O_{\partial M} \xrightarrow{\cong} i^*O_M$$

from the bundle map $n_v \oplus Ti\colon \mathbb{R} \oplus T\partial M \to TM$ covering $i\colon \partial M \to M$ and the isomorphism $O_{T\partial M} \xrightarrow{\cong} O_{\mathbb{R} \oplus T\partial M}$ of Lemma 5.23 (i), such that $\widehat{i}$ is independent of the choice of v. Moreover, the boundary map

$$\partial_n\colon H_n(M, \partial M; O_M) \to H_{n-1}(\partial M; O_{\partial M})$$

of the long exact sequence of the pair $(M, \partial M)$ with respect to $\widehat{i}$ sends $[[M, \partial M]]$ to $[[\partial M]]$.

Exercise 5.35 Show that in Lemma 5.34 the map

$$\partial_n\colon H_n(M, \partial M; O_M) \to H_{n-1}(\partial M; O_{\partial M})$$

is independent of the choice of the outward normal vector field.

Remark 5.36 (Recovering the classical fundamental class from the intrinsic one) Consider a compact manifold M. For simplicity assume that M is connected. Fix a choice (BP) for the universal covering, see Notation 5.6. Now we have a preferred isomorphism, see (5.14),

$$\beta_{p,x,o_x}\colon H_n(M, \partial M; \mathbb{Z}^w) \xrightarrow{\cong} H_n(M, \partial M; O_M).$$

We obtain a fundamental class $[M, \partial M] \in H_n(M, \partial M; \mathbb{Z}^w)$ by requiring that it is sent by the isomorphism above to the intrinsic fundamental class $[[M, \partial M]]$.

If we replace o_x by $-o_x$, then $[M, \partial M]$ is replaced by $-[M, \partial M]$. We see that $[M, \partial M]$ in contrast to $[[M, \partial M]]$ is not an intrinsic invariant of $(M, \partial M)$.

Remark 5.37 (Orientation conventions) Let us discuss our orientation conventions for manifolds. For simplicity we will only consider a connected compact orientable n-dimensional manifold M, possibly with boundary ∂M, where orientable means $w_1(M) = 0$, or, equivalently, that $H_n(M; \partial M)$ is infinite cyclic. Here is a list of desired properties or standard conventions.

(i) On the vector space $\mathbb{R}^n$ for $n \geq 1$ we use the standard orientation given by the ordered standard basis $\{e_1, e_2, \ldots, e_n\}$, where e_i is the vector

$$(0, 0, \ldots, 0, 1, 0, \ldots, 0)$$

whose only non-zero entry is at position i. If $n = 0$, an orientation on $\mathbb{R}^0$ is a choice of an element in $\{+, -\}$;

(ii) For $n \geq 1$ an orientation on TM is a choice of orientation on every $T_x M$ for $x \in M$ such that for every $x \in M$ there is an open neighbourhood U together with an isomorphism $TM|_U \xrightarrow{\cong} \underline{\mathbb{R}^n}$ of vector bundles over U with the property that for every $x \in U$ the isomorphism $T_x M \xrightarrow{\cong} \mathbb{R}^n$ respects the given orientation on $T_x M$ and the standard orientation of $\mathbb{R}^n$.
For $n = 0$ a choice of an orientation on TM is a choice of an element in $\{+, -\}$;
(iii) Since TD^n is $T\mathbb{R}^n|_{D^n}$ and we have the standard trivialisation $\underline{\mathbb{R}^n} \xrightarrow{\cong} T\mathbb{R}^n$, the standard orientation on the vector space $\mathbb{R}^n$ induces a standard orientation on TD^n. In particular on $D^1 = [-1, 1]$ we use the orientation on TD^1 coming from moving from -1 to 1;
(iv) An orientation on M is a choice of a generator $[M, \partial M]$ of the infinite cyclic group $H_n(M, \partial M)$;
(v) There is a preferred one-to-one-correspondence between the orientations on TM and the orientations on M which comes from the identification $H_n(T_x M, T_x M \setminus \{0\}) \xrightarrow{\cong} H_n(M, M \setminus \{x\})$ induced by the exponential map for $x \in M \setminus \partial M$;
(vi) The boundary homomorphism $H_n(M, \partial M) \to H_{n-1}(\partial M)$ sends $[M, \partial M]$ to a class $[\partial M]$ which induces for every path component $C \in \partial M$ a generator $[C] \in H_{n-1}(C)$. Thus an orientation on M induces an orientation on C.
(vii) We use the outward normal vector field and the canonical isomorphism $n_v \oplus Ti \colon \underline{\mathbb{R}} \oplus T\partial M \xrightarrow{\cong} TM|_{\partial M}$ in order to assign to an orientation on $T_x M$ an orientation on $T_x \partial M$ for $x \in X$. Thus an orientation on TM induces an orientation on TC for every path component C of ∂M;
(viii) The items (v), (vi), and (vii) are compatible with one another;
(ix) On TS^1 we use the anticlockwise orientation;
(x) With these conventions the standard orientation on $T[-1, 1]$ induces on $T\partial D^1 = T\partial[-1, 1] = T\{-1, 1\}$ the orientation which corresponds to $-$ on -1 and $+$ on 1.

We leave it to the reader to check that this can be arranged if and only if we use the outward normal field and the convention that in the identification $n_v \oplus Ti \colon \underline{\mathbb{R}} \oplus T\partial M \xrightarrow{\cong} TM|_{\partial M}$ we choose the order $\underline{\mathbb{R}} \oplus T\partial M$ and not the order $T\partial M \oplus \underline{\mathbb{R}}$. Namely (x) forces us to use the outward normal field and the order is determined by (iii) and (ix).

5.6 Poincaré Duality

5.6.1 Rings with Involutions

In order to state Poincaré duality on the level we will need, we have to introduce some algebra. Recall that a *ring with involution* R is an associative ring R with unit together with an *involution of rings*

5.6 Poincaré Duality

$$\bar{} \colon R \to R, \quad r \mapsto \bar{r},$$

i.e., a map satisfying $\bar{\bar{r}} = r, \overline{r+s} = \bar{r}+\bar{s}, \overline{r \cdot s} = \bar{s} \cdot \bar{r}$ and $\bar{1} = 1$ for $r, s \in R$. Our main example is the w-twisted involution on the group ring AG for some commutative associative ring A with unit and a group G together with a homomorphism $w \colon G \to \{\pm 1\}$, which we have defined by sending $\sum_{g \in G} a_g \cdot g$ to $\sum_{g \in G} w(g) \cdot a_g \cdot g^{-1}$ already in (3.26).

Let M be a left R-module. Then $M^* := \hom_R(M, R)$ carries a canonical right R-module structure given by $(fr)(m) = f(m) \cdot r$ for a homomorphism of left R-modules $f \colon M \to R$ and $m \in M$. The involution allows us to view $M^* = \hom_R(M; R)$ as a left R-module, namely, define rf for $r \in R$ and $f \in M^*$ by $(rf)(m) := f(m) \cdot \bar{r}$ for $m \in M$. Given an R-chain complex of left R-modules C_* and $n \in \mathbb{Z}$, we define its dual chain complex C^{n-*} to be the chain complex of left R-modules whose p-th chain module is $\hom_R(C_{n-p}, R)$ and whose p-th differential is given by

$$(-1)^{n-p+1} \cdot \hom_R(c_{n-p+1}, \mathrm{id}) \colon (C^{n-*})_p = \hom_R(C_{n-p}, R)$$
$$\to (C^{n-*})_{p-1} = \hom_R(C_{n-p+1}, R).$$

Remark 5.38 (Sign conventions) For a systematic treatment of chain complexes, rings with involutions and sign conventions, see Chapter 14.

5.6.2 Poincaré Complexes

Consider a connected finite CW-complex X with an infinite cyclic local coefficient system $O_X \colon \Pi(X) \to \mathbb{Z}\text{-MOD}$. Recall that it defines an element $w_1(O_X) \in H^1(X; \mathbb{F}_2)$. Now make the choice (BP) for the universal covering, see Notation 5.6. Abbreviate $\pi = \pi_1(X, x)$. Recall that $w_1(O_X)$ determines a group homomorphism $w \colon \pi \to \{\pm 1\}$, which in turn defines a $\mathbb{Z}\pi$-module $\mathbb{Z}^w$. In the sequel we use the w-twisted involution in $\mathbb{Z}\pi$. Denote by $C_*(\widetilde{X})$ the cellular $\mathbb{Z}\pi$-chain complex of the universal covering. Recall that this is a free $\mathbb{Z}\pi$-chain complex and the cellular structure on X determines a cellular $\mathbb{Z}\pi$-basis on it such that each basis element corresponds to a cell in X. This $\mathbb{Z}\pi$-basis is not quite unique, but its equivalence class depends only on the CW-structure of X, see Section 2.3. The product $\widetilde{X} \times \widetilde{X}$ equipped with the diagonal π-action is again a π-CW-complex. The diagonal map $D \colon \widetilde{X} \to \widetilde{X} \times \widetilde{X}$ sending $\tilde{x}$ to $(\tilde{x}, \tilde{x})$ is π-equivariant but not cellular. By the equivariant cellular Approximation Theorem, see for instance [250, Theorem 2.1 on page 32], there is up to cellular π-homotopy precisely one cellular π-map $\overline{D} \colon \widetilde{X} \to \widetilde{X} \times \widetilde{X}$, which is π-homotopic to D. It induces a $\mathbb{Z}\pi$-chain map unique up to $\mathbb{Z}\pi$-chain homotopy

$$C_*(\overline{D}) \colon C_*(\widetilde{X}) \to C_*(\widetilde{X} \times \widetilde{X}). \tag{5.39}$$

There is a natural isomorphism of based free $\mathbb{Z}\pi$-chain complexes

$$i_*: C_*(\widetilde{X}) \otimes_{\mathbb{Z}} C_*(\widetilde{X}) \xrightarrow{\cong} C_*(\widetilde{X} \times \widetilde{X}). \tag{5.40}$$

Given two finitely generated projective $\mathbb{Z}\pi$-chain complexes C_* and D_*, we can define the chain complex $\hom_{\mathbb{Z}\pi}(C^{-*}, D_*)$, see (14.13), and we obtain a natural $\mathbb{Z}$-chain map unique up to $\mathbb{Z}$-chain homotopy

$$s: \mathbb{Z}^w \otimes_{\mathbb{Z}\pi} (C_* \otimes_{\mathbb{Z}} D_*) \to \hom_{\mathbb{Z}\pi}(C^{-*}, D_*) \tag{5.41}$$

by sending $1 \otimes x \otimes y \in \mathbb{Z} \otimes C_p \otimes D_q$ to

$$s(1 \otimes x \otimes y): \hom_{\mathbb{Z}\pi}(C_p, \mathbb{Z}\pi) \to D_q, \quad (\phi: C_p \to \mathbb{Z}\pi) \mapsto (-1)^{pq+p}\overline{\phi(x)} \cdot y$$

where π acts diagonally on $C_* \otimes_{\mathbb{Z}} D_*$. The composite of the chain map (5.41) for $C_* = D_* = C_*(\widetilde{X})$, the inverse of the chain map (5.40), and the chain map (5.39) yields a $\mathbb{Z}$-chain map unique up to $\mathbb{Z}$-chain homotopy

$$\mathbb{Z}^w \otimes_{\mathbb{Z}\pi} C_*(\widetilde{X}) \to \hom_{\mathbb{Z}\pi}(C^{-*}(\widetilde{X}), C_*(\widetilde{X})).$$

Note that the n-th homology of $\hom_{\mathbb{Z}\pi}(C^{-*}(\widetilde{X}), C_*(\widetilde{X}))$ is the set of $\mathbb{Z}\pi$-chain homotopy classes $[C^{n-*}(\widetilde{X}), C_*(\widetilde{X})]_{\mathbb{Z}\pi}$ of $\mathbb{Z}\pi$-chain maps from $C^{n-*}(\widetilde{X})$ to $C_*(\widetilde{X})$, see Section 14.4.

Define the *homology with coefficients in* $\mathbb{Z}^w$

$$H_n(X; \mathbb{Z}^w) := H_n(\mathbb{Z}^w \otimes_{\mathbb{Z}\pi} C_*(\widetilde{X})).$$

Taking the n-th homology group yields a well-defined $\mathbb{Z}$-homomorphism

$$\cap: H_n(X; \mathbb{Z}^w) \to [C^{n-*}(\widetilde{X}), C_*(\widetilde{X})]_{\mathbb{Z}\pi}, \tag{5.42}$$

which sends a class $u \in H_n(X; \mathbb{Z}^w) = H_n(\mathbb{Z}^w \otimes_{\mathbb{Z}\pi} C_*(\widetilde{X}))$ to the $\mathbb{Z}\pi$-chain homotopy class of a $\mathbb{Z}\pi$-chain map denoted by $- \cap u: C^{n-*}(\widetilde{X}) \to C_*(\widetilde{X})$.

Definition 5.43 ((Connected simple) finite Poincaré complex) Let n be a natural number. A connected finite n-dimensional CW-complex X together with an infinite cyclic coefficient system O_X is called a *connected finite n-dimensional Poincaré complex* if there is an element $[[X]] \in H_n(X; O_X)$ with the following property: For one (and hence every) choice (BP) for the universal covering, see Notation 5.6, we get for the preimage $[X]$ of $[[X]]$ under the isomorphism

$$\beta_{p,\widetilde{x},o_x}: H_n(X; \mathbb{Z}^w) \xrightarrow{\cong} H_n(X; O_X)$$

of (5.14) that the $\mathbb{Z}\pi$-chain map well defined up to $\mathbb{Z}\pi$-chain homotopy

$$- \cap [X]: C^{n-*}(\widetilde{X}) \to C_*(\widetilde{X})$$

coming from (5.42) and $[X]$ is a $\mathbb{Z}\pi$-chain homotopy equivalence. We will call this chain map the *Poincaré $\mathbb{Z}\pi$-chain homotopy equivalence*.

5.6 Poincaré Duality

There are precisely two possible choices for $[[X]]$ which generate the infinite cyclic group $H_n(X; O_X) \cong H_n(X; \mathbb{Z}^w) \cong H^0(X; \mathbb{Z}) \cong \mathbb{Z}$. A triple $(X, O_X, [[X]])$ as above is called a *w-oriented connected finite n-dimensional Poincaré complex*. If we talk of an *oriented connected finite n-dimensional Poincaré complex*, we mean that X is w-oriented and satisfies $w_1(O_X) = 0$.

A connected finite n-dimensional Poincaré complex is called *simple* if the Poincaré $\mathbb{Z}\pi$-chain homotopy equivalence is simple for one (and hence every) choice (BP) for the universal covering and $[[X]]$.

The chain level Poincaré duality in the above definition leads to duality in the (co)homology groups of connected Poincaré complexes. Let B be a left $\mathbb{Z}\pi$-module and B' be a right $\mathbb{Z}\pi$-module. The *(co)homology groups with coefficients in B or B'* are the groups

$$H_m(X; B') := H_m(B' \otimes_{\mathbb{Z}\pi} C_*(\widetilde{X})),$$

and

$$H^m(X; B) := H^m(\hom_{\mathbb{Z}\pi}(C_*(\widetilde{X}), B)).$$

Let B^{op} be the right $\mathbb{Z}\pi$-module given by $b \cdot g := g^{-1} \cdot b$ for $g \in \pi$ and $b \in B$. We have the twisted left $\mathbb{Z}\pi$-module $B^w := B \otimes_{\mathbb{Z}} \mathbb{Z}^w$, where π acts diagonally. The following lemma was proven by Wall [408, Lemma 1.1].

Lemma 5.44 ((Co)homological Poincaré duality) *Let X be a w-oriented connected finite n-dimensional Poincaré complex. For any $\mathbb{Z}\pi$-module B, the cap product with $[X]$ induces an isomorphism*

$$- \cap [X] \colon H^{n-*}(X; B^w) \xrightarrow{\cong} H_*(X; B^{\mathrm{op}}).$$

Proof. The $\mathbb{Z}\pi$-chain homotopy equivalence $- \cap [X] \colon C^{n-*}(\widetilde{X}) \to C_*(\widetilde{X})$ induces a $\mathbb{Z}$-chain homotopy equivalence

$$B^{\mathrm{op}} \otimes_{\mathbb{Z}\pi} - \cap [X] \colon B^{\mathrm{op}} \otimes_{\mathbb{Z}\pi} C^{n-*}(\widetilde{X}) \to B^{\mathrm{op}} \otimes_{\mathbb{Z}\pi} C_*(\widetilde{X}).$$

Moreover, there is a canonical isomorphism of $\mathbb{Z}$-chain complexes

$$\hom_{\mathbb{Z}\pi}(C_{n-*}(\widetilde{X}), B^w) \xrightarrow{\cong} B^{\mathrm{op}} \otimes_{\mathbb{Z}\pi} C^{n-*}(\widetilde{X}),$$

and an obvious identification

$$H_m(\hom_{\mathbb{Z}\pi}(C_{n-*}(\widetilde{X}), B^w)) = H^{n-m}(\hom_{\mathbb{Z}\pi}(C_*(\widetilde{X}), B)).$$

□

Remark 5.45 (Not necessarily connected finite Poincaré complexes) Definition 5.43 extends in the obvious way to the non-connected case by considering each path component separately. More precisely, a *finite n-dimensional Poincaré complex* X is a finite n-dimensional CW-complex X with an infinite cyclic local coefficient system O_X such that for each path component C the pair $(C, O_X|_C)$ is a connected finite n-dimensional Poincaré complex.

A *w-oriented finite n-dimensional Poincaré complex* $(X, [[X]])$ is a finite n-dimensional Poincaré complex X together with a choice of an element $[[X]] \in H_n(X; O_X)$ such that the restriction of $[[X]]$ to any component C of X defines a fundamental class $[[C]] \in H_n(C; O_X|_C)$ for C. Often we shall suppress the choice of $[[X]]$ and O_X from the notation and write X instead of $(X, O_X, [[X]])$. With these conventions, X^- will denote the *w-oriented finite Poincaré complex with reversed orientation*; obtained from $X = (X, [[X]])$ by replacing $[[X]]$ by $-[[X]]$.

If $w_1(O_X) \in H^1(X; \mathbb{F}_2)$ vanishes, we call X *orientable* following the standard convention. (This is a little bit confusing since one can choose in any case a w-orientation, but it is the standard terminology.) When we say that X is *oriented*, we mean that X is orientable and w-oriented, or, equivalently, for every path component C of X the abelian group $H_n(C; \mathbb{Z})$ is infinite cyclic and we have specified a generator. A finite Poincaré complex is called *simple* if each of its components is simple.

One might wonder whether X admits the structure of a Poincaré complex for different choices of O_X. But in fact we have the following result.

Lemma 5.46 *Let X be a finite n-dimensional Poincaré complex. Then the isomorphism class of the infinite cyclic local coefficient system O_X and the element $w_1(O_X) \in H^1(X; \mathbb{F}_2)$ are uniquely determined by the homotopy type of X. If X is connected, we have $H_n(X; \mathbb{Z}^v) \neq 0$ for $v \colon \pi_1(X) \to \{\pm 1\}$ if and only if $v = w_1(X)$.*

Proof. Obviously we can assume without loss of generality that X is connected, otherwise treat each component separately. Put

$$w = w_1(O_X) \colon \pi = \pi_1(X) \to \{\pm 1\}.$$

Denote by $C^{n-*}(\widetilde{X})_{\mathrm{untw}}$ the $\mathbb{Z}\pi$-chain complex which is analogously defined as $C^{n-*}(\widetilde{X})$, but now with respect to the untwisted involution on $\mathbb{Z}\pi$. It can be identified with the $\mathbb{Z}\pi$-chain complex $\mathbb{Z}^w \otimes_{\mathbb{Z}} C^{n-*}(\widetilde{M})$, where π-acts diagonally. The 0-th homology $H_0(C^{n-*}(\widetilde{X})_{\mathrm{untw}})$ depends only on the homotopy type of X. The Poincaré $\mathbb{Z}\pi$-chain homotopy equivalence induces a $\mathbb{Z}\pi$-chain homotopy equivalence $C^{n-*}(\widetilde{X})_{\mathrm{untw}} \to \mathbb{Z}^w \otimes_{\mathbb{Z}} C_*(\widetilde{X})$. It yields an isomorphism of $\mathbb{Z}\pi$-modules

$$H_0(C^{n-*}(\widetilde{X})_{\mathrm{untw}}) \cong H_0(\mathbb{Z}^w \otimes_{\mathbb{Z}} C_*(\widetilde{X})) \cong \mathbb{Z}^w \otimes_{\mathbb{Z}} H_0(C_*(\widetilde{X})) \cong \mathbb{Z}^w \otimes_{\mathbb{Z}} \mathbb{Z} \cong \mathbb{Z}^w.$$

Thus we rediscover $w = w_1(O_X)$ from $H_0(C^{n-*}(\widetilde{X})_{\mathrm{untw}})$.

By Poincaré duality we have $H_n(X; \mathbb{Z}^v) \cong H^0(X; \mathbb{Z}^{v \cdot w})$. Since X is homotopy equivalent to a finite CW-complex with precisely one 0-cell, one can find a $\mathbb{Z}\pi$-chain complex D_* such that $C_*(\widetilde{X})$ is $\mathbb{Z}\pi$-chain homotopy equivalent to D_* and for some finite set of generators $\{s_1, s_2, \ldots s_m\}$ of π the first differential of D_* looks like $\bigoplus_{i=1}^m \mathbb{Z}\pi \xrightarrow{(s_1-1, s_2-1, \ldots, s_m-1)} \mathbb{Z}\pi$. Hence we get an exact sequence

5.6 Poincaré Duality

$$0 \to H^0(X; \mathbb{Z}^{v \cdot w}) \to \mathbb{Z} \xrightarrow{\begin{pmatrix} v(s_1) \cdot w(s_1) - 1 \\ v(s_2) \cdot w(s_2) - 1 \\ \vdots \\ v(s_m) \cdot w(s_m) - 1 \end{pmatrix}} \prod_{i=1}^{m} \mathbb{Z}.$$

This implies that $H^0(X; \mathbb{Z}^{v \cdot w})$ is non-trivial if and only if $v(s_i) \cdot w(s_i) - 1 = 0$ holds for $i = 1, 2, \ldots m$. Hence we get $v = w$.

Recall that two infinite cyclic coefficient systems O_X and O'_X are isomorphic if and only if $w_1(O_X) = w_1(O'_X)$ holds. □

Because of Lemma 5.46 we will abbreviate $w_1(O_X)$ by w and call it the *orientation homomorphism* or the *first Stiefel–Whitney class* of X.

Remark 5.47 (Formal dimension versus dimension) There is a weaker version of Definition 5.43 and its generalisation to the non-connected case version where one drops the condition that the number n is the dimension of X. Then one talks of a *finite Poincaré complex X of formal dimension n*. Its dimension as a CW-complex satisfies $\dim(X) \geq n$, but it is possible that $\dim(X) > n$ holds. For algebraic reasons it is sometimes more convenient to work with the formal dimension. Moreover, if $n \neq 2$, then a finite Poincaré complex of formal dimension n is homotopy equivalent to a finite n-dimensional Poincaré complex, see [408, Theorem 2.2].

Exercise 5.48 Show that for a finite Poincaré complex X of formal dimension n its dimension as a CW-complex is at least n. Decide when for a finite CW-complex Y its cone $\text{cone}(Y)$ is a finite simple Poincaré complex of formal dimension n. If it is, what is n?

Convention 5.49 (Convention about Poincaré complexes) Our exposition above is very precise in order to take care of certain inconsistencies appearing in the literature, but also more complicated than the one which usually appears in the literature. Indeed, we will often implicitly think of a connected finite Poincaré complex together with choice (BP) for the universal covering. Then the fundamental class becomes $[X] \in H_n(X; \mathbb{Z}^w)$ for the orientation homomorphism $w: \pi = \pi_1(X, x) \to \{\pm 1\}$, we can ignore the fundamental groupoid and work with the fundamental group π instead, and we always consider the Poincaré $\mathbb{Z}\pi$-chain homotopy equivalence

$$- \cap [X]: C^{n-*}(\widetilde{X}) \to C_*(\widetilde{X})$$

for the universal covering $\widetilde{X} \to X$.

Exercise 5.50 Compute $w = w_1(\mathbb{RP}^n): \pi_1(\mathbb{RP}^n) \to \{\pm 1\}$, the orientation homomorphism of n-dimensional real projective space $\mathbb{RP}^n$.

Theorem 5.51 (Closed manifolds are simple Poincaré complexes) *Let M be a closed manifold of dimension n. Then M admits the structure of a simple finite n-dimensional Poincaré complex where the infinite cyclic local coefficient system is*

O_M and the orientation homomorphism $w_1(M)\colon \pi_1(M) \to \{\pm 1\}$ *agrees with the first Stiefel–Whitney class $w_1(TM)$ in $H^1(M;\mathbb{Z}/2)$ of the tangent bundle of M.*

Proof. For a proof we refer for instance to [414, Theorem 2.1 on page 23]. The basic idea of the proof in the connected case is as follows.

Any closed manifold admits a smooth triangulation $h\colon K \to M$, see [317, 432], i.e., a finite simplicial complex together with a homeomorphism $h\colon K \to M$ such that h restricted to a simplex is a smooth embedding. Two such smooth triangulations have common subdivisions. Fix such a triangulation K. Denote by K' its barycentric subdivision. The vertices of K' are the barycentres $\widehat{\sigma^r}$ of simplices σ^r in K. A p-simplex in K' is given by a sequence $\widehat{\sigma^{i_0}}\widehat{\sigma^{i_1}}\ldots\widehat{\sigma^{i_p}}$ where σ^{i_j} is a face of $\sigma^{i_{j+1}}$. Equip M with the structure of a finite CW-complex coming from K'. Next we define the dual CW-complex K^* as follows. It is not a simplicial complex but shares the property of a simplicial complex that all attaching maps are embeddings. Each p-simplex σ in K determines a $(n-p)$-dimensional cell σ^* of K^* which is the union of all simplices in K' which begin with $\widehat{\sigma^p}$. So K has as many p-simplices as K^* has $(n-p)$-cells. See Figure 5.52 for illustration.

Figure 5.52 (Dual cell complex).

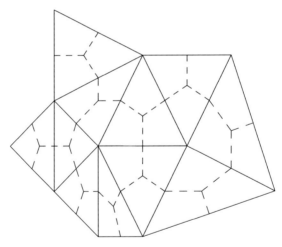

The cap product with the fundamental cycle, which is given by the sum of the n-dimensional simplices, yields an isomorphism of $\mathbb{Z}\pi$-chain complexes $C^{n-*}(\widetilde{K^*}) \to C_*(\widetilde{K})$. It preserves the cellular $\mathbb{Z}\pi$-bases and in particular its Whitehead torsion is trivial. Since K' is a subdivision of K and can also be viewed as a kind of subdivision of K^*, there are canonical $\mathbb{Z}\pi$-chain homotopy equivalences $C_*(\widetilde{K'}) \to C_*(\widetilde{K})$ and $C_*(\widetilde{K'}) \to C_*(\widetilde{K^*})$ that have trivial Whitehead torsion, see [302, Theorem 5.2 on page 372]. Thus we can write the $\mathbb{Z}\pi$-chain map $-\cap [M]\colon C^{n-*}(\widetilde{K'}) \to C_*(\widetilde{K'})$ as a composite of three simple $\mathbb{Z}\pi$-chain homotopy equivalences up to $\mathbb{Z}\pi$-chain homotopy. Hence it is a simple $\mathbb{Z}\pi$-chain homotopy equivalence. □

5.6 Poincaré Duality

Theorem 5.51 gives us the first obstruction for a topological space X to be homotopy equivalent to a closed n-dimensional manifold, see Problem 4.1. Namely, X must be homotopy equivalent to a finite simple n-dimensional Poincaré complex.

Let X be a connected finite n-dimensional Poincaré complex with fundamental group π and orientation homomorphism w. Recall that we have the Poincaré $\mathbb{Z}\pi$-chain homotopy equivalence $-\cap [X]\colon C^{n-*}(\widetilde{X}) \to C_*(\widetilde{X})$. Its dual

$$(-\cap [X])^{n-*}\colon C^{n-*}(\widetilde{X}) \to (C^{n-*}(\widetilde{X}))^{n-*} = C_*(\widetilde{X})$$

is $\mathbb{Z}\pi$-chain homotopic to $-\cap [X]$, as we will see in Construction 15.48 in Chapter 15. Let

$$*\colon \mathrm{Wh}(\pi) \to \mathrm{Wh}(\pi) \tag{5.53}$$

be the involution induced by the w-twisted involution on $\mathbb{Z}\pi$. It sends the class of an invertible matrix $A = (a_{i,j})$ to the class of the invertible matrix $A^* = (\overline{a_{j,i}})$. We get in $\mathrm{Wh}(\pi)$ from the homotopy invariance of Whitehead torsion

$$\tau(-\cap [X]) = (-1)^n \cdot *[(-\cap [X])]. \tag{5.54}$$

Let $f\colon X \to Y$ be a homotopy equivalence of connected finite Poincaré complexes. Then we can identify $\pi = \pi_1(X) = \pi_1(Y)$ and under this identification X and Y have the same orientation homomorphism $w\colon \pi \to \{\pm 1\}$ by Lemma 5.46. We can choose the fundamental classes so that they are respected by $H_n(X; \mathbb{Z}^w) \xrightarrow{\cong} H_n(Y; \mathbb{Z}^w)$. Hence the following diagram of $\mathbb{Z}\pi$-chain homotopy equivalences commutes

$$\begin{array}{ccc} C^{n-*}(X) & \xleftarrow{C^{n-*}(\widetilde{f})} & C^{n-*}(\widetilde{Y}) \\ {\scriptstyle -\cap [X]}\downarrow & & \downarrow{\scriptstyle -\cap [Y]} \\ C_*(\widetilde{X}) & \xrightarrow{C_*(\widetilde{f})} & C_*(\widetilde{Y}). \end{array}$$

We conclude from the homotopy invariance and the composition formula of Whitehead torsion that in $\mathrm{Wh}(\pi)$ we get

$$\tau(-\cap [X]) - \tau(-\cap [Y]) = \tau(f) + (-1)^n \cdot *(\tau(f)). \tag{5.55}$$

The n-th *Tate cohomology* of an abelian group A with involution $*\colon A \to A$ is defined to be the subquotient of A given by

$$\widehat{H}^n(\mathbb{Z}/2; A) := \ker(\mathrm{id} - (-1)^n \cdot *)/\mathrm{im}(\mathrm{id} + (-1)^n \cdot *).$$

This is well defined since the composite $(\mathrm{id} - (-1)^n \cdot *) \circ (\mathrm{id} + (-1)^n \cdot *)$ is trivial. Because of (5.54) we can assign to a connected finite n-dimensional Poincaré complex with fundamental group π its *Poincaré torsion*

$$\widehat{\tau}(X) \in \widehat{H}^n(\mathbb{Z}/2; \mathrm{Wh}(\pi)) \tag{5.56}$$

given by the class of $\tau(-\cap [X])$.

Lemma 5.57 *Let X and Y be connected finite Poincaré complexes of dimension $n \geq 3$.*

(i) *If $f\colon X \to Y$ is a homotopy equivalence, then the isomorphism*

$$\widehat{H}^n(\mathbb{Z}/2; \mathrm{Wh}(\pi_1(X))) \xrightarrow{\cong} \widehat{H}^n(\mathbb{Z}/2; \mathrm{Wh}(\pi_1(Y)))$$

induced by f sends $\widehat{\tau}(X)$ to $\widehat{\tau}(Y)$;

(ii) *We have $\widehat{\tau}(X) = 0$ if and only if X is homotopy equivalent to a simple finite Poincaré complex.*

Proof. (i) This follows from (5.55).

(ii) If X itself is simple, we get $\widehat{\tau}(X) = 0$ from the definitions. If X is homotopy equivalent to a simple finite Poincaré complex, we get $\widehat{\tau}(X) = 0$ from assertion (i).

Suppose that $\widehat{\tau}(X) = 0$. Then we can find an element $u \in \mathrm{Wh}(\pi)$ satisfying

$$\tau(-\cap [X]) = u + (-1)^n \cdot *(u).$$

Choose a finite n-dimensional CW-complex Y together with a homotopy equivalence $f\colon Y \to X$ such that $\tau(f) = -u$ holds. Then Y is a connected finite Poincaré complex, which is simple by (5.55). □

Remark 5.58 (Poincaré complexes with non-trivial Poincaré torsion) Note that a connected finite Poincaré complex X with $\widehat{\tau}(X) \neq 0$ cannot be homotopy equivalent to a closed manifold by Theorem 5.51 and Lemma 5.57. There exist such examples, but their construction is not easy and requires input from surgery theory and from the computation of L-groups of finite groups. The construction in [75, Theorem in Appendix] takes an appropriate bordism from $L^{2m+1} \times S^{2n-2}$ to itself for some $(2m + 1)$-dimensional lens space L^{2m+1} with fundamental group a cyclic 2-group and then glues $L^{2m+1} \times D^{2n-1}$ to the one end using the identity $\mathrm{id}_{L^{2m+1} \times S^{2n-2}}$ and to the other end using an appropriate self-homotopy equivalence of $L^{2m+1} \times S^{2n-2}$.

From a Morse-theoretic point of view, Poincaré duality and the dual cell decomposition correspond to the dual handlebody decomposition of a manifold, which we have already described in (2.39).

Remark 5.59 (Cohomology with compact supports) If Y is a space, its *singular cohomology with compact support* is defined to be the cohomology

$$H^n_c(Y) := H^n(C^*_c(Y))$$

of the cochain subcomplex $C^*_c(Y)$ of the singular cochain complex $C^*_{\mathrm{sing}}(Y)$ consisting of those maps $\phi\colon S_n(Y) \to \mathbb{Z}$ from the set of n-simplices to $\mathbb{Z}$, for which there exists a compact subset $K_\phi \subseteq Y$ such that for any simplex $\sigma\colon \Delta_n \to X$ with $\sigma(\Delta_n) \cap K_\phi = \emptyset$ we have $\phi(\sigma) = 0$. The various inclusions $Y = (Y; \emptyset) \to (Y, Y \setminus K)$ where K runs through the compact subsets of Y, induce an isomorphism, natural in Y,

5.6 Poincaré Duality

$$\mathrm{colim}_K H^n(Y, Y - K) \xrightarrow{\cong} H^n_c(Y).$$

If X is a finite CW-complex, then there is a natural isomorphism of $\mathbb{Z}\pi$-modules

$$H^n(C^*(\widetilde{X})_{\mathrm{untw}}) \xrightarrow{\cong} H^n_c(\widetilde{X})$$

where $C^*(\widetilde{X})_{\mathrm{untw}} = \hom_{\mathbb{Z}\pi}(C_*(\widetilde{X}), \mathbb{Z}\pi)$ is the $\mathbb{Z}\pi$-cochain complex which is defined with respect to the untwisted involution on $\mathbb{Z}\pi$ sending $\sum_{g \in g} \lambda_g \cdot g$ to $\sum_{g \in g} \lambda_g \cdot g^{-1}$. (See for example the proof of Proposition 3H.5 in [178]). If X is not finite, the statement above is not true anymore.

There is a version of Poincaré duality for a (not necessarily compact) orientable manifold N without boundary of dimension n predicting an isomorphism

$$H^{n-k}_c(N) \xrightarrow{\cong} H_k(N),$$

see for instance [178, Theorem 3.35]. Our version implies this version in the case when N is a covering of a closed orientable manifold M.

Remark 5.60 (No intrinsic fundamental class for Poincaré complexes) In contrast to a closed manifold, which has an intrinsic fundamental class, see Definition 5.29, we do not see for a finite Poincaré complex a preferred choice of its fundamental class $[[X]]$. This is the first hint that there is a fundamental difference between a closed manifold and a finite Poincaré complex. It has to do with the fact that being a manifold is a local structure which we have used for the definition of the intrinsic fundamental class and which is not present for a Poincaré complex.

Remark 5.61 (Poincaré duality on homology with R-coefficients) Suppose that X is an oriented finite Poincaré complex of dimension n. Recall that this implies $w_1(X) = 0$. Note that we have a canonical isomorphism of $\mathbb{Z}$-chain complexes $\mathbb{Z} \otimes_{\mathbb{Z}\pi} C_*(\widetilde{X}) \xrightarrow{\cong} C_*(X)$.

Now we obtain for any commutative ring R an R-isomorphism on (co-)homology with R-coefficients

$$- \cap [X]: H^{n-*}(X; R) \xrightarrow{\cong} H_*(X; R).$$

It suffices to explain this is in the connected case. Then Definition 5.43 of Poincaré duality implies that we get an R-chain homotopy equivalence

$$R \otimes_{\mathbb{Z}\pi} (- \cap [X]): R \otimes_{\mathbb{Z}\pi} C^{n-*}(\widetilde{X}) = \hom_{\mathbb{Z}}(C_{n-*}(X), R) =: C^{n-*}(X; R) \to$$
$$R \otimes_{\mathbb{Z}\pi} C_*(\widetilde{X}) = R \otimes_{\mathbb{Z}} C_*(X) =: C_*(X; R),$$

which induces the isomorphism above.

Let X be an oriented finite Poincaré complex of dimension n. For any commutative ring R with unit we have the *intersection pairing with coefficients in R*

$$I^p: H^p(X; R) \otimes_R H^{n-p}(X; R) \to R, \quad (x, y) \mapsto \langle x \cup y, i_*[X] \rangle \quad (5.62)$$

where i_* is the change of coefficients map associated to $\mathbb{Z} \to R$. This pairing is *non-degenerate*, i.e., if for $x \in H^p(X;R)$ we have $I^p(x,y) = 0$ for every $y \in H^{n-p}(X;R)$, then $x = 0$. If R is a field, this is equivalent to the bijectivity of the homomorphism $-\cap [X] \colon H^{n-*}(X;R) \to H_*(X;R)$ appearing in Remark 5.61.

There is a homological version of the intersection pairing

$$I_p \colon H_{n-p}(X;R) \otimes_R H_p(X;R) \to R, \quad (x,y) \mapsto x \cdot y \tag{5.63}$$

which is obtained from the cohomological version by applying Poincaré duality, namely, we require that the following diagram

$$\begin{array}{ccc} H^p(X;R) \otimes_R H^{n-p}(X;R) & \xrightarrow{I^p} & R \\ {\scriptstyle (-\cap[X]) \otimes_R (-\cap[X])} \downarrow \cong & & \downarrow \mathrm{id} \\ H_{n-p}(X;R) \otimes_R H_p(X;R) & \xrightarrow{I_p} & R \end{array}$$

commutes. If $R = \mathbb{Z}$, we just talk about the (homological) intersection pairing.

Remark 5.64 (Geometric interpretation of the intersection pairing) For a closed oriented manifold M, the homological intersection pairing has the following geometric interpretation, which justifies the name intersection pairing. Consider two closed oriented submanifolds N_1 and N_2 such that N_1 has dimension $n - p$, N_2 has dimension p and N_1 and N_2 intersect transversally. The intersection $N_1 \cap N_2$ consists of finitely many points since $\dim(N_1) + \dim(N_2) = \dim(M)$. For each $x \in N_1 \cap N_2$ we obtain a canonical isomorphism $T_x N_0 \oplus T_x N_2 \xrightarrow{\cong} T_x M$. The given orientations on N_1, N_2 and M induce orientations on the vector spaces $T_x N_1 \oplus T_x N_2$ and $T_x M$, see Remark 5.37 (v). Define a sign $\epsilon(x) \in \{\pm 1\}$ by requiring that $\epsilon(x) = 1$ if this isomorphism is orientation preserving, and $\epsilon(x) = -1$ otherwise. Let $i_k \colon N_k \to M$ be the inclusion for $k = 1, 2$. Then we get, see [45, Theorem 11.9 on page 372]:

$$(i_1)_*([N_1]) \cdot (i_2)_*([N_2]) = \sum_{x \in N_1 \cap N_2} \epsilon(x). \tag{5.65}$$

Every element in $H_p(M;\mathbb{Z})$ can be written as $i_*([N])$ for an embedded connected closed oriented submanifold N if $p \in \{0, 1, n-2, n-1, n\}$, see [45, Theorem 11.16 on page 377]. But in general it is not possible to find for an element $x \in H_p(M;\mathbb{Z})$ a connected closed oriented manifold N with some map $f \colon N \to M$ satisfying $x = f_*([N])$. If one considers singular chains, one can define the intersection pairing for singular homology for every two elements by intersecting singular simplices, see [375, Section 73].

Remark 5.66 (Analytic Poincaré duality) From the analytic point of view Poincaré duality can be explained as follows. Let M be a connected closed oriented Riemannian manifold. Let $(\Omega^*(M), d^*)$ be the de Rham complex of smooth p-forms on M. The p-th Laplacian is defined by

5.6 Poincaré Duality

$$\Delta_p = (d^p)^* d^p + d^{p-1}(d^{p-1})^* : \Omega^p(M) \to \Omega^p(M),$$

where $(d^p)^*$ is the adjoint of the p-th differential d^p. The kernel of the p-th Laplacian is the space $\mathcal{H}^p(M)$ of harmonic p-forms on M. The Hodge–de Rham Theorem yields an isomorphism

$$A^p : \mathcal{H}^p(M) \xrightarrow{\cong} H^p(M; \mathbb{R}) \tag{5.67}$$

from the space of harmonic p-forms to the singular cohomology of M with coefficients in $\mathbb{R}$. Let $[M]_\mathbb{R} \in H^n(M; \mathbb{R})$ be the fundamental cohomology class with $\mathbb{R}$-coefficients which is determined by the property $\langle [M]_\mathbb{R}, i_*([M]) \rangle = 1$ for $\langle \, , \, \rangle$ the Kronecker pairing, see (5.82), and $i_* : H_n(M; \mathbb{Z}) \to H_n(M; \mathbb{R})$ the change of rings homomorphism. Then A^n sends the volume form $d\mathrm{vol}$ to the class $\mathrm{vol}(M) \cdot [M]_\mathbb{R}$. The Hodge-star operator $* : \Omega^{n-p}(M) \to \Omega^p(M)$ induces isomorphisms

$$* : \mathcal{H}^{n-p}(M) \xrightarrow{\cong} \mathcal{H}^p(M). \tag{5.68}$$

We obtain from (5.67) and (5.68) isomorphisms

$$H^{n-p}(M; \mathbb{R}) \xrightarrow{\cong} H^p(M; \mathbb{R}).$$

This is the analytic version of Poincaré duality. It is equivalent to the claim that the $\mathbb{R}$-bilinear pairing

$$P^p : \mathcal{H}^p(M) \otimes_\mathbb{R} \mathcal{H}^{n-p}(M) \to \mathbb{R}, \quad (\omega, \eta) \mapsto \int_M \omega \wedge \eta \tag{5.69}$$

is non-degenerate. If we take $R = \mathbb{R}$, then the pairings (5.62) and (5.69) agree under the Hodge–de Rham isomorphism (5.67).

5.6.3 The Signature

One basic invariant of a finite CW-complex X is its *Euler characteristic* $\chi(X)$. It is defined by $\chi(X) := \sum_p (-1)^p \cdot n_p$, where n_p is the number of p-cells. Equivalently, it can be defined in terms of its homology by $\chi(X) = \sum_p (-1)^p \cdot \dim_\mathbb{Q} H_p(X; \mathbb{Q})$. Another basic invariant for Poincaré complexes is defined next.

Consider a symmetric $\mathbb{R}$-bilinear pairing $s : V \otimes_\mathbb{R} V \to \mathbb{R}$ for a finite-dimensional real vector space V. Choose a basis for V and let A be the square matrix describing s with respect to this basis. Since s is symmetric, A is symmetric. Hence A can be diagonalised by an orthogonal matrix U to a diagonal matrix whose entries on the diagonal are real numbers. Let n_0 be the number of zero entries, n_+ be the number of positive entries and n_- be the number of negative entries on the diagonal. These three numbers are independent of the choice of the basis and the orthogonal matrix U. Namely, n_0 is the dimension of the vector subspace called the *nullspace*

$V_0 = \{v \in V \mid s(v, w) = 0 \text{ for all } w \in V\}$ of V, n_+ is the maximum of the dimensions of vector subspaces $W \subset V$ on which s is positive-definite, i.e., $s(w, w) > 0$ for $w \neq 0$, and analogously for n_-. Obviously $n_+ + n_- + n_0 = \dim_{\mathbb{R}}(V)$. Define the *signature* of s to be the integer $n_+ - n_-$.

Note that s is non-degenerate if and only if $n_0 = 0$. Moreover, s induces a non-degenerate symmetric $\mathbb{R}$-bilinear pairing $\overline{s}: V/V_0 \otimes_{\mathbb{R}} V/V_0 \to \mathbb{R}$ satisfying $\text{sign}(\overline{s}) = \text{sign}(s)$.

Definition 5.70 (Signature) Let X be an oriented finite $4k$-dimensional Poincaré complex. We get from the intersection pairing with real coefficients for $p = 2k$ of (5.62) a symmetric non-degenerate $\mathbb{R}$-bilinear pairing

$$I^{2k}: H^{2k}(X; \mathbb{R}) \otimes_{\mathbb{R}} H^{2k}(X; \mathbb{R}) \to \mathbb{R}.$$

Define the *signature* $\text{sign}(X)$ of X to be the signature of I^{2k}. If the dimension X is not $4k$, put $\text{sign}(X) = 0$.

Exercise 5.71 For an oriented finite Poincaré complex, let X^- denote the oriented finite Poincaré complex obtained from X by replacing the fundamental class $[X]$ with $-[X]$. Show $\text{sign}(X^-) = -\text{sign}(X)$.

Exercise 5.72 Let X be a finite Poincaré complex of odd dimension. Show that its Euler characteristic $\chi(M)$ vanishes.

The notion of a Poincaré complex can be extended to a Poincaré pair. Just as Poincaré complexes give homotopy-theoretic models for closed manifolds, Poincaré pairs give homotopy-theoretic models for compact manifolds with boundary.

Definition 5.73 ((Simple) Poincaré pair) Let X be a connected finite CW-complex with a (not necessarily connected) subcomplex $A \subset X$. Suppose that X comes with an infinite cyclic local coefficient system $\mathcal{O}_X$. We call (X, A) a *connected finite n-dimensional Poincaré pair* if there is a so-called *fundamental class* $[[X, A]] \in H_n(X, A; \mathcal{O}_X)$ such that for one (and hence any) choice (BP) for the universal covering, see Notation 5.6, the following holds:

(i) Let $[X, A]$ be the preimage of $[[X, A]]$ under the version for pairs

$$\beta_{p, \widetilde{x}, o_x}: H_n(X, A; \mathbb{Z}^w) \xrightarrow{\cong} H_n(X, A; \mathcal{O}_X)$$

of the isomorphism (5.14). Denote by $\overline{A} \subset \widetilde{X}$ the preimage of A under the universal covering $\widetilde{X} \to X$. We require that the $\mathbb{Z}\pi$-chain maps

$$-\cap [X, A]: C^{n-*}(\widetilde{X}, \overline{A}) \to C_*(\widetilde{X});$$
$$-\cap [X, A]: C^{n-*}(\widetilde{X}) \to C_*(\widetilde{X}, \overline{A}),$$

are $\mathbb{Z}\pi$-chain equivalences;

(ii) Each component C of the space A inherits from (X, A) the structure of a connected finite $(n-1)$-dimensional Poincaré complex for the restriction $\mathcal{O}_X|_C$ to

5.6 Poincaré Duality

C as follows. The fundamental classes $[[C]]$ of the components $C \in \pi_0(A)$ are given by the image of the fundamental class $[[X, A]]$ under the boundary map $H_n(X, A; \mathcal{O}_X) \to H_{n-1}(A; \mathcal{O}_X|_A) \cong \bigoplus_{C \in \pi_0(A)} H_{n-1}(C; \mathcal{O}_X|_C)$.

We call (X, A) *simple* if the Whitehead torsion of the $\mathbb{Z}\pi$-chain maps $- \cap [X, A]\colon C^{n-*}(\widetilde{X}, \overline{A}) \to C_*(\widetilde{X})$ and $- \cap [X, A]\colon C^{n-*}(\widetilde{X}) \to C_*(\widetilde{X}, \overline{A})$ in $\mathrm{Wh}(\pi)$ vanishes and if for every component C of A the Whitehead torsion of the $\mathbb{Z}\pi_1(C)$-chain map $- \cap [C]\colon C^{n-1-*}(\widetilde{C}) \to C_*(\widetilde{C})$ vanishes in $\mathrm{Wh}(\pi_1(C))$.

In general, a finite n-dimensional Poincaré pair (X, A) for an infinite cyclic local coefficient system $\mathcal{O}_X$ is defined using the definition above for each connected pair $(C, C \cap A)$ for $\mathcal{O}_X|_C$ where C runs through the components of X.

Exercise 5.74 Consider a connected finite n-dimensional CW-pair (X, A). Suppose that for each component C of A the inclusion $C \to X$ induces an injection on the fundamental groups. Suppose that there is an infinite cyclic local coefficient system $\mathcal{O}_X$ and a class $[[X, A]] \in H_n(X, A; \mathcal{O}_X)$ such that condition (i) appearing in Definition 5.73 holds. Define $[[C]] \in H_{n-1}(C; \mathcal{O}_X|_C)$ as in condition (ii) appearing in Definition 5.73.

Show that then condition (ii) appearing in Definition 5.73 is automatically satisfied, and the same in the simple setting.

Theorem 5.75 (Compact manifolds are simple Poincaré pairs) *Let M be a compact manifold of dimension n with boundary ∂M. Then $(M, \partial M)$ admits the structure of a simple finite n-dimensional Poincaré pair.*

For the proof we refer to [414, Theorem 2.1].

Exercise 5.76 Let M be a closed manifold. Show that $(\mathrm{cone}(M), M)$ is a finite Poincaré pair of formal dimension n for some infinite cyclic local coefficient system $\mathcal{O}_{\mathrm{cone}(M)}$ if and only if M is $(n-1)$-dimensional, connected, and orientable and satisfies $H_k(M; \mathbb{Z}) \cong H_k(S^{n-1}; \mathbb{Z})$ for $k \in \mathbb{Z}$.

Exercise 5.77 Classify up to homeomorphism all simply connected closed manifolds M for which $(\mathrm{cone}(M), M)$ is a finite Poincaré complex of formal dimension n.

Remark 5.78 (Gluing along homotopy equivalences) If $(M, \partial M)$ and $(N, \partial N)$ are compact manifolds and $f\colon \partial M \to \partial N$ is a diffeomorphism, then $M \cup_f N$ is a closed manifold. This is no longer true in general if f is a homotopy equivalence.

If $(X, \partial X)$ and $(Y, \partial Y)$ are finite Poincaré pairs and $f\colon \partial X \to \partial Y$ is a homotopy equivalence, then $X \cup_f Y$ is a finite Poincaré complex. This flexibility for Poincaré complexes will be available and exploited when we deal with the chain complex version of surgery in Chapter 15, see Remark 15.325.

The signature will be the first and the most elementary surgery obstruction which we will encounter. This is due to the following result.

Theorem 5.79 (Bordism invariance of the signature) *Let (X, A) be a $(4k + 1)$-dimensional oriented finite Poincaré pair. Then*

$$\text{sign}(A) = \sum_{C \in \pi_0(A)} \text{sign}(C) = 0.$$

For the proof of Theorem 5.79 and later purposes we need the following notion.

Definition 5.80 (Lagrangian) Let $s \colon V \otimes_{\mathbb{R}} V \to \mathbb{R}$ be a symmetric bilinear non-degenerate pairing for a finite-dimensional real vector space V. A subspace $L \subset V$ for which $\dim_{\mathbb{R}}(V) = 2 \cdot \dim_{\mathbb{R}}(L)$ and $s(a, b) = 0$ for $a, b \in L$ holds is called a *lagrangian* for s.

Lemma 5.81 *Let $s \colon V \otimes_{\mathbb{R}} V \to \mathbb{R}$ be a symmetric bilinear non-degenerate pairing for a finite-dimensional real vector space V. Then $\text{sign}(s) = 0$ if and only if s admits a lagrangian L.*

Proof. Suppose that $\text{sign}(s) = 0$. One can find an orthonormal basis

$$\{b_1, b_2, \ldots, b_{n_+}, c_1, c_2, \ldots, c_{n_-}\}$$

(with respect to s) such that $s(b_i, b_i) = 1$ and $s(c_j, c_j) = -1$ holds. Since $0 = \text{sign}(s) = n_+ - n_-$, we can define L to be the vector subspace generated by $\{b_i + c_i \mid i = 1, 2, \ldots, n_+\}$. One easily checks that L has the desired properties.

Suppose such an $L \subset V$ exists. Choose vector subspaces V_+ and V_- of V such that s is positive-definite on V_+ and negative-definite on V_-, and that V_+ and V_- are maximal with respect to this property. Then $V_+ \cap V_- = \{0\}$ and $V = V_+ \oplus V_-$. Obviously $V_+ \cap L = V_- \cap L = \{0\}$. From

$$\dim_{\mathbb{R}}(V_\pm) + \dim_{\mathbb{R}}(L) - \dim_{\mathbb{R}}(V_\pm \cap L) \leq \dim_{\mathbb{R}}(V)$$

we conclude $\dim_{\mathbb{R}}(V_\pm) \leq \dim_{\mathbb{R}}(V) - \dim_{\mathbb{R}}(L)$. Since $2 \cdot \dim_{\mathbb{R}}(L) = \dim_{\mathbb{R}}(V) = \dim_{\mathbb{R}}(V_+) + \dim_{\mathbb{R}}(V_-)$ holds, we get $\dim_{\mathbb{R}}(V_\pm) = \dim_{\mathbb{R}}(L)$. This implies

$$\text{sign}(s) = \dim_{\mathbb{R}}(V_+) - \dim_{\mathbb{R}}(V_-) = \dim_{\mathbb{R}}(L) - \dim_{\mathbb{R}}(L) = 0.$$

$\square$

Proof of Theorem 5.79. Let $i \colon A \to X$ be the inclusion. Then the following diagram

$$\begin{CD}
H^{2k}(X; \mathbb{R}) @>{H^{2k}(i)}>> H^{2k}(A; \mathbb{R}) @>{\delta^{2k}}>> H^{2k+1}(X, A; \mathbb{R}) \\
@V{-\cap [X,A]}V{\cong}V @V{-\cap \partial_{4k+1}([X,A])}V{\cong}V @V{-\cap [X,A]}V{\cong}V \\
H_{2k+1}(X, A; \mathbb{R}) @>{\partial_{2k+1}}>> H_{2k}(A; \mathbb{R}) @>{H_{2k}(i)}>> H_{2k}(X; \mathbb{R})
\end{CD}$$

commutes for $n = 4k$. This implies $\dim_{\mathbb{R}}(\ker(H_{2k}(i))) = \dim_{\mathbb{R}}(\text{im}(H^{2k}(i)))$. Since $\mathbb{R}$ is a field, we get from the *Kronecker pairing* for a commutative ring R

5.6 Poincaré Duality

$$\langle -, - \rangle \colon H^i(X, R) \times H_i(X, R) \to R \tag{5.82}$$

an isomorphism $H^{2k}(X;\mathbb{R}) \xrightarrow{\cong} (H_{2k}(X;\mathbb{R}))^*$ and analogously for A. Under these identifications $H^{2k}(i)$ becomes $(H_{2k}(i))^*$. Hence

$$\dim_{\mathbb{R}}(\mathrm{im}(H_{2k}(i))) = \dim_{\mathbb{R}}(\mathrm{im}(H^{2k}(i))).$$

We conclude

$$\dim_{\mathbb{R}}(H^{2k}(A;\mathbb{R})) = 2 \cdot \dim_{\mathbb{R}}(\mathrm{im}(H^{2k}(i)))$$

from $\dim_{\mathbb{R}}(H_{2k}(A;\mathbb{R})) = \dim_{\mathbb{R}}(\mathrm{ker}(H_{2k}(i))) + \dim_{\mathbb{R}}(\mathrm{im}(H_{2k}(i)))$. We have for $x, y \in H^{2k}(X;\mathbb{R})$

$$\langle H^{2k}(i)(x) \cup H^{2k}(i)(y), \partial_{4k+1}([X, A]) \rangle = \langle H^{2k}(i)(x \cup y), \partial_{4k+1}([X, A]) \rangle$$
$$= \langle x \cup y, H_{2k}(i) \circ \partial_{4k+1}([X, A]) \rangle$$
$$= \langle x \cup y, 0 \rangle = 0.$$

If we apply Lemma 5.81 to the non-degenerate symmetric bilinear pairing

$$H^{2k}(A;\mathbb{R}) \otimes_{\mathbb{R}} H^{2k}(A;\mathbb{R}) \xrightarrow{\cup} H^{4k}(A;\mathbb{R}) \xrightarrow{\langle -, \partial_{4k+1}([X, A]) \rangle} \mathbb{R}$$

with L the image of $H^{2k}(i) \colon H^{2k}(X;\mathbb{R}) \to H^{2k}(A;\mathbb{R})$, we see that the signature of this pairing, which is $\mathrm{sign}(A)$, is zero. One easily checks that $\mathrm{sign}(A)$ is the sum of the signatures of the components of A. □

Exercise 5.83 Let M be a closed oriented $4k$-dimensional manifold. Let $\chi(M)$ be its Euler characteristic. Show $\mathrm{sign}(M) \equiv \chi(M) \mod 2$.

Definition 5.84 (Signature for finite Poincaré pairs) Let $(X, \partial X)$ be an oriented finite Poincaré pair of dimension $n = 4k$. Let $[X, \partial X]$ in $H_{4k}(X, \partial X)$ be its fundamental class. Define a symmetric $\mathbb{R}$-bilinear pairing, its *intersection pairing*, by

$$I^{2k} \colon H^{2k}(X, \partial X;\mathbb{R}) \times H^{2k}(X, \partial X;\mathbb{R}) \to \mathbb{R}, \quad (x, y) \mapsto \langle x \cup y, [X, \partial X]_{\mathbb{R}} \rangle,$$

where $[X, \partial X]_{\mathbb{R}} \in H^n(X, \partial X;\mathbb{R})$ is the image of $[X, \partial X]$ under the obvious change of coefficients map $H^n(X, \partial X;\mathbb{Z}) \to H^n(X, \partial X;\mathbb{R})$. The pairing I^{2k} is not in general non-degenerate since its null space is the kernel of the map $j_X \colon H^{2k}(X, \partial X;\mathbb{R}) \to H^{2k}(X;\mathbb{R})$ induced by the inclusion $X \to (X, \partial X)$. It follows that I^{2k} induces a non-degenerate bilinear pairing $\overline{I^{2k}}$ on the image of j_X where I^{2k} and $\overline{I^{2k}}$ have the same signature. The *signature* of $(X, \partial X)$, which we denote by $\mathrm{sign}(X, \partial X)$, is defined to be the signature of I^{2k}, equivalently of $\overline{I^{2k}}$. If n is not divisible by 4, we put $\mathrm{sign}(X, \partial X) = 0$.

Of course Definition 5.84 reduces to Definition 5.70 in the special case $\partial X = \emptyset$. The signature has the following further properties.

Lemma 5.85

(i) *Let M and N be compact oriented manifolds and $f\colon \partial M \to \partial N$ be an orientation reversing diffeomorphism. Then $M \cup_f N$ inherits an orientation from M and N and*
$$\mathrm{sign}(M \cup_f N) = \mathrm{sign}(M, \partial M) + \mathrm{sign}(N, \partial N);$$

(ii) *Let M and N be compact oriented manifolds. Then*
$$\mathrm{sign}(M \times N, \partial(M \times N)) = \mathrm{sign}(M, \partial M) \cdot \mathrm{sign}(N, \partial N);$$

(iii) *Let $p\colon \overline{M} \to M$ be a finite covering with d sheets of closed oriented manifolds. Then*
$$\mathrm{sign}(\overline{M}) = d \cdot \mathrm{sign}(M).$$

Proof. (i) This is due to Novikov. For a proof, see for instance [15, Proposition 7.1 on page 588].

(ii) This is well known when M and N are closed: see [388, pages 220 – 222] which we follow below. The additional argument for the proof in the general case was communicated to us by Andrew Ranicki.

Let $m = \dim(M)$, $n = \dim(N)$ and assume that $r = m + n = 4k$, otherwise there is nothing to prove. There is a homeomorphism of spaces $\partial(M \times N) \cong M \times \partial N \cup_{\partial M \times \partial N} \partial M \times N$, and so by the Künneth Theorem the exterior product in relative cohomology gives the horizontal isomorphisms of graded $\mathbb{R}$-algebras in the following commutative diagram.

$$\begin{array}{ccc} H^*(M, \partial M; \mathbb{R}) \otimes_{\mathbb{R}} H^*(N, \partial N; \mathbb{R}) & \xrightarrow{\cong} & H^*(M \times N, \partial(M \times N); \mathbb{R}) \\ \downarrow{\scriptstyle j_M \times j_N} & & \downarrow{\scriptstyle j_{M \times N}} \\ H^*(M; \mathbb{R}) \otimes_{\mathbb{R}} H^*(N; \mathbb{R}) & \xrightarrow{\cong} & H^*(M \times N; \mathbb{R}). \end{array}$$

We have to investigate the non-degenerate bilinear pairing defined on
$$j_{M \times N}(H^{2k}(M \times N, \partial(M \times N); \mathbb{R}))$$
by dividing out the nullspace that is $\ker(j_{M \times N})$. This form splits as an orthogonal sum of forms with the underlying $\mathbb{R}$-vector spaces the sums of

$$j_M(H^i(M, \partial M; \mathbb{R})) \otimes_{\mathbb{R}} j_N(H^{2k-i}(N, \partial N; \mathbb{R})); \tag{5.86}$$
$$j_M(H^{m-i}(M, \partial M; \mathbb{R})) \otimes_{\mathbb{R}} j_N(H^{2k+i-m}(N, \partial N; \mathbb{R})), \tag{5.87}$$

where $i < m/2$, and, if m and n are both even, there is an additional orthogonal summand
$$j_M(H^{m/2}(M, \partial M; \mathbb{R})) \otimes_{\mathbb{R}} j_N(H^{n/2}(N, \partial N; \mathbb{R})).$$

We claim that the signatures of the first summands are each zero: To see this, choose bases of $j_M(H^i(M, \partial M; \mathbb{R}))$ and $j_N(H^{2k-i}(N, \partial N; \mathbb{R}))$ and their duals in

5.6 Poincaré Duality

$j_M(H^{m-i}(M, \partial M; \mathbb{R}))$ and $j_N(H^{2k+i-m}(N, \partial N; \mathbb{R}))$ and observe that the form whose underlying $\mathbb{R}$-vector space is the sum of (5.86) and (5.87) contains the lagrangian given by (5.86), and so has signature zero by Lemma 5.81. Since the signature of an orthogonal sum of forms is the sum of their signatures, $\text{sign}(M \times N, \partial(M \times N)) = 0$ if m or n is odd. When m and n are both even, we have two cases. Either both m and n are congruent to 0 modulo 4 or both are congruent to 2 modulo 4. In the first case the forms on $j_M(H^{m/2}(M, \partial M; \mathbb{R}))$ and $j_N(H^{n/2}(N, \partial N; \mathbb{R}))$ are both symmetric and since the signature of the tensor product of forms is the product of the signatures, we have

$$\begin{aligned} &\text{sign}(M \times N, \partial(M \times N)) \\ &= \text{sign}\big(j_M(H^{m/2}(M, \partial M; \mathbb{R})) \otimes_{\mathbb{R}} j_N(H^{n/2}(N, \partial N; \mathbb{R}))\big) + 0 \\ &= \text{sign}\big(j_M(H^{m/2}(M, \partial M; \mathbb{R}))\big) \cdot \text{sign}\big(j_N(H^{n/2}(N, \partial N; \mathbb{R}))\big) \\ &= \text{sign}(M, \partial M) \cdot \text{sign}(N, \partial N). \end{aligned}$$

In the case when both m and n are congruent to 2 modulo 4, the forms on $j_M(H^{m/2}(M, \partial M; \mathbb{R}))$ and $j_N(H^{n/2}(N, \partial N; \mathbb{R}))$ are both skew symmetric and a special argument as at the bottom of the page 221 in [388] is needed to complete the argument.

(iii) This follows for (smooth) manifolds M from the Hirzebruch signature formula, see for instance [308, § 19], or from Atiyah's L^2-index theorem, see [12, (1.1)]. (Topological manifolds are treated in [372, Theorem 8].) □

Example 5.88 Wall has constructed a finite connected Poincaré space X together with a finite covering with d sheets $\overline{X} \to X$ such that the signature does not satisfy $\text{sign}(\overline{X}) = d \cdot \text{sign}(X)$, see [348, Example 22.28], [408, Corollary 5.4.1]. Hence X cannot be homotopy equivalent to a closed manifold by Lemma 5.85.

Exercise 5.89 Let M be an orientable compact manifold. Its double DM is the closed manifold $M \cup_{\text{id}: \partial M \to \partial M} M$. Show that DM is orientable and that for any choice of orientation the signature of DM is trivial.

Exercise 5.90 Compute for $n \geq 1$ the signature of the following manifolds:

(i) the complex projective space $\mathbb{CP}^n$;
(ii) the total space of the sphere bundle $S(TM \oplus \underline{\mathbb{R}})$ of $TM \oplus \underline{\mathbb{R}}$ for a closed oriented n-dimensional Riemannian manifold M;
(iii) a closed oriented n-dimensional manifold M admitting an orientation reversing self-homotopy equivalence.

Exercise 5.91 Let M and N be connected oriented n-manifolds with connected sum $M \sharp N$. Show that $\text{sign}(M \sharp N) = \text{sign}(M) + \text{sign}(N)$ holds and that $\mathbb{CP}^2 \sharp \mathbb{CP}^2$ is not homotopy equivalent to $(\mathbb{CP}^2)^- \sharp \mathbb{CP}^2$.

5.6.4 The Degree of a Map

We conclude this section by defining the degree of a map between w-oriented Poincaré pairs. Let $(X, \partial X)$ and $(Y, \partial Y)$ be w-oriented finite n-dimensional Poincaré complexes for infinite cyclic local coefficient systems O_X and O_Y and fundamental classes $[[X, \partial X]] \in H_n(X, \partial X; O_X)$ and $[[Y, \partial Y]] \in H_n(Y, \partial Y; O_Y)$. Consider a map $(f, \partial f) \colon (X, \partial X) \to (Y, \partial Y)$ and an isomorphism of infinite cyclic local coefficient systems $\overline{f} \colon O_X \to f^*O_Y$. We obtain a homomorphism $H_n(f, \partial f; \overline{f}) \colon H_n(X, \partial X; O_X) \to H_n(Y, \partial Y; O_Y)$. We have the canonical isomorphisms

$$\bigoplus_{C \in \pi_0(X)} H_n(C, C \cap \partial X; O_X|_C) \xrightarrow{\cong} H_n(X, \partial X; O_X);$$

$$\bigoplus_{D \in \pi_0(Y)} H_n(D, D \cap \partial Y; O_Y|_D) \xrightarrow{\cong} H_n(Y, \partial Y; O_Y).$$

Using these isomorphisms the classes $[[X, \partial X]]$ and $[[Y, \partial Y]]$ define classes

$$[[C, C \cap \partial X]] \in H_n(C, C \cap \partial X; O_X|_C);$$
$$[[D, D \cap \partial Y]] \in H_n(D, D \cap \partial Y; O_Y|_D),$$

which are in all cases generators of infinite cyclic groups. The homomorphism $H_n(f, \partial f; \overline{f})$ induces a map

$$H_n(f|_C, \partial f|_C; \overline{f}|_C) \colon H_n(C, C \cap \partial X; O_X|_C) \to H_n(D, D \cap \partial Y; O_Y|_D),$$

provided that $f(C) \subseteq D$. Let $\deg_{C,D}(f, \partial f; \overline{f})$ be the integer defined by

$$H_n(f|_C, \partial f|_C; \overline{f}|_C)([[C, C \cap \partial X]]) = \deg_{C,D}(f, \partial f; \overline{f}) \cdot [[D, D \cap \partial Y]].$$

Definition 5.92 (Degree of a map of w-oriented finite Poincaré pairs) Define the *degree* of $(f, \partial f; \overline{f})$ to be the element

$$\deg(f, \partial f; \overline{f}) \in H_0(Y) = \bigoplus_{D \in \pi_0(Y)} \mathbb{Z}$$

which assigns to $D \in \pi_0(Y)$ the integer $\sum_{\substack{C \in \pi_0(X) \\ f(C) \subseteq D}} \deg_{C,D}(f, \partial f; \overline{f})$.

We will show in Theorem 7.36 that the degree is a bordism invariant.

Example 5.93 (The degree in the connected orientable case) Suppose that $(X, \partial X)$ and $(Y, \partial Y)$ are oriented connected finite n-dimensional Poincaré pairs. Then we can choose O_X and O_Y to be the constant functors with value $\mathbb{Z}$ and obtain identifications

$$H_n(X, \partial X; O_X) \cong H_n(X, \partial X; \mathbb{Z});$$
$$H_n(Y, \partial Y; O_Y) \cong H_n(Y, \partial Y; \mathbb{Z}).$$

Denote by $[X, \partial X] \in H_n(X, \partial X; \mathbb{Z})$ and $[Y, \partial Y] \in H_n(Y, \partial Y; \mathbb{Z})$ the preimages of $[[X, \partial X]]$ and $[[Y, \partial Y]]$ under these isomorphism. Then $[X, \partial X]$ and $[Y, \partial Y]$ are generators of infinite cyclic groups. Let $\overline{f}: O_X \to f^*O_Y$ be given by the identity morphism $\text{id}_{\mathbb{Z}}$.

Then the definition of $\deg(f, \partial f; \overline{f})$ above reduces to the classical definition of the degree, namely, we have

$$H_n(f, \partial f; \mathbb{Z})([X, \partial X]) = \deg(f, \partial f; \overline{f}) \cdot [Y, \partial Y].$$

Note that the non-orientable case has to be more complicated since we cannot choose O_X and O_Y to be constant with value $\mathbb{Z}$.

5.7 Notes

In this chapter we defined Poincaré duality by a certain chain homotopy equivalence, which induces isomorphisms between cohomology and homology. This was then used in dimensions $n = 4k$ to define a symmetric bilinear form whose signature is an important invariant of manifolds. It turns out that at the level of chain complexes it is also possible to define what is called the symmetric signature. This much more refined invariant is due to Ranicki [344],[345]. We will come to this in more detail in Chapter 15, see Definition 15.312.

Chapter 6
The Spivak Normal Structure

6.1 Introduction

Recall that our goal is to find a solution to the following problem.

Problem 6.1 Let X be a finite Poincaré complex. When is X homotopy equivalent to a closed n-dimensional manifold?

In this chapter we work in the smooth category unless explicitly stated otherwise.

We explained in Chapter 5 why we have to strengthen the condition finite CW-complex in Problem 4.1 to finite Poincaré complex in Problem 6.1 above, see Theorem 5.51.

Recall from Chapter 4 that it will be important that bundle data are available. In this chapter we address the question whether we can achieve such bundle data and to what extent they are unique.

Recall that an n-dimensional manifold M can be embedded into an $(n + k)$-dimensional Euclidean space $\mathbb{R}^{n+k}$ for k large enough and such an embedding is unique up to isotopy. It possesses a well-defined normal bundle $\nu(M)$, which is a k-dimensional vector bundle. The Pontrjagin–Thom construction from Section 4.7 provides us with a collapse map $c_M : S^{n+k} \to \text{Th}(\nu_M)$.

In this chapter we investigate how to obtain analogous data when we start with a Poincaré complex X instead of a manifold M. It turns out that we have to relax the bundle statement in that we only obtain a spherical fibration and not a vector bundle.

For this reason we start with some background on spherical fibrations in Section 6.2. Section 6.3 is devoted to the Thom Isomorphism Theorem 6.54, which is an important tool for understanding cohomological properties of vector bundles and more generally of spherical fibrations. In Section 6.8 we introduce the notion of a Spivak normal structure, which is given by a spherical fibration and a collapse map, see Definition 6.65. We state the main result of this chapter, Theorem 6.68, which says that such a structure exists on a finite Poincaré complex and is unique up to stable strong fibre homotopy equivalence. Its existence is based on the construction of Section 6.4. If X happens to be a closed manifold, then a Spivak normal structure is given by the spherical fibration determined by the normal bundle, see Example 6.66.

Thus we obtain a new obstruction from algebraic topology for a positive answer to Problem 6.1 above, namely, the Spivak normal structure must have a reduction to a vector bundle, see Lemma 6.79 and Example 6.85.

Guide 6.2 It is crucial to understand the notion of the Spivak normal structure, see Definition 6.65, and the statement of its existence and uniqueness, see Theorem 6.68. The rather long proof of Theorem 6.68 is illuminating, but does not play a role in the rest of the book.

6.2 Spherical Fibrations

A $(k-1)$-*spherical fibration* is a (Hurewicz) fibration $p\colon E \to X$, i.e., a map having the homotopy lifting property whose fibres are all homotopy equivalent to S^{k-1}. The simplest example is the *trivial $(k-1)$-spherical fibration*,

$$p\colon X \times S^{k-1} \to X, \quad (x, y) \mapsto x,$$

which we denote by $\underline{S^{k-1}}$. If $k = 0$, then S^{-1} is defined to be the empty set and a (-1)-spherical fibration over X is the map $\emptyset \to X$. If $p_\xi \colon E \to X$ is the projection of a k-dimensional vector bundle ξ with Riemannian metric, its sphere bundle $p_{SE}\colon SE \to X$ is an example of a $(k-1)$-spherical fibration. Hence we regard $(k-1)$-spherical fibrations as homotopy-theoretic analogues of k-dimensional vector bundles.

The analogue of an isomorphism of vector bundles is a *strong fibre homotopy equivalence* of $(k-1)$-spherical fibrations, which we now define. Given two fibrations $p_0\colon E_0 \to X_0$ and $p_1\colon E_1 \to X_1$, a fibre map $(f, \overline{f})\colon p_0 \to p_1$ consists of maps $\overline{f}\colon E_0 \to E_1$ and $f\colon X_0 \to X_1$ satisfying $p_1 \circ \overline{f} = f \circ p_0$. There is an obvious notion of *fibre homotopy* $(h, \overline{h})$. A fibre homotopy is called a *strong fibre homotopy* if $h_t\colon X_0 \to X_1$ is stationary for $t \in [0,1]$. A fibre map $(\mathrm{id}, \overline{f})\colon p_0 \to p_1$ between two fibrations p_0 and p_1 over the same base space is a *strong fibre homotopy equivalence* if there is a fibre map $(\mathrm{id}, \overline{g})\colon p_1 \to p_0$ such that both composites are strongly fibre homotopic to the identity. In this case we say that p_0 and p_1 are *strongly fibre homotopy equivalent* and write $p_0 \simeq_{\mathrm{sfh}} p_1$.

The familiar constructions of the Thom space, pullbacks and Whitney sum for vector bundles have analogues for spherical fibrations, as we now describe.

To form the Thom space of a $(k-1)$-spherical fibration $p\colon E \to X$, we need the associated disk fibration that is defined by $Dp\colon DE := \mathrm{cyl}(p) \to X$, where $\mathrm{cyl}(p)$ is the mapping cylinder of p and Dp is the obvious map. A simple application of [97, Proposition 1.3] shows that Dp is a fibration.

Define the *Thom space* $\mathrm{Th}(p)$ of p to be the pointed space $\mathrm{cone}(p)$, where $\mathrm{cone}(p)$ is the mapping cone of p with its canonical base point, or, equivalently, put $\mathrm{Th}(p) = DE/E$. If $k = 0$, then $DE = X$ and $\mathrm{Th}(p) = X_+$. If $p_\xi \colon E \to X$ is the projection of a k-dimensional vector bundle ξ over X with sphere bundle $p_{SE}\colon SE \to X$, then we may identify the disk bundle of ξ with the mapping cylinder of p_{SE}, so that

6.2 Spherical Fibrations

$DE = \text{cyl}(p_{SE})$. Note that the canonical inclusion of X in $\text{cyl}(p_{SE})$ is a homotopy equivalence, which is analogous to the fact that the inclusion defined by the zero-section of ξ is a homotopy equivalence. The canonical inclusion of E into $\text{cyl}(p)$ corresponds to the inclusion of $SE \subset DE$. Hence the definition $\text{Th}(p) = DE/E = \text{cone}(p)$ for a $(k-1)$-spherical fibration p corresponds to $\text{Th}(\xi) = DE/SE$ for the k-dimensional vector bundle ξ.

To define pullbacks, let $f\colon X \to Y$ be a map and let $p\colon E \to Y$ be a $(k-1)$-spherical fibration. The *pullback* of p along f is the $(k-1)$-spherical fibration $f^*p\colon f^*E \to X$ with total space

$$f^*E := \{(x,e) \mid f(x) = p(e)\} \subset X \times E$$

and projection map $(f^*p)(x,e) = x$.

If one has two vector bundles ξ_0 over X_0 and ξ_1 over X_1, one can form the *exterior Whitney sum* $\xi_0 \times \xi_1$, which is a vector bundle over $X_1 \times X_2$. If $X = X_0 = X_1$, then the Whitney sum of ξ_0 and ξ_1 is the vector bundle over X defined by $\xi_0 \oplus \xi_1 := \Delta^*(\xi_0 \times \xi_1)$, where $\Delta\colon X \to X \times X$, $x \mapsto (x,x)$ is the diagonal map.

Note that $(SE_{\xi_0 \times \xi_1})_{(x_0,x_1)}$ for $(x_0,x_1) \in X_0 \times X_1$ is homeomorphic to the join $(SE_{\xi_0})_{x_0} * (SE_{\xi_1})_{x_1}$. This allows us to recover $SE_{\xi_1 \times \xi_2}$ as a spherical fibration from SE_{ξ_0} and SE_{ξ_1} by the following fibrewise join construction. Given a $(k-1)$-spherical fibration $p_0\colon E_0 \to X$ and an $(l-1)$-spherical fibration $p_1\colon E_1 \to Y$, define the $(k+l-1)$-spherical fibration

$$p_0 *_e p_1 \colon E_0 *_e E_1 \to X \times Y, \tag{6.3}$$

called the *exterior fibrewise join of p_0 and p_1*, as follows. Suppose that $k,l \geq 1$. Then the total space $E_0 *_e E_1$ is the quotient of the space $E_0 \times E_1 \times [0,1]$ under the equivalence relation generated by $(e_0, e_1, 0) \sim (e_0, e_1', 0)$, provided $p_1(e_1) = p_1(e_1')$, and $(e_0, e_1, 1) \sim (e_0', e_1, 1)$, provided $p_0(e_0) = p_0(e_0')$. The projection $p_0 *_e p_1$ sends $[e_0, e_1, t]$, the class of (e_0, e_1, t), to $(p_0(e_0), p_1(e_1))$. If $k = 0$, define the total space to be $X \times E_1$ and the projection $p_0 *_e p_1$ to be $\text{id}_X \times p_1$ and analogously for $l = 0$. We conclude from [97, Proposition 1.3] that $p_0 *_e p_1$ is a fibration. When $X = X_0 = X_1$, the *fibrewise join* of p_0 and p_1 is the $(k+l-1)$-spherical fibration

$$p_0 * p_1 := \Delta^*(p_0 *_e p_1).$$

Exercise 6.4 Show that

$$E_0 *_e E_1 \cong DE_0 \times E_1 \cup_{E_0 \times E_1} E_0 \times DE_1.$$

For the Thom space of the exterior fibrewise join, there is a natural homeomorphism

$$\text{Th}(p_0 *_e p_1) \cong \text{Th}(p_0) \wedge \text{Th}(p_1). \tag{6.5}$$

This is proven in [393, Proposition 12.28] for vector bundles; the proof for spherical fibrations is similar. This implies

$$\text{Th}(p * \underline{S^{k-1}}) \cong S^k \wedge \text{Th}(p). \tag{6.6}$$

We next define the orientation homomorphism of a spherical fibration. Given a fibration $p\colon E \to X$, the fibre transport along loops at $x \in X$ yields a functor from the fundamental groupoid $\Pi(X)$ to the homotopy category of topological spaces sending a point x to $p^{-1}(x)$, see for instance [393, 15.12 on page 343]. In particular we get for every $x \in X$ a map of monoids $t_x\colon \pi_1(X, x) \to [p^{-1}(x), p^{-1}(x)]$. We call p *orientable* if t_x is trivial for any base point $x \in X$. In the case of a $(k-1)$-spherical fibration over a CW-complex X, the fibre transport for $x \in X$ is the same as a homomorphism $t_x\colon \pi_1(X, x) \to \{\pm 1\}$ called *the orientation homomorphism* or *first Stiefel–Whitney class* since the degree defines a bijection $[S^{k-1}, S^{k-1}]^\times \to \{\pm 1\}$ for $k \geq 1$. This is the same as an element $w_1(p) \in H^1(X; \mathbb{Z}/2)$.

Remark 6.7 Recall from [308, §4] that every vector bundle ξ with base space X has Stiefel–Whitney classes $w_i(\xi) \in H^i(X; \mathbb{Z}/2)$. If $p_{S\xi}\colon SE_\xi \to X$ is the spherical fibration associated to ξ, then $w_1(p_{S\xi}) = w_1(\xi)$.

More generally, the Stiefel–Whitney classes are defined using the Thom-space $\text{Th}(\xi) = D(\xi)/S(\xi)$ only, and hence all the Stiefel–Whitney classes can be defined for any spherical fibration. In particular, $w_i(\xi)$ depends only on the strong fibre homotopy type of $p_{S\xi}$. This is not true for the Pontrjagin classes of a vector bundle.

Let X be a CW-complex and let $\text{VB}_k(X)$ be the set of isomorphism classes $[\xi]$ of k-dimensional (real) vector bundles ξ over X. There is a universal k-dimensional bundle γ_k over $\text{BO}(k)$ such that for any k-vector bundle ξ there is, up to homotopy, one map $f_\xi\colon X \to \text{BO}(k)$ such that ξ is isomorphic to $f_\xi^* \gamma_k$. Moreover, $[f_\xi]$ depends only on $[\xi]$. For the proof of the next theorem we refer, for instance, to [308, §5].

Theorem 6.8 (Classification of vector bundles) *If X is a CW-complex, then the map*

$$\text{VB}_k(X) \to [X, \text{BO}(k)], \quad [\xi] \mapsto [f_\xi],$$

is a bijection.

Next we explain the analogous statement for spherical $(k-1)$-fibrations. We write $[p]$ for the strong fibre homotopy equivalence class of a $(k-1)$-spherical fibration $p\colon E \to X$ and define

$$\text{SF}_k(X) := \{[p] \mid p\colon E \to X \text{ is a } (k-1)\text{-spherical fibration}\} \tag{6.9}$$

to be the set of strong fibre homotopy equivalence classes of $(k-1)$-spherical fibrations over a space X.

We have the topological monoid of self-homotopy equivalences of the $(k-1)$-sphere

$$G(k) := \{f\colon S^{k-1} \to S^{k-1} \mid \deg(f) = \pm 1\} \tag{6.10}$$

6.2 Spherical Fibrations

where we topologise $G(k)$ using the compact open topology. Stasheff proved that the topological monoid $G(k)$ plays the role in the theory of $(k-1)$-spherical fibrations which the topological group $O(k)$ plays in the theory of rank k vector bundles. In particular, there is a classifying space $BG(k)$ and a universal $(k-1)$-spherical fibration

$$q_{S^{k-1}} \colon E_{S^{k-1}} \to BG(k)$$

such that $BG(k)$ is a *CW*-complex and every $(k-1)$-spherical fibration $p \colon E \to X$ over a *CW*-complex X is classified by a map $f_p \colon X \to BG(k)$, i.e., there is a strong fibre homotopy equivalence $p \xrightarrow{\simeq} f_p^* q_{S^{k-1}}$. Moreover, the map f_p is uniquely determined up to homotopy by the strong fibre homotopy type of p. In other words, we have the following theorem of Stasheff [386, Classification Theorem].

Theorem 6.11 (Classification of spherical fibrations) *If X is a CW-complex, then the map*

$$SF_k(X) \to [X, BG(k)], \quad [p] \mapsto [f_p],$$

is a bijection.

Remark 6.12 Stasheff's theorem [386, Classification Theorem] is more general, and applies to all fibrations whose fibres have the homotopy type of a finite *CW*-complex. Given a finite *CW*-complex F, let $G(F)$ be the topological monoid of self-homotopy equivalences of F, for example $G(k) = G(S^{k-1})$. For such a monoid, Stasheff constructs a classifying space $BG(F)$ together with a fibration $q_F \colon E_F \to BG(F)$ with typical fibre F over a *CW*-complex $BG(F)$ such that the pullback construction yields a bijection between the homotopy classes of maps from X to $BG(F)$ and the set of strong fibre homotopy classes of fibrations over X with typical fibre F, see [386].

The case of more general fibres is studied in [289].

Lemma 6.13 *Let F be a finite CW-complex. The space $BG(F)$ is path connected and for any choice of base point $x \in BG(F)$ and for $n \geq 1$ there are isomorphisms of abelian groups*

$$\pi_n(BG(F), x) \cong \pi_{n-1}(G(F), \mathrm{id}_F).$$

Proof. There is a principal quasi-fibration $G(F) \to EG(F) \to BG(F)$ such that $EG(F)$ is contractible, see for instance [279, Theorem 1.2 on page 4 and Theorem 1.5 on page 7]. Since a quasi-fibration admits a long exact homotopy sequence, its boundary homomorphism induces the desired isomorphism $\pi_n(BG(F), x) \cong \pi_{n-1}(G(F), \mathrm{id}_F)$. □

Remark 6.14 We briefly describe the isomorphism appearing in Lemma 6.13 in more concrete terms. Consider a pointed map $f \colon (S^n, e) \to (BG(F), x)$ where e should be contained in $S^0 \subseteq S^n$. Choose a model for the universal F-fibration $q_F \colon E_F \to BG(F)$. Consider the pullback

$$\begin{array}{ccc} E & \xrightarrow{\overline{f}} & E_F \\ {\scriptstyle f^*q_F}\downarrow & & \downarrow{\scriptstyle q_F} \\ S^n & \xrightarrow{f} & \mathrm{BG}(F). \end{array}$$

Put $F := q_F^{-1}(x)$. Let $S_\pm^n$ be the upper and lower hemisphere. Note that the inclusion $\{e\} \to S_\pm^n$ is a homotopy equivalence. We obtain a strong fibre homotopy equivalence

$$\overline{h}_\pm \colon S_\pm^n \times F \to E|_{S_\pm^n}$$

such that $(\overline{h}_\pm)_e \colon F \to F$ is the identity. It is unique up to strong fibre homotopy relative to $e \in S^n$. Proofs of these claims can be derived from [393, Proposition 15.11 on page 342]. Let $\overline{h}_-^{-1} \colon E|_{S_-^n} \to S_-^n \times F$ be a strong fibre homotopy equivalence with $(\overline{h}_-^{-1})_e = \mathrm{id}_F$ such that $\overline{h}_- \circ \overline{h}_-^{-1}$ and $\overline{h}_-^{-1} \circ \overline{h}_-$ are strongly fibre homotopic relative e. Then

$$\overline{h}_-^{-1}|_{S^{n-1}} \circ \overline{h}_+|_{S^{n-1}} \colon S^{n-1} \times F \to S^{n-1} \times F$$

is a strong fibre homotopy equivalence with $(\overline{h}_-^{-1}|_{S^{n-1}} \circ \overline{h}_+|_{S^{n-1}})_e = \mathrm{id}_F$, which is unique up to strong fibre homotopy relative e and which we call *a characteristic map* for $f^*q_F \colon E \to S^n$. The characteristic map defines a unique element in $\pi_{n-1}(G(F), \mathrm{id}_F)$.

Now for $n \geq 1$ consider the bijection

$$\mathrm{char}_k(S^n) \colon \mathrm{SF}_k(S^n) \xrightarrow{\cong} [S^n, \mathrm{BG}(k)] \cong \pi_{n-1}(G(k), \mathrm{id}), \tag{6.15}$$

which, via the isomorphisms of Theorem 6.11 and Lemma 6.13, maps a $(k-1)$-spherical fibration $p \colon E \to S^n$ to the homotopy class of its characteristic map. Fix natural numbers k_0 and k_1. Let $s_0 \colon G(k_0) \to G(k_0 + k_1)$ and $s_1 \colon G(k_1) \to G(k_0 + k_1)$ respectively be the maps of monoids which send the class of $f_0 \colon S^{k_0-1} \to S^{k_0-1}$ and $f_1 \colon S^{k_1-1} \to S^{k_1-1}$ respectively to the class of $f_0 * \mathrm{id}_{S^{k_1-1}} \colon S^{k_0+k_1-1} \to S^{k_0+k_1-1}$ and $\mathrm{id}_{S^{k_0-1}} * f_1 \colon S^{k_0+k_1-1} \to S^{k_0+k_1-1}$ respectively (using the identification $S^{k_0-1} * S^{k_1-1} = S^{k_0+k_1-1}$). We get a bilinear map of abelian groups

$$\begin{aligned} \beta \colon \pi_{n-1}(G(k_0), \mathrm{id}) \times \pi_{n-1}(G(k_1), \mathrm{id}) &\to \pi_{n-1}(G(k_0+k_1), \mathrm{id}); \\ ([u_0], [u_1]) &\mapsto [s_0 \circ u_0] + [s_1 \circ u_1]. \end{aligned} \tag{6.16}$$

Let

$$* \colon \mathrm{SF}_{k_0}(S^n) \times \mathrm{SF}_{k_1}(S^n) \to \mathrm{SF}_{k_0+k_1}(S^n)$$

be the map sending $([p_0], [p_1])$ to $[p_0 * p_1]$.

6.2 Spherical Fibrations

Lemma 6.17 *The following diagram*

$$\begin{array}{ccc}
\mathrm{SF}_{k_0}(S^n) \times \mathrm{SF}_{k_1}(S^n) & \xrightarrow{\;\;*\;\;} & \mathrm{SF}_{k_0+k_1}(S^n) \\
{\scriptstyle \mathrm{char}_{k_0}(S^n) \times \mathrm{char}_{k_1}(S^n)} \downarrow \cong & & \cong \downarrow {\scriptstyle \mathrm{char}_{k_0+k_1}(S^n)} \\
\pi_{n-1}(G(k_0), \mathrm{id}) \times \pi_{n-1}(G(k_1), \mathrm{id}) & \xrightarrow{\;\;\beta\;\;} & \pi_{n-1}(G(k_0+k_1), \mathrm{id})
\end{array}$$

commutes for $n \geq 1$.

Proof. Consider the monoid map

$$*\colon G(k_0) \times G(k_1) \to G(k_0+k_1), \quad (g_0, g_1) \mapsto g_0 * g_1,$$

which maps a pair of homotopy equivalences to their join.

For $i = 0, 1$, let p_i be a $(k_i - 1)$-spherical fibration over S^n with characteristic function $f_i \colon S^{n-1} \to G(k)$. We conclude from Remark 6.14 that $p_0 * p_1$ has characteristic function $f_0 * f_1$ where for $x \in S^{n-1}$ we define

$$(f_0 * f_1)(x) = f_0(x) * f_1(x).$$

Obviously we have

$$f_0 * f_1 = (f_0 * c_{\mathrm{id}_{S^{k_1-1}}}) \circ (c_{\mathrm{id}_{S^{k_0-1}}} * f_1),$$

where $c_{\mathrm{id}_{S^{k_i-1}}}$ is the constant function $S^{n-1} \to G(k_i)$ with value $\mathrm{id}_{S^{k_i-1}}$. Suppose that $n \geq 2$. Then $\pi_{n-1}(G(k), \mathrm{id})$ has two group structures, one from the co-H-space structure on S^{n-1} and the other from the H-space structure on $G(k)$. These agree and $\pi_{n-1}(G(k), \mathrm{id})$ is an abelian group, see [393, Proposition 2.25 on page 21]. Hence we get in $\pi_{n-1}(G(k_0 + k_1), \mathrm{id})$

$$\begin{aligned}
[f_0 * f_1] &= [(f_0 * c_{\mathrm{id}_{S^{k_0-1}}}) \circ (c_{\mathrm{id}_{S^{k_0-1}}} * f_1)] \\
&= [(f_0 * c_{\mathrm{id}_{S^{k_0-1}}})] + [(c_{\mathrm{id}_{S^{k_0-1}}} * f_1)] \\
&= s_0([f_0]) + s_1([f_1]).
\end{aligned}$$

It remains to consider the case $n = 1$. Then the degree of a self-map of S^k gives a bijection $\pi_0(G(k)) \xrightarrow{\cong} \mathbb{Z}/2$ and the claim follows by a direct inspection using the fact that the degree of a join of two self-maps of spheres is the sum of their degrees. □

In general, the computation of $\mathrm{SF}_k(X)$ is a difficult problem. The classification problem becomes easier if we stabilise spherical fibrations with the trivial spherical fibration, as we now explain after having recalled the analogous construction for vector bundles. Let

$$\mathrm{VB}(X) := \mathrm{colim}_{k \to \infty} \mathrm{VB}_k(X) \tag{6.18}$$

be the set of stable isomorphism classes of vector bundles over X, where we call two bundles ξ_1 and ξ_2 over X *stably isomorphic* if for appropriate natural numbers j_1 and

j_2 the vector bundles $\mathbb{R}^{j_1} \oplus \xi_1$ and $\mathbb{R}^{j_2} \oplus \xi_2$ are isomorphic. The stable isomorphism class of a vector bundle ξ in this sense will be denoted $[\xi]_s$.

From now on suppose that X is finite. This assumption ensures that every vector bundle ξ over X has an inverse ξ^{-1}, i.e., another vector bundle such that $\xi \oplus \xi^{-1}$ is trivial. Hence the Whitney sum induces the structure of an abelian group on VB(X). Define BO as the classifying space of the topological group

$$O = \operatorname{colim}_{k \to \infty} O(k). \tag{6.19}$$

It has the structure of a homotopy associative and homotopy commutative H-space with homotopy inverse. Hence $[X, \mathrm{BO}]$ inherits the structure of an abelian group. Given an element $[\xi]_s \in \mathrm{VB}(X)$, we can choose a representative ξ, let us say of dimension k, choose for it a classifying map $f_\xi \colon X \to \mathrm{BO}(k)$ and define $f_{[\xi]}$ to be the composite of f_ξ with the map $\mathrm{BO}(k) \to \mathrm{BO}$. One easily checks that $f_{[\xi]}$ depends only on the stable isomorphism class $[\xi]_s$ of $[\xi]$ and not on the choice made above. For the proof of the next theorem we refer, for instance, to [393, Theorem 11.56 on page 210].

Theorem 6.20 (Classification of stable isomorphisms classes of vector bundles over finite CW-complexes) *Let X be a finite CW-complex. Then the map*

$$\mathrm{VB}(X) \to [X, \mathrm{BO}], \quad [\xi]_s \mapsto [f_{[\xi]}]$$

is an isomorphism of abelian groups.

Next we explain the analogous statement for spherical $(k-1)$-fibrations. Given a spherical $(k-1)$-fibration $p \colon E \to X$, its l-fold suspension $\underline{S^{l-1}} * p$ is the spherical $(k+l-1)$-fibration given by the fibrewise join of the trivial $(l-1)$-spherical fibration $\underline{S^{l-1}}$ and p. Since we may regard stabilisation as the exterior fibrewise join with the trivial fibration over a point, we note that (6.5) specialises to show that there is a natural homeomorphism for a spherical fibration p

$$\mathrm{Th}(\underline{S^{l-1}} * p) \cong S^l \wedge \mathrm{Th}(p). \tag{6.21}$$

Stabilisation defines maps

$$\mathrm{SF}_k(X) \to \mathrm{SF}_{k+j}(X), \quad [p] \mapsto [\underline{S^{j-1}} * p]$$

and we say that two spherical fibrations $p_1 \colon E_1 \to X$ and $p_2 \colon E_2 \to X$ are *stably equivalent* if there is a strong fibre homotopy equivalence

$$\underline{S^{j_1-1}} * p_1 \to \underline{S^{j_2-1}} * p_2$$

for some j_1, j_2. We define SF(X) to be the set of stable equivalence classes $[p]_s$ of spherical fibrations p over X so that

$$\mathrm{SF}(X) = \operatorname{colim}_{k \to \infty} \mathrm{SF}_k(X). \tag{6.22}$$

6.2 Spherical Fibrations

The fibrewise join of representatives endows $SF(X)$ with the structure of a commutative monoid with unit the equivalence class of the trivial spherical fibration. To show that $SF(X)$ has the structure of a group, we need to find stable inverses for a spherical fibration $p\colon E \to X$; i.e., we need to show that there is a spherical fibration p' where $p * p'$ is trivial.

Lemma 6.23 (Stable inverses for spherical fibrations) *For every spherical fibration p over a finite CW-complex X, there is a spherical fibration p^{-1} over X such that $p * p^{-1}$ is strongly fibre homotopy equivalent to the trivial spherical fibration.*

Proof. Let p be a $(k-1)$-spherical fibration over X. We use induction over the number of cells of X. The induction beginning with $X = \emptyset$ is trivial, the induction step is done as follows. Suppose that Y is obtained from X by attaching one cell, i.e., we have a pushout

$$\begin{array}{ccc} S^{d-1} & \xrightarrow{q} & X \\ \downarrow i & & \downarrow \\ D^d & \xrightarrow{Q} & Y. \end{array}$$

Given a $(k-1)$-spherical fibration $p_Y\colon E_Y \to Y$, denote by $p_X\colon E_X \to X, p_S\colon E_S \to S^{d-1}$ and $p_D\colon E_D \to D^d$ the $(k-1)$-spherical fibrations obtained by the pullback construction for p_Y applied to the obvious maps to Y. By the induction hypothesis there is a $(l-1)$-spherical fibration $(p_X^{-1})''\colon E_X^{-1} \to X$ over X such that

$$p_X * (p_X^{-1})'' \simeq_{\text{sfh}} \underline{S^{k+l-1}}.$$

Since D^d is contractible, we have $p_D \simeq_{\text{sfh}} \underline{S^{k-1}}$ and so $p_S \simeq_{\text{sfh}} \underline{S^{k-1}}$. Hence the pullback $q^*(p_X^{-1})''$ of $(p_X^{-1})''$ to S^{d-1} satisfies

$$\underline{S^{k-1}} * q^*(p_X^{-1})'' \simeq_{\text{sfh}} \underline{S^{k+l-1}}.$$

We now replace $(p_X^{-1})''$ by the $(k+l-1)$-fibration $(p_X^{-1})' := (p_X^{-1})'' * \underline{S^{k-1}}$, which is also a stable inverse of p_X and satisfies $q^*(p_X^{-1})' \simeq_{\text{sfh}} \underline{S^{k+l-1}}$. It follows using [97, Proposition 1.3] that $(p_X^{-1})'$ extends to a $(k+l-1)$-spherical fibration p_Y'. It remains to show, perhaps after further stabilisation, that p_Y' can be chosen so that $p_Y * p_Y'$ is trivial.

Let $c\colon Y \to S^d$ be the map collapsing $X \subset Y$ to a point. For any p_Y' we have that $(p_Y * p_Y')_X$ is trivial. Hence the classifying map for $(p_Y * p_Y')_X$ is nullhomotopic and so we can arrange by a cofibration argument that the classifying map for $p_Y * p_Y'$ factors over $c\colon Y \to S^d$. It follows that there is a spherical $(2k+l-1)$-fibration $p_{S^d}\colon E_{S^d} \to S^d$ such that $(p_Y * p_Y')_X \simeq_{\text{sfh}} c^* p_{S^d}$. Let $p'_{S^d}\colon E'_{S^d} \to S^d$ be a $(2k+l-1)$-spherical fibration with characteristic function $\text{char}_{2k+l}(S^d)(p'_{S^d}) = -\text{char}_{2k+l}(S^d)(p_{S^d})$, see (6.15).

Let
$$\beta\colon \pi_{d-1}(G(k)) \times \pi_{d-1}(G(k)) \to \pi_{d-1}(G(2k))$$
be the pairing defined in (6.16). Next we show that $\beta(x, -x) = 0$ holds for every $\pi_{d-1}(G(k))$. Unravelling the definition of β we see that it suffices to show $[s_0 \circ x] = [s_1 \circ x]$ for every $[x] \in \pi_{d-1}(G(k))$ for the homomorphisms induced by s_0 and s_1 appearing in the definition of β. In the proof of Lemma 6.104 we will construct a homotopy, see (6.105),
$$h\colon S^{k-1} * S^{k-1} \times [0,1] \to S^{k-1} * S^{k-1}$$
such that $h_0 = \text{id}$ and $h_1([v, w, t]) = [-w, v, 1-t]$. Since $(\text{id}_{S^{k-1}} * f) \circ h_1 = h_1 \circ (f * S^{k-1})$ holds for any self-homotopy equivalence $f\colon S^{k-1} \to S^{k-1}$, the equality $\beta(x, -x) = 0$ follows. Lemma 6.17 implies that p'_{S^d} is an inverse of p_{S^d}. We set $p_Y^{-1} := p'_Y * c^*(p'_{S^d})$. We get

$$\begin{aligned}
p_Y * p_Y^{-1} &\simeq_{\text{sfh}} p_Y * p'_Y * c^*(p'_{S^d}) \\
&\simeq_{\text{sfh}} c^* p_{S^d} * c^* p'_{S^d} \\
&\simeq_{\text{sfh}} c^*(p_{S^d} * p'_{S^d}) \\
&\simeq_{\text{sfh}} c^* \underline{S^{4k+2l-1}} \\
&\simeq_{\text{sfh}} \underline{S^{4k+2l-1}}
\end{aligned}$$

using the associativity and naturality of the fibrewise join of spherical fibrations. □

Remark 6.24 *A priori* the construction of p^{-1} in Lemma 6.23 need not be unique up to strong fibre homotopy equivalence. However, by Theorem 6.28 below, $[p^{-1}] \in \text{SF}(X)$ defines a unique stable equivalence class of spherical fibrations over X and by Lemma 6.30 below there is a unique fibre homotopy equivalence class p^{-1} of $(k-1)$-spherical fibrations over X for all $k > \dim(X)$.

From the perspective of the structure monoid $G(k)$, stabilising a spherical fibration corresponds to taking the suspension of a homotopy equivalence $S^{k-1} \to S^{k-1}$ to obtain a homotopy equivalence $S^k \to S^k$. Repeating this process j times defines a monoid map $S_k^j\colon G(k) \to G(k+j)$. It induces a map of classifying spaces
$$S_k^j\colon \text{BG}(k) \to \text{BG}(k+j)$$
that is compatible with stabilisation of spherical fibrations; i.e., we have $f_{S^{j-1}*p} = S_k^j \circ f_p$. We define
$$\text{BG} := \text{hocolim}_{k \to \infty} \text{BG}(k) \tag{6.25}$$
and note that, given $[p] \in \text{SF}(X)$, we get a map unique up to homotopy $f_{[p]}\colon X \to \text{BG}$ by taking $f_p\colon X \to \text{BG}(k)$ and composing with the inclusion $\text{BG}(k) \to \text{BG}$. Recall that for a sequence of maps $X_0 \xrightarrow{f_0} X_1 \xrightarrow{f_1} X_2 \xrightarrow{f_2} \cdots$ a model for the homotopy colimit $\text{hocolim}_{k \to \infty} X_k$ is the infinite mapping telescope. The canonical map

6.2 Spherical Fibrations

$\text{hocolim}_{k \to \infty} X_k \to \text{colim}_{k \to \infty} X_k$ is a homotopy equivalence if each map f_i is a cofibration. In contrast to the colimit the homotopy colimit is homotopy invariant in the sense that for a commutative diagram of the shape

$$\begin{array}{ccccccc} X_0 & \xrightarrow{f_0} & X_1 & \xrightarrow{f_1} & X_2 & \xrightarrow{f_2} & \cdots \\ {\scriptstyle u_0} \downarrow \simeq & & {\scriptstyle u_1} \downarrow \simeq & & {\scriptstyle u_1} \downarrow \simeq & & \\ Y_0 & \xrightarrow{g_0} & Y_1 & \xrightarrow{g_1} & Y_2 & \xrightarrow{g_2} & \cdots \end{array}$$

such that all vertical arrows are (weak) homotopy equivalences, the induced map

$$\text{hocolim}_{k \to \infty} u_k : \text{hocolim}_{k \to \infty} X_k \to \text{hocolim}_{k \to \infty} X_k$$

is a (weak) homotopy equivalence, and we have a canonical isomorphism

$$\text{colim}_{k \to \infty} \pi_i(X_k) \xrightarrow{\cong} \pi_i\big(\text{hocolim}_{k \to \infty} X_k\big). \tag{6.26}$$

Exercise 6.27 Give an example of a sequence $X_0 \xrightarrow{f_0} X_1 \xrightarrow{f_1} X_2 \xrightarrow{f_2} \cdots$ for which the canonical map $\text{hocolim}_{k \to \infty} X_k \to \text{colim}_{k \to \infty} X_k$ is not a homotopy equivalence and $\text{colim}_{k \to \infty} \pi_i(X_k) \to \pi_i(\text{colim}_{k \to \infty} X_k)$ is a not a bijection.

The exterior fibrewise join $q_{S^{k-1}} *_e q_{S^{k-1}} : E_{S^{k-1}} *_e E_{S^{k-1}} \to BG(k) \times BG(k)$ of the universal $(k-1)$-spherical fibration with itself is classified by a map

$$\mu_k : BG(k) \times BG(k) \to BG(2k).$$

The maps μ_k fit together to define a map $\mu: BG \times BG \to BG$ which turns BG into a homotopy associative and homotopy commutative H-space [303, page 60]. By construction this H-space structure reflects the fibrewise join of stable spherical fibrations. Moreover, up to homotopy $\mu: BG \times BG \to BG$ can be realised as the multiplication map for a loop space structure on BG, see [40, Example 6.39 (m)]. As a consequence $[X, BG]$ becomes an abelian group for any CW-complex X and there is a map $i: BG \to BG$ (which corresponds to reversing the parameter of a loop), such that $[f \circ i]$ is the inverse of $[f]$ for $f: X \to BG$ a map from a finite CW-complex to X. Hence we get the following theorem.

Theorem 6.28 (Classification of stable homotopy classes of spherical fibrations over finite CW complexes) Let X be a finite CW-complex. Then the map

$$SF(X) \to [X, BG], \quad [p]_s \mapsto [f_p]$$

is an isomorphism of abelian groups.

The following well-known result follows from the long homotopy sequence of the fibration $O(k) \to O(k+1) \to S^k$.

Lemma 6.29 *Fix a natural number $k \geq 1$. Let X be a CW-complex with $\dim(X) \leq k$. Then the natural stabilisation maps $\mathrm{VB}_k(X) \to \mathrm{VB}(X)$ and $\mathrm{VB}_{k+1}(X) \to \mathrm{VB}(X)$ are onto and bijective respectively, and the same holds for the natural maps $[X, \mathrm{BO}(k)] \to [X, \mathrm{BO}]$ and $[X, \mathrm{BO}(k+1)] \to [X, \mathrm{BO}]$.*

The analogue for spherical fibrations is the following lemma.

Lemma 6.30 *Fix a natural number $k \geq 3$. Let X be a CW-complex with $\dim(X) \leq k$. Then the natural stabilisation maps $\mathrm{SF}_k(X) \to \mathrm{SF}(X)$ and $\mathrm{SF}_{k+1}(X) \to \mathrm{SF}(X)$ are onto and bijective respectively, and the same holds for the natural maps $[X, \mathrm{BG}(k)] \to [X, \mathrm{BG}]$ and $[X, \mathrm{BG}(k+1)] \to [X, \mathrm{BG}]$.*

Proof. We conclude from Theorem 6.28 that it suffices to prove that the map $\mathrm{BG}(k) \to \mathrm{BG}(k+1)$ is k-connected. Lemma 6.13 implies that this is equivalent to the assertion that $G(k) \to G(k+1)$ is $(k-1)$-connected. This is proved as follows.

Let $y_0 \in S^k$ denote a base point and let $F(k) \subset G(k+1)$ be the submonoid of self-homotopy equivalences $f: S^k \to S^k$ with $f(y_0) = y_0$. There is a fibration sequence, see for example [279, Lemma 3.1 on page 46],

$$F(k) \to G(k+1) \xrightarrow{\mathrm{ev}} S^k,$$

where $\mathrm{ev}(f) = f(y_0)$. Hence the inclusion $F(k) \to G(k+1)$ is $(k-1)$-connected. Writing $S^k = \{\pm y_0\} * S^{k-1}$, the stabilisation map $G(k) \to G(k+1)$ factors through $F(k) \subset G(k+1)$. By [170, Remark 7.7] the map $G(k) \to F(k)$ is $(2k-3)$-connected, see also [201, Corollary 2.2], where the claim is proved for $(2k-4)$. Since $2k-4 \geq k-1$ for $k \geq 3$, the map $G(k) \to G(k+1)$ is $(k-1)$-connected. □

We now briefly discuss the computation of the homotopy groups of BG. Define the topological monoid

$$G := \mathrm{colim}_{k \to \infty} G(k). \tag{6.31}$$

Since $G(k) \to G(k+1)$ is a closed embedding, we get from (6.26) and Lemma 6.13 isomorphisms

$$\pi_i(\mathrm{BG}) \cong \mathrm{colim}_{k \to \infty} \pi_i(\mathrm{BG}(k));$$
$$\pi_i(G) \cong \mathrm{colim}_{k \to \infty} \pi_i(G(k));$$
$$\pi_{i+1}(\mathrm{BG}(k)) \cong \pi_i(G(k));$$
$$\pi_{i+1}(\mathrm{BG}) \cong \pi_i(G).$$

Note that we may regard $F(k)$ as a subspace of the space of loops $\Omega^k S^k$ and that we have taken $\mathrm{id}_{S^k}: S^k \to S^k$ as the base point of $G(k+1)$ and $F(k)$. For the computation of homotopy groups, it is easier to work with the base point of the constant loop $* \in \Omega^k S^k$. However, since $\Omega^k S^k$ is an H-space, all of its components have the same homotopy type and consequently isomorphic homotopy groups. We have for $i \geq 1$

$$\pi_{i+k}(S^k) \cong \pi_i(\Omega^k S^k, *) \cong \pi_i(\Omega^k S^k, \mathrm{id}_{S^k}) = \pi_i(F(k)), \tag{6.32}$$

where the first isomorphism is the standard property of the homotopy groups of the constant component of a loop space. Since the canonical map $F(k) \to G(k+1)$ is $(k-1)$-connected as explained in the proof of Lemma 6.29, we get for $1 \le i < k-1$ isomorphisms $\pi_{i+k}(S^k) \cong \pi_i(G(k+1), \mathrm{id}_{S^k})$. The stable i-stem is defined by

$$\pi_i^s := \pi_i^s(S^0) := \mathrm{colim}_{k \to \infty} \pi_{i+k}(S^k). \tag{6.33}$$

Taking the colimit for $k \to \infty$ gives an isomorphism for $i \ge 1$

$$\pi_{i+1}(\mathrm{BG}) \cong \pi_i(G) \cong \pi_i^s. \tag{6.34}$$

Exercise 6.35 Show $\pi_1(\mathrm{BG}) \cong \mathbb{Z}/2$ and $\pi_0^s \cong \mathbb{Z}$.

For example when $X = S^{i+1}$, we have $\mathrm{SF}(S^{i+1}) = [S^{i+1}, \mathrm{BG}] \cong \pi_i^s$, which, in general, is non-zero. Hence there are non-trivial spherical fibrations. However, Serre has proven that π_i^s is finite for $i > 0$ and so we have that $\pi_i(\mathrm{BG})$ is finite for all i. Elementary obstruction theory then gives the following

Corollary 6.36 *Let X be a finite CW-complex. Then $\mathrm{SF}(X) \cong [X, \mathrm{BG}]$ is a finite abelian group.*

6.3 The Thom Isomorphism

Recall that the classical Thom isomorphism describes a relation between the (co-)homology of a base space and of the Thom space of a vector bundle, see [308, Chapter 10]. In this section we will show that an analogous statement holds for spherical fibrations. We will relax the orientability condition at the cost of introducing local coefficients. We will also obtain a chain level version that is better suited for the proof of the existence and uniqueness of the Spivak normal fibration.

A reader who is not familiar with local coefficient systems and the associated homology groups may pass directly to Theorem 6.52 and from the very beginning make the choice (SBP), see Notation 6.47 and 6.49, and take the point of view analogous to that explained in Convention 5.49.

Let $p \colon E \to X$ be a $(k-1)$-spherical fibration. Put $E_x := p^{-1}(x)$. It comes with a preferred infinite cyclic local coefficient system

$$O_p \colon \Pi(X) \to \mathbb{Z}\text{-MOD}, \quad x \mapsto H_k(\mathrm{cone}(E_x), E_x; \mathbb{Z}). \tag{6.37}$$

It sends a morphism $[w] \colon x \to y$ to the isomorphism of infinite cyclic groups $H_k(\mathrm{cone}(t_{[w]}), t_{[w]}; \mathbb{Z}) \colon H_k(\mathrm{cone}(E_x), E_x; \mathbb{Z}) \to H_k(\mathrm{cone}(E_y), E_y; \mathbb{Z})$ induced by the homotopy class $[t_{[w]}]$ of maps $E_x \to E_y$ coming from the fibre transport along w, see for instance [393, 15.12 on page 343]. One easily checks that $w_1(O_p)$ agrees with the previously defined first Stiefel–Whitney class $w_1(p)$ of p and that for a k-dimensional vector bundle ξ there is a canonical isomorphism of infinite cyclic

154   6 The Spivak Normal Structure

local coefficient systems
$$O_\xi \xrightarrow{\cong} O_{ps\xi} \tag{6.38}$$
for the $(k-1)$-spherical fibration $ps\xi$ associated to ξ, where O_ξ has been defined in (5.21).

For $x \in X$ let $i(x): \{x\} \to X$ be the inclusion. Recall that $p_{DE}: DE \to X$ is the canonical projection. Hence we get an infinite local coefficient system $p_{DE}^* O_p$. We get from $i(x)$ a homomorphism

$$H^k(DE, E; p_{DE}^* O_p) \to H^k(\mathrm{cone}(E_x), E_x; i(x)^* p_{DE}^* O_p).$$

Since $i(x)^* p_{DE}^* O_p$ is the constant infinite cyclic local coefficient system on $\mathrm{cone}(E_x)$ with value $O_p(x)$, we get a canonical isomorphism

$$H^k(\mathrm{cone}(E_x), E_x; i(x)^* p_{DE}^* O_p) \xrightarrow{\cong} H^k(\mathrm{cone}(E_x), E_x; O_p(x)).$$

The universal coefficient theorem yields a natural isomorphism

$$H^k(\mathrm{cone}(E_x), E_x; O_p(x)) \xrightarrow{\cong} \hom_{\mathbb{Z}}(H_k(\mathrm{cone}(E_x), E_x), O_p(x))$$
$$\xrightarrow{\cong} \hom_{\mathbb{Z}}(O_p(x), O_p(x)).$$

Putting these isomorphisms together yields a homomorphism

$$\alpha_p(x): H^k(DE, E; p_{DE}^* O_p) \to \hom_{\mathbb{Z}}(O_p(x), O_p(x)). \tag{6.39}$$

Consider another point $y \in X$ and a path $v: [0, 1] \to X$ from x to y. Let h be a solution of the following homotopy lifting problem, where $j_0: E_x \to E_x \times [0, 1]$ sends x to $(x, 0)$ and $\mathrm{pr}: E_x \times [0, 1] \to [0, 1]$ is the projection

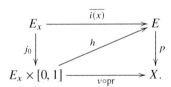

Then $h_1: E_x \to E_y$ is a representative of the fibre transport $t_{[v]}$ and using the homotopy h one shows that the following diagram commutes

6.3 The Thom Isomorphism

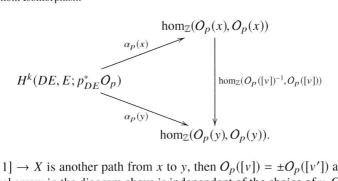

If $v' \colon [0, 1] \to X$ is another path from x to y, then $O_p([v]) = \pm O_p([v'])$ and hence the vertical arrow in the diagram above is independent of the choice of v. Obviously it sends $\mathrm{id}_{O_p(x)}$ to $\mathrm{id}_{O_p(y)}$. If we compose $\alpha_p(x)$ with the inverse of the isomorphism $\mathbb{Z} \xrightarrow{\cong} \hom_{\mathbb{Z}}(O_p(x), O_p(x))$ sending n to $n \cdot \mathrm{id}_{O_p(x)}$, we get a homomorphism

$$\alpha_p \colon H^k(DE, E; p_{DE}^* O_p) \to \mathbb{Z}, \tag{6.40}$$

which is, by the observations above, indeed independent of the choice of $x \in X$. By taking each path component separately, we obtain a homomorphism for a not necessarily connected X

$$\alpha_p \colon H^k(DE, E; p_{DE}^* O_p) \to H^0(X; \mathbb{Z}) = \prod_{\pi_0(X)} \mathbb{Z}. \tag{6.41}$$

Lemma 6.42 *Let $p \colon E \to X$ and $p' \colon E' \to X'$ be $(k-1)$-spherical fibrations over CW-complexes X and X'.*

(i) *Consider the following commutative diagram*

$$\begin{array}{ccc} E & \xrightarrow{\overline{f}} & E' \\ p \downarrow & & \downarrow p' \\ X & \xrightarrow{f} & X' \end{array}$$

such that $\overline{f}_x \colon E_x \to E'_{f(x)}$ is a homotopy equivalence for every $x \in X$. Then there is preferred and natural isomorphism $u \colon f^* O_{p'} \xrightarrow{\cong} O_p$ of infinite cyclic local coefficient systems over X. Moreover, $\overline{f}$ and u define a homomorphism

$$H^k(\mathrm{cone}(\overline{f}), \overline{f}; u) \colon H^k(DE', E'; p_{DE'}^* O_{p'}) \to H^k(DE, E; p_{DE}^* O_p)$$

such that the following diagram commutes

$$\begin{array}{ccc} H^k(DE', E'; p^*_{DE'} O_{p'}) & \xrightarrow{H^k(\operatorname{cone}(\overline{f}), \overline{f}; u)} & H^k(DE, E; p^*_{DE} O_p) \\ \alpha_{p'} \downarrow & & \downarrow \alpha_p \\ H^0(X'; \mathbb{Z}) & \xrightarrow{H^0(f; \mathbb{Z})} & H^0(X; \mathbb{Z}); \end{array}$$

(ii) *We have* $H^i(DE, E; p^*_{DE} O_p) = \{0\}$ *for* $i \leq k-1$;
(iii) *The homomorphism α_p is bijective;*

Proof. (i) In order to define u, we have to specify for every $x \in X$ an isomorphism $u_x : O_{p'}(f(x)) = H_k(\operatorname{cone}(E'_{f(x)}), E'_{f(x)}) \xrightarrow{\cong} O_p(x) = H_k(\operatorname{cone}(E_x), E_x)$. Since $(\operatorname{cone}(f_x), f_x) : (\operatorname{cone}(E_x), E_x) \to (\operatorname{cone}(E'_{f(x)}), E'_{f(x)})$ is a homotopy equivalence of pairs, we can take the inverse of the induced isomorphism $H_k(\operatorname{cone}(f_x), f_x)$. Now the commutativity of the diagram follows from the definitions using the fact that the homomorphism

$$\hom_{\mathbb{Z}}(O_{p'}(f(x)), O_{p'}(f(x))) \to \hom_{\mathbb{Z}}(O_p(x), O_p(x)), \quad v \mapsto u_x \circ v \circ u_x^{-1}$$

sends $\mathrm{id}_{O_{p'}(f(x))}$ to $\mathrm{id}_{O_p(x)}$.

(ii) and (iii) We first prove the claim for finite-dimensional X by induction over the dimension d. The induction beginning with $d = -1$ is trivial since then $X = \emptyset$. The induction step is done as follows. Choose a pushout

$$\begin{array}{ccc} \coprod_{i \in I} S^{d-1} & \longrightarrow & X_{d-1} \\ \downarrow & & \downarrow \\ \coprod_{i \in I} D^d & \longrightarrow & X_d. \end{array} \tag{6.43}$$

If we pullback the fibration over X to $\coprod_{i \in I} S^{d-1}$, $\coprod_{i \in I} D^d$, X_{d-1} and X_d, we obtain a pushout on the level of total spaces

$$\begin{array}{ccc} E_S & \longrightarrow & E_{d-1} \\ \downarrow & & \downarrow \\ E_D & \longrightarrow & E_d \end{array}$$

with the left vertical arrow a cofibration again, see [431, Theorem 7.14 in I.7 on page 33]. Denote the resulting $(k-1)$-spherical fibrations by p_S, p_D, p_{d-1}, and p_d. Up to homotopy one can arrange that this is a pushout of CW-complexes with the left vertical arrow an inclusion of CW-complexes, see for instance [152, Section 3]. Now one extends this to a pushout of pairs of finite CW-complexes

6.3 The Thom Isomorphism

$$\begin{array}{ccc} (DE_S, E_S) & \longrightarrow & (DE_{d-1}, E_{d-1}) \\ \downarrow & & \downarrow \\ (DE_D, E_D) & \longrightarrow & (DE_d, E_d). \end{array} \quad (6.44)$$

Since the inclusion $\{\bullet\} \to D^d$ is a homotopy equivalence, we conclude from assertion (i) that the claim holds for E_D if it holds for the trivial spherical fibration $S^{k-1} \to \{\bullet\}$ over $\{\bullet\}$, where the claim is obviously true. By induction hypothesis the claim is true for E_S and E_{d-1}. Using the Mayer–Vietoris sequence one easily checks that $H^i(DE_d, E_d; p^*_{DE_d}\mathcal{O}_{p_d}) = \{0\}$ for $i \leq k - 1$ and that we obtain a commutative diagram

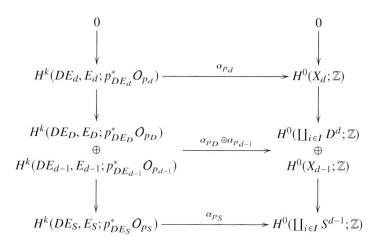

where the rows are exact and the middle and lowermost horizontal arrow are isomorphisms. Hence the uppermost horizontal arrow is an isomorphism.

Finally we treat arbitrary CW-complexes. We can write X as the union of its skeletons X_n. Let $p_n \colon E_n \to X_n$ be the restriction of p to X_n. Then the natural maps

$$H^k(DE, E; p^*_{DE}\mathcal{O}_p) \xrightarrow{\cong} \text{invlim}_{n\to\infty} H^k(DE_n, E_n; p^*_{DE_n}\mathcal{O}_{p_n});$$
$$H^0(X; \mathbb{Z}) \xrightarrow{\cong} \text{invlim}_{n\to\infty} H^0(X_n; \mathbb{Z}),$$

are isomorphisms. Since the inverse limit of isomorphisms is again an isomorphism, the proof of Lemma 6.42 is finished. □

Definition 6.45 Let $p \colon E \to X$ be a $(k-1)$-spherical fibration. Its *intrinsic Thom class*

$$\overline{U_p} \in H^k(DE, E; p^*_{DE}\mathcal{O}_p)$$

is defined to be the preimage of $1 \in H^0(X; \mathbb{Z})$ under the isomorphism

$$\alpha_p \colon H^k(DE, E; p_{DE}^* O_p) \to H^0(X; \mathbb{Z}) = \prod_{\pi_0(X)} \mathbb{Z}$$

of (6.41).

Exercise 6.46 Consider the situation of Lemma 6.42 (i). Show that the homomorphism

$$H^k(\mathrm{cone}(\overline{f}), \overline{f}; u) \colon H^k(DE', E'; p_{DE'}^* O_{p'}) \to H^k(DE, E; p_{DE}^* O_p)$$

sends $\overline{U}_{p'}$ to $\overline{U}_p$.

For further statements it is again convenient to choose base points and work over the group ring instead of the fundamental groupoid. Analogously to choice (BP) for the universal covering of Notation 5.6, we will often make the following choice

Notation 6.47 (Choice (SBP)) Suppose that the CW-complex X is connected. Choose a model for the universal covering $q_X \colon \widetilde{X} \to X$. Its pullback along $p_{DE} \colon DE = \mathrm{cyl}(p) \to X$

$$\begin{array}{ccc} \widetilde{DE} & \xrightarrow{\widetilde{p_{DE}}} & \widetilde{X} \\ {\scriptstyle q_{DE}}\downarrow & & \downarrow{\scriptstyle q_X} \\ DE & \xrightarrow{p_{DE}} & X \end{array}$$

is a model for the universal covering of DE which we will use. Let $\widetilde{E}$ be the preimage of E under q_{DE}. (Note that this is not necessarily the universal covering of E unless $k \geq 3$.) Choose base points $\widetilde{x} \in \widetilde{X}$ and $x \in X$ with $q_X(\widetilde{x}) = x$ and an isomorphism $o_x \colon O_p(x) \xrightarrow{\cong} \mathbb{Z}$.

With these choices put $\pi = \pi_1(DE, x) = \pi_1(X, x)$ and let $w \colon \pi \to \{\pm 1\}$ be the group homomorphism determined by $w_1(O_p)$.

With these choices we obtain an identification

$$H_\pi^k(DE, E; \mathbb{Z}^w) \xrightarrow{\cong} H^k(DE, E; p_{DE}^* O_p),$$

and we denote by

$$U_p \in H^k(DE, E; \mathbb{Z}^w) \tag{6.48}$$

the preimage of $\overline{U}_p$ under this identification. If we pass from o_x to $-o_x$, then we pass from U_p to $-U_p$.

Notation 6.49 For the remainder of this chapter, we denote by $C^{n-*}(\widetilde{X})$, $C^{n+k-*}(\widetilde{DE})$ and $C^{n+k-*}(\widetilde{DE}, \widetilde{E})$ the $\mathbb{Z}\pi$-chain complexes given by the dual of the $\mathbb{Z}\pi$-chain complexes $C_*(\widetilde{X})$, $C_*(\widetilde{DE})$ and $C_*(\widetilde{DE}, \widetilde{E})$ with respect to the *untwisted* involution $\mathbb{Z}\pi \to \mathbb{Z}\pi$, sending $\sum_{g \in \pi} \lambda_g \cdot g$ to $\sum_{g \in \pi} \lambda_g \cdot g^{-1}$. (This is in contrast to some other chapters in this book.)

Given any $\mathbb{Z}\pi$-chain complex D_*, we write D_*^w for the $\mathbb{Z}\pi$-chain complex $D_* \otimes_{\mathbb{Z}} \mathbb{Z}^w$ where π acts diagonally.

6.3 The Thom Isomorphism

Now, using a relative version of (5.42), we obtain the following composite of $\mathbb{Z}\pi$-chain maps, unique up to $\mathbb{Z}\pi$-chain homotopy,

$$C_*(\widetilde{DE}, \widetilde{E}) \xrightarrow{U_p \cap -} \Sigma^k C_*(\widetilde{DE})^w \xrightarrow{\Sigma^k C_*(\widetilde{Dp})^w} \Sigma^k C_*(\widetilde{X})^w. \quad (6.50)$$

Taking duals yields a $\mathbb{Z}\pi$-chain map

$$C^{n-*}(\widetilde{X})^w \xrightarrow{C^{n-*}(\widetilde{Dp})^w} C^{n-*}(\widetilde{DE})^w \xrightarrow{-\cup U_p} C^{n+k-*}(\widetilde{DE}, \widetilde{E}). \quad (6.51)$$

Theorem 6.52 (Thom chain homotopy equivalence) *Let $p\colon E \to X$ be a $(k-1)$-spherical fibration of connected finite CW-complexes with the first Stiefel–Whitney class $w\colon \pi = \pi_1(X) \to \{\pm 1\}$.*
Then the $\mathbb{Z}\pi$-chain maps (6.50) and (6.51) are $\mathbb{Z}\pi$-chain homotopy equivalences.

Proof. Since the $\mathbb{Z}\pi$-chain map (6.51) is essentially the $\mathbb{Z}\pi$-dual of the $\mathbb{Z}\pi$-chain map (6.50), it remains to show that the $\mathbb{Z}\pi$-chain map (6.50) is a $\mathbb{Z}\pi$-chain homotopy equivalence. This is obviously true if p is fibre homotopy equivalent to the trivial fibration $\underline{S^{k-1}}$. The general case is proved by induction over the number of cells of X, as carried out next.

The induction beginning with $X = \emptyset$ is trivial, the induction step is done as follows. As in the proof of Lemma 6.42, we obtain the pushout (6.44), where all spaces are pairs of finite CW-complexes and the left arrow is an inclusion of CW-pairs. Let $\widetilde{DE} \to DE$ be the universal covering. We can pull back $\widetilde{DE} \to DE$ to all of the pairs in the pushout (6.44), and we obtain a pushout

$$\begin{array}{ccc} (\overline{DE_S}, \overline{E_S}) & \longrightarrow & (\overline{DE_{d-1}}, \overline{E_{d-1}}) \\ \downarrow & & \downarrow \\ (\overline{DE_D}, \overline{E_D}) & \longrightarrow & (\overline{DE_d}, \overline{E_d}) \end{array}$$

of free π-CW-complexes. It projects to the pushout diagram

$$\begin{array}{ccc} \overline{S^{d-1}} & \longrightarrow & \overline{X_{d-1}} \\ \downarrow & & \downarrow \\ \overline{D^d} & \longrightarrow & \overline{X_d} \end{array}$$

obtained from the pushout (6.43) by pulling back $\widetilde{X} \to X$. The induction hypothesis applies for the fibration over X_{d-1} and the claim holds for the fibrations over D^l and S^{l-1} since these are trivial. Note that the claim in Theorem 6.52 for the universal covering automatically implies it for any covering of E. We obtain a commutative diagram of free $\mathbb{Z}\pi$-chain complexes

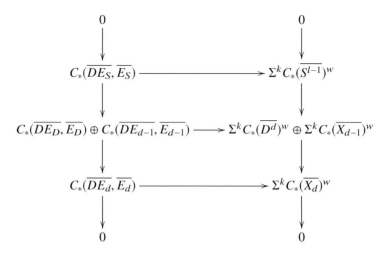

whose columns are exact and whose horizontal arrows are the appropriate versions of the maps (6.50). Since the upper two horizontal arrows are $\mathbb{Z}\pi$-chain homotopy equivalences, the lowermost horizontal arrow is a $\mathbb{Z}\pi$-chain homotopy equivalence. This finishes the proof of Theorem 6.52. □

Remark 6.53 One can actually show that the $\mathbb{Z}\pi$-chain homotopy equivalence appearing in Theorem 6.52 is simple. This makes sense and can be proved by the following argument. Since the Whitehead group of the trivial group and of $\mathbb{Z}$ is trivial, any space F which is homotopy equivalent to S^{k-1} has a preferred simple structure in the sense of [152, Section 3]. Hence the total space of a $(k-1)$-spherical fibration over a finite CW-complex inherits a simple structure. The diagram (6.44) is compatible with these simple structures. The details are left to the reader since we will not need this fact.

The advantage of the formulation in terms of a $\mathbb{Z}\pi$-chain homotopy equivalences is, just as in the case of Poincaré complexes, that this implies the corresponding versions of homology and cohomology with respect to coefficients in any $\mathbb{Z}\pi$-module M. Often we will take M to be $\mathbb{Z}\pi$ or to be $\mathbb{Z}^v$ for some homomorphism $v \colon \pi \to \{\pm 1\}$. We mention the following direct consequence of Theorem 6.52.

Theorem 6.54 (Thom Isomorphism Theorem) *Let $p \colon E \to X$ be a $(k-1)$-spherical fibration of connected finite CW-complexes with first Stiefel–Whitney class $w \colon \pi = \pi_1(X) = \pi_1(E) \to \{\pm 1\}$.*

Then the composites

$$H_p(DE, E; \mathbb{Z}) \xrightarrow{U_p \cap -} H^\pi_{p-k}(DE; \mathbb{Z}^w) \xrightarrow{H^\pi_{p-k}(pDE)} H^\pi_{p-k}(X; \mathbb{Z}^w); \qquad (6.55)$$

$$H^\pi_p(DE, E; \mathbb{Z}^w) \xrightarrow{U_p \cap -} H_{p-k}(DE; \mathbb{Z}) \xrightarrow{H_{p-k}(pDE)} H_{p-k}(X; \mathbb{Z}); \qquad (6.56)$$

$$H^p(X; \mathbb{Z}) \xrightarrow{H^p(pDE)} H^p(DE; \mathbb{Z}) \xrightarrow{-\cup U_p} H^{p+k}_\pi(DE, E; \mathbb{Z}^w); \qquad (6.57)$$

$$H^p_\pi(X; \mathbb{Z}^w) \xrightarrow{H^p_\pi(pDE)} H^p_\pi(DE; \mathbb{Z}^w) \xrightarrow{-\cup U_p} H^{p+k}(DE, E; \mathbb{Z}), \qquad (6.58)$$

6.3 The Thom Isomorphism

are bijective. These maps are called Thom isomorphisms.

Exercise 6.59 Let $f: S^k \to S^k$ be a self-homotopy equivalence of S^k for $k \geq 2$. Let T_f be the mapping torus of f and $p: T_f \to S^1$ be the projection. Let $q: E \to S^1$ be the fibration associated to p. Show that q is a k-spherical fibration with the first Stiefel–Whitney class $w(q): \pi_1(S^1) \to \{\pm 1\}$ given by the homomorphism sending a generator to the degree of f and compute $H_*(T_f; \mathbb{Z})$, $H^*(T_f; \mathbb{Z})$, $H_*^\pi(T_f; \mathbb{Z}^w)$, and $H_\pi^*(T_f; \mathbb{Z}^w)$.

Finally we prove a converse of the Thom Isomorphism Theorem.

Theorem 6.60 (Converse of the Thom Isomorphism Theorem) *Let* $p: E \to X$ *be a fibration of connected finite CW-complexes such that its fibre F is simply connected. Consider an integer $k \geq 3$. Suppose that there exists a homomorphism $w: \pi = \pi_1(X) \to \{\pm 1\}$ and an element $u \in H_\pi^{p+k}(DE, E; \mathbb{Z}^w)$ such that the $\mathbb{Z}\pi$-chain map of (6.50)*

$$\widetilde{T}_*: C_*(\widetilde{DE}, \widetilde{E}) \xrightarrow{U_p \cap -} \Sigma^k C_*(\widetilde{DE})^w \xrightarrow{\Sigma^k C_*(\widetilde{p})} \Sigma^k C_*(\widetilde{X})^w$$

is a $\mathbb{Z}\pi$-chain homotopy equivalence.

Then F is homotopy equivalent to S^{k-1} and w is the first Stiefel–Whitney class of the $(k-1)$-spherical fibration $p: E \to X$.

Proof. The covering $\widetilde{E} \to E$ can also be obtained by the following pullback

$$\begin{array}{ccc} \widetilde{E} & \xrightarrow{\widetilde{p}} & \widetilde{X} \\ \downarrow & & \downarrow \\ E & \xrightarrow{p} & X. \end{array}$$

The map $\widetilde{p}: \widetilde{E} \to \widetilde{X}$ is again a fibration with fibre F. The $\mathbb{Z}\pi$-chain homotopy equivalence $\widetilde{T}_*$ yields after forgetting the π-action a $\mathbb{Z}$-chain homotopy equivalence. Thus we have reduced the claim to the special case where X is a simply connected CW-complex, but is not necessarily a finite CW-complex anymore, and we have a $\mathbb{Z}$-chain homotopy equivalence

$$T_*: C_*(DE, E) \xrightarrow{\simeq} \Sigma^k C_*(X).$$

It is compatible with the two Leray–Serre spectral sequences, see for instance [292, Chapter 5] or [393, Theorem 15.27 on page 350], associated to the pair of fibrations $(DE, E) \to X$ and the trivial fibration $\mathrm{id}: X \to X$. Note that X is simply connected so that we will no longer encounter any twistings. Since X and E are not necessarily finite, we will work in the sequel with homology. The following argument is the homological version of Browder's argument [55, Lemma I.4.3 on page 17]. The chain homotopy equivalence T_* induces isomorphisms

$$H_n(DE, E) \xrightarrow{\cong} H_{n-k}(X)$$

for all $n \in \mathbb{Z}$. This implies $H_n(DE, E) = 0$ for $n < k$ and $H_n(DE, E) \cong \mathbb{Z}$. Since E and DE are simply connected, the relative Hurewicz theorem, see [178, Theorem 4.32 on page 366] or [431, Corollary IV.7.10 on page 181], implies that $\pi_n(DE, E) = 0$ for $n < k$ and $\pi_k(DE, E) \cong \mathbb{Z}$. Since the fibre of $p \colon E \to X$ is F, we conclude that F is $(k-2)$-connected and $\pi_{k-1}(F) \cong \mathbb{Z}$. Hence it remains to show by induction over $l = k, k+1, \ldots$ that $H_j(\text{cone}(F), F) \cong H_{j-1}(F) = 0$ holds for $j \in \mathbb{Z}$ with $k < j \le l$. The induction beginning with $l = k$ is trivial since there is no j with $l < j \le l$. The induction step from $l - 1 \ge k$ to l is done as follows.

Recall that the Leray–Serre spectral sequence converges to $H_{p+q}(DE, E)$ and its E^2-term is $E^2_{p,q} = H_p(X; H_q(\text{cone}(F), F))$. Consider the following commutative diagram

$$\begin{array}{ccc}
H_l(\text{cone}(F), F) & \xrightarrow{0} & H_{l-k}(\text{cone}(F)) \\
\cong \downarrow & & \downarrow \cong \\
E^2_{0,l}(DE, E) = H_0(X; H_l(\text{cone}(F), F)) & \longrightarrow & E^2_{0,l-k}(X) = H_0(X; H_{l-k}(\text{cone}(F))) \\
q \downarrow & & \downarrow = \\
E^\infty_{0,l}(DE, E) & \longrightarrow & E^\infty_{0,l-k}(X) \\
i \downarrow & & \downarrow j \\
H_l(DE, E) & \xrightarrow{\cong} & H_{l-k}(X)
\end{array}$$

where the left column and the right column come from the two spectral sequences mentioned above and the vertical maps are induced by T_*. The map q is the canonical epimorphism and the maps i and j are the canonical inclusions. They are due to the fact that the spectral sequence is a first quadrant spectral sequence and we are looking at the place $(0, l)$ on the vertical axis. The identity appearing on the right side is due to the fact that it belongs to the spectral sequence associated to the trivial fibration $\mathrm{id} \colon X \to X$. The uppermost two vertical arrows are isomorphisms since we are dealing here with untwisted homology groups and X is connected. The lowermost horizontal arrow is bijective by assumption. Hence it remains to show that $q \colon E^2_{0,l}(DE, E) \to E^\infty_{0,l}(DE, E)$ is injective.

Consider the following commutative diagram

$$\begin{array}{ccc}
E^\infty_{p,k}(DE, E) & \longrightarrow & E^\infty_{p,0}(X) \\
i_1 \downarrow & & \downarrow = \\
E^r_{p,k}(DE, E) & \longrightarrow & E^r_{p,0}(X) \\
i_2 \downarrow & & \downarrow = \\
E^2_{p,k}(DE, E) & \longrightarrow & E^2_{p,0}(X).
\end{array}$$

The left vertical arrows are the canonical inclusions, which come from the fact that every ingoing differential at (p, k) starts at zero because of $E^2_{p,q}(DE, E) = H_p(X; H_q(\text{cone}(F), F)) = 0$ for $q < k$. The maps on the right side are identities since they belong to the spectral sequence associated to the trivial fibration id: $X \to X$. As $E^2_{p,q}(DE, E) = H_p(X; H_q(\text{cone}(F), F)) = 0$ holds for $q < k$, we get for the filtration of $H_{p+k}(DE, E)$

$$F_{p,k} H_{p+k}(DE, E) = H_{p+k}(DE, E).$$

Since the map $H_{p+k}(DE, E) \to H_p(X)$ induced by T_* is surjective and compatible with the spectral sequences, the induced map $F_{p,k} H_{p+k}(DE, E) \to F_{p,0} H_p(X)$ and hence the induced map $E^{p,k}_\infty(DE, E) \to E^{p,0}_\infty(X)$, which is the uppermost horizontal arrow in the diagram above, are surjective. The lowermost horizontal arrow in the diagram above

$$E^2_{p,k}(DE, E) = H_p(X; H_k(\text{cone}(F), F)) \xrightarrow{\cong} E^2_{p,0}(X) = H_p(X; H_0(\{\bullet\}))$$

is bijective as it is induced by the isomorphism $H_k(\text{cone}(F), F) \xrightarrow{\cong} H_0(\{\bullet\})$. This implies that the map induced by T_*, which is the middle horizontal arrow in the diagram above,

$$E^r_{p,k}(DE, E) \xrightarrow{\cong} E^r_{p,0}(X)$$

is bijective for all $p \in \mathbb{Z}$ and $r \in \{2, 3, \ldots, \} \amalg \{\infty\}$. Since all differentials in the spectral sequence associated to the trivial fibration id: $X \to X$ are trivial, all differentials starting or ending at $E^r_{p,k}(DE, E)$ are trivial. Since $E^2_{p,q}(DE, E) = H_p(X; H_q(\text{cone}(F), F)) = 0$ for all $q \in \mathbb{Z}$ with $q < l$ and $q \neq k$, all differentials ending at $E^r_{0,l}(DE, E)$ are trivial and hence $q: E^2_{0,l}(DE, E) \to E^r_{0,l}(DE, E)$ is bijective. This finishes the proof that F is homotopy equivalent to S^{k-1}.

From the $\mathbb{Z}\pi$-chain homotopy equivalence $\widetilde{T}_*$, we obtain an isomorphism of $\mathbb{Z}\pi$-modules

$$H_k(\widetilde{DE}, \widetilde{E}) := H_k(C_*(\widetilde{DE}, \widetilde{E})) \xrightarrow{\cong} H_k(\Sigma^k C_*(\widetilde{X})^w) \cong H_0(\widetilde{X})^w \cong \mathbb{Z}^w.$$

Hence we can read off w from the π-action on the infinite cyclic group $H_k(\widetilde{DE}, \widetilde{E})$. Since we know that the first Stiefel–Whitney class of p also gives such a $\mathbb{Z}\pi$-chain homotopy equivalence $\widetilde{T}_*$ by Theorem 6.52, we conclude $w = w_1(p)$. This finishes the proof of Theorem 6.60. □

6.4 The Normal Fibration of a Connected Finite CW-Complex

Next we describe a construction which works for arbitrary finite CW-complexes, and will later give the Spivak fibration. Starting with a finite n-dimensional CW-complex X, we construct a fibration $p_N : S_N \to X$, see (6.61), and a class $c_N \in \pi_{n+k}(\text{Th}(p_N))$, see (6.64)), which are unique in a suitable sense explained in more detail in the text.

Consider a finite n-dimensional CW-complex X. One can always find an embedding $i_X \colon X \hookrightarrow \mathbb{R}^{n+k}$ for $k \geq n+1$ together with a regular neighbourhood $(N, \partial N)$ of the image if i_X. The regular neighbourhood is a compact manifold N with boundary ∂N such that there is a strong deformation retraction $r \colon N \times [0, 1] \to N$ of the inclusion $i_X \colon X \to N$, see [367, Chapter 3]. That is, r is a homotopy relative $i_X(X)$ from $r_0 = \mathrm{id}_N$ to a retraction $r_1 \colon N \to i_X(X)$ of $i_X(X)$. Consequently the inclusion $i_X \colon X \to N$ is a homotopy equivalence. In the case where X is a closed manifold M, such a regular neighbourhood is given by a tubular neighbourhood $(N(M), \partial N(M)) \cong (Dv(M), Sv(M))$.

Let $i_N \colon \partial N \to N$ be the inclusion. Denote by

$$q_N \colon E_N \to N$$

the associated fibration. An element in E_N consists of a pair (y, w) for an element $y \in \partial N$ and a path $w \colon [0, 1] \to N$ in N with $w(0) = y$. The map q_N sends (y, w) to $w(1)$. We have the embedding

$$j_N \colon \partial N \to E_N, \quad y \mapsto (y, c_y),$$

where c_y is the constant path c_y with value y. Define a map

$$r_N \colon E_N \to \partial N, \quad (y, w) \mapsto y.$$

We have $r_N \circ j_N = \mathrm{id}_{\partial N}$. The composite $j_N \circ r_N$ is homotopic to the identity on E_N; a homotopy is given by $E_N \times [0, 1] \to E_N$ sending $((x, w), t)$ to (x, w_t), where w_t is the path defined by $w_t(s) := w(ts)$. In particular, $r_N \colon E_N \to \partial N$ is a homotopy equivalence. We define the *normal fibration of* (N, X)

$$p_N \colon S_N \to X \qquad (6.61)$$

to be the pullback of $p_N \colon E_N \to N$ with i_X:

$$\begin{array}{ccc} S_N = i_X^* E_N & \xrightarrow{i_{X,N}} & E_N \\ {\scriptstyle p_N} \downarrow & & \downarrow {\scriptstyle q_N} \\ X & \xrightarrow{i_X} & N. \end{array}$$

Explicitly an element in S_N is a triple (y, w, x) consisting of $y \in \partial N$, $x \in X$, and a path $w \colon [0, 1] \to N$ with $w(0) = y$ and $w(1) = i_X(x)$. The map p_N sends (y, w, x) to x and the map $i_{X,N}$ sends (y, x, w) to (y, w). Since i_x is a homotopy equivalence and q_N is a fibration, $i_{X,N} \colon S_N \to E_N$ is a homotopy equivalence.

Consider the following diagram

6.4 The Normal Fibration of a Connected Finite CW-Complex

$$\begin{array}{ccc} S_N & \xrightarrow{p_N} & X \\ {\scriptstyle r_N \circ i_{X,N}} \downarrow & & \downarrow {\scriptstyle i_X} \\ \partial N & \xrightarrow{i_N} & N. \end{array}$$

It does not commute but there is a preferred homotopy from $i_N \circ r_N \circ i_{X,N}$ to $i_X \circ p_N$, namely

$$h \colon S_N \times [0,1] \to N, \quad ((y,w,x),t) \mapsto w(t).$$

Since the vertical arrows in the diagram above are homotopy equivalences, we obtain a homotopy equivalence of pairs

$$(u_N, p_N) \colon (\operatorname{cyl}(p_N), S_N) \to (N, \partial N), \tag{6.62}$$

and a pointed homotopy equivalence

$$v_N \colon \operatorname{Th}(p_N) := \operatorname{cone}(p_N \colon S_N \to X) \to N/\partial N. \tag{6.63}$$

If $(\mathbb{R}^{n+k})^c$ is the pointed space given by the one-point compactification of $\mathbb{R}^{n+k}$, then we have the canonical base point preserving *Thom collapse map*

$$c'_N \colon (\mathbb{R}^{n+k})^c \to N/\partial N$$

which is defined to be the identity on $N \setminus \partial N$ and maps any point outside $N \setminus \partial N$ to the base point $\partial N / \partial N$ in $N / \partial N$. Define an element

$$c_N \in \pi_{n+k}(\operatorname{Th}(p_N)) \tag{6.64}$$

by requiring that the isomorphism $\pi_{n+k}(v_N) \colon \pi_{n+k}(\operatorname{Th}(p_N)) \xrightarrow{\cong} \pi_{n+k}(N/\partial N)$ sends c_N to the class represented by $[c'_N]$.

It remains to check the dependency of the pair (p_N, c_N) on N. Suppose we have another embedding $i'_X \colon X \to \mathbb{R}^{n+k}$ and a regular neighbourhood $(N', \partial N')$ of $i_{X'}(X)$. Suppose that there is a compact subset $K \subseteq \mathbb{R}^{n+k}$ and a homeomorphism

$$F \colon \mathbb{R}^{n+k} \to \mathbb{R}^{n+k}$$

satisfying

- F is the identity outside K;
- $F(N) = N'$;
- $F \circ i_X = i'_X$.

Then F induces a fibre homotopy equivalence $S_F \colon p_N \to p'_N$

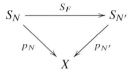

such that the induced map $\pi_{n+k}(\mathrm{Th}(S_F)) \colon \pi_{n+k}(\mathrm{Th}(p_N)) \xrightarrow{\cong} \pi_{n+k}(\mathrm{Th}(p_{N'}))$ sends c_N to $c_{N'}$. Namely, put $S_F(y, w, x) = (F(y), F \circ w, x)$. Since S_F itself is a homeomorphism and obviously compatible with the projections onto X, it is a fibre homotopy equivalence of fibrations over X. One easily checks that the following diagram

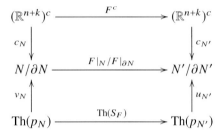

commutes. Since $\pi_{n+k}(F^c)$ is the identity on $\pi_{n+k}((\mathbb{R}^{n+k})^c)$, we conclude that the induced map $\pi_{n+k}(\mathrm{Th}(F)) \colon \pi_{n+k}(\mathrm{Th}(p_N)) \xrightarrow{\cong} \pi_{n+k}(\mathrm{Th}(p_{N'}))$ sends c_N to $c_{N'}$.

The standard uniqueness results about regular neighbourhoods, see [367, Chapter 3] will give the desired map F if $k \geq n + 2$.

We leave it to the reader to interpret and then check the statement that a homotopy F_t of two such maps F and F', which is the identity outside some compact subset L containing the compact subsets K and K' belonging to F and F', induces a fibre homotopy between S_F and $S_{F'}$.

6.5 The Spivak Normal Structure and Its Main Properties

In Section 4.7 we saw that every embedding $i \colon M^n \to \mathbb{R}^{n+k}$ of a closed n-manifold into Euclidean space gives rise to a unique homotopy class of collapse maps

$$c_M \colon S^{n+k} \to \mathrm{Th}(\nu(M))$$

where $\nu(M)$ is the normal k-disk bundle of the embedding i and $h([c_M])$, the image of $[c_M] \in \pi_{n+k}(\mathrm{Th}(\nu(M)))$ under the Hurewicz homomorphism $h \colon \pi_{n+k}(\mathrm{Th}(\nu(M))) \to H_{n+k}(\mathrm{Th}(\nu(M)))$, generates the infinite cyclic group $H_{n+k}(\mathrm{Th}(\nu(M)))$. We want to generalise this to finite Poincaré complexes.

6.5 The Spivak Normal Structure and Its Main Properties

Definition 6.65 (Spivak normal structure) A *Spivak normal $(k-1)$-structure* for a connected finite n-dimensional Poincaré complex (X, O_X) is a pair (p, c) where $p \colon E \to X$ is a $(k-1)$-spherical fibration satisfying $w_1(p) = w_1(X)$, called the *Spivak normal fibration*, and $c \colon S^{n+k} \to \text{Th}(p)$ is a map such that the Hurewicz homomorphism $h \colon \pi_{n+k}(\text{Th}(p)) \to H_{n+k}(\text{Th}(p))$ sends $[c]$ to a generator of the infinite cyclic group $H_{n+k}(\text{Th}(p))$.

This definition extends in the obvious way to the non-connected case by considering each path component separately.

Recall that any $(k-1)$-spherical fibration $p \colon E \to X$ comes with a preferred infinite cyclic local coefficient system O_p and intrinsic Thom class $\overline{U_p} \in H^k(DE, E; p_{DE}^* O_p)$, see Definition 6.45 whereas the choice of O_X is part of the structure of a finite Poincaré complex, see Definition 5.43. The fact that $H_{n+k}(\text{Th}(p))$ is an infinite cyclic group follows from the Thom isomorphism (6.55) appearing in Theorem 6.54. Note that $[c]$ is not determined by its image under the Hurewicz homomorphism $h \colon \pi_{n+k}(\text{Th}(p)) \to H_{n+k}(\text{Th}(p))$.

Example 6.66 (Closed manifolds have canonical Spivak normal structures) Let M be a closed n-manifold and let p_M be the spherical fibration underlying the normal bundle $\nu(M)$ of an embedding $i \colon M^n \to \mathbb{R}^{n+k}$. This together with the collapse map $c_M \colon S^{n+k} \to \text{Th}(\nu(M)) = \text{Th}(p_M)$ is a Spivak normal $(k-1)$-structure for M, where we regard M as a finite n-dimensional Poincaré complex as explained in Theorem 5.51.

We call the pair

$$(p_M, c_M)$$

the *Spivak normal structure of M*.

Lemma 6.67 (i) *In the situation of Definition 6.65 the so-called* Wu formula *holds, i.e., for any generator $[[X]]$ of the infinite cyclic group $H_n(X; O_X)$ there is precisely one isomorphism $o \colon O_p \xrightarrow{\cong} O_X$ of infinite cyclic local coefficient systems such that the element $h([c]) \cap \overline{U_p}$ is sent under the isomorphism*

$$H_n(DE; p_{DE}^* O_p) \xrightarrow{H_n(p_{DE})} H_n(X; O_p) \xrightarrow{H_n(\text{id}_X; o)} H_n(X; O_X)$$

to $[[X]]$;

(ii) *In the situation of Example 6.66 the Wu formula holds for the intrinsic fundamental class and the intrinsic Thom class, i.e., $h([c]) \cap \overline{U_{p_M}}$ is sent under the isomorphism*

$$H_n(DE; p_{DE}^* O_{p_M}) \xrightarrow{H_n(p_{DE})} H_n(M; O_{p_M}) \xrightarrow{H_n(\text{id}_M; o_M)} H_n(M; O_M)$$

to $[[M]]$, where $o_M \colon O_{p_M} \xrightarrow{\cong} O_M$ is the canonical isomorphism of infinite cyclic local coefficient systems coming from Lemma 5.23 (iv) and (6.38).

Proof. By assumption the infinite cyclic group $H_{n+k}(\mathrm{Th}(p)) \cong H_{n+k}(DE, E)$ is generated by $h([c])$. We conclude from Theorem 6.54 that $h([c]) \cap \overline{U_p}$ is a generator of the infinite cyclic group $H_n(DE; p_{DE}^* O_p)$. As the map $H_n(p_{DE}) \colon H_n(DE; p_{DE}^* O_p) \xrightarrow{\cong} H_n(X, O_p)$ is an isomorphism, $H_n(X, O_p)$ is infinite cyclic. Because of the condition $w_1(p) = w_1(X)$ there exists an isomorphism $\kappa \colon O_p \xrightarrow{\cong} O_X$. Now define the isomorphism $o \colon O_p \xrightarrow{\cong} O_X$ by $o = \pm \kappa$ if

$$H_n(DE; p_{DE}^* O_p) \xrightarrow{H_n(p_{DE})} H_n(X; O_p) \xrightarrow{H_n(\mathrm{id}_X; \kappa)} H_n(X; O_X)$$

sends $h([c]) \cap \overline{U_p}$ to $\pm[[X]]$.

(ii) One has to check this near a point $x \in M$ by unravelling the definitions. Namely, there is a small neighbourhood U of x in $\mathbb{R}^{n+k}$, which comes with a preferred identification by the various exponential maps with $DT_x M \times D\nu(M)_x$. Choose an orientation on $T_x M$, which is the same as a choice of a generator $[DT_x M, ST_x M] \in H_n(DT_x M, ST_x M)$, and an orientation on $\nu(M)_x$, which is the same as a choice of a generator $[D\nu(M)_x, S\nu(M)_x]$ in $H_n(D\nu(M)_x, S\nu(M)_x)$, such that the orientations fit with the standard orientation on $T_x \mathbb{R}^{n+k}$ under the isomorphism $T_x M \oplus \nu(M)_x \xrightarrow{\cong} T_x \mathbb{R}^{n+k}$. Then the collapse map $S^{n+k} \to \mathrm{Th}(p_M)$ composed with the projection $\mathrm{Th}(p_M) \to U/\partial U$ defines, using the Hurewicz homomorphism, an element in the abelian group $H_{n+k}(U/\partial U)$ whose image under the isomorphism $H_{n+k}(U/\partial U) \cong H_n(DT_x M, ST_x M) \otimes H_n(D\nu(M)_x, S\nu(M)_x)$ corresponds to $[DT_x M, ST_x M] \otimes [D\nu(M)_x, S\nu(M)_x]$. The details of the proof are left to the reader. □

The next theorem is one of the main results of this chapter.

Theorem 6.68 (Existence and uniqueness of Spivak normal fibrations) *Let X be a finite n-dimensional Poincaré complex.*

(i) *If k is a natural number satisfying $k \geq \dim(X) + 1$, then there exists a Spivak normal $(k-1)$-structure (p, c), namely, the normal fibration constructed in Section 6.4;*

(ii) *For $i = 0, 1$, let $p_i \colon E_i \to X$ and $c_i \colon S^{n+k_i} \to \mathrm{Th}(p_i)$ be Spivak normal $(k_i - 1)$-structures for X.*
Then there exists an integer k with $k \geq k_0, k_1$ such that there is up to strong fibre homotopy precisely one strong fibre homotopy equivalence

$$(\mathrm{id}, \overline{f}) \colon p_0 * \underline{S^{k-k_0}} \to p_1 * \underline{S^{k-k_1}}$$

for which $\pi_{n+k}(\mathrm{Th}(\overline{f}))(s^{k-k_0}([c_0])) = s^{k-k_1}([c_1])$ holds.

Definition 6.69 (Reduction of the Spivak normal fibration to a vector bundle)
Let X be a finite Poincaré complex. We say that the Spivak normal fibration of X has a *vector bundle reduction* if for one (and hence all) Spivak $(k-1)$-structure (p, c) there is a vector bundle ξ over X such that the underlying $(k-1)$-spherical fibration

6.5 The Spivak Normal Structure and Its Main Properties

p is stably fibre homotopy equivalent to the spherical fibration given by the sphere bundle associated to ξ.

Recall that we want to address Problem 6.1, which asks whether a Poincaré complex X is homotopy equivalent to a closed n-dimensional manifold. Next we explain a further necessary condition to a positive answer.

Let $J_k \colon \mathrm{BO}(k) \to \mathrm{BG}(k)$ be the classifying map for the universal k-dimensional vector bundle γ_k over $\mathrm{BO}(k)$ viewed as a spherical fibration. Taking the Whitney sum with $\mathbb{R}$ or the fibrewise join with $S\mathbb{R}$ yields stabilisation maps $\mathrm{BO}(k) \to \mathrm{BO}(k+1)$ and $\mathrm{BG}(k) \to \mathrm{BG}(k+1)$. Recall that we have introduced $\mathrm{BG} = \mathrm{hocolim}_{k \to \infty} \mathrm{BG}(k)$ in (6.25), and we have a natural homotopy equivalence $\mathrm{hocolim}_{k \to \infty} \mathrm{BO}(k) \xrightarrow{\simeq} \mathrm{BO}$. Define

$$J \colon \mathrm{BO} \to \mathrm{BG} \qquad (6.70)$$

by $\mathrm{hocolim}_{k \to \infty} J_k$. Let

$$\mathrm{G/O} \qquad (6.71)$$

be the homotopy fibre of $J \colon \mathrm{BO} \to \mathrm{BG}$. Actually, there is a homotopy fibration sequence

$$\mathrm{BO} \xrightarrow{J} \mathrm{BG} \xrightarrow{q} \mathrm{B(G/O)} \qquad (6.72)$$

such that $\Omega \mathrm{B(G/O)} = \mathrm{G/O}$ and in particular we have

$$\pi_{i+1}(\mathrm{B(G/O)}) \cong \pi_i(\mathrm{G/O}). \qquad (6.73)$$

Let M be a compact manifold of dimension n. Choose a classifying map $t_{M,n} \colon M \to \mathrm{BO}(n)$ for its tangent bundle TM, see Theorem 6.8. It defines an element

$$[t_{M,n}] \in [M, \mathrm{BO}(n)]. \qquad (6.74)$$

Let $i_n \colon \mathrm{BO}(n) \to \mathrm{BO}$ be the map induced by the inclusion $\mathrm{O}(n) \to \mathrm{O}$. So we get a map $t_M := i_n \circ t_{M,n} \colon M \to \mathrm{BO}$, which is well defined up to homotopy. Thus we get an element

$$[t_M] \in [M, \mathrm{BO}]. \qquad (6.75)$$

Given an embedding $i \colon M \to \mathbb{R}^{n+k}$, we can consider its normal bundle $\nu(i)$. It depends on the choice of the embedding. However, its stable isomorphism class is independent of the choice and defines an element

$$[\nu_M] \in \mathrm{VB}(M), \qquad (6.76)$$

which depends only on M. One easily checks that under the bijection of Theorem 6.20 the element $[\nu_M]$ corresponds to $-[t_M]$.

Let X be an n-dimensional finite Poincaré complex. Theorem 6.68 implies that we get from any choice of Spivak normal $(k-1)$-structure a well-defined element

$$[\nu_X] \in \mathrm{SF}(X). \qquad (6.77)$$

Its image under the bijection of Theorem 6.28 is denoted by

$$[s_X] \in [X, \mathrm{BG}]. \tag{6.78}$$

If $f \colon X \to Y$ is a homotopy equivalence of finite Poincaré complexes, we conclude from Theorem 6.68 that $f^* \colon \mathrm{SF}(Y) \to \mathrm{SF}(X)$ sends $[\nu_Y]$ to $[\nu_X]$, or, equivalently, that $[s_Y \circ f] = [s_X]$ holds.

If M is a compact manifold, then the map $S \colon \mathrm{VB}(X) \to \mathrm{SF}(X)$ given by assigning to a vector bundle the associated sphere bundle sends $[\nu_M]$ of (6.55) to the corresponding element $[\nu_M]$ of (6.77) given by considering M as a finite Poincaré complex.

Since we have the homotopy fibration sequence (6.72) we conclude for a finite Poincaré complex X that the element $[s_X] \in [X, \mathrm{BG}]$ lies in the image of $J_* \colon [X, \mathrm{BO}] \to [X; \mathrm{BG}]$ if and only if its image under the map $q_* \colon [X, \mathrm{BG}] \to [X, \mathrm{B}(G/O)]$ is the class of the constant map.

The upshot of the discussion above is the following lemma.

Lemma 6.79 *If the finite Poincaré complex X is homotopy equivalent to a closed manifold, then the following four equivalent conditions are satisfied:*

- *The Spivak normal fibration of X has a vector bundle reduction in the sense of Definition 6.69;*
- *The element $[\nu_X] \in \mathrm{SF}(X)$ lies in the image of $S \colon \mathrm{VB}(X) \to \mathrm{SF}(X)$;*
- *The element $[s_X] \in [X, \mathrm{BG}]$ lies in the image of $J_* \colon [X, \mathrm{BO}] \to [X; \mathrm{BG}]$, $[f] \mapsto [J \circ f]$;*
- *The element $[s_X] \in [X, \mathrm{BG}]$ is sent under $q_* \colon [X, \mathrm{BG}] \to [X, \mathrm{B}(G/O)]$ to the homotopy class of the constant map.*

6.6 Existence of Poincaré Complexes without Vector Bundle Reduction

There exists a finite Poincaré complex X for which the composite of maps $X \xrightarrow{s_X} \mathrm{BG} \xrightarrow{q} \mathrm{B}(G/O)$ is not nullhomotopic, and hence X cannot be homotopy equivalent to a closed manifold by Lemma 6.79. Examples were constructed by Gitler and Stasheff in [165, Theorem 5.2]. Madsen and Milgram [279] constructed a 5-dimensional example by attaching a 5-dimensional cell to $S^2 \vee S^3$ via the attaching map obtained using the Whitehead product and well-known non-trivial elements of the homotopy groups of low-dimensional spheres. Next we give our own construction.

Let $p \colon X \to B$ be a $(k-1)$-spherical fibration. If B is a finite n-dimensional Poincaré complex, then by a theorem of Gottlieb [167] the total space X is homotopy equivalent to a finite $(n + k - 1)$-dimensional Poincaré complex.

Suppose that ξ is a smooth vector bundle $p \colon E \to B$ over a closed smooth manifold B. Then E is a smooth manifold without boundary and there is an obvious

6.6 Existence of Poincaré Complexes without Vector Bundle Reduction

isomorphism $TB \oplus \xi \xrightarrow{\cong} TE|_B$. It induces an isomorphism $p^*TB \oplus p^*\xi \xrightarrow{\cong} TE$ of vector bundles over E. If $S\xi$ is the associated smooth sphere bundle $Sp \colon SE \to B$, then SE is a closed smooth manifold and we obtain an isomorphism of vector bundles over SE

$$(Sp)^*TB \oplus (Sp)^*\xi \xrightarrow{\cong} T(SE) \oplus \mathbb{R}.$$

If we choose embeddings $SE \subseteq \mathbb{R}^l$ and $B \subseteq \mathbb{R}^{l'}$ for large enough l and l', we obtain an isomorphism of vector bundles over SE

$$\nu(SE \subseteq \mathbb{R}^l) \oplus (Sp)^*\xi \oplus \underline{\mathbb{R}^{l'}} \xrightarrow{\cong} (Sp)^*\nu(B \subseteq \mathbb{R}^{l'}) \oplus \underline{\mathbb{R}^{l+1}}.$$

Hence we get in the abelian group $\mathrm{VB}(SE)$

$$[\nu_{SE}] = (Sp)^*([\nu_B]) - (Sp)^*([\xi]). \tag{6.80}$$

Next we prove that the analogous formula holds for the Spivak normal fibration.

Let X be a CW-complex that is homotopy equivalent to a finite Poincaré complex. Choose a homotopy equivalence $f \colon X \to Y$ to a finite Poincaré complex Y. Then we define $[\nu_X] \in \mathrm{SF}(X)$ and $[s_X] \in [X, \mathrm{BG}]$ to be the images of $[\nu_Y] \in \mathrm{SF}(Y)$ and $[s_Y] \in [Y, \mathrm{BG}]$ under the bijections $f^* \colon \mathrm{SF}(Y) \to \mathrm{SF}(X)$ and $f^* \colon [Y, \mathrm{BG}] \to [X, \mathrm{BG}]$ induced by f. This is independent of the choice of Y and $f \colon X \to Y$ by the following exercise.

Exercise 6.81 Let X be a finite n-dimensional CW-complex and Y be a finite n-dimensional Poincaré complex. Let $f \colon X \to Y$ be a homotopy equivalence. Then X is a finite n-dimensional Poincaré complex and the bijections $f^* \colon \mathrm{SF}(Y) \to \mathrm{SF}(X)$ and $f^* \colon [Y, \mathrm{BG}] \to [X, \mathrm{BG}]$ induced by f map $[\nu_Y]$ to $[\nu_X]$ and $[s_Y]$ to $[s_Y]$.

Lemma 6.82 Let $p \colon X \to B$ be a $(k-1)$-spherical fibration over a finite Poincaré complex for $k \geq 2$. Then we get in $\mathrm{SF}(X)$

$$[\nu_X] = p^*([\nu_B]) - p^*([p]).$$

To prove this we shall use the following lemma, whose statement for vector bundles and their Thom spectra can be found as equation (3.2) in [163].

Lemma 6.83 Let $p_i \colon X_i \to B$ be $(k_i - 1)$-spherical fibrations for $i = 1, 2$ over the same base space B.

Then there is a homotopy equivalence

$$\mathrm{cone}\bigl((\mathrm{Th}(p_1) \to \mathrm{Th}(p_1 * p_2)\bigr) \xrightarrow{\cong} \Sigma \mathrm{Th}(p_2^*(p_1))$$

where the map $\mathrm{Th}(p_1) \to \mathrm{Th}(p_1 * p_2))$ comes from the canonical map of spherical fibrations $p_1 \to p_1 * p_2$.

Proof. Let us write $X_1 * X_2$ for the total space of $p_1 * p_2$. We consider the following commutative diagram of spaces

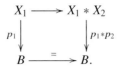

Recall that $\text{Th}(p_1)$ and $\text{Th}(p_1 * p_2)$ are by definition the mapping cones of the corresponding vertical maps. Let C_1 be the mapping cone of the upper horizontal arrow. Obviously $\text{cone}(B)$ is the mapping cone of the lower horizontal arrow. Let C_2 be the mapping cone of the induced map $\text{Th}(p_1) \to \text{Th}(p_1 * p_2)$ and C_3 be the mapping cone of the induced map $C_1 \to \text{cone}(B)$. Then C_2 and C_3 are homotopy equivalent. Since $\text{cone}(B)$ is contractible, there is a homotopy equivalence $\Sigma C_1 \xrightarrow{\simeq} C_2$. Hence it suffices to prove that there is a homotopy equivalence $C_1 \xrightarrow{\simeq} \text{Th}(p_2^*(p_1))$.

Recall the k-disk fibration associated to p_i is $D_i = \text{cyl}(p_i) \to B$. With this notation there is a homeomorphism

$$X_1 * X_2 = (X_1 \times_B D_2) \cup_{X_1 \times X_2} (D_1 \times_B X_2)$$

where $X_1 \times_B D_2$ deformation retracts to $X_1 = X_1 \times_B B$. It follows that there are homotopy equivalences

$$C_1 \simeq (X_1 * X_2)/X_1 \simeq (D_1 \times_B X_2)/(X_1 \times_B X_2).$$

But $(D_1 \times_B X_2)/(X_1 \times_B X_2) = \text{Th}(p_2^*(p_1))$ and so $C_1 \simeq \text{Th}(p_2^*(p_1))$ as required. □

Proof of Lemma 6.82. Let n be the dimension of the finite Poincaré complex B. Choose a Spivak normal $(l-1)$-structure (p_B, c_B) on B in the sense of Definition 6.65. Recall that $p_B \colon X \to B$ is a $(l-1)$-spherical fibration and c_B is an element in $\pi_{n+k}(\text{Th}(p_B))$. We can choose a $(k^- - 1)$-spherical fibration $p^- \colon X' \to B$ such that $p * p^-$ is strongly fibre homotopy equivalent to the trivial $(k + k^- - 1)$-spherical fibration $S^{k+k^- - 1}$ by Theorem 6.23.

We get from Lemma 6.83 applied to the spherical fibrations $p_1 = p_B * p^-$ and $p_2 = p$ over B an exact sequence

$$H_{n+k+k^-+l}(\text{Th}(p_B * p^-)) \to H_{n+k+k^-+l}(\text{Th}(p_B * p^- * p))$$
$$\to H_{n+k+k^-+l}(\Sigma \text{Th}(p^*(p_B * p^-))) \to H_{n+k+k^-+l-1}(\text{Th}(p_B * p^-)).$$

Since $\dim(\text{Th}(p_B * p^-)) = n + l + k^- < n + k + k' + l' - 1$ holds because of the assumption $k \geq 2$, we obtain a commutative diagram whose lower horizontal arrow is a bijection of infinite cyclic groups

$$\begin{array}{ccc} \pi_{n+k+k^-+l}(\text{Th}(p_B * p^- * p)) & \longrightarrow & \pi_{n+k+k^-+l}(\Sigma \text{Th}(p^*(p_B * p^-))) \\ \downarrow & & \downarrow \\ H_{n+k+k^-+l}(\text{Th}(p_B * p^- * p)) & \xrightarrow{\cong} & H_{n+k+k^-+l}(\Sigma \text{Th}(p^*(p_B * p^-))). \end{array}$$

6.6 Existence of Poincaré Complexes without Vector Bundle Reduction 173

There is an element in the left upper corner that is sent to a generator of the infinite cyclic groups appearing in the left lower corner. Its image under the horizontal upper arrow is an element in the right upper corner that is mapped under the right vertical arrow to a generator of the infinite cyclic group appearing in the right lower corner. Since $\Sigma \operatorname{Th}(p^*(p_B * p^-))$ is homeomorphic to $\operatorname{Th}(p^*(p_B * p^- * \underline{S^0}))$, see (6.6), we conclude that $p^*(p_B * p^- * \underline{S^0})$ is a normal Spivak structure on X. This implies the equality in $\operatorname{SF}(X)$

$$p^*([\nu_B]) - p^*([p]) = p^*([p_B * p^- * \underline{S^0}]) = [\nu_X].$$

This finishes the proof of Lemma 6.82. □

Another ingredient we need is some information about the space $B(G/O)$. Recall that we have the homotopy fibration $BO \to BG \to B(G/O)$. Homotopy groups of the spaces involved in low dimensions are depicted in the following table, which is taken from Remark 9.22 in the book [352].

n	1	2	3	4	5	6	7	8	9	10
$\pi_n(G/O)$	0	$\mathbb{Z}/2$	0	$\mathbb{Z}$	0	$\mathbb{Z}/2$	0	$\mathbb{Z} \oplus \mathbb{Z}/2$	$(\mathbb{Z}/2)^2$	$\mathbb{Z}/2 \oplus \mathbb{Z}/3$
$\pi_n(BO)$	$\mathbb{Z}/2$	$\mathbb{Z}/2$	0	$\mathbb{Z}$	0	0	0	$\mathbb{Z}$	$\mathbb{Z}/2$	$\mathbb{Z}/2$
$\pi_n(BG)$	$\mathbb{Z}/2$	$\mathbb{Z}/2$	$\mathbb{Z}/2$	$\mathbb{Z}/24$	0	0	$\mathbb{Z}/2$	$\mathbb{Z}/240$	$(\mathbb{Z}/2)^2$	$(\mathbb{Z}/2)^3$

(6.84)

Example 6.85 (Poincaré complexes without vector bundle reductions) Lemmas 6.79 and 6.82 allow us to construct examples of finite $(5 + l)$-dimensional Poincaré complexes Y_l that are up to homotopy total spaces of $(3 + l - 1)$-spherical fibrations over S^3 and do not have vector bundle reductions of their Spivak normal fibrations for all $l \geq 0$.

From the long exact sequence of homotopy groups associated to the homotopy fibration $BO \xrightarrow{J} BG \xrightarrow{q} B(G/O)$ of (6.72), the isomorphisms (6.73) and the vanishing of $\pi_3(BO)$ it follows that $\pi_3(q) \colon \pi_3(BG) \to \pi_3(B(G/O))$ is injective and its source and target are cyclic of order two. Hence it is actually an isomorphism and the induced map $q_* \colon [S^3, BG] \to [S^3, B(G/O)]$ is bijective.

It will be convenient to work with spherical fibrations $p \colon X \to S^3$ equipped with a section, that is, a map $s \colon S^3 \to X$ such that $p \circ s = \operatorname{id}_{S^3}$. If we take an element $c \in \pi_2(F(k), \operatorname{id}_{S^k})$ and map it to $\pi_2(G(k+1), \operatorname{id}_{S^k})$ via the inclusion $i \colon F(k) \to G(k+1)$, then the $(k-1)$-spherical fibration clutched by $i_*(c)$ as in Remark 6.14, $\operatorname{char}_k(S^3)^{-1}(i_*(c)) \in \operatorname{SF}(S^3)$, admits a section by construction. For $k = 2$, by (6.32) there is an isomorphism $\pi_2(F(2)) \to \pi_4(S^2)$ and by [397, Chapter V, Proposition 5.3] the stabilisation homomorphism $\pi_4(S^2) \to \pi_2^s$ is an isomorphism. The proof Lemma 6.30 then shows that there is a commutative square of isomorphisms

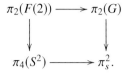

Here, the top horizontal arrow is induced by the inclusion $F(2) \to G(3) \to G$, the left vertical arrow is the isomorphism of (6.32), the bottom horizontal arrow is the stabilisation map, and the right vertical arrow is one of the isomorphisms of (6.34). Hence the non-trivial element in $\pi_3(BG) \cong \mathbb{Z}/2$ is realised by a fibration $S^2 \to X \xrightarrow{p} S^3$ that admits a section.

Consider any integer $l \geq 0$. By stabilising p with the trivial fibration $\underline{S^{l-1}}$, we obtain a fibration $S^{2+l} \to X_l \xrightarrow{p_l} S^3$ such that the stable class $[p_l]$ realises the non-trivial element in $\pi_3(BG) \cong \mathbb{Z}/2$. Since $[\nu_{S^3}] = 0$ holds in $\mathrm{SF}(S^3)$, we conclude from Lemma 6.82 that $[\nu_{X_l}] = -p_l^*[p_l]$ holds in $\mathrm{SF}(X_l)$. Hence the element $-[s_{X_l}] \in [X_l, BG]$ is given by $f_{p_l} \circ p_l$, where $f_{p_l} \colon S^3 \to BG(3+l)$ is the classifying map for p_l. By construction f_{p_l} represents the non-trivial element in $\pi_3(BG)$. Hence $[q \circ f_{p_l}] \in [S^3, B(G/O)]$ is non-trivial. Since p_l admits a section s_l, the image of the homotopy class $[s_l]$ under the map $q_* \colon [X_l, BG] \to [X_l, B(G/O)]$ is non-trivial. We conclude from Lemma 6.79 that there is a finite $(5+l)$-dimensional Poincaré complex Y_l that is homotopy equivalent to X_l and is not homotopy equivalent to a closed manifold.

Finally, the clutching analysis of spherical fibrations over spheres in Remark 6.14 shows that Y_l can be realised as a CW-complex $(S^{2+l} \vee S^3) \cup_f D^{5+l}$ where $f \colon S^{4+l} \to S^{2+l} \vee S^3$ is the attaching map of the single $(5+l)$-cell. In particular, Y_l is a finite $(5+l)$-dimensional Poincaré complex without vector bundle reduction, and it can be shown that Y_l is homotopy equivalent to the exotic example of Madsen and Milgram appearing in [279, page 32].

Remark 6.86 (A self-homotopy equivalence of $S^2 \times S^2$ that is not homotopic to a homeomorphism) We get from Example 6.85 a self-homotopy equivalence $g \colon S^2 \times S^2 \to S^2 \times S^2$ that is not homotopic to homeomorphism, as follows.

Namely, consider the fibration $S^2 \to X \xrightarrow{p} S^3$ appearing in Example 6.85, for which X is not homotopy equivalent to a closed manifold. We can decompose $S^3 = S^3_+ \cup_{S^2_0} S^3_-$ for $S^3_\pm$ the upper and lower hemisphere and $S^2_0 = S^3_+ \cap S^3_-$. Put $X_\pm = p^{-1}(S^3_\pm)$ and $X_0 = p^{-1}(S^2_0)$. Then the restrictions of p to $X_\pm$ and X_0 are again S^2-fibrations with base space $S^3_\pm$ and S^2_0.

The inclusions $i_\pm \colon X_0 \to X_\pm$ are cofibrations. Since $S^3_\pm$ is contractible, there are strong fibre homotopy equivalences $f_\pm \colon X_\pm \to S^3_\pm \times S^2$. We can find a strong fibre homotopy equivalence $g \colon S^2_0 \times S^2 \to S^2_0 \times S^2$ such that $g \circ f_+|_{S^2_0}$ is strongly fibre homotopic to $f_-|_{S^2_0}$. Hence we obtain a commutative diagram

6.7 Characterisation of Poincaré Duality in Terms of the Normal Fibration

$$
\begin{array}{ccc}
X_+ \xleftarrow{i_+} X_0 \xrightarrow{i_-} X_- \\
\downarrow f_+ \qquad \downarrow f_+|_{S_0^2} \qquad \downarrow f_- \\
S_+^3 \times S^2 \xleftarrow{j} S_0^2 \times S^2 \xrightarrow{j \circ g} S_-^3 \times S^2
\end{array}
$$

such that the left square commutes, the right square commutes up to homotopy, the upper left arrow, upper right arrow, and the lower left arrow, which is the inclusion, are cofibrations, and all vertical arrows are homotopy equivalences. By a cofibration argument we can change f_- up to homotopy so that also the right square commutes. Let $Y = S_+^3 \times S^2 \cup_g S_-^3 \times S^2$ be the pushout of the lower row. Then by the pushout property we obtain a map $u: X \to Y$ that is a homotopy equivalence. We conclude that Y is not homotopy equivalent to a closed manifold.

Suppose that g is homotopic to a homeomorphism g'. Then Y is homotopy equivalent to $M = S_+^3 \times S^2 \cup_{g'} S_-^3 \times S^2$, and M is a closed manifold, a contradiction. Hence $g: S^2 \times S^2 \to S^2 \times S^2$ is a self-homotopy equivalence that is not homotopic to a homeomorphism. Actually, g corresponds to the non-trivial element in $\pi_2(F(2))$ appearing in Example 6.85.

6.7 Characterisation of Poincaré Duality in Terms of the Normal Fibration

Theorem 6.87 (Characterisation of Poincaré duality in terms of the normal fibration) *Let X be an n-dimensional finite CW-complex. Consider any embedding $i_X : X \to \mathbb{R}^{n+k}$ together with a regular neighbourhood $(N, \partial N)$ for $k \geq 3$.*

Then X is a finite Poincaré complex if and only if the fibre $p_N^{-1}(x)$ of the normal fibration $p_N : S_N \to X$ of (6.61) is homotopy equivalent to S^{k-1} for every $x \in X$. If this is the case, we have $w_1(X) = w_1(p_N)$.

Proof. We will often implicitly make the choices (BP) and (SBP), see Notation 5.6 and 6.47. Also, let $s^l : \pi_i(X) \to \pi_{i+l}(\Sigma^l X)$ denote the suspension homomorphism.

We can assume without loss of generality that X is connected, otherwise treat each path component separately. Let F be $p_N^{-1}(x)$ for some $x \in X$.

We first show that the fibre F is simply connected. Since $j: X \to N$ is a homotopy equivalence, we have $\pi_i(F) \cong \pi_{i+1}(N, \partial N)$. Since N is a regular neighbourhood of X, the inclusion $\partial N \to N \setminus X$ is a homotopy equivalence. Hence it suffices to prove that $\pi_i(N, N \setminus X)$ vanishes for $i \leq 2$. Since $X \subset N$ is an n-dimensional subcomplex of an $(n + k)$-dimensional manifold, we conclude by a general position argument that for $i < k$ any map of pairs

$$\phi : (D^i, S^{i-1}) \to (N, N \setminus X)$$

is homotopic relative S^{i-1} to a map $\phi : (D^i, S^{i-1}) \to (N \setminus X, N \setminus X)$. Thus we have $\pi_i(N, N \setminus X) = 0$ for all $i < k$. Since $k \geq 3$, the fibre F is simply connected.

Note that in the sequel we can and will identify $\pi = \pi_1(N) = \pi_1(\partial N) = \pi_1(X) = \pi_1(S_N) = \pi_1(DS_N)$ using the isomorphisms induced by the inclusions $X \subseteq N$, $\partial N \subseteq N$, and $E_N \subseteq DE_N$ and the projection $DE_N \to X$. Moreover, the universal coverings $\widetilde{\partial N} \to \partial N$, $\widetilde{N} \to N$, $\widetilde{S_N} \to E_N$, and $\widetilde{DS_N} \to DE_N$ are obtained by restricting or pulling back the universal covering $\widetilde{X} \to X$ with the obvious inclusions or projections.

Choose a fundamental class $[N, \partial N] \in H_{n+k}(N, \partial N)$ for the orientable compact $(n+k)$-dimensional manifold $(N, \partial N)$. By Poincaré duality we obtain a $\mathbb{Z}\pi$-chain homotopy equivalence, unique up to $\mathbb{Z}\pi$-chain homotopy,

$$- \cap [N, \partial N] : C^{n+k-*}(\widetilde{N}, \widetilde{\partial N}) \xrightarrow{\simeq} C_*(\widetilde{N}).$$

The homotopy equivalence $(u_N, p_N) : (DS_N, S_N) \to (N, \partial N)$ of (6.62) yields a $\mathbb{Z}\pi$-chain homotopy equivalence

$$C^{n+k-*}(\widetilde{u_N}, \widetilde{p_N}) : C^{n+k-*}(\widetilde{N}, \widetilde{\partial N}) \xrightarrow{\simeq} C^{n+k-*}(\widetilde{DS_N}, \widetilde{S_N}).$$

The homotopy equivalence $i_X : X \to N$ induces a $\mathbb{Z}\pi$-chain homotopy equivalence

$$C_*(\widetilde{i_X}) : C_*(\widetilde{X}) \xrightarrow{\simeq} C_*(\widetilde{N}).$$

Recall that $C^{n-*}(\widetilde{X})$, $C^{n+k-*}(\widetilde{DE_N}, \widetilde{E_N})$, and $C^{n+k-*}(\widetilde{N}, \widetilde{\partial N})$ are the dual $\mathbb{Z}\pi$-chain complexes with respect to the untwisted involution $\mathbb{Z}\pi \to \mathbb{Z}\pi$ sending $\sum_{g \in \pi} \lambda_g \cdot g$ to $\sum_{g \in \pi} \lambda_g \cdot g^{-1}$, and that we can twist a $\mathbb{Z}\pi$-chain complex by w, i.e., by taking $- \otimes_{\mathbb{Z}} \mathbb{Z}^w$.

Now consider any homomorphism $w : \pi \to \{\pm 1\}$ and any class u in $H^k_\pi(DS_N, S_N; \mathbb{Z}^w)$. Define the element $u_X \in H^n_\pi(X; \mathbb{Z}^w)$ to be the image of $[N, \partial N]$ under the composite

$$H_{n+k}(N, \partial N) \xrightarrow[\cong]{H_n(u_N, p_N)^{-1}} H_{n+k}(DS_N, S_N)$$

$$\xrightarrow{-\cap u} H^\pi_n(DS_N; \mathbb{Z}^w) \xrightarrow[\cong]{H^\pi_n(Dp_N; \mathbb{Z}^w)} H^\pi_n(X; \mathbb{Z}^w).$$

Then we obtain $\mathbb{Z}\pi$-chain maps, unique up to $\mathbb{Z}\pi$-chain homotopy,

$$(- \cup u) \circ C^{n-*}(\widetilde{Dp})^w : C^{n-*}(\widetilde{X})^w \to C^{n+k-*}(\widetilde{DE_N}, \widetilde{E_N}); \qquad (6.88)$$

$$- \cap u_X : C^{n-*}(\widetilde{X})^w \to C_*(\widetilde{X}). \qquad (6.89)$$

All these $\mathbb{Z}\pi$-chain maps fit together to yield an up to $\mathbb{Z}\pi$-chain homotopy commutative diagram of up to $\mathbb{Z}\pi$-chain homotopy uniquely defined $\mathbb{Z}\pi$-chain maps

$$\begin{array}{ccc}
C^{n-*}(\widetilde{X})^w & \xrightarrow{-\cap u_X} & C_*(\widetilde{X}) \\
{\scriptstyle (-\cup u)\circ C^{n-*}(\widetilde{Dp})^w}\downarrow & & \\
C^{n+k-*}(\widetilde{DE_N},\widetilde{E_N}) & & \simeq \Big\downarrow C_*(\widetilde{i_X}) \\
{\scriptstyle C_*(\widetilde{u_N},\widetilde{p_N})}\Big\downarrow \simeq & & \\
C^{n+k-*}(\widetilde{N},\widetilde{\partial N}) & \xrightarrow[-\cap [N,\partial N]]{\simeq} & C_*(\widetilde{N}).
\end{array} \qquad (6.90)$$

Now suppose that X is a Poincaré complex with orientation homomorphism $w\colon \pi \to \{\pm 1\}$ and fundamental class $[X] \in H_n^\pi(X;\mathbb{Z}^w)$. We conclude from the diagram (6.90) that we have the composite of isomorphisms (or their inverses) of infinite cyclic groups

$$H_\pi^k(DS_N, S_N; \mathbb{Z}^w) \xrightarrow{H_\pi^k(u_N, p_N)} H^k(N, \partial N; \mathbb{Z}^w) \xrightarrow{-\cap [N,\partial N]} H_n(N; \mathbb{Z}^w)$$

$$\xleftarrow{H_n(i_X;\mathbb{Z}^w)} H_n(X;\mathbb{Z}^w) \xrightarrow{(-\cap [X])^-} H^0(X).$$

Let $u \in H_\pi^k(DS_N, S_N; \mathbb{Z}^w)$ be the preimage of the generator $1 \in H^0(X)$. Then the class u_X agrees with the fundamental class $[X]$. Again by diagram (6.90) we see that the $\mathbb{Z}\pi$-chain map (6.88) is a $\mathbb{Z}\pi$-chain homotopy equivalence since the $\mathbb{Z}\pi$-chain map (6.89) is a $\mathbb{Z}\pi$-chain homotopy equivalence because of Poincaré duality. The $\mathbb{Z}\pi$-dual of the $\mathbb{Z}\pi$-chain map (6.88) is a $\mathbb{Z}\pi$-chain homotopy equivalence and is up to $\mathbb{Z}\pi$-chain homotopy the $\mathbb{Z}\pi$-chain map appearing in Theorem 6.60. Hence Theorem 6.60 implies that the fibre F is homotopy equivalent to S^{k-1} and w is the first Stiefel–Whitney class of $p_N\colon S_N \to X$.

Now suppose that the fibre F of $p\colon S_N \to X$ is homotopy equivalent to S^{k-1}. Let $w\colon \pi = \pi_1(X) = \pi_1(E) \to \{\pm 1\}$ be the first Stiefel–Whitney class of p_N. We conclude from Theorem 6.52 that $H^k(DS_N, S_N; \mathbb{Z}^w)$ is an infinite cyclic group and for a choice of generator $u \in H^k(DE, E; \mathbb{Z}^w)$ the $\mathbb{Z}\pi$-chain map (6.88) is a $\mathbb{Z}\pi$-chain homotopy equivalence. Then the map (6.89) is a $\mathbb{Z}\pi$-chain homotopy equivalence. This implies that $H_n(X;\mathbb{Z}^w)$ is infinite cyclic with u_X a generator. Hence X is a Poincaré complex with respect to the first Stiefel–Whitney class of p and one can choose u_X as a fundamental class. This finishes the proof of Theorem 6.87. $\square$

6.8 The Existence of the Spivak Normal Fibration

This section is devoted to the proof of assertion (i) of Theorem 6.68, the existence of the Spivak normal fibration, which was originally proven by Spivak in [383]. Assertion (ii), the uniqueness of the Spivak normal fibrations, was proven by Wall in [408, Theorem 3.5], and will be covered in Section 6.10. Browder gives a detailed exposition of the proof of Theorem 6.68 in [55, I.4], but the existence part of the proof only covers the simply connected case.

Proof of Theorem 6.68 (i). We can assume without loss of generality that X is connected, otherwise treat each path component separately. Theorem 6.87 implies that the fibre of the normal fibration $p_N \colon S_N \to X$ is homotopy equivalent to S^{k-1} and $w_1(p_N) = w_1(X)$. Recall that by the Thom Isomorphism Theorem 6.54 the group $H_{n+k}(\mathrm{Th}(p_N))$ is isomorphic to $H_n(X; \mathbb{Z}^{w_1(X)})$ and hence infinite cyclic. It remains to show that the image of the class $c_N \in \pi_{n+k}(\mathrm{Th}(p_N))$ under the Hurewicz homomorphism is a generator of $H_{n+k}(\mathrm{Th}(p_N))$. This follows from the fact that c_N was given as a collapse map and for this collapse map each element in the interior of N is a regular value with precisely one preimage. $\square$

6.9 Spanier–Whitehead Duality

In this subsection we review another type of duality between homology and cohomology, namely the Spanier–Whitehead duality, or S-duality for short. It will crucially be used in the proof of the uniqueness of the Spivak normal structure of a Poincaré complex. Unlike Poincaré duality, which only makes sense for Poincaré complexes or manifolds, S-duality works for any finite CW-complex. However, if that CW-complex turns out to be a Poincaré complex, then we obtain an illuminating relationship between S-duality, the Thom isomorphism of the Spivak normal fibration, and Poincaré duality. This is a key observation that is needed later. We only state the properties of S-duality that we need for our purposes. For more details the reader may consult [55, Section I.4], [393, Chapter 14], or [399, Chapter 7].

In the sequel we will often work with pointed spaces. For a pointed space $X = (X, x)$ define its *reduced homology*

$$\widetilde{H}_n(X) := H_n(X, x). \tag{6.91}$$

Note that for an unpointed space Z we obtain an identification $\widetilde{H}_n(Z_+) = H_n(Z)$. For two pointed spaces X and Y we have the canonical isomorphism $\widetilde{H}_n(X \wedge Y) = H_n(X \times Y, X \vee Y)$. We will need the following versions of the so-called *slant product*:

$$-/-\colon \widetilde{H}_n(X \wedge Y) \otimes \widetilde{H}^k(Y) \to \widetilde{H}_{n-k}(X);$$
$$-/-\colon \widetilde{H}^n(X \wedge Y) \otimes \widetilde{H}_k(Y) \to \widetilde{H}^{n-k}(X). \tag{6.92}$$

In this section we work in the stable category of finite pointed CW-complexes. Let A and B be finite pointed CW-complexes. Two (pointed) maps $f_0 \colon \Sigma^{k_0} A \to \Sigma^{k_0} B$ and $f_1 \colon \Sigma^{k_1} A \to \Sigma^{k_1} B$ for some $k_0, k_1 \geq 0$ are *stably homotopic* if for some $l \geq k_0, k_1$ their (reduced) suspensions $\Sigma^l f, \Sigma^l g \colon \Sigma^l A \to \Sigma^l B$ are (pointed) homotopic. We write

$$\{A, B\} = \{[f] \mid f \colon \Sigma^k A \to \Sigma^k B, k \geq 0\} \tag{6.93}$$

6.9 Spanier–Whitehead Duality

for the *set of stable homotopy classes of pointed maps from A to B*. Note that a class $[f] \in \{A, B\}$ need not be realised by a map $f\colon A \to B$. A map $\alpha\colon A \wedge B \to S^m$ is called an *m-duality map* if for $[S^m]^* \in \widetilde{H}^m(S^m)$ a generator the slant product with $\alpha^*([S^m]^*) \in \widetilde{H}^*(A \wedge B)$ induces an isomorphism

$$\alpha^*([S^m]^*)/-\colon \widetilde{H}_i(A) \xrightarrow{\cong} \widetilde{H}^{m-i}(B)$$

for all i. Here and for this subsection we adopt the convention that, if coefficients are omitted from (co)homology groups, they are untwisted integral coefficients. For every pointed finite CW-complex A, there is a space B such that an m-duality map as above exists for m large enough, see [393, Theorem 14.34 on page 329], or [399, Proposition 7.5.1 on page 176]. It is given by the complement of the interior of a regular neighbourhood of some embedding of A in S^{m+1} for large enough m. Moreover, the stable homotopy type of B is unique, see [393, Corollary 14.25 on page 324]. We call B an *m-dual* of A or just the *S-dual* of A and denote it by A^*.

Exercise 6.94 Find an m-dual of the sphere S^n for $n \le m$.

There is a duality functor, which for a stable homotopy class of maps $[f]$ represented by $f\colon A \to B$ gives a stable homotopy class of maps $[g] = [Df]$ represented by $g\colon B^* \to A^*$. Indeed S-duality gives rise to a contravariant homotopy functor on the stable category of finite pointed CW-complexes

$$D\colon \{A, B\} \to \{B^*, A^*\}, \quad [f] \mapsto [Df].$$

The map D is a bijection, so that for every map $f\colon A \to B$ there is an *m-dual* $Df\colon B^* \to A^*$ whose stable homotopy class is uniquely defined by the stable homotopy class of f, see [393, 14.19 on page 321].

A basic property of S-duality is the following lemma, which follows easily from [393, Lemma 14.29 on page 326].

Lemma 6.95 *Let $\alpha\colon A \wedge A^* \to S^m$ and $\beta\colon B \wedge B^* \to S^m$ be m-duality maps and let $f\colon A \to B$ have the S-dual $Df\colon B^* \to A^*$.*
Then the following diagram commutes

$$\begin{array}{ccc} H_*(A) & \xrightarrow{H_*(f)} & H_*(B) \\ {\scriptstyle \alpha^*([S^m]^*)/-}\big\downarrow & & \big\downarrow{\scriptstyle \beta^*([S^m]^*)/-} \\ H^{m-*}(A^*) & \xleftarrow{H^*(Df)} & H^{m-*}(B^*). \end{array}$$

Following [55, page 24], Lemma 6.95 allows us to dualise and obtain an equivalent formulation of S-duality: An m-duality map $\beta\colon S^m \to A \wedge B$ is a map such that $\beta_*([S^m])/\colon H^i(A) \to H_{m-i}(B)$ is an isomorphism for all i with $[S^m] \in H_m(S^m)$ a generator.

The following theorem of Browder [55, Theorem I.4.14 on page 25], extending a theorem of Spanier [381], gives a criterion for recognising the S-dual Df of a map $f\colon A \to B$.

Theorem 6.96 (Recognising the S-dual of a map) *Let the maps $\alpha\colon S^m \to A \wedge A^*$ and $\beta\colon S^m \to B \wedge B^*$ be m-duality maps.*

Then two maps $f\colon A \to B$ and $g\colon B^* \to A^*$ are m-dual if the following diagram commutes up to homotopy:

$$\begin{array}{ccc} S^m & \xrightarrow{\alpha} & A \wedge A^* \\ \beta \downarrow & & \downarrow f \wedge \mathrm{id}_{A^*} \\ B \wedge B^* & \xrightarrow{\mathrm{id} \wedge g} & B \wedge A^*. \end{array}$$

Next we recall the promised illuminating relation between the Thom isomorphism, S-duality and Poincaré duality, which turns out to play a key role in surgery theory. It was first identified by Milnor and Spanier [307], see also [393, 14.41, 14.42, 14.43].

Theorem 6.97 (Poincaré duality and S-duality) *Let M be a closed manifold of dimension n. Let ν_M be the normal bundle of some embedding $M \to \mathbb{R}^{n+k}$. Let $\alpha\colon S^{n+k} \to \mathrm{Th}(\nu_M)$ be the associated collapse map onto the Thom space $\mathrm{Th}(\nu_M)$. Then $\mathrm{Th}(\nu_M)$ is an $(n+k)$-dual of M_+, and we have the commutative diagram*

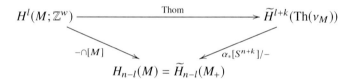

where Thom *comes from the Thom isomorphism, see Theorem 6.54.*

The next theorem gives a generalisation of the first part of Theorem 6.97 to the Poincaré category. This generalisation is due to Wall [408, Theorem 3.3], who in turn was building on Atiyah's extension of the results of Milnor and Spanier [11, Theorem 3.3].

Theorem 6.98 (Generalised Atiyah duality) *Consider a finite n-dimensional Poincaré complex X. Let $p_X\colon E_S \to X$ be a $(k-1)$-spherical fibration that is part of a Spivak normal structure on X in the sense of Definition 6.65. Let $p\colon E \to X$ be an (l_1-1)-spherical fibration and $p^{-1}\colon E \to X$ be an (l_2-1)-spherical fibration such that $p * p^{-1} \sim_{\mathrm{sfh}} S^{l_1+l_2-1}$.*

Then an $(n+k+l_1+l_2)$-dual of $\mathrm{Th}(p)$ is $\mathrm{Th}(p_X * p^{-1})$. The special case $l_1 = l_2 = 0$ shows that an $(n+k)$-dual of X_+ is $\mathrm{Th}(p_X)$.

Proof. Let $\Delta\colon X \to X \times X$ be the diagonal map and consider the spherical fibration $p_X * (p_X * p^{-1})$ over X. We note that $p * (p_X * p^{-1})$ is obtained from $p *_e (p_X * p^{-1})$ over $X \times X$ by pulling back along Δ. Hence there is a map of spherical fibrations

$(\Delta, \overline{\Delta})$: $p * (p_X * p^{-1}) \to p *_e (p_X * p^{-1})$. We fix a strong fibre homotopy equivalence $(\mathrm{id}, \overline{g})$: $\underline{S^{l_1+l_2-1}} * p_X \to p * p_X * p^{-1}$. Consider the map of Thom spaces

$$u \circ \mathrm{Th}(\overline{\Delta}) \circ \mathrm{Th}(\overline{g}) : \mathrm{Th}(\underline{S^{l_1+l_2-1}} * p_X) \xrightarrow{\mathrm{Th}(\overline{g})} \mathrm{Th}(p * p_X * p^{-1})$$
$$\xrightarrow{\mathrm{Th}(\overline{\Delta})} \mathrm{Th}(p *_e (p_X * p^{-1})) \xrightarrow{u} \mathrm{Th}(p) \wedge \mathrm{Th}(p_X * p^{-1})$$

where u is the homeomorphism of (6.5). Taking the canonical identification $\underline{S^{l_1+l_2-1}} * \mathrm{Th}(p_X) = \Sigma^{l_1+l_2} \mathrm{Th}(p_X)$ of (6.6) and precomposing $u \circ \mathrm{Th}(\overline{\Delta}) \circ \mathrm{Th}(\overline{g})$ with the map $\Sigma^{l_1+l_2} c_X : S^{n+k+l_1+l_2} = \Sigma^{l_1+l_2} S^{n+k} \to \Sigma^{l_1+l_2} \mathrm{Th}(p_X)$ yields a map

$$\alpha_X : S^{n+k+l_1+l_2} \to \mathrm{Th}(p) \wedge \mathrm{Th}(p_X * p^{-1}).$$

Next we show that α_X induces an $(n + k + l_1 + l_2)$-duality. Let $\mathcal{O}_p$, $\mathcal{O}_{p_X*p^{-1}}$, and $\mathcal{O}_{p_X}$ be the local coefficient systems induced by p, $p_X * p^{-1}$ and p_X respectively. The proof boils down to the following commutative diagram

$$\begin{array}{ccc}
H_n(X \times X; \mathcal{O}_{p_X}) & \xrightarrow{-/-} & \mathrm{hom}\big(H^i(X; \mathcal{O}_p), H_{n-i}(X; \mathcal{O}_{p_X*p^{-1}})\big) \\
\downarrow \scriptstyle{\mathrm{Thom}} & & \downarrow \scriptstyle{\mathrm{hom(Thom, Thom)}} \\
\widetilde{H}_{n+j}(\mathrm{Th}(p) \wedge \mathrm{Th}(p_X * p^{-1})) & \xrightarrow{-/-} & \mathrm{hom}\big(\widetilde{H}^{i+j}(\mathrm{Th}(p)), \widetilde{H}_{n-i}(\mathrm{Th}(p_X * p^{-1}))\big)
\end{array}$$

where we set $j := k + l_1 + l_2$ and the upper arrow is a twisted version of the slant product, which is well defined because of $\mathcal{O}_{p_X} = \mathcal{O}_p \otimes \mathcal{O}_{p_X*p^{-1}}$. The left hand vertical arrow sends the diagonal class $\Delta_*([X])$ to the class $\alpha_{X*}([S^{n+j}])$. On the other hand the upper horizontal arrow sends $\Delta_*([X])$ to the Poincaré duality isomorphism. Since the Thom maps in the right vertical map are isomorphisms, the slant product with $\alpha_{X*}([S^{n+j}])$ gives the required isomorphism from $\widetilde{H}^{i+j}(\mathrm{Th}(p))$ to $\widetilde{H}_{n-i}(\mathrm{Th}(p_X*p^{-1}))$ for all i. $\square$

6.10 The Uniqueness of the Spivak Normal Fibration

In this section we prove Theorem 6.68 (ii). Before giving the proof we need some basic facts about spherical fibrations and their automorphisms.

Lemma 6.99 *Let $p_i : E_i \to X$ be fibrations over the CW-complex X for $i = 0, 1$. Consider a map $\overline{f} : E_0 \to E_1$ satisfying $p_1 \circ \overline{f} = p_0$. Suppose that for each $x \in X$ the induced map $p_0^{-1}(x) \to p_1^{-1}(x)$ is a homotopy equivalence.*
Then $(\mathrm{id}, \overline{f})$ is a strong fibre homotopy equivalence.

Proof. See [132, Theorem 6.3 and Theorem 6.4]. $\square$

A feature of spherical fibrations, which distinguishes them from vector bundles, is that it is possible to give purely homotopy-theoretic conditions for their triviality and stable triviality. Let $p\colon E \to X$ be a $(k-1)$-dimensional spherical fibration over a connected finite CW-complex X, and let $i_x\colon p^{-1}(x) \to E$ be the inclusion of the fibre over x.

Lemma 6.100 *Let $t\colon E \to S^{k-1}$ be a map such that $t \circ i_{x_0}\colon p^{-1}(x_0) \to S^{k-1}$ is a homotopy equivalence for some $x_0 \in X$, i.e., $t \circ i_{x_0}$ has degree ± 1. Then p is trivial.*

Proof. Consider an $x \in X$ and any path u from x to x_0. Then the fibre transport gives a homotopy equivalence $t_u\colon p^{-1}(x) \to p^{-1}(x_0)$ together with a homotopy $i_{x_0} \circ t_u \simeq i_x$. Hence $t \circ i_x$ is a homotopy equivalence for all $x \in X$.

The product map $p \times t\colon E \to X \times S^{k-1}$ covers the identity $\mathrm{id}\colon X \to X$ and the paragraph above shows that it is fibrewise a homotopy equivalence. Hence it is a strong fibre homotopy equivalence by Lemma 6.99. □

If x is a point in X, the restriction of p to $\{x\}$ is a spherical fibration over a point, and we get a map $\mathrm{Th}(i_x)\colon \mathrm{Th}(p^{-1}(x)) \to \mathrm{Th}(p)$ that is the induced map of Thom spaces. The spherical fibration p is called *coreducible* if X is a connected CW-complex and for one $x \in X$ there is a map $r_x\colon \mathrm{Th}(p) \to \mathrm{Th}(p^{-1}(x))$ such that $r_x \circ \mathrm{Th}(i_x)$ is a homotopy equivalence. The following lemma is part of [11, Proposition 2.8].

Lemma 6.101 *If X is a connected finite CW-complex and $p\colon E \to X$ is coreducible, then $p * \underline{S^0}$ is trivial.*

Proof. Let $r\colon \mathrm{Th}(p) \to \mathrm{Th}(p^{-1}(x))$ be a map whose composite with the inclusion $\mathrm{Th}(i_x)\colon \mathrm{Th}(p^{-1}(x)) \to \mathrm{Th}(p)$ is a homotopy equivalence. Let E_0 be the total space of $p * \underline{S^0}$. Fix a base point $s_0 \in S^0$. We define the section $s\colon X \to E_0$ by $s_-(x) = [e, s_0, 1]$, where $p(e) = x$. This is well defined since by definition of the fibrewise join, $[e, s_0, 1] = [e', s_0, 1]$ holds for all $e, e' \in p^{-1}(x)$. Let $\mathrm{pr}\colon E_0 \to E_0/s_-(X)$ be the projection. There is a homeomorphism $f\colon E_0/s(X) \xrightarrow{\cong} \mathrm{Th}(p)$ such that the restriction of $t := r_x \circ f \circ \mathrm{pr}\colon E_0 \to \mathrm{Th}(p^{-1}(x))$ to the fibre of $p * \underline{S^0}$ over x is a homotopy equivalence. By Lemma 6.100, $p * \underline{S^0}$ is trivial. □

We now discuss the automorphisms of spherical fibrations. Let $p\colon E \to X$ be a $(k-1)$-spherical fibration over a finite CW-complex. We call a strong fibre homotopy equivalence $(\mathrm{id}, \overline{f})\colon p \to p$ a strong fibre homotopy automorphism of p. It is straightforward to verify that strong fibre homotopy defines an equivalence relation on strong fibre homotopy automorphisms of p, which is compatible with composition. We let $[\overline{f}]$ denote the homotopy class of $(\mathrm{id}, \overline{f})$ and define the group

$$\mathrm{aut}(p) := \{[\overline{f}] \mid (\mathrm{id}, \overline{f})\colon p \to p\}$$

with identity element the homotopy class of the identity $[\mathrm{id}_p]$ and inverses given by taking the homotopy class of the fibre homotopy inverse.

6.10 The Uniqueness of the Spivak Normal Fibration

If $\overline{f}$ is a strong fibre homotopy automorphism of p, then $\mathrm{id}_{S^0} * \overline{f}$ is a strong fibre homotopy automorphism of $\underline{S^0} * p$. In this way we obtain a well-defined stabilisation homomorphism

$$S\colon \mathrm{aut}(p) \to \mathrm{aut}(\underline{S^0} * p), \quad [\overline{f}] \mapsto [\overline{\mathrm{id}_{S^0} * \overline{f}}].$$

Lemma 6.102 *If $k \geq \dim(X) + 2$, then $S\colon \mathrm{aut}(p) \to \mathrm{aut}(\underline{S^0} * p)$ is a bijection.*

Proof. The same statement for $k \geq \dim(X) + 3$ is proven in [55, Theorem I.4.12 on page 23] using induction over the cells of X and the fact that $\pi_i(G(k) \to G(k+1)) = 0$ for $i \leq k - 2$. But in the proof of Lemma 6.30 we saw that $\pi_i(G(k) \to G(k+1)) = 0$ for $i \leq k - 1$ and so Browder's proof can be repeated with the dimension restriction improved by one. □

Define

$$\mathrm{aut}_\infty(p) := \mathrm{colim}_{k\to\infty} \mathrm{aut}(\underline{S^{k-1}} * p).$$

By Lemma 6.102, we can equivalently define $\mathrm{aut}_\infty(p)$ to be the set of stable homotopy classes of stable strong fibre homotopy automorphisms of p where two strong fibre homotopy automorphisms $\overline{f}_0\colon \underline{S^{k_0-1}} * p \to \underline{S^{k_0-1}} * p$ and $\overline{f}_1\colon \underline{S^{k_1-1}} * p \to \underline{S^{k_1-1}} * p$ are *stably fibre homotopic* if for some $k \geq k_0, k_1$, $\mathrm{id}_{\underline{S^{k-k_0-1}}} * \overline{f}_0$ and $\mathrm{id}_{\underline{S^{k-k_1-1}}} * \overline{f}_1$ are strongly fibre homotopic.

Example 6.103 Let $p = \underline{S^{k-1}}$ be the trivial $(k-1)$-spherical fibration. Restricting a strong fibre homotopy automorphism $\overline{f}\colon p \to p$ to a fibre $p^{-1}(x) = S^{k-1}$ gives a continuously varying element $\overline{f}_x$ of $G(k)$ for every point $x \in X$. Conversely, given any map $\alpha\colon X \to G(k)$, we define the strong fibre homotopy automorphism, see Lemma 6.99,

$$(\mathrm{id}, \overline{f}_\alpha)\colon X \times S^{k-1} \to X \times S^{k-1}, \quad (x, v) \mapsto (x, \alpha(x)v).$$

This gives a natural bijection between $\mathrm{map}(X, G(k))$ and the topological monoid of strong fibre homotopy automorphisms of $\underline{S^{k-1}}$, and so we get a bijection

$$[X, G(k)] \xrightarrow{\cong} \mathrm{aut}(\underline{S^{k-1}}).$$

Since it is compatible with stabilisation by the identity, there is a natural identification

$$\mathrm{aut}_\infty(\underline{S^{-1}}) = \mathrm{colim}_{k\to\infty} \mathrm{aut}(\underline{S^{k-1}}) = \mathrm{colim}_{k\to\infty}[X, G(k)]$$
$$= [X, \mathrm{hocolim}_{k\to\infty} G(k)] = [X, G].$$

Our next goal is to prove Lemma 6.109, which implies that the isomorphism class of $\mathrm{aut}_\infty(p)$ does not depend upon p but only on X. To prove this, we need the following result, which shows, as in other contexts, that stabilisation has the effect of abelianisation.

Lemma 6.104 *The group* $\mathrm{aut}_\infty(\underline{S^{-1}})$ *is abelian.*

Proof. Given $[\overline{f}_0], [\overline{f}_1] \in \mathrm{aut}_\infty(\underline{S^{-1}})$, we find l such that that $[\overline{f}_i]$ has a representative $\overline{f}_i \colon \underline{S^{l-1}} \to \underline{S^{l-1}}$ for $i = 0, 1$. We stabilise by $\mathrm{id}_{\underline{S^{l-1}}}$ to obtain the fibre homotopy equivalences
$$\mathrm{id}_{\underline{S^{l-1}}} * \overline{f}_i \colon \underline{S^{l-1}} * \underline{S^{l-1}} \to \underline{S^{l-1}} * \underline{S^{l-1}}.$$

We now rotate $\overline{f}_1$ to the first factor of the join preserving its fibre homotopy class as follows. Let V be a finite-dimensional Hilbert space. We have the standard homeomorphism
$$g \colon SV * SV \xrightarrow{\cong} S(V \oplus V), \quad [v, w, s] \mapsto \frac{1}{\|(sv, (1-s)w)\|} \cdot (sv, (1-s)w)$$

defined in terms of join coordinates $[v, w, t]$ for $v, w \in V$ and $t \in [0, 1]$ and the standard definition of $V * V$ as quotient for $V \times V \times [0, 1]$. For $t \in [0, 1]$ consider the homeomorphism

$$h'_t \colon S(V \oplus V) \xrightarrow{\cong} S(V \oplus V),$$
$$(v, w) \mapsto (\cos(t\pi/2)v + \sin(t\pi/2)w, -\sin(t\pi/2)v + \cos(t\pi/2)w).$$

Conjugating h'_t with g yields a homotopy

$$h \colon SV * SV \times [0, 1] \to SV * SV \tag{6.105}$$

such that $h_0 = \mathrm{id}$ and $h_1([v, w, t]) = [-w, v, 1-t]$. Take $V = \mathbb{R}^l$. Since this construction applies fibrewise, it yields a strong fibre homotopy equivalence

$$\overline{R} \colon \underline{S^{k-1}} * \underline{S^{k-1}} \to \underline{S^{k-1}} * \underline{S^{k-1}}, \quad [v, w, t] \mapsto [-w, v, 1-t], \tag{6.106}$$

which is strongly fibre homotopic to the identity $\mathrm{id}_{\underline{S^{k-1}} * \underline{S^{k-1}}}$

One easily checks $(\mathrm{id}_{\underline{S^{l-1}}} * \overline{f}_1) \circ \overline{R} = \overline{R} \circ (\overline{f}_1 * \mathrm{id}_{\underline{S^{l-1}}})$. This implies $\mathrm{id}_{\underline{S^{l-1}}} * \overline{f}_1 \simeq_\mathrm{sfh} \overline{f}_1 * \mathrm{id}_{\underline{S^{l-1}}}$. Since $\overline{f}_1 * \mathrm{id}_{\underline{S^{l-1}}}$ commutes with $\mathrm{id}_{\underline{S^{l-1}}} * \overline{f}_0$, we get $(\mathrm{id}_{\underline{S^{l-1}}} * \overline{f}_0) \circ (\mathrm{id}_{\underline{S^{l-1}}} * \overline{f}_1) \simeq_\mathrm{sfh} (\mathrm{id}_{\underline{S^{l-1}}} * \overline{f}_1) \circ (\mathrm{id}_{\underline{S^{l-1}}} * \overline{f}_0)$ and hence $[\overline{f}_0]_s \cdot [\overline{f}_1]_s = [\overline{f}_1]_s \cdot [\overline{f}_0]_s$ in $\mathrm{aut}_\infty(\underline{S^{-1}})$. $\square$

Lemma 6.107 *Let p be a $(k-1)$-spherical fibration over the finite CW-complex X. Then we obtain an isomorphism of abelian groups*

$$c_p \colon \mathrm{aut}_\infty(\underline{S^{-1}}) \xrightarrow{\cong} \mathrm{aut}_\infty(p)$$

by sending the class $[\overline{f}]_s$ in $\mathrm{aut}_\infty(\underline{S^{-1}})$ represented by the strong fibre homotopy equivalence $\overline{f} \colon \underline{S^{l-1}} \to \underline{S^{l-1}}$ to the class $[\overline{f} * \mathrm{id}_p]_s$ represented by the strong fibre homotopy equivalence $\overline{f} * \mathrm{id}_p \colon \underline{S^{l-1}} * p \to \underline{S^{l-1}} * p$.

Proof. One easily checks that c_p is a well-defined homomorphism of groups since stabilisation is done from the left and the map c_p is given by taking the join with p

6.10 The Uniqueness of the Spivak Normal Fibration

from the right. More generally we can consider a $(k-1)$-spherical fibration p and spherical $(l-1)$-fibration q over X and consider the group homomorphism

$$c_{q,p} \colon \mathrm{aut}_\infty(q) \to \mathrm{aut}_\infty(q*p), \quad [\overline{f}]_s \mapsto [\overline{f}*\mathrm{id}_p]_s. \tag{6.108}$$

By Lemma 6.23 we can choose a $(k'-1)$-spherical fibration p', a strong fibre homotopy equivalence $\overline{g} \colon p * p' \xrightarrow{\simeq} S^{k+k'-1}$, and a strong fibre homotopy inverse $\overline{g}^{-1}$ of $\overline{g}$. Define an isomorphism

$$\omega \colon \mathrm{aut}_\infty(q*p*p') \xrightarrow{\cong} \mathrm{aut}_\infty(q*S^{k+k'-1})$$

by sending the class $[\overline{f}]_s$ of the strong fibre homotopy equivalence

$$\overline{f} \colon q*p*p' \to q*p*p'$$

to the class of the strong fibre homotopy equivalence

$$(\mathrm{id}_{S^{l'-1}*q} * \overline{g}) \circ \overline{f} \circ (\mathrm{id}_{S^{l'-1}*q} * \overline{g}^{-1}) \colon S^{l'-1}*q*S^{k+k'-1} \to S^{l'-1}*q*S^{k+k'-1}.$$

One easily checks that the composite

$$\mathrm{aut}_\infty(q) \xrightarrow{c_{q,p}} \mathrm{aut}_\infty(q*p) \xrightarrow{c_{q*p,p'}} \mathrm{aut}_\infty(q*p*p') \xrightarrow{\omega} \mathrm{aut}_\infty(q*S^{k+k'-1})$$

agrees with $c_{q,S^{k+k'-1}}$. Hence it suffices to show for all natural numbers k that

$$c_{q,S^{k-1}} \colon \mathrm{aut}_\infty(q) \to \mathrm{aut}_\infty(q*S^{k-1})$$

is bijective because then one can consider the string of homomorphisms

$$\mathrm{aut}_\infty(S^{-1}) \xrightarrow{c_p} \mathrm{aut}_\infty(p) \xrightarrow{c_{p,p'}} \mathrm{aut}_\infty(p*p') \xrightarrow{c_{p*p',p}} \mathrm{aut}_\infty(p*p'*p)$$

and use the fact that $c_{p,p'} \circ c_p$ and $c_{p*p',p} \circ c_{p,p'}$ are isomorphisms to show that c_p is bijective. Note that this also implies that $c_{q,p}$ is bijective since $c_{q,p} \circ c_p = c_{q*p}$.

We begin with the surjectivity of the homomorphism $c_{q,S^{k-1}}$. Consider an element $y \in \mathrm{aut}_\infty(q*S^{k-1})$ represented by a strong fibre homotopy equivalence $\overline{f} \colon S^{l-1}*q*S^{k-1} \to S^{l-1}*q*S^{k-1}$. Then we get another representative of y by the strong fibre homotopy equivalence

$$\mathrm{id}_{S^{k-1}} * \overline{f} \colon S^{k-1}*S^{l-1}*q*S^{k-1} \to S^{k-1}*S^{l-1}*q*S^{k-1}.$$

In the sequel we use implicitly the fact that the join is associative and ignore the corresponding homeomorphism. The following diagram of strong fibre homotopy equivalences commutes

$$
\begin{array}{ccc}
\underline{S^{k-1}} * \underline{S^{l-1}} * q * \underline{S^{k-1}} & \xrightarrow{\overline{g} * \mathrm{id}_{S^{k-1}}} & \underline{S^{k-1}} * \underline{S^{l-1}} * q * \underline{S^{k-1}} \\
{\scriptstyle \mathrm{fl}_1 * \mathrm{id}_{S^{k-1}}} \Big\downarrow \cong & & \cong \Big\downarrow {\scriptstyle \mathrm{fl}_1 * \mathrm{id}_{S^{k-1}}} \\
\underline{S^{l-1}} * q * \underline{S^{k-1}} * \underline{S^{k-1}} & \xrightarrow{\overline{f} * \mathrm{id}_{S^{k-1}}} & \underline{S^{l-1}} * q * \underline{S^{k-1}} * \underline{S^{k-1}} \\
{\scriptstyle \mathrm{fl}_2} \Big\downarrow \cong & & \cong \Big\downarrow {\scriptstyle \mathrm{fl}_2} \\
\underline{S^{k-1}} * \underline{S^{l-1}} * q * \underline{S^{k-1}} & \xrightarrow{\mathrm{id}_{S^{k-1}} * \overline{f}} & \underline{S^{k-1}} * \underline{S^{l-1}} * q * \underline{S^{k-1}} \\
{\scriptstyle -\mathrm{id} * \mathrm{id} * \mathrm{id} * \mathrm{id}} \Big\downarrow \cong & & \cong \Big\downarrow {\scriptstyle -\mathrm{id} * \mathrm{id} * \mathrm{id} * \mathrm{id}} \\
\underline{S^{k-1}} * \underline{S^{l-1}} * q * \underline{S^{k-1}} & \xrightarrow{\mathrm{id}_{S^{k-1}} * \overline{f}} & \underline{S^{k-1}} * \underline{S^{l-1}} * q * \underline{S^{k-1}}
\end{array}
$$

where $\mathrm{fl}_1 \colon \underline{S^{k-1}} * \underline{S^{l-1}} * q \to \underline{S^{l-1}} * q * \underline{S^{k-1}}$ permutes the three factors in the obvious way, fl_2 puts the fourth factor in front of the others and $\overline{g} \colon \underline{S^{k-1}} * \underline{S^{l-1}} * q \to \underline{S^{k-1}} * \underline{S^{l-1}} * q$ is defined by $\mathrm{fl}_1^{-1} \circ \overline{f} \circ \mathrm{fl}_1$. The composite of the three vertical arrows on the left and on the right side is a map

$$\underline{S^{k-1}} * \underline{S^{l-1}} * q * \underline{S^{k-1}} \to \underline{S^{k-1}} * \underline{S^{l-1}} * q * \underline{S^{k-1}},$$

which becomes after conjugation with the map putting the last factor in the second position

$$\underline{S^{k-1}} * \underline{S^{l-1}} * q * \underline{S^{k-1}} \xrightarrow{\cong} \underline{S^{k-1}} * \underline{S^{k-1}} * \underline{S^{l-1}} * q$$

the join $\overline{R} * \mathrm{id}_{S^{l-1}} * \mathrm{id}_q$ for the strong fibre homotopy equivalence $\overline{R}$ defined in (6.106). Since R is fibre homotopic to the identity, the composition of the three vertical arrows on the left side and on the right side are fibre homotopic to the identity. Therefore $\overline{g} * \mathrm{id}_{S^{k-1}}$ and $\mathrm{id}_{S^{k-1}} * \overline{f}$ are fibre homotopic. Hence $[\overline{g} * \mathrm{id}_{S^{k-1}}]_s = [\mathrm{id}_{S^{k-1}} * \overline{f}]_s = y$ holds in $\mathrm{aut}_\infty(q * \underline{S^{k-1}})$. By definition $c_{q,S^{k-1}}$ sends $[\overline{g}]_s$ to $[\overline{g} * \mathrm{id}_{S^{k-1}}]_s$. This proves that $c_{q,S^{k-1}}$ is surjective.

Finally we show that $c_{q,S^{k-1}}$ is injective. Consider an element $x \in \mathrm{aut}_\infty(q)$ that is mapped under $c_{q,S^{k-1}}$ to the unit in $\mathrm{aut}_\infty(q * \underline{S^{k-1}})$. Choose a strong fibre homotopy equivalence $\overline{g} \colon \underline{S^{l-1}} * q \to \underline{S^{l-1}} * q$ with $[\overline{g}]_s = x$. Since $c_{q,S^{k-1}}(x)$ is the unit, we can arrange by possibly enlarging l that the strong fibre homotopy equivalence

$$\mathrm{id}_{S^{k-1}} * \overline{g} * \mathrm{id}_{S^{k-1}} \colon \underline{S^{k-1}} * \underline{S^{l-1}} * q * \underline{S^{k-1}} \to \underline{S^{k-1}} * \underline{S^{l-1}} * q * \underline{S^{k-1}}$$

is strongly fibre homotopic to the identity. Conjugating it with the permutation map

$$\underline{S^{k-1}} * \underline{S^{l-1}} * q * \underline{S^{k-1}} \xrightarrow{\cong} \underline{S^{k-1}} * \underline{S^{k-1}} * \underline{S^{l-1}} * q,$$

which puts the last factor in the second position, yields the strong fibre homotopy equivalence

$$\mathrm{id}_{S^{k-1}} * \mathrm{id}_{S^{k-1}} * \overline{g} \colon \underline{S^{k-1}} * \underline{S^{k-1}} * \underline{S^{l-1}} * q \to \underline{S^{k-1}} * \underline{S^{k-1}} * \underline{S^{l-1}} * q,$$

6.10 The Uniqueness of the Spivak Normal Fibration

which is strongly fibre homotopic to the identity. We conclude that x is the unit in $\mathrm{aut}_\infty(q)$. This finishes the proof of Lemma 6.107. □

Combining the above discussion with Lemmas 6.102, 6.104, and 6.107 yields the following result.

Lemma 6.109 *If $k \geq \dim(X) + 2$, then for any $(k-1)$-spherical fibration p over X there are natural isomorphisms of abelian groups*

$$\mathrm{aut}(p) \cong \mathrm{aut}_\infty(p) \cong \mathrm{aut}_\infty(S^{-1}).$$

Assume now that X is a connected w-oriented n-dimensional Poincaré complex and that $k \geq n + 2$. Define

$$\pi^s_{n+k}(\mathrm{Th}(p))_{\pm 1} \subset \pi^s_{n+k}(\mathrm{Th}(p))$$

to be the set of stable homotopy classes $c\colon S^{n+k} \to \mathrm{Th}(p)$ whose image under the Hurewicz map $h\colon \pi^s_{n+k}(\mathrm{Th}(p)) \to H_{n+k}(\mathrm{Th}(p))$ is a generator of the infinite cyclic group $H_{n+k}(\mathrm{Th}(p))$, or, equivalently, satisfy the Wu formula of Lemma 6.67. The group $\mathrm{aut}(p)$ acts on $\pi^s_{n+k}(\mathrm{Th}(p))_{\pm 1}$ via

$$\mathrm{aut}(p) \times \pi^s_{n+k}(\mathrm{Th}(p))_{\pm 1} \to \pi^s_{n+k}(\mathrm{Th}(p))_{\pm 1}, \quad ([\overline{f}], [c]) \mapsto \mathrm{Th}(\overline{f})_*([c]).$$

Lemma 6.110 *Let p_X be a Spivak normal fibration of X for $k \geq n+2$. Then the action of $\mathrm{aut}(p_X)$ on $\pi^s_{n+k}(\mathrm{Th}(p_X))_{\pm 1}$ is free and transitive.*

We defer the proof of Lemma 6.110 until the end of this subsection and first use it in the proof of the uniqueness of the Spivak normal fibration.

Proof of Theorem 6.68 (ii). We can assume without loss of generality that X is connected. Let p_1^{-1} be an inverse $(l-1)$-spherical fibration for p_1 and fix a strong fibre homotopy equivalence $p_1 * p_1^{-1} \simeq_{\mathrm{sfh}} S^{k_1+l-1}$.

Applying Theorem 6.98, we see that an $(n + k_0 + k_1 + l)$-dual of c_1 is given by a map

$$Dc_1 \colon \mathrm{Th}(p_0 * p_1^{-1}) \to S^{k_0+l}.$$

Since $H_{n+k_1}(c_1)\colon H_{n+k_1}(S^{n+k_1}) \to H_{n+k_1}(\mathrm{Th}(p_1))$ is an isomorphism, the same holds for $H_{n+k_0+l}(Dc_1)\colon H_{n+k_0+l}(\mathrm{Th}(p_0 * p_1^{-1})) \to H_{k_0+l}(S^{k_0+l})$ by Lemma 6.95 and the universal coefficient theorem. Consider $x \in X$. Then the inclusion $\mathrm{Th}(p_0 * p_1^{-1}(x)) \to \mathrm{Th}(p_0 * p_1^{-1})$ induces an isomorphism on H_{k_0+l} by Theorem 6.54. Hence $p_0 * p_1^{-1} * S^0$ is trivial by Lemma 6.101 applied to Dc_1. This implies that $p_0 * S^{k_1+l_1}$ and $p_1 * S^{k_0+l_1}$ are strongly fibre homotopy equivalent. Hence we can choose a natural number k large enough such that $k \geq \dim(X) + 2$ and $k \geq k_0, k_1$ hold and there is a strong fibre homotopy equivalence

$$\overline{g}'\colon p_0 * \underline{S^{k-k_0}} \to p_1 * \underline{S^{k-k_1}}.$$

Consider the homotopy classes

$$\mathrm{Th}(\overline{g}')_*(s^{k-k_0}([c_0])), \ s^{k-k_1}([c_1]) \in \pi_{n+k}(\mathrm{Th}(p_1 * \underline{S^{k-k_1}}))_{\pm 1}.$$

As $k \geq \dim(X)+2$, Lemma 6.110 applied to $p_1 * \underline{S^{k-k_1}}$ yields a strong fibre homotopy automorphism $\overline{f}\colon p_1 * \underline{S^{k-k_1}} \to p_1 * \underline{S^{k-k_1}}$ such that the induced map of Thom spaces satisfies $\mathrm{Th}(\overline{f})_*\big(\mathrm{Th}(\overline{g}')_*(s^{k-k_0}([c_0]))\big) = s^{k-k_1}([c_1])$. Hence $\overline{g} := \overline{f} \circ \overline{g}'\colon p_0 * \underline{S^{k-k_0}} \to p_1 * \underline{S^{k-k_1}}$ is the desired strong fibre homotopy equivalence satisfying

$$\mathrm{Th}(\overline{g})_*(s^{k-k_0}([c_0])) = s^{k-k_1}([c_1]) \in \pi_{n+k}(\mathrm{Th}(p_1 * \underline{S^{k-k_1}}))_{\pm 1}.$$

Moreover, the last equation determines $\overline{g}$ up to strong fibre homotopy by Lemma 6.110. □

We conclude this subsection with the proof of Lemma 6.110, which takes some time. We first consider an action that is S-dual to the action of Lemma 6.110. To define this S-dual action, we recall that the jth *cohomotopy set* of a space X, denoted $\pi^j(X)$, is the set of homotopy classes $[X, S^j]$ of maps from X to the j-sphere. If X is a co-H-space, e.g., if X is a suspension, then $\pi^j(X)$ has a natural group structure. The jth *stable cohomotopy group* of X is defined by

$$\pi_S^j(X) := \{X, S^j\}.$$

Fix a base point $x \in X$. Let $i_x\colon \{x\} \to X$ be the inclusion and denote by $\mathrm{Th}(i_x)\colon \mathrm{Th}(p^{-1}(x)) \to \mathrm{Th}(p)$ the induced map. Define

$$\pi_S^k(\mathrm{Th}(p))^{\pm 1} := \{[r\colon \mathrm{Th}(p) \to S^k] \mid r \circ \mathrm{Th}(i_x) \text{ is a homotopy equivalence}\}$$
$$\subset \pi_S^k(\mathrm{Th}(p)).$$

The group $\mathrm{aut}(p)$ acts on $\pi_S^k(\mathrm{Th}(p))$ and hence also on $\pi_S^k(\mathrm{Th}(p))^{\pm 1}$ via precomposition with the induced map of the Thom space:

$$\pi_S^k(\mathrm{Th}(p))^{\pm 1} \times \mathrm{aut}(p) \to \pi_S^k(\mathrm{Th}(p))^{\pm 1}, \quad ([r], [\overline{f}]) \mapsto [r \circ \mathrm{Th}(\overline{f})].$$

Lemma 6.111 *Let* $p\colon X \times S^{k-1} \to X$ *be the* $(k-1)$-*spherical fibration over a connected finite CW-complex satisfying* $k \geq \dim(X) + 2$. *Then the action of* $\mathrm{aut}(p)$ *on* $\pi_S^k(\mathrm{Th}(p))^{\pm 1}$ *is free and transitive.*

Proof. We have the adjunction homeomorphism of (unpointed) mapping spaces

$$\mathrm{Ad}\colon \mathrm{map}(X \times S^k, S^k) \xrightarrow{\cong} \mathrm{map}(X, \mathrm{map}(S^k, S^k))$$

defined by $\mathrm{Ad}(u)(x)\colon S^k \to S^k$, $s \mapsto u(x, s)$ for $u \in \mathrm{map}(X \times S^k, S^k)$ and $x \in X$. For $s_0 \in S^k$ the base point, $X_+ \wedge S^k$ is the quotient of $X \times S^k$ obtained by identifying $X \times \{s_0\}$ to a single point. Hence Ad induces a homeomorphism

6.10 The Uniqueness of the Spivak Normal Fibration

$$\overline{\mathrm{Ad}}\colon \mathrm{map}_*(X_+ \wedge S^k, S^k) \xrightarrow{\cong} \mathrm{map}(X, \mathrm{map}_*(S^k, S^k))$$

where map_* always denotes pointed mapping spaces. Let $i_x\colon \{x\} \to X$ be the inclusion of the base point. We get a commutative diagram

$$\begin{array}{ccc}
\mathrm{map}_*(X_+ \wedge S^k, S^k) & \xrightarrow{\overline{\mathrm{Ad}}} & \mathrm{map}(X, \mathrm{map}_*(S^k, S^k)) \\
& \searrow i_x^* \quad \swarrow & \\
& \mathrm{map}_*(S^k, S^k) &
\end{array}$$

where i_x^* is the obvious map induced by i_x and the right diagonal arrow is given by evaluating at x. Hence we obtain a homeomorphism

$$\mathrm{map}_*(X_+ \wedge S^k, S^k)_\pm \xrightarrow{\cong} \mathrm{map}(X, F(k))$$

where $\mathrm{map}_*(X_+ \wedge S^k, S^k)_\pm$ is defined to be the subspace of those elements in $\mathrm{map}_*(X_+ \wedge S^k, S^k)$ for which the image under i_x^* is given by a homotopy equivalence $S^k \to S^k$. Taking path components yields a bijection

$$\pi_0\left(\mathrm{map}_*(X_+ \wedge S^k, S^k)_\pm\right) \xrightarrow{\cong} [(X, F(k))].$$

Since $k \geq \dim(X) + 2$ and the inclusion $F(k) \to G(k+1)$ is $(k-1)$-connected, as explained in the proof of Lemma 6.30, the canonical maps

$$\pi_0\left(\mathrm{map}_*(X_+ \wedge S^k, S^k)_\pm\right) \xrightarrow{\cong} \pi_S^k(\mathrm{Th}(p))^{\pm 1}$$

and

$$[X, F(k)] \to [X, G(k+1)]$$

are bijective. Since we have the identification $\mathrm{Th}(\underline{S^{k-1}}) = X_+ \wedge S^k$, the three isomorphisms above yield an isomorphism

$$A\colon \pi_S^k(\mathrm{Th}(\underline{S^{k-1}}))^{\pm 1} \xrightarrow{\cong} [X, G(k+1)].$$

It is easy to check that the map A is compatible with the action of $\mathrm{aut}(\underline{S^{k-1}})$ on $\pi^k(\mathrm{Th}(\underline{S^{k-1}}))^{\pm 1}$ and the identification $\mathrm{aut}(\underline{S^{k-1}}) = [X, G(k)]$, namely, we have for any element α in $[X, G(k)]$ and the corresponding element $[f_\alpha]$ in $\mathrm{aut}(\underline{S^{k-1}})$ the equality

$$A \circ \pi_S^k(\mathrm{Th}([\overline{f_\alpha}])) = (S \circ \alpha)_* \circ A,$$

where $S\colon G(k) \to G(k+1)$ is given by suspension. This finishes the proof of Lemma 6.111. □

Now let p_X be the Spivak normal fibration of the Poincaré complex X defined in Section 6.8 above. Let p_X^{-1} be an $(l-1)$-spherical fibration that is a stable inverse for p_X. Recall Theorem 6.98 stating that $\mathrm{Th}(p_X)$ is $(n + 2k + l)$-dual to $\mathrm{Th}(\underline{S^{k+l-1}})$.

This implies that S-duality gives a bijection

$$D: \pi^s_{n+k}(\mathrm{Th}(p_X))_{\pm 1} \cong \pi^{k+l}_s(\mathrm{Th}(\underline{S^{k+l-1}}))^{\pm 1}.$$

Choose a strong fibre homotopy equivalence $\overline{u}: \underline{S^{k+l-1}} \simeq_{\mathrm{sfh}} p_X * p_X^{-1}$ and a homotopy inverse $\overline{u}^{-1}: p_X * p_X^{-1} \xrightarrow{\simeq_{\mathrm{sfh}}} \underline{S^{k+l-1}}$. Consider the isomorphism

$$\gamma: \mathrm{aut}(p_X) \xrightarrow{\cong} \mathrm{aut}(\underline{S^{k+l-1}}), \quad [\overline{f}] \mapsto [\overline{u} \circ (f * \mathrm{id}_{p_X^{-1}}) \circ \overline{u}^{-1}].$$

It is indeed an isomorphism by Lemmas 6.102 and 6.107, see also (6.108). Lemma 6.112 below states that the actions of $\mathrm{aut}(p_X)$ on $\pi^s_{n+k}(\mathrm{Th}(p_X))_{\pm 1}$ and $\mathrm{aut}(\underline{S^{k+l-1}})$ on $\pi^{k+l}_s(\mathrm{Th}(\underline{S^{k+l-1}}))^{\pm}$ are equivalent actions under the isomorphisms $\gamma: \mathrm{aut}(p_X) \cong \mathrm{aut}(\underline{S^{k+l-1}})$ and S-duality.

Lemma 6.112 *For all $[\overline{f}] \in \mathrm{aut}(p_X)$ and all $[c] \in \pi^s_{n+k}(\mathrm{Th}(p_X))_{\pm 1}$*

$$D(\mathrm{Th}(\overline{f}) \circ c) = D([c]) \circ \mathrm{Th}(\gamma([\overline{f}])) \in \pi^{k+l}_s(\mathrm{Th}(\underline{S^{k+l-1}}))^{\pm 1}.$$

Proof. For a spherical fibration p we define the homomorphism

$$\mathrm{Th}: \mathrm{aut}(p) \to \{\mathrm{Th}(p), \mathrm{Th}(p)\}, \quad [\overline{f}] \mapsto [\mathrm{Th}(\overline{f})].$$

Let $c \in \pi_{n+k}(\mathrm{Th}(p_X))_{\pm 1}$ and consider the following diagram,

$$\begin{array}{ccc}
\mathrm{aut}(p_X) & \xrightarrow{\gamma} & \mathrm{aut}(\underline{S^{k+l-1}}) \\
\downarrow \mathrm{Th} & & \downarrow \mathrm{Th} \\
\{\mathrm{Th}(p_X), \mathrm{Th}(p_X)\} & \xrightarrow{D} & \{\mathrm{Th}(\underline{S^{k+l-1}}), \mathrm{Th}(\underline{S^{k+l-1}})\} \\
\downarrow c_* & & \downarrow D(c)^* \\
\{S^n, \mathrm{Th}(p_X)\} & \xrightarrow{D} & \{\mathrm{Th}(\underline{S^{k+l-1}}), S^0\}
\end{array} \quad (6.113)$$

where c_* and $D(c)^*$ denote the maps induced by pre- and post-composition respectively. If Diagram (6.113) commutes, Lemma 6.112 follows immediately.

We now prove that the diagram commutes. For the lower square, this is just the functoriality of S-duality. The commutativity of the upper square is more involved: we shall need to apply the condition of Theorem 6.96 to prove it.

We consider the exterior fibrewise join $p_X *_e (p_X * p_X^{-1})$, which is a spherical fibration over $X \times X$ with

$$\mathrm{Th}(p_X *_e (p_X * p_X^{-1})) = \mathrm{Th}(p_X) \wedge \mathrm{Th}(p_X * p_X^{-1}) = \mathrm{Th}(p_X) \wedge \mathrm{Th}(\underline{S^{2+l-1}}).$$

Let $\Delta: X \to X \times X$ be the diagonal map and $(\Delta, \overline{\Delta})$ the map of spherical fibrations

$$(\Delta, \overline{\Delta}): p_X * p_X * p_X^{-1} \to p_X *_e (p_X * p_X^{-1}), \quad u * v * w = u *_e v * w.$$

The S-duality between $\text{Th}(p_X)$ and $\text{Th}(p_X * p_X^{-1})$ is defined via the composite

$$S^{n+2k+l} \xrightarrow{c'} \text{Th}(p_X * p_X * p_X^{-1}) \xrightarrow{\text{Th}(\overline{\Delta})} \text{Th}(p_X) \wedge \text{Th}(p_X * p_X^{-1}),$$

where c' belongs to $\pi_{n+2k+l}(\text{Th}(p_X * p_X * p_X^{-1}))_{\pm 1}$, see the proof of Theorem 6.98. Hence it suffices to show that the following diagram commutes up to homotopy

$$\begin{array}{ccc}
S^{n+2k+l} & \xrightarrow{\overline{\Delta} \circ c'} & p_X *_e (p_X * p_X^{-1}) \\
{\scriptstyle \overline{\Delta} \circ c'} \downarrow & & \downarrow {\scriptstyle \overline{f} *_e \text{id}} \\
p_X *_e (p_X * p_X^{-1}) & \xrightarrow{\overline{\text{id}} *_e \gamma(\overline{f})} & p_X *_e (p_X * p_X^{-1})
\end{array}$$

because then an application of Theorem 6.96 yields the desired equality $D(\text{Th}(\overline{f})) = \text{Th}(\gamma(\overline{f}))$. This will be achieved by proving that the following diagram of spherical fibrations commutes up to homotopy

$$\begin{array}{ccc}
p_X * p_X * p_X^{-1} & \xrightarrow{\overline{\Delta}} & p_X *_e (p_X * p_X^{-1}) \\
{\scriptstyle \overline{\Delta}} \downarrow & & \downarrow {\scriptstyle \overline{f} *_e \text{id}} \\
p_X *_e (p_X * p_X^{-1}) & \xrightarrow{\text{id} *_e \gamma(\overline{f})} & p_X *_e (p_X * p_X^{-1}).
\end{array}$$

Let $\overline{f}^{-1}$ be a strong fibre homotopy inverse of $\overline{f}$. Then a direct inspection shows that $(\overline{f} *_e \text{id}) \circ \overline{\Delta}$ and $(\text{id} *_e \gamma(\overline{f})) \circ \overline{\Delta} \circ (\overline{f} * \overline{f}^{-1} * \text{id}_{p_X^{-1}})$ are strongly fibre homotopic. Hence it suffices to show that $\overline{f} * \overline{f}^{-1} : p_X * p_X \to p_X * p_X$ is the identity element in $\text{aut}(p_X * p_X)$. For this purpose it suffices to show that $\overline{f} * \text{id}$ and $\text{id} * \overline{f}$ define the same element in $\text{aut}(p_X * p_X)$. Since they are conjugated by the obvious flip map $\text{fl}: p_X * p_X \to p_X * p_X$ and $\text{aut}(p_X * p_X)$ is abelian by Lemma 6.109 and the assumption $k \geq \dim(X) + 2$, the proof of Lemma 6.112 is finished. □

Finally, Lemma 6.110 is an immediate consequence of Lemmas 6.111 and 6.112.

6.11 Notes

Further references for spherical fibrations are [129, 300, 383, 381, 386].

Chapter 7
Normal Maps and the Surgery Problem

7.1 Introduction

Recall that we started in Chapter 4 with the following problem.

Problem 4.1. Let X be a finite CW-complex. When is X homotopy equivalent to a closed n-dimensional manifold?

By Theorem 5.51 we were forced to modify it to Problem 6.1 and by Lemma 6.79 we are forced to modify it even further to the following problem.

Problem 7.1 Let X be an n-dimensional finite Poincaré complex whose Spivak normal fibration has a reduction to a vector bundle in the sense of Definition 6.69. When is X homotopy equivalent to a closed n-dimensional manifold?

In this chapter we work in the smooth category unless explicitly stated otherwise.

In Section 7.3 we systematically study the set of reductions of the Spivak normal fibration to a vector bundle by analysing the set $\mathcal{NI}_n(X)$ of normal invariants of a finite n-dimensional Poincaré complex X. We have dealt with the question whether $\mathcal{NI}_n(X)$ is non-empty already in Lemma 6.79 and Example 6.85. In the case when it is non-empty, we construct a bijection, see (7.11),

$$[X, G/O] \xrightarrow{\cong} \mathcal{NI}_n(X).$$

In Section 7.4 we introduce the notion of a normal map of degree one. This is a map of degree one $f \colon M \to X$ from a closed n-dimensional manifold M to a w-oriented finite n-dimensional Poincaré complex X covered by some bundle data, which can be given in terms of normal bundles or, equivalently, tangent bundles. We show that elements in $\mathcal{NI}_n(X)$ correspond to normal bordism classes of normal maps of degree one, see Theorem 7.19 and Lemma 7.32.

The surgery problem is to modify the source manifold M and the map f to achieve a homotopy equivalence, while leaving the target fixed without losing the structure of a closed manifold on the source. In Section 4.6 we introduced the surgery step. This is the analogue for manifolds of attaching a cell to a CW-complex in order to kill an element in a homotopy group. Since the normal maps of degree one introduced in this chapter can be fed into the setup of Chapter 4, we will be able to show that we can make a normal map $f \colon M \to X$ highly connected, that is k-connected if $\dim(M) = 2k$ or $\dim(M) = 2k + 1$, by carrying out a finite number of surgery steps, see Theorem 7.41. The surgery obstructions will later arise as obstructions to increase the connectivity by one, namely to arrange f to be $(k + 1)$-connected. If f is $(k + 1)$-connected, then Poincaré duality implies that f is a homotopy equivalence from a closed manifold to X.

Finally we explain that our setting for normal maps contains a coupling of the bundle data and the degree one condition, which is often neglected in the literature, see Remark 7.45.

Recall that in surgery theory, besides Problem 1.1 of recognising manifolds, we also aim at Problem 1.2 of classifying manifolds. In Remarks 7.46 and 7.47 at the end of the chapter we describe in more detail the close links between the two problems and strategies for their solution.

Guide 7.2 It is crucial to understand the notion of a normal map (with respect to the tangent bundle) of degree one, see Definition 7.22, and to get the idea of the normal bordism relation (with cylindrical ends), see Definition 7.24 and Definition 7.25. One needs to know the mere facts that there is a bijection $[X, G/O] \xrightarrow{\cong} N\mathcal{I}_n(X)$, that every normal map of degree one can be made highly connected, see Theorem 7.41, and that the surgery obstruction arises, when raising this connectivity by one, which by Poincaré duality already implies that we have a homotopy equivalence, see Problem 7.42.

It may suffice go though Section 7.2 to get the basics of this chapter.

7.2 An Informal Preview

For technical reasons and for the sake of correctness, the forthcoming definitions are very complicated and hence not very accessible. Therefore we next give some informal definitions and facts, which shall motivate the forthcoming precise definitions, illustrate the main ideas, and should actually be used in the daily work of a "surgeon". There are some flaws with these informal definitions, which can lead to some inconsistencies and confusion. These will be repaired later. For simplicity we will only consider the orientable connected case in this Section 7.2.

Recall that the naive definition of a connected finite n-dimensional Poincaré complex is a connected finite CW-complex X of dimension n with fundamental group π and universal covering $\widetilde{X}$ such that $H_n(X; \mathbb{Z})$ is infinite cyclic and for a choice of some generator $[X] \in H_n(X; \mathbb{Z})$, called the fundamental class, the up to $\mathbb{Z}\pi$-chain homotopy unique $\mathbb{Z}\pi$-chain map given by the cap product with $[X]$ from the dual $\mathbb{Z}\pi$-chain complex $C^{n-*}(\widetilde{X})$ to $C_*(\widetilde{X})$ is a $\mathbb{Z}\pi$-chain homotopy equivalence. Note that it induces an isomorphism $H^{n-k}(X; \mathbb{Z}) \xrightarrow{\cong} H_k(X; \mathbb{Z})$ for every k. We call

7.2 An Informal Preview

X oriented if we have chosen a fundamental class $[X]$. A closed n-dimensional manifold M is a finite n-dimensional Poincaré complex.

Let X be an oriented connected finite CW-complex of dimension n and let ξ be a vector bundle over X. A normal map of degree one in the naive sense $(f, \bar{f})$ consists of an oriented connected closed manifold M of dimension n and a map $f\colon M \to X$ of degree one that is for some natural number a covered by a bundle map $\bar{f}\colon TM \oplus \mathbb{R}^a \to \xi$.

Consider two normal maps of degree one $(f_i, \bar{f_i})$ from M_i to X with the same target (X, ξ) for $i = 0, 1$ and $n = \dim(X)$. A normal bordism (with cylindrical target) between them in the naive sense consists of the following data. There is an oriented compact connected $(n + 1)$-dimensional manifold W with boundary ∂W and a map of degree one $(F, \partial F)\colon (W, \partial W) \to (X \times [0, 1], X \times \{0, 1\})$ covered by a bundle map $\bar{F}\colon TW \oplus \mathbb{R}^b \to \mathrm{pr}^*\xi \oplus \mathbb{R}^c$ for the projection $\mathrm{pr}\colon X \times [0, 1] \to X$. The boundary ∂W can be written as the disjoint union $\partial W = \partial_0 W \amalg \partial_1 W$. There are identifications of $(\bar{f_i}, f_i)$ with the restriction of $(\bar{F}, \partial F)$ to $\partial_i W$ for $i = 0, 1$ where one allows $\bar{f_i}$ to be stabilised by adding a trivial vector bundle $\mathbb{R}^d$ to $TM \oplus \mathbb{R}^a$ and ξ. By gluing such bordisms one can show that this is an equivalence relation. Surgery on a normal map does not change its normal bordism class.

One gets an analogous definition if one replaces $TM \oplus \mathbb{R}^a$ in the definition of a normal map of degree one by the normal bundle $\nu(M \hookrightarrow \mathbb{R}^{n+k})$ for some embedding of M into $\mathbb{R}^{n+k}$. One can pass from one setting to the other. The reason for this is that $TM \oplus \nu(M \hookrightarrow \mathbb{R}^{n+k})$ is trivial, and one does not change the normal bordism class of a normal map of degree one by stabilising the bundle data by adding trivial vector bundles.

If $f\colon M \to X$ is a homotopy equivalence, then one can choose fundamental classes so that it has degree one and, if we take $\xi = (f^{-1})^*TM$ for some homotopy inverse f^{-1} of f, we can cover f by a bundle map $\bar{f}$ from TM to ξ. Hence the existence of a normal map of degree one with X as target is a necessary condition for X to be homotopy equivalent to a closed manifold. Recall that one can make any normal map with target X highly connected by surgery and that the upcoming surgery obstruction will decide whether one can achieve by surgery that f is a homotopy equivalence. Moreover, we will later see for a normal map of degree one $(f, \bar{f})$ from M to X that, if X is simply connected with dimension $n = 4k$ for $k \geq 2$, the surgery obstruction is the integer given by the differences of the signatures of X and M.

It can happen that there are two different normal maps with target X such that the surgery obstruction for one of them is trivial and for the other not. Hence it is important to understand the set of normal bordism classes of maps of degree one with target X. This turns out to be the same as the set of vector bundle reductions of the Spivak normal bundle of X up to the obvious equivalence relation, which is called the set of normal invariants of X and denoted by $\mathcal{N}(X)$. More precisely, an element in $\mathcal{N}(X)$ is a pair (ξ, c) consisting of a k-dimensional vector bundle ξ and an element $c \in \pi_{n+k}(\mathrm{Th}(\xi))$ whose image under the Hurewicz map $\pi_{n+k}(\mathrm{Th}(\xi)) \to H_{n+k}(\mathrm{Th}(\xi))$ is a generator of the infinite cyclic group $H_{n+k}(\mathrm{Th}(\xi))$. Two such pairs (ξ, c) and (ξ', c') are called equivalent if, after stabilising by adding trivial vector bundles to both sides, there is a bundle isomorphism from ξ to ξ' such that the induced map $\pi_{n+k}(\mathrm{Th}(\xi)) \to \pi_{n+k}(\mathrm{Th}(\xi))$ sends c to c'.

Next we explain how one can assign to a normal map of degree one $(f, \bar{f})$ from M to X an element in $\mathcal{N}(X)$ where the bundle map is given in terms of the normal bundle ν of some embedding of M into $\mathbb{R}^{n+k}$. Let $c \in \pi_{n+k}(\mathrm{Th}(\xi))$ be the image of the canonical element in $\pi_{n+k}(\mathrm{Th}(\nu))$ coming from the collapse map from S^{n+k} to $\mathrm{Th}(\nu)$ under the map $\pi_{n+k}(\mathrm{Th}(\nu)) \to \pi_{n+k}(\mathrm{Th}(\xi))$ induced by $\bar{f}$. Since f has degree one, we conclude from the Thom isomorphism that (ξ, c) represents an element in $\mathcal{N}(X)$.

Consider $(\xi, c) \in \mathcal{N}(X)$. We can make the map $c \colon S^{n+k} \to \mathrm{Th}(\xi)$ transversal to the copy of X in $\mathrm{Th}(\xi)$. We obtain a map $f \colon M \to X$ covered by a bundle map $\nu(M \hookrightarrow S^{n+k}) \to \xi$. By surgery one can arrange that M is connected. One easily checks that the degree of f is one for the right choice of the fundamental class on M using the Thom isomorphism. Thus we obtain a bijection between set of normal bordism classes of maps of degree one with target X and $\mathcal{N}(X)$.

Since we have the homotopy fibration $G/O \to BO \to BG$, we can identify $\mathcal{N}(X)$ with $[X, G/O]$, provided that $\mathcal{N}(X)$ is non-empty and we have fixed a base point in it.

In order to indicate that all the definitions need to be worked out in more detail, the reader may wonder whether she or he can answer the following questions:

- For a normal map of degree one $(f, \bar{f})$ to (X, ξ), if we change the bundle map $\bar{f}$ by an automorphism of ξ covering id_X, do we stay in the same normal bordism class?
- Let M be an oriented closed manifold. Then we can consider the map $\mathrm{id}_M \amalg \mathrm{id}_M \amalg \mathrm{id}_M \colon M \amalg M \amalg M \to M$ which is covered by the bundle map $\mathrm{id}_{TM} \amalg \mathrm{id}_{TM} \amalg \mathrm{id}_{TM} \colon TM \amalg TM \amalg TM \to TM$. This is a normal map, but its degree is 3. Now change the orientation of one of the summands of its source. Then we get a normal map of degree one.
  Can we do surgery so that the result is a normal map of degree one whose source is connected?

The remainder of this section is devoted to giving the precise definitions of the notions above, full proofs of some of the claims above and of other assertions, and also to answering the two questions above.

7.3 Rank k Normal Invariants

In this section we introduce and study the set of rank k normal invariants, which is essentially the set of all possible reductions to a vector bundle of the Spivak normal structure.

Recall that we have defined the notion of a connected finite n-dimensional Poincaré complex $(X, \mathcal{O}_X)$ in Definition 5.43. We have also explained how it extends to the non-connected case, essentially by considering each path component

7.3 Rank k Normal Invariants

separately, see Remark 5.45. Recall that part of the structure is an infinite cyclic local coefficient system O_X over X.

A vector bundle ξ over X determines an infinite cyclic local coefficient system O_ξ, see (5.22). Recall that O_X and O_ξ are isomorphic if and only if $w_1(O_X) = w_1(O_\xi)$ holds in $H^1(X; \mathbb{F}_2)$, see (5.5).

We get isomorphisms

$$H_{n+k}(\mathrm{Th}(\xi)) \xrightarrow{\cong} H_n(X; O_X) \xrightarrow{\cong} H^0(X; \mathbb{Z}) \xrightarrow{\cong} \prod_{C \in \pi_0(X)} \mathbb{Z}$$

from Theorem 6.54 and from Poincaré duality, see Convention 5.49. The cup product induces a ring structure on $H^0(X)$, which corresponds under the canonical bijection $H^0(X) \xrightarrow{\cong} \prod_{C \in \pi_0(X)} \mathbb{Z}$ to the product ring structure on the target. Let $H^0(X)^\times$ be the abelian group of units in $H^0(X)$. It corresponds under the isomorphism above to $\prod_{C \in \pi_0(X)} \{\pm 1\}$.

Definition 7.3 (Componentwise generator) We will say that an element $x \in H_{n+k}(\mathrm{Th}(\xi))$ or an element $y \in H_n(X; O_X)$ is a *componentwise generator* if its image in $H^0(X)$ lies in $H^0(X)^\times$.

If X is connected, this means simply that x or y is a generator of the infinite cyclic group $H_{n+k}(\mathrm{Th}(\xi))$ or $H_n(X; O_X)$ respectively. Recall that a w-orientation on X is a choice of a componentwise generator $[[X]]$ in $H_n(X; O_X)$, which we will call a fundamental class, compare Remark 5.45.

Lemma 7.4 *Let $(X, O_X, [[X]])$ be a w-oriented finite n-dimensional Poincaré complex.*

(i) *There is an isomorphism of groups*

$$v \colon H^0(X)^\times = \prod_{\pi_0(X)} \{\pm 1\} \xrightarrow{\cong} \mathrm{aut}(O_X);$$

(ii) *For $\epsilon = \{\epsilon_C \in \{\pm 1\} \mid C \in \pi_0(X)\}$ the following diagram commutes*

$$\begin{array}{ccc}
\bigoplus_{C \in \pi_0(X)} H_n(C; O_X|_C) & \xrightarrow{\bigoplus_{C \in \pi_0(X)} \epsilon_C \cdot \mathrm{id}_{H_n(C; O_X|_C)}} & \bigoplus_{C \in \pi_0(X)} H_n(C; O_X|_C) \\
\cong \downarrow & & \downarrow \cong \\
H_n(X; O_X) & \xrightarrow{H_n(\mathrm{id}_X; v(\epsilon))} & H_n(X; O_X)
\end{array}$$

where the vertical isomorphisms are induced by the inclusions $C \to X$ for $C \in \pi_0(X)$;

(iii) *If $H_n(\mathrm{id}_X; o)([[X]]) = [[X]]$ holds for $o \in \mathrm{aut}(O_X)$, then $o = \mathrm{id}_{O_X}$.*

Proof. Obviously it suffices to treat the case where X is connected.

(i) Define v by sending $\epsilon \in H^0(X)^\times = \{\pm 1\}$ to $\epsilon \cdot \mathrm{id}_{O_X}$.

(ii) This follows from the obvious equation $H_n(\mathrm{id}_X; -o) = -H_n(\mathrm{id}_X, o)$ for $o \in \mathrm{aut}(O_X)$.

(iii) This follows directly from assertion (ii) since $H_n(C; O_C)$ is infinite cyclic for all $C \in \pi_0(X)$. □

Definition 7.5 (Rank k normal invariant) Let $X = (X, O_X, [[X]])$ be a w-oriented finite n-dimensional Poincaré complex. A *rank k normal invariant* (ξ, c, o) consists of

- A k-dimensional vector bundle ξ over X;
- An element $c \in \pi_{n+k}(\mathrm{Th}(\xi))$;
- An isomorphism $o \colon O_\xi \xrightarrow{\cong} O_X$ of infinite cyclic local coefficient systems over X such that the image of $h([c]) \cap \overline{U_{pS\xi}}$ for the intrinsic Thom class $\overline{U_{pS\xi}}$ of the spherical fibration $pS\xi$ associated to ξ in the sense of Definition 6.45 under the isomorphism

$$H_n(\mathrm{id}_X; o) \colon H_n(X; O_\xi) \xrightarrow{\cong} H_n(X; O_X)$$

is $[[X]]$.

We conclude from Lemma 7.4 (iii) that o is uniquely determined by the above equation

$$H_n(\mathrm{id}_X; o)(h([c]) \cap \overline{U_{pS\xi}}) = [[X]].$$

We call a rank k normal invariant (ξ_0, c_0, o_0) and a rank k normal invariant (ξ_1, c_1, o_1) *equivalent* if there is a bundle isomorphism $(\mathrm{id}, \overline{f}) \colon \xi_0 \xrightarrow{\cong} \xi_1$ such that

$$\pi_{n+k}(\mathrm{Th}(\overline{f})) \colon \pi_{n+k}(\mathrm{Th}(\xi_0)) \xrightarrow{\cong} \pi_{n+k}(\mathrm{Th}(\xi_1))$$

maps $[c_0]$ to $[c_1]$. Note that this already implies $o_1 \circ O_{\overline{f}} = o_0$ because of Lemma 7.4 (iii).

The *set of rank k normal invariants* $\mathcal{NI}_n(X, k)$ is the set of equivalence classes of rank k normal invariants of X.

Given a rank k normal invariant (ξ, c, o), we obtain a rank $(k+1)$ normal invariant $(\xi \oplus \mathbb{R}, \Sigma c, \Sigma o)$ where $\Sigma \colon \pi_k(\mathrm{Th}(\xi)) \to \pi_{k+1}(\Sigma \mathrm{Th}(\xi)) = \pi_{k+1}(\mathrm{Th}(\xi \oplus \mathbb{R}))$ is the suspension homomorphism and Σo comes from the canonical isomorphism $O_\xi \xrightarrow{\cong} O_{\xi \oplus \mathbb{R}}$ of Lemma 5.23 (i). Define

Definition 7.6 (Normal invariant) Let $X = (X, O_X, [[X]])$ be a w-oriented finite n-dimensional Poincaré complex X. Define the *set of normal invariants* $\mathcal{NI}_n(X)$ of X to be $\mathrm{colim}_{k \to \infty} \mathcal{NI}_n(X, k)$.

We have dealt with the question whether $\mathcal{NI}_n(X)$ is non-empty already in Lemma 6.79 and Example 6.85.

Let X be a finite CW-complex. Define an abelian group $\mathcal{G}/O(X)$ as follows. We consider pairs (ξ, t) consisting of a k-dimensional vector bundle ξ over X for some $k \geq 1$ and a strong fibre homotopy equivalence $t \colon S\xi \to \underline{S^{k-1}}$ from the associated

7.3 Rank k Normal Invariants

spherical fibration given by the sphere bundle $S\xi$ to the trivial spherical $(k-1)$-fibration over X. We call two such pairs (ξ_0, t_0) and (ξ_1, t_1) equivalent if for some k which is greater or equal to both $k_0 = \dim(\xi_0)$ and $k_1 = \dim(\xi_1)$ there is a bundle isomorphism $\overline{f} \colon \xi_0 \oplus \mathbb{R}^{k-k_0} \xrightarrow{\cong} \xi_1 \oplus \mathbb{R}^{k-k_1}$ covering the identity on X such that $\Sigma^{k-k_1} t_1 \circ S\overline{f}$ is fibre homotopic to $\Sigma^{k-k_0} t_0$. Let $\mathcal{G}/O(X)$ be the set of equivalence classes $[\xi, t]$ of such pairs (ξ, t). Note that $[\xi, t]$ for a pair (ξ, t) depends only on ξ and the fibre homotopy class of t. Addition is given by the Whitney sum. The neutral element is represented by $(\mathbb{R}^k, \mathrm{id})$ for some $k \geq 1$. The existence of inverses follows from the fact that for a vector bundle ξ over X we can find another vector bundle η such that $\xi \oplus \eta$ is trivial and from the next exercise.

Exercise 7.7 In the sequel all fibrations are assumed to live over the finite CW-complex X. Given $k \geq 1$, consider a strong fibre homotopy equivalence $f \colon S^{k-1} \to S^{k-1}$. Show that we can find a strong fibre homotopy equivalence $g \colon S^{k-1} \to S^{k-1}$ such that $f * g \colon S^{k-1} * S^{k-1} \to S^{k-1} * S^{k-1}$ is strongly fibre homotopic to the identity.

Let $X = (X, \mathcal{O}_X, [[X]])$ be a w-oriented finite n-dimensional Poincaré complex. Next we describe an action

$$\rho \colon \mathcal{G}/O(X) \times \mathcal{NI}_n(X) \to \mathcal{NI}_n(X). \tag{7.8}$$

Consider a pair (ξ, t) representing an element $[\xi, t] \in \mathcal{G}/O(X)$ and a rank k_1 normal invariant (η, c, o) representing an element $[\eta, c, o] \in \mathcal{NI}_n(X)$. If k_0 is the dimension of ξ, then $\xi \oplus \eta$ is a $(k_0 + k_1)$-dimensional vector bundle. Consider the composition

$$\mathrm{Th}(\xi \oplus \eta) \xrightarrow{\cong} \mathrm{Th}(S\xi * S\eta) \xrightarrow{t * \mathrm{id}} \mathrm{Th}(S^{k_0-1} * S\eta) \xrightarrow{\cong} \Sigma^{k_0}\mathrm{Th}(\eta).$$

This and the suspension $\Sigma^{k_0} c$ yield an element $d \in \pi_{n+k_0+k_1}(\mathrm{Th}(\xi \oplus \eta))$. Let $o' \colon \mathcal{O}_{\xi \oplus \eta} \to \mathcal{O}_X$ be the composite of $o \colon \mathcal{O}_\eta \xrightarrow{\cong} \mathcal{O}_X$ with the obvious isomorphism $\mathcal{O}_{\xi \oplus \eta} \xrightarrow{\cong} \mathcal{O}_\eta$ coming from t, Lemma 5.23 (i) and (6.38). One easily checks that $(\xi \oplus \eta, d, o')$ is a normal $(k_0 + k_1)$-invariant for X. Define

$$\rho([\xi, t], [\eta, c]) := [\xi \oplus \eta, d, o'].$$

We leave it to the reader to check that this is a well-defined group action.

Exercise 7.9 Let $X = (X, \mathcal{O}_X)$ be a finite n-dimensional Poincaré complex X. Let $[[X]]$ and $[[X]]'$ be two fundamental classes in $H_n(X; \mathcal{O}_X)$. Construct a bijection

$$\mathcal{NI}_n(X, \mathcal{O}_X, [[X]]) \xrightarrow{\cong} \mathcal{NI}_n(X, \mathcal{O}_X, [[X]]')$$

which is compatible with the $\mathcal{G}/O(X)$-actions of (7.8).

Since G/O is simply connected, it does not matter whether we consider pointed or unpointed homotopy classes when we deal with $[X, G/O]$, and analogously for G/PL and G/TOP instead of G/O, see Remark 11.12.

Theorem 7.10 (Normal invariants and G/O**)** *Let X be a w-oriented finite n-dimensional Poincaré complex.*

(i) *There is a bijection of sets*

$$\mu \colon [X, G/O] \xrightarrow{\cong} \mathcal{G}/\mathcal{O}(X);$$

(ii) *Suppose that $\mathcal{NI}_n(X)$ is non-empty. Then the action ρ of (7.8) is a transitive free operation of the abelian group $\mathcal{G}/\mathcal{O}(X)$ on the set $\mathcal{NI}_n(X)$.*

Proof. (i) Recall that G/O is the fibre of the fibration $\widehat{J} \colon E_J \to BG$ associated to $J \colon BO \to BG$ over some point $z \in BG$. By definition this is the space $\{(y, \lambda) \in BO \times \mathrm{map}([0,1], BG) \mid J(y) = \lambda(0), \lambda(1) = z\}$. Hence a map $F \colon X \to G/O$ is the same as a pair (f, h) consisting of a map $f \colon X \to BO$ and a homotopy $h \colon X \times [0, 1] \to BG$ such that $h_0 = J \circ f$ and h_1 is the constant map c_z. Since X is compact, we can find k such that the image of f and h lie in $BO(k)$ and $BG(k)$. Recall that $J_k \colon BO(k) \to BG(k)$ is covered by a fibre map $\overline{J_k} \colon S\mu_k \to p_k$ that is fibrewise a homotopy equivalence where μ_k is the universal k-dimensional bundle over $BO(k)$ and p_k is the universal spherical $(k-1)$-fibration. By the pullback construction applied to f and μ_k and h and p_k, we get a vector bundle ξ over X and a spherical $(k-1)$-fibration $p \colon E \to X \times [0, 1]$ over $X \times [0, 1]$ together with fibre homotopy equivalences $(\mathrm{pr}_0^{-1}, u) \colon S\xi \to p|_{X \times \{0\}}$ and $(\mathrm{pr}_1, v) \colon p|_{X \times \{1\}} \to X \times S^{k-1}$ for $\mathrm{pr}_i \colon X \times \{i\} \xrightarrow{\cong} X$ the obvious homeomorphism. Up to fibre homotopy there is precisely one fibre homotopy equivalence $(\mathrm{id}_{X \times [0,1]}, t) \colon p \to X \times [0, 1] \times S^{k-1}$ for $\mathrm{pr} \colon X \times [0, 1] \to X$ the projection whose restriction to $X \times \{1\} = X$ is v, see [393, Proposition 15.11 on page 342]. Thus we obtain a fibre homotopy equivalence $(\mathrm{pr}_0, t|_{X \times \{0\}}) \colon p|_{X \times \{0\}} \to X \times S^{k-1}$, which is unique up to fibre homotopy. Composing $t|_{X \times \{0\}}$ and u yields a fibre homotopy equivalence unique up to fibre homotopy $(\mathrm{id}, u \circ t|_{X \times \{0\}}) \colon S\xi \to X \times S^{k-1}$. Thus we can assign to a map $F \colon X \to G/O$ an element $[\xi, u \circ t|_{X \times \{0\}}] \in \mathcal{G}/\mathcal{O}(X)$. We leave it to the reader to check that this induces the desired bijection $\mu \colon [X, G/O] \xrightarrow{\cong} \mathcal{G}/\mathcal{O}(X)$.

(ii) Next we show transitivity. Consider two elements $[\xi_0, c_0, o_0]$ and $[\xi_1, c_1, o_1]$ in $\mathcal{NI}_n(X)$. Since we can stabilise the representatives without changing their classes, we can assume without loss of generality by Theorem 6.68 (ii) that for some natural number k we have $k = \dim(\xi_0) = \dim(\xi_1)$ and there is a strong strong fibre homotopy equivalence $\overline{t} \colon S\xi_0 \to S\xi_1$ such that $\pi_{n+k}(\mathrm{Th}(\overline{t})) \colon \pi_{n+k}(\mathrm{Th}(\xi_0)) \to \pi_{n+k}(\mathrm{Th}(\xi_1))$ sends $[c_0]$ to $[c_1]$. Since X is finite, we can choose a vector bundle η_1 of dimension l together with an isomorphism $\overline{u} \colon \xi_1 \oplus \eta_1 \xrightarrow{\cong} \underline{\mathbb{R}^{k+l}}$. Put $\mu = \xi_0 \oplus \eta_1$. Define a strong fibre homotopy equivalence

$$\overline{s} \colon S\mu = S(\xi_0 \oplus \eta_1) \xrightarrow{\cong} S\xi_0 * S\eta_1 \xrightarrow{\overline{t} * \mathrm{id}_{S\eta_1}} S\xi_1 * S\eta_1$$
$$\xrightarrow{\cong} S(\xi_1 \oplus \eta_1) \xrightarrow{S\overline{u}} S\underline{\mathbb{R}^{k+l}} = \underline{S^{k+l-1}}.$$

Thus we obtain an element $[\mu, \overline{s}]$ in $\mathcal{G}/\mathcal{O}(X)$. Using the bundle isomorphism

$$\xi_1 \oplus \mu = \xi_1 \oplus \xi_0 \oplus \eta_1 \xrightarrow{\text{flip}} \xi_0 \oplus \xi_1 \oplus \eta_1 \xrightarrow{\text{id}_{\xi_0} \oplus \overline{u}} \xi_0 \oplus \underline{\mathbb{R}^{k+l}}$$

one easily checks that $\rho([\mu, \overline{s}], [\xi_1, c_1, o_1]) = [\xi_0, c_0, o_o]$ holds.

The proof that the action is free is left to the reader. It uses the fact that the strong fibre homotopy equivalence appearing in Theorem 6.68 (ii) is unique up to stable strong fibre homotopy. This finishes the proof of Theorem 7.10. □

Note that, in the case when $\mathcal{NI}_n(X)$ is non-empty, Theorem 7.10 yields after a choice of an element in $\mathcal{NI}_n(X)$ a bijection of sets

$$[X, G/O] \xrightarrow{\cong} \mathcal{NI}_n(X). \tag{7.11}$$

Remark 7.12 Bijection (7.11) is important from the calculational point of view. However, to really use it, knowledge about the homotopy type of the space G/O is needed, which requires some deep results from algebraic topology. We will give more information in Chapters 12 and 17.

It is interesting to observe that, when we extend surgery theory to the PL category and the topological category, there are analogous bijections and in those cases we have a much better knowledge of the spaces G/PL and G/TOP, which are the analogues of G/O in those categories. This will be discussed in detail in Chapter 17 and then applied in the calculations in Chapters 18 and 19.

7.4 Normal Maps

The next definition is related to that of ξ-bordism for a vector bundle over a space X, as defined before Theorem 4.55. We will now additionally have a w-oriented finite n-dimensional Poincaré complex $X = (X, O_X, [[X]])$ as target, require an identification of O_ξ and O_X, impose a degree one condition, and will be more flexible concerning the vector bundle for a bordism. This is the setting that is needed if we want to apply the techniques of Chapter 4 to Problem 7.1.

Definition 7.13 (Rank k normal map (of degree one)) Let $X = (X, O_X, [[X]])$ be a w-oriented finite n-dimensional Poincaré complex.

(i) A *rank k normal map* $(M, i, f, \xi, \overline{f}, o)$ with target (X, O_X) consists of

- A closed manifold M of dimension n;
- An embedding $i \colon M \hookrightarrow \mathbb{R}^{n+k}$;
- A map $f \colon M \to X$;
- A k-dimensional vector bundle ξ over X;
- A bundle map $(f, \overline{f}) \colon \nu(i) \to \xi$ covering f where $\nu(i)$ is the normal bundle of the embedding i;
- An isomorphism $o \colon O_\xi \xrightarrow{\cong} O_X$ of infinite cyclic local coefficient systems;

(ii) The isomorphism
$$\alpha \colon O_M \to f^*O_X$$
of infinite cyclic local coefficient systems over M associated to the rank k normal map $(M, i, f, \xi, \overline{f}, o)$ is given by the canonical isomorphism $\kappa \colon O_M \xrightarrow{\cong} O_{\nu(i)}$ of Lemma 5.23 (iv), the isomorphisms $O_{\nu(i)} \xrightarrow{\cong} f^*O_\xi$ induced by $\overline{f}$ and the isomorphism $f^*O_\xi \xrightarrow{\cong} f^*O_X$ induced by o;

(iii) A *rank k normal map of degree one* with target $(X, O_X, [[X]])$ is a rank k normal map with target (X, O_X) such that the image of the intrinsic fundamental class $[[M]]$ of M, see Definition 5.29, under the map
$$H_n(f, \alpha) \colon H_n(M; O_M) \to H_n(X; O_X)$$
is $[[X]]$.

Remark 7.14 (Degree one) The name degree one makes sense since condition (iii) appearing Definition 7.13 is equivalent to the statement that the degree of $f \colon M \to X$ in the sense of Definition 5.92 with respect to $[[M]]$, $[[X]]$, and α is $1 \in H^0(X)$.

Next we define the relevant bordism relation. There will be two versions. We begin with the one where the targets of the two rank k normal maps under consideration are not necessarily the same and therefore the target of the bordism is not necessarily a cylinder.

Definition 7.15 (Bordism of rank k normal maps (of degree one)) For $m = 0, 1$, let $(X_m, O_{X_m}, [[X_m]])$ be a w-oriented finite n-dimensional Poincaré complex and let $(M_m, i_m, f_m, \xi_m, \overline{f}_m, o_m)$ be a rank k normal map with target $(X_m; O_{X_m}, [[X_m]])$.

(i) A *normal bordism of normal maps*
$$(W, I, F, \Xi, \overline{F}, O)$$
from $(M_0, i_0, f_0, \xi_0, \overline{f}_0, o_0)$ to $(M_1, i_1, f_1, \xi_1, \overline{f}_1, o_1)$ consists of:

- A compact manifold W of dimension $n+1$ whose boundary ∂W is the disjoint union $\partial_0 W \amalg \partial_1 W$;
- An embedding of manifolds with boundary $I \colon W \hookrightarrow \mathbb{R}^{n+k} \times [0, 1]$ sending $\partial_m W$ to $\mathbb{R}^{n+k} \times \{m\}$ for $m = 0, 1$;
- Diffeomorphisms $u_m \colon M_m \to \partial_m W$ and $U_m \colon \mathbb{R}^{n+k} \to \mathbb{R}^{n+k} \times \{m\}$ for $m = 0, 1$ satisfying $I \circ u_m = U_m \circ i_m$;
- A finite $(n + 1)$-dimensional Poincaré pair $(Y, \partial Y; O_Y)$ such that ∂Y is the disjoint union $\partial_0 Y \amalg \partial_1 Y$;
- Isomorphisms of CW-complexes $v_m \colon X_m \to \partial_m Y$ for $m = 0, 1$;
- A map $F \colon W \to Y$ satisfying $l_m \circ v_m \circ f_m = F \circ k_m \circ u_m$, where $k_m \colon \partial_m W \to W$ and $l_m \colon \partial_m Y \to Y$ are the obvious inclusions for $m = 0, 1$;
- A k-dimensional vector bundle Ξ over Y together with a bundle map $(F, \overline{F}) \colon \nu(I) \to \Xi$;

7.4 Normal Maps

- Bundle maps $(l_m \circ v_m, \overline{l_m \circ v_m}): \xi_m \to \Xi$ for $m = 0, 1$ satisfying $\overline{F} \circ \nu(u_m, U_m) = \overline{l_m \circ v_m} \circ \overline{f}_m$, where $(u_m, \nu(u_m, U_m)): \nu(i_m) \to \nu(I)$ is the composite of the bundle map $\nu(i_m) \to u_m^* \nu(I|_{\partial_m W})$ induced by Tu_m and TU_m and the pullback of the canonical bundle isomorphism $\kappa: \nu(I|_{\partial_m W}) \xrightarrow{\cong} \nu(I)|_{\partial_m W}$ of Lemma 5.23 (v) with u_m;

- Isomorphisms of infinite cyclic local coefficient systems $O: O_\Xi \xrightarrow{\cong} O_Y$ over Y and $O_{l_m \circ v_m}: O_{X_m} \to (l_m \circ v_m)^* O_Y$ over X_m for $m = 0, 1$ such that the isomorphisms $O_{\xi_m} \xrightarrow{o_m} O_{X_m} \xrightarrow{O_{l_m \circ v_m}} (l_m \circ v_m)^* O_Y$ and $O_{X_m} \xrightarrow{O_{\overline{l_m \circ v_m}}} (l_m \circ v_m)^* O_\Xi \xrightarrow{(l_m \circ v_m)^* O} (l_m \circ v_m)^* O_Y$ agree for $m = 0, 1$;

(ii) The isomorphism

$$\beta: O_W \to F^* O_Y$$

of infinite cyclic local coefficient systems over W associated to the bordism of rank k normal maps above is given by the canonical isomorphism $O_W \xrightarrow{\cong} O_{\nu(I)}$ of Lemma 5.23 (iv), the isomorphism $O_{\nu(I)} \xrightarrow{\cong} \overline{F}^* \Xi$ induced by $\overline{F}$ and the isomorphism $\overline{F}^* \Xi \xrightarrow{\cong} F^* O_Y$ induced by O;

(iii) If $(M_m, i_m, f_m, \overline{f}_m)$ is a rank k normal map of degree one for $m = 0, 1$, a bordism $(W, I, F, \Xi, \overline{F}, O)$ as above is called a *normal bordism of rank k normal maps of degree one* if the following holds for some fundamental class $[[Y, \partial Y]]$ of $(Y, \partial Y)$:

- The image of the intrinsic fundamental class of $(W, \partial W)$ under the map induced by F and β

$$H_{n+1}(W, \partial W; O_W) \to H_{n+1}(Y, \partial Y; O_Y)$$

is $[[Y, \partial Y]]$;

- The image of $[[Y, \partial Y]]$ under the composite

$$H_{n+1}(Y, \partial Y; O_Y) \xrightarrow{\partial} H_n(\partial Y; O_Y|_{\partial Y}) \xrightarrow{\cong} H_n(X_0; O_{X_0}) \oplus H_n(X_1; O_{X_1})$$

is $(-[[X_0]], [[X_1]])$, where the second arrow is the inverse of the isomorphism induced by $l_m \circ v_m: X_m \to \partial_m Y$ and $O_{X_m} \xrightarrow{o_m^{-1}} O_{\xi_m} \xrightarrow{O_{l_m \circ v_m}} (l_m \circ v_m)^* O_Y$ for $m = 0, 1$.

The second version of the relevant bordism relation only makes sense when the targets of the two rank k normal maps under consideration agree. The only difference is now that in this case we can demand the target of the bordism to be the cylinder.

Definition 7.16 (Bordism of a rank k normal map with cylindrical target (of degree one)) Let $X = (X, O_X, [[X]])$ be a w-oriented finite n-dimensional Poincaré complex. Let $(M_m, i_m, f_m, \xi_m, \overline{f}_m, o_m)$ be a rank k normal map with the same target $(X, O_X, [[X]])$ for $m = 0, 1$.

Then a normal bordism of normal maps (of degree one) $(W, I, F, \Xi, \overline{F}, O)$ from $(M_0, i_0, f_0, \xi_0, \overline{f}_0, o_0)$ to $(M_1, i_1, f_1, \xi_1, \overline{f}_1, o_1)$ in the sense of Definition 7.15 is called a *normal bordism with cylindrical target of normal maps (of degree one)* if

- $(Y, \partial Y)$ is the cylinder $(X \times [0, 1], X \times \{0, 1\})$, we have $\partial_m Y = X \times \{m\}$, and $v_m \colon X \to \partial_m Y$ sends x to (x, m) for $m = 0, 1$;
- The bundle Ξ over Y is $\Xi = \mathrm{pr}_X^* \xi$ for the projection $\mathrm{pr} \colon X \times [0, 1] \to X$;
- The isomorphism $O \colon O_\Xi := O_{\mathrm{pr}_X^* \xi} = \mathrm{pr}_X^* O_X \to O_Y := \mathrm{pr}_X^* O_X$ is $\mathrm{pr}_X^* o$.

Denote by $\mathcal{NM}_n(X, k)$ the set of normal bordism classes of rank k normal maps of degree one to X where the bordism relation is the one with cylindrical ends in the sense of Definition 7.16. Define $\mathcal{NM}_n(X, k) \to \mathcal{NM}_n(X, k+1)$ by sending the class of $(M, i, f, \xi, \overline{f}, o)$ to the class of $(M, i', f, \xi \oplus \mathbb{R}, \overline{f}', o')$ where $i' \colon M \hookrightarrow \mathbb{R}^{n+k+1}$ is the composite of i with the standard inclusion $\mathbb{R}^{n+k} \hookrightarrow \mathbb{R}^{n+k+1}$, $\overline{f}'$ is $\nu(i') = \nu(i) \oplus \mathbb{R} \xrightarrow{\nu(f) \oplus \mathrm{id}} \xi \oplus \mathbb{R}$, and o' is obtained from o using Lemma 5.23 (i).

Definition 7.17 (Set of normal maps) Let $X = (X, O_X, [[X]])$ be a w-oriented finite n-dimensional Poincaré complex. Define the *set of normal maps to X*

$$\mathcal{NM}_n(X) := \mathrm{colim}_{k \to \infty} \mathcal{NM}_n(X, k).$$

Exercise 7.18 Let $(M, i, f, \xi, \overline{f}, o)$ be a rank k normal map of degree one for k large. Show that there exists a fibre homotopy equivalence $S\xi \to \nu_X$ where ν_X is the Spivak normal k-fibration.

The proof of the next result is similar to the proof of Theorem 4.55 and uses Exercise 6.46 and Lemma 6.67 (ii).

Theorem 7.19 (Normal maps and normal invariants) Let $X = (X, O_X, [[X]])$ be a w-oriented finite n-dimensional Poincaré complex. Then the Pontrjagin–Thom construction yields for each $k \geq 1$ a bijection

$$P_k(X) \colon \mathcal{NM}_n(X, k) \xrightarrow{\cong} \mathcal{NI}_n(X, k)$$

and a bijection

$$P(X) \colon \mathcal{NM}_n(X) \xrightarrow{\cong} \mathcal{NI}_n(X).$$

Exercise 7.20 Let $X = (X, O_X, [[X]])$ be a w-oriented finite n-dimensional Poincaré complex. Let $\mathrm{VB}(X)$ be the set of stable isomorphism classes of vector bundles over X, see (6.18). Show that we get a well-defined map

$$\beta \colon \mathcal{NM}_n(X) \to \mathrm{VB}(X)$$

by sending the class of the normal map of degree one $(M, i, f, \xi, \overline{f}, o)$ to the class of ξ. Characterise its image in terms of the Spivak normal fibration.

7.4 Normal Maps

Remark 7.21 (Tangential versus normal data) In view of the Pontrjagin–Thom construction it is convenient to work with the normal bundle. On the other hand one always needs an embedding and one would prefer an intrinsic definition. This is possible if one defines the normal map in terms of the tangent bundle, which we will do below. Both approaches are equivalent, as is shown in Lemma 7.32. We will use in the sequel the one that is adequate for the concrete purpose. We mention that for a generalisation to the equivariant setting the approach using the tangent bundle is more useful, see [262, 263].

Definition 7.22 (Normal maps with respect to the tangent bundle (of degree one)) Let $X = (X, \mathcal{O}_X, [[X]])$ be a w-oriented finite n-dimensional Poincaré complex.

(i) A *normal map with respect to the tangent bundle* $(M, f, a, \xi, \overline{f}, o)$ with target $(X, \mathcal{O}_X)$ consists of the following data:

- A closed manifold M of dimension n;
- A map $f \colon M \to X$;
- A vector bundle ξ over X;
- A bundle map $(f, \overline{f}) \colon TM \oplus \mathbb{R}^a \to \xi$ for some natural number a;
- An isomorphism $o \colon \mathcal{O}_\xi \xrightarrow{\cong} \mathcal{O}_X$ of infinite cyclic local coefficient systems;

(ii) The isomorphism

$$\alpha \colon \mathcal{O}_M \to f^* \mathcal{O}_X$$

of infinite cyclic local coefficient systems over M associated to the normal map with respect to the tangent bundle $(M, i, f, \xi, \overline{f}, o)$ is given by the canonical isomorphism $\mathcal{O}_M = \mathcal{O}_{TM} \xrightarrow{\cong} \mathcal{O}_{TM \oplus \mathbb{R}^a}$ coming from Lemma 5.23 (i), the isomorphisms $\mathcal{O}_{TM \oplus \mathbb{R}^a} \xrightarrow{\cong} f^* \mathcal{O}_\xi$ induced by $\overline{f}$ and the isomorphism $f^* \mathcal{O}_\xi \xrightarrow{\cong} f^* \mathcal{O}_X$ induced by o;

(iii) A *normal map of degree one with respect to the tangent bundle* with target $(X, \mathcal{O}_X, [[X]])$ is a normal map with respect to the tangent bundle with target $(X, \mathcal{O}_X)$ such that the image of the intrinsic fundamental class $[[M]]$ of M, see Definition 5.29, under the map

$$H_n(f, \alpha) \colon H_n(M; \mathcal{O}_M) \to H_n(X; \mathcal{O}_X)$$

is $[[X]]$.

Remark 7.23 (Role of o) The class $[[X]]$ is determined by the other data because of condition (iii) appearing in Definition 7.22. However, when X is a manifold, we will have to take $[[X]]$ to be the intrinsic fundamental class. All this will fit nicely together when we will deal with the surgery exact sequence in Chapter 11.

Suppose that $\pi_0(f) \colon \pi_0(M) \to \pi_0(X)$ is bijective. Then the isomorphism $o \colon \mathcal{O}_\xi \xrightarrow{\cong} \mathcal{O}_X$ appearing in part (i) of Definition 7.22 is uniquely determined by the other data and the condition (iii). This is not true in general without the assumption that $\pi_0(f) \colon \pi_0(M) \to \pi_0(X)$ is bijective. In general o determines $[[X]]$, but not vice versa.

If $(M, f, a, \xi, \overline{f}, o)$ is a normal map of degree one with respect to the tangent bundle with target $(X, O_X, [[X]])$, then $(M, f, a, \xi, \overline{f}, -o)$ is a normal map of degree one with respect to the tangent bundle with target $(X, O_X, -[[X]])$. Note that the passage from $[[X]]$ to $-[[X]]$ requires o to change as well.

The notion of bordism is now rather obvious, but for the reader's convenience we state it explicitly. We will again have two versions.

Definition 7.24 (Bordism of normal maps with respect to the tangent bundle (of degree one)) For $m = 0, 1$, let $(X_m, O_{X_m}, [[X_m]])$ be a w-oriented finite n-dimensional Poincaré complex and let $(M_m, f_m, a_m, \xi_m, \overline{f}_m, o_m)$ be a normal map with respect to the tangent bundle (of degree one) with target $(X_m, O_{X_m}, [[X_m]])$.

(i) A *normal bordism of normal maps with respect to the tangent bundle* $(W, F, b, \Xi, \overline{F}, O)$ from $(M_0, f_0, a_0, \xi_0, \overline{f}_0, o_0)$ to $(M_1, f_1, a_1, \xi_1, \overline{f}_1, o_1)$ consists of:

- A compact manifold W of dimension $n+1$ whose boundary ∂W is the disjoint union $\partial_0 W \amalg \partial_1 W$;
- A finite $(n+1)$-dimensional Poincaré pair $(Y, \partial Y, O_Y)$ such that ∂Y is the disjoint union $\partial_0 Y \amalg \partial_1 Y$;
- Diffeomorphisms $u_m \colon M_m \to \partial_m W$ and isomorphisms of CW-complexes $v_m \colon X_m \to \partial_m Y$ for $m = 0, 1$;
- A map $F \colon W \to Y$ satisfying $l_m \circ v_m \circ f_m = F \circ k_m \circ u_m$, where $k_m \colon \partial_m W \to W$ and $l_m \colon \partial_m Y \to Y$ are the obvious inclusions for $m = 0, 1$;
- A vector bundle Ξ over Y and a bundle map $(F, \overline{F}) \colon TW \oplus \mathbb{R}^b \to \Xi$ for some natural number b satisfying $b \geq a_m$ for $m = 0, 1$;
- Bundle maps $(\overline{l_m \circ v_m}, \overline{l_m \circ v_m}) \colon \xi_m \oplus \mathbb{R}^{b+1-a_m} \to \Xi$ such that

$$(F, \overline{F}) \circ \left((u_m, \overline{u_m}) \oplus (u_m, \mathrm{id}_{\mathbb{R}^{b+1-a_m}})\right) \circ (\mathrm{id}_{TM} \oplus \mathrm{flip}_m, \mathrm{id}_M)$$
$$= (\overline{l_m \circ v_m}, \overline{l_m \circ v_m}) \circ \left((f_m, \overline{f}_m) \oplus (f_m, \mathrm{id}_{\mathbb{R}^{b+1-a_m}})\right)$$

holds where $(u_m, \overline{u_m}) \colon \mathbb{R} \oplus TM_m \to TW$ is the bundle map induced by Tu_m with respect to some inward normal vector field for $m = 0$ and for some outward normal vector field for $m = 1$, and $\mathrm{flip}_m \colon \mathbb{R}^{b+1} = \mathbb{R}^{a_m} \oplus \mathbb{R} \oplus \mathbb{R}^{b-a_m} \to \mathbb{R}^{b+1} = \mathbb{R} \oplus \mathbb{R}^{a_m} \oplus \mathbb{R}^{b-a_m}$ is the flip map;

- Isomorphisms of infinite cyclic local coefficient systems $O \colon O_\Xi \xrightarrow{\cong} O_Y$ over Y and $O_{l_m \circ v_m} \colon O_{X_m} \to (l_m \circ v_m)^* O_Y$ over X_m for $m = 0, 1$ such that the isomorphisms $O_{\xi_m} \xrightarrow{o_m} O_{X_m} \xrightarrow{O_{l_m \circ v_m}} (l_m \circ v_m)^* O_Y$ and $O_{\xi_m} \xrightarrow{\kappa} O_{\xi_m \oplus \mathbb{R}^{b+1-a_m}} \xrightarrow{\overline{l_m \circ v_m}} (l_m \circ v_m)^* O_\Xi \xrightarrow{(l_m \circ v_m)^* O} (l_m \circ v_m)^* O_Y$ agree for $m = 0, 1$, where κ comes from Lemma 5.23 (i);

7.4 Normal Maps

(ii) The isomorphism

$$\beta\colon O_W \to F^*O_Y$$

of infinite cyclic local coefficient systems over W associated to the bordism of rank k normal maps above is given by the canonical isomorphism $\kappa\colon O_W \xrightarrow{\cong} O_{TW \oplus \mathbb{R}^b}$ coming from Lemma 5.23 (i), the isomorphism $O_{TW \oplus \mathbb{R}^b} \xrightarrow{\cong} F^*\Xi$ induced by $\overline{F}$, and the isomorphism $F^*O\colon F^*O_\Xi \xrightarrow{\cong} F^*O_Y$;

(iii) If $(M_m, f_m, a_m, \xi_m, \overline{f}_m, o_m)$ is a normal map with respect to the tangent bundle of degree one with target $(X_m, O_{X_m}, [[X_m]])$ for $m = 0, 1$, a bordism $(W, F, b, \Xi, \overline{F}, O)$ as above is called a *normal bordism with respect to the tangent bundle of degree one* if the following holds for some fundamental class $[[Y, \partial Y]]$ of $(Y, \partial Y)$:

- The image of the intrinsic fundamental class of $(W, \partial W)$ under the map induced by F and β

$$H_{n+1}(W, \partial W; O_W) \to H_{n+1}(Y, \partial Y; O_Y)$$

is $[[Y, \partial Y]]$;
- The image of $[[Y, \partial Y]]$ under the composite

$$H_{n+1}(Y, \partial Y; O_Y) \xrightarrow{\partial} H_n(\partial Y; O_Y|_{\partial Y}) \xrightarrow{\cong} H_n(X_0; O_{X_0}) \oplus H_n(X_1; O_{X_1})$$

is $(-[[X_0]], [[X_1]])$, where the second arrow is the inverse of the isomorphism induced by $l_m \circ v_m \colon X_m \to \partial_m Y$ and $O_{X_m} \xrightarrow{o_m^{-1}} O_{\xi_m} \xrightarrow{O_{l_m \circ v_m}} (l_m \circ v_m)^* O_Y$ for $m = 0, 1$.

Definition 7.25 (Bordism of normal maps with cylindrical target with respect to the tangent bundle (of degree one)) Let $(X, O_X, [[X]])$ be a w-oriented finite n-dimensional Poincaré complex together with a bundle ξ over X and an isomorphism $o\colon O_\xi \xrightarrow{\cong} O_X$ of infinite cyclic coefficient systems. Let $(M_m, f_m, a_m, \xi_m, \overline{f}_m, o_m)$ be a normal map with respect to the tangent bundle with the same target $(X, O_X, [[X]])$ for $m = 0, 1$.

Then a bordism of normal maps with respect to the tangent bundle (of degree one) $(W, F, b, \Xi, \overline{F}, O)$ from $(M_0, f_0, a_0, \xi_0, \overline{f}_0, o_0)$ to $(M_1, f_1, a_1, \xi_1, \overline{f}_1, o_1)$ in the sense of Definition 7.24 is called a *normal bordism of normal maps with cylindrical target with respect to the tangent bundle (of degree one)* if

- $(Y, \partial Y)$ is the cylinder $(X \times [0, 1], X \times \{0, 1\})$, we have $\partial_m Y = X \times \{m\}$, and $v_m \colon X \to \partial_m Y$ sends x to (x, m) for $m = 0, 1$;
- The bundle Ξ over Y is $\Xi = \mathrm{pr}_X^* \xi$ for the projection $\mathrm{pr}\colon X \times [0, 1] \to X$;
- The isomorphism $O\colon O_\Xi := O_{\mathrm{pr}_X^* \xi} = \mathrm{pr}_X^* O_\xi \to O_{X \times [0,1]} := \mathrm{pr}_X^* O_X$ is $\mathrm{pr}_X^* o$.

We need the different choice of inward and outward normal vector field in Definition 7.24 (i) in order to be able to glue two normal bordism together when proving transitivity of the equivalence relation given by normal bordism. Moreover, it ensures that the bordism relation is reflexive and gives the right interpretation between bordism and nullbordism due to the following observations.

Example 7.26 (The role of the inward and the outward normal vector field) Let M be a closed manifold. Let $\mathrm{pr}_M \colon M \times [0,1] \to M$ and $\mathrm{pr}_{[0,1]} \colon M \times [0,1] \to [0,1]$ be the projections. Let $\overline{a} \colon T[0,1] \xrightarrow{\cong} \mathbb{R}$ be the standard trivialisation of $T[0,1]$ coming from the unit tangent vector pointing to the right at each point in $[0,1]$. We get a preferred isomorphism of vector bundles over $M \times [0,1]$

$$\overline{b} \colon T(M \times [0,1]) \xrightarrow{\cong} \mathrm{pr}_M^* TM \oplus \mathrm{pr}_{[0,1]}^* T[0,1] \xrightarrow{\mathrm{id}_{\mathrm{pr}_M^* TM} \oplus \mathrm{pr}_{[0,1]}^* \overline{a}} \mathrm{pr}_M^* TM \oplus \mathbb{R}.$$

For $m = 0, 1$ we have the inclusion $v_m \colon M \to M \times [0,1]$ sending x to (x, m). Its differential together with the inward normal vector field for $m = 0$ or the outward normal vector field for $m = 1$ induces an isomorphism $\overline{i_m} \colon TM \oplus \mathbb{R} \to T(M \times [0,1])$ covering i_m. For both $m = 0$ and $m = 1$ the composite $\overline{b} \circ \overline{i_m}$ induces an isomorphism of vector bundles over M from $TM \oplus \mathbb{R} \xrightarrow{\cong} v_m^* \mathrm{pr}_M^* TM \oplus \mathbb{R}$ that is under the obvious identification $v_m^* \mathrm{pr}_M^* TM = TM$ the identity.

Now it is easy to see that normal bordism is reflexive.

Example 7.27 (Bordism and nullbordism) Consider for $m = 0, 1$ a w-oriented finite n-dimensional Poincaré complex $(X_m, O_{X_m}, [[X_m]])$ and a rank k normal map $(M_m, i_m, f_m, \xi_m, \overline{f}_m, o_m)$ with respect to the tangent bundle of degree one with target $(X_m; O_{X_m}, [[X_m]])$. From $(M_1, i_1, f_1, \xi_1, \overline{f}_1, o_1)$ we obtain a rank k normal map with respect to the tangent bundle of degree one $(M_1, i_1, f_1, \xi_1, \overline{f}_1, -o_1)$ with target $(X_1; O_{X_1}, -[[X_1]])$, by inserting a sign in front of o_1 and $[[X_1]]$. Taking the disjoint union of $(M_0, i_0, f_0, \xi_0, \overline{f}_0, o_0)$ and $(M_1, i_1, f_1, \xi_1, \overline{f}_1, -o_1)$ we obtain a normal map with respect to the tangent bundle of degree one with target $(X_0; O_{X_0}, [[X_0]]) \sqcup (X_1; O_{X_1}, -[[X_1]])$. It is normally nullbordant, i.e., normally bordant to the empty normal map in the sense of Definition 7.24 if and only if $(M_0, i_0, f_0, \xi_0, \overline{f}_0, o_0)$ and $(M_1, i_1, f_1, \xi_1, \overline{f}_1, o_1)$ are bordant in the sense of Definition 7.24. The passage from o_1 to $-o_1$ and from $[[X_1]]$ to $-[[X_1]]$ comes from the fact that we have to distinguish between the inward and outward normal vector fields.

This already has an interesting consequence. Namely, consider the special case $(M_0, i_0, f_0, \xi_0, \overline{f}_0, o_0) = (M_1, i_1, f_1, \xi_1, \overline{f}_1, o_1)$. Then we conclude that the disjoint union of $(M_1, i_1, f_1, \xi_1, \overline{f}_1, o_1)$ and $(M_1, i_1, f_1, \xi_1, \overline{f}_1, -o_1)$ is normally nullbordant.

Remark 7.28 One reason for the complexity of Definitions 7.15 and 7.24 is that one has to take all the various isomorphisms, for instance between M_m and $\partial_m W$, into account, to ensure that the normal bordism relation is transitive. Transitivity is proved by gluing two normal bordisms together. The result of the gluing and the possibility to extend all the data to the new bordism depends on the gluing maps and a lot of compatibility conditions. Since normal bordism is obviously symmetric, it is an equivalence relation.

7.4 Normal Maps   209

Remark 7.29 In Definition 7.24 there is a sign appearing in $(-[[X_0]], [[X_1]])$, when dealing with the degree one condition. There are two explanations for this. Because we use an inward and an outward normal field in Definition 7.24, this is compatible with Lemma 5.34. Another argument is that we have for a w-oriented finite n-dimensional Poincaré complex $(X; O_X, [[X]])$ the exact sequence

$$H_{n+1}(X \times [0,1], X \times \{0,1\}; \mathrm{pr}_X^* O_X) \xrightarrow{\partial_{n+1}} H_n(X \times \{0,1\}; j^* \mathrm{pr}_X^* O_X)$$
$$\to H_n(X \times [0,1], \mathrm{pr}_M {}^*O_X)$$

where $\mathrm{pr}_X \colon X \times [0,1] \to X$ is the projection and $j \colon X \times \{0,1\} \to X \times [0,1]$ the inclusion. Under the obvious identifications

$$H_n(X \times \{0,1\}; j^* \mathrm{pr}_X O_X) = H_n(X; O_X) \oplus H_n(X; O_X);$$
$$H_n(X \times [0,1], \mathrm{pr}_M {}^*O_X) = H_n(X; O_X),$$

the latter map is $\mathrm{id} \oplus \mathrm{id}$. Hence the image of $[[X, \partial X]]$ under ∂_{n+1} cannot be $([[X]], [[X]])$, but with our conventions turns out to be $(-[[X]], [[X]])$.

Denote by $\mathcal{NT}_n(X)$ the set of bordism classes with cylindrical target of normal maps with respect to the tangent bundle of degree one.

Exercise 7.30 Let $(M, f, a, \xi, \overline{f}, o)$ be normal map with respect to the tangent bundle of degree one. Show that the following modifications do not change its class in $\mathcal{NT}_n(X)$:

(i) Let $u \colon \xi \xrightarrow{\cong} \xi$ be a bundle automorphism covering id_X. It induces an isomorphism $O_u \colon O_\xi \xrightarrow{\cong} O_\xi$. Replace $(M, f, a, \xi, \overline{f}, o)$ by $(M, f, a, \xi, u \circ \overline{f}, O_u \circ o)$;
(ii) Let $(h, \overline{h}) \colon TM \oplus \mathbb{R}^a \times [0,1] \to \xi$ be a homotopy of bundle maps from $(f, \overline{f})$ to $(g, \overline{g})$. Replace $(M, f, a, \xi, \overline{f}, o)$ by $(M, g, a, \xi, \overline{g}, o)$;
(iii) Let $g \colon M' \to M$ be a diffeomorphism. Replace $(M, f, a, \xi, \overline{f}, o)$ by $(M', f \circ g, a, \xi, \overline{f} \circ \overline{Tg} \oplus \mathrm{id}_{\mathbb{R}^a}, o)$;
(iv) Replace $(M, f, a, \xi, \overline{f}, o)$ by $(M, f, a+1, \xi \oplus \mathbb{R}, \overline{f} \oplus \mathrm{id}_{\mathbb{R}}, o \circ v)$, where the isomorphism $v \colon O_{\xi \oplus \mathbb{R}} \xrightarrow{\cong} O_\xi$ comes from Lemma 5.23 (i).

Remark 7.31 (ξ-bordism versus normal bordism) Note that there is one important difference between the notion of ξ-bordism and normal bordism. Consider a normal ξ-map $(M, i, f, \overline{f})$ in the sense of Definition 4.52. Let $u \colon \xi \xrightarrow{\cong} \xi$ be any bundle automorphism covering id_X such that the induced isomorphism $O_u \colon O_\xi \xrightarrow{\cong} O_\xi$ is the identity. Then $(M, i, f, \overline{f})$ and $(M, i, f, u \circ \overline{f})$ are not necessarily ξ-bordant in the sense of Definition 4.52. This is in sharp contrast to property (i) appearing in Exercise 7.30. See also Exercise 7.20.

Lemma 7.32 *Let X be a finite n-dimensional Poincaré complex. There is a natural bijection*

$$\mathcal{NM}_n(X, O_X) \cong \mathcal{NT}_n(X, O_X).$$

Proof. We define a map

$$\phi_n(k)\colon \mathcal{NM}_n(X,k) \to \mathcal{NT}_n(X)$$

as follows. Consider a rank k normal map $(M, i, f, \xi, \overline{f}, o)$. Since X is compact, we can find a bundle η together with an isomorphism $u\colon \mathbb{R}^a \xrightarrow{\cong} \xi \oplus \eta$. Choose a section of the canonical projection $\underline{\mathbb{R}^{n+k}} = i^*T\underline{\mathbb{R}^{n+k}} \to \nu(i)$. It yields together with $Ti\colon TM \to i^*T\underline{\mathbb{R}^{n+k}}$ an isomorphism $v\colon TM \oplus \nu(i) \xrightarrow{\cong} \underline{\mathbb{R}^{n+k}}$. We get from $\overline{f}, u$, and v an isomorphism of bundles covering the identity on M

$$TM \oplus \underline{\mathbb{R}^a} = TM \oplus f^*\underline{\mathbb{R}^a} \xrightarrow{\mathrm{id}_{TM} \oplus f^*u} TM \oplus f^*(\xi \oplus \eta) \xrightarrow{\mathrm{flip}} f^*\eta \oplus TM \oplus f^*\xi$$

$$\xrightarrow{\mathrm{id}_{f^*\eta} \oplus \mathrm{id}_{TM} \oplus \overline{f}^{-1}} f^*\eta \oplus TM \oplus \nu(i) \xrightarrow{\mathrm{id}_{f^*\eta} \oplus v} f^*\eta \oplus \underline{\mathbb{R}^{n+k}} = f^*(\eta \oplus \underline{\mathbb{R}^{n+k}}).$$

This is the same as a bundle map $(f, \overline{g})\colon TM \oplus \underline{\mathbb{R}^a} \to \eta \oplus \underline{\mathbb{R}^{n+k}}$ covering f. We obtain a preferred isomorphism $w\colon O_{\eta \oplus \mathbb{R}^{n+k}} \xrightarrow{\cong} O_\xi$ from u and Lemma 5.23 (iv). Define the image of the class of $(M, i, f, a, \xi, \overline{f}, o)$ under $\phi_n(k)$ to be the class of $(M, f, a, \eta \oplus \underline{\mathbb{R}^{n+k}}, \overline{g}, o \circ w)$. One easily checks that this is well defined and that the maps $\phi_n(k)$ fit together to yield a map $\phi_n\colon \mathcal{NM}_n(X) \to \mathcal{NT}_n(X)$. By the analogous construction one gets an inverse. □

Notation 7.33 In view of Theorem 7.19 and Lemma 7.32 we often identify for a w-oriented finite n-dimensional Poincaré complex $X = (X, O_X, [[X]])$ the sets $\mathcal{NI}_n(X) = \mathcal{NM}_n(X) = \mathcal{NT}_n(X)$ and write $\mathcal{N}(X)$ for short.

Theorem 7.34 (Normal Maps and G/O)

(i) Let X be a w-oriented finite n-dimensional Poincaré complex. If $\mathcal{N}(X)$ is non-empty, then there is a free transitive action of $[X, G/O]$ on $\mathcal{N}(X)$. Hence any choice of an element in $\mathcal{N}(X)$ gives a bijection

$$[X, G/O] \cong \mathcal{N}(X);$$

(ii) Let M be a closed n-dimensional manifold. Then the set $\mathcal{N}(M)$, which is to be understood with respect to O_M and the intrinsic fundamental class $[[M]]$, is non-empty and comes with a preferred base point, and there is a canonical bijection

$$[M, G/O] \cong \mathcal{N}(M).$$

Proof. (i) This follows from Theorem 7.10.

(ii) The preferred base point in $\mathcal{N}(M)$ is given by id_M, id_{TM}, and id_{O_M}. Now the claim follows from assertion (i). □

Exercise 7.35 Let S be a connected closed 2-dimensional manifold. Using the fact that G/O is simply connected and $\pi_2(\mathrm{G/O})$ is $\mathbb{Z}/2$ show that $\mathcal{N}_2(S)$ consists of precisely two elements.

7.4 Normal Maps

The next theorem essentially says that the degree is a bordism invariant. Actually, it contains the stronger statement that also the degree of the bordism (with cylindrical target) itself is fixed.

Theorem 7.36 (Bordism invariance of the degree) *Let (X, O_X) be a finite n-dimensional Poincaré complex. For $m = 0, 1$, let M_m be a closed n-dimensional manifold, $f : M_m \to X$ be a map, and $o_m : O_M \to f_m^* O_X$ be an isomorphism of infinite cyclic local coefficient systems over M_m. Suppose that (M_0, f_0, o_0) and a (M_1, f_1, o_1) are bordant in the following sense:*

- *There is a compact manifold W of dimension $n + 1$ whose boundary ∂W is the disjoint union $\partial_0 W \amalg \partial_1 W$;*
- *There are diffeomorphisms $u_m : M_m \to \partial_m W$ for $m = 0, 1$;*
- *There is a map $F : W \to X \times [0, 1]$ satisfying $l_m \circ f_m = F \circ k_m \circ u_m$ where $k_m : \partial_m W \to W$ is the inclusion and $l_m : X \to X \times [0, 1]$ sends x to (x, m) for $m = 0, 1$;*
- *Let $\mathrm{pr} : X \times [0, 1] \to X$ be the projection. Let $(u_m, \overline{u}_m) : \mathbb{R} \oplus TM_m \to TW$ be the bundle map induced by Tu_m with respect to some inward normal vector field for $m = 0$ and for some outward normal vector field for $m = 1$. Let $\kappa_m : O_{M_m} = O_{TM_m} \to O_{TM_m \oplus \mathbb{R}}$ be the isomorphism coming from Lemma 5.23 (i). Put $\widehat{l_m \circ u_m} = o_{\overline{u}_m} \circ \kappa_m : O_{M_m} \xrightarrow{\cong} u_m^* l_m^* O_W$.*

We require the existence of an isomorphism of infinite cyclic local coefficient systems $O : O_W \to F^* \mathrm{pr}^* O_X$ such that for $m = 0, 1$ the following diagram

$$
\begin{array}{ccc}
O_{M_m} & \xrightarrow{o_m} & f_m^* O_X \\
{\scriptstyle \widehat{l_m \circ u_m}} \downarrow & & \downarrow {\scriptstyle \mathrm{id}} \\
u_m^* l_m^* O_W & \xrightarrow{u_m^* l_m^* O} & u_m^* l_m^* F^* \mathrm{pr}^* O_X
\end{array}
$$

commutes.

Then we get in $H_n(X; O_X)$ the equality

$$H_n(f_0, o_0)([[M_0]]) = H_n(f_1, o_1)([[M_1]]) = \sigma \circ H_{n+1}(F, \partial F; O)([[W; \partial W]]),$$

where $\sigma : H_{n+1}(X \times [0, 1], X \times \{0, 1\}; \mathrm{pr}_X^* O_X) \xrightarrow{\cong} H_n(X; O_X)$ is the suspension isomorphism and $\partial F = F|_{\partial W} : \partial W \to X \times \{0, 1\}$.

Proof. The following diagram

$$
\begin{array}{ccc}
H_{n+1}(W, \partial W; O_W) & \xrightarrow{H_n(F, \partial F; O)} & H_{n+1}(X \times [0,1], X \times \{0,1\}; \mathrm{pr}_X^* O_X) \\
\downarrow \partial_{n+1} & & \downarrow \partial_{n+1} \\
H_n(\partial W; O_W|_{\partial W}) & \xrightarrow{H_n(\partial F; O|_{\partial W})} & H_n(X \times \{0,1\}; \mathrm{pr}_X^* O_X|_{X \times \{0,1\}}) \\
\end{array}
$$

$$
\begin{array}{c}
H_n(l_0 \circ u_0; \widehat{l_0 \circ u_0}) \\
\oplus \\
H_n(l_1 \circ u_1; \widehat{l_1 \circ u_1})
\end{array}
\cong
\quad
\begin{array}{c}
H_n(f_0; o_0) \\
\oplus \\
H_n(f_1; o_1)
\end{array}
\quad
\cong
\quad
\begin{array}{c}
H_n(l_0; \widehat{l_0}) \\
\oplus \\
H_n(l_1; \widehat{l_1})
\end{array}
$$

$$
\begin{array}{c}
H_n(M_0, O_{M_0}) \\
\oplus \\
H_n(M_1, O_{M_1})
\end{array}
\longrightarrow
\begin{array}{c}
H_n(X; O_X) \\
\oplus \\
H_n(X; O_X)
\end{array}
$$

commutes. Moreover, we get using Lemma 5.30 and Lemma 5.34

$$\partial_{n+1} \circ H_n(F; O)([[W]])$$
$$= H_n(\partial F; O|_{\partial W}) \circ \partial_{n+1}([[W]])$$
$$= H_n(\partial F; O|_{\partial W})([[\partial W]])$$
$$= H_n(\partial F; O|_{\partial W}) \circ \big(H_n(u_0; \widehat{u_0}) \oplus H_n(u_1; \widehat{u_1})\big)(-[[M_0]], [[M_1]])$$
$$= H_n(l_0; \widehat{l_0}) \oplus H_n(l_1; \widehat{l_1})\big(-H_n(f_0, o_0)([[M_0]]), H_n(f_1, o_1)([[M_1]])\big),$$

where we get two different signs in front of $[[M_i]]$ because of the different choice of inward and outward normal vector field. The sequence

$$H_{n+1}(X \times [0,1], X \times \{0,1\}; \mathrm{pr}_X^* O_X) \xrightarrow{\partial_{n+1}} H_n(X \times \{0,1\}; \mathrm{pr}_X^* O_X|_{\partial X \times [0,1]})$$
$$\xrightarrow{H_n(j;\bar{j})} H_n(X \times [0,1]; \mathrm{pr}_X^* O_X)$$

for $j \colon \partial(X \times [0,1]) \to X \times [0,1]$ the inclusion and $\bar{j} \colon \mathrm{pr}_X^* O_X|_{\partial X \times [0,1]} \xrightarrow{\cong} j^* \mathrm{pr}_X^* O_X$ the obvious isomorphism is exact. The composite

$$H_n(X; O_X) \oplus H_n(X; O_X) \xrightarrow{H_n(l_0;\widehat{l_0}) \oplus H_n(l_1;\widehat{l_1})} H_n(X \times \{0,1\}; \mathrm{pr}_X^* O_X|_{X \times \{0,1\}})$$
$$\xrightarrow{H_n(j;\bar{j})} H_n(X \times [0,1]; \mathrm{pr}_X^* O_X) \xrightarrow{H_n(\mathrm{pr}_X)} H_n(X; O_X)$$

agrees with $\mathrm{id}_{H_n(X;O_X)} \oplus \mathrm{id}_{H_n(X;O_X)}$. Since $H_n(l_0; \widehat{l_0}) \oplus H_n(l_1; \widehat{l_1})$ and $H_n(\mathrm{pr}_X)$ are bijective, the image of

7.4 Normal Maps

$$H_{n+1}(X \times [0,1], X \times \{0,1\}; \text{pr}_X^* O_X)$$

$$\xrightarrow{\partial_{n+1}} H_n(X \times \{0,1\}; \text{pr}_X^* O_X|_{X \times \{0,1\}})$$

$$\xrightarrow{(H_n(l_0;\widehat{l_0}) \oplus H_n(l_1;\widehat{l_1}))^{-1}} H_n(X; O_X) \oplus H_n(X; O_X) \quad (7.37)$$

is $\{(x, -x) \mid x \in H_n(X; O_X)\}$. This implies

$$H_n(f_0, o_0)([[M_0]]) = H_n(f_1, o_1)([[M_1]]).$$

Since the suspension isomorphism σ is the composite of the map (7.37) with the projection onto the second factor, we get

$$H_n(f_1, o_1)([[M_1]]) = \sigma \circ H_n(F; O)([[W; \partial W]]).$$

$\square$

Lemma 7.38 *Let $(X, O_X, [[X]])$ be a w-oriented finite n-dimensional Poincaré complex. Let $(M_0, f_0, a_0, \xi_0, \overline{f}_0, o_0)$ and $(M_1, f_1, a_1, \xi_1, \overline{f}_1, o_1)$ be normal maps with respect to the tangent bundle with target (X, O_X). Suppose that $(M_0, f_0, a_0, \xi_0, \overline{f}_0, o_0)$ has degree one and $(M_1, f_1, a_1, \xi_1, \overline{f}_1, o_1)$ is obtained from $(M_0, f_0, a_0, \xi_0, \overline{f}_0, o_0)$ by a finite sequence of surgery steps described in Definition 4.41.*

Then $(M_1, f_1, a_1, \xi_1, \overline{f}_1, o_1)$ also has degree one and there is a normal bordism with cylindrical target with respect to the tangent bundle of degree one from $(M_0, f_0, a_0, \xi_0, \overline{f}_0, o_0)$ to $(M_1, f_1, a_1, \xi_1, \overline{f}_1, o_1)$.

Proof. Obviously we can assume without loss of generality that we have performed precisely one surgery step. The desired normal bordism is described in Theorem 4.39 (iv). The normal map $(M_1, f_1, a_1, \xi_1, \overline{f}_1, o_1)$ and the normal bordism are of degree one by Theorem 7.36. $\square$

Lemma 7.39 *Let $X = (X, O_X, [[X]])$ be a w-oriented finite Poincaré complex of dimension n. Then there exists a closed manifold M homotopy equivalent to X only if there exists a normal map of degree one with target X.*

Proof. Suppose $f \colon M \to X$ is a homotopy equivalence with a closed manifold as source. Choose a homotopy inverse $f^{-1} \colon X \to M$. Put $\xi = (f^{-1})^* TM$. Then we can cover f by a bundle map $\overline{f} \colon TM \to \xi$. Since $w_1(O_X) = w_1(X) = w_1(M)$, we can choose an isomorphism $o \colon O_\xi \xrightarrow{\cong} O_X$. Thus we get a normal map with target X. By possibly composing o with some automorphism of O_X, we can arrange by Lemma 7.4 that it has degree one. $\square$

Recall that we started with Problem 4.1, then figured out that we have to pass to Problem 7.1 since the conditions appearing in Problem 7.1 are necessary. By Lemma 7.39 we come to the following problem.

Problem 7.40 (Surgery problem) Suppose we have some normal map of degree one from a closed n-dimensional manifold M to a w-oriented finite n-dimensional Poincaré complex X. Can we modify it via a normal bordism leaving X fixed to get a normal map such that the underlying map $M \to X$ is a homotopy equivalence?

This is an important step forward since, for the first time, we now have a candidate for a manifold, which hopefully can be changed so that we can solve Problem 7.40. Moreover, we know by Theorem 7.34 (i) how big the set of normal maps $\mathcal{N}(X)$ is.

We have explained the surgery step in Definition 4.41. We get from Theorem 4.44 and Lemma 7.39

Theorem 7.41 (Making a normal map highly connected) *Let X be a w-oriented connected finite n-dimensional Poincaré complex. Consider a normal map of degree one with respect to the tangent bundle $(M_0, f_0, a_0, \xi_0, \overline{f}_0, o_0)$ with target X of degree one.*

Then we can carry out a finite sequence of surgery steps to obtain a normal map of degree one and $(M_1, f_1, a_1, \xi_1, \overline{f}_1, o_1)$ such that $(M_0, f_0, a_0, \xi_0, \overline{f}_0, o_0)$ and $(M_1, f_1, a_1, \xi_1, \overline{f}_1, o_1)$ are normally bordant with cylindrical target and f_1 is k-connected where $n = 2k$ or $n = 2k + 1$.

So we are finally confronted with the following problem.

Problem 7.42 (Surgery problem for a highly connected map) Suppose we have some normal map of degree one with respect to the tangent bundle $(M, f, a, \xi, \overline{f}, o)$ from a closed n-dimensional manifold M to a w-oriented connected finite Poincaré complex X of dimension n such that the map f is k-connected where $n = 2k$ or $n = 2k + 1$.

Can we change M and f by finitely many surgery steps to get a normal map $(M', f', a', \xi', \overline{f}', o')$ from a closed manifold M' to X such that f' is a homotopy equivalence?

This problem cannot always be solved, since there may be surgery obstructions. Here is a first and very basic example of a surgery obstruction.

Remark 7.43 (The signature as a surgery obstruction) Suppose that X appearing in Problem 7.42 is orientable, i.e., $w_1(X) = 0$, connected, and of dimension $n = 4k$. Recall that we have the isomorphism $\alpha \colon \mathcal{O}_M \xrightarrow{\cong} f^*\mathcal{O}_X$. Fix an isomorphism $\omega \colon \mathcal{O}_X \xrightarrow{\cong} \underline{\mathbb{Z}}$ where $\underline{\mathbb{Z}}$ is the trivial local coefficient system on X with value $\mathbb{Z}$. Hence we also get an isomorphism $\omega_M \colon \mathcal{O}_M \xrightarrow{\cong} \underline{\mathbb{Z}}$ of local coefficient systems over M by the composite $f^*\omega \circ \alpha$. From ω_M and ω_X we obtain isomorphisms $H_n(M; \mathcal{O}_M) \xrightarrow{\cong} H_n(M; \mathbb{Z})$ and $H_n(X; \mathcal{O}_X) \xrightarrow{\cong} H_n(X; \mathbb{Z})$ and denote by $[M] \in H_n(M; \mathbb{Z})$ and $[X] \in H_n(X; \mathbb{Z})$ the images of $[[M]]$ and $[[X]]$. Then $[M]$ and $[X]$ are fundamental classes for M and X and $H_n(f; \mathbb{Z}) \colon H_n(M; \mathbb{Z}) \to H_n(X; \mathbb{Z})$ sends $[M]$ to $[X]$. Let $\mathrm{sign}(M; [M])$ and $\mathrm{sign}(X; [X])$ be the signatures of M and X with respect to these fundamental classes. We leave it to the reader to

7.4 Normal Maps   215

check that $\text{sign}(M;[M]) - \text{sign}(X;[X])$ is a normal bordism invariant using Theorem 5.79 and that $\text{sign}(M;[M]) - \text{sign}(X;[X])$ vanishes if f is a homotopy equivalence. Hence we see an obstruction to solve the Surgery Problem 7.42, namely $\text{sign}(M;[M]) - \text{sign}(X;[X])$ must be zero.

We will see that this is the only obstruction if X is a simply connected orientable $4k$-dimensional Poincaré complex for $k \geq 1$. If X is not simply connected, the vanishing of $\text{sign}(M;[M]) - \text{sign}(X;[X])$ will not be sufficient, more complicated surgery obstructions will occur.

Example 7.44 (The degree 3 Example) Let M be a closed connected n-dimensional manifold. Then we obtain a normal map (with respect to the tangent bundle) with target (M, O_M) by the map

$$\text{id}_M \amalg \text{id}_M \amalg \text{id}_M : N := M \amalg M \amalg M \to M$$

covered by the bundle map

$$\text{id}_{TM} \amalg \text{id}_{TM} \amalg \text{id}_{TM} : TM \amalg TM \amalg TM \to TM$$

and $o = \text{id}_{O_M}$. Obviously the image of the intrinsic fundamental class $[[M \amalg M \amalg M]]$ under the map $H_n(M \amalg M \amalg M; O_{M \amalg M \amalg M}) \to H_n(M; O_M)$ is $3 \cdot [[M]]$.

We conclude from Theorem 7.36 that the normal map above is not normally bordant to a normal map $(M', f', a', \xi', \overline{f}', o')$ of degree one with target (M, O_M) such that M' is connected.

Remark 7.45 (Coupling between the bundle data and the degree one condition) At first glance, Example 7.44 seems to contradict some of the folklore statements one can find in the literature. Namely, choose a fundamental class $[M] \in H_n(M; \mathbb{Z})$. Let M^+ be M with the orientation given by $[M]$ and let M^- be M with the orientation given by $-[M]$. Then $\text{id}_M \amalg \text{id}_M \amalg \text{id}_M : M^+ \amalg M^- \amalg M^+ \to M^+$ has degree one in the classical sense, and there is the claim that one can do surgery so that the resulting normal map with bundle map $\text{id}_{TM} \amalg \text{id}_{TM} \amalg \text{id}_{TM} : TM \amalg TM \amalg TM \to TM$ has a connected source and degree one. However, Example 7.44 predicts that this is not true.

So what is going wrong in the naive setting? Suppose that we want to do surgery to connect the first two summands in $M^+ \amalg M^- \amalg M^+ \to M^+$. This essentially means that we want to pass from $M^+ \amalg M^- \amalg M^+ \to M^+$ to $(M^+ \sharp M^-) \amalg M^+ \to M^+$. Note that the connected sum $M^+ \sharp M^-$ depends on the choice of orientation of the two summands since depending on this choice we have to use a specific embedding of D^n into M^+ and M^- to create $M^+ \sharp M^-$. It turns out that the bundle data cannot be extended in the case $M^+ \amalg M^-$, they can only be extended in the case $M^+ \amalg M^+$.

Roughly speaking, in our setting there is a coupling between the bundle data and the degree one condition since $\alpha : O_M \to O_X$ depends on the bundle data. This coupling is missing in the naive setting where we just changed the fundamental class of the second summand, and this lack causes the wrong conclusion mentioned above.

The necessary coupling can be seen in the naive setting as follows. Let M be a (not necessarily connected) closed manifold of dimension n with $w_1(M) = 0$ coming with a choice of a fundamental class $[M] \in H_n(M;\mathbb{Z})$. Let X be a connected finite Poincaré complex of dimension n with $w_1(X) = 0$ coming with a choice of fundamental class $[X] \in H_n(X;\mathbb{Z})$. Let $f \colon M \to X$ be a map of degree one covered by bundle data $\overline{f} \colon TM \oplus \underline{\mathbb{R}^a} \to \xi$ such that $w_1(\xi) = w_1(X)$. Consider any two points x_0 and x_1 in M and any path w from $f(x_0)$ to $f(x_1)$. The fundamental class $[M]$ determines fundamental classes $[C_m] \in H_n(C_m;\mathbb{Z})$ for $m = 0, 1$ where C_m is the component of M containing x_m. They determine a preferred orientation on $T_{x_m}M$ and hence also on $(TM \oplus \underline{\mathbb{R}^a})_{x_m}$ for $m = 0, 1$. We get from the fibre transport along w an isomorphism $\xi_{f(x_0)} \xrightarrow{\cong} \xi_{f(x_1)}$. This together with the isomorphism $\overline{f}_{x_m} \colon (TM \oplus \underline{\mathbb{R}^a})_{x_m} \xrightarrow{\cong} \xi_m$ for $m = 0, 1$ yield an isomorphism $u \colon T_{x_0}M \oplus \mathbb{R}^a \xrightarrow{\cong} T_{x_1}M \oplus \mathbb{R}^a$. The coupling is the requirement that u respects the orientations above coming from $[M]$. The requirement is independent of the choice of the path w because of the assumption $w_1(\xi) = w_1(X) = 0$.

This coupling is automatically satisfied if M is connected and it is compatible with surgery. However, this coupling is not satisfied in the case when we consider $\mathrm{id}_M \amalg \mathrm{id}_M \amalg \mathrm{id}_M \colon M^+ \amalg M^- \amalg M^+ \to M^+$ covered by the bundle map $\mathrm{id}_{TM} \amalg \mathrm{id}_{TM} \amalg \mathrm{id}_{TM} \colon TM \amalg TM \amalg TM \to TM$. Hence

$$(\mathrm{id}_M \amalg \mathrm{id}_M \amalg \mathrm{id}_M, \mathrm{id}_{TM} \amalg \mathrm{id}_{TM} \amalg \mathrm{id}_{TM})$$

cannot be viewed as a normal map of degree one in the sense of Definition 7.22.

Note that it also follows from the discussion above that, if $u \colon TM \to TM$ is an orientation reversing bundle map covering the identity on M, then

$$(\mathrm{id}_M \amalg \mathrm{id}_M \amalg \mathrm{id}_M, \mathrm{id}_{TM} \amalg (u \circ \mathrm{id}_{TM}) \amalg \mathrm{id}_{TM})$$

gives a degree one normal map in the sense of Definition 7.22.

We conclude with two remarks that interpret the development of the book from a bird's eye perspective.

Remark 7.46 (Problem 1.2 is a relative version of Problem 1.1) Suppose we developed a method for solving a relative version of Problem 1.1. That means, for a finite CW-pair (X, Y) and for a homotopy equivalence $g \colon N \to Y$ from some closed manifold N, we have a method of finding out whether there exists a homotopy equivalence $G \colon (W, \partial W) \to (X, Y)$ from a manifold with boundary $(W, \partial W)$, such that we have $\partial W = N$ and $g = G|_N \colon N \to Y$ on N. Given a homotopy equivalence $f \colon M_1 \to M_0$ between two closed manifolds, we may take $Y = M_0 \times \{0\} \sqcup M_0 \times \{1\}$, $X = M_0 \times I$ and $N = M_0 \sqcup M_1$. The method for solving a relative version of Problem 1.1 will then tell us whether we have a homotopy equivalence

$$F \colon (W, \partial W) \to (M_0 \times I, M_0 \times \{0\} \sqcup M_0 \times \{1\})$$

for some manifold W with $\partial W = M_0 \sqcup M_1$ such that $\mathrm{id}_{M_0} = F|_{M_0}$ and $f = F|_{M_1} \colon M_1 \simeq M_0$. If $(W, M_0 \sqcup M_1)$ can be further made into an s-cobordism, then the s-Cobordism Theorem 2.1 tells us that M_0 and M_1 are diffeomorphic.

The strategy for solving Problem 1.2 was to split it into steps as described in the Surgery Program 2.9. Looking back at the contents of Chapters 4 to 7, we see that we were in fact following an analogous strategy for solving Problem 1.1 as described in the next remark.

Remark 7.47 (Surgery Program for recognising manifolds) Let X be a finite CW-complex. Problem 1.1 asks when is X homotopy equivalent to a closed manifold M. This problem was addressed in the previous chapters by splitting it into the following steps, which we call the *Surgery Program for recognising manifolds*:

(i) Find necessary homotopical and homological conditions on X;
(ii) Construct a map $f \colon M \to X$ from some closed manifold M;
(iii) Modify M and f by *surgery* so that f becomes a homotopy equivalence.

Step (i) was addressed in Chapter 5, where we showed that X has to be a Poincaré complex. In the present chapter we showed that, in order to achieve Step (ii) with appropriate bundle data, we need that the Spivak normal fibration, which was studied in Chapter 6, has to have a vector bundle reduction. First improvements towards the last step were made via surgery below middle dimension in Chapter 4, while further considerations will be done in the next chapters.

We also see that steps from Surgery Program 2.9 and from the Surgery Program for recognising manifolds 7.47 correspond to each other. With respect to Remark 7.46 it indicates that not only can Problem 1.2 be viewed as a relative version of Problem 1.1, but also the methods for solving Problem 1.2 can be viewed as relative versions of methods for solving Problem 1.1. This heuristic will be completely fulfilled in Chapter 11, where we introduce the surgery exact sequence in Theorem 11.22 that finalises the Surgery Program 2.9.

7.5 Notes

We give more information about the space G/O in Section 12.7 and about its analogues G/PL and G/TOP in Chapter 17. This information will then be applied in Chapters 12, 18, and 19.

Chapter 8
The Even-Dimensional Surgery Obstruction

8.1 Introduction

In this chapter we shall give the solution to the surgery problem in even dimensions. We work in the smooth category unless explicitly stated otherwise. Let $(f, \overline{f}) \colon M \to X$ be a degree one normal map from a closed $2k$-dimensional manifold M to a Poincaré complex X. The surgery problem, stated as Problem 7.42, is to determine whether we can change the normal map $(f, \overline{f})$ by finitely many surgery steps to obtain a normal map $(g, \overline{g}) \colon N \to X$ from a closed manifold N to X such that g is a homotopy equivalence. We have already seen in Theorem 7.41 that we can arrange f to be k-connected. So it remains to achieve that f is $(k + 1)$-connected because then f is a homotopy equivalence by Poincaré duality, see Lemma 8.39 (iii) and (iv) below. We now want to do further surgery on elements in $\pi_{k+1}(f)$ to turn f into a $(k + 1)$-connected map. It will turn out that this is not possible in general and that we will encounter an obstruction, called the surgery obstruction,

$$\sigma(f, \overline{f}) \in L_{2k}(\mathbb{Z}\pi, w).$$

Here $(\mathbb{Z}\pi, w)$ is the group ring $\mathbb{Z}\pi$ with w-twisted involution defined in (3.26), and the even-dimensional L-groups $L_{2k}(\mathbb{Z}\pi, w)$ are defined in Section 8.5 as certain equivalence classes of non-singular quadratic forms over $(\mathbb{Z}\pi, w)$.

Theorem 8.112 and its intrinsic version Theorem 8.167 state that, if $k \geq 3$, then $\sigma(f, \overline{f})$ is a complete obstruction to converting $(f, \overline{f})$ into a normal map such that f is a homotopy equivalence, and that $\sigma(f, \overline{f})$ depends only on the normal bordism class with cylindrical target.

8.1.1 The Nature of the Surgery Obstruction

The surgery obstruction will arise from the following two facts. Firstly, given elements ω_0 and ω_1 in $\pi_{k+1}(f)$ defining immersions $q_i \colon S^k \looparrowright M$, $i = 0, 1$, we cannot in general find regular homotopies of the q_i so that their images in M are disjoint.

This is a necessary condition to kill both ω_0 and ω_1 via simultaneous surgeries. If we put q_0 and q_1 in general position, we may encounter intersection points, which are measured by the equivariant intersection pairing of M defined in Subsection 8.3.1. This is a pairing

$$\lambda \colon I_k(M) \times I_k(M) \to \mathbb{Z}\pi$$

where $I_k(M)$ is the $\mathbb{Z}\pi$-module of regular homotopy classes of immersions of the k-sphere in M, see Lemma 8.19. Secondly, even if all the elements $\omega \in \pi_{k+1}(f)$, on which we want to do surgery, are represented by disjoint immersions, we cannot in general do surgery on each element $\omega \in \pi_{k+1}(f)$. The main obstacle is that an immersion $q \colon S^k \hookrightarrow M$ associated to ω may not be regularly homotopic to an embedding. This assumption appears in Theorem 4.39 (iii). To handle this problem, we introduce in Subsection 8.3.2 equivariant self-intersection elements $\mu(q)$ for immersions $q \colon S^k \to M$. These define a function

$$\mu \colon I_k(M) \to Q_{(-1)^k}(\mathbb{Z}\pi, w)$$

where the abelian group $Q_{(-1)^k}(\mathbb{Z}\pi, w)$ is a quotient of $\mathbb{Z}\pi$, see (8.22). A fundamental theorem of Wall states that for $k \geq 3$ we have $\mu(q) = 0$ if and only f is regularly homotopic to an embedding, see Theorem 8.28. The main tool in the proof is the *Whitney trick*, which allows us to get rid of two of the double points under certain algebraic conditions. To place λ and μ in a wider algebraic setting, we introduce the kernel $\mathbb{Z}\pi$-modules $K_*(\widetilde{M})$ in Subsection 8.4.1. The kernel $\mathbb{Z}\pi$-module $K_i(\widetilde{M})$ is defined to be the kernel of the homomorphism $\widetilde{f}_* \colon H_i(\widetilde{M}) \to H_i(\widetilde{X})$ where $\widetilde{X} \to X$ is the universal covering of X and $\widetilde{M} \to M$ is the induced cover of M. When f is k-connected, there is a $\mathbb{Z}\pi$-module homomorphism

$$\alpha \colon K_k(\widetilde{M}) \to I_k(M)$$

defined in (8.58), which relates $K_k(\widetilde{M})$ and $I_k(M)$. Moreover, if the induced homomorphism $f_* \colon \pi_1(M) \to \pi_1(X)$ is an isomorphism, then Whitehead's Theorem implies that f is a homotopy equivalence if and only if all the kernel groups vanish, see Lemma 8.39 (iv).

In Subsection 8.4.3 we show that, if f is k-connected, then λ and μ induce via α the structure of a non-singular $(-1)^k$-quadratic form on the surgery kernel $K_k(\widetilde{M})$. This form is denoted by $(K_k(\widetilde{M}), s, t)$. We also show that we are able to kill the surgery kernel $K_k(\widetilde{M})$ by finitely many surgery steps if and only if, after perhaps adding a standard $(-1)^k$-hyperbolic form $H_{(-1)^k}(\mathbb{Z}\pi^a)$ for some a, the $(-1)^k$-quadratic form $(K_k(\widetilde{M}), s, t)$ is isomorphic to a standard $(-1)^k$-hyperbolic form $H_{(-1)^k}(\mathbb{Z}\pi^b)$ for some b, see Theorem 8.89. We point out that surgery on a standard generator of a hyperbolic summand not only kills that generator, or, equivalently, one direct summand $\mathbb{Z}\pi$ in the surgery kernel, but in fact such a surgery kills an entire hyperbolic summand $H_{(-1)^k}(\mathbb{Z}\pi)$; i.e., we kill two generators with such a surgery, or, equivalently, a summand $\mathbb{Z}\pi \oplus \mathbb{Z}\pi$.

8.1 Introduction

This leads naturally to the definition of the even-dimensional L-groups in Section 8.5 as the groups generated by isomorphism classes of non-singular $(-1)^k$-quadratic forms over finitely generated free $\mathbb{Z}\pi$-modules, modulo the relation defined by setting the standard hyperbolic form to zero. The surgery obstruction in even dimensions is then defined in Section 8.6 by

$$\sigma(f, \bar{f}) = [(K_k(\widetilde{M}), s, t)] \in L_{2k}(\mathbb{Z}\pi, w).$$

8.1.2 The Intrinsic Versions

The descriptions above make sense only after certain choices, including base points, which one has to get rid off. This is very easy, for instance, for the Whitehead group and the Whitehead torsion, compare Definition 3.22, whereas in L-theory this is a deeper problem that has to be taken seriously. One difficulty is that the conjugation $c_g \colon G \to G$ with $g \in G$ induces on $L_n(\mathbb{Z}G, w)$ the automorphisms $w(g) \cdot \mathrm{id}$ and not just the identity, see Lemma 8.152. To get the functoriality of the L-groups right, one might think that one can consider the category $\overline{\mathcal{R}}$ whose objects are pairs $(X, w_1(X))$ consisting of a path connected space X and an element $w_1(X) \in H^1(X; \mathbb{Z}/2)$ and whose morphisms $f \colon (X, w_1(X)) \to (Y, w_1(Y))$ are given by maps $f \colon X \to Y$ with $f^* w_1(Y) = w_1(X)$, and to define the associated functor $\overline{\mathcal{R}} \to \mathbb{Z}$-MOD by sending $(X, w_1(X))$ to $L_{2k}(\mathbb{Z}\pi_1(X, x), w_1(X))$ for some choice of base points $x \in X$. However, it turns out that one has to replace $\overline{\mathcal{R}}$ by the following larger category $\mathcal{R}$. Objects are pairs (X, O_X) consisting of a space X and an infinite cyclic coefficient system O_X over X, i.e., a covariant functor $O_X \colon \Pi(X) \to \mathbb{Z}$-MOD, such that its value at each object is an infinite cyclic group. A morphism $(u, \bar{u}) \colon (X_0, O_{X_0}) \to (X_1, O_{X_1})$ consists of a map $u \colon X_0 \to X_1$ and a natural transformation $\bar{u} \colon O_{X_0} \to O_{X_1} \circ \Pi(u)$. Note that $\overline{\mathcal{R}}$ is a quotient of $\mathcal{R}$. All of this is explained and carried out in Section 8.7 and leads to the intrinsic L-groups $L_{2k}(\mathbb{Z}\Pi(X); O_X)$ and the intrinsic surgery obstruction, see Definition 8.158 and Theorem 8.167.

8.1.3 The Simply Connected Case

The simply connected case is comparatively easy and fully understood. The values of the L-groups and the surgery obstruction in the simply connected case are presented in Subsections 8.5.2, 8.5.3, and 8.7.6. The L-groups vanish in odd dimensions, so there are no surgery obstructions in the simply connected case. In even dimensions $L_{4k}(\mathbb{Z}) = \mathbb{Z}$ and the surgery obstruction is given by the difference of signatures, and $L_{4k+2}(\mathbb{Z}) = \mathbb{Z}/2$ and the surgery obstruction is given by the Arf invariant. In Theorem 8.173 we describe the solution to Problem 7.1 (when is a finite Poincaré complex homotopy equivalent to a closed manifold?) in the simply connected case.

8.1.4 The Surgery Problem Relative Boundary and Wall's Realisation Theorem

In Section 8.8 we deal with the surgery problem for normal maps of degree one relative boundary. This means that we consider pairs $(f, \bar{f}) \colon (M, \partial M) \to (X, \partial X)$ where $(X, \partial X)$ is a $2k$-dimensional connected Poincaré pair, $(M, \partial M)$ is a compact $2k$-manifold with boundary ∂M, and the restriction of f to the boundary $\partial f \colon \partial M \to \partial X$ is a homotopy equivalence. The surgery problem is now to convert $(f, \bar{f})$ into a normal homotopy equivalence of pairs via a finite number of surgery steps in the interior of the domain manifold. As in the closed case, the techniques of surgery below the middle dimension ensure that we can arrange f to be k-connected. Again, however, we encounter a surgery obstruction

$$\sigma(f, \bar{f}) \in L_{2k}(\mathbb{Z}\pi, w).$$

Theorem 8.189 states that when $k \geq 3$, $\sigma(f, \bar{f})$ is a complete obstruction to converting $(f, \bar{f})$ into a homotopy equivalence of pairs by a finite sequence of surgeries in the interiors of the domain manifolds.

The surgery problem relative boundary plays a key role in the surgery classification of manifolds as suggested by the Surgery Program, see Remark 2.9. Here we are interested in deciding whether two homotopy equivalent manifolds M_0 and M_1 of dimension $(2k - 1)$ are diffeomorphic. Surgery theory first asks us to determine whether M_0 and M_1 are normally cobordant over $M_0 \times I$ and then analyses the normal cobordisms between M_0 and M_1 via their surgery obstructions.

In this setting a fundamental fact is Wall's Realisation Theorem 8.195. A consequence of Wall's Realisation Theorem in even dimensions is that, if X is a compact connected $(2k - 1)$-manifold, then every element $x \in L_{2k}(\mathbb{Z}\pi, w)$ can be realised as the surgery obstruction relative boundary of some degree one normal map of pairs relative boundary $(f, \bar{f}) \colon (M, \partial M) \to (X \times I, \partial(X \times I))$. An application of this result is then given in Chapter 11 where the geometric surgery exact sequence is constructed, providing us with the final solution of the Surgery Program.

Guide 8.1 This chapter is written in such a way that one can start reading it without knowing the previous chapters. For this purpose one should ignore the material about intrinsic versions and not worry about all the necessary choices such as the choice (NBP), see Notation 8.5. One can just start with the definition of a normal Γ-map and consider only the universal covering case, see Notation 8.34. The goal of this chapter is then to understand the definition of the quadratic L-groups $L_{2k}(\mathbb{Z}\pi, w)$ for the group ring $\mathbb{Z}\pi$ with the w-twisted involution and of the surgery obstruction, as it appears in Theorem 8.112.

One may also skip the detailed reading of Subsections 8.5.2, 8.5.3 and 8.6.2 and just believe the main statements. Subsections 8.7.7 and 8.8.4 can be skipped without harm.

An equivalent geometric approach to the surgery groups and the surgery obstruction is presented in Chapter 13.

8.2 Normal Γ-Maps

Recall that we have introduced the notion of a connected finite Poincaré complex in Definition 5.43 and could pass from the category of modules over the fundamental groupoid to the more familiar category of modules over a group ring, see Convention 5.49. We want to introduce a similar simplification for normal maps.

Let X be a finite n-dimensional Poincaré complex with Stiefel–Whitney class $w_1(X) \in H^1(X; \mathbb{Z}/2)$. For simplicity we suppose that X is connected. Let Γ be a group and $w\colon \Gamma \to \{\pm 1\}$ be a group homomorphism. Consider a Γ-covering $\widehat{p}_X\colon \widehat{X} \to X$ such that for the classifying map $u_{\widehat{p}_X}\colon X \to B\Gamma$ the induced map $u_{\widehat{p}_X}^*\colon H^1(B\Gamma; \mathbb{Z}/2) \to H^1(X; \mathbb{Z}/2)$ sends w to $w_1(X)$. Let $[X]$ be a generator of the infinite cyclic group $H_n^\Gamma(X; \mathbb{Z}^w) = H_n(\mathbb{Z}^w \otimes_{\mathbb{Z}\Gamma} C_*(\widehat{X}))$.

Exercise 8.2 Show that $H_n^\Gamma(X; \mathbb{Z}^w)$ is infinite cyclic.

Definition 8.3 (Normal Γ-map of degree one) A *normal Γ-map of degree one* $(M, [M], f, \widehat{f}, a, \xi, \overline{f})$ with target $(X, [X], \widehat{p}_X)$ consists of the following data:

- A connected closed manifold M of dimension n with a generator $[M]$ of the infinite cyclic group $H_n^\Gamma(M; \mathbb{Z}^w)$;
- A map $f\colon M \to X$ such that the map $H^1(f)\colon H^1(X; \mathbb{Z}/2) \to H^1(M; \mathbb{Z}/2)$ sends $w_1(X)$ to $w_1(M)$;
- A Γ-covering $\widehat{p}_M\colon \widehat{M} \to M$ and a Γ-map $\widehat{f}\colon \widehat{M} \to \widehat{X}$ covering f;
- We require that the map $H_n^\Gamma(M; \mathbb{Z}^w) \to H_n^\Gamma(X; \mathbb{Z}^w)$ induced by $\widehat{f}$ sends $[M]$ to $[X]$;
- A vector bundle ξ over X;
- A bundle map $(f, \overline{f})\colon TM \oplus \underline{\mathbb{R}^a} \to \xi$ covering f for some natural number a.

Sometimes we abbreviate $(M, [M], f, \widehat{f}, a, \xi, \overline{f})$ by $(f, \overline{f})$. The following diagram reflects all the data appearing in Definition 8.3 above

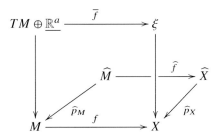

where the square involving the Γ-coverings is a pullback. Moreover, we have fundamental classes and the degree one condition.

Of course the most important choice for Γ and $\widehat{p}_X$ is $\pi_1(X, x)$ and the universal covering $\widetilde{p}_X\colon \widetilde{X} \to X$, see Notation 8.34. One can also consider $\Gamma = \{1\}$ and $\widehat{p} = \mathrm{id}_X$.

Remark 8.4 (Comparing Definition 7.22 and Definition 8.3) We recommend the reader to work with Definition 8.3 as long as possible. It is closer to the standard definition as it appears in the literature and is easier to handle than Definition 7.22. However, in the long run we will have to deal with Definition 7.22, namely, when we want to define the L-groups and the surgery obstructions intrinsically without choosing base points. We will see later that this is not only for cosmetic reasons, see Remarks 8.144, 8.160, and 8.162.

At this point we at least want to explain how the more general Definition 7.22 boils down to Definition 8.3 as long as we assume that M and X are connected and we make choices.

Consider a normal map with respect to the tangent bundle of degree one $(M, f, a, \xi, \overline{f}, o)$ in the sense of Definition 7.22 such that M and X are connected, and a Γ-covering $\widehat{p}_X \colon \widehat{X} \to X$.

Notation 8.5 (Choice (NBP))

- Base points $m \in M$, $\widehat{x} \in \widehat{X}$ and $x \in X$ with $p_X(\widehat{x}) = x = f(m)$;
- An isomorphism $o_x \colon O_X(x) \xrightarrow{\cong} \mathbb{Z}$;
- A group homomorphism $w \colon \Gamma \to \{\pm 1\}$ such that for the classifying map $u_{\widehat{p}_X} \colon X \to B\Gamma$ of $\widehat{p}$ the induced map $u^*_{\widehat{p}_X} \colon H^1(B\Gamma; \mathbb{Z}/2) \to H^1(X; \mathbb{Z}/2)$ sends w to $w_1(X)$.

We define $\widehat{f}$ and $\widehat{p}_M$ by the pullback

$$\begin{array}{ccc} \widehat{M} & \xrightarrow{\widehat{f}} & \widehat{X} \\ \widehat{p}_M \downarrow & & \downarrow \widehat{p}_X \\ M & \xrightarrow{f} & X. \end{array}$$

Let $\widehat{m} \in \widehat{M}$ be the point uniquely determined by $\widehat{p}_M(\widehat{m}) = m$ and $\widehat{f}(\widehat{m}) = \widehat{x}$. We get an isomorphism $O_{\overline{f}}(m) \colon O_M(m) \xrightarrow{\cong} O_X(x)$ from the bundle data and hence $o_x \colon O_X(x) \to \mathbb{Z}$ also defines an isomorphism

$$o_m \colon O_M(m) \xrightarrow{\cong} \mathbb{Z}.$$

Thus we have made a choice (BP), see Notation 5.6, for $\widehat{p}_M$ and $\widehat{p}_X$, namely $(m, \widehat{m}, o_m, w)$ and $(x, \widehat{x}, o_x, w)$. Recall that we obtain from the various maps (5.13) isomorphisms

$$\beta_M \colon H_n^{\Gamma}(\mathbb{Z}^w \otimes C_*(\widehat{M})) \xrightarrow{\cong} H_n(M; O_M);$$
$$\beta_X \colon H_n^{\Gamma}(\mathbb{Z}^w \otimes C_*(\widehat{X})) \xrightarrow{\cong} H_n(X; O_X),$$

and the following diagram commutes

8.2 Normal Γ-Maps

$$\begin{array}{ccc}
H_n^\Gamma(\mathbb{Z}^w \otimes C_*(\widehat{M})) & \xrightarrow{H_n^\Gamma(\widehat{f};\mathbb{Z}^w)} & H_n^\Gamma(\mathbb{Z}^w \otimes C_*(\widehat{X})) \\
\beta_M \downarrow \cong & & \cong \downarrow \beta_X \\
H_n(M;\mathcal{O}_M) & \xrightarrow{H_n(f;\alpha)} & H_n(X;\mathcal{O}_X).
\end{array}$$

Define $[M] \in H_n^\Gamma(\mathbb{Z}^w \otimes C_*(\widehat{M}))$ to be the image of the intrinsic fundamental class $[[M]]$ of Definition 5.29 and $[X]$ to be the image of $[M]$ under $H_n(\widehat{f};\mathbb{Z}^w)$.

Then $[M]$ and $[X]$ are generators and we get a normal Γ-map of degree one with target $(X, [X], \widehat{p}_X)$ in the sense of Definition 8.3 by $(M, [M], f, \widehat{f}, a, \xi, \overline{f})$.

Next we describe the relevant bordism relation where we allow the target of the bordism also to vary and not necessarily be the cylinder.

Namely, for $m = 0, 1$, let X_m be a finite n-dimensional Poincaré complex with Stiefel–Whitney class $w_1(X_m) \in H^1(X_m; \mathbb{Z}/2)$. Let Γ be a group, and $w \colon \Gamma \to \{\pm 1\}$ be a group homomorphism. Let $\widehat{p}_m \colon \widehat{X}_m \to X_m$ be a Γ-covering such that for the classifying map $u_{\widehat{p}_m} \colon X_m \to B\Gamma$ the induced map $u_{\widehat{p}_m}^* \colon H^1(B\Gamma; \mathbb{Z}/2) \to H^1(X_m; \mathbb{Z}/2)$ sends w to $w_1(X_m)$. Let $[X_m] \in H_\Gamma^n(X_m; \mathbb{Z}^w) = H_n(\mathbb{Z}^w \otimes_{\mathbb{Z}\Gamma} C_*(\widehat{X}_m))$ be a componentwise generator, see Definition 7.3.

Definition 8.6 (Bordism of normal Γ-maps of degree one) Consider two normal Γ-maps of degree one $(M_0, [M_0], f_0, \widehat{f}_0, a_0, \xi_0, \overline{f}_0)$ with target $(X_0, [X_0], \widehat{p}_0)$ and $(M_1, [M_1], f_1, \widehat{f}_1, a_1, \xi_1, \overline{f}_1)$ with target $(X_1, [X_1], \widehat{p}_1)$.

A *normal bordism of normal Γ-maps of degree one*

$$(W, [W, \partial W], Y, \partial Y, [Y, \partial Y], F, \widehat{F}, a, \Xi, \overline{F})$$

from $(M_0, [M_0], f_0, \widehat{f}_0, a_0, \xi_0, \overline{f}_0)$ to $(M_1, [M_1], f_1, \widehat{f}_1, a_1, \xi_1, \overline{f}_1)$ consists of:

- A compact manifold W of dimension $(n+1)$ whose boundary ∂W is the disjoint union $\partial_0 W \amalg \partial_1 W$;
- A finite $(n+1)$-dimensional Poincaré pair $(Y, \partial Y)$ such that ∂Y is the disjoint union $\partial_0 Y \amalg \partial_1 Y$;
- Diffeomorphisms $u_m \colon M_m \to \partial_m W$ and isomorphisms of CW-complexes $v_m \colon X_m \to \partial_m Y$ for $m = 0, 1$;
- A map $F \colon W \to Y$ satisfying $l_m \circ v_m \circ f_m = F \circ k_m \circ u_m$, where the maps $k_m \colon \partial_m W \to W$ and $l_m \colon \partial_m Y \to Y$ are the obvious inclusions for $m = 0, 1$;
- A Γ-covering $\widehat{p}_Y \colon \widehat{Y} \to Y$ and Γ-maps $\widehat{l_m \circ v_m} \colon \widehat{X}_m \to \widehat{Y}$ covering $l_m \circ v_m$ for $m = 0, 1$;
- A Γ-covering $\widehat{p}_W \colon \widehat{W} \to W$ and a Γ-map $\widehat{F} \colon \widehat{W} \to \widehat{Y}$ covering $F \colon W \to Y$;
- There are Γ-maps $\widehat{k_m \circ u_m} \colon \widehat{M}_m \to \widehat{W}$ covering $k_m \circ u_m$ for $m = 0, 1$ satisfying $\widehat{l_m \circ v_m} \circ \widehat{f}_m = \widehat{F} \circ \widehat{k_m \circ u_m}$;
- A choice of a componentwise generator $[Y, \partial Y] \in H_{n+1}^\Gamma(Y, \partial Y; \mathbb{Z}^w)$;
- A choice of a componentwise generator $[W, \partial W] \in H_{n+1}^\Gamma(W, \partial W; \mathbb{Z}^w)$;
- We require that the map $H_{n+1}^\Gamma(W; \partial W, \mathbb{Z}^w) \to H_{n+1}^\Gamma(Y; \partial Y; \mathbb{Z}^w)$ induced by $\widehat{f}$ sends $[W, \partial W]$ to $[Y, \partial Y]$ and that the image of $[W, \partial W]$ under the map

$$H^\Gamma_{n+1}(W, \partial W; \mathbb{Z}^w) \xrightarrow{\partial} H^\Gamma_n(\partial W; \mathbb{Z}^w)$$
$$\xrightarrow{\cong} H^\Gamma_n(\partial_0 W; \mathbb{Z}^w) \oplus H^\Gamma_n(\partial_1 W; \mathbb{Z}^w) \xrightarrow{\cong} H^\Gamma_n(M_0; \mathbb{Z}^w) \oplus H^\Gamma_n(M_1; \mathbb{Z}^w)$$

is $(-[M_0], [M_1])$ and analogously for $[Y, \partial Y]$, $-[X_0]$, and $[X_1]$;
- A vector bundle Ξ over Y and a bundle map $(F, \overline{F}) \colon TW \oplus \mathbb{R}^b \to \Xi$ for some natural number b satisfying $b \geq a_m$ for $m = 0, 1$;
- Bundle maps $(l_m \circ v_m, \overline{l_m \circ v_m}) \colon \xi_m \oplus \mathbb{R}^{b+1-a_m} \to \Xi$ such that

$$(F, \overline{F}) \circ \big((u_m, \overline{u_m}) \oplus (u_m, \mathrm{id}_{\mathbb{R}^{b+1-a_m}})\big) \circ (\mathrm{id}_{TM} \oplus \mathrm{flip}_m, \mathrm{id}_M)$$
$$= (l_m \circ v_m, \overline{l_m \circ v_m}) \circ \big((f_m, \overline{f}_m) \oplus (f_m, \mathrm{id}_{\mathbb{R}^{b+1-a_m}})\big)$$

holds, where $(u_m, \overline{u_m}) \colon TM_m \oplus \mathbb{R} \to TW$ is the bundle map induced by Tu_m with respect to some inward normal vector field for $m = 0$ and for some outward normal vector field for $m = 1$, and $\mathrm{flip}_m \colon \mathbb{R}^{b+1} = \mathbb{R}^{a_m} \oplus \mathbb{R} \oplus \mathbb{R}^{b-a_m} \to \mathbb{R}^{b+1} = \mathbb{R} \oplus \mathbb{R}^{a_m} \oplus \mathbb{R}^{b-a_m}$ is the flip map.

Let us illustrate this complicated definition. The following diagram reflects the data we have on the space level

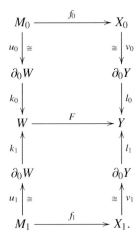

Moreover, there is a Γ-covering $p_Y \colon \widehat{Y} \to Y$ such that all other Γ-coverings are (up to Γ-isomorphism) given by the pullback with the various maps or composite of maps with target Y, namely, we have the commutative diagram of Γ-spaces

8.3 Intersection and Self-intersection Pairings

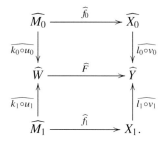

Moreover, everything is covered by appropriate bundle data and we have fundamental classes and the degree one condition.

If the normal Γ-maps of degree one have the same target, we will also consider the following more restrictive bordism relation.

Definition 8.7 (Bordism of normal Γ-maps with cylindrical target of degree one) Let $(M_0, [M_0], f_0, \widehat{f_0}, a_0, \xi_0, \overline{f_0})$ and $(M_1, [M_1], f_1, \widehat{f_1}, a_1, \xi_1, \overline{f_1})$ be two normal Γ-maps of degree one, which have the same target $(X, \widehat{p}, [X])$.

A *normal bordism of normal Γ-maps with cylindrical target of degree one*

$$(W, [W, \partial W], Y, \partial Y, [Y, \partial Y], F, \widehat{F}, a, \Xi, \overline{F})$$

from $(M_0, [M_0], f_0, \widehat{f_0}, a_0, \xi_0, \overline{f_0})$ to $(M_1, [M_1], f_1, \widehat{f_1}, a_1, \xi_1, \overline{f_1})$ is a normal bordism of normal Γ-maps of degree one in the sense of Definition 8.6 such that

- $(Y, \partial Y)$ is the cylinder $(X \times [0, 1], X \times \{0, 1\})$;
- We have $\partial_m Y = X \times \{m\}$, and $l_m \circ v_m \colon X \to Y = X \times [0, 1]$ sends x to (x, m) for $x \in X$ and $m = 0, 1$;
- The Γ-covering $\widehat{p}_Y \colon \widehat{Y} \to Y$ is the pullback of the Γ-covering $\widehat{p}_X \colon \widehat{X} \to X$ with the projection $Y = X \times [0, 1] \to X$;
- For $m = 0, 1$, the Γ-map $\overline{l_m \circ v_m} \colon (l_m \circ v_m)^* \widehat{p}_Y \to \widehat{p}_X$ covering the inclusion $l_m \circ v_m \colon X \to Y = X \times [0, 1]$ is the obvious one.

8.3 Intersection and Self-intersection Pairings

Let $(f, \overline{f}) \colon M \to X$ be a k-connected normal Γ-map of degree one where M and X have dimension $2k$. In order to solve the surgery problem for $(f, \overline{f})$, we have to understand the obstruction to representing a collection of elements $\omega_1, \omega_2, \ldots, \omega_a \in \pi_{k+1}(f)$ by disjoint embeddings $q_i \colon S^k \hookrightarrow M$ with trivial normal bundle (and with nullhomotopies of the maps $f \circ q_i$). By Theorem 4.39, the homotopy classes $\partial \omega_i \in \pi_k(M)$ can be represented by immersions $q_i \colon S^k \looparrowright M$, and by general position we can assume that the images of the q_i intersect transversally in isolated points. The key idea is to assign to each intersection point $d \in D := q_i(S^k) \cap q_j(S^k)$ both a sign $\epsilon(d) \in \{\pm 1\}$, and an element $g(d) \in \pi = \pi_1(M)$, which lies in the fundamental group of M. If the sum $\Sigma_{d \in D} \epsilon(d) g(d)$ vanishes in $\mathbb{Z}\pi$, then the *Whitney trick* leads to the existence of a regular homotopy of q_i, ending at an immersion q'_i that is disjoint from q_j. We describe these ideas in detail in Subsection 8.3.1.

In Subsection 8.3.2 we investigate the question when each q_i is regularly homotopic to an embedding. Here we find that the obstruction is more subtle and takes values in $Q_{(-1^k)}(\mathbb{Z}\pi, w)$, a certain quotient of $\mathbb{Z}\pi$ defined in (8.22).

8.3.1 Intersections of Immersions

We are facing the problem to decide whether we can change an immersion $q\colon S^k \looparrowright M$ within its regular homotopy class to an embedding where M is a compact manifold (possibly with non-empty boundary) of dimension $n = 2k$. This problem occurs when we want to carry out a surgery step in the middle dimension, see Theorem 4.39. We first deal with the necessary algebraic obstructions and then address the question whether their vanishing is also sufficient.

Notation 8.8 (Choice (GBP)) We fix base points $s \in S^k$ and $m \in M$ and an orientation of $T_m M$.

Assume that M is connected and $k \geq 2$. We will consider *pointed immersions* (q, w). These are immersions $q\colon S^k \looparrowright M$ together with a path w from m to $q(s)$. A *pointed regular homotopy* from (q_0, w_0) to (q_1, w_1) is a regular homotopy $h\colon S^k \times [0, 1] \to M$ from $h_0 = q_0$ to $h_1 = q_1$ such that $w_0 * h(s, -)$ and w_1 are homotopic paths relative end points. Here $h(s, -)$ is the path from $q_0(s)$ to $q_1(s)$ given by restricting h to $\{s\} \times [0, 1]$. We denote by $I_k(M)$ the set of pointed regular homotopy classes of pointed immersions from S^k to M:

$$I_k(M) := \{(q, w) \mid q\colon S^k \looparrowright M\}/\text{pointed regular homotopy}. \qquad (8.9)$$

We need the paths w to define the structure of an abelian group on $I_k(M)$. The sum of $[(q_0, w_0)]$ and $[(q_1, w_1)]$ is given by the connected sum along the path $w_0^- * w_1$ from $q_0(s)$ to $q_1(s)$, where one has to arrange $q_0(s) \neq q_1(s)$ by a general position argument. The zero element is given by the composition of the standard embedding $S^k \hookrightarrow \mathbb{R}^{k+1} \hookrightarrow \mathbb{R}^{k+1} \times \mathbb{R}^{k-1} = \mathbb{R}^n$ with some embedding $\mathbb{R}^n \hookrightarrow M$ and any path w from m to the image of s. The inverse of the class of (q, w) is the class of $(q \circ a, w)$ for any base point preserving diffeomorphism $a\colon S^k \to S^k$ of degree -1.

The fundamental group $\pi = \pi_1(M, m)$ operates on $I_k(M)$ by composing the path w with a loop at m. Thus $I_k(M)$ inherits the structure of a $\mathbb{Z}\pi$-module. Now suppose that $(f, \bar{f})\colon M \to X$ is a degree one normal Γ-map with a $2k$-dimensional manifold as source. The following lemma relates the relative homotopy group $\pi_{k+1}(f)$ of f to the group of pointed immersions $I_k(M)$.

8.3 Intersection and Self-intersection Pairings

Lemma 8.10 *There is a natural $\mathbb{Z}\pi$-homomorphism $t_k : \pi_{k+1}(f) \to I_k(M)$ for which the following diagram commutes*

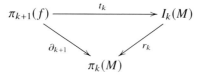

where r_k assigns to the regular homotopy class of an immersions $t_k(x)$ its underlying homotopy class.

Proof. The construction of t_k is analogous to the proof of Theorem 4.39 (ii), which was based on Theorem 4.30. Note that an element in $\pi_{k+1}(f, s)$ is given by a commutative diagram

$$\begin{array}{ccc} S^k & \xrightarrow{q} & M \\ \downarrow & & \downarrow \\ D^{k+1} & \xrightarrow{Q} & X \end{array}$$

together with a path w from m to $f(s)$. The commutativity of the diagram above is obvious. $\square$

Next we want to define the *geometric intersection pairing*

$$\lambda : I_k(M) \times I_k(M) \to \mathbb{Z}\pi. \tag{8.11}$$

Consider $\alpha_0 = [(q_0, w_0)]$ and $\alpha_1 = [(q_1, w_1)]$ in $I_k(M)$. Choose representatives (q_0, w_0) and (q_1, w_1). We can arrange without changing the pointed regular homotopy class that $D = \text{im}(q_0) \cap \text{im}(q_1)$ is finite, for any $y \in D$ both the preimage $f_0^{-1}(y)$ and the preimage $f_1^{-1}(y)$ consists of precisely one point, and for any two points x_0 and x_1 in S^k with $q_0(x_0) = q_1(x_1)$ we have

$$T_{x_0} q_0(T_{x_0} S^k) + T_{x_1} q_1(T_{x_1} S^k) = T_{q_0(x_0)} M.$$

The situation is illustrated in the following figure.

Figure 8.12 (Intersection pairing).

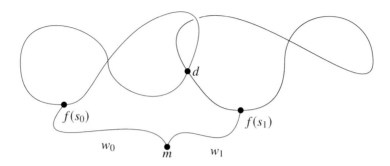

Consider $d \in D$. Let x_0 and x_1 in S^k be the points uniquely determined by $q_0(x_0) = q_1(x_1) = d$. Let u_i be a path in S^k from s to x_i. Then we obtain an element $g(d) \in \pi$ by the loop at m given by the composition $w_1 * q_1(u_1) * q_0(u_0)^- * w_0^-$. Recall that we have fixed an orientation of $T_m M$. The fibre transport along the path $w_0 * q_0(u_0)$ yields an isomorphism $T_m M \cong T_d M$ that is unique up to isotopy. Hence $T_d M$ inherits an orientation from the given orientation of $T_m M$. The standard orientation of S^k determines an orientation on $T_{x_0} S^k$ and $T_{x_1} S^k$. We have the isomorphism of oriented vector spaces
$$T_{x_0} q_0 \oplus T_{x_1} q_1 : T_{x_0} S^k \oplus T_{x_1} S^k \xrightarrow{\cong} T_d M.$$

Define $\epsilon(d) = 1$ if this isomorphism respects the orientations, and $\epsilon(d) = -1$ otherwise. The elements $g(d) \in \pi$ and $\epsilon(d) \in \{\pm 1\}$ are independent of the choices of u_0 and u_1 since S^k is simply connected for $k \geq 2$. Define
$$\lambda(\alpha_0, \alpha_1) := \sum_{d \in D} \epsilon(d) \cdot g(d).$$

Note that reversing the orientation of $T_m M$ has the effect that λ is replaced by $-\lambda$.

Now we want to consider the algebraic analogue of the geometric intersection pairing. For this we have to make the following choice:

Notation 8.13 (Choice (ABP)) We fix a model for the universal covering $p_M : \widetilde{M} \to M$, base points $\widetilde{m} \in \widetilde{M}$ and $m \in M$ with $p_M(\widetilde{m}) = m$ and an isomorphism $o_x : O_X(x) \xrightarrow{\cong} \mathbb{Z}$. (Note that the latter corresponds to a choice of a fundamental class $[M] \in H_n(\mathbb{Z}^w \otimes_{\mathbb{Z}\pi} C_*(\widetilde{M}))$.)

Recall that by the choice of $\widetilde{m} \in \widetilde{M}$ we have a preferred action of $\pi = \pi_1(M; m)$ on $\widetilde{M}$. We have the algebraic $\mathbb{Z}$-valued intersection pairing

$$s_{\mathbb{Z}} : H_k(\widetilde{M}; \mathbb{Z}) \otimes H_k(\widetilde{M}; \mathbb{Z}) \xrightarrow{\cong} H^k_c(\widetilde{M}, \widetilde{\partial M}; \mathbb{Z}) \otimes H^k_c(\widetilde{M}, \widetilde{\partial M}; \mathbb{Z})$$
$$\xrightarrow{\cup} H^{2k}_c(\widetilde{M}, \widetilde{\partial M}; \mathbb{Z}) \xrightarrow{\cong} H_0(\widetilde{M}; \mathbb{Z}) \xrightarrow{\cong} \mathbb{Z} \quad (8.14)$$

8.3 Intersection and Self-intersection Pairings

where the first and the third isomorphism come from Poincaré duality for open manifolds with respect to our choice of $[M]$, see Remark 5.59. It yields the $\mathbb{Z}\pi$-valued pairing

$$s\colon H_k(\widetilde{M};\mathbb{Z}) \otimes H_k(\widetilde{M};\mathbb{Z}) \to \mathbb{Z}\pi, \quad (u,v) \mapsto \sum_{g \in \pi} s_{\mathbb{Z}}(u, l_{g^{-1}}v) \cdot g, \quad (8.15)$$

where $l_{g^{-1}}$ denotes left multiplication by g^{-1}. If we replace $[M]$ by $-[M]$, then s is replaced by $-s$.

In order to relate λ and s, we need to relate our choices of orientation of $T_m M$ and of the fundamental class $[M]$, see also Remark 5.37 (v).

Definition 8.16 (Compatible local orientation of the tangent space) We call an orientation on $T_m M$, which is the same as a choice of an isomorphism $o_M(m)\colon O_M(m) \xrightarrow{\cong} \mathbb{Z}$, *compatible* with $[M] \in H_n(M;\mathbb{Z}^w)$ if the preferred isomorphism $\beta_{m,o_m}\colon H_n(M;\mathbb{Z}^w) \xrightarrow{\cong} H_n(M;O_M)$ given by $o_M(m)$ and $\widetilde{m} \in \widetilde{M}$, see (5.13), sends $[M]$ to the intrinsic fundamental class $[[M]]$, see Definition 5.29.

Exercise 8.17 Identify an open neighbourhood of $m \in M$ with $DT_m M$ via the exponential map. Since we have chosen a base point $\widetilde{m} \in \widetilde{M}$, we get an identification of the restriction of $p_M\colon \widetilde{M} \to M$ to $DT_m M$ with the trivial π-covering $\pi \times DT_m M \to DT_m M$. Thus we obtain an isomorphism of infinite cyclic groups

$$H_n(\mathbb{Z}^w \otimes_{\mathbb{Z}\pi} C_*(\widetilde{M})) \xrightarrow{\cong} H_n(\mathbb{Z}^w \otimes C_*(\widetilde{M}, p_M^{-1}(M \setminus \{m\})))$$
$$\xrightarrow{\cong} H_n(\mathbb{Z}^w \otimes_{\mathbb{Z}\pi} C_*(p_M^{-1}(T_m M), p_M^{-1}(T_m M \setminus \{0\})))$$
$$\xrightarrow{\cong} H_n(\mathbb{Z}^w \otimes_{\mathbb{Z}\pi} C_*(\pi \times (T_m M, T_m M \setminus \{0\})))$$
$$\xrightarrow{\cong} H_n(\mathbb{Z}^w \otimes_{\mathbb{Z}\pi} \mathbb{Z}\pi \otimes_{\mathbb{Z}} C_*(T_m M, T_m M \setminus \{0\}))$$
$$\xrightarrow{\cong} H_n(C_*(T_m M, T_m M \setminus \{0\}))$$
$$\xrightarrow{\cong} H_n(T_m M, T_m M \setminus \{0\})$$
$$\xrightarrow{\cong} O_M(m),$$

where the second isomorphism is given by excision and the last isomorphism comes from the fact that for a finite-dimensional vector space V a choice of an orientation on V is the same as a choice of a generator $H_n(V, V - \{0\})$, see Example 5.27.

Show that $[M]$ and the orientation on $T_m M$ are compatible if and only if the composite of the isomorphism above with the isomorphism $o_M(m)\colon O_M(m) \xrightarrow{\cong} \mathbb{Z}$ given by the orientation on $T_m M$ sends $[M]$ to 1.

Let $\widetilde{q}_i\colon S^k \to \widetilde{M}$ be the unique lift of q_i determined by w_i and $\widetilde{m}$ for $i = 0, 1$. Let $\lambda_{\mathbb{Z}}(\widetilde{q}_0, \widetilde{q}_1)$ be the geometric $\mathbb{Z}$-valued intersection number of $\widetilde{q}_0$ and $\widetilde{q}_1$, which is given by counting intersections with signs. Then $\lambda_{\mathbb{Z}}(\widetilde{q}_0, \widetilde{q}_1)$ is the same as the algebraic intersection number $s_{\mathbb{Z}}([\widetilde{q}_0], [\widetilde{q}_1])$ of the classes in the k-th homology defined by $\widetilde{q}_0$ and $\widetilde{q}_1$, which obviously depends only on the homotopy classes of $\widetilde{q}_0$ and $\widetilde{q}_1$;

the proof in [45, Theorem 11.9 in Chapter VI on page 372] can be extended to our setting. Since we have $\lambda(\alpha_0, \alpha_1) = \sum_{g \in \pi} \lambda_{\mathbb{Z}}(\widetilde{q}_0, l_{g^{-1}} \circ \widetilde{q}_1) \cdot g$, we get

$$\lambda(\alpha_0, \alpha_1) = s([\widetilde{q}_0], [\widetilde{q}_1]). \tag{8.18}$$

Note that for equation (8.18) we need the compatibility of the choice of the orientation of $T_m M$ with $[M]$, see Definition 8.16, since changing the orientation of $T_m M$ changes λ but not s, and changing the fundamental class changes s but not λ.

Recall that we have already treated equation (8.18) on the level of M instead of $\widetilde{M}$ in Remark 5.64. Equation (8.18) implies that $\lambda(\alpha_0, \alpha_1)$ depends only on the pointed regular homotopy classes of (q_0, w_0) and (q_1, w_1).

In the sequel we use the $w = w_1(M)$-twisted involution on $\mathbb{Z}\pi$ that sends $\sum_{g \in \pi} a_g \cdot g$ to $\sum_{g \in \pi} w(g) \cdot a_g \cdot g^{-1}$. The elementary proof of the next lemma is left to the reader.

Lemma 8.19 *For $\alpha, \beta, \beta_1, \beta_2 \in I_k(M)$ and $u_1, u_2 \in \mathbb{Z}\pi$ we have*

$$\lambda(\alpha, \beta) = \epsilon \cdot \overline{\lambda(\beta, \alpha)};$$
$$\lambda(\alpha, u_1 \cdot \beta_1 + u_2 \cdot \beta_2) = u_1 \cdot \lambda(\alpha, \beta_1) + u_2 \cdot \lambda(\alpha, \beta_2).$$

Remark 8.20 (Intersection pairings and embeddings) Suppose that the normal bundle of the immersion $q \colon S^k \hookrightarrow M$ has a nowhere vanishing section. (In our situation it actually will be trivial.) Suppose that q is regularly homotopic to an embedding r. Then the normal bundle of r has a nowhere vanishing section σ. Let r' be the embedding obtained by moving r a little in the direction of this normal vector field σ. Choose a path w_q from $q(s)$ to m. Then for appropriate paths w_r and $w_{r'}$ we get pointed embeddings (r, w_r) and $(r', w_{r'})$ such that the pointed regular homotopy classes of (q, w), (r, w_r) and $(r', w_{r'})$ agree. Since r and r' have disjoint images, we conclude

$$\lambda([q, w], [q, w]) = 0.$$

Hence the vanishing of $\lambda([q, w], [q, w])$ is a necessary condition for finding an embedding in the regular homotopy class of q, provided that the normal bundle of q has a nowhere vanishing section. It is not a sufficient condition. To get a sufficient condition, we have to consider self-intersections of immersions, which we do in the next subsection.

8.3.2 Self-intersections of Immersions

Let $\alpha \in I_k(M)$ be an element. Let (q, w) be a pointed immersion representing α. Recall that we have fixed base points $s \in S^k$, $m \in M$ and an orientation of $T_m M$ that has to be compatible with $[M]$. Since we can find arbitrarily close to q an immersion that is in general position, we can assume without loss of generality that q itself is in general position. This means that there is a finite subset D of $\text{im}(q)$ such that $q^{-1}(y)$ consists of precisely two points for $y \in D$ and of precisely one point for $y \in \text{im}(q) - D$ and for two points x_0 and x_1 in S^k with $x_0 \neq x_1$ and $q(x_0) = q(x_1)$

8.3 Intersection and Self-intersection Pairings

we have $T_{x_0}q(T_{x_0}S^k) + T_{x_1}q(T_{x_1}S^k) = T_{q_0(x_0)}M$. The situation is illustrated in the following figure.

Figure 8.21 (Self-intersections).

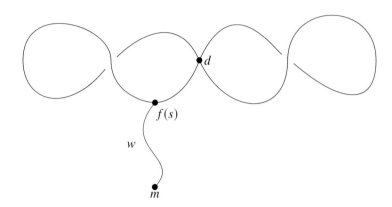

Now fix for any $d \in D$ an ordering $x_0(d), x_1(d)$ of $q^{-1}(d)$. Analogously to the construction above, one defines $\epsilon(x_0(d), x_1(d))$ in $\{\pm 1\}$ and $g(x_0(d), x_1(d))$ in $\pi = \pi_1(M, b)$. Consider the element $\sum_{d \in D} \epsilon(x_0(d), x_1(d)) \cdot g(x_0(d), x_1(d))$ of $\mathbb{Z}\pi$. It does not only depend on q but also on the choice of the ordering of $q^{-1}(d)$ for $d \in D$. One easily checks that the change of ordering of $q^{-1}(d)$ has the following effect for $w = w_1(M) \colon \pi \to \{\pm 1\}$

$$g(x_1(d), x_0(d)) = g(x_0(d), x_1(d))^{-1};$$
$$w(g(x_1(d), x_0(d))) = w(g(x_0(d), x_1(d)));$$
$$\epsilon(x_1(d), x_0(d)) = (-1)^k \cdot w(g(x_0(d), x_1(d))) \cdot \epsilon(x_0(d), x_1(d));$$
$$\epsilon(x_1(d), x_0(d)) \cdot g(x_1(d), x_0(d)) = (-1)^k \cdot \epsilon(x_0(d), x_1(d)) \cdot \overline{g(x_0(d), x_1(d))}.$$

Define an abelian group where we use the w-twisted involution on $\mathbb{Z}\pi$

$$Q_{(-1)^k}(\mathbb{Z}\pi, w) := \mathbb{Z}\pi/\{u - (-1)^k \cdot \overline{u} \mid u \in \mathbb{Z}\pi\}. \tag{8.22}$$

Define the *self-intersection element* for $\alpha \in I_k(M)$

$$\mu(\alpha) := [\sum_{d \in D} \epsilon(x_0(d), x_1(d)) \cdot g(x_0(d), x_1(d))] \in Q_{(-1)^k}(\mathbb{Z}\pi, w). \tag{8.23}$$

The passage from $\mathbb{Z}\pi$ to $Q_{(-1)^k}(\mathbb{Z}\pi, w)$ ensures that the definition is independent of the choice of the order on $q^{-1}(d)$ for $d \in D$. It remains to show that it depends only on the pointed regular homotopy class of (q, w). Let h be a pointed regular homotopy from (q, w) to (r, v). We can arrange that h is in general position, which means that the double point set of h is a compact one-dimensional manifold W consisting of circles and arcs joining points in $D_0 \cup D_1$ to points in $D_0 \cup D_1$.

Suppose that the point e and the point e' in $D_0 \cup D_1$ are joined by an arc. Then one easily checks that their contributions to

$$\mu(f, w) - \mu(r, w) := \left[\sum_{d_0 \in D_0} \epsilon(x_0(d_0), x_1(d_0)) \cdot g(x_0(d_0), x_1(d_0)) \right.$$

$$\left. - \sum_{d_1 \in D_1} \epsilon(x_0(d_1), x_1(d_1)) \cdot g(x_0(d_1), x_1(d_1)) \right]$$

cancel out. This implies that $\mu(q, w) = \mu(r, w)$. Hence there is a well-defined function, called the *self-intersection function*,

$$\mu \colon I_k(M) \to Q_{(-1)^k}(\mathbb{Z}\pi, w), \quad \alpha \mapsto \mu(\alpha). \tag{8.24}$$

In Lemma 8.26 below we establish some basic properties of the self-intersection function μ. Consider the pairing

$$\rho \colon \mathbb{Z}\pi \times Q_{(-1)^k}(\mathbb{Z}\pi, w) \to Q_{(-1)^k}(\mathbb{Z}\pi, w), \quad (u, [v]) \mapsto [uv\overline{u}]. \tag{8.25}$$

It is well defined since for $x, y, z \in \mathbb{Z}\pi$, if we put $z' = xz\overline{x}$, we get that

$$x\big(y + (z - (-1)^k \cdot \overline{z})\big)\overline{x} = xy\overline{x} + \big(z' - (-1)^k \cdot \overline{z'}\big).$$

It is additive in the second variable, i.e., $\rho(x, [y_1] - [y_2]) = \rho(x, [y_1]) - \rho(x, [y_2])$, but it is not additive in the first variable and in particular ρ does *not* give the structure of a left $\mathbb{Z}\pi$-module on $Q_{(-1)^k}(\mathbb{Z}\pi, w)$. Sometimes in the literature $\rho(x, [y])$ is denoted by $x[y]\overline{x}$, but this is a bit misleading since it might lead to the wrong impression that $Q_{(-1)^k}(\mathbb{Z}\pi, w)$ is a left or right $\mathbb{Z}\pi$-module.

Lemma 8.26 *Let* $\mu \colon I_k(M) \to Q_{(-1)^k}(\mathbb{Z}\pi, w)$ *be the self-intersection function of* (8.24), *and let* $\lambda \colon I_k(M) \times I_k(M) \to \mathbb{Z}\pi$ *be the intersection pairing of* (8.11). *Then the following statements hold:*

(i) *Let* $(1 + (-1)^k \cdot T) \colon Q_{(-1)^k}(\mathbb{Z}\pi, w) \to \mathbb{Z}\pi$ *be the homomorphism of abelian groups sending* $[u]$ *to* $u + (-1)^k \cdot \overline{u}$. *For* $\alpha \in I_k(M)$, *let* $\chi(\alpha) \in \mathbb{Z}$ *denote the Euler number of the normal bundle* $\nu(q)$ *for any representative* (q, w) *of* α *with respect to the orientation of* $\nu(q)$ *given by the standard orientation on* S^k *and the orientation on* q^*TM *coming from the fixed orientation on* T_mM *and* w. *Then*

$$\lambda(\alpha, \alpha) = (1 + (-1)^k \cdot T)(\mu(\alpha)) + \chi(\alpha) \cdot 1;$$

(ii) *We get for* $\mathrm{pr} \colon \mathbb{Z}\pi \to Q_{(-1)^k}(\mathbb{Z}\pi, w)$ *the canonical projection and* $\alpha, \beta \in I_k(M)$

$$\mu(\alpha + \beta) - \mu(\alpha) - \mu(\beta) = \mathrm{pr}(\lambda(\alpha, \beta));$$

(iii) *For $\alpha \in I_k(M)$ with trivial normal bundle and $x \in \mathbb{Z}\pi$, we have*

$$\mu(x \cdot \alpha) = \rho(x, \mu(\alpha)),$$

where ρ is defined in (8.25).

Note that there is a more complicated formula in assertion (iii) if one drops the assumption that the normal bundle is trivial, see [414, Theorem 5.2 on page 45] with the correction by Peter Teichner.

Proof. (i) Represent $\alpha \in I_k(M)$ by a pointed immersion (q, w) that is in general position. Choose a section σ of $\nu(q)$ that meets the zero section transversally. Note that then the Euler number satisfies

$$\chi(q) = \sum_{y \in N(\sigma)} \epsilon(y),$$

where $N(\sigma)$ is the (finite) set of zero points of σ and $\epsilon(y)$ is a sign, which depends on the local orientations. We can arrange that no zero of σ is the preimage of an element in the set of double points D_q of q. Now move q a little in the direction of this normal field σ. We obtain a new immersion $r \colon S^k \looparrowright M$ with a path ν from m to $r(s)$ such that (q, w) and (r, ν) are pointed regularly homotopic.

We want to compute $\lambda(\alpha, \alpha)$ using the representatives (q, w) and (r, ν). Note that any point in $d \in D_q$ corresponds to two distinct points $x_0(d)$ and $x_1(d)$ in the set $D = \operatorname{im}(q) \cap \operatorname{im}(r)$ and any element $n \in N(\sigma)$ corresponds to one point $x(n)$ in D. Moreover, any point in D occurs as $x_i(d)$ or $x(n)$ in a unique way. Now the contribution of d to $\lambda([(q, w)], [(r, \nu)])$ is $\epsilon(d) \cdot r(d) + (-1)^k \cdot \overline{\epsilon(d) \cdot r(d)}$ and the contribution of $n \in N(\sigma)$ is $\epsilon(n) \cdot 1$. The assertion (i) follows. The elementary proof of assertions (ii) and (iii) is left to the reader. This finishes the proof of Lemma 8.26. □

Proposition 8.27 *If $M = D^{2k}$, then $\mu \colon I_k(D^{2k}) \xrightarrow{\cong} Q_{(-1)^k}(\mathbb{Z})$ is an isomorphism of abelian groups.*

Proof. Since the intersection form λ on D^n is trivial, μ is a homomorphism of abelian groups.

Since D^{2k} is simply connected, we omit the paths w from the notation. By [352, Proposition 7.12] there is an immersion $q_1 \colon S^k \looparrowright D^{2k}$ such that $\mu(q_1) = [1]$ and the normal bundle of q_1 is given by $\nu(q_1) \cong -\tau_k$, where τ_k is the tangent bundle of the k-sphere. The construction of this immersion goes back to Whitney [434, §2].

When k is odd and $k \geq 3$, we deduce from Example 4.38 that the source and target of μ are cyclic groups of order two and μ maps q_1 to a generator. Actually, there are only two regular homotopy classes of immersions, which must be distinguished by the count of their double points modulo two.

When k is even or $k = 1$, we deduce from Example 4.38 that the source and target of μ are infinite cyclic groups and μ maps q_1 to a generator. Actually, we conclude from Example 4.36 that the normal bundle of an immersion $q_1 \colon S^k \looparrowright D^{2k}$ is a complete invariant of its regular homotopy class. □

Since $\mu(f) = 0$ for an embedding $f \colon S^k \hookrightarrow D^{2k}$, Proposition 8.27 implies that an immersion $g \colon S^k \looparrowright D^{2k}$ is regularly homotopic to an embedding if and only if $\mu(g) = 0$. The following theorem of Wall taken from [414, Theorem 5.2 on page 45] is a far-reaching and fundamental generalisation of this fact from $M = D^{2k}$ to any even-dimensional manifold M of dimension $2k \geq 6$.

Theorem 8.28 (Wall Embedding Theorem) *Let M be a compact connected manifold of dimension $n = 2k \geq 6$. Fix base points $s \in S^k$ and $m \in M$ and an orientation of $T_m M$. Let (q, w) be a pointed immersion of S^k in M. Suppose that $k \geq 3$. Then (q, w) is pointed regularly homotopic to a pointed immersion (r, v) with the property that $r \colon S^k \to M$ is an embedding if and only if $\mu(q, w) = 0$.*

Proof. If q is represented by an embedding, then $\mu(q, w) = 0$ by definition. Suppose that $\mu(q, w) = 0$. We can assume without loss of generality that q is in general position. Since $\mu(q, w) = 0$, we can find d and e in the set of double points D_q of q and a numeration $x_0(d), x_1(d)$ of $q^{-1}(d)$ and $x_0(e), x_1(e)$ of $q^{-1}(e)$ such that

$$\epsilon(x_0(d), x_1(d)) = -\epsilon(x_0(e), x_1(e));$$
$$g(x_0(d), x_1(d)) = g(x_0(e), x_1(e)).$$

Hence we can find arcs u_0 and u_1 in S^k such that $u_0(0) = x_0(d)$, $u_0(1) = x_0(e)$, $u_1(0) = x_1(d)$ and $u_1(1) = x_1(e)$, the paths u_0 and u_1 are disjoint from one another, $q(u_0((0, 1)))$ and $q(u_1((0, 1)))$ do not meet D_q, and $q(u_0)$ and $q(u_1)$ are homotopic relative endpoints. We can change u_0 and u_1 without destroying the properties above and find a smooth map $U \colon D^2 \to M$ such that: its restriction to S^1 is an embedding (ignoring corners on the boundary); it is given on the upper hemisphere S^1_+ by u_0 and on the lower hemisphere S^1_- by u_1; and the image of the interior of D^2 under U is disjoint from the image of q. There is a compact neighbourhood K of S^1 such that $U|_K$ is an embedding. Since $k \geq 3$ we can find arbitrarily close to U an embedding that agrees with U on K. Hence we can assume without loss of generality that U itself is an embedding. The Whitney trick, see [301, Theorem 6.6 on page 71], [435], makes it possible to change q within its pointed regular homotopy class to a new pointed immersion (r, v) such that $D_r = D_q - \{d, e\}$ and $\mu(r, v) = 0$. By iterating this process we achieve $D_q = \emptyset$. □

Figure 8.29 (Whitney trick).

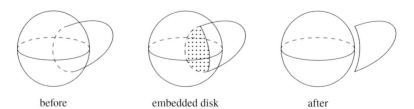

before   embedded disk   after

8.3 Intersection and Self-intersection Pairings

Remark 8.30 (The Whitney trick fails in low dimensions and good fundamental groups) The condition $\dim(M) \geq 5$, which arises in high-dimensional manifold theory, ensures in the proof of Theorem 8.28 that $k \geq 3$ and hence we can arrange U to be an embedding. If $k = 2$, one can achieve that U is an immersion but not necessarily an embedding. This is the technical reason why surgery in dimension 4 is much more complicated than in dimensions ≥ 5.

The problem with the Whitney trick can be solved and hence surgery can also be carried out in dimension 4, provided that we work in the topological category and the fundamental group π is good in the sense of Freedman, see [157, 158] or [37, Definition 12.12 on page 180]. If we work in the topological category, and if the fundamental group is good, then all the results presented in this book with the dimension condition ≥ 5 extend to dimension 4. The class of good groups is closed under taking subgroups, quotients, extensions, directed colimits, and passage to overgroups of finite index, see [158, page 660], [159, Exercise on page 44], [160, Lemma 1.2] or [37, Proposition 19.5 on page 279]. Groups with subexponential growth are good, see [160, 233]. All this implies that every elementary amenable group is good, see [37, Example 19.6 on page 280]. In [37, page 282] it is mentioned that it is likely that amenable groups are good. All groups are good if and only if $\mathbb{Z} * \mathbb{Z}$ is good, see [37, Proposition 19.7 on page 281]. No satisfactory characterisation of the class of good groups seems to be known. Our personal belief is that a group is good if and only if it is amenable.

We conclude this section with several examples of immersions with non-trivial self-intersection invariant.

Example 8.31 (The "diagonal" immersion in $S^k \times S^k$) Consider $M = S^k \times S^k$. For $i = 1, 2$ define $q_i \colon S^k \hookrightarrow M$ to be the coordinate embeddings $q_1(x) = (x, s)$ and $q_2(x) = (s, x)$ for a base point $s \in S^k$. Orient M so that q_1 and q_2 have intersection number $\lambda(q_1, q_2) = 1$. We assume $k \geq 2$ and that M is simply connected. We omit the paths w from the notation. The *"diagonal" immersion* $q_d := q_1 + q_2$ has self-intersection

$$\mu(q_d) = \mu(q_1) + \mu(q_2) + \mathrm{pr}(\lambda(q_1, q_2)) = [1] \in Q_{(-1)^k}(\mathbb{Z}).$$

Note that the *diagonal embedding*, $\Delta \colon S^k \to M, x \mapsto (x, x)$, has trivial self-intersection number and hence Δ and q_d are not regularly homotopic.

Example 8.32 (An elaboration of the diagonal immersion) The following example shows that every element of $Q_{(-1)^k}(\mathbb{Z}\pi, w)$ arises as the self-intersection of some immersion. It is taken from the proof of [345, Proposition 5.2 (v)].

Let X be a manifold with fundamental group π and M be the connected sum $M := X \sharp (S^k \times S^k)$. We let $q_i \colon S^k \to M$ be the composition of the coordinate embeddings from Example 8.31, which we assume are disjoint from the connected sum locus, composed with the inclusion into M. For any $u \in \mathbb{Z}\pi$ we set $q_{d,u} = u \cdot q_1 + q_2$ and then

$$\mu(q_{d,u}) = \mu(u \cdot q_1) + \mu(q_2) + \mathrm{pr}(\lambda(uq_1, q_2)) = 0 + 0 + \mathrm{pr}(u) = [u] \in Q_{(-1)^k}(\mathbb{Z}\pi, w).$$

The previous example gave examples of immersions q where the self-intersection invariant $\mu(q,u)$ could be any element of $Q_{(-1)}(\mathbb{Z}\pi, w)$, but this was an artefact of the choice of the element u. In the following example we give an immersion q such that the self-intersection number $\mu(q, w)$ for some $w \in \pi$ does not depend on the choice of w and where $\mu(q, w)$ maps non-trivially to $Q_{(-1)^k}(\mathbb{Z}\pi)/Q_{(-1)^k}(\mathbb{Z})$.

Example 8.33 (The covering immersion $S^k \to \mathbb{R}P^k$) The following example is due to [358, §5], see also [352, Ex. 11.33]. Denote points in $\mathbb{R}P^k$ by homogeneous coordinates $[x_0, \ldots, x_k]$ for $(x_0, \ldots, x_k) \in S^k$. We define an immersion $q \colon S^k \looparrowright \mathbb{R}P^k \times D^k$ by

$$q(x_0, \ldots, x_k) = \left([x_0, \ldots, x_k], (\frac{x_1}{2}, \ldots, \frac{x_k}{2})\right).$$

Note that q has one pair of double points $\{(1, 0, \ldots, 0), (-1, 0, \ldots, 0)\}$ and that the great circle in S^k connecting these double points maps to the generator of $\pi_1(\mathbb{R}P^k) = \mathbb{Z}/2$ under the canonical projection $S^k \to \mathbb{R}P^k$. For any choice of w we get

$$\mu([q, w]) = [0, T] \neq 0 \in Q_{(-1)^k}(\mathbb{Z}[\mathbb{Z}/2], w_k).$$

Here T is a generator of $\mathbb{Z}/2$, $w_k \colon \mathbb{Z}/2 \to \{\pm 1\}$ is trivial if k is odd and non-trivial if k is even and we conclude from the definition

$$Q_{(-1)^k}(\mathbb{Z}[\mathbb{Z}/2], w_k) = \begin{cases} \mathbb{Z}(e) \oplus \mathbb{Z}/2(T) & \text{if } k \text{ is even;} \\ \mathbb{Z}/2(e) \oplus \mathbb{Z}/2(T) & \text{if } k \text{ is odd,} \end{cases}$$

where $\mathbb{Z}(e)$, $\mathbb{Z}/2(e)$, and $\mathbb{Z}/2(T)$ denote $\mathbb{Z}$, $\mathbb{Z}/2$, and $\mathbb{Z}/2$ with the corresponding generator.

8.4 Surgery Kernels and Forms

Consider a normal Γ-map of degree one $(M, [M], f, \widehat{f}, a, \xi, \overline{f})$ with target $(X, \widehat{p}_X, [X])$ in the sense of Definition 8.3. Recall that X is a finite n-dimensional Poincaré complex with first Stiefel–Whitney class $w_1(X)$ in $H^1(X; \mathbb{Z}/2)$, we have a group homomorphism $w \colon \Gamma \to \{\pm 1\}$, and a Γ-covering $\widehat{p}_X \colon \widehat{X} \to X$ such that for the classifying map $u_{\widehat{p}_X} \colon X \to B\Gamma$ the induced map $u_{\widehat{p}_X}^* \colon H^1(B\Gamma; \mathbb{Z}/2) \to H^1(X; \mathbb{Z}/2)$ sends w to $w_1(X)$. Moreover, we have a componentwise generator $[X] \in H_n^\Gamma(\widehat{X}; \mathbb{Z}^w)$.

Notation 8.34 (Universal covering case) We say that we are looking at the *universal covering case* if

- M and X are connected;
- The map $\pi_1(f, m) \colon \pi_1(M, m) \to \pi_1(X, f(m))$ is an isomorphism for one (and hence every) base point $m \in M$;
- The space $\widehat{X}$ is simply connected;

We then write $\widetilde{p}_M \colon \widetilde{M} \to M$, $\widetilde{p}_X \colon \widetilde{X} \to X$ and $\widetilde{f}$ instead of $\widehat{p}_M \colon \widehat{M} \to M$, $\widehat{p}_X \colon \widehat{X} \to X$, and $\widehat{f}$ respectively.

8.4 Surgery Kernels and Forms

Moreover, we make choices of base points $m \in M$, $\widetilde{m} \in \widetilde{M}$, $x \in X$ and $\widetilde{x} \in \widetilde{X}$ with $\widetilde{p}_X(\widetilde{x}) = x$, $\widetilde{p}_M(\widetilde{m}) = m$, $f(m) = x$ and $\widetilde{f}(\widetilde{m}) = \widetilde{x}$ and thus get an identification

$$\pi_1(M, m) = \pi_1(X, x) = \Gamma.$$

We often abbreviate $\pi = \pi_1(X, x) = \pi_1(M, m)$.

Note that in the situation of Notation 8.34 the two maps $\widetilde{p}_M \colon \widetilde{M} \to M$ and $\widetilde{p}_X \colon \widetilde{X} \to X$ are models for the universal coverings of M and X and that $\widetilde{f}$ is π-equivariant.

The main aim of this section is to relate the topology of intersections and self-intersections of immersions from Section 8.3 to the homological algebra over the group ring $\mathbb{Z}\Gamma$ of the chain complexes of $\widehat{M}$ and $\widehat{X}$.

8.4.1 Surgery Kernels

For the remainder of this subsection, we fix a normal Γ-map of degree one $(M, [M], f, \widehat{f}, a, \xi, \overline{f})$ with target $(X, \widehat{p}_X, [X])$ in the sense of Definition 8.3.

In Chapter 7 we used the homotopy groups $\pi_{j+1}(f) = \pi_{j+1}(\mathrm{cyl}(f), M)$ to measure the failure of f to be a homotopy equivalence. In this subsection we look at the (co)homology groups of $\widehat{f}$, which are isomorphic to the surgery kernel groups of $\widehat{M}$.

We obtain maps of $\mathbb{Z}\Gamma$-chain complexes

$$C_*(\widehat{f}) \colon C_*(\widehat{M}) \to C_*(\widehat{X})$$

and

$$C^*(\widehat{f}) \colon C^*(\widehat{X}) \to C^*(\widehat{M}).$$

They induce $\mathbb{Z}\Gamma$-maps $\widehat{f}_* \colon H_*(\widehat{M}) \to H_*(\widehat{X})$ and $\widehat{f}^* \colon H^*(\widehat{X}) \to H^*(\widehat{M})$. The failure of $\widehat{f}_*$ or $\widehat{f}^*$ to be isomorphisms is measured by the $\mathbb{Z}\Gamma$-(co)homology modules

$$H_*(\widehat{f}) := H_*\bigl(\mathrm{cone}_*(C_*(\widehat{f}))\bigr);$$
$$H^*(\widehat{f}) := H^*\bigl(\mathrm{cone}^*(C^*(\widehat{f}))\bigr),$$

where $\mathrm{cone}_*(C_*(\widehat{f}))$ and $\mathrm{cone}^*(C^*(\widehat{f}))$ are the algebraic mapping cones of $C_*(\widehat{f})$ and $C^*(\widehat{f})$ respectively. The precise definition of the chain complex given by the mapping cone $\mathrm{cone}_*(f_*)$ of a map f_* of chain complexes has been given in (3.12). The definition of the cochain complex given by the mapping cone $\mathrm{cone}^*(f^*)$ of a map of cochain complexes $f^* \colon C^* \to D^*$ is analogous, namely,

$$\mathrm{cone}^{p+1}(f^*) := D^p \oplus C^{p+1} \xleftarrow{\begin{pmatrix} d^{p-1} & f^p \\ 0 & -c^p \end{pmatrix}} \mathrm{cone}^p(f^*) := D^{p-1} \oplus C^p. \quad (8.35)$$

If f_* is a chain map with dual cochain map f^*, then $\mathrm{cone}^*(f^*)$ is the dual of $\mathrm{cone}_*(f_*)$.

There are long exact sequences of $\mathbb{Z}\Gamma$-modules

$$\cdots \to H_{j+1}(\widehat{M}) \xrightarrow{\widehat{f}_*} H_{j+1}(\widehat{X}) \to H_{j+1}(\widehat{f}) \to H_j(\widehat{M}) \xrightarrow{\widehat{f}_*} H_j(\widehat{X}) \to \cdots,$$

and

$$\cdots \to H^j(\widehat{X}) \xrightarrow{\widehat{f}^*} H^j(\widehat{M}) \to H^{j+1}(\widehat{f}) \to H^{j+1}(\widehat{X}) \xrightarrow{\widehat{f}^*} H^{j+1}(\widehat{M}) \to \cdots.$$

Definition 8.36 (Surgery Kernel) Define the *surgery kernel*

$$K_j(\widehat{M}) := \ker\big(\widehat{f}_* \colon H_j(\widehat{M}) \to H_j(\widehat{X})\big),$$

and

$$K^j(\widehat{M}) := \operatorname{coker}\big(\widehat{f}^* \colon H^j(\widehat{X}) \to H^j(\widehat{M})\big).$$

Note that in the definition of the dual chain complex $C^*(\widehat{M})$ the orientation homomorphism $w \colon \pi \to \{\pm 1\}$ enters since the involution on $\mathbb{Z}\Gamma$ depends on it. Since we define $H^*(\widehat{M})$ by $H^*(C^*(\widehat{M}))$, the same remark applies to $H^*(\widehat{M})$ and $K^*(\widehat{M})$. Moreover, note that $H^*(\widehat{M})$ is cohomology with compact supports on $\widehat{M}$, see Remark 5.59. For the definition of the surgery kernels of f, the bundle data and the manifold structure on M play no role.

Exercise 8.37 Suppose that $K_i(\widehat{M}) = 0$ holds for $i \leq j$. Show that then $K^i(\widehat{M}) = 0$ holds for $i \leq j$.

Example 8.38 Suppose that X is a connected closed oriented manifold of dimension n, M is given by a connected sum $M = X \sharp N$ for some connected closed oriented manifold N of dimension n, and $f \colon M \to X$ is given by the collapse map. Then

$$K_j(\widehat{M}) = \begin{cases} H_j(N;\mathbb{Z}) \otimes_\mathbb{Z} \mathbb{Z}\Gamma & 0 < j < n; \\ \{0\} & \text{otherwise.} \end{cases}$$

The main result of this subsection is that the surgery kernels behave very much like the groups of a third space satisfying Poincaré duality.

Lemma 8.39 (i) *The $\mathbb{Z}\Gamma$-homomorphism $\widehat{f}_* \colon H_j(\widehat{M}) \to H_j(\widehat{X})$ is a split surjection whereas the $\mathbb{Z}\Gamma$-homomorphism $\widehat{f}^* \colon H^j(\widehat{X}) \to H^j(\widehat{M})$ is a split injection. There are natural $\mathbb{Z}\Gamma$-isomorphisms*

$$H_j(\widehat{M}) \xrightarrow{\cong} K_j(\widehat{M}) \oplus H_j(\widehat{X});$$
$$H^j(\widehat{M}) \xrightarrow{\cong} K^j(\widehat{M}) \oplus H^j(\widehat{X});$$
$$K_j(\widehat{M}) \xrightarrow{\cong} H_{j+1}(\widehat{f});$$
$$K^j(\widehat{M}) \xrightarrow{\cong} H^{j+1}(\widehat{f}) = H^{j+1}\big(\operatorname{cone}^*(C^*(\widehat{f}))\big);$$

(ii) *The cap product with* $[M]$ *induces a* $\mathbb{Z}\Gamma$*-isomorphism*

$$- \cap [M] \colon K^{n-j}(\widehat{M}) \xrightarrow{\cong} K_j(\widehat{M}),$$

which fits into a commutative diagram of split short exact sequences

$$\begin{array}{ccccccccc}
0 & \longleftarrow & K^{n-j}(\widehat{M}) & \longleftarrow & H^{n-j}(\widehat{M}) & \xleftarrow{\widehat{f}^*} & H^{n-j}(\widehat{X}) & \longleftarrow & 0 \\
& & \cong \downarrow -\cap[M] & & \cong \downarrow -\cap[M] & & \cong \downarrow -\cap[X] & & \\
0 & \longrightarrow & K_j(\widehat{M}) & \longrightarrow & H_j(\widehat{M}) & \xrightarrow{\widehat{f}_*} & H_j(\widehat{X}) & \longrightarrow & 0;
\end{array}$$

(iii) *Consider* $j \geq 1$. *If the map* f *is* j*-connected, then* $K_i(\widehat{M}) = 0$ *for all* $i < j$. *Suppose that we are in the universal covering case, see Notation 8.34. Then* f *is* j*-connected if and only if* $K_i(\widetilde{M}) = 0$ *for all* $i < j$. *In this case there is a composition of natural* $\mathbb{Z}\pi$*-isomorphisms*

$$h_j \colon \pi_{j+1}(f) \xrightarrow{\cong} \pi_{j+1}(\widetilde{f}) \xrightarrow{\cong} H_{j+1}(\widetilde{f}) \xrightarrow{\cong} K_j(\widetilde{M})$$

such that the following diagram of $\mathbb{Z}\pi$*-modules commutes*

$$\begin{array}{ccc}
\pi_{j+1}(f) & \xrightarrow{h_j} & K_j(\widetilde{M}) \\
\cong \uparrow & & \downarrow \\
\pi_{j+1}(\widetilde{f}) & & \\
\partial_{j+1} \downarrow & & \downarrow \\
\pi_j(\widetilde{M}) & \longrightarrow & H_j(\widetilde{M})
\end{array}$$

where the lower horizontal arrow is the Hurewicz homomorphism and the right vertical arrow is the inclusion;

(iv) *If* f *is a homotopy equivalence, then* $K_i(\widehat{M}) = 0$ *for all* i.
Suppose that we are in the universal covering case. Let k *be the natural number defined by* $n = \dim(X) = 2k$ *or* $n = \dim(X) = 2k + 1$. *Then* f *is a homotopy equivalence if and only if* $K_i(\widetilde{M}) = 0$ *for all* $i \leq k$.

Proof. (i) Now we come to a fundamental feature that the compatible Poincaré duality isomorphisms in the source and target allow us to define *Umkehr maps*, or "wrong way maps". For example, for $x \in H_j(\widehat{X})$, we define

$$\widehat{f_!} \colon H_j(\widehat{X}) \to H_j(\widehat{M}), \quad x \mapsto \widehat{f}^*\bigl((- \cap [X])^{-1}(x)\bigr) \cap [M]. \quad (8.40)$$

By standard properties of cap products, $\widehat{f}_* \widehat{f_!}(x) = x$ and so $\widehat{f}_*$ is a split surjection. Similarly we define the cohomology Umkehr map

$$\widehat{f}^!: H^j(\widehat{M}) \to H^j(\widehat{X}), \quad \alpha \mapsto (- \cap [X])^{-1}(\widehat{f}_*(\alpha \cap [M])),$$

which satisfies $\widehat{f}^! \widehat{f}^*(\beta) = \beta$ for all $\beta \in H^j(\widehat{X})$. This proves that $\widehat{f}^*$ is a split injection and that we obtain from the long exact sequences of $\widehat{f}$ split short exact sequences

$$0 \leftarrow H^{n-j+1}(\widehat{f}) \xleftarrow{\delta^{n-j}} H^{n-j}(\widehat{M}) \xleftarrow{\widehat{f}^*} H^{n-j}(\widehat{X}) \leftarrow 0;$$

$$0 \to H_{j+1}(\widehat{f}) \xrightarrow{\partial_{j+1}} H_j(\widehat{M}) \xrightarrow{\widehat{f}_*} H_j(\widehat{X}) \to 0.$$

This yields the first two isomorphisms of assertion (i). The other two isomorphisms are induced by the two boundary maps above.

(ii) Let $r: H_j(\widehat{M}) \to K_j(\widehat{M})$ be the retraction of the inclusion $K_j(\widehat{M}) \to H_j(\widehat{M})$ induced by the Umkehr map $f_!$, that is, $r(x) = x - f_! f_*(x)$. Capping with $[M]$ and composing with r yields a $\mathbb{Z}\Gamma$-homomorphism $H^{n-j}(\widehat{M}) \to K_j(M)$ sending $[\alpha]$ to $r(\alpha \cap [M])$. It induces a $\mathbb{Z}\Gamma$-isomorphism

$$- \cap [M]: K^{n-j}(\widehat{M}) \xrightarrow{\cong} K_j(\widehat{M}).$$

Hence we have constructed the maps and the diagram appearing in assertion (ii). Since the second and third vertical arrows are bijections, the first vertical arrow is also a bijection by the five lemma.

(iii) From assertion (ii) we get $H_i(\widehat{f}) \xrightarrow{\cong} K_{i-1}(\widehat{M})$. Since f and hence $\widehat{f}$ are j-connected, we conclude $H_i(\widehat{f}) = 0$ for $i \le j$ and therefore $K_i(\widehat{M}) = 0$ for all $i < j$.

Now suppose that we are in the universal covering case. Then f is j-connected if and only if $\widetilde{f}$ is j-connected, which is equivalent to $H_i(\widetilde{f}) = 0$ for $i \le j$, see [178, Corollary 4.33 on page 367] or [431, Theorem IV.7.13 on page 181]. Hence f is j-connected if and only if $K_i(\widetilde{M}) = 0$ for all $i < j$. Suppose that f is j-connected. Then the Hurewicz homomorphism

$$\pi_i(\widetilde{f}) \xrightarrow{\cong} H_i(\widetilde{f})$$

is bijective for $i \le j + 1$, see [178, Theorem 4.32 on page 366] or [431, Corollary IV.7.10 on page 181]. The obvious map $\pi_{j+1}(\widetilde{f}) \to \pi_{j+1}(f)$ is bijective. From assertion (ii) we get $H_i(\widetilde{f}) \cong K_{i-1}(\widetilde{M})$. The composition of the isomorphisms above or their inverses yields a natural $\mathbb{Z}\pi$-isomorphism $h_j: \pi_{j+1}(f) \to K_j(\widetilde{M})$.

(iv) If f is a homotopy equivalence, then clearly all surgery kernels vanish. Suppose that we are in the universal covering case and that $K_i(\widetilde{M}) = 0$ for $i \le k$. This implies that $K^i(\widetilde{M}) = 0$ for $i \le k$ by Exercise 8.37. Now by assertion (ii) $K_i(\widetilde{M}) \cong K^{n-i}(\widetilde{M})$ for all i. Hence $\widetilde{f}_*: H_i(\widetilde{M}) \to H_i(\widetilde{X})$ is an isomorphism for all i. We conclude from [178, Corollary 4.33 on page 367] or [431, Theorem IV.7.13 on page 181, Theorem V.3.1 and Theorem V.3.5 on page 230] that f is a homotopy equivalence. □

In the light of Lemma 8.39 we can now reformulate Theorem 7.41.

8.4 Surgery Kernels and Forms

Theorem 8.41 (Killing the surgery kernels of a normal map) *By carrying out a finite sequence of surgery steps, we can arrange that f is k-connected. In this case $K_i(\widehat{M}) = 0 = K^i(\widehat{M})$ unless $n = 2k$ and $i = k$, or $n = 2k + 1$ and $i = k, k + 1$.*

Proof. By Theorem 7.41 we may assume that $f \colon M \to X$ is k-connected. By Lemma 8.39 (iii), $K_i(\widehat{M}) = 0$ for $i < k$ and hence $K^i(\widehat{M}) = 0$ for $i < k$ by Exercise 8.37. From Lemma 8.39 (ii) we conclude $K_i(\widehat{M}) = 0 = K^i(\widehat{M})$ for $i > k$ if $n = 2k$ or $i > k + 1$ if $n = 2k + 1$. □

We conclude this subsection by introducing the *chain level Umkehr map*, which is a map of $\mathbb{Z}\Gamma$-chain complexes $C_*^!(\widehat{f}) \colon C_*(\widehat{X}) \to C_*(\widehat{M})$ inducing the homology Umkehr map $\widehat{f_!} \colon H_*(\widehat{X}) \to H_*(\widehat{M})$ of (8.40). The chain level Umkehr map leads to a deep explanation of the Poincaré duality isomorphism of surgery kernels in Lemma 8.39 (ii), plays an important role in the theory of algebraic surgery in Chapter 15, and is given by the following composition of $\mathbb{Z}\Gamma$-chain maps

$$C_*^!(\widehat{f}) \colon C_*(\widehat{X}) \xrightarrow{(-\cap [X])^{-1}} C^{n-*}(\widehat{X}) \xrightarrow{C^{n-*}(\widehat{f})} C^{n-*}(\widehat{M}) \xrightarrow{-\cap [M]} C_*(\widehat{M}). \quad (8.42)$$

Here $(-\cap [X])^{-1} \colon C_*(\widehat{X}) \to C^{n-*}(\widehat{X})$ is a $\mathbb{Z}\Gamma$-chain homotopy inverse of $-\cap [X] \colon C^{n-*}(\widehat{X}) \to C_*(\widehat{X})$. There are of course many choices for the chain map $(-\cap [X])^{-1}$, but they are all $\mathbb{Z}\Gamma$-chain homotopic and so the $\mathbb{Z}\Gamma$-chain homotopy class of $C_*^!(\widehat{f})$ is well defined. We write

$$C_!^{n-*}(\widehat{f}) \colon C^{n-*}(\widehat{M}) \to C^{n-*}(\widehat{X})$$

for the dual of $C_*^!(\widehat{f})$. From the definition of the chain level Umkehr map we see that

$$\widehat{f_!} = H_j(C_*^!(\widehat{f})) \colon H_j(\widehat{X}) \to H_j(\widehat{M}); \quad (8.43)$$
$$\widehat{f^!} = H_j(C_!^{n-*}(\widehat{f})) \colon H^j(\widehat{M}) \to H^j(\widehat{X}). \quad (8.44)$$

The mapping cone of the chain level Umkehr map $C_*^!(\widehat{f})$, see (3.12), plays an important role in surgery theory. It is the $\mathbb{Z}\Gamma$-chain complex

$$\mathrm{cone}_*(C_*^!(\widehat{f})).$$

Since $C_*(\widehat{M})$ is the target of $C_*^!(\widehat{f})$, we have the canonical chain map

$$e_* \colon C_*(\widehat{M}) \to \mathrm{cone}_*(C_*^!(\widehat{f})).$$

If we use e_* to push forward the fundamental class $[M] \in H_n(M; \mathbb{Z}^w)$ of M, we obtain $e_*([M]) \in H_n(\mathrm{cone}_*(C_*^!(\widehat{f})); \mathbb{Z}^w)$. Put

$$\mathrm{cone}^{n-*}(C_*^!(\widehat{f})) = \bigl(\mathrm{cone}_*(C_*^!(\widehat{f}))\bigr)^{n-*}. \quad (8.45)$$

This is the same as $\mathrm{cone}_*(C_!^{n-*}(\widehat{f}))$. We obtain a $\mathbb{Z}\Gamma$-chain map

$$- \cap e_*([M]) \colon \mathrm{cone}^{n-*}(C_*^!(\widehat{f})) \to \mathrm{cone}_*(C_*^!(\widehat{f})).$$

In the following lemma and its proof, we shall see that the fact that the map $f \colon M \to X$ has degree one implies that the surgery kernels $K_*(\widehat{M})$ are isomorphic to the homology groups of $\mathrm{cone}_*(C_*^!(\widehat{f}))$ and that the chain map $-\cap e_*([M])$ behaves very much like the cap product map of the fundamental class of a Poincaré complex. This gives a deeper explanation of the fact that the surgery kernels $K_*(\widehat{M})$ behave in many respects like the homology groups of a third space satisfying Poincaré duality.

Lemma 8.46 (i) *There are natural $\mathbb{Z}\Gamma$-isomorphisms*

$$K_j(\widehat{M}) \xrightarrow{\cong} H_j\big(\mathrm{cone}_*(C_*^!(\widehat{f}))\big);$$
$$K^j(\widehat{M}) \xrightarrow{\cong} H_{n-j}\big(\mathrm{cone}^{n-*}(C_*^!(\widehat{f}))\big);$$

(ii) *The cap product with $e_*([M])$ induces a $\mathbb{Z}\Gamma$-chain homotopy equivalence*

$$- \cap e_*([M]) \colon \mathrm{cone}^{n-*}(C_*^!(\widehat{f})) \xrightarrow{\simeq} \mathrm{cone}_*(C_*^!(\widehat{f})),$$

which fits into a commutative diagram of isomorphisms

$$\begin{array}{ccc} K^{n-j}(\widehat{M}) & \xrightarrow{\cong} & H_j\big(\mathrm{cone}^{n-*}(C_*^!(\widehat{f}))\big) \\ {\scriptstyle\cong}\Big\downarrow{\scriptstyle-\cap[M]} & & {\scriptstyle\cong}\Big\downarrow{\scriptstyle-\cap e_*([M])} \\ K_j(\widehat{M}) & \xrightarrow{\cong} & H_j\big(\mathrm{cone}_*(C_*^!(\widehat{f}))\big). \end{array}$$

Proof. By definition, the chain level Umkehr map fits into the following up to chain homotopy commutative diagram of chain maps:

$$\begin{array}{ccccccc} \mathrm{cone}^{n-*}(C_*^!(\widehat{f})) & \xrightarrow{e^{n-*}} & C^{n-*}(\widehat{M}) & \xleftarrow{C^{n-*}(\widehat{f})} & C^{n-*}(\widehat{X}) \\ {\scriptstyle -\cap e_*([M])}\Big\downarrow & & {\scriptstyle -\cap[M]}\Big\downarrow & {\scriptstyle C_*^!(\widehat{f})}\swarrow & \Big\downarrow{\scriptstyle -\cap[X]} \\ \mathrm{cone}_*(C_*^!(\widehat{f})) & \xleftarrow{e_*} & C_*(\widehat{M}) & \xrightarrow{C_*(\widehat{f})} & C_*(\widehat{X}). \end{array}$$

Applying (8.43) and (8.44), a diagram chase in the above diagram proves assertion (i). The commutativity up to chain homotopy of the diagram in assertion (ii) follows from the commutativity of the above diagram. Now assertion (i) and Lemma 8.39 (ii) show that the chain map $-\cap e_*([M])$ induces an isomorphism on homology groups and so $-\cap e_*([M])$ is a chain homotopy equivalence, see [361, Theorem 1.7.7]. This proves assertion (ii). □

8.4.2 Symmetric Forms and Surgery Kernels

For the remainder of this subsection we fix a normal Γ-map of degree one $(M, [M], f, \hat{f}, a, \xi, \overline{f})$ with target $(X, \hat{p}_X, [X])$ in the sense of Definition 8.3. Furthermore, we will assume that we are in the universal covering case, see Notation 8.34. In particular Γ will be identified with the fundamental group π of the connected CW-complex X.

Our aim in this subsection is to relate the topology of intersections of immersions from Section 8.3 with Poincaré duality for kernel groups from Subsection 8.4.1. Specifically, we shall show in the highly connected case that the kernel groups $K_k(\widetilde{M})$ defined in Definition 8.36 admit the structure of a symmetric pairing

$$s \colon K_k(\widetilde{M}) \times K_k(\widetilde{M}) \to \mathbb{Z}\pi,$$

called the *algebraic intersection form* on $K_k(\widetilde{M})$. Moreover, there is a homomorphism $\alpha \colon K_k(\widetilde{M}) \to I_k(M)$ so that the algebraic intersection form s and the geometric intersection pairing $\lambda \colon I_k(M) \times I_k(M) \to \mathbb{Z}\pi$ of (8.11) and Lemma 8.19 are compatible under α, see Lemma 8.59 below.

We begin by defining the purely algebraic notion of a symmetric form over a unital ring with involution R. Then we show how to define the symmetric form $(K_k(\widetilde{M}), s)$. Finally we relate the pairings $(K_k(\widetilde{M}), s)$ and $(I_k(M), \lambda)$.

Definition 8.47 (Symmetric forms) An ϵ-*symmetric form* (P, ϕ) over an associative ring R with unit and involution and $\epsilon = \pm 1$ is a finitely generated projective R-module P together with an R-homomorphism $\phi \colon P \to P^*$ such that the composition $P \xrightarrow{e(P)} (P^*)^* \xrightarrow{\phi^*} P^*$ agrees with $\epsilon \cdot \phi$. Here and elsewhere $e(P)$ is the canonical isomorphism sending $p \in P$ to the element in $(P^*)^*$ given by the homomorphism $P^* \to R$, $f \mapsto \overline{f(p)}$. We call (P, ϕ) *non-singular* if ϕ is an isomorphism.

Recall that non-degenerate stands for the weaker condition that ϕ is injective.

There are obvious notions of isomorphisms and direct sums of ϵ-symmetric forms.

We can rewrite (P, ϕ) as a pairing

$$\lambda \colon P \times P \to R, \quad (p, q) \mapsto \phi(p)(q). \tag{8.48}$$

Then the conditions that ϕ and $\phi(p)$ for each $p \in P$ are R-linear become the conditions

$$\lambda(p, r_1 \cdot q_1 + r_2 \cdot q_2,) = r_1 \cdot \lambda(p, q_1) + r_2 \cdot \lambda(p, q_2);$$
$$\lambda(r_1 \cdot p_1 + r_2 \cdot p_2, q) = \lambda(p_1, q) \cdot \overline{r_1} + \lambda(p_2, q) \cdot \overline{r_2}.$$

The condition $\phi = \epsilon \cdot \phi^* \circ e(P)$ translates to $\lambda(q, p) = \epsilon \cdot \overline{\lambda(p, q)}$.

Example 8.49 (Standard hyperbolic ϵ-symmetric form) Let P be a finitely generated projective R-module. The *standard hyperbolic ϵ-symmetric form* $H^\epsilon(P)$ is given by the R-module $P \oplus P^*$ and the R-isomorphism

$$\phi\colon (P\oplus P^*) \xrightarrow{\begin{pmatrix}0 & 1\\ \epsilon & 0\end{pmatrix}} P^*\oplus P \xrightarrow{\mathrm{id}\,\oplus e(P)} P^*\oplus (P^*)^* = (P\oplus P^*)^*.$$

If we write it as a pairing we obtain

$$(P\oplus P^*)\times (P\oplus P^*) \to R, \quad ((p,f),(p',f')) \mapsto f(p') + \epsilon\cdot \overline{f'(p)}.$$

Example 8.50 (Intersection form of a $2k$-dimensional Poincaré complex) Let X be an oriented connected finite $2k$-dimensional Poincaré complex with middle-dimensional integral homology group $H^k(X;\mathbb{Z})$ and let $\mathrm{tors}\, H^k(X;\mathbb{Z})$ be the torsion subgroup. The abelian group $H^k(X)_F := H^k(X;\mathbb{Z})/\mathrm{tors}\, H^k(X;\mathbb{Z})$ is a finitely generated free $\mathbb{Z}$-module, and the intersection pairing of X is the non-singular bilinear form over $\mathbb{Z}$

$$\lambda_X\colon H^k(X)_F \times H^k(X)_F \to \mathbb{Z}, \quad ([x],[y]) \mapsto \langle x\cup y,[X]\rangle.$$

The intersection form of $X = S^k\times S^k$ is canonically isomorphic to the standard $(-1)^k$-symmetric hyperbolic form over $\mathbb{Z}$

$$(H^k(S^k\times S^k;\mathbb{Z}), \lambda_{S^k\times S^k}) = H_{(-1)^k}(\mathbb{Z}).$$

Now we show how to equip the $\mathbb{Z}\pi$-module $K_k(\widetilde{M})$ with the structure of a symmetric form over the group ring $\mathbb{Z}\pi$ equipped with the $w = w_1(X)$-twisted involution. We first show that $K_k(\widetilde{M})$ is a stably finitely generated free module. Recall that an R-module V is called *stably finitely generated free* if for some non-negative integer l the R-module $V\oplus R^l$ is a finitely generated free R-module.

By Lemma 8.39 (i), the kernel group $K_k(\widetilde{M})$ is isomorphic to the homology group $H_{k+1}\bigl(\mathrm{cone}_*(C_*(\widetilde{f}))\bigr)$ of the algebraic mapping cone of the induced map of $\mathbb{Z}\pi$-chain complexes $C_*(\widetilde{f})\colon C_*(\widetilde{M}) \to C_*(\widetilde{X})$. Since f is k-connected, we have by Lemma 8.39 (iii) that the homology groups of $\mathrm{cone}_*(C_*(\widetilde{f}))$ vanish below dimension $k+1$. Hence we begin by establishing some purely algebraic facts about chain complexes with vanishing homology groups below a given degree.

Let D_* be a finite projective chain complex over a ring with involution R. The *Kronecker pairing*

$$\langle -,- \rangle\colon H^j(D_*) \times H_j(D_*) \to R$$

is induced by the evaluation pairing $D^j \times D_j \to R$ that sends a pair (ϕ,x) in $\hom_{\mathbb{Z}\pi}(D_j,R)\times D_j$ to $\phi(x)$. The Kronecker pairing is the same as a $\mathbb{Z}\pi$-homomorphism, see (8.48),

$$\kappa_j\colon H^j(D_*) \to H_j(D_*)^*. \tag{8.51}$$

8.4 Surgery Kernels and Forms

Lemma 8.52 *Let D_* be a finite projective R-chain complex. Consider an integer j and suppose that $H_i(D_*) = 0$ for $i < j$.*

(i) *Let E_* be the R-subchain complex of D_* given by*

$$\cdots \xrightarrow{d_{j+2}} D_{j+1} \xrightarrow{d_{j+1}} \ker(d_j) \to 0 \to 0 \to \cdots \to 0.$$

Then E_* is finite projective and the inclusion $i_* \colon E_* \to D_*$ is an R-chain homotopy equivalence;

(ii) *The homomorphism $\kappa_j \colon H^j(D_*) \to H_j(D_*)^*$ from (8.51) is an isomorphism;*

(iii) *Suppose $H^{j+1}(\hom_R(D_*, V)) = 0$ for any R-module V. Then $\operatorname{im}(d_{j+1})$ is a direct summand in D_j, and $H_j(D_*)$ is finitely generated projective;*

(iv) *Suppose that D_i is stably finitely generated free for $i \geq 0$, $H_i(D_*) = 0$ for $i \neq j$, and $H^{j+1}(\hom_R(D_*, V)) = 0$ for any R-module V. Then $H_j(D_*)$ is stably finitely generated free.*

Proof. (i) We have the long exact sequence

$$0 \to \operatorname{im}(d_j) \to D_{j-1} \xrightarrow{d_{j-1}} D_{j-2} \xrightarrow{d_{j-2}} \cdots \xrightarrow{d_1} D_0 \to 0.$$

Hence $\operatorname{im}(d_j)$ is finitely generated projective. From the short exact sequence $0 \to \ker(d_j) \to D_j \to \operatorname{im}(d_j) \to 0$ we conclude that $\ker(d_j)$ is finitely generated projective. Hence E_* is finite projective. The inclusion $i_* \colon E_* \to D_*$ induces isomorphisms on all homology modules. Since both D_* and E_* are finite projective, i_* is an R-chain homotopy equivalence.

(ii) Let E_* be the finite projective R-chain complex from part (i), which satisfies $E_i = 0$ for $i \leq j-1$, and let $i_* \colon E_* \to D_*$ be the R-chain homotopy equivalence given by the inclusion. We have the exact sequence $E_{j+1} \xrightarrow{e_{j+1}} E_j \to H_j(E_*) \to 0$. Dualising yields the short exact sequence of R-modules $0 \to H_j(E_*)^* \to E_j^* \xrightarrow{e_{j+1}^*} E_{j+1}^*$. Hence we obtain an isomorphism $H_j(E_*)^* \to H^j(E^*)$. Since $i_* \colon E_* \to D_*$ is an R-chain homotopy equivalence, it induces an R-cochain homotopy equivalence $i^* \colon D^* \to E^*$. Thus we obtain an isomorphism

$$H_j(D_*)^* \to H^j(D^*). \tag{8.53}$$

This yields an inverse to the homomorphism $\kappa_j \colon H^j(D_*) \to H_j(D_*)^*$ from (8.51).

(iii) If we apply the assumption $H^{j+1}(\hom_R(D_*, V)) = 0$ in the case $V = \operatorname{im}(d_{j+1})$, we obtain an exact sequence

$$\hom_{\mathbb{Z}\pi}(D_j, \operatorname{im}(d_{j+1})) \xrightarrow{\hom_{\mathbb{Z}\pi}(d_{j+1}, \operatorname{id})} \hom_{\mathbb{Z}\pi}(D_{j+1}, \operatorname{im}(d_{j+1}))$$

$$\xrightarrow{\hom_{\mathbb{Z}\pi}(d_{j+2}, \operatorname{id})} \hom_{\mathbb{Z}\pi}(D_{j+2}, \operatorname{im}(d_{j+1})).$$

Since $d_{j+1} \in \hom_{\mathbb{Z}\pi}(D_{j+1}, \mathrm{im}(d_{j+1}))$ is mapped to zero under $\hom_{\mathbb{Z}\pi}(d_{j+2}, \mathrm{id})$, we can find an R-homomorphism $\rho\colon D_j \to \mathrm{im}(d_{j+1})$ with $\rho \circ d_{j+1} = d_{j+1}$. Hence $\mathrm{im}(d_{j+1})$ is a direct summand in D_j and also in $\ker(d_j)$.

(iv) From the condition $H_i(D_*) = 0$ for $i > j$, we obtain an exact sequence

$$\ldots \to D_{j+3} \xrightarrow{d_{j+3}} D_{j+2} \xrightarrow{d_{j+2}} D_{j+1} \xrightarrow{d_{j+1}} \mathrm{im}(d_{j+1}) \to 0.$$

Since $\mathrm{im}(d_{j+1})$ is a direct summand of the projective R-module D_j by assertion (iii), $\mathrm{im}(d_{j+1})$ is projective. Hence we obtain an isomorphism

$$\mathrm{im}(d_{j+1}) \oplus \bigoplus_{i \geq 0} D_{j+2i+2} \cong \bigoplus_{i \geq 0} D_{j+2i+1}.$$

Since there exists a natural number N with $D_i = 0$ for $i \geq N$ and D_i is stably finitely generated free for $i > j$ by assumption, $\mathrm{im}(d_{j+1})$ is stably finitely generated free. From the long exact sequence

$$0 \to \ker(d_j) \to D_j \xrightarrow{d_j} D_{j-1} \xrightarrow{d_{j-1}} \cdots \xrightarrow{d_1} D_0 \to 0$$

we conclude that $\ker(d_j)$ is stably finitely generated free. Since we know already that $H_j(D_*)$ is finitely generated projective and we have the short exact sequence $0 \to \mathrm{im}(d_{j+1}) \to \ker(d_j) \to H_j(D_*) \to 0$, we conclude that $H_j(D_*)$ is stably finitely generated free. □

We now apply Lemma 8.52 to the chain complex $D_* = \mathrm{cone}_*(C_*(\widetilde{f}))$. Identifying $K_k(\widetilde{M}) = H_{k+1}\bigl(\mathrm{cone}_*(C_*(\widetilde{f}))\bigr)$, we have the Kronecker pairing

$$\langle\,,\,\rangle\colon K^k(\widetilde{M}) \times K_k(\widetilde{M}) \to \mathbb{Z}\pi$$

that is induced by the evaluation pairing

$$\hom_{\mathbb{Z}\pi}(\mathrm{cone}_{k+1}(C_*(\widetilde{f})), \mathbb{Z}\pi) \times \mathrm{cone}_{k+1}(C_*(\widetilde{f})) \to \mathbb{Z}\pi.$$

The Kronecker pairing is the same as a $\mathbb{Z}\pi$-homomorphism, see (8.48),

$$\kappa_k\colon K^k(\widetilde{M}) \to K_k(\widetilde{M})^*. \tag{8.54}$$

Lemma 8.55 *Suppose that f is k-connected. Then:*

(i) *The Kronecker homomorphism $\kappa_k\colon K^k(\widetilde{M}) \to K_k(\widetilde{M})^*$ of (8.54) is an isomorphism;*
(ii) *The kernel $K_k(\widetilde{M})$ is stably finitely generated free.*

Proof. (i) Let $D_* := \mathrm{cone}_*(C_*(\widetilde{f}))$. By Lemma 8.39 (i) we can identify $H_*(D_*) = K_{*-1}(\widetilde{M})$. Since f is k-connected, we conclude from Lemma 8.39 (iii) that $H_i(D_*) = 0$ for $i < k+1$. Now apply Lemma 8.52 (ii).

8.4 Surgery Kernels and Forms

(ii) We now set $D_* := \text{cone}_*(C_*^!(\widetilde{f}))$, the mapping cone of the chain level Umkehr map. By Lemma 8.46 (ii), there is a chain homotopy equivalence $-\cap e_*([M]): D^{n-*} \to D_*$, and by Lemma 8.46 (i), $H_*(D_*) \cong K_*(\widetilde{M})$ and $H_*(D^{n-*}) \cong K^{n-*}(\widetilde{M})$. Since we assume that the map $f: M \to X$ is k-connected, $K_i(\widetilde{M}) = 0 = K^i(\widetilde{M})$ for $i < k$ by Lemma 8.39 (iii). Hence $H_i(D_*) = 0$ for $i < k$ and $H^i(D^{n-*}) = 0$ for $i > k$. Now let $i_*: E_* \to D_*$ be as in the statement of Lemma 8.52 (i) so that i_* is a chain homotopy equivalence and $E_i = 0$ for $i < k$. As $E_* \to D_*$ is a chain homotopy equivalence, the dual chain map $D^{n-*} \to E^{n-*}$ is a chain homotopy equivalence with $E^{n-i} = 0$ for $i > k$. Since $-\cap e_*([M]): D^{n-*} \to D_*$ is a chain equivalence, $H_i(D_*) \cong H_i(D^{n-*}) = 0$ for $i > k$, and there are isomorphisms $H^i(\hom_R(D_*, V)) \cong H^i(\hom_R(D^{n-*}, V)) \cong H^i(\hom_R(E^{n-*}, V))$ for all i and for any R-module V. As $E^{n-i} = 0$ for $i > k$, we have

$$H^{n-i}(\hom_R(D_*, V)) \cong H^{n-i}(\hom_R(E^{n-*}, V)) = 0 \qquad \text{for } i \leq k-1.$$

In summary, we have

$$H_i(D_*) \cong \begin{cases} K_k(\widetilde{M}), & \text{if } i = k; \\ 0, & \text{if } i \neq k; \end{cases}$$

$$H^{k+1}(\hom_R(D_*, V)) = 0,$$

and now we apply Lemma 8.52 (iv) to D_*. $\square$

We can now give the main example from this subsection that is a crucial step in the definition of the even-dimensional surgery obstruction.

Example 8.56 (Symmetric form on $K_k(\widetilde{M})$) Let $(f, \overline{f}): M \to X$ be a k-connected normal Γ-map of degree one with $n = 2k$ in the sense of Definition 8.3, and suppose that we are in the universal covering case, see Notation 8.34. We take the inverse of the Poincaré duality isomorphism

$$(-\cap [M])^{-1}: K_k(\widetilde{M}) \xrightarrow{\cong} K^k(\widetilde{M})$$

of Lemma 8.39 (ii) and compose it with the Kronecker homomorphism $\kappa_k: K^k(\widetilde{M}) \xrightarrow{\cong} K_k(M)^*$ of (8.54), which is an isomorphism by Lemma 8.55 (i). This yields a non-singular $(-1)^k$-symmetric form

$$\kappa_k \circ (-\cap [M])^{-1}: K_k(\widetilde{M}) \xrightarrow{\cong} K^k(\widetilde{M}) \xrightarrow{\cong} K_k(\widetilde{M})^*.$$

This form can be rewritten using (8.48) as the *algebraic intersection form*

$$s: K_k(\widetilde{M}) \times K_k(\widetilde{M}) \to \mathbb{Z}\pi. \tag{8.57}$$

The pairing s of (8.57) is a non-singular $(-1)^k$-symmetric form over the group ring $\mathbb{Z}\pi$ with the $w_1(M)$-twisted involution. Note that the form s appearing in (8.57) is the restriction of the pairing s of (8.15).

By Lemma 8.39 (iii) there is an isomorphism $h_k \colon \pi_{k+1}(f) \xrightarrow{\cong} K_k(\widetilde{M})$, and from Lemma 8.10 we get the homomorphism $t_k \colon \pi_{k+1}(f) \to I_k(M)$. Hence we have the $\mathbb{Z}\pi$-homomorphism

$$\alpha \colon K_k(\widetilde{M}) \xrightarrow{h_k^{-1}} \pi_{k+1}(f) \xrightarrow{t_k} I_k(M). \tag{8.58}$$

We get from (8.18) the following result.

Lemma 8.59 *Equip $T_m M$ with the orientation that is compatible with $[M]$ in the sense of Definition 8.16.*

Then the algebraic intersection form and the topological intersection form agree under $\alpha \colon K_k(\widetilde{M}) \to I_k(M)$. That is, the following diagram commutes:

$$\begin{array}{ccc} K_k(\widetilde{M}) \times K_k(\widetilde{M}) & \xrightarrow{s} & \mathbb{Z}\pi \\ {\scriptstyle \alpha \times \alpha} \downarrow & & \downarrow {\scriptstyle \mathrm{id}} \\ I_k(M) \times I_k(M) & \xrightarrow{\lambda} & \mathbb{Z}\pi \end{array}$$

where s is defined in (8.57) and λ in (8.11).

We conclude this subsection with the following simple but important observation about the effect of trivial $(k-1)$-surgeries on the symmetric surgery kernel in even dimensions $n = 2k$.

Example 8.60 (Effect of trivial surgery) Let $(f, \overline{f}) \colon TM \oplus \underline{\mathbb{R}}^a \to \xi$ be a normal π-map covering a k-connected map of degree one $f \colon M \to X$ with target an n-dimensional Poincaré complex for $n = 2k$. If we do surgery on the zero element in $\pi_k(f)$, then the effect on M is that M is replaced by the connected sum $M' = M \sharp (S^k \times S^k)$. The effect on $K_k(\widetilde{M})$ is that it is replaced by $K_k(\widetilde{M'}) = K_k(\widetilde{M}) \oplus (\mathbb{Z}\pi \oplus \mathbb{Z}\pi)$. The intersection pairing on this new kernel is the orthogonal sum of the given intersection pairing on $K_k(\widetilde{M})$ together with the standard hyperbolic symmetric form $H^{(-1)^k}(\mathbb{Z}\pi)$. Since by Lemma 8.55 (ii) $K_k(\widetilde{M})$ is stably finitely generated free, we can arrange by finitely many surgery steps on the trivial element in $\pi_k(f)$ that $K_k(\widetilde{M})$ is a finitely generated free $\mathbb{Z}\pi$-module.

Remark 8.61 (The general case) So far we have only considered the universal covering case. We briefly explain what happens in general. Consider a normal Γ-map of degree one $(M, [M], f, \widehat{f}, a, \xi, \overline{f})$ with target $(X, \widehat{p}_X, [X])$ in the sense of Definition 8.3. We assume that X is connected, the dimension n satisfies $n = 2k$ for a natural number $k \geq 2$, and the map $f \colon M \to X$ is k-connected.

Let $\widetilde{p}_M \colon \widetilde{M} \to M$ and $\widetilde{p}_X \colon \widetilde{X} \to X$ be models for the universal coverings. Fix base points $\widetilde{m} \in \widetilde{M}$, $m \in M$, $\widetilde{x} \in \widetilde{X}$, and $x \in X$ such that $f(m) = x$. Let $\widetilde{f} \colon \widetilde{M} \to \widetilde{X}$ be the lift of f uniquely determined by $\widetilde{f}(\widetilde{m}) = \widetilde{x}$. Since f is k-connected for $k \geq 2$, we can identify $\pi = \pi_1(X, x)$ with $\pi_1(M, m)$ using the isomorphism $\pi_1(f, m) \colon \pi_1(M, m) \xrightarrow{\cong} \pi_1(X, x)$. We obtain from these choices π-actions on $\widetilde{M}$

8.4 Surgery Kernels and Forms

and $\widetilde{X}$. There is a preferred group homomorphism $u\colon \pi \to \Gamma$ and preferred Γ-homeomorphisms $\Gamma \times_u \widetilde{M} \xrightarrow{\cong} \widehat{M}$ and $\Gamma \times_u \widetilde{X} \xrightarrow{\cong} \widehat{X}$ such that $\mathrm{id}_\Gamma \times_u \widetilde{f}$ corresponds to $\widehat{f}$. There are isomorphisms of $\mathbb{Z}\Gamma$-chain complexes

$$\mathbb{Z}\Gamma \otimes_{\mathbb{Z}\pi} C_*(\widetilde{M}) \xrightarrow{\cong} C_*(\widehat{M});$$

$$\mathbb{Z}\Gamma \otimes_{\mathbb{Z}\pi} C_*(\widetilde{X}) \xrightarrow{\cong} C_*(\widehat{X}),$$

where we consider $\mathbb{Z}\Gamma$ as a right $\mathbb{Z}\pi$-module using u. They induce a $\mathbb{Z}\Gamma$-chain isomorphism $\mathbb{Z}\Gamma \otimes_{\mathbb{Z}\pi} \mathrm{cone}_*(\widetilde{f}) \xrightarrow{\cong} \mathrm{cone}_*(\widehat{f})$. The canonical $\mathbb{Z}\Gamma$-homomorphism $\mathbb{Z}\Gamma \otimes_{\mathbb{Z}\pi} H_{k+1}(\mathrm{cone}_*(\widetilde{f})) \xrightarrow{\cong} H_{k+1}(\mathbb{Z}\Gamma \otimes_{\mathbb{Z}\pi} \mathrm{cone}_*(\widetilde{f}))$ is bijective as $H_i(\mathrm{cone}_*(\widetilde{f}))$ vanishes for $i \leq k$. This follows from a spectral sequence argument or from Lemma 8.52 (ii) and the fact that the functor $\mathbb{Z}\Gamma \otimes_{\mathbb{Z}\pi} -$ is right exact. Hence we get a $\mathbb{Z}\Gamma$-isomorphism $\mathbb{Z}\Gamma \otimes_{\mathbb{Z}\pi} H_{k+1}(\mathrm{cone}_*(\widetilde{f})) \xrightarrow{\cong} H_{k+1}(\mathrm{cone}_*(\widehat{f}))$. Now Lemma 8.39 (ii) implies that we get a $\mathbb{Z}\Gamma$-isomorphism

$$\mathbb{Z}\Gamma \otimes_{\mathbb{Z}\pi} K_k(\widetilde{M}) \xrightarrow{\cong} K_k(\widehat{M}). \tag{8.62}$$

In particular we conclude from Lemma 8.55 that $K_k(\widehat{M})$ is a stably finitely generated free $\mathbb{Z}\Gamma$-module and the Kronecker homomorphism

$$\kappa_k \colon K^k(\widehat{M}) \xrightarrow{\cong} K_k(\widehat{M})^* \tag{8.63}$$

is a $\mathbb{Z}\Gamma$-isomorphism.

The pairing $s\colon K_k(\widetilde{M}) \times K_k(\widetilde{M}) \to \mathbb{Z}\pi$ of (8.57) descends to a pairing

$$s_\Gamma \colon K_k(\widehat{M}) \times K_k(\widehat{M}) \to \mathbb{Z}\Gamma, \tag{8.64}$$

the pairing λ of (8.11) descends to a pairing

$$\lambda_\Gamma \colon \mathbb{Z}\Gamma \otimes_{\mathbb{Z}\pi} I_k(M) \times \mathbb{Z}\Gamma \otimes_{\mathbb{Z}\pi} I_k(M) \to \mathbb{Z}\Gamma, \tag{8.65}$$

the $\mathbb{Z}\pi$-homomorphism $\alpha \colon K_k(\widetilde{M}) \to I_k(M)$ of (8.58) descends to the $\mathbb{Z}\Gamma$-homomorphism

$$\alpha_\Gamma \colon K_k(\widehat{M}) \to \mathbb{Z}\Gamma \otimes_{\mathbb{Z}\pi} I_k(M), \tag{8.66}$$

and the commutative diagram appearing in Lemma 8.59 yields the commutative diagram of $\mathbb{Z}\Gamma$-modules

$$\begin{array}{ccc} K_k(\widehat{M}) \times K_k(\widehat{M}) & \xrightarrow{s_\Gamma} & \mathbb{Z}\Gamma \\ {\scriptstyle \alpha_\Gamma \times \alpha_\Gamma} \downarrow & & \downarrow {\scriptstyle \mathrm{id}} \\ \mathbb{Z}\Gamma \otimes_{\mathbb{Z}\pi} I_k(M) \times \mathbb{Z}\Gamma \otimes_{\mathbb{Z}\pi} I_k(M) & \xrightarrow{\lambda_\Gamma} & \mathbb{Z}\Gamma. \end{array}$$

Remark 8.67 In this subsection we have *not* used the bundle data $\overline{f}$ at all. They come in when we deal with quadratic structures in the next subsection, see Example 8.86.

8.4.3 Quadratic Forms and Surgery Kernels

For the remainder of this subsection we fix a normal Γ-map of degree one $(M, [M], f, \widehat{f}, a, \xi, \overline{f})$ with target $(X, \widehat{p}_X, [X])$ in the sense of Definition 8.3. We often abbreviate it to $(f, \overline{f}): M \to X$. Furthermore, we will assume that we are in the universal covering case, see Notation 8.34. In particular, Γ will be identified with the fundamental group π of the connected CW-complex X.

We have already seen that it will not be enough to study intersections of different immersions, we must also deal with self-intersections of one immersion. We have seen in Lemma 8.26 that we can enrich the intersection pairing by the self-intersection function. This leads to the following algebraic analogue of the self-intersection function when R is an associative ring with unit and involution and $\epsilon \in \{\pm 1\}$. For a finitely generated projective R-module P, define an involution of R-modules

$$T: \hom_R(P, P^*) \to \hom_R(P, P^*), \quad \phi \mapsto \phi^* \circ e(P) \tag{8.68}$$

where $e(P): P \to (P^*)^*$ is the canonical isomorphism from Definition 8.47.

Let P be a finitely generated projective R-module. For $\epsilon \in \{\pm 1\}$, define abelian groups

$$Q^\epsilon(P) := \ker\left((1 - \epsilon \cdot T): \hom_R(P, P^*) \to \hom_R(P, P^*)\right); \tag{8.69}$$
$$Q_\epsilon(P) := \operatorname{coker}\left((1 - \epsilon \cdot T): \hom_R(P, P^*) \to \hom_R(P, P^*)\right). \tag{8.70}$$

Let

$$(1 + \epsilon \cdot T): Q_\epsilon(P) \to Q^\epsilon(P) \tag{8.71}$$

be the homomorphism that sends the class represented by $f: P \to P^*$ to the element $f + \epsilon \cdot T(f)$. An R-homomorphism $f: P \to Q$ of finitely generated projective R-modules induces homomorphisms of abelian groups $Q^\epsilon(f): Q^\epsilon(Q) \to Q^\epsilon(P)$ and $Q_\epsilon(f): Q_\epsilon(Q) \to Q_\epsilon(P)$ by sending the class of $\phi: P \to P^*$ to the class of $f^* \circ \phi \circ f$.

We have the isomorphism of R-modules $R^* \to R$, $\phi \mapsto \overline{\phi(1)}$. It yields an identification in the special case $P = R$

$$Q^\epsilon(R) = \ker\left(R \to R, \ r \mapsto r - \epsilon \overline{r}\right); \tag{8.72}$$
$$Q_\epsilon(R) = \operatorname{coker}\left(R \to R, \ r \mapsto r - \epsilon \overline{r}\right). \tag{8.73}$$

Under the identifications (8.72) and (8.73) the map $(1 + \epsilon \cdot T)$ of (8.71) is the map induced by $R \to R$, $r \mapsto r + \epsilon \cdot \overline{r}$.

Exercise 8.74 Let G be a group and let $w: G \to \{\pm 1\}$ be a group homomorphism. Equip $\mathbb{Z}G$ with the w-twisted involution. Show that there are canonical identifications between the abelian groups $Q_\epsilon(\mathbb{Z}G, w)$ of (8.22) and $Q_\epsilon(\mathbb{Z}G)$ of (8.70) for $P = \mathbb{Z}G$.

8.4 Surgery Kernels and Forms

Exercise 8.75 Let Λ be a commutative ring. Consider a group G with an orientation homomorphism $w\colon G \to \{\pm 1\}$ and an element $\epsilon \in \{\pm 1\}$. Equip ΛG with the involution sending $\sum_{g \in G} r_g \cdot g$ to $\sum_{g \in G} r_g \cdot w(g) \cdot g^{-1}$. Equip G with the $\mathbb{Z}/2$-action sending $g \to g^{-1}$. Let $S = \{g \in G \mid g = g^{-1}\}$ be the $\mathbb{Z}/2$-fixed point set. Let $T \subseteq G \setminus S$ be a subset such that each $\mathbb{Z}/2$-orbit of $G \setminus S$ meets T in exactly one element. Prove that we obtain isomorphisms of Λ-modules

$$Q_\epsilon(\Lambda G) = \bigoplus_{g \in S} \operatorname{coker}((1 - w_1(g) \cdot \epsilon) \cdot \operatorname{id}_\Lambda \colon \Lambda \to \Lambda) \oplus \bigoplus_{g \in T} \Lambda;$$

$$Q^\epsilon(\Lambda G) = \bigoplus_{g \in S} \ker((1 - w_1(g) \cdot \epsilon) \cdot \operatorname{id}_\Lambda \colon \Lambda \to \Lambda) \oplus \bigoplus_{g \in T} \Lambda.$$

Consider the pairing

$$\rho \colon R \times Q_{(-1)^k}(R) \to Q_{(-1)^k}(R), \quad (r, [s]) \mapsto [rs\bar{r}]. \tag{8.76}$$

It is well defined since for $r, s, t \in R$ we get if we put $t' = rt\bar{r}$,

$$r\big(s + (t - (-1)^k \cdot \bar{t})\big)\bar{r} = rs\bar{r} + \big(t' - (-1)^k \cdot \bar{t'}\big).$$

It is additive in the second variable, i.e., $\rho(r, [s_1] - [s_2]) = \rho(r, [s_1]) - \rho(r, [s_2])$, but it is not additive in the first variable and in particular ρ does *not* give the structure of a left R-module on $Q_{(-1)^k}(R)$. Sometimes in the literature $\rho(r, [s])$ is denoted by $r[s]\bar{r}$. As an aside, we point out that $Q_{(-1)^k}(R)$ does admit the structure of a quadratic R-module in the sense of [35, Definition 3.1].

Definition 8.77 (Quadratic form) An ϵ-*quadratic form* (P, ψ) is a finitely generated projective R-module P together with an element $\psi \in Q_\epsilon(P)$. It is called *non-singular* if the associated ϵ-symmetric form $(P, (1 + \epsilon \cdot T)(\psi))$ is non-singular, i.e., $(1 + \epsilon \cdot T)(\psi) \colon P \to P^*$ is bijective.

There is an obvious notion of the direct sum of two ϵ-quadratic forms. An isomorphism $f \colon (P, \psi) \to (P', \psi')$ of two ϵ-quadratic forms is an R-isomorphism $f \colon P \xrightarrow{\cong} P'$ such that the induced map $Q_\epsilon(f) \colon Q_\epsilon(P') \to Q_\epsilon(P)$ sends ψ' to ψ.

Remark 8.78 (Writing a quadratic form as a triple (P, λ, μ)) We can rewrite this as follows. An ϵ-quadratic form (P, ψ) is equivalent to a triple (P, λ, μ) consisting of a pairing

$$\lambda \colon P \times P \to R$$

satisfying

$$\lambda(p, r_1 \cdot q_1 + r_2 \cdot q_2) = r_1 \cdot \lambda(p, q_1) + r_2 \cdot \lambda(p, q_2);$$
$$\lambda(r_1 \cdot p_1 + r_2 \cdot p_2, q) = \lambda(p_1, q) \cdot \overline{r_1} + \lambda(p_2, q) \cdot \overline{r_2};$$
$$\lambda(q, p) = \epsilon \cdot \overline{\lambda(p, q)},$$

and a map

$$\mu\colon P \to Q_\epsilon(R) = R/\{r - \epsilon \cdot \overline{r} \mid r \in R\}$$

satisfying

$$\mu(rp) = \rho(r, \mu(p));$$
$$\mu(p + q) - \mu(p) - \mu(q) = \mathrm{pr}(\lambda(p, q));$$
$$\lambda(p, p) = (1 + \epsilon \cdot T)(\mu(p)),$$

where the pairing ρ was introduced in (8.76), $\mathrm{pr}\colon R \to Q_\epsilon(R)$ is the projection and $(1 + \epsilon \cdot T)\colon Q_\epsilon(R) \to R$ the map sending the class of r to $r + \epsilon \cdot \overline{r}$. Namely, put

$$\lambda(p, q) = \bigl((1 + \epsilon \cdot T)(\psi)(p)\bigr)(q);$$
$$\mu(p) = \mathrm{pr}(\psi(p)(p)).$$

These two descriptions of an ϵ-quadratic form are equivalent, see [410, Theorem 1].

Example 8.79 (Standard hyperbolic ϵ-quadratic form) Let P be a finitely generated projective R-module. The *standard hyperbolic ϵ-quadratic form $H_\epsilon(P)$* is given by the $\mathbb{Z}\pi$-module $P \oplus P^*$ and the class in $Q_\epsilon(P \oplus P^*)$ of the R-homomorphism

$$\phi\colon (P \oplus P^*) \xrightarrow{\begin{pmatrix} 0 & 1 \\ 0 & 0 \end{pmatrix}} P^* \oplus P \xrightarrow{\mathrm{id} \oplus e(P)} P^* \oplus (P^*)^* = (P \oplus P^*)^*.$$

The ϵ-symmetric form associated to $H_\epsilon(P)$ is $H^\epsilon(P)$. If we rewrite the standard hyperbolic ϵ-quadratic form as a triple (P, λ, μ) according to Remark 8.78, then we get

$$\lambda\colon (P \oplus P^*) \times (P \oplus P^*) \to R, \quad ((p, f), (p', f')) \mapsto f(p') + \epsilon \cdot f'(p);$$
$$\mu\colon P \oplus P^* \to Q_\epsilon(R), \quad (p, f) \mapsto [f(p)].$$

Example 8.80 (Linear ϵ-quadratic form) An ϵ-quadratic form (P, ψ) is called *linear* if the associated ϵ-symmetric form satisfies $(P, (1 + \epsilon \cdot T)\psi) = (P, 0)$. In this case $\psi(p + q) = \psi(p) + \psi(q)$ for all $p, q \in P$.

Note that ϵ-quadratic forms (P, ψ_0) and (P, ψ_1) have equal associated ϵ-symmetric forms, $(P, (1 + \epsilon \cdot T)\psi_0) = (P, (1 + \epsilon \cdot T)\psi_1)$, if and only if there is a linear ϵ-quadratic form (P, ψ) such that $\psi_1 = \psi_0 + \psi$. In this case we call (P, ψ_1) a *linear perturbation of ψ_0 by ψ*.

Remark 8.81 (Rings containing $1/2$) Suppose that $1/2 \in R$. Then the homomorphism

$$(1 + \epsilon \cdot T)\colon Q_\epsilon(P) \xrightarrow{\cong} Q^\epsilon(P), \quad [\psi] \mapsto [\psi + \epsilon \cdot T(\psi)]$$

is bijective. The inverse sends $[u]$ to $[u/2]$. Hence any ϵ-symmetric form carries a unique ϵ-quadratic structure and there is no difference between the symmetric and the quadratic setting if 2 is invertible in R.

8.4 Surgery Kernels and Forms

Definition 8.82 (Even ϵ-symmetric form) An ϵ-symmetric form (P, ϕ) is called *even* if the pairing λ associated to it in (8.48) satisfies for $p \in P$

$$\lambda(p, p) \in \operatorname{im}\big((1 + \epsilon T): Q_\epsilon(R) \to Q^\epsilon(R)\big).$$

The underlying ϵ-symmetric form of an ϵ-quadratic form is always even.

Lemma 8.83 *Suppose that the map $(1 + \epsilon T): Q_\epsilon(R) \to Q^\epsilon(R)$ is injective. Then the category of ϵ-quadratic forms and of even ϵ-symmetric forms are isomorphic.*

Proof. Consider an even ϵ-symmetric form (P, ϕ). Consider $p \in P$. The pairing λ associated to it in (8.48) satisfies $\lambda(p, p) \in \operatorname{im}\big((1 + \epsilon T): Q_\epsilon(R) \to Q^\epsilon(R)\big)$ by definition. Hence there is by assumption precisely one element $\mu(p) \in Q_\epsilon(R)$ which is sent under $(1 + \epsilon T): Q_\epsilon(R) \to Q^\epsilon(R)$ to $\lambda(p, p)$. One easily checks that (P, μ, λ) defines an ϵ-quadratic form (P, ψ) using Remark 8.78 such that its underlying ϵ-symmetric form is (P, ϕ). Since $(1 + \epsilon T): Q_\epsilon(R) \to Q^\epsilon(R)$ is injective, this is the only ϵ-quadratic form whose underlying ϵ-symmetric form is (P, ϕ). $\square$

Example 8.84 (Even ϵ-symmetric forms over $\mathbb{Z}$) Take $R = \mathbb{Z}$.

We begin with the case $\epsilon = +1$. Then $Q_{+1}(\mathbb{Z}) = \mathbb{Z}$ and $Q^{+1}(\mathbb{Z}) = \mathbb{Z}$ and the symmetrisation map $1 + T: Q_{+1}(\mathbb{Z}) \to Q^{+1}(\mathbb{Z})$ becomes under this identification the injective map $2 \cdot \operatorname{id}_\mathbb{Z}: \mathbb{Z} \to \mathbb{Z}$. Hence an $(+1)$-symmetric form (P, λ) is even if and only if $\lambda(p, p) \in \mathbb{Z}$ is even for all $p \in P$. Moreover, a $(+1)$-quadratic form over $\mathbb{Z}$ is the same as an even $(+1)$-symmetric form over $\mathbb{Z}$.

Now put $\epsilon = -1$. Then $Q_{-1}(\mathbb{Z}) = \mathbb{Z}/2$ and $Q^{-1}(\mathbb{Z}) = 0$. Hence a (-1)-symmetric form (P, λ) is always even and has a quadratic refinement. However, it can happen that there are two non-singular (-1)-quadratic forms that are not isomorphic as (-1)-quadratic forms but have the same underlying (-1)-symmetric form, see Theorem 8.111 and its proof.

Exercise 8.85 Show that for any group G equipped with the trivial orientation homomorphism $G \to \{\pm 1\}$ the map $(1 + T): Q_{+1}(\mathbb{Z}G) \to Q^{+1}(\mathbb{Z}G)$ defined in (8.71) is injective.

The following example is of fundamental importance in surgery and brings together all the main ideas of Sections 8.3 and 8.4.

Example 8.86 (Quadratic form coming from a surgery problem) An example of a non-singular $(-1)^k$-quadratic form over $\mathbb{Z}\pi$ with the w-twisted involution is given as follows for a normal Γ-map of degree one $(M, [M], f, \widehat{f}, a, \xi, \overline{f})$ with target $(X, \widehat{p}_X, [X])$ in the sense of Definition 8.3, for which f is k-connected, provided that we are in the universal covering case, see Notation 8.34, and have made choices (GBP) and (ABP) with the same $m \in M$, see Notations 8.8 and 8.13, that are compatible in the sense of Definition 8.16. Namely, take the triple $(K_k(\widetilde{M}), s, t)$ with the pairing s of (8.57) and quadratic refinement

$$t: K_k(\widetilde{M}) \xrightarrow{\alpha} I_k(M) \xrightarrow{\mu} Q_{(-1)^k}(\mathbb{Z}\pi, w), \qquad (8.87)$$

where $\mu\colon I_k(M) \to Q_{(-1)^k}(\mathbb{Z}\pi, w)$ is the self-intersection function defined in (8.23) and α is the map defined in (8.58). We need the compatibility of $[M]$ and the orientation $T_m M$ in order to be able to apply Lemma 8.59.

Note that the bundle data are not needed to define the non-singular $(-1)^k$-symmetric form $(K_k(\widetilde{M}), s)$ since s has a homological interpretation via Equation (8.18), but play a key role for the non-singular $(-1)^k$-quadratic form $(K_k(\widetilde{M}), s, t)$ since they appear in the definition of the map α.

The next result is the key step in translating the geometric question of whether we can change a normal map by a finite sequence of surgery steps to a homotopy equivalence to an algebraic question about quadratic forms. It will naturally lead to the definition of the surgery obstruction groups $L_{2k}(R)$.

Definition 8.88 (Normal homotopy equivalence) A normal map is called a *normal homotopy equivalence* if the underlying map of spaces is a homotopy equivalence.

Theorem 8.89 (Surgery and hyperbolic kernels) *Suppose that $k \geq 3$ holds and we are in the situation of Example 8.86. Moreover, assume that for the non-singular $(-1)^k$-quadratic form $(K_k(\widetilde{M}), s, t)$ there are integers $u, v \geq 0$ together with an isomorphism of non-singular $(-1)^k$-quadratic forms*

$$(K_k(\widetilde{M}), s, t) \oplus H_{(-1)^k}(\mathbb{Z}\pi^u) \cong H_{(-1)^k}(\mathbb{Z}\pi^v).$$

Then we can perform a finite number of surgery steps resulting in a normal homotopy equivalence $(f', \overline{f'})$.

Proof. If we do a surgery step on the trivial element in $\pi_k(f)$, we have explained the effect on $(K_k(\widetilde{M}), s)$ in Example 8.60. The effect on the quadratic form $(K_k(\widetilde{M}), s, t)$ is analogous, one adds a copy of $H_\epsilon(\mathbb{Z}\pi)$. Hence we can assume without loss of generality that the non-singular quadratic form $(K_k(\widetilde{M}), s, t)$ is isomorphic to $H_\epsilon(\mathbb{Z}\pi^v)$. Thus we can choose a $\mathbb{Z}\pi$-basis $\{b_1, b_2, \ldots, b_v, c_1, c_2, \ldots, c_v\}$ for $K_k(\widetilde{M})$ such that

$$\begin{aligned} s(b_i, c_i) &= 1 & &\text{for } i \in \{1, 2, \ldots, v\}; \\ s(b_i, c_j) &= 0 & &\text{for } i, j \in \{1, 2, \ldots, v\}, i \neq j; \\ s(b_i, b_j) &= 0 & &\text{for } i, j \in \{1, 2, \ldots, v\}; \\ s(c_i, c_j) &= 0 & &\text{for } i, j \in \{1, 2, \ldots, v\}; \\ t(b_i) &= 0 & &\text{for } i \in \{1, 2, \ldots, v\}. \end{aligned}$$

Note that f is a homotopy equivalence if and only if the number v is zero, see Lemma 8.39 (iii). Hence it suffices to explain how we can lower the number v to $(v-1)$ by a surgery step on an element in $\pi_{k+1}(f)$. Of course our candidate is the element ω in $\pi_{k+1}(f)$ that corresponds under the isomorphism $h_k\colon \pi_{k+1}(f) \to K_k(\widetilde{M})$, see Lemma 8.39 (iii), to the element b_v. By construction the composite

$$\pi_{k+1}(f) \xrightarrow{t_k} I_k(M) \xrightarrow{\mu} Q_\epsilon(\mathbb{Z}\pi, w)$$

of the maps defined in (8.23) and in Lemma 8.10 sends ω to zero. Now Theorem 4.39 and Theorem 8.28 ensure that we can perform surgery on ω. Note that the assumption

8.4 Surgery Kernels and Forms

$k \geq 3$ and the quadratic structure on the kernel become relevant exactly at this point. Finally it remains to check whether the effect on $K_k(\widetilde{M})$ is the desired one, namely, that we get rid of one of the hyperbolic summands $H_\epsilon(\mathbb{Z}\pi)$, or equivalently, ν is lowered to $\nu - 1$.

We have explained earlier that doing surgery yields not only a new manifold M' but also a bordism from M to M'. Namely, take

$$W := (M \times [0,1]) \cup_{S^k \times D^{n-k}} (D^{k+1} \times D^{n-k}),$$

where we attach $D^{k+1} \times D^{n-k}$ by an embedding $S^k \times D^{n-k} \hookrightarrow M \times \{1\}$, put $M' := \partial W - M$, and identify $M = M \times \{0\}$. The manifold W comes with a map $F: W \to X \times [0,1]$ whose restriction to M is the given map $f: M = M \times \{0\} \to X = X \times \{0\}$ and whose restriction to M' is a map $f': M' \to X \times \{1\}$.

We have introduced the surgery kernel for a normal map of closed manifolds in Subsection 8.4.1. We will treat their generalisations to pairs in detail in Subsections 8.6.2. Here we only need the following input,

$$K_j(\widetilde{W}) = \ker(H_j(\widetilde{W}) \to H_j(\widetilde{X} \times [0,1]));$$
$$K_j(\widetilde{M}) = \ker(H_j(\widetilde{M}) \to H_j(\widetilde{X} \times \{0\}));$$
$$K_j(\widetilde{M'}) = \ker(H_j(\widetilde{M'}) \to H_j(\widetilde{X} \times \{1\}));$$
$$K_j(\widetilde{W}, \widetilde{M}) = \ker(H_j(\widetilde{W}, \widetilde{M}) \to H_j(\widetilde{X} \times [0,1], \widetilde{X} \times \{0\}));$$
$$K_j(\widetilde{W}, \widetilde{M'}) = \ker(H_j(\widetilde{W}, \widetilde{M'}) \to H_j(\widetilde{X} \times [0,1], \widetilde{X} \times \{1\}));$$
$$K_j(\widetilde{W}, \widetilde{\partial W}) = \ker(H_j(\widetilde{W}, \widetilde{\partial W}) \to H_j(\widetilde{X} \times [0,1], \widetilde{X} \times \{0,1\})),$$

where $\widetilde{\partial W}$ is the restriction of the universal covering $\widetilde{W} \to W$ to ∂W, which is indeed over each component of ∂W the universal covering. Define $\widetilde{M'}$ analogously. The inclusion of the kernels into the homology groups, which they are submodules of, are always split injections. The kernels above vanish except in one dimension that is either k or $k + 1$. The various long exact sequences of pairs then yield the following exact braid. To get the full detailed proof of its existence, one has to combine Lemma 8.39 (ii) and Lemma 8.116 (ii)

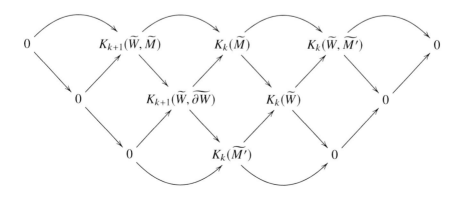

The $(k + 1)$-handle $D^{k+1} \times D^{n-k}$ defines an element ϕ^{k+1} in $K_{k+1}(\widetilde{W}, \widetilde{M})$ and the associated dual k-handle, see (2.38), defines an element ψ^k in $K_k(\widetilde{W}, \widetilde{M}')$. These elements constitute a $\mathbb{Z}\pi$-basis for $K_{k+1}(\widetilde{W}, \widetilde{M}) \cong \mathbb{Z}\pi$ and $K_k(\widetilde{W}, \widetilde{M}') \cong \mathbb{Z}\pi$. The $\mathbb{Z}\pi$-homomorphism $K_{k+1}(\widetilde{W}, \widetilde{M}) \to K_k(\widetilde{M})$ maps ϕ^{k+1} to b_v. The $\mathbb{Z}\pi$-homomorphism $K_k(\widetilde{M}) \to K_k(\widetilde{W}, \widetilde{M}')$ sends x to $s(b_v, x) \cdot \psi^k$. Hence we can find elements $b_1', b_2', \ldots, b_v'$ and $c_1', c_2', \ldots, c_{v-1}'$ in $K_{k+1}(\widetilde{W}, \widetilde{\partial W})$ uniquely determined by the property that b_i' is mapped to b_i and c_i' to c_i under the $\mathbb{Z}\pi$-homomorphism $K_{k+1}(\widetilde{W}, \widetilde{\partial W}) \to K_k(\widetilde{M})$. Moreover, these elements form a $\mathbb{Z}\pi$-basis for $K_{k+1}(\widetilde{W}, \widetilde{\partial W})$ and the element ϕ^{k+1} is mapped to b_v' under the $\mathbb{Z}\pi$-homomorphism $K_{k+1}(\widetilde{W}, \widetilde{M}) \to K_{k+1}(\widetilde{W}, \widetilde{\partial W})$. Define b_i'' and c_i'' for $i = 1, 2, \ldots, (v-1)$ to be the image of b_i' and c_i' under the $\mathbb{Z}\pi$-homomorphism $K_{k+1}(\widetilde{W}, \widetilde{\partial W}) \to K_k(\widetilde{M}')$. Then $\{b_i'' \mid i = 1, 2 \ldots, (v-1)\} \coprod \{c_i'' \mid i = 1, 2 \ldots, (v-1)\}$ is a $\mathbb{Z}\pi$-basis for $K_k(\widetilde{M}')$. Using the geometric definition of s and t via intersections and self-intersections of immersions, we observe that for the quadratic structure (s', t') on $K_k(\widetilde{M}')$ we get

$$\begin{aligned}
s'(b_i'', c_i'') &= s(b_i, c_i) = 1 && \text{for } i \in \{1, 2, \ldots, (v-1)\}; \\
s'(b_i'', c_j'') &= s(b_i, c_j) = 0 && \text{for } i, j \in \{1, 2, \ldots, (v-1)\}, i \neq j; \\
s'(b_i'', b_j'') &= s(b_i, b_j) = 0 && \text{for } i, j \in \{1, 2, \ldots, (v-1)\}; \\
s'(c_i'', c_j'') &= s(c_i, c_j) = 0 && \text{for } i, j \in \{1, 2, \ldots, (v-1)\}; \\
t'(b_i'') &= t(b_i) = 0 && \text{for } i \in \{1, 2, \ldots, (v-1)\}.
\end{aligned}$$

Via the correspondence from Remark 8.78 this is equivalent to the hyperbolic form $H_\epsilon(\mathbb{Z}\pi^{(v-1)})$. So we have succeeded in reducing the number v to $(v-1)$. This finishes the proof of Theorem 8.89. □

8.5 Even-Dimensional L-Groups

Let R be an associative unital ring with involution. In this section we define the even-dimensional L-groups $L_{2k}(R)$ of R for $k \in \mathbb{Z}$. When $R = (\mathbb{Z}\pi, w)$ is a group ring with w-twisted involution, then the surgery obstruction for even-dimensional surgery problems with fundamental group π and orientation homomorphism w will take values in $L_{2k}(\mathbb{Z}\pi, w)$.

The general definition of $L_{2k}(R)$ occurs in Subsection 8.5.1. In Subsections 8.5.2 and 8.5.3 we give the computation of $L_0(\mathbb{Z})$ and $L_2(\mathbb{Z})$.

8.5.1 The Definition of $L_{2k}(R)$ via Forms

Definition 8.90 (Quadratic L-groups in even dimensions) Let R be an associative unital ring with involution. For $k \in \mathbb{Z}$, define

$$L_{2k}(R),$$

8.5 Even-Dimensional L-Groups

called the *2k-th quadratic L-group* of R, to be the abelian group of equivalence classes $[(F, \psi)]$ of non-singular $(-1)^k$-quadratic forms (F, ψ) whose underlying R-module F is a finitely generated free R-module with respect to the following equivalence relation. We call (F, ψ) and (F', ψ') equivalent if and only if there exist integers $u, u' \geq 0$ and an isomorphism of non-singular ϵ-quadratic forms

$$(F, \psi) \oplus H_\epsilon(R)^u \cong (F', \psi') \oplus H_\epsilon(R)^{u'}.$$

Addition is given by the sum of two $(-1)^k$-quadratic forms. The zero element is represented by $[H_\epsilon(R)^u]$ for any integer $u \geq 0$. The inverse of $[(F, \psi)]$ is given by $[(F, -\psi)]$ because of Lemma 8.95.

Remark 8.91 (Category of modules) In the previous section we defined algebraic notions such as an ϵ-symmetric form (P, ϕ) and an ϵ-quadratic form (P, ψ) in the setting where P was a finitely generated projective module, but in Definition 8.90 we demand that the underlying module is finitely generated free. The reason for this is that we want the L-groups $L_{2k}(R)$ to be obstruction groups for modifying a surgery problem to a normal homotopy equivalence, see Theorem 8.112, and it turns out that for this purpose one needs to use the setting with finitely generated free modules. However, as we have already seen, the algebraic notions of forms can be defined and studied in a more general setting of finitely generated projective modules, see Definition 15.2, and in fact, the theory of forms in the setting of finitely generated projective modules is very pleasant and to a large extent similar to the setting of finitely generated free modules, Therefore we will continue to present it in the finitely generated projective setting, and whenever there is a place where special arguments are needed in the finitely generated free setting, we explicitly state them. On the other hand, there also exists a geometric setting that corresponds to the finitely generated projective modules; more on this will be mentioned in Chapter 16.

Remark 8.92 (L-groups via stably free R-modules) Suppose we only demanded in Definition 8.90 that the underlying modules of the $(-1)^k$-quadratic forms were stably finitely generated free R-modules and kept the rest of the definition unchanged. Then the resulting group would be isomorphic to the group $L_{2k}(R)$ from Definition 8.90, where we demanded the underlying modules to be finitely generated free. The reason for this is that if (P, ψ) is non-singular a $(-1)^k$-quadratic form whose underlying R-module P is only a stably finitely generated free R-module, then there exists an $a \geq 0$ such that $P \oplus R^a$ is finitely generated free, and therefore $(P, \psi) \oplus H_\epsilon(R)^a$ is a non-singular $(-1)^k$-quadratic form whose underlying module is finitely generated free. This modification of Definition 8.90 will be convenient at some places.

Remark 8.93 (Functoriality of L-groups in even dimensions) A morphism $u \colon R \to S$ of rings with involution induces homomorphisms $u_* \colon L_k(R) \to L_k(S)$ for $k = 0, 2$ by induction. Explicitly, given a finitely generated projective R-module P, we obtain a map

$$\mathrm{ind}_u \colon \hom_R(P, P^*) \to \hom_S(u_*P, (u_*P)^*)$$

by sending $f \in \hom_R(P, P^*)$ to $\ind_u(f) \in \hom_S(u_*P, (u_*P)^*)$ that is given by the S-homomorphism $S \otimes_u P \to \hom_S(S \otimes_u P, S)$ sending $s_0 \otimes p_0 \in S \otimes_u P$ to the following S-homomorphism

$$\ind_u(f)(s_0 \otimes p_0) \colon S \otimes_u P \to S, \quad (s_1 \otimes p_1) \mapsto s_1 \cdot u\big(f(p_0)(p_1)\big) \cdot \overline{s_0}.$$

One easily checks that the following diagram commutes

$$\begin{array}{ccc} \hom_R(P, P^*) & \xrightarrow{T_R} & \hom_R(P, P^*) \\ \downarrow{\ind_u} & & \downarrow{\ind_u} \\ \hom_S(S \otimes_u P, (S \otimes_u P)^*) & \xrightarrow{T_S} & \hom_S(S \otimes_u P, (S \otimes_u P)^*). \end{array}$$

Hence we can assign to a non-singular ϵ-quadratic form (P, ψ) over R a non-singular ϵ-quadratic form $(S \otimes_u P, \ind_u(\psi))$ over S. Obviously a standard hyperbolic form is mapped under $\ind_u$ to a standard hyperbolic form and everything is compatible with direct sums in the P variable. When P is a (stably) finitely generated free R-module, then u_*P is a (stably) finitely generated free S-module.

This shows that $\ind_u$ yields a well-defined homomorphism $u_* \colon L_{2k}(R) \to L_{2k}(S)$ for $k \in \mathbb{Z}$. One easily checks $(u \circ v)_* = u_* \circ v_*$ for homomorphisms of rings with involutions $u \colon R \to S$ and $v \colon S \to T$ and $(\id_R)_* = \id_{L_{2k}(R)}$ for $k \in \mathbb{Z}$.

Note that Remark 8.93 allows us to extend Remark 8.61 from the symmetric setting to the quadratic setting.

Next we present a criterion for an ϵ-quadratic form (P, ψ) to represent zero in $L_{2k}(R)$, which we will later need at several places. Let (P, ϕ) be an ϵ-symmetric form with $\epsilon = (-1)^k$. A *sublagrangian* $L \subset P$ is an R-submodule such that the inclusion $i \colon L \to P$ is split injective and the image of ϕ under the map $Q_\epsilon(i) \colon Q_\epsilon(P) \to Q_\epsilon(L)$ is zero. The second condition is equivalent to L being contained in its *annihilator*

$$L^\perp = \{p \in P \mid \lambda(p, q) = 0 \text{ for all } q \in L\},$$

which is the same as the kernel of $P \xrightarrow{\phi} P^* \xrightarrow{i^*} L^*$. A sublagrangian $L \subset P$ is called a *lagrangian* if $L = L^\perp$. Equivalently, a lagrangian $L \subset P$ is an R-submodule L such that the inclusion $i \colon L \to P$ has a retraction and the sequence $0 \to L \xrightarrow{i} P \xrightarrow{i^* \circ \phi} L^* \to 0$ is exact. An ϵ-symmetric form (P, ϕ) admitting a lagrangian is called *metabolic*.

In the quadratic case, let (P, ψ) be an ϵ-quadratic form. Then a *sublagrangian* $L \subset P$ of (P, ψ) is a sublagrangian of the induced symmetric form $(P, (1 + \epsilon T)\psi)$ such that the inclusion $i \colon L \to P$ satisfies $Q_\epsilon(i)(\psi) = 0$. A sublagrangian $L \subset P$ is called a *lagrangian* if $L = L^\perp$; i.e., it is a lagrangian of $(P, (1 + \epsilon T)\psi)$.

Exercise 8.94 Let (P, λ, μ) be ϵ-quadratic form given in the form described in Remark 8.78. Show that a submodule $L \subseteq P$ is a lagrangian if and only if it satisfies the following conditions:

8.5 Even-Dimensional L-Groups

- L is a direct summand in P;
- $\lambda(l, l') = 0$ for every $l, l' \in L$;
- If $p \in P$ satisfies $\lambda(p, l) = 0$ for all $l \in L$, then $p \in L$;
- $\mu \colon P \to Q_\epsilon(R)$ is trivial on L.

If P is a finitely generated projective R-module, then L is automatically also finitely generated projective since by Exercise 8.94 it is a direct summand of a projective R-module. If P is a (stably) finitely generated free R-module, then with the above definition L will still a priori only be finitely generated projective. Therefore in the setting of (stably) finitely generated free R-modules we demand in addition that a *lagrangian* L is a (stably) finitely generated free R-module.

The following lemma works in all three settings, that means for finitely generated projective, finitely generated free and stably finitely generated free R-modules with the corresponding definitions.

Lemma 8.95 *Let (P, ψ) be a non-singular ϵ-quadratic form. Let $L \subset P$ be a sublagrangian. Then L is a direct summand in $L^\perp$ and ψ induces the structure of a non-singular ϵ-quadratic form $(L^\perp/L, \psi^\perp/\psi)$ on $L^\perp/L$. Moreover, the inclusion $i \colon L \to P$ extends to an isomorphism of ϵ-quadratic forms*

$$H_\epsilon(L) \oplus (L^\perp/L, \psi^\perp/\psi) \xrightarrow{\cong} (P, \psi).$$

In particular a non-singular ϵ-quadratic form (P, ψ) is isomorphic to $H_\epsilon(Q)$ if and only if it contains a lagrangian $L \subset P$ such that L is isomorphic as an R-module to Q.

The analogous statement holds for ϵ-symmetric forms except that $H_\epsilon(L)$ is replaced by $H^\epsilon(L, \phi_1)$, the standard ϵ-symmetric hyperbolic form associated to some symmetric form (L^*, ϕ_1) as in (15.194), see also Lemma 15.195.

Proof. Let $s \colon L^* \to P$ be an R-map such that $i^* \circ (1 + \epsilon \cdot T)(\psi) \circ s$ is the identity on L^*. Our first attempt is the obvious split injection $i \oplus s \colon L \oplus L^* \to P$. The problem is that it is not necessarily compatible with the ϵ-quadratic structure. To be compatible with the ϵ-quadratic structure, it is necessary to be compatible with the ϵ-symmetric structure, i.e., the following diagram must commute

$$\begin{array}{ccc} L \oplus L^* & \xrightarrow{i \oplus s} & P \\ {\scriptsize\begin{pmatrix} 0 & 1 \\ \epsilon & 0 \end{pmatrix}}\Big\downarrow & & \Big\downarrow{\psi + \epsilon \cdot T(\psi)} \\ L^* \oplus L & \xleftarrow{(i^* \oplus s^*) \circ \Delta_{P^*}} & P^* \end{array}$$

where $\Delta_{P^*} \colon P^* \to P^* \oplus P^*$ is the diagonal map and we write s^* for the composition $P^* \xrightarrow{s^*} (L^*)^* \xrightarrow{e(L)^{-1}} L$. The diagram above commutes if and only if we have $s^* \circ (\psi + \epsilon \cdot T(\psi)) \circ s = 0$. Note that s is not unique, we can replace s by $s' = s + i \circ v$ for any R-map $v \colon L^* \to L$. For this new section s' the diagram above commutes if and only if $s^* \circ (\psi + \epsilon \cdot T(\psi)) \circ s + v^* + \epsilon \cdot v = 0$. This suggests to take $v = -\epsilon s^* \circ \psi \circ s$. Now one easily checks that

$$g := i \oplus (s - \epsilon \cdot i \circ s^* \circ \psi \circ s) \colon L \oplus L^* \to P$$

is split injective and compatible with the ϵ-quadratic structures and hence induces a morphism $g \colon H_\epsilon(L) \to (P, \psi)$ of ϵ-quadratic forms.

Let $\mathrm{im}(g)^\perp$ be the annihilator of $\mathrm{im}(g)$. Denote by $j \colon \mathrm{im}(g)^\perp \to P$ the inclusion. We obtain an isomorphism of ϵ-quadratic forms

$$g \oplus j \colon H_\epsilon(L) \oplus (\mathrm{im}(g)^\perp, j^* \circ \psi \circ j) \to (P, \psi).$$

The projection $L^\perp \to \mathrm{im}(g)^\perp$ induces an isomorphism $h \colon L^\perp/L \to \mathrm{im}(g)^\perp$. Let $\psi^\perp/\psi$ be the ϵ-quadratic structure on $L^\perp/L$, for which h becomes an isomorphism of ϵ-quadratic forms. This finishes the proof in the quadratic case.

The proof in the symmetric case is analogous. □

Remark 8.96 (Metabolic versus hyperbolic) We shall also call an ϵ-quadratic form (P, ψ) that admits a lagrangian *metabolic*. Then Lemma 8.95 states that every metabolic ϵ-quadratic form is hyperbolic, i.e., isomorphic to a standard hyperbolic ϵ-quadratic form.

The corresponding statement for symmetric forms is not true in general. Namely, the following $(+1)$-symmetric form over the integers

$$(P, \lambda) = \left(\mathbb{Z}^2, \begin{pmatrix} 0 & 1 \\ 1 & 1 \end{pmatrix} \right)$$

is metabolic, a lagrangian is given by $\mathbb{Z} \times \{0\}$, but is not hyperbolic since there are elements $x \in P$, e.g., $x = (0, 1)$ satisfying $\lambda(x, x) = 1$, which is not true for the hyperbolic form $H^+(\mathbb{Z})$.

This is the reason why the statement of Lemma 8.95 in the symmetric case uses the slightly modified notion of a standard symmetric hyperbolic form associated to a symmetric from, a notion treated more thoroughly in Chapter 15, see Lemma 15.195 and Remark 15.196.

8.5.2 The L-Groups of $\mathbb{Z}$ in Dimensions $n = 4k$

We work in the setting of non-singular symmetric bilinear forms (P, λ) over $\mathbb{Z}$. Recall that such a form is called *even* if the integer $\lambda(x, x)$ is even for all $x \in P$. An element $[P, \theta]$ in $L_0(\mathbb{Z})$ is represented by the $(+1)$-quadratic form $(P, \theta) = (P, \lambda, \mu)$ over $\mathbb{Z}$, which is nothing but the even symmetric form (P, λ) since $\mu(x)$ is determined by λ via the equation $\mu(x) = \frac{\lambda(x,x)}{2} \in \mathbb{Z}$.

Exercise 8.97 Show that a non-singular symmetric bilinear form (P, λ) over $\mathbb{Z}$ is metabolic if and only if there is a $\mathbb{Z}$-submodule $L \subset P$ with $\dim_\mathbb{Z}(P) = 2 \cdot \dim_\mathbb{Z}(L)$ such that λ vanishes on L, i.e., $\lambda(x, y) = 0$ for $x, y \in L$.

By Lemma 8.95, every metabolic $(+1)$-quadratic form is hyperbolic. The *Witt group*

8.5 Even-Dimensional L-Groups

$$W(\mathbb{Z}) \quad (8.98)$$

of $\mathbb{Z}$ is defined to be the abelian group of equivalence classes of non-singular symmetric forms over $\mathbb{Z}$ where we call two non-singular symmetric forms equivalent if they become isomorphic after adding some metabolic form. The addition is given by orthogonal sum of forms. There is an obvious homomorphism

$$L_0(\mathbb{Z}) \to W(\mathbb{Z}).$$

The crucial invariant of forms we shall need is the signature. By tensoring a symmetric form (P, λ) over $\mathbb{Z}$ with $\mathbb{R}$, we obtain a non-singular symmetric $\mathbb{R}$-bilinear pairing $\lambda_\mathbb{R} \colon (\mathbb{R} \otimes_\mathbb{Z} P) \times (\mathbb{R} \otimes_\mathbb{Z} P) \to \mathbb{R}$. Recall from the discussion just prior to Definition 5.70 that a symmetric bilinear form over $\mathbb{R}$ has a signature. It is defined to be the difference $n_+ - n_-$ where n_+ and n_- are respectively the number of positive entries and the number of negative entries on the diagonal of a diagonalisation of a real matrix representing the form.

Theorem 8.99 ($L_{4k}(\mathbb{Z})$) *The signature defines isomorphisms*

$$\operatorname{sign} \colon W(\mathbb{Z}) \xrightarrow{\cong} \mathbb{Z}, \qquad [P, \lambda] \mapsto \operatorname{sign}(\mathbb{R} \otimes_\mathbb{Z} P, \lambda_\mathbb{R});$$

$$\operatorname{sign} \colon L_0(\mathbb{Z}) \xrightarrow{\cong} 8 \cdot \mathbb{Z}, \qquad [P, \lambda, \mu] \mapsto \operatorname{sign}(\mathbb{R} \otimes_\mathbb{Z} P, \lambda_\mathbb{R}).$$

The proof of Theorem 8.99 rests on the following fact from the algebra of symmetric bilinear forms over $\mathbb{Z}$. For the proof we refer to [306, Lemma 4.1 in Chapter II on page 22]. Recall that a symmetric bilinear form (P, λ) is called *definite* if for all $x \in P$, $x \neq 0$ we have $\lambda(x, x) > 0$ or for all $x \in P$, $x \neq 0$ we have $\lambda(x, x) < 0$. This is equivalent to the condition $|\operatorname{sign}(P, \psi)| = \operatorname{rank}(P)$. It is called *indefinite* otherwise.

Theorem 8.100 *An indefinite non-singular symmetric bilinear form (P, λ) possesses an element $x \in P \setminus \{0\}$ such that $\lambda(x, x) = 0$.*

Proof of Theorem 8.99. We first show that $\operatorname{sign} \colon W(\mathbb{Z}) \xrightarrow{\cong} \mathbb{Z}$ is bijective and $\operatorname{sign} \colon L_0(\mathbb{Z}) \to \mathbb{Z}$ is injective. Note that, if (P, λ) is metabolic, then $\operatorname{sign}(P, \lambda) = 0$. It follows that for any (P, λ), $\operatorname{sign}(P, \lambda)$ only depends on $[P, \psi] \in W(\mathbb{Z})$. Since $\operatorname{sign}(\mathbb{Z}, (1)) = 1$, $\operatorname{sign}(W(\mathbb{Z})) \to \mathbb{Z}$ is onto.

We now suppose that $\operatorname{sign}(P, \lambda) = 0$. We show by induction on the rank of P that $[P, \lambda] = 0 \in W(\mathbb{Z})$; in addition, if λ is even we show that $[P, \lambda] = 0$ holds in $L_0(\mathbb{Z})$. If $\operatorname{rank}(P) = 0$ then trivially $[P, \lambda] = 0$ in either $W(\mathbb{Z})$ or $L_0(\mathbb{Z})$. If $\operatorname{rank}(P) > 0$, then (P, λ) is indefinite and so by Theorem 8.100 it has a non-trivial zero x. The submodule $L_x \subset P$ defined by

$$L_x := \{p \in P \mid kp = jx, \text{ for some } k, j \in \mathbb{Z}, k \neq 0\}$$

is a sublagrangian of (P, λ). By Lemma 8.95, there is an isomorphism of forms $(P, \lambda) \cong M(L_x) \oplus (P', \psi')$ where $M(L_x)$ is a metabolic form on L_x. If λ is even, then again by Lemma 8.95, $M(L_x) = H_+(L_x)$ is hyperbolic. Since $\operatorname{sign}(M(L_x)) = 0$ and $\operatorname{rank}(P') = \operatorname{rank}(P) - 2$, we have proven our inductive step and hence $\operatorname{sign} \colon W(\mathbb{Z}) \to \mathbb{Z}$

is injective. Moreover, in the case when λ is even, the above argument shows that there is an isomorphism $(P, \lambda) \cong H_+(\mathbb{Z}^r)$ where $\text{rank}(P) = 2r$, and so $\text{sign} \colon L_0(\mathbb{Z}) \to \mathbb{Z}$ is also injective.

It remains to prove that $\text{sign}(L_0(\mathbb{Z})) = 8 \cdot \mathbb{Z}$. The following non-singular even form is of fundamental importance in many areas in mathematics, it is the E_8-form, $(P, \lambda) = (\mathbb{Z}^8, \lambda_{E_8})$, where λ_{E_8} is represented by the matrix

$$\lambda_{E_8} = \begin{pmatrix} 2 & 1 & 0 & 0 & 0 & 0 & 0 & 0 \\ 1 & 2 & 1 & 0 & 0 & 0 & 0 & 0 \\ 0 & 1 & 2 & 1 & 0 & 0 & 0 & 0 \\ 0 & 0 & 1 & 2 & 1 & 0 & 0 & 0 \\ 0 & 0 & 0 & 1 & 2 & 1 & 0 & 1 \\ 0 & 0 & 0 & 0 & 1 & 2 & 1 & 0 \\ 0 & 0 & 0 & 0 & 0 & 1 & 2 & 0 \\ 0 & 0 & 0 & 0 & 1 & 0 & 0 & 2 \end{pmatrix}$$

The matrix λ_{E_8} may be understood as the incidence matrix of the following E_8-graph. From another point of view, this graph is the Dynkin diagram of the exceptional Lie group E_8, see for instance [161].

Figure 8.101 (The E_8-graph).

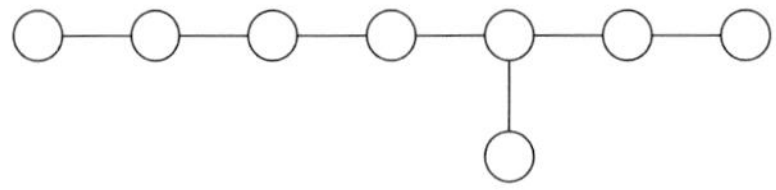

We have $\text{sign}(\mathbb{Z}^8, \lambda_{E_8}) = 8$. So it remains to show that $\text{sign}(L_0(\mathbb{Z})) \subset 8 \cdot \mathbb{Z}$. This follows from van der Blij's Lemma, which we prove next, see also [306, Lemma 5.2, Chapter II §5 on page 24]. Let (P, λ) be a non-singular symmetric bilinear form. Note that the map

$$\overline{\chi}^* \colon P \to \mathbb{Z}/2, \quad x \mapsto \lambda(x, x) \bmod 2$$

is a homomorphism of abelian groups. Since P is finitely generated free, $\overline{\chi}^*$ lifts (not uniquely) to a homomorphism of abelian groups $\chi^* \colon P \to \mathbb{Z}$, and we define $\chi \in P$ by

$$\chi^*(x) = \lambda(\chi, x), \quad \forall x \in P.$$

Any such $\chi \in P$ is called a *characteristic element for* (P, λ). Next we show that the map

$$\omega \colon W(\mathbb{Z}) \to \mathbb{Z}/8, \quad (P, \lambda) \mapsto \lambda(\chi, \chi) \bmod 8$$

is a well-defined homomorphism satisfying

$$\text{sign}(P, \lambda) = \omega(P, \lambda) \bmod 8.$$

To prove the claim, we note that any two lifts χ^* and χ'^* of $\overline{\chi}^*$ differ by a homomorphism $P \to 2 \cdot \mathbb{Z}$ and hence for the corresponding elements $\chi, \chi' \in P$ we have

8.5 Even-Dimensional L-Groups

$$\chi = \chi' + 2x$$

for some $x \in P$. We conclude

$$\lambda(\chi, \chi) - \lambda(\chi', \chi') = 2\lambda(\chi', 2x) + \lambda(2x, 2x) = 4(\lambda(\chi', x) + \lambda(x, x)).$$

Since χ' is characteristic for λ, we have $(\lambda(\chi', x) + \lambda(x, x)) \in 2\mathbb{Z}$. Hence $\omega(P, \lambda) \in \mathbb{Z}/8$ is well defined. It is clear that ω is additive under orthogonal sums and one checks that $\omega(P, \lambda) = 0$ for any metabolic form (P, λ). Hence $\omega \colon W(\mathbb{Z}) \to \mathbb{Z}/8$ is a well-defined homomorphism. To check that ω agrees with the signature homomorphism, we need only check on the generator $[\mathbb{Z}, (1)]$ of $W(\mathbb{Z})$, where the agreement is clear.

If (P, μ) is even, we can take χ to be zero and hence $\omega(P, \lambda)$ vanishes, which implies $\text{sign}(P, \lambda) \in 8\mathbb{Z}$. □

Remark 8.102 We point out that Lemma 5.81, Remark 8.81 and Lemma 8.95 imply that the signature defines an isomorphism $L_0(\mathbb{R}) \xrightarrow{\cong} \mathbb{Z}$.

8.5.3 The L-Groups of $\mathbb{Z}$ in Dimensions $n = 4k + 2$

We begin by classifying the non-singular (-1)-symmetric forms over R for $R = \mathbb{Z}, \mathbb{F}_2$.

Lemma 8.103 *Let R be $\mathbb{Z}$ or $\mathbb{F}_2$. Consider any non-singular (-1)-symmetric form (P, λ) over R. If R is $\mathbb{F}_2$, assume additionally that (P, λ) has a quadratic refinement.*
Then it is hyperbolic and admits a symplectic basis $\{e_1, f_1, \ldots, e_m, f_m\}$, which means

$$\lambda(e_i, f_i) = 1;$$
$$\lambda(e_i, f_j) = 0, \quad \text{if } i \neq j;$$
$$\lambda(e_i, e_j) = 0;$$
$$\lambda(f_i, f_j) = 0.$$

Proof. We use induction over the rank r of the finitely generated free $\mathbb{Z}$-module P. The induction beginning with $r = 0$, which is equivalent to $P = \{0\}$, is trivial. The induction step is done as follows.

We have $\lambda(x, x) = 0$ for any $x \in P$. If R is $\mathbb{Z}$, this is a direct consequence of λ being (-1)-symmetric. If R is $\mathbb{F}_2$, this follows from the existence of a quadratic refinement. Choose an element $e_1 \in P$ such that $e_1 = m \cdot e$ for $m \in R$ and $e \in P$ implies $m \in \{\pm 1\}$. Then $L = \{n \cdot e_1 \mid n \in R\}$ is a sublagrangian of (P, λ). We get from Lemma 8.95 an isomorphism of non-singular (-1)-symmetric forms

$$M^{-1}(L) \oplus (L^\perp/L, \phi') \cong (P, \lambda),$$

where $M^{-1}(L)$ is metabolic with lagrangian L. One easily checks that $M^{-1}(L) = H^{-1}(L)$ and possesses a symplectic basis. Since from the induction hypothesis applied to $(L^\perp/L, \phi')$ we conclude that the non-singular (-1)-symmetric form

$(L^\perp/L, \phi')$ is hyperbolic and has a symplectic basis, (P, λ) is hyperbolic and has a symplectic basis. □

Next we classify the non-singular (-1)-quadratic forms over R for $R = \mathbb{Z}, \mathbb{F}_2$. Recall that a non-singular (-1)-quadratic form (P, ψ) over R is equivalent to a triple (P, λ, μ) where (P, λ) is a non-singular (-1)-symmetric form over R and $\mu \colon P \to Q_-(R) = \mathbb{Z}/2$ is a function such that

$$\mu(x+y) = \mu(x) + \mu(y) + \mathrm{pr}(\lambda(x, y)) \in \mathbb{Z}/2.$$

We have the (-1)-quadratic hyperbolic form $H_{-1}(R)$ over R^2. There is also the following non-singular (-1)-quadratic form $A(R) = (R^2, \lambda_A, \mu_A)$ over R^2 whose underlying non-singular (-1)-symmetric form is $H^{-1}(R)$ and which is defined by

$$\lambda((m_1, m_2), (n_1, n_2)) = (m_1 n_2 - m_2 n_1);$$
$$\mu(m, n) = 0, \quad \text{if } m \text{ and } n \text{ are even};$$
$$\mu(m, n) = 1, \quad \text{if } m \text{ or } n \text{ is odd}.$$

Theorem 8.104 *Let R be $\mathbb{Z}$ or $\mathbb{F}_2$. Let (P, λ, μ) be a non-singular (-1)-quadratic form over the ring R such that P is different from $\{0\}$.*

(i) *Then P is isomorphic to R^{2k} for some natural number $k \geq 1$, and (P, λ, μ) is isomorphic to precisely one of the non-singular (-1)-quadratic forms over the ring R given by $H_{-1}(R^k)$ and $H_{-1}(R^{k-1}) \oplus A(R)$;*
(ii) *We have the isomorphism of non-degenerate (-1)-quadratic forms over the ring R*
$$A(R) \oplus A(R) \cong H_{-1}(R^2).$$

Proof. We first prove that there are precisely two isomorphism classes of non-singular (-1)-quadratic forms with underlying R-module R^2, namely, $H_-(R)$ and $A(R)$. By Lemma 8.103 the underlying non-singular (-1)-symmetric form (R^2, λ) over R^2 is isomorphic to $H^-(R)$. Now for any invertible matrix $X \in \mathrm{GL}(2, R)$ with $X = \begin{pmatrix} a & b \\ c & d \end{pmatrix}$, we have

$$\begin{pmatrix} a & b \\ c & d \end{pmatrix} \begin{pmatrix} 0 & 1 \\ -1 & 0 \end{pmatrix} \begin{pmatrix} a & b \\ c & d \end{pmatrix}^t = \det(X) \begin{pmatrix} 0 & 1 \\ -1 & 0 \end{pmatrix}.$$

Hence, using $\mathrm{aut}(D)$ to denote the group of automorphisms of a form D, we see that $\mathrm{aut}(H^-(R)) = \mathrm{SL}_2(R)$, the group of invertible 2×2 matrices over R with determinant one. It is also easy to see that $\mathrm{aut}(H_-(R)) \subset \mathrm{aut}(H^-(R))$ is given by

$$\mathrm{aut}(H_-(R)) = \left\{ \begin{pmatrix} a & b \\ c & d \end{pmatrix} \in \mathrm{SL}_2(R) \,\middle|\, a+b \equiv c+d \equiv 1 \bmod 2 \right\}$$

and that $\mathrm{aut}(H_-(R)) \subset \mathrm{aut}(H^-(R))$ is a subgroup of index three. Indeed there are precisely four quadratic refinements $\mu \colon R^2 \to \mathbb{Z}/2$ of $H^-(R)$, which we list for the

8.5 Even-Dimensional L-Groups

reader's convenience:

	$x = (0,0)$	$x = (1,0)$	$x = (0,1)$	$x = (1,1)$
$\mu_{H_1}(x)$	0	0	0	1
$\mu_{H_2}(x)$	0	0	1	0
$\mu_{H_3}(x)$	0	1	0	0
$\mu_A(x)$	0	1	1	1

(8.105)

These four quadratic refinements sit in two orbits under of the action of $\mathrm{aut}(H^-(R))$: the orbit $\{\mu_{H_1}, \mu_{H_2}, \mu_{H_3}\}$ corresponds to $H_-(R)$, and the orbit $\{\mu_A\}$ corresponds to $A(R)$. Hence every non-singular (-1)-quadratic form with underlying R-module R^2 is isomorphic to $H_-(R)$ or $A(R)$.

In particular, if $-A(R)$ is the (-1)-quadratic form $(R^2, -\lambda_{A(R)}, -\mu_{A(R)})$, then there is an isomorphism $f \colon A(R) \cong -A(R)$. Hence assertion (ii) follows since we obtain

$$A(R) \oplus A(R) \cong A(R) \oplus -A(R) \cong H_-(R^2)$$

by applying Lemma 8.95 to the diagonal lagrangian $\Delta \subset A(R) \oplus -A(R)$.

The following definition is taken from Browder [55, III.1.5-1.12]. Let (Q, λ, μ) be a non-singular (-1)-quadratic form over $\mathbb{F}_2$. We define the so-called *democratic invariant* of (Q, λ, μ) by

$$\mathrm{Arf}'(Q, \lambda, \mu) := \begin{cases} 0 & \text{if } |\mu^{-1}(0)| > |\mu^{-1}(1)|; \\ 1 & \text{if } |\mu^{-1}(1)| \geq |\mu^{-1}(0)|. \end{cases}$$

(8.106)

We note that $\mathrm{Arf}'(Q, \mu)$ is manifestly an isomorphism invariant. Since a direct computation shows $\mathrm{Arf}'(H_-(\mathbb{F}_2)) = 0$ and $\mathrm{Arf}'(A(\mathbb{F}_2)) = 1$, we conclude that $H_-(\mathbb{F}_2)$ and $A(\mathbb{F}_2)$ are not isomorphic. Since $H_{-1}(\mathbb{F}_2)$ and $A(\mathbb{F}_2)$ are obtained from $H_{-1}(\mathbb{Z})$ and $A(\mathbb{Z})$ by induction with the projection $\mathbb{Z} \to \mathbb{F}_2$, we conclude that $H_1(\mathbb{Z})$ and $A(\mathbb{Z})$ are not isomorphic. This finishes the proof of assertion (i) in the special case where $P = R^2$.

Consider any non-singular (-1)-quadratic form (Q, λ, μ) over the ring R with $Q \neq \{0\}$. The underlying non-singular (-1)-symmetric form (Q, λ) is isomorphic to $H^-(R^k)$ for some $k \geq 1$ by Lemma 8.103. Hence we can choose an isomorphism of non-singular (-1)-symmetric forms

$$\alpha \colon (Q, \lambda) \xrightarrow{\cong} \oplus_{i=1}^k H^-(R).$$

Now consider the quadratic form $(\alpha^{-1})^*\mu$, which is the pullback of μ along α^{-1}. This form is a quadratic refinement of the (-1)-symmetric form $\oplus_{i=1}^k H^-(R)$. From the equation $\mu(x+y) = \mu(x) + \mu(y) + \mathrm{pr}(\lambda(x,y)) \in \mathbb{Z}/2$, we deduce that $(\alpha^{-1})^*(\mu) = \oplus_{i=1}^k \mu_i$ is an orthogonal sum of quadratic refinements of $H^-(R)$. That is, α also defines an isomorphism of non-singular (-1)-quadratic forms

$$\alpha \colon (Q, \lambda, \mu) \cong \oplus_{i=1}^k (Q_i, \lambda_i, \mu_i)$$

where each R-module Q_i is isomorphic to R^2. Hence each (Q_i, λ_i, μ_i) is isomorphic to $H_-(R)$ or $A(R)$. Because of assertion (ii), we conclude that (Q, λ, μ) is isomorphic to $H_{-1}(R^k)$ or $H_{-1}(R^{k-1}) \oplus A(R)$.

It remains to show that $H_{-1}(R^k)$ and $H_{-1}(R^{k-1}) \oplus A(R)$ are not isomorphic. Since $H_{-1}(\mathbb{F}_2^k)$ and $H_{-1}(\mathbb{F}_2^{k-1}) \oplus A(\mathbb{F}_2)$ are obtained from $H_{-1}(\mathbb{Z}^k)$ and $H_{-1}(\mathbb{Z}^{k-1}) \oplus A(\mathbb{Z})$ by induction with the projection $\mathbb{Z} \to \mathbb{F}_2$, it suffices to do this for $R = \mathbb{F}_2$. A simple counting exercise shows that for a non-singular (-1)-quadratic form (Q, λ, μ) over the ring $\mathbb{F}_2$

$$|(\mu \oplus \mu_{H_{-1}(R)})^{-1}(1)| = 3|\mu^{-1}(1)| + |\mu^{-1}(0)|,$$

and that

$$|(\mu \oplus \mu_{A(R)})^{-1}(1)| = 3|\mu^{-1}(0)| + |\mu^{-1}(1)|.$$

We deduce that for $(Q_1, \lambda_1, \mu_1) = H_-(\mathbb{F}_2)$ or $A(\mathbb{F}_2)$

$$\mathrm{Arf}'((Q, \lambda, \mu) \oplus (Q_1, \lambda_1, \mu_1)) = \mathrm{Arf}'(Q, \lambda, \mu) + \mathrm{Arf}'(Q_1, \lambda_1, \mu_1). \tag{8.107}$$

Since $\mathrm{Arf}'(H_{-1}(\mathbb{F}_2)) = 0$ and $\mathrm{Arf}'(A(\mathbb{F}_2)) = 1$, we conclude that $\mathrm{Arf}'(H_{-1}(\mathbb{F}_2^k)) = 0$ and $\mathrm{Arf}'(H_{-1}(\mathbb{F}_2^{k-1}) \oplus A(\mathbb{F}_2)) = 1$. This finishes the proof of Theorem 8.104. □

Let $R = \mathbb{Z}$ or $R = \mathbb{F}_2$. Consider a non-singular (-1)-quadratic form (P, λ, μ) over R together with a symplectic basis $B = \{e_1, f_1, \ldots, e_k, f_k\}$ of the underlying non-singular (-1)-symmetric form (P, λ) over R. Define

$$\mathrm{Arf}(P, \lambda, \mu, B) := \sum_{i=1}^{k} \mu(e_i) \cdot \mu(f_i) \quad \in \mathbb{Z}/2. \tag{8.108}$$

We want to show that this definition is independent of the choice of the symplectic basis B. This follows directly from the following result.

Lemma 8.109 *Let (P, λ, μ) be a non-singular (-1)-quadratic form over R. Then we get for any choice of symplectic basis B of (P, λ)*

$$\mathrm{Arf}(P, \lambda, \mu, B) = \mathrm{Arf}'(P, \lambda, \mu),$$

where $\mathrm{Arf}'(P, \lambda, \mu)$ has been introduced in (8.106).

Proof. Let $\{e_1, f_1, \ldots, e_m, f_m\}$ be a symplectic basis for (P, λ, μ). It yields an isomorphism of non-singular (-1)-symmetric forms

$$\alpha \colon (Q, \lambda) \xrightarrow{\cong} \oplus_{i=1}^{k} H^-(R).$$

As explained in the proof of Theorem 8.104, we can find an isomorphism of non-singular (-1)-quadratic forms

$$\alpha \colon (Q, \lambda, \mu) \xrightarrow{\cong} \bigoplus_{i=1}^{k} (Q_i, \mu_i, \lambda_i)$$

8.5 Even-Dimensional L-Groups

that reduces to α if we only remember the (-1)-symmetric structure. Hence we get

$$\mathrm{Arf}(Q, \lambda, \mu, B) = \sum_{i=1}^{k} \mathrm{Arf}(Q_i, \lambda_i, \mu_i, B_i)$$

for the symplectic basis $B_i = \{e_i, f_i\}$ of (Q_i, λ_i, μ_i). We get from the proof of Theorem 8.104

$$\mathrm{Arf}'(Q, \lambda, \mu) = \sum_{i=1}^{k} \mathrm{Arf}'(Q_i, \lambda_i, \mu_i);$$
$$\mathrm{Arf}'(H_{-1}(R)) = 0;$$
$$\mathrm{Arf}'(A(R)) = 1.$$

Hence it suffices to show that for any choice of symplectic basis B we get

$$\mathrm{Arf}(H_{-1}(R), B) = 0;$$
$$\mathrm{Arf}(A(R); B) = 1.$$

We have

$$\mathrm{Arf}'(H_{-1}(\mathbb{Z}), B) = \mathrm{Arf}'(H_{-1}(\mathbb{F}_2), B');$$
$$\mathrm{Arf}'(A(\mathbb{Z}), B) = \mathrm{Arf}'(A(\mathbb{F}_2), B'),$$

where B' is the obvious symplectic basis obtained from B using the projection $\mathbb{Z} \to \mathbb{F}_2$. Therefore it suffices to show $\mathrm{Arf}(H_{-1}(\mathbb{F}_2), B) = 0$ and $\mathrm{Arf}(A(\mathbb{F}_2), B) = 1$ for any choice of a symplectic basis B, which is done by a direct calculation. □

In view of Lemma 8.109 the following definition makes sense.

Definition 8.110 (Arf invariant) Let $R = \mathbb{Z}$ or $R = \mathbb{F}_2$. Define the *Arf invariant* of a non-singular (-1)-quadratic form (P, λ, μ) over R for any choice of symplectic basis $B = \{e_1, f_1, \ldots, e_m, f_m\}$ of the non-singular (-1)-symmetric form (P, λ) over R by

$$\mathrm{Arf}(P, \lambda, \mu) := \mathrm{Arf}(P, \lambda, \mu, B) = \mathrm{Arf}'(P, \lambda, \mu) \quad \in \mathbb{Z}/2,$$

where $\mathrm{Arf}(P, \lambda, \mu, B)$ is defined in (8.108) and $\mathrm{Arf}'(P, \lambda, \mu)$ in (8.106).

Since we have shown above that for $R = \mathbb{Z}, \mathbb{F}_2$ the Arf invariant is compatible with direct sums, vanishes for hyperbolic non-singular (-1)-quadratic forms and two non-singular (-1)-quadratic forms of the same rank over R are isomorphic if and only their Arf invariants agree, we get the following result.

Theorem 8.111 ($L_{4k+2}(\mathbb{Z})$) *The Arf invariant defines isomorphisms*

$$\mathrm{Arf}: L_2(\mathbb{Z}) \xrightarrow{\cong} \mathbb{Z}/2, \quad [(P, \lambda, \mu)] \mapsto \mathrm{Arf}(P, \lambda, \mu) = \mathrm{Arf}(\mathbb{F}_2 \otimes_\mathbb{Z} (P, \lambda, \mu));$$
$$\mathrm{Arf}: L_2(\mathbb{F}_2) \xrightarrow{\cong} \mathbb{Z}/2, \quad [(P, \lambda, \mu)] \mapsto \mathrm{Arf}(P, \lambda, \mu),$$

and the change of rings homomorphism $\mathbb{Z} \to \mathbb{F}_2$ induces an isomorphism

$$L_2(\mathbb{Z}) \xrightarrow{\cong} L_2(\mathbb{F}_2).$$

8.6 The Even-Dimensional Surgery Obstruction in the Universal Covering Case

For the remainder of this subsection we fix a normal Γ-map of degree one $(M, [M], f, \widehat{f}, a, \xi, \overline{f})$ with target $(X, \widehat{p}_X, [X])$ in the sense of Definition 8.3, where the dimension $n = 2k$ is even. We will often abbreviate $(M, [M], f, \widehat{f}, a, \xi, \overline{f})$ by $(f, \overline{f})$. Moreover, we will assume that we are in the universal covering case, see Notation 8.34, and that $k \geq 3$.

The main result of this section is the following theorem.

Theorem 8.112 (Surgery obstruction in even dimension) *Suppose that $N = \dim(M) = \dim(X) = 2k$ for $k \geq 3$ and that we are in the universal covering case, see Notation 8.34.*

Then there is a well-defined surgery obstruction

$$\sigma(f, \overline{f}) \in L_{2k}(\mathbb{Z}\pi_1(X, x), w_1(X))$$

taking values in the even-dimensional algebraic L-group, in the sense of Definition 8.90, of the ring $\mathbb{Z}\pi_1(X, x)$ equipped with the $w_1(X)$-twisted involution sending $\sum_{g \in \Gamma} \lambda_g \cdot g$ to $\sum_{g \in \Gamma} \lambda_g \cdot w_1(X)(g) \cdot g^{-1}$. It has the following properties:

(i) *It depends only on the normal bordism with cylindrical ends class of $(f, \overline{f})$ in the sense of Definition 8.7;*
(ii) *We have $\sigma(f, \overline{f}) = 0 \in L_{2k}(\mathbb{Z}\pi_1(X, x), w_1(X))$ if and only if we can do a finite number of surgery steps to obtain a normal homotopy equivalence $(f', \overline{f'})$.*

Before we give its proof in Subsection 8.6.3, we consider as preparation for its proof normal maps of pairs and their surgery kernels in Subsections 8.6.1 and 8.6.2.

8.6.1 Normal Maps of Pairs

The remainder of this section is devoted to the proof of Theorem 8.112, which needs some preparation.

So far we have only considered normal Γ-maps $(M, [M], f, \widehat{f}, a, \xi, \overline{f})$ of degree one in the sense of Definition 8.3 for a closed manifold M and a finite Poincaré complex X. This definition also makes sense if we replace M by a compact manifold $(M, \partial M)$ with boundary ∂M and X by a Poincaré pair $(X, \partial X)$. The only difference is that f has to be replaced by a map of pairs $(f, \partial f) \colon (M, \partial M) \to (X, \partial X)$, and fundamental classes have to be replaced by the fundamental classes $[M, \partial M] \in H_n^\Gamma(M, \partial M; \mathbb{Z}^w)$

8.6 The Even-Dimensional Surgery Obstruction in the Universal Covering Case

and $[X, \partial X] \in H_n^\Gamma(X, \partial X; \mathbb{Z}^w)$. Moreover, we get Γ-coverings $p_{\partial M} \colon \widehat{\partial M} \to \partial M$ and $p_{\partial X} \colon \widehat{\partial X} \to \partial X$ from the Γ-coverings $p_X \colon \widehat{X} \to X$ and $p_M \colon \widehat{M} \to M$ by restriction to ∂X and ∂M and a Γ-map $\widehat{\partial f} \colon \widehat{\partial M} \to \widehat{\partial X}$ covering ∂f from the Γ-map $\widehat{f} \colon \widehat{M} \to \widehat{X}$.

The definitions concerning the bundle data are unchanged. This makes sense since TM is defined also for a manifold with boundary. The case of the universal covering appearing in Notation 8.34 is also unchanged. Note that we are not requiring anything about ∂M and ∂f and ∂X. It is possible that ∂M consists of more than one component and no connectivity assumptions about ∂f are made, unless explicitly stated otherwise. In particular, the restriction of $\widetilde{p}_M \colon \widetilde{M} \to M$ to ∂M is not necessarily the universal covering of ∂M and analogously for X.

8.6.2 Surgery Kernels for Maps of Pairs

For the remainder of this subsection we fix a normal Γ-map of pairs of degree one $(M, \partial M, [M, \partial M], f, \partial f, \widehat{f}, a, \xi, \overline{f})$ with target $(X, \partial X, [X, \partial X], \widehat{p}_X)$ in the sense of Subsection 8.6.1. We extend the material of Subsection 8.4.1 to this more general setting. This will be needed, for instance, when we deal with the bordism invariance of the surgery obstruction.

The Γ-map of pairs $(\widehat{f}, \widehat{\partial f})$ induces maps of $\mathbb{Z}\Gamma$-(co)chain complexes

$$C_*(\widehat{f}, \widehat{\partial f}) \colon C_*(\widehat{M}, \widehat{\partial M}) \to C_*(\widehat{X}, \widehat{\partial X})$$

and

$$C^*(\widehat{f}, \widehat{\partial f}) \colon C^*(\widehat{X}, \widehat{\partial X}) \to C^*(\widehat{M}, \widehat{\partial M}).$$

They induce $\mathbb{Z}\Gamma$-homomorphisms

$$(\widehat{f}, \widehat{\partial f})_* \colon H_*(\widehat{M}, \widehat{\partial M}) \to H_*(\widehat{X}, \widehat{\partial X}) \text{ and } (\widehat{f}, \widehat{\partial f})^* \colon H^*(\widehat{X}, \widehat{\partial X}) \to H^*(\widehat{M}, \widehat{\partial M}).$$

The failures of $(\widehat{f}, \widehat{\partial f})_*$ and $(\widehat{f}, \widehat{\partial f})^*$ to be isomorphisms are measured by the $\mathbb{Z}\Gamma$-(co)homology modules

$$H_*(\widehat{f}, \widehat{\partial f}) := H_*\big(\mathrm{cone}_*(C_*(\widehat{f}, \widehat{\partial f}))\big);$$
$$H^*(\widehat{f}, \widehat{\partial f}) := H^*\big(\mathrm{cone}^*(C^*(\widehat{f}, \widehat{\partial f}))\big),$$

where $\mathrm{cone}_*(C_*(\widehat{f}, \widehat{\partial f}))$ and $\mathrm{cone}^*(C^*(\widehat{f}, \widehat{\partial f}))$ are the algebraic mapping cones of $C_*(\widehat{f}, \widehat{\partial f})$ and $C^*(\widehat{f}, \widehat{\partial f})$ respectively. There are long exact sequences of $\mathbb{Z}\Gamma$-modules

$$\cdots \to H_{j+1}(\widehat{M}, \widehat{\partial M}) \xrightarrow{(\widehat{f}, \widehat{\partial f})_*} H_{j+1}(\widehat{X}, \widehat{\partial X}) \to H_{j+1}(\widehat{f}, \widehat{\partial f})$$
$$\to H_j(\widehat{M}, \widehat{\partial M}) \xrightarrow{(\widehat{f}, \widehat{\partial f})_*} H_j(\widehat{X}, \widehat{\partial X}) \to \cdots, \quad (8.113)$$

and

$$\cdots \to H^j(\widehat{X}, \widehat{\partial X}) \xrightarrow{(\widehat{f}, \widehat{\partial f})^*} H^j(\widehat{M}, \widehat{\partial M}) \to H^{j+1}(\widehat{f}, \widehat{\partial f})$$
$$\to H^{j+1}(\widehat{X}, \widehat{\partial X}) \xrightarrow{(\widehat{f}, \widehat{\partial f})^*} H^{j+1}(\widehat{M}, \widehat{\partial M}) \to \cdots . \quad (8.114)$$

Definition 8.115 (Kernels of degree one maps of pairs) Define the *kernels* by

$$K_j(\widehat{M}) := \ker(\widehat{f}_* \colon H_j(\widehat{M}) \to H_j(\widehat{X}));$$
$$K^j(\widehat{M}) := \operatorname{coker}(\widehat{f}^* \colon H^j(\widehat{X}) \to H^j(\widehat{M}));$$
$$K_j(\widehat{M}, \widehat{\partial M}) := \ker((\widehat{f}, \widehat{\partial f})_* \colon H_j(\widehat{M}, \widehat{\partial M}) \to H_j(\widehat{X}, \widehat{\partial X}));$$
$$K^j(\widehat{M}, \widehat{\partial M}) := \operatorname{coker}((\widehat{f}, \widehat{\partial f})^* \colon H^j(\widehat{X}, \widehat{\partial X}) \to H^j(\widehat{M}, \widehat{\partial M})).$$

The following lemma shows that the surgery kernels of a degree one map of pairs behave very much like the (co)homology groups of a third Poincaré pair.

Lemma 8.116 (i) *The $\mathbb{Z}\Gamma$-homomorphisms*

$$\widehat{f}_* \colon H_j(\widehat{M}) \to H_j(\widehat{X}) \quad \text{and} \quad (\widehat{f}, \widehat{\partial f})_* \colon H_j(\widehat{M}, \widehat{\partial M}) \to H_j(\widehat{X}, \widehat{\partial X})$$

are split surjections whereas the homomorphisms

$$\widehat{f}^* \colon H^j(\widehat{X}) \to H^j(\widehat{M}) \quad \text{and} \quad (\widehat{f}, \widehat{\partial f})^* \colon H^j(\widehat{X}, \widehat{\partial X}) \to H^j(\widehat{M}, \widehat{\partial M})$$

are split injections. Hence there are natural isomorphisms

$$H_j(\widehat{M}) \xrightarrow{\cong} K_j(\widehat{M}) \oplus H_j(\widehat{X});$$
$$H^j(\widehat{M}) \xrightarrow{\cong} K^j(\widehat{M}) \oplus H^j(\widehat{X});$$
$$H_j(\widehat{M}, \widehat{\partial M}) \xrightarrow{\cong} K_j(\widehat{M}, \widehat{\partial M}) \oplus H_j(\widehat{X}, \widehat{\partial X});$$
$$H^j(\widehat{M}, \widehat{\partial M}) \xrightarrow{\cong} K^j(\widehat{M}, \widehat{\partial M}) \oplus H^j(\widehat{X}, \widehat{\partial X}),$$

and also natural isomorphisms

$$K_j(\widehat{M}) \xrightarrow{\cong} H_{j+1}(\operatorname{cone}_*(C_*(\widehat{f})));$$
$$K^j(\widehat{M}) \xrightarrow{\cong} H^{j+1}(\operatorname{cone}^*(C^*(\widehat{f})));$$
$$K_j(\widehat{M}, \widehat{\partial M}) \xrightarrow{\cong} H_{j+1}(\operatorname{cone}_*(C_*(\widehat{f}, \widehat{\partial f})));$$
$$K^j(\widehat{M}, \widehat{\partial M}) \xrightarrow{\cong} H^{j+1}(\operatorname{cone}^*(C^*(\widehat{f}, \widehat{\partial f}))).$$

(ii) *The cap products with $[M, \partial M]$ and $[\partial M]$ induce isomorphisms*

8.6 The Even-Dimensional Surgery Obstruction in the Universal Covering Case   273

$$-\cap [M,\partial M]\colon K^{n-j}(\widehat{M}) \xrightarrow{\cong} K_j(\widehat{M},\widehat{\partial M});$$
$$-\cap [M,\partial M]\colon K^{n-j}(\widehat{M},\widehat{\partial M}) \xrightarrow{\cong} K_j(\widehat{M});$$
$$-\cap [\partial M]\colon K^{n-j-1}(\widehat{\partial M}) \xrightarrow{\cong} K_j(\widehat{\partial M}),$$

which fit into a commutative diagram of long exact sequences

$$\begin{array}{ccccccc}
\cdots \longrightarrow & K^{n-j-1}(\widehat{\partial M}) & \longrightarrow & K^{n-j}(\widehat{M},\widehat{\partial M}) & \longrightarrow & K^{n-j}(\widehat{M}) & \longrightarrow \cdots \\
& \cong \Big\downarrow{\scriptstyle -\cap [\partial M]} & & \cong \Big\downarrow{\scriptstyle -\cap [M,\partial M]} & & \cong \Big\downarrow{\scriptstyle -\cap [M,\partial M]} & \\
\cdots \longrightarrow & K_j(\widehat{\partial M}) & \longrightarrow & K_j(\widehat{M}) & \longrightarrow & K_j(\widehat{M},\widehat{\partial M}) & \longrightarrow \cdots\,;
\end{array}$$

(iii) *Consider $j \geq 1$. If f is j-connected, then $K_i(\widehat{M}) = 0$ for all $i < j$.
Suppose that we are in the universal covering case, see Notation 8.34. Then f is j-connected if only if $K_i(\widetilde{M}) = 0$ for all $i < j$. In this case there is a composition of natural $\mathbb{Z}\pi$-isomorphisms*

$$h_j\colon \pi_{j+1}(f) \xrightarrow{\cong} \pi_{j+1}(\widetilde{f}) \xrightarrow{\cong} H_{j+1}(\widetilde{f}) \xrightarrow{\cong} K_j(\widetilde{M})$$

such that the following diagram of $\mathbb{Z}\pi$-modules commutes

$$\begin{array}{ccc}
\pi_{j+1}(f) & \xrightarrow{h_j} & K_j(\widetilde{M}) \\
\cong \Big\uparrow & & \Big\downarrow \\
\pi_{j+1}(\widetilde{f}) & & \\
\partial_{j+1} \Big\downarrow & & \Big\downarrow \\
\pi_j(\widetilde{M}) & \longrightarrow & H_j(\widetilde{M})
\end{array}$$

where the lower horizontal arrow is the Hurewicz homomorphism and the right vertical arrow is the inclusion;

(iv) *If $(f,\partial f)$ is a homotopy equivalence of pairs, then $K_i(\widehat{M}) = 0$ for all i.
Suppose that we are in the universal covering case, see Notation 8.34, and that ∂f is a homotopy equivalence. Let k be the natural number defined by $n = \dim(X) = 2k$ or $n = \dim(X) = 2k+1$. Then $(f,\partial f)$ is a homotopy equivalence of pairs if and only if $K_i(\widetilde{M}) = 0$ for all $i \leq k$.*

Proof. (i) and (ii) The proof follows the same line of reasoning as the proof of Lemma 8.39 (i) and (ii), now for pairs.

For instance, the Umkehr maps for homology are now defined as follows. Given $x \in H_j(\widehat{X},\widehat{\partial X})$ and $y \in H_j(\widehat{X})$ we have

$(\widehat{f})^! : H_j(\widehat{X}) \to H_j(\widehat{M})$,

$$y \mapsto (\widehat{f, \partial f})^* ((- \cap [X, \partial X])^{-1}(x)) \cap [M, \partial M] \quad (8.117)$$

and

$(\widehat{f, \partial f})_! : H_j(\widehat{X}, \widehat{\partial X}) \to H_j(\widehat{M}, \widehat{\partial M})$

$$x \mapsto \widehat{f}^* ((- \cap [X, \partial X])^{-1}(x)) \cap [M, \partial M]. \quad (8.118)$$

As in the closed case, standard properties of cap products ensure that

$$(\widehat{f_*, \partial f})_* (\widehat{f, \partial f})_!(x) = x \text{ and } \widehat{f}_* \widehat{f}_!(y) = y,$$

and so $(\widehat{f, \partial f})_*$ and $\widehat{f}_*$ are split surjections. A similar argument works for cohomology.

Now one obtains the following commutative diagrams with split exact rows from the Umkehr maps and the long exact (co)homology sequences (8.113) and (8.114).

$$\begin{array}{ccccccccc}
0 & \longleftarrow & K^{n-j}(\widehat{M}, \widehat{\partial M}) & \longleftarrow & H^{n-j}(\widehat{M}, \widehat{\partial M}) & \stackrel{\widehat{f}^*}{\longleftarrow} & H^{n-j}(\widehat{X}, \widehat{\partial X}) & \longleftarrow & 0 \\
& & \cong \downarrow -\cap[M,\partial M] & & \cong \downarrow -\cap[M,\partial M] & & \cong \downarrow -\cap[X,\partial X] & & \\
0 & \longrightarrow & K_j(\widehat{M}) & \longrightarrow & H_j(\widehat{M}) & \stackrel{\widehat{f}_*}{\longrightarrow} & H_j(\widehat{X}) & \longrightarrow & 0
\end{array}$$
(8.119)

and

$$\begin{array}{ccccccccc}
0 & \longleftarrow & K^{n-j}(\widehat{M}) & \longleftarrow & H^{n-j}(\widehat{M}) & \stackrel{\widehat{f}^*}{\longleftarrow} & H^{n-j}(\widehat{X}) & \longleftarrow & 0 \\
& & \cong \downarrow -\cap[M,\partial M] & & \cong \downarrow -\cap[M,\partial M] & & \cong \downarrow -\cap[X,\partial X] & & \\
0 & \longrightarrow & K_j(\widehat{M}, \widehat{\partial M}) & \longrightarrow & H_j(\widehat{M}, \widehat{\partial M}) & \stackrel{\widehat{f}_*}{\longrightarrow} & H_j(\widehat{X}, \widehat{\partial X}) & \longrightarrow & 0.
\end{array}$$
(8.120)

The desired diagram appearing in assertion (ii) comes from the (co)homology long exact sequences (8.113) and (8.114) using (8.119) and (8.120).

(iii) The proof is identical to that of Lemma 8.39 (iii).

(iv) If $(f, \partial f)$ is a homotopy equivalence of pairs, then obviously all kernel groups vanish.

Now suppose that we are in the universal covering case, ∂f is a homotopy equivalence, and $K_i(\widetilde{M}) = 0$ for $i \leq k$. Then $K^i(\widetilde{M}) = 0$ for $i \leq k$. Now by assertion (ii)

$$K_i(\widetilde{M}) \cong K^{n-i}(\widetilde{M}, \widehat{\partial M}) \cong K^{n-i}(\widetilde{M})$$

since ∂f is a homotopy equivalence and hence $K^l(\widehat{\partial M}) = 0$ for all l. Therefore $K_i(\widetilde{M}) = 0$ for all i. So we have an isomorphism $\widetilde{f}_* : H_*(\widetilde{M}) \stackrel{\cong}{\to} H_*(\widetilde{X})$. Since f

8.6 The Even-Dimensional Surgery Obstruction in the Universal Covering Case 275

induces an isomorphism of fundamental groups, we conclude from [178, Corollary 4.33 on page 367] or [431, Theorem IV.7.13 on page 181, Theorem V.3.1 and Theorem V.3.5 on page 230] that f is a homotopy equivalence. Hence $(f, \partial f)$ is a homotopy equivalence of pairs. □

Note that all of the earlier arguments for performing surgery below the middle dimension in the closed case carry over to the degree one normal maps of pairs $(f, \bar{f}) \colon (M, \partial M) \to (X, \partial X)$ where we perform surgeries only in the interior of the source manifolds, leaving the normal map unchanged on the boundary. In the light of Lemma 8.116 we can now reformulate Theorem 7.41.

Theorem 8.121 (Killing the absolute kernels of a normal map of pairs) *Let k be the natural number satisfying $n = 2k + 1$ or $n = 2k$. By carrying out a finite sequence of surgery steps in the interior of the source manifolds we can arrange that f is k-connected and we have $K_i(\widehat{M}) = 0$ for $i < k$. If we additionally assume that ∂f is a homotopy equivalence, we have $K_i(\widehat{M}) = 0 = K^i(\widehat{M})$, unless $n = 2k$ and $i = k$, or $n = 2k + 1$ and $i = k, k+1$.*

We conclude this subsection by introducing the *chain level Umkehr maps for pairs*. These are $\mathbb{Z}\Gamma$-chain maps of $\mathbb{Z}\Gamma$-chain complexes

$$C_*^!(\widehat{f}) \colon C_*(\widehat{X}) \to C_*(\widehat{M}) \quad \text{and} \quad C_*^!(\widehat{f, \partial f}) \colon C_*(\widehat{X}, \widehat{\partial X}) \to C_*(\widehat{M}, \widehat{\partial M})$$

that induce the Umkehr maps on homology of (8.118). In fact, the chain map $C_*^!(\widehat{f})$ depends on the map of pairs $(\widehat{f, \partial f})$, but we suppress the ∂f from the notation so that it is clear that the source and target of $C_*^!(\widehat{f})$ are $C_*(\widehat{M})$ and $C_*(\widehat{X})$. As in the closed case, the chain level Umkehr maps lead to a deeper explanation of the Poincaré duality isomorphism of kernels in Lemma 8.116 (ii). They will also be used in the proof of Lemma 8.129 (ii) below and play an important role in the theory of algebraic surgery in Chapter 15. They are defined to be the following compositions of $\mathbb{Z}\Gamma$-chain maps

$$C_*^!(\widehat{f}) \colon C_*(\widehat{X}) \xrightarrow{(-\cap [X,\partial X])^{-1}} C^{n-*}(\widehat{X}, \widehat{\partial X}) \xrightarrow{C^{n-*}(\widehat{f, \partial f})} C^{n-*}(\widehat{M}, \widehat{\partial M}) \xrightarrow{-\cap [M,\partial M]} C_*(\widehat{M}), \quad (8.122)$$

and

$$C_*^!(\widehat{f, \partial f}) \colon C_*(\widehat{X}, \widehat{\partial X}) \xrightarrow{(-\cap [X,\partial X])^{-1}} C^{n-*}(\widehat{X}) \xrightarrow{C^{n-*}(\widehat{f})} C^{n-*}(\widehat{M}, \widehat{\partial M}) \xrightarrow{-\cap [M,\partial M]} C_*(\widehat{M}, \widehat{\partial M}). \quad (8.123)$$

Here the maps denoted $(-\cap [X, \partial X])^{-1}$ are $\mathbb{Z}\Gamma$-chain homotopy inverses to $-\cap [X, \partial X] \colon C^{n-*}(\widehat{X}, \widehat{\partial X}) \to C_*(\widehat{X})$ and also $-\cap [X, \partial X] \colon C^{n-*}(\widehat{X}) \to C_*(\widehat{X}, \widehat{\partial X})$. There are many choices for the chain maps $(-\cap [X, \partial X])^{-1}$, but they are all $\mathbb{Z}\Gamma$-chain

276   8 The Even-Dimensional Surgery Obstruction

homotopic, and so the $\mathbb{Z}\Gamma$-chain homotopy classes of $C_*^!(\widehat{f})$ and $C_*^!(\widehat{f}, \widehat{\partial f})$ are well defined.

The dual of $C_*^!(\widehat{f})$ is the $\mathbb{Z}\Gamma$-chain map $C^!(\widehat{f})^{n-*}: C^{n-*}(\widehat{M}) \to C^{n-*}(\widehat{X})$ and the dual of $C_*^!(\widehat{f}, \widehat{\partial f})$ is $C_!^{n-*}(\widehat{f}, \widehat{\partial f}): C^{n-*}(\widehat{M}, \widehat{\partial M}) \to C^{n-*}(\widehat{X}, \widehat{\partial X})$. From the definition of the chain level Umkehr map we see that

$$\widehat{f}_! = H_j(C_*^!(\widehat{f})): H_j(\widehat{X}) \to H_j(\widehat{M}); \tag{8.124}$$

$$(\widehat{f}, \widehat{\partial f})_! = H_*(C_*^!(\widehat{f}, \widehat{\partial f})): H_*(\widehat{X}, \widehat{\partial X}) \to H_*(\widehat{M}, \widehat{\partial M}); \tag{8.125}$$

$$\widehat{f}^! = H_j(C^!(\widehat{f})^{n-*}): H^j(\widehat{M}) \to H^j(\widehat{X}); \tag{8.126}$$

$$(\widehat{f}, \widehat{\partial f})^! = H_j(C_*^!(\widehat{f}, \widehat{\partial f})^{n-*}): H^j(\widehat{M}) \to H^j(\widehat{X}). \tag{8.127}$$

The mapping cone of $C_*^!(\widehat{f})$ is the $\mathbb{Z}\Gamma$-chain complex $\mathrm{cone}_*(C_*^!(\widehat{f}))$ and the mapping cone of $C_*^!(\widehat{f}, \widehat{\partial f})$ is $\mathrm{cone}_*(C_*^!(\widehat{f}, \widehat{\partial f}))$. Since the target of $C_*^!(\widehat{f}, \widehat{\partial f})$ is $C_*(\widehat{M}, \widehat{\partial M})$, we have the canonical chain map

$$e_*: C_*(\widehat{M}, \widehat{\partial M}) \to \mathrm{cone}_*(C_*^!(\widehat{f}, \widehat{\partial f})).$$

If we use e_* to push forward the fundamental class $[M, \partial M] \in H_n(M, \partial M; \mathbb{Z}^w)$ of $(M, \partial M)$, we obtain $e_*([M, \partial M]) \in H_n(\mathrm{cone}_*(C_*^!(\widehat{f}, \widehat{\partial f}); \mathbb{Z}^w))$ and hence also the $\mathbb{Z}\Gamma$-chain maps

$$- \cap e_*([M, \partial M]): \mathrm{cone}^{n-*}(C_*^!(\widehat{f})) \to \mathrm{cone}_*(C_*^!(\widehat{f}, \widehat{\partial f}))$$

and

$$- \cap e_*([M, \partial M]): \mathrm{cone}^{n-*}(C_*^!(\widehat{f}, \widehat{\partial f})) \to \mathrm{cone}_*(C_*^!(\widehat{f})).$$

In the following lemma and its proof, we shall see that the fact that the map $f: (M, \partial M) \to (X, \partial X)$ has degree one implies that the kernel groups of $K_*(\widehat{M})$ and $K_*(\widehat{M}, \widehat{\partial M})$ are isomorphic to the homology groups of $\mathrm{cone}_*(C_*^!(\widehat{f}))$ and $\mathrm{cone}_*(C_*^!(\widehat{f}, \widehat{\partial f}))$ respectively, and that the chain map $- \cap e_*([M])$ behaves very much like the cap product map of the fundamental class of a Poincaré pair. This gives a deeper explanation of the fact that the kernel groups $K_*(\widehat{M})$ and $K_*(\widehat{M}, \widehat{\partial M})$ behave in many respects like the homology groups of a third pair satisfying Poincaré duality.

Lemma 8.128 (i) *There are natural isomorphisms*

$$K_j(\widehat{M}) \xrightarrow{\cong} H_j\big(\mathrm{cone}_*(C_*^!(\widehat{f}))\big);$$

$$K_j(\widehat{M}, \widehat{\partial M}) \xrightarrow{\cong} H_j\big(\mathrm{cone}_*(C_*^!(\widehat{f}, \widehat{\partial f}))\big);$$

$$K^j(\widehat{M}) \xrightarrow{\cong} H_{n-j}\big(\mathrm{cone}^{n-*}(C_*^!(\widehat{f}))\big);$$

$$K^j(\widehat{M}, \widehat{\partial M}) \xrightarrow{\cong} H_{n-j}\big(\mathrm{cone}^{n-*}(C_*^!(\widehat{f}, \widehat{\partial f}))\big);$$

(ii) *The cap product with $e_*([M, \partial M])$ induces $\mathbb{Z}\Gamma$-chain homotopy equivalences*

8.6 The Even-Dimensional Surgery Obstruction in the Universal Covering Case

$$-\cap e_*([M,\partial M])\colon \operatorname{cone}^{n-*}(C_*^!(\widehat{f})) \xrightarrow{\simeq} \operatorname{cone}_*(C_*^!(\widehat{f},\widehat{\partial f}))$$

and

$$-\cap e_*([M,\partial M])\colon \operatorname{cone}^{n-*}(C_*^!(\widehat{f},\widehat{\partial f})) \xrightarrow{\simeq} \operatorname{cone}_*(C_*^!(\widehat{f})),$$

which in turn induce commutative diagrams of isomorphisms

$$\begin{array}{ccc}
K^{n-j}(\widehat{M}) & \xrightarrow{\cong} & H_j\bigl(\operatorname{cone}^{n-*}(C_*^!(\widehat{f}))\bigr) \\
{\scriptstyle \cong}\downarrow{\scriptstyle -\cap[M]} & & {\scriptstyle \cong}\downarrow{\scriptstyle -\cap e_*([M,\partial M])} \\
K_j(\widehat{M},\widehat{\partial M}) & \xrightarrow{\cong} & H_j\bigl(\operatorname{cone}_*(C_*^!(\widehat{f},\widehat{\partial f}))\bigr)
\end{array}$$

and

$$\begin{array}{ccc}
K^{n-j}(\widehat{M}) & \xrightarrow{\cong} & H_j\bigl(\operatorname{cone}^{n-*}(C_*^!(\widehat{f}))\bigr) \\
{\scriptstyle \cong}\downarrow{\scriptstyle -\cap[M]} & & {\scriptstyle \cong}\downarrow{\scriptstyle -\cap e_*([M,\partial M])} \\
K_j(\widehat{M},\widehat{\partial M}) & \xrightarrow{\cong} & H_j\bigl(\operatorname{cone}_*(C_*^!(\widehat{f},\widehat{\partial f}))\bigr).
\end{array}$$

Proof. By definition, the chain level Umkehr maps fit into the following up to chain homotopy commutative diagrams of chain maps:

$$\begin{array}{ccccc}
\operatorname{cone}^{n-*}(C_*^!(\widehat{f},\widehat{\partial f})) & \xrightarrow{e^{n-*}} & C^{n-*}(\widehat{M},\widehat{\partial M}) & \xleftarrow{C^{n-*}(\widehat{f},\widehat{\partial f})} & C^{n-*}(\widehat{X},\widehat{\partial X}) \\
\downarrow{\scriptstyle -\cap e_*([M,\partial M])} & & \downarrow{\scriptstyle -\cap[M,\partial M]} & {\scriptstyle C_*^!(\widehat{f})} & \downarrow{\scriptstyle -\cap[X,\partial X]} \\
\operatorname{cone}_*(C_*^!(\widehat{f})) & \xleftarrow{e_*} & C_*(\widehat{M}) & \xrightarrow{C_*(\widehat{f})} & C_*(\widehat{X})
\end{array}$$

and

$$\begin{array}{ccccc}
\operatorname{cone}^{n-*}(C_*^!(\widehat{f})) & \xrightarrow{e^{n-*}} & C^{n-*}(\widehat{M}) & \xleftarrow{C^{n-*}(\widehat{f})} & C^{n-*}(\widehat{X}) \\
\downarrow{\scriptstyle -\cap e_*([M,\partial M])} & & \downarrow{\scriptstyle -\cap[M,\partial M]} & {\scriptstyle C_*^!(\widehat{f},\widehat{\partial f})} & \downarrow{\scriptstyle -\cap[X,\partial X]} \\
\operatorname{cone}_*(C_*^!(\widehat{f},\widehat{\partial f})) & \xleftarrow{e_*} & C_*(\widehat{M},\widehat{\partial M}) & \xrightarrow{C_*(\widehat{f},\widehat{\partial f})} & C_*(\widehat{X},\widehat{\partial X}).
\end{array}$$

Diagram chases in the above diagrams prove assertion (i). The commutativity of the diagrams appearing in assertion (ii) follows from the commutativity up to chain homotopy of the above diagrams. Now part (i) and Lemma 8.116 (ii) show that the $\mathbb{Z}\Gamma$-chain maps $-\cap e_*([M,\partial M])$ induce an isomorphism on homology modules and so by [361, Theorem 1.7.7], $-\cap e_*([M,\partial M])$ is a $\mathbb{Z}\Gamma$-chain homotopy equivalence. This finishes the proof of Lemma 8.128. $\square$

The proof of the next result is analogous to the one of Lemma 8.55, see also Remark 8.61.

Lemma 8.129 *Suppose that f and ∂f are k-connected. Then:*

(i) *The Kronecker homomorphisms*

$$K^k(\widehat{M}) \xrightarrow{\cong} K_k(\widehat{M})^*;$$
$$K^k(\widehat{M}, \widehat{\partial M}) \xrightarrow{\cong} K_k(\widehat{M}, \widehat{\partial M})^*,$$

are isomorphisms;

(ii) *Suppose in addition that $K_k(\widehat{M}, \widehat{\partial M}) = 0$. Then the kernels $K_k(\widehat{M})$ and $K_{k+1}(\widehat{M}, \widehat{\partial M})$ are stably finitely generated free $\mathbb{Z}\Gamma$-modules and the Kronecker homomorphism*

$$K^{k+1}(\widehat{M}, \widehat{\partial M}) \xrightarrow{\cong} K_{k+1}(\widehat{M}, \widehat{\partial M})^*$$

is an isomorphism.

8.6.3 Proof of Theorem 8.112

Proof of Theorem 8.112. In this subsection we will only consider the universal covering case, see Notation 8.34, assume that the dimension n is of the form $n = 2k$ for some natural number $k \geq 3$, and all normal bordisms will have cylindrical target.

The proof is somewhat involved since the definition and the normal bordism invariance are linked to one another. We will proceed as follows.

(i) We first define the surgery obstruction for a highly connected normal map;
(ii) Then we show that it is invariant under highly connected normal bordism with cylindrical target;
(iii) We define the surgery obstruction for any normal map;
(iv) We prove invariance under any normal bordism with cylindrical target;
(v) We prove the obstruction property.

Consider a highly connected normal map $(f, \overline{f})$, where highly connected means that f is k-connected. Note that then we have identifications $\Gamma = \pi_1(X, x) = \pi_1(M, m)$ and the pullback of $\widetilde{p}_X \colon \widetilde{X} \to X$ to M is a model for the universal covering $\widetilde{p}_M \colon \widetilde{M} \to M$. In the sequel we abbreviate $\pi = \pi_1(X, x)$ and $w = w_1(X)$. The surgery kernel $K_k(\widetilde{M})$ is a stably free $\mathbb{Z}\pi$-module, see Lemma 8.55 (ii). Equip $T_m M$ with the orientation that is compatible with the fundamental class $[M]$ in the sense of Definition 8.16. Note that we have now made choices (GBP) and (ABP) as well, see Notations 8.8 and 8.13. The surgery kernel comes with the structure $(K_k(\widetilde{M}), s, t)$ of a non-singular $(-1)^k$-quadratic form over $\mathbb{Z}\pi$ by Example 8.86. Choose any natural number r such that $K_k(\widetilde{M}) \oplus \mathbb{Z}\pi^r$ is finitely generated free. Then $(K_k(\widetilde{M}), s, t) \oplus H_{(-1)^k}(\mathbb{Z}\pi^r)$ is a non-singular $(-1)^k$-quadratic form over $\mathbb{Z}\pi$ whose underlying $\mathbb{Z}\pi$-module is finitely generated free and hence it defines an element in $L_k(\mathbb{Z}\pi, w)$. We define

8.6 The Even-Dimensional Surgery Obstruction in the Universal Covering Case

$$\sigma(f, \overline{f}) \in L_{2k}(\mathbb{Z}\pi, w) \tag{8.130}$$

by this element. One easily checks that the choice of r does not matter.

Now consider highly connected normal π-maps of degree one $(f_i, \overline{f_i}) = (M_i, [M_i], f_i, \widehat{f_i}, a_i, \xi_i, \overline{f_i})$ for $i = 0, 1$ that have the same target $(X, \widehat{p}, [X])$. Moreover, let $(W, [W, \partial W], Y, \partial Y, [Y, \partial Y], F, \widehat{F}, a, \Xi, \overline{F})$, or sometimes for short $(F, \overline{F})$, be a highly connected normal bordism with cylindrical target in the sense of Definition 8.7 between them, where highly connected means that F is k-connected. We get from the projection $\mathrm{pr}\colon X \times [0,1] \to X$ and the maps F, f_0, and f_1 identifications

$$\pi = \pi_1(X) = \pi_1(X \times [0,1]) = \pi_1(X \times \{1\}) = \pi_1(X \times \{1\})$$
$$= \pi_1(W) = \pi_1(M_0) = \pi_1(M_1).$$

The pullbacks of the universal covering $p_X \colon \widetilde{X} \to X$ to $X \times [0,1]$, W, M_0 and M_1 are models for the universal coverings of these spaces. We have defined elements $\sigma(f_1, \overline{f_0})$ and $\sigma(f_1, \overline{f_1})$ in $L_{2k}(\mathbb{Z}\pi, w)$, see (8.130). We denote by $p_{\partial W}\colon \widetilde{\partial W} \to \partial W$ the restriction of $p_W \colon \widetilde{W} \to W$ to ∂W. Note that ∂W has precisely two boundary components and the restriction of $p_{\partial W}$ to each of these components is again a model for the universal covering, which justifies the notation $\widetilde{\partial W}$.

Lemma 8.131 *Consider the situation above. Additionally we suppose that $K_k(\widetilde{W}, \widetilde{\partial W}) = 0$. Then we get*

$$\sigma(f_0, \overline{f_0}) = \sigma(f_1, \overline{f_1}) \in L_{2k}(\mathbb{Z}\pi, w).$$

Proof. We can arrange by doing trivial surgery in dimension k on M_0 and M_1 that $K_k(\widetilde{M_0})$ and $K_k(\widetilde{M_1})$ are finitely generated free, see Example 8.60. This does not change $\sigma(f_0, \overline{f_0})$ and $\sigma(f_1, \overline{f_1})$ in $L_{(-1)^k}(\mathbb{Z}\pi)$, see Example 8.60, and does not change the normal bordism class with cylindrical target, see Theorem 4.39 (iv). Hence we can assume in the sequel without loss of generality that $K_k(\widetilde{M_0})$ and $K_k(\widetilde{M_1})$ are finitely generated free $\mathbb{Z}\pi$-modules and $\sigma(f_i, \overline{f_i})$ is given by the class $(K_k(\widetilde{M_i}), s_i, t_i)$ for $i = 0, 1$.

Recall that ∂W has two components, which we identified with M_0 and M_1. Hence we get a decomposition of stably finitely generated free $\mathbb{Z}\pi$-modules $K_k(\widetilde{\partial W}) = K_k(\widetilde{M_0}) \oplus K_k(\widetilde{M_1})$. The structures of non-singular $(-1)^k$-quadratic forms $(K_k(\widetilde{M_i}), s_i, t_i)$ on $K_k(\widetilde{M_i})$ for $i = 0, 1$ induce the structure of a non-singular $(-1)^k$-quadratic form $(K_k(\widetilde{\partial W}), s', t')$ on $K_k(\widetilde{\partial W})$. If we unravel the definitions, we get an identification of non-singular $(-1)^k$-quadratic forms

$$K_k(\widetilde{\partial W}, s', t') = K_k(\widetilde{M_0}, s_0, t_0) \oplus K_k(\widetilde{M_1}, -s_1, -t_1)$$

where the signs appearing in the second summand come from our conventions on how to relate the fundamental classes of M_0 and M_1 to that of ∂W, which in turn is induced by the fundamental class of W. Hence we get in $L_{2k}(\mathbb{Z}\pi, w)$

$$[(K_k(\widetilde{\partial W}), s', t')] = \sigma(f_0, \overline{f_0}) - \sigma(f_1, \overline{f_1}).$$

Hence we have to show $[(K_k(\widetilde{\partial W}), s', t')] = 0$. In view of Lemma 8.95, this boils down to constructing a lagrangian for $(K_k(\widetilde{\partial W}), s', t')$.

By assumption we have that $K_j(\widetilde{W}, \widetilde{\partial W}) = 0$ for $j \leq k$. Moreover, from Lemma 8.116 (ii) and Lemma 8.129 (i), we get

$$K_{k+1}(\widetilde{W}) \cong K^k(\widetilde{W}, \widetilde{\partial W}) \cong K_k(\widetilde{W}, \widetilde{\partial W})^* = 0$$

and from Lemma 8.129 (ii) that $K_{k+1}(\widetilde{W}, \widetilde{\partial W})$ and $K_k(\widetilde{W})$ are stably finitely generated free. From Lemma 8.116 (ii) we obtain a split short exact sequence of $\mathbb{Z}\pi$-modules

$$0 \to K_{k+1}(\widetilde{W}, \widetilde{\partial W}) \xrightarrow{\partial_{k+1}} K_k(\widetilde{\partial W}) \xrightarrow{K_k(\widetilde{i})} K_k(\widetilde{W}) \to 0 \tag{8.132}$$

where $\widetilde{i}\colon \widetilde{\partial W} \to \widetilde{W}$ is the inclusion. It splits since $K_k(\widetilde{W})$ is stably free. Our candidate for a lagrangian is $\mathrm{im}(\partial_{k+1})$.

Using Remark 8.92 and in view of the above, it is enough to check that the symmetric and quadratic forms on $\mathrm{im}(\partial_{k+1})$ induced by inclusion into $K_k(\widetilde{\partial W})$ vanish in the respective Q-groups.

For the symmetric structure, this is a generalisation of the fact that the signature of a boundary vanishes; cf. Theorem 5.79 and its proof. We first show for any $x, y \in K_{k+1}(\widetilde{W}, \widetilde{\partial W})$ that $s(\partial_{k+1}(x), \partial_{k+1}(y)) = 0$. From Lemma 8.116 (ii) and the exactness of (8.132) we deduce that the following diagram commutes, has exact rows, and its vertical arrows are isomorphisms.

$$\begin{array}{ccccccccc}
0 & \to & K^k(\widetilde{W}) & \xrightarrow{K^k(\widetilde{i})} & K^k(\widetilde{\partial W}) & \xrightarrow{\Delta^k} & K^{k+1}(\widetilde{W}, \widetilde{\partial W}) & \to & 0 \\
& & \downarrow \cong {\scriptstyle -\cap[W,\partial W]} & & \downarrow \cong {\scriptstyle -\cap[\partial W]} & & \downarrow \cong {\scriptstyle -\cap[W,\partial W]} & & \\
0 & \to & K_{k+1}(\widetilde{W}, \widetilde{\partial W}) & \xrightarrow{\partial_{k+1}} & K_k(\widetilde{\partial W}) & \xrightarrow{K_k(\widetilde{i})} & K_k(\widetilde{W}) & \to & 0.
\end{array} \tag{8.133}$$

We have for $x, y \in K_{k+1}(\widetilde{W}, \widetilde{\partial W})$

$$\begin{aligned}
s(\partial_{k+1}(x), \partial_{k+1}(y)) &= \langle (-\cap[\partial W])^{-1} \circ \partial_{k+1}(x), \partial_{k+1}(y) \rangle \\
&= \langle K^k(\widetilde{i}) \circ (-\cap[W, \partial W])^{-1}(x), \partial_{k+1}(y) \rangle \\
&= \langle \Delta^k \circ K^k(\widetilde{i}) \circ (-\cap[W, \partial W])^{-1}(x), y \rangle \\
&= \langle 0 \circ (-\cap[W, \partial W])^{-1}(x), y \rangle \\
&= 0.
\end{aligned}$$

Now suppose for $x \in K_k(\widetilde{\partial W})$ that $s(x, \partial_{k+1}(y)) = 0$ for all $y \in K_{k+1}(\widetilde{W}, \widetilde{\partial W})$. We want to show $x \in \mathrm{im}(\partial_{k+1})$. This is equivalent to showing that we have $(-\cap[W, \partial W])^{-1} \circ K_k(\widetilde{i})(x) = 0$. Since $K_j(\widetilde{W}, \widetilde{\partial W}) = 0$ for $j \leq k$, the canonical map

$$K^{k+1}(\widetilde{W}, \widetilde{\partial W}) \xrightarrow{\cong} \mathrm{hom}_{\mathbb{Z}\pi}(K_{k+1}(\widetilde{W}, \widetilde{\partial W}), \mathbb{Z}\pi), \quad \alpha \mapsto \langle \alpha, - \rangle$$

8.6 The Even-Dimensional Surgery Obstruction in the Universal Covering Case

is bijective by Lemma 8.52 (ii) and Lemma 8.128 (i). Hence the claim follows from the following calculation for $y \in K_{k+1}(\widetilde{W}, \widetilde{\partial W})$

$$\begin{aligned}\langle (-\cap [W,\partial W])^{-1} \circ K_k(\widetilde{i})(x), y \rangle &= \langle \Delta^k \circ (-\cap [\partial W])^{-1}(x), y \rangle \\ &= \langle (-\cap [\partial W])^{-1}(x), \partial_{k+1}(y) \rangle \\ &= s(x, \partial_{k+1}(y)) \\ &= 0.\end{aligned}$$

Hence we have shown that $\text{im}(\partial_{k+1})$ is a lagrangian for the non-singular $(-1)^k$-symmetric form $(K_k(\widetilde{\partial W}), s')$. It remains to show that it is also a lagrangian for the non-singular $(-1)^k$-quadratic form $(K_k(\widetilde{\partial W}), s', t')$. In other words, we must show that t' vanishes on $\text{im}(\partial_{k+1})$, and we prove this next.

We have the following commutative diagram of $\mathbb{Z}\pi$-modules

$$\begin{array}{ccccc} K_{k+1}(\widetilde{W}, \widetilde{\partial W}) & \xrightarrow{\partial_{k+1}} & K_k(\widetilde{\partial W}) & \xrightarrow{K_k(\widetilde{i})} & K_k(\widetilde{W}) \\ & & \cong \uparrow h_{\widetilde{\partial W}} & & \cong \uparrow h_{\widetilde{W}} \\ & & \pi_{k+1}(\widetilde{f_0}) \oplus \pi_{k+1}(\widetilde{f_1}) & \longrightarrow & \pi_k(\widetilde{F}) \\ & & \downarrow \partial_{\widetilde{\partial W}} & & \downarrow \partial_{\widetilde{W}} \\ & & \pi_k(\widetilde{M_0}) \oplus \pi_k(\widetilde{M_1}) & \longrightarrow & \pi_k(\widetilde{W}) \end{array} \quad (8.134)$$

where the right horizontal arrows are induced by the inclusions, the isomorphisms $h_{\widetilde{\partial W}}$ and $h_{\widetilde{W}}$ come from of Lemma 8.39 (iii), the identification $K_k(\widetilde{\partial W}) = K_k(\widetilde{M_0}) \oplus K_k(\widetilde{M_1})$, and Lemma 8.116 (iii), and $\partial_{\widetilde{\partial W}}$ and $\partial_{\widetilde{W}}$ are the boundary maps of the long exact homotopy sequences.

Consider $x \in K_{k+1}(\widetilde{W}, \widetilde{\partial W})$ and its image under ∂_{k+1}

$$\partial_{k+1}(x) \in K_k(\widetilde{\partial W}) = K_k(\widetilde{M_0}) \oplus K_k(\widetilde{M_1}).$$

By the construction of the map $t_k \colon \pi_{k+1}(f) \to I_k(M)$ of Lemma 8.10 and Lemma 8.39 (iii), we can find immersions $u_0 \colon S^k \looparrowright M_0$ and $u_1 \colon S^k \looparrowright M_1$ satisfying

$$\partial_{\widetilde{\partial W}} \circ h_{\widetilde{\partial W}}^{-1} \circ \partial_{k+1}(x) = ([u_0], -[u_1]). \quad (8.135)$$

Since the top row of the diagram (8.134) is exact, see (8.133), we conclude that the composition of u_0 with the inclusion $M_0 \to W$ and the composite of u_1 with the inclusion $M_1 \to W$ are homotopic. Hence we can find a smooth map

$$H \colon (S^k \times [0,1], S^k \times \{0,1\}) \to (W, \partial W)$$

such that $H|_{S^k \times \{i\}} = u_i$ for $i = 0, 1$ and H itself is in general position. We have

$$t'(\partial_{k+1}(x)) = \mu(u_0) + \mu(u_1), \quad (8.136)$$

where $\mu(u_i) \in Q_\epsilon(\mathbb{Z}\pi, w)$ is the self-intersection element defined in (8.23). The reader may have expected $t'(\partial_{k+1}(x)) = \mu(u_0) - \mu(u_1)$ in (8.136) in view of the minus sign appearing in (8.135), but one should realise that equation (8.136) depends on the convention on how to relate the fundamental classes of M_0 and M_1 to that of ∂W whereas (8.135) does not. Hence it suffices to show

$$\mu(u_0) + \mu(u_1) = 0 \in Q_\epsilon(\mathbb{Z}\pi, w). \tag{8.137}$$

Since H is in general position, the set of double points of H consists of circles in the interior of $\widetilde{W}$, which do not concern us, and arcs whose end points are on $\widetilde{\partial W}$. We claim that for each such arc the contributions of its two end points to $\mu(u_0) + \mu(u_1)$ cancel each other. This can be easily checked from the definition of μ in Subsection 8.3.2, using that fact that S^k is simply connected. Hence equation (8.137) is true. This finishes the proof of Lemma 8.131. $\square$

Next we explain that the condition $K_k(\widetilde{W}, \widetilde{\partial W}) = 0$ can be discarded in Lemma 8.131, namely, we will show the following result.

Lemma 8.138 *Consider the situation above. Then we get*

$$\sigma(f_0, \overline{f_0}) = \sigma(f_1, \overline{f_1}) \in L_{(-1)^k}(\mathbb{Z}\pi, w).$$

Proof. The idea is to manipulate the normal bordism $(F, \overline{F})$ and also $(f_1, \overline{f_1})$ without changing $(f_0, \overline{f_0})$ such that the condition $K_k(\widetilde{W}, \widetilde{\partial W}) = 0$ is satisfied, $\sigma(f_1, \overline{f_1})$ is unchanged, and then to apply Lemma 8.131.

We will achieve this by a handle subtraction argument, which appears for example in [414, Theorem 1.4 on page 15, proof of Theorem 5.6 on page 50]. In this book it is also used in Chapter 9 and in the proof of Lemma 13.15.

Recall that we have the obvious decomposition of $\mathbb{Z}\pi$-modules coming from the identification $\partial W = M_0 \amalg M_1$

$$K_j(\widetilde{\partial W}) = K_j(\widetilde{M_0}) \oplus K_j(\widetilde{M_1}).$$

Since f_0 and f_1 are k-connected, we get $K_{k-1}(\widetilde{\partial W}) = 0$ from Lemma 8.39 (iii). By Lemma 8.116 (ii) there is an exact sequence

$$K_k(\widetilde{\partial W}) \xrightarrow{i_*} K_k(\widetilde{W}) \to K_k(\widetilde{W}, \widetilde{\partial W}) \to 0. \tag{8.139}$$

So it will suffice to modify $(F, \overline{F})$ and $(f_1, \overline{f_1})$, leaving $(f_0, \overline{f_0})$ fixed, so that for the new normal bordism with cylindrical target the map $K_k(\widetilde{\partial W'}) \to K_k(\widetilde{W'})$ is onto, F and f_1 are k-connected and $\sigma(f_1', \overline{f_1'}) = \sigma(f_1, \overline{f_1})$ holds.

Since F is k-connected, we get from Lemma 4.4 and Lemma 8.116 (iii) that $K_k(\widetilde{W})$ is a finitely generated $\mathbb{Z}\pi$-module. We conclude from the exact sequence (8.139) that $K_k(\widetilde{W}, \widetilde{\partial W})$ is also a finitely generated $\mathbb{Z}\pi$-module.

We will now do induction over the minimal number of generators of the finitely generated $\mathbb{Z}\pi$-module $K_k(\widetilde{W}, \widetilde{\partial W})$. The beginning of the induction is immediate since

8.6 The Even-Dimensional Surgery Obstruction in the Universal Covering Case

$K_k(\widetilde{W}, \widetilde{\partial W})$ is trivial if its minimal number of generators is zero. The induction step is done as follows.

Choose a minimal finite set of generators S of $K_k(\widetilde{W}, \widetilde{\partial W})$. Fix $s \in S$. By general position s is represented by an embedding $q \colon S^k \hookrightarrow \text{Int}(W)$ and the bundle data ensure that this embedding extends to an embedding

$$Q \colon D^{k+1} \times S^k \hookrightarrow \text{int}(W).$$

We now choose an embedded path in W from a point on the boundary $Q(S^k \times S^k)$ of $Q(D^{k+1} \times S^k)$ to a point in M_1 that meets $Q(D^{k+1} \times S^k)$ and ∂W in exactly one point, namely, the starting point and the end point, and at these points the path is transversal to the boundary of $Q(D^{k+1} \times S^k)$ and to ∂W. We can thicken this path to obtain an embedding $I \times D^{2k} \hookrightarrow W$ such that $I \times D^{2k} \cap Q(D^{k+1} \times S^k) = I \times D^{2k} \cap Q(S^k \times S^k) = \{0\} \times D^{2k}$ and $I \times D^{2k} \cap \partial W = I \times D^{2k} \cap M_1 = \{1\} \times D^{2k}$ hold, and $I \times D^{2k}$ meets the boundary of $Q(D^{k+1} \times S^k)$ and M_1 in a transversal way. Define

$$V = I \times D^{2k} \cup Q(D^{k+1} \times S^k).$$

Since the composition $q \circ F \colon S^k \to X \times I$ is nullhomotopic, and V deformation retracts to $q(S^k)$, we can modify F and $\overline{F}$ (and also f_1 and $\overline{f_1}$), leaving $(f_0, \overline{f_0})$ fixed, by a homotopy such that the restriction of F to V is a constant map with image a point in $X \times \{1\}$ and $\sigma(f_0, \overline{f_0})$ and $\sigma(f_1, \overline{f_1})$ are unchanged. If we now remove the interior of V from W and set

$$W' := W - \text{int}(V),$$

then, after smoothing corners, we obtain a normal bordism $(F', \overline{F'})$ with domain $(W'; M_0, M_1')$, where $M_1' = M_1 \sharp (S^k \times S^k)$. The procedure of moving from W to W' is called a *handle subtraction*.

Figure 8.140 (Handle subtraction).

core of the cancelled handle

Since the passage from $(f_1, \overline{f_1})$ to $(f_1', \overline{f_1'})$ is just one trivial surgery step in dimension k, and we have not touched M_0, the normal map $(f_1', \overline{f_1})$ is highly connected, we have $\sigma(f_1, \overline{f_1}) = \sigma(f_1', \overline{f_1'})$, and there is a $\mathbb{Z}\pi$-isomorphism

$$i_1 \oplus i_2 \colon K_k(\widetilde{\partial W}) \oplus \mathbb{Z}\pi^2(b_1, b_2) \xrightarrow{\cong} K_k(\widetilde{\partial W'}) \tag{8.141}$$

where $\{b_1, b_2\}$ is a $\mathbb{Z}\pi$-basis for $\mathbb{Z}\pi^2$ coming from $Q(S^k \times \bullet\})$ and $Q(\{\bullet \times S^k)$. The composite

$$K_k(\widetilde{\partial W'}) \to K_k(\widetilde{W'}) \to K_k(\widetilde{W}) \to K_k(\widetilde{W}, \widetilde{\partial W})$$

sends b_2 to the generator $s \in S \subseteq K_k(\widetilde{W}, \partial W)$ we started with.

Since F is k-connected, W can be recovered from W' by attaching $Q(D^{k+1} \times S^k)$ along a map $Q(S^k \times S^{k+1})$ and F is constant on $Q(D^{k+1} \times S^k)$, the inclusion $W' \to W$ is k-connected, the map F' is k-connected, and there is an exact sequence of $\mathbb{Z}\pi$-modules

$$H_{k+1}(\widetilde{W}, \widetilde{W'}) \xrightarrow{\partial_{k+1}} K_k(\widetilde{W'}) \to K_k(\widetilde{W}) \to 0, \quad (8.142)$$

which comes from the long exact sequence of $\mathbb{Z}\pi$-homology modules of the pair $(\widetilde{W}, \widetilde{W'})$. In particular, the new normal bordism $(F', \overline{F'})$ is highly connected. The $\mathbb{Z}\pi$-module $H_{k+1}(\widetilde{W}, \widetilde{W'})$ is isomorphic to $\mathbb{Z}\pi$ and has a generator b coming from $Q(\{\bullet\} \times S^k)$. The image of b under $\partial_{k+1}: H_{k+1}(\widetilde{W}, \widetilde{W'}) \to K_k(\widetilde{W'})$ is the image of b_2 under $K_k(\widetilde{\partial W'}) \to K_k(\widetilde{W'})$. Hence the image of b under the composite

$$K_{k+1}(\widetilde{W}, \widetilde{W'}) \xrightarrow{\partial_{k+1}} K_k(\widetilde{W'}) \to K_k(\widetilde{W}) \to K_k(\widetilde{W}, \widetilde{\partial W})$$

is the generator $s \in S \subseteq K_k(\widetilde{W}, \partial W)$ we started with. Next we claim that we obtain an isomorphism

$$u: K_k(\widetilde{W'}, \widetilde{\partial W'}) \xrightarrow{\cong} K_k(\widetilde{W}, \widetilde{\partial W})/\langle s \rangle$$

where $\langle s \rangle \subseteq K_k(\widetilde{W}, \widetilde{\partial W})$ is the $\mathbb{Z}\pi$-module generated by s. It comes from the following commutative diagram of $\mathbb{Z}\pi$-modules

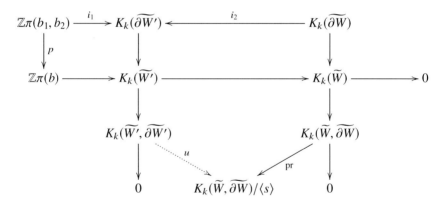

where p sends b_1 to zero and b_2 to b and pr is the canonical projection. Since the second row, which is (8.142), and the two columns are exact, $i_1 \oplus i_2$ is an isomorphism, see (8.141), and b is sent to s under the obvious composite of two horizontal arrows and one vertical arrow.

Hence the minimal number of generators of $K_k(\widetilde{W'}, \widetilde{\partial W'})$ is strictly smaller than the minimal number of generators of $K_k(\widetilde{W}, \widetilde{\partial W})$. Therefore the induction hypothesis applies and yields $\sigma(f_0, \overline{f_0}) = \sigma(f'_1, \overline{f'_1})$. Since we already know $\sigma(f'_1, \overline{f'_1}) = \sigma(f_1, \overline{f_1})$, this finishes the proof of the induction step of Lemma 8.138. □

Now we consider a normal map $(f, \overline{f})$, no longer assuming that it is highly connected. Choose any highly connected normal map $(f', \overline{f'})$ such that there is a normal bordism $(F, \overline{F})$ with cylindrical target between $(f, \overline{f})$ and $(f', \overline{f'})$. Such a choice is possible, actually one can construct $(f', \overline{f'})$ by a finite sequence of surgery steps in the interior, see Theorem 7.41. We have already defined $\sigma(f', \overline{f'})$ in (8.130). We want to define the surgery obstruction

$$\sigma(f, \overline{f}) \in L_k(\mathbb{Z}\pi, w) \tag{8.143}$$

by $\sigma(f', \overline{f'})$. We have to show that this definition is independent of the choice of $(f', \overline{f'})$. Suppose that $(f'', \overline{f''})$ is another such choice. Then there is a normal bordism $(F, \overline{F})$ with cylindrical target between $(f', \overline{f'})$ and $(f'', \overline{f''})$. Again by a finite sequence of surgery steps in the interior of the normal bordism we can change it into a highly connected normal bordism between $(f', \overline{f'})$ and $(f'', \overline{f''})$, see Theorem 8.121. Now Lemma 8.138 implies $\sigma(f', \overline{f'}) = \sigma(f'', \overline{f''})$. This finishes the construction of the surgery obstruction (8.143) for a normal map $(f, \overline{f})$ as long as we are in the universal covering case.

The same argument, namely making a normal bordism highly connected, shows that the surgery obstruction depends only on the normal bordism class with cylindrical target.

Finally we need to show that the vanishing of the surgery obstruction implies in the universal covering case that it can be transformed by finitely many surgery steps into a normal homotopy equivalence. We know already that we can convert $(f, \overline{f})$ by finitely many surgery steps into a highly connected normal map $(f', \overline{f'})$ and by definition of the surgery obstruction $\sigma(f', \overline{f'}) = \sigma(f, \overline{f})$ vanishes. Now the claim follows from Theorem 8.89. This finishes the proof of Theorem 8.112 in the universal covering case. □

8.7 The Intrinsic L-Group and Surgery Obstruction in Even Dimensions

So far we have defined the surgery obstruction only in the universal covering case. This requires in particular that the underlying spaces M and X are connected and the underlying map $f: M \to X$ induces an isomorphism $\pi_1(f, m): \pi_1(M, m) \to \pi_1(x, f(m))$ for one (and hence every) base point $m \in M$. Moreover, the obstruction was only well defined after a choice of model $\widetilde{p}_X: \widetilde{X} \to X$ of the universal covering and of base points in $x \in X$ and $\widetilde{x} \in \widetilde{X}$ with $\widetilde{p}(\widetilde{x}) = x$. We also want to get rid of these choices and this is more involved than expected.

Moreover, we have seen in Example 7.44 and Remark 7.45 that one cannot ignore the relation between the fundamental classes and the bundle data in the case where

$\pi_0(f)\colon \pi_0(M) \to \pi_0(X)$ is not necessarily bijective. We will take care of this problem by considering in the following intrinsic versions of the L-group and the intrinsic surgery obstruction.

We first discuss the base point issue for the obstruction groups.

Remark 8.144 (Base point issue for the obstruction groups) Consider a covariant functor $F\colon \text{GROUPS} \to \mathbb{Z}\text{-MOD}$ from the category of groups to the category of abelian groups, for instance $K_0(\mathbb{Z}G)$, $K_1(\mathbb{Z}G)$, or $\text{Wh}(G)$. We want to assign to X the value of F on the fundamental group, but this requires the choice of a base point. At first glance one may just make this choice and not worry because a different choice will certainly give isomorphic values. One problem is that a map $f\colon X \to Y$ may not respect the base points. Here one may use the fact that one can change f up to homotopy to obtain a base point preserving map. Now the problem occurs that within the homotopy class of such a map there can be many different pointed homotopy classes and the map induced on the fundamental groups depends on the pointed homotopy class and not only on the homotopy class.

In surgery theory another problem occurs. Consider two normal maps $(f_i, \overline{f_i})$ of orientable manifolds, for $i = 0, 1$. Assume that we have chosen the base points such that f_i preserves them. So we talk about the associated L-group $L_n(\mathbb{Z}\pi_1(X_i, x_i), w)$. But if we want to deal with bordism invariance, we have to consider a normal bordism $(F, \overline{F})$, and now we do not know which of the two possible choices of base points on the source of the bordism, namely, the one coming from $(f_0, \overline{f_0})$ or the one coming from $(f_1, \overline{f_1})$, we should take.

8.7.1 Conjugation Invariant Functors

The base point problem can easily be handled for a functor $F\colon \text{GROUPS} \to \mathbb{Z}\text{-MOD}$ under the assumption that F is *conjugation invariant*, i.e., for any group G and $g \in G$ the inner automorphism $c_g\colon G \to G$, $g' \mapsto gg'g^{-1}$ is sent under F to the identity. One can always assign to a space X its fundamental groupoid $\Pi(X)$. It has as objects the points in X and a morphism from $x \in X$ to $y \in X$ is a homotopy class $[v]$ relative endpoints of paths v in X from x to y. Composition is given by the concatenation of paths. A functor $F\colon \text{GROUPS} \to \mathbb{Z}\text{-MOD}$ induces a functor

$$F_X\colon \Pi(X) \to \mathbb{Z}\text{-MOD}$$

by sending an object x to $F(\pi_1(X, x))$ and a homotopy class relative endpoints of paths $[v]$ from x to y to $F(t_{[v]})$ for the group homomorphism

$$t_{[v]}\colon \pi_1(X, x) \to \pi_1(X, y), \quad [u] \mapsto [v^- * u * v]. \tag{8.145}$$

Now one can define an abelian group by the colimit

$$F_*(X) := \operatorname{colim}_{\Pi(X)} F_X. \tag{8.146}$$

8.7 The Intrinsic L-Group and Surgery Obstruction in Even Dimensions

An explicit model for $\mathrm{colim}_{\Pi(X)} F_X$ is $\coprod_{x \in X} F(x)/\sim$ for the equivalence relation $\sim$ given by identifying $a \in F(x)$ and $b \in F(y)$ if there exists a morphism $[v]\colon x \to y$ with $F([v])(a) = b$. For every $x \in X$ we have the obvious structure map

$$\psi_x \colon F(\pi_1(X, x)) \to F_*(X).$$

If $f\colon X \to Y$ is a map, it induces a functor $\Pi(f)\colon \Pi(X) \to \Pi(Y)$ and a natural transformation $F_X \to F_Y \circ \Pi(f)$ in the obvious way and hence a homomorphism

$$F_*(f)\colon F_*(X) \to F_*(Y). \tag{8.147}$$

Obviously $F_*(g) \circ F_*(f) = F_*(g \circ f)$ and $F(\mathrm{id}_X) = \mathrm{id}_{F[X]}$. Thus we obtain a covariant functor

$$F_* \colon \text{SPACES} \to \mathbb{Z}\text{-MOD}. \tag{8.148}$$

The following lemma shows that this definition has the desired properties. We leave its elementary proof to the reader.

Lemma 8.149 *Let $F\colon \text{GROUPS} \to \mathbb{Z}\text{-MOD}$ be a covariant functor that is conjugation invariant.*

(i) *Let X be a path connected space and $x \in X$. Then the structure map*

$$\psi_x \colon F(\pi_1(X, x)) \xrightarrow{\cong} F_*(X)$$

is a bijection;

(ii) *Let $u\colon X_0 \to X_1$ be a map of path connected spaces. Consider $x_i \in X_i$ for $i = 0, 1$. Choose a path v in X_1 from $u(x_0)$ to x_1. We have defined the isomorphism $t_{[v]}\colon \pi_1(X_1, u(x_0)) \xrightarrow{\cong} \pi_1(X_1, x_1)$ in (8.145). Then the diagram*

$$\begin{array}{ccc} F(\pi_1(X_0, x_0)) & \xrightarrow{\psi_{x_0}}_{\cong} & F_*(X_0) \\ {\scriptstyle F(t_{[v]} \circ \pi_1(u,x_0))}\Big\downarrow & & \Big\downarrow {\scriptstyle F_*(u)} \\ F(\pi_1(X_1, x_1)) & \xrightarrow[\psi_{x_1}]{\cong} & F_*(X_1) \end{array}$$

commutes;

(iii) *For a component $C \in \pi_0(X)$ let $i_C\colon C \to X$ be the inclusion. Then we obtain an isomorphism*

$$\bigoplus_{C \in \pi_0(X)} F_*(i_C)\colon F_*(C) \xrightarrow{\cong} F_*(X);$$

(iv) *Consider a map $u\colon X_0 \to X_1$ such that $\pi_0(u)\colon \pi_0(X_0) \to \pi_0(X_1)$ is bijective and the map $\pi_1(u, x_0)\colon \pi_1(X_0, x_0) \to \pi_1(X_1, u(x_0))$ is bijective for all base points $x_0 \in X_0$. Then*

$$F_*(u)\colon F_*(X_0) \xrightarrow{\cong} F_*(X_1)$$

is bijective;

(v) *If $u_i \colon X_0 \to X_1$ for $i = 0, 1$ are homotopic maps then we get an equality of homomorphisms*

$$F_*(u_0) = F_*(u_1) \colon F_*(X_0) \to F_*(X_1).$$

All of this applies to many interesting functors such as the ones sending a group G to $K_0(\mathbb{Z}G)$, $K_1(\mathbb{Z}G)$, and $\mathrm{Wh}(G)$, and also to $L_n(\mathbb{Z}G, w)$, provided that w is trivial. The case of non-trivial w is more complicated and is treated next. Moreover, even in the oriented case one has to do a more sophisticated construction for L-theory, see Remark 8.162.

Remark 8.150 (Groupoids) If we have a functor $F \colon \mathrm{GROUPOIDS} \to \mathbb{Z}\text{-MOD}$ which sends equivalences of groupoids to isomorphisms, then we can directly define

$$F_*(X) := F(\Pi(X))$$

and all the claims appearing in Lemma 8.149 carry directly over if one replaces ψ_x by the map $F(i_x) \colon F(\mathrm{aut}(x)) \to F(\Pi(X))$ induced by the obvious inclusion of groupoids $i_x \colon \mathrm{aut}(x) \to \Pi(X)$. Here we consider a group as a groupoid with precisely one object. Note that $\pi_1(X, x) = \mathrm{aut}(x)^{\mathrm{op}}$. Many functors $F \colon \mathrm{GROUPS} \to \mathbb{Z}\text{-MOD}$ which we are interested in, such as $K_n(\mathbb{Z}G)$, have natural extensions to functors $\widehat{F} \colon \mathrm{GROUPOIDS} \to \mathbb{Z}\text{-MOD}$, which send equivalences of groupoids to isomorphisms, see for instance [250, Chapter 10]. Note that L-theory is not such a functor since the algebraic L-groups depend on G plus a choice of an orientation homomorphism $w \colon G \to \{\pm 1\}$.

Exercise 8.151 Let $F \colon \mathrm{GROUPS} \to \mathbb{Z}\text{-MOD}$ be a covariant functor. Show that it is conjugation invariant if and only if it extends to a covariant functor $\widehat{F} \colon \mathrm{GROUPOIDS} \to \mathbb{Z}\text{-MOD}$ which sends equivalences of groupoids to isomorphisms.

8.7.2 Half-Conjugation Invariant Functors

Let $f \colon G_0 \to G_1$ be a group homomorphism. Suppose that we have orientation homomorphisms $w_0 \colon G_0 \to \{\pm 1\}$ and $w_1 \colon G_1 \to \{\pm 1\}$. Provided that $w_1 \circ f = w_0$ holds, we get an induced homomorphism of abelian groups

$$f_* \colon L_{2k}(\mathbb{Z}G_0, w_0) \to L_{2k}(\mathbb{Z}G_1, w_1)$$

since the ring homomorphism $\mathbb{Z}f \colon \mathbb{Z}G_0 \to \mathbb{Z}G_1$ is compatible with the w_0-twisted involution on the source and the w_1-twisted involution on the target, see Remark 8.93. Now consider a group G with an orientation homomorphism $w \colon G \to \{\pm 1\}$ and an element $g \in G$. Then $w \circ c_g = w$ and hence we get a homomorphism of abelian groups

8.7 The Intrinsic L-Group and Surgery Obstruction in Even Dimensions

$(c_g)_*\colon L_{2k}(\mathbb{Z}G, w) \to L_{2k}(\mathbb{Z}G, w)$. Next we see that $L_{2k}(\mathbb{Z}G, w)$ is not conjugation invariant in general, see also [395, Theorem 1]

Lemma 8.152 *The homomorphism $(c_g)_*\colon L_{2k}(\mathbb{Z}G, w) \to L_{2k}(\mathbb{Z}G, w)$ is equal to $w(g) \cdot \mathrm{id}_{L_{2k}(\mathbb{Z}G, w)}$.*

Proof. Consider a non-singular $(-1)^k$-quadratic form (P, ψ) over $\mathbb{Z}G$ representing an element $[(P, \psi)]$ in $L_{2k}(\mathbb{Z}G, w)$. According to Remark 8.93, the image of $[(P, \psi)]$ under $(c_g)_*$ is represented by the non-singular $(-1)^k$-quadratic form $((c_g)_*P, \overline{\mathrm{ind}_{c_g}}(\psi))$ over $\mathbb{Z}G$, where $(c_g)_*P$ is the induction of P with c_g, i.e., $\mathbb{Z}G \otimes_{\mathbb{Z}c_g} P$, and the homomorphism $\overline{\mathrm{ind}_{c_g}}\colon Q_{(-1)^k}(\mathbb{Z}G, w) \to Q_{(-1)^k}(\mathbb{Z}G, w)$ is induced by the homomorphism $\mathrm{ind}_{\mathbb{Z}c_g}\colon \hom_{\mathbb{Z}G}(P, P^*) \to \hom_{\mathbb{Z}G}(\mathbb{Z}G \otimes_{\mathbb{Z}c_g} P, (\mathbb{Z}G \otimes_{\mathbb{Z}c_g} P)^*)$, for which the following diagram commutes

$$\begin{array}{ccc}
\hom_{\mathbb{Z}G}(P, P^*) & \xrightarrow{T_{\mathbb{Z}G, w}} & \hom_{\mathbb{Z}G}(P, P^*) \\
{\scriptstyle \mathrm{ind}_{c_g}}\downarrow & & \downarrow{\scriptstyle \mathrm{ind}_{c_g}} \\
\hom_{\mathbb{Z}G}(\mathbb{Z}G \otimes_{\mathbb{Z}c_g} P, (\mathbb{Z}G \otimes_{\mathbb{Z}c_g} P)^*) & \xrightarrow{T_{\mathbb{Z}G, w}} & \hom_{\mathbb{Z}G}(\mathbb{Z}G \otimes_{\mathbb{Z}c_g} P, (\mathbb{Z}G \otimes_{\mathbb{Z}c_g} P)^*).
\end{array}$$

We have the $\mathbb{Z}G$-isomorphism $a\colon P \xrightarrow{\cong} (c_g)_*P := \mathbb{Z}G \otimes_{\mathbb{Z}c_g} P$ sending p to $g^{-1} \otimes p$, and its inverse sends $g' \otimes p$ to $g'gp$. Now a induces a map $Q_{(-1)^k}(a)\colon Q_{(-1)^k}((c_g)_*P) \to Q_{(-1)^k}(P)$ by sending the class of $\psi\colon P \to P^*$ to the class of $a^* \circ \psi \circ a$. We obtain an isomorphism of non-singular $(-1)^k$-quadratic forms

$$a\colon (P, Q_{(-1)^k}(a) \circ \overline{\mathrm{ind}_{c_g}}(\psi)) \to ((c_g)_*P, \overline{\mathrm{ind}_{c_g}}(\psi)).$$

This implies the equality in $L_{2k}(\mathbb{Z}G, w)$

$$(c_g)_*([(P, \psi)]) = [(P, Q_{(-1)^k}(a) \circ \overline{\mathrm{ind}_{c_g}}(\psi))].$$

The point is now the appearance of the sign $w(g)$ in the equality

$$Q_{(-1)^k}(a) \circ \overline{\mathrm{ind}_{c_g}}(\psi) = w(g) \cdot \psi. \tag{8.153}$$

This is proved as follows. We have the $\mathbb{Z}G$-isomorphism

$$\omega\colon \mathbb{Z}G \otimes_{\mathbb{Z}c_g} P^* \xrightarrow{\cong} (\mathbb{Z}G \otimes_{\mathbb{Z}c_g} P)^*$$

that sends $g' \otimes \alpha$ to the $\mathbb{Z}G$-homomorphism

$$\omega(g' \otimes \alpha)\colon \mathbb{Z}G \otimes_{\mathbb{Z}c_g} P \to \mathbb{Z}G, \quad (g'' \otimes p \mapsto g''g\alpha(p)g^{-1}(w_1(g') \cdot g'^{-1})).$$

If $\mu\colon P \to P^*$ is a representative of ψ, then a representative for $Q_{(-1)^k}(a) \circ \overline{\mathrm{ind}_{c_g}}(\psi)$ is given by the composite

$$P \xrightarrow{a} \mathbb{Z}G \otimes_{\mathbb{Z}c_g} P \xrightarrow{\mathrm{id}_{\mathbb{Z}G} \otimes \mu} \mathbb{Z}G \otimes_{\mathbb{Z}c_g} P^* \xrightarrow{\omega} (\mathbb{Z}G \otimes_{\mathbb{Z}c_g} P)^* \xrightarrow{a^*} P^*.$$

Now (8.153) follows by a direct calculation. We get from (8.153) in $L_{2k}(\mathbb{Z}G, w)$

$$(c_g)_*([(P, \psi)]) = [(P, Q_{(-1)^k}(a) \circ \overline{\mathrm{ind}_{c_g}}(\psi))] = [(P, w(g) \cdot \psi)] = w(g) \cdot [(P, \psi)].$$

This finishes the proof of Lemma 8.152. □

Lemma 8.152 shows that, in order to get a well-defined functor on spaces also for L-groups, we cannot proceed as in Subsection 8.7.1 as we have to take the transformation behaviour described in Lemma 8.152 into account.

Let GROUPS^w be the following category. Objects are pairs (G, w) consisting of a group G and a group homomorphism $w\colon G \to \{\pm 1\}$. A morphism $f\colon (G_0, w_0) \to (G_1, w_1)$ is a group homomorphism satisfying $w_1 \circ f = w_0$. Consider a covariant functor

$$F\colon \mathrm{GROUPS}^w \to \mathbb{Z}\text{-MOD}.$$

We call F *half-conjugation invariant* if for any object (G, w) and element $g \in G$ we have $F(c_g\colon G \to G) = w(g) \cdot \mathrm{id}_{F(G,w)}$.

Let $\mathcal{R}$ be the following category. Objects are pairs $(X, \mathcal{O}_X)$ consisting of a CW-complex X and an infinite cyclic coefficient system $\mathcal{O}_X$ over X, i.e., a covariant functor $\mathcal{O}_X\colon \Pi(X) \to \mathbb{Z}\text{-MOD}$ such that its value at each object is an infinite cyclic group. A morphism $(u, \overline{u})\colon (X_0, \mathcal{O}_{X_0}) \to (X_1, \mathcal{O}_{X_1})$ consists of a map $u\colon X_0 \to X_1$ and a natural transformation $\overline{u}\colon \mathcal{O}_{X_0} \to \mathcal{O}_{X_1} \circ \Pi(u)$. Note that $u^*\colon H^1(X_1; \mathbb{Z}/2) \to H^1(X_0; \mathbb{Z}/2)$ sends the first Stiefel–Whitney class $w_1(\mathcal{O}_{X_1})$ of $\mathcal{O}_{X_1}$, see (5.5), to the first Stiefel–Whitney class $w_1(\mathcal{O}_{X_0})$ of $\mathcal{O}_{X_0}$. Next we want to construct from a covariant functor $F\colon \mathrm{GROUPS}^w \to \mathbb{Z}\text{-MOD}$ a functor

$$F_*\colon \mathcal{R} \to \mathbb{Z}\text{-MOD}. \tag{8.154}$$

For this purpose we need the following variant of the fundamental groupoid $\Pi(X)$, the groupoid $\Pi(X, \mathcal{O}_X)$. Its objects are pairs $(x, \overline{x})$ consisting of a point $x \in X$ and a generator $\overline{x}$ of the infinite cyclic group $\mathcal{O}_X(x)$. A morphism $[v]\colon (x_0, \overline{x_0}) \to (x_1, \overline{x_1})$ is a morphism $[v]\colon x_0 \to x_1$ in $\Pi(X)$, i.e., a homotopy class relative endpoints $[v]$ of paths v from x_0 to x_1. The composition in $\Pi(X, \mathcal{O}_X)$ is given by the composition in $\Pi(X)$.

Consider a morphism $[v]\colon (x_0, \overline{x_0}) \to (x_1, \overline{x_1})$ in $\Pi(X, \mathcal{O}_X)$. Define

$$s([v]) \in \{\pm 1\} \tag{8.155}$$

by the equation $\mathcal{O}_X([v])(\overline{x_0}) = s([v]) \cdot \overline{x_1}$. Note that $s([v_1] \circ [v_0]) = s([v_1]) \cdot s([v_0])$ holds. Define a contravariant functor

$$F_{(X, \mathcal{O}_X)}\colon \Pi(X, \mathcal{O}_X) \to \mathbb{Z}\text{-MOD}$$

as follows. Let $w_1(\mathcal{O}_X) \in H^1(X; \mathbb{F}_2)$ be the first Stiefel–Whitney class of $\mathcal{O}_X$. It defines for every $x_0 \in X$ a homomorphism $w(x_0)\colon \pi_1(X, x_0) \to \{\pm 1\}$. Consider a morphism $[v]\colon (x_0, \overline{x_0}) \to (x_1, \overline{x_1})$ in $\Pi(X, \mathcal{O}_X)$. It is sent to the homomorphism of abelian groups

8.7 The Intrinsic L-Group and Surgery Obstruction in Even Dimensions 291

$$s([v]) \cdot F(t_{[v]}) \colon F(\pi_1(X, x_0), w(x_0)) \to F(\pi_1(X, x_1), w(x_1))$$

where the group homomorphism $t_{[v]} \colon \pi_1(X, x_0) \to \pi_1(X, x_1)$ has been defined in (8.145), and $s([v]) \in \{\pm 1\}$ has been defined in (8.155). So the object associated to $(x_0, \overline{x_0})$ only depends on x_0 but the morphisms depend on $\overline{x_0}$ and $\overline{x_1}$ and not only on x_0 and x_1. We define the abelian group

$$F_*(X, \mathcal{O}_X) := \operatorname{colim}_{\Pi(X, \mathcal{O}_X)} F_{(X, \mathcal{O}_X)}. \tag{8.156}$$

We leave it to the reader to check that a morphism $(u, \overline{u}) \colon (X_0, \mathcal{O}_{X_0}) \to (X_1, \mathcal{O}_{X_1})$ induces a functor $\Pi(u, \overline{u}) \colon \Pi(X_0, \mathcal{O}_{X_0}) \to \Pi(X_1, \mathcal{O}_{X_1})$, a natural transformation of functors $\Pi(X_0, \mathcal{O}_{X_0}) \to \mathbb{Z}\text{-MOD}$ from $F_{(X_0, \mathcal{O}_{X_0})}$ to $F_{(X_1, \mathcal{O}_{X_1})} \circ \Pi(u, \overline{u})$ and thus a homomorphism $F_*(u, \overline{u}) \colon F_*(X_0, \mathcal{O}_{X_0}) \to F_*(X_1, \mathcal{O}_{X_1})$. This finishes the construction of the covariant functor F_* announced in (8.154).

The following lemma shows that this definition has the desired properties. We call two morphisms $(u_m, \overline{u_m}) \colon (X, \mathcal{O}_X) \to (Y, \mathcal{O}_Y)$ homotopic if there exists a morphism $(h, \overline{h}) \colon (X \times [0, 1], \operatorname{pr}_X^* \mathcal{O}_X) \to (Y, \mathcal{O}_Y)$ for the projection $\operatorname{pr}_X \colon X \times [0, 1] \to X$ such that for $m = 0, 1$ its composite with the obvious morphism $(i_m, \overline{i_m}) \colon (X, \mathcal{O}_X) \to (X \times [0, 1], \operatorname{pr}_X^* \mathcal{O}_X)$, for which $i_m(x) = (x, m)$ holds for $x \in X$, is $(u_m, \overline{u_m})$.

Lemma 8.157 *Let $F \colon \text{GROUPS}^w \to \mathbb{Z}\text{-MOD}$ be a covariant functor that is half-conjugation invariant.*

(i) *Let $(X, \mathcal{O}_X)$ be an object in $\mathcal{R}$. Suppose that X is path connected. Consider an object $(x, \overline{x})$ in $\Pi(X, \mathcal{O}_X)$. Let $w \colon \pi_1(X, x) \to \{\pm 1\}$ be the group homomorphism given by $w_1(\mathcal{O}_X)$. Then the structure map*

$$\psi_{(x, \overline{x})} \colon F(\pi_1(X, x), w) \xrightarrow{\cong} F_*(X; \mathcal{O}_X)$$

is a bijection;

(ii) *Let $(u, \overline{u}) \colon (X_0, \mathcal{O}_{X_0}) \to (X_1, \mathcal{O}_{X_1})$ be a morphism in $\mathcal{R}$. Suppose that both X_0 and X_1 are path connected. Consider an object $(x_i, \overline{x_i})$ in $\Pi(X_i, \mathcal{O}_{X_i})$ for $i = 0, 1$. Choose a path v in X_1 from $u(x_0)$ to x_1. Let $w_i \colon \pi_1(X, x_i) \to \{\pm 1\}$ be the group homomorphism given by $w_1(\mathcal{O}_{X_i})$ for $i = 0, 1$. We have defined the isomorphism $t_{[v]} \colon \pi_1(Y, u(x_0)) \xrightarrow{\cong} \pi_1(X_1, x_1)$ in (8.145) and $s([v]) \in \{\pm 1\}$ in (8.155). Then the diagram*

$$\begin{CD}
F(\pi_1(X_0, x_0), w_0) @>{\psi_{(x_0, \overline{x_0})}}>{\cong}> F_*(X_0, \mathcal{O}_{X_0}) \\
@V{s([v]) \cdot F(t_{[v]} \circ \pi_1(u, x_0))}VV @VV{F_*(u, \overline{u})}V \\
F(\pi_1(X_1, x_1), w_1) @>>{\psi_{(x_1, \overline{x_1})}}^{\cong}> F_*(X_1, \mathcal{O}_{X_1})
\end{CD}$$

commutes;

(iii) *For a path component $C \in \pi_0(X)$, let $\mathcal{O}_C$ be the restriction of $\mathcal{O}_X$ to C. Let $(i_C, \overline{i_C}) \colon (C, \mathcal{O}_C) \to (X, \mathcal{O}_X)$ be the obvious morphism in $\mathcal{R}$. Then we obtain an* *isomorphism*

$$\bigoplus_{C\in\pi_0(X)} F_*(i_C,\overline{i_C})\colon \bigoplus_{C\in\pi_0(X)} F_*(C,O_C) \xrightarrow{\cong} F_*(X,O_X);$$

(iv) *Consider a morphism* $(u,\overline{u})\colon (X_0,O_{X_0}) \to (X_1,O_{X_1})$ *in* $\mathcal{R}$ *such that* $\pi_0(u)\colon \pi_0(X_0) \to \pi_0(X_1)$ *is bijective and the map* $\pi_1(u,x_0)\colon \pi_1(X_0,x_0) \to \pi_1(X_1,u(x_0))$ *is bijective for all base points* $x_0 \in X_0$. *Then*

$$F_*(u,\overline{u})\colon F_*(X_0,O_{X_0}) \xrightarrow{\cong} F_*(X_1,O_{X_1})$$

is bijective;

(v) *If* $(u_i,\overline{u_i})\colon (X_0,O_{X_0}) \to (X_1,O_{X_1})$ *for* $i = 0,1$ *are morphisms in* $\mathcal{R}$ *that are homotopic, then we get an equality of homomorphisms*

$$F_*(u_0,\overline{u_0}) = F_*(u_1,\overline{u_1})\colon F_*(X_0,O_{X_0}) \to F_*(X_1,O_{X_1}).$$

Proof. Assertion (i) follows from the observation that for an automorphism $[v]\colon (x,\overline{x}) \to (x,\overline{x})$ in $\Pi(X;O_X)$ we have $w([v]) = s([v])$ and hence half conjugation invariance of F implies $F_{(X,O_X)}([v]) = \mathrm{id}_{F_{(X,O_X)}(x,\overline{x})}$. The elementary remainder of the proof is left to the reader. $\square$

8.7.3 The Definition of Intrinsic L-Groups in Even Dimensions

By Lemma 8.152 we can apply the constructions of the previous section to the half-conjugation invariant functor

$$L_{2k}\colon \mathrm{GROUPS}^w \to \mathbb{Z}\text{-MOD}, \quad (G,w) \mapsto L_{2k}(\mathbb{Z}G,w).$$

Definition 8.158 (Intrinsic L-group in even dimensions) We denote by

$$(L_{2k})_*\colon \mathcal{R} \to \mathbb{Z}\text{-MOD}$$

the resulting functor and write

$$L_{2k}(\mathbb{Z}\Pi(X),O_X) := (L_{2k})_*(X,O_X).$$

The reason why we implemented the sign $s([v])$ of (8.155) is to enforce the following transformation behaviour.

Lemma 8.159 *Let* (X,O_X) *be an object in* $\mathcal{R}$. *Consider the morphism*

$$(\mathrm{id}_X, -\mathrm{id}_{O_X})\colon (X,O_X) \to (X,O_X)$$

in $\mathcal{R}$. *Then*

$$(L_{2k})_*(\mathrm{id}_X, -\mathrm{id}_{O_X}) = -\mathrm{id}_{(L_{2k})_*(X,O_X)}.$$

8.7 The Intrinsic L-Group and Surgery Obstruction in Even Dimensions

Proof. The functor $\Pi(\mathrm{id}_X, -\mathrm{id}_{O_X})\colon \Pi(X, O_X) \to \Pi(X, O_X)$ sends the object $(x, \overline{x})$ to the object $(x, -\overline{x})$ and the morphism $\mathrm{id}_x\colon (x, \overline{x}) \to (x, -\overline{x})$ in $\pi(X; O_X)$ given by the identity morphism $\mathrm{id}_x\colon x \to x$ in $\Pi(X)$ is mapped under the functor $L_{2k}\colon \mathrm{GROUPS}^w \to \mathbb{Z}\text{-MOD}$ to $-\mathrm{id}_{L_{2k}(\mathbb{Z}\pi_1(X, x), w_1(X))}$ because of the sign $s([c_x]) = -1$ of (8.155) for the constant path c_x. □

Remark 8.160 We conclude from Lemma 8.159 that the covariant functor $(L_{2k})_*\colon \mathcal{R} \to \mathbb{Z}\text{-MOD}$ does not factorise through the projection $\mathrm{pr}\colon \mathcal{R} \to \overline{\mathcal{R}}$, where $\overline{\mathcal{R}}$ is the category whose objects are pairs (X, v) consisting of a CW-complex X and an element $v \in H^1(X; \mathbb{Z}_2)$ and a morphism $f\colon (X_0, v_0) \to (X_1, v_1)$ is given by a map $f\colon X_0 \to X_1$ satisfying $f^*v_1 = v_0$, and the functor pr sends a morphism $(\overline{f}, f)\colon (X_0, O_{X_0}) \to (X_1, O_{X_1})$ to the morphism $f\colon (X_0, w_1(O_{X_1})) \to (X_1, w_1(O_{X_1}))$.

Note that here we follow [147, Paragraph after Theorem 3.1] where the bug in the construction by Wall [414, Chapter 9], who uses the quotient category $\overline{\mathcal{R}}$ of $\mathcal{R}$, is corrected.

Exercise 8.161 Check that the construction of the functor F_* of (8.154) above also makes sense if we replace the sign $s([v])$ of (8.155) by the trivial sign $s([v]) = 1$ everywhere, and we end up with a different covariant functor $F'_*\colon \mathcal{R} \to \mathbb{Z}\text{-MOD}$. Moreover, show that this functor factorises through the projection $\mathrm{pr}\colon \mathcal{R} \to \overline{\mathcal{R}}$ of Remark 8.160.

Remark 8.162 Note that Lemma 8.159 also holds in the case where $w_1(O_X)$ is trivial. In the orientable case, $L_n(\mathbb{Z}G)$ is a conjugation invariant functor and we get a functor $\mathrm{SPACES} \to \mathbb{Z}\text{-MOD}$ from Subsection 8.7.1 and thus by the obvious projection $\mathcal{R} \to \mathrm{SPACES}$ a functor $\mathcal{R} \to \mathbb{Z}\text{-MOD}$. But this functor does not have the transformation behaviour of Lemma 8.159 since it would send both $(\mathrm{id}_X, -\mathrm{id}_{O_X})$ and $(\mathrm{id}_X, \mathrm{id}_{O_X})$ to the identity. The specific transformation behaviour of the surgery obstruction for normal Γ-bordism will force us to implement the sign $s([v])$ of (8.155) in order to achieve the transformation behaviour of Lemma 8.159. We will see the same phenomenon when defining the surgery obstruction geometrically in Section 13.4.

8.7.4 The Intrinsic Even-Dimensional Surgery Obstruction

Next we make sense of the following definition.

Definition 8.163 (Intrinsic surgery obstruction in even dimension) Let $(X, O_X, [[X]])$ be a (not necessarily connected) w-oriented finite Poincaré complex in the sense of Definition 5.43 and Remark 5.45 of even dimension $n = 2k$ for $k \geq 3$. Consider a normal map with respect to the tangent bundle of degree one $(M, f, a, \xi, \overline{f}, o)$ with target $(X, O_X, [[X]])$ in the sense of Definition 7.22.

Its *intrinsic surgery obstruction in even dimension* is the element

$$\sigma(M, f, a, \xi, \overline{f}, o) \in L_{2k}(\mathbb{Z}\Pi(X), O_X)$$

defined below. We sometime abbreviate

$$(f, \overline{f}) := (M, f, a, \xi, \overline{f}, o);$$
$$\sigma(f, \overline{f}) = \sigma(M, f, a, \xi, \overline{f}, o).$$

We first explain the definition $\sigma(M, f, a, \xi, \overline{f}, o) \in L_{2k}(\mathbb{Z}\Pi(X), \mathcal{O}_X)$ in the special case where M and X are connected and $\pi_1(f, y) \colon \pi_1(M, y) \to \pi_1(X, f(y))$ is bijective for one (and hence all) base points $y \in M$. Make the choices (NBP), see Notation 8.5, namely,

- A model for the universal covering $p_X \colon \widetilde{X} \to X$;
- Base points $m \in M$, $\widetilde{x} \in \widetilde{X}$, and $x \in X$ and with $p_X(\widetilde{x}) = x = f(m)$;
- A generator $\overline{x} \in \mathcal{O}_X(x)$, or, equivalently, an isomorphism of abelian groups $o_X(x) \colon \mathcal{O}_X(x) \xrightarrow{\cong} \mathbb{Z}$.

We repeat the construction of Remark 8.4 to show that this defines a $\pi = \pi_1(X, x)$-normal map on the level of universal coverings. Let $w = w_1(X) \colon \pi \to \{\pm 1\}$ be the first Stiefel–Whitney class of $\mathcal{O}_X$. Consider the pullback of the π-covering $p_{\widetilde{X}}$

$$\begin{array}{ccc} \widetilde{M} & \xrightarrow{\widetilde{f}} & \widetilde{X} \\ {\scriptstyle p_M} \downarrow & & \downarrow {\scriptstyle p_X} \\ M & \xrightarrow{f} & X \end{array}$$

Let $\widetilde{m} \in \widetilde{M}$ be the point satisfying $p_M(\widetilde{m}) = m$ and $\widetilde{f}(\widetilde{m}) = \widetilde{x}$. Let $o_M(m) \colon \mathcal{O}_M(m) \xrightarrow{\cong} \mathbb{Z}$ be the composite $\mathcal{O}_M(m) \xrightarrow{\alpha(m)} \mathcal{O}_X(x) \xrightarrow{o_X(x)} \mathbb{Z}$ where $\alpha(m)$ is the evaluation of the morphisms α introduced in part (ii) appearing in Definition 7.22. Recall that we obtain from the various maps (5.13) isomorphisms

$$\beta_M \colon H_n^\pi(\mathbb{Z}^w \otimes C_*(\widetilde{M})) \xrightarrow{\cong} H_n(M; \mathcal{O}_M);$$
$$\beta_X \colon H_n^\pi(\mathbb{Z}^w \otimes C_*(\widetilde{X})) \xrightarrow{\cong} H_n(X; \mathcal{O}_X),$$

and the diagram

$$\begin{array}{ccc} H_n^\pi(\mathbb{Z}^w \otimes C_*(\widetilde{M})) & \xrightarrow{H_n^\pi(\widetilde{f}; \mathbb{Z}^w)} & H_n^\pi(\mathbb{Z}^w \otimes C_*(\widetilde{X})) \\ {\scriptstyle \beta_M} \downarrow {\scriptstyle \cong} & & {\scriptstyle \cong} \downarrow {\scriptstyle \beta_X} \\ H_n(M; \mathcal{O}_M) & \xrightarrow{H_n(f; \alpha)} & H_n(X; \mathcal{O}_X) \end{array}$$

commutes. Define $[M] \in H_n^\pi(\mathbb{Z}^w \otimes C_*(\widetilde{M}))$ to be the image of the intrinsic fundamental class $[[M]]$ of Definition 5.29 under β_M^{-1} and $[X]$ to be the image of $[M]$ under $H_n^\pi(\widetilde{f}; \mathbb{Z}^w)$. Then $[M]$ and $[X]$ are generators and we obtain a normal π-map

8.7 The Intrinsic L-Group and Surgery Obstruction in Even Dimensions

of degree one $(M, [M], f, \widetilde{f}, a, \xi, \overline{f})$ in the sense of Definition 8.3 in the universal covering case, see Notation 8.34. For this we have defined an element

$$\sigma(M, [M], f, \widetilde{f}, a, \xi, \overline{f}) \in L_{2k}(\mathbb{Z}\pi_1(X, x), w_1(X)),$$

see Theorem 8.112 (i). Recall that this involves the choice of an orientation of $T_m M$ compatible with the fundamental class $[M] \in H_n(\mathbb{Z}^w \otimes_{\mathbb{Z}\pi} C_*(\widetilde{M}))$, see Definition 8.16. It turns out from the construction that this orientation is just the one obtained by the isomorphism

$$O_M(m) \xrightarrow{\cong} O_{TM \oplus \mathbb{R}^a}(m) \xrightarrow{O_{\overline{f}_m}} O_\xi(\xi) \xrightarrow{o(x)} O_X(x) \xrightarrow{o_X(x)} \mathbb{Z},$$

where the first isomorphism comes from Lemma 5.23 (i), the second comes from the bundle data, the third comes from the isomorphism $o\colon O_\xi \xrightarrow{\cong} O_X$, and the last one is part of the choices.

Let $\psi_{x,\overline{x}}\colon L_{2k}(\mathbb{Z}\pi_1(X, x), w_1(X)) \to L_{2k}(\mathbb{Z}\Pi(X), O_X)$ be the structure map. We define

$$\sigma(M, f, a, \xi, \overline{f}, o) \in L_{2k}(\mathbb{Z}\Pi(X), O_X)$$

to be the image of

$$\sigma(M, [M], f, \widetilde{f}, a, \xi, \overline{f}) \in L_{2k}(\mathbb{Z}\pi_1(X, x), w_1(X))$$

under $\psi_{x,\overline{x}}$.

Next we explain why this definition is independent of the choices we made. Suppose that we have made a different set of choices

- A model for the universal covering $p'_X\colon \widetilde{X'} \to X$;
- Base points $\widetilde{x}' \in \widetilde{X}'$, $x' \in X$ and $m' \in M$ with $p_X(\widetilde{x}') = x' = f(m')$;
- A generator $\overline{x}' \in O_X(x')$, or, equivalently, an isomorphism $o_X(x')\colon O_X(x') \xrightarrow{\cong} \mathbb{Z}$.

Choose a homotopy $h\colon M \times [0, 1] \to M$ such that $h(y, 0) = y$ holds for $y \in M$ and $h(m, 1) = m'$. Let $a\colon M \to M$ be the map sending y to $h(y, 1)$. Note that $a(m) = m'$ holds. Let v be the path in X from x to x' given by $v(t) = f \circ h(x, t)$ for $t \in [0, 1]$. Next choose a homotopy $h'\colon X \times [0, 1] \to X$ such that $h(z, 0) = z$ holds for $z \in X$, $h(x, t) = v(t)$ for $t \in [0, 1]$ and $h(x, 1) = x'$. Let $b\colon X \to X$ be the map sending z to $h(z, 1)$ for $z \in X$. We have $b(x) = x'$. Thus we obtain the following diagram of pointed spaces, which commutes up to pointed homotopy, see [431, Section III.1]

$$\begin{array}{ccc} (M, m) & \xrightarrow{a} & (M, m') \\ {\scriptstyle f}\downarrow & & \downarrow{\scriptstyle f'} \\ (X, x) & \xrightarrow{b} & (X, x'). \end{array}$$

The map b induces the isomorphism $t_{[v]}\colon \pi_1(X, x) \xrightarrow{\cong} \pi_1(X, x')$. Now by taking lifts we obtain the following diagram

$$\begin{CD} (\widetilde{M}, \widetilde{m}) @>\widetilde{a}>> (\widetilde{M}', \widetilde{m}') \\ @V\widetilde{f}VV @VV\widetilde{f}'V \\ (\widetilde{X}, \widetilde{x}) @>>\widetilde{b}> (\widetilde{X}', \widetilde{x}') \end{CD}$$

where $\widetilde{f}$ is a $\pi_1(X, x)$-map, $\widetilde{f}'$ is a $\pi_1(X, x')$-map, the horizontal maps are $t_{[v]}$-equivariant and the diagram commutes up to $t_{[v]}$-equivariant homotopy. Hence it induces a $\mathbb{Z}\pi_1(X, x')$-isomorphism

$$(t_{[v]})_* \colon K_k(\widetilde{M}) \xrightarrow{\cong} K_k(\widetilde{M}'). \tag{8.164}$$

Let $s([v]) \in \{\pm 1\}$ be determined by the equation $O_X(t_{[v]})(\overline{x}) = s([v]) \cdot \overline{x'}$. Then the following diagram of $\mathbb{Z}$-isomorphisms commutes

$$\begin{CD} O_X(x) @>O_X(t_{[v]})>> O_X(x') \\ @Vo_X(x)VV @VVo_X(x')V \\ \mathbb{Z} @>>s([v])\cdot\mathrm{id}_{\mathbb{Z}}> \mathbb{Z}. \end{CD} \tag{8.165}$$

The following diagram commutes by Lemma 8.157 (ii)

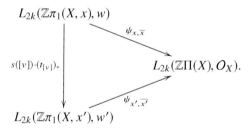

Hence it suffices to show for the two non-singular $(-1)^k$-quadratic forms $(K_k(\widetilde{M}), s, t)$ and $(K_k(\widetilde{M}'), s', t')$ whose classes represent

$$\sigma(M, [M], f, \widetilde{f}, a, \xi, \overline{f}) \in L_{2k}(\mathbb{Z}\pi_1(X, x), w_1(X));$$
$$\sigma(M, [M'], f, \widetilde{f}', a, \xi, \overline{f}) \in L_{2k}(\mathbb{Z}\pi_1(X, x'), w_1(X)),$$

that $(t_{[v]})_* \colon L_{2k}(\mathbb{Z}\pi_1(X, x), w) \to L_{2k}(\mathbb{Z}\pi_1(X, x'), w')$ sends $[(K_k(\widetilde{M}), s, t)]$ to $s([v]) \cdot [(K_k(\widetilde{M}'), s', t')]$. We have already constructed a $\mathbb{Z}\pi_1(X, x')$-isomorphism $(t_{[v]})_* \colon K_k(\widetilde{M}) \xrightarrow{\cong} K_k(\widetilde{M}')$ in (8.164). It remains to show that under this identification $s' = s([v]) \cdot s$ and $t' = s([v]) \cdot t$ hold. This follows for the symmetric structures s and s' from the appearance of the sign $\epsilon = s([v])$ in (5.16)

8.7 The Intrinsic L-Group and Surgery Obstruction in Even Dimensions 297

as it implies that the fundamental class $[M] \in H_n(\mathbb{Z}^w \otimes_{\mathbb{Z}\Pi_1(X,x)} C_*(\widetilde{M}))$ is mapped to $s([v]) \cdot [M'] \in H_n(\mathbb{Z}^w \otimes_{\mathbb{Z}\Pi_1(X,x')} C_*(\widetilde{M'}))$ under the isomorphism $H_n(\mathbb{Z}^w \otimes_{\mathbb{Z}\pi_1(X,x)} C_*(\widetilde{M})) \xrightarrow{\cong} H_n(\mathbb{Z}^w \otimes_{\mathbb{Z}\pi_1(X,x')} C_*(\widetilde{M'}))$ induced by $\widetilde{a}$. The claim for the quadratic structures t and t' follows from the appearance of $s([v])$ in the diagram (8.165) since replacing $o_X(x') \colon O_X(x') \xrightarrow{\cong} \mathbb{Z}$ by $-o_X(x') \colon O_X(x') \xrightarrow{\cong} \mathbb{Z}$ has the effect that for each self-intersection its contribution to the element in $Q_{(-1)^k}(\mathbb{Z}\pi_1(X, x'), w')$ given by a sign and an appropriate path is changed by multiplying the sign by -1 and leaving the path fixed, see Subsection 8.3.2.

This finishes the construction of the intrinsic surgery obstruction announced in Definition 8.163 in the case where M and X are connected and

$$\pi_1(f, m) \colon \pi_1(M, m) \to \pi_1(X, f(m))$$

is bijective for one (and hence all) base points $m \in M$.

Next we consider the special case where X is connected. Choose a normal map $(M', f', a', \xi, \overline{f'}, o)$ with target $(X, O_X, [[X]])$ such that M' is connected, $\pi_1(f')$ is an isomorphism, and $(M, [M], f, \widehat{f}, a, \xi, \overline{f})$ and $(M', f', a', \xi, \overline{f'}, o)$ are normal bordant with cylindrical ends. One can construct such $(M', f', a', \xi, \overline{f'}, o)$ by finitely many surgery steps on the normal map $(M, [M], f, \widehat{f}, a, \xi, \overline{f})$, see Theorem 7.41. We want to define the surgery obstruction $\sigma(M, f, a, \xi, \overline{f}, o)$ by $\sigma(M', f', a', \xi, \overline{f'}, o)$.

We have to show that this is independent of the choice of the surgery steps. Suppose that $(M'', f'', a'', \xi, \overline{f''}, o)$ is another such normal map. Then there exists a normal bordism $(W, I, F, \Xi, \overline{F}, O)$ with cylindrical ends between $(M', f', a, \xi, \overline{f'}, o)$ and $(M'', f'', a'', \xi, \overline{f''}, o)$ and we can arrange by surgery on the interior that W is connected and $\pi_1(F)$ an isomorphism, since the argument in the proof of Theorem 7.41 carries over to bordisms. Now it follows from unravelling the definitions and Theorem 8.112 (i) that $\sigma(M'', f'', a'', \xi, \overline{f''}, o)$ and $\sigma(M', f', a', \xi, \overline{f'}, o)$ agree.

The general case is now obtained by considering each path component C of X separately and using the isomorphism

$$\bigoplus_{C \in \pi_0(X)} L_{2k}(\mathbb{Z}\Pi(C); O_C) \xrightarrow{\cong} L_{2k}(\mathbb{Z}\Pi(X), O_X)$$

coming from Theorem 8.157 (iii).

Note that the intrinsic algebraic L-group $L_{2k}(\mathbb{Z}\Pi(X), O_X)$ is independent of $[[X]]$ whereas the next exercise shows that the intrinsic surgery obstruction $\sigma(M, f, a, \xi, \overline{f}, o)$ does depend on $[[X]]$.

Exercise 8.166 Consider a normal map with respect to the tangent bundle of degree one $(M, f, a, \xi, \overline{f}, o)$ of even dimension $n = 2k$ with target $(X, O_X, [[X]])$ for $k \geq 3$ in the sense of Definition 7.22.

(i) Then $(M, f, a, \xi, \overline{f}, -o)$ is a normal map with respect to the tangent bundle of degree one with target $(X, O_X, -[[X]])$. Show for its intrinsic surgery obstruction in $L_{2k}(\mathbb{Z}\Pi(X), w)$ the equality

$$\sigma(M, f, a, \xi, \overline{f}, -o) = -\sigma(M, f, a, \xi, \overline{f}, o);$$

(ii) Let θ be a bundle automorphism of ξ covering the identity. Suppose that for every $x \in X$ the linear automorphism θ_x of the fibre over x is orientation preserving or orientation reversing respectively, and let s be 1 or -1 respectively. Then $(M, f, a, \xi, \theta \circ \overline{f}, s \cdot o)$ is a normal map with respect to the tangent bundle of degree one with target $(X, O_X, s \cdot [[X]])$.
Show
$$\sigma(M, f, a, \xi, \theta \circ \overline{f}, s \cdot o) = \sigma(M, f, a, \xi, \overline{f}, o).$$

Theorem 8.167 (Intrinsic surgery obstruction in even dimensions) *Let $(X, O_X, [[X]])$ be a (not necessarily connected) w-oriented finite n-dimen-sional Poincaré complex in the sense of Definition 5.43 and Remark 5.45. Consider a normal map with respect to the tangent bundle of degree one $(M, f, a, \xi, \overline{f}, o)$ of even dimension $n = 2k$ with target $(X, O_X, [[X]])$ for $k \geq 3$ in the sense of Definition 7.22. Then:*

(i) *Its intrinsic surgery obstruction*
$$\sigma(f, \overline{f}) = \sigma(M, f, a, \xi, \overline{f}, o) \in L_{2k}(\mathbb{Z}\Pi(X), O_X)$$
depends only on the normal bordism with cylindrical ends class of $(\overline{f}, f)$;
(ii) *We have $\sigma(\overline{f}, f) = 0$ if and only if we can do a finite number of surgery steps to obtain a normal homotopy equivalence.*

Proof. Obviously it suffices to treat the case where X is connected. By finitely many surgery steps we can convert $(M, f, a, \xi, \overline{f}, o)$ to $(M', f', a, \xi, \overline{f}', o)$ such that such that M' is connected and $\pi_1(f')$ is an isomorphism, see Theorem 7.41. Above we have assigned to $(M', f', a, \xi, \overline{f}', o)$ after some choices a normal π-map of degree one $(M', [M'], f', \overline{f}', a, \xi, \overline{f}')$ in the sense of Definition 8.3 in the universal covering case, see 8.34 and defined
$$\sigma(M, f, a, \xi, \overline{f}, o) \in L_{2k}(\mathbb{Z}\Pi(X), O_X)$$
to be the image of
$$\sigma(M', [M'], f', \widetilde{f}', a, \xi, \overline{f}') \in L_{2k}(\mathbb{Z}\pi_1(X, x), w_1(X))$$
under $\psi_{x,\overline{x}}$. Since $\psi_{x,\overline{x}}$ is bijective, $\sigma(M, f, a, \xi, \overline{f}, o) \in L_{2k}(\mathbb{Z}\Pi(X), O_X)$ vanishes if and only if $\sigma(M', [M'], f', \widetilde{f}', a, \xi, \overline{f}')$ vanishes. Obviously, in finitely many surgery steps, we can turn $(M, f, a, \xi, \overline{f}, o)$ into a normal homotopy equivalence if, in finitely many surgery steps, we can turn $(M', f', a, \xi, \overline{f}', o)$ into a normal homotopy equivalence. Now the claim follows from Theorem 8.112 (ii). □

8.7.5 Bordism Invariance in Even Dimensions for a Not Necessarily Cylindrical Target

For $m = 0, 1$, let $(X_m, O_{X_m}, [[X_m]])$ be a w-oriented finite $2k$-dimensional Poincaré complex for $k \geq 3$ and let $(M_m, f_m, a_m, \xi_m, \overline{f}_m, o_m)$ be a normal map with respect to the tangent bundle of degree one with target $(X_m, O_{X_m}, [[X_m]])$ in the sense of Definition 7.22. Let $(W, F, b, \Xi, \overline{F}, O)$ be a normal bordism of normal maps with respect to the tangent bundle of degree one from $(M_0, f_0, a_0, \xi_0, \overline{f}_0, o_0)$ to $(M_1, f_1, a_1, \xi_1, \overline{f}_1, o_1)$ in the sense of Definition 7.24. Let $(Y, \partial Y, O_Y, [[Y]])$ be the Poincaré pair appearing in Definition 7.24. Moreover, we get from Definition 7.24 isomorphisms of CW-complexes $u_m \colon X_m \to \partial_m Y$, the inclusions $l_m \colon \partial_m Y \to Y$, and isomorphisms $O_{l_m \circ v_m} \colon O_{X_m} \to (l_m \circ v_m)^* O_Y$ over X_m of infinite cyclic local coefficient theorems for $m = 0, 1$. Thus we obtain for $m = 0, 1$ a morphism $(l_m \circ v_m, O_{l_m \circ v_m}) \colon (X_m, O_{Y_m}) \to (Y, O_Y)$ in $\mathcal{R}$ and hence a homomorphism of abelian groups, see Definition 8.158,

$$L_{2k}(l_m \circ v_m, O_{l_m \circ v_m}) \colon L_{2k}(\mathbb{Z}\Pi(X_m), O_{X_m}) \to L_{2k}(\mathbb{Z}\Pi(Y), O_Y).$$

Recall that we have defined the intrinsic surgery obstruction, see Theorem 8.167

$$\sigma(M_m, f_m, a_m, \xi_m, \overline{f}_m, o_m) \in L_{2k}(\mathbb{Z}\Pi(X_m), O_{X_m}).$$

Theorem 8.168 (Bordism invariance of the intrinsic surgery obstruction in even dimensions) *In $L_{2k}(\mathbb{Z}\Pi(Y), O_Y)$ we get the equality*

$$L_{2k}(l_0 \circ v_0, O_{l_0 \circ v_0})\bigl(\sigma(M_0, f_0, a_0, \xi_0, \overline{f}_0, o_0)\bigr)$$
$$= L_{2k}(l_1 \circ v_1, O_{l_1 \circ v_1})\bigl(\sigma(M_1, f_1, a_1, \xi_1, \overline{f}_1, o_1)\bigr).$$

Note that in Theorem 8.168 we do *not* require any of the spaces to be connected, nor do we require any of the maps induced on the fundamental groups to be injective or bijective, or the normal bordism to have cylindrical ends.

We will not give the proof of Theorem 8.168, which is more or less a variation of the proof of Theorem 8.167. One reason for omitting the proof is that we will later introduce a chain complex version of the surgery obstruction which will make the proof rather obvious. This material is discussed in Chapter 15 and the relevant statement is in Section 15.7.

If the normal bordism has cylindrical ends, then Theorem 8.168 boils down to the normal bordism invariance appearing in Theorem 8.167.

Remark 8.169 (Bordism invariance for nullbordism) The complications that appear concern the book-keeping of the various fundamental groups of the components and the behaviour of the obstructions when choosing different base points. Note that the proof of Theorem 8.167 proceeds in two stages. We first prove Theorem 8.112, which treats the case when X is connected and a base point $x \in X$ is fixed in the normal Γ-setting. In the second stage we take care of the behaviour of the resulting obstruction under the change of base points and this allows us to pass to the intrinsic setting. This is good for pedagogical reasons since we keep the two issues separated. The proof of the general bordism invariance for varying target would require us to incorporate the base point changes into the first stage, and hence into the proof of Theorem 8.112, but we think that this would be too complicated for the first-time reader.

On the other hand, there are special cases of bordism invariance that are more general than the case with cylindrical ends where the proof does go through without much additional effort. Consider for example the case when M_0 is connected, X_0 is connected, X_1 is empty (hence M_1 is also empty), and Y is connected. Then Theorem 8.168 says that the image of $\sigma(M_0, f_0, a_0, \xi_0, \overline{f}_0, o_0)$ is zero in $L_{2k}(\mathbb{Z}\Pi(Y), O_Y)$. Note that in this case we can choose a base point $x_0 \in X_0$ that can serve also as a base point in Y. Then we are in a setting that is close to that of Theorem 8.112. In fact, its proof goes through almost word for word, adjusting only for the fact that X_1 and hence M_1 are empty. There are two places where one needs to make a slightly bigger change. The first is the passage concerning the homotopy H after equation (8.135). The homotopy H here is replaced by a map $H: (D^{k+1}, S^k) \to (W, \partial W)$, and the rest of the proof should be modified with this in mind. If we later vary the base point in Y, no sign change appears since the obstruction is zero. The second place requiring change is the handle subtraction argument in the proof of Lemma 8.138. There one modifies the cobordism W between M_0 and M_1 in such a way that M_1 is also modified by a connected sum with $S^k \times S^k$, which is now not possible since M_1 is empty. But this change can equally well be made in M_0. Since a connected sum with $S^k \times S^k$ only adds the standard hyperbolic form to the surgery kernel of $(f_0, \overline{f}_0)$ the surgery obstruction is not affected.

Note that the proof of Theorem 8.112 not only gives that the image of $\sigma(M_0, f_0, a_0, \xi_0, \overline{f}_0, o_0)$ is zero in $L_{2k}(\mathbb{Z}\Pi(Y), O_Y)$, but it also gives the reason for it being zero, namely, it shows that the image of $K_{k+1}(\widetilde{W}, \widetilde{\partial W})$ in $K_k(\widetilde{M}_0)$ is a lagrangian. By the last lines in the paragraph above, this might be at the cost of modifying M_0 by a connected sum with $S^k \times S^k$. This special case will indeed be used in Chapter 9.

Exercise 8.170 Suppose in the situation of Theorem 8.168 that for $m = 0, 1$ the inclusions $\partial_m Y \to Y$ induce bijections $\pi_0(\partial_m Y) \to \pi_0(Y)$ and for every base point $y \in \partial_m Y$ bijections $\pi_1(\partial_m Y, y) \to \pi_1(Y, y)$. Show that then $\sigma(M_0, f_0, a_0, \xi_0, \overline{f}_0, o_0)$ vanishes if and only if $\sigma(M_1, f_1, a_1, \xi_1, \overline{f}_1, o_1)$ vanishes.

8.7.6 The Simply Connected Case in Even Dimensions

We are now in position to give a complete answer to the solution to the surgery problem in the even-dimensional simply connected case.

Since this answer in dimension $n = 4k$ is stated in terms of the signatures of the source and target of a degree one normal map $(f, \overline{f}): M \to X$, we first recall the computation of the signature of M via the Hirzebruch L-class of its tangent bundle.

8.7 The Intrinsic L-Group and Surgery Obstruction in Even Dimensions

Let ξ be a (real) vector bundle over a base space Y. Recall that the *Pontrjagin classes* of ξ, see [308, §15], are characteristic classes

$$p_j(\xi) \in H^{4j}(Y;\mathbb{Z})$$

satisfying $p_j(\xi) = p_j(\xi \oplus \mathbb{R})$. We shall also write $p_j(\xi) \in H^{4j}(Y;\mathbb{Q})$ for the image of the jth Pontrjagin class under the natural map $H^{4j}(Y;\mathbb{Z}) \to H^{4j}(Y;\mathbb{Q})$. Recall also that the *L-class* $\mathcal{L}(\xi)$ of ξ is a collection of homogeneous rational polynomials in the Pontrjagin classes of ξ

$$\mathcal{L}_j(\xi) = \mathcal{L}_j\big(p_1(\xi), p_2(\xi), p_3(\xi), \ldots, p_j(\xi)\big) \in H^{4j}(Y;\mathbb{Q}).$$

For the precise definition of the L-class, we refer the reader to [308, §19] but we at least mention the first three L-polynomials:

$$\mathcal{L}_1(\xi) = \frac{1}{3} \cdot p_1(\xi);$$

$$\mathcal{L}_2(\xi) = \frac{1}{45} \cdot \left(7 \cdot p_2(\xi) - p_1(\xi)^2\right);$$

$$\mathcal{L}_3(\xi) = \frac{1}{945} \cdot \left(62 \cdot p_3(\xi) - 13 \cdot p_1(\xi) \cup p_2(\xi) + 2 \cdot p_1(\xi)^3\right).$$

An important feature of the L-class is that in each L-polynomial $\mathcal{L}_k$, the coefficient s_k of p_k is non-zero. A fundamental fact about the L-class is the signature theorem of Hirzebruch, which we state next. Let M be a closed oriented smooth $4k$-manifold with tangent bundle TM and Pontrjagin classes

$$p_j(M) := p_j(TM) \in H^{4j}(M;\mathbb{Z}),$$

and recall that $\mathrm{sign}(M)$ is the signature of the intersection form of M as defined in (5.70).

Theorem 8.171 (Hirzebruch Signature Theorem) *If M is a closed smooth oriented $4j$-manifold with fundamental class $[M]$, then the signature of M is given by*

$$\mathrm{sign}(M) = \langle \mathcal{L}_j(p_1(M), \ldots, p_j(M)), [M] \rangle.$$

See [308, Theorem 19.4] for the proof.

Example 8.172 By [308, Example 15.6 on page 177] the total Pontrjagin class of complex $2j$-dimensional projective space $\mathbb{CP}^{2j}$ is given by

$$p(\mathbb{CP}^{2j}) = 1 + p_1(\mathbb{CP}^{2j}) + \cdots + p_j(\mathbb{CP}^{2j}) = (1 + a^2)^{2j+1} \in H^{4*}(\mathbb{CP}^{2j}),$$

where $a \in H^2(\mathbb{CP}^{2j};\mathbb{Z})$ is a generator. Consequently the signature theorem affirms that

$$\operatorname{sign}(\mathbb{CP}^2) = \langle \frac{1}{3} \cdot 3a^2, [\mathbb{CP}^2] \rangle = 1;$$

$$\operatorname{sign}(\mathbb{CP}^4) = \langle \frac{1}{45} \cdot (7 \cdot 10 - 25) \, a^4, [\mathbb{CP}^4] \rangle = 1;$$

$$\operatorname{sign}(\mathbb{CP}^6) = \langle \frac{1}{945} \cdot (62 \cdot 35 - 13 \cdot 7 \cdot 21 + 2 \cdot 343) \, a^6, [\mathbb{CP}^6] \rangle = 1.$$

We now state the solution to the simply connected surgery problem, see Problem 4.1 and Problem 7.42. Recall that for a simply connected finite Poincaré complex X of dimension $n = 4k$ for $k \geq 1$, we get after a choice of an orientation a preferred identification of the intrinsic L-group of $(X, \mathcal{O}_X)$ with $L_n(\mathbb{Z})$ and hence via the signature homomorphism with $8 \cdot \mathbb{Z}$. Since $L_n(\mathbb{Z})$ is isomorphic to $\mathbb{Z}/2$ if $n = 4k+2$, we get natural identifications of the intrinsic L-groups with $\mathbb{Z}/2$ if $n = 4k+2$ without choosing an orientation. Note that Theorem 8.173 also holds in dimension 4 because of Remark 8.30.

Theorem 8.173 (Solution to the simply connected surgery problem in even dimensions)

(i) Let $(f, \overline{f}) \colon M \to X$ be a degree one normal map from a closed oriented manifold M to a simply connected oriented finite Poincaré complex X of dimension $n = 4k$ for $k \geq 1$. Then

$$\sigma(f, \overline{f}) = \operatorname{sign}(M) - \operatorname{sign}(X) \in 8 \cdot \mathbb{Z} \cong L_{4k}(\mathbb{Z}).$$

Consequently, we can change M and f leaving X fixed by finitely many surgery steps to get a normal homotopy equivalence $(g, \overline{g})$ from a closed manifold N to X if and only if $\operatorname{sign}(M) = \operatorname{sign}(X)$;

(ii) Let $(f, \overline{f})$ be a degree one normal map from a closed manifold M to a simply connected finite Poincaré complex X of dimension $n = 4k+2 \geq 6$. Then we can change M and f leaving X fixed by finitely many surgery steps to get a normal homotopy equivalence $(g, \overline{g})$ from a closed manifold N to X if and only if the Arf invariant of the surgery obstruction, which takes values in $\mathbb{Z}/2$, vanishes;

(iii) Let X be a simply connected finite Poincaré complex of dimension $n = 4k$ for $k \geq 1$. Then X is homotopy equivalent to a closed manifold if and only if the Spivak normal fibration has a reduction to a vector bundle $\xi \colon E \to X$, i.e., such that for one (and hence all choices) of a fundamental class $[X] \in H_n(X; \mathbb{Z})$

$$\langle \mathcal{L}(\xi)^{-1}, [X] \rangle = \operatorname{sign}(X)$$

holds, where $\mathcal{L}(\xi)$ is the L-class of the bundle ξ;

(iv) Let X be a simply connected finite Poincaré complex of dimension $n = 4k+2 \geq 6$. Then X is homotopy equivalent to a closed manifold if and only if the Spivak normal fibration has a reduction to a vector bundle $\xi \colon E \to X$ such that the Arf invariant of the associated surgery problem, which takes values in $\mathbb{Z}/2$, vanishes.

8.7 The Intrinsic L-Group and Surgery Obstruction in Even Dimensions   303

Proof. (i) Because of Theorem 8.99 and Theorem 8.112 we have to show for a $2k$-connected normal map of degree one $f \colon M \to X$ from a closed simply connected oriented manifold M of dimension $n = 4k$ to a simply connected oriented Poincaré complex X of dimension $n = 4k$ that for the non-singular symmetric bilinear form $\mathbb{R} \otimes_{\mathbb{Z}} (K_{2k}(M), \lambda)$ induced by the intersection pairing we get

$$\mathrm{sign}\,(\mathbb{R} \otimes_{\mathbb{Z}} (K_{2k}(M), \lambda)) = \mathrm{sign}(M) - \mathrm{sign}(X).$$

This follows from elementary considerations about signatures and from the commutative diagram, see Lemma 8.39 (ii)

$$\begin{array}{ccccccccc}
0 & \longleftarrow & K^{2k}(M) & \longleftarrow & H^{2k}(M) & \stackrel{f^*}{\longleftarrow} & H^{2k}(X) & \longleftarrow & 0 \\
 & & \cong \downarrow {-\cap[M]} & & \cong \downarrow {-\cap[M]} & & \cong \downarrow {-\cap[X]} & & \\
0 & \longrightarrow & K_{2k}(M) & \longrightarrow & H_{2k}(M) & \stackrel{f_*}{\longrightarrow} & H_{2k}(X) & \longrightarrow & 0.
\end{array}$$

(ii) follows from Theorem 8.111 and Theorem 8.112.

(iii) follows from Theorem 7.19, assertion (i), the equality

$$\mathcal{L}(TM) \cdot \mathcal{L}(\nu(M) \oplus \underline{\mathbb{R}}^a) = \mathcal{L}(TM) \oplus \nu(M) \oplus \underline{\mathbb{R}}^a) = \mathcal{L}(\underline{\mathbb{R}}^b) = 1,$$

and the Hirzebruch signature formula that implies for a surgery problem $(f, \overline{f}) \colon \nu(M) \to \xi$ covering $f \colon M \to X$, which is obtained from a reduction ξ of the Spivak normal fibration of X,

$$\mathrm{sign}(M) = \langle \mathcal{L}(TM), [M] \rangle = \langle \mathcal{L}(\nu(M) \oplus \underline{\mathbb{R}}^a)^{-1}, [M] \rangle = \langle \mathcal{L}(\xi)^{-1}, [X] \rangle.$$

(iv) follows from Theorem 7.19 and assertion (ii). □

Exercise 8.174 Let $(f, \overline{f}) \colon M \to X$ be a normal map from a closed oriented manifold to a simply connected oriented finite Poincaré complex X of dimension $n \geq 5$. Show

(i) Crossing with $\mathbb{CP}^n$ yields another normal map of degree one denoted by $(f, \overline{f}) \times \mathbb{CP}^n$;
(ii) If n is divisible by four, then $(f, \overline{f})$ is normally bordant to a homotopy equivalence if and only if $(f, \overline{f}) \times \mathbb{CP}^2$ is normally bordant to a homotopy equivalence.

Remark 8.175 (Dependence on the bundle data in the simply connected case) Theorem 8.173 (iii) shows that the surgery obstruction $\sigma(f, \overline{f})$ of a simply connected surgery problem $(f, \overline{f}) \colon M \to X$ in dimension n for $n \equiv 0 \bmod 4$ is the difference of the signatures of the source and the target. We will later see that in the simply connected case the surgery obstruction is always trivial in odd dimensions. Hence the surgery obstruction in the simply connected case does not depend on the bundle data at all, provided that $n \not\equiv 2 \bmod 4$.

If $n \equiv 2 \mod 4$, then the surgery obstruction does depend on the bundle data. Consider the collapse map $f \colon T^2 \to S^2$, which has degree one. Since the tangent bundle TT^2 is trivial and $\pi_1(SO(3)) \cong \mathbb{Z}/2$, there are up to isotopy two bundle maps $\overline{f}_k \colon TT^2 \oplus \mathbb{R} \to \mathbb{R}^3$ for $k = 0, 1$ covering f. They satisfy $\sigma(f, \overline{f}_k) = k$ in $L_2(\mathbb{Z}) \cong \mathbb{Z}/2$.

Note that we have defined the surgery obstructions only for surgery problems of dimension ≥ 6. Therefore to make sense of the above example one has to employ another definition. There are two possibilities: a homotopy-theoretic definition due to Browder [55] or the chain complex definition due to Ranicki [345]. We review the latter in Chapter 15. Both of these definitions work in any dimension in the sense that they associate an element of the corresponding L-group to a surgery problem. However, one should not really call them surgery obstructions since we do not have an analogue of Theorem 8.173 and its relatives in the low dimensions. On the other hand one does have product formulas, see [55, Section III.5] or [344, Sec. 8] and [345, Sec. 8], which allow us to produce high-dimensional surgery problems with desired surgery obstructions from these low-dimensional problems. The above statement $\sigma(f, \overline{f}_k) = k \in L_2(\mathbb{Z}) \cong \mathbb{Z}/2$ about the surgery problems with the underlying map $f \colon T^2 \to S^2$ is due to Browder, see [54, Section 5].

In general the surgery obstruction depends on the bundle data, as the following example illustrates.

Example 8.176 (Dependence on the bundle data) Fix a normal Γ-map of degree one $(f, \overline{f}) = (M, [M], f, \widehat{f}, a, \xi, \overline{f})$. Let N be a closed oriented manifold of dimension n. Then we get a new normal Γ-map $(f, \overline{f}) \times N$ whose underlying map is $f \times \mathrm{id}_N \colon M \times N \to X \times N$ and whose bundle data are given by crossing with id_{TN}. We have the equality of surgery obstructions $\sigma(f, \overline{f}) = \sigma((f, \overline{f}) \times \mathbb{CP}^2)$ using the obvious identification $L_n(\mathbb{Z}\Gamma; w) = L_{n+4}(\mathbb{Z}\Gamma, w)$. This is a special case of a product formula for surgery obstructions, see [55, Section III.5] or [344, Sec. 8] and [345, Sec. 8]; we will also discuss it in Section 15.9. Hence we get from Remark 8.175, using the normal maps $(f, \overline{f}_k)$ in dimension two, examples of simply connected surgery problems $(f, \overline{f}_k) \times (\mathbb{CP}^2)^m$ for every natural number m such that the surgery obstructions of $(f, \overline{f}_0) \times (\mathbb{CP}^2)^m$ and $(f, \overline{f}_1) \times (\mathbb{CP}^2)^m$ are different.

Now consider any natural number l. By the fact that $\mathrm{Wh}(\mathbb{Z}^l)$ and $\widetilde{K}_n(\mathbb{Z}[\mathbb{Z}^l])$ vanish for $n \leq 0$ and using the Shaneson splitting, see Theorem 13.55, we conclude that the surgery obstructions of $(f, \overline{f}_0) \times (\mathbb{CP}^2)^m \times T^l$ and $(f, \overline{f}_1) \times (\mathbb{CP}^2)^m \times T^l$ in $L_{2+4k+l}(\mathbb{Z}[\mathbb{Z}^l])$ are different for $k = 0, 1$. Note that any natural number $n \geq 2$ can be written as $2 + 4k + l$ for appropriate k and l. So we get in every dimension ≥ 2 examples where the surgery obstruction can be changed by modifying the bundle data without changing the underlying map.

Exercise 8.177 Suppose in the situation of Theorem 8.168 that for $m = 0, 1$ the spaces X_m are simply connected and $w_1(W) = 0$. Show that then for $m = 0, 1$ there are isomorphisms $s_m \colon L_{2k}(\mathbb{Z}\Pi(X_m), O_{X_m}) \xrightarrow{\cong} L_{2k}(\mathbb{Z})$ such that

8.7 The Intrinsic L-Group and Surgery Obstruction in Even Dimensions   305

$$s_0\bigl(\sigma(M_0, f_0, a_0, \xi_0, \overline{f}_0, o_0)\bigr) = s_1\bigl(\sigma(M_1, f_1, a_1, \xi_1, \overline{f}_1, o_1)\bigr)$$

holds.

Exercise 8.178 Let $(f, \overline{f})\colon M \to X$ be a normal map from a closed oriented manifold to an oriented finite Poincaré complex X of dimension m. Consider a natural number $n \geq 2$. Show:

(i) Crossing with S^n yields another normal map of degree one, denoted by $(f, \overline{f}) \times \mathrm{id}_{S^n}$;
(ii) If $m + n \geq 5$, then $(f, \overline{f}) \times \mathrm{id}_{S^n}$ is normally bordant to a normal homotopy equivalence.

8.7.7 The Role of the Bundle Data in the Highly Connected Case

In this subsection we discuss of the role of the bundle data in the quadratic form coming from a surgery problem, after we have made it highly connected. More precisely, we investigate how the surgery obstruction of a highly connected surgery problem changes when we modify the bundle data leaving the underlying map fixed. Note that the results of this section do not apply to a surgery problem that is not highly connected since they do not carry over to normal bordisms and therefore do not contradict Example 8.176. Thus the material of this section is only relevant in the case where the normal map we are interested in is already highly connected.

For the remainder of this subsection, we fix two normal Γ-maps of degree one $(M, [M], f, \widehat{f}_i, a, \xi, \overline{f})$ with target $(X, \widehat{p}_X, [X])$ in the sense of Definition 8.3, where the dimension $n = 2k$ is even, the underlying maps $f\colon M \to X$ agree, and only the bundle maps $\overline{f}_i$ are different. We will often abbreviate $(M, [M], f, \widehat{f}_i, a, \xi, \overline{f})$ by $(f, \overline{f}_i)$. Moreover, we will assume that we are in the universal covering case, see Notation 8.34, $k \geq 3$ holds, and f is k-connected.

Let $(K_k(\widehat{M}), s, t_i)$ be their quadratic surgery kernels as defined in Example 8.86 above. We note that the definition of s in (8.57) means that the underlying symmetric form $(K_k(\widehat{M}), s)$ depends only on the map $f\colon M \to X$, and this explains the notation. The problem then is to determine the relationship between t_0 and t_1. We first develop the necessary notation and terminology to precisely answer this problem in Proposition 8.181 below.

There is a vector bundle isomorphism

$$\theta\colon TM \oplus \underline{\mathbb{R}^a} \xrightarrow{\cong} TM \oplus \underline{\mathbb{R}^a} \tag{8.179}$$

such that $\overline{f}_1 = \overline{f}_0 \circ \theta$, and compatibility between the bundle maps and orientations ensure that θ is orientation preserving. If we now add the identity morphism on the normal bundle $\nu(M)$ of M to θ, we obtain a bundle automorphism $\theta \oplus \mathrm{id}_{\nu(M)}$ of the trivial stable bundle over M. Since any orientation preserving bundle automorphism of the trivial stable bundle is equivalent to a map $M \to \mathrm{SO}$, where SO denotes the stable special orthogonal group as usual, this construction gives rise to a well-defined homotopy class also denoted $\theta \in [M, SO]$, see [55, Lemma I.4.6] for a related discussion of automorphisms of stable spherical fibrations.

Now we consider the usual integral surgery kernel where we *do not* pass to covering spaces. This is the group

$$K_k(M) := \ker(f_* : H_k(M; \mathbb{Z}) \to H_k(X; \mathbb{Z})).$$

From the canonical reduction homomorphism, see (8.62),

$$\mathrm{pr} : K_k(\widetilde{M}) \to \mathbb{Z} \otimes_{\mathbb{Z}\pi} K_k(\widetilde{M}) \cong K_k(M), \quad \omega \mapsto 1 \otimes \omega, \tag{8.180}$$

and the fact that $K_k(\widetilde{M})$ is stably finitely generated free, see Lemma 8.55, we conclude $K_k(M) \cong \mathbb{Z}^a$. Moreover, it follows from Lemma 8.39 (iii) that there is a map from a wedge of k-spheres $i \colon \vee_{i=1}^{a} S^k \to M$ representing a set of generators for $K_k(M)$. We have

$$i^*(\theta) \in [\vee_{i=1}^{a} S^k, SO] \cong H^k(\vee_{i=1}^{a} S^k; \pi_k(SO)) \cong \hom(K_k(M), \pi_k(SO)).$$

We regard $i^*(\theta)$ as an element of $\hom(K_k(M), \pi_k(SO))$ using the isomorphisms above. It will turn out, see Proposition 8.181 below, that the choice of bundle map can only affect the quadratic form t when $k = 3, 7$. If $k = 3, 7$, then $\pi_k(O) \cong \mathbb{Z}$, we identify

$$Q_{-1}(\mathbb{Z}) = \pi_k(SO)/2\pi_k(SO) = \mathbb{Z}/2,$$

and write $p_2(i^*(\theta)) \in \hom(K_k(M), Q_{-1}(\mathbb{Z}))$ for the projection $p_2 \colon \mathbb{Z} \to \mathbb{Z}/2$.

The inclusion of the trivial group $\{e\} \to \pi$ induces the inclusion of rings with involution $\mathbb{Z} \to (\mathbb{Z}\pi, w)$, which in turn induces an inclusion homomorphism $e \colon Q_{-1}(\mathbb{Z}) \to Q_{-1}(\mathbb{Z}\pi, w)$. We define the homomorphism

$$t_\theta \colon K_k(\widetilde{M}) \to Q_{-1}(\mathbb{Z}\pi, w), \quad \omega \mapsto e(p_2(i^*(\theta))(\mathrm{pr}(\omega)).$$

From the definition of t_θ we have that $(1 + (-1)^k \cdot T) t_\theta = 0$, so that $(K_k(\widetilde{M}), t_\theta)$ is a linear $(-1)^k$-quadratic form as in Example 8.80. Consequently $(K_k(\widetilde{M}), s, t_0 + t_\theta)$ is a non-singular $(-1)^k$-quadratic form on $K_k(\widetilde{M})$.

Proposition 8.181 *For $i = 0, 1$ let $(f, \overline{f}_i) \colon M \to X$ be k-connected degree one normal maps with quadratic surgery kernels $(K_k(\widetilde{M}), s, t_i)$.*

(i) *If $k \neq 3, 7$ then $(K_k(\widetilde{M}), s, t_1) = (K_k(\widetilde{M}), s, t_0)$;*
(ii) *If $k = 3, 7$, write $\overline{f}_1 = \overline{f}_0 \circ \theta$ as in (8.179) above. Then we have*

$$(K_k(\widetilde{M}), s, t_1) = (K_k(\widetilde{M}), s, t_0 + t_\theta).$$

Proof. To make the role of the bundle map $\overline{f}$ clear in the notation, let us write $t = t(\overline{f})$. We recall from (8.87) that the quadratic form $t(\overline{f})$ is given as the following composite

8.7 The Intrinsic L-Group and Surgery Obstruction in Even Dimensions

$$t(\overline{f}): K_k(\widetilde{M}) \xrightarrow{h_k^{-1}} \pi_{k+1}(f) \xrightarrow{t_k(\overline{f})} I_k(M) \xrightarrow{\mu} Q_{(-1)^k}(\mathbb{Z}\pi, w) \qquad (8.182)$$

where h_k is defined in Lemma 8.39, $t_k = t_k(\overline{f})$ is defined in Lemma 8.10, and the self-intersection function μ is defined in (8.23). Neither μ nor h_k^{-1} depend on the bundle map $\overline{f}$, so we must understand the role of the choice $\overline{f}$ in the definition of $t_k(\overline{f}): \pi_{k+1}(f) \to I_k(M)$. The homomorphism $t_k(\overline{f})$ is defined in Lemma 8.10, whose proof relies on Theorem 4.39, which uses Hirsch–Smale theory as described in Theorem 4.30. Since the relevant homotopies are always covered by corresponding bundle maps, Theorem 4.30 implies that there is an exact sequence

$$I_k(D^{2k}) \xrightarrow{i_M} I_k(M) \xrightarrow{F} \pi_k(M) \qquad (8.183)$$

where i_M is induced by a standard embedding $D^{2k} \hookrightarrow M$ and F is defined by taking the underlying homotopy class of a pointed immersion. For $x \in \pi_{k+1}(f)$, we see that the homotopy class $F(t_k(\overline{f})(x)) \in \pi_k(M)$ does not depend upon $\overline{f}$. Hence $t_k(\overline{f})(x)$ and $t_k(\overline{f} + \theta)(x)$ differ by an element of $i_M(I_k(D^{2k}))$. By Example 4.36, taking the derivative of an immersion defines an isomorphism

$$T: I_k(D^{2k}) = \pi_0(\mathrm{Imm}(S^k, D^{2k})) \xrightarrow{\cong} \pi_k(SO/SO(k))$$

and, when k is odd, $\pi_k(SO/SO(k)) = \mathbb{Z}/2$. Moreover, by Proposition 8.27, when k is odd, the two regular homotopy classes of immersions $S^k \hookrightarrow D^{2k}$ are distinguished by the number of double points, counted mod 2. Hence, if $(q_1, w_1) \in I_k(D^k)$ represents the regular homotopy class with an odd number of double points, so that $\mu(q_1, w_1) = 1$ holds in $\in Q_{-1}(\mathbb{Z}) = \mathbb{Z}/2$, then for any pointed immersion (q, w) with $q: S^k \hookrightarrow M$, we have

$$\mu\big((q, w) + (q_1, w_1)\big) = \mu(q, w) + e(1) \in Q_{-1}(\mathbb{Z}\pi, w). \qquad (8.184)$$

To compute $t_k(\overline{f} + \theta)(x) - t_k(\overline{f})(x)$, we let $\partial: \pi_{k+1}(f) \to \pi_k(M)$ be the boundary homomorphism and regard θ as a map $\theta: M \to SO$. Then $\theta_*(\partial(x))$ defines a homotopy class in $\pi_k(SO)$. To map this homotopy class to $I_k(D^{2k})$, we consider the quotient map $p: SO \to SO/SO(k)$, the induced homomorphism $p_*: \pi_k(SO) \to \pi_k(SO/SO(k))$ and finally the homomorphism

$$T^{-1} \circ p_*: \pi_k(SO) \to I_k(D^{2k}).$$

Using Theorem 4.30, the definition of $t_k(\overline{f})$ and (8.183) above, we deduce that for all $x \in \pi_{k+1}(f)$ we have

$$t_k(\overline{f}_0 + \theta)(x) = t_k(\overline{f}_0)(x) + i_M\big((T^{-1} \circ p_*)(\theta_*(\partial(x)))\big). \qquad (8.185)$$

Now by [246, Theorem 1.4 (2)], $p_* = 0$ unless $k = 3, 7$, in which case p_* is onto. Hence, if $k \neq 3, 7$, we conclude from (8.185) that $t_k(\overline{f}_0 + \theta) = t_k(\overline{f}_0)$, proving part (i).

When $k = 3, 7$, we already saw that $\pi_k(SO/SO(k)) \cong \mathbb{Z}/2$, and so (8.185) and the definition of $i^*(\theta)$ give the following formula for all $x \in \pi_{k+1}(f)$,

$$t_k(\overline{f}_0 + \theta)(x) = t_k(\overline{f}_0)(x) + i_M(i^*(\theta)(\partial(x))).$$

Here we regard $i^*(\theta(\partial(x)))$ as an element of $\mathbb{Z}/2 = I_k(D^{2k})$. Combining this formula with (8.184) above, we obtain that $t_1 = t(\overline{f}_0 + \theta) = t(\overline{f}_0) + t_\theta = t_0 + t_\theta$. This completes the proof of assertion (ii). □

Exercise 8.186 Let $k = 3, 7$. Put $M = S^k \times S^k$. Consider the Hopf collapse map $f: M \to S^{2k}$ of Example 4.8. Show that there are degree one normal maps $(f, \overline{f}_i): M \to S^{2k}$ for $i = 0, 1$ such that the surgery obstructions $\sigma(f, \overline{f}_i)$ in $L_{2k}(\mathbb{Z}) = \mathbb{Z}/2$ satisfy

$$\sigma(f, \overline{f}_0) = 0 \quad \text{and} \quad \sigma(f, \overline{f}_1) = 1.$$

8.8 The Even-Dimensional Surgery Obstruction Relative Boundary

In this section we extend the results of Subsection 8.7.4 from closed manifold to pairs, where we assume that on the boundary we already have a homotopy equivalence. This is analogous to the closed case since the surgery kernels of pairs look like the surgery kernel in the closed situation in this case. This will be important in Section 8.9, where we will consider the "Realisation problem": This is the problem of determining which elements of $L_{2k}(\mathbb{Z}\Pi(X), O_X)$ arise as the surgery obstructions to surgery problems. If we allow a boundary with a homotopy equivalence on the boundary, every element in $L_{2k}(\mathbb{Z}\Pi(X), O_X)$ occurs. This is not true if one considers only closed manifolds.

8.8.1 Normal Maps of Pairs

Next we explain how one can generalise the notions of rank k normal map of degree one introduced in Definition 7.13 and of normal maps with respect to the tangent bundle (of degree one) introduced in Definition 7.22 for closed manifolds to compact manifolds with boundary. We replace the w-oriented finite n-dimensional Poincaré complex $(X, O_X, [[X]])$ by a w-oriented finite n-dimensional Poincaré pair $(X, \partial X, O_X, [[X, \partial X]])$ in the sense of Definition 5.73. The closed manifold M is replaced by a compact manifold M with boundary ∂M. In this situation we have an intrinsic fundamental class $[[M, \partial M]] \in H_n(M, \partial M; O_M)$, see Definition 5.29. The map $f: M \to X$ is replaced by a map of pairs $(f, \partial f): (M, \partial M) \to (X, \partial X)$, where we require that ∂f is a homotopy equivalence.

8.8 The Even-Dimensional Surgery Obstruction Relative Boundary

As before, the bundle data $(\xi, \bar{f})$ include the choice of an isomorphism $o \colon O_\xi \to O_X$ of infinite cyclic local coefficient systems over X, and we have an isomorphism $\alpha \colon O_M \to f^*O_X$ of infinite cyclic coefficient systems. From $(f, \partial f)$ and α we get a homomorphism

$$H_n(f, \partial f; \alpha) \colon H_n(M, \partial M; O_M) \to H_n(X, \partial X, O_X).$$

We require that it sends $[[M, \partial M]]$ to $[[X, \partial X]]$. All in all the data above are encoded in the tuple $(M, \partial M, f, \partial f, a, \xi, \bar{f}, o)$. Sometimes we write $(f, \bar{f})$ for short. Since we require that $\partial f \colon \partial M \to \partial X$ is a homotopy equivalence, we call it a *normal map of degree one with respect to the tangent bundle relative boundary*.

Now it is also obvious how to extend the Definition 8.3 of a normal Γ-map of degree one to pairs.

The goal of doing surgery is to turn f into a homotopy equivalence without changing ∂f. It will be achieved by finitely many surgery steps on the interior of M, provided the appropriate surgery obstruction vanishes. The performance of a surgery step in the interior to kill a homotopy class in $\pi_k(f)$ for $(k-1)$-connected f can be carried out as in the closed case since one can always arrange that a representative of such a class and all of its later modifications do not meet the boundary. In particular we can always make f highly connected without changing ∂f.

8.8.2 Normal Bordism for Normal Maps of Pairs

Next we explain how the notion a bordism of normal maps with respect to the tangent bundle of degree one of Definition 7.24 carries over from the closed case to the case of pairs.

For $m = 0, 1$, let $(X_m, \partial X_m, O_{X_m}, [[X_m]])$ be a w-oriented finite n-dimensional Poincaré pair and let $(M_m, \partial M_m, f_m, \partial f_m, a_m, \xi_m, \bar{f}_m, o_m)$ be a normal map with respect to the tangent bundle of degree one relative boundary with target $(X_m, \partial X_m, O_{X_m}, [[X_m]])$. Next we define the notion of a *normal bordism of normal maps of degree one with respect to the tangent bundle relative boundary* $(W, F, b, \Xi, \bar{F}, O)$ from

$$(M_0, \partial M_0, f_0, \partial f_0, a_0, \xi_0, \bar{f}_0, o_0)$$

to

$$(M_1, \partial M_1, f_1, \partial f_1, a_1, \xi_1, \bar{f}_1, o_1).$$

As in the closed case W is a compact $(n+1)$-dimensional manifold with boundary ∂W, but now the boundary is the union of three pieces

$$\partial W = \partial_0 W \cup \partial_1 W \cup \partial_2 W$$

where $\partial_m W$ is a codimension zero submanifold of ∂W possibly with non-empty boundary $\partial\partial_m W$ for $m = 0, 1, 2$ satisfying

$$\partial_0 W \cap \partial_1 W = \emptyset;$$
$$\partial_2 W \cap \partial_m W = \partial\partial_m W \quad \text{for } m = 0, 1;$$
$$\partial\partial_2 W = \partial\partial_0 W \amalg \partial\partial_1 W.$$

Note that the local coefficient systems $O_{\partial W}$ and $O_{\partial_m W}$ for $m = 0, 1, 2$ can and will be identified with the coefficient systems O_W restricted to $\partial_m W$ using the outward normal field. The intrinsic fundamental classes $[[\partial W]] \in H_n(\partial W, O_{\partial W})$ and $[[\partial_m W, \partial\partial_m W]] \in H_n(\partial_m W, \partial\partial_m W; O_{\partial_m W})$ come from the intrinsic fundamental class $[[W, \partial W]] \in H_{n+1}(W, \partial W; O_W)$ by the boundary map

$$H_{n+1}(W, \partial W; O_W) \to H_n(\partial W; O_{\partial W})$$

and the composite

$$H_n(\partial W; O_{\partial W}) \to H_n(\partial W, \partial W - \partial W_m; O_{\partial W})) \xrightarrow{\cong} H_n(\partial_m W, \partial\partial_m W; O_{\partial_m W}),$$

see Lemma 5.30.

Figure 8.187 (Manifold W with boundary $\partial W = \partial_0 W \cup \partial_1 W \cup \partial_2 W$).

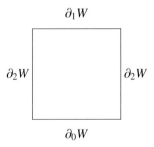

We have an $(n + 1)$-dimensional finite Poincaré pair $(Y, \partial Y, O_Y, [[Y, \partial Y]])$ with a decomposition of ∂Y into three n-dimensional finite CW-subcomplexes

$$\partial Y = \partial_0 Y \cup \partial_1 Y \cup \partial_2 Y$$

such that for appropriate $(n-1)$-dimensional finite CW-subcomplexes $\partial\partial_m Y \subseteq \partial_m Y$ for $m = 0, 1, 2$ we have

$$\partial_0 Y \cap \partial_1 Y = \emptyset;$$
$$\partial_2 Y \cap \partial_m Y = \partial\partial_m Y \quad \text{for } m = 0, 1;$$
$$\partial\partial_2 Y = \partial\partial_0 Y \amalg \partial\partial_1 Y.$$

Moreover, we demand that for $m = 0, 1, 2$ the CW-pairs $(\partial_m Y, \partial\partial_m Y)$ are w-oriented finite n-dimensional Poincaré pairs with fundamental classes $[[\partial_m Y, \partial\partial_m Y]]$ lying

8.8 The Even-Dimensional Surgery Obstruction Relative Boundary

in $H_n(\partial_m Y, \partial \partial_m Y; O_{\partial_m Y})$ for the restriction $O_{\partial_m Y}$ of the coefficient system O_Y to $\partial_m Y$ where the various fundamental classes above come from the fundamental class $[[Y, \partial Y]] \in H_{n+1}(Y, \partial Y; O_Y)$ by the boundary map

$$H_{n+1}(Y, \partial Y; O_Y) \to H_n(\partial Y; O_{\partial Y})$$

and the composite

$$H_n(\partial Y; O_{\partial Y}) \to H_n(\partial Y, \partial Y - \partial Y_m; O_{\partial Y}) \xrightarrow{\cong} H_n(\partial_m Y, \partial \partial_m Y; O_{\partial_m Y}).$$

The map $F \colon W \to Y$ is required to induce maps $\partial_m F \colon \partial_m W \to \partial_m Y$ for $m = 0, 1, 2$, and $\partial_2 F \colon \partial_2 W \to \partial_2 Y$ is required to be a homotopy equivalence. The various identifications $M_m \xrightarrow{\cong} \partial_m W$ and $X_m \xrightarrow{\cong} \partial_m Y$ for $m = 0, 1$ in the closed case are now required to be identifications $(M_m, \partial M_m) \xrightarrow{\cong} (\partial_m W, \partial \partial_m W)$ and $(X_m, \partial X_m) \xrightarrow{\cong} (\partial_m Y, \partial Y_m)$ for $m = 0, 1$. The bundle data $(\Xi, \overline{F})$, including the isomorphism $O \colon O_\Xi \to O_W$ over W, are unchanged. The obvious compatibility conditions concerning the bundle maps, the local coefficients systems, and the fundamental classes remain.

Finally we explain what it means if the bordism has *cylindrical target* in the special case where $(M_m, \partial M_m, f_m, \partial f_m, a_m, \xi_m, \overline{f}_m, o_m)$ for $m = 0, 1$ have the same target $(X, \partial X, O_X, [[X]])$. Namely, we require that $Y = X \times [0, 1]$ with $\partial_m Y = X \times \{m\}$ for $m = 0, 1$ and $\partial_2 Y = \partial X \times [0, 1]$, and that $\Xi = \text{pr}_Y^* \xi$ and $O = \text{pr}_Y^* o$ hold for the projection $\text{pr}_Y \colon Y \times [0, 1] \to Y$.

Definition 8.188 (Triads) If we take for W above $\partial_1 W = \emptyset$ and rename W as M, $\partial_0 W$ as $\partial_0 M$ and $\partial_2 W$ as $\partial_1 M$, we obtain a *manifold triad* $(M; \partial_0 M, \partial_1 M)$. We often use the notation

$$\partial_{0,1} M := \partial_0 M \cap \partial_1 M = \partial \partial_0 M = \partial \partial_1 M.$$

Analogously one defines a *Poincaré triad* $(X; \partial_0 X, \partial_1 X)$ by taking for Y above $\partial_1 Y = \emptyset$ and rename Y as X, $\partial_0 Y$ as $\partial_0 X$ and $\partial_2 Y$ as $\partial_1 X$. We often use the notation

$$\partial_{0,1} X := \partial_0 X \cap \partial_1 X = \partial \partial_0 X = \partial \partial_1 X.$$

8.8.3 The Intrinsic Even-Dimensional Surgery Obstruction Relative Boundary

Theorem 8.189 (Intrinsic surgery obstruction in even dimensions relative boundary) Let $(X, \partial X, O_X, [[X, \partial X]])$ be a (not necessarily connected) w-oriented finite n-dimensional Poincaré pair in the sense of Definition 5.73 of even dimension $n = 2k$ for $k \geq 3$. Consider a normal map with respect to the tangent bundle of degree one relative boundary $(f, \overline{f}) = (M, \partial M, f, \partial f, a, \xi, \overline{f}, o)$ with target $(X, \partial X, O_X, [[X, \partial X]])$ in the sense of Subsection 8.8.1. (Recall that this means that $\partial f \colon \partial M \to \partial X$ is a homotopy equivalence.) Then:

(i) *Its intrinsic surgery obstruction*

$$\sigma(f, \overline{f}) = \sigma(M, \partial M, f, \partial f, a, \xi, \overline{f}, o) \in L_{2k}(\mathbb{Z}\Pi(X), O_X)$$

depends only on the normal bordism with cylindrical ends class of $(\overline{f}, f)$;

(ii) *We have $\sigma(\overline{f}, f) = 0$ if and only if we can do a finite number of surgery steps on the interior of M without changing ∂f to obtain a normal homotopy equivalence, i.e., the underlying map f is a homotopy equivalence.*

Proof. We will only explain why the proof is essentially the same as that of Theorem 8.167.

Making a normal map highly connected can be done in the case relative boundary exactly as in the closed case, just do all surgery steps in the interior of M. The passage from the intrinsic setting to the setting of normal Γ-bordism, see Definitions 8.3 and 8.7 in the universal covering case, see Notation 8.34, goes through without changes. Suppose that f is highly connected. Now one has be careful concerning the following point only. We are not assuming any connectivity assumption on the inclusion of the boundary $\partial M \to M$. Therefore the restriction of the universal covering $\widetilde{M} \to M$ to ∂M is not necessarily the universal covering again and therefore is denoted by $\widehat{\partial M} \to \partial M$.

Next consider a highly connected normal bordism with cylindrical target. Then the restriction of the covering $\widetilde{W}$ to $\partial_m W$ for $m = 0, 1$ is the universal covering and will be denoted by $\widetilde{\partial_m W} \to \partial_m W$ for $m = 0, 1$ whereas the restriction of $\widetilde{W} \to W$ to $\partial_2 W$ is not necessarily the universal covering and therefore will be denoted by $\widehat{\partial_2 W} \to \partial_2 W$. The restriction of $\widetilde{W} \to W$ to ∂W will be denoted by $\widehat{\partial W} \to \partial W$ and not by $\widetilde{\partial W}$ as in the closed case. The same applies to Y and its subspaces $\partial_m Y$ for $m = 0, 1, 2$. All these coverings are pullbacks from the universal covering $\widetilde{X} \to X$ and are π-coverings for $\pi = \pi_1(X, x)$. Moreover, the homotopy equivalence $\partial_2 f : \partial_2 W \to \partial_2 Y$ lifts to a π-homotopy equivalence $\widehat{\partial_2 f} : \widehat{\partial_2 W} \to \widehat{\partial_2 Y}$. This will be crucial for the only extra argument that the kernels in the case relative boundary look like the kernels in the closed case, which we will explain next.

Suppose that f is highly connected. The definition of the kernels $K_j(\widetilde{M})$ and $K_j(\widetilde{M}, \widehat{\partial M})$ is unchanged, see Definition 8.115. Since $\widehat{\partial f}$ is a π-homotopy equivalence, $K_j(\widehat{\partial f}) = 0$ for $j \in \mathbb{Z}$. Hence we conclude from Lemma 8.116 (ii) that the natural map $K_j(\widetilde{M}) \to K_j(\widetilde{M}, \widehat{\partial M})$ is a $\mathbb{Z}\pi$-isomorphism. Hence the duality isomorphism of Lemma 8.116 (ii), which relate $K^{n-j}(\widetilde{M})$ to $K_j(\widetilde{M}, \widehat{\partial M})$, yield Poincaré duality isomorphisms $K^{n-j}(\widetilde{M}) \xrightarrow{\cong} K_j(\widetilde{M})$, which look like the ones we used in the closed case. Since the inclusion of $K_j(\widetilde{M}, \widehat{\partial M})$ into $H_j(\widetilde{M}, \widehat{\partial M})$ is split injective, the inclusion $K_j(\widetilde{M})$ into $H_j(\widetilde{M})$ is also split injective.

8.8 The Even-Dimensional Surgery Obstruction Relative Boundary 313

Next we consider the surgery kernel of a bordism. We define

$$K_j(\widetilde{W}) = \ker\bigl(\widetilde{F}_* \colon H_j(\widetilde{W}) \to H_j(\widetilde{Y})\bigr);$$
$$K_j(\widetilde{W}, \widehat{\partial W}) = \ker\bigl(\widetilde{F}_* \colon H_j(\widetilde{W}; \widehat{\partial W}) \to H_j(\widetilde{Y}, \widehat{\partial Y})\bigr);$$
$$K_j(\widetilde{W}, \widehat{\partial_0 W} \amalg \widehat{\partial_1 W}) = \ker\bigl(\widetilde{F}_* \colon H_j(\widetilde{W}; \widehat{\partial_0 W} \amalg \widehat{\partial_1 W}) \to H_j(\widetilde{Y}, \widehat{\partial_0 Y} \amalg \widehat{\partial_1 Y})\bigr);$$
$$K_j(\widetilde{W}, \widehat{\partial_2 W}) = \ker\bigl(\widetilde{F}_* \colon H_j(\widetilde{W}; \widehat{\partial_2 W}) \to H_j(\widetilde{Y}, \widehat{\partial_2 Y})\bigr);$$
$$K_j(\widetilde{W}, \widehat{\partial W}) = \ker\bigl(\widetilde{F}_* \colon H_j(\widetilde{W}; \widehat{\partial W}) \to H_j(\widetilde{Y}, \widehat{\partial Y})\bigr).$$

Since $\partial_2 F \colon (\partial_2 W, \partial \partial_2 W) \to (\partial_2 Y, \partial \partial_2 Y)$ is a homotopy equivalence of pairs, $\widetilde{\partial_2 F} \colon (\widetilde{\partial_2 W}, \widehat{\partial \partial_2 W}) \to (\widetilde{\partial_2 Y}, \widehat{\partial \partial_2 Y})$ is a π-homotopy equivalence of pairs.

The following two commutative diagrams with long exact homology sequences as columns, where the horizontal arrows marked with $\cong$ are isomorphisms

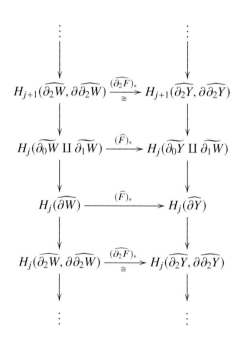

and

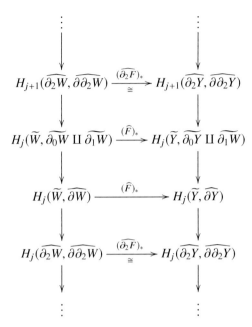

come from the excision isomorphisms

$$H_{j+1}(\widetilde{\partial W}, \widetilde{\partial_0 W} \amalg \widetilde{\partial_1 W}) \xrightarrow{\cong} H_{j+1}(\widehat{\partial_2 W}, \widehat{\partial\partial_2 W});$$
$$H_{j+1}(\widetilde{\partial Y}, \widetilde{\partial_0 Y} \amalg \widetilde{\partial_1 Y}) \xrightarrow{\cong} H_{j+1}(\widehat{\partial_2 Y}, \widehat{\partial\partial_2 Y}).$$

By an easy diagram chase one sees that the canonical $\mathbb{Z}\pi$-homomorphisms

$$K_j(\widetilde{\partial_0 W} \amalg \widetilde{\partial_1 W}) \xrightarrow{\cong} K_j(\widetilde{\partial W});$$
$$K_j(\widetilde{W}, \widetilde{\partial_0 W} \amalg \widetilde{\partial_1 W}) \xrightarrow{\cong} K_j(\widetilde{W}, \widetilde{\partial W}),$$

are bijective and the inclusions

$$K_j(\widetilde{\partial_0 W} \amalg \widetilde{\partial_1 W}) \to H_j(\widetilde{\partial_0 W} \amalg \widetilde{\partial_1 W});$$
$$K_j(\widetilde{W}, \widetilde{\partial_0 W} \amalg \widetilde{\partial_1 W}) \to H_j(\widetilde{W}, \widetilde{\partial_0 W} \amalg \widetilde{\partial_1 W}),$$

are split injective since the inclusions

$$K_j(\widetilde{\partial W}) \to H_j(\widetilde{\partial W});$$
$$K_j(\widetilde{W}, \widetilde{\partial W}) \to H_j(\widetilde{W}, \widetilde{\partial W}),$$

are split injective

This is all we need to carry over the proof for the closed case to the case with boundary. For instance, under the assumption that $K_k(\widetilde{W}, \widetilde{\partial W}) = 0$, the short exact sequence (8.132),

8.8 The Even-Dimensional Surgery Obstruction Relative Boundary

$$0 \to K_{k+1}(\widetilde{W}, \widetilde{\partial W}) \xrightarrow{\partial_{k+1}} K_k(\widetilde{\partial W}) \xrightarrow{K_k(\tilde{i})} K_k(\widetilde{W}) \to 0,$$

played a prominent role in the proof for the closed case. One easily checks that it is still available in the relative boundary case in the form

$$0 \to K_{k+1}(\widetilde{W}, \widetilde{\partial W}) \xrightarrow{\partial_{k+1}} K_k(\widetilde{\partial W}) \xrightarrow{K_k(\tilde{i})} K_k(\widetilde{W}) \to 0,$$

provided that $K_k(\widetilde{W}, \widetilde{\partial W}) = 0$. Another example is Lemma 8.138, whose proof was built on handle subtractions in order to lower the minimal numbers of generators of $K_k(\widetilde{W}, \widetilde{\partial W})$. We want to use the same argument to lower the minimal numbers of generators of $K_k(\widetilde{W}, \widetilde{\partial W})$, but now doing changes only in the interior of $\partial_0 W \amalg \partial_1 W$ and not touching $\partial_2 W$. This can be done successfully since we have the canonical isomorphism $K_j(\widetilde{W}, \widetilde{\partial_0 W} \amalg \widetilde{\partial_1 W}) \xrightarrow{\cong} K_j(\widetilde{W}, \widetilde{\partial W})$.

This finishes the proof of Theorem 8.189. □

Remark 8.190 Theorem 8.168 carries over to the case relative boundary in the obvious way.

Exercise 8.191 Let $(f, \overline{f}) \colon (M, \partial M) \to (X, \partial X)$ be a surgery problem relative boundary. Suppose that X is simply connected and $4k$-dimensional.

Show that one can carry out a finite number of surgery steps on the interior of M leaving the boundary fixed to turn f into a homotopy equivalence if and only if $\mathrm{sign}(M, \partial M) - \mathrm{sign}(X, \partial X) = 0$, where $\mathrm{sign}(M, \partial M)$ and $\mathrm{sign}(X, \partial X)$ denote the signature of a compact oriented manifold with boundary or a Poincaré pair respectively, see Definition 5.84.

8.8.4 Kernels for Maps of Triads

For later purposes we next consider triads in the sense of Definition 8.188. For the remainder of this section we fix a degree one map of Poincaré triads (without requiring bundle data)

$$(h; \partial_0 h, \partial_1 h) \colon (W; \partial_0 W, \partial_1 W) \to (X; \partial_0 X, \partial_1 X)$$

and a Γ-covering $\widehat{X} \to X$. By pulling back $\widehat{X} \to X$ along $h \colon W \to X$ and along the inclusions $\partial_0 X \to X$ and $\partial_1 X \to X$, we obtain the covering map of triads

$$(\widehat{h}; \widehat{\partial_0 h}, \widehat{\partial_1 h}) \colon (\widehat{W}; \widehat{\partial_0 W}, \widehat{\partial_1 W}) \to (\widehat{X}; \widehat{\partial_0 X}, \widehat{\partial_1 X}).$$

In particular, there are maps of pairs $(h, \partial_i h) \colon (W, \partial_i W) \to (X, \partial_i X)$.

Definition 8.192 (Kernels of degree one maps of triads) The kernel groups are the $\mathbb{Z}\Gamma$-modules

$$K_j(\widehat{W}, \widehat{\partial_i W}) := \ker\bigl((\widehat{h}, \widehat{\partial_i h})_* \colon H_j(\widehat{W}, \widehat{\partial_i W}) \to H_j(\widehat{X}, \widehat{\partial_i X})\bigr),$$

and
$$K^j(\widehat{W}, \widehat{\partial_i W}) := \mathrm{coker}\big((\widehat{h}, \widehat{\partial_i h})^* \colon H^j(\widehat{X}, \widehat{\partial_i X}) \to H^j(\widehat{W}, \widehat{\partial_i W})\big).$$

In order to state the properties of the kernel groups of a degree one map of Poincaré triads, we first recall the statement of Poincaré duality for Poincaré triads.

Lemma 8.193 (Poincaré duality for triads) *Given a Poincaré triad $(X; \partial_0 X, \partial_1 X)$ with fundamental class $[X, \partial X]$, we have the Poincaré $\mathbb{Z}\Gamma$-chain homotopy equivalence, unique up to $\mathbb{Z}\Gamma$-chain homotopy,*

$$- \cap [X, \partial X] \colon C^{n-*}(\widehat{X}, \widehat{\partial_0 X}) \xrightarrow{\simeq} C_*(\widehat{X}, \widehat{\partial_1 X}).$$

It induces Poincaré duality $\mathbb{Z}\Gamma$-isomorphisms:

$$- \cap [X, \partial X] \colon H^{n-j}(\widehat{X}, \widehat{\partial_0 X}) \xrightarrow{\cong} H_j(\widehat{X}, \widehat{\partial_1 X}).$$

Proof. We shall consider the cohomology long exact sequence of the pair $(\widehat{X}, \widehat{\partial_0 X})$ and the homology long exact sequence of the triple $(\widehat{X}, \widehat{\partial X}, \widehat{\partial_1 X})$. Using the excision isomorphism $e \colon H_j(\widehat{\partial_0 X}, \widehat{\partial_{01} X}) \cong H_j(\widehat{\partial X}, \widehat{\partial_1 X})$, we construct the following diagram of long exact sequences:

$$\cdots \to H^{n-j-1}(\widehat{\partial_0 X}) \xrightarrow{\delta} H^{n-j}(\widehat{X}, \widehat{\partial_0 X}) \xrightarrow{(j_0)^*} H^{n-j}(\widehat{X}) \to \cdots$$
$$\downarrow {-\cap[\partial_0 X]} \qquad \downarrow {-\cap[X,\partial X]} \qquad \downarrow {-\cap[X,\partial X]}$$
$$\cdots \to H_j(\widehat{\partial_0 X}, \widehat{\partial_{01} X}) \xrightarrow{i_* \circ e} H_j(\widehat{X}, \widehat{\partial_1 X}) \xrightarrow{(j_1)_*} H_j(\widehat{X}, \widehat{\partial X}) \to \cdots$$

Here δ is the boundary homomorphism, $i \colon (\widehat{\partial X}, \widehat{\partial_1 X}) \to (\widehat{X}, \widehat{\partial_1 X})$, $j_0 \colon \widehat{X} \to (\widehat{X}, \widehat{\partial_0 X})$ and $j_1 \colon (\widehat{X}, \widehat{\partial_1 X}) \to (\widehat{X}, \widehat{\partial X})$ are the inclusions, and the map $e \colon H_j(\widehat{\partial_0 X}, \widehat{\partial_{01} X}) \to H_j(\widehat{\partial X}, \widehat{\partial_1 X})$ is given by excision. We leave it as an exercise to verify that the diagram commutes. Since the first and third cap products are the Poincaré duality isomorphisms of the Poincaré pairs $(\partial_0 X, \partial_{01} X)$ and $(X, \partial X)$, the five lemma finishes the proof. $\square$

Just as the kernels of degree one maps of Poincaré pairs behave similarly to the (co)homology groups of a third Poincaré pair, the kernels of a map of Poincaré triads behave similarly to the (co)homology groups of a third Poincaré triad. We leave the proof of the next lemma to the reader since it is an easy variation of the one of Lemma 8.116.

Lemma 8.194

(i) *For $m \in \{0, 1\}$, the homomorphisms*

$$(\widehat{h}, \widehat{\partial_i h})_* \colon H_j(\widehat{W}, \widehat{\partial_m W}) \to H_j(\widehat{X}, \widehat{\partial_m X})$$

are split surjections whereas the homomorphisms

$$(\widehat{h}, \widehat{\partial_i h})^* : H^j(\widehat{X}, \widehat{\partial_m X}) \to H^j(\widehat{W}, \widehat{\partial_m W})$$

are split injections. Consequently there are natural isomorphisms

$$H_j(\widehat{W}, \widehat{\partial_m W}) \cong K_j(\widehat{W}, \widehat{\partial_m W}) \oplus H_j(\widehat{X}, \widehat{\partial_m X});$$
$$H^j(\widehat{W}, \widehat{\partial_m W}) \cong K^j(\widehat{W}, \widehat{\partial_m W}) \oplus H^j(\widehat{X}, \widehat{\partial_m X}),$$

and also natural isomorphisms

$$K_j(\widehat{W}, \widehat{\partial_m W}) \cong H_{j+1}(\operatorname{cone}_*(C_*(\widehat{h}, \widehat{\partial_m h})));$$
$$K^j(\widehat{W}, \widehat{\partial_m W}) \cong H^{j+1}(\operatorname{cone}^*(C^*(\widehat{h}, \widehat{\partial_m h})));$$

(ii) *For $m \in \{0, 1\}$, cap products with $[W, \partial W]$ and $[\partial_m W]$ induce isomorphisms*

$$- \cap [W, \partial W] : K^{n-j}(\widehat{W}, \widehat{\partial_0 W}) \xrightarrow{\cong} K_j(\widehat{W}, \widehat{\partial_1 W});$$
$$- \cap [\partial_m W] : K^{n-j-1}(\widehat{\partial_m W}) \xrightarrow{\cong} K_j(\widehat{\partial_m W}, \widehat{\partial_{01} W}),$$

which fit into a commutative diagram of long exact sequences

$$\begin{array}{ccccccc}
\cdots \to & K^{n-j-1}(\widehat{\partial_0 W}) & \to & K^{n-j}(\widehat{W}, \widehat{\partial_0 W}) & \to & K^{n-j}(\widehat{W}, \widehat{\partial W}) & \to \cdots \\
& \cong \downarrow {-\cap [\partial W_0]} & & \cong \downarrow {-\cap [W, \partial W]} & & \cong \downarrow {-\cap [W, \partial W]} & \\
\cdots \to & K_j(\widehat{\partial_0 W}, \widehat{\partial_{01} W}) & \to & K_j(\widehat{W}, \widehat{\partial_1 W}) & \to & K_j(\widehat{W}, \widehat{\partial W}) & \to \cdots.
\end{array}$$

8.9 Realisation of Even-Dimensional Surgery Obstructions

Theorem 8.167 does not identify which elements of $L_{2k}(\mathbb{Z}\Pi(X), O_X)$ arise as surgery obstructions of degree one normal maps $(f, \overline{f}) : (M, \partial M) \to (X, \partial X)$. In this subsection we show that, if N is a compact $(2k - 1)$-dimensional *manifold* possibly with boundary for $k \geq 3$, then all elements in the group $L_{2k}(\mathbb{Z}\Pi(N \times I), O_{N \times I})$ are realised as surgery obstructions of degree one normal maps with target $(N \times I, \partial(N \times I), O_{N \times [0,1]}, [[N \times I, \partial(N \times I)]])$ for $I = [0, 1]$. The term "Wall realisation" is often used to refer to the basic theorems for the realisation of surgery obstructions, after Wall [414, Theorems 5.8 and 6.5]. We now state a version of Wall realisation in even dimensions.

Theorem 8.195 (Realisation of even-dimensional surgery obstructions) *Suppose $n = 2k \geq 6$. Consider a compact oriented $(n - 1)$-manifold N with possibly empty boundary ∂N. Fix an element*

$$\omega \in L_{2k}(\mathbb{Z}\Pi(N \times I), O_{N \times I}).$$

Then there is a normal map of pairs of degree one relative boundary $(M, \partial M, f, \partial f, a, \xi, \overline{f}, o)$ *with target* $(N \times I, \partial(N \times I), O_{N \times I}, [[N \times I, \partial(N \times I)]])$ *for* $I = [0, 1]$ *with the following properties:*

(i) *The bundle* ξ *over* $N \times I$ *is* $T(N \times I) \oplus \mathbb{R}^a$ *and the isomorphism* o *of infinite cyclic local coefficient systems over* $N \times I$ *is the identity;*

(ii) *There is a decomposition* $\partial M = \partial_0 M \cup \partial_1 M \cup \partial_2 M$ *where* $\partial_m M$ *is a codimension zero submanifold of* ∂M *with possibly empty boundary* $\partial \partial_m M$ *for* $m = 0, 1, 2$ *satisfying*

$$\partial_0 M \cap \partial_1 M = \emptyset;$$
$$\partial_2 M \cap \partial_m M = \partial \partial_m M \quad \text{for } m = 0, 1;$$
$$\partial \partial_2 M = \partial \partial_0 M \amalg \partial \partial_1 M;$$
$$\partial_2 M = \partial N \times [0, 1];$$
$$\partial_0 M = N \times \{0\};$$

(iii) *The homotopy equivalence* $\partial f : \partial M \to \partial(N \times I)$ *induces the identity* $\partial_0 M = N \times \{0\} \to (N \times \{0\})$ *on* $\partial_0 M$ *and the identity* $\partial_2 M = \partial N \times [0, 1] \to \partial N \times I$ *on* $\partial_0 M$;

(iv) *The homotopy equivalence* $\partial f : \partial M \to \partial(N \times I)$ *maps* $\partial_1 M$ *to* $N \times \{1\}$ *and induces a homotopy equivalence*

$$\partial_1 f : \partial_1 M \to N \times \{1\};$$

(v) *The intrinsic surgery obstruction*

$$\sigma(M, \partial M, f, \partial f, a, \xi, \overline{f}, o) \in L_{2k}(\mathbb{Z}\Pi(N \times I), O_{N \times I}),$$

see Theorem 8.189, is the given element ω.

Proof. We can assume without loss of generality that N is connected. In the sequel fix a choice of a universal covering $p_N : \widetilde{N} \to N$ and choice (BP) for N, see Notation 5.6, where the base point $x \in N$ should not lie in ∂N. Put $\pi = \pi_1(N, x) = \pi_1(N \times I, (x, 0))$. If $w : \pi \to \{\pm 1\}$ is the first Stiefel–Whitney class of N, then we get an identification, see Lemma 8.157 (ii),

$$L_{2k}(\mathbb{Z}\Pi(N \times I), O_{N \times I}) = L_{2k}(\mathbb{Z}\pi, w). \tag{8.196}$$

We will consider ω as an element $L_{2k}(\mathbb{Z}\pi, w)$, and will work with normal π-maps of degree one in the sense of Definition 8.3.

Recall that the element $\omega \in L_{2k}(\mathbb{Z}\pi, w)$ is represented by a non-singular $(-1)^k$-quadratic form (P, ψ) with P a finitely generated free $\mathbb{Z}\pi$-module. We fix such a representative and write it as a triple (P, λ, μ), as explained in Subsection 8.4.3. Let $\{b_1, b_2, \ldots, b_r\}$ be a $\mathbb{Z}\pi$-basis for P.

Choose r disjoint embeddings $j_i : D^{2k-1} \hookrightarrow N \setminus \partial N$ into the interior of N for $i = 1, 2, \ldots, r$. Let $F^0 : S^{k-1} \times D^k \hookrightarrow D^{2k-1}$ be the standard embedding. Define embeddings $F_i^0 : S^{k-1} \times D^k \hookrightarrow N$ by the composite of the standard embedding F^0

8.9 Realisation of Even-Dimensional Surgery Obstructions

and j_i for $i = 1, 2, \ldots, r$. Let $f_i^0 \colon S^{k-1} \to N$ be the restriction of F_i to $S^{k-1} \times \{0\} \subset S^{k-1} \times D^k$. We have already fixed a base point x in $N \setminus \partial N$. Choose a base point $s \in S^{k-1}$ and paths w_i from x to $f_i^0(s)$ for $i = 1, 2, \ldots, r$. Thus each f_i^0 is a pointed immersion. Using the form (P, μ, λ) as a guide, we next construct regular homotopies $\eta_i \colon S^{k-1} \times I \to N$ from f_i^0 to a new embedding f_i^1. These regular homotopies define associated immersions $\eta_i' \colon S^{k-1} \times I \hookrightarrow N \times I$, $(x, t) \mapsto (\eta_i(x, t), t)$. Since these immersions η_i' are embeddings on the boundary, we can define their intersection and self-intersection number as in (8.11) and (8.23).

We want to achieve that the intersection number of η_i' and η_j' for $i < j$ is $\lambda(b_i, b_j)$ and the self-intersection number $\mu(\eta_i')$ is $\mu(b_i)$ for $i = 1, \ldots, r$. These numbers are additive under concatenating regular homotopies. Hence it suffices to explain how to introduce a single intersection with value $\pm g$ between η_i' and η_j' for $i < j$ or a self-intersection for η_i' with value $\pm g$ for $i \in \{1, 2, \ldots, r\}$ without introducing other intersections or self-intersections. We first explain the construction for introducing an intersection between η_i' and η_j' for $i < j$ and then the construction for a self-intersection of η_i'.

Join $f_i^0(s)$ and $f_j^0(s)$ by a path v such that v is an embedding, the interior of v does not intersect any $f_i^0(S^{k-1})$ and the composition $w_j * v * w_i^-$ represents the given element $g \in \pi$. Now there is an obvious regular homotopy from f_i^0 to another embedding along the path v inside a small tubular neighbourhood $N(v)$ of v that leaves f_i^0 fixed outside a small neighbourhood $U(s)$ of s and moves $f_i^0|_{U(s)}$ within this small neighbourhood $N(v)$ along v very close to $f_j^1(s)$. Now use a disk D^k that meets f_j^0 transversally at the origin to move f_i^0 further, thus introducing the desired intersection with value $\pm g$. Inside the disk the move looks like pushing the upper hemisphere of the boundary down to the lower hemisphere of the boundary, thus having no intersection with the origin at any point of time with one exception. We have to be careful with signs, but there is room to twist appropriately in a small tubular neighbourhood of v and realise either sign at the intersection point. To construct a self-intersection of η_i', we introduce a second base point $t \in S^{k-1}$ and fix a path u_i from x to $f_i^0(t)$. We then repeat the above argument replacing $f_j^0(s)$ and w_j with $f_i^0(t)$ and u_j. This can be done since the argument involves changes only on a small disk $D^{k-1} \subset S^{k-1}$ near s that we take disjoint from t.

Since f_i^1 is regularly homotopic to f_i^0, which is obtained from the trivial embedding F_i^0 and so has trivial normal bundle, f_i^1 has a trivial normal bundle. Hence we can extend f_i^1 to an embedding $F_i^1 \colon S^{k-1} \times D^k \hookrightarrow N \setminus \partial N$. Now for each $i \in \{1, 2, \ldots, r\}$ we attach a handle of index k $(\phi_i^k) = D^k \times D^k$ to $N \times [0, 1]$ using $\phi_i^k := F_i^1 \colon S^{k-1} \times D^k \to X \times \{1\}$, see Definition 2.12. Let M be the manifold obtained from all r of the handle attachments to $N \times I$:

$$M := (N \times I) + \sum_{i=1}^{r} (\phi_i^k).$$

The following figure gives a schematic representation of the situation when $r = 2$ and an intersection of $\pm g$, for some $g \in \pi$, is introduced between the two handles that are added to $N \times I$.

Figure 8.197 (Wall realisation).

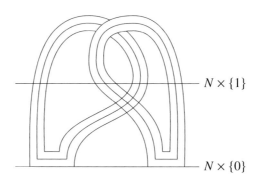

Since the attaching maps of the handles (ϕ_i^k) are each nullhomotopic, it follows that there is a homotopy equivalence $M \simeq N \vee (\vee_{i=1}^r S^k)$. We let $(M, \partial M)$ be the resulting manifold with boundary and we obtain a manifold triad structure $(M; \partial_0 M, \partial_1 M)$ on M by setting

$$\partial_0 M := (N \times \{0\}) \cup (\partial N \times [0, 1]) \quad \text{and} \quad \partial_1 M := \partial M - \text{int}(\partial_0 M).$$

By collapsing the handles (ϕ_i^k) inside a $(2k - 1)$-disk $D^{2k-1} \subset N \times \{1\}$, we can define an extension of the identity map $\text{id}\colon N \times [0, 1] \to N \times [0, 1]$ to a map $f\colon M \to N \times [0, 1]$ that induces a map of triads $(f; \partial_0 f, \partial_1 f)$ such that $\partial_0 f$ is a diffeomorphism. This map is covered by a bundle map $\overline{f}\colon TM \oplus \mathbb{R}^a \to TN \oplus \mathbb{R}^a$ that is given on $\partial_0 M$ by the differential of $\partial_0 f$. These claims follow from the fact that f_i^1 is regularly homotopic to f_i^0 and the f_i^0-s were trivial embeddings, so that we can regard this construction as the trace of surgeries on the identity map $N \to N$.

The kernel $K_k(\widetilde{M}) = K_k(\widetilde{M}, \widetilde{\partial_0 M})$ has a preferred basis corresponding to the cores $D^k \times \{0\} \subset D^k \times D^k = (\phi_i^k)$ of the new handles (ϕ_i^k). These cores can be completed to immersed spheres S_i by adjoining the images of the η_i' in $N \times [0, 1]$ and finally the obvious disks D^k in $N \times \{0\}$ whose boundaries are given by $f_i^0(S^{k-1})$. Then the intersection of S_i with S_j is the same as the intersection of η_i' with η_j' and the self-intersection of S_i is the self-intersection of η_i' with itself. Hence by construction

$$\lambda_M(S_i, S_j) = \lambda(b_i, b_j) \text{ for } i < j, \quad \text{and} \quad \mu_M(S_i) = \mu(b_i), \tag{8.198}$$

where λ_M is the geometric intersection pairing of Lemma 8.19, μ_M is the geometric self-intersection element of Lemma 8.26 and λ and μ come from the original quadratic form (P, λ, μ), see Subsection 8.4.3 and the proof of Theorem 8.167. The isomorphism $P \to P^*$ associated to the given non-singular $(-1)^k$-quadratic form can be identified with $K_k(\widetilde{M}) \to K_k(\widetilde{M}, \widetilde{\partial_1 M})$ if we use for $K_k(\widetilde{M}, \widetilde{\partial_1 M})$ the basis given by the cocores of the handles (ϕ_i^k). Hence $K_k(\widetilde{\partial_1 M}) = 0$. Since $k \geq 3$ and M

is obtained from $\partial_1 M$ by attaching k-handles, $\pi_1(\partial_1 M)$ is mapped isomorphically to $\pi_1(\partial_0 N)$ via $\partial_1 f$, and $K_*(\widetilde{\partial_1 M}) = 0$ for $* \leq k$. Lemma 8.116 (iv) implies that $\partial_1 f$ is a homotopy equivalence. Hence $\partial f = \partial_0 f \cup \partial_1 f \colon \partial M \to \partial N$ is also a homotopy equivalence and the normal map of pairs of degree one relative boundary $(f, \overline{f}) \colon (M, \partial M) \to (N, \partial N)$ has a surgery obstruction $\sigma(f, \overline{f}) \in L_{2k}(\mathbb{Z}\pi, w)$. From the definition of surgery obstruction relative boundary in the proof of Theorem 8.167, we get

$$\sigma(M, \partial M, f, \partial f, a, \xi, \overline{f}, o) = [(K_k(\widetilde{M}), s, t)] = \omega$$

under the identification (8.196). This finishes the proof of Theorem 8.195. □

Remark 8.199 (Non-realisability of surgery obstructions by closed manifolds)
It is not true that for any closed oriented manifold N of dimension $n = 2k \geq 6$ and any element $\omega \in L_{2k}(\mathbb{Z}\Pi(N), O_N)$, there is a normal map $(f, \overline{f}) \colon TM \oplus \mathbb{R}^a \to \xi$ covering a map of closed oriented manifolds $f \colon M \to N$ of degree one such that $\sigma(f, \overline{f}) = \omega$. Note that in Theorem 8.195 the target manifold $N \times [0, 1]$ is not closed.
If $N = S^{4k}$, then $\pi_1(N) = \{e\}$ and the signature gives an isomorphism $\sigma \colon L_{2k}(\mathbb{Z}\Pi(N), O_N) \xrightarrow{\cong} 8\mathbb{Z}$, see Theorem 8.99. By Theorem 8.173 (iii) the surgery obstruction to a degree one normal map $(f, \overline{f}) \colon M \to S^{4k}$ is $\text{sign}(M) \in 8\mathbb{Z}$. Since every stable vector bundle ξ over S^{4k} becomes trivial when restricted to $S^{4k} \setminus \text{pt}$ and the stable tangent bundle of M pulls back from ξ under f, it follows that M is a spin manifold. In dimension 4, Rokhlin's theorem [359] states that the signature of M is divisible by 16. Hence we see that not all elements of $L_4(\mathbb{Z})$ are realised as the surgery obstructions of degree one normal maps of closed smooth manifolds mapping to S^4. For $k \geq 2$, the signature of M becomes exponentially more divisible as k increases and fewer and fewer elements of $L_{4k}(\mathbb{Z})$ are realised as surgery obstructions of normal maps of closed manifolds to S^{4k}; for the precise statement, see Corollary 12.15 and Theorem 12.49 in Chapter 12 on homotopy spheres.
If $N = S^{4k+2}$, $4k + 2 \geq 6$, then $\pi_1(N) = \{e\}$ and the Arf invariant defines an isomorphism $\text{Arf} \colon L_{4k+2}(\mathbb{Z}) \xrightarrow{\cong} \mathbb{Z}/2$, see Theorem 8.111. If $(f, \overline{f}) \colon M \to S^{4k+2}$ is a $(2k + 1)$-connected degree one normal map, then it can be shown that M is stably parallelisable and that $\sigma(f, \overline{f})$ can be identified with the Arf invariant of any stable framing of M. It is known that the Arf invariant of a stably framed manifold is zero except in 5 or 6 special dimensions, and so except in these special dimensions $1 \in L_{4k+2}(\mathbb{Z})$ is not the surgery obstruction of a degree one normal map $(f, \overline{f}) \colon M \to S^{4k+2}$. For more information about the Arf invariant of framed manifolds in surgery, see Corollaries 12.15 and 12.53 in Chapter 12 on homotopy spheres.
For further information on which elements of $L_{2k}(\mathbb{Z}\pi, w)$ are realised as the surgery obstructions for closed manifold surgery problems, we refer to [171].

Consider w-oriented finite n-dimensional Poincaré pairs $(X, \partial X, [[X, \partial X]])$ and $(Y, \partial Y, [[Y, \partial Y]])$ for $n \geq 5$. Let $(u, \partial u) \colon (X, \partial X) \to (Y, \partial Y)$ be a map of pairs such that both u and ∂u are homotopy equivalences.
Since $(u, \partial u)$ is a homotopy equivalence of pairs, we have $w_1(X) = u^* w_1(Y)$. Hence there exists an isomorphism $O_X \to u^* O_Y$ of infinite cyclic local coefficients

systems over X. We conclude from the obvious generalisation of Lemma 7.4 to pairs that there is precisely one isomorphism of infinite cyclic local coefficients systems over X

$$o_u \colon O_X \to u^* O_Y \tag{8.200}$$

such that the isomorphism from $H_n(X, \partial X; O_X)$ to $H_n(Y, \partial Y; O_Y)$ induced by (u, o_u) sends $[[X, \partial X]]$ to $[[Y, \partial Y]]$.

Lemma 8.201 *Let $(M, \partial M, f, \partial f, a, \xi, \overline{f}, o)$ be a normal map of degree one relative boundary with target $(X, \partial X, O_X, [[X, \partial X]])$ in the sense of Subsection 8.8.1. Consider any vector bundle η over Y and a bundle map $(u, \overline{u}) \colon \xi \to \eta$ covering u. Then:*

(i) *There is precisely one isomorphism of infinite cyclic coefficients systems $o_{u,\overline{u}} \colon O_\eta \to O_Y$ over Y such that the isomorphism $o_u \circ o$ and $u^* o_{u,\overline{u}} \circ o_{\overline{u}}$ from O_ξ to $u^* O_Y$ agree;*
(ii) *We obtain a normal map $(M, \partial M, u \circ f, \partial u \circ \partial f, a, \eta, \overline{u} \circ \overline{f}, o_{u,\overline{u}})$ of degree one relative boundary with target $(Y, \partial Y, O_Y, [[Y, \partial Y]])$ in the sense of Subsection 8.8.1;*
(iii) *The isomorphism induced by the morphism $(u, o_u) \colon (X, O_X) \to (Y, O_Y)$ in $\mathcal{R}$*

$$L_n(\mathbb{Z}\Pi(X), O_X) \xrightarrow{\cong} L_n(\mathbb{Z}\Pi(Y), O_Y)$$

sends $\sigma(M, \partial M, f, \partial f, a, \xi, \overline{f}, o)$ to $\sigma(M, \partial M, u \circ f, \partial u \circ \partial f, a, \eta, \overline{u} \circ \overline{f}, o_{u,\overline{u}})$.

The analogous statement holds for the s-decoration where we additionally require that $(X, \partial X)$ and $(Y, \partial Y)$ are simple, u and ∂u are simple homotopy equivalences and one considers simple L-groups and simple surgery obstructions in the sense of Sections 10.2 and 10.4.

Proof. (i) This is obvious since u is a homotopy equivalence.

(ii) This follows from the definitions.

(iii) We consider for simplicity only the case where the decoration is h and $\partial X = \partial Y = \emptyset$ holds. Without loss of generality we can assume that Y and hence also M and X are connected. Choose base points $m \in M$, $x \in X$ and $y \in Y$ with $f(m) = x$ and $u(x) = y$. Choose a model for the universal covering $p_Y \colon \widetilde{Y} \to Y$ of Y and $\widetilde{y} \in \widetilde{Y}$ with $p_Y(\widetilde{y}) = y$. Let $p_X \colon \widetilde{X} \to X$ be the model for the universal covering of X and $p_M \colon \widetilde{M} \to M$ be the model for the universal covering of M obtained by the pullback construction applied to p_X with respect to u and $u \circ f$. Then we get unique base points $\widetilde{m} \in \widetilde{M}$ and $\widetilde{x} \in \widetilde{X}$ with $p_M(\widetilde{m}) = m$, $p_X(\widetilde{x}) = x$, and $\widetilde{u} \circ \widetilde{f}(\widetilde{m}) = \widetilde{u}(\widetilde{x}) = \widetilde{y}$. Choose an orientation of $T_m M$. Thus we have made a choice (NBP), see Notation 8.5, for both $(M, f, a, \xi, \overline{f}, o)$ and $(M, u \circ f, a, \eta, \overline{u} \circ \overline{f}, o_{u,\overline{u}})$ and we obtain an identification $\pi = \pi_1(M, m) = \pi_1(X, x) = \pi_1(Y, y)$. In view of the definition of the intrinsic surgery obstruction, we have to show that the (simple) surgery obstructions in the sense of Theorem 8.112 and Theorem 9.60 of the π-normal maps $(M, [M], f, \widetilde{f}, a, \xi, \overline{f})$ and $(M, [M], u \circ f, \widetilde{u} \circ \widetilde{f}, a, \eta, \overline{u} \circ \overline{f})$ in the sense of Definition 8.3 satisfy

$$\sigma(M, [M], f, \widetilde{f}, a, \xi, \overline{f}) = \sigma(M, [M], u \circ f, \widetilde{u} \circ \widetilde{f}, a, \eta, \overline{u} \circ \overline{f}) \quad \in L_n(\mathbb{Z}\pi, w).$$

8.9 Realisation of Even-Dimensional Surgery Obstructions   323

Firstly we treat the even-dimensional case $n = 2k$. We can assume without loss of generality that f is k-connected. Now the claim follows from the fact that the non-singular $(-1)^k$-quadratic forms $K_k(\widetilde{M}, \mu, \lambda)$ over $\mathbb{Z}\pi$ assigned to $(M, [M], f, \overline{f}, a, \xi, \overline{f})$ and $(M, [M], u \circ f, \widetilde{u} \circ \overline{f}, a, \eta, \overline{u} \circ \overline{f})$ using intersections and self-intersections agree.

Finally, we treat the odd-dimensional case $n = 2k + 1$. Again we can assume that the map f is k-connected. Now the claim follows since the two $(-1)^k$-quadratic formations assigned to the two maps in the proof of Theorem 9.60 agree. This finishes the proof of Lemma 8.201. □

Remark 8.202 Note that in Lemma 8.201 we are composing a given normal map with a (simple) homotopy equivalence and essentially get the same surgery obstruction if we identify the L-groups accordingly.

One may hope that the same is true if one precomposes with a (simple) homotopy equivalence. This is not true, as the following example shows.

Consider a normal map $(f, \overline{f})$ of degree one with underlying map $f \colon T^2 \to S^2$ covered by a bundle isomorphism $TT^2 \oplus \mathbb{R} \to \mathbb{R}^3$ such that $\sigma(f, \overline{f}) \in L_2(\mathbb{Z}) \cong \mathbb{Z}/2$ is non-trivial. There exists a normal map $(\mathrm{id}_{T^2}, \overline{g})$ of degree one with underlying map $\mathrm{id} \colon T^2 \to T^2$ covered by a bundle automorphism $\overline{g} \colon TT^2 \oplus \mathbb{R} \to TT^2 \oplus \mathbb{R}$ such that the surgery obstruction of the composite $(f \circ \mathrm{id}_{T^2}, \overline{f} \circ \overline{g})$ is trivial. Here the same ideas about dimensions apply as in Remark 8.175.

The following generalisation of Theorem 8.195, Theorem 9.115 and Theorem 10.31 will be needed in the construction of the geometric surgery exact sequence in Chapter 11.

Theorem 8.203 (Extended realisation of surgery obstructions) *Let $n \geq 5$. Let $(X, \partial X, O_X, [[X, \partial X]])$ be a w-oriented finite Poincaré pair of dimension $n - 1$. Consider a normal map $(M_0, \partial_0 M, f_0, \partial f_0, a_0, \xi, \overline{f_0}, o_0)$ of degree one relative boundary with target $(X, \partial X, O_X, [[X, \partial X]])$ in the sense of Subsection 8.8.1 such that f_0 is a homotopy equivalence and $\partial_0 f$ is a diffeomorphism or homotopy equivalence respectively. Consider*

$$\omega \in L_n^h(\mathbb{Z}\Pi(X \times [0, 1]), O_{X \times [0,1]}).$$

Then we can find a normal map of degree one relative boundary

$$(M_1, \partial M_1, f_1, \partial f_1, a_1, \xi, \overline{f_1}, o_1)$$

with target $(X, \partial X, O_X, [[X, \partial X]])$ such that f_1 is a homotopy equivalence and that ∂f_1 is a diffeomorphism or homotopy equivalence respectively, and a normal bordism

$$(W, \partial W, F, \partial F, b, \Xi, \overline{F}, O)$$

from $(M_0, f_0, a_0, \xi_0, \overline{f_0}, o_0)$ to $(M_1, f_1, a_1, \xi_1, \overline{f_1}, o_1)$ of degree one with cylindrical ends relative boundary in the sense of Subsection 8.8.2 such that $\partial_2 F$ is a diffeomorphism or homotopy equivalence respectively, and its surgery obstruction relative boundary, see Theorem 8.189, satisfies

$$\sigma(W, \partial W, F, \partial F, b, \Xi, \overline{F}, O) = \omega.$$

The analogous statement holds in the case where the decoration is s and we additionally require that X simple, u is a simple homotopy equivalence and one considers simple L-groups and simple surgery obstructions in the sense of Sections 10.2 and 10.4.

Proof. We conclude from Theorem 8.195, Theorem 9.115 that there exists a normal map of pairs of degree one relative boundary $(W, \partial W, G, \partial G, a, \xi, \overline{G}, o)$ with target $(M \times [0, 1], \partial(M \times [0, 1]), O_{M \times [0,1]}, [[M \times [0, 1], \partial(M \times [0, 1])]])$ satisfying:

(i) The bundle ξ over $M \times [0, 1]$ is $T(M \times [0, 1]) \oplus \mathbb{R}^a$ and the isomorphism o of infinite cyclic local coefficient systems over $M \times [0, 1]$ is the identity;

(ii) There is a decomposition $\partial W = \partial_0 W \cup \partial_1 W \cup \partial_2 W$ where $\partial_m W$ is a codimension zero submanifold of ∂W with possibly empty boundary $\partial \partial_m W$ for $m = 0, 1, 2$ satisfying

$$\partial_0 W \cap \partial_1 W = \emptyset;$$
$$\partial_2 W \cap \partial_m W = \partial \partial_m W \quad \text{for } m = 0, 1;$$
$$\partial \partial_2 W = \partial \partial_0 W \amalg \partial \partial_1 W.$$
$$\partial_2 W = \partial M \times [0, 1];$$
$$\partial_0 W = M \times \{0\};$$

(iii) The homotopy equivalence $\partial G \colon \partial W \to \partial(M \times [0, 1])$ induces the identity $\partial_0 W = M \times \{0\} \to (M \times \{0\})$ on $\partial_0 W$ and the identity $\partial_2 W = \partial M \times [0, 1] \to \partial M \times [0, 1]$ on $\partial_2 W$;

(iv) The homotopy equivalence $\partial G \colon \partial W \to \partial(M \times [0, 1])$ maps $\partial_1 M$ to $N \times \{1\}$ and induces a simple homotopy equivalence

$$\partial_1 G \colon \partial_1 W \to M \times \{1\};$$

(v) The intrinsic surgery obstruction

$$\sigma^h(W, \partial W, G, \partial G, a, \xi, \overline{G}, o) \in L_n^h(\mathbb{Z}\Pi(M \times [0, 1]), O_{M \times [0,1]})$$

in the sense of Section 10.4 is mapped to ω.

Now define the desired normal bordism $(W, F, b, \Xi, \overline{F}, O)$ of degree one with cylindrical ends to be the composite of $(W, \partial W, G, \partial G, a, \xi, \overline{G}, o)$ with the obvious normal map given by crossing $(M_0, f_0, a_0, \xi, \overline{f_0}, o_0)$ with $[0, 1]$ whose underlying map is $f_0 \times \mathrm{id}_{[0,1]} \colon M \times [0, 1] \to X \times [0, 1]$. Now apply Lemma 8.201 in the case $u = f_0 \times \mathrm{id}_{[0,1]}$. □

Exercise 8.204 Suppose $n = 2k \geq 6$. Let $(X; \partial_0 X, \partial_1 X)$ be a connected $2k$-dimensional manifold triad with $\partial_0 X$ connected and non-empty. Suppose that $\omega \in L_{2k}(\mathbb{Z}\Pi(X), O_X)$ lies in the image of the natural homomorphism

$$L_{2k}(\mathbb{Z}\Pi(\partial_0 X), O_{\partial_0 X}) \to L_{2k}(\mathbb{Z}\Pi(X), O_X).$$

Show that the underlying Poincaré pair $(X, \partial X)$ admits a degree one normal map relative boundary $(f, \overline{f}) \colon (M, \partial M) \to (X, \partial X)$ with $\sigma(f, \overline{f}) = \omega$.

8.10 Notes

Implicitly the fact that the signature is a complete invariant of $L_0(\mathbb{Z})$ is already in Milnor [297, Theorem 4] and implicitly the computation of $L_n(\mathbb{Z})$ is in Kervaire–Milnor [216]. The Arf invariant and its role in the classification of quadratic forms was identified by Arf [8]. For more information about bilinear forms over the integers and the Arf invariant we refer the reader to [55] and [306].

A guide for the calculation of the L-groups for finite groups will be given in Chapter 16.

Remark 8.205 (Browder's homotopy-theoretic formulation) We have seen that the underlying symmetric form of a surgery obstruction in even dimensions is given by the intersection numbers of the representing spheres and these numbers coincide with those obtained from the cup product pairing. The quadratic form is given by the self-intersection numbers. In the simply connected case these also have an algebraic topological interpretation, as described by Browder in [55, Chapter III].

His construction uses S-duality from stable homotopy theory. The main result used is that the S-dual of the Thom space $T(\nu_X)$ of the Spivak normal fibration of a Poincaré complex X is $T(\nu_X)^* = \Sigma^p X_+$. Hence a degree one normal map $(f, b) \colon M \to X$ induces a map of Thom spaces $T(b) \colon T(\nu_M) \to T(\nu_X)$, and this in turn induces $F := T(b)^* \colon T(\nu_X)^* \to T(\nu_M)^*$. Browder uses the map F and functional Steenrod squares to define a quadratic refinement of the mod 2 cup product pairing and shows that it coincides with the surgery obstruction as we defined it here.

An almost framed manifold possesses a degree one normal map to the sphere of the same dimension, and the Arf invariant of the associated quadratic form is called the Kervaire invariant. The formulation of Browder from the previous paragraph ultimately led to the formulation of the Kervaire Invariant 1 problem in terms of the Adams spectral sequence in [54]. More about this topic can be found in Chapter 12.

Browder's ideas were generalised by Ranicki. As we will see in Chapter 15, the generalisation is in terms of quadratic structures on chain complexes. Moreover, these constructions work also in the non-simply connected case and enable conceptual proofs of some properties of surgery obstructions such as product formulas and a unified treatment of both even and odd dimensions. The definition of the L-groups in this setting is presented in Definition 15.301 and the definition of the surgery obstructions in Definition 15.318.

Chapter 9
The Odd-Dimensional Surgery Obstruction

9.1 Introduction

In this chapter we shall give the solution to the surgery problem in odd dimensions. We work in the smooth category unless explicitly stated otherwise. Let $(f, \bar{f}) \colon M \to X$ be a normal map of degree one from a closed $(2k + 1)$-dimensional manifold M to a Poincaré complex X. The surgery problem, stated as Problem 7.42, is to determine whether we can change the normal map $(f, \bar{f})$ by finitely many surgery steps to obtain a normal homotopy equivalence $(g, \bar{g}) \colon N \to X$ from a closed manifold N to X.

It turns out that the problem in this case is both harder and easier than in the even-dimensional case. It is harder because the algebra that describes the surgery obstructions is more complicated. On the other hand it is easier, because the surgery obstruction groups turn out to vanish more often. This happens for example in the simply connected situation, so in that case there is no surgery obstruction and our problem always has a positive solution.

Here are some informal ideas about how one might approach our problem. As in the previous even-dimensional case, we can make the normal map $(f, \bar{f}) \colon M \to X$ be k-connected, which means that we now have two potentially non-vanishing surgery kernel groups, namely $K_k(\widehat{M})$ and $K_{k+1}(\widehat{M})$. Next one might attempt to kill $K_k(\widehat{M})$ by choosing a set of its generators, represent them by disjoint embedded spheres with trivial normal bundles (which in this case can always be done since $2k < n$) and do surgery on them. This may or may not be successful since it is not clear that one can find appropriate elements so that after doing surgery on them the surgery kernels are trivial. The problem is that surgery on such an element simultaneously affects $K_k(\widehat{M})$ and $K_{k+1}(\widehat{M})$ since these are related by Poincaré duality. So by killing a class we may at the same time create another one. Hence the question becomes: what does the success depend on?

Historically there were several approaches. We follow the approach of Ranicki, which builds on the work of Wall and Novikov and uses the algebra of so-called *formations*. In the above situation they arise as follows. The union of the embedded spheres representing generators of $K_k(\widehat{M})$ with the tubular neighbourhoods U induces a kind of Heegaard decomposition of the manifold $M = M_0 \cup U$. It turns out that the target can be similarly decomposed into $X = X_0 \cup D^n$ and the map $(f, \overline{f})$ can be made to respect the decomposition. Then the restriction of $(f, \overline{f})$ to $\partial M_0 = \partial U$ is an even-dimensional surgery problem that bounds in two different ways, namely via U and via M_0. Each of these coboundaries produces a lagrangian for $K_k(\widehat{\partial U})$. So we take the lagrangian, say F, coming from U, to obtain an isomorphism of the quadratic form on $K_k(\widehat{\partial U})$ with the standard hyperbolic form on F. The lagrangian, say G, from M_0 is then a submodule of $F \oplus F^*$. A formation is by definition a quadratic form with an ordered pair of lagrangians. Formations can be organised into an L-group as presented in Definition 9.15. In our geometric situation the quadratic form on the kernel $K_k(\widehat{\partial U})$ with the ordered pair of lagrangians (F, G) gives the formation that represents the surgery obstruction associated to $(f, \overline{f})$.

In the process of proving the main theorem, we work with the above Heegaard decomposition on one hand, this we call working "inside", but we also work with the trace W of the surgery on the elements that give U, this we call working "outside". It turns out that our formation can also be described via W and the two ends of it, M and M'. The equivalence between the two formations is proved in Lemma 9.88. This is very convenient since some properties are easier to see one way and some the other way. This gives us freedom to use whichever suits a particular purpose. For example the bordism formation gives an illuminating criterion for when the resulting map on M' is a homotopy equivalence, which is later used.

The L-group is constructed in such a way that the mutual positions of the lagrangians F and G decide whether a formation represents the zero element or not. The two lagrangians F and G are called *complementary* if we have $F \cap G = \{0\}$ and $F + G = P$. It turns out that there are two prominent classes of formations that represent the zero element. These are the *trivial formations*, i.e., formations with the property that they are isomorphic to formations with $G = F^*$, or, equivalently, that F and G are complementary, and the *boundary formations*, i.e., formations with the property that there is a lagrangian $L \subset P$ such that L is a complement of both F and G. That these two classes of formations should represent the zero element is nicely motivated by Lemma 9.75 and Lemma 9.96.

The main result of this chapter is Theorem 9.60 and its intrinsic version Theorem 9.108, where we show that the vanishing of the algebraic surgery obstruction is necessary and sufficient in dimensions ≥ 5 to do the desired surgery to achieve a normal homotopy equivalence and that the algebraic surgery obstruction is a normal bordism invariant. They are the odd-dimensional versions of Theorems 8.112 and 8.167.

The proof of the fact that the vanishing of the formation in the L-group suffices to give a solution to the surgery problem can be described as a two-attempts strategy. Namely, in the first attempt one does surgery on a set of generators of $K_k(\widehat{M})$ as described above. The result may not be a homotopy equivalence, but leads at least to a Heegaard decomposition as mentioned above. Moreover, the associated formation represents zero in the L-group. At this stage we will need two lemmas about the realisation of formations in the same class in the L-groups, namely Lemma 9.93

and Lemma 9.99. Then, since our formation represents zero in the L-group, we can realise a representative of the normal bordism class that gives a normal homotopy equivalence, see the proof of Theorem 9.60.

The intrinsic version, the version for manifolds with boundary, and the realisation theorem proceed analogously as in the even-dimensional case. Namely, Theorem 9.110 is the analogue of Theorem 8.173, Theorem 9.113 is the analogue of Theorem 8.189, and Theorem 9.115 is the analogue of Theorem 8.195.

Guide 9.1 The setup of the odd-dimensional surgery obstruction is rather complicated. The reader may skip this chapter completely just having in mind that the main results of Chapter 8 about the normal bordism invariance and the main obstruction property of the surgery obstruction in the even-dimensional cases also hold in the odd-dimensional case, and that in the simply connected case the L-groups vanish in odd dimensions.

The first time reader of this chapter may skip Sections 9.2.3 and 9.2.4 in the first round.

9.2 Odd-Dimensional L-Groups

In this section we explain the algebraic objects that describe the surgery obstruction in odd dimensions and which will represent the elements in the odd-dimensional surgery obstruction groups. Throughout this subsection R will be an associative unital ring with involution and $\epsilon \in \{\pm 1\}$. We also assume that R has the *invariant basis number (IBN) property*, i.e., the free R-modules R^n and R^m are R-isomorphic if and only if $n = m$. Fields and principal ideal domains obviously have property (IBN).

Exercise 9.2 Show for a ring R with property (IBN) that for any group G the group ring RG has property (IBN).

Recall that $K_1(R)$ is defined in terms of automorphisms of finitely generated free R-modules. Since L-groups are quadratic analogues of K-groups, it is plausible that the odd-dimensional L-groups, $L_{2k+1}(R)$, will be defined in terms of isometries of hyperbolic quadratic forms. Indeed, this was the case in Wall's original definition of these groups in [414]. Here we present a newer definition of $L_{2k+1}(R)$ via formations due to Ranicki, which first appeared in [341] and coincides with the one given in the book [352].

The first part of this section introduces formations and gives the first definition of $L_{2k+1}(R)$, which we use to define the surgery obstruction in Section 9.3. The second part establishes important properties of the groups $L_{2k+1}(R)$ and also gives Wall's original definition via automorphisms of hyperbolic quadratic forms. The two definitions are related in Lemma 9.6.

Ranicki states in [344] that formations first appeared in the paper of Wall [405], and were later used by Novikov in [323], where their importance in the present context became clearer. The definition of the L-group via formations first appears in [341], where their algebraic properties are investigated. The proof of the main theorem about surgery obstructions, which we present here, follows the same general strategy and also many details of the presentation in [352].

In this section all modules are finitely generated free R-modules.

9.2.1 The Definition of $L_{2k+1}(R)$ via Formations

Recall that an ϵ-quadratic form (P, ψ) on a module P is called non-singular if $(1 + \epsilon \cdot T)(\psi) \colon P \to P^*$ is an isomorphism and that a lagrangian $L \subset P$ is a summand with $L = L^\perp$ on which ψ vanishes.

Definition 9.3 (Quadratic formation) An ϵ-*quadratic formation*, denoted $(P, \psi; F, G)$, consists of a non-singular ϵ-quadratic form (P, ψ) together with an ordered pair of lagrangians F and G.

Remark 9.4 In some places, e.g., [344], formations, as we defined them, are called non-singular formations and the word "formation" is used for the case where one of F and G is a sublagrangian.

An isomorphism $f \colon (P, \psi; F, G) \to (P', \psi'; F', G')$ of ϵ-quadratic formations is an isomorphism $f \colon (P, \psi) \to (P', \psi')$ of non-singular ϵ-quadratic forms such that $f(F) = F'$ and $f(G) = G'$. The direct sum of ϵ-quadratic formations $(P_0, \psi_0; F_0, G_0)$ and $(P_1, \psi_1; F_1, G_1)$ is the ϵ-quadratic formation $(P_0 \oplus P_1, \psi_0 \oplus \psi_1; F_0 \oplus F_1, G_0 \oplus G_1)$.

For a module F, recall the standard ϵ-quadratic hyperbolic form $H_\epsilon(F)$ of Example 8.79. The underlying module of $H_\epsilon(F)$ is $F \oplus F^*$, where both $F := F \oplus \{0\}$ and $F^* := \{0\} \oplus F^*$ are lagrangians.

Lemma 9.5 *Given a formation* $(P, \psi; F, G)$, *there is a lagrangian* G' *of* $H_\epsilon(F)$ *and an isomorphism of formations* $f \colon (H_\epsilon(F); F, G') \xrightarrow{\cong} (P, \psi; F, G)$ *such that* $f \colon F \oplus F^* \xrightarrow{\cong} P$ *restricted to* F *is the identity* id_F.

Proof. By Lemma 8.95 the inclusion of ϵ-quadratic forms $(F, 0) \to (P, \psi)$ extends to an isomorphism $f \colon H_\epsilon(F) \xrightarrow{\cong} (P, \psi)$. Put $G' = f^{-1}(G)$. □

Example 9.6 Any automorphism $f \colon H_\epsilon(F) \xrightarrow{\cong} H_\epsilon(F)$ defines the formation $(H_\epsilon(F); F, f(F))$. In Lemma 9.33 below we prove that every formation is isomorphic to $(H_\epsilon(F); F, f(F))$ for some F and f.

Definition 9.7 (Trivial ϵ-quadratic formation) The *trivial ϵ-quadratic formation* associated to a module F is the formation $(H_\epsilon(F); F, F^*)$. A formation $(P, \psi; F, G)$ is called *trivial* if it isomorphic to the trivial ϵ-quadratic formation associated to some module.

9.2 Odd-Dimensional L-Groups

Definition 9.8 (Stable isomorphism of ϵ-quadratic formations) A *stable isomorphism* of ϵ-quadratic formations $(P, \psi; F, G)$ and $(P', \psi'; F', G')$ is an isomorphism

$$f: (P, \psi; F, G) \oplus (H_\epsilon(Q_0); Q_0, Q_0^*) \xrightarrow{\cong} (P', \psi'; F', G') \oplus (H_\epsilon(Q_1); Q_1, Q_1^*)$$

for trivial formations $(H_\epsilon(Q_0); Q_0, Q_0^*)$ and $(H_\epsilon(Q_1); Q_1, Q_1^*)$. In this case, $(P, \psi; F, G)$ and $(P', \psi'; F', G')$ are called *stably isomorphic* and we write $(P, \psi; F, G) \cong_s (P', \psi'; F', G')$.

Definition 9.9 (The boundary of a $(-\epsilon)$-quadratic form) Let (Q, μ) be *any* $(-\epsilon)$-quadratic form; in particular (Q, μ) can be singular. Define the *boundary* of (Q, μ), denoted $\partial(Q, \mu)$, to be the ϵ-quadratic formation $(H_\epsilon(Q); Q, \Gamma_\mu)$, where Γ_μ is the lagrangian given by the image of the R-homomorphism

$$Q \to Q \oplus Q^*, \quad x \mapsto (x, (1 - \epsilon \cdot T)(\mu)(x)).$$

Remark 9.10 Note that the ϵ-quadratic formation $\partial(Q, \mu)$ in Definition 9.9 only depends on the $(-\epsilon)$-symmetric form $(1 - \epsilon \cdot T)(\mu)$. The next exercise shows that Γ_μ is a lagrangian and explains why it is essential to require that we consider the graphs of $(-\epsilon)$-symmetric forms admitting $(-\epsilon)$-quadratic refinements.

Exercise 9.11 Let $\lambda: Q \to Q^*$ be an $(-\epsilon)$-symmetric form and $\Gamma_\lambda := \{(x, \lambda(x)) \mid x \in Q\} \subset Q \oplus Q^*$ be the graph of λ. Verify the following:

(i) The symmetric form of $H_\epsilon(Q)$ vanishes on $\Gamma_\lambda \times \Gamma_\lambda$;
(ii) Γ_λ is a lagrangian of $H_\epsilon(Q)$ if and only if λ admits a $(-\epsilon)$-quadratic refinement.

In particular, this shows that Γ_μ in Definition 9.9 is a lagrangian.

Definition 9.12 (Boundary formation) We call an ϵ-quadratic formation $(P, \psi; F, G)$ a *boundary formation* if it is isomorphic to $\partial(Q, \mu)$ for some (not necessarily non-singular) $(-\epsilon)$-quadratic form $\partial(Q, \mu)$.

The following lemma establishes the basic properties of formations needed for the definition of $L_{2k+1}(R)$. Two lagrangians F and G of a non-singular ϵ-quadratic form (P, ψ) are called *complementary* if we have $F \cap G = \{0\}$ and $F + G = P$, or equivalently, the obvious R-homomorphism $F \oplus G \to P$ induced by the inclusions of F and G in P is an isomorphism.

Lemma 9.13 Let $(P, \psi; F, G)$ be an ϵ-quadratic formation. Then:

(i) $(P, \psi; F, G)$ is trivial if and only F and G are complementary;
(ii) $(P, \psi; F, G)$ is isomorphic to a boundary if and only if there is a lagrangian $L \subset P$ such that L is a complement of both F and G;
(iii) There is an ϵ-quadratic formation $(P', \psi'; F', G')$ such that the sum of formations $(P, \psi; F, G) \oplus (P', \psi'; F', G')$ is a boundary;
(iv) An $(-\epsilon)$-quadratic form (Q, μ) is non-singular if and only if its boundary is a trivial formation.

Proof. (i) Obviously F and G are complementary if $(P, \psi; F, G)$ is isomorphic to $(H_\epsilon(F); F, F^*)$. It remains to show that $(P, \psi; F, G)$ is isomorphic to $(H_\epsilon(F); F, F^*)$, provided that F and G are complementary.

The inclusions of the lagrangians F and G into P induce an R-isomorphism $h: F \oplus G \to P$. Choose a representative $\psi: P \to P^*$ of $\psi \in Q_\epsilon(P)$. Then we can write
$$h^* \circ \psi \circ h = \begin{pmatrix} a & b \\ c & d \end{pmatrix} : F \oplus G \to (F \oplus G)^* = F^* \oplus G^*.$$

Let $\psi' \in Q_\epsilon(F \oplus G)$ be the class of $h^* \circ \psi \circ h$. Then h is an isomorphism of non-singular ϵ-quadratic forms $(F \oplus G, \psi') \to (P, \psi)$ and F and G are lagrangians in $(F \oplus G, \psi')$. Hence the isomorphism $(1 + \epsilon \cdot T)(\psi') : F \oplus G \to F^* \oplus G^*$ is given by
$$\begin{pmatrix} a & b \\ c & d \end{pmatrix} + \epsilon \cdot T \begin{pmatrix} a & b \\ c & d \end{pmatrix} = \begin{pmatrix} a + \epsilon a^* & b + \epsilon c^* \\ c + \epsilon b^* & d + \epsilon d^* \end{pmatrix} = \begin{pmatrix} 0 & e \\ f & 0 \end{pmatrix}$$
for some R-maps $e: G \to F^*$ and $f: F \to G^*$. Hence $e = (b + \epsilon \cdot c^*): G \to F^*$ is an isomorphism. Define an R-isomorphism
$$u = \begin{pmatrix} 1 & 0 \\ 0 & e \end{pmatrix} : F \oplus G \xrightarrow{\cong} F \oplus F^*.$$

One easily checks that u defines an isomorphism of formations
$$u: (F \oplus G, \psi'; F, G) \xrightarrow{\cong} (H_\epsilon(F); F, F^*). \tag{9.14}$$

(ii) One easily checks that for an $(-\epsilon)$-quadratic form (Q, μ) the lagrangian Q^* in its boundary $\partial(Q, \mu) = (H_\epsilon(Q); Q, \Gamma_\mu)$ is complementary to both Q and Γ_μ. Conversely, suppose that $(P, \psi; F, G)$ is an ϵ-quadratic formation such that there exists a lagrangian $L \subset P$ that is complementary to both F and G. By the argument appearing in the proof of assertion (i), we find an isomorphism of ϵ-quadratic formations
$$f: (H_\epsilon(F); F, F^*) \xrightarrow{\cong} (P, \psi; F, L)$$
that is the identity on F. The preimage $G' := f^{-1}(G)$ is a lagrangian in $H_\epsilon(F)$ that is complementary to F^*. The fact that G' is a complementary subspace to F^* means that
$$G' = \{(x, \lambda(x)) \in F \oplus F^* \mid x \in F\}$$
for some $\lambda \in \mathrm{Hom}_R(F, F^*)$. The fact that it is a lagrangian means that $\lambda = \psi' - \epsilon(\psi')^*$ for some element $\psi' \in Q_{-\epsilon}(F)$. The boundary of (F, ψ') is precisely $(H_\epsilon(F); F, G')$ and f induces an isomorphism of ϵ-quadratic formations
$$\partial(F, \psi') = (H_\epsilon(F); F, G') \xrightarrow{\cong} (P, \psi; F, G).$$

(iii) Because of Lemma 8.95, we can find lagrangians $\widehat{F}$ and $\widehat{G}$ such that F and $\widehat{F}$ are complementary and G and $\widehat{G}$ are complementary. Define the formation $(P', \psi'; F', G') := (P, -\psi; \widehat{F}, \widehat{G})$. Then $M := \{(p, p) \mid p \in P\} \subset P \oplus P$ is a la-

9.2 Odd-Dimensional L-Groups

grangian in the direct sum

$$(P, \psi; F, G) \oplus (P', \psi'; F', G') = (P \oplus P, \psi \oplus (-\psi), F \oplus \widehat{F}, G \oplus \widehat{G})$$

that is complementary to both $F \oplus \widehat{F}$ and $G \oplus \widehat{G}$. Hence the direct sum is isomorphic to a boundary by assertion (ii).

(iv) The lagrangian Γ_μ in the boundary $\partial(Q, \mu) := (H_\epsilon(Q); Q, \Gamma_\mu)$ is complementary to Q if and only if $(1 - \epsilon \cdot T)(\mu) \colon Q \to Q^*$ is an isomorphism. This finishes the proof of Lemma 9.13. □

Now we can define the odd-dimensional surgery groups.

Definition 9.15 (Quadratic L-groups in odd dimensions) Let R be an associative ring with unit and involution, which has property (IBN). For an odd integer $n = 2k + 1$ set $\epsilon := (-1)^k$ and define the abelian group $L_n(R)$, called the *n-th quadratic L-group* of R, to be the abelian group of equivalence classes $[P, \psi; F, G]$ of ϵ-quadratic formations $(P, \psi; F, G)$ such that P, F and G are finitely generated free R-modules with respect to the following equivalence relation. We call $(P, \psi; F, G)$ and $(P', \psi'; F', G')$ equivalent if and only if there exist $(-\epsilon)$-quadratic forms (Q, μ) and (Q', μ') for finitely generated free R-modules Q and Q' and a stable isomorphism of formations

$$(P, \psi; F, G) \oplus \partial(Q, \mu) \cong_s (P', \psi'; F', G') \oplus \partial(Q', \mu').$$

Addition is given by taking the orthogonal sum of two ϵ-quadratic formations. The zero element is represented by $\partial(Q, \mu) \oplus (H_\epsilon(S); S, S^*)$ for any $(-\epsilon)$-quadratic form (Q, μ) and any finitely generated free R-module S. The inverse of $[P, \psi; F, G]$ is represented by $(P, -\psi; \widehat{F}, \widehat{G})$ for any choice of lagrangians $\widehat{F}$ and $\widehat{G}$ in (P, ψ) such that F and $\widehat{F}$ are complementary and G and $\widehat{G}$ are complementary.

Exercise 9.16 If we replace in Definition 9.15 "finitely generated free" by "stably finitely generated free", we get another definition of the odd-dimensional L-group. There is a forgetful map from the L-group of Definition 9.15 to this modified definition. Show that it is an isomorphism.

The passage from finitely generated free to finitely generated projective modules will change the L-groups in general, whereas the algebraic K-groups $K_n(R)$ for $n \geq 1$ will not change under this passage. Some information on this topic can be found in Chapter 15.

Remark 9.17 (Functoriality of L-groups in odd dimensions) We have explained in Remark 8.93 that for $k = 0, 2$ a morphism $u \colon R \to S$ of rings with involution induces homomorphisms $u_* \colon L_k(R) \to L_k(S)$ by induction. Analogously one shows for $k = 1, 3$ that a morphism $u \colon R \to S$ of rings with involution induces homomorphisms $u_* \colon L_k(R) \to L_k(S)$ by induction and we have $(u \circ v)_* = u_* \circ v_*$ and $(\mathrm{id}_R)_* = \mathrm{id}_{L_k(R)}$.

In the previous Chapter 8, Lemma 8.152 about the fact that conjugation with an element $g \in \pi$ induces multiplication by $w(g)$ on $L_{2k}(\mathbb{Z}G)$ plays a major role. There is analogous lemma in the odd-dimensional case, which is formulated as Lemma 9.38 and proved at the end of Section 9.2.3.

In Section 9.2.4 below, we prove that the odd-dimensional L-groups of the ring of integers vanish.

Theorem 9.18 (Vanishing of $L_{2k+1}(\mathbb{Z})$) *We have $L_{2k+1}(\mathbb{Z}) = 0$.*

In this chapter we only deal with ϵ-quadratic formations. A notion of an ϵ-symmetric formation also exists and will be discussed in Chapter 15, see Definition 15.198.

9.2.2 Presentations of Formations

Presentations of formations are useful tools for understanding the structure of formations. Their refinement to split formations plays a key role in linking the geometric surgery obstruction to the surgery obstruction of algebraic surgery defined later in Chapter 15. The main result of this subsection is Lemma 9.29, which plays a key role in proving that the surgery obstruction is well defined in Section 9.3.

Definition 9.19 (Presentation of formation) Let $(P, \psi; F, G)$ be an ϵ-quadratic formation. A *presentation* $(f, (\gamma, \delta))$ of $(P, \psi; F, G)$, sometimes more briefly denoted by (γ, δ), consists of an isomorphisms of ϵ-quadratic formations

$$f \colon (P, \psi; F, G) \xrightarrow{\cong} (H_\epsilon(F); F, G')$$

such that $f|_F = \mathrm{id}_F$ and $f(G) = G'$ holds and the inclusion $G' \to F \oplus F'$ composed with the isomorphism $G \to G'$ induced by f is given by

$$\begin{pmatrix} \gamma \\ \delta \end{pmatrix} \colon G \to F \oplus F^*.$$

The maps $\gamma \colon G \to F$ and $\delta \colon G \to F^*$ are sometimes called *structure homomorphisms*.

Lemma 9.20 *Let $(P, \psi; F, G)$ be an ϵ-quadratic formation.*

(i) *There exists a presentation (γ, δ);*
(ii) *A presentation (γ, δ) yields the exact sequence*

$$0 \longrightarrow G \xrightarrow{\begin{pmatrix} \gamma \\ \delta \end{pmatrix}} F \oplus F^* \xrightarrow{\begin{pmatrix} \delta^* & \epsilon\gamma^* \end{pmatrix}} G^* \longrightarrow 0; \qquad (9.21)$$

(iii) *If (γ, δ) and (γ', δ') are two presentations, then there exists an $(-\epsilon)$-quadratic form (F^*, μ) and an automorphism $u \colon G \xrightarrow{\cong} G$ satisfying*

9.2 Odd-Dimensional L-Groups

$$\gamma' \circ u = \gamma + (\mu - \epsilon\mu^*) \circ \delta;$$
$$\delta' \circ u = \delta.$$

Proof. (i) We conclude from Lemma 9.5 that for every formation $(P, \psi; F, G)$ there is an isomorphism

$$f : (P, \psi; F, G) \xrightarrow{\cong} (H_\epsilon(F); F, G')$$

satisfying $f|_F = \mathrm{id}_F$ and $G = f(G')$.

(ii) The exactness of the sequence is equivalent to the condition that G' is a lagrangian in $H_\epsilon(F)$.

(iii) Consider isomorphisms of ϵ-quadratic formations

$$f : (P, \psi; F, G) \xrightarrow{\cong} (H_\epsilon(F); F, G');$$
$$f' : (P, \psi; F, G) \xrightarrow{\cong} (H_\epsilon(F); F, G''),$$

satisfying $f_F = \mathrm{id}_F$, $f'_F = \mathrm{id}_F$, $f(G) = G'$, and $f'(G) = G''$. Then

$$g := f' \circ f^{-1} : (H_\epsilon(F); F, G') \xrightarrow{\cong} (H_\epsilon(F); F, G'')$$

is an isomorphism of ϵ-quadratic formations satisfying $g|_F = \mathrm{id}_F$ and $g(G') = g(G'')$. A short calculation shows that the isomorphism g is expressed by a matrix

$$g = \begin{pmatrix} \mathrm{id} & \mu - \epsilon\mu^* \\ 0 & \mathrm{id} \end{pmatrix}$$

for some $(-\epsilon)$-quadratic form (F^*, μ). We get

$$g \circ \begin{pmatrix} \gamma \\ \delta \end{pmatrix} = \begin{pmatrix} \mathrm{id} & \mu - \epsilon\mu^* \\ 0 & \mathrm{id} \end{pmatrix} \circ \begin{pmatrix} \gamma \\ \delta \end{pmatrix} = \begin{pmatrix} \gamma + (\mu - \epsilon\mu^*) \circ \delta \\ \delta \end{pmatrix}.$$

Since $g \circ \begin{pmatrix} \gamma \\ \delta \end{pmatrix}$ and $\begin{pmatrix} \gamma' \\ \delta' \end{pmatrix}$ are injective and have the same image, we can take the automorphism $u : G \xrightarrow{\cong} G$ to be the composite of the three isomorphisms

$$G \xrightarrow{\begin{pmatrix} \gamma \\ \delta \end{pmatrix}} G' \xrightarrow{g|_{G'}} G'' \xrightarrow{\begin{pmatrix} \gamma' \\ \delta' \end{pmatrix}^{-1}} G.$$

This finishes the proof. □

Exercise 9.22 Consider finitely generated free R-modules F and G and morphisms $\gamma : G \to F$ and $\delta : G \to F^*$. Show that (γ, δ) is a presentation of some ϵ-quadratic formation if and only the sequence (9.21) is exact.

Consider a presentation (γ, δ) of the ϵ-quadratic formation $(P, \psi; F, G)$. Since the induced quadratic form on G is $(G, [\gamma^*\delta]) = (G, 0)$, we deduce that $[\gamma^*\delta] = 0$ in $Q_\epsilon(G)$. Hence $\gamma^*\delta \colon G \to G^*$ is a $(-\epsilon)$-symmetric form that admits a quadratic refinement, i.e., there is an $(-\epsilon)$-quadratic form $(G, \theta_{\gamma,\delta})$, called the *hessian* of the presentation $(f, (\gamma, \delta))$, satisfying

$$\theta_{\gamma,\delta} - \epsilon\theta^*_{\gamma,\delta} = \gamma^*\delta = -\epsilon\delta^*\gamma. \tag{9.23}$$

Hessians will play an important role in Chapter 15, which presents the definition of L-groups using chain complexes. Hessians are a tool that will enable us to relate ϵ-quadratic formations to 1-dimensional quadratic complexes in the sense of Chapter 15; see the material around Definition 15.208.

Lemma 9.24 *Let $(f, (\gamma, \delta))$ and $(f', (\gamma', \delta'))$ be presentations of ϵ-quadratic formations $(P, \psi; F, G)$ and $(P', \psi'; F', G')$. Then the following assertions hold:*

(i) *There are $(-\epsilon)$-quadratic forms $(G, \theta_{\gamma,\delta})$ and $(G', \theta'_{\gamma',\delta'})$ satisfying (9.23);*

(ii) *The ϵ-quadratic formations $(P, \psi; F, G)$ and $(P', \psi'; F', G')$ are isomorphic if and only if there are isomorphisms $\alpha \colon F \xrightarrow{\cong} F'$ and $\beta \colon G \xrightarrow{\cong} G'$ and an $(-\epsilon)$-quadratic form (F^*, μ) such that the following diagram commutes:*

$$\begin{array}{ccc} G & \xrightarrow{\beta} & G' \\ {\scriptsize \begin{pmatrix} \gamma \\ \delta \end{pmatrix}} \Big\downarrow & & \Big\downarrow {\scriptsize \begin{pmatrix} \gamma' \\ \delta' \end{pmatrix}} \\ F \oplus F^* & \xrightarrow{\begin{pmatrix} \alpha & \alpha(\mu - \epsilon\mu^*) \\ 0 & (\alpha^*)^{-1} \end{pmatrix}} & F' \oplus F'^* \end{array}$$

(iii) *$(P, \psi; F, G)$ is trivial if and only if for some presentation (γ, δ) of $(P, \psi; F, G)$ (and hence all such presentations), the map $\delta \colon G \to F^*$ is an isomorphism;*

(iv) *$(P, \psi; F, G)$ is a boundary if and only if for some presentation (γ, δ) of $(P, \psi; F, G)$, the map $\gamma \colon G \to F$ is an isomorphism.*

Proof. Parts (i), and (ii) follow from the definitions, Lemma 9.20 and its proof, and the discussion before Lemma 9.24.

(iii) Note that $F \cap G = \ker(\delta)$ and $P/(F + G) \cong \operatorname{coker}(\delta)$ hold. By Lemma 9.13 (i), $(P, \psi; F, G)$ is trivial if and only if $F \cap G = 0 = P/(F + G)$. Hence $(P, \psi; F, G)$ is trivial if and only if $\ker(\delta) = 0 = \operatorname{coker}(\delta)$, or, equivalently, δ is an isomorphism.

(iv) Suppose that γ is an isomorphism. Then $F^* \subset H_\epsilon(F)$ is a lagrangian complementary to both F and G. Hence $(P, \psi; F, G)$ is a boundary by Lemma 9.13 (ii).

Conversely, if $(P, \psi; F, G)$ is a boundary, then there is an isomorphism $f \colon (P, \psi; F, G) \xrightarrow{\cong} (H_\epsilon(F); F, \Gamma_\theta)$, where (F, θ) is a $(-\epsilon)$-quadratic form and Γ_θ is the graph of $\theta - \epsilon\theta^* \colon F \to F^*$. But then (P, ψ, F, G) has a presentation with $\gamma = \operatorname{id} \colon F = F$ and $\delta = \theta - \epsilon\theta^* \colon F \to F^*$; in particular, γ is an isomorphism. $\square$

9.2 Odd-Dimensional L-Groups

The following theorem of Ranicki [341, Theorem 2.3] gives an important criterion for recognising when a formation $(P, \psi; F, G)$ is isomorphic to the sum of a boundary formation and a trivial formation, in which case $[P, \psi; F, G] = 0 \in L_{2k+1}(R)$.

Theorem 9.25 (Zero criterion for formations) *A formation $(P, \psi; F, G)$ is isomorphic to the sum of a trivial formation and a boundary formation if and only if there is a lagrangian complement $\widehat{F}$ for F with the property that for the projection*

$$\pi \colon P = F \oplus \widehat{F} \xrightarrow{\begin{pmatrix} 1 & 0 \end{pmatrix}} F,$$

$\pi(G)$ is finitely generated free and there exists a finitely generated free submodule $L \subseteq F$ with $\pi(G) \oplus L = F$.

Moreover, the roles of F and G can be interchanged.

Proof. Suppose that $(P, \psi; F, G) \cong (H_\epsilon(Q); Q, \Gamma_\theta) \oplus (H_\epsilon(L); L, L^*)$, where $(H_\epsilon(Q); Q, \Gamma_\theta)$ is a boundary. Then we can take $\widehat{F} = Q^* \oplus L^*$ since $F \cong Q \oplus L$. Moreover, there is an isomorphism of pairs $(F, \pi(G)) \xrightarrow{\cong} (Q \oplus L, Q \oplus \{0\})$, and Q and L are finitely generated free.

Conversely, suppose we have $\widehat{F} \subset P$ as in the statement of the theorem. By the proof of Lemma 9.13 (i), there is an isomorphism $f \colon (P, \psi) \xrightarrow{\cong} H_\epsilon(F)$ that carries $\widehat{F}$ to F^* and is the identity on F, see (9.14). Let $(f, (\gamma, \delta))$ be the corresponding presentation of $(P, \psi; F, G)$ with homomorphisms

$$F \xleftarrow{\gamma} G \xrightarrow{\delta} F^*.$$

We can identify $\gamma \colon G \to F$ with the restriction of π to $G \subset P$. Let $Q := \pi(G) \subset F$ be the image of π. By assumption Q is finitely generated free and there is a finitely generated free submodule $L \subseteq F$ satisfying $F = Q \oplus L$. Dualising, we have $F^* = Q^* \oplus L^*$. Thus we get

$$F \oplus F^* = Q \oplus L \oplus Q^* \oplus L^*. \tag{9.26}$$

This induces an identification of ϵ-quadratic forms

$$H_\epsilon(Q) \oplus H_\epsilon(L) \xrightarrow{\cong} H_\epsilon(Q \oplus L) = H_\epsilon(F).$$

With respect to the decomposition (9.26), we see $G \subset Q \oplus \{0\} \oplus Q^* \oplus L^*$ since $\pi(G) = Q \subset Q \oplus L$. In the sequel we identify Q, L, Q^*, and L^* with the obvious summands in (9.26), for instance Q with $Q \oplus \{0\} \oplus \{0\} \oplus \{0\}$. Hence

$$L^* \subseteq (Q \oplus \{0\} \oplus Q^* \oplus L^*)^\perp \subseteq G^\perp = G$$

with respect to the form $H_\epsilon(F)$. If $p \colon G \to L^*$ is the restriction of the projection $F \oplus F^* = Q \oplus L \oplus Q^* \oplus L^* \to L^*$ to G, we get $p|_{L^*} = \mathrm{id}_{L^*}$. We define $M \subset G$ to be the submodule of G

$$M := G \cap (Q \oplus \{0\} \oplus Q^* \oplus \{0\}).$$

Then the kernel of $p \colon G \to L^*$ is M. Hence we get

$$G = M \oplus L^*.$$

We can write (γ, δ) in terms of the decompositions $G = M \oplus L^*$, $F = Q \oplus L$, and $F^* = Q^* \oplus L^*$,

$$Q \oplus L \xleftarrow{\begin{pmatrix} \gamma_M & 0 \\ 0 & 0 \end{pmatrix}} M \oplus L^* \xrightarrow{\begin{pmatrix} \delta_M & 0 \\ 0 & \mathrm{id} \end{pmatrix}} Q^* \oplus L^*$$

where the map γ_M is identified with $\pi|_M : M \to Q$ and hence is an isomorphism and δ_M is some homomorphism. We conclude from Lemma 9.24 (ii), by taking α to be the identification $F = Q \oplus L$, β to be the identification $G = M \oplus L^*$, and $\mu = 0$, that $(P, \psi; F, G)$ is isomorphic to the direct sum of formations $(H_\epsilon(Q); Q, M) \oplus (H_\epsilon(L); L, L^*)$. Since γ_M is an isomorphism, $(H_\epsilon(Q); Q, M)$ is a boundary by Lemma 9.24 (iv). Hence $(P, \psi; F, G)$ is isomorphic to the sum of a boundary formation and a trivial formation, as required.

To interchange the roles of F and G, we can repeat the above argument exchanging the places of F and G. The final step carries through since the properties of being a boundary formation and a trivial formation are symmetric in the order of the lagrangians by Lemma 9.13 (i) and (ii); e.g., $(H_\epsilon(Q); M, Q)$ is a boundary if and only if $(H_\epsilon(Q); Q, M)$ is a boundary. □

Corollary 9.27 *Let $\mathbb{F}$ be a field with any involution. Then $L_{2k+1}(\mathbb{F}) = 0$.*

Proof. Consider any formation $(P, \psi; F, G)$. We want to apply Theorem 9.25. By Lemma 9.5 there exists a complementary lagrangian $\widehat{F}$ for F. Since $\mathbb{F}$ is a field, any submodule of a finitely generated $\mathbb{F}$-module is a direct summand and finitely generated free. Hence $(P, \psi; F, G)$ represents zero in $L_{2k+1}(\mathbb{F})$. □

The following lemma establishes two key properties of formations, which we shall use in Section 9.3 to show that the surgery obstruction is well defined. In the sequel we put for a symmetric or quadratic form (P, ψ)

$$-(P, \psi) = (P, -\psi). \tag{9.28}$$

Lemma 9.29 *Let $(H_\epsilon(F); F, G)$ be a formation. Then the following hold:*

(i) *The formation $(H_\epsilon(F); F, G)$ is isomorphic to a boundary if and only if $(H_\epsilon(F); F^*, G)$ is trivial;*
(ii) *For every lagrangian $\widehat{G}$ complementary to G, we have*

$$[H_\epsilon(F); F, G] + [-H_\epsilon(F); F, \widehat{G}] = 0 \in L_{2k+1}(R).$$

9.2 Odd-Dimensional L-Groups

Proof. (i) Observe that $(H_\epsilon(F); F, G)$ is isomorphic to a boundary if and only if $G = \Gamma_\mu$ is the graph of a $(-\epsilon)$-quadratic form (F, μ). But this happens if and only if F^* and G are complementary and by Lemma 9.13 (i) this happens if and only if $(H_\epsilon(F); F^*, G)$ is trivial.

(ii) Let $\Delta = \{(x, \phi), (x, \phi) \mid x \in F, \phi \in F^*\} \subseteq (F \oplus F^*) \oplus (F \oplus F^*)$ be the diagonal. It is lagrangian complementary to $G \oplus \widehat{G}$ in $H_\epsilon(F) \oplus -H_\epsilon(F)$. The projection

$$\pi \colon (F \oplus F^*) \oplus (F \oplus F^*) = \Delta \oplus (G \oplus \widehat{G}) \xrightarrow{\begin{pmatrix} 0 & 1 \end{pmatrix}} G \oplus \widehat{G}$$

satisfies $\pi(F \oplus \{0\} \oplus F \oplus \{0\}) = F \subset G \oplus \widehat{G} = F \oplus F^*$. Hence we may apply Theorem 9.25 to conclude that $(H_\epsilon(F); F, G) \oplus (-H_\epsilon(F); F, \widehat{G})$ is isomorphic to the sum of a boundary formation and a trivial formation. This implies that it represents zero in $L_{2k+1}(R)$. □

9.2.3 The Definition of $L_{2k+1}(R)$ via Automorphisms

In this subsection we present Wall's original definition of the odd-dimensional L-groups. We also establish several useful identities for formations.

Recall that the L-groups are "quadratic analogues" of the K-groups. The group $K_1(R)$ is defined in terms of automorphisms of finitely generated free R-modules. There is an analogous definition of the odd-dimensional L-groups in terms of automorphisms of hyperbolic forms.

For the remainder of this subsection, we fix an integer k and put $\epsilon = (-1)^k$. Following [414, Chapter 6], we define the following groups of automorphisms of the standard hyperbolic form. Let $H_\epsilon(R^r) = (R^{2r}, \lambda, \mu)$ be the standard ϵ-hyperbolic form on R^r as defined in Example 8.79. We identify the underlying free R-module of $H_\epsilon(R^r)$ as

$$R^{2r} = R^r \oplus (R^r)^* = F_r \oplus F_r^*.$$

When we do not wish to specify the dimension of F_r, we shall simply write F. We define

$$\mathrm{SU}(r, R) := \{f \in \mathrm{GL}(2r, R) \mid f^*(\lambda, \mu) = (\lambda, \mu)\}$$

to be the group of automorphisms of $H_\epsilon(R^r)$. Extending by the identity defines an inclusion homomorphism $\mathrm{SU}(r, R) \to \mathrm{SU}(r+1, R)$ and we define

$$\mathrm{SU}(R) := \lim_{r \to \infty} \mathrm{SU}(r, R)$$

to be the stable unitary group over R. We recall from Example 9.6 that every automorphism $f \in \mathrm{SU}(r, R)$ defines the formation

$$\pi(f) := (H_\epsilon(F); F, f(F)).$$

Since $\pi(f \oplus \mathrm{id}) = \pi(f) \oplus \pi(\mathrm{id})$ and $[\pi(\mathrm{id})] = 0 \in L_{2k+1}(R)$, we obtain a well-defined map
$$\Pi \colon \mathrm{SU}(R) \to L_{2k+1}(R), \quad f \mapsto [\pi(f)].$$

The main result of this section, Theorem 9.31, states that Π is a surjective homomorphism. We now proceed to define the subgroup $\mathrm{RU}(R) \subset \mathrm{SU}(R)$, which we will later identify as the kernel of Π. Let $\mathrm{TU}(r, R) \subset \mathrm{SU}(r, R)$ be
$$\mathrm{TU}(r, R) := \{ f \in \mathrm{SU}(r, R) \mid f(F_r^*) = F_r^* \}.$$

Clearly the inclusion $\mathrm{SU}(r, R) \to \mathrm{SU}(r+1, R)$ maps $\mathrm{TU}(r, R)$ to $\mathrm{TU}(r+1, R)$ and we put
$$\mathrm{TU}(R) := \lim_{r \to \infty} \mathrm{TU}(r, R).$$

In Lemma 9.37 we prove that the subgroup $\mathrm{TU}(R)$ maps to boundary formations under π. To realise trivial formations under π, we start from the following basic automorphism $\sigma \in \mathrm{SU}(1, R)$
$$\sigma = \begin{pmatrix} 0 & 1 \\ \epsilon & 0 \end{pmatrix}. \tag{9.30}$$

Moreover, for $j \in \{1, \ldots, r\}$ we define $\sigma_{r,j} \in \mathrm{SU}(r, R)$ by identifying
$$H_\epsilon(R^r) = \bigoplus_{i=1}^{r} H_\epsilon(R)$$

and setting
$$\sigma_{r,j} = \mathrm{id} \oplus \cdots \oplus \mathrm{id} \oplus \sigma \oplus \mathrm{id} \oplus \cdots \oplus \mathrm{id} \colon \oplus_{i=1}^{r} H_\epsilon(R) \xrightarrow{\cong} \oplus_{i=1}^{r} H_\epsilon(R),$$

where σ lies in the jth place. We then define
$$\mathrm{RU}(r, R) := \langle \mathrm{TU}(r, R), \sigma_{r,1}, \ldots, \sigma_{r,r} \rangle$$

to be the subgroup of $\mathrm{SU}(r, R)$ generated by $\mathrm{TU}(r, R)$ and the automorphisms $\sigma_{r,j}$. As with the subgroups $\mathrm{TU}(r, R)$, the subgroups $\mathrm{RU}(r, R)$ are mapped to each other under stabilisation and so we define
$$\mathrm{RU}(R) := \lim_{r \to \infty} \mathrm{RU}(r, R).$$

Next we state the main result of this subsection.

Theorem 9.31 (Odd-dimensional L-groups via automorphisms) *The map $\Pi \colon \mathrm{SU}(R) \to L_{2k+1}(R)$ is a surjective homomorphism with kernel $\ker(\Pi) = \mathrm{RU}(R)$. In particular, $\mathrm{RU}(R) \subset \mathrm{SU}(R)$ is a normal subgroup and Π induces an isomorphism*
$$\mathrm{SU}(R)/\mathrm{RU}(R) \cong L_{2k+1}(R).$$

9.2 Odd-Dimensional L-Groups

The proof of Theorem 9.31 rests on the following identity in $L_{2k+1}(R)$. It is an important result related to many phenomena in topology, notably to the non-additivity of the signature for manifolds with boundary due to Wall [409] and to the Maslov index, as explained for example in the concluding remarks to Chapter 6 on pages 72–73 in [414]. The proof of Theorem 9.32 is taken from [352].

Theorem 9.32 (Maslov identity) *For any three lagrangians F, G, H in an ϵ-quadratic form (P, ψ) we have*

$$[P, \psi; F, G] + [P, \psi; G, H] = [P, \psi; F, H] \in L_{2k+1}(R).$$

Proof. Choose lagrangians $\widehat{F}, \widehat{G}$ and $\widehat{H}$ complementary to F, G and H respectively and define ϵ-quadratic formations $(P_i, \psi_i; F_i, G_i)$, $1 \le i \le 4$ by

$$(P_1, \psi_1; F_1, G_1) = (P, -\psi; \widehat{G}, \widehat{G});$$
$$(P_2, \psi_2; F_2, G_2) = (P \oplus P, \psi \oplus -\psi; F \oplus \widehat{F}, H \oplus \widehat{G})$$
$$\oplus (P \oplus P, -\psi \oplus \psi; \Delta P, \widehat{H} \oplus G);$$
$$(P_3, \psi_3; F_3, G_3) = (P \oplus P, \psi \oplus -\psi; F \oplus \widehat{F}, G \oplus \widehat{G});$$
$$(P_4, \psi_4; F_4, G_4) = (P \oplus P, \psi \oplus -\psi; G \oplus \widehat{G}, H \oplus \widehat{G})$$
$$\oplus (P \oplus P, -\psi \oplus \psi; \Delta P, \widehat{H} \oplus G),$$

where $\Delta P \subset P \oplus P$ is the diagonal copy of P. Then we have

$$(P, \psi; F, G) \oplus (P, \psi; G, H) \oplus (P_1, \psi_1; F_1, G_1) \oplus (P_2, \psi_2; F_2, G_2)$$
$$= (P, \psi; F, H) \oplus (P_3, \psi_3; F_3, G_3) \oplus (P_4, \psi_4; F_4, G_4).$$

Moreover, the formations $(P_i, \psi_i; F_i, G_i)$ are all boundaries since there are lagrangians H_i complementary to both F_i and G_i, see Lemma 9.13 (ii). For the lagrangians H_i we may take

$$H_1 = G \subset P_1 = P;$$
$$H_2 = \Delta_{P \oplus P} \subset P_2 = (P \oplus P) \oplus (P \oplus P);$$
$$H_3 = \Delta_P \subset P_3 = P \oplus P;$$
$$H_4 = \Delta_{P \oplus P} \subset P_4 = (P \oplus P) \oplus (P \oplus P).$$

By definition boundaries represent zero in $L_{2k+1}(R)$ and so we are done. □

The remainder of this subsection is devoted to the proof of Theorems 9.31. The proofs are organised as follows. We first show that Π is onto. Then we use Theorem 9.32 to show that Π is additive. Finally we show $\ker(\Pi) = RU(R)$.

Lemma 9.33 *Every formation $(P, \psi; F, G)$ satisfying $P \cong R^{2r}$ is isomorphic to $(H_\epsilon(F); F, f(F))$ for an automorphism $f \in \mathrm{SU}(r, R)$. Consequently, the map $\Pi: SU(R) \to L_{2k+1}(R)$ is onto.*

Proof. By Lemma 8.95 we can choose isomorphisms of non-singular ϵ-quadratic forms $f\colon H_\epsilon(F) \xrightarrow{\cong} (P, \psi)$ and $g\colon H_\epsilon(G) \xrightarrow{\cong} (P, \psi)$ such that $f(F) = F$ and $g(G) = G$. Choose a and b satisfying $F \cong R^a$ and $G \cong R^b$. Then $R^{2a} \cong F \oplus F^* \cong P \cong R^{2r} \cong G \oplus G^* \cong R^{2b}$. We conclude $a = b = r$ since R has property (IBN) by assumption. Hence we can choose an R-isomorphism $u\colon F \to G$. Then we obtain an isomorphism of non-singular ϵ-quadratic forms by the composite

$$v\colon H_\epsilon(F) \xrightarrow{H_\epsilon(u)} H_\epsilon(G) \xrightarrow{g} (P, \psi) \xrightarrow{f^{-1}} H_\epsilon(F)$$

and an isomorphism of ϵ-quadratic formations

$$f\colon (H_\epsilon(F); F, v(F)) \xrightarrow{\cong} (P, \psi; F, G).$$

This finishes the proof. □

Now we prove that Π is additive. Given $f, g \in SU(R)$, as a consequence of Theorem 9.32 we have the following equations in $L_{2k+1}(R)$:

$$\begin{aligned}\Pi(f) + \Pi(g) &= [H_\epsilon(F); F, f(F)] + [H_\epsilon(F); F, g(F)] \\ &= [H_\epsilon(F); F, f(F)] + [H_\epsilon(F); f(F), fg(F)] \\ &= [H_\epsilon(F); F, fg(F)] \\ &= \Pi(fg).\end{aligned}$$

We turn to the proof that $\ker(\Pi) = RU(R)$. Let us define $\Sigma_r \in RU(r, R)$ to be the "full flip automorphism",

$$\Sigma_r = \sigma \oplus \sigma \oplus \cdots \oplus \sigma \oplus \sigma,$$

where there are r copies of σ defined in (9.30). When we do not wish to emphasise the rank r, we shall simply write Σ. Recall that an automorphism $f \in SU(R)$ belongs to $TU(R)$ if and only if $f(F^*) = F^*$. Since $\Sigma(F) = F^*$, it follows that, if $f(F) = F$, then $\Sigma f \Sigma(F^*) = F^*$. Hence, if $f(F) = F$, then $\Sigma f \Sigma \in TU(R)$ and so $f \in RU(R)$.

Lemma 9.34 *Consider $f \in SU(R)$ and $g \in RU(R)$ such that there is a stable isomorphism $\pi(f) \cong_s \pi(g)$. Then $f \in RU(R)$.*

Proof. Let T_1 and T_2 be trivial formations such that

$$\pi(f) \oplus T_1 \cong \pi(g) \oplus T_2.$$

Now $T_i \cong \pi(\Sigma_i)$ and $\Sigma_i \in RU(R)$, $i = 1, 2$. Replacing f by $f \oplus \Sigma_1$ and g by $g \oplus \Sigma_2$, we can arrange that there is an isomorphism α from $\pi(f)$ to $\pi(g)$,

$$\alpha\colon (H_\epsilon(F); F, f(F)) \xrightarrow{\cong} (H_\epsilon(F); F, g(F)).$$

We see that $\alpha(F) = F$ and $\alpha f(F) = g(F)$. It follows that $\alpha \in RU(R)$ and that $g^{-1}\alpha f \in RU(R)$. But, since $g, \alpha \in RU(R)$, we also have $f \in RU(R)$. □

9.2 Odd-Dimensional L-Groups

Next we consider the subgroup $UU(r, R) \subset TU(r, R)$ defined by

$$UU(r, R) := \{f \mid f|_{F_r^*} = \mathrm{id}_{F_r^*}\}.$$

One easily checks that for every $f \in UU(r, R)$ there is a $(-\epsilon)$-quadratic form θ satisfying

$$f = f_\theta := \begin{pmatrix} \mathrm{id} & 0 \\ \theta - \epsilon\theta^* & \mathrm{id} \end{pmatrix}. \tag{9.35}$$

Stabilising as before, we set

$$UU(R) := \lim_{r \to \infty} UU(r, R).$$

Next we discuss the relationship between $UU(R)$ and $TU(R)$. Restricting $f \in TU(r, R)$ to F_r^* defines a homomorphism $TU(r, R) \to GL(r, R)$ that by definition is part of the following exact sequence

$$1 \to UU(r, R) \to TU(r, R) \to GL(r, R).$$

Indeed, $TU(r, R) \to GL(r, R)$ is onto and is split by the homomorphism,

$$s \colon GL(r, R) \to TU(r, R), \quad \alpha \mapsto f_\alpha := \begin{pmatrix} (\alpha^*)^{-1} & 0 \\ 0 & \alpha \end{pmatrix}$$

so that $TU(r, R)$ is identified with the semidirect product

$$TU(r, R) \cong UU(r, R) \rtimes GL(r, R). \tag{9.36}$$

Lemma 9.37 *A formation $(P, \psi; F, G)$ satisfying $P \cong R^{2r}$ is a boundary if and only if it isomorphic to $\pi(f)$ for some $f \in TU(R)$.*

Proof. By Definition 9.9, if $(P, \psi; F, G)$ is a boundary, then it is isomorphic to a formation $(H_\epsilon(F); F, \Gamma_\theta)$, where $\theta \colon F \to F^*$ is a $(-\epsilon)$-quadratic form and the lagrangian $\Gamma_\theta \subset F \oplus F^*$ is the graph of $\theta - \epsilon\theta^*$. Taking $f_\theta \in UU(R)$ as defined in 9.35, we have

$$\pi(f_\theta) = (H_\epsilon(F); F, \Gamma_\theta),$$

and so $(P, \psi; F, G)$ is isomorphic to $\pi(f_\theta)$. Conversely, given $f \in TU(r, R)$, we write $f = f_\theta \circ f_\alpha$ using (9.36). Since $f_\alpha(F) = F$, we get $f(F) = f_\theta(F)$ and hence

$$\pi(f) = \pi(f_\theta) = (H_\epsilon(F); F, \Gamma_\theta).$$

We conclude that $\pi(f)$ is a boundary for every $f \in TU(R)$. □

Next we prove $\ker(\Pi) = RU(R)$. Consider $f \in RU(R)$. We may assume $f \in RU(r, R)$ for some r and write f as a composite $f = f_1 \circ f_2 \circ \cdots \circ f_a$, where $f_i \in TU(r, R)$ or $f_i = \sigma_{r,j}$. Since Π is a homomorphism, we get

$$\Pi(f) = \sum_{i=1}^{a} \Pi(f_i).$$

We conclude from Lemma 9.37 that $\Pi(f_i) = 0$ holds if $f_i \in \mathrm{TU}(r, R)$ and $\Pi(\sigma_{r,j})$ is the sum of a trivial formation and the boundary of the zero formation. Hence we obtain $\mathrm{RU}(R) \subset \ker(\Pi)$. Now suppose $f \in \mathrm{SU}(R)$ and $\Pi(f) = 0 \in L_{2k+1}(R)$. By definition, there are boundary formations B_1 and B_2 and a stable isomorphism

$$\pi(f) \oplus B_1 \cong_s B_2.$$

Lemma 9.37 implies $B_i \cong \pi(g_i)$ for automorphisms $g_i \in \mathrm{UU}(R) \subset \mathrm{RU}(R)$. From Lemma 9.34 we conclude $f \oplus g_i \in \mathrm{RU}(R)$ and hence $f \in \mathrm{RU}(R)$. This implies $\ker(\Pi) \subset \mathrm{RU}(R)$ and so $\ker(\Pi) = \mathrm{RU}(R)$.

Next comes the promised odd-dimensional analogue of Lemma 8.152.

Lemma 9.38 *The automorphism* $(c_g)_* \colon L_{2k+1}(\mathbb{Z}\Gamma, w) \to L_{2k+1}(\mathbb{Z}\Gamma, w)$ *is* $w(g) \cdot \mathrm{id}$.

Proof. Choose a lagrangian $\widehat{G}$ that is complementary to G. We conclude from Lemma 9.13 (i) that $(-H_\epsilon(F); \widehat{G}, G)$ is trivial and hence represents zero in $L_{2k+1}(\mathbb{Z}\Gamma)$. From Lemma 9.29 (ii) and Theorem 9.32, we get in $L_{2k+1}(\mathbb{Z}\Gamma)$

$$-[(H_\epsilon(F); F, G)] = [(-H_\epsilon(F); F, \widehat{G})]$$
$$= [(-H_\epsilon(F); F, G)] + [(-H_\epsilon(F); \widehat{G}, G)] = [(-H_\epsilon(F); F, G)].$$

It was shown in the proof of Lemma 8.152 that conjugation by g transforms an ϵ-quadratic form (P, ψ) into $(P, w(g) \cdot \psi)$. Hence we get that conjugation by g transforms a formation $(H_\epsilon(F); F, G)$ to $(w(g) \cdot H_\epsilon(F); F, G)$. The above identity shows that c_g induces multiplication by $w(g)$ on $L_{2k+1}(\mathbb{Z}\Gamma, w)$. □

9.2.4 The Odd-Dimensional L-Groups of $\mathbb{Z}$

In this subsection we prove Theorem 9.18, which states that $L_{2k+1}(\mathbb{Z}) = 0$. Our aim is to give an elementary algebraic proof. We follow early work of Wall [405]. At the end of the subsection we review other approaches to proving $L_{2k+1}(\mathbb{Z}) = 0$.

We first introduce some notation and terminology. Let $\{P, \psi; F, G\}_s$ denote the stable isomorphism class of an ϵ-quadratic formation $(P, \psi; F, G)$. Let

$$\mathrm{Formation}_\epsilon(\mathbb{Z}) := \{\{P, \psi; F, G\}_s\}$$

denote the set of stable isomorphism classes of ϵ-quadratic formations over $\mathbb{Z}$. A formation $(P, \psi; F, G)$ is called $\mathbb{Q}$-*trivial* if the obvious $\mathbb{Z}$-homomorphism $F \oplus G \to Q$ induces a $\mathbb{Q}$-isomorphism

$$\mathbb{Q} \otimes_\mathbb{Z} (F \oplus G) \to \mathbb{Q} \otimes_\mathbb{Z} P.$$

9.2 Odd-Dimensional L-Groups

This is equivalent to saying that the formation $(P, \psi; F, G)$ is trivial after tensoring with $\mathbb{Q}$. We remark that in [346] $\mathbb{Q}$-trivial formations are called $(\mathbb{Z} - \{0\})$-non-degenerate formations.

We next identify two important subsets of $\text{Formation}_\epsilon(\mathbb{Z})$, namely, the set of stable isomorphism classes of $\mathbb{Q}$-trivial formations,

$$\text{Formation}_\epsilon^{\mathbb{Q}\text{-tr}}(\mathbb{Z}) := \{\{P, \psi; F, G\}_s \mid \mathbb{Q} \otimes_\mathbb{Z} (F \oplus G) \xrightarrow{\cong} \mathbb{Q} \otimes_\mathbb{Z} P\},$$

and the set of stable isomorphism classes of boundary formations

$$\text{Formation}_\epsilon^b(\mathbb{Z}) = \{\{P, \psi; F, G\}_s \mid (P, \psi; F, G) \cong \partial(F, \theta) = (H_\epsilon(F); F, \Gamma_\theta)\}.$$

Of course, all the stable isomorphism classes in $\text{Formation}_\epsilon^b(\mathbb{Z})$ map to zero in $L_{2k+1}(\mathbb{Z})$.

Theorem 9.18 is a direct consequence of the following two results; for the first one see [405, Lemma 3].

Lemma 9.39 *Let $(P, \psi; F, G)$ be a formation over $\mathbb{Z}$ and set $K := F \cap G$. Then there is an isomorphism*

$$(P, \psi; F, G) \xrightarrow{\cong} (H_\epsilon(K); K, K) \oplus (P_0, \psi_0; F_0, G_0)$$

where $(P_0, \psi_0; F_0, G_0)$ is $\mathbb{Q}$-trivial.

Theorem 9.40 ($\mathbb{Q}$-trivial formations versus boundary formations) *We have* $\text{Formation}_\epsilon^{\mathbb{Q}\text{-tr}}(\mathbb{Z}) \subset \text{Formation}_\epsilon^b(\mathbb{Z})$.

Proof of Lemma 9.39. Since $F, G \subset P$ are both lagrangians of (P, ψ), it follows that $K = F \cap G$ is a sublagrangian of (P, ψ). By Lemma 8.95 there is a quadratic form (P_0, ψ_0) and an isomorphism of ϵ-quadratic forms

$$f: (P, \psi) \xrightarrow{\cong} H_\epsilon(K) \oplus (P_0, \psi_0),$$

where $f(F \cap G) = K \subset K \oplus K^*$. In the sequel we identify K, K^* and P_0 with the obvious subspaces of $K \oplus K^* \oplus P_0$; e.g., $K = K \oplus \{0\} \oplus \{0\}$.

Next we show $f(F) = K \oplus (f(F) \cap P_0)$ and $f(G) = K \oplus (f(G) \cap P_0)$. Consider $x \in F$ and write

$$f(x) = (k_x, \alpha_x, p_x) \in K \oplus K^* \oplus P_0.$$

Using $\cdot$ to denote intersection with respect to the ϵ-symmetric form induced on $H_\epsilon(K) \oplus (P, \psi_0)$, we have $f(x) \cdot K = x \cdot (F \cap G) = 0$. Now for all $k \in K$, we have $(k_x, \alpha_x, p_x) \cdot (k, 0, 0) = \alpha_x(k)$. Hence $\alpha_x = 0$ and so $f(F) \subseteq K \oplus \{0\} \oplus P_0$. Since $K \subseteq f(F)$, we conclude $f(F) = K \oplus (f(F) \cap P_0)$. The proof for G is analogous.

Put $F_0 := f(F) \cap P_0$ and $G_0 := f(G) \cap P_0$. Then f defines an isomorphism of formations

$$f: (P, \psi; F, G) \xrightarrow{\cong} (H_\epsilon(K); K, K) \oplus (P_0, \psi_0; F_0, G_0).$$

Moreover, since $f(F) \cap f(G) = f(F \cap G) = K$, we conclude $F_0 \cap G_0 = 0$ and hence $(P_0, \psi_0; F, f(G)) \in \text{Formation}_\epsilon^{\mathbb{Q}\text{-tr}}(\mathbb{Z})$. □

Hence Theorem 9.18 is true if we can prove Theorem 9.40. To do this, we shall develop the theory of non-singular ϵ-symmetric and ϵ-quadratic linking forms on finite abelian groups. Let T be a finite abelian group. The *torsion dual of T* is the group
$$\widehat{T} := \hom_{\mathbb{Z}}(T, \mathbb{Q}/\mathbb{Z}).$$

An *ϵ-symmetric linking form on T* is a function
$$b \colon T \times T \to \mathbb{Q}/\mathbb{Z}$$
satisfying for all $x, y_1, y_2 \in T$

(i) $b(x, y_1 + y_2) = b(x, y_1) + b(x, y_2)$;
(ii) $b(x, y) = \epsilon b(y, x)$.

If in addition the adjoint homomorphism $\widehat{b}$ of b defined by
$$\widehat{b} \colon T \to \widehat{T}, \quad x \mapsto (y \mapsto b(x, y)) \tag{9.41}$$

is an isomorphism, then b is called *non-singular*. An *isomorphism* between linking forms (T_1, b_1) and (T_2, b_2) is a group isomorphism $A \colon T_1 \xrightarrow{\cong} T_2$ such that $b_1(x_1, x_2) = b_2(A(x_1), A(x_2))$ holds for all $x_1, x_2 \in T_1$. There is an obvious notion of the orthogonal sum of linking forms.

We now define ϵ-quadratic linking forms. For this we need the value group $Q_\epsilon(\mathbb{Q}/\mathbb{Z})$ defined by

$$Q_\epsilon(\mathbb{Q}/\mathbb{Z}) := (\mathbb{Q}/\mathbb{Z})/\{r - \epsilon r \mid r \in \mathbb{Q}/\mathbb{Z}\} = \begin{cases} \mathbb{Q}/\mathbb{Z} & \text{if } \epsilon = +1; \\ 0 & \text{if } \epsilon = -1. \end{cases}$$

An *ϵ-quadratic linking form* (T, b, q) on a finite abelian group T is an ϵ-symmetric linking form (T, b) together with a function
$$q \colon T \to Q_\epsilon(\mathbb{Q}/\mathbb{Z}),$$
satisfying

(i) $q(sx) = s^2 q(x), \forall s \in \mathbb{Z}$;
(ii) $q(x + y) - q(x) - q(y) = \mathrm{pr}(b(x, y))$;
(iii) $b(x, x) = (1 + \epsilon) q(x)$.

The ϵ-quadratic linking form q is called *non-singular* if and only if b is non-singular. If $\epsilon = -1$, then of course q is identically zero and hence a (-1)-quadratic linking form (T, b, q) is nothing but a (-1)-symmetric linking form (T, b) such that $b(x, x) = 0$ for all $x \in T$. If $\epsilon = +1$ and T has odd order, then the linking form b uniquely determines q, however, if T has even order, then q is not determined by b.

An *isomorphism* between two ϵ-quadratic linking forms (T_1, b_1, q_1) and (T_2, b_2, q_2) is an isomorphism of linking forms $A \colon (T_1, b_1) \xrightarrow{\cong} (T_2, b_2)$ such that $q_1(x) = q_2(A(x))$ holds for all $x \in T_1$. There is an obvious notion of the orthogonal

9.2 Odd-Dimensional L-Groups

sum of quadratic linking forms. We leave it to the reader to verify that a quadratic linking form (T, b, q) splits as an orthogonal sum if and only if the linking form (T, b) splits as an orthogonal sum and, moreover, every orthogonal splitting of b gives rise to an orthogonal splitting of q.

For later use, we let $\{T, b, q\}$ denote the isomorphism class of a non-singular ϵ-quadratic linking form (T, b, q) and define

$$Q_\epsilon(\mathbb{Z}) := \{\{T, b, q\}\}$$

to be the set of isomorphism classes of non-singular ϵ-quadratic linking forms over $\mathbb{Z}$.

Before proceeding, we digress somewhat to make some remarks about the geometric significance of linking forms, their historical role in the investigation of odd-dimensional L-groups, and their role in the proof of Theorem 9.40.

Remark 9.42 (Geometric significance of linking forms) An oriented $(2k + 1)$-dimensional Poincaré complex $(X, [X])$ with fundamental class $[X]$ in $H_{2k+1}(X; \mathbb{Z})$ possesses a non-singular $(-1)^{k+1}$-symmetric linking form b_X that is defined on tors $H^{k+1}(X; \mathbb{Z}) \subseteq H^{k+1}(X; \mathbb{Z})$, the torsion subgroup of $H^{k+1}(X; \mathbb{Z})$, as follows:

$$b_X \colon \text{tors } H^{k+1}(X; \mathbb{Z}) \times \text{tors } H^{k+1}(X; \mathbb{Z}) \to \mathbb{Q}/\mathbb{Z}, \quad (x, y) \mapsto \langle \beta^{-1}(x) \cup y, [X] \rangle.$$

Here $\beta \colon H^k(X; \mathbb{Q}/\mathbb{Z}) \to H^{k+1}(X; \mathbb{Z})$ is the Bockstein homomorphism and $\beta^{-1}(x)$ denotes any pre-image of x under β, see for example [3, §2] for further information. The linking form b_X is the odd-dimensional analogue of the intersection pairing $H^k(Y; \mathbb{Z}) \times H^k(Y; \mathbb{Z}) \to \mathbb{Z}$ of a $2k$-dimensional Poincaré complex Y given in (5.62). If $X = M$ is a $(2k+1)$-dimensional manifold, there is also a homological version of the linking form, which is geometrically defined via intersections and is described in [407, Section 12A].

Remark 9.43 (On the role of linking forms in the history of odd-dimensional surgery) When Kervaire and Milnor showed that the odd-dimensional simply connected obstruction vanishes [216, §5 and §6], they did so by delicately analysing the linking forms b_M of the sources of degree one normal maps $(f, \bar{f}) \colon M \to S^{2k+1}$. Specifically, they showed that every such normal map can be converted by a finite number of surgeries to one where tors $H^{k+1}(M; \mathbb{Z}) = 0$. This was sufficient since killing the free part of $H^{k+1}(M; \mathbb{Z}) \cong H_k(M; \mathbb{Z})$ was elementary.

Following Kervaire and Milnor, early attempts to analyse the odd dimensional surgery obstruction in the case of non-trivial finite fundamental groups used equivariant quadratic linking forms, see for example [105] and [406, §5]. In general, appropriate equivalence classes of equivariant non-singular $(-1)^{k+1}$-quadratic linking forms can be used to define invariants of $L_{2k+1}(\mathbb{Z}\pi)$ for finite groups π, but it turns out that these invariants are not complete for all integral group rings of finite groups.

In the simply connected setting, Kervaire and Milnor's discovery of the central role of the torsion group tors $H^{k+1}(M; \mathbb{Z})$ and the linking form on it corresponds to Lemma 9.39, which showed that every formation over $\mathbb{Z}$ is equivalent to a $\mathbb{Q}$-trivial formation in $L_{2k+1}(\mathbb{Z})$. The proof of Theorem 9.40 proceeds by showing that over $\mathbb{Z}$, equivalence classes of non-singular $(-1)^{k+1}$-quadratic linking forms can be used to define a complete invariant of $L_{2k+1}(\mathbb{Z})$, see Proposition 9.46, and that every such linking form bounds; i.e., is trivial, in the appropriate sense, see Proposition 9.53.

Let $C = (H_\epsilon(F); F, G)$ be a $\mathbb{Q}$-trivial ϵ-quadratic formation. We now explain how C defines a non-singular $(-\epsilon)$-quadratic linking form (T_C, b_C, q_C). Recall from Section 9.2.2 that $C = (H_\epsilon(F); F, G)$ can be presented by structure homomorphisms

$$\gamma \colon G \to F \quad \text{and} \quad \delta \colon G \to F^*$$

where we abuse notation and identify $G = \text{im}(\gamma \oplus \delta) \subset F \oplus F^*$. Since C is $\mathbb{Q}$-trivial, Lemma 9.24 (iii) implies that δ is injective. We define

$$T_C := \text{coker}(\delta).$$

This will be the group on which the quadratic linking form of C is defined. By definition there is a short exact sequence

$$0 \to G \xrightarrow{\delta} F^* \to T_C \to 0.$$

Recall from Section 9.2.2 that the homomorphism $\gamma^*\delta \colon G \to G^*$ defines a $(-\epsilon)$-symmetric form $(G, \gamma^*\delta)$ that admits a quadratic refinement θ. If $-\epsilon = 1$, then θ is uniquely determined by $\gamma^*\delta$. If $-\epsilon = -1$, there may be several choices of θ, but the choices will not matter for our purposes.

Following [352, Proposition 12.41], we make the following definition.

Definition 9.44 Let $C = (H_\epsilon(F); F, G)$ be a $\mathbb{Q}$-trivial ϵ-quadratic formation over $\mathbb{Z}$ with structure homomorphisms $\gamma \colon G \to F$ and $\delta \colon G \to F^*$. The $(-\epsilon)$-*quadratic linking form of* C is defined by

$$T_C := \text{coker}\bigl(\delta \colon G \to F^*\bigr);$$

$$b_C \colon T_C \times T_C \to \mathbb{Q}/\mathbb{Z}, \quad ([x],[y]) \mapsto \frac{y(\gamma(z))}{s};$$

$$q_C \colon T_C \to Q_\epsilon(\mathbb{Q}/\mathbb{Z}), \quad [x] \mapsto \frac{\theta(z)(z)}{s^2}.$$

Here $x, y \in F^*$ represent $[x], [y] \in T_C$, and $z \in G$ and $s \in \mathbb{Z} \setminus \{0\}$ satisfy $sx = \delta(z)$.

Exercise 9.45 Verify that (T_C, b_C, q_C) in Definition 9.44 is a well-defined non-singular $(-\epsilon)$-quadratic linking form.

The following proposition is due to Wall, see [405, Theorems 2 & 3]. It was later generalised by Ranicki in [346, Proposition 3.4.3].

Proposition 9.46 *The mapping that takes a $\mathbb{Q}$-trivial ϵ-quadratic formation to its associated $(-\epsilon)$-quadratic linking form gives rise to a well-defined bijective function*

$$q \colon \text{Formation}_\epsilon^{\mathbb{Q}\text{-tr}}(\mathbb{Z}) \xrightarrow{\cong} Q_{-\epsilon}(\mathbb{Z}), \quad [C] \mapsto [T_C, b_C, q_C].$$

9.2 Odd-Dimensional L-Groups

This means that every non-singular $(-\epsilon)$-quadratic linking form (T, b, q) is isomorphic to (T_C, b_C, q_C) for some $\mathbb{Q}$-trivial ϵ-quadratic formation C and that two $\mathbb{Q}$-trivial ϵ-quadratic formations C_1 and C_2 are stably isomorphic if and only if $(T_{C_1}, b_{C_1}, q_{C_1})$ and $(T_{C_2}, b_{C_2}, q_{C_2})$ are isomorphic.

The proof of Proposition 9.46 needs some preparation.

Starting from a non-singular $(-\epsilon)$-quadratic linking form (T, b, q) and a surjection $\pi \colon L \to T$ from a free abelian group L of rank r, we shall build an ϵ-quadratic formation $C_{(T,b,q),\pi} = (H_\epsilon(F); F, G)$ where F and G both have rank r. Later we discuss the properties of our construction and show that it defines a function

$$F \colon Q_{-\epsilon}(\mathbb{Z}) \to \text{Formation}_\epsilon^{\mathbb{Q}\text{-tr}}(\mathbb{Z}), \quad \{T, b, q\} \mapsto \{C_{(T,b,q),\pi}\}_s \tag{9.47}$$

that is an inverse for q.

Let

$$0 \to K \xrightarrow{i} L \xrightarrow{\pi} T \to 0 \tag{9.48}$$

be a short exact sequence of abelian groups where $K, L \cong \mathbb{Z}^r$. We define a $\mathbb{Q}$-valued bilinear pairing between L and K^* by

$$\phi \colon L \times K^* \to \mathbb{Q}, \quad (x, y) \mapsto \frac{y(z)}{s},$$

where $z \in K$ and $s \in \mathbb{Z}$ are elements satisfying $i(z) = sx$. The dual of the sequence (9.48) is the short exact sequence

$$0 \to L^* \xrightarrow{i^*} K^* \xrightarrow{\widehat{\pi}} \widehat{T} \to 0$$

where, for $x \in L$ with $[x] := \pi(x) \in T$ and $y \in K^*$, $\widehat{\pi}(y)$ is defined by

$$\widehat{\pi}(y)([x]) = \phi(x, y) \bmod \mathbb{Z}.$$

If $\widehat{b} \colon T \to \widehat{T}$ is the adjoint of b defined in (9.41), we set $[y] := (\widehat{b})^{-1}(\widehat{\pi}(y)) \in T$ for $y \in K^*$. Then we have

$$b([x], [y]) = \phi(x, y) \bmod \mathbb{Z}. \tag{9.49}$$

We define a $\mathbb{Q}$-valued ϵ-symmetric form on $L \oplus K^*$ by

$$\overline{\lambda} \colon (L \oplus K^*) \times (L \oplus K^*) \to \mathbb{Q}, \quad ((x_1, y_1), (x_2, y_2)) \mapsto \phi(x_1, y_2) + \epsilon\phi(x_2, y_1).$$

Let $H \subset L \oplus K^*$ be the subgroup

$$H := \{(x, y) \mid \widehat{b}(\pi(x)) = \widehat{\pi}(y)\} = \{(x, y) \mid [x] = [y] \in T\}. \tag{9.50}$$

Put

$$\lambda := \overline{\lambda}|_{H \times H} \colon H \times H \to \mathbb{Q}.$$

We claim that (H, λ) is an ϵ-symmetric hyperbolic form over $\mathbb{Z}$. We first verify that $\lambda(H, H) \subset \mathbb{Z}$. For $(x_1, y_1), (x_2, y_2) \in H$, we conclude from (9.49)

$$\lambda\bigl((x_1, y_1), (x_2, y_2)\bigr) = \phi(x_1, y_2) + \epsilon\phi(x_2, y_1)$$
$$= b([x_1], [y_2]) + \epsilon b([x_2], [y_1]) \bmod \mathbb{Z}$$
$$= b([x_1], [x_2]) + \epsilon b([x_2], [x_1]) = 0 \bmod \mathbb{Z}.$$

Hence (H, λ) is an ϵ-symmetric bilinear form over $\mathbb{Z}$.

Define an ϵ-symmetric bilinear form over $\mathbb{Q}$ by

$$\lambda_\mathbb{Q} : \mathbb{Q} \otimes_\mathbb{Z} H \times \mathbb{Q} \otimes_\mathbb{Z} H \to \mathbb{Q}, \quad (r_1 \otimes h_1, r_2 \otimes h_2) \mapsto r_1 r_2 \lambda(h_1, h_2).$$

It is clear from the definitions that there is an isomorphism $(\mathbb{Q} \otimes_\mathbb{Z} H, \lambda_\mathbb{Q}) \xrightarrow{\cong} H^\epsilon(\mathbb{Q}^r)$ and hence the adjoint $\widehat{\lambda} : H \to H^*$ of λ is injective. Next we show that

$$\widehat{\lambda} : H \to H^*$$

is onto. For $f \in H^*$, let $\overline{f} : L \oplus K^* \to \mathbb{Q}$ be the unique linear extension of f to the abelian group $L \oplus K^*$. One easily checks that the form $(\mathbb{Q} \otimes_\mathbb{Z} (L \oplus K^*), \overline{\lambda}_\mathbb{Q})$ is non-singular where $\overline{\lambda}_\mathbb{Q}$ is defined analogously to $\lambda_\mathbb{Q}$. Hence there is an element $\widehat{f} \in \mathbb{Q} \otimes_\mathbb{Z} (L \oplus K^*)$ that is uniquely determined by the property

$$\overline{\lambda}_\mathbb{Q}(\widehat{f}, r \otimes (x, y)) = r\overline{f}(x, y) \quad \text{for all } r \in \mathbb{Q}, (x, y) \in L \oplus K^*.$$

Next we show that $\widehat{f}$ lies in $L \oplus K^*$. Under the canonical isomorphism

$$\mathbb{Q} \otimes_\mathbb{Z} (L \oplus K^*) \xrightarrow{\cong} (\mathbb{Q} \otimes_\mathbb{Z} L) \oplus (\mathbb{Q} \otimes_\mathbb{Z} K^*)$$

the element $\widehat{f}$ becomes $(s_1 \otimes x_f, s_2 \otimes y_f)$ for $s_1, s_2 \in \mathbb{Q}$, $x_f \in L$, and $y_f \in K^*$. Let $k \in K$ and $\phi \in L^*$, so that $(i(k), 0)$ and $(0, i^*(\phi))$ belong to H. Hence $\overline{\lambda}_\mathbb{Q}(\widehat{f}, (i(x), 0))$ and $\overline{\lambda}_\mathbb{Q}(\widehat{f}, (0, i^*(\phi)))$ lie in $\mathbb{Z}$ because of $\overline{f}(H) \subset \mathbb{Z}$. A direct computation shows

$$\overline{\lambda}_\mathbb{Q}(\widehat{f}, (i(k), 0)) = s_2 y_f(k);$$
$$\overline{\lambda}_\mathbb{Q}(\widehat{f}, (0, i^*(\phi))) = s_1 \phi(x_f).$$

We conclude that $s_2 y_f(k) \in \mathbb{Z}$ and $s_1 \cdot \phi(x_f)$ holds for every $k \in K$ and $\phi \in L^*$. This implies $s_1, s_2 \in \mathbb{Z}$. Hence $\widehat{f}$ lies in $L \oplus K^*$ and so can be written as $\widehat{f} = (x_f, y_f)$ for $x_f \in L$ and $y_f \in K^*$.

Next we show $\widehat{f} \in H$. For $(x, y) \in L \oplus K^*$ we have

$$\overline{\lambda}((x, y), (x_f, y_f)) = b([x], [y_f]) + \epsilon b([x_f], [y]) \bmod \mathbb{Z}.$$

Since (x, y) is arbitrary, we get $b([x_f], [x]) = b([y_f], [x])$ for all $[x] \in T$. Since b is non-singular, this implies $[x_f] = [y_f]$. Hence $\widehat{f} = (x_f, y_f) \in H$. Therefore $\widehat{\lambda}(\widehat{f}) = f$. This finishes the proof that $\widehat{\lambda}$ is bijective and that (H, λ) is a non-singular form over $\mathbb{Z}$.

9.2 Odd-Dimensional L-Groups

Finally, if $\epsilon = +1$, note that for $(x, y) \in H$, we have

$$\phi(x, y) = b([x], [y]) = b([x], [x]) = 0 \mod \mathbb{Z},$$

where the last equality holds since (T, b, q) is a (-1)-quadratic linking form. Hence

$$\lambda\big((x, y), (x, y)\big) = \phi(x, y) + \phi(x, y) \in 2\mathbb{Z},$$

and (H, λ) is even.

The above shows that (H, λ) is a non-singular ϵ-symmetric form over $\mathbb{Z}$, which becomes hyperbolic over $\mathbb{Q}$, and which is even if $\epsilon = +1$. It follows that (H, λ) admits a quadratic refinement (H, λ, μ). If $\epsilon = +1$, μ is determined by λ. If $\epsilon = -1$, we use $q \colon T \to \mathbb{Q}/\mathbb{Z}$ to define μ as follows. Firstly, we define the isomorphism of abelian groups

$$\tau \colon \mathbb{Q}/\mathbb{Z} \to \mathbb{Q}/2\mathbb{Z}, \quad r \cdot \mathbb{Z} \mapsto 2r \cdot (2\mathbb{Z}).$$

Note that $\mathbb{Z}/2$ is a subgroup of $\mathbb{Q}/2\mathbb{Z}$. For an element $r \in \mathbb{Q}$ we denote by $\bar{r}$ its class in $\mathbb{Q}/2\mathbb{Z}$. We define the quadratic refinement $\mu \colon H \to Q_-(\mathbb{Z}) = \mathbb{Z}/2$ by the equation

$$\mu(x, y) := \tau \circ q([x]) + \overline{\phi(x, y)} \in \mathbb{Z}/2\mathbb{Z}.$$

A priori $\tau \circ q([x]) + \overline{\phi(x, y)}$ is an element in $\mathbb{Q}/2\mathbb{Z}$ but turns out to lie in $\mathbb{Z}/2$ by the following observation. If $(x, y) \in H$, then by definition, see (9.50), $[x] = [y]$ and so we get $\phi(x, y) = b([x], [y]) = b([x], [x]) = 0 \mod \mathbb{Z}$ from (9.49) and $2q([x]) = b([x], [x])$ from the definitions. For $(x_1, y_1), (x_2, y_2) \in H$ we have the following equations in $\mathbb{Q}/2\mathbb{Z}$

$$\begin{aligned}
\mu\big((x_1, y_1) + (x_2, y_2)\big) &= \tau \circ q([x_1 + x_2]) + \overline{\phi(x_1 + x_2, y_1 + y_2)} \\
&= \tau\big(q([x_1]) + q([x_2]) + b([x_1], [x_2])\big) + \overline{\phi(x_1 + x_2, y_1 + y_2)} \\
&= \tau \circ q([x_1]) + \overline{\phi(x_1, y_1)} + \tau \circ q([x_2]) + \overline{\phi(x_2, y_2)} \\
&\quad + \tau \circ b([x_1], [x_2]) + \overline{\phi(x_1, y_2)} + \overline{\phi(x_2, y_1)} \\
&= \mu(x_1, y_1) + \mu(x_2, y_2) + \mathrm{pr} \circ \lambda\big((x_1, y_1), (x_2, y_2)\big),
\end{aligned}$$

where the last equality holds since in $\mathbb{Q}/2\mathbb{Z}$ we have

$$\begin{aligned}
\overline{\lambda\big((x_1, y_1), (x_2, y_2)\big)} &= \overline{\phi(x_1, y_2)} - \overline{\phi(x_2, y_1)} \\
&= \overline{\phi(x_1, y_2)} + \overline{\phi(x_2, y_1)} - \overline{2\phi(x_2, y_1)} \\
&\stackrel{(9.49)}{=} \overline{\phi(x_1, y_2)} + \overline{\phi(x_2, y_1)} - \tau(b([x_2], [y_1])) \\
&\stackrel{[x_1]=[y_1]}{=} \overline{\phi(x_1, y_2)} + \overline{\phi(x_2, y_1)} - \tau(b([x_2], [x_1])) \\
&= \overline{\phi(x_1, y_2)} + \overline{\phi(x_2, y_1)} + \tau(b([x_1], [x_2])).
\end{aligned}$$

The above shows that (H, λ, μ) is an ϵ-quadratic form. Moreover, it is easy to see that the submodules $F, G \subset H \subset L \oplus K^*$ defined by

$$F := (\{0\} \oplus K^*) \cap H \quad \text{and} \quad G := (L \oplus \{0\}) \cap H$$

are lagrangians of the form (H, λ, μ). Hence the form is hyperbolic and we identify it with $H_\epsilon(F)$. We define the formation

$$C_{(T,b,q),\pi} := (H_\epsilon(F); F, G).$$

The following properties of $C_{(T,b,q),\pi}$ are easily verified:

(i) A commutative diagram

$$\begin{array}{ccc} L_1 & \xrightarrow{\pi_1} & T_1 \\ \alpha \downarrow & & \downarrow A \\ L_2 & \xrightarrow{\pi_2} & T_2 \end{array}$$

where $A \colon (T_1, b_1, q_1) \xrightarrow{\cong} (T_2, b_2, q_2)$ is an isomorphism of $(-\epsilon)$-quadratic linking forms and $\alpha \colon L_1 \xrightarrow{\cong} L_2$ is an isomorphism of abelian groups, induces an isomorphism of ϵ-quadratic formations

$$f_\alpha \colon C_{(T_1,b_1,q_1),\pi_1} \xrightarrow{\cong} C_{(T_2,b_2,q_2),\pi_2};$$

(ii) Given a surjection $\pi_1 \colon L_1 \to T$, then for any free abelian group L_2, the surjection $\pi_1 \oplus 0 \colon L_1 \oplus L_2 \to T$ induces an isomorphism of ϵ-quadratic formations

$$C_{(T,b,q),\pi_1 \oplus 0} \xrightarrow{\cong} C_{(T,b,q),\pi_1} \oplus (H_\epsilon(L_2); L_2, L_2^*);$$

(iii) Consider $C = (H_\epsilon(F); F, G)$ with $T_C = \operatorname{coker}(\delta)$ and $\delta \colon G \to F^*$. If we take $(L, \pi) = (F^*, \pi_\delta)$, where $\pi_\delta \colon F^* \to T_C$ is the canonical projection, then there is a canonical isomorphism of ϵ-quadratic formations

$$C_{(T_C,b_C,q_C),\pi_\delta} \xrightarrow{\cong} C.$$

To make use of the above properties of $C_{(T,b,q),\pi}$, we shall need the following result.

Lemma 9.51 *Let $\pi_1 \colon L_1 \to T_1$ and $\pi_2 \colon L_2 \to T_2$ be surjections from finitely generated free abelian groups to torsion groups and suppose that $A \colon T_1 \to T_2$ is an isomorphism. Then there are finitely generated free abelian groups F_1, F_2 and an isomorphism $\alpha \colon F_1 \oplus L_1 \to F_2 \oplus L_2$ such that the following diagram commutes:*

$$\begin{array}{ccc} F_1 \oplus L_1 & \xrightarrow{0 \oplus \pi_1} & T_1 \\ \alpha \downarrow & & \downarrow A \\ F_2 \oplus L_2 & \xrightarrow{0 \oplus \pi_2} & T_2. \end{array}$$

9.2 Odd-Dimensional L-Groups

Proof. Let us call a diagram as above an equivalence of $(0 \oplus \pi_1)$ and $(0 \oplus \pi_2)$ over A. Let r be the minimal number of generators of T. Then any $\pi: L \to T$ is equivalent over the identity of T to $(0 \oplus \pi')$, where $\pi': \mathbb{Z}^r \to T$ is a surjection with $\ker(\pi') \cong \mathbb{Z}^r$. To see this, choose an $\mathbb{Z}$-epimorphism $\pi'': \mathbb{Z}^r \to T$ and a $\mathbb{Z}$-map $u: L \to \mathbb{Z}$ with $\pi'' \circ u = \pi$, and take $\pi' := \pi''|_{\mathrm{im}(u)}: \mathrm{im}(u) \to T$. Hence we can reduce to the case where $L_1 \cong L_2$.

We have the short exact sequence $0 \to \ker(\pi_2) \to L_2 \xrightarrow{\pi_2} T_2 \to 0$ and there exists an isomorphism $\ker(\pi_2) \cong L_2$ since both these torsionfree groups have the same rank. Hence we can find an exact sequence of the form

$$0 \to L_2 \xrightarrow{j_2} L_2 \xrightarrow{\pi_2} T_2 \to 0.$$

Since L_1 is free, there is a homomorphism $\bar{A}: L_1 \to L_2$ such that the following diagram commutes:

$$\begin{array}{ccc} L_1 & \xrightarrow{\pi_1} & T_1 \\ \bar{A} \downarrow & & \downarrow \cong A \\ L_2 & \xrightarrow{\pi_2} & T_2. \end{array}$$

Choose an isomorphism $u: L_1 \xrightarrow{\cong} L_2$. The map

$$(j_2 \circ u) \oplus \bar{A}: L_1 \oplus L_1 \to L_2$$

is onto since $\pi_2 \circ \bar{A} = A \circ \pi_1$ is onto and $0 \to L_2 \xrightarrow{j_2} L_2 \xrightarrow{\pi_2} T_2 \to 0$ is exact. Its kernel is isomorphic to L_2. Hence we can choose an exact sequence

$$0 \to L_2 \xrightarrow{i} L_1 \oplus L_1 \xrightarrow{(j_2 \circ u) \oplus \bar{A}} L_2 \to 0.$$

We may consider it as an acyclic free $\mathbb{Z}$-chain complex. Hence there is a retraction $(a, b): L_1 \oplus L_1 \to L_2$ of i. Moreover, if we define the homomorphism $\alpha: L_1 \oplus L_1 \to L_2 \oplus L_2$ via the matrix

$$\alpha = \begin{pmatrix} a & b \\ j_2 \circ u & \bar{A} \end{pmatrix} : L_1 \oplus L_1 \to L_2 \oplus L_2,$$

then α is an isomorphism. One easily checks that we obtain a commutative diagram

$$\begin{array}{ccc} L_1 \oplus L_1 & \xrightarrow{0 \oplus \pi_1} & T_1 \\ \alpha \downarrow & & \downarrow A \\ L_2 \oplus L_2 & \xrightarrow{0 \oplus \pi_2} & T_2. \end{array}$$

This completes the proof of Lemma 9.51 □

Proof of Proposition 9.46. The properties (i), (ii), and (iii) of $C_{(T,b,q),\pi}$ above together with Lemma 9.51 show that the desired function F of (9.47) is well defined and an inverse to q. This finishes the proof of Proposition 9.46. □

Let $(F, \theta) = (F, \lambda, \mu)$ be an $(-\epsilon)$-quadratic form over the free abelian group F with symmetrisation $(F, \widehat{\lambda})$. We say that (F, λ) is $\mathbb{Q}$-*non-singular* if $(\mathbb{Q} \otimes_{\mathbb{Z}} F, \text{id}_{\mathbb{Q}} \otimes_{\mathbb{Z}} \lambda)$ is non-singular and that (F, θ) is $\mathbb{Q}$-*non-singular* if (F, λ) is. Suppose that (F, θ) is $\mathbb{Q}$-non-singular and recall the boundary formation of (F, θ), that is, the formation

$$\partial(F, \theta) = (H_\epsilon(F); F, \Gamma_{\widehat{\lambda}}),$$

where $\Gamma_{\widehat{\lambda}} = \{(x, \widehat{\lambda}(x)) \mid x \in F\}$ is the graph of λ. The assumption that (F, λ, μ) is $\mathbb{Q}$-non-singular is equivalent to $\partial(F, \theta)$ being a $\mathbb{Q}$-trivial formation. In this situation we show how to express the $(-\epsilon)$-quadratic linking form $(T_{\partial(F,\theta)}, b_{\partial(F,\theta)}, q_{\partial(F,\theta)})$ introduced in Definition 9.44 directly from (F, λ). Since (F, λ) is $\mathbb{Q}$-non-singular, there is a short exact sequence

$$0 \to F \xrightarrow{\widehat{\lambda}} F^* \to T_\lambda \to 0 \tag{9.52}$$

where $T_\lambda := \text{coker}(\widehat{\lambda})$ is finite. If we tensor the exact sequence (9.52) with $\mathbb{Q}$, then, since $T_{\widehat{\lambda}}$ is finite, we see that

$$\text{id}_\mathbb{Q} \otimes_\mathbb{Z} \widehat{\lambda} \colon \mathbb{Q} \otimes_\mathbb{Z} F \to \mathbb{Q} \otimes_\mathbb{Z} F^*$$

is an isomorphism. The inverse $(\text{id}_\mathbb{Q} \otimes_\mathbb{Z} \widehat{\lambda})^{-1}$ defines a $(-\epsilon)$-symmetric bilinear form on $\mathbb{Q} \otimes_\mathbb{Z} F^*$. We denote the restriction of this bilinear form to F^* by

$$\lambda^{-1} \colon F^* \times F^* \to \mathbb{Q}.$$

Given $x, y \in F^*$, we write $[x], [y]$ for their images in T_λ and define the $(-\epsilon)$-quadratic linking form $\overline{\partial}(F, \theta) = (T_\lambda, b_\lambda, q_\lambda)$ by

$$b_\lambda([x], [y]) := \lambda^{-1}(x, y) \mod \mathbb{Z};$$

$$q_\lambda([x]) := \frac{\lambda^{-1}(x, x)}{2} \mod \mathbb{Z}.$$

From the definitions of $\partial(F, \theta)$ we directly conclude

$$\overline{\partial}(F, \theta) = q(\partial(F, \theta)),$$

where the map q is the bijection appearing in Proposition 9.46. Moreover, Proposition 9.46 shows that $\overline{\partial}(F, \theta)$ is a well-defined non-singular $(-\epsilon)$-quadratic linking form and that there is a commutative diagram

9.2 Odd-Dimensional L-Groups

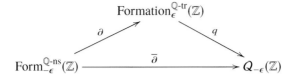

where $\text{Form}_{-\epsilon}^{\text{Q-ns}}(\mathbb{Z})$ is the set of isomorphism classes of $\mathbb{Q}$-non-singular $(-\epsilon)$-quadratic forms (F, θ) over $\mathbb{Z}$ and $\overline{\partial}$ is the function

$$\overline{\partial}\colon \text{Form}_{-\epsilon}^{\text{Q-ns}}(\mathbb{Z}) \to Q_{-\epsilon}(\mathbb{Z}), \quad [F, \theta] \mapsto [\overline{\partial}(F, \theta)].$$

Since the map q is a bijection, see Proposition 9.46, the map ∂ is surjective if and only the map $\overline{\partial}$ is surjective. Recall from Lemma 9.13 (iv) that (F, θ) is $\mathbb{Q}$-non-singular if and only if $\partial(F, \theta)$ is a $\mathbb{Q}$-trivial. Hence Theorem 9.40 is equivalent to the following result of Wall [405, Theorem 6].

Proposition 9.53 *The boundary map*

$$\overline{\partial}\colon \text{Form}_{-\epsilon}^{\text{Q-ns}}(\mathbb{Z}) \to Q_{-\epsilon}(\mathbb{Z})$$

is onto.

Hence it remains to give the proof of Proposition 9.53. It is based on the fact that $\overline{\partial}$ is additive with respect to orthogonal sum of forms and linking forms, i.e.,

$$\overline{\partial}([F_1, \theta_1] \oplus [F_2, \theta_2]) = \overline{\partial}([F_1, \theta_1]) \oplus \overline{\partial}([F_2, \theta_2]).$$

With this observation in mind, we call a quadratic linking form (T, b, q) *decomposable* if it can be written as a non-trivial orthogonal sum and *indecomposable* otherwise; and similarly for linking forms (T, b). Note that (T, b, q) is indecomposable if and only if (T, b) is indecomposable. It is sufficient to express each isometry class of an indecomposable linking form as the boundary of a non-degenerate form. We now list all indecomposable non-singular ϵ-quadratic linking forms. The proof is due to Wall [405, Theorems 4, 5].

Theorem 9.54 (Indecomposable linking forms over $\mathbb{Z}$) *Consider any decomposable non-singular ϵ-quadratic linking form. Then we can find a prime p, an integer θ coprime to p, a natural number k satisfying $-p^k < \theta < p^k$, a generator x of $\mathbb{Z}/p^k$, and generators x_1, x_2 of $(\mathbb{Z}/p^k)^2$ such that the given decomposable non-singular ϵ-quadratic linking form is isomorphic to one of the following linking forms:*

- *ϵ-Hyperbolic:* $H_\epsilon(\mathbb{Z}/p^k) \colon \left((\mathbb{Z}/p^k)^2, \begin{pmatrix} 0 & p^{-k} \\ \epsilon p^{-k} & 0 \end{pmatrix}, q(x_1) = 0 = q(x_2)\right);$
- *Pseudo-hyperbolic:* $F_{2,k} \colon \left((\mathbb{Z}/2^k)^2, \begin{pmatrix} 2^{1-k} & 2^{-k} \\ 2^{-k} & 2^{1-k} \end{pmatrix}, q(x_1) = 2^{-k} = q(x_2)\right);$
- *Cyclic:* $A_{p,k}(\theta) \colon \left(\mathbb{Z}/p^k, \left(\frac{\theta}{p^k}\right), q(x) = \frac{\theta}{2p^k}\right)$

where $\epsilon = 1$ in the pseudo-hyperbolic and cyclic case.

Remark 9.55 The standard ϵ-quadratic hyperbolic linking form $H_\epsilon(\mathbb{Z}/p^k)$ is a torsion analogue of the standard ϵ-quadratic hyperbolic form $H_\epsilon(\mathbb{Z})$ from Definition 8.79.

Now we list the $\mathbb{Q}$-non-singular ϵ-quadratic forms $(F, \theta) = (F, \lambda, \mu)$ that give the above quadratic linking forms as their boundaries. Define the weighted ϵ-quadratic hyperbolic form by

$$H_\epsilon^{p^k}(\mathbb{Z}) = \left(\mathbb{Z}^2, \begin{pmatrix} 0 & p^k \\ \epsilon p^k & 0 \end{pmatrix}, \mu(x) = 0 = \mu(y)\right).$$

Then we have $\overline{\partial}(H_\epsilon^{p^k}(\mathbb{Z})) = H_\epsilon(\mathbb{Z}/p^k)$, and it remains to consider pseudo-hyperbolic and cyclic linking forms where $\epsilon = +1$.

Recall that, when $\epsilon = +1$, a quadratic form $(F, \theta) = (F, \lambda, \mu)$ is determined by λ since μ is uniquely given by

$$\mu(x) = \frac{\lambda(x, x)}{2} \in Q_+(\mathbb{Z}) = \mathbb{Z}.$$

Hence we shall list the required even symmetric bilinear forms, see Wall [405, Proof of Theorem 6]. For the pseudo-hyperbolic $(+1)$-quadratic linking forms, we define the integers

$$a_k := \frac{1}{3}(2^k - (-1)^k) \quad \text{and} \quad b_k := (-1)^{k-1},$$

and let $(\mathbb{Z}^4, \lambda_{F_{2,k}})$ be represented by the following matrix:

$$\lambda_{F_{2,k}} := \begin{pmatrix} 2^{k+1}(4a_k b_k - 1 - 2^k b_k) & -2^k(4a_k b_k - 1) & 2^{k+1} b_k & -2^k \\ -2^k(4a_k b_k - 1) & 2^{k+1}(4a_k b_k - 1) & -2^{k+2} b_k & 2^{k+1} \\ 2^{k+1} b_k & -2^{k+2} b_k & 6 b_k & -12 \\ -2^k & 2^{k+1} & -12 & 12 a_k - 2^{k+1} \end{pmatrix}.$$

A calculation shows that the determinant of $\lambda_{F_{2,k}}$ is 2^{2k} and that the inverse of $\lambda_{F_{2,k}}$ contains the submatrix $\begin{pmatrix} 2^{1-k} & 2^{-k} \\ 2^{-k} & 2^{1-k} \end{pmatrix}$. As a consequence we have $\overline{\partial}(\mathbb{Z}^4, \lambda_{F_{2,k}}) = F_{2,k}$.

For the cyclic $(+1)$-quadratic linking forms, the situation is more complex and we shall need the following general symmetric bilinear form. Given an n-tuple of even integers $\underline{a} = \{a_1, \ldots, a_n\}$, we define the even symmetric bilinear form on $\mathbb{Z}^n$ using the matrix

$$\lambda_{\underline{a}} := \begin{pmatrix} a_1 & 1 & 0 & \cdots & 0 & 0 & 0 \\ 1 & a_2 & 1 & \cdots & 0 & 0 & 0 \\ 0 & 1 & a_3 & \cdots & 0 & 0 & 0 \\ \vdots & \vdots & \vdots & \ddots & \vdots & \vdots & \vdots \\ 0 & 0 & 0 & \cdots & a_{n-2} & 1 & 0 \\ 0 & 0 & 0 & \cdots & 1 & a_{n-1} & 1 \\ 0 & 0 & 0 & \cdots & 0 & 1 & a_n \end{pmatrix}.$$

9.2 Odd-Dimensional L-Groups

The following lemma is adapted from Wall's ideas in the proof of [405, Theorem 6], and its proof is taken from [114, §6.2].

Lemma 9.56 *Let (U, b, q) be a $(+1)$-quadratic linking form on a cyclic group U of order p^k. Then there is a string of even integers $\underline{a} = \{a_1, \ldots, a_n\}$ such that*

$$\overline{\partial}(\mathbb{Z}^n, \lambda_{\underline{a}}) \cong (U, b, q).$$

Proof. We first show that U has a generator x satisfying $q(x) = \frac{\theta}{2p^k}$, and θ has the opposite parity from p. If p is even, this is necessary for (b, q) to be non-singular. If p is odd, then by [405, §5] the isometry class of (b, q) is determined by whether θ is a quadratic non-residue or not. Since 1 and 4 are both quadratic residues mod p and the number of quadratic residues equals the number of quadratic non-residues, we see that for $p > 3$ there is an even quadratic non-residue. For $p = 3$ we can take $\theta = 2$ and $\theta = -2$. Hence we can assume that θ has the opposite parity from p.

To find the even integers a_i, we need an even version of the Euclidean algorithm. Let d_1 and d_2 be co-prime integers of opposite parity and suppose that $|d_1| > |d_2|$. We are interested in the case where $(d_1, d_2) = (p^k, \theta)$. Since $\frac{-\theta}{-p^k} = \frac{\theta}{p^k}$, we have the freedom to change the sign of *both* d_1 and d_2. Starting from $i = 1$, we define even integers $a_i \in 2\mathbb{Z}$ and integers $d_i \in \mathbb{Z}$ as follows. Given d_i and d_{i+1}, we find the even integer multiple $a_i \cdot d_{i+1}$ of d_{i+1} which is closest to d_i, and we take d_{i+2} to be the difference of $a_i \cdot d_{i+1}$ and d_i, i.e.,

$$d_i = a_i \cdot d_{i+1} - d_{i+2}, \tag{9.57}$$

where $0 \leq |d_{i+2}| < |d_{i+1}|$. As in the standard Euclidean algorithm, if d_{i+2} divides d_{i+1} then d_{i+2} divides d_1 and d_2. But d_1 and d_2 are co-prime, so the numbers $|d_i|$ form a strictly decreasing sequence ending with 1 or 0. Choose n so that $d_{n+2} = 0$ and $d_{n+1} = \pm 1$. Since below we shall want $d_{n+1} = 1$, we define a new sequence $\{d'_1, d'_2, \ldots, d'_{n+1}\}$ by

$$d'_i := d_{n+1} \cdot d_i.$$

The d'_i-s also satisfy (9.57) for the given even a_i and setting $i = n-1, n$ in (9.57) we have

$$d'_{n-1} = a_{n-1} \cdot d'_n - 1 \quad \text{and} \quad d'_n = a_n.$$

We must now show that $\overline{\partial}(\mathbb{Z}^n, \lambda_{\underline{a}}) \cong (U, b, q)$. By induction, the last i rows and columns of the matrix $\lambda_{\underline{a}}$ have determinant d'_{n-i+1}. Hence $\lambda_{\underline{a}}$ has determinant $d'_1 = d_{n+1} \cdot p^k$. Moreover, the submatrix $M_{1,1}(\lambda_{\underline{a}})$ obtained by deleting the first row and column of $\lambda_{\underline{a}}$ has determinant $d'_2 = d_{n+1} \cdot \theta$. This implies that the $(1, 1)$-entry of $\lambda_{\underline{a}}^{-1}$ is

$$\lambda_{\underline{a}}^{-1}(1, 1) = \frac{d'_2}{d'_1} = \frac{\theta}{p^k}.$$

Hence $\overline{\partial}(\mathbb{Z}^n, \lambda_{\underline{a}})$ contains an element y of order p^k satisfying $q_{\lambda_{\underline{a}}}(y) = \frac{\theta}{2p^k}$. Since $\det(\lambda_{\underline{a}}) = p^k$, it follows that $\overline{\partial}(\mathbb{Z}^n, \lambda_{\underline{a}}) \cong (U, b, q)$. □

If we apply Lemma 9.56 to the cyclic $(+1)$-quadratic linking form, we see that Proposition 9.53 holds if we can prove Theorem 9.54, which we do next.

Given a linking form (T, b) and a subgroup $U \subset T$, we define the *annihilator* of U by
$$U^\perp := \{x \in T \mid b(x, y) = 0, \forall y \in U\}.$$

We shall repeatedly use the following lemma to split off non-singular linking forms from larger linking forms.

Lemma 9.58 *Let (T, b) be a non-singular linking form with $U \subset T$ a subgroup such that $b|_{U \times U}$ is non-singular. Then $(U^\perp, b|_{U^\perp \times U^\perp})$ is non-singular and*
$$(T, b) = (U, b|_{U \times U}) \oplus (U^\perp, b|_{U^\perp \times U^\perp}).$$

Proof. Let $R_U \colon \widehat{T} \to \widehat{U}$ be the homomorphism defined by restricting the domain of an element in the torsion dual of T to U. By definition, there is a short exact sequence
$$0 \to U^\perp \to T \xrightarrow{R_U \circ \widehat{b}} \widehat{U} \to 0,$$
and the assumption that $b|_{U \times U}$ is non-singular means that $(R_U \circ \widehat{b})|_U$ is an isomorphism. It follows that $U \cap U^\perp = \{e\}$ and that U and $U^\perp$ generate T. Hence $T = U \oplus U^\perp$ and, again by definition, this is an orthogonal splitting for the linking form b. As $\widehat{b}$ is an isomorphism, $\widehat{b}|_{U^\perp} \colon U^\perp \xrightarrow{\cong} \widehat{U^\perp}$ is an isomorphism, and hence $b_{U^\perp \times U^\perp}$ is non-singular. □

If (T, b, q) is quadratic linking form and $T = \bigoplus_p T_p$ is the decomposition of T into its p-primary components, then Lemma 9.58 implies that (T, b, q) splits into p-primary components
$$(T, b, q) = \bigoplus_p (T_p, b_p, q_p),$$
where $b_p := b|_{T_p \times T_p}$ and $q_p := q|_{T_p}$. We therefore focus on one prime at a time and assume that $T = T_p$ is p-primary. Recall that a group U is called an elementary abelian p-group if U is isomorphic to $(\mathbb{Z}/p^k)^j$ for some j, k. The group T has a decomposition into elementary abelian groups U_k,
$$T \cong \bigoplus_k U_k.$$

Lemma 9.58 implies that every splitting of T into elementary abelian groups gives rise to splitting of quadratic linking forms
$$(T, b, q) \cong \bigoplus_k (U_k, b_k, q_k)$$
where $b_k := b|_{U_k \times U_k}$ and $q_k := q|_{U_k}$.

We now suppose that (U, b, q) is an ϵ-quadratic linking form over an elementary abelian p-group, $U \cong (\mathbb{Z}/p^k)^j$. Let $x \in U$ generate a $\mathbb{Z}/p^k$ summand. Since b is

9.2 Odd-Dimensional L-Groups

non-singular, there is a $y \in U$ such that

$$b(x, y) = \frac{1}{p^k}.$$

If $\epsilon = -1$, then $b(x,x) = b(y,y) = 0$ and setting $V \subset U$ to be the subgroup generated by x and y, we see that $b|_{V \times V} \colon V \times V \xrightarrow{\cong} H_-(\mathbb{Z}/p^k)$ is an isomorphism. Applying Lemma 9.58 (U, b, q) splits off $H_-(\mathbb{Z}/p^k)$ as an orthogonal summand and proceeding by induction, we see that $j = 2i$ is even and $(U, b, q) \cong \bigoplus_i H_-(\mathbb{Z}/p^k)$.

If $\epsilon = +1$, then the argument is somewhat more complex. Suppose that $b(x, x) = \frac{r_x}{p^k}$ and $b(y, y) = \frac{r_y}{p^k}$. If r_x is prime to p, then b restricted to the summand of U generated by x is non-singular and so by Lemma 9.58 splits off (U, b, q), and we proceed inductively. The same holds for r_y. Hence we may assume that p divides both r_x and r_y. In the case where p is odd, we have $b(x+y, x+y) = \frac{r_x + r_y + 2}{p^k}$. Hence $x + y$ generates a summand of U on which b is non-singular, and so we proceed as before.

If $p = 2$, let $q(x) = \frac{s_x}{2^k}$ and $q(y) = \frac{s_y}{2^k}$. If $s_x = 0$, then $q(-s_y x + y) = 0$ and hence $\{x, -s_y x + y\}$ forms a basis for a summand $V \subset U$. As $b|_{V \times V} \colon V \times V \xrightarrow{\cong} H_+(\mathbb{Z}/2^k)$ is an isomorphism, we can split off $(V, b|_{V \times V}, q|_V)$ from (U, b, q). If $s_x = 2s'$ is even, then $q(-s_y x + y) = \frac{2s' s_y^2}{2^k}$. Inductively replacing y by $-s_y x + y$, we can eventually achieve that $s_y = 0$, in which case we have already shown that (U, b, q) splits off a copy of $H_+(\mathbb{Z}/2^k)$.

This leaves the final case, in which both s_x and s_y are odd. Recall that the abelian group of multiplicative units $(\mathbb{Z}/2^k)^\times$ consists of elements $\overline{m}$ for odd m and is isomorphic to $\mathbb{Z}/2 \oplus \mathbb{Z}/2^{k-2}$. In particular, it has exponent $\max\{2, 2^{k-2}\}$. As before, we note that $q(-s_y x + y) = \frac{s_x s_y^2}{2^k}$. Hence, replacing y by $-s_y x + y$, we may replace s_y by $s_x s_y^2$ without changing s_x. Iterating this we can replace s_y by $s_x^{2^j-1} s_y^{2^j}$ for every $j \geq 0$ without changing s_x. By choosing j large enough, we can arrange $s_y = 1$, leaving s_x unchanged. Reversing the roles of x and y we can achieve that $s_x = 1 = s_y$. Now $\{x, y\}$ generates a summand $V \subset U$ such that $b|_{V \times V} V \times V \xrightarrow{\cong} F_{2,k}$ is an isomorphism. Hence we can split off a copy of $F_{2,k}$ from (U, b, q). This finishes the proof of Theorem 9.54 and hence also of Proposition 9.53, Theorem 9.40, and Theorem 9.18.

Remark 9.59 Note that the above argument shows that every non-singular $(+1)$-quadratic linking form (T, b, q) on an odd order group T is isomorphic to a sum of cyclic linking forms. Hence in the list of Theorem 9.54 we only need the linking form $H_+(\mathbb{Z}/p^k)$ for $p = 2$.

Much more is known about the classification of linking forms. Wall already proved more than we have used in [405, §5]. A complete classification of $(+1)$-symmetric linking forms over $\mathbb{Z}$ is given in [213], and combined with [405, Theorem 5] this gives a complete classification of $(+1)$-quadratic linking forms over $\mathbb{Z}$.

9.3 The Odd-Dimensional Surgery Obstruction in the Universal Covering Case

In this section we fix a normal Γ-map of degree one $(M, [M], f, \widehat{f}, a, \xi, \overline{f})$ with target $(X, \widehat{p}_X, [X])$ in the sense of Definition 8.3, where the dimension $n = 2k + 1$ is odd. We will often abbreviate $(M, [M], f, \widehat{f}, a, \xi, \overline{f})$ by $(f, \overline{f})$. Moreover, we will assume that we are in the universal covering case, see Notation 8.34 and that $k \geq 2$. We identify $\Gamma = \pi = \pi_1(X, x)$ and write $w = w_1(X)$. Next we state the main result of this section.

Theorem 9.60 (Surgery obstruction in odd dimensions) *Consider a normal π-map of degree one $(f, \overline{f}) \colon M \to X$ with $\dim(M) = 2k + 1$.*
Then there is a well-defined surgery obstruction

$$\sigma(f, \overline{f}) \in L_{2k+1}(\mathbb{Z}\pi_1(X, x), w_1(X))$$

taking values in the odd-dimensional algebraic L-group (in the sense of Definition 9.15) of the ring $\mathbb{Z}\pi_1(X, x)$ equipped with the $w_1(X)$-twisted involution. It has the following properties:

(i) *It depends only on the normal bordism with cylindrical target class of $(f, \overline{f})$ in the sense of Definition 8.7;*
(ii) *Suppose $k \geq 2$. Then we have $\sigma(f, \overline{f}) = 0 \in L_{2k+1}(\mathbb{Z}\pi_1(X, x), w_1(X))$ if and only if we can do a finite number of surgery steps to obtain a normal homotopy equivalence $(f', \overline{f'})$.*

Given a k-connected degree one normal π-map $(f, \overline{f}) \colon M \to X$, there are two ways to define an ϵ-quadratic formation $(H_\epsilon(F); F, G)$ representing the surgery obstruction $\sigma(f, \overline{f}) \in L_{2k+1}(\mathbb{Z}\pi, w)$ of $(f, \overline{f})$. The first of these is the *kernel formation* and is defined working "inside M" using a Heegaard splitting of $(f, \overline{f})$ in Definition 9.70. The second produces what we call the *bordism formation*: it is defined working "outside M" using certain normal bordisms $(F, \overline{F})$ between $(f, \overline{f})$ and some other k-connected degree one normal map $(f', \overline{f'})$ in Definition 9.85. An important insight of Ranicki, see for example [352, Proposition 12.2], is that the kernel formation is isomorphic to the bordism formation.

We begin by defining the kernel formation of $(f, \overline{f})$ in Subsection 9.3.1. We next define the bordism formation and prove that the kernel formation is isomorphic to the bordism formation in Subsection 9.3.2. Moreover, we prove some auxiliary properties to be used later. Finally, in Subsection 9.3.3, we first prove that $\sigma(f, \overline{f})$ gives a well-defined element in $L_{2k+1}(\mathbb{Z}\pi, w)$ using the bordism formation, and then we use all we know to show that the vanishing of $\sigma(f, \overline{f})$ is both a necessary and sufficient condition for a positive solution to the surgery problem $(f, \overline{f})$.

9.3.1 The Kernel Formation

In this subsection we define the kernel formation of a k-connected normal π-map of degree one $(f, \bar{f}): M \to X$ with target a $(2k+1)$-dimensional Poincaré complex with choices (ABP) and (GBP) in Definition 9.70 and prove some of its properties.

To achieve a homotopy equivalence, it remains to kill the kernel groups $K_k(\widetilde{M})$ and $K_{k+1}(\widetilde{M})$. By Poincaré duality $K_{k+1}(\widetilde{M}) \cong K^k(\widetilde{M}) \cong K_{k+1}(\widetilde{M})^*$. So it is enough to kill $K_k(\widetilde{M})$. Now by Lemma 4.4 and Lemma 8.55 the $\mathbb{Z}\pi$-module $K_k(\widetilde{M})$ is finitely generated, and so we can choose a set of generators $\underline{\omega} = \{x_1, \ldots, x_a\}$. By Lemma 8.39 (iii) there is a canonical isomorphism $K_k(\widetilde{M}) \xrightarrow{\cong} \pi_{k+1}(f)$. Hence, by Lemma 8.10 and Theorem 4.39, each generator x_i is represented by a triple $(\bar{q}_i, w_i, h_i)$, where

$$\bar{q}_i : D^{k+1} \times S_i^k \hookrightarrow M$$

is an embedding, w_i is a path from $\bar{q}_i(0, s)$ to the base point $m \in M$, and h_i is a null homotopy of $f \circ \bar{q}_i$ compatible with w_i. For later use, we note that by Theorem 4.39 (ii) and Remark 4.40 the regular homotopy class of $\bar{q}_i$, considered as an immersion, is determined by $x_i \in K_k(\widetilde{M})$ and the bundle map $\bar{f}$. By general position, we can arrange that the images of the $\bar{q}_i$ are pairwise disjoint, and so there is an embedding

$$\coprod_{i=1}^{a} \bar{q}_i : \coprod_{i=1}^{a} (D^{k+1} \times S_i^k) \hookrightarrow M.$$

The paths w_i can be arranged to be embedded and to lie in the interior of thick tubes $D^{2k} \times w_i(I) \subset M$ that meet the boundary of $(D^{k+1} \times S_i^k)$ in an appropriate transversal way. Collecting these data we obtain an embedding

$$\bar{q} : \overline{U} \hookrightarrow M,$$

where $\overline{U} := \natural_{i=1}^{a}(D^{k+1} \times S_i^k)$ is the indicated boundary connected sum. In this case we say that the embedding $\bar{q} : \overline{U} \hookrightarrow M$ represents $\underline{\omega}$ and that the pair $(\underline{\omega}, \bar{q})$ is a *representation of the surgery kernel* $K_k(\widetilde{M})$. We can also arrange that the base point $m \in M$ lies in the image of the boundary of $\overline{U}$ and take $x = f(m)$ as the base point for X.

Remark 9.61 A k-connected degree one normal map $(f, \bar{f}): M \to X$ admits many representations of its surgery kernel $K_k(\widetilde{M})$. Firstly there are many choices for $\underline{\omega}$, and, even once the choice $\underline{\omega}$ is made, there are many choices of isotopy classes of embeddings $\bar{q} : \overline{U} \hookrightarrow M$ representing $\underline{\omega}$.

We now explain the notion of a Heegaard splitting of $(f, \bar{f})$ as defined by Ranicki in [352, Chapter 12]. Heegaard splittings illuminate the topology of k-connected normal maps and are essential for defining the kernel formation.

Henceforth we shall always write $U := \bar{q}(\overline{U}) \subset M$ for the image of $\bar{q}$. We may then decompose M as the union

$$M = U \cup M_0,$$

where M_0 is the closure of the complement of U and with $m \in \partial U$. This decomposition of the source of $(f, \overline{f})$ can be mirrored in the target X. By a theorem of Wall [408, Theorem 2.4], X is homotopy equivalent to a *CW* complex with precisely one cell in dimension $2k + 1$. So, perhaps after replacing X with a homotopy equivalent *CW* complex, we may assume that

$$X = X_0 \cup D^{2k+1},$$

where X_0 is a *CW* complex of dimension at most $2k$ and D^{2k+1} is the top cell of X. We regard $S^{2k} = \partial D^{2k+1}$ as a subspace of X_0 so that (X_0, S^{2k}) is a $(2k + 1)$-dimensional Poincaré pair. The fundamental class of (X_0, S^{2k}), $[X_0, S^{2k}] \in H_{2k+1}(X_0, S^{2k}; \mathbb{Z}^w)$, is defined by requiring that it maps to the fundamental class $[X] \in H_{2k+1}(X; \mathbb{Z}^w)$ under the isomorphism

$$H_{2k+1}(X_0, S^{2k}; \mathbb{Z}^w) \xrightarrow{\cong} H_{2k+1}(X, D^{2k+1}; \mathbb{Z}^w) \xrightarrow{\cong} H_{2k+1}(X; \mathbb{Z}^w).$$

Definition 9.62 Let $(f, \overline{f}) \colon M \to X$ be a k-connected normal π-map of degree one and $(\underline{\omega}, \overline{q})$ be a representation of $K_k(\widetilde{M})$ so that $M = M_0 \cup U$, where $U = \overline{q}(U)$. A *Heegaard splitting* of $(f, \overline{f})$ based on $(\underline{\omega}, \overline{q})$ is the expression of $(f, \overline{f})$ as a union

$$(f, \overline{f}) = (f_0, \overline{f}_0) \cup (e, \overline{e}) \colon M_0 \cup U \to X_0 \cup D^{2k+1},$$

where $(f_0, \overline{f}_0) \colon (M_0, \partial U) \to (X_0, S^{2k})$ and $(e, \overline{e}) \colon (U, \partial U) \to (D^{2k+1}, S^{2k})$ are degree one normal maps of pairs.

Figure 9.63 (Schematic drawing of a Heegaard splitting).

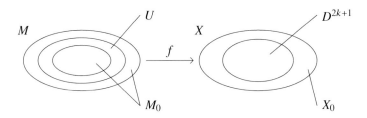

Lemma 9.64 *For every representation $(\underline{\omega}, \overline{q})$ of the surgery kernel $K_k(\widetilde{M})$ of a k-connected degree one normal map $(f, \overline{f}) \colon M \to X$, there is a Heegaard splitting of $(f, \overline{f})$ based on $(\underline{\omega}, \overline{q})$.*

Proof. Using the embedding $\overline{q} \colon U \hookrightarrow M$, we decompose $M = M_0 \cup U$ and $X = X_0 \cup D^{2k+1}$ as above. We show that $f \colon M_0 \cup U \to X_0 \cup D^{2k+1}$ is homotopic to a union of degree one maps of pairs; the existence of the bundle data follows easily and we leave the reader to verify this.

Since $f|_U$ is nullhomotopic and $U \subset M$ satisfies the homotopy extension property, we may assume after a homotopy that $f(U) \subset D^{2k+1}$. After another homo-

9.3 The Odd-Dimensional Surgery Obstruction in the Universal Covering Case

topy, we may further assume that $f|_U$ gives rise to a degree one map of pairs $(f|_U, \partial f|_U) \colon (U, \partial U) \to (D^{2k+1}, S^{2k})$. Keeping f fixed on U, we now look for a homotopy relative U to a map where $f(M_0) \subset X_0$. The obstructions to finding such a homotopy lie in $H^*_\pi(M_0, \partial U; \pi_*(X, X_0))$, where $\pi_*(X, X_0)$ is viewed as a $\mathbb{Z}\pi$-module in the obvious way. The only value of $*$ where this group is non-zero is $* = 2k+1$, where $H^{2k+1}_\pi(M_0, \partial U; \pi_{2k+1}(X, X_0)) \xrightarrow{\cong} \mathbb{Z}$ is given by the fundamental class of $(M_0, \partial U)$ and the obstruction is identified with the difference between the degree of f and $(f|_U, \partial f|_U)$, which is zero. □

We now explain how a Heegaard splitting $(f, \overline{f}) = (f_0, \overline{f}_0) \cup (e, \overline{e})$ based on a representation $(\underline{\omega}, \overline{q})$ of the surgery kernel $K_k(M)$ of a k-connected degree one normal map $(f, \overline{f}) \colon M \to X$ defines a $(-1)^k$-quadratic formation. Let $\widehat{D}^{2k+1} \subset \widetilde{X}$ denote the cover of D^{2k+1} induced by $\widetilde{X} \to X$ and similarly for $\widehat{S}^{2k} = \partial \widehat{D}^{2k+1}$. Let $\widehat{U} \subset \widetilde{M}$ and $\widehat{\partial U} \subset \widetilde{M}$ denote respectively the pre-image of U and ∂U under the covering map $\widetilde{M} \to M$. Then $\widehat{D}^{2k+1}$ is a disjoint union of copies of D^{2k+1} indexed by the elements of π and similarly for $\widehat{S}^{2k}$, $\widehat{U}$ and $\widehat{\partial U}$:

$$\widehat{D}^{2k+1} = \coprod_{g \in \pi} D^{2k+1}_g, \quad \widehat{S}^{2k} = \coprod_{g \in \pi} S^{2k}_g, \quad \widehat{U} = \coprod_{g \in \pi} U_g \quad \text{and} \quad \widehat{\partial U} = \coprod_{g \in \pi} \partial U_g.$$

In the sequel we use the notation of Section 8.2. Note that in the Heegaard splitting

$$(f, \overline{f}) = (f_0, \overline{f}_0) \cup (e, \overline{e}) \colon M_0 \cup U \to X_0 \cup D^{2k+1},$$

from Definition 9.62 the fundamental group in the target is π for X_0, and it is the trivial group for D^{2k+1} and S^{2k}. But we would like to consider the normal maps obtained via the π-covering coming from the universal covering $\widetilde{X}$. This is exactly the setting of normal Γ-maps with $\Gamma = \pi$.

The normal π-map $(e, \overline{e}) \colon (U, \partial U) \to (D^{2k+1}, S^{2k})$ has surgery kernels

$$K_k(\widehat{U}), \quad K_k(\widehat{\partial U}) \quad \text{and} \quad K_{k+1}(\widehat{U}, \widehat{\partial U}),$$

see Definition 8.115. The kernels $K_{k+1}(\widehat{U}, \widehat{\partial U})$, $K_k(\widehat{\partial U})$ and $K_k(\widehat{U})$ are free $\mathbb{Z}\pi$-modules of rank a, $2a$ and a respectively, where we recall that $\underline{\omega} = \{x_1, \ldots, x_a\}$ has a elements. Clearly there are isomorphisms

$$K_k(\widehat{\partial U}) \xrightarrow{\cong} H_k(\widehat{\partial U}), \quad K_k(\widehat{U}) \xrightarrow{\cong} H_k(\widehat{U}), \quad K_{k+1}(\widehat{U}, \widehat{\partial U}) \xrightarrow{\cong} H_{k+1}(\widehat{U}, \widehat{\partial U}), \quad (9.65)$$

which we shall use freely below. Our choices yield preferred lifts of the canonical k-spheres in U, which in turn give a basis $\{\widetilde{x}'_1, \ldots, \widetilde{x}'_a\}$ for $K_k(\widehat{U})$. Write $\{\widetilde{x}_1, \ldots, \widetilde{x}_a\} \subset K_k(\widehat{\partial U})$ for the set of lifts given by the identification $U = \overline{q}(\natural_{i=1}^a(D^{k+1} \times S^k))$. Taking the corresponding lifts of the canonical $(k+1)$-disks in U, we obtain a basis $\{\widetilde{y}'_1, \ldots \widetilde{y}'_a\}$ of $K_{k+1}(\widehat{U}, \widehat{\partial U})$. Define $\widetilde{y}_i := \partial(\widetilde{y}'_i)$ for the boundary map $\partial \colon K_{k+1}(\widehat{U}, \widehat{\partial U}) \to K_k(\widehat{\partial U})$. Then $\{\widetilde{x}_1, \ldots, \widetilde{x}_a, \widetilde{y}_1, \ldots, \widetilde{y}_a\}$ is a $\mathbb{Z}\pi$-basis for $K_k(\widehat{\partial U})$ that is symplectic. Actually, the k-spheres corresponding to $\{\widetilde{x}_1, \ldots, \widetilde{x}_a, \widetilde{y}_1, \ldots, \widetilde{y}_a\}$

intersect in $\widehat{\partial U}$ via the equations

$$\widetilde{x}_i \cap \widetilde{x}_j = \emptyset = \widetilde{y}_i \cap \widetilde{y}_j \quad \text{and} \quad \widetilde{x}_i \cap \widetilde{y}_j = \begin{cases} \emptyset & i \neq j; \\ \{\bullet\} & i = j. \end{cases}$$

With the symplectic basis $\{\widetilde{x}_1, \ldots, \widetilde{x}_a, \widetilde{y}_1, \ldots, \widetilde{y}_a\}$ the $\mathbb{Z}\pi$-module $K_k(\widehat{\partial U})$ carries the structure of the standard hyperbolic ϵ-quadratic form $H_\epsilon((\mathbb{Z}\pi, w)^a)$ with respect to this basis. We shall often use $K_k(\widehat{\partial U})$ to denote not only the kernel module, but also the standard hyperbolic ϵ-quadratic form on this module.

Given a Heegaard splitting based on $(\underline{\omega}, \overline{q})$, thanks to the last paragraph in Remark 8.169, the coboundaries $\widehat{U}$ and $\widetilde{M}_0$ of $\widehat{\partial U}$ give rise to a pair of lagrangians $F, G \subset K_k(\widehat{\partial U})$ defined by

$$F := \mathrm{im}\bigl(K_{k+1}(\widehat{U}, \widehat{\partial U}) \to K_k(\widehat{\partial U})\bigr) = \mathrm{im}\bigl(H_{k+1}(\widehat{U}, \widehat{\partial U}) \to H_k(\widehat{\partial U})\bigr), \qquad (9.66)$$

and

$$G := \mathrm{im}\bigl(K_{k+1}(\widetilde{M}_0, \widehat{\partial U}) \to K_k(\widehat{\partial U})\bigr) = \mathrm{im}\bigl(H_{k+1}(\widetilde{M}_0, \widehat{\partial U}) \to H_k(\widehat{\partial U})\bigr). \qquad (9.67)$$

That we can equally well use the homology groups H_k or the kernel groups K_k above is clear for the lagrangian F, and for G it follows from a simple diagram chase in the following commutative diagram with exact rows and exact columns, where we recall that $K_k(\widetilde{M}_0) \to H_k(\widetilde{M}_0)$ is by definition injective:

$$\begin{array}{ccccc} K_{k+1}(\widetilde{M}_0, \widehat{\partial U}) & \longrightarrow & K_k(\widehat{\partial U}) & \longrightarrow & K_k(\widetilde{M}_0) \\ \downarrow & & \downarrow \cong & & \downarrow \\ H_{k+1}(\widetilde{M}_0, \widehat{\partial U}) & \longrightarrow & H_k(\widehat{\partial U}) & \longrightarrow & H_k(\widetilde{M}_0). \end{array}$$

Lemma 9.68 *The $\mathbb{Z}\pi$-submodules $F, G \subset K_k(\widehat{\partial U})$ of (9.66) and (9.67) are lagrangians of $K_k(\widehat{\partial U})$ that depend only on the representation $(\underline{w}, \overline{q})$ of the surgery kernel $K_k(\widetilde{M})$ of a k-connected degree one normal map $(f, \overline{f}): M \to X$. Moreover, if $(f, \overline{f}) = (f_0, \overline{f}_0) \cup (e, \overline{e})$ is a Heegaard splitting with $(f_0, \overline{f}_0): M_0 \to X_0$, then we have $K_{k+1}(\widetilde{M}_0) = 0$.*

Proof. That F and G are lagrangians of $K_k(\widehat{\partial U})$ follows by the arguments from Remark 8.169. Since by (9.65) we can equally well use the homology groups H_* instead of the kernel groups K_* in (9.66) and (9.67), the kernel formation depends only on the inclusion $U \subset M$ and hence only on the representation $(\underline{\omega}, \overline{q})$ of $K_k(\widetilde{M})$.

To see that $K_{k+1}(\widetilde{M}_0) = 0$, we argue as follows. By Poincaré duality for surgery kernels of pairs, see Lemma 8.116 (ii), there is an isomorphism $K_{k+1}(\widetilde{M}_0) \xrightarrow{\cong} K^k(\widetilde{M}_0, \widehat{\partial U})$. Hence by Lemma 8.52 (ii) it is sufficient to prove that $K_j(\widetilde{M}_0, \widehat{\partial U}) = 0$ for $j \leq k$. For the case $j < k$, the natural map $K_j(\widetilde{M}_0) \to K_j(\widetilde{M}) = 0$ is an isomorphism since by general position $(j+1)$-cycles in M can be homotoped relative

9.3 The Odd-Dimensional Surgery Obstruction in the Universal Covering Case

boundary into M_0 and similarly the natural map $K_k(\widetilde{M}_0) \to K_k(\widetilde{M})$ is onto. Because of the exact sequence

$$\cdots \to K_j(\widehat{\partial U}) \to K_j(\widetilde{M}_0) \to K_j(\widetilde{M}_0, \widehat{\partial U}) \to \cdots,$$

it remains to prove that $K_k(\widehat{\partial U}) \to K_k(\widetilde{M}_0)$ is onto. It is easy to prove a version of excision for surgery kernels that gives the following exact sequence

$$K_{k+1}(\widehat{U}, \widehat{\partial U}) \to K_k(\widetilde{M}_0) \to K_k(\widetilde{M}).$$

Since $K_k(\widehat{U}) \to K_k(\widetilde{M})$ is onto and $K_k(\widehat{\partial U}) \xrightarrow{\cong} K_k(\widehat{U}) \oplus K_{k+1}(\widehat{U}, \widehat{\partial U})$, it follows that $K_k(\widehat{\partial U}) \to K_k(\widetilde{M}_0)$ is onto and this completes the proof. □

Remark 9.69 Since $K_{k+1}(\widehat{U}) = 0$ and by Lemma 9.68 $K_{k+1}(\widetilde{M}_0) = 0$, it follows that $K_{k+1}(\widehat{U}, \widehat{\partial U})$ and $K_{k+1}(\widetilde{M}_0, \widehat{\partial U})$ inject into $K_k(\widehat{\partial U})$ and so in the sequel we shall identify $K_{k+1}(\widehat{U}, \widehat{\partial U})$ and $K_{k+1}(\widetilde{M}_0, \widehat{\partial U})$ with their isomorphic images in $K_k(\widehat{\partial U})$.

Definition 9.70 (Kernel formation) Let $(f, \overline{f}) \colon M \to X$ be a k-connected degree one normal map. The *kernel formation* defined by a representation $(\underline{\omega}, \overline{q})$ of $K_k(\widetilde{M})$ is the formation

$$\Sigma(f, \overline{f}, \underline{\omega}, \overline{q}) = (H_\epsilon(F); F, G) = \bigl(K_k(\widehat{\partial U}); K_{k+1}(\widehat{U}, \widehat{\partial U}), K_{k+1}(\widetilde{M}_0, \widehat{\partial U})\bigr).$$

Remark 9.71 The isomorphism class of the kernel formation $\Sigma(f, \overline{f}, \underline{\omega}, \overline{q})$ depends upon the choice of the representation $(\underline{\omega}, \overline{q})$, and this is before we consider the kernel formations of different but normally bordant k-connected degree one normal maps. However, the equivalence class of $\Sigma(f, \overline{f}, \underline{\omega}, \overline{q})$ will be shown to be well defined in $L_{2k+1}(\mathbb{Z}\pi, w)$ and a normal bordism invariant. It will give the definition of the odd-dimensional surgery obstruction.

The kernel formation $\Sigma(f, \overline{f}, \underline{\omega}, \overline{q})$ has an *interior presentation* defined as follows. Recall that $\overline{q} \colon \overline{U} \hookrightarrow M$ is an embedding with image U. Fix the standard embedding $\overline{U} \hookrightarrow S^{2k+1}$ and then write

$$S^{2k+1} = \overline{U} \cup_{\partial U = \partial U'} U',$$

where $U' \cong \natural_{i=1}^a (S^k \times D^{k+1})$ is the closure of the complement of $\overline{U}$ in S^{2k+1}. Here and elsewhere we shall identify $\partial U = \partial U'$ via $(\overline{q}|_{\partial \overline{U}})^{-1}$. Using this convention, let $i_U \colon \partial U \to U$ and $i'_U \colon \partial U \to U'$ be the inclusions. We also write $K_k(\widehat{U}') := H_k(\widehat{U}')$ and $K_{k+1}(\widehat{U}', \widehat{\partial U}) := H_{k+1}(\widehat{U}', \widehat{\partial U})$, where $\widehat{U}' = \coprod_{g \in \pi} U'$ and $\widehat{\partial U}'$ is identified with the boundary of $\widehat{U}$. The notation is convenient now and will be justified further when we consider doing surgery on $(\overline{q}, \underline{w})$ below, see Definition 9.79. Now consider the composites

$$K_{k+1}(\widehat{U}, \widehat{\partial U}) \xrightarrow{\partial_U} K_k(\widehat{\partial U}) \xrightarrow{i'_U} K_k(\widehat{U}')$$

and

$$K_{k+1}(\widehat{U}', \widehat{\partial U}) \xrightarrow{\partial_{U'}} K_k(\widehat{\partial U}) \xrightarrow{i_U} K_k(\widehat{U}),$$

where ∂_U and $\partial_{U'}$ are the boundary maps in the homology long exact sequences of the pairs $(\widehat{U}, \widehat{\partial U})$ and $(\widehat{U}', \widehat{\partial U})$ respectively. These composites define isomorphisms of free based $\mathbb{Z}\pi$-modules that yield the isomorphism

$$H_\epsilon(F) = F \oplus F^* = K_{k+1}(\widehat{U}, \widehat{\partial U}) \oplus K_{k+1}(\widehat{U}', \widehat{\partial U}) \xrightarrow{\cong} K_k(\widehat{U}') \oplus K_k(\widehat{U}). \quad (9.72)$$

Note that the generating set $\underline{\omega}$ provides a basis for $K_k(\widehat{U}')$. The structure maps of the kernel formation $\gamma \colon G \to F$ and $\delta \colon G \to F^*$ are given by the composites

$$\gamma \colon G = K_{k+1}(\widetilde{M}_0, \widehat{\partial U}) \xrightarrow{\partial_{M_0}} K_k(\widehat{\partial U}) \xrightarrow{i'_U} K_k(\widehat{U}') \cong F \quad (9.73)$$

and

$$\delta \colon G = K_{k+1}(\widetilde{M}_0, \widehat{\partial U}) \xrightarrow{\partial_{M_0}} K_k(\widehat{\partial U}) \xrightarrow{i_U} K_k(\widehat{U}) \cong F^*, \quad (9.74)$$

where ∂_{M_0} is the boundary homomorphism in the homology long exact sequence of the pair $(\widetilde{M}_0, \widehat{\partial U})$.

We now identify some key properties of the kernel formation.

Lemma 9.75 *Let $\Sigma(f, \overline{f}, \underline{\omega}, \overline{q}) = (H_\epsilon(F); F, G)$ be the kernel formation of a representation $(\underline{\omega}, \overline{q})$ of the surgery kernel of a k-connected degree one normal map $(f, \overline{f})$. Then the following hold:*

(i) *There are identifications*

$$K_k(\widetilde{M}) = H_\epsilon(F)/(F+G) \quad \text{and} \quad K_{k+1}(\widetilde{M}) = F \cap G;$$

(ii) *The map f is a homotopy equivalence if and only if $\Sigma(f, \overline{f}, \underline{\omega}, \overline{q})$ is a trivial formation.*

Proof. (i) The short exact sequence of kernel modules

$$0 \to K_{k+1}(\widehat{U}, \widehat{\partial U}) \to K_k(\widehat{\partial U}) \to K_k(\widehat{U}) \to 0$$

identifies $K_k(\widehat{U}) = H_\epsilon(F)/F = F^*$. Using a Heegaard splitting of the normal map $(f, \overline{f}) = (f_0, \overline{f}_0) \cup (e, \overline{e})$, it is easy to show that the kernel groups $K_*(\widehat{U}), K_*(\widetilde{M})$ and $K_*(\widetilde{M}, \widehat{U})$ behave like ordinary homology groups in that there is an excision isomorphism $K_{k+1}(\widetilde{M}_0, \widehat{\partial U}) \xrightarrow{\cong} K_{k+1}(\widetilde{M}, \widehat{U})$ and a long exact sequence of the pair $(\widetilde{M}, \widehat{U})$

$$0 \to K_{k+1}(\widetilde{M}) \to K_{k+1}(\widetilde{M}, \widehat{U}) \xrightarrow{\partial} K_k(\widehat{U}) \to K_k(\widetilde{M}) \to 0.$$

Now the structure map

$$\delta \colon G = K_{k+1}(\widetilde{M}_0, \widehat{\partial U}) \to K_k(\widehat{\partial U}) \to K_k(\widehat{U}) = F^*$$

of the kernel formation $(H_\epsilon(F); F, G)$ is identified with the boundary homomorphism $\partial \colon K_{k+1}(\widetilde{M}, \widehat{U}) \to K_k(\widehat{U})$ in the long exact sequence above. Hence we have isomor-

9.3 The Odd-Dimensional Surgery Obstruction in the Universal Covering Case

phisms $K_k(\widetilde{M}) \xrightarrow{\cong} \mathrm{coker}(\delta) \xrightarrow{\cong} H_\epsilon(F)/(F+G)$ and $K_{k+1}(\widetilde{M}) = \ker(\delta) = F \cap G$.

(ii) By Lemma 9.24 (iii) a formation is trivial if and only if $\delta \colon G \to F^*$ is an isomorphism. From the proof of part (i), we see that $\Sigma(f, \bar{f}, \omega, \bar{q})$ is trivial if and only if $K_k(\widetilde{M}) = K_{k+1}(\widetilde{M}) = 0$, or, equivalently, f is a homotopy equivalence. □

Exercise 9.76 Consider the identity map $\mathrm{id} \colon S^{2k+1} \to S^{2k+1}$ covered by the identity on the tangent bundles as a surgery problem. Identify $S^{2k+1} = \partial D^{2k+2} = \partial(D^{k+1} \times D^{k+1})$. Using this identification consider the embedding $S^k \times D^{k+1} \hookrightarrow \partial(D^{k+1} \times D^{k+1}) = S^{2k+1}$ and denote its image by U so that there is a decomposition $S^{2k+1} = U \cup S_0^{2k+1}$.

Modify the surgery problem so that it respects this Heegaard decomposition and show that the associated kernel formation is trivial in the sense of Definition 9.7. What is the result of the corresponding surgery?

Exercise 9.77 Consider a degree one normal $(f, \bar{f}) \colon S^k \times S^{k+1} \to S^{2k+1}$ obtained by collapsing the complement of a neighbourhood of some point $(x, s) \in S^k \times S^{k+1}$ so that we have $K_k(S^k \times S^{k+1}) \cong K_{k+1}(S^k \times S^{k+1}) \cong \mathbb{Z}$. Consider the embedding $S^k \times \{s\} \hookrightarrow S^k \times S^{k+1}$ with trivial normal bundle and denote its tubular neighbourhood by U so that there is a decomposition $S^k \times S^{k+1} = U \cup (S^k \times S^{k+1})_0$.

Modify the surgery problem so that it respects this Heegaard decomposition and show that the associated kernel formation is a boundary in the sense of Definition 9.9. What is the result of the corresponding surgery?

We now take the first step towards understanding how the kernel formation $\Sigma(f, \bar{f}, \omega, \bar{q})$ can vary as $(f, \bar{f}) \colon M \to X$ varies amongst π-normally bordant k-connected degree one normal maps. Let $(F, \overline{F}) \colon (W; M, M') \to X \times (I, \partial I)$ be a normal π-s-cobordism between $(f, \bar{f}) \colon M \to X$ and $(f', \bar{f'}) \colon M' \to X$. From s-cobordism Theorem 2.1 we conclude that W determines a diffeomorphism $h \colon M \xrightarrow{\cong} M'$ and the map $F \colon W \to X \times I$ determines a homotopy $f \simeq f' \circ h$. If $(\underline{\omega} = \{x_1, \ldots, x_a\}, \bar{q})$ is a representation of $K_k(\widetilde{M})$, then $(h_*(\underline{\omega}), h \circ \bar{q})$ is a representation of $K_k(\widetilde{M}')$ where $h_*(\underline{\omega}) = \{h_*(x_1), \ldots, h_*(x_a)\}$. Setting $U' := h(U)$, and letting M'_0 be the closure of the complement of U' in M', then the diffeomorphism h gives a diffeomorphism of the unions

$$h \colon M_0 \cup U \xrightarrow{\cong} M'_0 \cup U'.$$

It induces an isomorphism $h_* \colon K_k(\widehat{\partial U}) \xrightarrow{\cong} K_k(\widehat{\partial U'})$ of ϵ-quadratic forms as well as isomorphisms

$$h_* \colon K_{k+1}(\widehat{U}, \widehat{\partial U}) \xrightarrow{\cong} K_{k+1}(\widehat{U'}, \widehat{\partial U'});$$
$$h_* \colon K_{k+1}(\widetilde{M}_0, \widehat{\partial U}) \xrightarrow{\cong} K_{k+1}(\widetilde{M}'_0, \widehat{\partial U'}).$$

Hence we have proven the following lemma.

Lemma 9.78 *With the notation above, a normal π-s-cobordism*

$$(F, \overline{F}) \colon (W; M, M') \to X \times (I, \partial I)$$

induces an isomorphism of kernel formations

$$h_* \colon \Sigma(f, \overline{f}, \underline{\omega}, \overline{q}) \xrightarrow{\cong} \Sigma(f', \overline{f'}, h_*(\underline{\omega}), h \circ \overline{q}).$$

We next consider how the kernel formation $\Sigma(f, \overline{f}, \underline{w}, \overline{q})$ varies, as we perform surgery on the embedding $\overline{q} \colon U \hookrightarrow M$. Lemma 9.75 tells us that the triviality of the kernel formation recognises whether $(f, \overline{f}) \colon M \to X$ is a normal homotopy equivalence. If $(f, \overline{f})$ is not a homotopy equivalence, then note that the embedding $\overline{q} \colon U \hookrightarrow M$ is a reasonable candidate for doing surgery to achieve a homotopy equivalence: after all $\overline{q}$ represents a set of generators $\underline{\omega}$ for $K_k(\widetilde{M})$. Hence we consider the effect of surgery on the embedding $\overline{q} \colon U \hookrightarrow M$. The trace of this surgery is the bordism

$$W := (M \times I) \cup_U D^{2k+2}.$$

By Theorem 4.39 (iv) the normal map $(f, \overline{f}) \colon M \to X$ extends to a normal map

$$(F, \overline{F}) \colon (W; M, M') \to X \times (I; \partial I)$$

where $M' := M_0 \cup U'$ is the outcome of the surgery on $\overline{q}$ and there is a normal π-map $(f', \overline{f'}) \colon M' \to X$, which is the restriction of $(F, \overline{F})$ to M'. Along with π-s-cobordisms, normal π-bordisms $(F, \overline{F})$ obtained as the trace of surgery on a representation of the kernel $K_k(\widetilde{M})$ play a key role in what follows and this leads to the following definition.

Definition 9.79 Let $(f, \overline{f}) \colon M \to X$ be a k-connected degree one normal map. A normal π-bordism,

$$(F, \overline{F}) \colon (W; M, M') \to X \times (I, \partial I),$$

obtained from surgery on an embedding $\overline{q} \colon U \hookrightarrow M$ representing a generating set $\underline{\omega} = \{x_1, \dots, x_a\}$ for $K_k(\widetilde{M})$ will be called a $(k+1)$-*trace normal π-bordism*.

More generally, a $(k+1)$-*normal π-bordism* is a degree one normal map

$$(F, \overline{F}) \colon (W; M, M') \to X \times (I, \partial I),$$

which can be written as a union

$$(F, \overline{F}) = (F_0, \overline{F}_0) \cup (F_1, \overline{F}_1) \cup (F_2, \overline{F}_2),$$

where $(F_1, \overline{F}_1) \colon (W_1; M'_0, M_2) \to X \times (I, \partial I)$ is a $(k+1)$-trace normal π-bordism and where for $i = 0, 2$, $(F_i, \overline{F}_i) \colon (W_i; M_i, M'_i) \to X \times (I, \partial I)$ are normal π-s-cobordisms.

9.3 The Odd-Dimensional Surgery Obstruction in the Universal Covering Case

Remark 9.80 While $(k+1)$-normal π-bordisms are somewhat more complicated than $(k+1)$-trace normal π-bordisms, note that, by Lemma 9.78, the difference between these types of normal π-bordisms does not effect the isomorphism class of kernel formations of the normal maps at either end of the normal bordisms.

We also point out that $(k+1)$-normal π-bordisms are ubiquitous, in that every normal bordism between k-connected degree on normal maps can be modified by interior surgeries to become a $(k+1)$-normal π-bordism, see Lemma 9.100.

The surgery kernels of $(f, \overline{f}), (f', \overline{f'})$ and $(F, \overline{F})$ found in a $(k+1)$-normal π-bordism $(F, \overline{F}): (W; M, M') \to X \times (I, \partial I)$ fit into a braid, a fragment of which runs as follows:

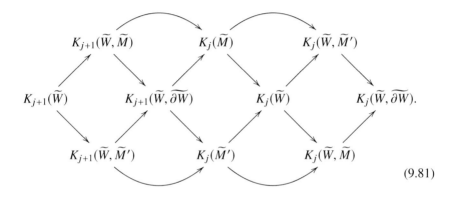

(9.81)

Since $K_j(\widetilde{W}, \widetilde{M})$ and $K_j(\widetilde{W}, \widetilde{M}')$ are non-zero if and only if $j = k+1$, we see from the braid (9.81) that $f': M' \to X$ is k-connected. Indeed, if W is a $(k+1)$-trace normal π-bordism, then W can equally well be presented as

$$W := (M' \times I) \cup_{U'} D^{2k+2},$$

where $\overline{q}': U' \to M'$ results from the surgery decomposition $M' = M_0 \cup U'$. We also see that $K_k(\widehat{U}') \to K_k(\widetilde{M}')$ is onto and so the canonical generating set for $K_k(\widehat{U}')$ gives a generating set $\underline{\omega}'$ for $K_k(\widetilde{M}')$ that is represented by $\overline{q}'$, i.e., $(\underline{\omega}', \overline{q}')$ represents $K_k(\widetilde{M}')$. Since we assumed that the base point $m \in \partial U \subset M$ and $\partial U = \partial U'$, we also obtain the corresponding base point $m' \in M'$.

We now consider the kernel formation $\Sigma(f', \overline{f'}, \underline{\omega}', \overline{q}')$ for the outcome of surgery $(f', \overline{f'})$, that is defined by the representation $(\underline{\omega}', \overline{q}')$, and we compare it to the original kernel formation $\Sigma(f, \overline{f}, \underline{\omega}, \overline{q})$, which we denote by $(H_\epsilon(F); F, G)$. Since $\partial U' = \partial U$, the ambient hyperbolic form is unchanged, i.e., $K_k(\widehat{\partial U}') = K_k(\widehat{\partial U}) = H_\epsilon(F)$. The lagrangian $K_{k+1}(\widetilde{M}_0, \widehat{\partial U})$ is unchanged and the lagrangian $K_{k+1}(\widehat{U}, \widehat{\partial U})$ is replaced by the lagrangian $K_{k+1}(\widehat{U}', \widehat{\partial U}')$. By (9.72), we see that $K_{k+1}(\widehat{U}, \widehat{\partial U})$ and $K_{k+1}(\widehat{U}', \widehat{\partial U})$ respectively define the complementary lagrangians F and F^* in $K_k(\widehat{\partial U})$. It follows that $\Sigma(f', \overline{f'}, \underline{\omega}', \overline{q}') = (H_\epsilon(F); F^*, G)$. Applying the first part of

Remark 9.80 to extend the above to any $(k+1)$-normal π-bordism, we have therefore proven the next lemma.

Lemma 9.82 *Let $(F, \overline{F}) \colon (W; M, M') \to X \times (I, \partial I)$ be a $(k+1)$-normal π-bordism where $(f, \overline{f})$ has a kernel formation $\Sigma(f, \overline{f}, \underline{\omega}, \overline{q}) = (H_\epsilon(F); F, G)$. Then $(f', \overline{f'})$ is a k-connected normal π-map of degree one with a kernel formation*

$$\Sigma(f', \overline{f'}, \underline{\omega}', \overline{q}') = (H_\epsilon(F); F^*, G).$$

The representation $(\underline{\omega}', \overline{q}')$ of $K_k(\widetilde{M}')$ from Lemma 9.82 is called *the dual representation* of the representation $(\underline{\omega}, \overline{q})$ of $K_k(\widetilde{M})$.

In the next section we will use the following exercise, which is a useful general statement about commutative braids.

Exercise 9.83 Consider the commutative braid of R-modules:

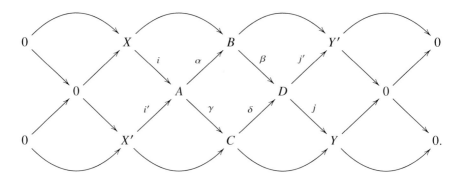

Show that then we have an exact sequence of R-modules

$$0 \to A \xrightarrow{\alpha \oplus \gamma} B \oplus C \xrightarrow{\delta - \beta} D \to 0.$$

9.3.2 The Bordism Formation

In this subsection we show how a $(k+1)$-normal π-bordism $(F, \overline{F})$ with domain $(W; M, M')$ defines an ϵ-quadratic formation, which we denote by $\Sigma(F, \overline{F}, M) = (H_\epsilon(F); F, G)$. The main result of the subsection is Lemma 9.88, which identifies the bordism formation $\Sigma(F, \overline{F}, M)$ with a particular kernel formation $\Sigma(f, \overline{f}, \underline{\omega}, \overline{q})$ for $(f, \overline{f}) = (F|_M, \overline{F}|_M)$ in the case when $(W; M, M')$ is a $(k+1)$-trace normal π-bordism.

Assume now that $(W; M, M')$ is the source of a $(k+1)$-normal bordism $(F, \overline{F})$ as defined in Definition 9.79. Recalling the notation of the previous subsection, we have $\natural_{i=1}^a (D^{k+1} \times S^k) \cong U \subset M$ and homotopy equivalences

9.3 The Odd-Dimensional Surgery Obstruction in the Universal Covering Case

$$M' \cup_{U'} D^{2k+2} \simeq W \simeq M \cup_U D^{2k+2},$$

where $\partial D^{2k+2} = S^{2k+1} \cong U \cup U'$ and $U' \cong \natural_{i=1}^a (S^k \times D^{k+1})$. The $\mathbb{Z}\pi$-modules $K_{k+1}(\widetilde{W}, \widetilde{M}') \cong K_{k+1}(\widetilde{D}^{2k+2}, \widehat{U}')$ and $K_{k+1}(\widetilde{W}, \widetilde{M}) \cong K_{k+1}(\widetilde{D}^{2k+2}, \widehat{U})$ are free of rank a. Put $F := K_{k+1}(\widetilde{W}, \widetilde{M}')$. Using Poincaré duality for the triad $(W; M, M')$, see Lemma 8.194 (ii), we identify

$$K_{k+1}(\widetilde{W}, \widetilde{M}) \cong K^{k+1}(\widetilde{W}, \widetilde{M}') \cong K_{k+1}(\widetilde{W}, \widetilde{M}')^* = F^*$$

where

$$K^{k+1}(\widetilde{W}, \widetilde{M}') \xrightarrow{\cong} K_{k+1}(\widetilde{W}, \widetilde{M}')^*, \qquad \alpha \mapsto \langle \alpha, ? \rangle$$

is bijective by Lemma 8.52 (ii).

The bordism formation is a formation $(H_\epsilon(F); F, G)$ where the second lagrangian of the formation, $G \cong K_{k+1}(\widetilde{W})$, is defined to be the image of the homomorphism

$$j' \oplus j \colon K_{k+1}(\widetilde{W}) \to K_{k+1}(\widetilde{W}, \widetilde{M}') \oplus K_{k+1}(\widetilde{W}, \widetilde{M})$$

for $j' \colon K_{k+1}(\widetilde{W}) \to K_{k+1}(\widetilde{W}, \widetilde{M})$ and $j \colon K_{k+1}(\widetilde{W}) \to K_{k+1}(\widetilde{W}, \widetilde{M})$ the canonical homomorphisms in the long exact sequences of the pairs $(\widetilde{W}, \widetilde{M}')$ and $(\widetilde{W}, \widetilde{M})$.

To see that G is indeed a lagrangian in the sense of Section 8.5.2, we need to verify:

- The map $j' \oplus j \colon K_{k+1}(\widetilde{W}) \to K_{k+1}(\widetilde{W}, \widetilde{M}') \oplus K_{k+1}(\widetilde{W}, \widetilde{M})$ is split injective with $K_{k+1}(\widetilde{W})$ stably free;
- The restriction of the standard hyperbolic ϵ-quadratic form to G via $j' \oplus j$ is zero in $Q_\epsilon(G)$;
- The sequence

$$0 \to G \xrightarrow{j' \oplus j} K_{k+1}(\widetilde{W}, \widetilde{M}') \oplus K_{k+1}(\widetilde{W}, \widetilde{M}) \xrightarrow{\epsilon j^* \oplus (j')^*} G^* \to 0$$

is exact.

To see the second condition, we observe that there exists a *hessian* for G in the sense of Section 9.2.2. Namely, it is the quadratic form $(K_{k+1}(\widetilde{W}), \theta_W)$ defined just as in Example 8.86. Our setting is more general than Example 8.86, and so $(K_{k+1}(\widetilde{W}), \theta_W)$ is not necessarily non-singular, but it is a $(-\epsilon)$-quadratic form (the minus sign comes in since W has dimension $2(k+1)$). The commutative diagram

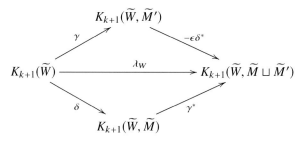

implies that the hessian form satisfies

$$\lambda_W = \theta_W - \epsilon\theta_W^* = \gamma^*\delta = -\epsilon\delta^*\gamma.$$

To see the first and the third condition, we need a preliminary calculation. We show that there is an isomorphism

$$k\colon K_{k+1}(\widetilde{M}_0, \widehat{\partial U}) \xrightarrow{\cong} K_{k+1}(\widetilde{W}), \qquad (9.84)$$

constructed as follows. Consider the following sequence of inclusions of pairs that have the property that each inclusion induces an isomorphism on H_{k+1} either by excision or by homotopy invariance

$$(M_0, \partial U) \hookrightarrow (M, U) \hookrightarrow (M \times I, U \times I) \hookrightarrow (W, T).$$

This shows that $k_T \colon H_{k+1}(\widetilde{M}_0, \widehat{\partial U}) \to H_{k+1}(\widetilde{W})$ is an isomorphism, and as a consequence the map k from (9.84) is an isomorphism as well. Furthermore, $K_{k+1}(\widetilde{M}_0, \widehat{\partial U})$ is a stably free $\mathbb{Z}\pi$-module. Hence $G = K_{k+1}(\widetilde{W})$ is a stably free $\mathbb{Z}\pi$-module.

Now, Exercise 9.83 applied to our situation yields the third point. Since G is a stably finitely generated free $\mathbb{Z}\pi$-module, its dual G^* is as well and therefore the exact sequence splits. This also gives the desired splitting from the first bullet point.

The module underlying $H_\epsilon(F)$ is

$$F \oplus F^* = K_{k+1}(\widetilde{W}, \widetilde{M}') \oplus K_{k+1}(\widetilde{W}, \widetilde{M}),$$

and we use the $(k+1)$-normal π-bordism $(F, \overline{F})$ to give a geometric description of the standard hyperbolic ϵ-quadratic form $H_\epsilon(F)$ of Example 8.79.

Definition 9.85 (Bordism formation) The *bordism formation* for M of a $(k+1)$-normal bordism $(F, \overline{F}) \colon (W; M, M') \to X \times (I, \partial I)$ is the formation

$$\Sigma(F, \overline{F}, M) := \big(K_{k+1}(\widetilde{W}, \widetilde{M}') \oplus K_{k+1}(\widetilde{W}, \widetilde{M}); K_{k+1}(\widetilde{W}, \widetilde{M}'), K_{k+1}(\widetilde{W})\big)$$

with the structure homomorphisms $\gamma = j'\colon K_{k+1}(\widetilde{W}) \to K_{k+1}(\widetilde{W}, \widetilde{M}')$ and $\delta = j\colon K_{k+1}(\widetilde{W}) \to K_{k+1}(\widetilde{W}, \widetilde{M})$.

9.3 The Odd-Dimensional Surgery Obstruction in the Universal Covering Case

Remark 9.86 If $\Sigma(F, \overline{F}, M) = (H_\epsilon(F); F, G)$ is the bordism formation for M of a $(k+1)$-normal π-bordism $(F, \overline{F}): (W; M, M') \to X \times (I, \partial I)$, then we see immediately from the definitions that

$$\Sigma(F, \overline{F}, M') = (H_\epsilon(F); F^*, G)$$

is the bordism formation for M' of $(F, -\overline{F})$. Once we have established that the kernel formation and the bordism formation are isomorphic in Lemma 9.88 below, this observation is equivalent to Lemma 9.82 above.

Remark 9.87 The hyperbolic form on $K_{k+1}(\widetilde{W}, \widetilde{M}') \oplus K_{k+1}(\widetilde{W}, \widetilde{M})$ that appears in the bordism formation above is not geometric in the sense that it is not the quadratic form associated to any normal map. However, Poincaré duality in the bordism $(W; M, M')$ between $K_{k+1}(\widetilde{W}, \widetilde{M}')$ and $K^{k+1}(\widetilde{W}, \widetilde{M})$ plays a key role in the definition of the bordism formation.

The following key lemma identifies the kernel formation $\Sigma(f, \overline{f}, \omega, \overline{q})$ with the bordism formation relative to M of a $(k+1)$-trace normal π-bordism given by surgery on $\overline{q}: U \to M$.

Lemma 9.88 Let $(F, \overline{F}): (W; M, M') \to X \times (I, \partial I)$ be a $(k+1)$-trace normal π-bordism obtained from surgery on a representation $(w, \overline{q})$ of the surgery kernel of a degree one normal map $(f, \overline{f}): M \to X$. Then

$$\Sigma(F, \overline{F}, M) \cong \Sigma(f, \overline{f}, \omega, \overline{q}).$$

Proof. Recall that $\partial D^{2k+2} = S^{2k+1} = \overline{U} \cup_{\partial U} U'$, where there are identifications $\overline{U} = \bigsqcup_{i=1}^{j} (D^{k+1} \times S^k)$ and $U' = \bigsqcup_{i=1}^{j} (S^k \times D^{k+1})$. The trace bordism W is formed from $M \times I$ by attaching D^{2k+2} to $M \times \{1\}$ using the embedding $\overline{q}: U \hookrightarrow M \times \{1\}$:

$$W = (M \times I) \cup_U D^{2k+2}. \tag{9.89}$$

Dually, the trace W is formed from $M' \times I$ by attaching D^{2k+2} to $M' \times \{0\}$ along $U' \subset M'$:

$$W = (M' \times I) \cup_{U'} D^{2k+2}. \tag{9.90}$$

We define $T \subset W$ to be the union of D^{2k+2} and $U \times I \subset M \times I$ along $U \times \{1\}$. There is a homeomorphism $T \xrightarrow{\cong} D^{2k+2}$ and for the boundary we have $\partial T \cong S^{2k+1} \cong U \cup_{\partial U \times \{0\}} (\partial U \times I) \cup_{\partial U \times \{1\}} U'$.

Figure 9.91 (Bordism W).

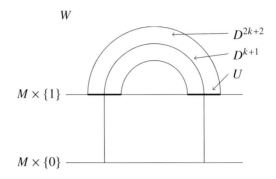

Now let $(H_\epsilon(F_B); F_B, G_B)$ denote the bordism formation $\Sigma(F, \overline{F}, M)$ and let $(H_\epsilon(F_K); F_K, G_K)$ denote the kernel formation $\Sigma(f, \overline{f}, \omega, \overline{q})$. We shall define an isomorphism of formations

$$A \colon (H_\epsilon(F_B); F_B, G_B) \xrightarrow{\cong} (H_\epsilon(F_K); F_K, G_K).$$

Let $\{y_1, \ldots, y_a\}$ and $\{y'_1, \ldots, y'_a\}$ be $\mathbb{Z}\pi$-bases for the kernels $K_{k+1}(\widehat{U}, \widehat{\partial U})$ and $K_{k+1}(\widehat{U'}, \widehat{\partial U})$ respectively that are represented by the canonical embeddings $q_i \colon (D^{k+1}, S^k) \hookrightarrow (U, \partial U)$ and $q'_i \colon (D^{k+1}, S^k) \hookrightarrow (U', \partial U)$ for $i = 1, \ldots, a$. From the structure of W displayed in (9.89) and (9.90), we see that there are maps of pairs $i \colon (U, \partial U) \to (W, M')$ and $i' \colon (U', \partial U') \to (W, M)$ such that $K_{k+1}(\widetilde{W}, \widetilde{M}')$ is a free $\mathbb{Z}\pi$-module with generators $i_*(y_i)$ and $K_{k+1}(\widetilde{W}, \widetilde{M})$ is a free $\mathbb{Z}\pi$-module with generators $i'_*(y'_i)$. Hence the maps of pairs i and i' induce isomorphisms

$$i_* \colon F_K = K_{k+1}(\widehat{U}, \widehat{\partial U}) \xrightarrow{\cong} K_{k+1}(\widetilde{W}, \widetilde{M}') = F_B$$

and

$$i'_* \colon F_K^* = K_{k+1}(\widehat{U'}, \widehat{\partial U}) \xrightarrow{\cong} K_{k+1}(\widetilde{W}, \widetilde{M}) = F_B^*.$$

The sum of i_* and i'_* defines an isomorphism of ϵ-quadratic forms

$$A := (i_* \oplus i'_*) \colon H_\epsilon(F_B) \xrightarrow{\cong} H_\epsilon(F_K).$$

By construction $A(F_K) = F_B$, and it remains to verify that $A(G_K) = G_B$, where $G_K = K_{k+1}(\widetilde{M}_0, \widehat{\partial U})$ and $G_B = K_{k+1}(\widetilde{W})$. Firstly, we recall that we have the isomorphism $k \colon K_{k+1}(\widetilde{M}_0, \widehat{\partial U}) \to K_{k+1}(\widetilde{W})$ from (9.84). Next, we study the following diagram:

9.3 The Odd-Dimensional Surgery Obstruction in the Universal Covering Case

$$\begin{array}{ccccc}
K_{k+1}(\widetilde{W}, \widetilde{M}') & \xleftarrow{j'} & K_{k+1}(\widetilde{W}) & \xrightarrow{j} & K_{k+1}(\widetilde{W}, \widetilde{M}) \\
\uparrow i_* & & \uparrow k & & \uparrow i'_* \\
K_{k+1}(\widehat{U}, \widehat{\partial U}) & \xleftarrow{\gamma_K} & K_{k+1}(\widehat{M_0}, \widehat{\partial U}) & \xrightarrow{\delta_K} & K_{k+1}(\widehat{U'}, \widehat{\partial U}) \\
\uparrow (i'_{U*} \circ \partial_{U})^{-1} & & \downarrow \partial_{M_0} & & \uparrow (i_{U*} \circ \partial_{U'})^{-1} \\
K_k(\widehat{U'}) & \xleftarrow{i'_{U*}} & K_k(\widehat{\partial U}) & \xrightarrow{i_{U*}} & K_k(\widehat{U}).
\end{array} \quad (9.92)$$

Here $i_U \colon \partial U \to U$ and $i_{U'} \colon \partial U \to U'$ are the inclusions, ∂_U and $\partial_{U'}$ are the boundary homomorphisms in the homology long exact sequences of the pairs $(U, \partial U)$ and $(U', \partial U)$ respectively, and j and j' are the maps in the exact sequences of the pairs (W, M) and (W, M'). From (9.73) and (9.74) we see that the arrows labelled γ_K and δ_K are indeed the structure homomorphisms of the canonical presentation of the kernel formation of $(\omega, \overline{q})$ and by definition $j' = \gamma_B$ and $j = \delta_B$ are the structure homomorphisms of the canonical presentation of the bordism formation. Assuming that the diagram (9.92) commutes, we see that $A(G_K) \subseteq G_B$ is a lagrangian in $H_\epsilon(F_B)$ and hence $A(G_K) = G_B$.

We complete the proof by showing that (9.92) commutes. The lower part commutes by definitions of the maps involved, so it remains to prove that the upper two squares commute. We must show $j' \circ k = i_* \circ \gamma_K$ and $j \circ k = i'_* \circ \delta_K$. We only prove the first equality, the argument for the second is analogous. The Poincaré duality isomorphism $K_{k+1}(\widetilde{W}, \widetilde{M}') \xrightarrow{\cong} K_{k+1}(\widetilde{W}, \widetilde{M})^*$ shows that any element $u \in K_{k+1}(\widetilde{W}, \widetilde{M}')$ is determined by its $\mathbb{Z}\pi$-intersection numbers $\lambda_W(u, i'_*(y'_i))$ with the $\mathbb{Z}\pi$-basis elements $i'_*(y'_i) \in K_{k+1}(\widetilde{W}, \widetilde{M})$. Here we add the subscript W in λ_W to indicate the ambient manifold since later we will also consider other ambient manifolds. Given $z \in K_{k+1}(\widetilde{M_0}, \widehat{\partial U})$, the element $k(z)$ is constructed by writing

$$\partial_{M_0}(z) = \Sigma_{i=1}^a a_i \partial_U(y_i) + a'_i \partial_{U'}(y'_i)$$

and then adding $\Sigma_{i=1}^a a_i y_i + a'_i y'_i$ to z. After passing to $K_{k+1}(\widetilde{W}, \widetilde{M}')$, the elements y'_i vanish and hence $j'(k(z))$ is represented by adding just $\Sigma_{i=1}^a a_i y_i$ to z. These classes can be represented by immersions as follows. Let $q_z \colon (D^{k+1}, S^k) \looparrowright (M_0, \partial U)$ be an immersion representing the element $z \in K_{k+1}(\widetilde{M_0}, \widehat{\partial U})$. Then an immersion $q_{j'(k(z))} \colon (D^{k+1}, S^k) \looparrowright (\widetilde{W}, \widetilde{M}')$ representing $j'(k(z))$ can be constructed by adjoining to the image of q_z the images of the immersions $a_i q_i \colon (D^{k+1}, S^k) \looparrowright (U, \partial U)$, where we remind the reader that q_i is the representative of y_i chosen at the beginning of the proof. Given $1 \leq j \leq a$, we take the class $i'_*(y'_j)$ and consider its $\mathbb{Z}\pi$-intersection number $\lambda_W(j'(k(z)), i'_*(y'_j))$. Since q_z lies in M_0, which is disjoint from U', we see that the immersion q_z plays no role, and the $\mathbb{Z}\pi$-intersection number $\lambda_W(j'(k(z)), i'_*(y'_j))$ is only determined by how the immersions q_i and q_j intersect. Their contribution is in turn given by the $\mathbb{Z}\pi$-intersection numbers $\lambda_T(y_i, y_j)$ with the ambient manifold T since the images of q_i and q_j lie in T. Altogether we have

$$\lambda_W(j'(k(z)), i'_*(y'_j)) = \lambda_T(\Sigma_{i=1}^a a_i y_i, y'_j) = a_j.$$

By the definition of γ_K, this is precisely the intersection of $i_*(\gamma_K(z))$ with $i'_*(y'_j)$. Hence $j'(k(z)) = i_*(\gamma_K(z))$. This finishes the proof of Lemma 9.88. □

Lemma 9.93 (i) *The stable isomorphism class of the formation $\Sigma(f, \overline{f}, \underline{\omega}, \overline{q})$ depends only on $(f, \overline{f})$, i.e., is independent of the choices of $\underline{\omega}$ and $\overline{q}$;*
(ii) *Every formation in the stable isomorphism class of $(H_\epsilon(F); F, G)$ can be realised as $\Sigma(f, \overline{f}, \underline{\omega}, \overline{q})$ for certain choices of $\underline{\omega}$ and $\overline{q}$.*

Proof. (i) Let $(\underline{\omega}_1, \overline{q}_1)$ and $(\underline{\omega}_2, \overline{q}_2)$ be two choices of representation for $K_k(\widetilde{M})$ with associated kernel formations $\Sigma(f, \overline{f}, \underline{\omega}_1, \overline{q}_1)$ and $\Sigma(f, \overline{f}, \underline{\omega}_2, \overline{q}_2)$. We let H_i be the free $\mathbb{Z}\pi$-module generated by $\underline{\omega}_i$, so that there are surjections of $\mathbb{Z}\pi$-modules $H_i \to K_k(\widetilde{M})$ for $i = 1, 2$. To begin we assume that $\underline{\omega}_1$ and $\underline{\omega}_2$ have the same number of elements and that there is an isomorphism $\alpha \colon H_1^* \xrightarrow{\cong} H_2^*$ such that the following diagram commutes

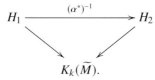

Suppose that the normal maps

$$(F_1, \overline{F}_1) \colon (W; M, M') \to X \times (I, \partial I) \text{ and } (F_2, \overline{F}_2) \colon (W; M, M'') \to X \times (I, \partial I)$$

are $(k+1)$-trace normal π-bordisms obtained from surgery on $\overline{q}_1$ and $\overline{q}_2$. By Lemma 9.88, there are isomorphisms

$$\Sigma(f, \overline{f}, \underline{\omega}_i, \overline{q}_i) \xrightarrow{\cong} \Sigma(F_i, \overline{F}_i, M)$$

for $i = 1, 2$. We shall show that there is an isomorphism between the π-bordism formations $\Sigma(F_1, \overline{F}_1, M) \xrightarrow{\cong} \Sigma(F_2, \overline{F}_2, M)$. Recall that $\underline{\omega}_i$ defines a basis for $K_k(\widetilde{W}_i, \widetilde{M}) = (F_i)_B^*$. Hence the isomorphism $(\alpha^*)^{-1}$ above defines an isomorphism

$$(\alpha^*)^{-1} \colon K_{k+1}(\widetilde{W}_1, \widetilde{M}) \xrightarrow{\cong} K_{k+1}(\widetilde{W}_2, \widetilde{M}).$$

By Theorem 4.39 (ii), the regular homotopy class of $\overline{q}_i$ is determined by the $x_i \in \underline{\omega}_1$ and the bundle map $\overline{f}$. Hence there is a regular homotopy

$$Q \colon \left(\sqcup_{i=1}^a (D^{k+1} \times S_i^k)\right) \times I \hookrightarrow M \times I$$

between the embeddings $\overline{q}_1$ and $\overline{q}_2$. From the canonical homotopy equivalences $W_i \xrightarrow{\simeq} M_i \cup_U D^{2k+2}$ for $i = 1, 2$, we see that the homotopy Q gives a homotopy equivalence of pairs,

$$f_Q \colon (W_1, M) \xrightarrow{\simeq} (W_2, M),$$

9.3 The Odd-Dimensional Surgery Obstruction in the Universal Covering Case

where $f_Q|_M = \mathrm{id}_M$ and the induced homomorphism $f_{Q*}\colon K_{k+1}(\widetilde{W}_1, \widetilde{M}) \xrightarrow{\cong} K_{k+1}(\widetilde{W}_2, \widetilde{M})$ is α by construction. If we use the canonical dual basis $\underline{\omega}_1^*$ for $K_{k+1}(\widetilde{W}_i, M)$, then, by the arguments about $\mathbb{Z}\pi$-intersection numbers obtained via regular homotopies as in the proof of Theorem 8.195 for Wall realisation, the intersection forms

$$\lambda_i \colon K_{k+1}(\widetilde{W}_i, \widetilde{M}) \to K_{k+1}(\widetilde{W}_i, \widetilde{M})^*$$

satisfy the relation

$$\lambda_1 = \alpha^* \lambda_2 \alpha + \lambda_Q, \tag{9.94}$$

where $\lambda_Q \colon K_{k+1}(\widetilde{W}_1, \widetilde{M}) \to K_{k+1}(\widetilde{W}_1, \widetilde{M}')$ is the symmetric form defined by the regular homotopy Q and the basis $\underline{\omega}_1$. Consequently the matrix

$$A = \begin{bmatrix} \alpha & \lambda_Q \alpha \\ 0 & (\alpha^*)^{-1} \end{bmatrix}$$

gives an isomorphism of $\mathbb{Z}\pi$-modules

$$A \colon K_{k+1}(\widetilde{W}_1, \widetilde{M}') \oplus K_{k+1}(\widetilde{W}_1, \widetilde{M}) \xrightarrow{\cong} K_{k+1}(\widetilde{W}_2, \widetilde{M}') \oplus K_{k+1}(\widetilde{W}_2, \widetilde{M}).$$

Moreover, the homotopy equivalence $f_Q \colon W_1 \to W_2$ induces a $\mathbb{Z}\pi$-homo-morphism $\beta \colon K_{k+1}(\widetilde{W}_1) \to K_{k+1}(\widetilde{W}_2)$ such that, thanks to (9.94), the following diagram commutes,

$$\begin{array}{ccc} K_{k+1}(\widetilde{W}_1) & \xrightarrow{\beta} & K_{k+1}(\widetilde{W}_2) \\ \downarrow & & \downarrow \\ K_{k+1}(\widetilde{W}_1, \widetilde{M}') \oplus K_{k+1}(\widetilde{W}_1, \widetilde{M}) & \xrightarrow{A} & K_{k+1}(\widetilde{W}_2, \widetilde{M}') \oplus K_{k+1}(\widetilde{W}_2, \widetilde{M}). \end{array}$$

This ensures that β and A define an isomorphism of the π-bordism formations

$$(\beta, A) \colon \Sigma(F_1, \overline{F}_1, M) \xrightarrow{\cong} \Sigma(F_2, \overline{F}_2, M).$$

Now we consider choosing different generating sets $\underline{\omega}_1 = \{x_1, \ldots, x_a\}$ and $\underline{\omega}_2 = \{y_1, \ldots, y_b\}$. We will use the standard fact [414, page 61] that any two sets of generators are related by the following elementary operations:

$$\{x_1, \ldots, x_a\} \to \{x_1, \ldots, x_a, 0\} \to \{x_1, \ldots, x_a, y_1\} \to$$
$$\to \{x_1, \ldots, x_a, y_1, 0\} \to \cdots \to \{x_1, \ldots, x_a, y_1, \ldots, y_b\} \to$$
$$\to \{y_1, \ldots, y_b, x_1, \ldots, x_a\} \to \cdots \to \{y_1, \ldots, y_b\}.$$

Each of these basic operations takes one of the following forms:

(a) add or delete a zero;
(b) add a linear combination of the preceding elements to the last element;
(c) permute the elements.

We consider first adding or removing zero: suppose that $\overline{q}_1$ represents $\underline{\omega}_1 = \{x_1, \ldots, x_a\}$. To represent $\underline{\omega}_{1,1} = \{x_1, \ldots, x_a, 0\}$, we can take $\overline{q}_{1,1}$, where $\overline{q}_{1,1}$ is formed from $\overline{q}_1$ by tubing an embedding of $D^{q+1} \times S^q \hookrightarrow M$ that lies in a small disk $D^{2q+1} \subset M$ disjoint from $\overline{q}_1(U)$. We see that passing from $(\underline{\omega}_1, \overline{q}_1)$ to $(\underline{\omega}_{1,1}, \overline{q}_{1,1})$ has the effect of adding the trivial formation $(H_\epsilon(\mathbb{Z}\pi); \mathbb{Z}\pi, \mathbb{Z}\pi^*)$ and so does not change the stable isomorphism class of the kernel formation. Similarly, passing from $(\underline{\omega}_{1,1}, \overline{q}_{1,1})$ to $(\underline{\omega}_1, \overline{q}_1)$ has the effect of removing the trivial formation $(H_\epsilon(\mathbb{Z}\pi); \mathbb{Z}\pi, \mathbb{Z}\pi^*)$ and so does not change the stable isomorphism class of the kernel formation.

The cases (b) and (c) are covered by the argument above since in both of these cases the different choices of generating set are related by an isomorphism α as in the assumption above.

(ii) Suppose that $\Sigma(f, \overline{f}, \underline{\omega}, \overline{q}) = (H_\epsilon(F_1); F_1, G_1)$ and that there is an isomorphism $A: (H_\epsilon(F_1); F_1, G) \xrightarrow{\cong} (H_\epsilon(F_2), F_2, G_2)$. Then A induces an isomorphism $\overline{A}: H_\epsilon(F_1)/F_1 \xrightarrow{\cong} H_\epsilon(F_2)/F_2$. As there is a canonical isomorphism $H_\epsilon(F_i)/F_i \xrightarrow{\cong} F_i^*$, we have an isomorphism $\overline{A}: F_1^* \xrightarrow{\cong} F_2^*$, where F_1^* is based by $\underline{\omega} = \{x_1, \ldots, x_a\}$. Choosing a basis for F_2^*, we regard $\overline{A} = (a_{ij})_{i,j=1}^a$ as an invertible matrix with $a_{ij} \in \mathbb{Z}\pi$. We then define a new generating set for $K_k(\widetilde{M})$ by

$$\underline{\omega}_2 = \overline{A} \cdot \underline{\omega}_1 = \{y_1, \ldots, y_a\}$$

where

$$y_i = \sum_{j=1}^a a_{ij} x_j.$$

The argument above shows that we can choose $\overline{q}_2$ representing $\underline{\omega}_2$ such that $\Sigma(f, \overline{f}, \underline{\omega}_2, \overline{q}_2) = (H_\epsilon(F_2); F_2, G_2)$.

Finally, suppose that

$$\Sigma(f, \overline{f}, \underline{\omega}, \overline{q}) = (H_\epsilon(F); F, G) = (H_\epsilon(F_1); F_1, G_1) \oplus (H_\epsilon(F_2); F_2, G_2),$$

where $(H_\epsilon(F_2); F_2, G_2)$ is trivial. Then by Lemma 9.75 (i),

$$K_k(\widetilde{M}) = H_\epsilon(F)/(F+G) \cong H_\epsilon(F_1)/F_1.$$

It follows that we may choose a representation $(\underline{\omega}_1, \overline{q}_1)$ for $K_k(\widetilde{M})$ where $\Sigma(f, \overline{f}, \underline{\omega}_1, \overline{q}_1) = (H_\epsilon(F_1); F_1, G_1)$. Hence we can remove a trivial formation from a kernel formation and still realise the resulting formation. A similar argument shows that we can add a trivial formation to a kernel formation and realise the resulting larger formation. Thus Lemma 9.93 is proven. □

As a consequence of Lemma 9.93 (i), any kernel formation $\Sigma(f, \overline{f}, \underline{\omega}, \overline{q})$ of a k-connected degree one normal π-map $(f, \overline{f}): M \to X$ gives rise to a well-defined *stable isomorphism class of kernel formations*, which we denote by

$$\Sigma(f, \overline{f}) := \{\Sigma(f, \overline{f}, \underline{\omega}, \overline{q})\}_s. \tag{9.95}$$

9.3.3 Proof of Theorem 9.60

In this subsection we define the odd-dimensional surgery obstruction in (9.104) and prove Theorem 9.60. As usual, $(f, \overline{f}) \colon M \to X$ is a k-connected normal π-map of degree one and we make choices (ABP) and (GBP). A choice of the representation $(\omega, \overline{q})$ yields an ϵ-quadratic kernel formation $\Sigma(f, \overline{f}, \omega, \overline{q})$ with the stable isomorphism class $\Sigma(f, \overline{f})$, see (9.95). A $(k+1)$-normal π-bordism between k-connected degree one normal π-maps $(f, \overline{f})$ and $(f', \overline{f'})$ yields an ϵ-quadratic bordism formation $\Sigma(F, \overline{F}; M)$. If the bordism is obtained as the trace of the surgery done using choices $(\omega, \overline{q})$, then these formations are isomorphic by Lemma 9.88.

Lemma 9.96 *Let $(F, \overline{F})$ be a $(k+1)$-normal bordism between k-connected degree one normal π-maps $(f, \overline{f})$ and $(f', \overline{f'})$. Then f' is a homotopy equivalence if and only if $\Sigma(F, \overline{F}; M)$ is a boundary.*

Proof. Recall from Lemma 9.88 that for an appropriate choice of $(\omega, \overline{q})$ we have $\Sigma(f, \overline{f}, \omega, \overline{q}) \cong \Sigma(F, \overline{F}, M)$. Let $\Sigma(F, \overline{F}, M) = (H_\epsilon(F); F, G)$. By Lemma 9.82, for an appropriate choice of $(\omega', \overline{q}')$, the kernel formation of $(f', \overline{f'})$ satisfies $\Sigma(f', \overline{f'}, \omega', \overline{q}') \cong (H_\epsilon(F); F^*, G)$. We conclude from Lemma 9.75 that f' is a homotopy equivalence if and only if $(H_\epsilon(F); F^*, G)$ is trivial. Lemma 9.29 (i) implies that $(H_\epsilon(F); F^*, G)$ is trivial if and only if $(H_\epsilon(F); F, G)$ is a boundary. □

The following Lemma 9.97 is a special case of Lemma 9.99. We state it separately because we find it and its proof illuminating.

Lemma 9.97 *Let Y be a $(2k+1)$-dimensional manifold with $\pi = \pi_1(Y)$. Let (P, θ) be a (possibly singular) $(-1)^k$-quadratic form on a finitely generated free $\mathbb{Z}[\pi]$-module P.*

Then there a k-connected degree one normal π-map $(g, \overline{g}) \colon N_0 \to Y$ from some $(2k+1)$-dimensional manifold N_0 and a $(k+1)$-normal π-bordism $(G, \overline{G}) \colon (W; N_0, N_1) \to (Y \times I; Y \times \{0\}, Y \times \{1\})$ such that $(G, \overline{G})|_{N_0} = (g, \overline{g})$, the map $G|_{N_1}$ is a diffeomorphism, and

$$\Sigma(G, \overline{G}, N_0) = \partial(P, \theta).$$

Proof. The proof is essentially the same as the proof of the realisation theorem for even-dimensional surgery obstructions, which is our Theorem 8.195, with the following two differences. Firstly, the form (P, θ) that we are realising, is possibly singular. This has the effect that the degree one normal π-map that we obtain "on the top end" is possibly not a homotopy equivalence, it might have non-trivial K_{k+1} and K_k. Secondly, the order of the two boundary components of W is switched compared to the proof of Theorem 8.195. This means that we start with the identity map on Y, then we realise the form (P, θ), and at the "top end" we obtain a new manifold that will be our N_0. Then we look at the resulting bordism "from the other side" to obtain the desired equation.

More precisely, consider id: $Y \to Y$ and let $P = \mathbb{Z}[\pi]^r$ and, just as in the proof of Theorem 8.195, choose r disjoint embeddings $j_i \colon D^{2k+1} \hookrightarrow Y$. Composing with the standard embedding $F^0 \colon S^k \times D^{k+1} \hookrightarrow D^{2k+1}$, we obtain r disjoint embeddings $F_i^0 \colon S^k \times D^{k+1} \hookrightarrow Y$. Write the form (P, θ) as a triple (P, λ, μ) as in Subsection 8.4.3. Using the maps λ and μ as a guide, we modify these embeddings by regular homotopies to new embeddings denoted by $F_i^1 \colon S^k \times D^{k+1} \hookrightarrow Y$ so that, when we consider the resulting maps denoted by $F_i^t \colon S^k \times D^{k+1} \times I \to Y \times I$, we can attach handles $D^{k+1} \times D^{k+1}$ to $Y \times \{1\}$ along the embeddings F_i^1 and call the resulting bordism W'. Denote its bottom end by $N_0' = Y \times \{0\}$ and the top end by N_1'. The degree one normal π-map $(G', \overline{G}') \colon (W'; N_0', N_1') \to (Y \times I, Y \times \{0\}, Y \times \{1\})$ is constructed analogously as in the proof of Theorem 8.195. It has the property that the $\mathbb{Z}\pi$-intersection and $\mathbb{Z}\pi$-self-intersection numbers on the kernel $K_{k+1}(W') \cong P$ are given by

$$\lambda_{W'} = \lambda \quad \text{and} \quad \mu_{W'} = \mu.$$

Next comes the promised switching of the two ends of the cobordism. Let $W = W'$, $N_0 = N_1'$, $N_1 = N_0'$. Let $\tau \colon Y \times I \to Y \times I$ be the map $\tau(y, t) = (y, 1 - t)$ and let $\overline{\tau} \colon T_{Y \times I} \to T_{Y \times I}$ be its differential. Denote $(\tau, \overline{\tau}) \circ (G', \overline{G}')$ by $(G, \overline{G})$. Then we obtain the desired degree one normal map $(G, \overline{G}) \colon (W; N_0, N_1) \to (Y \times I; Y \times \{0\}, Y \times \{1\})$.

To prove the claim, we study the following commutative braid:

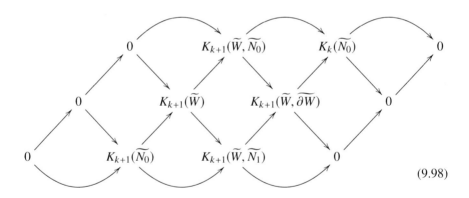

(9.98)

By choosing the cores of the attaching handles as the basis for the module $K_{k+1}(\widetilde{W}, \widetilde{N_1}) \cong K_{k+1}(\widetilde{W})$ and the cocores as the basis for $K_{k+1}(\widetilde{W}, \widetilde{N_0})$, the map $K_{k+1}(\widetilde{W}) \to K_{k+1}(\widetilde{W}, \widetilde{N_0})$ can be identified with the symmetrisation map $\theta + \epsilon \theta^* \colon P \to P^*$. By Definition 9.85 of the bordism formation, we obtain the desired formula. □

Lemma 9.99 *Let $(F_1, \overline{F}_1)$ be a $(k + 1)$-normal π-bordism between k-connected degree one normal π-maps $(f, \overline{f}) \colon M \to X$ and $(f_1, \overline{f}_1) \colon M_1 \to X$ and let $B = \partial(P, \theta)$ be a boundary formation.*

9.3 The Odd-Dimensional Surgery Obstruction in the Universal Covering Case

Then there is a $(k+1)$-normal π-bordism from $(F_B, \overline{F}_B)$ from some k-connected degree one normal π-map $(f_B, \overline{f}_B) \colon N \to X$ to $(f_1, \overline{f}_1)$ such that

$$\Sigma(F_B, \overline{F}_B, N) \cong \Sigma(F_1, \overline{F}_1, M) \oplus \partial(P, \theta).$$

Proof. The idea is to combine $(F_1, \overline{F}_1)$ with the construction from the previous lemma. In order to do that, we may assume that $(F_1, \overline{F}_1)$ is a $(k+1)$-trace normal π-bordism. In that case, it is obtained from $(f, \overline{f})$ by surgery on some representation $(\underline{\omega}, \overline{q})$. Looked at from the other side, we may see it as obtained from $(f_1, \overline{f}_1)$ by surgery on the dual representation $(\underline{\omega}_1, \overline{q}_1)$ in the sense described below Lemma 9.82.

We suppose that the $(-\epsilon)$-quadratic form (P, θ) has $P \cong (\mathbb{Z}\pi)^l$ and that $\underline{\omega}_1 = \{x_1, \ldots x_a\}$. We extend the representation $(\underline{\omega}_1, \overline{q}_1)$ of $K_k(\widetilde{M}_1)$ as follows. Set $\underline{\omega}_2 := \underline{\omega}_1 \sqcup \{x_{a+1}, \ldots, x_{a+l}\}$, where $x_{a+1} = \cdots = x_{a+l} = 0$, and set $\overline{q}_B := \overline{q}_1 \sqcup \overline{q}_2$, where $\overline{q}_2 \colon \sqcup_{i=1}^l (D^{k+1} \times S^k) \to D^{2k+1} \subset M$, with D^{2k+1} disjoint from $\operatorname{im}(\overline{q}_1)$. Now we proceed as in the proof of Theorem 8.195. We use the quadratic form (P, θ) as a guide to construct a regular homotopy of the embedding $\overline{q}_2$ to another embedding $\overline{q}_2'$ such that, when we attach handles to the cylinder using these new embeddings $\overline{q}_2'$, the resulting cobordism W_2 will map to the cylinder $X \times I$ by a k-connected degree one normal map whose intersection form on $K_{k+1}(\widetilde{W}_2)$ will be (P, θ). By general position this can be done in such a way that the regular homotopies from $\overline{q}_2$ to $\overline{q}_2'$ stay away from $\overline{q}_1$.

If we do the above procedure simultaneously with the surgeries prescribed by $(\underline{\omega}_1, \overline{q}_1)$, we obtain the desired cobordism W_B. From the construction we see that we may view it as

$$W_B = W_1 \cup_M W_2.$$

Such a decomposition can be obtained via an appropriate Morse function corresponding to first surgeries via $(\underline{\omega}_1, \overline{q}_1)$ and then handle attachments via $\overline{q}_2'$. The top end of the cobordism is the desired N, the corresponding normal maps $(F_1, \overline{F}_1)$ and $(F_2, \overline{F}_2)$ are obtained in an obvious way, and $(F_B, \overline{F}_B)$ is their union. Then we look at the result from the other side.

We claim that with these definitions the desired equation holds. This follows from the isomorphisms

$$A_G \colon K_{k+1}(\widetilde{W}_B) \xrightarrow{\cong} K_{k+1}(\widetilde{W}_1) \oplus P;$$

$$A_F \colon K_{k+1}(\widetilde{W}_B, \widetilde{M}_1) \xrightarrow{\cong} K_{k+1}(\widetilde{W}_1, \widetilde{M}_1) \oplus P,$$

and

$$A_F^* \colon K_{k+1}(\widetilde{W}_B, \widetilde{N}) \xrightarrow{\cong} K^{k+1}(\widetilde{W}_B, \widetilde{M}_1) \xrightarrow{\cong} K_{k+1}(\widetilde{W}_B, \widetilde{M}_1)^*$$

$$\cong K_{k+1}(\widetilde{W}_1, \widetilde{M}_1)^* \oplus P^* \xrightarrow{\cong} K_{k+1}(\widetilde{W}_1, \widetilde{M}) \oplus P^*.$$

The required isomorphism of formations A is given by $A = (A_F, A_G)$:

$$K_{k+1}(\widetilde{W}_B, \widetilde{M}_1) \xleftarrow{j_B} K_{k+1}(\widetilde{W}_B) \xrightarrow{j'_B} K_{k+1}(\widetilde{W}_B, \widetilde{N})$$
$$\downarrow A_F \qquad\qquad \downarrow A_G \qquad\qquad \downarrow A_F^*$$
$$K_{k+1}(\widetilde{W}_1, \widetilde{M}_1) \oplus P \xleftarrow{j_1 \oplus \mathrm{id}} K_{k+1}(\widetilde{W}_1) \oplus P \xrightarrow{j'_1 \oplus \lambda_2} K_{k+1}(\widetilde{W}_1, \widetilde{M}) \oplus P^*$$

where j_B, j'_B, j_1, and j'_1 are the connecting maps for the pairs (W_B, M_1), (W_B, N), (W_1, M_1) and (W_1, M) respectively and where $\lambda_2 = \theta - \epsilon\theta^*$. □

The following lemma is a key step in proving that the surgery obstruction is a well defined invariant of normal bordism. Its proof uses three further lemmas, stated below the proof, about handlebody decompositions of cobordisms, which use arguments from Chapter 2.

Lemma 9.100 *Let* $(G, \overline{G}) \colon (V; M, M') \to (X \times [0,1]; X \times \{0\}, X \times \{1\})$ *be a normal π-bordism between two k-connected degree one normal π-maps* $(f, \overline{f}) \colon M \to X$ *and* $(f', \overline{f'}) \colon M' \to X$.

Then by a finite sequence of surgeries on the interior of normal π-bordisms starting from $(G, \overline{G})$, *we can replace* $(G, \overline{G})$ *by a $(k+1)$-normal π-bordism* $(F, \overline{F}) \colon (W; M, M') \to (X \times [0,1]; X \times \{0\}, X \times \{1\})$ *between* $(f, \overline{f})$ *and* $(f', \overline{f'})$.

Proof. We start by performing surgeries below the middle dimension on the interior of V to obtain a $(k+1)$-connected normal π-bordism $(F', \overline{F}')$ with domain $(V'; M, M')$. Then the surgery kernels $K_i(\widetilde{V}', \widetilde{M})$ and $K_i(\widetilde{V}', \widetilde{M}')$ vanish for all $i \neq k+1$.

Since the map $M \to V'$ is k-connected, by handle cancellation arguments as in Section 2.4, the cobordism V' has a handlebody decomposition relative M consisting of handles of index $(k+1)$ and higher, see Lemma 9.101 below. The inclusion of the other end $M' \to V'$ is also k-connected, and so by the same arguments we may assume that V' has a handlebody decomposition relative M' with only handles of index $(k+1)$ and higher. This second handlebody decomposition can also be viewed as the dual decomposition relative M in the sense of Section 2.4, and as such it has handles of index $(k+1)$ and smaller. So we have two handlebody decompositions of V' relative M, one with handles of index $(k+1)$ and higher and the other with handles of index $(k+1)$ and smaller. What we are seeking is a handlebody decomposition relative M with only handles of index $(k+1)$.

In fact, taking the second handlebody decomposition of V' relative M, i.e., the one with handles of index $(k+1)$ and smaller, the same handle cancellation arguments can be employed to obtain a handle decomposition of V' relative M with only handles of index k and $(k+1)$. Now we need to get rid of the k-handles and we know that this can be done if we are able to introduce new $(k+1)$-handles with trivial attaching embeddings that would cancel them. In the proof of the s-cobordism theorem, given a k-handle (ϕ^k), using the assumption $H_k(\widetilde{V}', \widetilde{M}) = 0$, which is fulfilled here as well because of $K_k(\widetilde{V}', \widetilde{M}) \cong H_k(\widetilde{V}', \widetilde{M})$, we were able to achieve such an introduction of a $(k+1)$-handle with trivial attaching embeddings (ϕ^{k+1}) that would cancel (ϕ^k). However, it was at the expense of also introducing a $(k+2)$-handle, which we now

9.3 The Odd-Dimensional Surgery Obstruction in the Universal Covering Case

do not want to do. But in the present proof we have more freedom than in the proof of the s-cobordism theorem since we can modify the cobordism by more drastic operations. In particular, we can perform the operation of the connected sum with $S^{k+1} \times S^{k+1}$ on the interior of V'. By Lemma 9.102 such a connected sum introduces two new $(k+1)$-handles with trivial attaching embeddings and no handles of another index. If we do this sufficiently many times, we obtain enough $(k+1)$-handles to cancel all k-handles, see Lemma 9.103.

Altogether we obtain a cobordism W from M to M' with only $(k+1)$-handles, which is what we wanted. □

Lemma 9.101 (Normal Form Lemma for highly connected cobordisms) *Let $(W; \partial_0 W, \partial_1 W)$ be a compact cobordism of dimension $n \geq 6$. Let q, s be integers with $2 \leq q < s \leq n - 3$ and such that the inclusion $\partial_0 W \to W$ is q-connected and the inclusion $\partial_1 W \to W$ is $(n - s)$-connected.*

Then there is a handlebody decomposition that has only handles of index r for $q \leq r \leq s$, i.e., there is a diffeomorphism relative $\partial_0 W$

$$W \cong \partial_0 W \times [0, 1] + \sum_{i=1}^{p_q} (\phi_i^q) + \cdots + \sum_{i=1}^{p_s} (\phi_i^s).$$

Proof. By a straightforward modification of the proof of Lemma 2.37 adapted to our situation. □

Lemma 9.102 (Handlebody decompositions and connected sums with products of spheres) *Let $(W; \partial_0 W, \partial_1 W)$ be a compact cobordism of dimension $n = 2k+2 \geq 6$. Consider the compact cobordism $(V; \partial_0 W, \partial_1 W)$ obtained from $(W; \partial_0 W, \partial_1 W)$ by connected sum with $S^{k+1} \times S^{k+1}$ in the interior of W, i.e., we have*

$$V = W \# (S^{k+1} \times S^{k+1}).$$

Let

$$W \cong \partial_0 W \times [0, 1] + \sum_{i=1}^{p_0} (\phi_i^0) + \cdots + \sum_{i=1}^{p_n} (\phi_i^n)$$

be a handlebody decomposition relative $\partial_0 W$.

Then the cobordism $(V; \partial_0 W, \partial_1 W)$ has a handlebody decomposition relative $\partial_0 W$ that differs from the above decomposition only by adding two handles of index $(k+1)$ along trivial attaching embeddings, i.e., we have

$$V \cong \partial_0 W \times [0, 1] + \sum_{i=1}^{p_0} (\phi_i^0) + \cdots + \sum_{i=1}^{p_{k+1}} (\phi_i^{k+1}) + (\psi_1^{k+1}) + (\psi_2^{k+1})$$
$$+ \cdots + \sum_{i=1}^{p_n} (\phi_i^n),$$

where ψ_j^{k+1} for $j = 1, 2$ are the two k-handles corresponding to the two spheres of $S^{k+1} \times S^{k+1}$ and the attaching embeddings are trivial in $\partial_1 W_k$.

Proof. The connected sum with the product of spheres $(S^{k+1} \times S^{k+1})$ can be thought of as follows. Introduce an n-handle and an $(n-1)$-handle that cancel each other. Then cut out of W the n-handle, thus introducing a new part of the boundary diffeomorphic to S^{2k+1}. Attach the product of the spheres with a removed disk $(S^{k+1} \times S^{k+1}) \setminus D^{2k+2}$ that has three handles, two of index $(k + 1)$ with trivial attaching embeddings and one of index $n = 2k + 2$. This handle of index n cancels with the $(n - 1)$-handle that was introduced at the beginning. □

Lemma 9.103 (Removing k-handles of a k-connected $(2k + 2)$-dimensional cobordism) *Let $(W; \partial_0 W, \partial_1 W)$ be a compact cobordism of dimension $n = 2k+2 \geq 6$. Suppose that the inclusion $\partial_0 W \to W$ is k-connected.*

Let

$$W \cong \partial_0 W \times [0, 1] + \sum_{i=1}^{p_k} (\phi_i^k) + \sum_{i=1}^{p_{k+1}} (\phi_i^{k+1}) \cdots + \sum_{i=1}^{p_n} (\phi_i^n)$$

be a handlebody decomposition relative $\partial_0 W$ with $p_k \geq 1$ and let $i_0 \leq p_k$. (We know that such a decomposition exists by Lemma 9.101.)

Then the cobordism $(V; \partial_0 W, \partial_1 W)$ obtained from $(W; \partial_0 W, \partial_1 W)$ by connected sum with $S^{k+1} \times S^{k+1}$ in the interior of W so that we have $V = W \# (S^{k+1} \times S^{k+1})$ has a handlebody decomposition where the handles $\phi_{i_0}^k$ and ψ_1^{k+1} cancel, i.e., we have

$$V \cong \partial_0 W \times [0, 1] + \sum_{i=1,\ldots p_k, i \neq i_0} (\phi_i^k) + (\psi_2^{k+1}) + \sum_{i=1}^{p_{k+1}} (\phi_i^{k+1}) \cdots + \sum_{i=1}^{p_n} (\phi_i^n).$$

Proof. Use Lemma 9.102 to modify the proof of Lemma 2.37 according to the new situation and the new goal. □

Proof of Theorem 9.60. We define the surgery obstruction of a k-connected degree one normal π-map $(f, \overline{f}): M \to X$ to be the element of $L_{2k+1}(\mathbb{Z}, w)$ represented by a kernel formation of $(f, \overline{f})$:

$$\sigma(f, \overline{f}) := [\Sigma(f, \overline{f})] \in L_{2k+1}(\mathbb{Z}\pi, w). \tag{9.104}$$

If $(g, \overline{g}): N \to X$ is any degree one normal π-map, then by Theorem 7.41, $(g, \overline{g})$ is normally π-bordant to a k-connected degree one normal π-map $(f, \overline{f}): M \to X$, and we define the surgery obstruction of $(g, \overline{g})$ to be the surgery obstruction of $(f, \overline{f})$, i.e., we put

$$\sigma(g, \overline{g}) := \sigma(f, \overline{f}). \tag{9.105}$$

We first show that the surgery obstruction is a well-defined invariant of the normal π-bordism class of $(g, \overline{g})$. From the definition, it suffices to do this for a k-connected

9.3 The Odd-Dimensional Surgery Obstruction in the Universal Covering Case

degree one normal π-map $(f, \overline{f}) \colon M \to X$. Suppose that $(G, \overline{G}) \colon (V; M, M') \to (X \times [0, 1]; X \times \{0\}, X \times \{1\})$ is a normal π-bordism between $(f, \overline{f})$ and another k-connected degree one normal π-map $(f', \overline{f'}) \colon M' \to X$. Applying Lemma 9.100, we replace $(G, \overline{G})$ with a $(k + 1)$-normal π-bordism $(F, \overline{F}) \colon (W; M, M') \to (X \times [0, 1]; X \times \{0\}, X \times \{1\})$ satisfying $(F, \overline{F})|_M = (f, \overline{f})$ and $(F, \overline{F})|_{M'} = (f', \overline{f'})$. If $\Sigma(f, \overline{f}) = (H_\epsilon(F); F, G)$, then by Lemma 9.82 there is an isomorphism $\Sigma(f', \overline{f'}) \xrightarrow{\cong} (H_\epsilon(F); F^*, G)$. Now, by definition of minus in $L_{2k+1}(\mathbb{Z}\pi)$,

$$-[\Sigma(f', \overline{f'})] = -[H_\epsilon(F); F^*, G] = [-H_\epsilon(F); F, \widehat{G}],$$

where $\widehat{G} \subset H_\epsilon(F)$ is a complementary lagrangian to G and we used the fact that F is complementary to F^*. Lemma 9.29 (ii) implies

$$\sigma(f, \overline{f}) - \sigma(f', \overline{f'}) = [\Sigma(f, \overline{f})] - [\Sigma(f', \overline{f'})]$$
$$= [H_\epsilon(F); F, G] \oplus [-H_\epsilon(F); F, \widehat{G}]$$
$$= 0 \in L_{2k+1}(\mathbb{Z}\pi, w).$$

This shows that the surgery obstruction is a well-defined invariant of normal bordism.

Now we show that $\sigma(f, \overline{f}) = 0 \in L_{2k+1}(\mathbb{Z}, w)$ if and only if $(f, \overline{f})$ is normal π-bordant to a normal homotopy equivalence. If $\sigma(f, \overline{f}) = 0$, then there are boundary formations $\partial(P_0, \theta_0)$ and $\partial(P_1, \theta_1)$ and a stable isomorphism

$$\Sigma(f, \overline{f}) \oplus \partial(P_0, \theta_0) \cong_s \partial(P_1, \theta_1),$$

where $\Sigma(f, \overline{f})$ is the kernel formation of $(f, \overline{f})$. Consider some choice of a presentation $(\omega, \overline{q})$ for $(f, \overline{f})$ and its associated $(k+1)$-normal π-bordism $(F_1, \overline{F}_1)$ with domain $(W_0; M, M_1)$ from $(f, \overline{f})$ to some $(f_1, \overline{f}_1)$. Because of Lemma 9.88 we can identify $\Sigma(f, \overline{f})$ and $\Sigma(F_1, \overline{F}_1, M)$. We deduce from Lemma 9.99 that there is a $(k+1)$-normal π-bordism $(F_2, \overline{F}_2)$ with domain $(W_1; M_2, M_1)$ from some $(f_2, \overline{f}_2)$ to $(f_1, \overline{f}_1)$ above such that $\Sigma(F_2, \overline{F}_2, M_2) = \Sigma(F_1, \overline{F}_1, M) \oplus \partial(P_0, \theta_0)$. By Lemma 9.93 (ii) there is a normal bordism $(F_3, \overline{F}_3)$ with domain $(W_2; M_2, M')$ from $(f_2, \overline{f}_2)$ to $(f', \overline{f'})$ with the associated formation $\Sigma(F_3, \overline{F}_3, M_2) = \partial(P_1, \theta_1)$ and by Lemma 9.96 the map $(f', \overline{f'})$ is a normal homotopy equivalence. The union of normal π-bordisms

$$(F_1, \overline{F}_1) \cup_{(f_1, \overline{f}_1)} (F_2, \overline{F}_2) \cup_{(f_2, \overline{f}_2)} (F_3, \overline{F}_3)$$

is then a normal π-bordism from $(f, \overline{f})$ to the normal homotopy equivalence $(f', \overline{f'})$.

Conversely, if $(f, \overline{f})$ is normally π-bordant to a normal homotopy equivalence, then by Lemma 9.100, we may assume that the normal π-bordism is a $(k+1)$-normal π-bordism and then by Lemmas 9.78, 9.88 and 9.96, $\Sigma(f, \overline{f})$ is stably isomorphic to a boundary and so $\sigma(f, \overline{f}) = 0 \in L_{2k+1}(\mathbb{Z}\pi, w)$. This finishes the proof of Theorem 9.60. $\square$

9.4 The Intrinsic Surgery Obstruction in Odd Dimensions

This section is built analogously to Section 8.7. Since the setup there is mostly independent of the parity of the dimension, the proofs will mostly be the same, just needing the input from this chapter to make the odd-dimensional claim.

9.4.1 The Definition of Intrinsic L-Group in Odd Dimensions

Now we will apply the technology of Section 8.7 to the odd-dimensional L-groups. By Lemma 9.38 we have the half-conjugation invariant functor

$$L_{2k+1} \colon \mathrm{GROUPS}^w \to \mathbb{Z}\text{-MOD}, \quad (G, w) \mapsto L_{2k+1}(\mathbb{Z}G, w),$$

and we can apply the construction from the end of Section 8.7.2.

Definition 9.106 (Intrinsic L-group in odd dimensions) We denote by

$$(L_{2k+1})_* \colon \mathcal{R} \to \mathbb{Z}\text{-MOD}$$

the resulting functor and write

$$L_{2k+1}(\mathbb{Z}\Pi(X), \mathcal{O}_X) := (L_{2k+1})_*(X, \mathcal{O}_X).$$

9.4.2 The Intrinsic Odd-Dimensional Surgery Obstruction

Definition 9.107 (Intrinsic surgery obstruction in odd dimension) Let $(X, \mathcal{O}_X, [[X]])$ be a (not necessarily connected) w-oriented finite Poincaré complex in the sense of Definition 5.43 and Remark 5.45 of odd dimension $n = 2k + 1$ for $k \geq 2$. Consider a normal map with respect to the tangent bundle of degree one $(M, f, a, \xi, \overline{f}, o)$ with target $(X, \mathcal{O}_X, [[X]])$ in the sense of Definition 7.22.

Its *intrinsic surgery obstruction in odd dimension* is the element

$$\sigma(M, f, a, \xi, \overline{f}, o) \in L_{2k+1}(\mathbb{Z}\Pi(X), \mathcal{O}_X),$$

defined below. We sometime abbreviate

$$(f, \overline{f}) := (M, f, a, \xi, \overline{f}, o);$$
$$\sigma(f, \overline{f}) = \sigma(M, f, a, \xi, \overline{f}, o).$$

The definition proceeds analogously as in Chapter 8. We first consider the case when M and X are connected and $\pi_1(f, m) \colon \pi_1(M, m) \to \pi_1(X, f(m))$ is bijective for one (and hence all) base points $m \in M$. We make the choices (NBP), see Notation 8.5. Then we have the structure map

9.4 The Intrinsic Surgery Obstruction in Odd Dimensions

$$\psi_{x,\bar{x}} \colon L_{2k+1}(\mathbb{Z}\pi_1(X,x), w_1(X)) \to L_{2k+1}(\mathbb{Z}\Pi(X), \mathcal{O}_X).$$

We define $\sigma(M, f, a, \xi, \overline{f}, o) \in L_{2k+1}(\mathbb{Z}\Pi(X), \mathcal{O}_X)$ to be the image of

$$\sigma(M, [M], f, \widetilde{f}, a, \xi, \overline{f}) \in L_{2k+1}(\mathbb{Z}\pi_1(X,x), w_1(X))$$

from Theorem 9.60 under $\psi_{x,\bar{x}}$. Next we verify that the surgery obstruction of normal π-maps of degree one $\sigma(f, \overline{f}) \in L_{2k+1}(\mathbb{Z}\pi_1(X,x), w_1(X))$ behaves well with respect to all the choices we have made. Given two choices (NBP) for a normal π-map of degree one $(f, \overline{f})$, a choice of a homotopy h with a path v as in Section 8.7.4 provides us with an isomorphism of the underlying $\mathbb{Z}\pi$-modules of the two formations coming from the two choices of (NBP) analogously to the isomorphism (8.164). We observe that the underlying quadratic form of the formation is replaced by the form obtained by multiplication by $s([v])$ for the same reasons as in Chapter 8 since it is in fact a quadratic form associated to an even-dimensional surgery problem. Then we pass to the case when X is connected, and finally to the general case exactly as in Chapter 8.

Theorem 9.108 (The Intrinsic surgery obstruction in odd dimensions) *Let $(X, \mathcal{O}_X, [[X]])$ be a (not necessarily connected) w-oriented finite n-dimensional Poincaré complex in the sense of Definition 5.43 and Remark 5.45. Consider a normal map with respect to the tangent bundle of degree one $(M, f, a, \xi, \overline{f}, o)$ of odd dimension $n = 2k + 1$ with target $(X, \mathcal{O}_X, [[X]])$ for $k \geq 2$ in the sense of Definition 7.22. Then:*

(i) *Its intrinsic surgery obstruction*

$$\sigma(f, \overline{f}) = \sigma(M, f, a, \xi, \overline{f}, o) \in L_{2k+1}(\mathbb{Z}\Pi(X), \mathcal{O}_X)$$

depends only on the normal bordism with cylindrical ends class of $(\overline{f}, f)$;

(ii) *We have $\sigma(\overline{f}, f) = 0$ if and only if we can do a finite number of surgery steps to obtain a normal homotopy equivalence.*

Proof. The theorem follows from Theorem 9.60 in the same way as Theorem 8.167 follows from Theorem 8.112. □

9.4.3 Bordism Invariance in Odd Dimensions for Not Necessarily Cylindrical Target

Assume the same setup and the same notation as at the beginning of Section 8.7.5, but for dimension $n = 2k + 1$ with $k \geq 2$.

Theorem 9.109 (Bordism invariance of the intrinsic surgery obstruction in odd dimensions) *In $L_{2k+1}(\mathbb{Z}\Pi(Y), \mathcal{O}_Y)$ we get the equality*

$$L_{2k+1}(l_0 \circ v_0, \mathcal{O}_{l_0 \circ v_0})\big(\sigma(M_0, f_0, a_0, \xi_0, \overline{f}_0, o_0)\big)$$
$$= L_{2k+1}(l_1 \circ v_1, \mathcal{O}_{l_1 \circ v_1})\big(\sigma(M_1, f_1, a_1, \xi_1, \overline{f}_1, o_1)\big).$$

We note that the proof of Theorem 9.109 is significantly harder than the proof of Theorem 9.108. One difficulty is that in the proof of Theorem 9.60, which was its main ingredient, we have used special handlebody decompositions of the bordism W relative to M that very well relate to the surgery kernel groups. This relation becomes more complicated when we deal with varying target. In fact in Wall's book [414] it is the contents of the difficult Chapter 8. Therefore we postpone the proof to Chapter 15 where we will give the proof for both even and odd dimensions.

9.4.4 The Simply Connected Case in Odd Dimensions

Now we can give a rather complete answer to Problem 4.1 and Problem 7.42 for odd dimensions in the simply connected case.

Theorem 9.110 (Solution to the simply connected surgery problem in odd dimensions)

(i) *Suppose we have normal map of degree one $(f, \overline{f})$ from a closed manifold M to a simply connected finite Poincaré complex X of odd dimension $n = 2k + 1 \geq 5$. Then we can always change M and $(f, \overline{f})$ leaving X fixed by finitely many surgery steps to get a normal map $(g, \overline{g})$ from a closed manifold N to X such that g is a homotopy equivalence;*
(ii) *Let X be a simply connected finite connected Poincaré complex of odd dimension $n = 2k + 1 \geq 5$. Then X is homotopy equivalent to a closed manifold if and only if the Spivak normal fibration has a reduction to a vector bundle $\xi \colon E \to X$, i.e., the set of normal invariants $\mathcal{T}_n(X)$ is non-empty.*

Proof. Part (i) follows from Theorem 9.60 and Theorem 9.18.

Part (ii) follows from assertion (i) together with Theorem 7.19. ⊔⊓

Note that the above arguments only used the algebra from Sections 9.2.1 and 9.2.2.

Exercise 9.111 Consider a degree one normal map $(f, \overline{f}) \colon M \to X$ with $(2k + 1)$-dimensional M and X and with $\pi_1(M) \cong \pi_1(X) \cong \{1\}$. Suppose it is k-connected and that $K_k(M)$ is finitely generated free.

Show that surgery on the elements of any $\mathbb{Z}$-basis of $K_k(M)$ produces a homotopy equivalence.

Exercise 9.112 Consider a degree one normal map $(f, \overline{f}) \colon M \to X$ with $(2k + 1)$-dimensional M and X and with $\pi_1(M) \cong \pi_1(X) = \{1\}$. Suppose it is k-connected and that $K_k(M)$ is a finite cyclic group $\mathbb{Z}/s$ for some $q \geq 2$ and that k is even.

Show that surgery on any generator of $K_k(M) = \mathbb{Z}/s$ produces a k-connected degree one normal map $(f', \overline{f}') \colon M' \to X$ with $K_k(M')$ infinite cyclic.

9.5 The Odd-Dimensional Surgery Obstruction Relative Boundary

Analogously to the even-dimensional situation in Section 8.8, we now extend in the odd-dimensional situation the results of Subsection 8.7.4 from closed manifold to pairs, where we assume that on the boundary we already have a homotopy equivalence. This is again analogous to the closed case since the surgery kernels of pairs look like the surgery kernel in the closed situation in this case. Similarly, this will be important in Section 9.6, where we will consider the "Realisation problem": This is the problem of determining, which elements of $L_{2k+1}(\mathbb{Z}\Pi(X), O_X)$ arise as the surgery obstructions to surgery problems. If we allow boundary with a homotopy equivalence on the boundary, every element in $L_{2k+1}(\mathbb{Z}\Pi(X), O_X)$ occurs. This is not true if one considers only closed manifolds. Hence in this aspect the situation is perfectly parallel to the even dimensional case.

Theorem 9.113 (The intrinsic surgery obstruction in odd dimensions relative boundary) *Let $(X, \partial X, O_X, [[X, \partial X]])$ be a (not necessarily connected) w-oriented finite n-dimensional Poincaré pair (in the sense of Definition 5.73) of odd dimension $n = 2k + 1$ for $k \geq 2$. Consider a normal map with respect to the tangent bundle of degree one relative boundary $(f, \overline{f}) = (M, \partial M, f, \partial f, a, \xi, \overline{f}, o)$ with target $(X, \partial X, O_X, [[X, \partial X]])$ in the sense of Subsection 8.8.1. (Recall that this means that $\partial f \colon \partial M \to \partial X$ is a homotopy equivalence.) Then:*

(i) *Its intrinsic surgery obstruction*

$$\sigma(f, \overline{f}) = \sigma(M, \partial M, f, \partial f, a, \xi, \overline{f}, o) \in L_{2k+1}(\mathbb{Z}\Pi(X), O_X)$$

  depends only on the normal bordism with cylindrical ends class of $(\overline{f}, f)$;

(ii) *We have $\sigma(\overline{f}, f) = 0$ if and only if we can do a finite number of surgery steps on the interior of M without changing ∂f to obtain a normal homotopy equivalence, i.e., the underlying map f is a homotopy equivalence.*

Proof. The proof is analogous to the proof of Theorem 9.108. Similarly as in the proof of Theorem 8.189, one has to make appropriate modifications using relative kernels. In this case we leave the details to the reader. □

Remark 9.114 Theorem 9.109 carries over to the case relative boundary in the obvious way.

9.6 Realisation of Odd-Dimensional Surgery Obstructions

Again analogously to the even-dimensional situation, we see that Theorem 9.108 does not identify which elements of $L_{2k+1}(\mathbb{Z}\Pi(X), O_X)$ arise as surgery obstructions of degree one normal maps $(f, \overline{f}) \colon (M, \partial M) \to (X, \partial X)$. In this subsection we show that, if N is a compact $(2k)$-dimensional *manifold*, possibly with boundary for

$k \geq 2$, then all elements in the group $L_{2k+1}(\mathbb{Z}\Pi(N \times I), \mathcal{O}_{N \times I})$ are realised as surgery obstructions of degree one normal maps with target $(N \times I, \partial(N \times I), \mathcal{O}_{N \times [0,1]}, [[N \times I, \partial(N \times I)]])$ for $I = [0, 1]$. We now state a version of Wall realisation in odd dimensions.

Theorem 9.115 (Realisation of odd-dimensional surgery obstructions) *Suppose $k \geq 2$. Consider a compact oriented $(2k)$-manifold N with possibly empty boundary ∂N. Fix an element*

$$\omega \in L_{2k+1}(\mathbb{Z}\Pi(N \times I), \mathcal{O}_{N \times I}).$$

Then there is a normal map of pairs of degree one relative boundary $(M, \partial M, f, \partial f, a, \xi, \overline{f}, o)$ with target $(N \times I, \partial(N \times I), \mathcal{O}_{N \times I}, [[N \times I, \partial(N \times I)]])$ for $I = [0, 1]$ with the following properties:

(i) *The bundle ξ over $N \times I$ is $T(N \times I) \oplus \mathbb{R}^a$ and the isomorphism o of infinite cyclic local coefficient systems over $N \times I$ is the identity;*

(ii) *There is a decomposition $\partial M = \partial_0 M \cup \partial_1 M \cup \partial_2 M$ where $\partial_m M$ is a codimension zero submanifold of ∂M with possibly empty boundary $\partial \partial_m M$ for $m = 0, 1, 2$ satisfying*

$$\partial_0 M \cap \partial_1 M = \emptyset;$$
$$\partial_2 M \cap \partial_m M = \partial \partial_m M \quad \text{for } m = 0, 1;$$
$$\partial \partial_2 M = \partial \partial_0 M \amalg \partial \partial_1 M;$$
$$\partial_2 M = \partial N \times [0, 1];$$
$$\partial_0 M = N \times \{0\};$$

(iii) *The homotopy equivalence $\partial f \colon \partial M \to \partial(N \times I)$ induces the identity $\partial_0 M = N \times \{0\} \to (N \times \{0\})$ and the identity $\partial_2 M = \partial N \times [0, 1] \to \partial N \times I$;*

(iv) *The homotopy equivalence $\partial f \colon \partial M \to \partial(N \times I)$ maps $\partial_1 M$ to $N \times \{1\}$ and induces a homotopy equivalence*

$$\partial_1 f \colon \partial_1 M \to N \times \{1\};$$

(v) *The intrinsic surgery obstruction*

$$\sigma(M, \partial M, f, \partial f, a, \xi, \overline{f}, o) \in L_{2k+1}(\mathbb{Z}\Pi(N \times I), \mathcal{O}_{N \times I})$$

from Theorem 9.113 is the given element ω.

Proof. As in the proof of Theorem 8.195, we fix a choice of a universal covering $p_N \colon \widetilde{N} \to N$ and choice (BP) for N, see Notation 5.6, where the base point $x \in N$ should not lie in ∂N. Put $\pi = \pi_1(N, x) = \pi_1(N \times I, (x, 0))$ and let $w \colon \pi \to \{\pm 1\}$ be the orientation homomorphism of N. Then we get an identification, see Lemma 8.157 (ii)

$$L_{2k+1}(\mathbb{Z}\Pi(N \times I), \mathcal{O}_{N \times I}) = L_{2k+1}(\mathbb{Z}\pi, w). \tag{9.116}$$

We will consider ω as an element $L_{2k+1}(\mathbb{Z}\pi, w)$ and will work with normal π-maps of degree one in the sense of Definition 8.3.

9.6 Realisation of Odd-Dimensional Surgery Obstructions

Recall that an element $\omega \in L_{2k+1}(\mathbb{Z}\pi, w)$ is represented by an ϵ-quadratic formation $(H_\epsilon(F); F, G)$, where $\epsilon = (-1)^k$ and F and G are free $\mathbb{Z}\pi$-modules of the same rank, r say. We form $(f, \overline{f}) \colon M \to N \times I$ in three stages. We start by taking $M_1 = N \times I$ and $(f_1, \overline{f}_1) = (\mathrm{id}, \overline{\mathrm{id}})$. Then we fix a $(2k)$-disk $D^{2k} \subset \mathrm{int}(N \times \{1\})$ in the interior of $N \times \{1\}$, choose r disjoint trivially framed embeddings $S^{k-1} \times D^{k+1} \hookrightarrow D^{2k}$ as in Definition 4.22 and paths w_i from the image of $s \times \{0\}$ to the origin of D^{2k}. Now attach r copies of a k-handle $D^k \times D^{k+1}$ to $N \times I$ along these embeddings to obtain M_2:

$$M_2 := N \times I + \sum_{i=1}^{r} (\phi_i^k).$$

Its boundary now decomposes into the union of $\partial_0 M_2 = N$, $\partial_2 M_2 = \partial N \times I$ and $\partial_1 M_2$, that is, the complement of $\partial_0 M_2 \cup \partial_2 M_2$ in ∂M_2 or, in other words, the top end of the resulting cobordism M_2. We remark that the cobordism M_2 can equally well be described as the trace of surgery on r disjoint embeddings $S^{k-1} \times D^{k+1} \hookrightarrow D^{2k}$. Hence we can extend the identity map on N to a degree one normal map

$$(f_2, \overline{f}_2) \colon (M_2, \partial_0 M_2 \cup \partial_2 M_2, \partial_1 M_2) \to (N \times I, N \times \{0\} \cup \partial N \times I, N \times \{1\}).$$

It also tells us that $\partial_1 M_2 = N \natural_i^r S^k \times S^k$.

For future purposes we denote by $U' \subset M_2$ the union of the handles with the product of the images of the attaching maps $S^k \times D^{k+1}$ with the interval and the product of the images of the thickened paths w_i with the interval. Moreover let $\partial_0 U' = U' \cap \partial_0 M_2$.

Now observe that $(\partial_1 f_2, \overline{\partial_1 f_2}) \colon \partial_1 M_2 \to N \times \{1\}$ is a degree one normal map with kernel form $H_\epsilon(F)$. Let $\{b_1, \ldots, b_r\}$ be a basis for $G \subset H_\epsilon(F)$. By Theorems 4.39 and 8.28 we may assume that each element b_i is represented by an embedding $g_i \colon S^k \times D^k \hookrightarrow \partial_1 M_2$ and we may assume that the images of the g_i are pairwise disjoint. We define $M = M_3$ to be the manifold obtained by attaching r copies of a $(k+1)$-handle to M_2 along the embeddings g_i:

$$M = M_3 := M_2 + \sum_{i=1}^{r} (\phi_i^{k+1}) = N \times I + \sum_{i=1}^{r} (\phi_i^k) + \sum_{i=1}^{r} (\phi_i^{k+1}).$$

Applying Theorem 4.39, the normal map $(f_2, \overline{f}_2)$ extends to a degree one normal map of triads $(f, \overline{f})$:

$$(f, \partial_0 f, \partial_1 f) \colon (M; \partial_0 M, \partial_1 M) \to (N \times [0, 1]; N \times \{0\} \cup \partial N \times [0, 1], N \times \{1\})$$

where $\partial_0 f = \mathrm{id}$ is a diffeomorphism and $\partial_1 f$ is a homotopy equivalence since it is the outcome of surgery on a lagrangian on a hyperbolic kernel, see Theorem 8.89.

We complete the proof by showing that $(f, \overline{f})$ has surgery obstruction $\omega = [H_\epsilon(F); F, G]$. By construction, we see that there is a homotopy equivalence

$$M \simeq N \cup (\cup_{i=1}^{r} e_i^k) \cup (\cup_{i=1}^{r} e_i^{k+1})$$

and that $f|_N = \text{id}_N$. If $\{x_1, \ldots, x_r\}$ is the canonical basis $\pi_k(U', \partial_0 U') \cong \mathbb{Z}^r$, then under the inclusion $U \to M_2 \subset M$ the set $\{x_1, \ldots, x_r\}$ maps to a generating set ω for $K_k(\widetilde{M})$. Taking $U \subset U'$ to be a copy of U' in the interior of M, we obtain $\overline{q}\colon U \to U' \to M$ so that $(\omega, \overline{q})$ is a representation of $K_k(\widetilde{M})$. The kernel formation defined by $(\omega, \overline{q})$ is given by

$$\Sigma(f, \overline{f}, \omega, \overline{q}) = (K_k(\widetilde{\partial U}); K_{k+1}(\widetilde{U}, \widetilde{\partial U}), K_{k+1}(\widetilde{M}_0, \widetilde{\partial U})),$$

where M_0 is the complement of U in M. Now by construction there is an equality of pairs $(K_k(\widetilde{\partial U}), K_{k+1}(\widetilde{U}, \widetilde{\partial U})) = (H_\epsilon(F), F)$ and, since $M = M_3$ was formed from M_2 by attaching $(k+1)$-handles along a basis for G, it follows that $K_{k+1}(\widetilde{M}_0, \widetilde{\partial U}) = G$. Hence

$$\sigma(f, \overline{f}) = [\Sigma(f, \overline{f}, \omega, \overline{q})] = [H_\epsilon(F); F, G] = \omega,$$

which is what we wanted to prove. □

Remark 9.117 (Non-realisability of surgery obstructions by odd-dimensional closed manifolds) It is not true that for any closed oriented manifold N of dimension $n = 2k+1$ with fundamental group π and orientation homomorphism $w\colon \pi \to \{\pm 1\}$ and any element $\omega \in L_n(\mathbb{Z}\pi, w)$ there is a normal map $(f, \overline{f})\colon TM \oplus \underline{\mathbb{R}^a} \to \xi$ covering a map of closed oriented manifolds $f\colon M \to N$ of degree one such that $\sigma(f, \overline{f}) = \omega$. Note that in Theorem 8.195 the target manifold $N \times [0, 1]$ is not closed. We refer the reader to [171] for more information.

Remark 9.118 We remind the reader about the generalisation of Theorem 8.195 to Theorem 8.203. We will later use the evident fact that the analogous extension of Theorem 9.115 is true.

9.7 Notes

Surgery in odd dimensions was studied by Kervaire and Milnor in [216], where the target was a sphere and it was implicitly shown that $L_{2k+1}(\mathbb{Z}) = 0$ by mostly geometric means. Another geometric treatment in the simply connected case appears in [55]. As already mentioned in the introduction, the treatment of surgery obstructions using automorphisms of quadratic forms appeared in [414, Chapters 6,8] and the treatment of surgery obstructions via formations in [352].

An equivalent geometric approach to the surgery groups and the surgery obstruction is presented in Chapter 13.

Chapter 10
Decorations and the Simple Surgery Obstruction

10.1 Introduction

The goal of this chapter is to modify the L-groups and the surgery obstructions so that the surgery obstruction is the obstruction to achieving a *simple* homotopy equivalence and not just a homotopy equivalence. This will be important since at one point we want to apply the s-Cobordism Theorem 2.1. In this chapter we work in the smooth category unless explicitly stated otherwise.

Guide 10.1 For this chapter one needs to understand the notion of Whitehead torsion as presented in Chapter 3 and the L-groups and the surgery obstructions as presented in Chapters 8 and 9.

For the remainder of this book it would suffice to be aware of the fact, without knowing any details, that there are versions of the L-groups and the surgery obstructions to converting a normal map with a simple Poincaré complex as target to a simple homotopy equivalence.

10.2 Decorated L-Groups

We begin with the L-groups. Let R be a ring with involution. Suppose that we have specified a subgroup $U \subset K_1(R)$ such that U is closed under the involution (3.25) on $K_1(R)$ coming from the involution of R and contains the image of the change of ring homomorphism $K_1(\mathbb{Z}) \to K_1(R)$.

Two bases B and B' for the same finitely generated free R-module V are called *U-equivalent* if the change of bases matrix defines an element in $K_1(R)$ which belongs to U. Note that the U-equivalence class of a basis B is unchanged if we permute the order of elements of B. An R-module V is called *U-based* if V is finitely generated free and we have chosen a U-equivalence class of bases.

Let V be a *stably finitely generated free* R-module, i.e, an R-module V such that there exist natural numbers a, b satisfying $V \oplus R^a \cong_R R^b$. Note that there are stably finitely generated free R-modules that are not finitely generated free.

Exercise 10.2 Let R be the commutative ring $C^\infty(S^2)$ given by the smooth functions from S^2 to $\mathbb{R}$. The real vector space of smooth sections $V := \Gamma^\infty(TS^2 \to S^2)$ of the tangent bundle TS^2 has an obvious R-module structure. Show that V is not finitely generated free, but $V \oplus R \cong_R R^3$.

A *stable basis* for a stably finitely generated free R-module V is a basis B for $V \oplus R^u$ for some integer $u \geq 0$. Denote for any integers a, u the direct sum of the basis B and the standard basis S^a for R^a by $B \coprod S^a$, which is a basis for $V \oplus R^{u+a}$. Let C be a basis for $V \oplus R^v$ for some integer $v \geq 0$. We call the stable bases B and C of V *stably U-equivalent* if there is an integer $w \geq u, v$ such that $B \coprod S^{w-u}$ and $C \coprod S^{w-v}$ are U-equivalent bases for $V \oplus R^w$. We call an R-module V *stably U-based* if V is stably finitely generated free and we have specified a stable U-equivalence class of stable bases for V.

Let V and W be stably U-based R-modules. Let $f: V \xrightarrow{\cong} W$ be an R-isomorphism. Choose a non-negative integer c together with bases for $V \oplus R^c$ and $W \oplus R^c$ that represent the given stable U-equivalence classes of bases for V and W. Let A be the matrix of $f \oplus \mathrm{id}_{R^c}: V \oplus R^c \xrightarrow{\cong} W \oplus R^c$ with respect to these bases. It defines an element $[A]$ in $K_1(R)$. Define the *U-torsion*

$$\tau^U(f) \in K_1(R)/U \qquad (10.3)$$

by the class represented by $[A]$. One easily checks that $\tau^U(f)$ is independent of the choices of c and the basis representing the stable U-equivalence classes of stable bases and depends only on f and the stable U-equivalence class of bases for V and W. Moreover, one easily verifies

$$\tau^U(g \circ f) = \tau^U(g) + \tau^U(f); \qquad (10.4)$$

$$\tau^U \begin{pmatrix} f & 0 \\ u & v \end{pmatrix} = \tau^U(f) + \tau^U(v); \qquad (10.5)$$

$$\tau^U(\mathrm{id}_V) = 0, \qquad (10.6)$$

for R-isomorphisms $f: V_0 \xrightarrow{\cong} V_1$, $g: V_1 \xrightarrow{\cong} V_2$, and $v: V_3 \xrightarrow{\cong} V_4$, and an R-homomorphism $u: V_0 \to V_4$ for stably U-based R-modules V_i.

Exercise 10.7 Let V be a stably U-based R-module. Show that the dual R-module V^* is stably finitely generated free and inherits a preferred stable U-equivalence class of stable bases.

Show for two U-stably based R-modules V and W and an R-isomorphism $f: V \to W$ that $\tau(f^*: W^* \to V^*)$ is the image of $\tau^U(f: V \to W)$ under the involution $*: K_1(R)/U \to K_1(R)/U$.

10.2 Decorated L-Groups

A chain complex C_* is called *stably U-based finite* if it is finite and each chain module is stably U-based. Let C_* be a contractible stably U-based finite R-chain complex. The definition of *Reidemeister torsion* in (3.16) carries over to the definition of the *Reidemeister U-torsion*

$$\Delta^U(C_*) = [A] \quad \in K_1(R)/U. \tag{10.8}$$

Similarly we can associate to an R-chain homotopy equivalence $f \colon C_* \to D_*$ of stably U-based finite R-chain complexes its *Whitehead U-torsion*, cf. Definition 3.17,

$$\tau^U(f_*) := \Delta^U(\mathrm{cone}_*(f_*)) \quad \in K_1(R)/U. \tag{10.9}$$

Lemma 3.18 carries over to U-torsion in the obvious way.

We will consider *stably U-based non-singular ϵ-quadratic forms* (P, ψ), i.e., non-singular ϵ-quadratic forms whose underlying R-module P is a stably U-based R-module such that the U-torsion of the isomorphism $(1 + \epsilon \cdot T)(\psi) \colon P \xrightarrow{\cong} P^*$ is zero in $K_1(R)/U$, where P^* inherits a stable U-equivalence class of stable bases from the one on P in the obvious way. An isomorphism $f \colon (P, \psi) \to (P', \psi')$ of stably U-based non-singular ϵ-quadratic forms is *U-simple* if the U-torsion of $f \colon P \to P'$ vanishes in $K_1(R)/U$. Given a stably U-based R-module P, the standard hyperbolic ϵ-quadratic form $H_\epsilon(P)$ inherits the structure of a stably U-based non-singular ϵ-quadratic form. The sum of two stably U-based non-singular ϵ-quadratic forms is again a stably U-based non-singular ϵ-quadratic form. It is worthwhile to mention the following U-simple version of Lemma 8.95.

Lemma 10.10 *Let (P, ψ) be a stably U-based non-singular ϵ-quadratic form. Let $L \subset P$ be a lagrangian such that L is stably U-based and the U-torsion of the following 2-dimensional stably U-based finite R-chain complex*

$$0 \to L \xrightarrow{i} P \xrightarrow{i^* \circ (1 + \epsilon \cdot T)(\psi)} L^* \to 0$$

vanishes in $K_1(R)/U$.

Then the inclusion $i \colon L \to P$ extends to a U-simple isomorphism of stably U-based non-singular ϵ-quadratic forms

$$H_\epsilon(L) \xrightarrow{\cong} (P, \psi).$$

Next we give the simple version of the even-dimensional L-groups.

Definition 10.11 (Decorated quadratic L-groups in even dimensions) Let R be an associative ring with unit and involution. Let k be an integer. Define the $2k$-th U-decorated quadratic L-group of R

$$L^U_{2k}(R)$$

to be the abelian group of equivalence classes $[(P, \psi)]$ of stably U-based non-singular $(-1)^k$-quadratic forms (P, ψ) with respect to the following equivalence relation.

We call (P, ψ) and (P', ψ') equivalent if and only if there exist stably U-based R-modules M and M' and a U-simple isomorphism of stably U-based non-singular $(-1)^k$-quadratic forms

$$(P, \psi) \oplus H_{(-1)^k}(M) \xrightarrow{\cong} (P', \psi') \oplus H_{(-1)^k}(M').$$

Addition is given by the sum of two stably U-based non-singular $(-1)^k$-quadratic forms. The zero element is represented by $[H_\epsilon(M)]$ for any stably U-based R-module M. The inverse of $[F, \psi]$ is given by $[F, -\psi]$.

Next we define the odd-dimensional version. For $\epsilon \in \{\pm 1\}$, define a *stably U-based ϵ-quadratic formation* $(P, \psi; F, G)$ to be an ϵ-quadratic formation $(P, \psi; F, G)$ such that (P, ψ) is a stably U-based non-singular ϵ-quadratic form, the lagrangians F and G are stably U-based R-modules and the U-torsion of the following two contractible stably U-based finite R-chain complexes

$$0 \to F \xrightarrow{i} P \xrightarrow{i^* \circ (1 + \epsilon \cdot T)(\psi)} F^* \to 0$$

and

$$0 \to G \xrightarrow{j} P \xrightarrow{j^* \circ (1 + \epsilon \cdot T)(\psi)} G^* \to 0$$

vanish in $K_1(R)/U$, where $i: F \to P$ and $j: G \to P$ denote the inclusions. An isomorphism $f: (P, \psi; F, G) \to (P', \psi'; F', G')$ of stably U-based ϵ-quadratic formations is U-simple if the U-torsion of the induced R-isomorphisms $P \xrightarrow{\cong} P'$, $F \xrightarrow{\cong} F'$ and $G \xrightarrow{\cong} G'$ vanishes in $K_1(R)/U$. Given a stably U-based R-module P, the trivial ϵ-quadratic formation $(H_\epsilon(P); P, P^*)$ inherits a U-equivalence of bases. Given a stably U-based $(-\epsilon)$-quadratic form (Q, ψ), its boundary $\partial(Q, \psi)$ is a stably U-based ϵ-quadratic formation. Obviously the sum of two stably U-based ϵ-quadratic formations is again a stably U-based ϵ-quadratic formation. Next we give the simple version of the odd-dimensional L-groups.

Definition 10.12 (Decorated quadratic L-groups in odd dimensions) Let R be an associative ring with unit and involution. Let k be an integer. Define the $(2k + 1)$-th U-decorated quadratic L-group of R

$$L^U_{2k+1}(R)$$

to be the abelian group of equivalence classes $[(P, \psi; F, G)]$ of stably U-based $(-1)^k$-quadratic formations $(P, \psi; F, G)$ with respect to the following equivalence relation. We call two stably U-based ϵ-quadratic formations $(P, \psi; F, G)$ and $(P', \psi'; F', G')$ equivalent if there exist stably U-based $(-(-1)^k)$-quadratic forms (Q, μ) and (Q', μ') and stably U-based R-modules M and M', with a U-simple isomorphism of stably U-based ϵ-quadratic formations

$$(P, \psi; F, G) \oplus \partial(Q, \mu) \oplus (H_\epsilon(M); M, M^*)$$
$$\cong (P', \psi'; F', G') \oplus \partial(Q', \mu') \oplus (H_\epsilon(M'); M', (M')^*).$$

10.2 Decorated L-Groups

Addition is given by the sum of two stably U-based $(-1)^k$-quadratic formations. The zero element is represented by $\partial(Q, \mu) \oplus (H_\epsilon(M); M, M^*)$ for any U-based $(-(-1)^k)$-quadratic form (Q, μ) and stably U-based R-module M. The inverse of $[(P, \psi; F, G)]$ is represented by $(P, -\psi; F', G')$ for any choice of stably U-based lagrangians F' and G' in $H_\epsilon(P)$ such that F and F' are complementary and G and G' are complementary and the U-torsion of the obvious isomorphism $F \oplus F' \xrightarrow{\cong} P$ and $G \oplus G' \xrightarrow{\cong} P$ vanishes in $K_1(R)/U$.

Notation 10.13 If $R = \mathbb{Z}\pi$ with the w-twisted involution and $T \subset K_1(\mathbb{Z}\pi)$ is the abelian group of elements of the form $(\pm g)$ for $g \in \pi$, then we write

$$L_n^s(\mathbb{Z}\pi, w) := L_n^T(\mathbb{Z}\pi);$$
$$L_n^h(\mathbb{Z}\pi, w) := L_n(\mathbb{Z}\pi),$$

where $L_n(\mathbb{Z}\pi)$ is the L-group introduced in Definitions 8.90 and 9.15 and $L_n^T(\mathbb{Z}\pi)$ is the L-group introduced in Definitions 10.11 and 10.12 for $R = \mathbb{Z}\pi$ with the w-twisted involution. The L-groups $L_n^s(\mathbb{Z}\pi, w)$ are called the *simple quadratic L-groups*.

Lemma 8.152 and Lemma 9.38 carry directly over to $L_n^s(\mathbb{Z}\pi, w)$. Thus we can define analogues $L_n^s(\mathbb{Z}\Pi(X); O_X)$ of $L_n(\mathbb{Z}\Pi(X); O_X)$ exactly as explained in Subsection 8.7.3.

Remark 10.14 Let $\pi_{ab} = \pi/[\pi, \pi]$ be the abelianisation of π. The subgroup $T \subseteq K_1(\mathbb{Z}\pi)$ agrees with the image of the map

$$A: \pi_{ab} \times \{\pm 1\} \to K_1(\mathbb{Z}\pi)$$

which sends $([g], \epsilon) \in \pi_{ab} \times \{\pm 1\}$ to the class in $K_1(\mathbb{Z}\pi)$ given by the trivial unit $\epsilon \cdot g$. The map A is injective, see [266, Lemma 2 in Section 2.2.1 on page 709].

Definition 10.15 (Intrinsic simple L-groups) We denote by

$$(L_n^s)_*: \mathcal{R} \to \mathbb{Z}\text{-MOD}$$

the covariant functor coming from the simply quadratic L-groups and write

$$L_n^s(\mathbb{Z}\Pi(X), O_X) := (L_n^s)_*(X, O_X).$$

We also sometimes write

$$L_n^h(\mathbb{Z}\Pi(X), O_X) := L_n(\mathbb{Z}\Pi(X), O_X)$$

for the intrinsic L-groups $L_n(\mathbb{Z}\Pi(X), O_X)$ of Definitions 8.158 and 9.106.

Exercise 10.16 Show that the obvious homomorphism $L_n(\mathbb{Z}) \to L_n^s(\mathbb{Z})$ is a bijection.

10.3 Reidemeister U-Torsion

We will need the following modification of torsion invariants.

Let C_* be a projective R-chain complex such that $H_i(C_*)$ is projective for $i \in \mathbb{Z}$. Denote by $H_*(C_*)$ the R-chain complex whose i-th chain module is $H_i(C_*)$ and whose differentials are all trivial. Then there is up to R-chain homotopy precisely one R-chain map

$$j(C_*)_* \colon H_*(C_*) \to C_* \qquad (10.17)$$

such that $H_i(j(C_*))$ is the identity on $H_i(C_*)$. Namely, for each $i \in \mathbb{Z}$ we can choose an R-map $j(C_*)_i \colon H_i(C_*) \to \ker(c_i)$ whose composite with the projection $\ker(c_i) \to H_i(C_*)$ is the identity. Then the collection of the $j(C_*)_i$ yields the desired R-chain map $j(C_*)_*$. Let $j(C_*)'_* \colon H_*(C_*) \to C_*$ be another R-chain map with $H_i(j'_*) = \mathrm{id}_{H_i(C_*)}$. Then $j(C_*)_i - j(C_*)'_i \colon H_i(C_*) \to C_i$ has image in $\mathrm{im}(c_{i+1})$. Since $H_i(C_*)$ is projective, we can find an R-map $\gamma(C_*)_i \colon H_i(C_*) \to C_{i+1}$ with $c_{i+1} \circ \gamma(C_*)_i = j(C_*)_i - j(C_*)'_i$. The collection of the $\gamma(C_*)_i$ defines an R-chain homotopy $\gamma(C_*)_* \colon j(C_*)_i \simeq j(C_*)'_i$.

Given an R-chain complex C_*, we say that its homology $H_*(C_*)$ is *stably U-based* if $H_i(C_*)$ is stably U-based, i.e., is finitely generated stably free and comes with a preferred stable U-equivalence class of stable bases, for each $i \geq 0$.

Definition 10.18 (Reidemeister U-torsion of a (not necessarily acyclic) chain complex) Let C_* be a stably U-based finite R-chain complex. Suppose that its homology is stably U-based. Let $j(C_*)_* \colon H_*(C_*) \to C_*$ be an R-chain map inducing the identity on the homology, see (10.17). Define the *Reidemeister U-torsion*

$$\Delta^U(C_*) := \tau^U(j(C)_*)$$

to be the Whitehead U-torsion of the R-chain homotopy equivalence $j(C_*)_*$ in the sense of (10.9).

Obviously Definition 10.18 extends (10.8).

Let $0 \to C_* \xrightarrow{i_*} D_* \xrightarrow{q_*} E_* \to 0$ be an exact sequence of stably U-based finite R-chain complexes. Let $\Delta^U(C_p, D_p, E_p)$ be the Reidemeister U-torsion in the sense of (10.8) of the contractible stably U-based finite R-chain complex concentrated in dimension 2, 1 and 0 given by $\cdots \to 0 \to C_p \xrightarrow{i_p} D_p \xrightarrow{q_p} E_p \to 0$. Define

$$\Delta^U(C_*, D_*, E_*) := \sum_{p \geq 0} (-1)^p \cdot \Delta^U(C_p, D_p, E_p). \qquad (10.19)$$

The long exact homology sequence associated to the short exact sequence $0 \to C_* \xrightarrow{i_*} D_* \xrightarrow{q_*} E_* \to 0$ yields a contractible stable U-based finite R-chain complex $LHS(C_*, D_*, E_*)$

10.3 Reidemeister U-Torsion

$$\cdots \to H_1(C_*) \xrightarrow{H_1(i_*)} H_1(D_*) \xrightarrow{H_1(q_*)} H_1(E_*) \xrightarrow{\partial_1} H_0(C_*)$$
$$\xrightarrow{H_0(i_*)} H_0(D_*) \xrightarrow{H_0(q_*)} H_0(E_*) \to 0 \to 0 \to \cdots \quad (10.20)$$

where $H_0(E_*)$ sits in dimension 0. Define

$$\Delta^U(LHS(C_*, D_*, E_*)) := \tau^U(LHS(C_*, D_*, E_*)) \quad (10.21)$$

to be its Reidemeister U-torsion in the sense of (10.8).

Lemma 10.22 *Let C_* and D_* be stably U-based R-chain complexes with stably U-based homology. Let $f_* : C_* \to D_*$ be an R-chain homotopy equivalence. Let $\tau^U(f_*)$ and $\tau^U(H_*(f_*))$ be the Whitehead U-torsion of f_* and $H_*(f_*)$ in the sense of (10.9). Let $\Delta^U(C_*)$ and $\Delta^U(D_*)$ be the Reidemeister U-torsion of C_* and D_* in the sense of Definition 10.18.*

Then we get in $K_1(R)/U$

$$\Delta^U(D_*) - \Delta^U(C_*) = \tau^U(f_*) - \tau^U(H_*(f_*));$$
$$\tau^U(H_*(f_*)) = \sum_{i \geq 0} (-1)^i \cdot \tau^U\big(H_i(f_*) \colon H_i(C_*) \to H_i(D_*)\big).$$

Proof. Consider the following up to R-chain homotopy commutative diagram of R-chain homotopy equivalences of stably U-based finite R-chain complexes

$$\begin{array}{ccc} H_*(C_*) & \xrightarrow{H_*(f_*)} & H_*(D_*) \\ {\scriptstyle j(C_*)_*}\downarrow & & \downarrow {\scriptstyle j(D_*)_*} \\ C_* & \xrightarrow{f_*} & D_*. \end{array}$$

We get from the analogue of Lemma 3.18 (ii) and (iii)

$$\tau^U(f_*) + \tau^U(j(C_*)_*) = \tau^U(j(D_*)_*) + \tau^U(H_*(f_*))$$

and from the analogue of Lemma 3.18 (i)

$$\tau^U(H_*(f_*)) = \sum_{i \geq 0} (-1) \cdot \tau^U\big(H_i(f_*) \colon H_i(C_*) \to H_i(D_*)\big).$$

This completes the proof. □

Lemma 10.23 *Let $0 \to C_* \xrightarrow{i_*} D_* \xrightarrow{q_*} E_* \to 0$ be an exact sequence of stably U-based finite R-chain complexes with stably U-based homology. Then we get in $K_1(R)/U$*

$$\Delta^U(C_*) - \Delta^U(D_*) + \Delta^U(E_*) = \Delta^U(C_*, D_*, E_*) - \Delta^U(LHS(C_*, D_*, E_*)).$$

Proof. We only give the proof in what is for us the most important special case, where the boundary map $\partial_i \colon H_i(E_*) \to H_{i-1}(C_*)$ vanishes for all $i \geq 0$. The proofs in [254, Theorem 3.35 on page 142] and in [302, Theorem 3.2] carry over to our setting in the general case.

We begin with the case when C_*, D_* and E_* are contractible. By taking the direct sum with *elementary chain complexes*, i.e., chain complexes concentrated in two consecutive dimensions and the identity as the only non-trivial differential, we can reduce the claim to the case where C_*, D_* and E_* are finite free R-chain complexes and the stable U-equivalence classes of stable bases are represented by actual bases. Moreover, we can construct a commutative diagram of based finite free R-chain complexes

$$\begin{array}{ccccccccc} 0 & \longrightarrow & C_* & \longrightarrow & D_* & \longrightarrow & E_* & \longrightarrow & 0 \\ & & \downarrow \mathrm{id} & & \downarrow \mathrm{id} & & \downarrow \mathrm{id} & & \\ 0 & \longrightarrow & C_* & \longrightarrow & D'_* & \longrightarrow & E_* & \longrightarrow & 0 \end{array}$$

such that the lower exact sequence is based exact where D'_* has D_* as underlying R-chain complex, but comes with a different basis. Lemma 3.18 implies

$$\Delta^U(D'_*) = \Delta^U(C_*) + \Delta^U(E_*);$$
$$\Delta^U(D'_*) - \Delta^U(D_*) = \sum_{i \geq 0}(-1)^i \cdot \tau^U(\mathrm{id} \colon D_i \to D'_i);$$
$$\Delta^U(C_*, D_*, E_*) = \sum_{i \geq 0}(-1)^i \cdot \tau^U(\mathrm{id} \colon D_i \to D'_i),$$

and the claim follows in the case when C_*, D_* and E_* are acyclic.

Now suppose that $\partial_i \colon H_i(E_*) \to H_{i-1}(C_*)$ vanishes for all $i \geq 0$. Then we can arrange the chain maps $i(C_*)$, $i(D_*)$ and $i(E_*)$ such that we obtain a commutative diagram of stably U-based finite R-chain complexes with exact rows and R-chain homotopy equivalences as vertical maps:

$$\begin{array}{ccccccccc} 0 & \longrightarrow & H_*(C_*) & \xrightarrow{H_*(i_*)} & H_*(D_*) & \xrightarrow{H_*(q_*)} & H_*(E_*) & \longrightarrow & 0 \\ & & \downarrow j(C_*)_* & & \downarrow j(D_*)_* & & \downarrow j(E_*)_* & & \\ 0 & \longrightarrow & C_* & \xrightarrow{i_*} & D_* & \xrightarrow{q_*} & E_* & \longrightarrow & 0. \end{array}$$

Taking mapping cones yields the short exact sequence of contractible stably U-based finite R-chain complexes

$$0 \to \mathrm{cone}(j(C_*)) \xrightarrow{j'_*} \mathrm{cone}(j_*(D_*)) \xrightarrow{q'_*} \mathrm{cone}(j_*(E_*)) \to 0,$$

for which we already know

10.3 Reidemeister U-Torsion

$$\Delta^U(\text{cone}(j(C_*)_*)) - \Delta^U(j(D_*)_*) + \Delta^U(\text{cone}(j(E_*)_*))$$
$$= \Delta^U(\text{cone}(j(C_*)_*), \text{cone}(j(D_*)_*, \text{cone}(j(E_*)_*)).$$

One easily checks

$$\Delta^U(\text{cone}(j(C_*)_*), \text{cone}(j(D_*)_*, \text{cone}(j(E_*)_*))$$
$$= \Delta^U(C_*, D_*, E_*) - \Delta^U(LHS(C_*, D_*, E_*)).$$

Since by definition

$$\Delta^U(C_*) = \tau^U(\text{cone}(j(C_*)_*));$$
$$\Delta^U(D_*) = \tau^U(\text{cone}(j(D_*)_*));$$
$$\Delta^U(E_*) = \tau^U(\text{cone}(j(E_*)_*)),$$

the claim follows in the case where $\partial_i \colon H_i(E_*) \to H_{i-1}(C_*)$ vanishes for all $i \geq 0$. □

Exercise 10.24 Let C_* be a finite $\mathbb{Z}$-chain complex. Choose for any $i \in \mathbb{Z}$ a $\mathbb{Z}$-basis for $C_i/\text{tors}(C_i)$ and $H_i(C_*)/\text{tors}(H_i(C_*))$. They induce $\mathbb{Q}$-bases on $\mathbb{Q} \otimes_{\mathbb{Z}} C_i$ and $H_i(\mathbb{Q} \otimes_{\mathbb{Z}} C_*)$ by the canonical $\mathbb{Q}$-isomorphisms

$$\mathbb{Q} \otimes_{\mathbb{Z}} C_i \xrightarrow{\cong} \mathbb{Q} \otimes_{\mathbb{Z}} C_i/\text{tors}(C_i);$$
$$\mathbb{Q} \otimes_{\mathbb{Z}} H_i(C_*) \xrightarrow{\cong} \mathbb{Q} \otimes_{\mathbb{Z}} H_i(C_*)/\text{tors}(H_i(C_*));$$
$$\mathbb{Q} \otimes_{\mathbb{Z}} H_i(C_*) \xrightarrow{\cong} H_i(\mathbb{Q} \otimes_{\mathbb{Z}} C_i).$$

Put $U = \{\pm 1\}$. Show that these $\mathbb{Q}$-bases are unique up to stable U-equivalence. Hence $\mathbb{Q} \otimes_{\mathbb{Z}} C_*$ is a stably U-based finite $\mathbb{Q}$-chain complex with stably U-based homology and $\Delta^U(\mathbb{Q} \otimes_{\mathbb{Z}} C_*) \in K_1(\mathbb{Q})/U$ is defined, see Definition 10.18.

Show that we obtain an isomorphism

$$D \colon K_1(\mathbb{Q})/U \xrightarrow{\cong} \mathbb{Q}^{>0}, \quad [A] \mapsto |\det(A)|$$

where $\mathbb{Q}^{>0}$ is the multiplicative abelian group of positive rational numbers and that we get

$$D(\Delta^U(\mathbb{Q} \otimes_{\mathbb{Z}} C_*)) = \prod_{i \geq 0} |\text{tors}(H_i(C_*))|^{(-1)^i}.$$

Note that, for any stably U-based R-module V and element $x \in K_1(R)/U$, we can find another stable U-equivalence class of stable bases C for V with the property that the U-torsion $\tau^U(\text{id} \colon (V, B) \to (V, C))$ is x. This is not true in the unstable setting. For instance, there exists a ring R with an element $x \in K_1(R)/U$ for U the image of $K_1(\mathbb{Z}) \to K_1(R)$ such that x cannot be represented by a unit in R, in other words, x is not the U-torsion of any R-automorphism of R, see for instance [324, Example 5.1 on page 127].

Now Lemma 8.52 (iv) has the following version in the simple homotopy setting.

Lemma 10.25 *Let D_* be a stably U-based finite R-chain complex. Suppose for a fixed integer r that $H_i(D_*) = 0$ for $i \neq r$ and $H^{r+1}(\hom_R(D_*, V)) = 0$ for any R-module V.*

(i) *Then $H_r(D_*)$ is stably finitely generated free and inherits a preferred stable U-equivalence class of stable bases;*
(ii) *Let C_* be a stably U-based finite R-chain complex and $f_*: C_* \to D_*$ be an R-chain homotopy equivalence with $\tau^U(f_*) = 0$. Then $H_i(C_*) = 0$ for $i \neq r$ and $H^{r+1}(\hom_R(C_*, V)) = 0$ for any R-module V and hence C_* inherits a preferred U-equivalence class of stable bases by assertion (i). Moreover, the R-isomorphism $H_r(f_*): H_r(C_*) \to H_r(D_*)$ respects the preferred stable U-equivalence class of stable bases, i.e., $\tau^U(H_r(f_*)) = 0$ holds in $K_1(R)/U$.*

Proof. (i) We know from Lemma 8.52 (iv) that $H_r(D_*)$ is stably finitely generated free. Now equip it with the stable U-equivalence class of stable bases, for which $\Delta^U(D_*) = 0$ holds in $K_1(R)/U$, where we equip for $i \neq r$ the trivial R-module $H_i(D_*) = 0$ with the trivial basis. This is well defined by Lemma 10.22.
(ii) This follows from Lemma 10.22. □

10.4 The Simple Surgery Obstruction

Let $(X, \partial X, O_X, [[X, \partial X]])$ be a (not necessarily connected) simple w-oriented finite n-dimensional Poincaré pair in the sense of Definition 5.73 for $n \geq 5$. Consider a normal map $(f, \overline{f}) = (M, \partial M, f, \partial f, a, \xi, \overline{f}, o)$ of degree one relative boundary with target $(X, \partial X, O_X, [[X, \partial X]])$ in the sense of Subsection 8.8.1. Suppose additionally that $\partial f: \partial M \to \partial X$ is a simple homotopy equivalence.

Next we want to explain how the definition of the intrinsic surgery obstruction relative boundary of Definitions 8.189 and 9.113 can be modified to the simple setting. Note that the difference between the L-groups $L_n^h(\mathbb{Z}\Pi(X), O_X)$ and the simple L-groups $L_n^s(\mathbb{Z}\Pi(X), O_X)$ is the additional structure of a stable U-equivalence class of stable bases. We will explain below that the definition of the simple surgery obstruction

$$\sigma^s(f, \overline{f}) \in L_n^s(\mathbb{Z}\Pi(X), O_X) \tag{10.26}$$

is the same as in Section 8.8 and 9.5 except that we must specify how the various surgery kernels inherit a stable U-equivalence class of stable bases.

For the remainder of this section, we take $R = \mathbb{Z}\pi$ with the w-twisted involution and put

$$U = \{\pm g \mid g \in \pi\}.$$

We need at least $\{\pm g \mid g \in \pi\} \subseteq U$ since the Whitehead torsion of a homotopy equivalence is only well defined modulo the trivial units because the cellular structure of a CW-complex X with fundamental group π defines on the cellular $\mathbb{Z}\pi$-chain complex of its universal covering only a $\{\pm g \mid g \in \pi\}$-equivalence class of bases.

10.4 The Simple Surgery Obstruction

10.4.1 The Even-Dimensional Case

Next we can prove the following version of Lemma 8.55 (ii).

Lemma 10.27 *Suppose that X is connected with fundamental group π, the Poincaré pair $(X, \partial X)$ is simple, $f \colon M \to X$ is k-connected for $n = 2k$ and $\partial f \colon \partial M \to \partial X$ is a simple homotopy equivalence. Assume that we are in the universal covering case, see Notation 8.34. Then:*

(i) *The kernels $K_k(\widetilde{M})$, $K_k(\widetilde{M}, \widehat{\partial M})$, $K^k(\widetilde{M})$, and $K^k(\widetilde{M}, \widehat{\partial M})$ are stably finitely generated free $\mathbb{Z}\pi$-modules and inherit a preferred stable U-equivalence class of stable bases from the identifications of Lemma 8.128 (i) and from Lemma 10.25 applied to the based finite $\mathbb{Z}\pi$-chain complexes $\operatorname{cone}_*(C^!_*(\widetilde{f}))$, $\operatorname{cone}_*(C^!_*(\widetilde{f}, \widehat{\partial f}))$, $\operatorname{cone}^{n-*}(C^!_*(\widetilde{f}))$, and $\operatorname{cone}^{n-*} C^!_*(\widetilde{f}, \widehat{\partial f}))$;*

(ii) *The canonical $\mathbb{Z}\pi$-isomorphisms*

$$K_k(\widetilde{M}) \xrightarrow{\cong} K_k(\widetilde{M}, \widehat{\partial M});$$
$$K^k(\widetilde{M}) \xrightarrow{\cong} K_k(\widetilde{M})^*;$$
$$K^k(\widetilde{M}, \widehat{\partial M}) \xrightarrow{\cong} K_k(\widetilde{M}, \widehat{\partial M})^*;$$
$$K^k(\widetilde{M}, \widehat{\partial M}) \xrightarrow{\cong} K_k(\widetilde{M})^*;$$
$$K^k(\widetilde{M}) \xrightarrow{\cong} K_k(\widetilde{M}, \widehat{\partial M})^*,$$

given by the inclusion $\widetilde{M} \to (\widetilde{M}, \widehat{\partial M})$, the Kronecker pairings, and Poincaré duality have trivial U-torsion.

Proof. (i) The four $\mathbb{Z}\pi$-chain complexes $\operatorname{cone}_*(C^!_*(\widetilde{f}))$, $\operatorname{cone}_*(C^!_*(\widetilde{f}, \widehat{\partial f}))$, $\operatorname{cone}^{n-*}(C^!_*(\widetilde{f}))$, and $\operatorname{cone}^{n-*} C^!_*(\widetilde{f}, \widehat{\partial f}))$ inherit U-equivalence classes of bases from the structures of CW-pairs on $(M, \partial M)$ and $(X, \partial X)$ using additionally the fact that a homeomorphism is a simple homotopy equivalence, see [94]. (In the smooth case it suffices to use the easier to prove fact that two smooth triangulations of a closed smooth manifold determine the same simple homotopy type, see [302, Theorem 7.1 on page 378 and Section 9]). Since $K_j(\widetilde{M})$, $K_j(\widetilde{M}, \widehat{\partial M})$, $K^j(\widetilde{M})$, and $K^j(\widetilde{M}, \widehat{\partial M})$ vanish for $j \neq k$, one can indeed apply Lemma 10.25 to obtain stable U-bases on $H_k(\operatorname{cone}_*(C^!_*(\widetilde{f})))$, $H_k(\operatorname{cone}_*(C^!_*(\widetilde{f}, \widehat{\partial f})))$, $H^k(\operatorname{cone}^{n-*}(C^!_*(\widetilde{f})))$, and $H^k(\operatorname{cone}^{n-*}(C^!_*(\widetilde{f}, \widehat{\partial f})))$. Now equip $K_k(\widetilde{M})$, $K_k(\widetilde{M}, \widehat{\partial M})$, $K^k(\widetilde{M})$, and $K^k(\widetilde{M}, \widehat{\partial M})$ with the stable U-bases for which the canonical isomorphisms of Lemma 8.128 (i) have vanishing U-torsion.

(ii) This assertion follows essentially from Lemma 10.22 and Lemma 10.23, and the assumption that ∂f is a simple homotopy equivalence and $(X, \partial X)$ is a simple Poincaré pair. □

Remark 10.28 (Simple surgery obstruction in even dimensions) We can define the surgery obstruction against achieving a simple homotopy equivalence as before in Subsection 8.8.3, but now taking the preferred stable U-equivalence class of stable bases on the surgery kernels coming from Lemma 10.27 (i) into account. The proof that the simple surgery obstruction is well defined is analogous if one uses Lemma 10.27 (ii).

10.4.2 The Odd-Dimensional Case

The odd-dimensional surgery obstructions are defined in Chapter 9 using formations. The underlying modules of a formation associated to a k-connected degree one normal map $(f, \overline{f}) \colon (M, \partial M) \to (X, \partial X)$ in dimension $(2k + 1)$ inherit stable U-bases from the CW-structures of $(M, \partial M)$ and $(X, \partial X)$ in the same way as described in the previous section since they are in fact obtained either as the surgery kernel groups of bordisms of even-dimensional problems (the kernel formation) or as the surgery kernel groups of even-dimensional maps with boundary (the bordism formation). Moreover, the structure maps of these formations are such that with these choices of stable U-bases these formations are stably U-based quadratic formations. In the process of constructing these formations several choices are made, so it needs to be checked that the stable U-equivalence class of the associated stable U-based formations do not depend on these choices. This is again done using the bordism invariance as in Section 9.5 and Lemma 10.27 (ii).

Remark 10.29 (Simple surgery obstruction in odd dimensions) We can define the surgery obstruction against achieving a simple homotopy equivalence as before in Section 9.5, but now taking the preferred stable U-equivalence class of stable bases on the surgery kernels coming from Lemma 10.27 (i) into account, as described in the previous paragraph. The proof that the simple surgery obstruction is well defined is analogous if one uses Lemma 10.27 (ii).

10.4.3 Main Properties of the Simple Surgery Obstruction

Next we can give the simple version of the surgery obstruction theorem. Its proof is analogous to that of Theorems 8.189 and 9.113 if we take Lemma 10.27 (ii) into account.

Theorem 10.30 (Simple surgery obstruction for manifolds with boundary) *Let $(X, \partial X, O_X, [[X, \partial X]])$ be a (not necessarily connected) simple w-oriented finite n-dimensional Poincaré pair in the sense of Definition 5.73 for $n \geq 5$. Consider a normal map with respect to the tangent bundle of degree one relative boundary $(f, \overline{f}) = (M, \partial M, f, \partial f, a, \xi, \overline{f}, o)$ with target $(X, \partial X, O_X, [[X, \partial X]])$ in the sense of Subsection 8.8.1. Suppose additionally that $\partial f \colon \partial M \to \partial X$ is a simple homotopy equivalence. Then:*

10.4 The Simple Surgery Obstruction

(i) *The simple surgery obstruction*

$$\sigma^s(f,\overline{f}) = \sigma^s(M,\partial M, f, \partial f, a, \xi, \overline{f}, o) \in L_n^s(\mathbb{Z}\Pi(X), O_X)$$

depends only on the simple normal bordism with cylindrical ends class of $(f, \overline{f})$, where simple means that in the definition of normal bordism of Subsection 8.8.2 we require additionally that the map $\partial_2 F$ is a simple homotopy equivalence and $(X, \partial X)$ is a simple Poincaré pair;

(ii) *We have $\sigma^s(f, \overline{f}) = 0$ in $L_n^s(\mathbb{Z}\Pi(X), O_X)$ if and only if we can do a finite number of surgery steps on the interior of M without changing ∂f to arrange that f is a simple homotopy equivalence.*

Theorem 8.195 and 9.115 carry over to the simple setting as follows.

Theorem 10.31 (Realisation of simple-dimensional surgery obstructions) *Suppose $n \geq 5$. Consider a connected compact oriented $(n-1)$-manifold N with possibly empty boundary ∂N. Fix an element*

$$\omega \in L_n^s(\mathbb{Z}\Pi(N \times I), O_{N \times I}).$$

Then there is a normal map of pairs of degree one relative boundary $(M, \partial M, f, \partial f, a, \xi, \overline{f}, o)$ with target $(N \times I, \partial(N \times I), O_{N \times I}, [[N \times I, \partial(N \times I)]])$ for $I = [0, 1]$ with the following properties:

(i) *The bundle ξ over $N \times I$ is $T(N \times I) \oplus \underline{\mathbb{R}^a}$ and the isomorphism o of infinite cyclic local coefficient systems over $N \times I$ is the identity;*

(ii) *There is a decomposition $\partial M = \partial_0 M \cup \partial_1 M \cup \partial_2 M$ where $\partial_m M$ is a codimension zero submanifold of ∂M with possibly empty boundary $\partial \partial_m M$ for $m = 0, 1, 2$ satisfying*

$$\partial_0 M \cap \partial_1 M = \emptyset;$$
$$\partial_2 M \cap \partial_m M = \partial \partial_m M \quad \text{for } m = 0, 1;$$
$$\partial \partial_2 M = \partial \partial_0 M \amalg \partial \partial_1 M;$$
$$\partial_2 M = \partial N \times [0, 1];$$
$$\partial_0 M = N \times \{0\};$$

(iii) *The simple homotopy equivalence $\partial f \colon \partial M \to \partial(N \times I)$ induces the identity $\partial_0 M = N \times \{0\} \to N \times \{0\}$ on $\partial_0 M$ and the identity $\partial_2 M = \partial N \times [0, 1] \to \partial N \times I$ on $\partial_2 M$;*

(iv) *The simple homotopy equivalence $\partial f \colon \partial M \to \partial(N \times I)$ maps $\partial_1 M$ to $N \times \{1\}$ and induces a simple homotopy equivalence*

$$\partial_1 f \colon \partial_1 M \to N \times \{1\};$$

(v) *The simple intrinsic surgery obstruction*

$$\sigma^s(M, \partial M, f, \partial f, a, \xi, \overline{f}, o) \in L^s_{2k}(\mathbb{Z}\Pi(N \times I), \mathcal{O}_{N \times I})$$

in the sense of Theorem 10.30 is the given element ω.

Remark 10.32 Theorem 8.168 and Theorem 9.109 about the bordism invariance for not necessarily cylindrical targets carry over to the simple setting in the obvious way. This is also true for their versions relative boundary.

10.5 Notes

We will later compare the L-groups $L^h_n(\mathbb{Z}\pi, w)$ and $L^s_n(\mathbb{Z}\pi, W)$ when we deal with the Rothenberg sequence in Section 13.5.

Chapter 11
The Geometric Surgery Exact Sequence

11.1 Introduction

In this chapter we introduce the surgery exact sequence, see Theorem 11.22 and Remark 11.23. It is the realisation of the Surgery Program, which we have explained in Remark 2.9. The surgery exact sequence is the main theoretical tool in solving the classification problem of manifolds of dimensions greater than or equal to five. To get a first impression of the potential of the surgery exact sequence, we recommend studying Chapter 12. Other calculations and applications can be found in Chapters 18 and 19.

In this chapter we work in the smooth category unless explicitly stated otherwise.

Guide 11.1 One may well start reading this book with this chapter and then work back through some of the material of the previous chapters when needed.

11.2 The Geometric Structure Set

Definition 11.2 (The simple geometric structure set) Let X be a compact manifold of dimension n with possibly empty boundary ∂X.

(i) A *simple manifold structure relative boundary on X*, or briefly a *simple manifold structure on X*, is a map of pairs $(f, \partial f) \colon (M, \partial M) \to (X, \partial X)$ from a compact manifold $(M, \partial M)$ such that f is a simple homotopy equivalence and ∂f is a diffeomorphism;

(ii) The *simple structure set* $\mathcal{S}^s(X, \partial X)$ of $(X, \partial X)$ is the set of equivalence classes of simple manifold structures on X where for $i = 0, 1$ two simple manifold structures $(f_i, \partial f_i) \colon (M_i, \partial M_i) \to (X, \partial X)$ are *equivalent* if there exists a diffeomorphism $(g, \partial g) \colon (M_0, \partial M_0) \to (M_1, \partial M_1)$ such that $f_1 \circ g$ is homotopic to f_0 relative ∂M_0.

The simple structure set has a preferred base point, namely, the structure defined by the identity map id: $(X, \partial X) \to (X, \partial X)$. If ∂X is empty, we abbreviate $\mathcal{S}^s(X) := \mathcal{S}^s(X, \emptyset)$.

The simple structure set $\mathcal{S}^s(X, \partial X)$ is the basic object in the study of manifolds that are simple homotopy equivalent to a compact manifold X relative to ∂X. Next we briefly discuss its role in the classification of manifolds. For simplicity, we shall assume that X is closed in this discussion, leaving the relative boundary version of this discussion to the reader.

Note that a simple homotopy equivalence $f \colon M \to X$ between closed manifolds is homotopic to a diffeomorphism if and only if it represents the base point in $\mathcal{S}^s(X)$. Hence a manifold M is diffeomorphic to X if and only if for some simple homotopy equivalence $f \colon M \to X$ the class of $[f]$ in $\mathcal{S}^s(X)$ agrees with the preferred base point. However, care is necessary if one wants to use $\mathcal{S}^s(X)$ to classify manifolds that are simple homotopy equivalent to X up to diffeomorphism since it is possible that a simple homotopy equivalence $f \colon M \to X$ is not homotopic to a diffeomorphism, although M and X are diffeomorphic.

Definition 11.3 (The simple manifold set) Let X be a closed manifold of dimension n. The *simple manifold set* $\mathcal{M}^s(X)$ of X is the set of diffeomorphism classes of closed manifolds simple homotopy equivalent to X.

As the following fundamental exercise shows, in order to pass from $\mathcal{S}^s(X)$ to $\mathcal{M}^s(X)$ one has to understand the operation of ho-aut$^s(X)$, the group of homotopy classes of simple self-homotopy equivalences of X, on $\mathcal{S}^s(X)$.

Exercise 11.4 The group ho-aut$^s(X)$ acts from the left of $\mathcal{S}^s(X)$ by post-composition ho-aut$^s(X) \times \mathcal{S}^s(X) \to \mathcal{S}^s(X)$, $([g], [f]) \mapsto [g \circ f]$.

Show that the natural map $\mathcal{S}^s(X) \to \mathcal{M}^s(X)$, $[f \colon M \to X] \mapsto [M]$, induces a bijection ho-aut$^s(X) \backslash \mathcal{S}^s(X) \xrightarrow{\cong} \mathcal{M}^s(X)$. Deduce that the action of ho-aut$^s(X)$ on $\mathcal{S}^s(X)$ is transitive if and only if every closed manifold that is simple homotopy equivalent to X is already diffeomorphic to X.

The action of ho-aut$^s(X)$ on $\mathcal{S}^s(X)$ can be complicated to analyse in general, see [111]. However in many cases, for example when $X = S^n$, the action can be computed, see Example 18.4. More on this topic can be found in Section 18.2 and further calculations in later sections of Chapter 18.

There is also a version of the structure set that considers homotopy equivalences, simple or not, and so does not take Whitehead torsion into account, as defined below.

Definition 11.5 (The geometric structure set) Let X be a compact manifold of dimension n with possibly empty boundary ∂X.

(i) A *manifold structure relative boundary on X*, or briefly *manifold structure on X*, is a map of pairs $(f, \partial f) \colon (M, \partial M) \to (X, \partial X)$ from a compact manifold $(M, \partial M)$ such that f is a homotopy equivalence and ∂f is a diffeomorphism;

(ii) The *structure set* $\mathcal{S}^h(X, \partial X)$ of $(X, \partial X)$ is the set of equivalence classes of manifold structures on X where for $i = 0, 1$ two manifold structures $(f_i, \partial f_i) \colon (M_i, \partial M_i) \to (X, \partial X)$ are *equivalent* if there exists:

11.2 The Geometric Structure Set

- A compact manifold $(W, \partial W)$ of dimension $(n+1)$ with a homotopy equivalence of pairs

$$(F, \partial F): (W, \partial W) \to (X \times [0,1], \partial(X \times [0,1]));$$

- The boundary ∂W is the union of three pieces

$$\partial W = \partial_0 W \cup \partial_1 W \cup \partial_2 W,$$

where $\partial_i W$ is a codimension zero submanifold of ∂W with possibly empty boundary $\partial \partial_i W$ for $i = 0, 1, 2$ satisfying

$$\partial_0 W \cap \partial_1 W = \emptyset;$$
$$\partial_2 W \cap \partial_i W = \partial \partial_i W \quad \text{for } i = 0, 1;$$
$$\partial \partial_2 W = \partial \partial_0 W \amalg \partial \partial_1 W;$$

- The map $\partial F: \partial W \to \partial(X \times [0,1]) = (X \times \{0,1\}) \cup (\partial X \times [0,1])$, which is a homotopy equivalence, is required to induce a homotopy equivalence $\partial_i F: \partial_i W \to X \times \{i\}$ for $i = 0, 1$ and in addition a diffeomorphism $\partial_2 F: \partial_2 W \to \partial X \times [0, 1]$;
- There are diffeomorphisms $(u_i, \partial u_i): (M_i, \partial M_i) \to (\partial_i W, \partial \partial_i W)$ for $i = 0, 1$ satisfying $\partial_i F \circ u_i = j_i \circ f_i$, where $j_i: X \to X \times \{i\}$ sends x to (x, i).

The class of the identity $\mathrm{id}: (X, \partial X) \to (X, \partial X)$ is a preferred base point for the structure set. If ∂X is empty, we abbreviate $\mathcal{S}^h(X) := \mathcal{S}^h(X, \emptyset)$.

Remark 11.6 (Simple structure sets and the s-Cobordism Theorem) If we additionally require in Definition 11.5 that the homotopy equivalences f_0, f_1, F, $\partial_0 F$, and $\partial_1 F$ are simple homotopy equivalences, we get back the simple structure set $\mathcal{S}^s(X)$ of Definition 11.2, provided that $n \geq 5$. We prove this only in the case $\partial X = \emptyset$. Then the claim follows from the s-Cobordism Theorem 2.1. Namely, consider two manifold structures $f_i: M_i \to X$ for $i = 0, 1$ that are equivalent in the sense above. Then the h-cobordism W appearing in Definition 11.5 is actually an s-cobordism by Theorem 3.1 since we require F, f_0 and f_1 to be simple homotopy equivalences. Hence there is a diffeomorphism $\Phi: \partial_0 W \times [0,1] \to W$ inducing the obvious identification $\partial_0 W \times \{0\} \to \partial_0 W$ and some diffeomorphism $g': (\partial_0 W) = (\partial_0 W \times \{1\}) \xrightarrow{\cong} \partial_1 W$. If $g: M_0 \to M_1$ is the diffeomorphism $u_1^{-1} \circ g' \circ u_0$, then $f_1 \circ g$ is homotopic to f_0 and hence f_0 and f_1 are equivalent in the sense of Definition 11.2. The other implication is obvious.

If we allow a non-empty boundary ∂X, one has to use the obvious version of the s-Cobordism Theorem 2.1 for manifolds with boundary.

11.3 The Set of Normal Maps

Recall from Definition 5.29 that a compact n-dimensional manifold X with possibly empty boundary ∂X has an intrinsic fundamental class $[[X, \partial X]]$ in $H_n(X, \partial X; O_X)$.

Definition 11.7 (The set of normal maps) Let X be a compact manifold of dimension n with possibly empty boundary ∂X. Define the *set of normal maps to* $(X, \partial X)$,

$$\mathcal{N}(X, \partial X),$$

to be the set of equivalence classes of normal maps of degree one with respect to the tangent bundle relative boundary $(M, \partial M, f, \partial f, a, \xi, \overline{f}, o)$ with target $(X, \partial X, [[X, \partial X]])$ in the sense of Subsection 8.8.1, where we additionally assume that $\partial f \colon \partial M \to \partial X$ is a diffeomorphism, under the equivalence relation given by normal bordism with cylindrical ends of degree one in the sense of Subsection 8.8.2, where we additionally assume that $\partial_2 F \colon \partial_2 W \to \partial_2 Y = \partial X \times [0, 1]$ is a diffeomorphism.

This set has a preferred base point, namely, the class given by the identity normal map on $(X, \partial X, [[X, \partial X]])$.

If ∂X is empty, then X is a closed manifold, hence a Poincaré complex of dimension n, and the definition of $\mathcal{N}(X, \emptyset)$ agrees with the definition of $\mathcal{N}(X)$ from Section 7.4. Hence we identify $\mathcal{N}(X)$ and $\mathcal{N}(X, \emptyset)$.

We emphasise that in Definition 11.7 we require that ∂f and $\partial_2 F$ are diffeomorphisms whereas in Subsections 8.8.1 and 8.8.2 we have only demanded that ∂f and $\partial_2 F$ are homotopy equivalences and in Section 10.4 that ∂f and $\partial_2 F$ are simple homotopy equivalences.

Let X be a compact manifold of dimension n with possibly empty boundary ∂X. Then we define maps of pointed sets

$$\eta_n^h \colon \mathcal{S}^h(X, \partial X) \to \mathcal{N}(X, \partial X); \tag{11.8}$$

$$\eta_n^s \colon \mathcal{S}^s(X, \partial X) \to \mathcal{N}(X, \partial X), \tag{11.9}$$

as follows. Consider a manifold structure $[f, \partial f] \in \mathcal{S}^h(X, \partial X)$ given by a map of pairs $(f, \partial f) \colon (M, \partial M) \to (X, \partial X)$. Under η_n^h it is sent to the normal bordism class of a normal map $(M, \partial M, f, \partial f, 0, \xi, \overline{f}, o_{f,\overline{f}})$ with target $(X, \partial X, [[X, \partial X]])$ defined as follows. The underlying map is the original map $(f, \partial f) \colon (M, \partial M) \to (X, \partial X)$. To define a bundle map $\overline{f}$, we choose a homotopy inverse $f^{-1} \colon X \to M$ of f. Let ξ be the bundle $(f^{-1})^*TM$ over X. Since $f^{-1} \circ f \simeq \mathrm{id}_M$, we can find a bundle isomorphism $f^*(f^{-1})^*TM \xrightarrow{\cong} TM$ covering id_M. This is the same as a bundle map $\overline{f} \colon TM \to \xi$ covering f. Since $(f, \partial f)$ is a homotopy equivalence of pairs, we have $w_1(M) = f^*w_1(X)$. Hence there exists an isomorphism of infinite cyclic local coefficients systems $O_\xi \to O_X$. The obvious extension of Lemma 7.4 to pairs implies that there is precisely one isomorphism of infinite cyclic local coefficients systems

$$o_{f,\overline{f}} \colon O_\xi \to O_X \tag{11.10}$$

such that the isomorphism from $H_n(M, \partial M; O_M)$ to $H_n(X, \partial X; O_X)$ induced by $(f, \overline{f}, o_{f,\overline{f}})$ sends $[[M, \partial M]]$ to $[[X, \partial X]]$. Since a change of $\overline{f}$ causes an appropriate change of $o_{f,\overline{f}}$, one easily checks that the normal bordism class of $(M, \partial M, f, \partial f, 0, \xi, \overline{f}, o_{f,\overline{f}})$, regarded as a degree one normal map with cylindrical ends relative boundary, is independent of the choice of $\overline{f}$ and depends only on the manifold structure $[f, \partial f] \in \mathcal{S}^h(X, \partial)$. The map η^s is defined to be the composite of η^h with the forgetful map $\mathcal{S}^s(X) \to \mathcal{S}^h(X)$.

A key ingredient in the successful applications of surgery is the ability to calculate the set of normal maps $\mathcal{N}(X, \partial X)$. In the closed case, we have already seen in Theorem 7.34(ii) that the set of normal maps $\mathcal{N}(X)$ is in bijection with the set of homotopy classes of maps $[X, G/O]$. This generalises for normal maps relative boundary to a bijection

$$\mathcal{N}(X, \partial X) \xrightarrow{\cong} [X/\partial X, G/O], \qquad (11.11)$$

which maps the base point $[\mathrm{id}_{(X, \partial X)}]$ to the class of the constant map.

Remark 11.12 Recall that we use the convention $X/\emptyset = X_+$. If X consists of several path components, $X/\partial X$ becomes $\bigvee_{C \in \pi_0(X)} C/\partial C$. Since G/O is simply connected, it does not matter whether we consider pointed or unpointed homotopy classes when we deal with $[X/\partial X, G/O]$. Moreover, the map $[X_+, G/O] \xrightarrow{\cong} [X, G/O]$ given by the inclusion $X \to X_+$ is a bijection. The same is true for G/PL and G/TOP (see Section 11.6) instead of G/O.

Exercise 11.13 Show that $\mathcal{N}(D^p \times S^q, S^{p-1} \times S^q) \xrightarrow{\cong} \pi_p(G/O) \oplus \pi_{p+q}(G/O)$ and also that $\mathcal{N}(S^p \times S^q) \xrightarrow{\cong} \pi_p(G/O) \oplus \pi_q(G/O) \oplus \pi_{p+q}(G/O)$.

11.4 The Surgery Obstruction Groups

Let X be a compact n-dimensional manifold with possibly empty boundary ∂X for $n \geq 5$. We have introduced the intrinsic surgery obstruction group $L_n^h(\mathbb{Z}\Pi(X), O_X) := L_n(\mathbb{Z}\Pi(X), O_X)$ in Subsections 8.7.3 and 9.4.1 and its simple version $L_n^s(\mathbb{Z}\Pi(X), O_X)$ in Section 10.2. If X is connected, π is the fundamental group of X, and $w \colon \pi \to \{\pm 1\}$ is the first Stiefel–Whitney class of X, then $L_n^h(\mathbb{Z}\Pi(X), O_X)$ and $L_n^s(\mathbb{Z}\Pi(X), O_X)$ are isomorphic to the L-groups $L_n^h(\mathbb{Z}\pi, w) := L_n(\mathbb{Z}\pi, w)$ and $L_n^s(\mathbb{Z}\pi, w)$ that are associated to the group ring $\mathbb{Z}\pi$ with the w-twisted involution, see Subsections 8.5.1 and 9.2.1, and Section 10.2. The intrinsic surgery obstruction relative boundary, see Section 8.8 and 9.4, and its simple version, see Section 10.4, define maps

$$\sigma_n^h \colon \mathcal{N}(X, \partial X) \to L_n^h(\mathbb{Z}\Pi(X), O_X); \quad (11.14)$$

$$\sigma_n^s \colon \mathcal{N}(X, \partial X) \to L_n^s(\mathbb{Z}\Pi(X), O_X); \quad (11.15)$$

$$\sigma_{n+1}^h \colon \mathcal{N}(X \times [0,1], \partial(X \times [0,1])) \to L_{n+1}^h(\mathbb{Z}\Pi(X), O_X); \quad (11.16)$$

$$\sigma_{n+1}^s \colon \mathcal{N}(X \times [0,1], \partial(X \times [0,1])) \to L_{n+1}^s(\mathbb{Z}\Pi(X), O_X), \quad (11.17)$$

using the identifications

$$L_m^h(\mathbb{Z}\Pi(X \times [0,1]), O_{X \times [0,1]}) \xrightarrow{\cong} L_m^h(\mathbb{Z}\Pi(X), O_X); \quad (11.18)$$

$$L_m^s(\mathbb{Z}\Pi(X \times [0,1]), O_{X \times [0,1]}) \xrightarrow{\cong} L_m^s(\mathbb{Z}\Pi(X), O_X), \quad (11.19)$$

coming from the projection $X \times [0,1] \to X$ and the standard isomorphism $\overline{\mathrm{pr}} \colon O_{X \times [0,1]} \xrightarrow{\cong} \mathrm{pr}^* O_M$.

Next we define an operation of the abelian group $L_{n+1}^s(\mathbb{Z}\Pi(X), O_X)$ on $\mathcal{S}^s(X, \partial X)$:

$$\rho_{n+1}^s \colon L_{n+1}^s(\mathbb{Z}\Pi(X), O_X) \times \mathcal{S}^s(X, \partial X) \to \mathcal{S}^s(X, \partial X). \quad (11.20)$$

Consider a simple manifold structure $[f, \partial f] \in \mathcal{S}^s(X, \partial X)$ that is represented by the map of pairs $(f, \partial f) \colon (M, \partial M) \to (X, \partial X)$ such that f is a simple homotopy equivalence and ∂f is a diffeomorphism. As explained in the construction of the map η_n^s of (11.9), we get from $(f, \partial f) \colon (M, \partial M) \to (X, \partial X)$ a normal map of degree one relative boundary $(M, \partial M, f, \partial f, a, \xi, \overline{f}, o)$ with target $(X, \partial X, O_X, [[X, \partial X]])$. Fix $\omega \in L_{n+1}^s(\mathbb{Z}\Pi(X), O_X)$. From Theorem 10.31 applied to ω we get a normal map of degree one relative boundary

$$(M_1, \partial M_1, f_1, \partial f_1, a_1, \xi, \overline{f_1}, o_1)$$

with target $(X, \partial X, O_X, [[X, \partial X]])$ such that f_1 is a simple homotopy equivalence and ∂f_1 is a diffeomorphism, and a normal bordism of degree one with cylindrical ends relative boundary in the sense of Subsection 8.8.2

$$(W, \partial W, F, \partial F, b, \Xi, \overline{F}, O)$$

from $(M, \partial M, f, \partial f, a, \xi, \overline{f}, o)$ to $(M_1, \partial M_1, f_1, \partial f_1, a_1, \xi, \overline{f_1}, o_1)$ with the property that $\sigma^s(W, \partial W, F, \partial F, b, \Xi, \overline{F}, O) \in L_{n+1}^s(\mathbb{Z}\Pi(X \times [0,1]), O_{X \times [0,1]})$ is mapped under the isomorphism (11.19) to ω. Now define $\rho_{n+1}^s((\omega, [f, \partial f]))$ to be the simple manifold structure $[(f_1, \partial f_1) \colon (M_1, \partial M_1) \to (X, \partial X)]$.

We have to show that the class of $(f_1, \partial f_1)$ in $\mathcal{S}^s(X, \partial X)$ is independent of the choices of $(M_1, \partial M_1, f_1, \partial f_1, a_1, \xi, \overline{f_1}, o_1)$ and $(W, \partial W, F, \partial F, b, \Xi, \overline{F}, O)$. Let $(M_1', \partial M_1', f_1', \partial f_1', a_1', \xi', \overline{f_1'}, o_1')$ and $(W', \partial W', F', \partial F', b', \Xi', \overline{F'}, O')$ be a second choice. We can glue the normal bordisms $(W, \partial W, F, \partial F, b, \Xi, \overline{F}, O)$ and $(W', \partial W', F', \partial F', b', \Xi', \overline{F'}, O')$ together along $(M, \partial M, f, \partial f, a, \xi, \overline{f}, o)$ so that we obtain a normal bordism from $(M_1, \partial M_1, f_1, \partial f_1, a_1, \xi, \overline{f_1}, o_1)$ to $(M_1', \partial M_1', f_1', \partial f_1', a_1', \xi', \overline{f_1'}, o_1')$ whose simple surgery obstruction vanishes. Because of Theorem 10.30 (ii) we can arrange that its underlying map to $X \times [0,1]$ is a simple homotopy equivalence. Remark 11.6 implies that $(f_1, \partial f_1) \colon (M_1, \partial M_1) \to$

$(X, \partial X)$ and $(f'_1, \partial f'_1) \colon (M'_1, \partial M'_1) \to (X, \partial X)$ define the same manifold structure in $\mathcal{S}^s(X, \partial X)$.

This indeed defines a group action since the surgery obstruction is additive under stacking normal bordisms together.

Analogously, using Theorem 8.203, we define an operation of the abelian group $L^h_{n+1}(\mathbb{Z}\Pi(X), O_X)$ on $\mathcal{S}^h(X)$

$$\rho^h_{n+1} \colon L^h_{n+1}(\mathbb{Z}\Pi(X), O_X) \times \mathcal{S}^h(X, \partial X) \to \mathcal{S}^h(X, \partial X). \tag{11.21}$$

11.5 The Geometric Surgery Exact Sequence

Now we can establish one of the main tools in the classification of manifolds, the surgery exact sequence.

Theorem 11.22 (Geometric surgery exact sequence) *Let X be an n-dimensional compact smooth manifold with possibly empty boundary ∂X. Suppose $n \geq 5$. Then:*

(i) *There is the so-called* simple geometric surgery exact sequence

$$\mathcal{N}(X \times [0,1], \partial(X \times [0,1])) \xrightarrow{\sigma^s_{n+1}} L^s_{n+1}(\mathbb{Z}\Pi(X), O_X) \xrightarrow{\rho^s_{n+1}} \mathcal{S}^s(X, \partial X)$$
$$\xrightarrow{\eta^s_n} \mathcal{N}(X, \partial X) \xrightarrow{\sigma^s_n} L^s_n(\mathbb{Z}\Pi(X), O_X)$$

which is exact in the following sense:

(a) *An element $\alpha \in \mathcal{N}(X, \partial X)$ lies in the image of the map η^s_n defined in (11.9) if and only if $\sigma^s_n(\alpha) = 0$ holds for the map σ^s_n defined in (11.15);*

(b) *Two elements β_1 and β_2 in $\mathcal{S}^s(X, \partial X)$ have the same image under η^s_n if and only if there is an $\omega \in L^s_{n+1}(\mathbb{Z}\Pi(X), O_X)$ with $\rho^s_{n+1}(\omega, \beta_1) = \beta_2$ for the operation ρ^s_{n+1} defined in (11.20);*

(c) *For $\omega \in L^s_{n+1}(\mathbb{Z}\Pi(X), O_X)$, we have $\rho^s_{n+1}(\omega, [\mathrm{id}_{(X, \partial X)}]) = [\mathrm{id}_{(X, \partial X)}]$ in $\mathcal{S}^s(X, \partial X)$ if and only if there is a normal bordism class γ in $\mathcal{N}_{n+1}(X \times [0,1], \partial(X \times [0,1]))$ with $\sigma^s_{n+1}(\gamma) = \omega$ for the map σ^s_{n+1} defined in (11.17).*

(ii) *There is the so-called* geometric surgery exact sequence

$$\mathcal{N}(X \times [0,1], \partial(X \times [0,1])) \xrightarrow{\sigma^h_{n+1}} L^h_{n+1}(\mathbb{Z}\Pi(X), O_X) \xrightarrow{\rho^h_{n+1}} \mathcal{S}^h(X, \partial X)$$
$$\xrightarrow{\eta^h_n} \mathcal{N}(X, \partial X) \xrightarrow{\sigma^h_n} L^h_n(\mathbb{Z}\Pi(X), O_X)$$

that is exact in the sense analogous to the one in assertion (i).

Proof. (i). We have to prove exactness:

(ia) This follows directly from Theorem 10.30.

(ib) By definition, two elements of $\mathcal{S}^s(X)$ that lie in the same orbit of the $L^s_{n+1}(\mathbb{Z}\Pi(X), O_X)$-action have the same image under η^s_n.

If two elements $\beta_1, \beta_2 \in \mathcal{S}^s(X)$ have the same image under η^s_n, then there is a normal bordism between the representatives of their images under η^s_n in $\mathcal{N}(X, \partial X)$ whose simple surgery obstruction σ is an element in $L^s_{n+1}(\mathbb{Z}\Pi(X), O_X)$. By construction σ satisfies the desired equality in $\mathcal{S}^s(X)$

$$\rho^s_{n+1}(\sigma, \beta_1) = \beta_2.$$

(ic) Consider an element $\omega \in L^s_{n+1}(\mathbb{Z}\Pi(X), O_X)$ such that we have the equation $\rho^s_{n+1}(\omega, [\mathrm{id}_{(X, \partial X)}]) = [\mathrm{id}_{(X, \partial X)}]$. In the sequel we specify only the underlying maps of manifolds when considering a normal map or normal bordism. By the definition of the action ρ^s_{n+1}, there is a normal bordism $(F, \partial F)\colon (W, \partial W) \to (X \times [0, 1], \partial(X \times [0, 1]))$ of degree one with cylindrical ends relative boundary between $\mathrm{id}\colon (X, \partial X) \to (X, \partial X)$ and $\mathrm{id}\colon (X, \partial X) \to (X, \partial X)$ whose surgery obstruction is ω. Since $\mathrm{id}_{(X, \partial X)}$ is a diffeomorphism, ∂F is a diffeomorphism. Hence $(F, \partial F)$ defines an element $[F, \partial F]$ in $\mathcal{N}(X \times [0, 1], \partial(X \times [0, 1]))$ satisfying

$$\omega = \sigma^s_{n+1}([F, \partial F]).$$

Consider an element $[(F, \partial F)\colon (W, \partial W) \to (X \times [0, 1], \partial(X \times [0, 1]))]$ in the set $\mathcal{N}(X \times [0, 1], \partial(X \times [0, 1]))$. Since ∂F is a diffeomorphism, we can interpret $(F, \partial F)$ as a normal bordism of degree one with cylindrical ends relative boundary between $\mathrm{id}\colon (X, \partial X) \to (X, \partial X)$ and $\mathrm{id}\colon (X, \partial X) \to (X, \partial X)$. Hence we conclude from the definition of the action ρ^s_{n+1} that

$$\rho^s_{n+1}(\sigma^s_{n+1}([F, \partial F]), [\mathrm{id}_{(X, \partial X)}]) = [\mathrm{id}_{(X, \partial X)}].$$

This finishes the proof of assertion (i).

(ii) The proof of assertion (ii) is completely analogous, just ignore everything about the condition of simplicity and replace Theorem 10.30 by Theorems 8.189 and 9.113. This finishes the proof of Theorem 11.22. □

Remark 11.23 (Extension to the left) Let X be a compact manifold of dimension $n \geq 5$ with possibly empty boundary ∂X. Recall that we always use the angle straightening procedure to give a product of two smooth manifolds with possibly non-empty boundary a smooth structure, see [316, 7.5] or [415, 2.6]. Using the obvious diffeomorphisms of compact manifolds with boundary for $i \geq 0$,

$$([0, 1]^i, \partial[0, 1]^i) \cong (D^{i-1}, S^{i-2}) \times ([0, 1], \partial[0, 1]) \cong (D^i, S^{i-1}),$$

one can iterate the surgery exact sequence appearing in Theorem 11.22 (i) and obtain an exact sequence that is infinite to the left:

11.6 The Piecewise Linear and Topological Categories

$$\cdots \xrightarrow{\sigma^s_{n+i+1}} L^s_{n+i+1}(\mathbb{Z}\Pi(X), \mathcal{O}_X) \xrightarrow{\rho^s_{n+i+1}} \mathcal{S}^s(X \times D^i, \partial(X \times D^i))$$
$$\xrightarrow{\eta^s_{n+i}} \mathcal{N}(X \times D^i, \partial(X \times D^i)) \xrightarrow{\sigma^s_{n+i}} L^s_{n+i}(\mathbb{Z}\Pi(X), \mathcal{O}_X) \xrightarrow{\rho^s_{n+i}} \cdots$$
$$\cdots \xrightarrow{\sigma^s_{n+1}} \mathcal{N}(X \times D^1, \partial(X \times D^1)) \xrightarrow{\sigma^s_{n+1}} L^s_{n+1}(\mathbb{Z}\Pi(X), \mathcal{O}_X)$$
$$\xrightarrow{\rho^s_{n+1}} \mathcal{S}^s(X, \partial X) \xrightarrow{\eta^s_n} \mathcal{N}(X, \partial X) \xrightarrow{\sigma^s_n} L^s_n(\mathbb{Z}\Pi(X), \mathcal{O}_X).$$

The corresponding statement is also true for the h decoration.

See also Remark 11.36 below, which discusses the group structures on the terms in the part where $i \geq 1$ and explains that the exactness in that part can be understood in the usual way.

11.6 The Piecewise Linear and Topological Categories

In addition to the smooth category DIFF one can also develop surgery theory in the piecewise linear category PL or topological category TOP. This requires finding analogues of the notions of vector bundle and tangent bundle and also developing transversality and Hirsch–Smale theory in these categories. In the PL category this was done earlier, see for example [197] or [367]. In the topological category this was done later, see the foundational book [219] of Kirby and Siebenmann, and also [159, 9.6C on page 160], which removes some dimension restrictions from [219]. Note that in [219] the results from surgery in the smooth category and in the PL category are used as an input to develop these techniques in the topological category.

There are analogues of the structure sets

$$\mathcal{S}^{\text{PL},h}(X, \partial X), \ \mathcal{S}^{\text{PL},s}(X, \partial X), \ \mathcal{S}^{\text{TOP},h}(X, \partial X), \text{ and } \mathcal{S}^{\text{TOP},s}(X, \partial X)$$

and also analogues of the sets of normal maps $\mathcal{N}^{\text{PL}}(X, \partial X)$ and $\mathcal{N}^{\text{TOP}}(X, \partial X)$. To compute these sets of normal maps, we start with analogues PL and TOP of the group O. The topological group TOP is the limit of the groups TOP(k) that are the groups of homeomorphisms of $\mathbb{R}^k$ fixing the origin:

$$\text{TOP} = \text{colim}_{k \to \infty} \text{TOP}(k).$$

The definition of PL is more elaborate. There are classifying spaces BPL (resp. BTOP) that classify stable isomorphism classes of PL (resp. TOP) $\mathbb{R}^k$-bundles and are infinite loop spaces with multiplication corresponding to the Whitney sum of bundles. There are also canonical maps BPL → BG (resp. BTOP → BG) which classify the passage to strong fibre homotopy equivalence classes of stable spherical fibrations. The homotopy fibres of these maps are denoted G/PL (resp. G/TOP) and have infinite loop space structures so that the canonical maps G/PL → BPL and G/TOP → BTOP are maps of infinite loop spaces. Theorem 7.19, Remark 7.21,

Theorem 7.34, and (11.11), carry over to the piecewise linear category PL and the topological category TOP.

Theorem 11.24 (**Normal maps,** G/PL **and** G/TOP)

(i) *Let X be a finite n-dimensional Poincaré complex. Suppose that $\mathcal{N}^{\mathrm{PL}}(X)$ is non-empty. Then there is a canonical group structure on the set $[X, \mathrm{G/PL}]$ of homotopy classes of maps from X to G/PL and a transitive free operation of this group on $\mathcal{N}^{\mathrm{PL}}(X)$. The analogous statement holds in the topological category with* TOP *in place of* PL *and* G/TOP *instead of* G/PL*;*

(ii) *Let X be a compact n-dimensional* PL *manifold. Then there is a canonical group structure on the set $[X/\partial X, \mathrm{G/PL}]$ of homotopy classes of maps from $X/\partial X$ to G/PL and a transitive free operation of this group on $\mathcal{N}^{\mathrm{PL}}(X, \partial X)$. The analogous statement holds in the topological category with* TOP *in place of* PL *and* G/TOP *instead of* G/PL.

There are analogues of the surgery exact sequence in the smooth category presented in Theorem 11.22, see also Remark 11.23, for the categories PL and TOP.

Theorem 11.25 (**Surgery exact sequence for** PL **and** TOP) *Let X be a compact* PL *manifold of dimension $n \geq 5$.*

Then there is a surgery exact sequence, infinite to the left,

$$\cdots \xrightarrow{\rho_{n+2}^{\mathrm{PL},s}} \mathcal{N}^{\mathrm{PL}}(X \times D^1, \partial(X \times D^1)) \xrightarrow{\sigma_{n+1}^{\mathrm{PL},s}} L_{n+1}^s(\mathbb{Z}\Pi(X), O_X)$$

$$\xrightarrow{\rho_{n+1}^{\mathrm{PL},s}} \mathcal{S}^{\mathrm{PL},s}(X, \partial X) \xrightarrow{\eta_n^{\mathrm{PL},s}} \mathcal{N}^{\mathrm{PL}}(X, \partial X) \xrightarrow{\sigma_n^{\mathrm{PL},s}} L_n^s(\mathbb{Z}\Pi(X), O_X)$$

that is exact for $n \geq 5$ in the sense of Theorem 11.22.

There is an analogous surgery exact sequence, infinite to the left,

$$\cdots \xrightarrow{\rho_{n+2}^{\mathrm{PL},h}} \mathcal{N}^{\mathrm{PL}}(X \times D^1, \partial(X \times D^1)) \xrightarrow{\sigma_{n+1}^{\mathrm{PL},h}} L_{n+1}^h(\mathbb{Z}\Pi(X), O_X)$$

$$\xrightarrow{\rho_{n+1}^{\mathrm{PL},h}} \mathcal{S}^{\mathrm{PL},h}(X, \partial X) \xrightarrow{\eta_n^{\mathrm{PL},h}} \mathcal{N}^{\mathrm{PL}}(X, \partial X) \xrightarrow{\sigma_n^{\mathrm{PL},h}} L_n^h(\mathbb{Z}\Pi(X), O_X).$$

The analogous exact sequences exist in the topological category if X is instead a compact topological manifold of dimension $n \geq 5$.

Note that the surgery obstruction groups are independent of the category, i.e, they are the same in the smooth category, the PL category and the smooth category, but they depend on the decoration h or s. The set of normal invariants is independent of the decoration h or s, but depends on the category. The structure sets depend on both the decoration h and s and the category.

A feature of the categories PL and TOP is that all the surgery obstruction maps are surjective for simply connected manifolds X. Since X is simply connected, the map $L_{n+1}^s(\mathbb{Z}\Pi(X), O_X) \to L_{n+1}^h(\mathbb{Z}\Pi(X), O_X)$ is an isomorphism. In fact an even more general statement holds as follows.

11.6 The Piecewise Linear and Topological Categories

Theorem 11.26 (Surjectivity of the simply connected surgery obstruction maps for PL **and** TOP**)** *Let X be a simply connected finite Poincaré complex of dimension $n \geq 5$ such that the Spivak normal fibration of X admits a* PL *reduction. Then the surgery obstruction map*

$$\mathcal{N}^{\mathrm{PL}}(X, \partial X) \xrightarrow{\sigma_n^{\mathrm{PL},s}} L_n^s(\mathbb{Z}\Pi(X), \mathcal{O}_X)$$

is onto.

The analogous statement holds in the topological category if the Spivak normal fibration of X admits a topological reduction.

We will prove Theorem 11.26 in Section 17.8, see Theorem 17.30. Here we record some important consequences for simply connected PL and topological compact manifolds. The first of these is due independently to Browder and Novikov, see [52], [321] and also [56], [322].

Theorem 11.27 (Browder–Novikov Theorem) *Let X be a simply connected finite Poincaré complex of dimension $n \geq 5$. Then X is homotopy equivalent to a* PL *manifold if and only if the Spivak normal fibration of X admits a* PL *reduction.*

Proof. If the Spivak normal fibration of X does not have a PL reduction then X cannot be homotopy equivalent to a PL manifold. Conversely, if the Spivak normal fibration has PL reduction, then $\mathcal{N}^{\mathrm{PL}}(X)$ is non-empty. By Theorem 11.26 there is a degree one normal map $(f, \overline{f}) \colon M \to X$ with vanishing surgery obstruction and so $(f, \overline{f})$ is normally bordant to $(g, \overline{g}) \colon N \to X$, where $g \colon N \to X$ is a homotopy equivalence. □

Following the work of Kirby and Siebenmann, the topological version of the Browder–Novikov theorem can be proven analogously:

Theorem 11.28 (Topological Browder–Novikov Theorem) *Let X be a simply connected finite Poincaré complex of dimension $n \geq 5$. Then X is homotopy equivalent to a topological manifold if and only if the Spivak normal fibration of X admits a topological reduction.*

Moving from the existence problem to the classification problem for simply connected PL and topological manifolds, the exactness of surgery exact sequences in the categories PL and TOP means that Theorem 11.26 applied to $X \times I$ has the following consequence.

Theorem 11.29 (PL **and topological structure sets in the simply connected case)** *Let X be a simply connected compact* PL *manifold of dimension $n \geq 5$. Then the map*

$$\mathcal{S}^{\mathrm{PL},s}(X, \partial X) \xrightarrow{\eta_n^{\mathrm{PL},s}} \mathcal{N}^{\mathrm{PL}}(X, \partial X)$$

is injective.

The analogous statement holds in the topological category if X is instead a simply connected compact topological manifold of dimension $n \geq 5$.

Note that Theorem 11.29 does not hold in the smooth category. Counterexamples will be constructed in Chapter 12, see Lemma 12.9 and Theorem 12.49.

Remark 11.30 (Surgery obstruction map is not a homomorphism with respect to the Whitney sum) In this remark we explain that the surgery obstruction map is not a homomorphism for the abelian group structure on normal invariants given by the Whitney sum.

Namely, consider the composite

$$\alpha \colon \pi_4(G/O) \oplus \pi_4(G/O) \oplus \pi_8(G/O) \xrightarrow{\beta} \mathcal{N}(S^4 \times S^4) \xrightarrow{\sigma} L_8(\mathbb{Z}) \xrightarrow{\text{sign}} 8 \cdot \mathbb{Z}$$

where β is the bijection appearing in Exercise 11.13, the map σ is given by taking the surgery obstruction, and sign is the isomorphism of abelian group of Theorem 8.99. Obviously the source and target of α are abelian groups. We want to show that α is *not* a homomorphism of abelian groups.

Let $\text{pr}_1 \colon S^4 \times S^4 \to S^4$ be the projection to the first factor and let $\text{pr}_2 \colon S^4 \times S^4 \to S^4$ be the projection to the second factor. Then the subgroup $\pi_4(G/O) \oplus \{0\} \oplus \{0\}$ agrees with $\text{pr}_1^*(\pi_4(G/O))$ for the map $\text{pr}_1^* \colon [S^4, G/O] \to [S^4 \times S^4, G/O]$ and the subgroup $\{0\} \oplus \pi_4(G/O) \oplus \{0\}$ agrees with $\text{pr}_2^*(\pi_4(G/O))$. In the sequel we use the fact that $\pi_4(G/O)$ and $\pi_4(BO)$ are infinite cyclic groups, see (6.84). Since BG has finite homotopy groups and we have the fibration $G/O \xrightarrow{i} BO \xrightarrow{J} BG$, see (6.71), the map $\pi_4(i) \colon \pi_4(G/O) \to \pi_4(BO)$ is injective.

We conclude from Theorem 8.173 (i) and Theorem 8.171 that the following diagram commutes

$$\begin{array}{ccc} [S^4 \times S^4, G/O] & \xrightarrow{\alpha} & \mathbb{Z} \\ \downarrow{i_*} & \nearrow{\gamma} & \\ [S^4 \times S^4, BO] & & \end{array}$$

where γ sends the homotopy class $[f]$ of $f \colon S^4 \times S^4 \to BO$ to the integer $\frac{1}{45} \cdot \langle 7 \cdot p_2(\xi_f) - p_1(\xi_f)^2, [S^4 \times S^4] \rangle$ for ξ_f any vector bundle over $S^4 \times S^4$ representing f, see Theorem 6.20.

Now consider elements a_1 in $\pi_p(S^4) = \pi_4(G/O) \oplus \{0\} \oplus \{0\}$ and a_2 in $\pi_p(S^4) = \{0\} \oplus \pi_4(G/O) \oplus \{0\}$ satisfying $a_1 \neq 0$ and $a_2 \neq 0$. Since $H^8(S^4)$ is trivial, we get $p_2(\xi_{a_i}) = p_1(\xi_{a_i})^2 = 0$ for $i = 1, 2$, where ξ_{a_i} is any vector bundle over S^4 representing the homotopy class of the image of a_i under $i_* \colon [S^4, G/O] \to [S^4, BO]$. We conclude

$$\alpha(a_1) = \alpha(a_2) = 0.$$

Now suppose that α is a homomorphism of abelian groups. This implies $\alpha(a_1 + a_2) = \alpha(a_1) + \alpha(a_2) = 0$. Since Pontrjagin classes are modulo 2-torsion multiplicative, see [308, Theorem 15.3 on page 175], we conclude

$$0 = \sigma_8(a_1 + a_2) = \frac{1}{45} \cdot \langle -\text{pr}_1^*(p_1(\xi_{a_1})) \cup \text{pr}_1^*(p_1(\xi_{a_2})), [S^4 \times S^4] \rangle.$$

Hence the element $\mathrm{pr}_1^*(p_1(\xi_{a_1})) \cup \mathrm{pr}_1^*(p_1(\xi_{a_2}))$ in the infinite cyclic group $H^8(S^4 \times S^4)$ vanishes. We conclude from the Künneth Theorem that one of the elements $p_1(\xi_{a_1})$ or $p_1(\xi_{a_2})$ in the infinite cyclic group $H^4(S^4)$ vanish. Since the map $\pi_4(\mathrm{BO}) \to H^4(S^4)$ given by taking the first Pontrjagin class and the map $\pi_4(i): \pi_4(\mathrm{G/O}) \to \pi_4(\mathrm{BO})$ are injective, we conclude that a_1 or a_2 is trivial. This is a contradiction. Hence α is *not* a homomorphism of abelian groups.

The same argument works in the categories PL and TOP, see [340, Example 3.6]. However, a major difference in PL and TOP is that it is possible to endow the normal invariants with a different structure of an abelian group, such that the surgery obstruction map becomes a homomorphism, see Remark 11.38 below.

The failure of the surgery map to be a homomorphism in the smooth category for the case when $X = \mathbb{CP}^{2k}$ is explained in [352, Example 13.3].

11.7 Rigidity

Definition 11.31 (Topological rigidity) A closed topological manifold N is called *topologically rigid* if any homotopy equivalence $M \to N$ from a closed topological manifold M to N is homotopic to a homeomorphism.

Note that for any closed topological manifold N we have a forgetful map $\mathcal{S}^{\mathrm{TOP},s}(N) \to \mathcal{S}^{\mathrm{TOP},h}(N)$. If $\mathrm{Wh}(\pi_1(N)) = \{1\}$ and $\dim(N) \geq 5$, then this map is a bijection, but in general it is neither injective nor surjective. More information about the forgetful map can found in Section 13.5. Still, one gets the following lemma directly from the definitions.

Lemma 11.32 *If a closed topological manifold N is topologically rigid, then $\mathcal{S}^{\mathrm{TOP},s}(N) \to \mathcal{S}^{\mathrm{TOP},h}(N)$ is a bijection and $\mathcal{S}^{\mathrm{TOP},s}(N) \cong \mathcal{S}^{\mathrm{TOP},h}(N)$ consists precisely of one element.*

If a closed topological manifold N of dimension $n \geq 5$ has the properties that $\mathrm{Wh}(\pi_1(N)) = \{1\}$ and $\mathcal{S}^{\mathrm{TOP},s}(N)$ consists precisely of one element, then X is topologically rigid.

Exercise 11.33 Prove that the n-sphere S^n is topologically rigid if and only if the Poincaré Conjecture in dimension n is true.

Remark 11.34 (Smoothly rigid) A closed smooth manifold N is *smoothly rigid* if any homotopy equivalence $M \to N$ with a closed smooth manifold M as source and N as target is homotopic to a diffeomorphism. This phenomenon occurs only very rarely in dimensions ≥ 5, and therefore we will not pursue this notion further. For instance, we will see in Remark 12.36 that S^n is smoothly rigid in very few dimensions whereas S^n is topologically rigid in all dimensions.

We will systematically treat the question of which closed topological manifolds are topologically rigid in Chapter 19.

11.8 Group Structures in the Surgery Exact Sequences

In this subsection by "surgery exact sequence" we mean the extended surgery exact sequence of Remark 11.23, which is infinite to the left, or its analogues from Theorem 11.25. *A priori* the surgery exact sequences are exact sequences of pointed sets. However, computations with these sequences are significantly easier to make and summarise when the sets involved have natural group structures and the maps between them are homomorphisms. This is frequently, but not always, the case for the surgery exact sequences of a compact n-dimensional manifold X with $n \geq 5$.

Definition 11.35 (Additive and abelian exact sequences) We say that an exact sequence of pointed sets is *additive* if the sets admit group structures such that the base points are the identity elements and the maps are homomorphisms. An additive exact sequence is called *abelian* if all the groups are abelian.

In the surgery exact sequences, the L-group terms $L_{n+i}^s(\mathbb{Z}\Pi(X), O_X)$ and $L_{n+i}^h(\mathbb{Z}\Pi(X), O_X)$ of course come with group structures and by (11.11) and Theorem 11.24 the sets of normal maps also admit groups structures since the classifying spaces are infinite loop spaces via the Whitney sum operation. However, for these group structures the surgery obstruction maps

$$\mathcal{N}(X, \partial X) \xrightarrow{\sigma_n^s} L_n^s(\mathbb{Z}\Pi(X), O_X) \quad \text{and} \quad \mathcal{N}(X, \partial X) \xrightarrow{\sigma_n^h} L_n^h(\mathbb{Z}\Pi(X), O_X)$$

are not in general homomorphisms and the same holds in the categories PL and TOP, see Remark 11.30. On the other hand, this is not the case if we move at least three terms to the left in the surgery exact sequence, and indeed we can make the following observation.

Remark 11.36 (Existence of group structures in high enough degrees) There are group structures on all sets appearing in the simple geometric surgery exact sequence that lie on the left side of or are equal to $L_{n+1}^s(\mathbb{Z}\Pi(X), O_X)$, and all these group structures are abelian, except possibly for the structure set of $(X \times D^1, \partial(X \times D^1))$, see [93, Remark 4.89]. Moreover, all maps between these sets are homomorphisms of groups. Hence the simple surgery exact sequence is additive until $L_{n+1}^s(\mathbb{Z}\Pi(X), O_X)$ and it is abelian until $L_{n+2}^s(\mathbb{Z}\Pi(X), O_X)$.

This is due to the fact that one can stack normal maps and structures with target $X \times [0, 1]$ together, as one can glue $[0, 1]$ and $[1, 2]$ along $\{1\}$ together to obtain $[0, 2]$, which can be identified with $[0, 1]$ again.

The bijection of Remark (11.11)

$$\mathcal{N}(X \times D^i, \partial(X \times D^i)) \xrightarrow{\cong} [(X \times D^i)/\partial(X \times D^i), G/O]$$

is for $i \geq 1$ an isomorphism of abelian groups. The abelian group structure in the target comes from the Whitney-sum infinite loop space structure on G/O and from the fact that $(X \times D^i)/\partial(X \times D^i)$ is the i-fold suspension of $X/\partial X$.

11.8 Group Structures in the Surgery Exact Sequences

The same remark applies for the h-decoration and also in the categories PL and TOP.

Exercise 11.37 Using the fact that G/PL and G/TOP are simply connected show that for $n \geq 6$ the surgery obstruction map yields isomorphisms of abelian groups $\pi_n(\text{G/PL}) \cong L_n(\mathbb{Z})$ and $\pi_n(\text{G/TOP}) \cong L_n(\mathbb{Z})$.

Remark 11.38 (The surgery exact sequences in the categories PL and TOP are abelian) We start with the category TOP. The discussions in this remark apply for both the s and h decoration and so we shall not mention decorations again.

The first statement that the topological surgery exact sequence is abelian is due to Siebenmann [219, Periodicity Theorem for Structures C.5 on page 283], where it appeared as an application of Quinn's then recent construction of surgery spaces [337], which we discuss in the following section. Quinn's work produced a new infinite loop space structure on G/TOP, different from the Whitney-sum infinite loop space structure discussed to date. In particular, this gave a new group structure on $[X/\partial X, \text{G/TOP}] \equiv \mathcal{N}^{\text{TOP}}(X)$ such that the surgery obstruction map is a homomorphism. However, Siebenmann's proofs relied on his mistaken version of what can now be called "Siebenmann near periodicity". The correct statement of Siebenmann periodicity, along with a complete proof that the topological surgery exact sequence is abelian for triangulable topological manifolds X, can be found in §5 of [320] together with Remark 1 on page 81, which explains how to obtain this statement when X is not triangulable. Nicas [320, pp. 81–83] also shows that the PL surgery exact sequence is abelian. We come back to this point in Section 17.6.

Another major development in topological surgery, parallel to Quinn's surgery spaces, was Ranicki's theory of algebraic surgery. An algebraic surgery exact sequence of abelian groups was constructed in [343] and also [348, § 14, § 18]. This sequence is identified with whole of the geometric surgery sequence, infinite to the left, from Theorem 11.25 in the topological category for both the h and s decoration, see Theorem 15.482. In particular in the topological category one can find abelian group structures on $\mathcal{S}^{\text{TOP},s}(X, \partial X)$, $\mathcal{S}^{\text{TOP},h}(X, \partial X)$ and $\mathcal{N}^{\text{TOP}}(X, \partial X)$ such that whole of the surgery exact sequence, infinite to the left, becomes an exact sequence of abelian groups that agrees with the given abelian group structures coming from Remark 11.36. Here the main point is also to find the right addition on G/TOP. We will say more about how to do this in Section 17.6.

Finally, we note that the abelian structures on the topological surgery exact sequence defined by Quinn–Siebenmann and Ranicki are equal and they are also equal to the stacking group structures of Remark 11.36 in the part where these are defined, see Remark 17.19.

The following remark explains that in the smooth category it is not in general possible to do better than Remark 11.36.

Remark 11.39 (The smooth surgery exact sequence is not additive) Consider the start of the simple surgery exact sequence for a compact smooth manifold X of dimension n

$$L^s_{n+1}(\mathbb{Z}\pi) \xrightarrow{\rho} \mathcal{S}^s(X) \xrightarrow{\eta} [X, \text{G/O}] \xrightarrow{\sigma^s} L^s_n(\mathbb{Z}\pi),$$

where we have identified $\mathcal{N}(X) \equiv [X, G/O]$. In [93, Proposition 4.87] Chang and Weinberger calculate that for $X = T^8$ the image of σ^s in $L_8^s(\mathbb{Z}[\mathbb{Z}^8])$ is not a subgroup. Assuming we keep the given group structure on the L-groups, it follows that the smooth surgery exact sequence cannot be abelian for any H-space structure on G/O. In particular, there is no map of H-spaces from G/O to G/TOP where G/TOP has the Quinn H-space structure.

Another counterexample to the smooth surgery exact sequence being abelian appears when $X = S^3 \times S^4$. By Exercise 11.13, we have $\mathcal{N}(S^3 \times S^4) \cong \pi_4(G/O)$ and $\pi_4(G/O) \cong \mathbb{Z}$. In [108] is was shown that, for each integer $l \in \pi_4(G/O) \cong \mathbb{Z}$, the size of the finite set $\eta^{-1}(l) \subset \mathcal{S}(S^3 \times S^4)$ varies; specifically

$$|\eta^{-1}(l)| = \begin{cases} 4 & (l,7) = 1; \\ 28 & l = 7k. \end{cases}$$

If $\eta \colon \mathcal{S}(S^3 \times S^4) \to \mathcal{N}(S^3 \times S^4)$ were a homomorphism, then it would have a finite kernel and each of its pre-images would be the same size. Hence there are no group structures on $\mathcal{S}(S^3 \times S^4)$ and $\mathcal{N}(S^3 \times S^4)$ such that η is a homomorphism. In particular, the smooth surgery exact sequence is not additive in general.

11.9 Some Information about G/O, G/PL and G/TOP

In this section we shall assume that the reader is familiar with the localisation of spaces. For background we refer either to our Section 17.2 or to the notes [389, Chapter 2]. We write $X[1/p]$ for the result of inverting a prime p, the symbol $X_{(p)}$ stands for the localisation of a space X at a prime p, i.e., all primes except p are inverted, and the symbol $X_\mathbb{Q}$ means the rationalisation of X, i.e., all primes are inverted.

Since BG has finite homotopy groups, the rationalisation $BG_\mathbb{Q}$ is contractible. Hence the canonical map G/O $\to$ BSO is a rational equivalence. By Theorem 12.1 and by (12.61), the rationalisation PL/O$_\mathbb{Q}$ is contractible. By Theorem 11.40 below, the rationalisation TOP/PL$_\mathbb{Q}$ is contractible. Hence the canonical maps G/O $\to$ G/PL $\to$ G/TOP after rationalisation fit into the following zigzag of equivalences:

$$\text{BSO}_\mathbb{Q} \xleftarrow{\simeq} \text{G/O}_\mathbb{Q} \xrightarrow{\simeq} \text{G/PL}_\mathbb{Q} \xrightarrow{\simeq} \text{G/TOP}_\mathbb{Q}.$$

Of the spaces G/O, G/PL and G/TOP, the homotopy types of G/PL and G/TOP are known whereas the determination of the homotopy type of G/O would entail the computation of the stable homotopy groups of spheres and so is currently out of reach, but more on this below.

The spaces G/PL and G/TOP are connected with homotopy groups

$$\pi_n(\text{G/PL}) \cong \pi_n(\text{G/TOP}) \cong \mathbb{Z}, 0, \mathbb{Z}/2, 0 \text{ for } n \equiv 0, 1, 2, 3 \bmod 4$$

when $n \geq 1$, see Exercise 11.37 for the case $n \geq 6$ and Section 17.1 for $n \leq 5$. The determination of the homotopy type of G/PL is due to Sullivan and is explained in detail in [279, Chapter 4]. The homotopy fibre of the map G/PL $\to$ G/TOP is the space TOP/PL that is defined to be the homotopy fibre of the canonical map BPL $\to$ BTOP. Kirby and Siebenmann [219, Theorem 5.5 in Essay V on page 251], see also [369], determined the homotopy type of TOP/PL. To state their result, we let $K(A, l)$ denote the *Eilenberg–MacLane space* of type (A, l), i.e., a *CW*-complex such that $\pi_n(K(A, l))$ is trivial for $n \neq l$ and is isomorphic to A for $n = l$.

Theorem 11.40 (TOP/PL is an Eilenberg–MacLane space) *The space* TOP/PL *has the homotopy type of* $K(\mathbb{Z}/2, 3)$.

The results of Sullivan combined with the above give the following homotopy equivalences:

$$\text{G/PL}\left[\frac{1}{2}\right] \simeq \text{G/TOP}\left[\frac{1}{2}\right] \simeq \text{BO}\left[\frac{1}{2}\right];$$

$$\text{G/PL}_{(2)} \simeq E \times \prod_{j \geq 2} K(\mathbb{Z}_{(2)}, 4j) \times \prod_{j \geq 2} K(\mathbb{Z}/2, 4j - 2);$$

$$\text{G/TOP}_{(2)} \simeq \prod_{j \geq 1} K(\mathbb{Z}_{(2)}, 4j) \times \prod_{j \geq 1} K(\mathbb{Z}/2, 4j - 2).$$

Here E is the total space of the fibration $K(\mathbb{Z}, 4) \to E \to K(\mathbb{Z}/2, 2)$ with k-invariant the element of order 2 in $H^5(K(\mathbb{Z}/2, 2); \mathbb{Z}) \cong \mathbb{Z}/4$ and $\mathbb{Z}_{(2)} \subset \mathbb{Q}$ is the subgroup of all fractions with odd denominators when expressed in lowest terms. The proof of these equivalences will be discussed in more detail in Chapter 17.

With the group structure on $\mathcal{N}^{\text{TOP}}(X) \cong [X, \text{G/TOP}]$ given by the Quinn–Siebenmann–Ranicki H-space structure on G/TOP, see Remark 11.38, we obtain for a finite *CW*-complex X isomorphisms of abelian groups:

$$\mathcal{N}^{\text{TOP}}(X)_{(2)} \cong \prod_{j \geq 1} H^{4j}(X; \mathbb{Z}_{(2)}) \times \prod_{j \geq 1} H^{4j-2}(X; \mathbb{Z}/2);$$

$$\mathcal{N}^{\text{TOP}}(X)\left[\frac{1}{2}\right] \cong \widetilde{\text{KO}}^0(X)\left[\frac{1}{2}\right],$$

(11.41)

where $\widetilde{\text{KO}}^0$ is the reduced K-theory of real vector bundles. For more information, see Section 17.6.

We have already encountered the space G/O in Chapter 6, see (6.84) for its low-dimensional homotopy groups. The following theorem of Sullivan, see [279, Theorem 5.18], isolates the mysterious part of the homotopy type of G/O in certain spaces coker$(J)_{(p)}$ related to the J-homomorphism, a classical object in homotopy theory, which we study in more detail in Section 12.4.

Theorem 11.42 (Sullivan splitting of $G/O_{(p)}$) *For every prime p there are spaces* $\text{coker}(J)_{(p)}$ *and equivalences*

$$G/O_{(p)} \simeq BSO_{(p)} \times \text{coker}(J)_{(p)}.$$

Remark 11.43 (The homotopy groups of the space $\text{coker}(J)_{(p)}$) The homotopy groups of $\text{coker}(J)_{(p)}$ are finite p-torsion groups. If p is odd, then $\pi_n(\text{coker}(J)_{(p)})$ is isomorphic to $\text{coker}(J_n)_{(p)}$, for the J-homomorphism $J_n\colon \pi_n(SO) \to \pi_n^s$ of Definition 12.29. If $p = 2$, then $\pi_n(\text{coker}(J)_{(p)})$ is isomorphic to $\text{coker}(J_n)_{(2)}$, unless $n \equiv 1, 2 \bmod 8$ when $\pi_n(\text{coker}(J)_{(p)})$ is isomorphic to an index two summand of $\text{coker}(J_n)_{(2)}$.

Remark 11.44 (Comparison of smooth and topological surgery) A consequence of the above formulae is that the surgery exact sequence is easier to calculate in the topological category than in the smooth category. This might seem paradoxical at first glance since, as already mentioned in Section 11.6, it is much harder to construct the surgery exact sequence in the topological category than in the smooth category. Nevertheless, once it is set up, the calculations are easier and we will see examples of this in Chapter 18 and 19. As we will see later in Chapter 17, more specifically display (17.1), one reason for this situation is that the generalised Poincaré Conjecture holds in the topological category, but not in the smooth category.

More information about the spaces O, PL, TOP, G, G/O, G/PL and G/TOP, in particular about their homotopy groups, will be given in Section 12.7. The work of Sullivan on G/PL, with its consequences for the topology of G/TOP and G/O, will be reviewed in Chapter 17. Theorem 17.17 and Remark 19.8 describe rational calculation of normal invariants using characteristic classes.

Exercise 11.45 Let M be a simply connected closed 5-dimensional topological manifold. Show that $\mathcal{S}^{\text{TOP},h}(M)$ is finite.

Exercise 11.46 Let M be a simply connected closed 6-dimensional topological manifold. Show that $\mathcal{S}^{\text{TOP},s}(M)$ is finite if and only if $\pi_2(M)$ is finite.

11.10 Functorial Properties of the Geometric Surgery Exact Sequence

We briefly discuss the functorial properties of the geometric surgery exact sequence. Fix a natural number n.

We begin with simple finite Poincaré complexes. Let $\mathcal{P}$ be the following category. An object is a w-oriented finite simple n-dimensional Poincaré complex $X = (X, \mathcal{O}_X, [[X]])$. A morphism $(g, u)\colon X = (X_0, \mathcal{O}_{X_0}, [[X_0]]) \to X_1 = (X_1, \mathcal{O}_{X_1}, [[X_1]])$ consists of a simple homotopy equivalence $g\colon X_0 \xrightarrow{\simeq} X_1$ and a natural equivalence $u\colon \mathcal{O}_{X_0} \to g^*\mathcal{O}(X_1) = \mathcal{O}_{X_1} \circ \Pi(g)$ such that the induced isomorphism $H_n(g; u)\colon H_n(X_0; \mathcal{O}_{X_0}) \xrightarrow{\cong} H_n(X_1; \mathcal{O}_{X_1})$ sends $[[X_0]]$ to $[[X_1]]$.

11.10 Functorial Properties of the Geometric Surgery Exact Sequence

Given a morphism $(g, u) \colon X = (X_0, \mathcal{O}_{X_0}, [[X_0]]) \to X_1 = (X_1, \mathcal{O}_{X_1}, [[X_1]])$, we obtain a map of sets
$$\mathcal{S}^s(g, u) \colon \mathcal{S}^s(X_0) \to \mathcal{S}^s(X_1)$$
by sending the class $[f]$ of a simple homotopy equivalence $f \colon M \to X_0$ with a closed manifold M as source to the class of $g \circ f \colon M \to X_1$. Note that $\mathcal{S}^s(g, u)$ depends only on the homotopy class of g and is independent of u.

We also get a map of sets
$$\mathcal{N}(g, u) \colon \mathcal{N}(X_0) \to \mathcal{N}(X_1)$$
as follows. An element in $\mathcal{N}(X_0)$ can be thought of as a rank k-normal invariant (ξ, c, o) in the sense of Definition 7.5 because of Theorem 7.19 and Remark 7.21. Choose a vector bundle ξ' over X_1 together with a bundle isomorphism $\overline{g} \colon \xi \to g^*\xi'$. Define $c' \in \pi_{n+k}(\mathrm{Th}(\xi'))$ to be the image of c under the isomorphism $\pi_{n+k}(\mathrm{Th}(\xi)) \xrightarrow{\cong} \pi_{n+k}(\mathrm{Th}(\xi'))$ induced by $\overline{g}$ and define the natural equivalence $o' \colon \mathcal{O}_{\xi'} \xrightarrow{\cong} \mathcal{O}_{X_1}$ by requiring that the composite $\mathcal{O}_\xi \xrightarrow{\mathcal{O}_{\overline{g}}} \mathcal{O}_{g^*\xi'} = g^*\mathcal{O}_{\xi'} \xrightarrow{g^*o'} g^*\mathcal{O}_{X_1}$ agrees with the composite $\mathcal{O}_\xi \xrightarrow{o} \mathcal{O}_{X_0} \xrightarrow{u} g^*\mathcal{O}_{X_1}$. Then we define the image of the class of (ξ, c, o) to be the class of (ξ', c', o'). We leave it to the reader to check that this definition makes sense. We have the obvious forgetful functor $\mathcal{P} \to \mathcal{R}$, where $\mathcal{R}$ has been defined in the paragraph before (8.154). Thus we get from Definitions 8.158 and 9.106 a homomorphism of abelian groups
$$L_n^s(g, u) \colon L_n^s(\mathbb{Z}\Pi(X_0), \mathcal{O}_{X_0}) \to L_n^s(\mathbb{Z}\Pi(X_1), \mathcal{O}_{X_1}).$$

One easily checks that the following diagram commutes

$$\begin{array}{ccccc}
\mathcal{S}^s(X_0) & \xrightarrow{\eta_n^s(X_0)} & \mathcal{N}(X_0) & \xrightarrow{\sigma_n^s(X_0)} & L_n^s(\mathbb{Z}\Pi(X_0), \mathcal{O}_{X_0}) \\
{\scriptstyle \mathcal{S}^s(g,u)} \downarrow \cong & & {\scriptstyle \mathcal{N}(g,u)} \downarrow \cong & & {\scriptstyle L_n^s(g,u)} \downarrow \cong \\
\mathcal{S}^s(X_1) & \xrightarrow{\eta_n^s(X_1)} & \mathcal{N}(X_1) & \xrightarrow{\sigma_n^s(X_1)} & L_n^s(\mathbb{Z}\Pi(X_1), \mathcal{O}_{X_1}).
\end{array} \quad (11.47)$$

Hence we get functors $\mathcal{S}^s$, $\mathcal{N}$, and L_n^s from $\mathcal{P}$ to the category whose objects are sets and whose morphisms are bijections, by sending $(X, \mathcal{O}_X, [[X]])$ to $\mathcal{S}^s(X)$, $\mathcal{N}(X)$, and $L_n^s(\mathbb{Z}\Pi(X, \mathcal{O}_X))$ and obtain a sequence of natural transformations $\mathcal{S}^s \xrightarrow{\eta_n^s} \mathcal{N} \xrightarrow{\sigma_n^s} L_n$ which is exact.

Now we turn to closed manifolds. Let $\mathcal{M}$ be the following category. Objects are closed manifolds of dimension n. A morphism $g \colon X_0 \to X_1$ is a simple homotopy equivalence $g \colon X_0 \xrightarrow{\cong} X_1$.

Recall that a closed manifold X comes with a preferred infinite cyclic coefficient system $\mathcal{O}_X$, see (5.25), and a preferred class $[[X]] \in H_n(X; \mathcal{O}_X)$, its intrinsic fundamental class, see Definition 5.29. These will be used in the sequel and actually yield the structure of a simple Poincaré complex on $X = (X, \mathcal{O}_X, [[X]])$, see Theorem 5.51.

Now we define a functor of categories

$$I \colon \mathcal{M} \to \mathcal{P}. \tag{11.48}$$

It sends a closed manifold X to the simple finite Poincaré complex $X = (X, O_X, [[X]])$. Consider a morphism $g \colon X_0 \to X_1$ in $\mathcal{M}$. There is precisely one natural equivalence $u_g \colon O_{X_0} \to g^*O(X_1) = O_{X_1} \circ \Pi(g)$ such that the induced isomorphism $H_n(g; u_g) \colon H_n(X_0; O_{X_0}) \xrightarrow{\cong} H_n(X_1; O_{X_1})$ sends $[[X_0]]$ to $[[X_1]]$. We define $I(g)$ to be (g, u_g). Using I, we obtain functors $\mathcal{S}^s$, $\mathcal{N}$, and L_n^s from $\mathcal{M}$ to the category whose objects are sets and whose morphisms are bijections. These functors depend only on the homotopy class of g. If we abbreviate $L_m^s(X_k) = L_m^s(\mathbb{Z}\Pi(X_k), O_{X_k})$, we obtain the following diagram

$$\begin{array}{ccccccc}
L_{n+1}^s(X_0) & \xrightarrow{\rho_{n+1}^s(X_0)} & \mathcal{S}^s(X_0) & \xrightarrow{\eta_n^s(X_0)} & \mathcal{N}(X_0) & \xrightarrow{\sigma_n^s(X_0)} & L_n^s(X_0) \\
\cong \big\downarrow L_{n+1}^s(I(g)) & & \cong \big\downarrow \mathcal{S}^s(I(g)) & & \mathcal{N}(I(g)) \big\downarrow \cong & & L_n^s(I(g)) \big\downarrow \cong \\
L_{n+1}^s(X_1) & \xrightarrow{\rho_{n+1}^s(X_1)} & \mathcal{S}^s(X_1) & \xrightarrow{\eta_n^s(X_1)} & \mathcal{N}(X_0) & \xrightarrow{\sigma_n^s(X_1)} & L_n^s(X_1).
\end{array} \tag{11.49}$$

Thus we obtain a sequence $L_{n+1}^s \xrightarrow{\rho_{n+1}^s} \mathcal{S}^s \xrightarrow{\eta_n^s} \mathcal{N} \xrightarrow{\sigma_n^s} L_n^s$ of natural transformations of such functors that is exact.

If we additionally assume that g is a diffeomorphism, then we can extend Diagram (11.49) infinitely to the left using the surgery exact sequences of X_0 and X_1 as rows.

Note that there are no requirements about X_0, X_1, or g concerning orientations thanks to the use of the intrinsic fundamental class.

All this works also in the PL category and in the topological category and for the decoration h instead of s.

Remark 11.50 (Non-commutative diagram) Note that the following diagram of bijections does *not* commute in general for a homotopy equivalence $g \colon X_0 \xrightarrow{\simeq} X_1$ of closed manifolds

$$\begin{array}{ccc}
[X_0, G/O] & \xrightarrow[\cong]{\alpha(X_0)} & \mathcal{N}_n(X_0) \\
g^* \big\uparrow \cong & & \cong \big\downarrow \mathcal{N}(g) \\
[X_1, G/O] & \xrightarrow[\cong]{\alpha(X_1)} & \mathcal{N}_n(X_1)
\end{array} \tag{11.51}$$

where g^* is given by precomposition with g and the bijections $\alpha(X_0)$ and $\alpha(X_1)$ come from Theorem 7.34 (ii). Namely, start with the constant map in the left lower corner. Its image under the lower vertical arrow $\alpha(x_1)$ is represented by the normal map given by id_{X_1} and id_{TX_1} whereas its image under the composite $\alpha(X_0) \circ g^*$ is the normal map given by id_{X_0} and id_{TX_0}. The latter element is sent under the right vertical arrow $\mathcal{N}(g)$ to the normal map whose underlying map is given by $g \colon X_0 \to X_1$ covered by bundle data of the form $\overline{g} \colon TX_0 \to (g^{-1})^*TX_0$ for some homotopy inverse g^{-1} of g. It is not necessarily true that the latter normal map and the one given by id_{X_1} and id_{TX_1} define the same element in $\mathcal{N}_n(X_1)$. Namely, there are examples of orientation preserving homotopy equivalences $g \colon X_0 \to X_1$ of oriented closed manifolds X_0 and X_1 such that X_0 and X_1 are not oriented cobordant, see Example 12.66. In view of Remark 19.8, Diagram (11.51) is not commutative since a homotopy equivalence g does not necessarily respect the L-classes.

Note that Diagram (11.51) does commute if g is a diffeomorphism.

All of this carries over directly to the PL category and the topological category. In particular, we get a for a homeomorphism $g \colon X_0 \to X_1$ a commutative diagram

$$\begin{array}{ccc} [X_0, G/\mathrm{TOP}] & \xrightarrow[\cong]{\alpha(X_0)} & \mathcal{N}_n^{\mathrm{TOP}}(X_0) \\ g^* \uparrow \cong & & \cong \downarrow \mathcal{N}(g) \\ [X_1, G/\mathrm{TOP}] & \xrightarrow[\cong]{\alpha(X_1)} & \mathcal{N}_n^{\mathrm{TOP}}(X_1). \end{array} \qquad (11.52)$$

It is not commutative in general if we only assume that g is a homotopy equivalence. This phenomenon is analysed in [66, Proposition 2.2],[281, Lemma 2.5] and [340, Section 3], and is also related to Theorem 18.7.

11.11 Notes

Given a finite Poincaré complex X of dimension ≥ 5, a single obstruction, the so-called *total surgery obstruction*, was constructed by Ranicki in [348, § 17], see also [234] and Subsection 15.9.6. It vanishes if and only if X is homotopy equivalent to a closed topological manifold. It combines the two stages of the obstruction we have seen before, namely, the problem whether the Spivak normal fibration has a reduction to a topological bundle and whether the surgery obstruction of the associated normal map is trivial. More information about it is given in Section 15.9.6 in the chapter about algebraic surgery, which is one of the key technical ingredients in the construction.

A generalisation of the ideas from this chapter with applications to automorphism spaces of manifolds was given by Quinn in [337]. Given a closed n-dimensional manifold M, he constructed spaces $\widetilde{\mathcal{S}}(M)$, $\widetilde{\mathcal{N}}(M)$ and $\widetilde{\mathcal{L}}_n(M)$ and maps such that the sequence

$$\widetilde{\mathcal{S}}(M) \to \widetilde{\mathcal{N}}(M) \to \widetilde{\mathcal{L}}_n(M) \qquad (11.53)$$

turns out to be a homotopy fibration sequence whose long exact sequence of homotopy groups is the geometric surgery exact sequence, in any category with any decoration. Moreover, the additive structure on the surgery exact sequence of Remark 11.36 corresponds to the additive structure on homotopy groups π_i for $i \geq 1$. In the smooth category, it is only an exact sequence of sets on π_0 (as holds in general for the long exact sequence of homotopy groups of a fibration), and this cannot be improved in view of Remark 11.39. On the other hand, discussion from Remark 11.38 says that in the categories PL and TOP it is also possible to endow the sets of components with abelian group structures so that the sequence becomes an exact sequence of abelian groups, but this is a special feature of these categories that requires ideas discussed in Remark 11.38. More information on surgery spaces and on the relation of $\widetilde{S}(M)$ to automorphism spaces of M can be found in [337], [414, Chapter 17A] and [430]. We also touch this topic again briefly in the Notes section of Chapter 13.

Finally, we mention the following version of a braid appearing in [414, page 116], whose notation is explained below and agrees with the one in [414, page 116]. Namely, let $(M, \partial M)$ be a compact manifold with boundary of dimension $n \geq 6$. Then we get the following braid for the decoration h that can be extended infinitely to the left

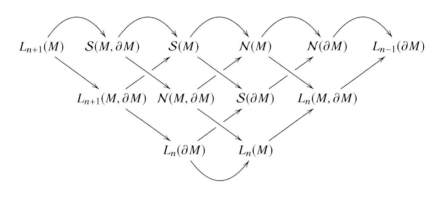

(11.54)

Here $L_*(M)$ and $L_*(\partial M)$ denote the intrinsic L-groups defined in Subsections 8.7.3 and 9.4.1, and $L_*(M, \partial M)$ denotes a certain relative version that depends only on the map of the groupoids $\Pi(\partial M) \to \Pi(M)$ and a choice of an infinite cyclic coefficient system O_M over M, which is suppressed from the notation. The structure sets $S(M, \partial M)$ and $S(\partial M)$ and the set of normal invariants $N(M, \partial M)$ and $N(\partial M)$ have been defined in Definition 11.5 and Definition 11.7. We denote by $S(M)$ and $N(M)$ the structure set and the set of normal invariants where on the boundary one requires only homotopy equivalences for $S(M)$, no conditions on the boundary occur for $N(M)$, and surgery on the boundary is allowed when dealing with the map $N(M) \to L_n(M, \partial M)$. There is also a version for the decoration s.

If $\Pi(\partial M) \to \Pi(M)$ is an equivalence of groupoids, then $L_*(M, \partial M)$ is trivial and hence $S(M) \to N(M)$ is bijective. This is the statement appearing in the π-π-Theorem 13.4.

11.11 Notes

More information on this topic can be found for instance in [89]. See also [348, page 207].

Remark 11.55 Our notation for $\mathcal{N}(M, \partial M)$ and $\mathcal{S}(M, \partial M)$, which agrees with the one in [414, page 116], deviates from the notation one may find elsewhere. Sometimes in the literature $\mathcal{N}(M, \text{rel } \partial M)$ and $\mathcal{S}(M, \text{rel } \partial M)$ or $\mathcal{N}^\partial(M, \partial M)$ and $\mathcal{S}^\partial(M, \partial M)$ are used instead of $\mathcal{N}(M, \partial M)$ and $\mathcal{S}(M, \partial M)$ in order to emphasise that surgery is only done on the interior. The symbols $\mathcal{S}(M, \partial M)$ and $\mathcal{N}(M, \partial M)$ are sometimes used for the versions denoted by $\mathcal{S}(M)$ and $\mathcal{N}(M)$ above, where on the boundary one requires only (simple) homotopy equivalences for the structure sets and no condition on the boundary for the set of normal invariants.

Chapter 12
Homotopy Spheres

12.1 Introduction

A *homotopy n-sphere* Σ is a closed n-dimensional smooth manifold that is homotopy equivalent to S^n. The (Generalised) Poincaré Conjecture says that any homotopy n-sphere Σ is homeomorphic to S^n and is known to be true in all dimensions. It was proven by Smale for $n \geq 5$ in [378], by Freedman for $n = 4$ in [157], by Perelman for $n = 3$ in [333], see also [310] for a published account, and for $n \leq 2$ it follows from the classification of surfaces and curves.

In this chapter we want to solve the problem of how many oriented diffeomorphism classes of oriented homotopy n-spheres exist for a given n, which is the same as determining the smooth structure set $\mathcal{S}(S^n)$. This is a beautiful and illuminating example. It shows how the general surgery methods, which we have developed so far, apply to a concrete problem, and illustrates what kind of input from homotopy theory and algebra is needed for the classification of smooth manifolds via the surgery exact sequence.

Recall that S^n is the standard sphere $S^n = \{x \in \mathbb{R}^{n+1} \mid ||x|| = 1\}$. We will equip it with the structure of a smooth manifold for which the canonical inclusion $S^n \hookrightarrow \mathbb{R}^{n+1}$ is an embedding of smooth manifolds. We use the orientation on S^n that is for each $x \in S^n$ compatible with the isomorphisms $\nu_x(S^n, \mathbb{R}^{n+1}) \oplus T_x S^n \xrightarrow{\cong} T_x \mathbb{R}^{n+1}$ where on $\nu_x(S^n, \mathbb{R}^n)$ we use the orientation coming from the normal vector field pointing from the origin and on $T_x \mathbb{R}^n$ the standard orientation. This agrees with the convention that $S^n = \partial D^{n+1}$ inherits its orientation from D^{n+1} in the sense of Remark 5.37.

Let Θ_n be the abelian group of oriented h-cobordism classes of oriented homotopy n-spheres under the operation of connected sum. We consider the subgroup $bP_{n+1} \subseteq \Theta_n$ of those homotopy n-spheres that bound a stably parallelisable compact manifold, the J-homomorphism $J_n \colon \pi_n(O) \to \pi_n^s$, and the n-th Bernoulli number B_n. These notions will be presented in this chapter, whose main result is Theorem 12.1. Since π_n^s is finite for $n \geq 1$, the same holds for $\mathrm{coker}(J_n)$, the cokernel of the J-homomorphism.

Theorem 12.1 (Classification of Homotopy Spheres) *The group Θ_n is finite. Moreover the following statements hold:*

(i) *Set $a_k := (3 - (-1)^k)/2$. For all $k \geq 2$, bP_{4k} is a finite cyclic group of order*
$$a_k \cdot 2^{2k-2} \cdot (2^{2k-1} - 1) \cdot \text{numerator}(B_k/4k);$$

(ii) *For all $k \geq 1$, the group bP_{4k+2} is either trivial or isomorphic to $\mathbb{Z}/2$. We have*
$$bP_{4k+2} = \begin{cases} \mathbb{Z}/2 & 4k+2 \neq 2^l - 2, \text{ or } 4k+2 = 2^l - 2 \geq 254; \\ \{0\} & 4k+2 \in \{6, 14, 30, 62\}. \end{cases}$$

(In particular, the only unresolved case is bP_{126});

(iii) *For any natural number $k \geq 2$, there is an exact sequence*
$$0 \to \Theta_{4k+2} \to \text{coker}(J_{4k+2}) \to \mathbb{Z}/2$$
where the map $\text{coker}(J_{4k+2}) \to \mathbb{Z}/2$ is non-trivial if and only if bP_{4k+2} is trivial;

(iv) *We have for $k \geq 1$*
$$\Theta_{4k} \cong \text{coker}(J_{4k});$$

(v) *For any natural number $k \geq 2$, there is a short exact sequence*
$$0 \to bP_{2k+2} \to \Theta_{2k+1} \to \text{coker}(J_{2k+1}) \to 0;$$

(vi) *For all $k \geq 3$, the group bP_{2k+1} is trivial.*

The following table, largely taken from [216, pages 504 and 512], gives the orders of the finite groups Θ_n, bP_{n+1}, and Θ_n/bP_{n+1} up to dimension 15. The values in dimension 1 and 2 come from obvious ad hoc computations. The only change in the table since 1963 corresponds to Perelman's solution of the Poincaré Conjecture, see [310] for details of the proof and the relevant literature that gives the values for $n = 3$. Note that $|\Theta_4| = 1$ does *not* mean that any homotopy 4-sphere is diffeomorphic to S^4, see Lemma 12.5.

n	1	2	3	4	5	6	7	8	9	10	11	12	13	14	15
$|bP_{n+1}|$	1	1	1	1	1	1	28	1	2	1	992	1	1	1	8128
$|\Theta_n|$	1	1	1	1	1	1	28	2	8	6	992	1	3	2	16256
$|\Theta_n/bP_{n+1}|$	1	1	1	1	1	1	1	2	4	6	1	1	3	2	2

(12.2)

More information about $|\Theta_n|$ up to dimension 90 can be found in the recent paper [199, Table 1].

Most of Theorem 12.1 was proven by Kervaire and Milnor in their seminal 1963 paper [216] with the title "Groups of homotopy spheres. I". It contains a systematic study of surgery and can be viewed as the beginning of surgery theory. Nearly all the results presented here are taken from [216]. However, a feature of the following presentation using the surgery exact sequence is that in dimensions $n \geq 5$ we only use that homotopy spheres are almost parallelisable, see Definition 12.10, whereas the proof that homotopy spheres are stably parallelisable, see Definition 12.7, was a first basic step in the line of argument in [216].

Since 1963, two major advances in stable homotopy theory have contributed to the statement of Theorem 12.1. The first advance was the final determination of the order of the image of the J-homomorphism, see Theorem 12.33 and discussion preceding it, which is an input to assertion (i). The second advance was progress on the Kervaire invariant problem, see Corollary 12.53 and the discussion preceding it, which is an input to the second sentence of assertion (ii). The proof of Theorem 12.1 is distributed across Theorem 12.49, Corollary 12.53, Theorem 12.54, and Theorem 12.55.

The paper "Groups of homotopy spheres. II" never appeared, but a paper by Levine [246] is often regarded as the intended sequel. It presents another exposition of the arguments from [216] and also the Kervaire–Milnor Braid, see Theorem 12.59. There is also a detailed survey article about homotopy spheres by Lance [239], and for more recent information about Θ_n we refer to [36].

Guide 12.3 It is best to read this chapter linearly.

12.2 The Group of Homotopy Spheres

Define the *n-th group of homotopy spheres* Θ_n as follows. Elements are oriented h-cobordism classes $[\Sigma]$ of oriented homotopy n-spheres Σ where Σ and Σ' are called oriented h-cobordant if there is an oriented h-cobordism $(W, \partial_0 W, \partial_1 W)$ and orientation preserving diffeomorphisms $\Sigma \to \partial_0 W$ and $(\Sigma')^- \to \partial_1 W$. The addition is given by connected sum. The zero element is represented by S^n. The inverse of $[\Sigma]$ is given by $[\Sigma^-]$, where Σ^- is obtained from Σ by reversing the orientation. Obviously Θ_n is abelian.

We at least check that $[\Sigma^-]$ is indeed an inverse of $[\Sigma]$. We have to show that for a homotopy n-sphere Σ there is an h-cobordism W from the connected sum $\Sigma \sharp \Sigma^-$ to S^n. Let $D^n \subset \Sigma$ be an embedded disk. Let $\Sigma \setminus \text{int}(D^n)$ be the manifold with boundary obtained by removing this disk. Delete the interior of an embedded disk $D^{n+1} \subset (\Sigma \setminus \text{int}(D^n)) \times [0, 1]$. The result is the desired h-cobordism W.

Exercise 12.4 For $n \geq 6$, prove that every homotopy n-sphere Σ is oriented diffeomorphic to $D^n \cup_f (D^n)^-$ for some orientation preserving diffeomorphism $f \colon S^{n-1} \to S^{n-1}$.

Lemma 12.5 *Let $\overline{\Theta}_n$ be the set of oriented diffeomorphism classes $[\Sigma]$ of oriented homotopy n-spheres Σ. The forgetful map*

$$f \colon \overline{\Theta}_n \to \Theta_n$$

is bijective for $n \neq 4$.

Proof. By definition f is onto. For $n \geq 5$, f is injective by the s-cobordism Theorem 2.1. For $n = 2, 3$, the set $\overline{\Theta}_n$ contains just one element, by the classification of surfaces for $n = 2$ and by Perelman's proof of the Poincaré Conjecture for $n = 3$ combined with the fact that 3-dimensional manifolds have unique smooth structures [309]. □

Lemma 12.6 *There is a natural bijection*

$$\alpha \colon \mathcal{S}(S^n) \xrightarrow{\cong} \overline{\Theta}_n, \quad [f \colon M \to S^n] \mapsto [M]$$

where we choose the orientation on M such that f has degree one with respect to the standard orientation on S^n. In addition, there is an obvious bijection

$$\{\text{smooth oriented structures on } S^n\}/\text{o.p. diffeomorphism} \xrightarrow{\cong} \overline{\Theta}_n$$

where "o.p." abbreviates "orientation preserving".

Proof. For any oriented homotopy n-sphere Σ, there is up to homotopy precisely one map $f \colon \Sigma \to S^n$ of degree one. Recall that the (Generalised) Poincaré Conjecture, which is known to be true, says that any homotopy n-sphere is homeomorphic to S^n. □

Definition 12.7 (Stably parallelisable) A manifold M is called *stably parallelisable* if $TM \oplus \mathbb{R}^a$ is trivial for some $a \geq 0$.

Definition 12.8 (The set bP_{n+1}) Let $bP_{n+1} \subseteq \Theta_n$ be the subset of elements $[\Sigma]$ for which Σ is oriented diffeomorphic to the boundary ∂M of a stably parallelisable compact manifold M.

Lemma 12.9 *The subset $bP_{n+1} \subseteq \Theta_n$ is a subgroup of Θ_n. For $n \neq 4$, bP_{n+1} is the preimage under the composite*

$$\Theta_n \xrightarrow{(f \circ \alpha)^{-1}} \mathcal{S}(S^n) \xrightarrow{\eta} \mathcal{N}(S^n)$$

of the base point $[\mathrm{id} \colon TS^n \to TS^n]$ in $\mathcal{N}(S^n)$, where f is the bijection of Lemma 12.5 and α is the bijection of Lemma 12.6.

Proof. Suppose that Σ bounds W and Σ' bounds W' for stably parallelisable manifolds W and W'. Then the boundary connected sum $W \natural W'$ is stably parallelisable and has $\Sigma \sharp \Sigma'$ as boundary. This shows that $bP_{n+1} \subseteq \Theta_n$ is a subgroup.

Consider an element $[f \colon \Sigma \to S^n]$ in $\eta^{-1}([\mathrm{id} \colon TS^n \to TS^n])$. Then there is a normal bordism from a normal map $(f, \overline{f}) \colon T\Sigma \oplus \mathbb{R}^a \to \xi$ covering $f \colon \Sigma \to S^n$ to the normal map $(\mathrm{id}, \overline{\mathrm{id}}) \colon TS^n \to TS^n$. This normal bordism is given by a bundle map $(F, \overline{F}) \colon TW \oplus \mathbb{R}^{a+b} \to \eta$ covering a map of triads $(F; \partial_0 F, \partial_1 F) \colon (W; \partial_0 W, \partial_1 W) \to (S^n \times [0,1]; S^n \times \{0\}, S^n \times \{1\})$, as well as two bundle isomorphisms $(u, \overline{u}) \colon T\Sigma \oplus \mathbb{R}^{a+b+1} \to TW \oplus \mathbb{R}^b$ and $(u', \overline{u'}) \colon TS^n \oplus \mathbb{R}^{a+b+1} \to TW \oplus \mathbb{R}^b$ covering, respectively, orientation preserving diffeomorphisms $u \colon \Sigma \to \partial_0 W$ and

12.3 The Surgery Exact Sequence for Homotopy Spheres

$u'\colon (S^n)^- \to \partial_1 W$, and finally two further bundle isomorphisms $(\nu, \overline{\nu})\colon \xi \oplus \mathbb{R}^{b+1} \to \eta$ and $(\nu', \overline{\nu'})\colon TS^n \oplus \mathbb{R}^{a+b+1} \to \eta$, covering, respectively, the obvious maps $v\colon S^n \to S^n \times \{0\}$ and $v'\colon S^n \to S^n \times \{1\}$, such that $(f, \overline{f}) \circ (u, \overline{u}) = (\nu, \overline{\nu}) \circ (\overline{f} \oplus \mathrm{id}_{\mathbb{R}^{b+1}})$ and $(f, \overline{f}) \circ (u', \overline{u'}) = (\nu', \overline{\nu'})$ hold. Then $D^{n+1} \cup_{u'} W$ is a manifold whose boundary is oriented diffeomorphic to Σ by u and for which the bundle data above yield a stable isomorphism $TW \oplus \mathbb{R}^{a+b} \to \mathbb{R}^{n+1+a+b}$. Hence $[\Sigma]$ lies in bP_{n+1}.

Conversely, consider $[\Sigma]$ such that there exists a stably parallelisable manifold W together with an orientation preserving diffeomorphism $u\colon \Sigma \to \partial W$. We can assume without loss of generality that W is connected. Choose an orientation preserving homotopy equivalence $f\colon \Sigma \to S^n$. We can extend $\partial_0 F := f \circ u^{-1}\colon \partial W \to S^n$ to a smooth map $F\colon W \to D^{n+1}$. Since f has degree one, the map $(F, \partial F)\colon (W, \partial W) \to (D^{n+1}, S^n)$ has degree one. Let $y \in D^{n+1} \setminus S^n$ be a regular value. Then the degree of F is the finite sum $\sum_{x \in F^{-1}(y)} \epsilon(x)$ where $\epsilon(x) = 1$ if $T_x F\colon T_x W \to T_y D^n$ preserves the induced orientations and $\epsilon(x) = -1$ otherwise. If two points x_1 and x_2 in $F^{-1}(y)$ satisfy $\epsilon(x_1) \neq \epsilon(x_2)$, one can change F up to homotopy relative ∂W so that $F^{-1}(y)$ contains two points less than before. Thus one can arrange that $F^{-1}(y)$ consists of precisely one point x and that $T_x F\colon T_x W \to T_x D^{n+1}$ is orientation preserving. Then one can change F up to homotopy in a small neighbourhood of x so that there is an embedded disk $D_0^{n+1} \subset W \setminus \partial W$ such that F induces a diffeomorphism $D_0^{n+1} \to F(D_0^{n+1})$ and no point outside D_0^{n+1} is mapped to $F(D_0^{n+1})$. Define $V = W \setminus \mathrm{int}(D_0^{n+1})$. Then F induces a map, also denoted by $F\colon V \to D^{n+1} \setminus F(D_0^{n+1})$. If we identify $D^{n+1} \setminus F(D_0^{n+1})$ with $S^n \times [0, 1]$ by an orientation preserving diffeomorphism, we get a map of triads $(F; \partial_0 F, \partial_1 F)\colon (V; \partial_0 V, \partial_1 V) \to (S^{n+1} \times [0, 1], S^n \times \{0\}, S^n \times \{1\})$ together with diffeomorphisms $u\colon \Sigma \to \partial_0 V$, $v\colon S^n \to S^n \times \{0\}$, $u'\colon S^n \to \partial_0 V$, and $v'\colon S^n \to S^n \times \{1\}$ such that $\partial_1 F$ is an orientation preserving diffeomorphism and $F \circ u = v \circ f$ and $F \circ u' = v'$. Now one covers everything with appropriate bundle data to obtain a normal bordism from $(f, \overline{f})\colon T\Sigma \oplus \mathbb{R}^a \to \xi$ to $\mathrm{id}\colon TS^n \to TS^n$. This shows $\eta \circ \alpha^{-1} \circ f^{-1}([\Sigma]) = [\mathrm{id}\colon TS^n \to TS^n]$. □

By Theorem 12.17 below, $\Theta_4 = \{0\}$ and so $bP_5 = \Theta_4$.

12.3 The Surgery Exact Sequence for Homotopy Spheres

In this section we examine the surgery sequence, see Theorem 11.22, in the case of the sphere. In contrast to the general case, we will obtain a long exact sequence of abelian groups. We have to introduce the following bordism groups.

Definition 12.10 (Stable and almost stable framing) A *stable framing* of a compact oriented manifold M of dimension n is a strong bundle isomorphism $\overline{u}\colon TM \oplus \mathbb{R}^a \xrightarrow{\cong} \mathbb{R}^{n+a}$ for some $a \geq 0$ that is compatible with the given orientation. (Recall that strong means that $\overline{u}$ covers the identity.) A *stably framed manifold* is a pair $(M, \overline{u})$ as above.

An *almost stable framing* of a closed oriented manifold M of dimension n is a choice of a point $x \in M$ together with a strong bundle isomorphism $\overline{u} \colon TM|_{M \setminus \{x\}} \oplus \mathbb{R}^a \xrightarrow{\cong} \mathbb{R}^{n+a}$ for some $a \geq 0$ that is compatible with the given orientation on $M \setminus \{x\}$. An *almost stably framed manifold* is a triple $(M, x, \overline{u})$ as above.

Of course any stably framed manifold $(M, \overline{u})$ induces in particular an almost stably framed manifold for every $x \in M$ by restricting $\overline{u}$. A homotopy n-sphere Σ admits an almost stable framing since for any point $x \in \Sigma$ the complement $\Sigma \setminus \{x\}$ is contractible and hence $T\Sigma|_{\Sigma \setminus \{x\}}$ is trivial and, up to homotopy of bundle isomorphisms, admits one homotopy class of trivialisation. We will later show the non-trivial fact that any homotopy n-sphere is stably parallelisable, i.e., admits a stable framing. The standard sphere S^n inherits its standard stable framing from its embedding in $\mathbb{R}^{n+1}$.

A *stably framed nullbordism* for a stably framed manifold $(M, \overline{u})$ is a compact manifold W with a stable framing $\overline{U} \colon TW \oplus \mathbb{R}^{a+b} \xrightarrow{\cong} \mathbb{R}^{n+1+a+b}$ and a bundle isomorphism $(v, \overline{v}) \colon TM \oplus \mathbb{R}^{a+1+b} \xrightarrow{\cong} TW \oplus \mathbb{R}^{a+b}$ coming from the differential of an orientation preserving diffeomorphism $v \colon M \to \partial W$ such that $\overline{U} \circ \overline{v} = \overline{u} \oplus \mathrm{id}_{\mathbb{R}^{b+1}}$. Now define a *stably framed bordism* from a stably framed manifold $(M, \overline{u})$ to another stably framed manifold $(M', \overline{u'})$ to be a stably framed nullbordism for the disjoint union of $(M^-, \overline{u}^-)$ and $(M', \overline{u'})$, where M^- is obtained from M by reversing the orientation and $\overline{u}^-$ is the composition

$$\overline{u}^- \colon TM \oplus \mathbb{R}^a \xrightarrow{\overline{u}} \mathbb{R}^{a+n} = \mathbb{R} \oplus \mathbb{R}^{a+n-1} \xrightarrow{-\mathrm{id}_{\mathbb{R}} \oplus \mathrm{id}_{\mathbb{R}^{a+n-1}}} \mathbb{R} \oplus \mathbb{R}^{a+n-1} = \mathbb{R}^{a+n}.$$

Consider almost stably framed manifolds $(M, x, \overline{u})$ and $(M', x', \overline{u'})$. An almost stably framed bordism from $(M, x, \overline{u})$ to $(M', x', \overline{u'})$ consists of the following data. A compact oriented $(n+1)$-dimensional manifold triad $(W; \partial_0 W, \partial_1 W)$ satisfying $\partial_0 W \cap \partial_1 W = \emptyset$ together with an embedding $j \colon ([0,1]; \{0\}, \{1\}) \hookrightarrow (W; \partial_0 W, \partial_1 W)$ such that j is transversal at the boundary and a strong bundle isomorphism $\overline{U} \colon TW_{W - \mathrm{im}(j)} \oplus \mathbb{R}^b \xrightarrow{\cong} \mathbb{R}^{n+1+b}$ for some $b \geq a, a'$. Furthermore, we require the existence of a bundle isomorphism $(v, \overline{v}) \colon TM \oplus \mathbb{R}^{b+1} \xrightarrow{\cong} TW \oplus \mathbb{R}^b$ coming from the differential of an orientation preserving diffeomorphism $v \colon M^- \to \partial_0 W$ with $v(x) = j(0)$ and of a bundle isomorphism $(v', \overline{v'}) \colon TM' \oplus \mathbb{R}^{b+1} \xrightarrow{\cong} TW \oplus \mathbb{R}^b$ coming from the differential of an orientation preserving diffeomorphism $v' \colon M' \to \partial_1 W$ with $v'(x') = j(1)$ such that $\overline{U} \circ \overline{v} = \overline{u}^- \oplus \mathrm{id}_{\mathbb{R}^{b-a+1}}$ and $\overline{U} \circ \overline{v'} = \overline{u'} \oplus \mathrm{id}_{\mathbb{R}^{b+1-a'}}$ hold.

Definition 12.11 (Framed and almost framed bordism) Let Ω_n^{fr} be the abelian group of stably framed bordism classes of stably framed closed oriented manifolds of dimension n. This becomes an abelian group under disjoint union. The zero element is represented by S^n with its standard stable framing. The inverse of the class of $(M, \overline{u})$ is represented by the class of $(M^-, \overline{u} \oplus (-\mathrm{id}_{\mathbb{R}}))$.

Let Ω_n^{alm} be the abelian group of almost stably framed bordism classes of almost stably framed closed oriented manifolds of dimension n. This becomes an abelian group under the connected sum at the base points. The zero element is represented

12.3 The Surgery Exact Sequence for Homotopy Spheres 437

by S^n with the base point $s = (1, 0, \ldots, 0)$ with its standard stable framing restricted to $S^n \setminus \{s\}$. The inverse of the class of $(M, x, \bar{u})$ is represented by the class of $(M^-, x, \bar{u} \oplus (-\mathrm{id}_\mathbb{R}))$.

Lemma 12.12 *There are canonical bijections of pointed sets*

$$\beta \colon \mathcal{N}(S^n) \xrightarrow{\cong} \Omega_n^{\mathrm{alm}};$$

$$\gamma \colon \mathcal{N}(S^n \times [0,1], S^n \times \{0,1\}) \xrightarrow{\cong} \mathcal{N}(S^{n+1}).$$

Proof. Let $r \in \mathcal{N}(S^n)$ be represented by a normal map $(f, \bar{f}) \colon TM \oplus \mathbb{R}^a \to \xi$ covering a map of degree one $f \colon M \to S^n$. Since f has degree one, one can change $(f, \bar{f})$ by a homotopy such that $f^{-1}(s)$ consists of one point $x \in M$ for a fixed point $s \in S^n$. Since $S^n \setminus \{s\}$ is contractible, $\xi|_{S^n \setminus \{s\}}$ admits a trivialisation that is unique up to homotopy. It induces together with $\bar{f}$ an almost stable framing on (M, x). The class of (M, x) with this stable framing in Ω_n^{alm} is defined to be the image of r under β.

Next we define the inverse β^{-1} of β. Let $r \in \Omega_n^{\mathrm{alm}}$ be represented by the almost framed manifold $(M, x, \bar{u})$. Let $c \colon M \to S^n$ be the Hopf collapse map for a small embedded disk $D^n \subset M$ with origin x. By construction c induces a diffeomorphism $c|_{\mathrm{int}(D^n)} \colon \mathrm{int}(D^n) \to S^n \setminus \{s\}$ and obviously maps $M \setminus \mathrm{int}(D^n)$ to $\{s\}$ for fixed $s \in S^n$. The almost stable framing $\bar{u}$ yields a bundle map $\overline{c'} \colon TM|_{M \setminus \{x\}} \oplus \mathbb{R}^a \to \mathbb{R}^{n+a}$ covering $c|_{M \setminus \{x\}} \colon M \setminus \{x\} \to S^n \setminus \{c(x)\}$. Since D^n is contractible, we obtain a bundle map unique up to homotopy $\overline{c''} \colon TD^n \oplus \mathbb{R}^a \to \mathbb{R}^{n+a}$ covering $c|_{D^n} \colon D^n \to S^n$. The composition of the inverse of the restriction of $\overline{c''}$ to $\mathrm{int}(D^n) \setminus \{x\}$ and of the restriction of $\overline{c'}$ to $\mathrm{int}(D^n) \setminus \{x\}$ yields a strong bundle automorphism of the trivial bundle $\mathbb{R}^{n+a}$ over $S^n \setminus \{s, c(x)\}$. Let ξ be the vector bundle obtained by gluing the trivial bundle $\mathbb{R}^{a+n}$ over $S^n \setminus \{s\}$ and the trivial bundle $\mathbb{R}^{a+n}$ over $S^n \setminus \{c(x)\}$ together using this bundle automorphism over $S^n \setminus \{s, c(x)\}$. Then $\overline{c'}$ and $\overline{c''}$ fit together to a bundle map $\bar{c} \colon TM \oplus \mathbb{R}^{n+a} \to \xi$ covering c. Define the image of r under β^{-1} to be the class of $(c, \bar{c})$.

Consider $r \in \mathcal{N}(S^n \times [0, 1], S^n \times \{0, 1\})$ represented by a normal map $(f, \bar{f}) \colon TM \oplus \mathbb{R}^a \to \xi$ covering $(f, \partial f) \colon (M, \partial M) \to (S^n \times [0, 1], S^n \times \{0, 1\})$. Recall that ∂f is a diffeomorphism. Hence one can form the closed manifold $N = M \cup_{\partial f} D^{n+1} \times \{0, 1\}$. The map f and the identity on $D^{n+1} \times \{0, 1\}$ induce $g \colon N \to (S^n \times [0, 1]) \cup_{S^n \times \{0,1\}} (D^{n+1} \times \{0, 1\}) \cong S^{n+1}$, which is a map of degree one. We define the bundle η over the manifold $(S^n \times [0, 1]) \cup_{S^n \times \{0,1\}} (D^{n+1} \times \{0, 1\}) \cong S^{n+1}$ by gluing the bundles ξ and $T(D^{n+1} \times \{0, 1\}) \oplus \mathbb{R}^{a-1}$ together over $S^n \times \{0, 1\}$ by the strong bundle isomorphism

$$\xi|_{S^{n-1} \times \{0,1\}} \xrightarrow{\left(\bar{f}|_{S^{n-1} \times \{0,1\}}\right)^{-1}} TM|_{\partial M} \oplus \mathbb{R}^a = T(\partial M) \oplus \mathbb{R}^{a+1}$$

$$\xrightarrow{T\partial f \oplus \mathrm{id}_{\mathbb{R}^{a+1}}} T(S^{n-1} \times \{0, 1\}) \oplus \mathbb{R}^{a+1} = T(D^{n+1} \times \{0, 1\}) \oplus \mathbb{R}^a.$$

Then $\bar{f}$ and id: $T(D^{n+1} \times \{0,1\}) \oplus \underline{\mathbb{R}^a} \to T(D^{n+1} \times \{0,1\}) \oplus \underline{\mathbb{R}^a}$ fit together to give a bundle map $\bar{g} \colon TN \oplus \underline{\mathbb{R}^a} \to \eta$ covering g. Define the image of r under γ by the class of $(g, \bar{g})$. We leave it to the reader to construct the inverse of γ, which is similar to the construction in Lemma 12.9, but now two embedded disks instead of one embedded disk are removed. □

Next we wish to place Θ_n in a long exact sequence containing L-groups. Since the fundamental group π of a homotopy n-sphere is trivial for $n \geq 2$, the orientation homomorphism w is always trivial and $\mathrm{Wh}(\pi) = \{1\}$. Hence it does not matter whether we work with simple homotopy equivalences or homotopy equivalence since $\mathcal{S}^s(S^n) = \mathcal{S}^h(S^n)$ for $n \geq 5$ and $L_n^s(\mathbb{Z}\pi) = L_n^h(\mathbb{Z}\pi) = L_n(\mathbb{Z})$ holds. Therefore we will omit the decoration h or s for the remainder of this chapter.

For $n \geq 5$ we want to construct a long exact sequence of abelian groups

$$\cdots \to \Omega_{n+1}^{\mathrm{alm}} \xrightarrow{\sigma} L_{n+1}(\mathbb{Z}) \xrightarrow{\rho} \Theta_n \xrightarrow{\eta} \Omega_n^{\mathrm{alm}} \xrightarrow{\sigma} L_n(\mathbb{Z}) \to \cdots.$$

The map

$$\sigma \colon \Omega_{n+1}^{\mathrm{alm}} \to L_{n+1}(\mathbb{Z})$$

is given by the composition

$$\Omega_{n+1}^{\mathrm{alm}} \xrightarrow{\beta^{-1}} \mathcal{N}(S^{n+1}) \xrightarrow{\sigma} L_{n+1}(\mathbb{Z})$$

where β is the bijection of Lemma 12.12 and $\sigma \colon \mathcal{N}(S^{n+1}) \to L_{n+1}(\mathbb{Z})$ is given by the surgery obstruction and has already appeared in the surgery sequence, see Theorem 11.22. The map

$$\rho \colon L_{n+1}(\mathbb{Z}) \to \Theta_n$$

is the composite of the inverse of the bijection $\alpha \circ f \colon \mathcal{S}(S^n) \xrightarrow{\cong} \Theta_n$ coming from Lemma 12.5 and Lemma 12.6 and the map $\rho \colon L_{n+1}(\mathbb{Z}) \to \mathcal{S}(S^n)$ of the surgery sequence, see Theorem 11.22. The map

$$\eta \colon \Theta_n \to \Omega_n^{\mathrm{alm}} \tag{12.13}$$

sends the class of a homotopy sphere Σ to the class of $(\Sigma, x, \bar{u})$, where x is any point in Σ and the stable framing of $T\Sigma|_{\Sigma \setminus \{x\}}$ comes from the fact that $\Sigma \setminus \{x\}$ is contractible. This map η corresponds to the map η appearing in the surgery exact sequence, see Theorem 11.22, under the identification $\alpha \circ f \colon \mathcal{S}(S^n) \xrightarrow{\cong} \Theta_n$ coming from Lemma 12.5 and Lemma 12.6.

We leave it to the reader to check that all these maps are homomorphisms of abelian groups. The surgery exact sequence, see Theorem 11.22, implies the following theorem.

12.3 The Surgery Exact Sequence for Homotopy Spheres

Theorem 12.14 (The surgery sequence for homotopy spheres) *The sequence of abelian groups that extends infinitely to the left*

$$\cdots \to \Omega^{\text{alm}}_{n+1} \xrightarrow{\sigma} L_{n+1}(\mathbb{Z}) \xrightarrow{\rho} \Theta_n \xrightarrow{\eta} \Omega^{\text{alm}}_n \xrightarrow{\sigma} L_n(\mathbb{Z}) \xrightarrow{\rho} \cdots \xrightarrow{\eta} \Omega^{\text{alm}}_5 \xrightarrow{\sigma} L_5(\mathbb{Z})$$

is exact.

We shall next cut the long exact sequence of Theorem 12.14 into shorter sequences that lead to the proof of Theorem 12.1. Recall that we have shown in Theorem 8.99, Theorem 8.111 and Theorem 9.18 that $L_{2i+1}(\mathbb{Z}) = 0$ for $i \in \mathbb{N}$ and there are isomorphisms

$$\frac{1}{8} \cdot \text{sign} \colon L_0(\mathbb{Z}) \xrightarrow{\cong} \mathbb{Z}$$

and

$$\text{Arf} \colon L_2(\mathbb{Z}) \xrightarrow{\cong} \mathbb{Z}/2.$$

Recall also from Theorem 8.173 (i) and (ii) that the surgery obstruction of a degree one normal map $(f, \overline{f}) \colon M \to S^n$ from a closed n-manifold to the n-sphere is $\frac{1}{8} \text{sign}(M)$ when $n = 4i$, and the Arf invariant of the quadratic form on the surgery kernel when $n = 4i - 2$. In the latter case, this is called the *Kervaire invariant* of the degree one normal map $(f, \overline{f})$, denoted $\text{KI}(f, \overline{f})$. Hence Theorem 12.14 and Lemma 12.9 imply the following result.

Corollary 12.15 *For $i \geq 2$ and $j \geq 3$ there are short exact sequences of abelian groups*

$$0 \to \Theta_{4i} \xrightarrow{\eta} \Omega^{\text{alm}}_{4i} \xrightarrow{\frac{\text{sign}}{8}} \mathbb{Z} \xrightarrow{\rho} bP_{4i} \to 0$$

and

$$0 \to \Theta_{4i-2} \xrightarrow{\eta} \Omega^{\text{alm}}_{4i-2} \xrightarrow{\text{KI}} \mathbb{Z}/2 \xrightarrow{\rho} bP_{4i-2} \to 0$$

and

$$0 \to bP_{2j} \to \Theta_{2j-1} \xrightarrow{\eta} \Omega^{\text{alm}}_{2j-1} \to 0.$$

Here the map

$$\frac{\text{sign}}{8} \colon \Omega^{\text{alm}}_{4i} \to \mathbb{Z}$$

sends a bordism class $[M, x, \overline{u}]$ to $\frac{1}{8} \cdot \text{sign}(M)$ and $\text{KI} \colon \Omega^{\text{alm}}_{4i-2} \to \mathbb{Z}/2$ sends $[M, x, \overline{u}]$ to the Arf invariant of the normal map $\beta^{-1}([M, x, \overline{u}]) \in \mathcal{N}(S^{4i-2})$ where β is the bijection appearing in Lemma 12.12.

We conclude this section by showing that $\Theta_4 = \{0\}$ and that the long exact sequence of Theorem 12.14 extends three terms to the right. For $n = 4$ (but only for $n = 4$), we shall need the following fundamental fact about homotopy spheres.

Theorem 12.16 (Stable parallelisability of homotopy spheres) *For $n \geq 1$ any homotopy n-sphere Σ is stably parallelisable.*

Proof. For an almost parallelisable manifold $(M, \overline{u})$ the image of its class $[M, \overline{u}] \in \Omega_n^{\text{alm}}$ under the homomorphism $\partial \colon \Omega_n^{\text{alm}} \to \pi_n(\text{SO}(n-1))$ of (12.21) is exactly the obstruction to extending the almost stable framing to a stable framing. Recall that any homotopy n-sphere is almost stably parallelisable. The map ∂ is trivial for $n \neq 0 \bmod 4$ by Lemma 12.23, Lemma 12.30, and Theorem 12.33 (i). If $n = 0 \bmod 4$, the claim follows from Lemma 12.23, Lemma 12.30 and Lemma 12.46 since the signature of a homotopy n-sphere is trivial. $\square$

Theorem 12.17 (The long exact sequence for homotopy spheres) *We have $\Theta_4 = \{0\}$ and the sequence of Theorem 12.14 extends three terms to the right as an exact sequence:*

$$\cdots \xrightarrow{\eta} \Omega_5^{\text{alm}} \xrightarrow{\sigma} L_5(\mathbb{Z}) \xrightarrow{\rho} \Theta_4 \xrightarrow{\eta} \Omega_4^{\text{alm}} \xrightarrow{\sigma} L_4(\mathbb{Z}).$$

Proof. Let Σ be a homotopy 4-sphere. Theorem 12.16 implies that Σ is stably parallelisable and so we choose a stable framing $\overline{u}$ of Σ. By (12.27) and by (12.31) below, $(\Sigma, \overline{u})$ bounds a stably framed manifold $(W, \overline{F})$. We fix a degree one map $c \colon \Sigma \to S^4$ and extend it to a degree one map of pairs $(C, c) \colon (W, \Sigma) \to (D^5, S^4)$. Since $TD^5 = \underline{\mathbb{R}^5}$, the stable framing $\overline{F}$ is equivalent to the bundle data of a degree one normal map of pairs $(C, \overline{C}) \colon (W, \Sigma) \to (D^5, S^4)$. Since c is a homotopy equivalence and $L_5(\mathbb{Z}) = \{0\}$, Theorem 9.113 applies to show that we can perform surgery on the interior of W to obtain a homotopy equivalence of pairs $(C', c) \colon (W', \Sigma) \to (D^5, S^4)$. Now delete the interior of an embedded disk $D^5 \subset W'$. The result is an h-cobordism from Σ to S^4. Hence we have shown $\Theta_4 = \{0\}$.

Since $L_5(\mathbb{Z}) = \Theta_4 = \{0\}$, the exactness of the extension is equivalent to the surgery obstruction map $\Omega_4^{\text{alm}} \xrightarrow{\sigma} L_4(\mathbb{Z})$ being injective. We have seen above that $\sigma(M, x, \overline{u}) = \frac{\text{sign}(M)}{8}$. Now since $\pi_i(O) = \{0\}$ for $i = 2$, it follows that Ω_4^{alm} is identified with the spin bordism group Ω_4^{Spin} that is known to be isomorphic to $16 \cdot \mathbb{Z}$, detected by the signature [299]. $\square$

We conclude this section by considering the tangent bundle $T\Sigma$ of a homotopy n-sphere Σ. Since this topic is not needed for the surgery classification of homotopy spheres, it is a digression from the main line of argument in this chapter. We have seen that there is an orientation preserving homeomorphism $f \colon \Sigma \to S^n$ and Theorem 12.16 implies $T\Sigma \oplus \underline{\mathbb{R}^a} \cong \underline{\mathbb{R}^{n+a}}$. Since $TS^n \oplus \underline{\mathbb{R}} \cong \underline{\mathbb{R}^{n+1}}$, it follows that $f^*(TS^n \oplus \underline{\mathbb{R}^a}) \cong (T\Sigma \oplus \underline{\mathbb{R}^a})$. It is natural to ask whether this isomorphism destabilises to an isomorphism $TS^n \cong f^*(T\Sigma)$.

Proposition 12.18 (The tangent bundle of a homotopy sphere is standard) *Let $f \colon S^n \to \Sigma$ be an orientation preserving homotopy equivalence. Then there is an orientation preserving vector bundle isomorphism $TS^n \to f^*(T\Sigma)$.*

12.3 The Surgery Exact Sequence for Homotopy Spheres

Proof. A short elegant proof is given by Pedersen and Ray in [357, Lemma 1.1]. We give an elementary proof, but only for even dimensions $n = 2k$. An elementary proof when $n = 2k + 1$ seems to be significantly more challenging.

Oriented n-dimensional vector bundles like TS^n and $f^*(T\Sigma)$ over S^n are classified by the elements c_{TS^n} and $c_{f^*(T\Sigma)}$ of the group $\pi_{n-1}(SO(n))$ given by their clutching functions. Obstruction theory for the fibrations $S^{n+b-1} \to BO(n+b-1) \to BO(n+b)$, $1 \leq b \leq a$, shows that the bundle isomorphism $TS^n \oplus \mathbb{R}^a \cong f^*(T\Sigma) \oplus \mathbb{R}^a$ destabilises to an isomorphism $TS^n \oplus \mathbb{R} \cong f^*(T\Sigma) \oplus \mathbb{R}$. It follows that the homotopy classes of the clutching functions of TS^n and $f^*(T\Sigma)$ agree when mapped to $\pi_{n-1}(SO(n+1))$. Hence we consider the long exact homotopy sequence of the fibre bundle

$$SO(n) \to SO(n+1) \to S^n \qquad (12.19)$$

that contains the exact sequence

$$\pi_n(S^n) \xrightarrow{\partial} \pi_{n-1}(SO(n)) \xrightarrow{i_*} \pi_{n-1}(SO(n+1))$$

where $i: SO(n) \to SO(n+1)$ is the standard inclusion. This sequence shows that

$$c_{TS^n} = c_{f^*(T\Sigma)} + \partial(a)$$

for some $a \in \pi_n(S^n)$. Since (12.19) is the oriented orthogonal frame bundle of TS^n, which is a principal $SO(n)$-bundle, $\partial([\mathrm{id}_{S_n}]) = c_{TS^n}$ by [387, 18.5 Classification Theorem].

When $n = 2k$, the Euler class of an oriented n-dimensional bundle induces a homomorphism

$$\mathrm{eul}: \pi_n(BSO(n)) \to \mathbb{Z}.$$

Define the homomorphism

$$e: \pi_{n-1}(SO(n)) \xrightarrow{\cong} \pi_n(BSO(n)) \xrightarrow{\mathrm{eul}} H^n(S^n) \xrightarrow{\kappa} \mathbb{Z}$$

where the first isomorphism is given by the inverse of the boundary map of the fibration $SO(n) \to ESO(n) \to BSO(n)$ and sends the class c_ξ given by the clutching function of an oriented n-dimensional vector bundle ξ over S^n to the class of the classifying map for ξ and the isomorphism κ sends x to $\langle x, [S^n] \rangle$.

Since $\partial([\mathrm{id}_{S_n}]) = c_{TS^n}$ and the Euler class of the tangent bundle of an oriented closed manifold evaluated against its fundamental class yields the Euler characteristic of the manifold [308, Corollary 11.12], we get the equality $e \circ \partial([\mathrm{id}_{S^n}]) = 2$. Hence $e \circ \partial$ is an injective homomorphism between infinite cyclic groups whose image has index two. Since S^n and Σ both have Euler characteristic 2, we get $e(c_{TS^n}) = e(c_{f^*(T\Sigma)})$ and hence $e \circ \partial(a) = 0$. This implies $a = 0$ and hence $c_{TS^n} = c_{f^*(T\Sigma)}$. We conclude that TS^n and $f^*(T\Sigma)$ are isomorphic. □

12.4 The J-Homomorphism and Stably Framed Bordism

By Theorem 12.14 we have reduced the computation of Θ_n to computations about Ω_n^{alm} and certain maps to $\mathbb{Z}$ and $\mathbb{Z}/2$ given by the signature divided by 8 and the Arf invariant. This reduction is essentially due to the machinery of surgery. The rest of the computation will essentially rely on homotopy theory. First we try to understand Ω_*^{fr} geometrically. There is an obvious forgetful map

$$f \colon \Omega_n^{\text{fr}} \to \Omega_n^{\text{alm}}. \tag{12.20}$$

We next define the group homomorphism

$$\partial \colon \Omega_n^{\text{alm}} \to \pi_{n-1}(SO). \tag{12.21}$$

Given $r \in \Omega_n^{\text{alm}}$, choose a representative $(M, x, \bar{u})$. Let $D^n \subset M$ be an embedded disk with origin x. Since D^n is contractible, we obtain a strong bundle isomorphism unique up to isotopy $\bar{v} \colon TM|_{D^n} \oplus \mathbb{R}^a \xrightarrow{\cong} \mathbb{R}^{a+n}$. The composition of the inverse of the restriction of $\bar{u}$ to $S^{n-1} = \partial D^n$ and of the restriction of $\bar{v}$ to S^{n-1} is an orientation preserving bundle automorphism of the trivial bundle $\mathbb{R}^{a+n}$ over S^{n-1}. This is the same as a map $S^{n-1} \to SO(n+a)$. Its composite with the canonical map $SO(n+a) \to SO$ represents an element in $\pi_{n-1}(SO)$ that is defined to be the image of r under $\partial \colon \Omega_n^{\text{alm}} \to \pi_{n-1}(SO)$. Let

$$\bar{J} \colon \pi_n(SO) \to \Omega_n^{\text{fr}} \tag{12.22}$$

be the group homomorphism that assigns to the element $r \in \pi_n(SO)$ represented by a map $\bar{u} \colon S^n \to SO(n+a)$ the class of S^n with the stable framing $TS^n \oplus \mathbb{R}^a \xrightarrow{\cong} \mathbb{R}^{a+n}$ that is induced by the standard trivialisation

$$TS^n \oplus \mathbb{R} \cong T(\partial D^{n+1}) \oplus \nu(\partial D^{n+1}, D^{n+1}) \cong TD^{n+1}|_{\partial D^{n+1}} \cong \mathbb{R}^{n+1}$$

and the strong bundle automorphism of the trivial bundle $\mathbb{R}^{a+n}$ over S^n given by $\bar{u}$. One easily checks the following result.

Lemma 12.23 *The following sequence is a long exact sequence of abelian groups*

$$\cdots \xrightarrow{\partial} \pi_n(SO) \xrightarrow{\bar{J}} \Omega_n^{\text{fr}} \xrightarrow{f} \Omega_n^{\text{alm}} \xrightarrow{\partial} \pi_{n-1}(SO) \xrightarrow{\bar{J}} \Omega_{n-1}^{\text{fr}} \xrightarrow{f} \cdots.$$

Next we want to interpret the exact sequence appearing in Lemma 12.23 homotopy theoretically. We begin with the homomorphism $\bar{J}$. Note that there is a natural bijection

$$\tau' \colon \operatorname{colim}_{k \to \infty} \Omega_n(\mathbb{R}^k \to \{*\}) \xrightarrow{\cong} \Omega_n^{\text{fr}} \tag{12.24}$$

that is defined as follows. Consider an element

12.4 The J-Homomorphism and Stably Framed Bordism

$$x = [(M, i: M \to \mathbb{R}^{n+k}, \mathrm{pr}: M \to \{*\}, \overline{u}: \nu(i) \to \underline{\mathbb{R}^k})] \in \Omega_n(\underline{\mathbb{R}^k} \to \{*\}).$$

There is a canonical strong isomorphism $\overline{u}': TM \oplus \nu(i) \xrightarrow{\cong} \underline{\mathbb{R}^{n+k}}$. From $\overline{u}$ and $\overline{u}'$ we get an isomorphism $\overline{v}: TM \oplus \underline{\mathbb{R}^k} \to \underline{\mathbb{R}^{n+k}}$ covering the projection $M \to \{*\}$. Define $\tau'(x)$ by the class of $(M, \overline{v})$. The Thom space $\mathrm{Th}(\underline{\mathbb{R}^{n+k}})$ is S^{n+k}. Hence the Pontrjagin–Thom construction, see Theorem 4.55, induces an isomorphism

$$P: \mathrm{colim}_{k \to \infty} \Omega_n(\underline{\mathbb{R}^k} \to \{*\}) \cong \mathrm{colim}_{k \to \infty} \pi_{n+k}(S^k). \tag{12.25}$$

Note that the *stable n-th homotopy group* of a space X is defined by

$$\pi_n^s(X) := \mathrm{colim}_{k \to \infty} \pi_{n+k}(S^k \wedge X). \tag{12.26}$$

Recall that stable *n-stem* is defined by

$$\pi_n^s := \pi_n^s(S^0) := \mathrm{colim}_{k \to \infty} \pi_{n+k}(S^k).$$

Thus the isomorphism τ' of (12.24) and the isomorphism P of (12.25) yield an isomorphism

$$\tau: \Omega_n^{\mathrm{fr}} \xrightarrow{\cong} \pi_n^s. \tag{12.27}$$

Next we explain the so-called *Hopf construction* that defines for spaces X, Y and Z a map

$$H: [X \times Y, Z] \to [X * Y, \Sigma Z] \tag{12.28}$$

as follows. Recall that the join $X * Y$ is defined by $X \times Y \times [0, 1]/\sim$, where $\sim$ is given by $(x, y, 0) \sim (x', y, 0)$ and $(x, y, 1) \sim (x, y', 1)$, and that the (unreduced) suspension ΣZ is defined by $Z \times [0, 1]/\sim$, where $\sim$ is given by $(z, 0) \sim (z', 0)$ and $(z, 1) \sim (z', 1)$. Given $f: X \times Y \to Z$, let $H(f): X * Y \to \Sigma Z$ be the map induced by $f \times \mathrm{id}: X \times Y \times [0, 1] \to Z \times [0, 1]$. Consider the following composite

$$[S^n, \mathrm{SO}(k)] \to [S^n, \mathrm{aut}(S^{k-1})] \to [S^n \times S^{k-1}, S^{k-1}]$$
$$\xrightarrow{H} [S^n * S^{k-1}, \Sigma S^{k-1}] = [S^{n+k}, S^k].$$

Note that $\pi_1(\mathrm{SO}(k))$ acts trivially on $\pi_n(\mathrm{SO}(k))$ and $\pi_1(S^k)$ acts trivially on $\pi_{n+k}(S^k)$ for $k, n \geq 1$. Hence no base point questions arise in the next definition.

Definition 12.29 (J-homomorphism) For $n, k \geq 1$ the composition above induces homomorphisms of abelian groups

$$J_{n,k}: \pi_n(\mathrm{SO}(k)) \to \pi_{n+k}(S^k).$$

Taking the colimit for $k \to \infty$ induces the so-called *J-homomorphism*

$$J_n: \pi_n(\mathrm{SO}) \to \pi_n^s.$$

With some care one can check the following lemma.

Lemma 12.30 *The composite of the homomorphism* $\bar{J} \colon \pi_n(SO) \to \Omega_n^{\mathrm{fr}}$ *of* (12.22) *with the isomorphism* $\tau \colon \Omega_n^{\mathrm{fr}} \xrightarrow{\cong} \pi_n^s$ *of* (12.27) *is the J-homomorphism* $J_n \colon \pi_n(SO) \to \pi_n^s$ *of Definition 12.29.*

As discovered by Bott [42], the homotopy groups of O are 8-periodic and are listed in the following table:

$i \bmod 8$	0	1	2	3	4	5	6	7
$\pi_i(O)$	$\mathbb{Z}/2$	$\mathbb{Z}/2$	0	$\mathbb{Z}$	0	0	0	$\mathbb{Z}$

Note that $\pi_i(SO) = \pi_i(O)$ for $i \geq 1$ and $\pi_0(SO) = 1$. The first 10 stable stems can be found in the book of Toda [397] and are listed in the following table:

$$
\begin{array}{|c||c|c|c|c|c|c|c|c|c|c|}
\hline
n & 0 & 1 & 2 & 3 & 4 & 5 & 6 & 7 & 8 & 9 \\
\hline
\pi_n^s & \mathbb{Z} & \mathbb{Z}/2 & \mathbb{Z}/2 & \mathbb{Z}/24 & 0 & 0 & \mathbb{Z}/2 & \mathbb{Z}/240 & \mathbb{Z}/2 \oplus \mathbb{Z}/2 & \mathbb{Z}/2 \oplus \mathbb{Z}/2 \oplus \mathbb{Z}/2 \\
\hline
\end{array}
\tag{12.31}
$$

More information about the stable stems up to dimension 90 can be found in [199, Table 1]. The *Bernoulli numbers* B_n for $n \geq 1$ are defined by

$$\frac{z}{e^z - 1} = 1 - \frac{z}{2} + \sum_{n \geq 1} \frac{(-1)^{n+1} \cdot B_n}{(2n)!} \cdot z^{2n}. \tag{12.32}$$

The first values are given by

n	1	2	3	4	5	6	7	8
B_n	$\frac{1}{6}$	$\frac{1}{30}$	$\frac{1}{42}$	$\frac{1}{30}$	$\frac{5}{66}$	$\frac{691}{2730}$	$\frac{7}{6}$	$\frac{3617}{510}$

The next deep result that we will need in the sequel is due to Adams [1, Theorem 1.1, Theorem 1.3 and Theorem 1.5], modulo a factor of two in the case $n = 7$ mod 8. This mod 2 ambiguity was resolved by Mahowald [284] and also follows from the solution of the Adams Conjecture [336, 390].[1]

Theorem 12.33 (Computation of the J-homomorphism)

(i) *If $n \not\equiv 3 \bmod 4$, then the J-homomorphism $J_n \colon \pi_n(SO) \to \pi_n^s$ is injective;*
(ii) *The image of the J-homomorphism $J_{4k-1} \colon \pi_{4k-1}(SO) \to \pi_{4k-1}^s$ has order* denominator$(B_k/4k)$ *where B_k is the k-th Bernoulli number.*

Exercise 12.34 Let n be an odd natural number. Show that Θ_n vanishes if and only if one of the following conditions is satisfied:

---

[1] See [336] and [390] for statements of the Adams Conjecture and also [279, Ch. 5] for a discussion of the role of the Adams Conjecture and its solution in surgery theory.

12.4 The J-Homomorphism and Stably Framed Bordism

(i) $n \in \{1, 3, 5\}$;
(ii) $n = 29$ and $\pi_n^s = 0$;
(iii) $n = 61$ and $\pi_n^s = 0$;
(iv) $n = 125$, $bP_{n+1} = 0$, and $\pi_n^s = 0$.

Exercise 12.35 Let n be an even natural number. Show that Θ_n vanishes if and only if one of the following conditions is satisfied:

(i) $n \in \{2, 4, 6, 12\}$;
(ii) $n = 8k$ for $k \geq 2$ and $\pi_n^s = \mathbb{Z}/2$;
(iii) $n = 8k + 4$ for $k \geq 2$ and $\pi_n^s = \{0\}$;
(iv) $n = 4k + 2$ for $k \geq 3$, $n \notin \{30, 162, 126\}$, and $\pi_n^s = 0$;
(v) $n = 30, 62$ and $\pi_n^s = \mathbb{Z}/2$;
(vi) $n = 126$ and either ($\pi_n^s = \mathbb{Z}/2$ and $bP_n = 0$) or ($\pi_n^s = 0$ and $bP_n = \mathbb{Z}/2$) hold.

Remark 12.36 (Smoothly rigid spheres) Next we want to discuss the following question: For which n is S^n smoothly rigid in the sense of Remark 11.34? If $n \leq 14$ and $n \neq 4$, then S^n is smoothly rigid if and only if $n \in \{1, 2, 3, 5, 6, 12\}$. If n is odd, Wang and Xu [417] have proven that S^n is smoothly rigid if and only if $n \in \{1, 3, 5, 61\}$. If n is even, $n \neq 4$ and $16 \leq n \leq 138$, Behrens, Hill, Hopkins and Mahowald [36] have proven that S^n is smoothly rigid if and only if $n = 56$. The general belief is that there are only finitely many natural numbers for which S^n is smoothly rigid. Conjecture 1.7 in [199] predicts that a sphere S^n for $n \geq 5$ is smoothly rigid if and only if $n = 5, 6, 12, 56, 61$.

We now review the history of the question above and how surgery theory reduces the above question to questions in stable homotopy theory for $n \geq 5$. Recall that S^n is topologically rigid in the sense of Definition 11.31 in all dimensions since the Poincaré Conjecture holds in all dimensions, see Theorem 2.5 and Subsection 2.6. Since in dimensions ≤ 3 there is no difference between the topologically and smooth category, S^n is smoothly rigid in dimensions ≤ 3.

It is unknown whether the smooth 4-dimensional Poincaré Conjecture is true. It is equivalent to the statement that S^4 is smoothly rigid. Note that the smooth s-Cobordism Theorem is known to fail in dimension 4 by the work of Donaldson [135].

In dimensions ≥ 5 a closed smooth manifold M is smoothly rigid if and only if its smooth structure set consists of precisely one structure. We conclude from Lemma 12.5 and Lemma 12.6 that S^n is smoothly rigid in dimension $n \geq 5$ if and only if Θ_n vanishes.

We conclude from Exercise 12.34 that, if S^n is smoothly rigid, then $n = 1, 3$ or $n \in \{29, 61, 125\}$ and $|\pi_n^s| = 1$. Now $\pi_{29}^s \cong \mathbb{Z}/3$, see e.g., [356, Appendix A3], is non-trivial and the computations of Wang and Xu show that $\pi_{61}^s = \{e\}$ but that π_{125}^s is non-trivial.

We conclude from Exercise 12.35 that S^{2k} is smoothly rigid for $2k \geq 16$ if and only if $|\operatorname{coker}(J_{2k})| = 1$. Behrens, Hill, Hopkins and Mahowald have shown that for $16 \leq 2k \leq 138$ the only even integer $2k$ with $\operatorname{coker}(J_{2k})$ trivial is $2k = 56$.

12.5 The Computation of bP_{n+1}

In this section we want to compute the subgroups $bP_{n+1} \subseteq \Theta_n$, see Definition 12.8.

We have introduced the bijection $\beta \colon \mathcal{N}(S^n) \xrightarrow{\cong} \Omega_n^{\mathrm{alm}}$ in Lemma 12.12 and the map $\partial \colon \Omega_n^{\mathrm{alm}} \to \pi_{n-1}(\mathrm{SO})$ in (12.21). Let

$$\delta(k) \colon \pi_n(\mathrm{BSO}(k)) \xrightarrow{\cong} \pi_{n-1}(\mathrm{SO}(k)) \qquad (12.37)$$

be the boundary map in the long exact homotopy sequence associated to the fibration $\mathrm{SO}(k) \to \mathrm{ESO}(k) \to \mathrm{BSO}(k)$. It is an isomorphism since $\mathrm{ESO}(k)$ is contractible. It can be described as follows. Consider $x \in \pi_n(\mathrm{BSO}(k))$. Choose a representative $f \colon S^n \to \mathrm{BSO}(k)$. If $\gamma_k \to \mathrm{BSO}(k)$ is the universal k-dimensional oriented vector bundle, $f^*\gamma_k$ is a k-dimensional oriented vector bundle over S^n. Let S_-^n be the lower and S_+^n be the upper hemisphere and $S^{n-1} = S_-^n \cap S_+^n$. Since the hemispheres are contractible, we obtain up to isotopy unique strong bundle isomorphisms $\bar{u}_- \colon f^*\gamma_k|_{S_-^n} \xrightarrow{\cong} \underline{\mathbb{R}}^k$ and $\bar{u}_+ \colon f^*\gamma_k|_{S_+^n} \xrightarrow{\cong} \underline{\mathbb{R}}^k$. The composition of the inverse of the restriction of $\bar{u}_-$ to S^{n-1} with the restriction of $\bar{u}_+$ to S^{n-1} is a bundle automorphism of the trivial bundle $\underline{\mathbb{R}}^k$ over S^{n-1}, which is the same as the map $S^{n-1} \to \mathrm{SO}(k)$, often called the *clutching function*. Its class in $\pi_{n-1}(\mathrm{SO}(k))$ is the image of x under $\delta(k)^{-1} \colon \pi_n(\mathrm{BSO}(k)) \to \pi_{n-1}(\mathrm{SO}(k))$. Analogously we get an isomorphism

$$\delta \colon \pi_n(\mathrm{BSO}) \xrightarrow{\cong} \pi_{n-1}(\mathrm{SO}). \qquad (12.38)$$

Define a map

$$\gamma \colon \mathcal{N}(S^n) \to \pi_n(\mathrm{BSO}) \qquad (12.39)$$

by sending the class of the normal map of degree one $(f, \bar{f}) \colon TM \oplus \underline{\mathbb{R}}^a \to \xi$ covering a map $f \colon M \to S^n$ to the class represented by the classifying map $f_\xi \colon S^n \to \mathrm{BSO}(n+k)$ of ξ. One easily checks the following result.

Lemma 12.40 *The following diagram commutes*

$$\begin{array}{ccc}
\Omega_n^{\mathrm{alm}} & \xrightarrow{\partial} & \pi_{n-1}(\mathrm{SO}) \\
{\scriptstyle \beta^{-1}}\downarrow & & \downarrow {\scriptstyle \delta^{-1}} \\
\mathcal{N}(S^n) & \xrightarrow{\gamma} & \pi_n(\mathrm{BSO}).
\end{array}$$

The next ingredient is the Hirzebruch Signature Theorem 8.171. Recall that it says that for a closed oriented manifold M of dimension $n = 4k$ its signature can be computed in terms of the L-class $\mathcal{L}(M)$ by

$$\mathrm{sign}(M) = \langle \mathcal{L}_k(M), [M] \rangle, \qquad (12.41)$$

12.5 The Computation of bP_{n+1}

where $\mathcal{L}_k(M) = \mathcal{L}_k(M)(p_1(TM), \ldots, p_k(TM))$ is a polynomial in the Pontrjagin classes of the tangent bundle of M. The coefficient s_k of $p_k(TM)$ is given in terms of the Bernoulli number B_k by the following formula, see [192, Theorem 8.2.2]:

$$s_k := \frac{2^{2k} \cdot (2^{2k-1} - 1) \cdot B_k}{(2k)!}. \tag{12.42}$$

Assume that M is almost stably parallelisable. Then for some point $x \in M$ the restriction of the tangent bundle TM to $M \setminus \{x\}$ is stably trivial and hence has trivial Pontrjagin classes. Since the inclusion induces an isomorphism $H^p(M; \mathbb{Z}) \xrightarrow{\cong} H^p(M \setminus \{x\}; \mathbb{Z})$ for $p \le n - 2$, we obtain $p_i(M) = 0$ for $i \le k - 1$. Hence (12.41) implies for a closed oriented almost stably parallelisable manifold M of dimension $4k$

$$\mathrm{sign}(M) = s_k \cdot \langle p_k(TM), [M] \rangle. \tag{12.43}$$

We omit the proof of the next lemma, which is based on certain homotopy theoretical computations, see for instance [246, Theorem 3.8 on page 76].

Lemma 12.44 *Let $n = 4k$. Then there is an isomorphism*

$$\phi \colon \pi_{n-1}(SO) \xrightarrow{\cong} \mathbb{Z}.$$

Define a map

$$p_k \colon \pi_n(BSO) \to \mathbb{Z}$$

by sending the element $x \in \pi_n(BSO)$ represented by a map $f \colon S^n \to BSO(m)$ to $\langle p_k(f^*\gamma_m), [S^n] \rangle$ for γ_m the universal m-dimensional vector bundle over $BSO(m)$. Let $\delta \colon \pi_n(BSO) \to \pi_{n-1}(SO)$ be the isomorphism of (12.38). Put

$$t_k := a_k \cdot (2k - 1)! \tag{12.45}$$

where we recall that $a_k = (3 - (-1)^k)/2$. Then

$$t_k \cdot \phi = p_k \circ \delta^{-1}.$$

Lemma 12.46 *For $n = 4k$ the following diagram commutes:*

$$\begin{array}{ccc}
\Omega_n^{\mathrm{alm}} & \xrightarrow{\frac{\mathrm{sign}}{8}} & \mathbb{Z} \\
\partial \downarrow & & \uparrow \frac{s_k \cdot t_k}{8} \cdot \mathrm{id} \\
\pi_{n-1}(SO) & \xrightarrow[\phi]{\cong} & \mathbb{Z}
\end{array}$$

Here $\frac{\mathrm{sign}}{8}$ is the homomorphism appearing in Corollary 12.15, the homomorphism ∂ has been defined in (12.21), and the isomorphism ϕ is taken from (12.44).

Proof. Consider an almost stably parallelisable manifold M of dimension $n = 4k$. We conclude from Lemma 12.40 and Lemma 12.44

$$\frac{s_k \cdot t_k}{8} \cdot \phi \circ \partial([M]) = \frac{s_k}{8} \cdot p_k \circ \delta^{-1} \circ \partial([M]) = \frac{s_k}{8} \cdot p_k \circ \gamma \circ \beta^{-1}([M]). \quad (12.47)$$

By definition the composite

$$\Omega_n^{\mathrm{alm}} \xrightarrow{\beta^{-1}} \mathcal{N}(S^n) \xrightarrow{\gamma} \pi_n(\mathrm{SO}) \xrightarrow{p_k} \mathbb{Z}$$

sends the class of $(M, x, \overline{u})$ to $\langle p_k(\xi), [S^n] \rangle$ for a bundle ξ over S^n for which there exists a bundle map $(c, \overline{c}) \colon TM \oplus \underline{\mathbb{R}^a} \to \xi$ covering a map $c \colon M \to S^n$ of degree one. This implies

$$\begin{aligned}
\frac{s_k}{8} \cdot p_k \circ \gamma \circ \beta^{-1}([M]) &= \frac{s_k}{8} \cdot \langle p_k(\xi), [S^n] \rangle \\
&= \frac{s_k}{8} \cdot \langle p_k(\xi), c_*([M]) \rangle \\
&= \frac{s_k}{8} \cdot \langle c^*(p_k(\xi)), [M] \rangle \\
&= \frac{s_k}{8} \cdot \langle p_k(TM), [M] \rangle.
\end{aligned} \quad (12.48)$$

Now the claim follows from (12.43), (12.47) and (12.48). $\square$

Theorem 12.49 (Computation of bP_{4k}) *Let $k \geq 2$ be an integer and recall that $a_k = (3 - (-1)^k)/2$. Then bP_{4k} is a finite cyclic group of order*

$$\begin{aligned}
&\frac{s_k \cdot t_k}{8} \cdot \left| \mathrm{im}\left(J_{4k-1} \colon \pi_{4k-1}(\mathrm{SO}) \to \pi^s_{4k-1} \right) \right| \\
&= \frac{1}{8} \cdot \frac{2^{2k} \cdot (2^{2k-1} - 1) \cdot B_k}{(2k)!} \cdot a_k \cdot (2k-1)! \cdot \mathrm{denominator}(B_k/4k) \\
&= a_k \cdot 2^{2k-2} \cdot (2^{2k-1} - 1) \cdot \mathrm{numerator}(B_k/4k).
\end{aligned}$$

Proof. We conclude from Lemma 12.23 and Lemma 12.30 that the cokernel of the map $\partial \colon \Omega_{4k}^{\mathrm{alm}} \to \pi_{4k-1}(\mathrm{SO})$ is isomorphic to the image of the homomorphism $J_{4k-1} \colon \pi_{4k-1}(\mathrm{SO}) \to \pi^s_{4k-1}$. Now the claim follows from Corollary 12.15, Theorem 12.33 (ii) and Lemma 12.46. $\square$

Next we treat the case $n = 4k + 2$ for $k \geq 1$. The Kervaire invariant in stable homotopy

$$\mathrm{KI} \colon \pi^s_{4k+2} \to \mathbb{Z}/2 \quad (12.50)$$

is defined to be the composition of $\tau^{-1} \colon \pi^s_n \xrightarrow{\cong} \Omega_n^{\mathrm{fr}}$, the inverse of the Pontrjagin–Thom isomorphism of (12.27), $f \colon \Omega_{4k+2}^{\mathrm{fr}} \to \Omega_{4k+2}^{\mathrm{alm}}$, the forgetful homomorphism of (12.20), and the map $\mathrm{KI} \colon \Omega_{4k+2}^{\mathrm{alm}} \to \mathbb{Z}/2$ appearing in Corollary 12.15.

12.5 The Computation of bP_{n+1}

The determination of the Kervaire invariant is called the *Kervaire invariant problem* and it has proven to be one of the most challenging problems in stable homotopy theory. Before summarising progress on this problem, we summarise its relevance for the group bP_{4k+2}.

Theorem 12.51 (Computation of bP_{4k+2} in terms of the Kervaire invariant problem) *Let $k \geq 3$. Then bP_{4k+2} is a trivial group if the homomorphism* KI: $\pi^s_{4k+2} \to \mathbb{Z}/2$ *of* (12.50) *is surjective and is $\mathbb{Z}/2$ if the homomorphism* KI: $\pi^s_{4k+2} \to \mathbb{Z}/2$ *of* (12.50) *is trivial.*

Proof. We conclude from Lemma 12.23, Lemma 12.30 and Theorem 12.33 (i) that the forgetful map $f \colon \Omega^{\text{fr}}_{4k+2} \to \Omega^{\text{alm}}_{4k+2}$ is surjective. Now the claim follows from Corollary 12.15. □

We conclude from Exercise 8.186 that for $k = 3, 7$ there are framings $\bar{u}_k$ on $S^k \times S^k$ satisfying KI$([S^k \times S^k, \bar{u}_k]) = 1$. Hence $bP_6 = bP_{14} = 0$, by Theorem 12.51. This was known to Kervaire and Milnor who wondered [216, Remark on page 536] whether the Kervaire invariant might vanish in all dimensions besides 2, 6 and 14.

In 1969 Browder proved the following result.

Theorem 12.52 (Browder's Theorem on the Kervaire invariant) *The homomorphism* KI: $\pi^s_{4k+2} \to \mathbb{Z}/2$ *of* (12.50) *is trivial if $2k + 1 \neq 2^l - 1$.*

Indeed Browder proved more than this, giving purely homotopic-theoretic condition determining precisely when KI $\neq 0$. Results of Mahowald and Tangora in dimension 30 [285] and Barratt, Jones and Mahowald in dimension 62 [21] imply the Kervaire invariant is non-zero in dimensions 30 and 62. Recently the spectacular work of Hill, Hopkins and Ravenel [187], built on the work of Browder, shows that the Kervaire invariant vanishes in dimensions $4k + 2 \geq 254$, leaving $4k + 2 = 126$ the only dimension in which the Kervaire invariant problem remains unsolved. Theorem 12.51 and the above discussion imply the following statement.

Corollary 12.53 *The group bP_{4k+2} is trivial or isomorphic to $\mathbb{Z}/2$. We have*

$$bP_{4k+2} = \begin{cases} \mathbb{Z}/2 & 4k+2 \neq 2^l - 2, k \geq 1, \text{ or } 4k+2 = 2^l - 2 \geq 254; \\ \{0\} & 4k+2 \in \{6, 14, 30, 62\}. \end{cases}$$

In particular, the only unresolved case is bP_{126}.

In odd degrees we get the statements as follows.

Theorem 12.54 (Vanishing of bP_{2k+1}) *We have for $k \geq 3$*

$$bP_{2k+1} = 0.$$

Proof. We can identify bP_{2k+1} with the kernel of the map $\eta \colon \Theta_{2k} \xrightarrow{\eta} \Omega^{\text{alm}}_{2k}$ defined in (12.13) by Lemma 12.9, and Lemma 12.12. Now apply Corollary 12.15. □

12.6 The Group Θ_n/bP_{n+1}

In this section we relate the group Θ_n/bP_{n+1} to the stable homotopy groups of spheres.

Theorem 12.55 (The group Θ_n/bP_{n+1})

(i) *If $n = 4k + 2$, then there is an exact sequence*

$$0 \to \Theta_n/bP_{n+1} \to \operatorname{coker}\left(J_n \colon \pi_n(SO) \to \pi_n^s\right) \to \mathbb{Z}/2;$$

(ii) *If $n \not\equiv 2 \bmod 4$ or if $n = 4k + 2$ with $2k + 1 \neq 2^l - 1$, then*

$$\Theta_n/bP_{n+1} \cong \operatorname{coker}\left(J_n \colon \pi_n(SO) \to \pi_n^s\right).$$

Proof. Lemma 12.23, Lemma 12.30, Theorem 12.33 (i) and Lemma 12.46 imply

$$\ker\left(\partial \colon \Omega_n^{\text{alm}} \to \pi_{n-1}(SO)\right) = \Omega_n^{\text{alm}} \qquad n \not\equiv 0 \bmod 4;$$
$$\ker\left(\partial \colon \Omega_n^{\text{alm}} \to \pi_{n-1}(SO)\right) = \ker\left(\tfrac{\text{sign}}{8} \colon \Omega_n^{\text{alm}} \to \mathbb{Z}\right) \qquad n \equiv 0 \bmod 4;$$
$$\ker\left(\partial \colon \Omega_n^{\text{alm}} \to \pi_{n-1}(SO)\right) = \operatorname{coker}\left(J_n \colon \pi_n(SO) \to \pi_n^{\text{fr}}\right).$$

Now the claim follows from Corollary 12.15, Theorem 12.52, and Theorem 12.54. $\square$

12.7 The Kervaire–Milnor Braid

The Kervaire–Milnor Braid of Theorem 12.59 below relates the groups we have studied in this chapter. It is comprised of the following four long exact sequences. In Lemma 12.23 we established the long exact sequence

$$\cdots \xrightarrow{\partial} \pi_n(SO) \xrightarrow{\overline{J}} \Omega_n^{\text{fr}} \xrightarrow{f} \Omega_n^{\text{alm}} \xrightarrow{\partial} \pi_{n-1}(SO) \xrightarrow{\overline{J}} \Omega_{n-1}^{\text{fr}} \xrightarrow{f} \cdots.$$

The long exact sequence

$$\cdots \to L_{n+1}(\mathbb{Z}) \to \Theta_n \to \Omega_n^{\text{alm}} \to L_n(\mathbb{Z}) \to \Theta_{n-1} \to \cdots \qquad (12.56)$$

is taken from Theorem 12.14. Denote by Θ_n^{fr} the abelian group of stably framed h-cobordism classes of stably framed homotopy n-spheres. There is a long exact sequence

$$\cdots \to \Theta_{n+1} \xrightarrow{0} \pi_n(SO) \to \Theta_n^{\text{fr}} \to \Theta_n \xrightarrow{0} \pi_{n-1}(SO) \to \cdots \qquad (12.57)$$

where the map $\pi_n(SO) \to \Theta_n^{\text{fr}}$ is the map defined by reframing the standard stable framing of S^n and $\Theta_n^{\text{fr}} \to \Theta_n$ is the forgetful map. There is also a long exact sequence

12.7 The Kervaire–Milnor Braid

$$\ldots \to L_{n+1}(\mathbb{Z}) \to \Theta_n^{\mathrm{fr}} \to \Omega_n^{\mathrm{fr}} \to L_n(\mathbb{Z}) \to \Theta_{n-1}^{\mathrm{fr}} \to \ldots \tag{12.58}$$

that is defined as follows. The map $\Theta_n^{\mathrm{fr}} \to \Omega_n^{\mathrm{fr}}$ assigns to the class of a framed homotopy sphere its class in Ω_n^{fr}. The map $\Omega_n^{\mathrm{fr}} \to L_n(\mathbb{Z})$ assigns to a bordism class $[M, \overline{u}]$ in Ω_n^{fr} the surgery obstruction of the degree one normal map $(c, \overline{c}) \colon TM \oplus \mathbb{R}^a \to \mathbb{R}^{n+a}$ for any map of degree one $c \colon M \to S^n$, where $\overline{c}$ is given by the framing $\overline{u}$. The map $L_{n+1}(\mathbb{Z}) \to \Theta_n^{\mathrm{fr}}$ assigns to $x \in L_{n+1}(\mathbb{Z})$ the class of the framed homotopy n-sphere $(\Sigma, \overline{v})$ for which there is a normal map of degree one $(U, \overline{U}) \colon TW \oplus \mathbb{R}^{a+b} \to \mathbb{R}^{n+a+b}$ covering a map of triads $U \colon (W; \partial_0 W, \partial_1 W) \to (S^n \times [0, 1]; S^n \times \{0\}, S^n \times \{1\})$ and a bundle map $(u_0, \overline{u_0}) \colon T\Sigma \oplus \mathbb{R}^{a+b+1} \to TW_0 \oplus \mathbb{R}^{a+b}$ covering the orientation preserving diffeomorphism $u_0 \colon \Sigma \to \partial_0 W$ satisfying $\overline{U} \circ \overline{u_0} = \overline{v} \oplus \mathrm{id}_{\mathbb{R}^{b+1}}$, U induces a diffeomorphism $\partial_1 W \to S^n$, and the surgery obstruction associated to $(U, \overline{U})$ is the given element $x \in L_{n+1}(\mathbb{Z})$.

Theorem 12.59 (The Kervaire–Milnor braid) *For $n \geq 5$ the long exact sequences (12.7), (12.56), (12.57) and (12.58) above fit together to form the following exact braid:*

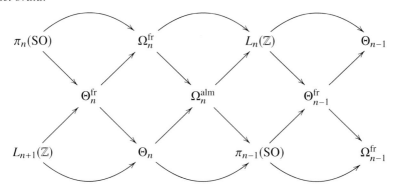

Next we interpret the braid of Theorem 12.59 as a braid of homotopy long exact sequences of fibrations of classifying spaces. Recall from Section 7.3 that the space G/O is the homotopy fibre of the canonical map BO → BG and similarly from Section 11.6, G/PL is the homotopy fibre of the canonical map BPL → BG. We introduce the space PL/O, which is the homotopy fibre of the canonical map BPL → BO. Since $\Omega\,\mathrm{BO} \simeq \mathrm{O}$, $\Omega\,\mathrm{BPL} \simeq \mathrm{PL}$ and $\Omega\,\mathrm{BG} \simeq \mathrm{G}$, we get fibrations

$$\mathrm{O} \to \mathrm{G} \to \mathrm{G/O}, \quad \mathrm{PL} \to \mathrm{G} \to \mathrm{G/PL} \quad \text{and} \quad \mathrm{O} \to \mathrm{PL} \to \mathrm{PL/O}\,.$$

For an inclusion of topological groups $H \subset K$, there is an obvious fibration $H \to K \to K/H$ and the fibrations above are in this spirit. But we have to use the classifying spaces since for instance G is not a group and we cannot talk about the homogeneous space G/PL. More information about these spaces, their homotopy-theoretic properties, and their role in manifold theory can be found, for instance, in [279].

Theorem 12.60 (Homotopy-theoretic interpretation of the Kervaire–Milnor braid) *The long exact homotopy sequences of these three fibrations above yield an exact braid*

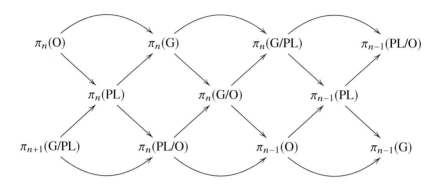

which for $n \geq 5$ is isomorphic to the Kervaire–Milnor braid of Theorem 12.59.

Proof. We at least explain how the two braids are related by isomorphisms. Since O/SO is the discrete group $\{\pm 1\}$, the inclusion induces an isomorphism

$$\pi_n(\mathrm{SO}) \xrightarrow{\cong} \pi_n(\mathrm{O}) \quad \text{for } n \geq 1.$$

The Pontrjagin–Thom isomorphism $\tau \colon \Omega_n^{\mathrm{fr}} \xrightarrow{\cong} \pi_n^s$ of (12.27) and the canonical isomorphism $\pi_n(\mathrm{G}) \cong \pi_n^s$ of (6.34) yield an isomorphism

$$\Omega_n^{\mathrm{fr}} \xrightarrow{\cong} \pi_n(\mathrm{G}).$$

The isomorphism $\beta \colon \mathcal{N}(S^n) \xrightarrow{\cong} \Omega_n^{\mathrm{alm}}$ of Lemma 12.12 and the bijection $\pi_n(\mathrm{G/O}) \xrightarrow{\cong} \mathcal{N}(S^n)$, see Theorem 7.34, induce a bijection

$$\Omega_n^{\mathrm{alm}} \xrightarrow{\cong} \pi_n(\mathrm{G/O}).$$

As we saw in Section 11.6, one can also develop surgery theory in the PL category instead of the smooth category. In Exercise 11.37 we saw that this leads to the isomorphism $\pi_n(\mathrm{G/PL}) \cong L_n(\mathbb{Z})$ for $n \geq 6$. The low-dimensional cases follow from direct computations, see Remark 12.72, and we obtain isomorphisms of abelian groups

$$\pi_n(\mathrm{G/PL}) \xrightarrow{\cong} L_n(\mathbb{Z}) \quad n \geq 1.$$

There is an isomorphism

$$\Theta_n \xrightarrow{\cong} \pi_n(\mathrm{PL/O}) \quad n \geq 0 \qquad (12.61)$$

that is defined as follows. Let Σ be a homotopy n-sphere. As explained above, there is an orientation preserving PL homeomorphism $h\colon \Sigma \to S^n$. Fix a classifying map $f_{S^n}\colon S^n \to \mathrm{BPL}$ for the PL tangent bundle of S^n. We obtain a classifying map $f_\Sigma\colon \Sigma \to \mathrm{BO}$ of the smooth tangent bundle of Σ together with a homotopy $h\colon Bi \circ f_\Sigma \simeq f_{S^n}$, where $Bi\colon \mathrm{BO} \to \mathrm{BPL}$ is the canonical map. The pair (f_Σ, h) yields a map $S^n \to \mathrm{PL/O}$ since PL/O is the homotopy fibre of $Bi\colon \mathrm{BO} \to \mathrm{BPL}$. The bijectivity of this map for $n \geq 5$ follows from the five lemma and the comparison of the surgery exact sequence with the long homotopy sequence associated to the fibration $\mathrm{PL/O} \to \mathrm{G/O} \to \mathrm{G/PL}$. There is an isomorphism

$$\Theta_n^{\mathrm{fr}} \xrightarrow{\cong} \pi_n(\mathrm{PL}) \quad n \geq 1$$

that is defined as follows. As above we get a pair (f_Σ, h). The framing also yields a homotopy $g\colon f_\Sigma \simeq c$ for c the constant map. Since PL can be viewed as the homotopy fibre of the obvious map $\mathrm{PL/O} \to \mathrm{BO}$, these data yield a map $S^n \to \mathrm{PL}$. The bijectivity of this map follows from the five lemma, comparison with the long exact sequence homotopy sequence associated to the fibration $\mathrm{O} \to \mathrm{PL} \to \mathrm{PL/O}$, and the bijection $\Theta_n \cong \pi_n(\mathrm{PL/O})$. □

Exercise 12.62 Use the Kervaire–Milnor braid, the fact that $\pi_8(G) \cong \pi_8^s \cong (\mathbb{Z}/2)^2$, the fact that $J_7\colon \pi_7(\mathrm{O}) \to \pi_7(G)$ is onto, and computations reported in this chapter, to determine the isomorphism class of the abelian group $\pi_7(\mathrm{PL})$.

12.8 Notes

Remark 12.63 (Generators of bP_{4k}) We have shown in Theorem 12.49 that bP_{4k} is a finite cyclic group for $k \geq 2$. An explicit generator can be constructed as follows. Recall that $\frac{\mathrm{sign}}{8}\colon L_{4k}(\mathbb{Z}) \to \mathbb{Z}$ is an isomorphism. Let W be any oriented stably parallelisable manifold of dimension $4k$ whose boundary ∂W is a homotopy sphere and whose intersection pairing on $H_{2k}(W)$ has signature 8. Then ∂W is an exotic sphere representing a generator of bP_{4k}. Such manifolds W can be explicitly constructed by a plumbing construction, see for instance [55, Theorem V.2.9 on page 122].

Example 12.64 (Milnor's exotic spheres) The first example of exotic spheres, i.e., closed manifolds that are homeomorphic but not diffeomorphic to S^n, were constructed by Milnor [294] in a paper published in 1956. See also [305]. The construction and the proof that these are exotic spheres are summarised below.

There is an isomorphism

$$\mathbb{Z} \oplus \mathbb{Z} \xrightarrow{\cong} \pi_3(\mathrm{SO}(4))$$
$$(h, j) \mapsto [S^3 \to \mathrm{SO}(4), \quad x \mapsto (y \mapsto x^h \cdot y \cdot x^j)]$$

where we identify $\mathbb{R}^4$ with the quaternions $\mathbb{H}$ by $(a, b, c, d) \mapsto a + bi + cj + dk$ and $x^h \cdot y \cdot x^j$ is to be understood with respect to the multiplication in $\mathbb{H}$. Since $\pi_3(SO(4)) \cong \pi_4(BSO(4))$, each pair $(h, j) \in \mathbb{Z} \oplus \mathbb{Z}$ determines an oriented vector bundle $E(h, j)$ with Riemannian metric over S^4 unique up to orientation and Riemannian metric preserving isomorphism. The Euler number and the first Pontrjagin class of $E(j, h)$ are given by

$$\chi(E(h, j)) = h + j;$$
$$\langle p_1(E(h, j)), [S^4] \rangle = 2(h - j).$$

The Gysin sequence shows that the sphere bundle $SE(h, j)$ is a homotopy 7-sphere if and only if $\chi(E(h, j)) = h + j = \pm 1$.[2]

For an odd integer k, let $W(k)$ be the disk bundle $DE((1 + k)/2, (1 - k)/2)$ and $\Sigma(k)$ be $\partial W(k) = SE((1 + k)/2, (1 - k)/2)$. Then $\Sigma(k)$ is a homotopy 7-sphere. Next we recall Milnor's argument why $\Sigma(k)$ cannot be diffeomorphic to S^7 for certain values of k.

The embedding $i\colon S^4 \hookrightarrow W(k)$ given by the zero section is a homotopy equivalence and $i^*TW(k)$ is isomorphic to $TS^4 \oplus E((1 + k)/2, (1 - k)/2)$. Hence

$$\langle i^* p_1(TW(k)), [S^4] \rangle = 2k.$$

Suppose that there exists a diffeomorphism $\Sigma(k) \to S^7$. Then we can form the closed oriented smooth 8-dimensional manifold $M(k) = W(k) \cup_f D^8$. Let $j\colon W(k) \to M(k)$ be the inclusion. Since the inclusion $j \circ i\colon S^4 \to M(k)$ induces an isomorphism on H_4, the signature of $M(k)$ is one. The Hirzebruch Signature Theorem 8.171 states that $1 = \text{sign}(M) = \langle \mathcal{L}_2(M), [M] \rangle$. From

$$\mathcal{L}_2(M) = \frac{7}{45} p_2(TM) - \frac{1}{45} p_1(TM)^2$$

we conclude

$$1 = \langle \frac{7}{45} p_2(TM) - \frac{1}{45} p_1(TM)^2, [M] \rangle = \frac{7}{45} \langle p_2(TM), [M] \rangle - \frac{1}{45} \langle p_1(TM)^2, [M] \rangle$$

$$= \frac{7}{45} \langle p_2(TM), [M] \rangle - \frac{1}{45} \langle i^* j^* p_1(TM), [S^4] \rangle^2 = \frac{7}{45} \langle p_2(TM), [M] \rangle - \frac{4k^2}{45}.$$

Since $\langle p_2(TM), [M] \rangle$ is an integer, we conclude $k^2 = 1 \mod 7$. Hence $\Sigma(k)$ is an exotic 7-sphere if $k^2 \neq 1 \mod 7$.

Remark 12.65 (Milnor's examples in the general framework) In Example 12.64 we only used the integrality of the signature and obtained an invariant mod 7. Eells and Kuiper [141] exploited the integrality of the $\widehat{A}$-genus of spin manifolds to define

---

[2] In 1956 the (Generalised) Poincaré Conjecture was not known and Milnor proved that $SE\left(\frac{1+k}{2}, \frac{1-k}{2}\right)$ is homeomorphic to S^7 by explicitly constructing a Morse function on $SE\left(\frac{1+k}{2}, \frac{1-k}{2}\right)$ with just two critical points.

12.8 Notes

a more refined invariant that detects all 28 homotopy 7-spheres. Using the Eells–Kuiper invariant, see [141, §6], one can show that Milnor's Example 12.64 above fits into the general context as follows. Let $\rho_{28}\colon \mathbb{Z} \to \mathbb{Z}/28$ be the canonical projection and recall that $bP_8 = \mathbb{Z}/28$ and that we have an isomorphism

$$\rho_{28} \circ \frac{\text{sign}}{8}\colon \Theta_7 \to \mathbb{Z}/28$$

that sends $[\Sigma]$ to $\text{sign}(W)/8$ mod 28 for any stably parallelisable manifold W whose boundary is oriented diffeomorphic to Σ. If $\Sigma(k)$ is the oriented homotopy 8-sphere of Example 12.64, then the isomorphism above sends $[\Sigma(k)]$ to the residue $\rho_{28}\left(\frac{1-k^2}{8}\right) \in \mathbb{Z}/28$.

Example 12.66 (Manifolds oriented homotopy equivalent and not oriented cobordant) Pairs of smooth manifolds that are oriented homotopy equivalent, but not oriented cobordant can be constructed using Example 12.64 and Remark 12.65 by the following argument, which was communicated to us by Matthias Kreck.

Given an odd integer $k \in \mathbb{Z}$, let $W(k)$ be the oriented D^4-bundle over S^4 constructed in Example 12.64 with the boundary $\Sigma(k)$. From Remark 12.65 it follows that if $(k^2 - 1)$ is divisible by 56, then $\Sigma(k)$ is diffeomorphic to the standard sphere S^7. Hence, after choosing a diffeomorphism $f_k\colon \Sigma(k) \to S^7$ (there is up to isotopy only one that preserves orientations), we can form the closed smooth oriented manifold $M(k) = W(k) \cup_{f_k} D^8$.

Let us recall from [308, Lemma 17.3] that two oriented cobordant manifolds must have all Pontrjagin numbers equal. We claim that it is possible to find two odd integers k, k' as above such that $M(k)$ and $M(k')$ do not have the same Pontrjagin numbers, but are oriented homotopy equivalent, thus yielding the desired example.

To see this, observe that the calculations in Example 12.64 show that the Pontrjagin number associated to p_1^2 is

$$\langle p_1^2(M(k)), [M(k)] \rangle = 4k^2.$$

It remains to find a suitable oriented homotopy equivalence for some pair of distinct k and k' as above. For this purpose we view the manifolds $W(k)$ as total spaces of oriented disk fibrations and their boundaries $\Sigma(k)$ as the total spaces of the associated 3-dimensional oriented spherical fibrations. If we can find pairs k and k' such that these spherical fibrations are oriented fibre homotopy equivalent, then the fibre homotopy equivalence will induce an oriented homotopy equivalence between $M(k)$ and $M(k')$.

Recall Theorem 6.11 from Chapter 6 about the classification of spherical fibrations up to fibre homotopy equivalence, or rather its version for oriented fibrations. It says that the oriented S^3-fibrations over S^4 up to oriented fibre homotopy equivalence are in bijection with the homotopy classes of maps from S^3 to the monoid SG(4) of homotopy equivalences S^3 to S^3 of degree one. This monoid contains a submonoid SF(4) of those elements that preserve the base point. Then SF(4) is the fibre of the fibration projection SG(4) $\to S^3$ given by evaluation at the base point of S^3.

Moreover, recall the calculation $\pi_3(\mathrm{SF}(4)) \cong \pi_6(S^3) \cong \mathbb{Z}/12$. The J-homomorphism $J_{3,3}\colon \pi_3(\mathrm{SO}(3)) \to \pi_6(S^3)$ from Definition 12.29 is known to be a surjective homomorphism $\mathbb{Z} \to \mathbb{Z}/12$, see [202]. Since the fibration $\mathrm{SG}(4) \to S^3$ has a section (because $\mathrm{SO}(4) \to S^3$ has one via quaternion multiplication), the associated long exact sequence of the homotopy groups splits into short exact sequences and the above discussion tells us that we have the isomorphism of short exact sequences

It can be shown that $\Sigma(k)$ viewed as an oriented spherical fibration corresponds to the element $((k-1)/2, 1)$ in the middle group, see the discussion on pages 376–377 and Remark 1.1 in [109] for full details. Hence, if we can pick k and k' so that their difference is divisible by 24, then we have that $\Sigma(k)$ and $\Sigma(k')$ are oriented fibre homotopy equivalent and hence also $M(k)$ and $M(k')$ are oriented homotopy equivalent.

Let $a, b \in \mathbb{Z}$ be any two positive integers and consider $k = 28a + 1$ and $k' = 28a + 6 \cdot 28b + 1$. Then elementary calculations show that the manifolds $M(k)$ and $M(k')$ satisfy all the above requirements and thus provide us with the example that we wanted.

Other examples of pairs of such manifolds are given by the so-called fake complex projective spaces, see Theorems 8.10 and 8.31 in [279]. (We also discuss fake complex projective spaces later in Section 18.4 of Chapter 18.)

Example 12.67 (Brieskorn spheres) Let $W^{2n-1}(d)$ be the subset of $\mathbb{C}^{n+1}$ consisting of those points $(z_0, z_1, \ldots, z_n)$ that satisfy the equations $z_0^d + z_1^2 + \cdots + z_n^2 = 0$ and $||z_0||^2 + ||z_1||^2 + \cdots + ||z_n||^2 = 1$. These turn out to be smooth submanifolds and are called *Brieskorn varieties*, see [47, 193]. Suppose that d and n are odd. Then $W^{2n-1}(d)$ is a homotopy $(2n-1)$-sphere. It is diffeomorphic to the standard sphere S^{2n-1} if $d = \pm 1 \bmod 8$, and it is an exotic sphere representing the generator of bP_{2n} if $d = \pm 3 \bmod 8$, see [47, page 11]. In general one can study the intersection $K = f^{-1}(0) \cap \{z \in \mathbb{C}^{n+1} \mid ||z|| = \epsilon\}$ for a polynomial $f(z_0, z_1, \ldots, z_n)$ with an isolated singularity at the origin and examine when K is a homotopy sphere and when K is an exotic sphere, see [304, § 8, § 9].

Remark 12.68 (Self-diffeomorphisms of spheres) Let Σ be a homotopy n-sphere for $n \geq 6$. In Exercise 12.4 we saw that there is an orientation preserving diffeomorphism $f\colon S^{n-1} \to S^{n-1}$ such that Σ is oriented diffeomorphic to $D^n \cup_f (D^n)^-$. If f is isotopic to the identity, the Σ is oriented diffeomorphic to S^n. Hence the existence of exotic spheres shows the existence of self-diffeomorphisms of S^{n-1} that are homotopic but not isotopic to the identity.

The following result is due to Brendle–Schoen [46].

12.8 Notes

Theorem 12.69 (Differentiable Sphere Theorem) *Let M be a simply connected closed Riemannian manifold whose sectional curvature is pinched by $1 \geq \sec(M) > \frac{1}{4}$. Then M is diffeomorphic to the standard sphere.*

Note that complex projective spaces are simply connected closed Riemannian manifolds coming with a Riemannian metric whose sectional curvature is pinched by $1 \geq \sec(M) \geq \frac{1}{4}$.

The following theorem is due to Brumfiel [63, 64, 65] and independently to Frank [155] in the special case of $n = 4k - 1$.

Theorem 12.70 (Brumfiel splitting) *For $n \neq 2^k - 3$ the sequence*

$$0 \to bP_{n+1} \to \Theta_n \to \mathrm{coker}(J_n) \to 0$$

splits.

Remark 12.71 (The Kervaire–Milnor extension) Following the work of Hill, Hopkins and Ravenel [187], the group bP_{254} is known to be $\mathbb{Z}/2$. At the time of writing, it is not known whether the short exact sequence

$$0 \to bP_{254} \to \Theta_{253} \to \mathrm{coker}(J_{253}) \to 0$$

splits or not. Indeed this is the situation for all the groups Θ_{2^k-3} with $k \geq 8$.

Remark 12.72 (Smoothing theory) The isomorphism $\Theta_n \cong \pi_n(\mathrm{PL/O})$ of (12.61) is a basic result of PL-to-DIFF smoothing theory. It studies when a PL manifold admits a smooth structure and, if so, how many such structures there are up to concordance. Smoothing theory, along with surgery theory, is one of the basic tools of manifold theory and can be used to prove the low-dimensional isomorphisms $\pi_n(\mathrm{PL/O}) \cong \{0\}$ for $n \leq 6$. For further information we refer the reader to [190].

Other examples of manifolds that look close to standard models, but are not equal to them, include fake projective spaces, fake lens spaces and fake tori. These have been investigated by Wall [414, 14D, 14E, 15A] and we will say something about them in Chapter 18.

Chapter 13
The Geometric Surgery Obstruction Group and Surgery Obstruction

13.1 Introduction

This chapter is the equivalent to Chapter 9 in Wall's book [414]. We give a geometric approach to the L-groups and the surgery obstruction based on bordism theory in Section 13.4. We prove that this geometric surgery obstruction is a normal bordism with cylindrical ends invariant and vanishes in high dimensions if and only if we can transform a normal map relative boundary to a (simple) homotopy equivalence by surgery on the interior, see Theorem 13.45. This illuminating geometric approach will only require the notion of surgery kernels, but not the notions of quadratic forms and algebraic L-groups. The geometric approach to the surgery obstruction is easier than the algebraic one, which we have presented in Chapters 8 and 9. We will show that both are equivalent, see Theorem 13.46. However, the algebraic description is needed when one wants to compute the L-groups.

There are more general situations, e.g., equivariant surgery where the geometric approach can be extended quite easily, but the algebra becomes very difficult, see for instance [139, 262, 263].

This geometric approach is based on the π-π-Theorem, which is formulated and proved in Section 13.3 and is the main technical result of this chapter. Consider a surgery problem with underlying map $f \colon (M, \partial M) \to (X, \partial X)$ where we do *not* demand that ∂f is a (simple) homotopy equivalence. Assume that the inclusion $\partial X \to X$ induces bijections on π_0 and on π_1 for any choice of base point in ∂X. Then the π-π-Theorem says that we can always arrange by surgery on the boundary and on the interior of M that both f and ∂f are (simple) homotopy equivalences, no surgery obstructions occur. Surgery on the boundary, which we have not performed beforehand, is explained in Subsection 13.2.2.

We will use the geometric approach to establish a geometric Rothenberg sequence in Section 13.5, which relates the two L-groups associated with $\mathbb{Z}G$ for decorations s and h in terms of the Tate cohomology of the Whitehead group of G, and the geometric Shaneson splitting in Section 13.6, which computes the L-groups of $\mathbb{Z}[G \times \mathbb{Z}]$ in terms of that of $\mathbb{Z}G$.

Guide 13.1 The material of this chapter is not needed for the rest of the book if one is familiar with the basic results of Chapters 8 and 9.

13.2 Surgery on and Normal Bordisms for Pairs

13.2.1 Short Review of Normal Bordism for Manifolds Relative Boundary

Recall that we have introduced in Subsection 8.8.1 the notion of a normal map $(M, \partial M, f, \partial f, a, \xi, \overline{f}, o)$ of degree one of manifolds relative boundary with target a w-oriented finite n-dimensional Poincaré pair $(X, \partial X, O_X, [[X, \partial X]])$. It was essential to require that $\partial f \colon \partial M \to \partial X$ is a (simple) homotopy equivalence.

In Subsection 8.8.2 we have explained the notion of a normal bordism $(W, F, b, \Xi, \overline{F}, O)$ of normal maps of degree one with cylindrical ends relative boundary between two normal maps $(M_0, \partial M_0, f_0, \partial f_0, a_0, \xi_0, \overline{f}_0, o_0)$ and $(M_1, \partial M_1, f_1, \partial f_1, a_1, \xi_1, \overline{f}_1, o_1)$ of degree one relative boundary with the same target $(X, \partial X, O_X, [[X, \partial X]])$. Recall that we had an appropriate decomposition of $\partial W = \partial_0 W \cup \partial_1 W \cup \partial_2 W$ and demanded that $\partial_2 F \colon \partial_2 W \to \partial X \times [0, 1]$ is a (simple) homotopy equivalence.

Notation 13.2 To keep the notation simple, we will often only denote the underlying maps of spaces $(f, \partial f) \colon (M, \partial M) \to (X, \partial X)$ instead of $(M, \partial M, f, \partial f, a, \xi, \overline{f}, o)$ and analogously only $(F, \partial F) \colon (W, \partial W) \to (X \times [0, 1], \partial(X \times [0, 1]))$ instead of $(W, F, b, \Xi, \overline{F}, O)$ keeping the decomposition $\partial F = \partial_0 F \cup \partial_1 F \cup \partial_2 F$ in mind. Moreover, when considering a normal bordism, we will often think of the diffeomorphism $u_i \colon M_i \xrightarrow{\cong} \partial_i W$ as the identity and just identify $M_i = \partial_i W$ for $i = 0, 1$.

In Subsections 8.8.3 and 9.4.2, for $n \geq 5$, we have assigned to such a normal map $(f, \partial f) \colon (M, \partial M) \to (X, \partial X)$ its intrinsic surgery obstruction $\sigma(f, \partial f)$ in the intrinsic surgery obstruction group $L_n(\mathbb{Z}\Pi(X), O_X)$. It is a normal bordism with cylindrical ends invariant and vanishes if and only if we can do surgery on the interior of M without changing ∂f such that f becomes a homotopy equivalence. The simple versions of these notions and results were treated in Chapter 10.

Next we want to drop the conditions that ∂f is a (simple) homotopy equivalence and that $\partial_2 F$ is a (simple) homotopy equivalence and also allow us to do surgery on the boundary as explained below, in other words, we will also modify ∂f. Then it turns out that for $n \geq 5$ there are no surgery obstructions at all to achieve by surgery on the interior of M and on ∂M that both f and ∂f are (simple) homotopy equivalences, provided that the π-π condition explained below is satisfied. For this purpose we have to explain what surgery on the boundary means.

13.2.2 Surgery on the Boundary

The main difference is that in the case of Subsections 8.8.3 and 9.4.2 we did not change the boundary, but we will allow the boundary to change in this chapter. Namely, we can also do *surgery on the boundary* for a normal map of degree one of manifolds with boundary $(f, \partial f) \colon (M, \partial M) \to (X, \partial X)$, as described next. The idea is to consider the surgery problem $\partial f \colon \partial M \to \partial X$. Note that the bundle data for $(f, \partial f)$ restrict to the boundary. Moreover, ∂f is automatically of degree one by the following argument. The diagram below commutes

$$\begin{array}{ccc} H_n(M, \partial M; O_M) & \xrightarrow{H_n(F, \partial F; \alpha)} & H_n(X, \partial X; O_X) \\ \partial_n \downarrow & & \partial_n \downarrow \\ H_{n-1}(\partial M; O_{\partial M}) & \xrightarrow{H_n(\partial F; \partial \alpha)} & H_{n-1}(\partial X; O_{\partial X}) \end{array}$$

where $\alpha \colon O_M \to F^* O_X$ and $\partial \alpha \colon O_{\partial M} \to \partial F^* O_{\partial X}$ are the isomorphisms of infinite cyclic coefficients systems over M and ∂M coming from the bundle data, see Definition 7.13. The vertical maps are the boundary maps. The left one sends $[[M, \partial M]]$ to $[[\partial M]]$ by Lemma 5.34, the right one sends $[[X, \partial X]]$ to $[[\partial X]]$ by Definition 5.73.

We know how to do surgery on $\partial f \colon \partial M \to \partial X$, see Section 4.6, and we intend to just glue this surgery step back to the given surgery problem $(f, \partial f) \colon (M, \partial M) \to (X, \partial X)$, as explained in the next paragraph. In applications one usually does surgery on the boundary first to arrange the map on the boundary and then does surgery on the interior to arrange everything on the whole manifold without changing the boundary.

Consider $\omega \in \pi_{k+1}(\partial f)$. Recall that we obtain from Theorem 4.39 (iv) a manifold W by attaching an appropriate handle to $\partial M \times [0, 1]$ and a map $F \colon W \to \partial X$. The boundary of W is the disjoint union of a copy of ∂M and a new closed manifold N such that F restricts on ∂M to ∂f and on N to a map $g \colon N \to \partial X$. Everything is covered by bundle maps so that $F \colon W \to \partial X$ is a normal bordism of degree one between the normal map of degree one $\partial f \colon \partial M \to \partial X$ and the new normal map of degree one $g \colon N \to \partial X$ that is the result of surgery on ∂f and ω, see Definition 4.41. Now we can form $M' = M \cup_{\partial M} W$ and obtain a map $f' \colon M' \to X$ that is uniquely determined by the properties that its restriction to M is f and to W is F. Note that $N = \partial M'$ and F restricted to $\partial M' = N$ is the map $g \colon N \to \partial X$. Everything is covered by bundle data so that we obtain a normal map of degree one of pairs $(f', \partial f') \colon (M', \partial M') \to (X, \partial X)$. We call it the result of *surgery on the boundary* on $(f, \partial f)$ and $\omega \in \pi_{k+1}(\partial f)$.

Note that $(f, \partial f) \colon (M, \partial M) \to (X, \partial X)$ and $(f', \partial f') \colon (M', \partial M') \to (X, \partial X)$ are normally bordant with cylindrical ends. The underlying map $(E, \partial E) \colon (V, \partial V) \to (X \times [0, 1], \partial(X \times [0, 1]))$ is given by $V = M' \times [0, 1]$ and $E = f' \times \mathrm{id}_{[0,1]}$. One easily checks that E maps ∂V to $\partial(X \times [0, 1])$ and therefore we can define $\partial E = E|_{\partial V}$. We decompose

$$\partial V = \partial_0 V \cup \partial_1 V \cup \partial_2 V$$

by putting

$$\partial_0 V = M \times \{0\};$$
$$\partial_1 V = M' \times \{1\};$$
$$\partial_2 V = W \times \{0\} \cup \partial M' \times [0, 1],$$

where we consider $M \times \{0\}$ and $W \times \{0\}$ as subsets of $M' \times \{0\}$ in the obvious way. One easily checks

$$\partial_0 V \cap \partial_1 V = \emptyset;$$
$$\partial \partial_2 V = \partial \partial_0 V \amalg \partial \partial_1 V;$$
$$\partial \partial_0 V = \partial M \times \{0\};$$
$$\partial \partial_1 V = \partial M' \times \{1\}.$$

Moreover, ∂E induces maps

$$\partial_0 E \colon \partial_0 V \to X \times \{0\};$$
$$\partial_1 E \colon \partial_1 V \to X \times \{1\};$$
$$\partial_2 E \colon \partial_2 V \to \partial(X \times [0, 1]),$$

such that $\partial_0 E = f$ and $\partial_1 E = f'$ hold.

We leave to the reader to check that everything is covered by appropriate bundle data and E has degree one. Note that in contrast to Subsection 8.8.2 we do not demand that $\partial_2 E$ is a (simple) homotopy equivalence.

Exercise 13.3 Check that the normal map $(f', \partial f') \colon (M', \partial M') \to (X, \partial X)$ and the normal bordism $(E, \partial E) \colon (V, \partial V) \to (X \times [0, 1], \partial(X \times [0, 1]))$ have degree one.

13.3 The π-π-Theorem

The main result of this section is the following theorem.

Theorem 13.4 (π-π-Theorem) *Let $(X, \partial X, O_X, [[X, \partial X]])$ be a w-oriented finite n-dimensional (simple) Poincaré pair in the sense of Definition 5.73. Suppose that the inclusion $\partial X \to X$ induces a bijection on the set of path components and an isomorphism on the fundamental groups for every choice of base point in ∂X, or, equivalently, induces an equivalence $\Pi(\partial X) \to \Pi(X)$ between the fundamental groupoids. Assume that $n \geq 6$.*

Then every normal map of pairs of degree one $(f, \partial f) \colon (M, \partial M) \to (X, \partial X)$ with target $(X, \partial X)$ in the sense of Subsection 8.8.1 can be changed by applying a sequence of homotopies, surgeries on the interior, and surgeries on the boundary to a normal map of pairs $(g, \partial g) \colon (N, \partial N) \to (X, \partial X)$ such that g and ∂g are (simple) homotopy equivalences. In particular, $(f, \partial f)$ and $(g, \partial g)$ are normally bordant with cylindrical ends.

13.3 The π-π-Theorem

Exercise 13.5 Let N be a connected closed smooth manifold of dimension $4k$ for $k \geq 2$. Suppose that ∂N is connected and non-empty and the inclusion $\partial N \to N$ induces an isomorphism $\pi_1(\partial N) \xrightarrow{\cong} \pi_1(N)$. Construct a normal map of degree one $(f, \partial f): (M, \partial M) \to (N, \partial N)$ with the following properties:

(i) The map $\partial f: \partial M \to \partial N$ is a diffeomorphism;
(ii) The normal map $(f, \partial f)$ is not normally bordant relative boundary to a normal map $(f', \partial f): (M', \partial M) \to (N, \partial N)$ such that f' is a homotopy equivalence;
(iii) The normal map $(f, \partial f): (M, \partial M) \to (N, \partial N)$ can be changed by applying a sequence of homotopies, surgeries on the interior, and surgeries on the boundary to a normal map of pairs $(f', \partial f'): (M', \partial M') \to (N, \partial N)$ such that f' and $\partial f'$ are simple homotopy equivalences.

There is a slightly more general version of Theorem 13.4, whose proof is completely analogous.

Theorem 13.6 (π-π-Theorem for triads) *Let $(X; \partial X_0, \partial_1 X)$ be an n-dimensional finite (simple) Poincaré triad. Suppose that the inclusion $\partial_0 X \to X$ induces a bijection on the set of path components and an isomorphism on the fundamental groups for every choice of base point in $\partial_0 X$, or, equivalently, induces an equivalence $\Pi(\partial_0 X) \to \Pi(X)$ between the fundamental groupoids. Assume that $n \geq 6$.*

Consider a manifold triad $(M; \partial_0 M, \partial_1 M)$. Let

$$(f; \partial_0 f, \partial_1 f): (M; \partial_0 M, \partial_1 M) \to (X, \partial_0 X_0, \partial_1 X)$$

be a map such that $\partial_1 f: \partial_1 M \to \partial_1 X$ is a (simple) homotopy equivalence and f has degree one. Suppose that f is covered by appropriate bundle data so that we obtain a normal map of degree one of pairs $(f, \partial f): (M, \partial M) \to (X, \partial X)$.

Then this normal map can be changed by applying a sequence of homotopies, surgeries on the interior of M, and surgeries on the boundary to a normal map of triads $(g; \partial_0 g, \partial_1 g): (N; \partial_0 N, \partial_1 N) \to (X; \partial_0 X, \partial_1 X)$ such that no modifications take place at $\partial_1 M$ and hence $\partial_1 M = \partial_1 N$ and $\partial_1 g = \partial_1 f$, and both g and ∂g are (simple) homotopy equivalences.

In particular, $(f, \partial f)$ and $(g, \partial g)$ are normally bordant.

Subsections 13.3.2 and 13.3.3 are devoted to proving the π-π-Theorem 13.4. We will follow [414, part 1, Sections 5 and 6] and give a purely geometric proof that does not use the algebra or the theory of surgery obstructions that we have already established. The reason for this is that the proof itself is quite illuminating, and we can finally give a geometric approach to the surgery obstruction that is independent of (but of course equivalent to) the already developed algebraic approach. Moreover, there are more general situations, e.g., equivariant surgery where the geometric approach can be extended quite easily, but the algebra becomes very difficult, see for instance [139, 262, 263].

13.3.1 Algebraic Proof of the π-π-Theorem 13.4

Nevertheless, we outline a quick proof of the π-π-Theorem 13.4 using the material of the previous sections to convince the reader of its validity.

Consider the situation of Theorem 13.4 assuming that ∂X and X are path connected. We consider only the simple case. Then $f \colon M \to X$ is a normal nullbordism of degree one for the normal map of degree one $\partial f \colon \partial M \to \partial X$. Hence the surgery obstruction of $\partial f \colon \partial M \to \partial X$ in $L^s_{n-1}(\mathbb{Z}\Pi(\partial X))$ is mapped to zero under the map $L^s_{n-1}(\mathbb{Z}\Pi(\partial X)) \to L^s_{n-1}(\mathbb{Z}\Pi(X))$ induced by the inclusion $\partial X \to X$, see Theorem 8.168. Since this inclusion induces an equivalence on the fundamental groupoids by assumption, the map $L^s_{n-1}(\mathbb{Z}\Pi(\partial X)) \to L^s_{n-1}(\mathbb{Z}\Pi(X))$ is bijective. Hence the surgery obstruction of $\partial f \colon \partial M \to \partial X$ vanishes. By Theorem 10.30, we can do finitely many surgery steps on $\partial f \colon \partial M \to \partial X$ to obtain a simple homotopy equivalence $\partial f' \colon \partial M' \to \partial X$. Let $g \colon (V; \partial M, \partial M') \to X$ be the corresponding normal bordism of degree one between $\partial f \colon \partial M \to \partial X$ and $\partial f' \colon \partial M' \to \partial X$. Define $M' = M \cup_{\partial M} V$ and $f' \colon M' \to X$ by $f \cup_{\partial f} g$. Then $f' \colon M' \to X$ is obtained from $f \colon M \to X$ by finitely many surgeries on the boundary and induces a simple homotopy equivalence $\partial f' \colon \partial M' \to \partial X$ on the boundary. Hence it remains to treat the case where $\partial f \colon \partial M \to \partial X$ is already a simple homotopy equivalence, otherwise replace f by f'.

The normal map of degree one $(f, \partial f) \colon (M, \partial M) \to (X, \partial X)$ has a surgery obstruction in $L^s_n(\mathbb{Z}\Pi(X))$. By Theorem 10.31, there is a normal map of degree one of the shape $(g; \partial_0 g, \partial_1 g) \colon (V; \partial_0 V, \partial_1 V) \to (\partial M \times [0,1]; \partial M \times \{0\}, \partial M \times \{1\})$ such that $\partial_0 g$ is an orientation preserving diffeomorphism, $\partial_1 g$ is a simple homotopy equivalence, and its surgery obstruction in $L^s_n(\mathbb{Z}\Pi(\partial M \times [0,1]))$ is mapped under the composite of the isomorphisms $L^s_n(\mathbb{Z}\Pi(\partial M \times [0,1])) \to L^s_n(\mathbb{Z}\Pi(\partial X))$, induced by ∂f and the projection $\partial M \times [0,1] \to \partial M$, and $L^s_n(\mathbb{Z}\Pi(\partial X)) \to L^s_n(\mathbb{Z}\Pi(X))$, induced by the inclusion to the negative of the surgery obstruction of $(f, \partial f) \colon (M; \partial M) \to (X, \partial X)$. Define $N = M \cup_{(\partial_0 g)^{-1}} V$ and $F \colon N \to X$ by requiring that $F|_M$ is f and $F|_V$ is the composite $V \xrightarrow{g} \partial M \times [0,1] \xrightarrow{pr} \partial M \xrightarrow{\partial f} \partial X \to X$. Then F yields a normal map of degree one $(F, \partial F) \colon (N, \partial N) \to (X, \partial X)$ that is obtained by finitely many surgery steps on the boundary of $(f, \partial f) \colon (M, \partial M) \to (X, \partial X)$ such that ∂F is a simple homotopy equivalence and the surgery obstruction of $(F, \partial F)$ in $L^s_n(\mathbb{Z}\Pi(X))$ is trivial. Hence by Theorem 10.30 we can do finitely many surgery steps on the interior of N to obtained a simple homotopy equivalence $F' \colon N' \to X$. This finishes the algebraic proof of Theorem 13.4 in the simple case.

13.3.2 Preliminaries for the Proof of the π-π-Theorem 13.4

To give a purely geometric proof of the π-π-Theorem 13.4, we need a little bit of input about the homotopy groups of squares and only a small portion of what we have developed so far about surgery kernels and simple structures on them. We collect the

13.3 The π-π-Theorem

basic input here for the reader's convenience so that the reader does not necessarily have to go through the material of some of the previous chapters.

Without loss of generality we can assume for the n-dimensional finite Poincaré pair $(X, \partial X)$ that ∂X and X are path connected and the inclusion $\pi_1(\partial X) \to \pi := \pi_1(X)$ is an isomorphism since in the general case we can treat each path component of X separately.

Consider a normal map $(f, \partial f): (M, \partial M) \to (X, \partial X)$ of degree one of pairs. By doing surgery on the boundary and then surgery in the interior, we can arrange that the maps f and ∂f are highly connected, see Theorem 7.41. This means that f is k-connected and ∂f is $(k-1)$-connected if $n = 2k$, and that f and ∂f are k-connected if $n = 2k + 1$.

Note that by assumption $k \geq 3$. Hence ∂M, M, ∂X, and X are path connected and in the following commutative diagram all arrows are isomorphisms

$$\begin{array}{ccc} \pi_1(\partial M) & \xrightarrow[\cong]{\pi_1(\partial f)} & \pi_1(\partial X) \\ \pi_1(i_M) \downarrow \cong & & \cong \downarrow \pi_1(i_X) \\ \pi_1(M) & \xrightarrow[\pi_1(f)]{\cong} & \pi_1(X) \end{array}$$

where the vertical maps are induced by the inclusions. We will identify all these fundamental groups and denote them by π. By passing to the universal covering we obtain a commutative square of π-equivariant maps

$$\begin{array}{ccc} \widetilde{\partial M} & \xrightarrow{\widetilde{\partial f}} & \widetilde{\partial X} \\ \widetilde{i_M} \downarrow & & \downarrow \widetilde{i_X} \\ \widetilde{M} & \xrightarrow{\widetilde{f}} & \widetilde{X}. \end{array} \qquad (13.7)$$

Define the surgery kernel $K_l(\widetilde{M})$ to be the $\mathbb{Z}\pi$-module given by the kernel of $\widetilde{f}_*: H_l(\widetilde{M}) \to H_l(\widetilde{X})$. Define analogously $K_l(\widetilde{M}, \widetilde{\partial M})$ and $K_l(\widetilde{\partial M})$ to be the kernels of the maps $(\widetilde{f, \partial f})_*$ and $(\widetilde{\partial f})_*$. Define $K^l(\widetilde{M})$ to be the $\mathbb{Z}\pi$-module given by the cokernel of $\widetilde{f}^*: H^l(\widetilde{X}) \to H^l(\widetilde{M})$. Define analogously $K^l(\widetilde{M}, \widetilde{\partial M})$ and $K^l(\widetilde{\partial M})$ to be the cokernels of the maps $(\widetilde{f, \partial f})^*$ and $(\widetilde{\partial f})^*$.

We will frequently use Poincaré duality for kernels in the form

$$K^l(\widetilde{M}, \widetilde{\partial M}) \cong K_{n-l}(\widetilde{M});$$
$$K^l(\widetilde{M}) \cong K_{n-l}(\widetilde{M}, \widetilde{\partial M});$$
$$K^l(\widetilde{\partial M}) \cong K_{n-1-l}(\widetilde{\partial M}),$$

which we have explained in Theorem 8.116 (ii).

We only treat the more complicated case where we want to achieve a simple homotopy equivalence of pairs. The proof of the case where we only want to achieve a homotopy equivalence is then given by ignoring all arguments concerning stable U-equivalence classes of stable bases and torsion invariants.

We will often use the fact that for highly connected $(f, \partial f)$ the surgery kernels are stably finitely generated free and inherit a preferred U-equivalence class of stable $\mathbb{Z}\pi$-bases, as explained in Lemma 10.25 and Lemma 10.27. The point is that the kernel is stably finitely generated free and can be identified with the only non-trivial homology group of a U-based finite $\mathbb{Z}\pi$-chain complex associated to a manifold or a map of manifolds and therefore there exists a preferred U-equivalence class of stable bases on this homology group for which the Reidemeister U-torsion in the sense of Definition 10.18 vanishes. In this chapter we will call this stable U-equivalence class of stable $\mathbb{Z}\pi$-bases just the *Reidemeister stable U-equivalence class of stable $\mathbb{Z}\pi$-bases*.

More generally, consider a stably U-based finite $\mathbb{Z}\pi$-chain complex C_* such that each homology group $H_i(C_*)$ is stably finitely generated free and, for some natural number j, each $H_i(C_*)$ for $i \neq j$ comes with a preferred stable U-equivalence class of $\mathbb{Z}\pi$-bases. Then $H_j(C_*)$ inherits a unique stable U-equivalence class of stable $\mathbb{Z}\pi$-bases, also called the Reidemeister stable U-equivalence class of stable $\mathbb{Z}\pi$-bases, for which the Reidemeister U-torsion in the sense of Definition 10.18 vanishes. We use the convention that a trivial module is always equipped with the trivial stable U-equivalence class of stable $\mathbb{Z}\pi$-bases. In all applications the $\mathbb{Z}\pi$-bases for the chain modules of C_* come from a CW-structure of a manifold and we will use the fact that the Whitehead torsion of a homeomorphism of finite CW-complexes vanishes.

When we say that an isomorphisms of stably based stably finitely generated free modules is simple, we mean that it respects the Reidemeister stable U-equivalence class of stable $\mathbb{Z}\pi$-bases.

The reader can assume without harm that all stable bases are actually bases since this can always be achieved by trivial surgery or by adding elementary chain complexes.

In the proof of the π-π-Theorem 13.4 we take $U = \{\pm g \mid g \in \pi\}$.

13.3.3 Proof of the π-π-Theorem 13.4 in the Even-Dimensional Case

In this subsection we give the proof Theorem 13.4 in the even-dimensional case $n = 2k$ for $k \geq 3$.

Consider a normal map of manifolds with boundary $(f, \partial f) \colon (M, \partial M) \to (X, \partial X)$. Recall that, by doing surgery on the boundary and then surgery on the interior, we can arrange that the map ∂f is $(k-1)$-connected and the map f is k-connected, see Theorem 7.41. In [140, Appendix 9] and [318, Section 2] one can find the definition of the homotopy groups of a square and the homology groups of a square and the proof that there exists a Hurewicz isomorphism

$$\pi_{k+1}(\widetilde{f}, \widetilde{\partial f}) \xrightarrow{\cong} H_{k+1}(\widetilde{f}, \widetilde{\partial f}). \tag{13.8}$$

13.3 The π-π-Theorem

An element in $\pi_{k+1}(\widetilde{f}, \widetilde{\partial f})$ is given by pointed homotopy classes of pointed maps from the square

$$\begin{array}{ccc} S^{k-1} & \longrightarrow & S^k_+ \\ \downarrow & & \downarrow \\ S^k_- & \longrightarrow & D^{k+1} \end{array}$$

where S^k_+ and S^k_- are the upper and lower hemispheres in $S^k = \partial D^{k+1}$, we have $S^{k-1} = S^k_+ \cap S^k_+$, and the maps are the inclusions, to the square

$$\begin{array}{ccc} \widetilde{\partial M} & \xrightarrow{\widetilde{\partial f}} & \widetilde{\partial X} \\ \widetilde{i_M} \downarrow & & \downarrow \widetilde{i_X} \\ \widetilde{M} & \xrightarrow{\widetilde{f}} & \widetilde{X}. \end{array}$$

The homology group $H_{k+1}(\widetilde{f}, \widetilde{\partial f})$ is defined to be the $(k+1)$-th homology of the mapping cone of the $\mathbb{Z}\pi$-chain map $C_*(\widetilde{f}, \widetilde{\partial f}) \colon C_*(\widetilde{M}, \widetilde{\partial M}) \to C_*(\widetilde{X}, \widetilde{\partial X})$. There are long exact sequences

$$\cdots \to \pi_{l+1}(\widetilde{M}, \widetilde{\partial M}) \xrightarrow{(\widetilde{f}, \widetilde{\partial f})_*} \pi_{l+1}(\widetilde{X}, \widetilde{\partial X}) \to \pi_{l+1}(\widetilde{f}, \widetilde{\partial f})$$
$$\to \pi_l(\widetilde{M}, \widetilde{\partial M}) \xrightarrow{(\widetilde{f}, \widetilde{\partial f})_*} \pi_l(\widetilde{X}, \widetilde{\partial X}) \to \cdots$$

and

$$\cdots \to H_{l+1}(\widetilde{M}, \widetilde{\partial M}) \xrightarrow{(\widetilde{f}, \widetilde{\partial f})_*} H_{l+1}(\widetilde{X}, \widetilde{\partial X}) \to H_{l+1}(\widetilde{f}, \widetilde{\partial f})$$
$$\to H_l(\widetilde{M}, \widetilde{\partial M}) \xrightarrow{(\widetilde{f}, \widetilde{\partial f})_*} H_l(\widetilde{X}, \widetilde{\partial X}) \to \cdots.$$

We define the surgery kernel $K_k(\widetilde{M}, \widetilde{\partial M})$ to be the kernel of the map $(\widetilde{f}, \widetilde{\partial f})_* \colon H_k(\widetilde{M}, \widetilde{\partial M}) \to H_k(\widetilde{X}, \widetilde{\partial X})$, see Definition 8.115. We obtain from Lemma 8.116 (i) an isomorphism $H_{k+1}(\widetilde{f}, \widetilde{\partial f}) \cong K_k(\widetilde{M}, \widetilde{\partial M})$. The obvious map $\pi_{k+1}(\widetilde{f}, \widetilde{\partial f}) \to \pi_{k+1}(f, \partial f)$ is a bijection. Hence we obtain from (13.8) an explicit isomorphism

$$\pi_{k+1}(f, \partial f) \xrightarrow{\cong} K_k(\widetilde{M}, \widetilde{\partial M}). \tag{13.9}$$

In the sequel we put

$$U = \{\pm g \mid g \in \pi\}.$$

The kernel $K_k(\widetilde{M}, \widetilde{\partial M})$ is a stably finitely generated free $\mathbb{Z}\pi$-module. We equip $K_k(\widetilde{M}, \widetilde{\partial M})$ with the Reidemeister stable U-equivalence class of stable $\mathbb{Z}\pi$-bases, see Subsection 13.3.2. By trivial surgery we can arrange that $K_k(\widetilde{M}, \widetilde{\partial M})$ is a finitely generated free $\mathbb{Z}\pi$-module, compare Example 8.60, and that we can choose a $\mathbb{Z}\pi$-basis $B = \{b_1, b_2, \ldots, b_r\}$ for $K_k(\widetilde{M}, \widetilde{\partial M})$ representing the Reidemeister stable U-equivalence class of stable $\mathbb{Z}\pi$-bases. Let $A = \{a_1, a_2, \ldots, a_r\}$ be the $\mathbb{Z}\pi$-basis on $\pi_{k+1}(f, \partial f)$ that corresponds under the isomorphism (13.9) to B. Analogously to Theorem 4.39 (i) and (ii) one shows that each $a_i \in A$ determines an up to regular homotopy unique immersion

$$u_i \colon (D^k \times D^k, \partial D^k \times D^k) \to (M, \partial M).$$

Lemma 13.10 *We can change the immersions u_i within their regular homotopy class so that they are disjoint embeddings.*

Proof. It is enough to show this by a tubular neighbourhood argument for the restricted immersions $\overline{u_i} \colon (D^k, \partial D^k) \to (M, \partial M)$. The strategy of proof is analogous to the one in Theorem 8.28.

We can arrange that the system of immersions $\overline{u_i}$ is in general position. Then the set of intersection and self-intersection points consists of isolated points in the interior of M, the preimage of each self-intersection point of a map $\overline{u_j}$ consists of precisely two points, and the preimage of an intersection point of two different maps $\overline{u_i}$ and $\overline{u_j}$ under each of the maps $\overline{u_i}$ and $\overline{u_j}$ consist of precisely one point since the dimension of M is $2k$. We have to explain how we get rid of each intersection and self-intersection point without changing the regular homotopy class of the $\overline{u_i}$-s and without introducing new intersection or self-intersection points.

Consider an intersection point P between two different embeddings $\overline{u_i}$ and $\overline{u_j}$. We can choose embedded arcs α_i and α_j in D^k such that the image of α_i under $\overline{u_i}$ and the image of α_j under $\overline{u_j}$ is still an embedded arc in M that starts in ∂M and ends in P and does not meet any other intersection or self intersection point of the $\overline{u_l}$-s. The union of these images is an embedded arc whose endpoints lie on ∂M. Recall that ∂M and M are path connected and the inclusion $\partial M \to M$ induces an isomorphism on the fundamental groups and $k \geq 3$. Hence we can find an embedded disk $D^2 \in M$ whose boundary ∂S^1 consist of three segments, the first one corresponds to $\overline{u_i}(\alpha_i)$, the second one to $\overline{u_j}(\alpha_j)$, and the third segment lies on ∂M, and the intersection of this disk D^2 with the union of the images of the embeddings $\overline{u_l}$ for $l = 1, 2, \ldots, r$ consists of the union of the first and the second segment on ∂D^2. Now apply the Whitney trick, see [301, Theorem 6.6 on page 71], [435] and Figure 8.29, to get rid of the intersection point P. Essentially one pulls the arc $\overline{u_i}(\alpha_i)$ along the disk D^2 to the other side of the arc $\overline{u_j}(\alpha_j)$ to get rid of the intersection point P. Everything takes place in a small enough tubular neighbourhood of D^2 so that one does not introduce new intersection or self-intersection points.

In the case of a self intersection point P for the immersion $\overline{u_l}$, the preimage $\overline{u_l}^{-1}(P)$ consist of two points x and x' in D^k, and we choose disjoint arcs α and α' in D^k such that they do not meet any further preimages of self-intersection points or intersection points and they connect a point on ∂D^k with x and x' respectively. Now proceed as above, replacing $\overline{u_i}(\alpha_i)$ and $\overline{u_j}(\alpha_j)$ by $\overline{u_l}(\alpha)$ and $\overline{u_l}(\alpha')$. This finishes the proof of Lemma 13.10. □

13.3 The π-π-Theorem

Let $H \subseteq M$ be the union of the images of the embeddings u_i for $i = 1, 2, \ldots, r$. Since each u_i corresponds to a class $a_i \in \pi_{k+1}(f, \partial f)$, the restriction of $(f, \partial f)$ to $(H, H \cap \partial M)$ is homotopic as a map of pairs to a map $(H, H \cap \partial M) \to (X, \partial X)$ whose image lies in $(\partial X, \partial X)$. By a cofibration argument we can change $(f, \partial f): (M, \partial M) \to (X, \partial X)$ up to homotopy such that $f(H) \subseteq \partial X$ holds. Changing the underlying map of a normal map of pairs up to homotopy results in a normal map that is normally bordant to the old one. So we can assume in the sequel without loss of generality that $f(H) \subseteq \partial X$ holds.

Next we perform a handle subtraction. Namely, let N be obtained from M by deleting H. Let $\widetilde{N} \to N$ and $\widehat{H} \to H$ be the restriction of the universal covering $\widetilde{M} \to M$ of M to N and H. Then N is path connected, the inclusion induces an isomorphism $\pi_1(N) \to \pi_1(M)$, which we will use to identify $\pi = \pi_1(N)$, the covering $\widetilde{N} \to N$ is the universal covering of N, and $\widehat{H}$ is π-diffeomorphic to $H \times \pi$. Let $(g, \partial g): (N, \partial N) \to (X, \partial X)$ be the normal map obtained by restricting $(f, \partial f): (M, \partial M) \to (X, \partial X)$ to $(N, \partial N)$. One easily checks that $(g, \partial g)$ and $(f, \partial f)$ are normally bordant. Let $\widetilde{\partial N} \to \partial N$ and $\widetilde{\partial X} \to \partial X$ be the restrictions of the universal coverings $\widetilde{N} \to N$ and $\widetilde{X} \to X$ to ∂N and ∂X. Let $(\widetilde{f}, \widetilde{\partial f}): (\widetilde{M}, \widetilde{\partial M}) \to (\widetilde{X}, \widetilde{\partial X})$ and $(\widetilde{g}, \widetilde{\partial g}): (\widetilde{N}, \widetilde{\partial N}) \to (\widetilde{X}, \widetilde{\partial X})$ be the π-maps obtained by lifting f and g to the universal coverings.

Lemma 13.11 *The $\mathbb{Z}\pi$-chain map $C_*(\widetilde{g}, \widetilde{\partial g}): C_*(\widetilde{N}, \widetilde{\partial N}) \to C_*(\widetilde{X}, \widetilde{\partial X})$ is a simple $\mathbb{Z}\pi$-chain homotopy equivalence.*

Proof. In the sequel of the proof we will use Lemma 10.22 and Lemma 10.23 over and over again without mentioning it explicitly.

Excision gives an explicit $\mathbb{Z}\pi$-isomorphism

$$H_l(\widehat{H}, \widehat{H} \cap \widetilde{\partial M}) \cong \bigoplus_{i=1}^{r} H_l(\pi \times D^k \times D^k, \pi \times S^{k-1} \times D^k) \cong \begin{cases} \mathbb{Z}\pi^r & l = k; \\ 0 & l \neq k. \end{cases}$$

We equip $H_k(\widehat{H}, \widehat{H} \cap \widetilde{\partial M})$ with the stable U-equivalence class of stable $\mathbb{Z}\pi$-bases coming from the standard basis on $\mathbb{Z}\pi^r$ and the isomorphism above. One easily checks that it agrees with the Reidemeister stable U-equivalence class of stable $\mathbb{Z}\pi$-bases on $H_k(\widehat{H}, \widehat{H} \cap \widetilde{\partial M})$. Recall that we have already chosen a $\mathbb{Z}\pi$-basis B for $K_k(\widetilde{M}, \widetilde{\partial M})$ above, whose choice we demand to be compatible with the Reidemeister stable U-equivalence class of stable $\mathbb{Z}\pi$-bases. The obvious composite of $\mathbb{Z}\pi$-isomorphisms

$$H_k(\widehat{H}, \widehat{H} \cap \widetilde{\partial M}) \xrightarrow{\cong} H_k(\widehat{H} \cup \widetilde{\partial M}, \widetilde{\partial M}) \xrightarrow{\cong} K_k(\widetilde{M}, \widetilde{\partial M}) \qquad (13.12)$$

is simple. This implies that the obvious composite of $\mathbb{Z}\pi$-chain maps

$$h_*: \Sigma C_*(\widehat{H}, \widehat{H} \cap \widetilde{\partial M}) \to \Sigma C_*(\widetilde{M}, \widetilde{\partial M}) \to \mathrm{cone}(C_*(\widetilde{f}, \widetilde{\partial f})) \qquad (13.13)$$

is a simple $\mathbb{Z}\pi$-chain homotopy equivalence since the homology groups of the source and target are concentrated in the same dimensions and in this dimension it is given by the simple isomorphism (13.12). We have the following short exact sequence of finite $\mathbb{Z}\pi$-chain complexes compatible with the preferred stable U-equivalence classes of $\mathbb{Z}\pi$-bases

$$0 \to C_*(\widehat{H}, \widehat{H} \cap \widetilde{\partial M}) \to C_*(\widetilde{M}, \widetilde{\partial M}) \to C_*(\widetilde{M}, \widehat{H} \cup \widetilde{\partial M}) \to 0.$$

It yields a short exact sequence of finite free $\mathbb{Z}\pi$-chain complexes respecting the preferred stable U-equivalence classes of stable $\mathbb{Z}\pi$-bases

$$0 \to \Sigma C_*(\widehat{H}, \widehat{H} \cap \widetilde{\partial M}) \xrightarrow{h_*} \mathrm{cone}(C_*(\widetilde{f}, \widetilde{\partial f})) \to \mathrm{cone}(C_*(\widetilde{f}, \widetilde{f}_{\widehat{H} \cup \widetilde{\partial M}})) \to 0.$$

Since the $\mathbb{Z}\pi$-chain map h_* of (13.13) is a simple $\mathbb{Z}\pi$-chain homotopy equivalence, the $\mathbb{Z}\pi$-chain map

$$C_*(\widetilde{f}, \widetilde{f}_{\widehat{H} \cup \widetilde{\partial M}}) \colon C_*(\widetilde{M}, \widehat{H} \cup \widetilde{\partial M}) \to C_*(\widetilde{X}, \widetilde{\partial X})$$

is a simple $\mathbb{Z}\pi$-chain homotopy equivalence. The $\mathbb{Z}\pi$-chain map

$$C_*(\widetilde{N}, \widetilde{\partial N}) \xrightarrow{\cong} C_*(\widetilde{M}, \widehat{H} \cup \widetilde{\partial M})$$

induced by the inclusion is a base preserving isomorphism of $\mathbb{Z}\pi$-chain complexes and in particular a simple $\mathbb{Z}\pi$-chain homotopy equivalence. Since the composite of two simple $\mathbb{Z}\pi$-chain homotopy equivalences is again a simple $\mathbb{Z}\pi$-chain homotopy equivalence,

$$C_*(\widetilde{g}, \widetilde{\partial g}) \colon C_*(\widetilde{N}, \widetilde{\partial N}) \to C_*(\widetilde{X}, \widetilde{\partial X})$$

is a simple $\mathbb{Z}\pi$-chain homotopy equivalence. □

Lemma 13.14 *Let $(f, \partial f) \colon (X, \partial X) \to (Y, \partial Y)$ be a map of (simple) finite Poincaré pairs of dimension n. Suppose that X, Y, ∂X, and ∂Y are path connected, the maps induced by the inclusions $\pi_1(\partial X) \to \pi_1(X)$ and $\pi_1(\partial Y) \to \pi_1(Y)$ and the map $\pi_1(f) \colon \pi_1(X) \to \pi_1(Y)$ are bijections, and $(f, \partial f)$ has degree one. Let $\widetilde{f} \colon \widetilde{X} \to \widetilde{Y}$ be the $\pi = \pi_1(X) = \pi_1(Y)$-equivariant lift of f to the universal coverings. Let $\widetilde{\partial X} \to \partial X$ and $\widetilde{\partial Y} \to \partial Y$ be the restriction of the universal coverings $\widetilde{X} \to X$ and $\widetilde{Y} \to Y$ to ∂X and ∂Y.*

Consider the following assertions:

(i) *$f \colon X \to Y$ is a (simple) homotopy equivalence;*
(ii) *The $\mathbb{Z}\pi$-chain map $C_*(\widetilde{f}, \widetilde{f}|_{\widetilde{\partial X}}) \colon C_*(\widetilde{X}, \widetilde{\partial X}) \to C_*(\widetilde{Y}, \widetilde{\partial Y})$ is a (simple) $\mathbb{Z}\pi$-chain homotopy equivalence;*
(iii) *$\partial f \colon \partial X \to \partial Y$ is a (simple) homotopy equivalence.*

Then all of them are true if and only if one of the assertions (i) and assertion (ii) is true.

13.3 The π-π-Theorem

Proof. We treat the case of simple homotopy equivalences only, the case of a homotopy equivalence is obtained by ignoring all remarks about torsion invariants. Since $\pi_1(f)$ is a bijection by assumption, the classical Whitehead Theorem implies that assertion (i) is equivalent to the statement that $C_*(\widetilde{f})$ is a simple $\mathbb{Z}\pi$-chain homotopy equivalence. The maps $\widetilde{\partial X} \to \partial X$ and $\widetilde{\partial Y} \to \partial Y$ are the universal coverings and $\widetilde{f}|_{\widetilde{\partial X}}: \widetilde{\partial X} \to \widetilde{\partial Y}$ covers $\partial f: \partial X \to \partial Y$. Hence assertion (iii) is equivalent to the statement that $C_*(\widetilde{\partial f}): C_*(\widetilde{\partial X}) \to C_*(\widetilde{\partial Y})$ is a simple $\mathbb{Z}\pi$-chain complex. In the sequel of the proof we will use Lemma 3.18 over and over again without mentioning it explicitly.

We have the following commutative diagram of finite free $\mathbb{Z}\pi$-chain complexes

$$\begin{array}{ccccccccc} 0 & \to & C_*(\widetilde{\partial X}) & \to & C_*(\widetilde{X}) & \to & C_*(\widetilde{X}, \widetilde{\partial X}) & \to & 0 \\ & & \downarrow C_*(\widetilde{\partial f}) & & \downarrow C_*(\widetilde{f}) & & \downarrow C_*(\widetilde{f}, \widetilde{\partial f}) & & \\ 0 & \to & C_*(\widetilde{\partial Y}) & \to & C_*(\widetilde{Y}) & \to & C_*(\widetilde{Y}, \widetilde{\partial Y}) & \to & 0 \end{array}$$

where the rows are U-based exact. Hence all of the three vertical arrows are simple $\mathbb{Z}\pi$-chain homotopy equivalences if two of them are. Since $(f, \partial f)$ has degree one, the following diagram of finite based free $\mathbb{Z}\pi$-chain complexes

$$\begin{array}{ccc} C^{n-*}(\widetilde{X}, \widetilde{\partial X}) & \xleftarrow{C^{n-*}(\widetilde{f}, \widetilde{\partial f})} & C^{n-*}(\widetilde{Y}, \widetilde{\partial Y}) \\ \downarrow -\cap [X, \partial X] & & \downarrow -\cap [Y, \partial Y] \\ C_*(\widetilde{X}) & \xrightarrow{C_*(\widetilde{f})} & C_*(\widetilde{Y}) \end{array}$$

commutes. The vertical arrows are simple $\mathbb{Z}\pi$-chain homotopy equivalences since $(X, \partial X)$ and $(Y, \partial Y)$ are simple finite Poincaré pairs by assumption. One easily checks that $C_*(\widetilde{f}, \widetilde{\partial f})$ is a simple $\mathbb{Z}\pi$-chain homotopy equivalence if and only if $C^{n-*}(\widetilde{f}, \widetilde{\partial f})$ is a simple $\mathbb{Z}\pi$-chain homotopy since their Whitehead torsions are mapped to one another under the involution on $\mathrm{Wh}(\pi)$ of (3.27).

Hence $C_*(\widetilde{f}, \widetilde{\partial f})$ is a simple $\mathbb{Z}\pi$-chain homotopy equivalence if and only if $C_*(\widetilde{f})$ is a simple $\mathbb{Z}\pi$-chain homotopy equivalence. This finishes the proof of Lemma 13.14. $\square$

We conclude from Lemma 13.11 and Lemma 13.14 that both $g: N \to X$ and $\partial g: X \to Y$ are simple homotopy equivalences. Since $(g, \partial g)$ is obtained from $(f, \partial f)$ by a sequence of homotopies, surgeries on the interior, surgeries on the boundary, and handle subtractions, and a handle subtraction can be interpreted as the inverse of a surgery on the boundary, this finishes the proof of the π-π-Theorem 13.4 in the even-dimensional case.

13.3.4 Proof of the π-π-Theorem 13.4 in the Odd-Dimensional Case

In this subsection we give the proof of Theorem 13.4 in the odd-dimensional case $n = 2k + 1$ for $k \geq 3$.

Lemma 13.15 *We can perform a sequences of homotopies, surgeries on the interior, and surgeries on the boundary on the normal map $(f, \partial f) \colon (M, \partial M) \to (X, \partial X)$ to arrange that f and ∂f are k-connected and $K_k(\widetilde{M}, \widetilde{\partial M}) = 0$.*

Proof. By doing surgery on the boundary and then surgery in the interior, we can arrange that the maps f and ∂f are k-connected, see Theorem 7.41. It remains to explain how we can perform additionally a handle subtraction to arrange $K_k(\widetilde{M}, \widetilde{\partial M}) = 0$ without destroying the property that f and ∂f are k-connected.

Recall that $\pi = \pi_1(M) = \pi_1(X) = \pi_1(\partial M) = \pi_1(\partial X)$. We obtain from Lemma 8.116 (iii) a sequence of $\mathbb{Z}\pi$-isomorphisms

$$h_k \colon \pi_{k+1}(f) \xrightarrow{\cong} \pi_{k+1}(\widetilde{f}) \xrightarrow{\cong} H_{k+1}(\widetilde{f}) \xrightarrow{\cong} K_k(\widetilde{M}).$$

Since $\widetilde{f}$ is k-connected, $H_l(\mathrm{cone}(C_*(\widetilde{f}))) = 0$ for $l \leq k$. Denote by c_{k+1} the $(k+1)$th differential of $\mathrm{cone}_* := \mathrm{cone}(C_*(\widetilde{f}))$. We obtain the exact sequence of $\mathbb{Z}\pi$-modules

$$0 \to \ker(c_{k+1}) \to \mathrm{cone}_{k+1} \xrightarrow{c_{k+1}} \mathrm{cone}_k \xrightarrow{c_k} \mathrm{cone}_{k-1} \xrightarrow{c_{k-1}} \cdots \xrightarrow{c_1} \mathrm{cone}_0 \to 0.$$

Since each $\mathbb{Z}\pi$-module cone_l is finitely generated free, $\ker(c_{k+1})$ is a finitely generated projective $\mathbb{Z}\pi$-module. Since $H_{k+1}(\widetilde{f}) = H_{k+1}(\mathrm{cone}_*)$ is a quotient of $\ker(c_{k+1})$, the $\mathbb{Z}\pi$-module $H_{k+1}(\widetilde{f})$ is finitely generated. Hence also $\pi_{k+1}(f)$ is a finitely generated $\mathbb{Z}\pi$-module. Fix a finite subset $\{x_1, x_2, \ldots, x_r\}$ of $\pi_{k+1}(f)$ that generates the $\mathbb{Z}\pi$-module $\pi_{k+1}(f)$. Choose for $i = 1, 2, \ldots, r$ commutative diagrams

$$\begin{array}{ccc} S^k \times D^{k+1} & \xrightarrow{u_i} & M \\ \downarrow & & \downarrow f \\ D^{k+1} \times D^{k+1} & \longrightarrow & X \end{array}$$

such that the corresponding class in $\pi_{k+1}(\widetilde{f})$ is x_i and u_i is an embedding whose image is disjoint from ∂M. This is possible by Theorem 4.39. By a general position argument, we can additionally arrange that $u_i(S^k \times \{0\})$ and $u_j(S^k \times \{0\})$ are disjoint for $i \neq j$ and then by possibly shrinking D^{k+1} that the images of u_i and u_j are disjoint for $i \neq j$.

Next we describe how we can connect each u_i with a tube to the boundary ∂M so that we obtain a commutative diagram

$$\begin{array}{ccc} S^{k-1} \times D^{k+1} & \xrightarrow{v_i} & \partial M \\ \downarrow & & \downarrow \\ D^k \times D^{k+1} & \xrightarrow{V_i} & M \end{array}$$

13.3 The π-π-Theorem

such that it still represents x_i, each map $(V_i, v_i) \colon (D^k \times D^{k+1}, S^{k-1} \times D^{k+1}) \to (M, \partial M)$ is an embedding, and the images of two different embeddings (V_i, v_i) and (V_j, v_j) are disjoint. Namely, choose an embedded closed disk $B_k \subseteq S^k$, which one should think of as being very small. Consider $X_k = S^k \setminus \mathrm{int}(B_k) \cup_{\partial B_k} \partial B_k \times [0, 1]$, where we identify ∂B_k with $\partial B_k \times \{0\}$ in $\partial B_k \times [0, 1]$. Now we can find an embedding $V_i \colon X_k \times D^{k+1} \hookrightarrow M$ such that $V_i|_{S^k \setminus B_k \times D^{k+1}}$ agrees with the restriction of u_i to $S^k \setminus B_k \times D^{k+1} \subseteq S^k \times D^{k+1}$ and the intersection of the image of V_i with ∂M is $V_i(\partial B_k \times \{1\} \times D^{k+1}) = V_i(\partial X_k \times D^{k+1})$. Since $(X_k, \partial X_k)$ is isomorphic to (D^k, S^{k-1}), we obtain the desired embedding V_i. Since we can choose the disks B_k as small as we want, the images of u_i and u_j are disjoint for $i \neq j$, and two embedded arcs in M between two different endpoints can be changed by a homotopy relative endpoint to be disjoint, we can additionally arrange that the images of V_i and V_j are disjoint for $i \neq j$.

Let $H \subseteq M$ be the union of the images of the maps (V_i, v_i) for $i = 1, 2, \ldots, r$. We can change $(f, \partial f)$ up to homotopy to arrange that $f(H) \subseteq \partial X$ holds. Let N be obtained from M by deleting the interior of H. Let $(g, \partial g) \colon (N, \partial N) \to (X, \partial X)$ be obtained by restricting f to N. Note that N and ∂N are path connected and the inclusions $\partial N \to N \to M$ induce isomorphisms on π_1. Hence we can identify $\pi = \pi_1(N) = \pi_1(\partial N)$, and the coverings $\widetilde{N} \to N$ and $\widetilde{\partial N} \to \partial N$ obtained by restricting the universal covering $\widetilde{M} \to M$ to N and ∂N are the universal coverings of N and ∂N. Let $\widehat{H} \to H$ be the restriction $\widetilde{M} \to M$ to H. Then $\widehat{H} = \pi \times D^k \times D^{k+1}$.

It remains to prove that g and ∂g are k-connected and $K_k(\widetilde{N}, \widetilde{\partial N}) \cong H_{k+1}(\widetilde{g}, \widetilde{\partial g})$ vanishes.

We first show that $H_l(\widetilde{g}, \widetilde{\partial g})$ vanishes for $l \leq k + 1$. We have the exact sequence of finite free $\mathbb{Z}\pi$-chain complexes

$$0 \to C_*(\widehat{H} \cup \widetilde{\partial M}, \widetilde{\partial M}) \to C_*(\widetilde{M}, \widetilde{\partial M}) \to C_*(\widetilde{M}, \widehat{H} \cup \widetilde{\partial M}) \to 0$$

and the $\mathbb{Z}\pi$-chain isomorphism $C_*(\widetilde{N}, \widetilde{\partial N}) \xrightarrow{\cong} C_*(\widetilde{M}, \widehat{H} \cup \widetilde{\partial M})$. Hence we obtain an exact sequence of $\mathbb{Z}\pi$-chain complexes

$$0 \to \Sigma C_*(\widehat{H} \cup \widetilde{\partial M}, \widetilde{\partial M}) \to \mathrm{cone}(C_*(\widetilde{f}, \widetilde{\partial f})) \to \mathrm{cone}(C_*(\widetilde{g}, \widetilde{\partial g})) \to 0.$$

The associated long exact homology sequence is

$$\cdots \to H_l(\widehat{H} \cup \widetilde{\partial M}, \widetilde{\partial M}) \to H_{l+1}(\widetilde{f}, \widetilde{\partial f}) \to H_{l+1}(\widetilde{g}, \widetilde{\partial g})$$
$$\to H_{l-1}(\widehat{H} \cup \widetilde{\partial M}, \widetilde{\partial M}) \to \cdots.$$

Since $H_l(\pi \times D^k \times D^{k+1}, \pi \times S^{k-1} \times D^{k+1}) = 0$ holds for $l \leq (k-1)$, we get $H_l(\widehat{H} \cup \widetilde{\partial M}, \widetilde{\partial M}) = 0$ for $l \leq (k-1)$. Hence we obtain for $l \leq k$ an isomorphism

$$H_l(\widetilde{f}, \widetilde{\partial f}) \to H_l(\widetilde{g}, \widetilde{\partial g}),$$

and we have the exact sequence

$$\cdots \to H_k(\widehat{H} \cup \widetilde{\partial M}, \widetilde{\partial M}) \to H_{k+1}(\widetilde{f}, \widetilde{\partial f}) \to H_{k+1}(\widetilde{g}, \widetilde{\partial g}) \to 0.$$

Since $\widetilde{f}$ and $\widetilde{\partial f}$ are k-connected, $H_l(\widetilde{f}, \widetilde{\partial f})$ and hence $H_l(\widetilde{g}, \widetilde{\partial g})$ vanish for $l \le k$. Next we show that $H_{k+1}(\widetilde{g}, \widetilde{\partial g})$ is trivial. It suffices to show that the map

$$H_k(\widehat{H} \cup \widetilde{\partial M}, \widetilde{\partial M}) \to H_{k+1}(\widetilde{f}, \widetilde{\partial f})$$

is surjective. Since ∂f is k-connected and hence $H_k(\widetilde{\partial f}) = 0$, the map $H_{k+1}(\widetilde{f}) \to H_{k+1}(\widetilde{f}, \widetilde{\partial f})$ is surjective. The Hurewicz homomorphism $\pi_{k+1}(\widetilde{f}) \to H_{k+1}(\widetilde{f})$ is surjective as $\widetilde{f}$ is k-connected. Hence the composite

$$\pi_{k+1}(\widetilde{f}) \to H_{k+1}(\widetilde{f}) \to H_{k+1}(\widetilde{f}, \widetilde{\partial f})$$

is surjective. Let $y_i \in H_{k+1}(\widetilde{f}, \widetilde{\partial f})$ be the image of the class x_i under this composite. Since the x_i-s generate $\pi_{k+1}(\widetilde{f})$, the y_i-s generate $H_{k+1}(\widetilde{f}, \widetilde{\partial f})$. The element y_i lies in the image of $H_k(\widehat{H} \cup \widetilde{\partial M}, \widetilde{\partial M}) \to H_{k+1}(\widetilde{f}, \widetilde{\partial f})$, a preimage is given by the class of the i-th handle in

$$H_k(\widehat{H} \cup \widetilde{\partial M}, \widetilde{\partial M}) \cong \bigoplus_{i=1}^{r} H_k(\pi \times D^k \times D^{k+1}, \pi \times S^{k-1} \times D^{k+1}) \cong \bigoplus_{i=1}^{r} \mathbb{Z}\pi.$$

This finishes the proof that $H_l(\widetilde{g}, \widetilde{\partial g})$ vanishes for $l \le k+1$.

Since we have the pushout

$$\begin{array}{ccc} D^k \times S^k & \longrightarrow & N \\ \downarrow & & \downarrow \\ D^k \times D^{k+1} & \longrightarrow & M \end{array}$$

the inclusion $N \to M$ is k-connected. Since $f: M \to X$ is k-connected and g is $f|_N$, the map g is k-connected.

It remains to show that $\partial g: \partial M \to \partial X$ is k-connected. This is equivalent to showing that $H_l(\widetilde{\partial g})$ vanishes for $l \le k$. This follows from the facts that $H_l(\widetilde{g}) = 0$ vanishes for $l \le k$ and $H_l(\widetilde{g}, \widetilde{\partial g}) = 0$ vanishes for $l \le k+1$. This finishes the proof of Lemma 13.15. □

In the sequel we can assume that the conclusions of Lemma 13.15 hold.

Lemma 13.16 *We have*

$$\begin{aligned} K_i(\widetilde{M}) &= 0 \quad \text{for } i \ne k; \\ K_i(\widetilde{\partial M}) &= 0 \quad \text{for } i \ne k; \\ K_i(\widetilde{M}, \widetilde{\partial M}) &= 0 \quad \text{for } i \ne k+1. \end{aligned}$$

Proof. Because of Lemma 13.15 we can assume that $\widetilde{f}$ and $\widetilde{\partial f}$ are k-connected. Hence $K_i(\widetilde{M}) = H_{i+1}(\widetilde{f})$ and $K_i(\widetilde{\partial M}) = H_{i+1}(\widetilde{\partial f})$ vanish for $i \le k-1$. We have the short exact sequence of finite free $\mathbb{Z}\pi$-chain complexes

13.3 The π-π-Theorem

$$0 \to C_*(\widetilde{\partial f}) \to C_*(\widetilde{f}) \to C_*(\widetilde{f}, \widetilde{\partial f}) \to 0. \tag{13.17}$$

We conclude from its long exact homology sequence that $K_i(\widetilde{M}, \widetilde{\partial M}) \cong H_{i+1}(\widetilde{f}, \widetilde{\partial f})$ vanish for $i \leq k - 1$. We know already from Lemma 13.15 that $K_k(\widetilde{M}, \widetilde{\partial M})$ vanishes. Hence we have proven so far

$$K_i(\widetilde{\partial M}) = 0 \quad \text{for } i \leq k - 1;$$
$$K_i(\widetilde{\partial M}) = 0 \quad \text{for } i \leq k - 1;$$
$$K_i(\widetilde{M}, \widetilde{\partial M}) = 0 \quad \text{for } i \leq k.$$

Since $H_i(\widetilde{f})$ and $H_i(\widetilde{\partial f})$ vanish for $i \leq k$, the chain complexes $\text{cone}(C_*(\widetilde{f}))$ and $\text{cone}(C_*(\widetilde{\partial f}))$ are $\mathbb{Z}\pi$-chain homotopy equivalent to $\mathbb{Z}\pi$-chain complexes whose chain modules are trivial in dimensions $\leq k$. Hence the cochain complexes $\text{cone}(C^*(\widetilde{f}))$ and $\text{cone}(C^*(\widetilde{\partial f}))^*$ respectively are $\mathbb{Z}\pi$-chain homotopy equivalent to cochain complexes whose cochain modules vanish in dimensions $\leq k$. This implies $K^i(\widetilde{M}) = H^{i+1}(\text{cone}(C^*(\widetilde{f}))) = 0$ and $K^i(\widetilde{\partial M}) = H^{i+1}(\text{cone}(C^*(\widetilde{\partial f}))) = 0$ for $i \geq k - 1$. Poincaré duality yields isomorphisms, see Theorem 8.116 (ii),

$$K_i(\widetilde{\partial M}) \cong K^{2k-i}(\widetilde{\partial M});$$
$$K_i(\widetilde{M}, \widetilde{\partial M}) \cong K^{2k+1-i}(\widetilde{M}).$$

Hence $K_i(\widetilde{\partial M}) = 0$ for $i \geq k + 1$ and $K_i(\widetilde{M}, \widetilde{\partial M}) = 0$ for $i \geq k + 2$. The long homology sequence associated to (13.17) implies $K_i(\widetilde{\partial M}) = 0$ for $i \geq k + 1$. This finishes the proof of Lemma 13.16. $\square$

We have $\pi_{k+2}(\widetilde{f}, \widetilde{\partial f}) \cong K_{k+1}(\widetilde{M}, \widetilde{\partial M})$. We can find for each $i \in \{1, 2, \ldots, r\}$ immersions $u_i : (D^{k+1}, S^k) \to (M, \partial M)$ that are framed, i.e., come with explicit trivialisations of their normal bundles, such that b_i is given by a map of squares

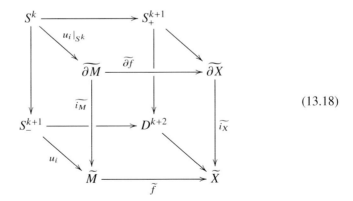

(13.18)

if we identify $S^{k+1}_- = D^{k+1}$. These u_i are unique up to regular homotopy.

Lemma 13.19 *We can change the framed immersions u_i so that their restrictions to S^k form a system of disjoint embeddings S^k to ∂M.*

Proof. We can put the framed immersions u_i into general position. The set of intersections and self-intersections form a system of 1-dimensional submanifolds of M, which intersect transversely and meet ∂M transversely and for which there are no triple intersections, see for instance [9, 363, 364, 365, 366, 436]. These 1-dimensional submanifolds are either circles that do not touch the boundary and therefore do not concern us here, or they are embedded arcs that meet the boundary transversely and in precisely two points. Fix such an embedded arc α. Next we explain how we can change the framed embeddings in their regular homotopy class so that the two endpoints of α disappear and hence we get rid of two of the intersection or self-intersection points on the boundary without introducing new intersection or self-intersection points on the boundary. The arc then becomes a circle within the interior of M.

We only treat the case where the arc consists of intersection points of two different framed embeddings u_i and u_j. The case of self-intersections is analogous. For the immersion u_i we can find an embedded 2-disks δ_i in S^k whose boundary consist of two segments, one segment is mapped under u_i to the arc α, the second segment is mapped under u_i to an arc β_i in ∂M, and analogously we get δ_j and β_j for u_j. Moreover, the arcs β_i and β_j meet in precisely two points, namely, the endpoints of α on ∂M. Obviously $\beta_i * \beta_j^-$ is a loop in ∂M. If we regard $\beta_i * \beta_j^-$ in M, then it is nullhomotopic, thanks to the disks δ_i and δ_j. Since $\partial M \to M$ induces an isomorphism on π_1, the loop $\beta_i * \beta_j^-$ is nullhomotopic in ∂M. Since $k \geq 3$ by assumption, we can find an embedded disk $D^2 \subseteq \partial M$ whose boundary is just the loop $\beta_i * \beta_j^-$. Now we can use the Whitney trick mentioned earlier to get rid of the two intersection points of $u_i|_{S^k}: S^k \to \partial M$ and $u_j|_{S^k}: S^k \to \partial M$. This move extends to a move in M in the obvious way. The result is that we have two intersection points on the boundary less than before the move. □

Since each u_i is framed, we can extend each u_i to an immersion

$$\overline{u_i}: (D^{k+1} \times D^k, S^k \times D^k) \to (M, \partial M)$$

such that $\overline{u_i}|_{S^k \times D^k}: S^k \times D^k \hookrightarrow \partial M$ is an embedding and the images of $\overline{u_i}|_{S^k \times D^k}$ and $\overline{u_j}|_{S^k \times D^k}$ are disjoint for $i \neq j$. For $i \in \{1, 2, \ldots, r\}$ we attach a handle $D^{k+1} \times D^k$ to M using the embedding $\overline{u_i}|_{S^k \times D^k}: S^k \times D^k \hookrightarrow \partial M$. Denote by N the resulting $(2k+1)$-dimensional compact manifold. Let $H \subseteq N$ be the union of these r attached handles.

By a cofibration argument using (13.18), we can change $f: (M, \partial M) \to (X, \partial X)$ up to homotopy with the following properties:

- The map f is the trivial map on $\mathrm{im}(\overline{u_i}) \cap \partial M$ for $i \in \{1, 2, \ldots, r\}$;
- There is a uniquely determined map $(g, \partial g): (N, \partial N) \to (X, \partial X)$ which agrees with f on M and is constant on H;

13.3 The π-π-Theorem

- There is a simple homotopy equivalence $h\colon M \vee \bigvee_{i=1}^{r} S^{k+1} \xrightarrow{\simeq} N$ such that the composite of g with h is homotopic to the wedge of $f\colon M \to X$ with the constant map $\bigvee_{i=1}^{r} S^{k+1} \to X$.

Hence the simple homotopy equivalence h together with the inclusion $M \to M \vee \bigvee_{i=1}^{r} S^{k+1}$ induce for $l \leq k$ an isomorphism

$$\xi_l \colon K_l(\widetilde{M}) \xrightarrow{\cong} K_l(\widetilde{N}) \tag{13.20}$$

and for $l = k + 1$ an isomorphism

$$\xi_{k+1} \colon K_{k+1}(\widetilde{M}) \oplus \mathbb{Z}\pi^r \xrightarrow{\cong} K_{k+1}(\widetilde{N}). \tag{13.21}$$

The long exact homology sequence of the triple $(\partial N \subseteq H \cup \partial N \subseteq N)$ yields the long exact sequence

$$\cdots \to K_{l+1}(\widetilde{N}, \widetilde{\partial N}) \to K_{l+1}(\widetilde{N}, \widehat{H} \cup \widetilde{\partial N})$$
$$\to K_l(\widehat{H} \cup \widetilde{\partial N}, \widetilde{\partial N}) \to K_l(\widetilde{N}, \widetilde{\partial N}) \to \cdots \tag{13.22}$$

where the π-spaces $\widetilde{N}, \widetilde{\partial N}$, and $\widehat{H}$ are obtained by pulling back the universal covering $\widetilde{M} \to M$. The excision isomorphisms associated to the inclusions $(\widehat{H}, \widehat{H} \cap \widetilde{\partial N}) \to (\widehat{H} \cup \widetilde{\partial N}, \widetilde{\partial N})$ and $(\widetilde{M}, \widetilde{\partial M}) \to (\widetilde{N}, \widehat{H} \cup \widetilde{\partial N})$ yield $\mathbb{Z}\pi$-isomorphisms

$$K_l(\widehat{H}, \widehat{H} \cap \widetilde{\partial N}) \xrightarrow{\cong} K_l(\widehat{H} \cup \widetilde{\partial N}, \widetilde{\partial N}); \tag{13.23}$$

$$K_l(\widetilde{M}, \widetilde{\partial M}) \xrightarrow{\cong} K_l(\widetilde{N}, \widehat{H} \cup \widetilde{\partial N}). \tag{13.24}$$

Next we show

$$K_l(\widetilde{M}, \widetilde{\partial M}) = 0 \quad \text{for } l \neq k + 1; \tag{13.25}$$

$$K_k(\widehat{H}, \widehat{H} \cap \widetilde{\partial N}) = 0 \quad \text{for } l \neq k; \tag{13.26}$$

$$K_{k+1}(\widetilde{N}, \widetilde{\partial N}) = 0 \quad \text{for } l \neq k, k + 1. \tag{13.27}$$

We have

$$K_l(\widehat{H}, \widehat{H} \cap \widetilde{\partial N}) \cong H_l(\widehat{H}, \widehat{H} \cap \widetilde{\partial N}) \cong$$
$$\cong \bigoplus_{i=1}^{k} H_l(\pi \times D^{k+1} \times D^k, \pi \times D^{k+1} \times S^{k-1}) \cong \begin{cases} 0 & l \neq k; \\ \mathbb{Z}\pi^r & l = k. \end{cases} \tag{13.28}$$

Recall that f and ∂f are k-connected. Hence $K_l(\widetilde{M}) = K_l(\widetilde{\partial M}) = K_l(\widetilde{M}, \widetilde{\partial M}) = 0$ for $l \leq k - 1$. This implies $K_l(\widetilde{N}) \cong K_l(\widetilde{M}) = 0$ for $l \leq k - 1$. Since ∂N is obtained from ∂M by surgery in dimension k, we have $K_l(\widetilde{\partial N}) = 0$ and hence $K_l(\widetilde{N}, \widetilde{\partial N}) = 0$ for $l \leq k-2$. We conclude that $K^l(\widetilde{M}) \cong K_l(\widetilde{M})^*$ and $K^l(\widetilde{N}) \cong K_l(\widetilde{N})^*$ vanish for $l \leq k - 1$. Poincaré duality implies that $K_l(\widetilde{M}, \widetilde{\partial M}) \cong K^{2k+1-l}(\widetilde{M})$ and $K_l(\widetilde{N}, \widetilde{\partial N}) \cong K^{2k+1-l}(\widetilde{N}) = 0$ vanish for $l \geq k + 2$. This finishes the verification of (13.25), (13.26), and (13.27). Combining these with the isomorphisms (13.23) and (13.24) implies that the long exact sequence (13.22) reduces to the exact sequence of finitely generated free $\mathbb{Z}\pi$-modules

$$0 \to K_{k+1}(\widetilde{N}, \widetilde{\partial N}) \to K_{k+1}(\widetilde{M}, \widetilde{\partial M}) \xrightarrow{\delta} K_k(\widehat{H}, \widehat{H} \cap \widetilde{\partial N})$$
$$\to K_k(\widetilde{N}, \widetilde{\partial N}) \to 0. \quad (13.29)$$

Consider the composite $\mu \colon K_{k+1}(\widehat{H}, \widehat{H} \cap \widetilde{M}) \to K_k(\widehat{H} \cap \widetilde{M}) \to K_k(\widetilde{M})$ where the first map comes from the boundary map in the long homology sequence of the pair $(\widehat{H}, \widehat{H} \cap \widetilde{M})$ and the second map comes from the inclusion $\widehat{H} \cap \widetilde{M} \to \widetilde{M}$. We have the isomorphism of $\mathbb{Z}\pi$-modules

$$K_{k+1}(\widehat{H}, \widehat{H} \cap \widetilde{M}) \xrightarrow{\cong} H_{k+1}(\widehat{H}, \widehat{H} \cap \widetilde{M})$$
$$\xleftarrow{\cong} \bigoplus_{i=1}^{r} H_{k+1}(\pi \times D^{k+1} \times D^k, \pi \times S^k \times D^k) \cong \mathbb{Z}\pi^r.$$

The composite μ sends the base element corresponding to the i-th handle to the class represented by the lift $\widetilde{u}_i \colon S^k \to \widetilde{M}$ of the attaching map $u_i \colon S^k \to M$. Since u_i is nullhomotopic, the map μ is trivial. Hence the following composite is also trivial

$$\nu \colon K_k(\widehat{H}, \widehat{H} \cap \widetilde{\partial N})^* \xrightarrow{\cong} K^k(\widehat{H}, \widehat{H} \cap \widetilde{\partial N}) \xrightarrow{\cong} K_{k+1}(\widehat{H}, \widehat{H} \cap \widetilde{M})$$
$$\xrightarrow{\mu} K_k(\widetilde{M}) \xrightarrow{\cong} K^{k+1}(\widetilde{M}, \widetilde{\partial M}) \xrightarrow{\cong} K_{k+1}(\widetilde{M}, \widetilde{\partial M})^*$$

where the maps except μ are given by Poincaré duality or the Kronecker pairing. The following diagram commutes

$$\begin{array}{ccc}
K_{k+1}(\widetilde{M}, \widetilde{\partial M}) & \xrightarrow{\delta} & K_k(\widehat{H}, \widehat{H} \cap \widetilde{\partial N}) \\
\Big\downarrow \cong & & \Big\downarrow \cong \\
\left(K_{k+1}(\widetilde{M}, \widetilde{\partial M})^*\right)^* & \xrightarrow{\nu^*} & \left(K_k(\widehat{H}, \widehat{H} \cap \widetilde{\partial N})^*\right)^*
\end{array}$$

where δ is the map appearing in (13.29). Hence δ is trivial and the short exact sequence (13.29) reduces to the $\mathbb{Z}\pi$-isomorphisms

$$K_{k+1}(\widetilde{N}, \widetilde{\partial N}) \xrightarrow{\cong} K_{k+1}(\widetilde{M}, \widetilde{\partial M}); \quad (13.30)$$
$$K_k(\widehat{H}, \widehat{H} \cap \widetilde{\partial N}) \xrightarrow{\cong} K_k(\widetilde{N}, \widetilde{\partial N}). \quad (13.31)$$

The handles attached using the u_i-s give the preferred stable U-equivalence class of $\mathbb{Z}\pi$-basis $\{b_1, b_2, \ldots, b_r\}$ for $K_{k+1}(\widetilde{M}, \widetilde{\partial M})$ by construction. In view of the isomorphisms (13.21) and (13.30), the canonical map

$$K_{k+1}(\widetilde{N}) \to K_{k+1}(\widetilde{N}, \widetilde{\partial N})$$

13.3 The π-π-Theorem

is surjective since the composite

$$K_{k+1}(\widetilde{M}) \oplus \mathbb{Z}\pi^r \xrightarrow{\cong} K_{k+1}(\widetilde{N}) \to K_{k+1}(\widetilde{N}, \widetilde{\partial N}) \xrightarrow{\cong} K_{k+1}(\widetilde{M}, \widetilde{\partial M})$$

sends the standard base of $\mathbb{Z}\pi^r$ to this $\mathbb{Z}\pi$-base of $K_{k+1}(\widetilde{M}, \widetilde{\partial M})$. If we combine this with the isomorphisms (13.20) and inspect the long exact sequences of kernels associated to $(\widetilde{N}, \widetilde{\partial N})$, we conclude $K_l(\widetilde{\partial N}) = 0$ for $l \leq k$. Since ∂N is $2k$-dimensional, Poincaré duality implies $K_l(\widetilde{N}) = 0$ for all $l \geq 0$. In particular $K_{k+1}(\widetilde{N}) \xrightarrow{\cong} K_{k+1}(\widetilde{N}, \widetilde{\partial N})$ is an isomorphism. Hence $K_{k+1}(\widetilde{M}) = 0$. In the sequel we equip $K_{k+1}(\widetilde{N})$ with the $\mathbb{Z}\pi$-basis from the isomorphism $\mathbb{Z}\pi^r \xrightarrow{\cong} K_{k+1}(\widetilde{N})$ of (13.21). Then the composite of $\mathbb{Z}\pi$-isomorphisms

$$K_{k+1}(\widetilde{N}) \xrightarrow{\cong} K_{k+1}(\widetilde{N}, \widetilde{\partial N}) \xrightarrow{\cong} K_{k+1}(\widetilde{M}, \widetilde{\partial M})$$

is simple.

Since $K_l(\widetilde{N}) = 0$ for all $l \geq 0$, the map ∂g is a homotopy equivalence. Actually, it is a simple homotopy equivalence, as we will see below. It remains now to achieve by surgery on the interior that g is also a simple homotopy equivalence.

Recall that we have already equipped $K_{k+1}(\widetilde{N})$ with a $\mathbb{Z}\pi$-basis. Recall that we have $H_i(\text{cone}(C_*(\widetilde{g}))) = 0$ for $i \neq k+1, k+2$ and canonical identifications $H_{i+1}(\text{cone}(C_*(\widetilde{g}))) = K_i(\widetilde{N})$ for $i = k, k+1$. Hence $K_k(\widetilde{N})$ inherits a Reidemeister stable U-equivalence class of stable $\mathbb{Z}\pi$-bases as explained in Subsection 13.3.2. After possible trivial surgery, we can choose a $\mathbb{Z}\pi$-basis $\{a_1, a_2, \ldots, a_r\}$ for $K_k(\widetilde{N})$ representing the Reidemeister stable U-equivalence class of stable $\mathbb{Z}\pi$-bases. Since $\dim(N) = 2k + 1$, we can perform surgery on the interior of N according to this basis. Thus we get a new normal map that will be the one we are aiming for,

$$(h, \partial h) \colon (P, \partial P) \to (X, \partial X),$$

with $\partial P = \partial N$ and $\partial h = \partial g$ and a normal bordism

$$(L; \partial_0 L, \partial_1 L) \colon (W; \partial_0 W, \partial_1 W) \to (X \times [0, 1]; X \times 0, \partial X \times [0, 1] \cup X \times \{1\})$$

such that $\partial_0 W = N$, $\partial_0 L = g$, $\partial_1 W = P \cup_{\partial P \times \{1\}} \partial P \times [0, 1]$, and $\partial_1 L = h \cup_{\partial h \times \text{id}_{\{1\}}} \partial h \times \text{id}_{\partial P \times [0,1]}$ hold. The bordism W is obtained from $P \times [0, 1]$ by attaching a handle $D^{k+1} \times D^k$ to $P \times \{1\}$ for each $i \in \{1, 2, \ldots, r\}$. We have

$$K_l(\widetilde{W}, \widetilde{\partial_0 W}) \cong \bigoplus_{i=1}^{r} H_l(D^{k+1} \times D^k \times \pi, S^k \times D^k \times \pi) \cong \begin{cases} \mathbb{Z}\pi^r & l = k+1; \\ 0 & l \neq k+1. \end{cases}$$

We will equip $K_{k+1}(\widetilde{W}, \widetilde{\partial_0 W})$ with the preferred stable U-equivalence class of stable $\mathbb{Z}\pi$-bases coming from the $\mathbb{Z}\pi$-isomorphism above. It agrees with the Reidemeister stable U-equivalence class of $\mathbb{Z}\pi$-bases on $K_{k+1}(\widetilde{W}, \widetilde{\partial_0 W})$. We obtain from the pair $(\widetilde{W}, \widetilde{\partial_0 W})$ the exact sequence of $\mathbb{Z}\pi$-modules

$$0 \to K_{k+1}(\widetilde{\partial_0 W}) \xrightarrow{j} K_{k+1}(\widetilde{W}) \to K_{k+1}(\widetilde{W}, \widetilde{\partial_0 W})$$
$$\xrightarrow{\partial} K_k(\widetilde{\partial_0 W}) \to K_k(\widetilde{W}) \to 0.$$

The map ∂ sends the $\mathbb{Z}\pi$-basis of $K_{k+1}(\widetilde{W}, \widetilde{\partial_0 W})$ determined by the attached handles to the $\mathbb{Z}\pi$-basis $\{a_1, a_2, \ldots, a_r\}$ of $K_k(\widetilde{\partial_0 W}) = K_k(\widetilde{N})$. Hence it is a simple isomorphism. We conclude that $K_k(\widetilde{W})$ vanishes. Hence the sequence above reduces to the exact sequence

$$0 \to K_{k+1}(\widetilde{\partial_0 W}) \xrightarrow{j} K_{k+1}(\widetilde{W}) \to K_{k+1}(\widetilde{W}, \widetilde{\partial_0 W}) \xrightarrow{\partial} K_k(\widetilde{\partial_0 W}) \to 0, \qquad (13.32)$$

and we have $K_i(\widetilde{W}) = 0$ for $i \neq k+1$. So $K_{k+1}(\widetilde{W})$ comes with a Reidemeister stable U-equivalence class of $\mathbb{Z}\pi$-bases. We have specified for each term in the short exact sequence (13.32) a preferred stable U-equivalence class of $\mathbb{Z}\pi$-bases. The Reidemeister U-torsion in the sense of Definition 10.8 of the sequence (13.32) is trivial because of Lemma 10.23. Since ∂ is a simple isomorphism, the map $j \colon K_{k+1}(\widetilde{\partial_0 W}) \xrightarrow{\cong} K_{k+1}(\widetilde{W})$ is a simple isomorphism.

Let $J \colon K_{k+1}(\widetilde{W}) \to K_{k+1}(\widetilde{W}, \widetilde{\partial_1 W})$ be induced by the map $\widetilde{W} \to (\widetilde{W}, \widetilde{\partial W})$. The following diagram commutes

$$\begin{array}{ccc}
K_k(\widetilde{\partial_0 W})^* & \xrightarrow[\cong]{\partial^*} & K_{k+1}(\widetilde{W}, \widetilde{\partial_0 W})^* \\
\cong \uparrow & & \cong \uparrow \\
K^k(\widetilde{\partial_0 W}) & & K^{k+1}(\widetilde{W}, \widetilde{\partial_0 W}) \\
\cong \downarrow & & \cong \downarrow \\
K_{k+1}(\widetilde{\partial_0 W}) & \xrightarrow[j]{\cong} K_{k+1}(\widetilde{W}) \xrightarrow{J} & K_{k+1}(\widetilde{W}, \widetilde{\partial_1 W}).
\end{array}$$

Since all maps in the diagram above different from J are simple isomorphisms, the isomorphisms $J \colon K_{k+1}(\widetilde{W}) \to K_{k+1}(\widetilde{W}, \widetilde{\partial_1 W})$ is a simple isomorphism. We conclude from Lemma 10.22 that the obvious chain map $\mathrm{cone}(C_*(\widetilde{L})) \to \mathrm{cone}(C_*(\widetilde{L}, \widetilde{\partial_1 L}))$ is a simple $\mathbb{Z}\pi$-homotopy equivalence. Since we have the based exact short exact sequence of $\mathbb{Z}\pi$-chain complexes

$$0 \to \mathrm{cone}(C_*(\widetilde{\partial_1 L})) \to \mathrm{cone}(C_*(\widetilde{L})) \to \mathrm{cone}(C_*(\widetilde{L}, \widetilde{\partial_1 L})) \to 0,$$

the Reidemeister U-torsion of the contractible based free $\mathbb{Z}\pi$-chain complex $\mathrm{cone}(C_*(\widetilde{\partial_1 L}))$ vanishes. Hence $\partial_1 L \colon \partial_1 W \to \partial X \times [0,1] \cup X \times \{1\}$ is a simple homotopy equivalence. We conclude from Lemma 13.14 that the induced map on the boundary $\partial_1 L|_{\partial(\partial_1 W)} \colon \partial(\partial_1 W) \to \partial X$ is also a simple homotopy equivalence. Hence the maps $h \colon P \to X$ and $\partial h \colon \partial P \to \partial X$ are simple homotopy equivalences. Since $(h, \partial h) \colon (P, \partial P) \to (X, \partial X)$ is obtained from $(f, \partial f) \colon (M, \partial M) \to (X, \partial X)$ by a sequence of homotopies, surgeries on the interior, surgeries on the boundary,

13.4 The Geometric Surgery Obstruction Group and Surgery Obstruction

In this section we want to give a geometric construction of the (simple) surgery obstruction group that will be identified with the algebraic L-group, and of the geometric (simple) surgery obstruction that will be identified with the algebraic surgery obstruction.

13.4.1 The Construction of the Geometric Surgery Groups

We begin with the construction of the geometric surgery obstruction group in terms of normal bordisms of normal maps of degree one with reference.

Recall the following category $\mathcal{R}$ of Subsection 8.7.2. Objects are pairs (K, O_K) consisting of a space K and an infinite cyclic coefficient system O_K over K, i.e., a covariant functor $O_K \colon \Pi(K) \to \mathbb{Z}\text{-MOD}$ such that its value at each object is an infinite cyclic group. A morphism $(u, \overline{u}) \colon (K_0, O_{K_0}) \to (K_1, O_{K_1})$ consists of a map $u \colon K_0 \to K_1$ and a natural transformation $\overline{u} \colon O_{K_0} \to O_{K_1} \circ \Pi(u)$.

The goal of this section is to define certain covariant functors

$$\mathcal{L}_n^h, \mathcal{L}_n^s \colon \mathcal{R} \to \mathbb{Z}\text{-MOD}. \tag{13.33}$$

They will be defined as bordism groups of *normal maps of degree one with reference to* (K, O_K) where a normal map of degree one with reference to (K, O_K) is a normal map $(M, \partial M, f, \partial f, a, \xi, \overline{f}, o)$ with target some w-oriented finite n-dimensional Poincaré pair $(X, \partial X, O_X, [[X, \partial X]])$ in the sense of Subsection 8.8.1 together with a morphism $(u, \overline{u}) \colon (X, O_X) \to (K, O_K)$ in $\mathcal{R}$. Recall that here we demand that $\partial f \colon \partial M \to \partial X$ is a (simple) homotopy equivalence.

Note that we here follow [147, Paragraph after Theorem 3.1] where an apparent bug in Wall's construction from [414, Chapter 9] is corrected. It leads us to consider the category $\mathcal{R}$ above whereas in [414, Chapter 9] a quotient category of $\mathcal{R}$ is used. This is related to the appearance of the term $w(g)$ in Lemma 8.152.

If we have two such normal maps $(M_m, \partial M_m, f_m, \partial f_m, a_m, \xi_m, \overline{f_m}, o_m)$ with target $(X_m, \partial X_m, O_{X_m}, [[X_m, \partial X_m]])$ and with reference maps $(u_m, \overline{u_m}) \colon (X_m, O_{X_m}) \to (K, O_K)$ for $m = 0, 1$, we call them *normally bordant* if there exists a normal bordism of degree one (*not necessarily with cylindrical ends*) $(W, F, b, \Xi, \overline{F}, O)$ with target $(Y, \partial Y, O_Y, [[Y, \partial Y]])$ from $(M_0, \partial M_0, f_0, \partial f_0, a_0, \xi_0, \overline{f}_0, o_0)$ to $(M_1, \partial M_1, f_1, \partial f_1, a_1, \xi_1, \overline{f}_1, o_1)$ in the sense of Subsection 8.8.2 together with a reference map $(U, \overline{U}) \colon (Y, O_Y) \to (K, O_K)$ such that the composite of $(U, \overline{U})$ with the morphism $(X_m, O_{X_m}) \to (Y, O_Y)$ in $\mathcal{R}$ induced by the inclusion agrees with $(u_m, \overline{u}_m)$ for $m = 0, 1$. Recall that here we demand that $\partial_2 F \colon \partial_2 W \to \partial_2 Y$ is a (simple) homotopy equivalence.

The addition is given by taking the disjoint union. The inverse of the bordism class of the normal map of degree one with reference to (K, O_K) given by $(M, \partial M, f, \partial f, a, \xi, \overline{f}, o)$ with target some w-oriented finite n-dimensional Poincaré pair $(X, \partial X, O_X, [[X, \partial X]])$ and reference map $(u, \overline{u}) \colon (X, O_X) \to (K, O_K)$ is given by the same data, except that we replace $\overline{u}$ by $-\overline{u}$.

By taking compositions of reference maps to (K_0, O_{K_0}) with morphisms $(u, \overline{u}) \colon (K_0, O_{K_0}) \to (K_1, O_{K_1})$ in $\mathcal{R}$, we get the desired covariant functors (13.33).

Definition 13.34 ((Simple) geometric surgery obstruction group) We call $\mathcal{L}_n^s(K, O_K)$ the *n*-th simple geometric obstruction group and $\mathcal{L}_n^h(K, O_K)$ the *n*-th geometric obstruction group.

Lemma 13.35 *Let $(u_i, \overline{u}_i) \colon (K, O_K) \to (K', O'_{K'})$ for $i = 0, 1$ be two morphisms in $\mathcal{R}$. Suppose that they are homotopic, see the paragraph before Lemma 8.157. Then we get*

$$\mathcal{L}_n^s(u_0, \overline{u}_0) = \mathcal{L}_n^s(u_1, \overline{u}_1);$$
$$\mathcal{L}_n^h(u_0, \overline{u}_0) = \mathcal{L}_n^h(u_1, \overline{u}_1).$$

Proof. The idea is to construct from the homotopy between $(u_0, \overline{u}_0)$ and $(u_1, \overline{u}_1)$ a normal bordism between the representatives of the image under $\mathcal{L}_n^s(u_0, \overline{u}_0)$ and $\mathcal{L}_n^s(u_1, \overline{u}_1)$ of a given element in $\mathcal{L}_n^s(K_0, O_{K_0})$, and analogously for the h-decoration. $\square$

Exercise 13.36 Let (K, O_K) be an object in $\mathcal{R}$ for a *CW*-complex K. Let $\mathcal{K}(K)$ be the direct set of finite sub-*CW*-complexes of K directed by inclusion. For $L \in \mathcal{K}(K)$ let $i_L \colon L \to K$ be the inclusion, $i_L^* O_K$ be the pullback of O_K with i_L and $\overline{i_L} \colon i_L^* O_K \to O_K$ be the canonical bundle map covering i_L. For $L, L' \subseteq \mathcal{K}(K)$ let $i_{L \subseteq L'} \colon L \to L'$ be the inclusion and let $\overline{i_{L \subseteq L'}} \colon i_L^* O_K \to i_{L \subseteq L'}^* i_{L'} * O_K = i_L^* O_K$ be the identity. Let ϵ stand for h or s. Thus we obtain a directed system of abelian groups $\{\mathcal{L}_n^\epsilon(L, i_L^* O_K) \mid L \in \mathcal{K}(K)\}$ and a canonical homomorphism

$$\psi \colon \operatorname{colim}_{L \subseteq \mathcal{K}(K)} \mathcal{L}_n^\epsilon(L, i_L^* O_K) \to \mathcal{L}_n^\epsilon(K, O_K).$$

Show that ψ is an isomorphism.

13.4.2 Making the Reference Map a π_0- and π_1-Isomorphism

In order to apply the π-π-Theorem 13.4 later, the following result will be crucial. It roughly says that we can arrange that the reference map $u \colon X \to K$ induces an equivalence on the fundamental groupoids.

Lemma 13.37 Let (K, O_K) be an object in $\mathcal{R}$. Consider a natural number $n \geq 5$. Let $(M, \partial M, f, \partial f, a, \xi, \overline{f}, o)$ be a normal map of degree one with target some w-oriented finite n-dimensional Poincaré pair $(X, \partial X, O_X, [[X, \partial X]])$ coming with a reference map $(u, \overline{u}): (X, O_X) \to (K, O_K)$ such that ∂f is a homotopy equivalence.

Then it can be modified by surgery without changing $\partial M, \partial X, \partial f, \overline{f}|_{\partial M}, u|_{\partial X}$, $\overline{u}|_{\partial X}$, and its class in $\mathcal{L}_n^h(K, O_K)$ so that the reference map $u \colon X \to K$ induces a bijection on π_0 and for every base point in X an isomorphism on π_1.

The analogous statement holds with the decoration s.

Its proof needs some preparation. The next lemma will ensure that we have a "manifold part" in a Poincaré pair $(X, \partial X)$ on which we can perform surgery on the reference to K.

Definition 13.38 (Special Poincaré pair) Let $(X, \partial X)$ be a finite (simple) Poincaré pair of dimension $n \geq 5$. It is called *special* if there exists:

- An n-dimensional compact smooth manifold H with boundary ∂H such that the inclusion $i \colon \partial H \to H$ induces an epimorphism $\pi_0(\partial H) \to \pi_0(H)$ and for every component D of the manifold H it induces an epimorphism $*_{C \in \pi_0(\partial H), i(C) \subseteq D} \pi_1(C) \to \pi_1(D)$;
- A finite CW-subcomplex $\widehat{X}$ of X containing ∂X such that $\widehat{X} \setminus \partial X$ contains only cells of dimension $\leq (n-2)$;
- A cellular map $z \colon \partial H \to \widehat{X}$ that induces a bijection on π_0 and for every choice of base point in ∂H an epimorphism on π_1, where we consider $(H, \partial H)$ as a simple CW-pair by a smooth triangulation,

such that X is the pushout

$$\begin{array}{ccc} \partial H & \xrightarrow{z} & \widehat{X} \\ \downarrow & & \downarrow \\ H & \longrightarrow & X = H \cup_z \widehat{X}. \end{array}$$

Lemma 13.39 Let $(X, \partial X)$ be an n-dimensional finite (simple) Poincaré pair of dimension $n \geq 5$. Then there exists a special Poincaré pair $(X', \partial X')$ with $\partial X = \partial X'$ together with a (simple) homotopy equivalence $g \colon X \to X'$ inducing the identity on ∂X.

Proof. See [414, Lemma 2.8 on page 30 and the following paragraph]. □

Lemma 13.40 Let $X = H \cup_z \widehat{X}$ be a special Poincaré pair. Then the inclusion $H \to X$ induces bijections on π_0 and for every choice of base points of H an epimorphism on π_1.

Proof. Recall from the definition of a special Poincaré pair that $\partial H \to \widehat{X}$ induces bijections on π_0 and for every choice of base point in ∂H a surjection on π_1, and the inclusion $i \colon \partial H \to H$ induces an epimorphism $\pi_0(\partial H) \to \pi_0(H)$, and for every component D of H an epimorphism $*_{C \in \pi_0(\partial H), i(C) \subseteq D} \pi_1(C) \to \pi_1(D)$. This implies that $H \to X$ induces a bijection on π_0. The Seifert–van Kampen Theorem implies that the inclusion of $H \to X = H \cup_z \widehat{X}$ induces a surjection on π_1 for every choice of base point in H. □

In the sequel we often only mention the underlying maps when dealing with normal maps, normal bordisms or reference maps.

Lemma 13.41 *Let $(f, \partial f) \colon (M, \partial M) \to (X, \partial X)$ be the underlying map of a normal map of degree one such that f is 2-connected. Suppose that $(X, \partial X)$ is special in the sense of Definition 13.38, i.e., $X = H \cup_z \widehat{X}$. Consider an embedded $S^l \subseteq H \setminus \partial H$ for $l \in \{0, 1\}$ such that the normal bundle $\nu(S^l \subseteq H \setminus \partial H)$ is trivial.*

Then we can change f up to homotopy relative to the boundary ∂M so that $f^{-1}(S^l) \subseteq M$ is an embedded submanifold and f induces a diffeomorphism $f|_{f^{-1}(S^l)} \colon f^{-1}(S^l) \to S^l$ that is covered by a bundle map between the normal bundles $\nu(f^{-1}(S^l) \subseteq M)$ and $\nu(S^l \subseteq H)$.

Proof. We begin with the case $l = 0$. Obviously it suffices to treat each point of S^l separately. Hence it suffices to show for $y \in H$ that we can change f up to homotopy so that $f^{-1}(y)$ consists of one point. We can change f up to homotopy so that f is transverse to the submanifold $\{y\} \subseteq H$. Hence $f^{-1}(y)$ is a finite set and for every $x \in f^{-1}$ the differential $T_x f \colon T_x M \to T_y H$ is an isomorphism. We fix an orientation on the fibre ξ_y of the bundle ξ over $y \in H$. Then the bundle data yield preferred orientations on $T_x M$ for every $x \in f^{-1}(y)$. Define the local degree $\deg(f, x)$ for $x \in f^{-1}(y)$ to be 1 if $T_x f$ is orientation preserving and to be -1 if $T_x f$ is orientation reversing. Let D be the path component of X containing y. Since $\pi_0(f)$ is bijective by assumption, there is precisely one path component C of M with $f(C) \subseteq D$. The degree one condition on the normal map $(f, \partial f) \colon (M, \partial M) \to (X, \partial X)$ implies that the sum of the local degrees $\sum_{x \in f^{-1}(y)} \deg(f, x)$ is ± 1. Hence $f^{-1}(y)$ can be written as $\{x_0, x_1, \ldots, x_{2m}\}$ for some integer $m \geq 0$ such that $\deg(f, x_{2i-1}) = -\deg(f, x_{2i})$ holds for $i = 1, 2, \ldots m$. Hence it suffices to describe a procedure that modifies f so that $f^{-1}(y)$ is reduced to $\{x_0, x_1, \ldots, x_{2m-2}\}$ provided that $m \geq 1$.

Choose an embedded arc in M joining x_{2m-1} and x_{2m} that does not meet any of the other points $x_0, x_1, \ldots, x_{2(m-1)}$. Let U be an open neighbourhood of x_{2m} that is diffeomorphic to $\mathbb{R}^n$. Now perform a local homotopy of f along this arc to move x_{2m-1} so close to x_{2m} such that x_{2m-1} lies in U. Hence it suffices to prove the following: Given a map $f \colon \mathbb{R}^n \to \mathbb{R}^n$ such that f is transversal to $0 \in \mathbb{R}^n$, the preimage $f^{-1}(0)$ consists of precisely two points x_0 and x_1 belonging to the interior of the disk $D^n \subseteq \mathbb{R}^n$, the differential $T_{x_0} f \colon T_{x_0} \mathbb{R}^n \to T_0 \mathbb{R}^n$ is bijective and reverses the standard orientations, and the differential $T_{x_1} f \colon T_{x_1} \mathbb{R}^n \to T_0 \mathbb{R}^n$ is bijective and preserves the standard orientations, then we can change f up to homotopy relative $\mathbb{R}^n \setminus D^n$ so that $f^{-1}(0)$ is empty.

Choose $\epsilon > 0$ so small that the image of $S^{n-1} \subseteq \mathbb{R}^n$ under f does not meet the interior of $\epsilon \cdot D^n$. Let $\mathrm{pr}_\epsilon \colon \mathbb{R}^n \to \epsilon \cdot D_n$ be the retraction that sends $x \in \mathbb{R}^n$ to $\frac{\epsilon}{||x||} \cdot x$ if $||x|| \geq \epsilon$, and to x if $||x|| \geq \epsilon$. Then $\mathrm{pr}_\epsilon \circ f$ induces a map $(D^n, S^{n-1}) \to$

13.4 The Geometric Surgery Obstruction Group and Surgery Obstruction

$(\epsilon \cdot D^n, \epsilon \cdot S^{n-1})$ of compact oriented manifolds. By inspecting the preimage of $0 \in \epsilon \cdot D^n$ we see that its degree is zero. Hence also the induced map $S^{n-1} \to \epsilon \cdot S^{n-1}$ has degree zero and hence is nullhomotopic. This implies that the map $f_0 \colon S^{n-1} \to \mathbb{R}^n \setminus \{0\}$ induced by f is nullhomotopic and hence extends to a map $f_1 \colon D^n \to \mathbb{R}^n \setminus \{0\}$. Let $f' \colon \mathbb{R}^n \to \mathbb{R}^n \setminus \{0\}$ be the map whose restriction to D^n is f_1 and whose restriction to $\mathbb{R}^n \setminus D^n$ agrees with the restriction of f to $\mathbb{R}^n \setminus D^n$. We obtain a homotopy $h \colon f \simeq f'$ of maps $\mathbb{R}^n \to \mathbb{R}^n$ by $h(x, t) = t \cdot f'(x) + (1 - t) \cdot f$ that is stationary outside the interior of D^n. Since the image of f' does not contain zero, the claim follows in the case $l = 0$.

Next we treat the case $l = 1$. Again we can make f transversal to $S^1 \subseteq M$. Then the preimage $f^{-1}(S^1)$ is an embedded 1-dimensional submanifold of M. Hence each component C of $f^{-1}(S^1)$ is an embedded $S^1 \subseteq M$ with trivial normal bundle. We want to reduce inductively the number of components of $f^{-1}(S^1)$ to 1. If there is one component C in $f^{-1}(S^1)$ such that $f(C) \neq S^1$, then $f|_C \colon C \to S^1$ is nullhomotopic and one easily changes f up to homotopy in a small neighbourhood of C so that the component C in $f^{-1}(S^1)$ disappears and no new components in $f^{-1}(S^1)$ are generated. Hence it remains to treat the case where $f(C) = S^1$ holds for all components C of $f^{-1}(S^1)$.

We next describe a procedure to change f within its homotopy type so that f is still transversal to $S^1 \subseteq H$, but two previously different components of $f^{-1}(S^1)$ are now connected to one component. Consider two different components C and D of $f^{-1}(S^1)$. Choose points $x_C \in C$ and $x_D \in D$ with $f(x_C) = f(X_D) = y$ for some $y \in S^1$. We can arrange that in a neighbourhood of x_C and x_D the differentials $T_{x_C}(f|C) \colon T_{x_C}C \to T_y S^1$ and $T_{x_D}(f|D) \colon T_{x_D}D \to T_y S^1$ are bijective. Choose an arc $v \colon [0, 1] \to M$ in M joining x_C and x_D such that the loop $f \circ v$ in X is nullhomotopic. This can be done since f is assumed to be 1-connected. We can furthermore arrange that the arc v meets $f^{-1}(S^1)$ only at x_C and x_D (transversely) and that $f \circ v$ is an embedding that on $[0, 1/3]$ moves a small distance along the normal fibre of y away from S^1, stays constant on $[1/3, 2/3]$, and then on $[2/3, 1]$ moves back to y just reversing the part on $[0, 1/3]$. We leave it to the reader to check that one can perform a homotopy of f in a small neighbourhood of the arc v so that for the resulting map the components C and D become their connected sum along this arc and f is still transversal to $S^1 \subseteq H$. Hence we have achieved that $f^{-1}(S^1)$ has one component less. Finally, we can arrange by this procedure that $f^{-1}(S^1)$ is connected and f is transversal to $S^1 \subseteq H$. In particular, $f^{-1}(S^1)$ is diffeomorphic to S^1.

The map f induces a map $f|_{f^{-1}(S^1)} \colon f^{-1}(S^1) \to S^1$ whose degree is $\pm - 1$ because of the degree one condition on the normal map $(f, \partial f) \colon (M, \partial M) \to (X, \partial X)$. Hence $f|_{f^{-1}(S^1)} \colon f^{-1}(S^1) \to S^1$ is homotopic to a diffeomorphism. Since f is covered by a bundle map between the normal bundles $\nu(f^{-1} \subseteq M)$ and $\nu(S^1 \subseteq H)$, we can change f itself up to homotopy so that $f|_{f^{-1}(S^1)} \colon f^{-1}(S^1) \to S^1$ is a diffeomorphism and f is still traversal to $S^1 \subseteq H$. The latter implies that $f|_{f^{-1}(S^1)}$ is covered by a bundle map between the normal bundles $\nu(f^{-1}(S^l) \subseteq M)$ and $\nu(S^l \subseteq H)$. This finishes the proof of Lemma 13.41. □

Lemma 13.42 Let (K, O_K) be an object in $\mathcal{R}$. Consider a normal map of degree one whose underlying map is $(f, \partial f)\colon (M, \partial M) \to (X, \partial X)$. Suppose that it comes with a reference map $(u, \overline{u})\colon (X, O_X) \to (K, O_K)$. Let $(g, \partial g)\colon (X, \partial X) \to (X', \partial X')$ be a (simple) homotopy equivalence of (simple) finite Poincaré pairs.

Then we can find another normal map of degree one with reference to (K, O_K) whose underlying map is $(g, \partial g) \circ (f, \partial f)\colon (M, \partial M) \to (X', \partial X')$ that is normally bordant to the given one as a normal map of degree one with reference to (K, O_K) and hence defines the same element in the geometric L-group of (K, O_K).

Proof. Choose a map $(h, \partial h)\colon (X', \partial X') \to (X, \partial X)$ that is as a map of pairs a homotopy inverse of $(g, \partial g)$. If ξ is the vector bundle over X coming from the bundle data, let ξ' over X' be the pullback $h^*\xi$. Then $g\colon X \to X'$ is covered by a bundle map $\overline{g}\colon \xi \to g^*\xi'$. Since g is a homotopy equivalence, we can find an isomorphism $o_g\colon O_X \to g^*O_{X'}$ of infinite cyclic coefficient systems over X. Let $o'\colon O_{\xi'} \xrightarrow{\cong} O_{X'}$ be an isomorphism such that the following diagram of isomorphisms of infinite cyclic coefficient systems over X

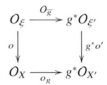

commutes. By composition with $(g, \overline{g})$ and using o', we obtain a new normal map of degree one with target $(X', \partial X')$ whose underlying map of spaces is the composite $(g, \partial g) \circ (f, \partial f)\colon (M, \partial M) \to (X', \partial X')$. We equip it with the reference map to (K, O_K) given by the composite $(u, \overline{u}) \circ (h, \overline{h})$ where $\overline{h}\colon O_{X'} \to h^*O_X$ comes from the inverse of $h^*o_g\colon h^*o_X \to h^*g^*O_{X'}$ and an isomorphism $O_{X'} \xrightarrow{\cong} h^*g^*O_{X'}$ induced by a homotopy between $g \circ h$ and id_X.

We leave it to the reader to check that the new and the old normal map with reference to (K, O_K) are normally bordant and hence give the same element in the geometric L-group of (K, O_K). □

Exercise 13.43 Let X and Y be CW-complexes that are homotopy equivalent to 1-dimensional CW-complexes. Let $u\colon X \to Y$ be a map. Let ξ and ξ' be n-dimensional vector bundles over X and let η be an n-dimensional vector bundle over Y for $n \geq 1$. Show:

(i) The vector bundles ξ and ξ' over X are isomorphic if and only if the infinite cyclic coefficient systems O_ξ and $O_{\xi'}$ are isomorphic;
(ii) For every morphisms of infinite cyclic coefficients systems $\alpha\colon O_\xi \to u^*O_\eta$, there is a bundle map $\overline{u}\colon \xi \to \eta$ covering u such that $O_{\overline{u}} = \alpha$ holds.

Now we are ready to prove Lemma 13.37.

Proof of Lemma 13.37. We can assume without loss of generality that K is connected since the image of $u\colon X \to K$ meets only finitely many of the components of K and we can treat any of these components of K separately.

13.4 The Geometric Surgery Obstruction Group and Surgery Obstruction

We want to perform surgery on the reference map $u: X \to K$ although X is not a manifold and K is not a Poincaré complex. Just to do surgery to improve the connectivity of a normal map itself does not use Poincaré duality for the target, but of course we have to deal with the problem that X itself is not a compact manifold. This problem is taken care of by the conclusion from Lemma 13.39 and Lemma 13.42 that we can assume that $(X, \partial X)$ is special in the sense of Definition 13.38, i.e., $X = H \cup_z \widehat{X}$, so that we can perform surgery within H. Moreover, since we want to do surgeries only in dimensions 0 and 1, the data about the infinite cyclic coefficient systems are enough to get the bundle data necessary to carry out the surgery step.

From the very beginning we can do surgery on $f: M \to X$ to arrange that f is 2-connected, compare Theorem 4.44.

Consider an element ω in $\pi_{l+1}(u|_H)$ for $l = 0, 1$. We want to kill it by surgery on the interior $H^\circ := H \setminus \partial H$ of H. Essentially we proceed now as in Theorem 4.44; the proofs of some of the claims below are analogous to the one appearing there.

We can represent ω by a diagram

$$\begin{array}{ccc} S^l \times D^{n-l} & \xrightarrow{q} & H^\circ \\ \downarrow & & \downarrow u|_{H^\circ} \\ D^{l+1} \times D^{n-l} & \xrightarrow{Q} & K. \end{array}$$

We can find an isomorphism $\overline{Q}: O_{D^{l+1} \times D^{n-l}} \xrightarrow{\cong} Q^* O_K$ of infinite cyclic coefficient systems since $D^{l+1} \times D^{n-l}$ is contractible. There is precisely one isomorphism $\overline{q}: O_{S^l \times D^{n-l}} \to q^* O_\xi|_{H^\circ}$ of infinite cyclic coefficient systems such that the following diagram

$$\begin{array}{ccc} O_{S^l \times D^{n-l}} & \xrightarrow{\overline{q}} & q^* O_X|_{H^\circ} \\ \downarrow & & \downarrow \overline{u}|_{H^\circ} \\ O_{D^{l+1} \times D^{n-l}} & \xrightarrow{\overline{Q}} & Q^* O_K \end{array}$$

commutes. The group homomorphism $\det: \mathrm{GL}(n, \mathbb{R}) \to \mathrm{GL}(1, \mathbb{R})$ given by the determinant is 1-connected. Hence the map $\mathrm{B}\det: \mathrm{BGL}(n, \mathbb{R}) \to \mathrm{BGL}(1, \mathbb{R})$ is 2-connected. Since we assume that $l \leq 1$ and $S^l \times D^{n-l}$ is homotopic S^l and $w_1(H) = w_1(\xi)$, we can find a bundle map

$$\overline{\overline{q}}: TS^l \times D^{n-l} \to TH^\circ$$

covering q and an isomorphism $k: O_H \to O_X|_H$ of infinite cyclic coefficient systems over H such that the composite $k \circ O_{\overline{\overline{q}}}$ and $O_{\overline{q}}$ agree.

The bundle map $\overline{\overline{q}}$ determines an embedding $q': S^l \times D^{n-l} \hookrightarrow H^\circ$ such that Tq' and $\overline{\overline{q}}$ are homotopic bundle maps, after possible stabilisation with a trivial bundle. Now we can perform surgery on q' and obtain a new map $u_{H'}: H' \to K$ in the usual way, delete the interior of the image of q' and attach $D^{l+1} \times D^{n-l-1}$ along the common boundary $S^l \times S^{n-l-1}$. A bordism $R_W: W \to K$ between $u|_H: H \to K$ and

$u_{H'} : H' \to K$ is given by attaching a handle $D^{l+1} \times D^{n-l}$ to $H \times [0, 1]$ where R_W is determined by Q and the map $H \times [0, 1] \to K$ sending (x, t) to $u(x)$. Everything is covered by appropriate bundle data. The effect of this surgery on u is that we have killed $\omega \in \pi_{l+1}(u|_H)$.

This surgery is done relative boundary of H and hence $\partial H = \partial H'$ and W contains a copy of $\partial H \times [0, 1]$. So we can perform a gluing construction with $\widehat{X}$ to get back to X instead of H. Namely, we can define a new Poincaré pair $(X', \partial X')$ with $\partial X' = \partial X$ by putting $X' = H' \cup_z \widehat{X}$. Moreover, we get a Poincaré triad $(Z; \partial_0 Z, \partial_1 Z)$ by putting $Z = W \cup_{z \times \mathrm{id}_{[0,1]}} \widehat{X} \times [0, 1]$, $\partial_0 Z = X \times \{0\}$, and $\partial_1 Z = X' \times \{1\} \cup_{\partial X \times \{1\}} \cup \partial X \times [0, 1]$. The reference map $u : X' = X' \times \{0\} \to K$ extends to a reference map $U : Z \to K$ inducing a reference map $u_1 : \partial_1 Z \to K$. Moreover, everything can be covered by maps of infinite cyclic coefficient systems as well. One easily checks that all this data factorises through the quotient of Z that is obtained by collapsing $\partial X \times [0, 1]$ to ∂X so that $\partial_1 Z$ becomes X'.

Note that X' is again a special Poincaré pair, the conditions about maps on π_0 and π_1 appearing in Definition 13.38 survive the surgery step, essentially because of $\partial H = \partial H'$ and the Seifert–van Kampen Theorem.

Of course we have to take care of the surgery problem $f : M \to X$ as well. This is no problem since, because of Lemma 13.41, we can arrange that $f^{-1}(S^l \times D^{n-l})$ is an embedded submanifold of M and f induces a diffeomorphism

$$f|_{f^{-1}(S^l \times D^{n-l})} : f^{-1}(S^l \times D^{n-l}) \to q(S^l \times D^{n-l})$$

for the embedding $q : S^l \times D^{n-l} \hookrightarrow H^\circ$. Namely, then we can perform a simultaneous surgery step on the interior of M according to the surgery on X so that we also obtain a new normal map $f' : M' \to X'$ together with an appropriate normal bordism to the given surgery problem $f : M \to X$.

So with these kind of surgery steps we do not change the bordism class of the normal map of degree one with reference to (K, O_K). We cannot kill arbitrary elements in $\pi_{l+1}(u : X \to K)$, but we can kill elements in $\pi_{l+1}(u : X \to K)$ that lie in the image of $\pi_{l+1}(u|_H : H \to K)$, provided that $l = 0, 1$. This is enough for our purposes by the following argument.

Recall that we can assume that K is connected. We can arrange that H is connected and $H \to K$ is 1-connected by doing this surgery step on the reference map finitely many times for $l = 0$. Hence X is also connected by Lemma 13.40.

Now we perform these kind of surgery steps for $l = 1$ in order to arrange that $u|_H : H \to K$ is 2-connected and in particular that $\pi_1(u|_H) : \pi_1(H) \to \pi_1(K)$ is a bijection. The inclusion $H \to X = H \cup_z \widehat{X}$ induces a surjection on π_1 by Lemma 13.40. Hence $u : X \to K$ induces an isomorphism on π_1. This finishes the proof of Lemma 13.37. □

13.4.3 The Geometric Surgery Obstruction

Definition 13.44 (Geometric surgery obstruction) Consider a w-oriented finite n-dimensional Poincaré pair $(X, \partial X, O_X, [[X, \partial X]])$ in the sense of Definition 5.73 of dimension $n \geq 5$. Consider a normal map with respect to the tangent bundle of degree one relative boundary $(f, \overline{f}) = (M, \partial M, f, \partial f, a, \xi, \overline{f}, o)$ with target $(X, \partial X, O_X, [[X, \partial X]])$ in the sense of Subsection 8.8.1. (Recall that this means that $\partial f \colon \partial M \to \partial X$ is a (simple) homotopy equivalence.) Define its *geometric simple surgery obstruction*

$$\sigma^s_{\text{geo}}(f, \overline{f}) \in \mathcal{L}^s_n(X, O_X)$$

to be the class of the normal map of degree one with reference to (X, O_X) given by $(f, \overline{f})$ and the reference map $\text{id} \colon X \to X$ covered by the identity $\text{id} \colon O_X \to O_X$.

If we require only that ∂f is a homotopy equivalence and $(X, \partial X)$ a finite Poincaré pair, we can define analogously its *geometric surgery obstruction*

$$\sigma^h_{\text{geo}}(f, \overline{f}) \in \mathcal{L}^h_n(X, O_X).$$

Theorem 13.45 (Geometric Surgery Obstruction Theorem) Let $(X, \partial X, O_X, [[X, \partial X]])$ be a w-oriented finite n-dimensional Poincaré pair (in the sense of Definition 5.73) of dimension $n \geq 5$. Consider a normal map with respect to the tangent bundle of degree one relative boundary $(f, \overline{f}) = (M, \partial M, f, \partial f, a, \xi, \overline{f}, o)$ with target $(X, \partial X, O_X, [[X, \partial X]])$ in the sense of Subsection 8.8.1. (Recall that this means that $\partial f \colon \partial M \to \partial X$ is a (simple) homotopy equivalence.) Then:

(i) *Its geometric (simple) surgery obstruction*

$$\sigma^h_{\text{geo}}(f, \overline{f}) \in \mathcal{L}^h_n(X, O_X);$$
$$\sigma^s_{\text{geo}}(f, \overline{f}) \in \mathcal{L}^s_n(X, O_X),$$

depends only on the normal bordism with cylindrical ends class of $(f, \overline{f})$;

(ii) *We have $\sigma^s_{\text{geo}}(f, \overline{f}) = 0$ if and only if we can do a finite number of surgery steps on the interior of M without changing ∂f to obtain a normal simple homotopy equivalence, i.e., the underlying map f is a simple homotopy equivalence. The analogous statement holds for the decoration h.*

Proof. (i) This follows directly from the definition of $\mathcal{L}^s_n(X, O_X)$ in terms of a normal bordism with reference map to (X, O_X). Since the normal bordism has cylindrical ends, we can extend the identity reference map to (X, O_X) using the projection $X \times [0, 1] \to X$ to the target of the bordism.

(ii) This follows from the π-π-Theorem 13.6 for triads applied to a nullbordism with reference map to (X, O_X) since we can use Lemma 13.37 to arrange that the π-π-condition is satisfied. □

13.4.4 Identifying the Geometric Surgery Obstruction Groups with the Algebraic L-Groups

Theorem 13.46 (Identifying the geometric surgery obstruction groups with the algebraic L-groups)

(i) *Consider $n \geq 5$. Let (K, O_K) be an object in $\mathcal{R}$.*
   Then there are isomorphisms, natural in (K, O_K),

$$\zeta_n^s \colon \mathcal{L}_n^s(K, O_K) \xrightarrow{\cong} L_n^s(\mathbb{Z}\Pi(K), O_K);$$
$$\zeta_n^h \colon \mathcal{L}_n^h(K, O_K) \xrightarrow{\cong} L_n^h(\mathbb{Z}\Pi(K), O_K),$$

   from the geometric L-groups in the sense of Definition 13.44 to the algebraic L-groups in the sense of Definitions 8.158, 9.106 and 10.15;

(ii) *Let $(X, \partial X, O_X, [[X, \partial X]])$ be a w-oriented finite n-dimensional Poincaré pair in the sense of Definition 5.73 of dimension $n \geq 5$. Consider a normal map with respect to the tangent bundle of degree one relative boundary $(f, \overline{f}) = (M, \partial M, f, \partial f, a, \xi, \overline{f}, o)$ with target $(X, \partial X, O_X, [[X, \partial X]])$ in the sense of Subsection 8.8.1.*
   Then the isomorphism $\zeta_n^h \colon \mathcal{L}_n^h(X, O_X) \xrightarrow{\cong} L_n^h(\mathbb{Z}\Pi(X), O_X)$ sends the geometric surgery obstruction $\sigma_{\mathrm{geo}}^h(f, \overline{f})$ of Definition 13.44 to the algebraic surgery obstruction $\sigma(f, \overline{f})$ in the sense of Definitions 8.163 and 9.107.
   The analogous statement holds for the decoration s.

Proof. We only treat the simple case, the other one is obtained from it by deleting the word simple everywhere. We define the map

$$\zeta_n^s \colon \mathcal{L}_n^s(K, O_K) \xrightarrow{\cong} L_n^s(\mathbb{Z}\Pi(K), O_K)$$

as follows. Consider a normal map of degree one $(M, \partial M, f, \partial f, a, \xi, \overline{f}, o)$ with target a w-oriented finite n-dimensional Poincaré pair $(X, \partial X, O_X, [[X, \partial X]])$ in the sense of Section 10.4 together with a morphism $(u, \overline{u}) \colon (X, O_X) \to (K, O_K)$ in $\mathcal{R}$. Its class in $\mathcal{L}_n^s(K, O_K)$ is sent to the image of its algebraic surgery obstruction $\sigma^s(f, \overline{f})$, see (10.26), under the homomorphism $L_n^s(\mathbb{Z}\Pi(X), O_X) \to L_n^s(\mathbb{Z}\Pi(K), O_K)$ induced by $(u, \overline{u})$. This is well defined by Remark 10.32.

One easily checks that ζ_n^s is a homomorphism. Injectivity follows from Theorem 10.30 (ii) and Lemma 13.37 whereas surjectivity follows from Theorem 10.31. Hence assertion (i) is proved. Assertion (ii) holds by construction. □

13.5 The Geometric Rothenberg Sequence

In this section we establish a geometric version of the Rothenberg sequence, which measures the difference between the decorations s and h.

Firstly, we want to compare for an object $(K, \mathcal{O}_K)$ in $\mathcal{R}$ the two geometric L-groups $\mathcal{L}_n^s(K, \mathcal{O}_K)$ and $\mathcal{L}_n^h(K, \mathcal{O}_K)$ of Definition 13.34. For this purpose we define a third geometric L-group

$$\mathcal{L}_n^{s,h}(K, \mathcal{O}_K), \tag{13.47}$$

construct a long exact sequence involving these three types of geometric L-groups, and finally compute $\mathcal{L}_n^{s,h}(K, \mathcal{O}_K)$ in terms of the Tate cohomology of the Whitehead group of $(K, \mathcal{O}_K)$.

The definition of $\mathcal{L}_n^{s,h}(K, \mathcal{O}_K)$ is the same as that of $\mathcal{L}_n^s(K, \mathcal{O}_K)$, but we additionally assume for a normal map that the underlying map $f \colon M \to X$ is a homotopy equivalence. Recall that the underlying map $\partial f \colon M \to \partial X$ is required to be a simple homotopy equivalence and $(X, \partial X)$ to be a simple finite Poincaré pair. In the definition of a bordism we additionally require that the underlying map $F \colon W \to Y$ is a homotopy equivalence, and keep the condition that $\partial_2 F \colon \partial_2 W \to \partial_2 Y$ is a simple homotopy equivalence and $(Y, \partial Y)$ is a simple Poincaré pair.

Lemma 13.48 *There is a long exact sequence of abelian groups*

$$\cdots \xrightarrow{\partial_{n+2}} \mathcal{L}_{n+1}^{s,h}(K, \mathcal{O}_K) \xrightarrow{i_{n+1}} \mathcal{L}_{n+1}^s(K, \mathcal{O}_K) \xrightarrow{j_{n+1}} \mathcal{L}_{n+1}^h(K, \mathcal{O}_K)$$
$$\xrightarrow{\partial_{n+1}} \mathcal{L}_n^{s,h}(K, \mathcal{O}_K) \xrightarrow{i_n} \mathcal{L}_n^s(K, \mathcal{O}_K) \xrightarrow{j_n} \mathcal{L}_n^h(K, \mathcal{O}_K) \xrightarrow{\partial_n} \cdots.$$

Proof. The maps i_n and j_n are given by forgetting conditions on the underlying maps of spaces of the normal maps. The map ∂_n is given by restricting to ∂M and ∂X. One easily checks that we obtain homomorphisms of abelian groups since the addition is given by taking disjoint unions.

It remains to show the exactness of the sequence above. In the following arguments we only explain what happens for the underlying maps of the normal map and ignore both the bundle data and the reference to $(K, \mathcal{O}_K)$ since these are always given in the obvious way.

We start with exactness at $\mathcal{L}_{n+1}^h(K, \mathcal{O}_K)$. Consider an element in the geometric simple L-group $\mathcal{L}_{n+1}^s(K, \mathcal{O}_K)$ given by a map $(f, \partial f) \colon (M, \partial M) \to (X, \partial X)$ with ∂f a simple homotopy equivalence. Its image under the composite $i_n \circ \partial_{n+1}$ is the class of $\partial f \colon \partial M \to \partial N$, which is zero since the map $(f, \partial f) \colon (M, \partial M) \to (X, \partial X)$ gives the desired nullbordism. This shows $\mathrm{im}(j_{n+1}) \subseteq \ker(\partial_{n+1})$.

Next we show $\ker(\partial_{n+1}) \subseteq \mathrm{im}(j_{n+1})$. Consider $[(f, \partial f)]$ in $\mathcal{L}_{n+1}^h(K, \mathcal{O}_K)$ represented by the map $(f, \partial f) \colon (M, \partial M) \to (X, \partial X)$. Suppose that $[(f, \partial f)]$ lies in the kernel of ∂_{n+1}. Then there is a map $(F; \partial_0 F, \partial_2 F) \colon (W; \partial_0 W, \partial_2 W) \to (Y; \partial_0 Y, \partial_2 Y)$ such that $F \colon W \to Y$ is a homotopy equivalence, $\partial_2 F \colon \partial_2 W \to \partial_2 Y$ is a simple homotopy equivalence, and $\partial_0 F \colon \partial_0 W \to Y_0$ can be identified with $\partial f \colon M \to X$. Now we can glue $(f, \partial f) \colon (M, \partial M) \to (X, \partial X)$ and $(F, \partial_0 F, \partial_2 F) \colon (W; \partial_0 W, \partial_2 W) \to$

$(Y, \partial_0 Y, \partial_2 Y)$ along $\partial f\colon \partial M \to \partial X$ and obtain a map $f \cup_{\partial f} F\colon M\cup_{\partial M} W \to X\cup_{\partial X} Y$. One easily checks that it defines an element in $\mathcal{L}^s_{n+1}(K, O_K)$. Its image under j_{n+1} is $[(f, \partial f)]$ since we can construct from $(f \cup_{\partial f} F) \times \mathrm{id}_{[0,1]}\colon M \times [0,1] \cup_{\partial M} W \to X \cup_{\partial X} Y \times [0,1]$ the desired bordism between $f \cup_{\partial f} F\colon M \cup_{\partial M} W \to X \cup_{\partial X} Y$ and $(f, \partial f)\colon (M, \partial M) \to (X, \partial X)$. Hence the sequence is exact at $\mathcal{L}^h_{n+1}(K, O_K)$.

Next we show exactness at $\mathcal{L}^{s,h}_n(K, O_K)$. The composite $i_n \circ \partial_{n+1}$ sends a normal map $(f, \partial f)\colon (M, \partial M) \to (X, \partial X)$ representing an equivalence class in $\mathcal{L}^h_{n+1}(K, O_K)$ to the class given by $\partial f\colon \partial M \to \partial X$ in $\mathcal{L}^s_n(K, O_K)$. The latter is zero, a nullbordism is given by $(f, \partial f)\colon (M, \partial M) \to (X, \partial X)$. This shows $\mathrm{im}(\partial_{n+1}) \subseteq \ker(i_n)$.

Next we show $\ker(i_n) \subseteq \mathrm{im}(\partial_{n+1})$. Consider an element $[(f, \partial f)]$ in $\mathcal{L}^{s,h}_n(K, O_K)$ given by a map $(f, \partial f)\colon (M, \partial M) \to (X, \partial X)$ with $f\colon M \to X$ a homotopy equivalence and $\partial f\colon \partial M \to \partial X$ a simple homotopy equivalence. Suppose that $[(f, \partial f)]$ is sent to zero under the map i_n. Then we can find a nullbordism $(F; \partial_0 F, \partial_1 F)\colon (W; \partial_0 W, \partial_1 W) \to (Y; \partial_0 Y, \partial_1 Y)$ such that $\partial_1 F\colon \partial_1 W \to \partial_1 Y$ is a simple homotopy equivalence and the map $(\partial_0 F, \partial\partial_0 F)\colon (\partial_0 W, \partial\partial_0 W) \to (\partial_0 Y, \partial\partial_0 Y)$ can be identified with the map $(f, \partial f)\colon (M, \partial M) \to (X, \partial X)$. This nullbordism $(F; \partial_0 F, \partial_1 F)$ defines an element in $\mathcal{L}^h_{n+1}(K, O_K)$. Its image under ∂_{n+1} is $[(f, \partial f)]$ since $\partial F\colon \partial W \to \partial Y$ and $(f, \partial f)$ define the same class in $\mathcal{L}^{s,h}_n(K, O_K)$, and a desired bordism is given by $\partial F \times [0,1]\colon \partial W \times [0,1] \to \partial Y \times [0,1]$.

Finally we show exactness at $\mathcal{L}^s_n(K, O_K)$. Given an element in $\mathcal{L}^s_n(K, O_K)$ represented by $(f, \partial f)\colon (M, \partial M) \to (X, \partial X)$, its class under the composite $j_n \circ i_n$ is zero since $(f, \partial f) \times \mathrm{id}_{[0,1]}\colon (M \times [0,1], \partial(M \times [0,1])) \to (X \times [0,1], \partial(X \times [0,1]))$ gives the desired nullbordism. This shows $\mathrm{im}(i_n) \subseteq \ker(j_n)$.

Next we show $\ker(j_n) \subseteq \mathrm{im}(i_n)$. Consider an element $[(f, \partial f)]$ in $\mathcal{L}^s_n(K, O_K)$ represented by $(f, \partial f)\colon (M, \partial M) \to (X, \partial X)$ that is mapped to zero under j_n. Then we can find a map $(F; \partial_0 F, \partial_1 F)\colon (W; \partial_0 W, \partial_1 W) \to (Y; \partial_0 Y, \partial_1 Y)$ such that $(\partial_0 F, \partial\partial_0 F)\colon (\partial_0 W, \partial\partial_0 W) \to (\partial_0 Y, \partial\partial_0 Y)$ can be identified with $(f, \partial f)\colon (M, \partial M) \to (X, \partial X)$ and $\partial_1 F\colon \partial_1 W \to \partial_1 Y$ is a homotopy equivalence. Then $\partial_1 F\colon \partial_1 W \to \partial_1 Y$ defines an element in $\mathcal{L}^{s,h}_n(K, O_K)$ since $\partial\partial_1 F = \partial f$ is a simple homotopy equivalence. Its image under i_n is $[(f, \partial f)]$, a desired bordism between $\partial_1 F\colon \partial_1 W \to \partial_1 Y$ and $(f, \partial f)$ is given by $(F; \partial_0 F, \partial_1 F)$. This finishes the proof of Lemma 13.48. $\square$

Consider an object (K, O_K) in $\mathcal{R}$. Then we have defined the Whitehead group $\mathrm{Wh}(\Pi(K))$ in Definition 3.22 and an involution on $\mathrm{Wh}(\Pi(K))$ in (3.27). Recall from (5.56) that the n-th Tate cohomology of an abelian group A with involution $*\colon A \to A$ is defined by

$$\widehat{H}^n(\mathbb{Z}/2; A) := \ker(\mathrm{id} - (-1)^n \cdot *)/\mathrm{im}(\mathrm{id} + (-1)^n \cdot *).$$

Exercise 13.49 Show that for every abelian group A with involution each element in the Tate cohomology $\widehat{H}^n(\mathbb{Z}/2; A)$ is annihilated by multiplication by 2.

13.5 The Geometric Rothenberg Sequence

Next we want to define a homomorphism

$$\tau \colon \mathcal{L}_n^{s,h}(K, O_K) \to \widehat{H}^{n+1}(\mathbb{Z}/2; \mathrm{Wh}(\Pi(K))). \tag{13.50}$$

Consider a class $[(\underline{f}, \partial f)]$ in $\mathcal{L}_n^{s,h}(K, O_K)$ represented by a normal map $(f, \overline{f}) = (M, \partial M, f, \partial f, a, \xi, \overline{f}, o)$ with target a w-oriented finite n-dimensional Poincaré pair $(X, \partial X, O_X, [[X, \partial X]])$ and reference map $(u, \overline{u}) \colon O_X \to O_K$. Then the underlying map $f \colon M \to X$ is a homotopy equivalence. Let $\tau(f) \in \mathrm{Wh}(\Pi(X))$ be its Whitehead torsion. We want to show that $\tau(f)$ lies in the kernel of $\mathrm{id} - (-1)^{n+1} \cdot * \colon \mathrm{Wh}(\Pi(X)) \to \mathrm{Wh}(\Pi(X))$. Without loss of generality we can assume that X is connected, otherwise treat each component of X separately. Make the choice (NBP), see Notation 8.5 (in the universal covering case, see Notation 8.34).

The following diagram of finite based free $\mathbb{Z}[\pi_1(X)]$-chain complexes commutes up to $\mathbb{Z}[\pi_1(X)]$-chain homotopy

$$\begin{array}{ccc}
C^{n-*}(\widetilde{M}, \widehat{\partial M}) & \xleftarrow{C^{n-*}(\widetilde{f}, \widehat{\partial f})} & C^{n-*}(\widetilde{X}, \widehat{\partial X}) \\
{\scriptstyle ?\cap [M,\partial M]} \downarrow & & \downarrow {\scriptstyle ?\cap [X,\partial X]} \\
C_*(\widetilde{M}) & \xrightarrow{C_*(\widetilde{f})} & C_*(\widetilde{X})
\end{array}$$

where $\widehat{\partial M}$ is the preimage of ∂M under the universal covering $\widetilde{M} \to M$, analogously for $\widehat{\partial X}$, and $\widehat{\partial f}$ is the restriction of $\widetilde{f} \colon \widetilde{M} \to \widetilde{X}$. Here the dual chain complexes are to be understood with respect to the $w_1(X)$-twisted involution on $\mathbb{Z}[\pi_1(X)]$. The vertical arrows are simple homotopy equivalences since M is a manifold and X is assumed to be a simple Poincaré complex. We conclude from Lemma 3.18 (i) and (iii) and the definitions

$$\tau(C^{n-*}(\widetilde{f}, \widehat{\partial f})) = -\tau(C_*(\widetilde{f}));$$
$$\tau(C_*(\widetilde{f}, \widehat{\partial f})) = \tau(C_*(\widetilde{f})) - \tau(C_*(\widehat{\partial f}));$$
$$\tau(C^{n-*}(\widetilde{f}, \widehat{\partial f})) = (-1)^n \cdot *\big(\tau(C_*(\widetilde{f}, \widehat{\partial f}))\big).$$

Since $\partial f \colon \partial M \to \partial X$ is by definition a simple homotopy equivalence, this implies

$$\begin{aligned}
\big(\mathrm{id} - (-1)^{n+1} \cdot *\big)(\tau(f)) &= \tau(C_*(\widetilde{f})) - (-1)^{n+1} \cdot *(\tau(C_*(\widetilde{f}))) \\
&= \tau(C_*(\widetilde{f})) - (-1)^n \cdot *\big(\tau(C^{n-*}(\widetilde{f}, \widehat{\partial f}))\big) \\
&= \tau(C_*(\widetilde{f})) - \tau(C_*(\widetilde{f}, \widehat{\partial f})) \\
&= \tau(C_*(\widetilde{f})) - \big(\tau(C_*(\widetilde{f}, \widehat{\partial f})) - \tau(C_*(\widehat{\partial f}))\big) \\
&= \tau(C_*(\widehat{\partial f})) \\
&= 0.
\end{aligned}$$

Hence $\tau(f)$ defines an element in $\widehat{H}^{n+1}(\mathbb{Z}/2; \mathrm{Wh}(\Pi(X)))$. We want to define the desired homomorphism $\tau \colon \mathcal{L}_n^{s,h}(K, O_K) \to \widehat{H}^{n+1}(\mathbb{Z}/2; \mathrm{Wh}(\Pi(K)))$ by sending the class of $[(f,\partial f)]$ to the image of $\tau(f)$ under the homomorphism $\widehat{H}^{n+1}(\mathbb{Z}/2;\mathrm{Wh}(\Pi(X))) \to \widehat{H}^{n+1}(\mathbb{Z}/2;\mathrm{Wh}(\Pi(K)))$ induced by the map $u\colon X \to K$. The latter homomorphism is well defined since $u^*w_1(O_K) = w_1(O_X)$ holds.

We have to show that this is compatible with the bordism relation. Consider a null-bordism whose underlying map is $(F;\partial_0 F, \partial_1 F)\colon (W;\partial_0 W, \partial_2 W) \to (Y;\partial_0 Y, \partial_2 Y)$. Recall that F and $\partial_0 F$ are homotopy equivalences and $\partial_2 F$ and $\partial \partial_0 F = \partial \partial_2 F$ are simple homotopy equivalences. Let $j\colon \partial X \to X$, $j_i\colon \partial_i X \to X$ for $i = 0, 2$, and $j'\colon \partial_0 Y \cap \partial_2 Y \to Y$ be the inclusions. Poincaré duality implies $\tau(F, \partial F) = -(-1)^{n+1} \cdot *(\tau(F))$ as explained earlier. We conclude from Theorem 3.1 that the following holds in $\mathrm{Wh}(\Pi(X))$

$$\begin{aligned}(j_0)_*(\tau(\partial_0 F)) &= (j_0)_*(\tau(\partial_0 F)) - j'_*(\tau(\partial \partial_0 F)) + (j_2)_*(\tau(\partial_2 F))\\ &= j_*(\tau(\partial F))\\ &= \tau(F) - \tau(F, \partial F)\\ &= \tau(F) + (-1)^{n+1} \cdot *(\tau(F)).\end{aligned}$$

Hence $(j_0)_*(\tau(\partial_0 F))$ lies in the image of $\mathrm{id} + (-1)^{n+1} \cdot *$. This implies that the class of $(j_0)_*(\tau(\partial_0 F))$ in $\widehat{H}^{n+1}(\mathbb{Z}/2;\mathrm{Wh}(\Pi(Y)))$ vanishes. We conclude that the map $\tau\colon \mathcal{L}_n^{s,h}(K, O_K) \to \widehat{H}^{n+1}(\mathbb{Z}/2;\mathrm{Wh}(\Pi(K)))$ announced in (13.50) is well defined. It is a $\mathbb{Z}$-homomorphism since the addition on the source is given by the disjoint union and $\widehat{H}^{n+1}(\mathbb{Z}/2;\mathrm{Wh}(\Pi(K)))$ is 2-torsion.

Lemma 13.51 *Consider an object (K, O_K) in $\mathcal{R}$. Then the homomorphism of (13.50)*

$$\tau\colon \mathcal{L}_n^{s,h}(K, O_K) \to \widehat{H}^{n+1}(\mathbb{Z}/2;\mathrm{Wh}(\Pi(K)))$$

is injective if $n \geq 5$, and bijective if $n \geq 6$.

Proof. Obviously we can assume without loss of generality that K is connected, otherwise treat each component of K separately.

We begin with injectivity. Fix an element in the kernel of $\tau\colon \mathcal{L}_n^{s,h}(K, O_K) \to \widehat{H}^{n+1}(\mathbb{Z}/2;\mathrm{Wh}(\Pi(K)))$. It can be represented by a normal map $(f, \overline{f}) = (M, \partial M, f, \partial f, a, \xi, \overline{f}, o)$ with target a w-oriented finite n-dimensional Poincaré pair $(X, \partial X, O_X, [[X, \partial X]])$ and reference map $(u, \overline{u})\colon O_X \to O_K$. Recall that the underling map $f\colon M \to X$ is a homotopy equivalence and the restriction $\partial f\colon M \to \partial X$ is a simple homotopy equivalence. We have explained in Lemma 13.42 how to arrange for an element in $\mathcal{L}_n^s(K, O_K)$ that X is connected and the map $u\colon X \to K$ induces a bijection $\pi_1(X) \to \pi_1(K)$. Going through the procedure, we see that it also applies to elements in $\mathcal{L}_n^{s,h}(K, O_K)$, the main observation is that the simultaneous surgery to adjust the normal map results again in a homotopy equivalence and we have never changed ∂f. Hence we can assume without loss of generality that X and K are connected and u induces an isomorphism $\pi_1(X) \xrightarrow{\cong} \pi_1(K)$. In the sequel we will consider all torsion invariants in $\mathrm{Wh}(\Pi(K))$ by pushing them forward by the obvious maps of Whitehead groups to $\mathrm{Wh}(\Pi(K))$. By assumption there is an element $u \in \mathrm{Wh}(\Pi(K))$ such that

13.5 The Geometric Rothenberg Sequence

$$\tau(f) = u + (-1)^{n+1} \cdot *(u)$$

holds. Since $n \geq 5$, we can choose by the obvious version of Theorem 2.1 (ii) for manifolds with boundaries an h-cobordism (W, M, N) over $(M, \partial M)$ such that $\tau(i) = \tau(f) - u$ holds for the inclusion $i \colon M \to W$. We can choose a map $(F, f, \partial_1 F) \colon (W; M, \partial_1 W) \to (X \times [0, 1], X \times \{0\}, \partial X \times [0, 1] \cup X \times \{1\})$ such that F and $\partial_1 F$ are a homotopy equivalences and $\partial f \colon \partial M \to X$ is a simple homotopy equivalence. Poincaré duality implies $\tau(F, \partial F) = -(-1)^n \cdot *(\tau(F))$, as explained earlier. We conclude from Theorem 3.1 and Lemma 3.28 (ii)

$$\begin{aligned}
\tau(\partial_1 F) &= \tau(\partial F) - \tau(f) + \tau(\partial f) \\
&= \tau(\partial F) - \tau(f) \\
&= \tau(F) - \tau(F, \partial F) - \tau(f) \\
&= \bigl(\mathrm{id} + (-1)^{n+1} \cdot *\bigr)(\tau(F)) - \tau(f) \\
&= \bigl(\mathrm{id} + (-1)^{n+1} \cdot *\bigr)\bigl(\tau(F \circ i) - \tau(i)\bigr) - \tau(f) \\
&= \bigl(\mathrm{id} + (-1)^{n+1} \cdot *\bigr)\bigl(\tau(f) - (\tau(f) - u)\bigr) - \tau(f) \\
&= \bigl(\mathrm{id} + (-1)^{n+1} \cdot *\bigr)(u) - \bigl(u + (-1)^{n+1} \cdot *(u)\bigr) \\
&= 0.
\end{aligned}$$

Hence $\partial_1 F$ is a simple homotopy equivalence. Since F is a homotopy equivalence, one can find the necessary bundle data and reference to $(K, \mathcal{O}_K)$ such that $(F, f, \partial_1 F)$ defines the desired nullbordism for the representative of the class in the kernel of $\tau \colon \mathcal{L}_n^{s,h}(K, \mathcal{O}_K) \to \widehat{H}^{n+1}(\mathbb{Z}/2; \mathrm{Wh}(\Pi(K)))$ we have started with. This proves injectivity.

Finally we show surjectivity. Consider a class $[u] \in \widehat{H}^{n+1}(\mathbb{Z}/2, \mathrm{Wh}(K)))$ represented by $u \in \mathrm{Wh}(\Pi(K))$ satisfying $u - (-1)^{n+1} * (u) = 0$. Choose a connected closed $(n-1)$-dimensional manifold M with $w_1(M) = w_1(\mathcal{O}_K)$ and reference map $(u, \bar{u}) \colon \mathcal{O}_M \to \mathcal{O}_K$ such that $\pi_1(u) \colon \pi_1(M) \to \pi_1(K)$ is an isomorphism. This is possible since $(n-1) \geq 5$. In the sequel we will push all Whitehead torsions to $\mathrm{Wh}(\Pi(K))$ with the appropriate map to K. Since $(n-1) \geq 5$, we can choose by Theorem 2.1 (ii) an h-cobordism $(W; M, N)$ over M such $\tau(i) = -u$ holds for the inclusion $i \colon M \to W$. Choose a map $(F; \mathrm{id}; \partial_1 F) \colon (W; M, N) \to (M \times [0, 1], M \times \{0\}, M \times \{1\})$ such that F is a homotopy equivalence. Poincaré duality implies $\tau(F, \partial F) = -(-1)^n \cdot *(\tau(F))$, as explained earlier. We conclude from Theorem 3.1 and Lemma 3.28 (ii), if $j_0 \colon M \times \{0\} \to M \times [0, 1]$ is the inclusion,

$$\tau(F) = \tau(F \circ i) - \tau(i) = \tau(j_0) - \tau(i) = u,$$

and

$$\begin{aligned}
\tau(\partial_1 F) &= \tau(\partial F) = \tau(F) - \tau(F, \partial F) \\
&= \tau(F) - (-1)^{n+1} \cdot *(\tau(F)) = \bigl(\mathrm{id} - (-1)^{n+1} \cdot *\bigr)(u) = 0.
\end{aligned}$$

Therefore $\partial F\colon \partial W \to \partial W$ is a simple homotopy equivalence. We cover id: $M \to M$ with the bundle map id: $TM \to TM$. Since F is a homotopy equivalence, we can cover F and the map $U\colon M \times [0,1] \to R$, $(x,t) \mapsto u(x)$ with appropriate bundle maps so that we obtain an element in $\mathcal{L}_n^{s,h}(K;O_K)$ represented by a normal map of degree one with reference to (K,O_K), whose underlying maps are $F\colon W \to M \times [0,1]$ and $U\colon M \times [0,1] \to K$. It is mapped under $\tau\colon \mathcal{L}_n^{s,h}(K,O_K) \to \widehat{H}^{n+1}(\mathbb{Z}/2; \mathrm{Wh}(\Pi(K)))$ to $[u]$. This finishes the proof of Lemma 13.51. □

Lemma 13.48 and Lemma 13.51 imply

Theorem 13.52 (Geometric Rothenberg sequence) *Consider an object (K,O_K). There is a long exact sequence of abelian groups*

$$\cdots \xrightarrow{\partial_{n+2}} \widehat{H}^{n+2}(\mathbb{Z}/2, \mathrm{Wh}(\Pi(K))) \xrightarrow{i_{n+2}} \mathcal{L}_{n+1}^s(K,O_K) \xrightarrow{j_{n+1}} \mathcal{L}_{n+1}^h(K,O_K)$$

$$\xrightarrow{\partial_{n+1}} \widehat{H}^{n+1}(\mathbb{Z}/2, \mathrm{Wh}(\Pi(K))) \xrightarrow{i_n} \mathcal{L}_n^s(K,O_K) \xrightarrow{j_n} \mathcal{L}_n^h(K,O_K) \xrightarrow{\partial_{n-1}} \cdots$$

which is infinite to the left and ends at $\widehat{H}^6(\mathbb{Z}/2, \mathrm{Wh}(\Pi(K)))$.

Exercise 13.53 Let $(v,\overline{v})\colon (K,O_K) \to (L,O_L)$ be a morphism in $\mathcal{R}$. Let ϵ be the decoration h or s. Construct an abelian group $\mathcal{L}_n^\epsilon(K \xrightarrow{(v,\overline{v})} L)$ for $n \geq 0$ and a long exact sequence, natural in $(v,\overline{v})$,

$$\cdots \xrightarrow{\partial_{n+2}} \mathcal{L}_{n+1}^\epsilon(K,O_K) \xrightarrow{\mathcal{L}_{n+1}^\epsilon(v,\overline{v})} \mathcal{L}_{n+1}^\epsilon(L,O_L) \xrightarrow{j_{n+1}} \mathcal{L}_{n+1}^\epsilon(K \xrightarrow{(v,\overline{v})} L)$$

$$\xrightarrow{\partial_{n+1}} \mathcal{L}_n^\epsilon(K,O_K) \xrightarrow{\mathcal{L}_n^\epsilon(v,\overline{v})} \mathcal{L}_n^\epsilon(L,O_L) \xrightarrow{j_n} \mathcal{L}_n^\epsilon(K \xrightarrow{(v,\overline{v})} L) \xrightarrow{\partial_n} \cdots .$$

Once we have established the relationship between the L-groups with s and h decorations, it is natural to ask what the relationship is between the structure sets for a closed manifold X with decorations s and h. Recall the surgery exact sequences in both cases established in Theorem 11.22, see also Remark 11.23 for the extension to the left. Note that the normal invariants do not depend on the decoration. Thus Theorem 13.52 leads to the following theorem for structure sets, see [424, Section 2.4.A].

Theorem 13.54 (Geometric Rothenberg sequence for structure sets) *Let X be a closed n-dimensional manifold with $n \geq 5$ and $\Pi = \Pi(X)$. There is a long exact sequence*

$$\cdots \xrightarrow{i_{n+2}} \mathcal{S}^s(X \times I, \partial(X \times I)) \xrightarrow{j_{n+1}} \mathcal{S}^h(X \times I, \partial(X \times I)) \xrightarrow{\partial_{n+1}}$$

$$\widehat{H}^{n+2}(\mathbb{Z}/2, \mathrm{Wh}(\Pi)) \xrightarrow{i_{n+2}} \mathcal{S}^s(X) \xrightarrow{j_{n+1}} \mathcal{S}^h(X) \xrightarrow{\partial_{n+1}} \widehat{H}^{n+1}(\mathbb{Z}/2, \mathrm{Wh}(\Pi))$$

which is infinite to the left.

The group $\widehat{H}^{n+2}(\mathbb{Z}/2, \mathrm{Wh}(\Pi))$ acts on $\mathcal{S}^s(X)$ via the action ρ_{n+1}^s of $\mathcal{L}_{n+1}^s(X, \mathcal{O}_X)$ $\cong \mathcal{L}_{n+1}^s(\mathbb{Z}\Pi, \mathcal{O}_X)$ on $\mathcal{S}^s(X)$, see (11.20), and the map i_{n+2} from Theorem 13.52, and the meaning of exactness in the last four terms is analogous to the meaning of exactness in Theorem 11.22.

Using methods of algebraic surgery from Chapter 15, the sequence from Theorem 13.52 and the topological version of the sequence from Theorem 13.54 can be extended infinitely to the right, see Section 15.9.3 and (15.484) in Section 15.9.5 respectively.

13.6 The Geometric Shaneson Splitting

Consider an object $(K, \mathcal{O}_K)$ in $\mathcal{R}$. Let $i \colon K \to K \times S^1$ be the inclusion sending x to (x, s) for a fixed base point $s \in S^1$ and let $\mathrm{pr} \colon K \times S^1 \to K$ be the projection. Define the following homomorphism of abelian groups

$$i_n \colon \mathcal{L}_n^s(K, \mathcal{O}_K) \to \mathcal{L}_n^s(K \times S^1, \mathrm{pr}^* \mathcal{O}_K);$$
$$s_n \colon \mathcal{L}_n^h(K, \mathcal{O}_K) \to \mathcal{L}_{n+1}^s(K \times S^1, \mathrm{pr}^* \mathcal{O}_K).$$

The map i_n is induced by the morphisms $(i, \overline{i}) \colon (K, \mathcal{O}_K) \to (K \times S^1, \mathrm{pr}^* \mathcal{O}_K)$ for the obvious bundle map $\overline{i}$.

Consider an element $y \in \mathcal{L}_n^h(K, \mathcal{O}_K)$. It can be represented by a normal map $(f, \overline{f}) = (M, \partial M, f, \partial f, a, \xi, \overline{f}, o)$ with target a w-oriented finite n-dimensional Poincaré pair $(X, \partial X, \mathcal{O}_X, [[X, \partial X]])$ and reference map $(u, \overline{u}) \colon \mathcal{O}_X \to \mathcal{O}_K$. By taking the cartesian product with S^1, we obtain maps $f \times \mathrm{id}_{S^1} \colon M \times S^1 \to X \times S^1$ and $u \times \mathrm{id}_{S^1} \colon X \times S^1 \to K \times S^1$. Using the obvious identification $T(M \times S^1) = \mathrm{pr}^* TM \oplus \mathbb{R}$, we can cover $f \times \mathrm{id}_{S^1}$ by the bundle map $\mathrm{pr}^* \overline{f} \oplus \mathrm{id}_\mathbb{R} \to \mathrm{pr}^* \xi \oplus \mathrm{id}_\mathbb{R}$. We can extend $u \times \mathrm{id}_{S^1}$ to a reference map $(u \times \mathrm{id}_{S^1}, \overline{u} \times \mathrm{id}_{S^1}) \colon \mathcal{O}_{X \times S^1} \to \mathrm{pr}^* \mathcal{O}_K$ in the obvious way. Note that $\partial(M \times S^1) = \partial M \times S^1$ and $\partial f \times \mathrm{id}_{S^1} \colon \partial M \times S^1 \to X \times S^1$ is a simple homotopy equivalence by Theorem 3.1 (iv). Hence we have defined, essentially by taking the cartesian product with S^1, a normal map of dimension $n + 1$ in the simple setting with reference to $(K \times S^1, \mathrm{pr}^* \mathcal{O}_K)$. It determines a class in $\mathcal{L}_{n+1}^s(K \times S^1, \mathrm{pr}^* \mathcal{O}_K)$ that we define to be $s_n(y)$. Obviously this is compatible with the bordism relation and the structures of abelian groups.

Theorem 13.55 (Geometric Shaneson splitting) *For $n \geq 6$ the map*

$$i_n \oplus s_{n-1} \colon \mathcal{L}_n^s(K, \mathcal{O}_K) \oplus \mathcal{L}_{n-1}^h(K, \mathcal{O}_K) \xrightarrow{\cong} \mathcal{L}_n^s(K \times S^1, \mathrm{pr}^* \mathcal{O}_K)$$

is a bijection.

Proof. We only sketch the proof. Details can be found in [376]. Moreover, we confine ourselves to describing the inverse. The obvious morphism $(\mathrm{pr}, \overline{\mathrm{pr}}) \colon \mathrm{pr}^* \mathcal{O}_K \to \mathcal{O}_K$ in $\mathcal{R}$ induces a map

$$p_n \colon \mathcal{L}_n^s(K \times S^1, \mathrm{pr}^* \mathcal{O}_K) \to \mathcal{L}_n^s(K, \mathcal{O}_K).$$

The main difficulty is to construct using transversality a map

$$t_n \colon \mathcal{L}_n^s(K \times S^1, \mathrm{pr}^* \mathcal{O}_K) \to \mathcal{L}_{n-1}^h(K, \mathcal{O}_K).$$

We do this only in the case where K is a connected compact manifold with non-empty boundary ∂K. Consider an element z in $\mathcal{L}_n^s(K \times S^1, \mathrm{pr}^* \mathcal{O}_K)$. It can be represented by a normal map of degree one, with underlying map

$$(f; \mathrm{id}_{K \times S^1}, \partial_1 f) \colon (W; K \times S^1, \partial_1 W) \to$$
$$(K \times S^1 \times [0,1]; K \times S^1 \times \{0\}, K \times S^1 \times \{1\} \cup \partial K \times S^1 \times [0,1])$$

such that $\partial_1 f \colon \partial_1 W \to K \times \{1\}$ is a simple homotopy equivalence, and a reference to $(K \times S^1, \mathrm{pr}^* \mathcal{O}_K)$ whose underlying map is given by the projection $K \times S^1 \times [0,1] \to K \times S^1$, see Theorem 10.31. One can additionally arrange that f is transversal to $K \times \{s\} \times [0,1]$. Put $V = f^{-1}(K \times \{s\} \times [0,1])$. Then the boundary of V is the disjoint union of $\partial f^{-1}(K \times \{s\} \times \{1\})$ and $K \times \{0\}$, and we obtain by restriction a normal map of degree one $(V; K \times \{0\}, \partial_1 V) \to (K \times [0,1]; K \times \{0\}, K \times \{1\})$. One can additionally arrange that the induced map $\partial_1 V \to K \times \{1\}$ is a (not necessarily simple) homotopy equivalence. (Here we use that $\partial_1 f$ is a simple homotopy equivalence, the arguments are related to the topic discussed in Remark 17.29 below.) It comes with an obvious reference map to $(K, \mathcal{O}_K)$ whose underlying map is the projection $K \times [0,1] \to K$. Hence we obtain a class in $\mathcal{L}_{n-1}^h(K, \mathcal{O}_K)$ that we define to be the image of z under t_n. It turns out that t_n is a well-defined homomorphism of abelian groups. Now it is easy to check that an inverse of the map $i_n \oplus s_{n-1} \colon \mathcal{L}_n^s(K, \mathcal{O}_K) \oplus \mathcal{L}_{n-1}^h(K, \mathcal{O}_K) \xrightarrow{\cong} \mathcal{L}_n^s(K \times S^1, \mathrm{pr}^* \mathcal{O}_K)$ is given by $p_n \oplus t_n \colon \mathcal{L}_n^s(K \times S^1, \mathrm{pr}^* \mathcal{O}_K) \xrightarrow{\cong} \mathcal{L}_n^s(K, \mathcal{O}_K) \oplus \mathcal{L}_{n-1}^h(K, \mathcal{O}_K)$. □

Exercise 13.56 Using the fact that $\mathrm{Wh}(\mathbb{Z}^m)$ vanishes for $m \geq 0$, compute $L_n^s(\mathbb{Z}[\mathbb{Z}^m])$ and $L_n^h(\mathbb{Z}[\mathbb{Z}^m])$ for $m, n \in \mathbb{Z}$ with $m \geq 0$.

There is also a version of the Shaneson splitting for the structure sets. However, the argument uses group structures on structure sets, so we only obtain statements in the categories PL and TOP or in the category DIFF for structures sets of $X \times D^k$ for $k \geq 2$. The maps analogous to i_n and s_n are obtained as follows. Let $(f, \partial f) \colon (M, \partial M) \to (X \times I, \partial(X \times I))$ be a simple homotopy equivalence whose restriction to the boundary is a diffeomorphism. It represents an element in $\mathcal{S}^s(X \times I, \partial(X \times I))$. Gluing the two boundary copies of X in the target, and the corresponding two copies of the boundary in the source, produces a simple homotopy equivalence $g \colon W \to X \times S^1$ that represents an element in $\mathcal{S}^s(X \times S^1)$. This gives a well-defined map which we denote i_n. The map s_n is defined simply by taking the product with the identity on S^1. The same construction produces such maps in the categories PL and TOP, and it can be shown that these maps are also homomorphisms of abelian groups. This is also true in the DIFF category when we replace X by $X \times D^k$ for $k \geq 2$. We can now formulate the promised version that follows from Shaneson's thesis, as explained, for instance, in [424, page 97].

Theorem 13.57 (Geometric Shaneson splitting for structure sets) *Let X be a closed n-dimensional topological manifold with $n \geq 5$. Then the map*

$$i_{n+1} \oplus s_n \colon \mathcal{S}^{\mathrm{TOP},s}(X \times I, \partial(X \times I)) \oplus \mathcal{S}^{\mathrm{TOP},h}(X) \xrightarrow{\cong} \mathcal{S}^{\mathrm{TOP},s}(X \times S^1)$$

is a bijection. An analogous statement holds in the category PL.

Let X be a closed n-dimensional DIFF *manifold and let $k \geq 2$ be such that $(n+k) \geq 5$. Then the map*

$$i_{n+1} \oplus s_n \colon \mathcal{S}^{\mathrm{DIFF},s}(X \times D^{k+1}, \partial(X \times D^{k+1})) \oplus \mathcal{S}^h(X \times D^k, \partial(X \times D^k))$$
$$\xrightarrow{\cong} \mathcal{S}^s(X \times S^1 \times D^k, \partial(X \times S^1 \times D^k))$$

is a bijection.

Algebraic surgery version of the Shaneson splitting for L-groups is discussed in Section 15.9.4 and for the structure sets in (15.485).

13.7 Notes

All the material of this section that we have explained for the smooth category extends to the PL category and the topological category.

The geometric definition of the L-groups from Subsection 13.4.1 can be modified to obtain the so-called geometric formulation of the surgery exact sequence as in [337].

Note that the groups $\mathcal{L}_n^h$ and $\mathcal{L}_n^s$ were defined using as representatives surgery problems of manifolds with boundary modulo the relation, which used the notion of a triad of manifolds. Quinn generalised this idea further using the notion of manifold k-ads for any $k \geq 2$. (The standard k-dimensional simplex Δ^k with the usual decomposition of its boundary into faces is a manifold $(k+2)$-ad in the language of Chapter 0 of [414].) Given a space K and an integer n, he obtained via k-ads not only a geometric definition of the L-groups, but he also constructed spaces $\widetilde{\mathcal{L}}_n(K)$ whose homotopy groups $\pi_k(\widetilde{\mathcal{L}}_n(K))$ are the geometric L-groups $\mathcal{L}_{n+k}(K)$ (in his model these are Δ-sets, which is a mild generalisation of simplicial sets). Similar considerations for the other terms of the surgery exact sequence provided him with the homotopy fibration sequence (11.53), which we already mentioned in Chapter 11.

Another nice feature of the geometric model is that the sequence of the L-spaces $\widetilde{\mathcal{L}}_n(K)$ constructed in this way for a given K with varying dimension naturally form an Ω-spectrum in the sense of homotopy theory. Beware that the notation convention here is the opposite to the standard one in homotopy theory, so that we have homotopy equivalences $\widetilde{\mathcal{L}}_n(K) \xrightarrow{\cong} \Omega \widetilde{\mathcal{L}}_{n-1}(K)$. Both the L-spaces and L-spectra have algebraic versions that can be constructed using the technology of algebraic surgery from Chapter 15, see Section 15.9.5. The algebraic versions are often denoted by **L** or $\mathbb{L}$ with various decorations.

It is a distinct feature of the topological category that the normal invariants in the topological category form a generalised (co-)homology theory with the coefficients in the appropriate version of this spectrum. More on this is discussed in Section 15.9.5 and in Remark 17.20, here we just mention that this fact has many further applications, some of which we touch on in Chapter 19.

There is an algebraic version of the Geometric Rothenberg sequence, see Theorem 15.475, and of the Geometric Shaneson splitting, see Theorem 15.478.

Chapter 14
Chain Complexes

14.1 Introduction

This chapter has two goals. Firstly, we summarise the sign conventions that we use in the subsequent chapter about algebraic surgery and in fact throughout the book. We use those that are standard in textbooks on algebraic topology. Section IV.9 of the textbook [134] in particular contains what is the standard definition of hom of chain complexes as a chain complex and also the tensor product of chain complexes as a chain complex. These conventions follow the Koszul sign convention [134, Remark 9.8], which is also standard. Therefore we choose to have the sign conventions from this source. They are in conflict with the foundational papers in algebraic surgery, including [344] and [427]. As a result the formulas we obtain in Chapter 15 look slightly different than the corresponding formulas in [344] and [427], but we think that, in view of the benefit of the compatibility with many other texts, this is worthwhile. With some effort it is possible to translate between some of the conventions, and in many cases the difference is just in signs. When we feel it is appropriate, we comment on the differences more specifically in the text. Instead of just confronting the reader with our sign conventions, we give a rather axiomatic approach in the sense that we fix a few reasonable familiar basic conventions and then all other sign conventions are consequences of taking certain algebraic or geometric considerations into account.

The second goal of this chapter is to review some basic homotopy theory of chain complexes. This is scattered around many places in the existing literature, for example in [134], [290], and [421]. There are also some technical propositions that are certainly folklore, but we could not explicitly find them in any of the above sources, for example those in Sections 14.6.9 or 14.6.11. Moreover, coherent choices for signs and notation are also needed. Hence, for the benefit of the reader, we summarise those conventions and results we use in one place at hand. The notions we are interested in include mapping cones, suspensions, homotopy pushouts, homotopy pullbacks, sequential homotopy colimits, and homotopy cofibration sequences. For a reader who is familiar with triangulated categories, higher categories, or the stable homotopy category of spectra, this section contains no new information since the homotopy category of chain complexes is a very easy and elementary example or analogue. However, for someone who is not familiar with these concepts and wants to learn them, this chapter is a good entry.

Guide 14.1 The reader who is familiar with homotopy theory of chain complexes may just skim through this chapter and come back to it whenever something needs to be clarified while reading the next chapter.

Contrary to our conventions in the previous chapters, we denote a chain complex by a letter without a subscript $*$. This is to avoid too much repetition.

14.2 Modules over Associative Rings

Let R, T, and S be associative rings with units. Given bimodules $_TM_R$ and $_RN_S$, their tensor product $M \otimes_R N$ inherits the structure of a T-S-bimodule. Given bimodules $_SM_R$ and $_TN_R$, the abelian group $\hom_R(M, N)$ of homomorphisms of right R-modules inherits the structure of a T-S-bimodule. Given bimodules $_RM_S$ and $_RN_T$, the abelian group $\hom_R(M, N)$ of homomorphisms of left R-modules inherits the structure of an S-T-bimodule. For bimodules $_TL_R$, $_RM_S$ and $_TN_S$, we have the isomorphism of abelian groups

$$\eta \colon \hom_{T\text{-}S}(L \otimes_R M, N) \to \hom_{T\text{-}R}(L, \hom_S(M, N)) \qquad (14.2)$$
$$f \mapsto (l \mapsto (m \mapsto f(l \otimes m)))$$

where $\hom_{T\text{-}S}$ denotes the abelian group of T-S-bimodules and analogously for $\hom_{T\text{-}R}$.

14.3 Modules over Rings with Involution

Recall that a *ring with involution* R is an associative ring R with unit and with an anti-automorphism $r \mapsto \bar{r}$ satisfying $\bar{\bar{r}} = r$.

A right R-module M can be made into a left R-module M^t via

$$R \times M^t \to M^t, \quad (r, m) \mapsto m \cdot \bar{r}, \qquad (14.3)$$

and similarly a left R-module can be made into a right R-module. Under this procedure homomorphisms of left R-modules automatically become homomorphisms of right R-modules and vice versa.

For a left R-module M, its *dual* is the a priori right R-module denoted $M^* = \hom_R(M, R)$ with the right R-module structure given by the formula for $f \in M^*$ and $r \in R$

$$(f \cdot r) \colon M \to R, \quad m \mapsto f(m) \cdot r. \qquad (14.4)$$

14.3 Modules over Rings with Involution

Via construction (14.3), it can be made into a left R-module M^{*t} by the formula for $f \in M^*$ and $r \in R$

$$(r \cdot f) \colon M \to R, \quad m \mapsto \overline{f(m) \cdot \overline{r}}. \tag{14.5}$$

The evaluation map $M \otimes_{\mathbb{Z}} M^* \to R$ given by $m \otimes f \mapsto f(m)$ is an R-R-bimodule homomorphism with the standard R-R-bimodule structure on R in the target. If M is a finitely generated free or finitely generated projective R-module, then its dual M^* is respectively also finitely generated free or finitely generated projective. Suppose that M is finitely generated projective. Then the adjoint of the evaluation map under (14.2) provides us with the identification of left R-modules $M \to (M^*)^*$ given by $x \mapsto (f \mapsto f(x))$. However $f \mapsto f(x)$ is a homomorphism of right R-modules. Since we would like to work with the dual as a left R-module most of the time, we will not use the isomorphism above, but

$$e(M) \colon M \to (M^{*t})^{*t} \quad x \mapsto (f \mapsto \overline{f(x)}). \tag{14.6}$$

Now $f \mapsto \overline{f(x)}$ is a homomorphism of left R-modules and $e(M)$ itself is a homomorphism of left R-modules.

Exercise 14.7 Let M be a finitely generated projective R-module. Show that the evaluation map $e(M)$ from (14.6) is an isomorphism of R-modules. Also show that both hypotheses on M are necessary.

Consider an R-homomorphism of left R-modules $f \colon M \to N$. We denote by $f^{*t} \colon N^{*t} \to M^{*t}$ the homomorphisms of left R-modules given by precomposition with f. One easily checks for a finitely generated projective R-module M that the map $e(M) \colon M \to (M^{*t})^{*t}$ is an isomorphism and we have the equality of maps $M^{*t} \to M^{*t}$

$$\mathrm{id}_{M^{*t}} \colon M^{*t} \xrightarrow{e(M^{*t})} ((M^{*t})^{*t})^{*t} \xrightarrow{e(M)^{*t}} M^{*t}. \tag{14.8}$$

Moreover, for an R-homomorphism of left R-modules $f \colon M \to N$, the following diagram commutes

$$\begin{array}{ccc} M & \xrightarrow{f} & N \\ {\scriptstyle e(M)}\downarrow & & \downarrow{\scriptstyle e(N)} \\ (M^{*t})^{*t} & \xrightarrow{(f^{*t})^{*t}} & (N^{*t})^{*t} \end{array} \tag{14.9}$$

For finitely generated projective left R-modules M and N, we obtain the *slant isomorphisms* of abelian groups

$$M^* \otimes_R N \to \mathrm{hom}_R(M, N), \quad f \otimes n \mapsto (m \mapsto f(m) \cdot n);$$
$$M^t \otimes_R N \to \mathrm{hom}_R(M^{*t}, N), \quad m \otimes n \mapsto (f \mapsto \overline{f(m)} \cdot n).$$

Let M be a finitely generated projective R-module. Then the abelian group $M^t \otimes_R M$ inherits the structure of a $\mathbb{Z}[\mathbb{Z}/2]$-module from the involution of abelian groups $M^t \otimes_R M \xrightarrow{\cong} M^t \otimes_R M$ sending $x \otimes y$ to $y \otimes x$. This is well defined because $xr \otimes y$ and $x \otimes ry$ are both sent to $y \otimes \bar{r}x = y\bar{r} \otimes x$. We obtain an involution of abelian groups on $\hom_R(M^{*,t}, M)$ by sending $f \colon M^{*t} \to M$ to $e(M)^{-1} \circ f^{*t}$, where $e(M)^{-1}$ is the inverse $e(M)^{-1} \colon (M^{*t})^{*t} \to M$ of the isomorphism $e(M)$ of (14.6). This is indeed an involution because of (14.8) and (14.9).

Exercise 14.10 Let M be a finitely generated projective R-module. Show that the slant isomorphism above $s \colon M^t \otimes_R M \xrightarrow{\cong} \hom_R(M^{*t}, M)$ is an isomorphism of $\mathbb{Z}[\mathbb{Z}/2]$-modules.

Later in the book we very often drop the superscript t to increase readability, so a right R-module M in the place of a left module means M^t by default. We also adopt the following notation.

Notation 14.11 (Categories of modules). Let R be a ring with involution. Denote by

R-MOD the category of left R-modules;

R-MOD$_{\mathrm{fgf}}$ the full subcat. of R-MOD of fin. gen. free R-modules;

R-MOD$_{\mathrm{fgp}}$ the full subcat. of R-MOD of fin. gen. projective R-modules.

14.4 Some Basic Chain Complex Constructions

We will give some basic definitions concerning the construction of chain complexes or chain maps, and some sign conventions. We will justify our choices. Then all other relevant constructions and sign conventions will be consequences of these.

Let C and D be chain complexes of left R-modules. Then we have chain complexes of abelian groups $(C^t \otimes_R D, d_\otimes)$ and $(\hom_R(C, D), d_{\hom})$ defined in [134, VI.9.1] and [134, VI.10.1] by

$$(C^t \otimes_R D)_n = \bigoplus_{p+q=n} C_p^t \otimes_R D_q, \tag{14.12}$$

$$d_\otimes(x \otimes y) = c(x) \otimes y + (-1)^{|x|} x \otimes d(y),$$

and

$$\hom_R(C, D)_n = \prod_{q-p=n} \hom_R(C_p, D_q), \tag{14.13}$$

$$d_{\hom}(f) = d \circ f + (-1)^{n+1} f \circ c.$$

Let us spell out how these formulas for the differentials have to be interpreted. The formula for $d_\otimes$ explains what the image under the differential of the element $x \otimes y$ in the summand $C_p^t \otimes_R D_q$ of $(C^t \otimes_R D)_n$ is. The image takes possibly non-zero values only in the summands $C_{p-1} \otimes_R D_q$ and $C_p \otimes_R D_{q-1}$ of $(C^t \otimes_R D)_{n-1}$ and these are

14.4 Some Basic Chain Complex Constructions

$c_p(x) \otimes y$ and $(-1)^p \cdot x \otimes d_q(y)$. The formula for d_{hom} explains what the image under the differential of the element f in the summand $\hom_R(C_p, D_q)$ of $\hom_R(C, D)_n$ is. The image takes possibly non-zero values only in the summands $\hom_R(C_p, D_{q-1})$ and $\hom_R(C_{p+1}, D_q)$ of $\hom_R(C, D)_{n-1}$, and these are $d_q \circ f$ and $(-1)^{n+1} \cdot f \circ c_{p+1}$. In the sequel we define for an element $x \in C_p$ its degree $|x|$ by $|x| = p$.

Besides [134] this also agrees with [44, 85, 178, 272, 382]. The formula $d_\otimes$ differs from both [344] and [427]. The d_{hom} differs from [344], but coincides with [427]. As a result of this, almost all formulas from [344] and [427] will be changed. The notes [240] are also very helpful.

Remark 14.14 We can ask why we need signs in the definitions of the above complexes. A purely algebraic answer is that these complexes are total complexes of double complexes and in order to obtain a chain complex, that means for $d \circ d = 0$ to hold, we need some signs. For the tensor product there is also a geometric motivation for this particular choice. Namely, if X and Y are CW-complexes, then the cellular chain complex of $X \times Y$ is the tensor product of the cellular chain complexes of X and cellular chain complexes of Y. The particular choice of the signs for the hom complex is motivated by the discussion below about chain maps. These choices are compatible in the sense that we obtain the pleasant adjunction (14.16).

The conventions about the hom-complex are convenient since they lead to the following interpretation of the homology groups.

Given $n \in \mathbb{Z}$, an *R-chain map of degree n* of chain complexes of left R-modules $f: C \to D$ is a collection of R-homomorphisms $f_i: C_i \to D_{i+n}$ for $i \in \mathbb{Z}$ satisfying $d_{i+n} \circ f_i = (-1)^n \cdot f_{i-1} \circ c_i$. If n is 0, we just talk about an *R-chain map*. (If $R = \mathbb{Z}$ or if the ring with involution is obvious, we sometimes drop the letter R and only talk about a chain map.)

An *R-chain homotopy* $g: f \simeq f'$ from an R-chain map f of degree n to an R-chain map f' of degree n is an R-map $g: C \to D$ of degree $n+1$ satisfying $d_{i+n+1} \circ g_i + (-1)^n \cdot g_{i-1} \circ c_i = f'_i - f_i$ for $i \in \mathbb{Z}$. We obtain the identification of abelian groups

$H_n(\hom_R(C, D))$
$\qquad = \{\text{homotopy classes of chain maps of degree } n \text{ from } C \text{ to } D\}.$

For the case $n = 0$, we use the notation

$$[C, D] := H_0(\hom_R(C, D)). \qquad (14.15)$$

An R-chain map $f: C \to D$ is called an *R-chain homotopy equivalence* if there exists an R-chain map $g: D \to C$ and R-chain homotopies $h: \text{id}_C \simeq g \circ f$ and $k: \text{id}_D \simeq f \circ g$.

Next we want to introduce the conventions for relations between $\hom_R$ and $\otimes_R$ complexes.

The functors $\hom_R$ and $\otimes_R$ are *adjoint* via the $\mathbb{Z}$-chain isomorphism [134, VI.10.29.2] and [272, V.3.2]

$$\Phi\colon \hom_{\mathbb{Z}}(C^t \otimes_R D, E) \to \hom_R(C, \hom_{\mathbb{Z}}(D, E)),$$
$$f = \{f_s\}_{s \in \mathbb{Z}} \mapsto (x \mapsto (y \mapsto f_{|x|+|y|}(x \otimes y))), \tag{14.16}$$

and, more generally, for chain complexes of bimodules $_T C_R$, $_R D_S$ and $_T E_S$, we have the isomorphism

$$\Phi\colon \hom_{T-S}(C \otimes_R D, E) \to \hom_{T-R}(C, \hom_S(D, E)),$$
$$f = \{f_s\}_{s \in \mathbb{Z}} \mapsto (x \mapsto (y \mapsto f_{|x|+|y|}(x \otimes y))). \tag{14.17}$$

It is convenient that no signs appear in (14.16) and (14.17).

There is the *switch chain map* [134, VI.9.5]

$$\mathrm{sw}\colon C^t \otimes_R D \xrightarrow{\cong} D^t \otimes_R C, \quad x \otimes y \mapsto (-1)^{|x| \cdot |y|} \cdot y \otimes x \tag{14.18}$$

which is a $\mathbb{Z}$-chain isomorphism. The sign is forced here because otherwise we would not have a chain map. This sign is also the source of the convention in homological algebra which says that, whenever you permute tensor product expressions, you introduce the sign as in this definition.

Next we have the *composition chain map* [134, VI.10.29]

$$-\circ-\colon \hom_R(C', D) \otimes_{\mathbb{Z}} \hom_R(C, C') \to \hom_R(C, D), \quad f \otimes g \mapsto f \circ g. \tag{14.19}$$

It is convenient that no signs appear in (14.19).

We have the *tensoring chain map* [134, VI.10.23]

$$\otimes\colon \hom_R(C, D) \otimes_{\mathbb{Z}} \hom_R(C', D') \to \hom_R(C \otimes_{\mathbb{Z}} C', D \otimes_{\mathbb{Z}} D'),$$
$$f \otimes g \mapsto (x \otimes y \mapsto (-1)^{|x| \cdot |g|} \cdot f(x) \otimes g(y)). \tag{14.20}$$

The signs are forced here because otherwise we would not have a chain map.

Given R-chain maps $\alpha\colon D \to C$ and $\beta\colon C' \to D'$, we have the *hom chain map*

$$\hom(\alpha, \beta)\colon \hom_R(C, C') \to \hom_R(D, D'),$$
$$\gamma \mapsto (-1)^{|\alpha| \cdot (|\beta|+|\gamma|)} \beta \circ \gamma \circ \alpha. \tag{14.21}$$

The sign appears here since this map is obtained as the adjoint of

$$\hom_R(D, C) \otimes \hom_R(C', D') \otimes \hom_R(C, C') \to \hom_R(D, D')$$

which itself is obtained as a composition of permuting the factors and applying the composition map (14.19).

14.5 Further Chain Complex Constructions over Rings with Involution

Notation 14.22 (Categories of chain complexes). Let R be a ring with involution. Denote by

R-CH the category of chain complexes over R-MOD;

R-CH$_{fgf}$ the category of chain complexes over R-MOD$_{fgf}$;

R-CH$_{fgp}$ the category of chain complexes over R-MOD$_{fgp}$;

bR-CH the full subcat. of R-CH of bounded chain complexes;

bR-CH$_{fgf}$ the full subcat. of R-CH$_{fgf}$ of bounded chain complexes;

bR-CH$_{fgp}$ the full subcat. of R-CH$_{fgp}$ of bounded chain complexes,

where *bounded* means that there exists a natural number N such that $C_n = 0$ for $|n| > N$. Moreover, denote by

hR-CH$_{fgf}$ the full subcat. of R-CH of chain compl.
R-chain homotopy equiv. to R-chain compl. in bR-CH$_{fgf}$;

hR-CH$_{fgp}$ the full subcat. of R-CH of chain compl.
R-chain homotopy equiv. to R-chain compl. in bR-CH$_{fgp}$.

14.5 Further Chain Complex Constructions over Rings with Involution

Note that the composite of (14.19) with the corresponding switch map (14.18) involves a sign, namely, it looks like

$$\hom_R(C, C') \otimes_{\mathbb{Z}} \hom_R(C', D) \to \hom_R(C, D), \quad g \otimes f \mapsto (-1)^{|f| \cdot |g|} \cdot f \circ g.$$

We come to other useful chain maps that can be derived from the chain maps (14.16), (14.18), (14.19), and (14.20). In the sequel, for an integer n and an R-module M, we denote by $n[M]$ the R-chain complex concentrated in dimension n with M as n-th chain module. We start with the *dual chain complex* C^{-*} of a chain complex C, defined by $C^{-*} = \hom_R(C, 0[R])$. This means that we have

$$C^{-r} = \hom_R(C, 0[R])_r = \hom_R(C_{-r}, R) = C^*_{-r},$$
$$(d_{C^{-*}})_r = (-1)^{r+1} \cdot \hom_R(c_{-r+1}, \mathrm{id}_R) = (-1)^{r+1} \cdot c^*_{-r+1}. \tag{14.23}$$

Note that, with these conventions, the usual cohomology groups of a cochain complex associated to a chain complex C are expressed as follows:

$$H^k(C) = H_{-k}(C^{-*}). \tag{14.24}$$

As it stands, the dual chain complex C^{-*} is a priori a chain complex of right R-modules. A special case of (14.19) is the map

$$C^{-*} \otimes_{\mathbb{Z}} C \to 0[R], \quad f \otimes x \mapsto f(x), \qquad (14.25)$$

namely, take $D = C = 0[R]$, replace C' by C, and use the obvious identification of chain complexes $\hom_R(0[R], C) \xrightarrow{\cong} C$. It is compatible with the obvious R-R-bimodule structures. The evaluation map (14.25) composed with the switch map (14.18) is a chain map of chain complexes of R-R-bimodules

$$C \otimes_{\mathbb{Z}} C^{-*} \to 0[R], \quad x \otimes f \mapsto (-1)^{|x| \cdot |f|} \cdot f(x) = (-1)^{|x|} \cdot f(x).$$

Its adjoint under the adjunction (14.17) is a map of chain complexes of left R-modules

$$C \to (C^{-*})^{-*}, \quad x \mapsto (f \mapsto (-1)^{|x| \cdot |f|} \cdot f(x) = (-1)^{|x|} \cdot f(x))$$

where the switch map is responsible for the sign $(-1)^{|x|}$.

A drawback of this map is, as noted above, that C^{-*} is a chain complex of right R-modules. Since we want to work entirely in the category of left R-modules, we replace C^{-*} by C^{-*t} and proceed as in the case of R-modules. We will use the modified version of (14.25) called the *evaluation chain map*

$$e(C) \colon C \to (C^{-*t})^{-*t}, \quad x \mapsto (f \mapsto (-1)^{|x|} \cdot \overline{f(x)}). \qquad (14.26)$$

The convention above means that, when taking the dual of a chain map, we also need to introduce signs. Namely, we define a $\mathbb{Z}$-chain isomorphism

$$I \colon \hom_R(C, D) \to \hom_R(D^{-*t}, C^{-*t}) \qquad (14.27)$$

by associating to the element $f \in \hom_R(C_p, D_{p+n})$ the element given as $I(f)_{-(p+n)} = (-1)^{pn+n} \cdot (f_p)^{*t} \colon D^{*t}_{p+n} \to C^{*t}_p$ in $\hom_R((D^{-*t})_{-p-n}, (C^{-*t})_{-p})$. It is sometimes convenient to phrase this identity as

$$I(f)_{-k} = (-1)^{kn} \cdot (f_{k-n})^{*t} \colon D^{*t}_k \to C^{*t}_{k-n}.$$

This follows since the map I can be seen as the adjoint of the map

$$\hom_R(C, D) \otimes_{\mathbb{Z}} \hom_R(D, 0[R]) \to \hom_R(C, 0[R]), \quad f \otimes \alpha \mapsto (-1)^{|\alpha| \cdot |f|} \cdot \alpha \circ f.$$

If C belongs to bR-CH$_{\text{fgp}}$, then the map (14.26) is an isomorphism of chain complexes of left R-modules, and we have the equality of R-chain maps from C^{-*t} to C^{-*t}

$$\text{id}_{C^{-*t}} \colon C^{-*t} \xrightarrow{e(C^{-*t})} ((C^{-*t})^{-*t})^{-*t} \xrightarrow{I(e(C))} C^{-*t}. \qquad (14.28)$$

Incidentally, in this case we have $I(e(C)) = (e(C))^{-*t}$ since the degree of the map $e(C)$ is zero. Moreover, for an R-chain map $f \colon C \to D$, the following diagram commutes

14.5 Further Chain Complex Constructions over Rings with Involution 509

$$\begin{array}{ccc} C & \xrightarrow{f} & D \\ e(C) \downarrow & & \downarrow e(D) \\ (C^{-*t})^{*t} & \xrightarrow{I^2(f)} & (D^{-*t})^{-*t}. \end{array} \qquad (14.29)$$

Hence, upon the identification (14.26) the map I of (14.27) becomes an involution in the sense that the above diagram commutes. Note that on elements we have for $\alpha \in (C^{-*t})^{*t}$ that $I^2(f)(\alpha) = (-1)^{|f|}(f^{-*t})^{-*t}(\alpha)$.

Finally we obtain the *slant maps* (only the second one will really be used):

$$\begin{aligned} -\backslash - \colon (C^{-*} \otimes_R D) &\to \hom_R(C, D), \\ f \otimes y &\mapsto (x \mapsto (-1)^{|x| \cdot |y|} \cdot f(x) \cdot y) \end{aligned} \qquad (14.30)$$

and

$$\begin{aligned} -\backslash - \colon (C^t \otimes_R D) &\to \hom_R(C^{-*t}, D), \\ (x \otimes y) &\mapsto (f \mapsto (-1)^{|x| \cdot |y| + |x|} \cdot \overline{f(x)} \cdot y). \end{aligned} \qquad (14.31)$$

Because of the adjunction (14.17) we have to specify a chain map of left R-modules $C^{-*} \otimes_R D \otimes_\mathbb{Z} C \to D$ in order to define the $\mathbb{Z}$-chain map (14.30). This is done by the composite

$$C^{-*} \otimes_R D \otimes_\mathbb{Z} C \xrightarrow{\text{sw}} C^{-*} \otimes_\mathbb{Z} C \otimes_R D \xrightarrow{\text{ev} \otimes_R \text{id}_D} 0[R] \otimes_R D \xrightarrow{\cong} D.$$

The switch map is responsible for the sign $(-1)^{|x| \cdot |y|}$ in (14.30).

Because of the adjunction (14.17) we have to specify a chain map of left R-modules $C^t \otimes_R D \otimes_\mathbb{Z} C^{-*t} \to D$ in order to define the $\mathbb{Z}$-chain map (14.31). This is done by the composite

$$C^t \otimes_R D \otimes_\mathbb{Z} C^{-*t} \xrightarrow{\text{sw}} C^{-*t} \otimes_\mathbb{Z} C^t \otimes_R D \xrightarrow{\text{ev} \otimes_R \text{id}_D} 0[R] \otimes_R D \xrightarrow{\cong} D,$$

where ev: $C^{-*t} \otimes_\mathbb{Z} C^t \to 0[R]$ is the chain map, which comes from the R-R-bimodule map sending $f \otimes x$ to $\overline{f(x)}$, and is motivated by (14.25). The switch map is responsible for the sign $(-1)^{(|x|+|y|) \cdot |f|} = (-1)^{|x| \cdot |y| + |x|}$ in (14.31).

If C and D belong to bR-CH$_{\text{fgp}}$, then the slant maps (14.30) and (14.31) are $\mathbb{Z}$-chain isomorphisms. If C and D belong to hR-CH$_{\text{fgp}}$, then the slant maps (14.30) and (14.31) are $\mathbb{Z}$-chain homotopy equivalences.

Ranicki in [344] and Weiss in [427] do not have the sign in (14.31), so here we pay a price for our conventions (14.12) and (14.13).

Given two R-chain maps $\alpha \colon C \to C'$ and $\beta \colon D \to D'$, we may consider the $\mathbb{Z}$-chain map $\alpha \otimes \beta \colon C^t \otimes_R D \to C'^t \otimes_R D'$, as well as the $\mathbb{Z}$-chain map $\hom(I(\alpha), \beta) \colon \hom_R(C^{-*t}, D) \to \hom_R((C')^{-*t}, D')$ obtained by combining the map I from (14.27) and $\hom(-, -)$ from (14.21). These two maps turn out to be compatible via the following commutative diagram:

$$\begin{array}{ccc} C^t \otimes_R D & \xrightarrow{\alpha \otimes \beta} & C'^t \otimes_R D' \\ {\scriptstyle -\backslash-}\Big\downarrow & & \Big\downarrow{\scriptstyle -\backslash-} \\ \hom_R(C^{-*t}, D) & \xrightarrow[\hom(I(\alpha), \beta)]{} & \hom_R((C')^{-*t}, D'). \end{array} \qquad (14.32)$$

Note that given $|\alpha| = m$, $|\beta| = n$, $\varphi \in \hom_R(C_p^{*t}, D_q)$, and $g \in (C')_{p+m}^{*t}$ we have

$$\hom(I(\alpha), \beta)(\varphi)(g) = (-1)^{m(n+q+m)}(\beta \circ \varphi \circ \alpha^*)(g).$$

14.6 Homotopy Theory of Chain Complexes

In this section we describe several chain complex constructions. We are mainly interested in the notions of homotopy pushout and homotopy pullback for chain maps and their special cases, mapping cone alias homotopy cofibre and mapping cocone alias homotopy fibre. Recall that a pushout of spaces can be thought of as a gluing construction. However, it suffers from the fact that it is not homotopy invariant and therefore we prefer to work with the homotopy pushout alias double mapping cylinder which is homotopy invariant. The situation with chain complexes is analogous. We want to do gluing constructions (most prominently gluing cobordisms along a common piece of boundary, see 15.5) and these ought to be homotopy invariant. The situation for chain complexes is better than for spaces due to the facts that the notions of homotopy cofibration sequence and homotopy fibration sequence, see Theorem 14.78, as well as the notions of a homotopy cocartesian square and homotopy cartesian square, see Theorem 14.95, coincide. The same situation happens in spectra, we deal with a "stable category".

The situation with references is similar to the previous section. There is no single place where one can find all conventions. Therefore one has to combine several conventions or come up with one's own. We have decided to use mainly the conventions from [130], which seem compatible with [134]. The introductions to [290] and [427] are also very useful. Further inspiration comes from the first chapter of [421].

14.6.1 Chain Homotopy

Recall that $0[\mathbb{Z}]$ denotes the chain complex concentrated in dimension 0 with the value $\mathbb{Z}$, in other words the cellular chain complex of the one-point space. Let I denote the 1-dimensional cellular $\mathbb{Z}$-chain complex of the unit interval $[0, 1]$, alias the 1-simplex. Explicitly it is given by

$$I_1 = \mathbb{Z}\{\sigma_{01}\}, \quad I_0 = \mathbb{Z}\{\sigma_0\} \oplus \mathbb{Z}\{\sigma_1\} \quad d_1(\sigma_{01}) = \sigma_1 - \sigma_0. \qquad (14.33)$$

14.6 Homotopy Theory of Chain Complexes

or, equivalently, by specifying its first differential by $\mathbb{Z} \xrightarrow{\begin{pmatrix} -1 \\ +1 \end{pmatrix}} \mathbb{Z} \oplus \mathbb{Z}$. The two inclusions of the end points define two chain maps $i_0, i_1: 0[\mathbb{Z}] \to I$, and projection to the point defines a chain map $\mathrm{pr}: I \to 0[\mathbb{Z}]$. They satisfy $\mathrm{pr} \circ i_k = \mathrm{id}_{0[\mathbb{Z}]}$ for $k = 0, 1$. There is a preferred chain homotopy $h_k: i_k \circ \mathrm{pr} \simeq \mathrm{id}_I$ for $k = 0, 1$. Explicitly $h_k: \mathbb{Z} \oplus \mathbb{Z} \to \mathbb{Z}$ is given by $(0, 1)$ for $k = 0$ and by $(-1, 0)$ for $k = 1$.

Consider two R-chain maps $f_0, f_1: C \to D$ and an R-chain homotopy $h: f_0 \simeq f_1$ between them. Recall that the n-th-differential of $I \otimes C$ is given by

$$\begin{pmatrix} -c_{n-1} & 0 & 0 \\ -\mathrm{id} & c_n & 0 \\ +\mathrm{id} & 0 & c_n \end{pmatrix}: C_{n-1} \oplus C_n \oplus C_n = (I_1 \otimes C_{n-1}) \oplus (I_0 \otimes C_n)$$

$$\to C_{n-2} \oplus C_{n-1} \oplus C_{n-1} = (I_1 \otimes C_{n-2}) \oplus (I_0 \otimes C_{n-1}). \quad (14.34)$$

To get closer to standard conventions, we have changed the order of $(I_0 \otimes C_n) \oplus (I_1 \otimes C_{n-1})$ to $(I_1 \otimes C_{n-1}) \oplus (I_0 \otimes C_n)$. We can associate to this data the R-chain map

$$\overline{h}: I \otimes_{\mathbb{Z}} C \to D, \quad \sigma_j \otimes x \mapsto f_j(x) \quad \text{and} \quad \sigma_{01} \otimes x \mapsto h(x) \quad (14.35)$$

satisfying $\overline{h} \circ i_j = f_j$ for $j = 0, 1$. Equivalently, it is defined in degree n by $\overline{h}_n = (h_{n-1} \, (f_0)_n \, (f_1)_n): C_{n-1} \oplus C_n \oplus C_n \to D_n$.

Vice versa, given an R-chain map $\overline{h}: I \otimes C \to D$ satisfying $\overline{h} \circ i_j = f_j$ for $j = 0, 1$, we get an R-chain homotopy $h: f_0 \simeq f_1$ by the composite of $\overline{h}_n$ with

$$\begin{pmatrix} \mathrm{id} \\ 0 \\ 0 \end{pmatrix}: C_{n-1} \to C_{n-1} \oplus C_n \oplus C_n.$$

Of course all of this is motivated by the observation that, for a CW-complex X and a cellular map $h: [0, 1] \times X \to Y$, we get an R-chain map $I \otimes_{\mathbb{Z}} C_*^c(X; R) = C_*^c([0, 1] \times X; R) \xrightarrow{C_*^c(h)} C_*^c(Y; R)$.

Note that we have the switch $\mathbb{Z}$-chain isomorphism $C \otimes_{\mathbb{Z}} I \xrightarrow{\cong} I \otimes_{\mathbb{Z}} C$ of (14.18). Hence we can also think of such a chain homotopy as a chain map of the shape $C \otimes I \to D$, but signs creep into formula (14.35). Explicitly, when we define

$$\overline{\overline{h}}(x \otimes \sigma_j) = f_j(x) \quad \text{and} \quad \overline{\overline{h}}(x \otimes \sigma_{01}) = (-1)^{|x|} \cdot h(x), \quad (14.36)$$

we obtain an R-chain map

$$\overline{\overline{h}}: C \otimes I \to D \quad (14.37)$$

such that $\overline{\overline{h}} \circ i_j = f_j$ holds for $j = 0, 1$. To avoid these extra signs and to get nicer sign conventions for mapping cones and mapping cylinders, we use $I \otimes C$ instead of $C \otimes I$, although at the space level one usually writes a homotopy with $X \times [0, 1]$ as source.

14.6.2 Pushouts

Pushouts exist in the category of R-chain complexes. Namely, given a diagram of R-chain complexes $E \xleftarrow{g} C \xrightarrow{f} D$, we can complete it to a commutative square of R-chain complexes called a *pushout*

$$\begin{array}{ccc} C & \xrightarrow{f} & D \\ g \downarrow & & \downarrow \bar{g} \\ E & \xrightarrow{\bar{f}} & D \cup_C E \end{array}$$

with the following universal property: Given R-chain maps $u \colon C \to Z$, $v \colon D \to Z$ and $w \colon E \to Z$ satisfying $v \circ f = u = w \circ g$, there exists precisely one R-chain map $z \colon D \cup_C E \to Z$ with $z \circ \bar{f} = w$ and $z \circ \bar{g} = v$. Recall that the category of R-chain complexes inherits the structure of an abelian category from the structure of an abelian category on the category of R-modules by a degreewise construction. Therefore we can construct the pushout to be the cokernel of the R-chain map $\begin{pmatrix} f \\ -g \end{pmatrix} \colon C \to D \oplus E$. In particular a commutative square of R-chain complexes

$$\begin{array}{ccc} C & \xrightarrow{f} & D \\ g \downarrow & & \downarrow \bar{g} \\ E & \xrightarrow{\bar{f}} & F \end{array}$$

is a pushout if and only if the following sequence of R-chain complexes

$$C \xrightarrow{\begin{pmatrix} f \\ -g \end{pmatrix}} D \oplus E \xrightarrow{(\bar{g}\ \bar{f})} F \to 0.$$

is exact.

The drawback of the pushout is that it is not homotopy invariant, i.e., the following statement is *not* true: If we have two pushouts of R-chain complexes

$$\begin{array}{ccc} C & \xrightarrow{f} & D \\ g \downarrow & & \downarrow \bar{g} \\ E & \xrightarrow{\bar{f}} & F \end{array} \qquad \begin{array}{ccc} C' & \xrightarrow{f'} & D' \\ g' \downarrow & & \downarrow \bar{g}' \\ E' & \xrightarrow{\bar{f}'} & F' \end{array}$$

14.6 Homotopy Theory of Chain Complexes

and three R-chain homotopy equivalences $u\colon C \xrightarrow{\simeq} C'$, $v\colon D \xrightarrow{\simeq} D'$, and $w\colon E \xrightarrow{\simeq} E'$ satisfying $v \circ f = f' \circ u$ and $w \circ g = g' \circ w$ and $z\colon F \to F'$ is the unique R-chain map coming from the pushout property, then z is an R-chain homotopy equivalence.

Exercise 14.38 Give a counterexample to the statement that the pushout of R-chain complexes is homotopy invariant.

The deficiency of the pushout not being homotopy invariant will lead to the notions of homotopy pushout. Before we explain this, we introduce the notion of a mapping cylinder and of a mapping cone.

14.6.3 Mapping Cylinders, Mapping Cones and Suspensions

In view of the definition of $\mathrm{cyl}(f)$ as a pushout, see (3.34), we define the mapping cylinder of an R-chain map $f\colon C \to D$ by the following pushout

$$\begin{array}{ccc} 0[\mathbb{Z}] \otimes_{\mathbb{Z}} C = C & \xrightarrow{i_1 \otimes_{\mathbb{Z}} \mathrm{id}_C} & I \otimes_{\mathbb{Z}} C \\ {\scriptstyle f}\downarrow & & \downarrow {\scriptstyle \bar{f}} \\ D & \xrightarrow{j} & \mathrm{cyl}(f). \end{array} \qquad (14.39)$$

Explicitly, the n-th differential of the mapping cylinder is given by

$$\begin{pmatrix} -c_{n-1} & 0 & 0 \\ -\mathrm{id} & c_n & 0 \\ f_{n-1} & 0 & d_n \end{pmatrix} \colon C_{n-1} \oplus C_n \oplus D_n \to C_{n-2} \oplus C_{n-1} \oplus D_{n-1}. \qquad (14.40)$$

The R-chain map $j\colon D \to \mathrm{cyl}(f)$ is given in terms of (14.40) in degree n by the canonical inclusion $D_n \to C_{n-1} \oplus C_n \oplus D_n$, and the R-chain map $\bar{f}\colon I \otimes_{\mathbb{Z}} C \to \mathrm{cyl}(f)$ is given in terms of (14.34) and (14.39) by

$$\begin{pmatrix} \mathrm{id} & 0 & 0 \\ 0 & \mathrm{id} & 0 \\ 0 & 0 & f_n \end{pmatrix} \colon C_{n-1} \oplus C_n \oplus C_n \to C_{n-1} \oplus C_n \oplus D_n.$$

Define the R-chain map

$$k\colon C \to \mathrm{cyl}(f)$$

to be the composite $C = 0[\mathbb{Z}] \otimes_{\mathbb{Z}} C \xrightarrow{i_0 \otimes_{\mathbb{Z}} \mathrm{id}} I \otimes_{\mathbb{Z}} C \xrightarrow{\bar{f}} \mathrm{cyl}(f)$. It is explicitly given by the canonical inclusion $C_n \to C_{n-1} \oplus C_n \oplus D_n$.

Thanks to the pushout property of (14.39), the map $\mathrm{id}_D\colon D \to D$ and the composite $I \otimes_{\mathbb{Z}} C \xrightarrow{\mathrm{pr} \otimes_{\mathbb{Z}} \mathrm{id}} 0[\mathbb{Z}] \otimes_{\mathbb{Z}} C = C \xrightarrow{f} D$ together yield the R-chain map

$$p\colon \mathrm{cyl}(f) \to D$$

satisfying $p \circ j = \mathrm{id}_D$ and $p \circ k = f$. It is explicitly given by the formula $\begin{pmatrix} 0 & f_n & \mathrm{id} \end{pmatrix} : C_{n-1} \oplus C_n \oplus D_n \to D_n$.

Remark 14.41 (Universal property of the mapping cylinder) The mapping cylinder $\mathrm{cyl}(f)$ of an R-chain map $f \colon C \to D$ has the following universal property. Consider R-chain maps $u \colon C \to E$ and $v \colon D \to E$ and a chain homotopy $h \colon u \simeq v \circ f$. We can and will think of h as an R-chain map $\overline{h} \colon I \otimes_{\mathbb{Z}} C \to E$ such that the composite $C = 0[\mathbb{Z}] \otimes_{\mathbb{Z}} C \xrightarrow{i_k \otimes_{\mathbb{Z}} \mathrm{id}_C} I \otimes_{\mathbb{Z}} C \xrightarrow{\overline{h}} E$ agrees with u for $k = 0$ and with $v \circ f$ for $k = 1$, see (14.35). By the pushout property (14.39), we obtain precisely one R-chain map $z(u, v, h) \colon \mathrm{cyl}(f) \to E$ such that $z(u, v, h) \circ j = v$ and $z(u, v, h) \circ \overline{f} = \overline{h}$ hold. It is explicitly given by $\begin{pmatrix} h_{n-1} & u_n & v_n \end{pmatrix} : C_{n-1} \oplus C_n \oplus D_n \to E_n$.

Conversely, if $z \colon \mathrm{cyl}(f) \to E$ is an R-chain map, then we get R-chain maps $u \colon C \to E$ by $u = z \circ \overline{f} \circ i_0$, and $v \colon D \to E$ by $z \circ j$ together with an R-chain homotopy $h \colon v \circ f \simeq u$ defined by $h = z \circ \overline{f}$. These two constructions are inverse to one another.

Obviously $p \circ j = \mathrm{id}_D$. There is an R-chain homotopy $h \colon \mathrm{id}_{\mathrm{cyl}(f)} \simeq j \circ p$ given by

$$h_n = \begin{pmatrix} 0 & \mathrm{id} & 0 \\ 0 & 0 & 0 \\ 0 & 0 & 0 \end{pmatrix} : C_{n-1} \oplus C_n \oplus D_n \to C_n \oplus C_{n+1} \oplus D_{n+1}$$

in degree n. In particular $p \colon \mathrm{cyl}(f) \to D$ is an R-chain homotopy equivalence. To summarise, we get a diagram of R-chain complexes

$$\begin{array}{c}
 & & D \\
 & \nearrow^{f} & \uparrow p \simeq \\
C & \xrightarrow{k} & \mathrm{cyl}(f) \\
 & \searrow_{f} \; \simeq & \uparrow j \simeq \\
 & & D
\end{array} \qquad (14.42)$$

where the upper triangle commutes strictly, the lower triangle commutes up to a preferred R-chain homotopy, and j and p are R-chain homotopy equivalences satisfying $p \circ j = \mathrm{id}_D$.

The *mapping cone* $\mathrm{cone}(f)$ of an R-chain map $f \colon C \to D$ is the cokernel of the R-chain map $k \colon C \to \mathrm{cyl}(f)$. By construction we obtain a canonical exact sequence of R-chain complexes

$$0 \to C \xrightarrow{k} \mathrm{cyl}(f) \xrightarrow{q} \mathrm{cone}(f) \to 0 \qquad (14.43)$$

that is degreewise split exact and in degree n given by the canonical exact sequence $0 \to C_n \to C_{n-1} \oplus C_n \oplus D_n \to C_{n-1} \oplus D_n \to 0$. Explicitly, the nth differential of the mapping cone is given by

14.6 Homotopy Theory of Chain Complexes

$$\begin{pmatrix} -c_{n-1} & 0 \\ f_{n-1} & d_n \end{pmatrix} : C_{n-1} \oplus D_n \to C_{n-2} \oplus D_{n-1}. \tag{14.44}$$

The R-chain map

$$l := q \circ j \colon D \to \mathrm{cone}(f)$$

is in degree n given by the canonical inclusion $D_n \to C_{n-1} \oplus D_n$.

Remark 14.45 (Universal property of the mapping cone) There is a preferred homotopy $h \colon 0 \simeq l \circ f$ for the composite of $f \colon C \to D$ with the R-chain map $l \colon D \to \mathrm{cone}(f)$, namely, $h_n \colon C_n \to \mathrm{cone}(f)_{n+1} = C_n \oplus D_{n+1}$ is the canonical inclusion. Consider an R-chain map $v \colon D \to E$ together with an R-chain homotopy $h' \colon 0 \simeq v \circ f$. Then there is precisely one R-chain map $z(v, h') \colon \mathrm{cone}(f) \to E$ such that the composite of z with the canonical inclusion $D \to \mathrm{cone}(f)$ is v and the composite of the preferred chain homotopy h with z is h', namely, put $z_n = (h'_{n-1}, v_n) \colon \mathrm{cone}(f)_n = C_{n-1} \oplus D_n \to E_n$.

Conversely, let $z \colon \mathrm{cone}(f) \to E$ be an R-chain map. Let $v \colon D \to E$ be the composite $z \circ l$. Let h' be the R-chain homotopy given by the composite of the preferred R-chain homotopy h with z. Then h' is an R-chain homotopy $h' \colon 0 \simeq v \circ f$ and $z = z(v, h')$.

Given an R-chain complex C, its *suspension* ΣC is the R-chain complex whose n-th chain module $(\Sigma C)_n$ is C_{n-1} and whose n-differential is $-c_{n-1} \colon C_{n-1} \to C_{n-2}$. Given an R-chain map $f \colon C \to D$, we define $\Sigma f \colon \Sigma C \to \Sigma D$ by $(\Sigma f)_n = f_{n-1}$.

For an R-chain map $f \colon C \to D$ consider the composite

$$l \colon D \xrightarrow{j} \mathrm{cyl}(f) \xrightarrow{q} \mathrm{cone}(f)$$

that is in degree n given by the canonical inclusion $D_n \to C_{n-1} \oplus D_n$. We obtain a canonical exact sequence of R-chain complexes

$$0 \to D \xrightarrow{l} \mathrm{cone}(f) \xrightarrow{r} \Sigma C \to 0 \tag{14.46}$$

that is degreewise split exact and given in degree n by the canonical sequence $0 \to D_n \to C_{n-1} \oplus D_n \to C_{n-1} \to 0$.

The suspension ΣC can be viewed as the tensor product $I/\partial I \otimes_{\mathbb{Z}} C$ where $\partial I = C_*^c(\partial [0,1]) = C_*^c(\{0,1\}) = 0[\mathbb{Z}] \oplus 0[\mathbb{Z}]$, and $I/\partial I$ denotes the cokernel of the inclusion $\partial I \to I$. Actually, $I/\partial I = 1[\mathbb{Z}] = \Sigma 0[1]$. In particular we see that for a pointed CW-complex (X, x) and its suspension $S^1 \wedge X = [0,1] \times X/[0,1] \times \{x\} \cup 0 \times X$ with its preferred base point y we get an identification of cellular $\mathbb{Z}$-chain complexes

$$C_*^c(S^1 \wedge X, \{y\}) = \Sigma C_*^c(X, \{x\}). \tag{14.47}$$

We will also use the inverse operation, namely the *desuspension* $\Sigma^{-1} C$, explicitly it is

$$(\Sigma^{-1} C)_n = C_{n+1} \quad (d_{\Sigma^{-1} C})_n = -c_{n+1}. \tag{14.48}$$

This can be viewed as the tensor product $(I/\partial I)^{-*} \otimes_{\mathbb{Z}} C = (-1)[\mathbb{Z}] \otimes_{\mathbb{Z}} C$. Note that Ranicki uses Ω instead of Σ^{-1} in [344], but we prefer Σ^{-1}.

Chain homotopy behaves with respect to the suspension in the following manner. Let $h \colon f \simeq g$ be an R-chain homotopy between R-chain maps $f, g \colon C \to D$, so that we have $d \circ h + h \circ c = g - f$. Denote by $\Sigma h \colon \Sigma C \to \Sigma D$ a map of degree 1 given by $(\Sigma h)_n(x) = -h_{n-1}(x)$. Then Σh is an R-chain homotopy between Σf and Σg via the following calculation for $x \in (\Sigma C)_n = C_{n-1}$:

$$\begin{aligned}(d_{\Sigma D} \circ (\Sigma h) + (\Sigma h) \circ c_{\Sigma C})(x) &= \bigl((-d_n) \circ (-h_{n-1}) + (-h_{n-2}) \circ (-c_{n-1})\bigr)(x) \\ &= (d_n \circ h_{n-1} + h_{n-2} \circ c_{n-1})(x) \\ &= g_{n-1}(x) - f_{n-1}(x) = (\Sigma g)_n(x) - (\Sigma f)_n(x).\end{aligned}$$

Example 14.49 (Mapping cone versus suspension) Note that our conventions imply the equality

$$\Sigma \operatorname{cone}(f) = \operatorname{cone}(-\Sigma f),$$

since by our definitions $(d_{\Sigma \operatorname{cone}(f)})_n = -(d_{\operatorname{cone}(f)})_{n-1}$,

$$(d_{\operatorname{cone}(f)})_{n-1} = \begin{pmatrix} -c_{n-2} & 0 \\ f_{n-2} & d_{n-1} \end{pmatrix} \text{ and } (d_{\operatorname{cone}(-\Sigma f)})_n = \begin{pmatrix} c_{n-2} & 0 \\ -(\Sigma f)_{n-1} & -d_{n-1} \end{pmatrix}.$$

An elementary calculation reveals $\Sigma l_f = \tau \circ l_{\Sigma f}$ and $-r_{\Sigma f} = \Sigma r_f \circ \tau$ where $\tau \colon \operatorname{cone}(\Sigma f) \to \Sigma(\operatorname{cone}(f))$ is given by $\tau(x, y) = (-x, y)$ and l_- and r_- are the maps associated to R-chain maps as in (14.46).

Example 14.50 Take f in Remark 14.45 to be the zero map. Then $\operatorname{cone}(f) = \Sigma C \oplus D$, and we get an obvious bijection of abelian groups

$$[\Sigma C, E] \times [D, E] \xrightarrow{\cong} [\operatorname{cone}(f), E] = [\Sigma C \oplus D, E], \quad ([h], [v]) \mapsto [h \oplus v],$$

where $[D, E]$ denotes the abelian group of R-chain homotopy classes of R-chain maps and $[v]$ the R-chain homotopy class of an R-chain map v. Note that an R-chain map $h \colon \Sigma C \to D$ is the same as a homotopy $h \colon 0 \simeq 0$ of R-chain maps $C \to D$. If we fix an R-chain map $v \colon D \to E$, then an R-chain map $h \colon \Sigma C \to E$ is an R-chain homotopy of R-chain maps $C \to E$ from 0 to $v \circ f$ and the R-chain map $z(v, h)$ of Remark 14.45 represents $[h \oplus v]$. In particular, for fixed v, we see that the R-chain homotopy class of $z(v, h)$ depends on h.

Exercise 14.51 Let $f \colon C \to D$ and $v \colon D \to E$ be R-chain maps. Consider two R-chain homotopies $h', h'' \colon 0 \simeq v \circ f$. Show that the R-chain maps $z(v, h')$ and $z(v, h'')$ defined in Remark 14.45 are R-chain homotopic if and only if there exists an R-homotopy $b \colon 0 \simeq 0$ of R-chain maps $D \to E$ together with an R-chain homotopy $a \colon h' + b \circ f \simeq h''$ of R-chain homotopies of R-chain maps $C \to E$, i.e., a map $a \colon C \to E$ of degree 2 satisfying $e \circ a - a \circ c = h'' - (h' + b \circ f)$.

Finally we prove two propositions about R-chain homotopy equivalences needed later.

14.6 Homotopy Theory of Chain Complexes

Proposition 14.52 *Consider an R-chain map $f\colon C \to D$.*

(i) *The R-chain map f is an R-chain homotopy equivalence if and only if its mapping cone $\mathrm{cone}(f)$ is contractible;*

(ii) *If the R-chain complexes C and D are projective and bounded from below, i.e., there exists a natural number N with $C_n = 0$ for $n \leq N$, then the following assertions are equivalent:*

   (a) *The R-chain map f is an R-chain homotopy equivalence;*
   (b) *The map $H_n(f)$ is an isomorphism for all $n \in \mathbb{Z}$;*
   (c) *The mapping cone $\mathrm{cone}(f)$ is acyclic.*

Proof. (i) Let $\gamma\colon \mathrm{cone}(f) \to \mathrm{cone}(f)$ be an R-chain contraction given in degree n by

$$\begin{pmatrix} h_{n-1} & g_n \\ l_{n-1} & k_n \end{pmatrix} \colon C_{n-1} \oplus D_n \to C_n \oplus D_{n+1}.$$

Since it is an R-chain contraction, we obtain the equality of maps of the form $C_{n-1} \oplus D_n \to C_{n-1} \oplus D_n$

$$\begin{pmatrix} -c_n & 0 \\ f_n & d_{n+1} \end{pmatrix} \circ \begin{pmatrix} h_{n-1} & g_n \\ l_{n-1} & k_n \end{pmatrix} + \begin{pmatrix} h_{n-2} & g_{n-1} \\ l_{n-2} & k_{n-1} \end{pmatrix} \circ \begin{pmatrix} -c_{n-1} & 0 \\ f_{n-1} & d_n \end{pmatrix} = \begin{pmatrix} \mathrm{id}_{C_{n-1}} & 0 \\ 0 & \mathrm{id}_{D_n} \end{pmatrix}.$$

This is equivalent to

$$-c_n \circ h_{n-1} - h_{n-2} \circ c_{n-1} + g_{n-1} \circ f_{n-1} = \mathrm{id}_{C_{n-1}};$$
$$-c_n \circ g_n + g_{n-1} \circ d_n = 0;$$
$$f_n \circ h_{n-1} + d_{n+1} \circ l_{n-1} - l_{n-2} \circ c_{n-1} + k_{n-1} \circ f_{n-1} = 0;$$
$$f_n \circ g_n + d_{n+1} \circ k_n + k_{n-1} \circ d_n = \mathrm{id}_{D_n}.$$

Hence the collection of the g_n-s, h_n-s, and k_n-s define an R-chain map $g\colon D \to C$ and R-chain homotopies $h\colon g \circ f \simeq \mathrm{id}_C$ and $k\colon f \circ g \simeq \mathrm{id}_D$, and thus f is an R-chain homotopy equivalence.

Given an R-chain map $g\colon D \to C$ and R-chain homotopies $h\colon g \circ f \simeq \mathrm{id}_C$ and $k\colon f \circ g \simeq \mathrm{id}_D$, define an R-chain isomorphism $u\colon \mathrm{cone}(f) \xrightarrow{\cong} \mathrm{cone}(f)$ by $u_n = \begin{pmatrix} \mathrm{id}_{C_{n-1}} & 0 \\ f_n \circ h_{n-1} - k_{n-1} \circ f_{n-1} & \mathrm{id}_{D_n} \end{pmatrix}$ and an R-chain homotopy $\delta\colon u \simeq 0$ by $\delta_n = \begin{pmatrix} h_{n-1} & g_n \\ 0 & -k_n \end{pmatrix}$. Then we obtain a chain contraction γ by $\gamma = u^{-1} \circ \delta$.

(ii) The implication (iia) $\Longrightarrow$ (iib) is obvious. We get implication (iib) $\Longrightarrow$ (iic) from the long exact homology sequence of the short exact sequence of R-chain complexes $0 \to D \to \mathrm{cone}(f) \to \Sigma C \to 0$. It remains to prove the implication (iic) $\Longrightarrow$ (iia). Because of assertion (i) it suffices to show for an acyclic projective R-chain complex (P, c) bounded from below that it admits an R-chain contraction γ. Fix a natural number N with $P_n = 0$ for $n \leq N$. We define the desired chain contraction γ as follows. Put $\gamma_n = 0$ for $n \leq N - 1$. Now we define inductively for $n = N-1, N+1, \ldots$ R-maps $\gamma_n\colon P_n \to P_{n+1}$ satisfying $c_{m+1} \circ \gamma_m + \gamma_{m-1} \circ c_m = \mathrm{id}_{P_m}$

for $m \in \mathbb{Z}, m \leq n$. The induction step from n to $n+1$ is done as follows. We have

$$c_{n+1} \circ \left(\mathrm{id}_{P_{n+1}} - \gamma_n \circ c_{n+1}\right) = c_{n+1} - c_{n+1} \circ \gamma_n \circ c_{n+1}$$
$$= c_{n+1} - \left(\mathrm{id}_{P_n} - \gamma_{n-1} \circ c_n\right) \circ c_{n+1}$$
$$= c_{n+1} - \mathrm{id}_{P_n} \circ c_{n+1}$$
$$= 0.$$

Since P is acyclic and hence $\mathrm{im}(c_{n+2}) = \ker(c_{n+1})$, the image of the R-homomorphism $\mathrm{id}_{P_{n+1}} - \gamma_n \circ c_{n+1} \colon P_{n+1} \to P_{n+1}$ is contained in $\mathrm{im}(c_{n+2})$. Since P_{n+1} is projective, we can find an R-homomorphism $\gamma_{n+1} \colon P_{n+1} \to P_{n+2}$ satisfying $c_{n+1} \circ \gamma_{n+1} + \gamma_n \circ c_{n+1} = \mathrm{id}_{P_{n+1}}$. □

Exercise 14.53 Show that assertion (ii) of Proposition 14.52 is no longer true if we drop the assumption that the chain complexes are projective or that they are bounded from below.

Lemma 14.54 *Let* $0 \to C \xrightarrow{i} D \xrightarrow{p} E \to 0$ *be an exact sequence of R-chain complexes such that the exact sequence* $0 \to C_n \xrightarrow{i_n} D_n \xrightarrow{p_n} E_n \to 0$ *is split exact for each $n \in \mathbb{Z}$ and E is contractible.*

Then there exists an R-chain map $s \colon E \to D$ such that $p \circ s = \mathrm{id}_E$ holds and $i \oplus s \colon C \oplus E \xrightarrow{\cong} D$ is an R-chain isomorphism.

Proof. By assumption we can choose for each $n \in \mathbb{Z}$ an R-map $t_n \colon E_n \to D_n$ with $p_n \circ t_n = \mathrm{id}_{E_n}$. Let $\epsilon \colon E \to E$ be a chain contraction of E. Define $s_n \colon E_n \to D_n$ by $s_n = d_{n+1} \circ t_{n+1} \circ \epsilon_n + t_n \circ \epsilon_{n-1} \circ e_n$. The collection of the s_n-s defines an R-chain map $s \colon E \to D$ satisfying $p \circ s = \mathrm{id}_E$ by the following calculations

$$d_n \circ s_n = d_n \circ d_{n+1} \circ t_{n+1} \circ \epsilon_n + d_n \circ t_n \circ \epsilon_{n-1} \circ e_n$$
$$= d_n \circ t_n \circ \epsilon_{n-1} \circ e_n$$
$$= d_n \circ t_n \circ \epsilon_{n-1} \circ e_n + t_{n-1} \circ \epsilon_{n-2} \circ e_{n-1} \circ e_n$$
$$= s_{n-1} \circ e_n$$

and

$$p_n \circ s_n = p_n \circ d_{n+1} \circ t_{n+1} \circ \epsilon_n + p_n \circ t_n \circ \epsilon_{n-1} \circ e_n$$
$$= d_n \circ p_{n+1} \circ t_{n+1} \circ \epsilon_n + p_n \circ t_n \circ \epsilon_{n-1} \circ e_n$$
$$= d_n \circ \mathrm{id}_{E_{n+1}} \circ \epsilon_n + \mathrm{id}_{E_n} \circ \epsilon_{n-1} \circ e_n$$
$$= d_n \circ \epsilon_n + \epsilon_{n-1} \circ e_n$$
$$= \mathrm{id}_{E_n}.$$

Obviously we obtain an R-chain isomorphism $i \oplus s \colon C \oplus E \xrightarrow{\cong} D$. □

Proposition 14.55 *Consider the commutative diagram of R-chain complexes*

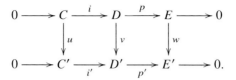

Assume that the rows are in each degree short split exact sequences of R-modules. Suppose that two of the R-chain maps u, v, and w are R-chain homotopy equivalences. Then all three R-chain maps u, v, and w are R-chain homotopy equivalences.

Proof. Because of Proposition 14.52 (i) it suffices to show for a sequence of R-chain complexes $0 \to C \xrightarrow{i} D \xrightarrow{p} E \to 0$ that is in each degree a split exact sequence of R-modules that all three R-chain complexes C, D, and E are contractible if two of them are.

The case where C and E are contractible follows from Lemma 14.54.

Now suppose that C and D are contractible. We have to show that E is contractible. We have the two short exact sequences of R-chain complexes $0 \to \Sigma C \to \text{cone}(p) \to \text{cone}(\text{id}_E) \to 0$ and $0 \to D \to \text{cyl}(p) \xrightarrow{q} \text{cone}(p) \to 0$. There is an obvious chain contraction for $\text{cone}(\text{id}_E)$. Since C and $\text{cone}(\text{id}_E)$ are contractible, $\text{cone}(p)$ is contractible by the case we have already dealt with. Since D and $\text{cone}(p)$ are contractible, $\text{cyl}(p)$ is contractible by the case we have already dealt with. Since $\text{cyl}(p)$ and E are R-chain homotopy equivalent, E is contractible.

It remains to treat the case where D and E are contractible and we must show that C is contractible. If we apply the second case, which we have already proved, to $0 \to D \to \text{cyl}(p) \xrightarrow{q} \text{cone}(p) \to 0$ and observe that D and $\text{cyl}(p) \simeq E$ are contractible, we conclude that $\text{cone}(p)$ is contractible. Since $0 \to \Sigma C \to \text{cone}(p) \to \text{cone}(\text{id}_E) \to 0$ is degreewise split exact and $\text{cone}(\text{id}_E)$ is contractible, we get from Lemma 14.54 an R-chain isomorphism $\Sigma C \oplus \text{cone}(\text{id}_E) \to \text{cone}(p)$ and hence an R-chain homotopy equivalence $\Sigma C \to \text{cone}(p)$. Hence C is contractible. This finishes the proof of Proposition 14.55. □

Exercise 14.56 Let $f, f' \colon C \to D$ be R-chain maps and let $g \colon f \simeq f'$ be an R-chain homotopy from f to f'. Then the induced map

$$z = z(e', -g \circ e' + h') \colon \text{cone}(f) \to \text{cone}(f')$$

is an isomorphism of R-chain complexes such that $e' \circ z = e$ holds, where $e \colon D \to \text{cone}(f)$ and $e' \colon D \to \text{cone}(f')$ are the canonical chain maps and $h' \colon C \to \text{cone}(f')$ is the canonical chain homotopy $h' \colon 0 \simeq e' \circ f'$.

14.6.4 Cofibrations and Homotopy Cofibrations

An R-chain map $f \colon C \to D$ is a *cofibration* if it is degreewise a split monomorphism.

Exercise 14.57 Consider a pushout of R-chain complexes

$$\begin{array}{ccc} C & \xrightarrow{f} & D \\ {\scriptstyle g}\downarrow & & \downarrow{\scriptstyle \bar{g}} \\ E & \xrightarrow[\bar{f}]{} & F \end{array}$$

such that f is a cofibration. Show:

(i) The R-chain map $\bar{f}$ is a cofibration;
(ii) The R-chain map g is an R-chain homotopy equivalence if and only if the R-chain map $\bar{g}$ is an R-chain homotopy equivalence.

Exercise 14.58 Let $f \colon C \to D$ be an R-chain map. Show that f is a cofibration if and only if it has the following homotopy extension property: If $u \colon C \to E$ and $v \colon D \to E$ are R-chain maps and $h \colon v \circ f \simeq u$ is an R-chain homotopy of R-chain maps $C \to E$, then there exist an R-chain map $u' \colon D \to E$ and an R-chain homotopy $h' \colon v \simeq u'$ of R-chain maps $D \to E$ such that $u' \circ f = u$ and $h' \circ f = h$ hold. (Compare with the homotopy extension property for cofibrations of spaces, e.g., in [45, Chapter VII.1].)

Lemma 14.59 *Consider two pushouts of R-chain complexes*

$$\begin{array}{ccc} C & \xrightarrow{f} & D \\ {\scriptstyle g}\downarrow & & \downarrow{\scriptstyle \bar{g}} \\ E & \xrightarrow[\bar{f}]{} & F \end{array} \qquad \begin{array}{ccc} C' & \xrightarrow{f'} & D' \\ {\scriptstyle g'}\downarrow & & \downarrow{\scriptstyle \bar{g}'} \\ E' & \xrightarrow[\bar{f}']{} & F' \end{array}$$

and three R-chain homotopy equivalences $u \colon C \xrightarrow{\simeq} C'$, $v \colon D \xrightarrow{\simeq} D'$, and $w \colon E \xrightarrow{\simeq} E'$ satisfying $v \circ f = f' \circ u$ and $w \circ g = g' \circ u$. Let $z \colon F \to F'$ be the unique R-chain map coming from the pushout property. Suppose that f and f' are cofibrations.

Then z is an R-chain homotopy equivalence.

Proof. Since f and f' are cofibrations, we get a commutative diagram of R-chain complexes such that the rows are degreewise split exact

14.6 Homotopy Theory of Chain Complexes

$$
\begin{array}{ccccccccc}
0 & \longrightarrow & C & \xrightarrow{\binom{f}{-g}} & D \oplus E & \xrightarrow{(\bar{g}\ \bar{f})} & F & \longrightarrow & 0 \\
& & \downarrow u & & \downarrow v \oplus w & & \downarrow z & & \\
0 & \longrightarrow & C' & \xrightarrow{\binom{f'}{-g'}} & D' \oplus E & \xrightarrow{(\bar{g'}\ \bar{f'})} & F' & \longrightarrow & 0.
\end{array}
$$

Now apply Proposition 14.55. $\qquad\square$

The mapping cylinder can be used to replace an arbitrary chain map by a cofibration, see (14.42), since $k: C \to \mathrm{cyl}(f)$ is a cofibration.

Lemma 14.60 *Consider a diagram of R-chain complexes*

$$
\begin{array}{ccc}
C & \xrightarrow{f} & D \\
\downarrow u & & \downarrow v \\
C' & \xrightarrow{f'} & D'
\end{array}
$$

together with an R-chain homotopy $h: f' \circ u \simeq v \circ f$.

(i) *By the universal property of the mapping cone, see Remark 14.45, we get an R-chain map $z(u, v, h): \mathrm{cone}(f) \to \mathrm{cone}(f')$;*
(ii) *If two of the three R-chain maps u, v, and $z(u, v, h)$ are R-chain homotopy equivalences, then all three are R-chain homotopy equivalences.*

Proof. (i) We have the R-chain map $v': D \xrightarrow{v} D' \xrightarrow{l'} \mathrm{cone}(f')$ for the canonical inclusion l'. The R-chain homotopy $h: f' \circ u \simeq v \circ f$ and the canonical R-chain homotopy $0 \simeq l' \circ f'$ together yield an R-chain homotopy $h': 0 \simeq v' \circ f$. Let $z = z(u, v, h): \mathrm{cone}(f) \to \mathrm{cone}(f')$ be the R-chain map associated to (v', h') by the universal property of $\mathrm{cone}(f')$ of Remark 14.45. It is explicitly given by $\begin{pmatrix} u_{n-1} & 0 \\ h_{n-1} & v_n \end{pmatrix}: C_{n-1} \oplus D_n \to C'_{n-1} \oplus D'_n$.

(ii) There is an obvious diagram of R-chain complexes

$$
\begin{array}{ccccccccc}
0 & \longrightarrow & D & \longrightarrow & \mathrm{cone}(f) & \longrightarrow & \Sigma C & \longrightarrow & 0 \\
& & \downarrow v & & \downarrow z & & \downarrow \Sigma u & & \\
0 & \longrightarrow & D' & \longrightarrow & \mathrm{cone}(f') & \longrightarrow & \Sigma C' & \longrightarrow & 0.
\end{array}
$$

Now the claim follows from Proposition 14.55. $\qquad\square$

Definition 14.61 (Cofibration and homotopy cofibration sequence) We call a sequence of R-chain complexes $C \xrightarrow{f} D \xrightarrow{g} E$ a *cofibration sequence* if f is a cofibration and the sequence $0 \to C \xrightarrow{f} D \xrightarrow{g} E \to 0$ is exact.

A *homotopy cofibration sequence* consists of a sequence of R-chain maps $C \xrightarrow{f} D \xrightarrow{g} E$ together with a nullhomotopy $h \colon 0 \simeq g \circ f$ such that the induced map $z \colon \mathrm{cone}(f) \to E$, see Remark 14.45, is an R-chain homotopy equivalence.

The mapping cone $\mathrm{cone}(f)$ is sometimes also called the *homotopy cofibre* of f.

Exercise 14.62 Show that a cofibration sequence $0 \to C \xrightarrow{f} D \xrightarrow{g} E \to 0$ is a homotopy cofibration sequence with respect to the trivial R-chain homotopy $0 \colon 0 \simeq g \circ f$.

Exercise 14.63 Construct an example of a sequence of R-chain complexes $C \xrightarrow{f} D \xrightarrow{g} E$ together with two R-chain homotopies $h, h' \colon 0 \simeq g \circ f$ such that the sequence is a homotopy cofibration sequence for h but not for h'.

Lemma 14.64 (i) *Consider the (not necessarily strictly commutative) diagram of R-chain maps*

$$\begin{array}{ccccc} C & \xrightarrow{f} & D & \xrightarrow{g} & E \\ \simeq \downarrow u & & \simeq \downarrow v & & \simeq \downarrow w \\ C' & \xrightarrow{f'} & D' & \xrightarrow{g'} & E'. \end{array}$$

Suppose that u, v and w are R-chain homotopy equivalences. Consider R-chain homotopies $h \colon 0 \simeq g \circ f$, $h' \colon 0 \simeq g' \circ f'$, $k \colon f' \circ u \simeq v \circ f$ and $k' \colon g' \circ v \simeq w \circ g$. Suppose that these are compatible in the sense that

$$w \circ h = h' \circ u + g' \circ k + k' \circ f$$

holds.

Then $C \xrightarrow{f} D \xrightarrow{g} E$ together with h is a homotopy cofibration sequence if and only if $C' \xrightarrow{f'} D' \xrightarrow{g'} E'$ together with h' is a homotopy cofibration sequence;

(ii) *Consider the sequence of R-chain maps $C \xrightarrow{f} D \xrightarrow{g} E$ together with two R-chain homotopies $h, h' \colon 0 \simeq g \circ f$. Suppose that h and h' are R-chain homotopic, i.e., there exists a map $k \colon C \to E$ of degree 2 satisfying $e_{n+2} \circ k_n - k_{n-1} \circ c_n = h'_n - h_n$. Then $C \xrightarrow{f} D \xrightarrow{g} E$ together with h is a homotopy cofibration sequence if and only if $C \xrightarrow{f} D \xrightarrow{g} E$ together with h' is a homotopy cofibration sequence.*

Proof. (i) Let $z(g, h) \colon \mathrm{cone}(f) \to E$ and $z(g', h') \colon \mathrm{cone}(f') \to E'$ be the R-chain maps given by the universal property of the mapping cone, see Remark 14.45. Note that we have to show that $z(g, h)$ is an R-chain homotopy equivalence if and only if $z(g', h')$ is an R-chain homotopy equivalence. Let $z(u, v, k) \colon \mathrm{cone}(f) \to \mathrm{cone}(f')$ be the R-chain map associated to the (not necessarily strictly commutative) diagram of R-chain complexes

14.6 Homotopy Theory of Chain Complexes

$$\begin{array}{ccc} C & \xrightarrow{f} & D \\ u \downarrow & & \downarrow v \\ C' & \xrightarrow{f'} & D' \end{array}$$

together with an R-chain homotopy $k \colon f' \circ u \simeq v \circ f$, see Lemma 14.60 (i). Since u and v are R-chain homotopy equivalences, $z(u, v, h)$ is an R-chain homotopy equivalence by Lemma 14.60 (ii). We get explicitly in degree n

$$\begin{array}{ccc} \mathrm{cone}(f)_n = C_{n-1} \oplus D_n & \xrightarrow{z(g,h)_n = (h_{n-1}\ g_n)} & E_n \\ z(u,v,k)_n = \begin{pmatrix} u_{n-1} & 0 \\ k_{n-1} & v_n \end{pmatrix} \downarrow & & \downarrow w_n \\ \mathrm{cone}(f')_n = C'_{n-1} \oplus D'_n & \xrightarrow{z(g',h')_n = (h'_{n-1}\ g'_n)} & E'_n. \end{array}$$

We get an R-chain homotopy $l \colon z(g', h') \circ z(u, v, k) \simeq w \circ z(g, h)$ if we put $l_n = (0 \ k'_n) \colon \mathrm{cone}(f)_n = C_{n-1} \oplus D_n \to E'_{n+1}$ because of the following calculation

$$w_n \circ z(g, h)_n - z(g', h')_n \circ z(u, v, k)_n$$
$$= w_n \circ (h_{n-1}\ g_n) - (h'_{n-1}\ g'_n) \circ \begin{pmatrix} u_{n-1} & 0 \\ k_{n-1} & v_n \end{pmatrix}$$
$$= ((w_n \circ h_{n-1} - h'_{n-1} \circ u_{n-1} - g'_n \circ k_{n-1})\ (w_n \circ g_n - g'_n \circ v_n))$$
$$= ((k'_{n-1} \circ f_{n-1})\ (k'_{n-1} \circ d_n + e'_n \circ k'_{n-1}))$$
$$= (0\ k'_{n-1}) \circ \begin{pmatrix} -c_{n-1} & 0 \\ f_{n-1} & d_n \end{pmatrix} + e'_n \circ (0\ k'_{n-1}).$$

Since w and $z(u, v, k)_n$ are R-chain homotopy equivalences, $z(g, h)$ is an R-chain homotopy equivalence if and only if $z(g', h')$ is an R-chain homotopy equivalence.

(ii) The difference $z(g, h') - z(g, h) \colon \mathrm{cone}(f) \to E$ is given in degree n by $(h'_{n-1} - h_{n-1}\ g_n) \colon C_{n-1} \oplus D_n \to E_n$. Hence $z(g, h')$ and $z(g, h)$ are R-chain homotopic. We conclude that the desired R-chain homotopy is given by $(k_{n-1}\ 0) \colon C_{n-1} \oplus D_n \to E_{n+1}$. $\square$

Exercise 14.65 Consider a sequence of R-chain complexes $C \xrightarrow{f} D \xrightarrow{g} E$. Show that the following statements are equivalent:

(i) There exists a nullhomotopy $h \colon 0 \simeq g \circ f$ such that $C \xrightarrow{f} D \xrightarrow{g} E$ together with h is a homotopy cofibration;

(ii) There exists a cofibration sequence $C' \xrightarrow{f'} D' \xrightarrow{g'} E'$ and a strictly commutative square of R-chain complexes

$$
\begin{array}{ccccc}
C' & \xrightarrow{f'} & D' & \xrightarrow{g'} & E' \\
{\scriptsize \simeq}\downarrow{\scriptstyle u} & & {\scriptsize \simeq}\downarrow{\scriptstyle v} & & {\scriptsize \simeq}\downarrow{\scriptstyle w} \\
C & \xrightarrow{f} & D & \xrightarrow{g} & E
\end{array}
$$

whose vertical arrows are R-chain homotopy equivalences.

14.6.5 Homotopy Pushouts

Consider the diagram of R-chain maps

$$ E \xleftarrow{g} C \xrightarrow{f} D. $$

The *homotopy pushout* $D\widetilde{\cup}_C E$ is defined to be the mapping cone of the R-chain map $\begin{pmatrix} f \\ -g \end{pmatrix} \colon C \to D \oplus E$. The n-differential of the homotopy pushout $D\widetilde{\cup}_C E$ is
$$\begin{pmatrix} -c_{n-1} & 0 & 0 \\ f_{n-1} & d_n & 0 \\ -g_{n-1} & 0 & e_n \end{pmatrix} \colon C_{n-1} \oplus D_n \oplus E_n \to C_{n-2} \oplus D_{n-1} \oplus E_{n-1}.$$ We obtain a (not necessarily strictly commutative) square of R-chain complexes

$$
\begin{array}{ccc}
C & \xrightarrow{f} & D \\
g \downarrow & & \downarrow \overline{g} \\
E & \xrightarrow{\overline{f}} & D\widetilde{\cup}_C E
\end{array}
$$

together with a preferred R-chain homotopy $h \colon \overline{f} \circ g \simeq \overline{g} \circ f$ from Remark 14.45. Explicitly the R-maps $\overline{g}_n \colon D \to C_{n-1} \oplus D_n \oplus E_n$, $\overline{f}_n \colon E \to C_{n-1} \oplus D_n \oplus E_n$, and $h_n \colon C_{n-1} \to C_{n-1} \oplus D_n \oplus E_n$ are given by the canonical inclusions for $n \in \mathbb{Z}$.

Remark 14.66 (Universal property of the homotopy pushout) The homotopy pushout has the following universal property by Remark 14.45. Let $u \colon D \to Z$ and $v \colon E \to Z$ be R-chain maps and let $h' \colon v \circ g \simeq u \circ f$ be an R-chain homotopy. Then there exists precisely one R-chain map $z = z(u, v, h') \colon D\widetilde{\cup}_C E \to Z$ such that $u = z \circ \overline{g}$, $v = z \circ \overline{f}$ and $h' = z \circ h$ hold. Explicitly z_n is given by $(h'_{n-1}\ u_n\ v_n) \colon C_{n-1} \oplus D_n \oplus E_n \to Z_n$.

Conversely, given an R-chain map $z \colon D\widetilde{\cup}_C E \to Z$, we obtain an R-chain map $u \colon D \to Z$ by $z \circ \overline{f}$, an R-chain map $v \colon E \to Z$ by $z \circ \overline{g}$, and an R-chain homotopy $h' \colon v \circ g \simeq u \circ f$ by $h' = z \circ h$ satisfying $z = z(u, v, h')$.

14.6.6 Homotopy Cocartesian Squares

Consider a diagram of R-chain complexes

$$\begin{array}{ccc} C & \xrightarrow{u} & D \\ v \downarrow & & \downarrow v' \\ E & \xrightarrow{u'} & F \end{array} \qquad (14.67)$$

together with an R-chain homotopy $h \colon u' \circ v \simeq v' \circ u$.

We have the R-chain map $l' \circ v' \colon D \to \mathrm{cone}(u')$ for the canonical inclusion $l' \colon F \to \mathrm{cone}(u')$. The R-chain homotopy $h \colon u' \circ v \simeq v' \circ u$ and the canonical R-chain homotopy $0 \simeq l' \circ u'$ together yield an R-chain homotopy $h' \colon 0 \simeq (l' \circ v') \circ u$. Let

$$a \colon \mathrm{cone}(u) \to \mathrm{cone}(u')$$

be the R-chain map associated to $(l' \circ v', h')$ by the universal property of $\mathrm{cone}(u')$ of Remark 14.45. It is explicitly given by $\begin{pmatrix} v_{n-1} & 0 \\ h_{n-1} & v'_n \end{pmatrix} \colon C_{n-1} \oplus D_n \to E_{n-1} \oplus F_n$. Analogously we get an R-chain map

$$b \colon \mathrm{cone}(v) \to \mathrm{cone}(v')$$

by $\begin{pmatrix} u_{n-1} & 0 \\ -h_{n-1} & u'_n \end{pmatrix} \colon C_{n-1} \oplus E_n \to D_{n-1} \oplus F_n$. Moreover, there is an R-chain map

$$z \colon D\tilde{\cup}_C E = \mathrm{cone}\left(\begin{pmatrix} u \\ -v \end{pmatrix} \colon C \to D \oplus E\right) \to F$$

given by $(h_{n-1}, v'_n, u'_n) \colon C_{n-1} \oplus D_n \oplus E_n \to F_n$.

Lemma 14.68 *The following statements are equivalent:*

(i) *The sequence* $C \xrightarrow{\begin{pmatrix} u \\ -v \end{pmatrix}} D \oplus E \xrightarrow{(v'\ u')} F$ *together with the R-chain nullhomotopy* $0 \simeq (v'\ u') \circ \begin{pmatrix} u \\ -v \end{pmatrix}$ *induced by h is a homotopy cofibration sequence;*

(ii) *The canonical R-chain map $z \colon D\tilde{\cup}_C E \to F$ is an R-chain homotopy equivalence;*

(iii) *The canonical map $a \colon \mathrm{cone}(u) \to \mathrm{cone}(u')$ is an R-chain homotopy equivalence;*

(iv) *The canonical map $b \colon \mathrm{cone}(v) \to \mathrm{cone}(v')$ is an R-chain homotopy equivalence.*

Proof. Assertions (i) and (ii) are equivalent by definition. It remains to show that assertions (ii), (iii), and (iv) are equivalent.

The n-th differential of cone(a) is explicitly given by

$$\begin{pmatrix} c_{n-2} & 0 & 0 & 0 \\ -u_{n-2} & -d_{n-1} & 0 & 0 \\ v_{n-2} & 0 & -e_{n-1} & 0 \\ h_{n-2} & v'_{n-1} & u'_{n-1} & f_n \end{pmatrix} : C_{n-2} \oplus D_{n-1} \oplus E_{n-1} \oplus F_n$$
$$\to C_{n-2} \oplus D_{n-2} \oplus E_{n-2} \oplus F_{n-1}.$$

The n-th differential of cone(b) is explicitly given by

$$\begin{pmatrix} c_{n-2} & 0 & 0 & 0 \\ -v_{n-2} & -e_{n-1} & 0 & 0 \\ u_{n-2} & 0 & -d_{n-1} & 0 \\ -h_{n-2} & u'_{n-1} & v'_{n-1} & f_n \end{pmatrix} : C_{n-2} \oplus E_{n-1} \oplus D_{n-1} \oplus F_n$$
$$\to C_{n-2} \oplus E_{n-2} \oplus D_{n-2} \oplus F_{n-1}.$$

The n-th differential of cone(z) is explicitly given by

$$\begin{pmatrix} c_{n-2} & 0 & 0 & 0 \\ -u_{n-2} & -d_{n-1} & 0 & 0 \\ v_{n-2} & 0 & -e_{n-1} & 0 \\ h_{n-2} & v'_{n-1} & u'_{n-1} & f_n \end{pmatrix} : C_{n-2} \oplus D_{n-1} \oplus E_{n-1} \oplus F_n$$
$$\to C_{n-2} \oplus D_{n-2} \oplus E_{n-2} \oplus F_{n-1}.$$

Obviously we have cone(a) = cone(z). There is an R-chain isomorphism cone(b) $\xrightarrow{\cong}$ cone(z) that is given in degree n by

$$\begin{pmatrix} -\mathrm{id} & 0 & 0 & 0 \\ 0 & 0 & \mathrm{id} & 0 \\ 0 & \mathrm{id} & 0 & 0 \\ 0 & 0 & 0 & \mathrm{id} \end{pmatrix} : C_{n-2} \oplus E_{n-1} \oplus D_{n-1} \oplus F_n \xrightarrow{\cong} C_{n-2} \oplus D_{n-1} \oplus E_{n-1} \oplus F_n.$$

Now Lemma 14.68 follows from Proposition 14.52 (i). □

We call the diagram (14.67) together with the R-chain homotopy h *homotopy cocartesian* if one of the equivalent conditions appearing in Proposition 14.52 is satisfied.

Exercise 14.69 Give an example of a square (14.67) together with two R-chain homotopies $h, h' : u' \circ v \simeq v' \circ u$ such that it is homotopy cocartesian for h but not for h'.

14.6 Homotopy Theory of Chain Complexes

Exercise 14.70 Let

$$\begin{array}{ccc} C & \xrightarrow{f} & D \\ g \downarrow & & \downarrow \overline{g} \\ E & \xrightarrow{\overline{f}} & F \end{array}$$

be a homotopy cocartesian square. Show that then

$$\begin{array}{ccc} C & \xrightarrow{f \oplus g} & D \oplus E \\ \overline{g} \circ f \downarrow & & \downarrow \overline{f} \oplus \overline{g} \\ F & \xrightarrow{\Delta} & F \oplus F \end{array}$$

where Δ is the diagonal map, is also a homotopy cocartesian square.

14.6.7 Pullbacks

Next we introduce the dual notions of pushout, homotopy pushout, cofibration, homotopy cofibration, and homotopy cofibre, namely of pullback, homotopy pullback, fibration, homotopy cofibration, and homotopy fibre.

Given a diagram of R-chain complexes $D \xrightarrow{f} F \xleftarrow{g} E$, we can complete it to a commutative square of R-chain complexes, the so-called *pullback*

$$\begin{array}{ccc} D \times_F E & \xrightarrow{\overline{g}} & D \\ \overline{f} \downarrow & & \downarrow f \\ E & \xrightarrow{g} & F \end{array}$$

with the following universal property: Given R-chain maps $u \colon A \to D$, $v \colon A \to E$ and $w \colon A \to F$ satisfying $f \circ u = g \circ v = w$, then there exists precisely one R-chain map $z \colon A \to D \times_F E$ satisfying $\overline{f} \circ z = v$ and $\overline{g} \circ z = u$. A commutative square of R-chain complexes

$$\begin{array}{ccc} C & \xrightarrow{\overline{g}} & D \\ \overline{f} \downarrow & & \downarrow f \\ E & \xrightarrow{g} & F \end{array}$$

is a pullback if and only if the sequence of R-chain complexes

$$0 \to C \xrightarrow{\binom{\bar{g}}{f}} D \times E \xrightarrow{(-f\ g)} F$$

is exact. The pullback can be constructed to be the kernel of the R-chain map $(-f\ g) : D \oplus E \to F$.

The pullback is not homotopy invariant. Therefore we will introduce the notion of a homotopy pullback later.

14.6.8 Homotopy Fibres

The homotopy fibre $\operatorname{hofib}(f)$ of a map of topological spaces $f : X \to Y$ over $y \in Y$ is defined to be the pullback of spaces

$$\begin{array}{ccc} \operatorname{hofib}(f) & \xrightarrow{\bar{f}} & \operatorname{map}([0,1], Y) \\ {\scriptstyle \bar{e}}\downarrow & & \downarrow{\scriptstyle e} \\ X & \xrightarrow[c_y \times f]{} & Y \times Y \end{array}$$

where $\operatorname{ev}_k : \operatorname{map}([0,1], Y) \to Y$ is the evaluation at k, i.e., $w \mapsto w(k)$, $e = \operatorname{ev}_0 \times \operatorname{ev}_1$, and c_y is the constant map with value y. This leads to the definition of the *homotopy fibre* $\operatorname{hofib}(f)$ of an R-chain map $f : C \to D$ to be the pullback of R-chain complexes

$$\begin{array}{ccc} \operatorname{hofib}(f) & \xrightarrow{\bar{f}} & \hom_{\mathbb{Z}}(I, D) \\ {\scriptstyle \bar{e}}\downarrow & & \downarrow{\scriptstyle e} \\ C & \xrightarrow[0 \times f]{} & D \times D \end{array} \qquad (14.71)$$

where $\operatorname{ev}_k : \hom_{\mathbb{Z}}(I, D) \to D$ is given by composition with $i_k : 0[\mathbb{Z}] \to I$ and the canonical identification $\hom_{\mathbb{Z}}(0[\mathbb{Z}], C) = C$ for $k = 0, 1$ and $e = \operatorname{ev}_0 \times \operatorname{ev}_1$.

The n-th differential of $\hom_{\mathbb{Z}}(I, D)$ can be identified with

$$\begin{pmatrix} d_n & 0 & 0 \\ 0 & d_n & 0 \\ (-1)^n \cdot \operatorname{id} & (-1)^{n+1} \cdot \operatorname{id} & d_{n+1} \end{pmatrix} : D_n \oplus D_n \oplus D_{n+1} \to D_{n-1} \oplus D_{n-1} \oplus D_n.$$

The n-differential of $\operatorname{hofib}(f)$ is given by

$$\begin{pmatrix} c_n & 0 \\ (-1)^{n+1} f_n & d_{n+1} \end{pmatrix} : C_n \oplus D_{n+1} \to C_{n-1} \oplus D_n. \qquad (14.72)$$

14.6 Homotopy Theory of Chain Complexes

In degree n Diagram (14.71) becomes

$$\begin{array}{ccc} C_n \oplus D_{n+1} & \xrightarrow{(1)} & D_n \oplus D_n \oplus D_{n+1} \\ {\scriptsize \begin{pmatrix} \mathrm{id} & 0 \end{pmatrix}} \downarrow & & \downarrow {\scriptsize \begin{pmatrix} \mathrm{id} & 0 & 0 \\ 0 & \mathrm{id} & 0 \end{pmatrix}} \\ C_n & \xrightarrow{(2)} & D_n \times D_n \end{array}$$

where the top and the bottom horizontal arrows are given by the matrices $(1) = \begin{pmatrix} 0 & 0 \\ f_n & 0 \\ 0 & \mathrm{id} \end{pmatrix}$ and $(2) = \begin{pmatrix} 0 \\ f_n \end{pmatrix}$.

Remark 14.73 (Universal property of the homotopy fibre) The homotopy fibre has the following universal property. The R-chain map $\overline{e}\colon \mathrm{hofib}(f) \to C$ is explicitly given in degree n by the canonical projection $C_n \oplus D_{n+1} \to C_n$. There is a preferred nullhomotopy $h\colon 0 \simeq f \circ \overline{e}$ given in degree n by $(0 \; (-1)^n \cdot \mathrm{id})\colon C_n \oplus D_{n+1} \to D_{n+1}$.

Let $u\colon A \to C$ be an R-chain map and let $h'\colon 0 \simeq f \circ u$ be an R-chain homotopy. Then there is precisely one R-chain map $z = z(u, h')\colon A \to \mathrm{hofib}(f)$ such that $u = \overline{e} \circ z$ and $h' = h \circ z$ holds, namely, define $z_n = \begin{pmatrix} u_n \\ (-1)^n \cdot h'_n \end{pmatrix}\colon A_n \to C_n \oplus D_{n+1}$.

Conversely, given an R-chain map $z\colon A \to \mathrm{hofib}(f)$, we get an R-chain map $u\colon A \to C$ by $u = \overline{e} \circ z$ and an R-chain homotopy $h'\colon 0 \simeq f \circ u$ by $h' = h \circ z$ such that $z = z(u, h')$.

The underlying modules in dimension n of $\Sigma^{-1}\mathrm{cone}(f)$ and $\mathrm{hofib}(f)$ agree, but the differentials differ by signs. This is due to our convention which defines the mapping cone in terms of $I \otimes C$ instead of $C \otimes I$. We obtain an isomorphism of R chain complexes

$$w\colon \Sigma^{-1}\mathrm{cone}(f) \to \mathrm{hofib}(f) \tag{14.74}$$

by $w_n = \begin{pmatrix} \mathrm{id} & 0 \\ 0 & (-1)^{n+1} \cdot \mathrm{id} \end{pmatrix}\colon C_n \oplus D_{n+1} \to C_n \oplus D_{n+1}$.

Definition 14.75 (Fibration and homotopy fibration sequence) An R-chain map $f\colon C \to D$ is a *fibration* if it is degreewise split surjective.

We call a sequence of R-chain complexes $C \xrightarrow{f} D \xrightarrow{g} E$ a *fibration sequence* if g is a fibration and the sequence $0 \to C \xrightarrow{f} D \xrightarrow{g} E \to 0$ is exact.

A *homotopy fibration sequence* consists of a sequence of R-chain maps $C \xrightarrow{f} D \xrightarrow{g} E$ together with a nullhomotopy $h\colon 0 \simeq g \circ f$ such that the induced map $z\colon C \to \mathrm{hofib}(g)$, see Remark 14.73, is an R-chain homotopy equivalence.

Every fibration sequence is a homotopy fibration sequence. Every cofibration sequence is a fibration sequence and vice versa. There exists a sequence of R-chain complexes $C \xrightarrow{f} D \xrightarrow{g} E$ together with a nullhomotopies $h, h'\colon 0 \simeq g \circ f$ such that the sequence is a homotopy fibration sequence with respect to h but not with respect to h'.

We have already mentioned that the pullback of R-chain complexes is not homotopy invariant, i.e., the following statement is *not* true: If we have two pullbacks of R-chain complexes

$$\begin{array}{ccc} C & \xrightarrow{\overline{g}} & D \\ \overline{f} \downarrow & & \downarrow f \\ E & \xrightarrow{g} & F \end{array} \qquad \begin{array}{ccc} C' & \xrightarrow{\overline{g'}} & D' \\ \overline{f'} \downarrow & & \downarrow f' \\ E' & \xrightarrow{g'} & F' \end{array}$$

and three R-chain homotopy equivalences $u\colon D \xrightarrow{\simeq} D'$, $v\colon E \xrightarrow{\simeq} E'$, and $w\colon F \xrightarrow{\simeq} F'$ satisfying $w \circ f = f' \circ u$ and $w \circ g = g' \circ v$, and $z\colon C \to C'$ is the unique R-chain map coming from the pullback property, then z is an R-chain homotopy equivalence. However, this statement is true, provided that f and f' are fibrations. We leave the elementary proof to the reader.

Exercise 14.76 Consider the commutative square of R-chain complexes

$$\begin{array}{ccc} C & \xrightarrow{f} & D \\ \kappa \downarrow & & \downarrow \lambda \\ C' & \xrightarrow{f'} & D'. \end{array}$$

Let A and A' be R-chain complexes with an R-chain map $\mu\colon A \to A'$ and let $\alpha\colon A \to C$ and $\alpha'\colon A' \to C'$ be R-chain maps with R-chain nullhomotopies $\beta\colon 0 \simeq f \circ \alpha\colon A \to D$ and $\beta'\colon 0 \simeq f' \circ \alpha'\colon A' \to D'$ so that there are induced R-chain maps $z(\alpha, \beta)\colon A \to \mathrm{hofib}(f)$ and $z(\alpha', \beta')\colon A' \to \mathrm{hofib}(f')$. Assume further that
$$\kappa \circ \alpha = \alpha' \circ \mu \quad \text{and} \quad \overline{\lambda}\colon \lambda \circ \beta \simeq \beta' \circ \mu.$$

Then the square

$$\begin{array}{ccc} A & \xrightarrow{z(\alpha,\beta)} & \mathrm{hofib}(f) \\ \mu \downarrow & & \downarrow \mathrm{hofib}(\kappa,\lambda) \\ A' & \xrightarrow{z(\alpha',\beta')} & \mathrm{hofib}(f') \end{array}$$

commutes up to an R-chain homotopy $z(\overline{\lambda})$ induced by $\overline{\lambda}$. If $\overline{\lambda}$ can be chosen to be the constant homotopy, i.e., if $\lambda \circ \beta = \beta' \circ \mu$, then the square commutes strictly.

Exercise 14.77 Consider two commutative squares of R-chain complexes with the same horizontal maps

$$\begin{array}{ccc} C & \xrightarrow{f} & D \\ \kappa_i \downarrow & & \downarrow \lambda_i \\ C' & \xrightarrow{f'} & D' \end{array}$$

for $i = 0, 1$. Assume there are chain homotopies $\mu \colon \kappa_0 \simeq \kappa_1$ and $\eta \colon \lambda_0 \simeq \lambda_1$ such that $\eta \circ f = f' \circ \mu$. Show that these data induce a chain homotopy

$$z(\mu, \eta) \colon z(\kappa_0, \lambda_0) \simeq z(\kappa_1, \lambda_1) \colon \mathrm{hofib}(f) \to \mathrm{hofib}(f').$$

14.6.9 Relating the Homotopy Cofibre and the Homotopy Fibre

Consider an R-chain map $f \colon C \to D$. Recall that we have associated to it the homotopy cofibration sequence $C \xrightarrow{f} D \xrightarrow{l} \mathrm{cone}(f)$ and the homotopy fibration sequence $\mathrm{hofib}(f) \xrightarrow{\overline{e}} C \xrightarrow{f} D$.

The n-th differential of the homotopy fibre $\mathrm{hofib}(l)$ of l is

$$d_{\mathrm{hofib}(l)} = \begin{pmatrix} d_n & 0 & 0 \\ 0 & -c_n & 0 \\ (-1)^{n+1} \cdot \mathrm{id} & f_n & d_{n+1} \end{pmatrix} \colon D_n \oplus C_n \oplus D_{n+1} \to D_{n-1} \oplus C_{n-1} \oplus D_n.$$

We obtain R-chain maps $a \colon C \to \mathrm{hofib}(l)$ and $b \colon \mathrm{hofib}(l) \to C$ by

$$a_n = \begin{pmatrix} f_n \\ (-1)^n \, \mathrm{id} \\ 0 \end{pmatrix} \colon C_n \to D_n \oplus C_n \oplus D_{n+1};$$

$$b_n = \begin{pmatrix} 0 & (-1)^n \, \mathrm{id} & 0 \end{pmatrix} \colon D_n \oplus C_n \oplus D_{n+1} \to C_n.$$

They are R-chain homotopy equivalences, R-chain homotopy inverse to one another, since $b_n \circ a_n = \mathrm{id}_C$ holds and there exists an R-chain homotopy $h \colon \mathrm{id}_{\mathrm{hofib}(l)} \simeq a_n \circ b_n$ given by

$$h_n = \begin{pmatrix} 0 & 0 & (-1)^n \cdot \mathrm{id} \\ 0 & 0 & 0 \\ 0 & 0 & 0 \end{pmatrix} \colon D_n \oplus C_n \oplus D_{n+1} \to D_{n+1} \oplus C_{n+1} \oplus D_{n+2}.$$

Conversely, the n-th differential of $\mathrm{cone}(\overline{e})$ is

$$\begin{pmatrix} -c_{n-1} & 0 & 0 \\ (-1)^{n-1} f_{n-1} & -d_n & 0 \\ \mathrm{id} & 0 & c_n \end{pmatrix} \colon C_{n-1} \oplus D_n \oplus C_n \to C_{n-2} \oplus D_{n-1} \oplus C_{n-1}.$$

We get R-chain maps $u\colon \operatorname{cone}(\overline{e}) \to D$ and $v\colon D \to \operatorname{cone}(\overline{e})$ by

$$u_n = \begin{pmatrix} 0 & (-1)^{n+1}\operatorname{id} & f_n \end{pmatrix} \colon C_{n-1} \oplus D_n \oplus C_n \to D_n;$$

$$v_n = \begin{pmatrix} 0 \\ (-1)^{n+1}\cdot \operatorname{id} \\ 0 \end{pmatrix} v_n \colon D_n \to C_{n-1} \oplus D_n \oplus C_n.$$

They are R-chain homotopy equivalences, R-chain homotopy inverse to one another, since $v_n \circ u_n = \operatorname{id}_D$ holds and there exists an R-chain homotopy $h\colon \operatorname{id}_{\operatorname{hofib}(l)} \simeq u_n \circ v_n$ given by

$$h_n = \begin{pmatrix} 0 & 0 & \operatorname{id} \\ 0 & 0 & 0 \\ 0 & 0 & 0 \end{pmatrix} \colon C_{n-1} \oplus D_n \oplus C_n \to C_n \oplus D_{n-1} \oplus C_{n+1}.$$

Theorem 14.78 (Homotopy cofibrations and homotopy fibrations agree for chain complexes) *Consider R-chain maps $f\colon C \to D$ and $g\colon D \to E$ together with an R-chain homotopy $h\colon 0 \simeq g \circ f$.*

Then $C \xrightarrow{f} D \xrightarrow{g} E$ is a homotopy cofibration if and only if it is a homotopy fibration.

Proof. Let the R-chain map $z(f,h)\colon C \to \operatorname{hofib}(g)$ be given by the universal property of the homotopy fibre, see Remark 14.73, and let the R-chain map $z(g,h)\colon E \to \operatorname{cone}(f)$ be given by the universal property of the mapping cone, see Remark 14.45. We have to show that $z(f,h)$ is an R-chain homotopy equivalence if and only if $z(g,h)$ is an R-chain homotopy equivalence.

Next we construct a commutative diagram of R-chain complexes

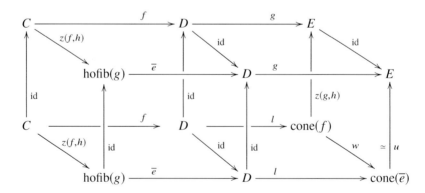

The R-chain map $u\colon \operatorname{cone}(\overline{e}) \to E$ has been constructed above. The R-chain map $w\colon \operatorname{cone}(f) \to \operatorname{cone}(\overline{e})$ is given by

$$w_n = \begin{pmatrix} f_{n-1} & 0 \\ (-1)^{n-1}\cdot h_{n-1} & 0 \\ 0 & \operatorname{id} \end{pmatrix} \colon C_{n-1} \oplus D_n \to D_{n-1} \oplus E_n \oplus D_n.$$

It agrees with the R-chain map appearing in Lemma 14.60 (i) applied to the commutative square

$$\begin{array}{ccc} C & \xrightarrow{f} & D \\ {\scriptstyle z(f,h)}\downarrow & & \downarrow{\scriptstyle \mathrm{id}} \\ \mathrm{hofib}(g) & \xrightarrow[e]{} & D \end{array}$$

and trivial R-chain homotopy from $0 \circ z(f, h) \simeq f$. Lemma 14.60 (ii) implies that $z(f, h)\colon C \to \mathrm{hofib}(g)$ is an R-chain homotopy equivalence if and only if w is an R-chain homotopy equivalence. Recall that u is an R-chain homotopy equivalence satisfying $u = z(g, h) \circ w$. Hence $z(g, h)\colon C \to \mathrm{hofib}(g)$ is an R-chain homotopy equivalence if and only if w is an R-chain homotopy equivalence. We conclude that $z(f, h)$ is an R-chain homotopy equivalence if and only if $z(g, h)$ is an R-chain homotopy equivalence. This finishes the proof of Theorem 14.78. □

Theorem 14.79 (Puppe sequences for chain complexes) *Consider the sequence of R-chain maps $C \xrightarrow{f} D \xrightarrow{g} E$ together with an R-chain homotopy $0 \simeq g \circ f$. Suppose that this gives a homotopy cofibration sequence, or, equivalently, a homotopy fibre sequence.*

Then there is an R-chain map $t\colon E \to \Sigma C$ such that for any R-chain complex Z we have in both directions infinite natural long exact sequences of abelian groups given by homotopy classes of R-chain maps

$$\cdots \xrightarrow{(\Sigma^2 f)^*} [\Sigma^2 C, Z] \xrightarrow{(-\Sigma t)^*} [\Sigma E, Z] \xrightarrow{(-\Sigma g)^*} [\Sigma D, Z]$$
$$\xrightarrow{(-\Sigma f)^*} [\Sigma C, Z] \xrightarrow{t^*} [E, Z] \xrightarrow{g^*} [D, Z] \xrightarrow{f^*} [C, Z] \xrightarrow{(-\Sigma^{-1} t)^*} [\Sigma^{-1} E, Z]$$
$$\xrightarrow{(-\Sigma^{-1} g)^*} [\Sigma^{-1} D, Z] \xrightarrow{(-\Sigma^{-1} f)^*} [\Sigma^{-1} C, Z] \xrightarrow{(\Sigma^{-2} t)^*} [\Sigma^{-2} E, Z] \xrightarrow{(\Sigma^{-2} g)^*} \cdots,$$

and

$$\cdots [Z, \Sigma^{-2} E] \xrightarrow{(\Sigma^{-2} t)_*} [Z, \Sigma^{-1} C] \xrightarrow{(-\Sigma^{-1} f)_*} [Z, \Sigma^{-1} D] \xrightarrow{(-\Sigma^{-1} g)_*}$$
$$[Z, \Sigma^{-1} E] \xrightarrow{(-\Sigma^{-1} t)_*} [Z, C] \xrightarrow{f_*} [Z, D] \xrightarrow{g_*} [Z, E] \xrightarrow{t_*} [Z, \Sigma C]$$
$$\xrightarrow{(-\Sigma f)_*} [Z, \Sigma D] \xrightarrow{(-\Sigma g)_*} [Z, \Sigma E] \xrightarrow{(-\Sigma t)_*} [Z, \Sigma^2 C] \xrightarrow{(\Sigma^2 f)_*} \cdots.$$

Proof. Since $C \xrightarrow{f} D \xrightarrow{g} E$ is a homotopy cofibration, we obtain a commutative diagram of R-chain complexes with R-chain homotopy equivalences as vertical arrows

$$\begin{array}{ccccc} C & \xrightarrow{f} & D & \xrightarrow{l} & \mathrm{cone}(f) \\ {\scriptstyle \mathrm{id}}\downarrow & & {\scriptstyle \mathrm{id}}\downarrow & & \simeq \downarrow{\scriptstyle z(g,h)} \\ C & \xrightarrow{f} & D & \xrightarrow{g} & E. \end{array}$$

The upper row induces by the universal property of the mapping cone, see Remark 14.45, a sequence $[\mathrm{cone}(f), Z] \xrightarrow{l^*} [D, Z] \xrightarrow{f^*} [C, Z]$ which is exact (in the middle). Hence we obtain an exact sequence

$$[E, Z] \xrightarrow{g^*} [D, Z] \xrightarrow{f^*} [C, Z]. \tag{14.80}$$

Next we construct a diagram of R-chain complexes where the squares are either strictly commutative or commutative up to R-chain homotopy:

$$\begin{array}{ccccccccc}
C & \xrightarrow{f} & D & \xrightarrow{g} & E & \xrightarrow{t} & \Sigma C & \xrightarrow{-\Sigma f} & \Sigma D \\
& & \uparrow \mathrm{id} & & \uparrow \simeq z(g,h) & & \uparrow \mathrm{id} & & \uparrow \mathrm{id} \\
& & D & \xrightarrow{l} & \mathrm{cone}(f) & \xrightarrow{r} & \Sigma C & & \\
& & \uparrow \mathrm{id} & & \uparrow \simeq z(r,0) & & & & \\
& & \mathrm{cone}(f) & \xrightarrow{l'} & \mathrm{cone}(l) & \xrightarrow{r'} & \Sigma D. & &
\end{array}$$

Since the R-chain map $z(g, h)$ is an R-chain homotopy equivalence, we can choose an R-chain homotopy inverse $z(g, h)^{-1}$ and define $t = r \circ z(g, h)^{-1}$.

The R-chain map

$$z(r, 0) \colon \mathrm{cone}(l) \to \Sigma C \tag{14.81}$$

is the induced map associated to the cofibration sequence $0 \to D \xrightarrow{l} \mathrm{cone}(f) \xrightarrow{r} \Sigma C \to 0$ associated to f of (14.46) and the trivial R-chain homotopy $0 \simeq r \circ l$. It is an R-chain homotopy equivalence either by Exercise 14.62, or by observing that it has a homotopy inverse

$$u \colon \Sigma C \to \mathrm{cone}(l), \quad u(x) = (-f(x), x, 0). \tag{14.82}$$

That u is indeed an R-chain homotopy inverse of $z(r, 0)$ is seen by observing that $z(r, 0) \circ u = \mathrm{id}_{\Sigma C}$ and that there is an R-chain homotopy $v \colon u \circ z(r, 0) \simeq \mathrm{id}_{\mathrm{cone}(l)}$ given by $v_n(x, y, z) = (z, 0, 0)$.

We also have the cofibration sequence $0 \to \mathrm{cone}(f) \xrightarrow{l'} \mathrm{cone}(l) \xrightarrow{r'} \Sigma D \to 0$ associated to l of (14.46).

It remains to show that the rightmost square commutes up to chain homotopy. The R-chain map $\mathrm{id} \circ r' - (-\Sigma f) \circ \mathrm{id} \circ z(r, 0) \colon \mathrm{cone}(l) \to \Sigma D$ is in degree n explicitly given by $(\mathrm{id}\ f_{n-1}\ 0) \colon D_{n-1} \oplus C_{n-1} \oplus D_n \to D_{n-1}$. Hence we get an R-chain homotopy $k \colon (-\Sigma f) \circ \mathrm{id} \simeq \mathrm{id} \circ r'$ given in degree n by $k_n = (0\ 0\ \mathrm{id}) \colon D_{n-1} \oplus C_{n-1} \oplus D_n \to D_n$. From the diagram above together with Lemma 14.64 (i), we conclude that

$$D \xrightarrow{g} E \xrightarrow{t} \Sigma C \tag{14.83}$$

and $E \xrightarrow{t} \Sigma C \xrightarrow{-\Sigma f} \Sigma D$ with appropriate choices of homotopies $0 \simeq t \circ g$ and $0 \simeq (-\Sigma f) \circ t$ are homotopy cofibration sequences. This is sometimes called the *rota-*

14.6 Homotopy Theory of Chain Complexes

tion trick. Hence we get from (14.80) the short exact sequences $[\Sigma C, Z] \xrightarrow{t^*} [E, Z] \xrightarrow{g^*} [D, Z]$ and $[\Sigma D, Z] \xrightarrow{(-\Sigma f)^*} [\Sigma C, Z] \xrightarrow{t^*} [E, Z]$. They give together with (14.80) the exact sequence

$$[\Sigma D, Z] \xrightarrow{(-\Sigma f)^*} [\Sigma C, Z] \xrightarrow{t^*} [E, Z] \xrightarrow{g^*} [D, Z] \xrightarrow{f^*} [C, Z].$$

Since $\Sigma^n C \xrightarrow{(-1)^n \Sigma^n f} \Sigma^n D \xrightarrow{(-1)^n \Sigma^n g} \Sigma^n E$ together with $\Sigma^n h$ is for every n a homotopy cofibration, we get by the same method an exact sequence

$$[\Sigma^{n+1} D, Z] \xrightarrow{((-1)^{n+1}\Sigma^{n+1} f)^*} [\Sigma^{n+1} C, Z] \xrightarrow{(t')^*} [\Sigma^n E, Z]$$

$$\xrightarrow{((-1)^n \Sigma^n g)^*} [\Sigma^n D, Z] \xrightarrow{((-1)^n \Sigma^n f)^*} [\Sigma^n C, Z]$$

where t' is defined analogously as t, but taking into account that we now work in the suspended situation. We need to show that $t' = (-1)^n \Sigma t$. This is seen for $n = 1$ by observing that using the equations from Example 14.49 we have

$$t' = r_{-\Sigma f} \circ z(-\Sigma g, \Sigma h)^{-1} = (\Sigma r_f) \circ (-\Sigma(z(g, h)))^{-1} =$$

$$- \Sigma(r_f \circ z(g, h)^{-1}) = -\Sigma t,$$

and for other $n \in \mathbb{Z}$ by iterating. The collection of these exact sequences can be spliced together to the infinite (in both directions) long exact sequence appearing as the first sequence in the statement of the theorem.

Since $C \xrightarrow{f} D \xrightarrow{g} E$ is also homotopy fibration by Theorem 14.78, we obtain a commutative diagram of R-chain complexes with R-chain homotopy equivalences as vertical arrows

$$\begin{array}{ccccc} C & \xrightarrow{f} & D & \xrightarrow{g} & E \\ {\scriptstyle z}\uparrow {\scriptstyle \simeq} & & {\scriptstyle id}\uparrow & & {\scriptstyle id}\uparrow \\ \mathrm{hofib}(g) & \xrightarrow{\bar{e}} & D & \xrightarrow{g} & E. \end{array}$$

The lower row induces by the universal property of the homotopy fibre, see Remark 14.73, an exact sequence $[Z, \mathrm{hofib}(g)] \xrightarrow{\bar{e}_*} [D, Z] \xrightarrow{g_*} [E, Z]$. Hence we obtain an exact sequence

$$[Z, C] \xrightarrow{f_*} [Z, D] \xrightarrow{g_*} [Z, E]. \tag{14.84}$$

Now one proceeds as above, replacing (14.80) by (14.84) to get the second desired sequence. □

Remark 14.85 (Extended homotopy cofibration sequence) Consider an R-chain map $f: C \to D$. We have associated to it the canonical homotopy cofibration sequence $D \xrightarrow{l} \mathrm{cone}(f) \xrightarrow{r} \Sigma C$ in (14.46). The construction in the proof of Theorem 14.79 yields a sequence of chain maps, the so-called *signed extended homotopy* *cofibration sequence*

$$\cdots \xrightarrow{\Sigma^{-2}r} \Sigma^{-1}C \xrightarrow{-\Sigma^{-1}f} \Sigma^{-1}D \xrightarrow{-\Sigma^{-1}l} \Sigma\operatorname{cone}(f) \xrightarrow{-\Sigma^{-1}r}$$
$$C \xrightarrow{f} D \xrightarrow{l} \operatorname{cone}(f) \xrightarrow{r} \Sigma C \xrightarrow{-\Sigma f} \Sigma D \xrightarrow{-\Sigma l} \Sigma\operatorname{cone}(f) \xrightarrow{-\Sigma r} \cdots \quad (14.86)$$

which is infinite in both directions and has the property that each pair of consecutive arrows form a homotopy cofibration sequence.

This sequence appears in the literature sometimes as in (14.86), see for example in [2, page 153] or [421, page 413] in the version with spectra, and sometimes without the minus signs in front of the suspensions, see for example [130] in the version with chain complexes, or [399, page 94–95] in the version with spectra.

In fact, the construction in the proof of Theorem 14.79 can be modified to yield a sequence of chain maps, the so-called *not signed extended homotopy cofibration sequence*

$$\cdots \xrightarrow{\Sigma^{-2}r} \Sigma^{-1}C \xrightarrow{\Sigma^{-1}f} \Sigma^{-1}D \xrightarrow{\Sigma^{-1}l} \Sigma\operatorname{cone}(f) \xrightarrow{\Sigma^{-1}r}$$
$$C \xrightarrow{f} D \xrightarrow{l} \operatorname{cone}(f) \xrightarrow{r} \Sigma C \xrightarrow{\Sigma f} \Sigma D \xrightarrow{\Sigma l} \Sigma\operatorname{cone}(f) \xrightarrow{\Sigma r} \cdots \quad (14.87)$$

which is also infinite in both directions and has the property that each pair of consecutive arrows form a homotopy cofibration sequence.

The construction of sequence (14.87) is achieved by employing the isomorphism of chain complexes τ from Example 14.49 (as is done in [130]) as follows. In the large diagram in the proof of Theorem 14.79 we put $-\operatorname{id}$ as the right vertical arrow and Σf as the rightmost arrow in the top row. However, if we do this, we have to check what happens with the connecting map three terms to the right. What appears there is the top row of the last diagram in Example 14.49, which can by the indicated isomorphism be replaced by the bottom row in that diagram, and this in turn gives the desired sequence.

The whole sign issue is related to the fact that an exact sequence of R-modules remains an exact sequence of R-modules if the maps are multiplied by -1. We have stated the Puppe sequence in both versions since both appear in the literature.

Lemma 14.88 *Let $f: C \to D$ be an R-chain map. Let $r: \operatorname{cone}(f) \to \Sigma C$ be the canonical R-chain map, see (14.46). Then there is a natural R-chain map*

$$v: \operatorname{cone}(r) \xrightarrow{\simeq} \Sigma D$$

that is an R-chain homotopy equivalence.

Proof. We have the canonical cofibration sequence $C \xrightarrow{f} D \xrightarrow{l} \operatorname{cone}(f)$ of (14.46). After desuspending once, we obtain the cofibration sequence

14.6 Homotopy Theory of Chain Complexes

$$\Sigma^{-1}C \xrightarrow{-\Sigma^{-1}f} \Sigma^{-1}D \xrightarrow{-\Sigma^{-1}l} \Sigma^{-1}\operatorname{cone}(f).$$

The associated extended homotopy cofibration sequence contains the homotopy cofibration sequence

$$\Sigma^{-1}\operatorname{cone}(f) \xrightarrow{r} C \xrightarrow{f} D.$$

This sequence induces the desired canonical R-chain homotopy equivalence $v \colon \operatorname{cone}(r) \xrightarrow{\simeq} \Sigma D$. □

In Chapter 15 we will also need the two propositions stated below. In their proofs the following observation is used. Given an R-chain map $f \colon C \to D$ the $\mathbb{Z}$-chain maps

$$\iota_{f,V} = z(l \otimes \operatorname{id}_V, h \otimes \operatorname{id}_V) \colon \operatorname{cone}(f \otimes \operatorname{id}_V) \xrightarrow{\cong} \operatorname{cone}(f) \otimes_R V; \tag{14.89}$$

$$\iota_{V,f} = z(\operatorname{id}_V \otimes l, \operatorname{id}_V \otimes h) \colon \operatorname{cone}(\operatorname{id}_V \otimes f) \xrightarrow{\cong} V \otimes_R \operatorname{cone}(f), \tag{14.90}$$

are isomorphisms of $\mathbb{Z}$-chain complexes. In degree n the map

$$(\iota_{f,V})_n \colon \operatorname{cone}(f \otimes \operatorname{id}_V)_n = (C \otimes_R V)_{n-1} \oplus (D \otimes_R V)_n \to (\operatorname{cone}(f) \otimes_R V)_n$$

is explicitly given by

$$(\iota_{f,V})_n = (h \otimes \operatorname{id}_V)_{n-1} \oplus (l \otimes \operatorname{id}_V)_n, \tag{14.91}$$

and $(\iota_{V,f})_n$ has an analogous formula.

Proposition 14.92 (Behaviour of homotopy cofibration sequences with respect to tensoring) *Let*

$$C \xrightarrow{f} D \xrightarrow{g} E \quad \text{with} \quad h \colon 0 \simeq g \circ f$$

be a homotopy cofibration sequence of R-chain complexes and let V be an R-chain complex. Then

$$C \otimes_R V \xrightarrow{f \otimes \operatorname{id}_V} D \otimes_R V \xrightarrow{g \otimes \operatorname{id}_V} E \otimes_R V \quad \text{with} \quad h \otimes \operatorname{id}_V \colon 0 \simeq (g \otimes \operatorname{id}_V) \circ (f \otimes \operatorname{id}_V)$$

is a homotopy cofibration sequence.

Proof. Using Remark 14.45 and (14.89), the composition

$$\operatorname{cone}(f \otimes \operatorname{id}_V) \xrightarrow[\cong]{\iota_{f,V}} \operatorname{cone}(f) \otimes_R V \xrightarrow[\simeq]{z(g,h) \otimes \operatorname{id}_V} E \otimes_R V$$

is the desired $\mathbb{Z}$-chain homotopy equivalence. □

Proposition 14.93 (Behaviour of homotopy fibration sequences with respect to hom-functor) *Let*

$$C \xrightarrow{f} D \xrightarrow{g} E \quad \text{with} \quad h \colon 0 \simeq g \circ f$$

be a homotopy fibration sequence of R-chain complexes and let V be an R-chain complex. Then

$$\hom_R(V,C) \xrightarrow{\hom(\mathrm{id}_V,f)} \hom_R(V,D) \xrightarrow{\hom(\mathrm{id}_V,g)} \hom_R(V,E),$$

$$\text{with}\quad \hom_R(\mathrm{id}_V,h)\colon 0 \simeq \hom_R(\mathrm{id}_V,g) \circ \hom_R(V,f),$$

is a homotopy fibration sequence.

Proof. Given an R-chain map $g\colon D \to E$ there is an isomorphism of $\mathbb{Z}$-chain complexes

$$\mathrm{hofib}(\hom_R(\mathrm{id}_V, g)) \xrightarrow{\cong} \hom_R(\mathrm{id}_V, \mathrm{hofib}(g)).$$

Given our situation, we have the desired $\mathbb{Z}$-chain homotopy equivalence

$$\hom_R(\mathrm{id}_V, C) \xrightarrow{\simeq} \hom_R(\mathrm{id}_V, \mathrm{hofib}(g))$$

since applying $\hom_R(\mathrm{id}_V, -)$ to an R-chain homotopy equivalence produces a chain homotopy equivalence. □

14.6.10 Homotopy Pullbacks

Consider the diagram of R-chain maps

$$C \xrightarrow{f} E \xleftarrow{g} D.$$

The *homotopy pullback* $C \tilde{\times}_E D$ is defined to be the homotopy fibre of the R-chain map $C \oplus D \xrightarrow{(-f+g)} E$. Explicitly, its n-th differential is

$$\begin{pmatrix} c_n & 0 & 0 \\ 0 & d_n & 0 \\ (-1)^n f_n & (-1)^{n+1} g_n & e_{n+1} \end{pmatrix} \colon C_n \oplus D_n \oplus E_{n+1} \to C_{n-1} \oplus D_{n-1} \oplus E_n.$$

We obtain a homotopy commutative square

$$\begin{array}{ccc} C\tilde{\times}_E D & \xrightarrow{\overline{f}} & D \\ {\scriptstyle \overline{g}}\downarrow & & \downarrow {\scriptstyle g} \\ C & \xrightarrow{f} & E \end{array}$$

and we have the corresponding universal property analogous to the universal property of the homotopy pushout from Remark 14.66 as follows.

14.6 Homotopy Theory of Chain Complexes

Remark 14.94 (Universal property of the homotopy pullback) The homotopy pullback has the following universal property by Remark 14.73. Let $u\colon Z \to C$ and $v\colon Z \to D$ be R-chain maps and let $h'\colon g \circ v \simeq f \circ u$ be an R-chain homotopy. Then there exists precisely one R-chain map $z = z(u, v, h')\colon Z \to C\widetilde{\times}_E D$ such that $u = \overline{g} \circ z$, $v = \overline{f} \circ z$ and $h' = h \circ z$ hold. Explicitly $z_n\colon Z_n \to C_n \oplus D_n \oplus E_{n+1}$ is given by $z_n(\gamma) = (u_n(\gamma), v_n(\gamma), h'_n(\gamma))$.

Conversely, given an R-chain map $z\colon Z \to C\widetilde{\times}_E D$, we obtain an R-chain map $u\colon Z \to C$ by $\overline{f} \circ z$, an R-chain map $v\colon Z \to E$ by $\overline{g} \circ z$, and an R-chain homotopy $h'\colon g \circ v \simeq f \circ u$ by $h' = h \circ z$ satisfying $z = z(u, v, h')$.

Moreover, the homotopy pullback is homotopy invariant in the obvious sense.

We call the diagram (14.67) together with an R-chain homotopy $h\colon u' \circ v \simeq v' \circ u$ *homotopy cartesian* if the canonical map $z\colon C \to D\widetilde{\times}_F E$ is an R-chain homotopy equivalence.

One easily checks that the next theorem follows from Theorem 14.78.

Theorem 14.95 (Homotopy cartesian and homotopy cocartesian agree for chain complexes) *Consider the diagram* (14.67) *together with an R-chain homotopy* $h\colon u' \circ v \simeq v' \circ u$.

Then it is homotopy cocartesian if and only if it is homotopy cartesian.

Next we state and prove the following Lemma 14.98, which will be needed in Chapter 15. Consider a commutative diagram of R-chain complexes

$$\begin{array}{ccc} C & \xrightarrow{u} & D \\ {\scriptstyle v}\downarrow & & \downarrow{\scriptstyle v'} \\ E & \xrightarrow{u'} & F \end{array} \qquad (14.96)$$

and let f denote $u' \circ v = v' \circ u$. This is a special case of Diagram (14.67) from the beginning of Section 14.6.6 when the homotopy h is zero. Taking cones and induced maps as done there, we obtain the 3×3 part of the diagram below, not involving $\mathrm{cone}(f)$, where the maps l_- denote various canonical inclusions and the two wobbly arrows starting from the top left entry indicate the R-chain nullhomotopy $h_{u'} \circ v = a \circ h_u \colon 0 \simeq a \circ l_u \circ u = l_{u'} \circ u' \circ v$, with $h_u\colon 0 \simeq l_u \circ u$ and $h_{u'}\colon 0 \simeq l'_{u} \circ u'$ the canonical R-chain homotopies, and analogously $\widetilde{h}_{v'} \circ u = b \circ h_v$:

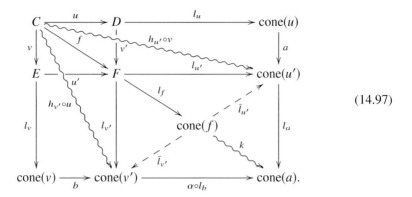
(14.97)

The map $\alpha \colon \mathrm{cone}(b) \to \mathrm{cone}(a)$ is the R-chain isomorphism given by the matrix

$$\alpha = \begin{pmatrix} -\mathrm{id} & 0 & 0 & 0 \\ 0 & 0 & \mathrm{id} & 0 \\ 0 & \mathrm{id} & 0 & 0 \\ 0 & 0 & 0 & \mathrm{id} \end{pmatrix} \colon C_{n-2} \oplus E_{n-1} \oplus D_{n-1} \oplus F_n \xrightarrow{\cong} C_{n-2} \oplus D_{n-1} \oplus E_{n-1} \oplus F_n.$$

Next we explain the part of the diagram involving $\mathrm{cone}(f)$. Consider the canonical inclusion $l_f \colon F \to \mathrm{cone}(f)$. Let $\bar{l}_{u'} \colon \mathrm{cone}(f) \to \mathrm{cone}(u')$ be the R-chain map $z(l_f, h_{u'} \circ v)$ obtained by the universal property of $\mathrm{cone}(f)$ from Remark 14.45, and define $\bar{l}_{v'} \colon \mathrm{cone}(f) \to \mathrm{cone}(v')$ analogously. Consider the canonical inclusions $l_b \colon \mathrm{cone}(v') \to \mathrm{cone}(b)$ and $l_a \colon \mathrm{cone}(u') \to \mathrm{cone}(a)$. A direct calculation shows that the map $k \colon \mathrm{cone}(f) \to \mathrm{cone}(a)$ of degree one that sends the element (x, y) in $C_{n-1} \oplus D_n = \mathrm{cone}(f)_n$ to the element $(x, 0, 0, 0)$ in $C_{n-1} \oplus D_n \oplus E_n \oplus F_{n+1} = \mathrm{cone}(b)_{n+1}$ defines an R-chain homotopy $\alpha \circ l_b \circ \bar{l}_{v'} \simeq l_a \circ \bar{l}_{u'}$. In this situation we have the following lemma.

Lemma 14.98 *In the above situation, the square*

$$\begin{array}{ccc} \mathrm{cone}(f) & \xrightarrow{\bar{l}_{u'}} & \mathrm{cone}(u') \\ {\scriptstyle \bar{l}_{v'}}\downarrow & & \downarrow{\scriptstyle l_a} \\ \mathrm{cone}(v') & \xrightarrow[\alpha \circ l_a]{} & \mathrm{cone}(a) \end{array}$$

together with the R-chain homotopy $k \colon \alpha \circ l_b \circ \bar{l}_{v'} \simeq l_a \circ \bar{l}_{u'}$ is homotopy cartesian.

Proof. The idea of the proof is that we will consider the induced map $\Phi \colon \mathrm{cone}(\bar{l}_{v'}) \to \mathrm{cone}(l_a)$ of mapping cones of the vertical maps and we will show that there exist R-chain homotopy equivalences $\pi_1 \colon \mathrm{cone}(\bar{l}_{v'}) \to \Sigma \mathrm{cone}(u)$ and $\pi_2 \colon \mathrm{cone}(l_a) \to \Sigma \mathrm{cone}(u)$ satisfying $\pi_2 \circ \Phi = \pi_1$. This implies that the map Φ is an R-chain homotopy equivalence. Hence by Lemma 14.68 (iv) the square under investigation is homotopy

14.6 Homotopy Theory of Chain Complexes   541

cocartesian, and by Theorem 14.95 it is also homotopy cartesian. In the proof we will use some techniques from the proof of Theorem 14.79.

We start with the map $\pi_2\colon \operatorname{cone}(l_a) \to \Sigma\operatorname{cone}(u)$. Observe that we have the canonical projection map $r_a\colon \operatorname{cone}(a) \to \Sigma\operatorname{cone}(u)$, see (14.46), such that $r_a \circ l_a = 0$. This together with the zero R-chain homotopy induces the R-chain homotopy equivalence $\pi_2 := z(r_a, 0)\colon \operatorname{cone}(l_a) \to \Sigma\operatorname{cone}(u)$, see (14.81). Explicitly, it is given in degree n by the 2×6 matrix

$$\begin{pmatrix} 0 & 0 & \mathrm{id} & 0 & 0 & 0 \\ 0 & 0 & 0 & \mathrm{id} & 0 & 0 \end{pmatrix} \colon E_{n-2} \oplus F_{n-1} \oplus C_{n-2} \oplus D_{n-1} \oplus E_{n-1} \oplus F_n \to C_{n-2} \oplus D_{n-1}.$$

Next we discuss the map π_1. Consider the homotopy commutative square

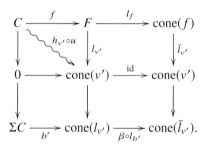

Taking cones, it induces the following homotopy commutative diagram with all rows and columns homotopy cofibration sequences:

$$\begin{array}{ccccc}
C & \xrightarrow{f} & F & \xrightarrow{l_f} & \operatorname{cone}(f) \\
\downarrow & \searrow^{h_{v'} \circ u} \downarrow^{l_{v'}} & & \downarrow^{\bar{l}_{v'}} \\
0 & \longrightarrow & \operatorname{cone}(v') & \xrightarrow{\mathrm{id}} & \operatorname{cone}(v') \\
\downarrow & & \downarrow & & \downarrow \\
\Sigma C & \xrightarrow{b'} & \operatorname{cone}(l_{v'}) & \xrightarrow{\beta \circ l_{b'}} & \operatorname{cone}(\bar{l}_{v'}).
\end{array}$$

Here b' is the induced map, and $\beta\colon \operatorname{cone}(b') \to \operatorname{cone}(\bar{l}_{v'})$ is the canonical R-chain isomorphism from the proof of Lemma 14.68 whose matrix formula in degree n is the analogue of the formula for α below Diagram (14.97) in the present situation. In degree n the map $b'_n\colon C_{n-1} \to F_{n-1} \oplus \operatorname{cone}(v')_n$ is explicitly given by $b'_n\colon x \mapsto (f_{n-1}(x), -(h_{v'} \circ u)_{n-1}(x))$.

Next we recall the canonical projection $r_{v'}\colon \operatorname{cone}(v') \to \Sigma D$. It induces the R-chain homotopy equivalence $z(r_{v'}, 0)\colon \operatorname{cone}(v') \to \Sigma D$ that is in degree n given by the canonical projection $F_{n-1} \oplus D_{n-1} \oplus F_n \to D_{n-1}$. We have $z(r_{v'}, 0)_n \circ b'_n(x) = -u_{n-1}(x) = -\Sigma(u)_n(x)$. Recall from Example 14.49 that $\operatorname{cone}(-\Sigma u) = \Sigma\operatorname{cone}(u)$. Hence we obtain the commutative diagram with rows homotopy cofibration sequences

$$\begin{CD}
\Sigma C @>b'>> \mathrm{cone}(l_{v'}) @>>> \mathrm{cone}(b') \\
@VV\mathrm{id}V @VVz(r_{v'},0)V @VVz'V \\
\Sigma C @>>-\Sigma u> \Sigma D @>>> \Sigma\,\mathrm{cone}(u)
\end{CD}$$

where z' is the induced map, which is an R-chain homotopy equivalence since $z(r_{v'}, 0)$ is an R-chain homotopy equivalence. Hence we obtain an R-chain homotopy equivalence

$$\pi_1 := z' \circ \beta^{-1} \colon \mathrm{cone}(l_{v'}) \to \Sigma\,\mathrm{cone}(u).$$

Explicitly it is given in degree n by the 2×4 matrix

$$\begin{pmatrix} -\mathrm{id} & 0 & 0 & 0 \\ 0 & 0 & \mathrm{id} & 0 \end{pmatrix} \colon C_{n-2} \oplus F_{n-1} \oplus D_{n-1} \oplus F_n \to C_{n-2} \oplus D_{n-1}.$$

Next the induced map $\Phi \colon \mathrm{cone}(\bar{l}_{v'}) \to \mathrm{cone}(l_a)$ is given in degree n by the 6×4 matrix

$$\begin{pmatrix} v & 0 & 0 & 0 \\ 0 & \mathrm{id} & 0 & 0 \\ -\mathrm{id} & 0 & 0 & 0 \\ 0 & 0 & \mathrm{id} & 0 \\ 0 & 0 & 0 & 0 \\ 0 & 0 & 0 & \mathrm{id} \end{pmatrix} \colon C_{n-2} \oplus F_{n-1} \oplus D_{n-1} \oplus F_n \to$$

$$\to E_{n-2} \oplus F_{n-1} \oplus C_{n-2} \oplus D_{n-1} \oplus E_{n-1} \oplus F_n.$$

Multiplying matrices we see that $\pi_2 \circ \Phi = \pi_1$ and we are done. $\square$

14.6.11 Homotopy Colimits of Sequences

The homotopy pushout is an example of a homotopy colimit and the homotopy pullback is an example of a homotopy limit in an appropriate category of chain complexes. We will need one more example of a homotopy colimit, namely, the homotopy colimit of a sequence, alias infinite mapping telescope.

Let $(C\{i\}, f\{i\})_{i \in \mathbb{N}}$ be a sequence of R-chain maps, i.e., a sequence indexed by $\mathbb{N}$,

$$C\{0\} \xrightarrow{f\{0\}} C\{1\} \xrightarrow{f\{1\}} C\{2\} \to \cdots \to C\{i\} \xrightarrow{f\{i\}} C\{i+1\} \to \cdots \quad (14.99)$$

of R-chain complexes and R-chain maps. There always exists the colimit R-chain complex of such a sequence (sometimes also called a direct limit)

$$\mathrm{colim}_{i \to \infty} C\{i\}.$$

14.6 Homotopy Theory of Chain Complexes

However, this notion is not homotopy invariant. It is homotopy invariant under the assumption that each map $f\{i\}$ is a cofibration by the following lemma.

Proposition 14.100 *Consider two sequences of R-chain maps $(C\{i\}, f\{i\})_{i\in\mathbb{N}}$ and $(D\{i\}, g\{i\})_{i\in\mathbb{N}}$ and a morphism $(u\{i\})_{n\in\mathbb{N}}$ between them, i.e., a sequence of R-chain maps $u\{i\}\colon C\{i\} \to D\{i\}$ satisfying $g\{i\} \circ u\{i\} = u\{i+1\} \circ f\{i\}$ for every $i \in \mathbb{N}$. Suppose that $u\{i\}$ is an R-chain homotopy equivalence and both $f\{i\}$ and $g\{i\}$ are cofibrations for every $i \in \mathbb{N}$.*

Then $\mathrm{colim}_{i\to\infty} g\{i\}\colon \mathrm{colim}_{i\to\infty} C\{i\} \to \mathrm{colim}_{i\to\infty} D\{i\}$ is an R-chain homotopy equivalence.

Proof. We get an obvious exact sequence of sequences of R-chain maps

$$0 \to (C\{i\}, f\{i\})_{i\in\mathbb{N}} \xrightarrow{(k\{i\})_{n\in\mathbb{N}}} (\mathrm{cyl}(u\{i\}), f'\{i\})_{i\in\mathbb{N}}$$
$$\xrightarrow{(q\{i\})_{i\in\mathbb{N}}} (\mathrm{cone}(u\{i\}), f''\{i\})_{i\in\mathbb{N}} \to 0$$

where for every $i \in \mathbb{N}$ and $n \in \mathbb{Z}$ the induced sequence of R-modules $0 \to C\{i\}_n \to \mathrm{cyl}(u\{i\})_n \to \mathrm{cone}(u\{i\})_n \to 0$ is split exact, see (14.43). Taking the colimit yields an exact sequence of R-chain complexes

$$0 \to \mathrm{colim}_{i\to\infty} C\{i\} \xrightarrow{\mathrm{colim}_{i\to\infty} k\{i\}} \mathrm{colim}_{i\to\infty} \mathrm{cyl}(u\{i\})$$
$$\xrightarrow{\mathrm{colim}_{i\to\infty} q\{i\}} \mathrm{colim}_{i\to\infty} \mathrm{cone}(u\{i\}) \to 0 \quad (14.101)$$

that is degreewise split exact. The canonical inclusion $j\{i\}\colon D\{i\} \to \mathrm{cyl}(u\{i\})$ and the canonical projection $p\{i\}\colon \mathrm{cyl}(u\{i\}) \to D\{i\}$, see (14.42), yield morphisms of sequences of R-chain complexes

$$(j\{i\})_{n\in\mathbb{N}}\colon (D\{i\}, g\{i\})_{i\in\mathbb{N}} \to (\mathrm{cyl}(u\{i\}), f'\{i\})_{i\in\mathbb{N}};$$
$$(p\{i\})_{n\in\mathbb{N}}\colon (\mathrm{cyl}(u\{i\}), f'\{i\})_{i\in\mathbb{N}} \to (D\{i\}, g\{i\})_{i\in\mathbb{N}},$$

satisfying $(p\{i\})_{n\in\mathbb{N}} \circ (j\{i\})_{n\in\mathbb{N}} = \mathrm{id}_{(D\{i\},g\{i\})_{i\in\mathbb{N}}}$. Hence we get R-chain maps

$$\mathrm{colim}_{i\to\infty} j\{i\}\colon \mathrm{colim}_{i\to\infty} D\{i\} \to \mathrm{colim}_{i\to\infty} \mathrm{cyl}(u\{i\});$$
$$\mathrm{colim}_{i\to\infty} p\{i\}\colon \mathrm{colim}_{i\to\infty} \mathrm{cyl}(u\{i\}) \to \mathrm{colim}_{i\to\infty} D\{i\},$$

satisfying $\mathrm{colim}_{i\to\infty} p\{i\} \circ \mathrm{colim}_{i\to\infty} j\{i\} = \mathrm{id}_{\mathrm{colim}_{i\to\infty} D\{i\}}$. Recall that for every i there is a preferred R-chain homotopy $j\{i\} \circ p\{i\} \simeq \mathrm{id}_{\mathrm{cyl}(u\{i\})}$, see (14.42). One easily checks that they fit together to an R-chain homotopy $\mathrm{colim}_{i\to\infty} j\{i\} \circ \mathrm{colim}_{i\to\infty} p\{i\} \simeq \mathrm{id}_{\mathrm{colim}_{i\to\infty} \mathrm{cyl}(u\{i\})}$. Hence $\mathrm{colim}_{i\to\infty} j\{i\}$ and $\mathrm{colim}_{i\to\infty} p\{i\}$ are R-chain homotopy equivalences. Therefore it suffices to show that $\mathrm{colim}_{i\to\infty} k\{i\}\colon \mathrm{colim}_{i\to\infty} C\{i\} \to \mathrm{colim}_{i\to\infty} \mathrm{cyl}(u\{i\})$ is an R-chain homotopy equivalence. Since we can apply Lemma 14.54 to (14.101), it remains to show that $\mathrm{colim}_{i\to\infty} \mathrm{cone}(u\{i\})$ is contractible.

Since the canonical R-chain map

$$\mathrm{cone}(\mathrm{colim}_{i\to\infty} u\{i\}\colon \mathrm{colim}_{i\to\infty} C\{i\} \to \mathrm{colim}_{i\to\infty} D\{i\})$$
$$\xrightarrow{\cong} \mathrm{colim}_{i\to\infty} \mathrm{cone}(u\{i\})$$

is an R-chain isomorphism and $\mathrm{cone}(u\{i\})$ is contractible for every $i \in \mathbb{N}$ by Proposition 14.52 (i), it suffices to prove that $\mathrm{colim}_{i\to\infty} C\{i\}$ is contractible, provided that each R-chain complex $C\{i\}$ is contractible and each R-chain map $f\{i\}\colon C\{i\} \to C\{i+1\}$ is a cofibration.

Proposition 14.55 implies that $\mathrm{coker}(f\{i\})$ is contractible. By Lemma 14.54 there is an R-chain isomorphism of the form $f\{i\} \oplus s\{i\}\colon C\{i\} \oplus \mathrm{coker}(f\{i\}) \xrightarrow{\cong} C\{i+1\}$. Hence the sequence of R-chain complexes $(C\{i\}, f\{i\})_{i\in\mathbb{N}}$ is isomorphic to a sequence of R-chain complexes

$$E\{0\} \to E\{0\} \oplus E\{1\} \to E\{0\} \oplus E\{1\} \oplus E\{2\} \to \cdots$$

where the structure maps are the obvious inclusions and each R-chain complex $E\{i\}$ is contractible. Just take $E\{0\} = C\{0\}$ and $E\{i+1\} = \mathrm{coker}(f\{i\})$ for $i \in \mathbb{N}$. This implies

$$\mathrm{colim}_{i\to\infty} C\{i\} \cong \bigoplus_{i\in\mathbb{N}} E\{i\}.$$

Since each $E\{i\}$ is contractible, $\mathrm{colim}_{i\to\infty} C\{i\}$ is contractible. This finishes the proof of Proposition 14.100. □

Similarly as in the previous cases, there exists a notion that replaces the colimit of the sequence (14.99) and is homotopy invariant, namely its homotopy colimit. The basic idea, motivated by Proposition 14.100, is to replace each R-chain map $f\{i\}$ by a cofibration using the mapping cylinder. It is defined as follows. Each R-chain map $f\{i\}\colon C\{i\} \to C\{i+1\}$ can be replaced up to chain homotopy by a cofibration $k\{i\}\colon C\{i\} \to \mathrm{cyl}(f\{i\})$ by (14.42). Next we construct inductively the *mapping telescope* $\mathrm{tel}(f\{i\})$ for each $i \geq -1$ as follows. We define $\mathrm{tel}(f\{-1\}) := C\{0\}$, $\mathrm{tel}(f\{0\}) = \mathrm{cyl}(f\{0\})$ and $l\{-1\}$ by $k\{0\}$. We have the canonical cofibration $\bar{j}\{0\} = j\{0\}\colon C\{1\} \to \mathrm{tel}(f\{0\})$, which is an R-chain homotopy equivalence, see (14.42). Suppose that $\mathrm{tel}(f\{i\})$ has been constructed together with a cofibration $\bar{j}\{i\}\colon C\{i+1\} \to \mathrm{tel}(f\{i\})$ which is an R-chain homotopy equivalence, for some $i \in \mathbb{N}$. Then $\mathrm{tel}(f\{i+1\})$ is constructed as the pushout

$$\begin{array}{ccc} C\{i+1\} & \xrightarrow{k\{i+1\}} & \mathrm{cyl}(f\{i+1\}) \\ {\scriptstyle \bar{j}\{i\}}\downarrow & & \downarrow \\ \mathrm{tel}(f\{i\}) & \xrightarrow{l\{i\}} & \mathrm{tel}(f\{i+1\}). \end{array} \quad (14.102)$$

We also obtain the cofibration $\bar{j}\{i+1\}\colon C\{i+2\} \to \mathrm{tel}(f\{i+1\})$, namely the composite of $j\{i+1\}\colon C\{i+1\} \to \mathrm{cyl}(f\{i+1\})$ and the right vertical map in Dia-

14.6 Homotopy Theory of Chain Complexes

gram (14.102). It is an R-chain homotopy equivalence, see Exercise 14.57. In this way the sequence (14.99) induces another sequence

$$C\{0\} = \mathrm{tel}(f\{-1\}) \xrightarrow{l\{-1\}} \mathrm{tel}(f\{0\}) \xrightarrow{l\{0\}} \cdots \xrightarrow{l\{i-1\}} \mathrm{tel}(f\{i\})$$
$$\xrightarrow{l\{i\}} \mathrm{tel}(f\{i+1\}) \xrightarrow{l\{i+1\}} \cdots . \quad (14.103)$$

The difference now is that all maps $l\{i\}$ are cofibrations. The *homotopy colimit of the sequence* (14.99) (also sometimes called the (infinite) mapping telescope) is then defined as

$$\mathrm{hocolim}_{i\to\infty} C\{i\} := \mathrm{colim}_{i\to\infty} \mathrm{tel}(f\{i\}). \quad (14.104)$$

The homotopy colimit has the following universal property. Denote by $s\{i\}\colon \mathrm{tel}(f\{i\}) \to \mathrm{hocolim}_{i\to\infty} C\{i\}$ the structure map of $\mathrm{colim}_{i\to\infty} \mathrm{tel}(f\{i\})$. There exists a preferred R-chain homotopy $h\{i\}\colon s\{i\} \circ \overline{j}\{i\} \circ f\{i\} \simeq s\{i-1\} \circ \overline{j}\{i-1\}$. For an R-chain complex D, R-chain maps $g\{i\}\colon C\{i\} \to D$, and R-chain homotopies $h'\{i\}\colon g\{i+1\} \circ f\{i\} \simeq g\{i\}$ for $i \in \{-1, 0, 1, \ldots\}$, there is precisely one R-chain map $g\colon \mathrm{hocolim}_{i\to\infty} C\{i\} \to D$ such that $g \circ s\{i\} \circ \overline{j}\{i\} = g\{i+1\}$ and $h'\{i\} = g \circ h\{i\}$ holds for all $i \in \mathbb{N}$.

For every $i \in \{-1, 0, 1, \ldots\}$, we define R-chain homotopy equivalences $\mathrm{pr}\{i\}\colon \mathrm{tel}(f\{i\}) \to C\{i+1\}$ satisfying $f\{i+1\} \circ \mathrm{pr}\{i\} = \mathrm{pr}\{i+1\} \circ l\{i\}$ and $\mathrm{pr}\{i\} \circ j\{i\} = \mathrm{id}_{C\{i+1\}}$ as follows. We define $\mathrm{pr}\{-1\} = \mathrm{id}_{C\{0\}}$. Define $\mathrm{pr}\{0\}$ to be the canonical projection $p\colon \mathrm{tel}(f\{0\}) = \mathrm{cyl}(f\{0\}) \to C\{1\}$. If we have defined $\mathrm{pr}\{i\}\colon \mathrm{tel}(f\{i\}) \to C\{i+1\}$, define $\mathrm{pr}\{i+1\}\colon \mathrm{tel}(f\{i+1\}) \to C\{i+2\}$ by the pushout diagram (14.102) and the R-chain map $f\{i+1\} \circ \mathrm{pr}\{i\}$ and the canonical projection $\mathrm{cyl}(f\{i+1\}) \to C\{i+2\}$. Since each R-chain map $j\{i+1\}$ is an R-chain homotopy equivalence, each R-chain map $\mathrm{pr}\{i\}$ is an R-chain homotopy equivalence.

The collection of the R-chain maps $(\mathrm{pr}\{i\})_{i\in\mathbb{N}}$ defines a natural R-chain map

$$\mathrm{pr}\colon \mathrm{hocolim}_{i\to\infty} C\{i\} \to \mathrm{colim}_{i\to\infty} C\{i\} \quad (14.105)$$

which is in general not an R-chain homotopy equivalence, see Example 14.112. We conclude from Proposition 14.100 that it is an R-chain homotopy equivalence if each R-chain map $f\{i\}$ is a cofibration.

The discussion so far tells us that the homotopy colimit of a sequence is natural for maps between such sequences. That means that, if we have two sequences of R-chain maps $(C\{i\}, f\{i\})_{i\in\mathbb{N}}$ and $(D\{i\}, g\{i\})_{i\in\mathbb{N}}$ and a morphism $(u\{i\})_{n\in\mathbb{N}}$ between them, i.e., a sequence of R-chain maps $u\{i\}\colon C\{i\} \to D\{i\}$ satisfying $g\{i\} \circ u\{i\} = u\{i+1\} \circ f\{i\}$ for every $i \in \mathbb{N}$, then we obtain an induced map

$$\mathrm{hocolim}_{i\to\infty} u\{i\}\colon \mathrm{hocolim}_{i\to\infty} C\{i\} \to \mathrm{hocolim}_{i\to\infty} D\{i\}.$$

This naturality can be generalised to systems of chain maps $(u\{i\})_{n\in\mathbb{N}}$ such that the corresponding squares commute only up to chain homotopy, which means there exist $h\{i\}$ such that $h\{i\}\colon g\{i\} \circ u\{i\} \simeq u\{i+1\} \circ f\{i\}$. Then one obtains an induced map

$$\mathrm{hocolim}_{i\to\infty} z(u\{i\}, h\{i\}) : \mathrm{hocolim}_{i\to\infty} C\{i\} \to \mathrm{hocolim}_{i\to\infty} D\{i\}. \quad (14.106)$$

by the following argument. The basic idea is that the system with only homotopy commutative squares is replaced by the system with strictly commutative squares, by iteratively applying the next Exercise 14.107.

Exercise 14.107 Let
$$\begin{array}{ccc} C\{1\} & \xrightarrow{f\{1\}} & C\{2\} \\ {\scriptstyle u\{1\}}\downarrow & {\scriptstyle h\{1\}} \nwarrow & \downarrow {\scriptstyle u\{2\}} \\ D\{1\} & \xrightarrow[g\{1\}]{} & D\{2\} \end{array}$$

be a homotopy commutative square, i.e., $h\{1\}: g\{1\} \circ u\{1\} \simeq u\{2\} \circ f\{1\}$. Then there exists a map $\overline{u\{2\}}: \mathrm{cyl}(f\{1\}) \to \mathrm{cyl}(g\{1\})$ such that the following diagram commutes strictly

$$\begin{array}{ccc} C\{1\} & \xrightarrow{i_{C\{1\}}} & \mathrm{cyl}(f\{1\}) \\ {\scriptstyle u\{1\}}\downarrow & & \downarrow {\scriptstyle \overline{u\{2\}}} \\ D\{1\} & \xrightarrow[i_{D\{1\}}]{} & \mathrm{cyl}(g\{1\}). \end{array}$$

Since $\mathrm{colim}_{i\in\mathbb{N}}$ is an exact functor, the canonical maps

$$\mathrm{colim}_{i\in\mathbb{N}} H_n(C\{i\}) \xrightarrow{\cong} H_n(\mathrm{colim}_{i\in\mathbb{N}} C\{i\});$$
$$\mathrm{colim}_{i\in\mathbb{N}} H_n(\mathrm{tel}(f\{i\})) \xrightarrow{\cong} H_n(\mathrm{hocolim}_{i\in\mathbb{N}} C\{i\}),$$

are isomorphisms for every $n \in \mathbb{N}$. Since each R-chain map $\mathrm{pr}\{i\}$ is an R-chain homotopy equivalence satisfying $f\{i+1\} \circ \mathrm{pr}\{i\} = \mathrm{pr}\{i+1\} \circ l\{i\}$, the map induced by pr

$$H_n(\mathrm{pr}): H_n(\mathrm{hocolim}_{i\in\mathbb{N}} C\{i\}) \xrightarrow{\cong} H_n(\mathrm{colim}_{i\in\mathbb{N}} C\{i\}) \quad (14.108)$$

is bijective for all $n \in \mathbb{Z}$. Hence we get for every $n \in \mathbb{Z}$ a natural isomorphism

$$H_n(\mathrm{hocolim}_{i\to\infty} C\{i\}) \xrightarrow{\cong} \mathrm{colim}_{i\to\infty} H_n(C\{i\}), \quad (14.109)$$

cf. [134, Proposition VIII.5.20].

Explicitly, the homotopy colimit $\mathrm{hocolim}_{i\to\infty} C\{i\}$ is given by the R-chain complex whose n-th R-chain module $\left(\mathrm{hocolim}_{i\to\infty} C\{i\}\right)_n$ is

$$C\{0\}_{n-1} \oplus C\{0\}_n \oplus C\{1\}_{n-1} \oplus C\{1\}_n \oplus C\{2\}_{n-1} \oplus C\{2\}_n \oplus \cdots \quad (14.110)$$

and whose nth-differential is given by

14.6 Homotopy Theory of Chain Complexes

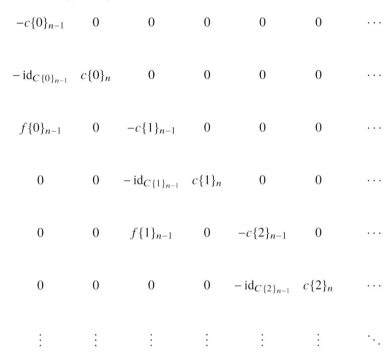

An alternative construction of the homotopy colimit of the sequence (14.99) is via the mapping cone

$$\operatorname{hocolim}_{i \to \infty} C\{i\} := \operatorname{cone}(\operatorname{id} - \operatorname{Shift}(f)) \tag{14.111}$$

of the R-chain map $\operatorname{id} - \operatorname{Shift}(f) \colon \bigoplus_{i \geq 0} C\{i\} \longrightarrow \bigoplus_{i \geq 0} C\{i\}$ where $\operatorname{Shift}(f)$ sends the summand $C\{i\}$ to the summand $C\{i+1\}$ by the map $f\{i\}$, cf. [41].

Example 14.112 Consider the sequence of finitely generated free $\mathbb{Z}$-chain complexes that are all concentrated in dimension 0

$$0[\mathbb{Z}] \xrightarrow{p \cdot \operatorname{id}_{\mathbb{Z}}} 0[\mathbb{Z}] \xrightarrow{p \cdot \operatorname{id}_{\mathbb{Z}}} 0[\mathbb{Z}] \xrightarrow{p \cdot \operatorname{id}_{\mathbb{Z}}} \cdots .$$

Its colimit is the $\mathbb{Z}$-chain complex $0[\mathbb{Z}[1/p]]$. Obviously the homotopy colimit is a free $\mathbb{Z}$-chain complex concentrated in dimension 0 and 1. The canonical $\mathbb{Z}$-chain map from the homotopy colimit to the colimit induces an isomorphism on homology by (14.108). It cannot be a $\mathbb{Z}$-chain homotopy equivalence as $\mathbb{Z}[1/p]$ is not projective as a $\mathbb{Z}$-module.

Proposition 14.113 *Suppose that we have a commutative ladder of R-chain complexes*

$$\begin{array}{ccccccccc} C\{0\} & \xrightarrow{f\{0\}} & C\{1\} & \longrightarrow & \cdots & \longrightarrow & C\{i\} & \xrightarrow{f\{i\}} & C\{i+1\} & \longrightarrow & \cdots \\ {\scriptstyle u\{0\}}\downarrow & & {\scriptstyle u\{1\}}\downarrow & & & & {\scriptstyle u\{i\}}\downarrow & & {\scriptstyle u\{i+1\}}\downarrow & & \\ D\{0\} & \xrightarrow[g\{0\}]{} & D\{1\} & \longrightarrow & \cdots & \longrightarrow & D\{i\} & \xrightarrow[g\{i\}]{} & D\{i+1\} & \longrightarrow & \cdots \end{array}$$

such that all maps $u\{i\}$ are R-chain homotopy equivalences. Then there is an induced R-chain map

$$\operatorname{hocolim}_{i \to \infty} u\{i\} \colon \operatorname{hocolim}_{i \to \infty} C\{i\} \to \operatorname{hocolim}_{i \to \infty} D\{i\}$$

which is also an R-chain homotopy equivalence.

Proof. This follows from Proposition 14.100. □

Proposition 14.114 *Suppose that we have for each $i \geq 0$ a commutative diagram of R-chain complexes*

$$\begin{array}{ccccc} C\{i\} & \xrightarrow{\alpha\{i\}} & D\{i\} & \xrightarrow{\beta\{i\}} & E\{i\} \\ {\scriptstyle f\{i\}}\downarrow & & {\scriptstyle g\{i\}}\downarrow & & {\scriptstyle h\{i\}}\downarrow \\ C\{i+1\} & \xrightarrow[\alpha\{i+1\}]{} & D\{i+1\} & \xrightarrow[\beta\{i+1\}]{} & E\{i+1\} \end{array}$$

whose rows are homotopy cofibration sequences such that the corresponding chain homotopies (which are not depicted) are compatible in the obvious sense.

Then the induced maps and homotopies provide us with a homotopy cofibration sequence of the homotopy colimits

$$\operatorname{hocolim}_{i \to \infty} C\{i\} \xrightarrow{\alpha} \operatorname{hocolim}_{i \to \infty} D\{i\} \xrightarrow{\beta} \operatorname{hocolim}_{i \to \infty} E\{i\}$$

where $\alpha := \operatorname{hocolim}_{i \to \infty} \alpha\{i\}$ and $\beta := \operatorname{hocolim}_{i \to \infty} \beta\{i\}$ and the required chain homotopy of the composition is induced by the chain homotopies in each level indexed by i that we suppressed above.

Proof. The induced maps are constructed using the universal properties of homotopy pushouts and colimits. We have to show that the canonical R-chain map

$$\operatorname{cone}(\alpha \colon \operatorname{hocolim}_{i \to \infty} C\{i\} \to \operatorname{hocolim}_{i \to \infty} D\{i\}) \to \operatorname{hocolim}_{i \to \infty} E\{i\}$$

14.6 Homotopy Theory of Chain Complexes

is an R-chain homotopy equivalence. There is an obvious R-chain isomorphism

$$\mathrm{hocolim}_{i\to\infty} \mathrm{cone}(\alpha\{i\}) \xrightarrow{\cong} \mathrm{cone}(\alpha\colon \mathrm{hocolim}_{i\to\infty} C\{i\} \to \mathrm{hocolim}_{i\to\infty} D\{i\}).$$

Hence it suffices to show that the canonical R-chain map

$$\mathrm{hocolim}_{i\to\infty} \mathrm{cone}(\alpha\{i\}) \to \mathrm{hocolim}_{i\to\infty} E\{i\}$$

is an R-chain homotopy equivalence. Since for every $i \geq 0$ the R-chain map $\mathrm{cone}(\alpha\{i\}) \to E\{i\}$ is an R-chain homotopy equivalence, the claim follows from Proposition 14.113. $\square$

Proposition 14.115 *Let C be an R-chain complex such that for every $p \in \mathbb{N}$ there exists an R-chain complex $C\{p\}$ with the property that $C\{p\}_n = 0$ for $n < p$ and an R-chain homotopy equivalence $\omega\{p\}\colon C \to C\{p\}$. Then C is R-chain contractible.*

Proof. We construct inductively over $p = 0, 1, 2, \ldots$ maps $\gamma_n\colon C_n \to C_{n+1}$ for $n < p$ such that for every $n < p$ we have $c_{n+1} \circ \gamma_n + \gamma_{n-1} \circ c_n = \mathrm{id}_{C_n}$. Then the collection of the maps $\{\gamma_n \mid n \in \mathbb{Z}\}$ gives the desired chain contraction.

The existence of the R-chain homotopy equivalence $\omega\{p\}\colon C \to C\{p\}$ together with the assumption that $C\{p\}_n = 0$ for $n < p$ imply that there exists a sequence of R-maps $s\{p\}_n\colon C_n \to C_{n+1}$ for $n < p$ such that for every $n < p$ we have $c_{n+1} \circ s\{p\}_n + s\{p\}_{n-1} \circ c_n = \mathrm{id}_{C_n}$. So the induction beginning with $p = 0$ is given by applying the observation above in the case $p = 0$ and putting $\gamma_n = s\{0\}_n$ for $n < 0$.

The induction step from p to $p+1$ is done as follows. By applying the observation above for $p+1$ we get a sequence of R-maps $s\{p+1\}_n\colon C_n \to C_{n+1}$ for $n < p+1$ such that we have $c_{n+1} \circ s\{p+1\}_n + s\{p+1\}_{n-1} \circ c_n = \mathrm{id}_{C_n}$ for every $n < p+1$. Now define

$$\gamma_p = s\{p+1\}_p \circ \bigl(\mathrm{id}_{C_p} + s\{p+1\}_{p-1} \circ c_p - \gamma_{p-1} \circ c_p\bigr).$$

We compute

$$c_p \circ \bigl(s\{p+1\}_{p-1} \circ c_p - \gamma_{p-1} \circ c_p\bigr)$$
$$= c_p \circ s\{p+1\}_{p-1} \circ c_p - c_p \circ \gamma_{p-1} \circ c_p$$
$$= \bigl(c_p \circ s\{p+1\}_{p-1} + s\{p+1\}_{p-2} \circ c_{p-1}\bigr) \circ c_p$$
$$\quad - \bigl(c_p \circ \gamma_{p-1} + \gamma_{p-2} \circ c_{p-1}\bigr) \circ c_p$$
$$= \mathrm{id}_{C_{p-1}} \circ c_p - \mathrm{id}_{C_{p-1}} \circ c_p$$
$$= 0.$$

This implies

$$c_{p+1} \circ \gamma_p + \gamma_{p-1} \circ c_p$$
$$= c_{p+1} \circ s\{p+1\}_p \circ \left(\mathrm{id}_{C_p} + s\{p+1\}_{p-1} \circ c_p - \gamma_{p-1} \circ c_p\right) + \gamma_{p-1} \circ c_p$$
$$= c_{p+1} \circ s\{p+1\}_p + c_{p+1} \circ s\{p+1\}_p \circ \left(s\{p+1\}_{p-1} \circ c_p - \gamma_{p-1} \circ c_p\right)$$
$$+ \gamma_{p-1} \circ c_p$$
$$= \left(c_{p+1} \circ s\{p+1\}_p + s\{p+1\}_{p-1} \circ c_p\right) \circ \left(s\{p+1\}_{p-1} \circ c_p - \gamma_{p-1} \circ c_p\right)$$
$$+ c_{p+1} \circ s\{p+1\}_p + \gamma_{p-1} \circ c_p$$
$$= \mathrm{id}_{C_p} \circ \left(s\{p+1\}_{p-1} \circ c_p - \gamma_{p-1} \circ c_p\right) + c_{p+1} \circ s\{p+1\}_p + \gamma_{p-1} \circ c_p$$
$$= s\{p+1\}_{p-1} \circ c_p - \gamma_{p-1} \circ c_p + c_{p+1} \circ s\{p+1\}_p + \gamma_{p-1} \circ c_p$$
$$= \mathrm{id}_{C_p}.$$

This finishes the proof of Proposition 14.115. □

Let $p \in \mathbb{Z}^{\geq 0}$ and $(C\{i\}, f\{i\})_{i \in \mathbb{N}}$ be a sequence of R-chain complexes and R-chain maps $f\{i\}: C\{i\} \to C\{i+1\}$. We may consider the homotopy colimit of this sequence and its truncated version $\mathrm{trun}_{\leq p} C\{i\}$ where we replace $C\{i\}$ by $\{0\}$ for $i \leq p$ and leave $f\{i\}$ unchanged for $i \geq p+1$. There is an obvious inclusion $j[p]\{i\}: \mathrm{trun}_{\leq p} C\{i\} \to C\{i\}$ that induces an R-chain map

$$\mathrm{hocolim}_{i \to \infty} j[p]\{i\}: \mathrm{hocolim}_{i \to \infty} \mathrm{trun}_{\leq p} C\{i\} \to \mathrm{hocolim}_{i \to \infty} C\{i\}. \quad (14.116)$$

Lemma 14.117 *The R-chain map $\mathrm{hocolim}_{i \to \infty} j[p]\{i\}$ of (14.116) is an R-chain homotopy equivalence.*

Proof. We get an exact sequence of such sequences of R-chain complexes $0 \to \mathrm{trun}_{\leq p} C\{i\} \to C\{i\} \to \mathrm{coker}(j[p]\{i\}) \to 0$ that is for a fixed i in each degree split exact. This implies that we obtain a cofibration sequence of R-chain complexes

$$\mathrm{hocolim}_{i \to \infty} \mathrm{trun}_{\leq p} C\{i\} \to \mathrm{hocolim}_{i \to \infty} C\{i\} \to \mathrm{hocolim}_{i \to \infty} \mathrm{coker}(j[p]\{i\})$$

by inspecting the formulas for the homotopy colimit described in (14.110) or (14.111). We conclude from Lemma 14.59 that it suffices to show that $\mathrm{hocolim}_{i \to \infty} \mathrm{coker}(j[p]\{i\})$ is contractible. As $\mathrm{coker}(j[p]\{i\}) = \{0\}$ for $i > p$, the canonical R-chain map $\mathrm{tel}(f\{p\}) \to \mathrm{hocolim}_{i \to \infty} \mathrm{coker}(j[p]\{i\})$ is an R-chain isomorphism. The canonical inclusion $C\{i+1\} \to \mathrm{tel}(f\{i\})$ is an R-chain homotopy equivalence for every i and every sequence $(C\{i\}, f\{i\})$. This follows by induction from the statement that for an R-chain map $f: C \to D$ the canonical inclusion $j: D \to \mathrm{cyl}(f)$ is an R-chain homotopy equivalence, see Remark 14.41 and Exercise 14.57. Hence $\mathrm{coker}(j[p]\{p+1\}) = \{0\}$ is R-chain homotopy equivalent to $\mathrm{hocolim}_{i \to \infty} \mathrm{coker}(j[p]\{i\})$. □

14.6 Homotopy Theory of Chain Complexes

Proposition 14.118 *Let $(C\{i\}, f\{i\})_{i \in \mathbb{N}}$ be a sequence R-chain complexes and R-chain maps $f\{i\}: C\{i\} \to C\{i+1\}$. Suppose that there is a function $\rho: \mathbb{N} \to \mathbb{N}$ with $\lim_{i \to \infty} \rho(i) = \infty$ such that we have $C\{i\}_n = 0$ for $n < \rho(i)$.*

Then $\operatorname{hocolim}_{i \to \infty} C\{i\}$ is R-chain contractible.

Proof. Consider any natural number p. Choose a natural number q such that $\rho(i) \geq p$ holds for $i \geq q$. Then $\operatorname{trun}_{\leq q} C\{i\}$ has the property that $\left(\operatorname{trun}_{\leq q} C\{i\}\right)_n$ vanishes for $n < p$. The formulas for the homotopy colimit described in (14.110) or (14.111) imply $\left(\operatorname{hocolim}_{i \to \infty} \operatorname{trun}_{\leq q} C\{i\}\right)_n = \{0\}$ for $n < p$. Now we conclude from Proposition 14.115 and Lemma 14.117 that $\operatorname{hocolim}_{i \to \infty} C\{i\}$ is contractible. $\square$

14.6.12 Homotopy Limits of Inverse Systems of Maps

Dually to the previous section we will also need the example of the homotopy limit of an inverse system of maps. Let $(C\{i\}, f\{i\})_{i \in \mathbb{N}}$ be an inverse systems of R-chain maps, i.e., a sequence indexed by $\mathbb{N}$

$$C\{0\} \xleftarrow{f\{0\}} C\{1\} \xleftarrow{f\{1\}} C\{2\} \leftarrow \cdots \leftarrow C\{i\} \xleftarrow{f\{i\}} C\{i+1\} \leftarrow \cdots \quad (14.119)$$

of R-chain complexes and R-chain maps. There always exists the limit R-chain complex of such a sequence (sometimes also called an inverse limit)

$$\lim_{i \to \infty} C\{i\}.$$

However, this notion is not homotopy invariant. It is homotopy invariant under the assumption that each map $f\{i\}$ is a fibration by the following proposition, dual to Proposition 14.100.

Proposition 14.120 *Consider two inverse systems of R-chain maps denoted by $(C\{i\}, f\{i\})_{i \in \mathbb{N}}$ and $(D\{i\}, g\{i\})_{i \in \mathbb{N}}$ and a morphism $(u\{i\})_{n \in \mathbb{N}}$ between them, i.e., a sequence of R-chain maps $u\{i\}: C\{i\} \to D\{i\}$ satisfying $g\{i\} \circ u\{i+1\} = u\{i\} \circ f\{i\}$ for every $i \in \mathbb{N}$. Suppose that $u\{i\}$ is an R-chain homotopy equivalence and both $f\{i\}$ and $g\{i\}$ are fibrations for every $i \in \mathbb{N}$.*

Then $\lim_{i \to \infty} g\{i\}: \lim_{i \to \infty} C\{i\} \to \lim_{i \to \infty} D\{i\}$ is an R-chain homotopy equivalence.

The proof is by dualising the proof of Proposition 14.100. It is also possible to dualise the rest of the previous section, but we skip this since we will not need it.

14.6.13 Chain Complexes of Projective Modules

In this section we recall a proposition from [250], stated here as Proposition 14.122, which enables us in Chapter 15 under certain assumptions on homology and cohomology to replace a chain complex by a chain homotopic chain complex concentrated in some dimensions. The proposition and its proof are also related to Lemma 8.52 from Chapter 8 where similar reasoning was used. In fact, for us only a part of the proposition will be of direct use and this part is distilled below as Corollary 14.123. The statements use the following terminology, which will also be used in the next chapter.

Definition 14.121 (Chain complex terminology for dimension and connectedness) An R-chain complex C is called:

- *positive* if $C_n = 0$ holds for $n \leq -1$;
- *bounded below* if there exists an integer j such that $C_n = 0$ holds for $n < j$;
- *d-dimensional* for the integer d if $C_n = 0$ for $n > d$ holds;
- *k-connected* if $H_i(C) = 0$ for $i \leq k$.

Proposition 14.122 *Let C be a projective R-chain complex. Then the following assertions are equivalent:*

(i) *Let d be a natural number. Let $B_d(C)$ be the image of $c_{d+1} \colon C_{d+1} \to C_d$ and $j \colon B_d(C) \to C_d$ be the inclusion. There is an R-submodule $C_d^\perp$ such that for the inclusion $i \colon C_d^\perp \to C_d$ the map $i \oplus j \colon C_d^\perp \oplus B_d(C) \to C_d$ is an isomorphism. Moreover, the following chain map from a d-dimensional projective R-chain complex to C is an R-chain homotopy equivalence*

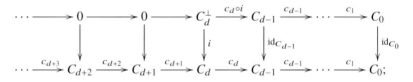

(ii) *The R-chain complex C is R-chain homotopy equivalent to a d-dimensional projective R-chain complex;*
(iii) *The R-chain complex C is R-chain homotopy equivalent to a d-dimensional R-chain complex;*
(iv) *The R-chain complex C is dominated by an d-dimensional R-chain complex D, i.e., there are R-chain maps $i \colon C \to D$ and $r \colon D \to C$ satisfying $r \circ i \simeq \mathrm{id}_C$;*
(v) *The R-module $B_d(C)$ is a direct summand in C_d and $H_i(C) = 0$ for $i \geq d+1$;*
(vi) *The homology group $H_R^{d+1}(C; M) := H^{d+1}(\hom_R(C, M))$ vanishes for every R-module M and $H_i(C) = 0$ for all $i \geq d + 1$.*

Proof. The proof of [250, Proposition 11.10 on page 221] carries over to our setting. □

Note that, if we assume additionally that C is positive or bounded below, then also the R-chain complex appearing in assertion (ii) of Theorem 14.122 can be assumed to have this property.

The promised, more directly useful, corollary is as follows.

14.6 Homotopy Theory of Chain Complexes

Corollary 14.123 (Criterion for finite dimensionality)

(i) Let C be a bounded below projective R-chain complex, let j be an integer, and suppose that C is $(j-1)$-connected. Then there exists a projective R-chain complex D with $D_i = 0$ for $i < j$ and a chain homotopy equivalence $D \to C$;

(ii) Let $d \geq j$ be another integer. Suppose in addition to the above that $H_i(C) = 0$ for $i > d$ and that $H^{d+1}(C, V) = 0$ for any R-module V. Then there exists a projective R-chain complex E with $E_i = 0$ unless $j \leq i \leq d$, and a chain homotopy equivalence $E \to C$;

(iii) If in addition to the conditions appearing in assertion (ii) C is a bounded R-chain complex of finitely generated free R-modules and $j < d$, then the R-chain complexes D and E can also be chosen to be R-chain complexes of finitely generated free R-modules;

(iv) If in addition to the conditions appearing in assertion (ii) C is a bounded R-chain complex of finitely generated free R-modules and $j = d$, then the chain complexes D and E can be chosen to be R-chain complexes of stably free finitely generated R-modules.

Proof. The proof can be obtained by combining Lemma 8.52 and Proposition 14.122. For the reader's convenience we present the proof here in one place at the expense of repeating ourselves to some extent.

(i) By the assumptions we have the long exact sequence

$$0 \to \mathrm{im}(c_j) \to C_{j-1} \xrightarrow{c_{j-1}} C_{j-2} \xrightarrow{c_{j-2}} \cdots \to \cdots \to 0.$$

Therefore $\mathrm{im}(c_j)$ is projective. From the short exact sequence $0 \to \ker(c_j) \to C_j \to \mathrm{im}(c_j) \to 0$ we obtain that $\ker(c_j)$ is projective. Hence, by Proposition 14.52, the inclusion provides us with an R-chain homotopy equivalence

$$\begin{array}{ccccccccc}
\cdots & \longrightarrow & C_{j+1} & \xrightarrow{c_{j+1}} & \ker(c_j) & \longrightarrow & 0 & \longrightarrow \cdots \longrightarrow & 0 \\
& & \downarrow{\mathrm{id}} & & \downarrow{i} & & \downarrow & & \downarrow \\
\cdots & \xrightarrow{c_{j+2}} & C_{j+1} & \xrightarrow{c_{j+1}} & C_j & \xrightarrow{c_j} & C_{j-1} & \xrightarrow{c_{j-1}} \cdots \longrightarrow & 0.
\end{array}$$

The top row is the desired chain complex D and the inclusion is the desired R-chain homotopy equivalence.

(ii) Now consider $V := \mathrm{im}(c_{d+1}) \subset C_d$. The assumption $H^{d+1}(C, V) = 0$ means that the following sequence is exact

$$\hom(C_d, V) \to \hom(C_{d+1}, V) \to \hom(C_{d+2}, V).$$

Taking $c_{d+1} \in \hom(C_{d+1}, V)$, we see that there exists an $\alpha \colon C_d \to \mathrm{im}(c_{d+1})$ such that $\alpha \circ c_{d+1} = c_{d+1}$. Therefore α splits the inclusion of $\mathrm{im}(c_{d+1})$ into C_d, and hence $\mathrm{im}(c_{d+1})$ is a summand in C_d. Denoting by $C_d^\perp$ its direct complement and using Proposition 14.52, we obtain an R-chain equivalence

$$\begin{array}{ccccccccc}
\cdots \longrightarrow & 0 & \longrightarrow & C_d^\perp & \xrightarrow{c_d \circ i} & C_{d-1} & \xrightarrow{c_{d-1}} & \cdots \xrightarrow{c_{j+1}} & \ker(c_j) & \longrightarrow & 0 & \longrightarrow \cdots \\
& \downarrow & & \downarrow i & & \downarrow \text{id} & & & \downarrow i & & \downarrow \\
\cdots \xrightarrow{c_{d+2}} & C_{d+1} & \xrightarrow{c_{d+1}} & C_d & \xrightarrow{c_d} & C_{d-1} & \xrightarrow{c_{d-1}} & \cdots \xrightarrow{c_{j+1}} & C_j & \xrightarrow{c_j} & C_{j-1} & \longrightarrow \cdots .
\end{array}$$

The top row is the desired chain complex E and the inclusion is the desired R-chain homotopy equivalence.

(iii) We describe the additional arguments. In part (i) the R-module $\text{im}(c_j)$ is stably finitely generated free by the exact sequence and hence $\ker(c_j)$ is as well. Thus we can make D free by adding an elementary R-chain complex, that is concentrated in dimensions j and $j+1$ and associated to a free R-module. In part (ii) we have that $\text{im}(c_{d+1})$ is stably finitely generated free by the exact sequence

$$0 \to \cdots \to C_{d+2} \xrightarrow{c_{d+2}} C_{d+1} \to \text{im}(c_{d+1}) \to 0,$$

and hence $C_d^\perp$ is as well. Thus we can make E free by adding an elementary R-chain complex that is concentrated in dimensions d and $d-1$ and associated to a free R-module.

(iv) If $j = d$ we cannot add free modules to two consecutive dimensions in the above argument. But by the above arguments we still achieve stably free R-chain complexes D and E. □

Definition 14.124 (k-connected chain map) An R-chain map $f: C \to D$ is called *k-connected* if $H_i(\text{cone}(f)) = 0$ for $i \leq k$.

Given an R-chain complex C with the differentials $c_n: C_n \to C_{n-1}$ and a cycle $x \in C_n$, we will sometimes seek an $(n+1)$-dimensional chain $y \in C_{n+1}$ (if it exists), such that $c_{n+1}(y) = x$. The chain y will be called a *nullhomology* of x since it guarantees that $0 = [x] \in H_n(C)$.

Exercise 14.125 Show that an R-chain map is k-connected if and only if the induced map $H_i(f): H_i(C) \to H_i(D)$ is an isomorphism for $i < k$ and an epimorphism for $i = k$.

14.7 Notes

Given a ring R, one can consider the derived category of the category of R-chain complexes as described, for example, in [421, Chapter 10]. The ideas presented in Section 14.6.9 are important in showing that the derived category is a *triangulated category* in the sense of Verdier. For the definition of a triangulated category, see [421, Definition 10.2.1], and for the proof of the result, see [421, Section 10.4].

Homotopy colimits and limits over arbitrary indexing categories are treated, for instance, in [43], [116, Section 3], and [194].

Chapter 15
Algebraic Surgery

15.1 Introduction

The main purpose of this chapter is to introduce a chain complex version of the L-groups and of the surgery obstructions and to identify them with the L-groups and surgery obstructions from Chapters 8 and 9, see Theorem 15.3 and Theorem 15.4.

To motivate the need for such versions, let us discuss the theory of surgery obstructions presented in Chapters 8 and 9 from a more conceptual point of view. To a given normal map of degree one $(f, \overline{f}) \colon M \to X$, we associated an element $\sigma(f, \overline{f}) \in L_n(\mathbb{Z}\pi, w)$ with the property that, provided $n \geq 5$, $\sigma(f, \overline{f}) = 0$ if and only if $(f, \overline{f})$ is normally cobordant to a homotopy equivalence, see Theorem 8.112 and Theorem 9.60.

As we have seen, this is a very powerful result. However, it suffers from two conceptual drawbacks:

(i) The structures representing $\sigma(f, \overline{f})$ are of different natures depending on the parity of n. We have a quadratic form when $n = 2k$ and a quadratic formation when $n = 2k + 1$;
(ii) In order to read off either the quadratic form or the quadratic formation representing $\sigma(f, \overline{f})$, we first need to do surgery below middle dimension.

Ranicki's algebraic theory of surgery was primarily invented to rectify these deficiencies, see [344], [345], [348] following a request of Wall [414, 17G]. However, it also turned out to allow for far reaching generalisations. Some early examples are localisation sequences, see [346], the construction of L-groups of additive categories with chain duality, see [347], the construction of L-spectra, the total surgery obstruction, see [343, 348], and algebraic transfer for fibre bundles, see [265]. We will say more about these in Section 15.9.

The algebraic theory of surgery uses chain complexes instead of modules. The first obstacle in developing such a theory is to find out what the analogue of a symmetric bilinear form or a quadratic form is when we replace the underlying module by a chain complex. These analogues come under the names of symmetric and quadratic structures on chain complexes and the definition is inspired by geometry. Recall that the intersection form of a manifold is given by the cup product pairing. The symmetric structure is a straightforward generalisation of the cup product map on the level of chain complexes. Recall that at this level the cup product is only well defined up to chain homotopy. Hence we are lead to the "derived" point of view. As a result, the symmetric and quadratic structures on chain complexes will be well defined only up to chain homotopy but they will be chain homotopy invariant.

The next question is how to organise such structured chain complexes into an L-group. Here the inspiration comes from Chapter 13, where the L-groups were redefined as certain complicated cobordism groups. The structured chain complexes can also be organised in cobordism groups and so we obtain yet another definition of L-groups. It will be shown that these groups are isomorphic to the L-groups from Chapters 8 and 9. The main tool will be algebraic surgery on structured chain complexes. The hyperbolic forms, respectively the boundary formations, will correspond to nullcobordant complexes (= boundaries). So at the end Wall's group $L_n(\mathbb{Z}\pi, w)$ becomes the cobordism group of n-dimensional quadratic chain complexes over $\mathbb{Z}\pi$. Note that this solves our problem (i) above.

Furthermore we explain how such a quadratic chain complex comes from a normal map of degree one. The construction will rely on stable homotopy theory. The use of stable homotopy theory is inspired by the construction of Browder in [55, Chapter III], as mentioned in Remark 8.205, and utilises the application of S-duality to the Spivak normal structures, which we already dealt with in Chapter 6. The resulting construction also provides us with a non-simply connected generalisation of Browder's construction and a solution to problem (ii).

The chapter is organised as follows. Section 15.2 contains a non-technical overview of the whole chapter, including the statement of the main Theorems 15.3 and 15.4. In Section 15.3 we introduce the symmetric and quadratic structures on chain complexes and describe the relationships between these structures on a fixed chain complex C. We also discuss the geometric situations that yield such structured chain complexes. The symmetric case is standard algebraic topology, while the quadratic case requires some stable homotopy theory. In Section 15.4 we show that 0-dimensional complexes correspond to forms and 1-dimensional complexes correspond to (split) formations. In Section 15.5 we introduce relative versions of structured chain complexes and obtain the notion of a cobordism. It takes some work to see that cobordism yields an equivalence relation, but, after this is settled, we obtain a chain complex version of L-groups in Definition 15.301. Sections 15.5.3 to 15.5.5 introduce technical tools needed for establishing the isomorphism between the quadratic L-groups from Chapters 8 and 9 and the quadratic L-groups from Definition 15.301, including the notion of algebraic surgery. The proof of the isomorphism from Theorem 15.3 is obtained in Section 15.6. Section 15.7 contains the identification of the surgery obstructions from Theorem 15.4. Section 15.8 treats the simple versions. Finally, Section 15.9 contains a brief overview of some of the applications.

Guide 15.1 The chapter can be read linearly, but, in case the reader is interested only in certain parts of the story, some sections may be skipped. We recommend to all readers to go through the overview of this chapter in Section 15.2 to get some insight without going through all the technical details. Below we offer two possibilities how to proceed afterwards for the readers only interested in specific parts.

For the reader who is interested in the *algebraic part*, it is possible to reach the purely algebraic Theorem 15.3 without studying stable homotopy theory by going through the following sections:

- Definition of $L_n(R)_{\text{chain}}$: Subsections 15.3.1, 15.3.7, 15.5.1 and 15.5.2;
- $L_n(R)_{\text{chain}}$ versus $L_n(R)_{\text{forms}}$: Section 15.4 for how to view forms and formations as structured chain complexes, then for certain tools Subsections 15.5.3, 15.5.4, 15.5.5, and finally Section 15.6 for the identification itself.

On the other hand, the reader who wishes to understand the definition of the *surgery obstruction* in the chain complex version and is willing to take the identification of the L-groups as a black box may proceed by going through:

- Chain complex definition of $\sigma(f, \overline{f})$: Subsections 15.3.1 to 15.3.9, 15.5.1, and 15.5.2;
- Identification with the previously defined surgery obstructions in terms of forms and formations: Section 15.7.

One drawback of Ranicki's presentation in the original sources is that he very often describes certain constructions, such as the algebraic surgery, only by writing down formulas without giving any structural insight. In our exposition we try to give certain general chain complex constructions which shall motivate the outcome and lead finally to explicit formulas. Moreover, Ranicki switches between a (co)cycle version and a version where only the relevant (co)homology class in the Q-group matters. We try to make clear where the differences are and why one sometimes should consider (co)cycles, although the basic invariants involve only their (co)homology class.

This chapter also contains many remarks, which are only meant as a motivation. Their name often explains the contents, so if the reader feels they are not necessary, they can be skipped.

The chain complex conventions that we use and that differ from [344] and [427] are summarised in the previous Chapter 14.

15.2 Overview

In this section we give an overview of the material of this chapter, which contains the main results, the basic ideas, and the strategies of proofs. It is designed for a reader for whom the introduction is not detailed enough, but who does not want to go through all the technical details, which will be presented in the remainder of this chapter.

15.2.1 Statement of the Main Results

There are two main theorems appearing in this chapter. The first one identifies the versions of the L-groups of Chapters 8 and 9 in terms of quadratic forms and quadratic formations with their chain complex analogues, which we will develop below. Here we only state the version where the underlying modules are finitely generated free, which corresponds to the decoration h, and the version where the underlying modules are finitely generated projective, which corresponds to the decoration p. The simple version, i.e., the decoration s, will be stated in Theorem 15.464.

Definition 15.2 (Projective L-groups) If we replace in Definitions 8.90 and 9.15 finitely generated free by finitely generated projective everywhere, we obtain the *projective quadratic L-groups $L_n^p(R)$*.

Theorem 15.3 (Identification of the form/formation and the chain complex version of quadratic L-groups) *Let R be a ring with involution. Then there are for all $n \in \mathbb{Z}$ natural isomorphisms*

$$L_n^p(R)_{\text{chain}} \xrightarrow{\cong} L_n^p(R);$$
$$L_n(R)_{\text{chain}} = L_n^h(R)_{\text{chain}} \xrightarrow{\cong} L_n(R) = L_n^h(R)$$

between the L-groups defined via chain complexes, see Definitions 15.301 and 15.302, and the L-groups defined via quadratic forms and formations, see Definitions 8.90, Definition 9.15, and Definition 15.2.

We will define the quadratic signature $\text{qsign}_\pi(f, \overline{f})$ of a normal map of degree one $(f, \overline{f}) \colon M \to X$ in Definition 15.318. It is a chain complex version of the surgery obstructions defined in Chapters 8 and 9. We will identify them for the decoration h by proving the following Theorem 15.4, which is the second main theorem of this chapter. The simple case will be stated in Theorem 15.464.

Theorem 15.4 (Identification of the surgery obstruction with the quadratic signature) *Let $(f, \overline{f}) \colon M \to X$ be normal map of degree one of dimension $n \geq 5$. Then under the identification of L-groups from Theorem 15.3 we get the identity*

$$\sigma(f, \overline{f}) = \text{qsign}_\pi(f, \overline{f}) \in L_n(\mathbb{Z}\pi, w)$$

between the surgery obstruction of Theorem 8.112 and 9.60 and the quadratic signature of Definition 15.318.

15.2.2 Some Basic Notions and Explanations

Recall from Chapter 8 that an ε-symmetric form φ on a finitely generated projective R-module P is an element

$$\varphi \in Q^\varepsilon(P) = \ker((1 - \varepsilon \cdot T)\colon \hom_R(P, P^*) \to \hom_R(P, P^*)) \cong$$
$$\cong \hom_{\mathbb{Z}[\mathbb{Z}/2]}(\mathbb{Z}, \hom_R(P, P^*))$$

where $\mathbb{Z}$ in the last expression is considered as the trivial $\mathbb{Z}[\mathbb{Z}/2]$-module and $\hom_R(P, P^*)$ becomes a $\mathbb{Z}[\mathbb{Z}/2]$-module via the involution $\varepsilon \cdot T$.

An n-dimensional *symmetric structure* on a bounded R-chain complex C of finitely generated projective modules is a cohomology class

$$[\varphi] \in Q^n(C) := H^n(\mathbb{Z}/2, C \otimes_R C) := H_n(\hom_{\mathbb{Z}[\mathbb{Z}/2]}(W, C \otimes_R C))$$

where W denotes the standard free $\mathbb{Z}[\mathbb{Z}/2]$-resolution of $\mathbb{Z}$. It is represented by a sequence $(\varphi_s)_{s \geq 0}$, with $\varphi_s \in (C \otimes_R C)_{n+s}$, such that, if we use the slant isomorphism $C \otimes_R C \cong \hom_R(C^{-*}, C)$, then $\varphi_0 \colon C^{n-*} \to C$ is an R-chain map and $\varphi_{s+1}\colon \varphi_s \simeq \varphi_s^*$ for all $s \geq 0$. It turns out that it is both necessary and convenient to keep all these maps φ_s in consideration. Observe that this definition is made independently of the parity of n and that, if we pick $C = 0[P^*]$, then we have $Q^0(0[P^*]) = Q^{+1}(P)$. A symmetric complex $(C, [\varphi])$ is called *Poincaré* if φ_0 is an R-chain homotopy equivalence, which is a condition that corresponds to the non-singularity of a symmetric form.

In Chapter 8 the key example of a symmetric bilinear form was the algebraic intersection form (8.57) of a $(2k)$-dimensional degree one normal map $(f, \bar{f})\colon M \to X$ defined on its surgery kernel $K_k(\widetilde{M})$. In its definition the Poincaré duality isomorphism on surgery kernels was used, which was derived from the Poincaré duality isomorphisms of M and X. These are obtained by the cap products. For X it is on the chain level given by

$$- \cap [X]\colon C^{n-*}(\widetilde{X}) \to C(\widetilde{X})$$

for some representative of the fundamental class $[X] \in H_n(X; \mathbb{Z}^w)$ and analogously for M.

Abbreviate $R = \mathbb{Z}\pi$, $C = C(\widetilde{X})$, and for any class $[\gamma] \in H_n(X; \mathbb{Z}^w)$ take $\varphi_0 = -\cap \gamma$ for some representative γ. Recall that this cap product is not self-dual at the level of chain complexes, meaning that in general $\varphi_0 \neq \varphi_0^*$. However, it is a well-known fact in algebraic topology that it is self-dual on homology, namely, there exists a chain homotopy $\varphi_1\colon \varphi_0 \simeq \varphi_0^*$. This homotopy is again self-dual only up to homotopy and so on, so there exists a sequence of higher homotopies $\varphi_{s+1}\colon \varphi_s \simeq \varphi_s^*$ and we obtain a symmetric structure, which we denote by $\overline{\varphi}_{\widetilde{X}, \pi, w}([\gamma]) \in Q^n(C)$. For X a Poincaré complex (and hence also for a closed manifold) the resulting n-dimensional symmetric complex

$$(C(\widetilde{X}), \overline{\varphi}_{\widetilde{X}, \pi, w}([X]))$$

is Poincaré.

In the above construction there is the technical issue, suppressed in this overview, that we want to obtain a construction that is natural in X. For this it is necessary to work with the singular chain complex of X, rather than with the cellular chain complex. However, once all necessary constructions are done, we may pass to the cellular complex, which has better finiteness properties.

In analogy, an n-dimensional *quadratic structure* on a bounded R-chain complex C of finitely generated projective modules is a homology class

$$[\psi] \in Q_n(C) := H_n(\mathbb{Z}/2, C \otimes_R C) := H_n(W \otimes_{\mathbb{Z}[\mathbb{Z}/2]} (C \otimes_R C)).$$

It is also represented by a sequence $(\psi_s)_{s \geq 0}$, with $\psi_s \in (C \otimes_R C)_{n-s}$, satisfying some relations, but this time it is somewhat less intuitive to explain what they mean, so we refrain from doing so here.

A standard fact from homological algebra is that the norm map $1 + T$ from group homology to group cohomology exists, and that the difference is measured by the Tate cohomology via the long exact sequence:

$$\cdots \to Q_n(C) \xrightarrow{1+T} Q^n(C) \to \widehat{Q}^n(C) \to Q_{n-1}(C) \to \cdots$$

where $\widehat{Q}^n(C) := \widehat{H}^n(\mathbb{Z}/2, C \otimes_R C)$. A quadratic complex $(C, [\psi])$ is called *Poincaré* if its symmetrisation $(C, (1 + T)([\psi]))$ is Poincaré. The exact sequence also suggests a method for producing a quadratic structure, namely to obtain a symmetric structure and a reason for its vanishing in the Tate cohomology.

At this point stable homotopy can be employed via the following idea. Given a pointed space Y and a homology class $[\gamma] \in \widetilde{H}_n(Y)$, we can consider the reduced suspension ΣY and the corresponding homology class $\Sigma([\gamma])$ in $\widetilde{H}_{n+1}(\Sigma Y) \cong \widetilde{H}_n(Y)$. The symmetric construction φ_Y applied to $[\gamma]$ gives rise to an n-dimensional symmetric structure on $C(Y)$ and $\varphi_{\Sigma Y}$ applied to $\Sigma([\gamma])$ gives rise to an $(n + 1)$-dimensional symmetric structure on the reduced chain complex $\widetilde{C}(\Sigma Y) \simeq \Sigma \widetilde{C}(Y)$. In fact, generalising to the setting with arbitrary R, it is possible to define for any R-chain complex C the suspension map

$$S \colon Q^n(C) = H^n(\mathbb{Z}/2; C \otimes_R C) \to Q^{n+1}(\Sigma C) = H^{n+1}(\mathbb{Z}/2; \Sigma C \otimes_R \Sigma C)$$

such that, given $C = C(\widetilde{X})$ and $[\gamma] \in \widetilde{H}_n(X; \mathbb{Z}^w)$ as above, the application of S to the symmetric structure associated via $\overline{\varphi}_{\widetilde{X}, \pi, w}$ to $[\gamma]$ produces the symmetric structure associated via $\overline{\varphi}_{\Sigma \widetilde{X}_+, \pi, w}$ to $\Sigma([\gamma])$. Moreover, it turns out that for the Tate cohomology group we have

$$\widehat{Q}^n(C) \cong \operatorname{colim}_k Q^{n+k}(\Sigma^k C),$$

where the structure maps in the colimit are obtained by iterating the above suspension map S. Hence to produce a quadratic structure, it is enough to obtain a symmetric structure and a reason for its vanishing in the colimit of the suspension maps.

This is utilised in the context of surgery as follows. Recall that a degree one normal map $(f, \overline{f}) \colon M \to X$ with $\pi = \pi_1(X)$ and $w \colon \pi \to \{\pm 1\}$ induces a map of Thom spaces $\mathrm{Th}(\overline{f}) \colon \mathrm{Th}(\nu_M) \to \mathrm{Th}(\nu_X)$, which gives via S-duality a homotopy

15.2 Overview

class of maps of the form $F: \Sigma^p(\widetilde{X}_+) \to \Sigma^p(\widetilde{M}_+)$ for some p. This map induces the Umkehr map $C^!(f): C(\widetilde{X}) \to C(\widetilde{M})$ at the level of chain complexes whose mapping cone $\mathrm{cone}(C^!(f))$ has homology groups isomorphic to the surgery kernel groups $K_i(\widetilde{M})$. We obtain the following symmetric structure using the induced map $C^!(f)^{Q,n}: Q^n(C(\widetilde{X})) \to Q^n(C(\widetilde{M}))$ by

$$C^!(f)^{Q,n}(\overline{\varphi}_{\widetilde{X},\pi,w}([X])) - \overline{\varphi}_{\widetilde{M},\pi,w}([M]) \in Q^n(C(\widetilde{M})).$$

It can be pushed forward by the canonical map $e: C(\widetilde{M}) \to \mathrm{cone}(C^!(f))$ to $\mathrm{cone}(C^!(f))$, and this class will be the underlying symmetric structure of the desired quadratic structure.

It remains to find a reason for its vanishing in the colimit of suspension maps. By the naturality of the symmetric construction with respect to maps of spaces we obtain that

$$C(F)^{Q,n+p}(\overline{\varphi}_{\Sigma^p\widetilde{X}_+,\pi,w}(\Sigma^p[X])) - \overline{\varphi}_{\Sigma^p\widetilde{M}_+,\pi,w}(\Sigma^p[M]) = 0 \in Q^{n+p}(\widetilde{C}(\Sigma^p\widetilde{M}_+)),$$

and certain quite complicated technical arguments provide us with a well-defined *reason* for this equation, and hence we obtain a quadratic structure denoted $\overline{\psi}_{\{F\},\pi,w}[X] \in Q_n(C(\widetilde{M}))$. The pushforward to $\mathrm{cone}(C^!(f))$ yields the n-dimensional quadratic complex

$$(\mathrm{cone}(C^!(f)), e_{Q,n} \circ \overline{\psi}_{\{F\},\pi,w}[X]),$$

which is Poincaré. In the above reasoning it is important that the symmetric construction φ_- is natural for maps induced by maps between spaces, such as $C(F)$, but not for arbitrary chain maps, such as $C^!(f)$. This has the consequence that in the above displays we get a possibly non-zero element in $Q^n(C(\widetilde{M}))$, but a zero element in $Q^{n+p}(\widetilde{C}(\Sigma^p\widetilde{M}_+))$.

Section 15.4 contains a discussion of the relation between symmetric and quadratic forms and 0-dimensional symmetric and quadratic complexes, which is quite obvious, and between symmetric and quadratic formations and 1-dimensional symmetric and quadratic complexes, where the basic idea is that the two non-zero modules of a 1-dimensional complex correspond to the two lagrangians of a formation. However, there are some technical details, which mean that in the quadratic case we need to consider a certain refinement of formations, called *split formations*. In the L-groups the difference between formations and split formations will not play a role.

In Section 15.5 we turn to cobordisms of structured chain complexes. The basic idea is as follows. First, an $(n+1)$-dimensional *symmetric pair* is an R-chain map $f: C \to D$ together with an element $[\varphi, \delta\varphi]$ in $Q^{n+1}(f)$, which is a group that fits into a long exact sequence

$$\cdots \to Q^{n+1}(C) \xrightarrow{f^{Q,n+1}} Q^{n+1}(D) \to Q^{n+1}(f) \xrightarrow{\partial^Q_{n+1}} Q^n(C) \xrightarrow{f^{Q,n}} Q^n(D) \to \cdots.$$

In particular, $[\varphi] = \partial^Q_{n+1}[\varphi, \delta\varphi] \in Q^n(C)$ is an n-dimensional symmetric structure on C. Here $\delta\varphi$ also consists of components $\delta\varphi_s \in (D \otimes_R D)_{n+1+s}$, and certain relations between φ_s and $\delta\varphi_s$ hold, which we again suppress here. Geometrically this corresponds to the situation where we have a pair of spaces (X, A) and a class $[\gamma] \in H_{n+1}(X, A)$ and we take $f\colon C = C(A) \to D = C(X)$ induced by the inclusion. Then the map

$$\mathrm{ev}_r(\varphi, \delta\varphi) := \begin{pmatrix} \varphi_0 \circ f^* \\ \delta\varphi_0 \end{pmatrix} \colon D^{n+1-*} \to \mathrm{cone}(f)$$

corresponds to the relative cap product $-\cap \gamma\colon C^{n+1-*}(X) \to C(X, A)$. A symmetric pair is called Poincaré if $\mathrm{ev}_r(\varphi, \delta\varphi)$ is an R-chain homotopy equivalence. This is a condition that holds in the above geometric example if (X, A) is a geometric Poincaré pair (e.g., a manifold with boundary) and $[\gamma] \in H_{n+1}(X, A)$ is its fundamental class. A *cobordism* between Poincaré complexes $(C, [\varphi])$ and $(C', [\varphi'])$ is a symmetric Poincaré pair of the shape

$$(f \oplus f'\colon C \oplus C' \to D, [-\varphi \oplus \varphi', \delta\varphi]).$$

Cobordism is shown to be an equivalence relation, the hardest part being the proof of transitivity, where the key technical part is the definition of gluing of cobordisms with common ends. Analogously one obtains the quadratic cobordisms, which produces an equivalence relation. Here, the higher components of the structures are used, see the formulas in Remarks 15.294 and 15.296. Note that we need the whole sequence of higher homotopies since the s-th component of the glued structure contains the $(s-1)$-th component of the input and we want to be able to iterate the gluing construction. The chain complex L-groups $L^n(R)_{\text{chain}}$ and $L_n(R)_{\text{chain}}$ appearing in Theorem 15.3 are groups of cobordism classes of n-dimensional structured Poincaré complexes with addition given by direct sum in both symmetric and quadratic case. The parity of n plays no role in the definition.

Given an n-dimensional geometric Poincaré complex X with the fundamental class $[X] \in H_n(X; \mathbb{Z}^w)$ and $\pi = \pi_1(X)$, the symmetric construction discussed at the beginning of this section provides us with a well-defined element, called the *symmetric signature* of X, see Definition 15.312,

$$\mathrm{ssign}_\pi(X) = [(C(\widetilde{X}), \overline{\varphi}_{\widetilde{X}, \pi, w}([X]))] \in L^n(\mathbb{Z}\pi)_{\text{chain}}.$$

Given a degree one normal map $(f, \overline{f})\colon M \to X$ from an n-dimensional manifold to an n-dimensional Poincaré complex X, the so-called quadratic construction provides us with a well-defined element, called the *quadratic signature* of $(f, \overline{f})$

$$\mathrm{qsign}_\pi(f, \overline{f}) = [(\mathrm{cone}(C^!(f)), e_{Q,n} \circ \overline{\psi}_{\{F\}, \pi, w}([X]))] \in L_n(\mathbb{Z}\pi)_{\text{chain}}$$

appearing in Theorem 15.4. By the relative version of the quadratic construction, it only depends on the normal bordism class of $(f, \overline{f})$.

15.2 Overview

The task of the identification in Theorem 15.3 requires a substantial technical preparation, involving the three related notions of the algebraic Thom Construction, the algebraic boundary and the algebraic surgery.

The *algebraic Thom Construction* associates to an $(n + 1)$-dimensional symmetric pair $(f\colon C \to D, [\varphi, \delta\varphi])$ and $(n + 1)$-dimensional symmetric complex $(\operatorname{cone}(f), [\delta\varphi/\varphi])$, using a certain homomorphism of the shape

$$z\colon Q^{n+1}(f) \to Q^{n+1}(\operatorname{cone}(f)), \quad z\colon [\varphi, \delta\varphi] \mapsto [\delta\varphi/\varphi].$$

It corresponds to the geometric situation where we again have a pair of spaces (X, A) and a class $[\gamma] \in H_{n+1}(X, A)$ and take $f\colon C = C(A) \to D = C(X)$ induced by the inclusion. As explained above, in this situation we have a symmetric pair $[\varphi, \delta\varphi]$, which encodes the relative cap products with γ. However, γ also induces absolute cap products of the shape $-\cap \gamma\colon C^{n+1-*}(X, A) \simeq \widetilde{C}^{n+1-*}(X/A) \to C(X, A) \simeq \widetilde{C}(X/A)$. The algebraic Thom Construction $[\delta\varphi/\varphi]$ encodes these absolute products, which also motivates its name.

Note that, in accordance with the geometric motivation, when we start with a Poincaré pair we in general end up with a complex that is not Poincaré, just as passing from a manifold with boundary $(W, \partial W)$ to the quotient $W/\partial W$ in general produces a space that is not Poincaré.

The *algebraic boundary* of an $(n + 1)$-dimensional symmetric complex $(E, [\varphi])$ is a certain n-dimensional symmetric Poincaré complex $(\partial E, [\partial\varphi])$ where

$$\partial E := \Sigma^{-1} \operatorname{cone}(\varphi_0\colon E^{n+1-*} \to E),$$

and that is defined together with a certain $(n + 1)$-dimensional Poincaré pair $(i\colon \partial E \to E^{n+1-*}, [\partial\varphi, \overline{\varphi}])$, where i is the canonical map. It reverses the algebraic Thom Construction in the sense that the algebraic Thom Construction of this pair is homotopy equivalent to $(E, [\varphi])$. The geometric significance is as follows. Assume that we have an $(n+1)$-dimensional manifold with boundary $(W, \partial W)$. Then we have a cofibration sequence $C = C(\partial W) \to D = C(W) \to E = C(W, \partial W)$. Moreover, recall that cap products with $\gamma = [W, \partial W] \in H_{n+1}(W, \partial W)$ induce the commutative ladder

$$\begin{array}{ccccccc}
H^{k-1}(\partial W) & \longrightarrow & H^k(W, \partial W) & \longrightarrow & H^k(W) & \longrightarrow & H^k(\partial W) \\
\cong \Big\downarrow {-\cap\partial_{n+1}(\gamma)} & & \cong \Big\downarrow {-\cap\gamma} & & \Big\downarrow {-\cap\gamma} \cong & & \Big\downarrow {-\cap\partial_{n+1}(\gamma)} \cong \\
H_{n-(k-1)}(\partial W) & \longrightarrow & H_{n+1-k}(W) & \longrightarrow & H_{n+1-k}(W, \partial W) & \longrightarrow & H_{n-k}(\partial W).
\end{array}$$

The isomorphism in the second column is induced by a chain homotopy equivalence $E^{n+1-*} \to D$, and it turns out that the diagonal composition in the second square is induced by φ_0. This tells us that we can recover the chain complex C from φ_0 since C is chain homotopy equivalent to the homotopy fibre of $D \to E$, and it is also possible to recover the symmetric structure on C. The algebraic boundary of a quadratic complex is a straightforward analogue.

For us the algebraic boundary will primarily be a device for producing nullcobordant complexes: observe that $(\partial E, [\partial \varphi])$ is nullcobordant since it is the source in a Poincaré pair with the underlying map i. We note that the higher components of the structures also play a crucial role here, as can be seen from the formulas in Remark 15.370 and 15.414.

The *algebraic surgery* is defined in the following way. Let $(C, [\varphi])$ be an n-dimensional symmetric complex (in general it does not have to be Poincaré, but we will only be interested in this case). The data for algebraic surgery on $(C, [\varphi])$ consist of an $(n + 1)$-dimensional symmetric pair $(f : C \to D, [\varphi, \delta\varphi])$ (this pair will typically not be Poincaré). The outcome of algebraic surgery on these data will be a certain n-dimensional symmetric complex $(C', [\varphi'])$ with

$$C' := \Sigma^{-1} \operatorname{cone}\left(\operatorname{ev}_r(\varphi, \delta\varphi) = \begin{pmatrix} \varphi_0 \circ f^* \\ \delta\varphi_0 \end{pmatrix} : D^{n+1-*} \to \operatorname{cone}(f)\right).$$

Just as in the geometric situation, algebraic surgery also produces cobordant complexes via certain $(n + 1)$-dimensional cobordism of the shape

$$(g \oplus g' : C \oplus C' \to F, [-\varphi \oplus \varphi', \delta\varphi'']),$$

where $F = \operatorname{cone}(f)^{n+1-*}$.

The above setup corresponds to the geometric situation with a cobordism $(W; M, M')$ by assigning $f : C = C(M) \to D = C(W, M')$ induced by the inclusion. Note that then $\operatorname{cone}(f) \simeq C(W, \partial W)$, and $D^{n+1-*} \simeq C^{n+1-*}(W, M') \simeq C(W, M)$. It turns out that the defining map $\operatorname{ev}_r(\varphi, \delta\varphi)$ for C' corresponds to the canonical map $C(W, M) \to C(W, \partial W)$ so that we obtain an identification $C(M') \simeq C'$. Also $F = \operatorname{cone}(f)^{n+1-*} \simeq C^{n+1-*}(W, \partial W) \simeq C(W)$. A convenient summary of the whole situation is provided in Diagram (15.391) in Section 15.5.5. Again there is a straightforward quadratic analogue. We warn the reader here that, although the definitions in the symmetric and quadratic case are analogous, some properties will be different.

The setup of geometric surgery is such that, given an embedding of a thickened sphere with compatible bundle data into M, we perform the surgery to obtain M' and the trace W. The inclusion of the sphere defines a homology class, say $x \in H_k(M)$. The starting data of algebraic surgery include a map f from C to D. This corresponds to the dual $x^* \in H^{n-k}(M)$, so in algebraic surgery we a priori kill cohomology classes rather than homology classes.[1] The benefit of the algebraic version is that we do not need to do surgery step-by-step, we might attempt to kill all the cohomology above the middle dimension in one step by taking D to be the quotient of C given by factoring out everything below and including in the middle dimension. It turns out that this is always possible in the quadratic case (but in the symmetric case there are obstructions to this).

---

[1] Of course, by duality in both cases homology and cohomology classes are killed simultaneously, but in terms of presentation it makes a difference whether we define surgery using homological or cohomological data.

15.2.3 Outline of the Proofs

Armed with all the tools from the previous section, we proceed with an outline of the proofs of the main theorems. Diagram (15.5) below summarises the whole identification process leading to Theorem 15.3, and we augment it with the following explanations.

$$
\begin{array}{c}
n\text{-dim. quad. compl.}\big/\text{cobordism} \\
\Big\downarrow \text{skew suspension} \left\{ \text{Prop. 15.427} \right. \\
0/1\text{-dim. quad. compl.}\big/\text{cobordism} \\
\Big\downarrow \text{algebraic surgery} \left\{ \text{Prop. 15.430} \right. \\
\text{highly ctd } 0/1\text{-dim. quad. compl.}\big/\text{highly ctd cobordism} \\
\Big\downarrow \text{algebraic surgery} \left\{ \text{Prop. 15.433} \right. \\
\text{highly ctd } 0/1\text{-dim. quad. compl.}\big/\text{well ctd cobordism} \\
\Big\downarrow \text{homological algebra} \left\{ \text{Prop. 15.434} \right. \\
\text{strictly } 0/1\text{-dim. quad. compl.}\big/\text{strictly well ctd cob.} \\
\Big\downarrow \text{identification of low dimensions and boundaries} \left\{ \text{Prop. 15.202, 15.214, 15.439, 15.440} \right. \\
\text{forms/formations}\big/\text{hyperbolics/boundaries.}
\end{array}
\tag{15.5}
$$

Before we start, we note that the chain complex version of the L-groups is defined as the cobordism group of n-dimensional quadratic Poincaré complexes $(C, [\psi])$ where the dimension refers to the dimension of $[\psi] \in Q_n(C)$ and for C itself it is only demanded that it is a bounded chain complex of finitely generated projective modules, in particular it is allowed that it is non-zero in negative dimensions. In the first step of the desired identification the skew-suspension map, which induces 4-periodicity, is used to shift our attention from dimension n to dimension 0 or 1, bearing in mind the above fact about the underlying chain complexes. Next, algebraic surgery is used to kill the homology and cohomology groups of C in dimensions other than 0 when n is even, or 0 and 1 when n is odd. This can always be done in the quadratic case, and in addition it can be done in one step as explained above. It is also necessary to do algebraic surgery on cobordisms. This is done in two steps. We first achieve what we call highly connected cobordisms, so that their homology is contained in at most three dimensions by a direct analogue of surgery on complexes.

In the second step we achieve what we call well-connected cobordisms, which is an improvement ensuring that we only need to consider cobordisms whose homology is restricted to at most two dimensions. Once we have this, some additional homological algebra allows us to assume that the complexes themselves are concentrated in one or two degrees, and similarly for cobordisms. At this point, the identification of forms and formations with low-dimensional complexes from Section 15.4 is applied, together with some additional arguments leading to the identification of those forms and formations that were factored out in their L-groups (i.e., hyperbolic forms and boundary formations), with nullcobordant low-dimensional complexes that are obtained via the algebraic boundary construction.

For the proof of Theorem 15.4 in Section 15.7 we only offer the following rough ideas. Recall that at this point we already have bordism invariance of the quadratic signature with varying target and bordism invariance of the surgery obstruction with cylindrical target. Hence it suffices to provide an identification for highly connected degree one normal maps. Since the surgery obstruction in odd dimensions is defined in terms of surgery kernels for even-dimensional maps and their cobordisms, the crucial part is the identification in the even-dimensional case.

In the highly connected even-dimensional situation the quadratic chain complex corresponds to a quadratic form, and we need to look at its values on (co)homology classes representing framed immersed spheres of the surgery kernel. The surgery obstruction form is on these classes given by a $\mathbb{Z}\pi$-valued self-intersection number. The values of the quadratic from corresponding to the quadratic signature are extracted from it by the so-called Wu classes from Section 15.3.9. The key observation is a relation, for a (co)homology class representing an immersed sphere, between this Wu class, the framing obstruction, and the self-intersection number, which is obtained by carefully studying the quadratic construction in Proposition 15.456. Since the immersions in the surgery kernel are all framed, the framing obstruction vanishes. This enters into the identification of the two quadratic forms in question in Proposition 15.455. This technical proposition is then used to prove Propositions 15.460 (even-dimensional case) and 15.461 (odd-dimensional case), which together yield the desired Theorem 15.4.

15.3 Structured Chain Complexes

In this section we introduce structured chain complexes, meaning symmetric, quadratic and hyperquadratic chain complexes. The symmetric chain complexes generalise the symmetric bilinear forms on modules and the quadratic chain complexes generalise quadratic forms on modules. The primary purpose of hyperquadratic complexes is to measure the difference between the symmetric and quadratic structures on a fixed chain complex.

We note that, strictly speaking, one should view the structured chain complexes as analogues of co-forms, rather than forms. Recall that a form on a finitely generated projective R-module P is an element in $\hom_R(P, P^*) \cong (P \otimes_R P)^*$ whereas a co-form

15.3 Structured Chain Complexes

is an element in $\hom_R(P^*, P) \cong P \otimes_R P$. In the case of finitely generated projective modules, we have a non-canonical bijection $P \otimes_R P \to (P \otimes_R P)^*$ between forms and co-forms. We will indeed work with bounded chain complexes of finitely generated projective modules or at least with chain complexes chain homotopy equivalent to bounded chain complexes of finitely generated projective modules. The use of co-forms is convenient at certain stages, as explained in more detail in Remark 15.34.

15.3.1 Basic Definitions

Let R be a ring with involution and let $\varepsilon = \pm 1$. Several conventions for working with R-modules and with chain complexes of R-modules were collected in Chapter 14. Recall in particular various categories of such chain complexes introduced in Notation 14.22, and observe that we have obvious inclusions of full subcategories

$$\mathrm{b}R\text{-CH}_{\mathrm{fgp}} \subseteq \mathrm{h}R\text{-CH}_{\mathrm{fgp}} \subseteq R\text{-CH}.$$

Moreover, for chain complexes C, D in R-CH, the definition of the tensor product $\mathbb{Z}$-chain complex $C^t \otimes_R D$ is stated in (14.12) and the definition of the hom-complex $\hom_R(C, D)$, which is also a $\mathbb{Z}$-chain complex, is stated in (14.13); see especially the sign conventions for the differentials.

The cyclic group of order 2 is denoted $\mathbb{Z}/2 = \{1, T\}$. When dealing with several copies of $\mathbb{Z}/2$, we might index the elements as $1_s, T_s$ to distinguish the copies.

Let P be an R-module in $R\text{-MOD}_{\mathrm{fgp}}$. Recall Definitions 8.47 and 8.77 of an ε-*symmetric bilinear form* on P and of an ε-*quadratic form* on P. In (8.68) we introduced the involution $T: \varphi \mapsto \varphi^* \circ e(P)$ on the abelian group of forms $(P \otimes P)^* \cong \hom_R(P, P^*)$. An ε-symmetric bilinear form is a *fixed point* of this involution, in other words, an element in the abelian group

$$Q^\varepsilon(P) := \ker(1 - \varepsilon \cdot T) = \hom_R(P, P^*)^{\mathbb{Z}/2} \cong \hom_{\mathbb{Z}[\mathbb{Z}/2]}(\mathbb{Z}, \hom_R(P, P^*)).$$

An ε-quadratic form is an *orbit* of the involution T, in other words, an element in the abelian group

$$Q_\varepsilon(P) := \mathrm{coker}(1 - \varepsilon \cdot T) \cong \hom_R(P, P^*)_{\mathbb{Z}/2} := \mathbb{Z} \otimes_{\mathbb{Z}[\mathbb{Z}/2]} \hom_R(P, P^*).$$

The definition of a symmetric and of a quadratic chain complex in R-CH is directly inspired by these definitions. However, as already pointed out, we need homotopy invariant notions. Therefore the fixed points have to be replaced by *homotopy fixed points* and orbits have to be replaced by *homotopy orbits*.

Let C be a chain complex in R-CH. The tensor product $C^t \otimes_R C$ becomes a chain complex of $\mathbb{Z}[\mathbb{Z}/2]$-modules through the involution

$$T = T_\varepsilon: C_p^t \otimes_R C_q \to C_q^t \otimes_R C_p,$$
$$x \otimes y \mapsto (-1)^{pq} y \otimes \varepsilon x. \tag{15.6}$$

Remark 15.7 (Dropping "t") In the above definition and in the previous chapter, we used the notation $C^t \otimes_R C$ to emphasise that one has to be careful about the left and right modules and use the involution appropriately in this context. However, to ease the notation, we will drop the superscript "t" in the sequel. We implicitly assume that, whenever we have a left module in a place where only a right module fits, then we use the involution to make it a right module via the "t" procedure. For example, we simply write $C \otimes_R C$.

Remark 15.8 Let $C \in bR\text{-}\mathrm{CH}_{\mathrm{fgp}}$. Using the evaluation map e from (14.26) and the duality map I from (14.27), one can define the involution

$$T_{\mathrm{hom}}\colon \hom_R(C^{-*}, C) \to \hom_R(C^{-*}, C)$$

by $T_{\mathrm{hom}}(\varphi) = \hom(\mathrm{id}_{C^{-*}}, e(C))^{-1} \circ I(\varphi) = e(C)^{-1} \circ I(\varphi)$. On the component $\hom_R(C^p, C_q)$ it is given by the formula

$$T_{\mathrm{hom}}\colon \varphi \mapsto \left((-1)^{pq+p+q} e(C_p)^{-1} \circ \varphi^*\right).$$

The involutions T from (15.6) and T_{hom} are compatible with respect to the slant product isomorphism $-\backslash-\colon (C \otimes_R C) \to \hom_R(C^{-*}, C)$ from (14.31).

Definition 15.9 (Standard resolutions) The standard free $\mathbb{Z}[\mathbb{Z}/2]$-resolution of $\mathbb{Z}$ is the chain complex W given by

$$W_s = \begin{cases} \mathbb{Z}[\mathbb{Z}/2] = \mathbb{Z}\{1_s, T_s\} & \text{for } s \geq 0; \\ 0 & \text{otherwise,} \end{cases}$$

and

$$d_W = (1 + (-1)^s T)\colon W_s \to W_{s-1},$$

so that we have

$$W := \cdots \xrightarrow{1-T} \mathbb{Z}[\mathbb{Z}/2] \xrightarrow{1+T} \mathbb{Z}[\mathbb{Z}/2] \xrightarrow{1-T} \mathbb{Z}[\mathbb{Z}/2] \longrightarrow 0.$$

The complete standard free $\mathbb{Z}[\mathbb{Z}/2]$-resolution of $\mathbb{Z}$ is the chain complex $\widehat{W}$ given by

$$\widehat{W}_s = \mathbb{Z}[\mathbb{Z}/2] = \mathbb{Z}\{1_s, T_s\} \text{ for } s \in \mathbb{Z}$$

and

$$d_{\widehat{W}} = (1 + (-1)^s T)\colon \widehat{W}_s \to \widehat{W}_{s-1}$$

so that we have

$$\widehat{W} := \cdots \xrightarrow{1-T} \mathbb{Z}[\mathbb{Z}/2] \xrightarrow{1+T} \mathbb{Z}[\mathbb{Z}/2] \xrightarrow{1-T} \mathbb{Z}[\mathbb{Z}/2] \xrightarrow{1+T} \mathbb{Z}[\mathbb{Z}/2] \xrightarrow{1-T} \cdots.$$

We have a degreewise split short exact sequence of chain complexes

$$0 \to \Sigma^{-1} W^{-*} \to \widehat{W} \to W \to 0 \tag{15.10}$$

15.3 Structured Chain Complexes

where

$$(\Sigma^{-1}W^{-*})_s \to \widehat{W}_s \quad \text{is} \quad \begin{cases} +\mathrm{id}_{\mathbb{Z}[\mathbb{Z}/2]} & s \equiv 1,2 \ (4), s \leq -1; \\ -\mathrm{id}_{\mathbb{Z}[\mathbb{Z}/2]} & s \equiv 0,3 \ (4), s \leq -1; \\ 0 & s \geq 0, \end{cases} \tag{15.11}$$

and the map $\widehat{W} \to W$ is the projection. See equations (14.23) and (14.48) in Chapter 14 for the sign conventions in $\Sigma^{-1}W^{-*}$, which force the signs in (15.11).

Definition 15.12 (W-complexes) Let C be a chain complex in R-CH and let T be the involution on $C \otimes_R C$ from (15.6). The complex $W^\%(C)$ of the "homotopy fixed points" of the involution T, the complex $W_\%(C)$ of "homotopy orbits" of T, and the "Tate" complex $\widehat{W}^\%(C)$ of T are the following chain complexes of abelian groups:

$$W_\%(C) := W \otimes_{\mathbb{Z}[\mathbb{Z}_2]} (C \otimes_R C) \cong \hom_{\mathbb{Z}[\mathbb{Z}_2]}(W^{-*}, C \otimes_R C);$$
$$W^\%(C) := \hom_{\mathbb{Z}[\mathbb{Z}_2]}(W, C \otimes_R C);$$
$$\widehat{W}^\%(C) := \hom_{\mathbb{Z}[\mathbb{Z}_2]}(\widehat{W}, C \otimes_R C).$$

For C, D in R-CH and an R-chain map $f \colon C \to D$, the map of $\mathbb{Z}[\mathbb{Z}/2]$-chain complexes $f \otimes f \colon C \otimes_R C \to D \otimes_R D$ induces chain maps

$$f_\% \colon W_\%(C) \to W_\%(D), \quad f^\% \colon W^\%(C) \to W^\%(D), \quad \widehat{f}^\% \colon \widehat{W}^\%(C) \to \widehat{W}^\%(D).$$

Observe that the ring with involution R and $\varepsilon = \pm 1$ play a role in the definition of the chain complexes $W_\%(C)$, $W^\%(C)$, and $\widehat{W}^\%(C)$, although they do not appear in the notation. Most of the time, when we work with these complexes, the ring with involution R and the sign ε are clear so that this does not lead to confusion. However, there are a small number of instances where it is convenient to specify either R or ε or both. On those occasions we will introduce temporary notations such as $W^\%_{R,\varepsilon} C$ or similar and explicitly inform the reader about their meaning.

Remark 15.13 The functors $C \mapsto W^\%(C)$, $C \mapsto W_\%(C)$ and $C \mapsto \widehat{W}^\%(C)$ are homotopy functors from R-CH to $\mathbb{Z}$-CH in the sense that they map chain homotopy equivalent complexes to chain homotopy equivalent complexes. This and some of their other properties are studied in detail in Subsections 15.3.4 and 15.3.5. The functor $C \mapsto W_\%(C)$ has the useful property that for $C \in bR\text{-CH}_{\text{fgp}}$ the complex $W_\%(C)$ is bounded from below.

We come straight to the definition of structures on chain complexes that generalise ε-symmetric bilinear forms and ε-quadratic forms. As already indicated, we will need one more auxiliary type of structure that does not have a straightforward analogue on modules. These hyperquadratic structures will be used to study the difference between the symmetric and quadratic structures.

Definition 15.14 (Q-groups) For C in R-CH, the Q-groups of C are defined by

$$Q_n(C) = H_n(W_\%(C)), \quad Q^n(C) = H_n(W^\%(C)), \quad \widehat{Q}^n(C) = H_n(\widehat{W}^\%(C)).$$

Remark 15.15 (Q-groups are group (co-)homology groups) It might be helpful to think about these groups in terms of group co-homology since we have

$$Q_n(C) \cong H_n(\mathbb{Z}/2; C \otimes_R C);$$
$$Q^n(C) \cong H^n(\mathbb{Z}/2; C \otimes_R C);$$
$$\widehat{Q}^n(C) \cong \widehat{H}^n(\mathbb{Z}/2; C \otimes_R C).$$

Definition 15.16 (Structured chain complexes) Let C be a chain complex in bR-CH$_{\text{fgp}}$.
 (i) An n-dimensional *symmetric structure* on the chain complex C is an element $[\varphi] \in Q^n(C) := H_n(W^\%(C))$;
 (ii) An n-dimensional *quadratic structure* on the chain complex C is an element $[\psi] \in Q_n(C) := H_n(W_\%(C))$;
 (iii) An n-dimensional *hyperquadratic structure* on the chain complex C is an element $[\theta] \in \widehat{Q}^n(C) := \widehat{W}^\%(C)$;
 (iv) The pair $(C, [\varphi])$ is called an n-dimensional *symmetric complex*;
 (v) The pair $(C, [\psi])$ is called an n-dimensional *quadratic complex*;
 (vi) The pair $(C, [\theta])$ is called an n-dimensional *hyperquadratic complex*.

Note that the dimension n refers only to the degree of the element $[\varphi]$, $[\psi]$, or $[\theta]$ and does not mean that the chain complex C has to be concentrated between dimensions 0 and n.

We will also need a version where we do not consider structures as homology classes, but as actual cycles. We call the choice of a representing cycle within a given homology class a *polarisation*.

Definition 15.17 (Polarisation of structured chain complexes) Let $(C, [\varphi])$ be an n-dimensional symmetric chain complex. A *polarisation* is a choice of a cycle $\varphi \in W^\%(C)_n$ representing the class $[\varphi] \in Q^n(C)$. When we write (C, φ), this means that we have chosen a polarisation for $(C, [\varphi])$.

The definition and notation (C, ψ), and (C, θ), in the quadratic and hyperquadratic setting are analogous.

Remark 15.18 (Various versions of structured chain complexes) Our Definition 15.16 differs from definitions given by Ranicki in [344] since only chain complexes that are homotopy equivalent to chain complexes concentrated between dimensions 0 and n are considered in [344]. It agrees with Ranicki in [344] in the sense that structures are homology classes.

On the other hand, Ranicki later modifies the definitions in his book [348, Definition 1.6] where structures are cycles and the underlying chain complexes are from the category bR-CH$_{\text{fgp}}$ (there bR-CH$_{\text{fgp}}$ is denoted $\mathbb{B}(R)$). This fits with Definition 15.17.

15.3 Structured Chain Complexes

The definition including polarisation is more convenient from certain points of view, for example in the theory of normal complexes presented in [348, §3] and in [427], [428]. Moreover, certain constructions such as algebraic surgery require input on the level of a polarised structure. However, we will show that the choice of a different polarisation does not change the cobordism class and hence the element in the L-group represented by a structured chain complex. Therefore the difference between polarised and not polarised structured chain complexes plays no significant role for us in the end.

The dimension restrictions do make a difference in the symmetric case, as the resulting theories have different periodicity properties, see Remark 15.310 or [348, Example 1.11] for more explanation. In the quadratic case the resulting L-groups are the same in both settings.

Remark 15.19 (The role of $\varepsilon = \pm 1$.) Although we suppress it from the notation, the unit $\varepsilon = \pm 1$ plays a role in Definitions 15.16 and 15.17 because it appears in (15.6). Recall that ε was used in Chapters 8 and 9 to treat the groups $L_n(R)$ simultaneously for $n \equiv 0, 2$ (4) and for $n \equiv 1, 3$ (4). In the chain complex treatment it is possible to work with $\varepsilon = 1$ all the time. However, by including $\varepsilon = \pm 1$ we are able to treat the groups $L_n(R)$ simultaneously for n and $n+2$ via the skew-suspension map of Section 15.6. In particular, when identifying the groups defined by forms or formations with those defined by chain complexes, we only need to consider two cases, see also Section 15.4.

Remark 15.20 (Polarised Structures in terms of components) When we write out the structures on C from Definition 15.17 in terms of their components, we obtain the following equations.

Let $\varphi \in W^{\%}(C)_n$ be an n-dimensional symmetric structure on C. Denoting $\varphi(1_s)$ by φ_s, where $1_s \in W_s = \mathbb{Z}[\mathbb{Z}/2] = \mathbb{Z}\{1_s, T_s\}$, we see that φ is a collection of chains

$$\varphi = (\varphi_s)_{s \geq 0}, \quad \text{where} \quad \varphi_s \in (C \otimes_R C)_{n+s},$$

satisfying

$$d_{C \otimes_R C}(\varphi_s) = (-1)^n (\varphi_{s-1} + (-1)^s T(\varphi_{s-1})) \in (C \otimes_R C)_{n+s-1}. \quad (15.21)$$

These describe φ_0 as a cycle and each φ_s for $s \geq 1$ as an explicit reason why φ_{s-1} and $T(\varphi_{s-1})$ (or their negatives) are homologous.

Let $\psi \in W_{\%}(C)_n$ be an n-dimensional quadratic structure on C. If we write out $\psi = \Sigma_{s \geq 0} (1_s \otimes \psi_s)$, then ψ becomes a collection of chains

$$\psi = (\psi_s)_{s \geq 0}, \quad \text{where} \quad \psi_s \in (C \otimes_R C)_{n-s},$$

satisfying

$$d_{C \otimes_R C}(\psi_s) = T(\psi_{s+1}) + (-1)^{s+1} \psi_{s+1} \in (C \otimes_R C)_{n-s-1}. \quad (15.22)$$

These identities describe each ψ_s as an explicit reason why ψ_{s+1} and $T(\psi_{s+1})$ (or their negatives) are homologous.

Let $\theta \in \widehat{W}^{\%}(C)_n$ be an n-dimensional hyperquadratic structure on C. If we denote $\theta(1_s)$ by θ_s, we see that θ is a collection of chains $\{\theta_s \in (C \otimes_R C)_{n+s} | s \in \mathbb{Z}\}$,

$$\theta = (\theta_s)_{s \in \mathbb{Z}}, \quad \text{where} \quad \theta_s \in (C \otimes_R C)_{n+s},$$

satisfying

$$d_{C \otimes_R C}(\theta_s) = (-1)^n (\theta_{s-1} + (-1)^s T(\theta)_{s-1}) \in (C \otimes_R C)_{n+s-1}. \tag{15.23}$$

These identities describe each θ_s as an explicit reason why θ_{s-1} and $T(\theta)_{s-1}$ (or their negatives) are homologous for all $s \in \mathbb{Z}$.

Note that, when $C \in bR\text{-CH}_{\text{fgp}}$, in each case only a finite number of components is non-zero since then we have $(C \otimes_R C) \in b\mathbb{Z}[\mathbb{Z}/2]\text{-CH}_{\text{fgp}}$.

Note that, in view of the definition of $C \otimes_R C$, each of $\varphi_s, \psi_s, \theta_s$ also consists of components, which we will later index by r. As an exercise the reader may try to write out the corresponding formulas in terms of these components.

Recall that for $C \in R\text{-CH}$ we have the slant chain map (14.31)

$$-\backslash-: C \otimes_R C \to \hom_R(C^{-*}, C),$$

which is an isomorphism if $C \in bR\text{-CH}_{\text{fgp}}$, but only a chain homotopy equivalence if $C \in hR\text{-CH}_{\text{fgp}}$. Therefore, when $C \in bR\text{-CH}_{\text{fgp}}$, the identities from Remark 15.20 can be rephrased as identities with chains $\varphi_s, \psi_s, \theta_s$ represented by collections of R-module morphisms. Before we do that, we need to recall what our sign conventions from Chapter 14 tell us about the relationship between the functors hom and Σ.

Observe first that we have isomorphisms

$$\Sigma \hom_R(C, D) \to \hom_R(C, \Sigma D) \quad f \mapsto f \tag{15.24}$$

and

$$\Sigma \hom_R(C, D) \to \hom_R(\Sigma^{-1} C, D) \quad f \mapsto (-1)^{|f|} f. \tag{15.25}$$

The sign conventions for chain maps and chain homotopies guarantee that for $n \in \mathbb{Z}$ we have:

$$H_n(\hom_R(C, D)) = \{\text{htpy classes of chain maps of degree } n \text{ from } C \text{ to } D\}$$
$$= H_0(\hom_R(\Sigma^n C, D))$$
$$= \{\text{htpy classes of chain maps from } \Sigma^n C \text{ to } D\}.$$

Recall the dual R-chain complex C^{-*} from (14.23), see also Remark 15.7. We will often work with the R-chain complex $C^{n-*} = \Sigma^n C^{-*}$. By our conventions[2] we have

$$(C^{n-*})_p = (C_{n-p})^* = \hom_R(C_{n-p}, R)$$

---

[2] Here we differ from the original treatment of Ranicki. From pages 98, 104 and 105 in [344] we see that there the identification $C^{n-*} = \Sigma^n C^{-*}$ does not hold because of signs.

15.3 Structured Chain Complexes

and
$$(d_{C^{n-*}})_p = (-1)^{p+1} \cdot (c_{n-p+1})^* : (C_{n-p})^* \to (C_{n-p+1})^*.$$

Our conventions imply that an R-chain map $g: C^{n-*} \to D$ is a collection $g = \{g_p : (C_{n-p})^* \to D_p \mid p \in \mathbb{Z}\}$ satisfying

$$d_p \circ g_p = (-1)^{p+1} \cdot g_{p-1} \circ (c^*_{n-p+1}) : (C_{n-p})^* \to D_{p-1}.$$

Remark 15.26 (Polarised structures in terms of components as R-module morphisms) By the above conventions, for $C \in bR\text{-CH}_{\text{fgp}}$, the promised identities from Remark 15.20 with chains $\varphi_s, \psi_s, \theta_s$ represented as collections of R-module morphisms look as follows.

A cycle $\varphi \in W^{\%}(C)_n$ gives a collection $\{\varphi_s : C^{n+s-*} \to C \mid s \in \mathbb{N}\}$ of maps satisfying

$$d_C \circ \varphi_s - (-1)^{n+s} \varphi_s \circ d_{C^{-*}} = (-1)^n (\varphi_{s-1} + (-1)^s T(\varphi_{s-1})). \tag{15.27}$$

These identities describe φ_0 as a chain map and each φ_s for $s \geq 1$ as a chain homotopy between φ_{s-1} and $T(\varphi)_{s-1}$ (or their negatives).

Recall again that each φ_s is itself a collection of R-module homomorphisms. In terms of the components $\varphi_{s,r} : C^{n+s-r} \to C_r$, the formula above becomes

$$(d_C)_{r+1} \circ \varphi_{s,r+1} - (-1)^r \varphi_{s,r} \circ (d_C)^*_{n-r+s}$$
$$= (-1)^n \varphi_{s-1,r} + (-1)^{t(n,s,r)} \varepsilon e(C_r)^{-1} \circ \varphi^*_{s-1,n-1+s-r} : C^{n-1-r+s} \to C_r \tag{15.28}$$

where $t(n, r, s) = r \cdot (n + s) + 1$ and $e(C_r) : C_r \to (C_r^*)^*$ is the map appearing in the r-th component of the isomorphism (14.26).

A cycle $\psi \in W_{\%}(C)_n$ gives a collection $\{\psi_s : C^{n-s-*} \to C \mid s \in \mathbb{N}\}$ of maps satisfying

$$d_C \circ \psi_s - (-1)^{n-s} \psi_s \circ d_{C^{-*}} = T(\psi)_{s+1} + (-1)^{s+1} \psi_{s+1}. \tag{15.29}$$

These identities describe $T(\psi)_{s+1} + (-1)^{s+1} \psi_{s+1}$ as a measure of failure for ψ_s to be chain map of a certain degree. Another interpretation is to observe that, since C is bounded, there exists an $s' \geq 0$ satisfying $\psi_s = 0$ for each $s \geq s'$. So there will be an $s'' \geq 0$ such that $\psi_{s''}$ is a chain map of appropriate degree, $\psi_{s''-1}$ is a chain homotopy between $\psi_{s''}$ and $T(\psi)_{s''}$, and ψ_s for $s < s''$ are higher homotopies.

Similarly as above, in terms of the components $\psi_{s,r} : C^{n-s-r} \to C_r$, the formula above becomes

$$(d_C)_{r+1} \circ \psi_{s,r+1} - (-1)^r \psi_{s,r} \circ (d_C)^*_{n-r+s}$$
$$= (-1)^{t'(n,r,s)} \varepsilon e(C_r)^{-1} \circ \psi^*_{s+1,n-s-r-1} + (-1)^{s+1} \psi_{s+1,r} : C^{n-1-s-r} \to C_r \tag{15.30}$$

where now $t'(n, r, s) = (r + 1) \cdot (n + s) + 1$.

Finally a cycle $\theta \in \widehat{W}^{\%}(C)_n$ gives a collection $\{\theta_s : C^{n+s-*} \to C \mid s \in \mathbb{Z}\}$ of maps satisfying

$$d_C \circ \theta_s - (-1)^{n+s}\theta_s \circ d_{C^{-*}} = (-1)^n(\theta_{s-1} + (-1)^s T(\theta)_{s-1}). \tag{15.31}$$

These identities describe $T(\theta)_{s-1} + (-1)^s \theta_{s-1}$ as a measure of failure for θ_s to be chain map of a certain degree.

In terms of the components $\theta_{s,r}: C^{n+s-r} \to C_r$ we have

$$(d_C)_{r+1} \circ \theta_{s,r+1} - (-1)^r \theta_{s,r} \circ (d_C)^*_{n-r+s}$$
$$= (-1)^n \theta_{s-1,r} + (-1)^{t(n,s,r)} \varepsilon e(C_r)^{-1} \circ \theta^*_{s-1,n-1+s-r} : C^{n-1-r+s} \to C_r \tag{15.32}$$

with $t(n,r,s)$ as in (15.28). Since $C \in bR\text{-CH}_{\text{fgp}}$, in the symmetric and quadratic case only finitely many of the components $\varphi_{s,r}, \psi_{s,r}$ are non-zero.

Diagram (15.33) below depicts a useful way of thinking about the components of the maps φ_s of an n-dimensional symmetric complex (C, φ):

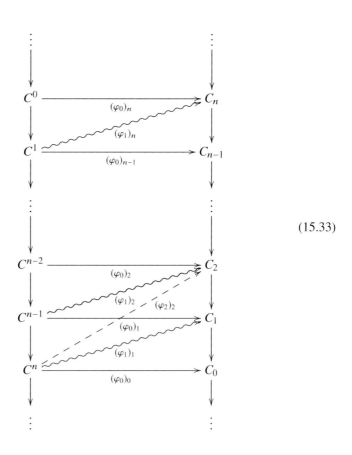

(15.33)

15.3 Structured Chain Complexes

For the sake of clarity only some components of φ_0, φ_1, and φ_2 are contained in the picture, but it should be clear where the further components fit. A similar picture could be made for an n-dimensional quadratic complex (C, ψ), the difference would be that the maps ψ_1, ψ_2 and so on would point downwards rather than upwards. We recommend to the reader to draw the picture for himself or herself. In the hyperquadratic case one would have maps going both down and up, again we leave the details to the reader.

Remark 15.34 (Forms versus co-forms) The reason why we use co-forms rather than forms in algebraic surgery is as follows. When we worked with forms on modules in Chapters 8 and 9, we used the canonical map $e \colon P \to (P^*)^*$ with P as source and $(P^*)^*$ as target, and it (and not its inverse) appears when dualising a form $\alpha \colon P \to P^*$ to $P \xrightarrow{e} (P^*)^* \xrightarrow{\varphi^*} P^*$. On the chain complex side Poincaré duality goes naturally from the cohomology to the homology as the cap product goes in this way. The slant isomorphism (14.31) then links co-forms and cap products.

Recall that for an R-chain map $f \colon C \to D$ we have defined the induced maps $f^{\%}$, $f_{\%}$, and $\widehat{f}^{\%}$ in Definition 15.12 above. Taking homology yields maps of abelian groups

$$f^{Q,n} \colon Q^n(C) \to Q^n(D); \tag{15.35}$$

$$f_{Q,n} \colon Q_n(C) \to Q_n(D); \tag{15.36}$$

$$\widehat{f}^{Q,n} \colon \widehat{Q}^n(C) \to \widehat{Q}^n(D). \tag{15.37}$$

Definition 15.38 (Maps of structured chain complexes)

(i) A *map* of n-dimensional symmetric complexes $f \colon (C, [\varphi]) \to (C', [\varphi'])$ is an R-chain map $f \colon C \to C'$ such that $f^{Q,n}([\varphi]) = [\varphi'] \in Q^n(C')$;
(ii) A *map* of n-dimensional quadratic complexes $f \colon (C, [\psi]) \to (C', [\psi'])$ is an R-chain map $f \colon C \to C'$ such that $f_{Q,n}([\psi]) = [\psi'] \in Q_n(C')$.

In both cases the map f is called an *isomorphism/homotopy equivalence* if the underlying R-chain map is an isomorphism/chain homotopy equivalence.

Exercise 15.39 Let $\varphi \in W^{\%}(C)_n$ be a polarisation of the n-dimensional symmetric structure on C. Then $T(\varphi) = T \circ \varphi \in W^{\%}(C)_n$ is also an n-dimensional polarised symmetric structure on C.

Show that $[\varphi] = [T(\varphi)] \in Q^n(C)$ by considering the chain $\widetilde{\varphi} \in W^{\%}(C)_{n+1}$ given by $\widetilde{\varphi}_s = (-1)^n \varphi_{s+1}$ for s even and $\widetilde{\varphi}_s = 0$ for s odd, and proving that $d_{W^{\%}(C)}(\widetilde{\varphi}) = \varphi - T(\varphi)$.

The relationship between polarised symmetric and polarised quadratic structures is described by Proposition 15.43 below. It will be further elaborated on in Section 15.3.5. As a preparation we note that there is a cofibration sequence in the sense of Definition 14.61

$$W^{\%}(C) \xrightarrow{J} \widehat{W}^{\%}(C) \longrightarrow \Sigma W_{\%}(C), \tag{15.40}$$

which comes from the degreewise split short exact sequence (15.10) by applying the functor $\hom_{\mathbb{Z}[\mathbb{Z}/2]}(-, C \otimes_R C)$ and using the identification $W \otimes_{\mathbb{Z}[\mathbb{Z}_2]} (C \otimes_R C) \cong \hom_{\mathbb{Z}[\mathbb{Z}_2]}(W^{-*}, C \otimes_R C)$. Here

$$J \colon W^{\%}(C) \to \widehat{W}^{\%}(C) \qquad (15.41)$$

is the chain map induced by the projection $\widehat{W} \to W$. We call the chain map

$$1 + T \colon W_{\%}(C) \to W^{\%}(C) \qquad (1+T)(\psi)_s = \begin{cases} (1+T)(\psi_0) & \text{if } s = 0; \\ 0 & \text{if } s \geq 1, \end{cases} \qquad (15.42)$$

the *symmetrisation map*.

Proposition 15.43 (The long exact sequence of Q-groups)

(i) We have the homotopy cofibration sequence

$$W_{\%}(C) \xrightarrow{1+T} W^{\%}(C) \xrightarrow{J} \widehat{W}^{\%}(C);$$

(ii) For a chain complex C in R-CH, we have a long exact sequence of Q-groups

$$\cdots \longrightarrow Q_n(C) \xrightarrow{1+T} Q^n(C) \xrightarrow{J} \widehat{Q}^n(C) \xrightarrow{H} Q_{n-1}(C) \longrightarrow \cdots.$$

Proof. (i) By applying Σ^{-1} to (15.40) we obtain the cofibration sequence $\Sigma^{-1} W^{\%}(C) \to \Sigma^{-1} \widehat{W}^{\%}(C) \to W_{\%}(C)$. This is in particular a homotopy cofibration sequence in the sense of Definition 14.61 by taking the trivial nullhomotopy. By employing the sequence (14.87) from Remark 14.85, we see that the sequence $\Sigma^{-1} \widehat{W}^{\%}(C) \to W_{\%}(C) \to W^{\%}(C)$ and then also the sequence $W_{\%}(C) \to W^{\%}(C) \to \widehat{W}^{\%}(C)$ are homotopy cofibration sequences.

(ii) This is the long exact homology sequence associated to (15.40). $\square$

Remark 15.44 (Recipe for a quadratic structure, version 1) Proposition 15.43 (i) is the most convenient formulation of the relationship between polarised symmetric and polarised quadratic structures on a fixed chain complex C in bR-CH$_{\text{fgp}}$. Namely, it tells us that, in order to produce a polarised quadratic structure on C, it is enough to produce a polarised symmetric structure together with a hyperquadratic nullhomology of this polarised symmetric structure in the chain complex of hyperquadratic structures. This follows from the universal property of a homotopy fibration, which is the same as homotopy cofibration in the category of chain complexes, and the fact that a cycle in the n-th chain group of some R-chain complex E is the same as a chain map $n[R] \to E$. The present remark will be further simplified in the following section, see Remark 15.117.

15.3 Structured Chain Complexes

Remark 15.45 (Recipe for a polarised quadratic structure in components, version 1) Let us state concretely what it means for an n-dimensional polarised symmetric structure $\varphi \in W^{\%}(C)_n$ to have a nullhomology $\theta \in \widehat{W}^{\%}(C)_{n+1}$. It means that we have the equality

$$d(\theta) = J(\varphi) \in \widehat{W}^{\%}(C)_n. \tag{15.46}$$

Written out in coordinates we get

$$d_{C \otimes_R C}(\theta_s) - (-1)^n(\theta_{s-1} + (-1)^s T(\theta_{s-1})) = \varphi_s \in (C \otimes_R C)_{n+s} \tag{15.47}$$

for all $s \geq 0$. In particular, for $s = 0$ we have $(-1)^n(1+T)(\theta_{-1}) \sim \varphi_0$ in $(C \otimes_R C)_n$, where the symbol $\sim$ means "homologous". Moreover, we see that defining

$$\psi_s := (-1)^n \theta_{-1-s} \quad \text{for} \quad s \geq 0$$

provides us with a polarised quadratic structure that symmetrises to a polarised structure homologous to φ.

15.3.2 The Symmetric Construction

In this section we present a geometric construction that produces a symmetric chain complex.

Below $C^s(X)$ will denote the *singular chain complex* of a space X. Recall that the cellular chain complex is denoted by $C(X)$. Remark 15.62 below contains an explanation why we use $C^s(X)$ in this section. If X is a pointed space, we denote by $\widetilde{C}^s(X)$ the *reduced singular chain complex*, that is, $\widetilde{C}^s(X) := C^s(X, x)$ for $x \in X$ the base point. Note that for an unpointed space X we have $\widetilde{C}^s(X_+) = C^s(X)$, where the pointed space X_+ is obtained from X by adding a disjoint base point. Analogously we define the *reduced cellular chain complex* $\widetilde{C}(X)$ for a pointed CW-complex X.

Since below we have different versions of various constructions where the ground ring with involution varies, we temporarily adopt the following notations to make clear which ring is being used. If C is a $\mathbb{Z}$-chain complex, $w \colon \Gamma \to \{\pm 1\}$ is a group homomorphism and D is a $\mathbb{Z}\Gamma$-chain complex, we put

$$W_{\mathbb{Z}}^{\%}(C) = \hom_{\mathbb{Z}[\mathbb{Z}/2]}(W, C \otimes_{\mathbb{Z}} C);$$
$$W_{\mathbb{Z}\Gamma,w}^{\%}(D) = \hom_{\mathbb{Z}[\mathbb{Z}/2]}(W, D \otimes_{\mathbb{Z}\Gamma} D).$$

In the second line the involution comes into play on the right-hand side since in the tensor product $D \otimes_{\mathbb{Z}\Gamma} D$ we need to use both the left and the right $\mathbb{Z}\Gamma$-module structure on D and the right one is obtained from the left one via w.

Construction 15.48 (The symmetric construction) The *symmetric construction of X* is a $\mathbb{Z}$-chain map of the shape

$$\varphi_X \colon C^s(X) \to W_{\mathbb{Z}}^{\%}(C^s(X)) \tag{15.49}$$

explained below.

For every n-dimensional cycle $\gamma \in C^s(X)$, after applying the slant map from (14.31), the component $\varphi_X(\gamma)_0 \colon C_s^{n-*}(X) \to C^s(X)$ is the cap product with the cycle γ. After taking (co)homology, it induces the cap-product $-\cap [\gamma] \colon H^{n-*}(X) \to H_*(X)$ on the singular (co)homology with $[\gamma] \in H_n(X)$. The $\mathbb{Z}$-chain homotopy class of the $\mathbb{Z}$-chain map $\varphi_X(\gamma)_0 \colon C_s^{n-*}(X) \to C^s(X)$ depends only on the homology class $[\gamma] \in H_n(X)$.

The construction of φ_X is via the standard theory of acyclic models. It is the same construction that underlies the construction of Steenrod squares. A general setup for constructions via this method is explained in [393, Theorem 13.26 on page 246]. An application of the method that yields Steenrod squares is described in [393, Theorem 18.1 on page 441]. Another reference is in the textbook [45, Sections IV.16, VI.1, VI.4 and VI.16].

We give an informal description of φ_X. Recall that an Alexander–Whitney diagonal map is a chain map of degree 0 of the shape

$$\Delta_0 \colon C^s(X) \to C^s(X) \otimes_{\mathbb{Z}} C^s(X) \tag{15.50}$$

that equals the actual diagonal on the 0-chains. Such a map exists by acyclic models. On the other hand, precisely because it is constructed by acyclic models, the choices that one has to make can be made independently of X and hence there exists a choice such that Δ_0 is natural in X. Furthermore, all the choices of Δ_0 are naturally chain homotopic.

However, the chain map Δ_0 is not equivariant with respect to the trivial $\mathbb{Z}/2$-action on the source and the switch involution on the target. On the other hand, the composite $T \circ \Delta_0$ is another choice of an Alexander–Whitney diagonal map and so there is a chain homotopy $\Delta_1 \colon \Delta_0 \simeq T \circ \Delta_0$. This process can be iterated and at the end we obtain a sequence of maps of degree s

$$\Delta_s \colon C^s(X) \to C^s(X) \otimes_{\mathbb{Z}} C^s(X) \tag{15.51}$$

for $s \geq 0$, satisfying

$$d_{C^s(X)} \circ \Delta_{s+1} + (-1)^s \Delta_{s+1} \circ d_{C_s^{-*}(X)} = (T + (-1)^{s+1} \mathrm{id}) \circ \Delta_s. \tag{15.52}$$

These can be put together to form a $\mathbb{Z}$-chain map of degree 0

$$\Delta_X \colon W \otimes C^s(X) \to C^s(X) \otimes_{\mathbb{Z}} C^s(X), \quad 1_s \otimes x \mapsto \Delta_s(x). \tag{15.53}$$

The symmetric construction map φ_X in (15.49) is defined to be the adjoint of Δ_X. By the above discussion it can be chosen to be natural in X and is unique up to chain homotopy.

15.3 Structured Chain Complexes

Furthermore, since Δ_0 is the diagonal approximation in the sense of [45, Chapter VI.4], it defines the cup product on the cochain complex of X and via the slant homomorphism also the cap product. An explicit formula for one choice of such Δ_0 is given by the well-known "front face" and "back face" maps, as presented in [45, Definition 4.4. in Section VI.4]. To see how the maps Δ_s for $s \geq 1$ are used to construct the Steenrod squares, we refer to [45, Chapter VI.16]. We will use this interpretation explicitly in Section 15.3.9. This finishes the Construction 15.48.

Construction 15.54 (The symmetric construction – equivariant version) The symmetric construction has an equivariant version as follows. Let Γ be a group and let Y be a free Γ-space. The symmetric construction (15.49) over $\mathbb{Z}$ applied to Y yields a $\mathbb{Z}\Gamma$-chain map

$$\varphi_{Y,\Gamma} \colon C^s(Y) \to W_{\mathbb{Z}}^{\%}(C^s(Y))$$

where $C^s(Y) \otimes_{\mathbb{Z}} C^s(Y)$ is considered as a $\mathbb{Z}\Gamma$-chain complex via the diagonal Γ-action. This is indeed a $\mathbb{Z}\Gamma$-chain map by the naturality of (15.49).

Now fix a homomorphism $w \colon \Gamma \to \{\pm 1\}$. Applying $\mathbb{Z}^w \otimes_{\mathbb{Z}\Gamma} -$, we obtain a chain map of chain complexes of abelian groups

$$\varphi_{Y,\Gamma,w} \colon \mathbb{Z}^w \otimes_{\mathbb{Z}\Gamma} C^s(Y) \to W_{\mathbb{Z}\Gamma,w}^{\%}(C^s(Y)), \tag{15.55}$$

called the *equivariant symmetric construction of* (Y, Γ, w). For an n-dimensional cycle $\gamma \in \mathbb{Z}^w \otimes_{\mathbb{Z}\Gamma} C^s(Y)$, after applying the slant map (14.31), the component

$$\varphi(\gamma)_0 \colon C_s^{n-*}(Y) \to C^s(Y) \tag{15.56}$$

is called the cap product with the cycle γ, but now $\varphi(\gamma)_0$ is a $\mathbb{Z}\Gamma$-chain map with $C_s^{n-*}(Y) = \Sigma^n \hom_{\mathbb{Z}\Gamma}(C_*^s(Y), \mathbb{Z}\Gamma)$, a $\mathbb{Z}\Gamma$-chain complex obtained via the twisted involution from w. The $\mathbb{Z}\Gamma$-chain homotopy class of the $\mathbb{Z}\Gamma$-chain map $\varphi(\gamma)_0 \colon C_s^{n-*}(Y) \to C^s(Y)$ depends only on the class $[\gamma]$ in $H_n^\Gamma(Y; \mathbb{Z}^w) = H_n(\mathbb{Z}^w \otimes_{\mathbb{Z}\Gamma} C^s(Y))$ represented by γ.

If X is a path connected space, we can take $\Gamma = \pi_1(X)$ and $Y = \widetilde{X}$.

Construction 15.57 (The symmetric construction – pointed version) There is also a version of the symmetric construction for pointed spaces. For a pointed space X we obtain the *pointed symmetric construction*, see [345, page 198],

$$\widetilde{\varphi}_X \colon \widetilde{C}^s(X) \to W_{\mathbb{Z}}^{\%}(\widetilde{C}^s(X)). \tag{15.58}$$

Construction 15.59 (The symmetric construction – pointed equivariant version) Let Γ be a group. A *pointed free Γ-space* Y is a pointed space with a Γ-action that preserves the base point and is free outside the base point. In this setup we obtain the pointed equivariant symmetric construction

$$\widetilde{\varphi}_{Y,\Gamma} \colon \widetilde{C}^s(Y) \to W_{\mathbb{Z}}^{\%}(\widetilde{C}^s(Y)), \tag{15.60}$$

cf. [345, page 198], which is a $\mathbb{Z}\Gamma$-chain map by the naturality of $\widetilde{\varphi}_X$ of (15.58).

Let $w\colon \Gamma \to \{\pm 1\}$ be a group homomorphism. Applying $\mathbb{Z}^w \otimes_{\mathbb{Z}\Gamma} -$ we finally obtain the $\mathbb{Z}$-chain map

$$\widetilde{\varphi}_{Y,\Gamma,w} \colon \mathbb{Z}^w \otimes_{\mathbb{Z}\Gamma} \widetilde{C}^s(Y) \to W^{\%}_{\mathbb{Z}\Gamma,w}(\widetilde{C}^s(Y)), \tag{15.61}$$

called the *pointed equivariant symmetric construction of* (Y, Γ, w).

An example relevant in our context is the reduced suspension of a universal cover $\widetilde{X}$ of a path connected space X with added base point $Y = \Sigma^p(\widetilde{X}_+)$ for any $p \geq 0$.

Moreover, Construction 15.59 is natural in Y in the following sense. Let $f\colon Y \to Y'$ be a map of pointed free Γ-spaces. Assume further that we have a homomorphism $w\colon \Gamma \to \{\pm 1\}$. Then the following diagram strictly commutes:

$$\begin{array}{ccc}
\mathbb{Z}^w \otimes_{\mathbb{Z}\Gamma} \widetilde{C}^s(Y) & \xrightarrow{\widetilde{\varphi}_{Y,\Gamma,w}} & W^{\%}_{\mathbb{Z}\Gamma,w}(\widetilde{C}^s(Y)) \\
\mathrm{id}_{\mathbb{Z}} \otimes \widetilde{C}^s(f) \downarrow & & \downarrow (\widetilde{C}^s(f))^{\%} \\
\mathbb{Z}^w \otimes_{\mathbb{Z}\Gamma} \widetilde{C}^s(Y') & \xrightarrow{\widetilde{\varphi}_{Y',\Gamma,w}} & W^{\%}_{\mathbb{Z}\Gamma,w}(\widetilde{C}^s(Y')).
\end{array}$$

This finishes the Construction 15.59.

The examples of quadratic and hyperquadratic structures arising from geometry are relegated to Section 15.3.6.

We conclude this section with a remark explaining why we use the singular chain complex instead of the cellular chain complex. Recall that there are constructions of Steenrod squares that do use the cellular chain complex and thus avoid the need for acyclic models, see for example [178, Chapter 4.L]. The remark below explains why such an approach is not good enough for our purposes.

Remark 15.62 (Cellular version) Let $w\colon \Gamma \to \{\pm 1\}$ be a group homomorphism and let Y be a free Γ-CW-complex. Then $Y \times Y$ with the diagonal Γ-action is a Γ-CW-complex. The trivial $\mathbb{Z}/2$-action on Y and the flip $\mathbb{Z}/2$-action on $Y \times Y$ commute with these Γ-actions. Note that $Y \times Y$ is not a $\mathbb{Z}/2$-CW-complex. The diagonal map $\Delta\colon Y \to Y \times Y$ is a Γ-map. However, it is not cellular, not even after forgetting the Γ-action. By the Equivariant Approximation Theorem, see for instance [250, Theorem 2.1 on page 32], it is Γ-homotopic to a cellular Γ-map $\overline{\Delta}\colon Y \to Y \times Y$. Using the canonical $\mathbb{Z}\Gamma$-chain isomorphism $C(Y) \otimes_{\mathbb{Z}} C(Y) \xrightarrow{\cong} C(Y \times Y)$, we get from $\overline{\Delta}$ a $\mathbb{Z}\Gamma$-chain map, unique up to $\mathbb{Z}\Gamma$-chain homotopy,

$$C(Y) \to C(Y) \otimes_{\mathbb{Z}} C(Y),$$

which is the equivariant cellular analogue of the map (15.50), or equivalently, of the map (15.51) for $s = 0$. The process can be iterated to obtain a cellular analogue

$$\varphi^c_{Y,\Gamma,w} \colon \mathbb{Z}^w \otimes_{\mathbb{Z}\Gamma} C(Y) \to W^{\%}_{\mathbb{Z}\Gamma,w}(C(Y))$$

of the symmetric construction map $\varphi_{Y,\Gamma,w}$ from (15.55). It has the pleasant feature that $C(Y) \in \mathrm{b}\mathbb{Z}\Gamma\text{-}\mathrm{CH}_{\mathrm{fgf}}$ whereas we only have $C^s(Y) \in \mathrm{h}\mathbb{Z}\Gamma\text{-}\mathrm{CH}_{\mathrm{fgf}}$ in (15.55).

15.3 Structured Chain Complexes

However, the chain map $\varphi^c_{Y,\Gamma}$ will *not be natural* in Y since choices were made and unlike in the singular case these choices depend on Y. This is the reason why we use the singular chain complex $C^s(Y)$.

Note that there are natural zigzags of $\mathbb{Z}\Gamma$-chain homotopy equivalences

$$C(Y) \xleftarrow{\simeq} C^{\text{in}}(Y) \xrightarrow{\simeq} C^s(Y) \tag{15.63}$$

and

$$C(Y \times Y) \xleftarrow{\simeq} C^{\text{in}}(Y \times Y) \xrightarrow{\simeq} C^s(Y \times Y), \tag{15.64}$$

where $C^{\text{in}}(Y)$ denotes a certain "intermediate" chain complex, see [250, Proposition 13.12 on page 264]. These can be used to relate the singular and cellular symmetric constructions and hence singular and cellular cup products and Steenrod squares in cohomology, but the correspondence will not be natural in Y at the level of chain complexes by the remarks above.

On the other hand, on the level of Q-groups we will get a well-defined isomorphism between the Q-groups associated to the singular and the cellular chain complexes, and we will use it to transport symmetric and quadratic structures from the singular chain complex to the cellular chain complex.

Remark 15.65 By applying homology to (15.55) we get a map

$$\overline{\varphi}_{Y,\Gamma,w} \colon H^\Gamma_n(Y; \mathbb{Z}^w) \to Q^n_{\mathbb{Z}\Gamma,w}(C^s(Y)) := H_n(W^\%_{\mathbb{Z}\Gamma,w}(C^s(Y))). \tag{15.66}$$

Composing it with the map $H_n(W^\%_{\mathbb{Z}\Gamma,w}(C^s(Y))) \to [C^{n-*}_s(Y), C^s_*(Y)]_{\mathbb{Z}\Gamma}$ coming from the passage to (15.56) yields a map

$$\cap \colon H^\Gamma_n(Y; \mathbb{Z}^w) \to [C^{n-*}_s(Y), C^s_*(Y)]_{\mathbb{Z}\Gamma}. \tag{15.67}$$

Recall that in (15.67) we consider singular chain complexes whereas in (5.42) we consider cellular chain complexes. The maps (5.42) and (15.67) agree under the obvious identification of the targets coming from the $\mathbb{Z}\Gamma$-chain homotopy equivalences (15.63).

15.3.3 Products

Next we would like to define operations on structured chain complexes that correspond to products of topological spaces. The products are needed in Section 15.3.4 to establish the homotopy invariance of the W-functors from Section 15.3.1.

Recall that Section 15.3.2 tells us that we can informally think of an n-dimensional polarised symmetric structure on a chain complex C as a topological space together with a choice of an n-dimensional cycle in the singular chain complex, the choice of a cup product on the chain level, the choice of a homotopy between the cup product and its composite with the switch map, and choices of higher homotopies. Another way of thinking about these data is as a $\mathbb{Z}/2$-equivariant map from $\Sigma^n S^\infty$ to $X \times X$.

As already mentioned, these data are also used in the standard construction of Steenrod squares. Recall that for Steenrod squares one has the Cartan formula that describes the Steenrod squares on a product, see [45, Chapter VI, Proposition 15.1]. A crucial ingredient in the proof is the diagonal map for the resolution W, which can be thought of as a $\mathbb{Z}/2$-equivariant diagonal approximation for S^∞. In [45] a mod 2 formula for such a diagonal approximation is implicit on page 419 (the map D_n). An explicit formula from the paper [344, Section 8] is extracted from the more general case described in [85, Chapter XII] together with an explicit formula for a diagonal map of the periodic resolution $\widehat{W}$.

Another way of thinking about the products below is, in view of Remark 15.15, as about the products in group co-homology.

Notation 15.68 So far we have always specified the ring in the notation of the tensor product, e.g. $C \otimes_R D$. To ease the notation slightly, we keep this convention except in the case $R = \mathbb{Z}$, when we leave it out, so henceforth $C \otimes D$ means $C \otimes_\mathbb{Z} D$. Also we omit the subscript from the tensor notation of elements, so we write $x \otimes y$ for an element in $C \otimes_R D$. Since we know the group in which the element lives, the meaning should be clear.

Now we recall the formulas for the diagonal maps. The r-th module W_r, respectively $\widehat{W}_r$, is identified in the notation as $\mathbb{Z}[\mathbb{Z}/2] \cong \mathbb{Z}\{1_r, T_r\}$. In some cases we also use the completed tensor product $C \widehat{\otimes}_R D_n = \prod_{p+q=n} C_p \otimes_R D_q$ so that we work with "infinite chains". Define

$$\Delta \colon W \to W \otimes W \qquad 1_s \mapsto \sum_{r=0}^{s} (-1)^{r(s-r)} 1_r \otimes T^r_{s-r}; \qquad (15.69)$$

$$\Delta \colon \widehat{W} \to \widehat{W} \widehat{\otimes} \widehat{W} \qquad 1_s \mapsto \sum_{r=-\infty}^{\infty} (-1)^{r(s-r)} 1_r \otimes T^r_{s-r}. \qquad (15.70)$$

We note that the formulas look slightly different than those in the paper [344, Section 8] due to the fact that we have a different convention for the differential in the tensor product of chain complexes. We omit the proof that Δ is $\mathbb{Z}/2$-equivariant (w.r.t. the diagonal $\mathbb{Z}[\mathbb{Z}/2]$-chain complex structure on the target) and that $(\Delta \otimes \mathrm{id}_W) \circ \Delta = (\mathrm{id}_W \otimes \Delta) \circ \Delta$ holds.

The diagonal (15.70) restricts to another diagonal

$$\Delta \colon W^{-*} \to W \widehat{\otimes} W^{-*} \qquad 1_{-s} \mapsto \sum_{r=0}^{\infty} (-1)^{r(-s-r)} 1_r \otimes T^r_{-s-r}, \qquad (15.71)$$

with $s \geq 0$, which will be useful when dealing with quadratic structures, in view of $W_{\%}(C) = W \otimes_{\mathbb{Z}[\mathbb{Z}_2]} (C \otimes_R C) \cong \hom_{\mathbb{Z}[\mathbb{Z}_2]}(W^{-*}, C \otimes_R C)$.

A general setup for constructing products in algebraic surgery is presented in [344, Section 8]. There one starts with two structured chain complexes over two rings with involution, say R and S, and produces a structured chain complex on the tensor product (over $\mathbb{Z}$) of the underlying chain complexes considered as a chain complex

15.3 Structured Chain Complexes

over $R \otimes S$. We will not need this generality, so to keep things simple we will consider only the case $S = \mathbb{Z}$ and hence $R \otimes \mathbb{Z} = R$.

Definition 15.72 (Products of structures on chain complexes)
For $C \in h\mathbb{Z}\text{-CH}_{\text{fgp}}$, $D \in hR\text{-CH}_{\text{fgp}}$, such that both C and D are also bounded from below, the products

$$- \otimes -\colon W^{\%}(C) \otimes W^{\%}(D) \to W^{\%}(C \otimes D); \tag{15.73}$$

$$- \otimes -\colon \widehat{W}^{\%}(C) \otimes \widehat{W}^{\%}(D) \to \widehat{W}^{\%}(C \otimes D); \tag{15.74}$$

$$- \otimes -\colon W^{\%}(C) \otimes W_{\%}(D) \to W_{\%}(C \otimes D), \tag{15.75}$$

are defined via the formula (the "same" one in all cases):

$$\varphi_C \otimes \varphi_D \mapsto (\text{id}_C \otimes \text{sw} \otimes \text{id}_D) \circ (\varphi_C \otimes \varphi_D) \circ \Delta, \tag{15.76}$$

where $\text{sw}\colon C \otimes D \to D \otimes C$ is the switch map (14.18). So the image of $\varphi_C \otimes \varphi_D$ in the target in (15.76) is the composite

$$W \xrightarrow{\Delta} W \otimes_{\mathbb{Z}} W \xrightarrow{(\varphi_C \otimes \varphi_D)} (C \otimes_{\mathbb{Z}} C) \otimes_{\mathbb{Z}} (D \otimes_R D) \xrightarrow{\text{sw}} (C \otimes_{\mathbb{Z}} D) \otimes_R (C \otimes_{\mathbb{Z}} D).$$

Note that these products admit a kind of unit. It is the 0-dimensional symmetric chain complex $0[\mathbb{Z}]$ that is by a chain complex concentrated in dimension 0 with the value $\mathbb{Z}$ and with the symmetric structure defined by $v = 1 \otimes 1 \in (0[\mathbb{Z}] \otimes 0[\mathbb{Z}])_0 \cong \mathbb{Z}$.

15.3.4 Homotopy Invariance

As the first application of these products we show the next result.

Lemma 15.77 *The functors $W^{\%}(-)$, $W_{\%}(-)$, and $\widehat{W}^{\%}(-)$ are homotopy functors, i.e., chain homotopic chain maps are sent to chain homotopic chain maps. In particular, they send chain homotopy equivalences to chain homotopy equivalences.*

Proof. Consider first an R-chain homotopy $h\colon f \simeq g\colon C \to D$. The R-chain maps f and g induce chain maps

$$f \otimes f\colon C \otimes_R C \to D \otimes_R D \quad \text{and} \quad g \otimes g\colon C \otimes_R C \to D \otimes_R D,$$

and h can be used to obtain a chain homotopy between them, for example $k = (h \otimes f) - (g \otimes h)$. However, this chain homotopy is not $\mathbb{Z}[\mathbb{Z}/2]$-equivariant. In fact, such a $\mathbb{Z}[\mathbb{Z}/2]$-chain homotopy does not exist, and hence we do not obtain a canonical homotopy between $f^{\%}$ and $g^{\%}$ just by simple concatenation with such a hypothetical chain homotopy.

Therefore we proceed as follows. We use notation from Chapter 14, in particular, we denote by I the cellular chain complex of the 1-simplex Δ^1 and by $i_j\colon 0[\mathbb{Z}] \to I$ for $j = 0, 1$ the inclusions of the two ends. Recall from Subsection 14.6.1 that a chain homotopy $h\colon f_0 \simeq f_1\colon C \to D$ between chain maps f_0, f_1 can be seen as a chain map $\bar{h}\colon I \otimes C \to D$ that restricts to f_0 and f_1 upon precomposing with the inclusions i_0 and i_1.

Now choose once and for all a 1-chain $\omega \in W^{\%}(I)$ satisfying

$$d_{W^{\%}(I)}(\omega) = i_1^{\%}(v) - i_0^{\%}(v).$$

The choice of ω is equivalent to choosing a chain map $\bar{\omega}\colon I \to W^{\%}(I)$ whose restrictions to $0[I_0]$ coincide with the pushforwards of v. Then observe that a chain homotopy $h\colon f_0 \simeq f_1$ induces the chain homotopies $h^{\%}, h_{\%}$ and $\widehat{h}^{\%}$:

$$I \otimes W^{\%}(C) \xrightarrow{\bar{\omega} \otimes -} W^{\%}(I) \otimes W^{\%}(C) \to W^{\%}(I \otimes C) \xrightarrow{\bar{h}^{\%}} W^{\%}(D); \qquad (15.78)$$

$$I \otimes W_{\%}(C) \xrightarrow{\bar{\omega} \otimes -} W^{\%}(I) \otimes W_{\%}(C) \to W_{\%}(I \otimes C) \xrightarrow{\bar{h}^{\%}} W_{\%}(D); \qquad (15.79)$$

$$I \otimes \widehat{W}^{\%}(C) \xrightarrow{\bar{\omega} \otimes -} W^{\%}(I) \otimes \widehat{W}^{\%}(C) \to \widehat{W}^{\%}(I \otimes C) \xrightarrow{\bar{h}^{\%}} \widehat{W}^{\%}(D). \qquad (15.80)$$

This concludes the proof of the lemma. □

Remark 15.81 (A choice of ω) One choice of ω is as follows. Denote the non-degenerate simplices of Δ^1 by σ_{01} for the 1-simplex and σ_1, σ_0 for the two 0-simplices so that we have $d(\sigma_{01}) = \sigma_1 - \sigma_0$. Define

$$\omega_0 := \sigma_{01} \otimes \sigma_0 + \sigma_1 \otimes \sigma_{01} \in (I \otimes I)_1;$$
$$\omega_1 := -\sigma_{01} \otimes \sigma_{01} \in (I \otimes I)_2.$$

Then we get

$$d_\otimes(\omega_0) = \sigma_1 \otimes \sigma_1 - \sigma_0 \otimes \sigma_0 = i_1^{\%}(v) - i_0^{\%}(v);$$
$$d_\otimes(\omega_1) = -[\sigma_1 \otimes \sigma_{01} - \sigma_0 \otimes \sigma_{01} - \sigma_{01} \otimes \sigma_1 + \sigma_{01} \otimes \sigma_0] = T(\omega_0) - \omega_0.$$

To be able to write down explicit formulas let us add the following observation. The R-chain homotopy $h\colon f_0 \simeq f_1\colon C \to D$ and the corresponding R-chain map $\bar{h}\colon I \otimes C \to D$ allow us to consider the tensor product maps:

$$\bar{h} \otimes \bar{h}\colon (I \otimes C) \otimes_R (I \otimes C) \to D \otimes_R D;$$
$$h \otimes h\colon C \otimes_R C \to D \otimes_R D,$$

where the first map is an R-chain map of degree 0 and the second map is of degree 2, it is a 2-chain in $\hom_{\mathbb{Z}}(C \otimes_R C, D \otimes_R D)$. Given $x \in C_p$, $y \in C_q$, and taking $\sigma_{01} \in I_1$ as above, one obtains the formula:

$$(\bar{h} \otimes \bar{h})(\mathrm{id}_I \otimes \mathrm{sw} \otimes \mathrm{id}_C)(\sigma_{01} \otimes \sigma_{01} \otimes x \otimes y) = (-1)^p h(x) \otimes h(y) = (h \otimes h)(x \otimes y)$$

where in the right equation we use (14.20).

15.3 Structured Chain Complexes

Remark 15.82 (Formulas for $h^{\%}$, $h_{\%}$ and $\widehat{h}^{\%}$) Given n-dimensional structures $\varphi \in W^{\%}(C)_n$, $\psi \in W_{\%}(C)_n$, and $\theta \in \widehat{W}^{\%}(C)_n$, taking the particular choice of ω from Remark 15.81 and using the above observation we obtain the formulas:

$$h^{\%}(\varphi)_s = (h \otimes f_0)(\varphi_s) + (f_1 \otimes h)(\varphi_s) + (-1)^{n+s}(h \otimes h)(T(\varphi_{s-1})); \quad (15.83)$$

$$h_{\%}(\psi)_s = (h \otimes f_0)(\psi_s) + (f_1 \otimes h)(\psi_s) + (-1)^{n+s}(h \otimes h)(T(\psi_{s+1})); \quad (15.84)$$

$$\widehat{h}^{\%}(\theta)_s = (h \otimes f_0)(\theta_s) + (f_1 \otimes h)(\theta_s) + (-1)^{n+s}(h \otimes h)(T(\theta_{s-1})). \quad (15.85)$$

Exercise 15.86 Note that another choice in Remark 15.81 would be ω' with $\omega'_s = T(\omega_s)$ for $s = 0, 1$. Find the corresponding modification of the formulas from Remark 15.82.

15.3.5 The Suspension Map

In this section we define the suspension maps for structured chain complexes. The definition is designed to be compatible with the suspension of topological spaces, see Proposition 15.107. It is needed in order to bring stable homotopy theory into the game, which in turn will be used to obtain quadratic complexes from geometric situations in Section 15.3.6. As we mentioned in the introduction to this chapter, this was motivated by the relation of the Spivak normal fibration to suspensions via S-duality and by Browder's idea of employing suspensions to obtain quadratic refinements as in [55, Chapter III]. However, his definition of the quadratic refinement in terms of functional Steenrod squares is involved.

On the other hand, in algebraic surgery we have two relatively straightforward ways to see the relation between the suspension and quadratic refinements. One way is to study abstractly the behaviour of the functors $W^{\%}(-)$, $W_{\%}(-)$ and $\widehat{W}^{\%}(-)$ with respect to the suspension maps, this leads us to Corollary 15.116, which reformulates Proposition 15.43 entirely in terms of quadratic and symmetric structures. The second way is to look at the concrete formulas in terms of components for the suspension maps as given in Remark 15.97. Using these formulas, one can directly construct a quadratic refinement of a symmetric structure whose suspension is nullhomologous as described in Remark 15.120.

Definition 15.87 (The suspension maps) For $C \in \mathrm{h}R\text{-}\mathrm{CH}_{\mathrm{fgp}}$ the *suspension maps*

$$S = S_C : \Sigma W^{\%}(C) \to W^{\%}(\Sigma C); \quad (15.88)$$

$$S = S_C : \Sigma W_{\%}(C) \to W_{\%}(\Sigma C); \quad (15.89)$$

$$S = S_C : \Sigma \widehat{W}^{\%}(C) \to \widehat{W}^{\%}(\Sigma C), \quad (15.90)$$

are defined as follows. Consider the collapse map $\overline{\mathrm{col}} : I \otimes C \to \Sigma C$ as a chain homotopy $\mathrm{col} : 0 \simeq 0 : C \to \Sigma C$, given by $\mathrm{col}(x) = x$. Then the product map (15.78) uniquely factors through a map $\Sigma W^{\%}(C) \to W^{\%}(\Sigma C)$. This is the desired suspension map. The two other cases are similar. These maps can be iterated and we get for $p \geq 0$

$$S^p = S^p_C : \Sigma^p W^{\%}(C) \to W^{\%}(\Sigma^p C); \qquad (15.91)$$
$$S^p = S^p_C : \Sigma^p W_{\%}(C) \to W_{\%}(\Sigma^p C); \qquad (15.92)$$
$$S^p = S^p_C : \Sigma^p \widehat{W}^{\%}(C) \to \widehat{W}^{\%}(\Sigma^p C), \qquad (15.93)$$

via the formula

$$S^p_C := S_{\Sigma^{p-1}C} \circ \Sigma S^{p-1}_C \qquad (15.94)$$

in all three cases.

To be able to write down explicit formulas, consider the chain complex $C \otimes_R C$ and define μ_1 and μ_2 to be the degreewise homomorphisms (not chain maps)

$$\mu_1 : C \otimes_R C \to C \otimes_R C \quad x \otimes y \mapsto (-1)^{|x|} \cdot x \otimes y, \qquad (15.95)$$

and

$$\mu_2 : C \otimes_R C \to C \otimes_R C \quad x \otimes y \mapsto (-1)^{|y|} \cdot x \otimes y. \qquad (15.96)$$

Using this notation observe that by inserting col for h into the formula just above Remark 15.82 we obtain

$$\overline{(\mathrm{col} \otimes \mathrm{col})}(\mathrm{id}_I \otimes \mathrm{sw} \otimes \mathrm{id}_C)(\sigma_{01} \otimes \sigma_{01} \otimes x \otimes y) = \mu_1(\mathrm{col}(x) \otimes \mathrm{col}(y)) = \mu_1(x \otimes y).$$

Remark 15.97 (Formulas for the suspension maps) For the choice of ω from Remark 15.81, the formula for the suspension $S(\varphi)$ is the $(n+1)$-dimensional polarised symmetric structure on ΣC given by

$$(S(\varphi))_0 = 0;$$
$$(S(\varphi))_{s+1} = (-1)^{n+s+1} \mu_1 T(\varphi_s) \quad s \geq 0.$$

We note that the fact that the 0-th component is zero corresponds to the well-known observation from algebraic topology that cup products vanish on suspensions. It also shows that $S \colon \Sigma W^{\%}(C) \to W^{\%}(\Sigma C)$ is not surjective, but it is degreewise split injective and hence is a cofibration.

For polarised quadratic structures we obtain

$$(S(\psi))_{s-1} = (-1)^{n+s-1} \mu_1 T(\psi_s) \quad s \geq 1.$$

From the formula we see that the map $S \colon \Sigma W_{\%}(C) \to W_{\%}(\Sigma C)$ is not injective. For polarised hyperquadratic structures we obtain

$$(S(\theta))_{s+1} = (-1)^{n+s+1} \mu_1 T(\theta_s) \quad s \in \mathbb{Z}.$$

15.3 Structured Chain Complexes

The occurrence of T is caused by the occurrence of T in the formulas of Remark 15.82, which is in turn forced by the choices of Δ and ω. It is possible to make these choices so that T does not occur in the suspension formulas, but then we would pay for this at another place. Hence we consider the occurrence of T in them to be acceptable. In [344] the corresponding formulas are to be found on page 106 and in [427] on page 159.

Exercise 15.98 Check that $S(\varphi)$ is indeed an $(n+1)$-dimensional polarised symmetric structure on ΣC, that means, check that it is a cycle in $W^{\%}(\Sigma C)$.

Proposition 15.99 (Suspension is a bijection for hyperquadratic structures) *For $C \in$ hR-$\mathrm{CH}_{\mathrm{fgp}}$ the map $S \colon \Sigma \widehat{W}^{\%}(C) \to \widehat{W}^{\%}(\Sigma C)$ is bijective.*

Proof. This follows directly from the formula in Remark 15.97. □

In Proposition 15.113 we will give a conceptual proof that on $\widehat{Q}$-groups the map $\widehat{Q}^n(C) \to \widehat{Q}^{n+1}(\Sigma C)$ induced by the map S above is an isomorphism without using explicit formulas.

The suspension maps from Definition 15.87 are natural in the following sense. Let $f \colon C \to D$ be an R-chain map. Then we have the induced map $f^{\%} \colon W^{\%}(C) \to W^{\%}(D)$ and also $(\Sigma f)^{\%} \colon W^{\%}(\Sigma C) \to W^{\%}(\Sigma D)$. Inspecting the definitions shows that we have the equation

$$(\Sigma f)^{\%} \circ S_C = S_D \circ \Sigma(f^{\%}). \tag{15.100}$$

Similarly in the quadratic and hyperquadratic case.

Exercise 15.101 Consider the product from (15.73) and use it to obtain from the 1-chain $\omega \in W^{\%}(I)_1$ the p-chain $\omega^{\otimes p} \in W^{\%}(I^{\otimes p})_p$. Show that, repeating the same process as in the definition of S_C in (15.88), we obtain a map that equals S_C^p from (15.94).

Exercise 15.102 Show that the iterated suspension has the following pleasant feature. The map $S_C^2 \colon \Sigma^2 W^{\%}(C) \to W^{\%}(\Sigma^2 C)$ is given by the formulas

$$(S_C^2(\varphi))_s = 0 \quad s = 0, 1;$$
$$(S_C^2(\varphi))_{s+1} = \varphi_{s-1} \quad s \geq 1,$$

so that there are no signs and the map equals just a shift by two degrees.

Exercise 15.103 Given an R-chain homotopy $h \colon f \simeq g \colon C \to D$ show the equation $(\Sigma h)^{\%} \circ S_C = S_D \circ \Sigma(h^{\%})$.

We have already mentioned that the suspension map (15.88) corresponds to the suspension of topological spaces. This is made precise by the following lemma that links the symmetric construction on a pointed topological space X with the symmetric construction on its reduced suspension $\Sigma X = S^1 \wedge X$. To prepare for it, recall that the acyclic models technology yields for any pointed space X a $\mathbb{Z}$-chain homotopy equivalence $\Sigma_X \colon \Sigma \widetilde{C}^s(X) \to \widetilde{C}^s(\Sigma X)$, natural in X, where $\widetilde{C}^s(-)$ denotes the reduced singular chain complex. Iterating it by the formula $\Sigma_X^{p+1} = \Sigma_{\Sigma^p X} \circ \Sigma \Sigma_X^p$, we get for $p \geq 0$ a $\mathbb{Z}$-chain homotopy equivalence

$$\Sigma_X^p \colon \Sigma^p \widetilde{C}^s(X) \to \widetilde{C}^s(\Sigma^p X), \tag{15.104}$$

which is natural in X and where $\Sigma^p X = S^p \wedge X$, so that $\Sigma^0 \wedge X = X$ and $\Sigma^p X = \Sigma(\Sigma^{p-1} X)$ for $p \geq 1$.

If Γ is a group and Y is a pointed free Γ-space, we obtain a $\mathbb{Z}\Gamma$-chain homotopy equivalence

$$\Sigma_{Y,\Gamma}^p \colon \Sigma^p \widetilde{C}^s(Y) \to \widetilde{C}^s(\Sigma^p Y), \tag{15.105}$$

which is natural in Y. A special case relevant to us is that for a path connected unpointed space X, we can take $\Gamma = \pi_1(X,x)$ for some $x \in X$ and $Y = \widetilde{X}_+$ and obtain a $\mathbb{Z}\Gamma$-chain homotopy equivalence

$$\Sigma_{\widetilde{X}}^p \colon \Sigma^p C^s(\widetilde{X}) = \Sigma^p \widetilde{C}^s(\widetilde{X}_+) \to \widetilde{C}^s(\Sigma^p \widetilde{X}_+), \tag{15.106}$$

which is natural in X.

The next result, which is taken from [345, page 201], describes the behaviour of the symmetric construction with respect to the suspension of pointed spaces.

Proposition 15.107 (Suspension maps for symmetric complexes versus suspensions of spaces) *For every pointed space X there exists a natural $\mathbb{Z}$-chain homotopy Γ_X as follows:*

$$\begin{array}{ccc}
\Sigma \widetilde{C}^s(X) & \xrightarrow{S_{\widetilde{C}^s(X)} \circ \Sigma(\widetilde{\varphi}_X)} & W_{\mathbb{Z}}^{\%}(\Sigma \widetilde{C}^s(X)) \\
{\scriptstyle \Sigma_X} \downarrow & \overset{\Gamma_X}{\rightsquigarrow} & \downarrow {\scriptstyle \Sigma_X^{\%}} \\
\widetilde{C}^s(\Sigma X) & \xrightarrow[\widetilde{\varphi}_{\Sigma X}]{} & W_{\mathbb{Z}}^{\%}(\widetilde{C}^s(\Sigma X))
\end{array}$$

where $\widetilde{\varphi}_X$ and $\widetilde{\varphi}_{\Sigma X}$ are the $\mathbb{Z}$-chain maps from (15.58), the chain map $S_{\widetilde{C}^s(X)}$ has been defined in (15.88) and the chain map Σ_X in (15.104) for $p = 1$.

There also exists a pointed equivariant version. It says that for any group homomorphism $w \colon \Gamma \to \{\pm 1\}$ and pointed free Γ-space Y there exists a natural $\mathbb{Z}$-chain homotopy $\Gamma_{Y,\Gamma,w}$ as follows:

$$\begin{array}{ccc}
\mathbb{Z}^w \otimes_{\mathbb{Z}\Gamma} \Sigma \widetilde{C}^s(Y) & \xrightarrow{S_{\widetilde{C}^s(Y)} \circ \Sigma(\widetilde{\varphi}_{Y,\Gamma,w})} & W_{\mathbb{Z}\Gamma,w}^{\%}(\Sigma \widetilde{C}^s(Y)) \\
{\scriptstyle \mathrm{id}_{\mathbb{Z}^w} \otimes \Sigma_{Y,\Gamma}} \downarrow & \overset{\Gamma_{Y,\Gamma,w}}{\rightsquigarrow} & \downarrow {\scriptstyle (\Sigma_{Y,\Gamma})^{\%}} \\
\mathbb{Z}^w \otimes_{\mathbb{Z}\Gamma} \widetilde{C}^s(\Sigma Y) & \xrightarrow[\widetilde{\varphi}_{\Sigma Y,\Gamma,w}]{} & W_{\mathbb{Z}\Gamma,w}^{\%}(\widetilde{C}^s(\Sigma Y))
\end{array}$$

15.3 Structured Chain Complexes

where $\widetilde{\varphi}_{Y,\Gamma,w}$ and $\widetilde{\varphi}_{\Sigma Y,\Gamma,w}$ have been defined in (15.61) and $\Sigma_{Y,\Gamma}$ in (15.105) for $p = 1$.

The proof is also by acyclic models. The proposition can be viewed as a special case of commuting between products and symmetric constructions, and as such the proof is analogous to the proof of the Cartan formula for Steenrod squares. See for example [393, Proposition 18.12] or [45, Section VI.16].

On the other hand we have one more relation between the algebraic and topological suspension as follows.

Proposition 15.108 *For any pointed space X the following diagram of $\mathbb{Z}$-chain maps*

$$\begin{array}{ccc}
\Sigma W^{\%}(\widetilde{C}^s(X)) & \xrightarrow{\Sigma(\Sigma^{-1}\Sigma_X)^{\%}} & \Sigma W^{\%}(\Sigma^{-1}\widetilde{C}^s(\Sigma X)) \\
{\scriptstyle S_{\widetilde{C}^s(X)}}\downarrow & & \downarrow{\scriptstyle S_{\Sigma^{-1}\widetilde{C}^s(\Sigma X)}} \\
W^{\%}(\Sigma\widetilde{C}^s(X)) & \xrightarrow{(\Sigma_X)^{\%}} & W^{\%}(\widetilde{C}^s(\Sigma X))
\end{array}$$

commutes strictly.

There is also a corresponding pointed equivariant version. It says that for any group homomorphism $w\colon \Gamma \to \{\pm 1\}$ and pointed free Γ-space Y the following diagram of $\mathbb{Z}$-chain maps

$$\begin{array}{ccc}
\Sigma W^{\%}_{\mathbb{Z}\Gamma,w}(\widetilde{C}^s(Y)) & \xrightarrow{\Sigma(\Sigma^{-1}\Sigma_{Y,\Gamma})^{\%}} & \Sigma W^{\%}_{\mathbb{Z}\Gamma,w}(\Sigma^{-1}\widetilde{C}^s(\Sigma Y)) \\
{\scriptstyle S_{\widetilde{C}^s(Y)}}\downarrow & & \downarrow{\scriptstyle S_{\Sigma^{-1}\widetilde{C}^s(\Sigma Y)}} \\
W^{\%}_{\mathbb{Z}\Gamma,w}(\Sigma\widetilde{C}^s(Y)) & \xrightarrow{(\Sigma_{Y,\Gamma})^{\%}} & W^{\%}_{\mathbb{Z}\Gamma,w}(\widetilde{C}^s(\Sigma Y))
\end{array}$$

commutes strictly.

Proof. Observe that the following diagram is commutative

$$\begin{array}{ccc}
I \otimes I \otimes \widetilde{C}^s(X) \otimes \widetilde{C}^s(X) & \longrightarrow & I \otimes I \otimes \Sigma^{-1}\widetilde{C}^s(\Sigma X) \otimes \Sigma^{-1}\widetilde{C}^s(\Sigma X) \\
{\scriptstyle \mathrm{id} \otimes \mathrm{sw} \otimes \mathrm{id}}\downarrow & & \downarrow{\scriptstyle \mathrm{id} \otimes \mathrm{sw} \otimes \mathrm{id}} \\
I \otimes \widetilde{C}^s(X) \otimes I \otimes \widetilde{C}^s(X) & \longrightarrow & I \otimes \Sigma^{-1}\widetilde{C}^s(\Sigma X) \otimes I \otimes \Sigma^{-1}\widetilde{C}^s(\Sigma X).
\end{array}$$

The passage first right then down in the desired diagram is obtained from the same passage in this diagram and likewise for the passage first down then right.

In the pointed equivariant version we start with an analogous diagram of universal covers, in which all the entries are considered as a $\mathbb{Z}\Gamma$-chain complex via the diagonal $\mathbb{Z}\Gamma$-structure, and all maps are $\mathbb{Z}\Gamma$-chain maps, and apply $\mathbb{Z}^w \otimes_{\mathbb{Z}\Gamma} -$. □

Now we describe how the construction of Proposition 15.107 is iterated. For this we consider the following diagram:

$$\begin{CD}
\Sigma^{p+1}\widetilde{C}^s(X) @>{\Sigma S^p_{\widetilde{C}^s(X)} \circ \Sigma^{p+1}\widetilde{\varphi}_X}>> \Sigma W^{\%}_{\mathbb{Z}}(\Sigma^p \widetilde{C}^s(X)) @>{S_{\Sigma^p(\bar{C}^s X)}}>> W^{\%}_{\mathbb{Z}}(\Sigma^{p+1}\widetilde{C}^s(X)) \\
@V{\Sigma\Sigma^p_X}VV @VV{\Sigma(\Sigma^p_X)^{\%}}V @VV{(\Sigma\Sigma^p_X)^{\%}}V \\
\Sigma\widetilde{C}^s(\Sigma^p X) @>{\Sigma\widetilde{\varphi}_{\Sigma^p X}}>> \Sigma W^{\%}_{\mathbb{Z}}(\widetilde{C}^s(\Sigma^p X)) @>{S_{\widetilde{C}^s(\Sigma^p X)}}>> W^{\%}_{\mathbb{Z}}(\Sigma \widetilde{C}^s(\Sigma^p X)) \\
@V{\Sigma_{\Sigma^p X}}VV @. @VV{\Sigma^{\%}_{\Sigma^p X}}V \\
\widetilde{C}^s(\Sigma^{p+1}X) @>>{\widetilde{\varphi}_{\Sigma^{p+1}X}}> W^{\%}_{\mathbb{Z}}(\widetilde{C}^s(\Sigma^{p+1}X)).
\end{CD}$$

The upper left diagram commutes up to the preferred natural in X homotopy $\Sigma\Gamma_{p,X}$ given by the induction assumption, the upper right square commutes by Proposition 15.108, and the lower square commutes up to the preferred natural in X homotopy $\Gamma_{\Sigma^p X}$ given by Proposition 15.107. The diagram tells us that we can "paste" the chain homotopies $\Gamma_{\Sigma^p X}$ and $\Sigma\Gamma_{p,X}$ by the formula

$$\Gamma_{p+1,X} := \Gamma_{\Sigma^p X} \circ \Sigma\Sigma^p_X + \Sigma^{\%}_{\Sigma^p X} \circ S_{\widetilde{C}^s(\Sigma^p X)} \circ \Sigma\Gamma_{p,X}, \tag{15.109}$$

to obtain a preferred $\mathbb{Z}$-chain homotopy that is natural in X

$$\Gamma_{p+1,X} : \widetilde{\varphi}_{\Sigma^{p+1}X} \circ \Sigma^{p+1}_X \simeq (\Sigma^{p+1}_X)^{\%} \circ S^{p+1}_{\widetilde{C}^s(X)} \circ \Sigma^{p+1}\widetilde{\varphi}_X.$$

The actual application of these chain homotopies will be in the situation where the above diagram is desuspended appropriately. For the convenience of the reader we state the suitable version here explicitly

$$\begin{CD}
\widetilde{C}^s(X) @>{\Sigma^{-p} S^p_{\widetilde{C}^s(X)} \circ \widetilde{\varphi}_X}>> \Sigma^{-p} W^{\%}_{\mathbb{Z}}(\Sigma^p \widetilde{C}^s(X)) @>{\Sigma^{-p-1} S_{\Sigma^p(\bar{C}^s X)}}>> \Sigma^{-p-1} W^{\%}_{\mathbb{Z}}(\Sigma^{p+1}\widetilde{C}^s(X)) \\
@V{\Sigma^{-p}\Sigma^p_X}VV @VV{\Sigma^{-p}(\Sigma^p_X)^{\%}}V @VV{\Sigma^{-p-1}(\Sigma\Sigma^p_X)^{\%}}V \\
\Sigma^{-p}\widetilde{C}^s(\Sigma^p X) @>{\Sigma^{-p}\widetilde{\varphi}_{\Sigma^p X}}>> \Sigma^{-p} W^{\%}_{\mathbb{Z}}(\widetilde{C}^s(\Sigma^p X)) @>{\Sigma^{-p-1}S_{\widetilde{C}^s(\Sigma^p X)}}>> \Sigma^{-p-1} W^{\%}_{\mathbb{Z}}(\Sigma\widetilde{C}^s(\Sigma^p X)) \\
@V{\Sigma^{-p-1}\Sigma_{\Sigma^p X}}VV @. @VV{\Sigma^{-p-1}\Sigma^{\%}_{\Sigma^p X}}V \\
\Sigma^{-p-1}\widetilde{C}^s(\Sigma^{p+1}X) @>>{\Sigma^{-p-1}\widetilde{\varphi}_{\Sigma^{p+1}X}}> \Sigma^{-p-1} W^{\%}_{\mathbb{Z}}(\widetilde{C}^s(\Sigma^{p+1}X)).
\end{CD}$$

Lemma 15.110 *Let $C \xrightarrow{i} D \xrightarrow{p} E$ be a cofibration sequence of chain complexes in* $bR\text{-}CH_{\mathrm{fgp}}$. *It induces a long exact sequence of* $\widehat{Q}^*$-*groups*

$$\cdots \to \widehat{Q}^n(C) \xrightarrow{\widehat{i}^{Q,n}} \widehat{Q}^n(D) \xrightarrow{\widehat{p}^{Q,n}} \widehat{Q}^n(E) \to \widehat{Q}^{n-1}(C) \to \cdots.$$

15.3 Structured Chain Complexes

Proof. We follow closely [427, Theorem 1.1]. Recall that the assumption means that we have an exact sequence of bounded chain complexes $0 \to C \xrightarrow{i} D \xrightarrow{p} E \to 0$ that yields in each degree a split short exact sequence of finitely generated projective modules. The induced sequence of $\mathbb{Z}[\mathbb{Z}/2]$-chain complexes

$$C \otimes_R C \xrightarrow{i \otimes i} D \otimes_R D \xrightarrow{p \otimes p} E \otimes_R E \tag{15.111}$$

is not short exact because we might have $Q := \ker(p \otimes p)/\mathrm{im}(i \otimes i) \neq 0$. (This is the reason why we do not get a similar sequence for $Q_*(-)$ and $Q^*(-)$, see Proposition 15.119.) The sequence for $\widehat{Q}^*(-)$ is obtained by the following additional observations, which work only for $\widehat{Q}^*(-)$.

At least $i \otimes i$ is split injective and $p \otimes p$ is split surjective. Hence we get cofibration sequences of bounded finitely generated projective $\mathbb{Z}[\mathbb{Z}/2]$-chain complexes $C \otimes_R C \xrightarrow{i \otimes i} \ker(p \otimes p) \xrightarrow{q} Q$ and $\ker(p \otimes p) \to D \otimes_R D \xrightarrow{p \otimes p} E \otimes_R E$. They induce cofibration sequences of $\mathbb{Z}$-chain complexes

$$\widehat{W}^{\%}(C) \to \hom_{\mathbb{Z}[\mathbb{Z}/2]}(\widehat{W}, \ker(p \otimes p)) \to \hom_{\mathbb{Z}[\mathbb{Z}/2]}(\widehat{W}, Q);$$
$$\hom_{\mathbb{Z}[\mathbb{Z}/2]}(\widehat{W}, \ker(p \otimes p)) \to \widehat{W}^{\%}(D) \to \widehat{W}^{\%}(E).$$

Hence the sequence $\widehat{W}^{\%}(C) \xrightarrow{\widehat{W}^{\%}(i)} \widehat{W}^{\%}(D) \xrightarrow{\widehat{W}^{\%}(p)} \widehat{W}^{\%}(E)$ induces a long exact sequence in homology if we can show that the chain map $\widehat{W}^{\%}(C) \to \hom_{\mathbb{Z}[\mathbb{Z}/2]}(\widehat{W}, \ker(p \otimes p))$ induces an isomorphism in homology, and this is equivalent to the assertion that $H_n(\hom_{\mathbb{Z}[\mathbb{Z}/2]}(\widehat{W}, Q)) = \widehat{H}^n(Q)$ is trivial for all $n \in \mathbb{Z}$.

We will use a variant of Shapiro's lemma from homological algebra. Let G be a finite group with a subgroup $H < G$. A $\mathbb{Z}[G]$-module N is called induced if $N \cong \mathbb{Z}[G] \otimes_{\mathbb{Z}[H]} M$ for some $\mathbb{Z}[H]$-module M. Shapiro's lemma says that then $\widehat{H}^*(H, M) \cong \widehat{H}^*(G, N)$ [58, (5.2)]. If $H = \{1\}$, then we have $\widehat{H}^*(G, N) = 0$ for all $* \in \mathbb{Z}$.

Let $H = \{1\}$ and $G = \mathbb{Z}/2$ and let A be a chain complex in $b\mathbb{Z}[\mathbb{Z}/2]$-CH such that each A_r is a coinduced $\mathbb{Z}[\mathbb{Z}/2]$-module. Then an inductive argument as in [133, Anhang Proposition 2.1] shows that $\widehat{H}^*(\mathbb{Z}/2; A)$ vanishes for all $* \in \mathbb{Z}$. Now the claim follows since $Q := \ker(p \otimes p)/\mathrm{im}(i \otimes i)$ belongs to $b\mathbb{Z}[\mathbb{Z}/2]$-CH and is degreewise coinduced over $\mathbb{Z}[\mathbb{Z}/2]$. □

One may ask whether the arguments in the above proof might be improved to show that the functor $\widehat{W}^{\%}(-)$ sends a cofibration sequence of chain complexes in bR-CH$_{\mathrm{fgp}}$ to a homotopy cofibration sequence. We were not able to obtain such an improvement since the arguments above use homology and the chain complex $\widehat{W}^{\%}(A)$ is unbounded even if A is in bR-CH$_{\mathrm{fgp}}$.

Exercise 15.112 Show that applying the functor $\widehat{Q}^*(-)$ on a homotopy cofibration sequence of bounded finitely generated projective chain complexes induces a long exact sequence of abelian groups.

From the previous remark we can get the following proposition, whose proof is independent of the proof of Proposition 15.99.

Proposition 15.113 (Suspension is an isomorphism on $\widehat{Q}^*$-groups) *For C in* $hR\text{-}CH_{fgp}$ *the suspension map*

$$S\colon \widehat{Q}^n(C) \to \widehat{Q}^{n+1}(\Sigma C)$$

is an isomorphism for all $n \in \mathbb{Z}$.

Proof. We can assume without loss of generality $C \in bR\text{-}CH_{fgp}$ because of homotopy invariance, see Subsection 15.3.4. We get from Lemma 15.110 applied to the canonical cofibration sequence $C \to \text{cone}(C) \to \Sigma C$ a long exact sequence $\cdots \to \widehat{Q}^n(C) \to \widehat{Q}^n(\text{cone}(C)) \to \widehat{Q}^n(\Sigma C) \to \cdots$. We conclude from homotopy invariance, see Subsection 15.3.4, that $\widehat{W}^{\%}(\text{cone}(C))$ is contractible as the mapping cone $\text{cone}(C)$ is contractible, and hence $\widehat{Q}^n(\text{cone}(C)) = 0$ for all $n \in \mathbb{Z}$. Therefore the connecting map in the long exact sequence $\widehat{Q}^{n+1}(\Sigma C) \to \widehat{Q}^n(C)$ is an isomorphism for all $n \in \mathbb{Z}$. We leave it to the reader to check that this is an inverse of S. □

The analogous propositions for the symmetric and quadratic structures do not hold. The failure of the suspension map to be chain equivalence in the symmetric case is described in more detail in Proposition 15.119 below.

Finally, there is an important corollary that is the main device for producing quadratic structures. In the next subsection it will be used in a geometric situation.

Corollary 15.114 (Hyperquadratic structures are stable symmetric structures) *For a chain complex $C \in hR\text{-}CH_{fgp}$ we have a zigzag of chain homotopy equivalences*

$$\widehat{W}^{\%}(C) \xrightarrow{\simeq} \text{hocolim}_{p\to\infty} \Sigma^{-p}\widehat{W}^{\%}(\Sigma^p C) \xleftarrow{\simeq} \text{hocolim}_{p\to\infty} \Sigma^{-p}W^{\%}(\Sigma^p C)$$

where the homotopy colimits are taken over the maps given by applying Σ^{-1} to the suspension maps appearing in Definition 15.87.

Proof. We can assume without loss of generality $C \in bR\text{-}CH_{fgp}$ because of homotopy invariance, see Subsection 15.3.4. We have for any $p \geq 0$ the commutative diagram

$$\begin{array}{ccccc}
W^{\%}(C) & \longrightarrow & \widehat{W}^{\%}(C) & \longrightarrow & \Sigma W_{\%}(C) \\
{\scriptstyle S^p}\downarrow & & {\scriptstyle S^p}\downarrow & & {\scriptstyle S^p}\downarrow \\
\Sigma^{-p}W^{\%}(\Sigma^p C) & \longrightarrow & \Sigma^{-p}\widehat{W}^{\%}(\Sigma^p C) & \longrightarrow & \Sigma^{-p+1}W_{\%}(\Sigma^p C).
\end{array}$$

The upper row is a cofibration sequence, which means split short exact sequences in each degree, see (15.40). The bottom sequence is a cofibration sequence because it is obtained just by reindexing another cofibration sequence. By Proposition 14.114 these cofibration sequences induce a homotopy cofibration sequence of homotopy colimits.

15.3 Structured Chain Complexes

So we get a commutative diagram of chain complexes whose columns are homotopy cofibration sequences, depicted in (15.115) below. The middle arrow in the diagram is a homotopy equivalence by Proposition 15.99.

Because $C \in bR\text{-CH}_{\text{fgp}}$, it is concentrated between dimensions, say, l and m, with $l < m$. Then $C \otimes_R C$ is concentrated between dimensions $2l$ and $2m$, and $\Sigma^p C \otimes_R \Sigma^p C$ is concentrated between dimensions $2(l+p)$ and $2(m+p)$. This also implies that $\Sigma^{-p+1} W^{\%}_{\%}(\Sigma^p C)$ is bounded from below by $2l + p + 1$. Hence we can apply Proposition 14.118 to the homotopy colimit in the bottom right and we obtain that it is a contractible $\mathbb{Z}$-chain complex.

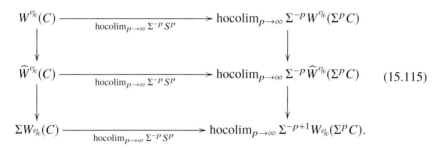

(15.115)

This yields the desired statement. □

Corollary 15.116 (Quadratic versus symmetric versus stable symmetric structures in the polarised setting) *For a chain complex $C \in hR\text{-CH}_{\text{fgp}}$ we have a homotopy cofibration sequence of chain complexes*

$$W_{\%}^{\%}(C) \xrightarrow{1+T} W^{\%}(C) \xrightarrow{S^{\infty}} \operatorname{hocolim}_{p \to \infty} \Sigma^{-p} W^{\%}(\Sigma^p C)$$

where the homotopy colimit is taken over the maps given by applying Σ^{-p} to the suspension maps appearing in Definition 15.87 and where the notation S^{∞} is the zeroth structure map associated to $\operatorname{hocolim}_{p \to \infty}$.

Proof. The sequence is obtained from Proposition 15.43 by replacing the polarised hyperquadratic structures with stable polarised symmetric structures using Corollary 15.114. From the second diagram in the proof of Corollary 15.114, we conclude that the zigzag of chain homotopy equivalences identifying the polarised hyperquadratic structures as stable polarised symmetric structures fits into a commutative square with the maps $J: W^{\%}(C) \to \widehat{W}^{\%}(C)$ and $S^{\infty}: W^{\%}(C) \to \operatorname{hocolim}_{p \to \infty} \Sigma^{-p} W^{\%}(\Sigma^p C)$. Now the statement follows from Lemma 14.64 (i). □

We note that since the suspension maps $S: \Sigma W^{\%}(C) \to W^{\%}(\Sigma C)$ are cofibrations for any $C \in hR\text{-CH}_{\text{fgp}}$, as pointed out in Remark 15.97, the homotopy colimit in Corollary 15.116 can be replaced by the colimit.

Remark 15.117 (Recipe for a polarised quadratic structure, version 2) The usefulness of Corollary 15.116 is best seen together with the statement of Proposition 15.43. It tells us that, in order to produce a polarised quadratic structure on a chain complex C, it is up to homotopy enough to produce a polarised symmetric structure together with a nullhomology of some suspension of this polarised symmetric structure. Such a nullhomology might not exist, in which case the polarised symmetric structure does not come from a polarised quadratic structure. However, as we will see in Subsection 15.3.6, there are geometric situations where this procedure can be used effectively.

Corollary 15.118 (Quadratic versus symmetric versus stable symmetric structures in Q-groups) *For a chain complex $C \in$ hR-CH$_{\text{fgp}}$ we have a long exact sequence of Q-groups*

$$\cdots \to Q_n(C) \xrightarrow{1+T} Q^n(C) \xrightarrow{J} \operatorname{colim}_p Q^{n+p}(\Sigma^p C) \xrightarrow{H} Q_{n-1}(C) \to \cdots .$$

We close this subsection with a brief discussion of a modification of the above situation where the map S^∞ whose target is a homotopy colimit is replaced by the p-th suspension map S_C^p for some specific $p \geq 1$. It will not be directly used in the proofs in the subsequent sections, but it is closely related to some constructions, see Remark 15.143. It might also be of independent interest and is closer to the original treatment of Ranicki, see [344, Proposition 1.3], so we mention it.

We need the following notation. For $-\infty \leq i \leq j \leq \infty$ let $\widehat{W}[i, j]$ be the truncation of $\widehat{W}$ given by $\widehat{W}[i, j]_s = \widehat{W}_s$ for $i \leq s \leq j$ and $\widehat{W}[i, j]_s = 0$ otherwise. We put for a chain complex C

$$W_\%^{[0,p]}(C) := \widehat{W}[0, p] \otimes_{\mathbb{Z}[\mathbb{Z}/2]} (C \otimes_R C).$$

Proposition 15.119 (Truncated quadratic versus symmetric versus suspended symmetric structures in the polarised setting) *For a chain complex $C \in$ hR-CH$_{\text{fgp}}$ and $p \geq 1$ we have a homotopy cofibration sequence*

$$W_\%^{[0,p-1]}(C) \xrightarrow{1+T} W^\%(C) \xrightarrow{\Sigma^{-p} S_C^p} \Sigma^{-p} W^\%(\Sigma^p C).$$

Remark 15.120 (Recipe for a polarised quadratic structure in components, version 2) Let us state concretely what it means for an n-dimensional polarised symmetric structure $\varphi \in W^\%(C)_n$ of $C \in$ bR-CH$_{\text{fgp}}$ to possess a nullhomology of its suspension

$$S^p(\varphi) \sim 0 \in W^\%(\Sigma^p C)_{n+p}.$$

We start with the case $p = 0$. Then $\varphi \sim 0 \in W^\%(C)_n$ means that there exists a $\theta \in W^\%(C)_{n+1}$ such that $\varphi = d_{W^\%(C)}(\theta)$. The latter equation implies

$$\varphi_0 = d_C \theta_0 - (-1)^n \theta_0 d_{C^{-*}},$$

which means that the component φ_0 is nullhomologous.

Next consider the case $p = 1$. Then $S(\varphi) \sim 0 \in W^\%(\Sigma C)_{n+1}$ means that there exists a $\theta \in W^\%(\Sigma C)_{n+2}$ satisfying $\varphi = d_{W^\%(C)}(\theta)$. The latter equation implies

15.3 Structured Chain Complexes

$$\varphi_0 = d_C \circ \theta_0 - (-1)^n \theta_0 \circ d_{C^{-*}} - (-1)^n(\theta_{-1} + T(\theta)_{-1});$$
$$0 = d_C \circ \theta_{-1} - (-1)^{n-1} \theta_{-1} \circ d_{C^{-*}},$$

which means that φ_0 is homologous to a symmetrisation of a chain map. (The chain map is θ_{-1} and the homology is θ_0.)

In the general case $p \geq 0$ a nullhomology $S^p(\varphi) \sim 0 \in W^{\%}(\Sigma^p C)_{n+p}$ consists of a sequence maps θ_i satisfying

$$\varphi_0 = d_C \circ \theta_0 - (-1)^n \theta_0 \circ d_{C^{-*}} - (-1)^n(\theta_{-1} + T(\theta)_{-1});$$
$$(-1)^n(\theta_{s-1} + (-1)^s T(\theta)_{s-1}) = d_C \circ \theta_s - (-1)^{n+s} \theta_s \circ d_{C^{-*}} \quad 0 > s > -p;$$
$$0 = d_C \circ \theta_{-p} - (-1)^{n-p} \theta_{-p} \circ d_{C^{-*}},$$

which means that φ_0 is homologous to a symmetrisation of a map that is the p-th level chain homotopy between maps and their duals where the sequence starts with a chain map (namely θ_{-p}).

This gives a "recipe" in the sense that, in order to show that the p-th symmetrisation of φ vanishes in $Q^{n+p}(\Sigma^p C)$, it is enough to find a sequence of maps θ_i satisfying the above equations.

In the above verification we used a certain trick. Namely, according to the formulas in Remark 15.97 the signs, the homomorphism μ_1, and the involution T should interfere. What we have done here is to consider φ and θ as chains in $\widehat{W}^{\%}(C)$. Note that the map $W^{\%}(C) \to \widehat{W}^{\%}(C)$ is injective. Moreover, the suspension map is bijective for $\widehat{W}^{\%}(-)$ by Proposition 15.99 and so we can uniquely desuspend θ. The desuspended formulas are precisely those given here.

15.3.6 The Quadratic Construction

In this section we describe the geometric situation that gives rise to quadratic structures. The idea is to employ Corollary 15.116, which tells us that, in order to construct a quadratic structure, it is enough to construct a symmetric structure together with a nullhomology of some suspension of this symmetric structure.

Such a situation arises when we have a map of pointed topological spaces of the form $F: \Sigma^p X \to \Sigma^p Y$ for some fixed $p \geq 0$, where we remind the reader that $\Sigma^p X = S^p \wedge X$ is the reduced suspension of a pointed space X. In Construction 15.161 we present geometric problems in the context of surgery theory that lead to maps of such form. Now recall that the version of the symmetric construction from (15.58) for pointed spaces

$$\widetilde{\varphi}_X : \widetilde{C}^s(X) \to W^{\%}_{\mathbb{Z}}(\widetilde{C}^s(X))$$

is natural with respect to maps between pointed spaces (keep in mind that $\widetilde{C}^s(-)$ denotes the reduced singular chain complex). However, the symmetric constructions on the pointed spaces X and Y are *not* natural with respect to maps induced by the maps of the form $F: \Sigma^p X \to \Sigma^p Y$ for some $p > 0$. This means that we have the

equation
$$\widetilde{C}^s(F)^{\%} \circ \widetilde{\varphi}_{\Sigma^p X} - \widetilde{\varphi}_{\Sigma^p Y} \circ \widetilde{C}^s(F) = 0, \tag{15.121}$$
but, when we consider the composition chain map
$$f: \widetilde{C}^s(X) \xrightarrow{\Sigma^{-p}\Sigma_X^p} \Sigma^{-p}\widetilde{C}^s(\Sigma^p X) \xrightarrow{\Sigma^{-p}\widetilde{C}^s(F)} \Sigma^{-p}\widetilde{C}^s(\Sigma^p Y) \xrightarrow{\Sigma^{-p}\Sigma_Y^{-p}} \widetilde{C}^s(Y), \tag{15.122}$$
where $\Sigma_Y^{-p}: \widetilde{C}^s(\Sigma^p Y) \to \Sigma^p \widetilde{C}^s(Y)$ is some chain homotopy inverse[3] of the $\mathbb{Z}$-homotopy equivalence $\Sigma_Y^p: \Sigma^p \widetilde{C}^s(Y) \to \widetilde{C}^s(\Sigma^p Y)$ of (15.104), we have in general
$$f^{\%} \circ \widetilde{\varphi}_X - \widetilde{\varphi}_Y \circ f \neq 0. \tag{15.123}$$

Therefore, when we pick some n-dimensional cycle γ in $\widetilde{C}_n^s(X)$ and apply the left-hand side of (15.123) to it, we may obtain some non-zero symmetric structure on $\widetilde{C}^s(Y)$. On the other hand, the p-th suspension of this structure can be identified with the structure given by applying the left-hand side of (15.121) to $\Sigma_X^p(\gamma)$ and hence it is zero. This is realised by some nullhomologies and so, in order to describe the resulting quadratic structure, we have to describe these nullhomologies. We will do this now, but we remark that the situation is perhaps more technically demanding than one would expect at first glance. One of the problems is that we aim for a construction that is natural in F in a specific sense to be explained later, and in the description above we have one map that is not natural, namely, the chain homotopy inverse Σ_Y^{-p} in the composition (15.122). To remedy this we follow [346, page 30].

Our ultimate goal in this section is to produce the quadratic construction map $\overline{\psi}_{\{F\}}$ in (15.149) as a map from homology groups to quadratic Q-groups with good naturality and additivity properties, see Lemma 15.148, and Proposition 15.152, and its equivariant version $\overline{\psi}_{\{F\}_{\Gamma,w}}$ in (15.154), also with good properties, see Lemma 15.155 and Proposition 15.156. To achieve this goal we first have to make some constructions at the chain level and prove that they behave well when we vary p and F in the above discussion. In general we follow [345, Section 1] and [346, page 30], but we give additional explanations. As was the case with the symmetric construction, we start with the non-equivariant version and later discuss the equivariant one.

Construction 15.124 (The quadratic construction)
Let $C_{p,F}(X)$ be the mapping cone of the following natural chain map
$$C_{p,F}(X) := \mathrm{cone}\bigl(\Sigma^{-1}\widetilde{C}^s(X) \oplus \Sigma^{-1}\widetilde{C}^s(Y) \xrightarrow{\Sigma^{-p-1}((F \circ \Sigma_X^p) \oplus (-\Sigma_Y^p))} \Sigma^{-p-1}\widetilde{C}^s(\Sigma^p Y)\bigr). \tag{15.125}$$

We have the homotopy pullback diagram

---

[3] Note that there is no natural in Y construction of such an inverse since the acyclic models cannot be used. See [346, page 30] where this issue plays a role.

15.3 Structured Chain Complexes

$$\begin{array}{ccc} C_{p,F}(X) & \xrightarrow{\pi_{p,F,Y}} & \widetilde{C}^s(Y) \\ \pi_{p,F,X} \downarrow \simeq & & \simeq \downarrow \Sigma^{-p}\Sigma_Y^p \\ \widetilde{C}^s(X) & \xrightarrow{\Sigma^{-p}(F \circ \Sigma_X^p)} & \Sigma^{-p}\widetilde{C}^s(\Sigma^p Y). \end{array} \qquad (15.126)$$

The projection map $\pi_{p,F,X}: C_{p,F}(X) \to \widetilde{C}^s(X)$ is a chain homotopy equivalence as Σ_Y^p is. Therefore we can replace $\widetilde{C}^s(X)$ by $C_{p,F}(X)$. The advantage of this passage is that we obtain a natural chain map to $\widetilde{C}^s(Y)$, namely, the upper horizontal map $\pi_{p,F,Y}$ in the above diagram.

By its definition, the diagram commutes up to a preferred natural chain homotopy

$$\Upsilon_{p,F}: \Sigma^{-p}(F \circ \Sigma_X^p) \circ \pi_{p,F,X} \simeq \Sigma^{-p}\Sigma_Y^p \circ \pi_{p,F,Y}. \qquad (15.127)$$

An explicit formula is given as follows. Recall first that an n-dimensional chain in $C_{p,F}(X)$ is a triple (x, y, z) such that $x \in \widetilde{C}_n^s(X)$, $y \in \widetilde{C}_n^s(Y)$ and $z \in (\Sigma^{-p}\widetilde{C}^s(\Sigma^p Y))_{n+1}$. It is a cycle if x and y are cycles and the boundary of z equals $(\Sigma^{-p}\Sigma_Y^p)(y) - (\Sigma^{-p}\widetilde{C}^s(F) \circ \Sigma_X^p)(x)$. The desired formula is

$$\Upsilon_{p,F}(x, y, z) = -z.$$

Now the whole situation can be summarised in Diagram (15.132) below. Its inner square does not commute in general, cf., (15.123). On the other hand, all the other squares in the diagram commute either strictly or up to preferred natural chain homotopy, the outer square by the equation (15.121), the left and the right square by chain homotopies $\Gamma_{p,X}$ and $\Gamma_{p,Y}$ obtained in (15.109) by iterating Proposition 15.107, the upper square by the above chain homotopy $\Upsilon_{p,F}$, and the lower square by Proposition 15.108.

These chain homotopies can be put together to obtain the desired *quadratic construction map*

$$\psi_{p,F}: C_{p,F}(X) \to W_{p,Y}, \qquad (15.128)$$

as described below, where

$$W_{p,Y} := \mathrm{hofib}\left(W^{\%}(\Sigma^{-p}\widetilde{C}^s(\Sigma^p Y)) \xrightarrow{\Sigma^{-p}S^p} \Sigma^{-p}W^{\%}(\widetilde{C}^s(\Sigma^p Y))\right). \qquad (15.129)$$

Consider the map given by the (non-commuting) inner square

$$\alpha_{p,F}: C_{p,F}(X) \to W^{\%}(\Sigma^{-p}\widetilde{C}^s(\Sigma^p Y)) \qquad (15.130)$$

defined by

$$\alpha_{p,F} := (\Sigma^{-p}\widetilde{C}^s(F))^{\%} \circ (\Sigma^{-p}\Sigma_X^p)^{\%} \circ \widetilde{\varphi}_X \circ \pi_{p,F,X} - (\Sigma^{-p}\Sigma_Y^p)^{\%} \circ \widetilde{\varphi}_Y \circ \pi_{p,F,Y}.$$

The fact that the outer square commutes together with the preferred natural chain homotopies of the remaining four inner squares provide us with a preferred natural nullhomotopy

$$\beta_{p,F} := 0 \simeq \Sigma^{-p} S^p \circ \alpha_{p,F} : C_{p,F}(X) \to \Sigma^{-p} W^{\%}(\widetilde{C}^s(\Sigma^p Y)). \quad (15.131)$$

The desired $\mathbb{Z}$-chain map $\psi_{p,F}$ is obtained using the universal property of the homotopy fibre from Remark 14.73 as the $\mathbb{Z}$-chain map $\psi_{p,F} := z(\alpha_{p,F}, \beta_{p,F})$.

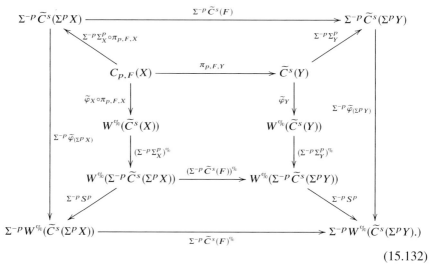

(15.132)

An explicit formula is obtained as follows. Consider a triple (x, y, z) representing an n-dimensional chain in $C_{p,F}(X)$. The image $\psi_{\{p,F\}}(x, y, z)$ has two components, one in $\Sigma^{-p}\widetilde{C}^s(\Sigma^p Y)_n$ and one in $\Sigma^{-p}\widetilde{C}^s(\Sigma^p Y)_{n+1}$. The first component $\alpha_{p,F}(x, y, z) \in \Sigma^{-p}\widetilde{C}^s(\Sigma^p Y)_n$ is

$$(\Sigma^{-p}\widetilde{C}^s(F))^{\%} \circ (\Sigma^{-p}\Sigma_X^p)^{\%} \circ \widetilde{\varphi}_X(x) - (\Sigma^{-p}\Sigma_Y^p)^{\%} \circ \widetilde{\varphi}_Y(y). \quad (15.133)$$

The second component $\beta_{p,F}(x, y, z) \in \Sigma^{-p} W^{\%}(C(\Sigma^p Y))_{n+1}$ is given in terms of the chain homotopies $\Gamma_{p,X}$ and $\Gamma_{p,Y}$ of Proposition 15.107 and $\Upsilon_{p,F}$ of (15.127) as

$$\Sigma^{-p} C^s(F)^{\%} \circ \Sigma^{-p}\Gamma_{p,X}(x) - \Sigma^{-p}\Gamma_{p,Y}(y) + \Sigma^{-p}\widetilde{\varphi}_{\Sigma^p Y}(z). \quad (15.134)$$

The map $\psi_{p,F}$ of (15.128) is natural in the following sense.

Lemma 15.135 *Let $f: X \to X'$ and $g: Y \to Y'$ be pointed maps. Consider $p \geq 0$ and pointed maps $F: \Sigma^p X \to \Sigma^p Y$ and $F': \Sigma^p X' \to \Sigma^p Y'$ such that the following square commutes*

$$\begin{array}{ccc} \Sigma^p X & \xrightarrow{F} & \Sigma^p Y \\ {\scriptstyle \Sigma^p f} \downarrow & & \downarrow {\scriptstyle \Sigma^p g} \\ \Sigma^p X' & \xrightarrow{F'} & \Sigma^p Y'. \end{array}$$

15.3 Structured Chain Complexes

Then the following diagram commutes

$$\begin{array}{ccc} C_{p,F}(X) & \xrightarrow{\psi_{p,F}} & W_{p,Y} \\ {\scriptstyle C_{p,F,F'}(f,g)} \downarrow & & \downarrow {\scriptstyle W_p(g)} \\ C_{p,F'}(X') & \xrightarrow{\psi_{p,F'}} & W_{p,Y'} \end{array}$$

where the vertical arrows are the obvious maps induced by f and g.

Note that there is no reasonable formulation of Lemma 15.135 where $f\colon X \to X'$ and $g\colon Y \to Y'$ are replaced by maps of the shape $f\colon \Sigma^p X \to \Sigma^p X'$ and $g\colon \Sigma^p Y \to \Sigma^p Y'$.

Proof. Let us first explain in more detail the vertical arrows in the desired diagram. The left arrow $C_{p,F,F'}(f,g)$ is the dashed arrow in the following diagram

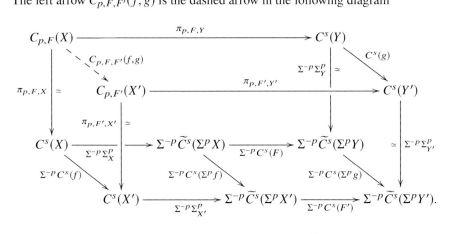

The bottom left square commutes strictly by naturality of Σ^p_X, the bottom right square commutes strictly by the assumption of the lemma, and the right vertical square commutes strictly by naturality of Σ^p_Y. By the universal property of the homotopy pullback from Remark 14.94, we obtain the dashed arrow, and the top and the left squares commute strictly.

The right vertical arrow is the induced map between homotopy fibres obtained via the universal property of the homotopy fibre using the diagram

$$\begin{array}{ccc} W_{p,Y} & \longrightarrow W^{\%}(\Sigma^{-p}\widetilde{C}^s(\Sigma^p Y)) & \xrightarrow{\Sigma^{-p} S^p} \Sigma^{-p} W^{\%}(\widetilde{C}^s(\Sigma^p Y)) \\ {\scriptstyle W_p(g)} \downarrow & \downarrow {\scriptstyle (\Sigma^{-p}\widetilde{C}^s(\Sigma^p g))^{\%}} & \downarrow {\scriptstyle \Sigma^{-p}(\widetilde{C}^s(\Sigma^p g))^{\%}} \\ W_{p,Y'} & \longrightarrow W^{\%}(\Sigma^{-p}\widetilde{C}^s(\Sigma^p Y')) & \xrightarrow{\Sigma^{-p} S^p} \Sigma^{-p} W^{\%}(\widetilde{C}^s(\Sigma^p Y')) \end{array}$$

whose right square commutes by the naturality of the suspension map (15.88), as stated in (15.100), applied to the chain map $\Sigma^{-p}\widetilde{C}^s(\Sigma^p g)$.

The horizontal arrows $\psi_{p,F}$ and $\psi_{p,F'}$ in the desired square are obtained via the universal property of the homotopy fibre using the chain maps $\alpha_{p,F}$ and $\alpha_{p,F'}$ from (15.130) and chain homotopies $\beta_{p,F}\colon 0 \simeq \Sigma^{-p}S^p \circ \alpha_{p,F}$ and $\beta_{p,F'}\colon 0 \simeq \Sigma^{-p}S^p \circ \alpha_{p,F'}$ from (15.131). The proof of the commutativity is achieved by studying the following diagram

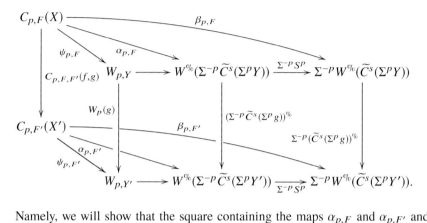

Namely, we will show that the square containing the maps $\alpha_{p,F}$ and $\alpha_{p,F'}$ and the square containing chain homotopies $\beta_{p,F}$ and $\beta_{p,F'}$ commute. Since the maps $\psi_{p,F}$ and $\psi_{p,F'}$ are obtained via the universal property of the homotopy fibre, the desired commutativity is acquired by the naturality of such induced maps from Exercise 14.76.

To get the statement about the square containing the maps $\alpha_{p,F}$ and $\alpha_{p,F'}$, we look at how our situation applies to the (non-commuting) inner square of Diagram (15.132). We obtain the following diagram:

15.3 Structured Chain Complexes

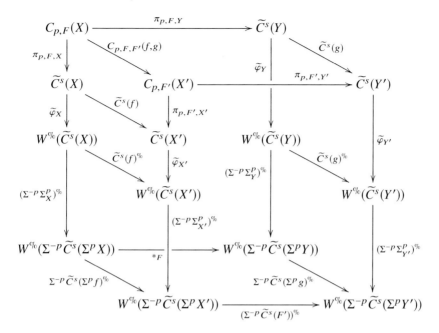

with the arrow $*_F = (\Sigma^{-P}\widetilde{C}^s(F))^{\%}$. We claim that in this diagram all squares except the front and the back commute.

The first square in the left vertical face commutes since it is the left square in the first diagram in this proof, the second by the naturality of the symmetric construction $\widetilde{\varphi}_X$ from (15.55), and the third by the functoriality of $W^{\%}(-)$ and naturality of the map Σ^P_- from (15.104).

The top square commutes since it is the top square in the first diagram in this proof, the first square in the right vertical face by the naturality of the symmetric construction $\widetilde{\varphi}_Y$ from (15.55), the second and the bottom square by the functoriality of $W^{\%}(-)$ and naturality of the map Σ^P_- from (15.104).

Now the map $\alpha_{p,F}$ is obtained as the difference between the two compositions of the non-commuting back face, and the map $\alpha_{p,F'}$ is the difference between the two compositions of the non-commuting front face. Let us write $\alpha_{p,F}$ as $\alpha_{p,F} = \alpha_{p,F,1} - \alpha_{p,F,2}$, where

$$\alpha_{p,F,1} := (\Sigma^{-P}\widetilde{C}^s(F))^{\%} \circ (\Sigma^{-P}\Sigma^P_X)^{\%} \circ \widetilde{\varphi}_X \circ \pi_{p,F,X};$$
$$\alpha_{p,F,2} := (\Sigma^{-P}\Sigma^P_Y)^{\%} \circ \widetilde{\varphi}_Y \circ \pi_{p,F,Y},$$

and similarly for $\alpha_{p,F'} = \alpha_{p,F',1} - \alpha_{p,F',2}$. The commutativity of the other faces implies

$$\Sigma^{-P}\widetilde{C}^s(\Sigma^P g)^{\%} \circ \alpha_{p,F,i} = \alpha_{p,F',i} \circ C_{p,F,F'}(f,g)$$

for both $i = 1, 2$. Subtracting the two identities tells us that the square containing maps $\alpha_{p,F}$ and $\alpha_{p,F'}$ in the desired diagram commutes.

To get the statement about the square containing the chain homotopies $\beta_{p,F}$ and $\beta_{p,F'}$, we also look at how our situation applies to the remaining squares of Diagram (15.132). Recall that all of them either consist of natural maps and commute strictly, or they consist of natural maps and commute up to a preferred natural chain homotopy. The chain homotopies $\beta_{p,F}$ and $\beta_{p,F'}$ are obtained by pasting these preferred natural chain homotopies.

The desired commutativity is achieved as follows. Let us first recall that we have by (15.134) and by the definitions of $\pi_{p,F,X}$, $\pi_{p,F,Y}$ from (15.126) and $\Upsilon_{p,F}$ from (15.127)

$$\beta_{p,F} = \Sigma^{-P} C^s(F)^{\%} \circ \Sigma^{-P} \Gamma_{p,X} \circ \pi_{p,F,X} - \Sigma^{-P} \Gamma_{p,Y} \circ \pi_{p,F,Y} - \Sigma^{-P} \widetilde{\varphi}_{\Sigma^P Y} \circ \Upsilon_{p,F}.$$

Next we calculate

$$\begin{aligned}
\Sigma^{-P} \widetilde{C}^s(\Sigma^P g)^{\%} &\circ \beta_{p,F} = \\
&= \Sigma^{-P} \widetilde{C}^s(\Sigma^P g)^{\%} \circ \Sigma^{-P} C^s(F)^{\%} \circ \Sigma^{-P} \Gamma_{p,X} \circ \pi_{p,F,X} - \\
&\quad - \Sigma^{-P} \widetilde{C}^s(\Sigma^P g)^{\%} \circ \Sigma^{-P} \Gamma_{p,Y} \circ \pi_{p,F,Y} - \\
&\quad - \Sigma^{-P} \widetilde{C}^s(\Sigma^P g)^{\%} \circ \Sigma^{-P} \widetilde{\varphi}_{\Sigma^P Y} \circ \Upsilon_{p,F} = \\
&= \Sigma^{-P} C^s(F')^{\%} \circ \Sigma^{-P} \widetilde{C}^s(\Sigma^P f)^{\%} \circ \Sigma^{-P} \Gamma_{p,X} \circ \pi_{p,F,X} - \\
&\quad - \Sigma^{-P} \widetilde{C}^s(\Sigma^P g)^{\%} \circ \Sigma^{-P} \Gamma_{p,Y} \circ \pi_{p,F,Y} - \\
&\quad - \Sigma^{-P} \widetilde{C}^s(\Sigma^P g)^{\%} \circ \Sigma^{-P} \widetilde{\varphi}_{\Sigma^P Y} \circ \Upsilon_{p,F} = \\
&= \Sigma^{-P} C^s(F')^{\%} \circ \Sigma^{-P} \Gamma_{p,X'} \circ \widetilde{C}^s(f) \circ \pi_{p,F,X} - \\
&\quad - \Sigma^{-P} \Gamma_{p,Y'} \circ \widetilde{C}^s(g) \circ \pi_{p,F,Y} - \\
&\quad - \Sigma^{-P} \widetilde{\varphi}_{\Sigma^P Y'} \circ \Sigma^{-P} \widetilde{C}^s(\Sigma^P g) \circ \Upsilon_{p,F} = \\
&= \Sigma^{-P} C^s(F')^{\%} \circ \Sigma^{-P} \Gamma_{p,X'} \circ \pi_{p,F',X'} \circ C_{p,F,F'}(f,g) - \\
&\quad - \Sigma^{-P} \Gamma_{p,Y'} \circ \pi_{p,F',Y'} \circ C_{p,F,F'}(f,g) - \\
&\quad - \Sigma^{-P} \widetilde{\varphi}_{\Sigma^P Y'} \circ \Upsilon_{p,F'} \circ C_{p,F,F'}(f,g) = \\
&= \beta_{p,F'} \circ C_{p,F,F'}(f,g).
\end{aligned}$$

The first equation is the definition, in the second only the first of the three terms has changed by the functoriality of $W^{\%}(-)$ and the assumption of the desired lemma. In the first two terms of the third equation we used naturality of $\Gamma_{p,-}$ from (15.109), see also Proposition 15.107, and in the third term naturality of the symmetric construction $\widetilde{\varphi}_-$ applied to $\Sigma^P g$. In the fourth equation we have used in the first two terms the commutativity of the left and top square of the first diagram in this proof and in the third term naturality of the chain homotopy $\Upsilon_{p,-}$ from (15.127). The last equation is the definition.

In summary we have obtained the commutativity of the square containing $\beta_{p,F}$ and $\beta_{p,F'}$. This finishes the proof of Lemma 15.135. □

15.3 Structured Chain Complexes

To proceed, we need to produce a certain zigzag of chain maps containing the map $\psi_{p,F}$ from (15.128) so that the zigzag will at the end induce the map $\overline{\psi}_{\{F\}}$ in (15.149). The zigzag is depicted as the left column in Diagram (15.136) below and what follows is a sequence of descriptions of all vertical maps in that column from the top to the bottom. The horizontal maps in the diagram are needed to study the behaviour of the zigzag when we vary p. Their description from the top to the bottom is also included below.

We claim that there is a diagram

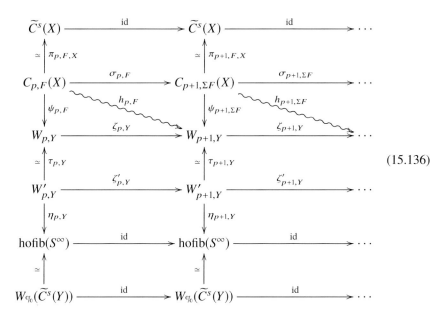

(15.136)

where all squares commute strictly or up to preferred chain homotopy indicated by a wobbly arrow and all chain maps pointing upwards are chain homotopy equivalences. The vertical arrows stem from (15.126), (15.128), (15.138), and (15.139). The horizontal arrows are constructed in (15.140), (15.141), and (15.142). The chain homotopy indicated by the wobbly arrow is constructed in (15.144) inside the proof of the (homotopy) commutativity of the diagram, which follows after the descriptions of the maps.

The maps $\pi_{p,F,X}$ and $\psi_{p,F}$ were described in the previous construction.

The map $\tau_{p,Y}$: The chain complex $W'_{p,Y}$ is defined as the homotopy fibre

$$W'_{p,Y} := \mathrm{hofib}\left(W^{c/_0}(\widetilde{C}^s(Y)) \xrightarrow{\Sigma^{-p} S^p_{\widetilde{C}^s(Y)}} \Sigma^{-p} W^{c/_0}(\Sigma^p \widetilde{C}^s(Y))\right). \tag{15.137}$$

The desired chain map is obtained as the induced map between the homotopy fibres in the following diagram:

$$\begin{CD}
W'_{p,Y} @>>> W^{\%}(\widetilde{C}^s(Y)) @>\Sigma^{-p}S^p_{\widetilde{C}^s(Y)}>> \Sigma^{-p}W^{\%}(\Sigma^p\widetilde{C}^s(Y)) \\
@VV\tau_{p,Y}V @VV(\Sigma^{-p}\Sigma^p_Y)^{\%}V @VV\Sigma^{-p}(\Sigma^p_Y)^{\%}V \\
W_{p,Y} @>>> W^{\%}(\Sigma^{-p}\widetilde{C}^s(\Sigma^pY)) @>\Sigma^{-p}S^p_{\Sigma^{-p}\widetilde{C}^s(\Sigma^pY)}>> \Sigma^{-p}W^{\%}(\widetilde{C}^s(\Sigma^pY)).
\end{CD} \quad (15.138)$$

The map $\eta_{p,Y}$: Consider the map S^∞ from Corollary 15.116 for $C = \widetilde{C}^s(Y)$. Denote by $i_p \colon \Sigma^{-p}W^{\%}(\Sigma^p\widetilde{C}^s(Y)) \to \text{hocolim}_p \Sigma^{-p}W^{\%}(\Sigma^p\widetilde{C}^s(Y))$ the map induced by the inclusion of the p-th entry into the homotopy colimit. The desired map $\eta_{p,Y}$ is obtained as the induced map between the homotopy fibres in the following diagram:

$$\begin{CD}
W'_{p,Y} @>>> W^{\%}(\widetilde{C}^s(Y)) @>\Sigma^{-p}S^p_{\widetilde{C}^s(Y)}>> \Sigma^{-p}W^{\%}(\Sigma^p\widetilde{C}^s(Y)) \\
@VV\eta_{p,Y}V @VV\text{id}V @VVi_pV \\
\text{hofib}(S^\infty) @>>> W^{\%}(\widetilde{C}^s(Y)) @>S^\infty>> \text{hocolim}_p \Sigma^{-p}W^{\%}(\Sigma^p\widetilde{C}^s(Y)).
\end{CD} \quad (15.139)$$

The last map is the chain homotopy equivalence

$$W^{\%}(\widetilde{C}^s(Y)) \xrightarrow{\simeq} \text{hofib}(S^\infty),$$

obtained from Corollary 15.116 by the universal property of the homotopy fibre.

We have now defined all the maps in the desired zigzag from the left column of Diagram (15.136). Next we study the behaviour of all these maps when we vary p, which includes producing the horizontal maps in Diagram (15.136), which we now do, again from the top to the bottom.

The map $\sigma_{p,F}$: The commutative diagram of chain complexes

$$\begin{CD}
\widetilde{C}^s(X) @>\Sigma^{-p}(F\circ\Sigma^p_X)>> \Sigma^{-p}\widetilde{C}^s(\Sigma^pY) @<\Sigma^{-p}\Sigma^p_Y<\simeq< \widetilde{C}^s(Y) \\
@VV\text{id}V @VV\Sigma^{-(p+1)}\Sigma_{\Sigma^pY}V @VV\text{id}V \\
\widetilde{C}^s(X) @>\Sigma^{-(p+1)}(\Sigma F\circ\Sigma^{p+1}_X)>> \Sigma^{-(p+1)}\widetilde{C}^s(\Sigma^{p+1}Y) @<\Sigma^{-(p+1)}\Sigma^{p+1}_Y<\simeq< \widetilde{C}^s(Y)
\end{CD}$$

induces in view of (15.126) a chain map

$$\sigma_{p,F} \colon C_{p,F}(X) \to C_{p+1,\Sigma F}(X). \quad (15.140)$$

15.3 Structured Chain Complexes

The map $\zeta_{p,Y}$: The commutative diagram of chain complexes

$$\begin{array}{ccc}
W^{\%}(\Sigma^{-p}\widetilde{C}^s(\Sigma^p Y)) & \xrightarrow{\Sigma^{-p} S^p} & \Sigma^{-p} W^{\%}(\widetilde{C}^s(\Sigma^p Y)) \\
{\scriptstyle (\Sigma^{-(p+1)}\Sigma_{\Sigma^p Y})^{\%}} \downarrow & & \downarrow {\scriptstyle \Sigma^{-(p+1)}((\Sigma_{\Sigma^p Y})^{\%} \circ S_{\widetilde{C}^s(\Sigma^p Y)})} \\
W^{\%}(\Sigma^{-(p+1)}\widetilde{C}^s(\Sigma^{p+1} Y)) & \xrightarrow{\Sigma^{-(p+1)} S^{p+1}} & \Sigma^{-(p+1)} W^{\%}(\widetilde{C}^s(\Sigma^{p+1} Y))
\end{array}$$

induces in view of (15.129) a chain map between the homotopy fibres of the horizontal maps

$$\zeta_{p,Y} : W_{p,Y} \to W_{p+1,Y}. \tag{15.141}$$

The map $\zeta'_{p,Y}$: The commutative diagram of chain complexes

$$\begin{array}{ccc}
W^{\%}(\widetilde{C}^s(Y)) & \xrightarrow{\Sigma^{-p} S^p} & \Sigma^{-p} W^{\%}(\Sigma^p \widetilde{C}^s(Y)) \\
{\scriptstyle \mathrm{id}} \downarrow & & \downarrow {\scriptstyle \Sigma^{-(p+1)} \circ S_{\Sigma^p \widetilde{C}^s(Y)}} \\
W^{\%}(\widetilde{C}^s(Y)) & \xrightarrow{\Sigma^{-(p+1)} S^{p+1}} & \Sigma^{-(p+1)} W^{\%}(\Sigma^{p+1} \widetilde{C}^s(Y))
\end{array}$$

induces in view of (15.137) a chain map between the homotopy fibres of the horizontal maps

$$\zeta'_{p,Y} : W'_{p,Y} \to W'_{p+1,Y}. \tag{15.142}$$

Remark 15.143 Proposition 15.119 tells us that we have a natural chain homotopy equivalence $W'_{p,Y} \simeq W^{[0,p-1]}_{\%}(\widetilde{C}^s(Y))$ and similarly for $W'_{p+1,Y}$, and the corresponding natural square commutes, but we will not use this.

This finishes the construction of all the maps in Diagram (15.136). We proceed with the proof of the (homotopy) commutativity of the respective squares, which includes the construction of the chain homotopy $h_{p,F}$ appearing in the second square of the diagram.

Proof of the (homotopy) commutativity of Diagram (15.136). We will proceed from the top to the bottom of Diagram (15.136). This means we have to prove the (homotopy) commutativity of five squares. Each square is marked below.

Square (1).

Recall in more detail how the map $\sigma_{p,F}$ from (15.140) is defined. It is the dashed map in the diagram below obtained from the universal property of the homotopy pullback applied to (15.126). In this diagram everything is so defined that it commutes strictly, except the back and the front face, which commute up to preferred natural homotopies $\Upsilon_{p,F}$ and $\Upsilon_{p+1,\Sigma F}$ from (15.127) respectively.

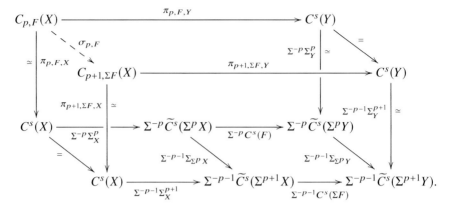

The left square takes care of the first square in the desired Diagram (15.136).

Square (2).

The map $\sigma_{p,F} \colon C_{p,F}(X) \to C_{p+1,\Sigma F}$ has just been discussed above, while the map $\zeta_{p,Y} \colon W_{p,Y} \to W_{p+1,Y}$ is defined in (15.141) via the universal property of the homotopy fibre as the induced map between the horizontal homotopy fibres of the strictly commutative square depicted just above (15.141).

The maps $\psi_{p,F}$ and $\psi_{p+1,\Sigma F}$ in the desired square are obtained via the universal property of the homotopy fibre using the chain maps $\alpha_{p,F}$ and $\alpha_{p+1,\Sigma F}$ from (15.130) and chain homotopies $\beta_{p,F} \colon 0 \simeq \Sigma^{-p} S^p \circ \alpha_{p,F}$ and $\beta_{p+1,\Sigma F} \colon 0 \simeq \Sigma^{-p-1} S^{p+1} \circ \alpha_{p+1,\Sigma F}$ from (15.131). The proof of the commutativity up to a preferred natural homotopy of the desired square from Diagram (15.136) is achieved by studying the following diagram:

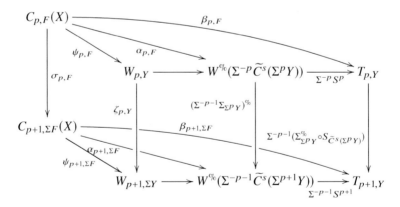

where $T_{l,Y} = \Sigma^{-l} W^{\%}(\widetilde{C}^s(\Sigma^l Y))$.

The square involving the maps $\alpha_{-,-}$ commutes strictly, which follows from the naturality of $W^{\%}(-)$ as follows. Recall from the proof of Lemma 15.135 that $\alpha_{p,F}$ is equal to the difference of two maps $\alpha_{p,F,1} - \alpha_{p,F,2}$. Immediately from definitions we observe that

15.3 Structured Chain Complexes

$$\alpha_{p+1,\Sigma F,2} \circ \sigma_{p,F} = (\Sigma^{-p-1}\Sigma_{\Sigma^p Y})^{\%} \circ \alpha_{p,F,2},$$

and similarly for $\alpha_{p,F,1}$. Taking the difference shows the commutativity of the desired square.

We show below that the square involving the maps $\beta_{-,-}$ commutes up to a preferred natural homotopy induced by the preferred natural chain homotopies $\Gamma_{-,-}$ from (15.109), and $\Upsilon_{-,-}$ from (15.127). As a result we obtain via Exercise 14.76 that the desired square commutes up to a preferred natural homotopy.

As a preparation we recall some definitions and properties of the relevant maps and chain homotopies:

$$\beta_{p,F} \stackrel{(a)}{=} \Sigma^{-p} C^s(F)^{\%} \circ \Sigma^{-p}\Gamma_{p,X} \circ \pi_{p,F,X} - \Sigma^{-p}\Gamma_{p,Y} \circ \pi_{p,F,Y} - \Sigma^{-p}\widetilde{\varphi}_{\Sigma^p Y} \circ \Upsilon_{p,F};$$

$$\Gamma_{p+1,X} \stackrel{(b)}{=} \Gamma_{\Sigma^p X} \circ \Sigma\Sigma^p_X + \Sigma^{\%}_{\Sigma^p X} \circ S_{\widetilde{C}^s(\Sigma^p X)} \circ \Sigma\Gamma_{p,X};$$

$$\Upsilon_{p+1,\Sigma F} \circ \sigma_{p,F} \stackrel{(c)}{=} \Sigma^{-p-1}\Sigma_{\Sigma^p Y} \circ \Upsilon_{p,F}, \quad \pi_{p+1,\Sigma F,X} \circ \sigma_{p,F} \stackrel{(d)}{=} \pi_{p,F,X}.$$

The desired preferred and natural chain homotopy

$$\beta'_{p,F} : \beta_{p+1,\Sigma F} \circ \sigma_{p,F} \simeq \Sigma^{-p-1}(\Sigma^{\%}_{\Sigma^p Y} \circ S_{\widetilde{C}^s(\Sigma^p Y)}) \circ \beta_{p,F}$$

is obtained by the calculation (explanations follow)

$$\beta'_{p,F} : \beta_{p+1,\Sigma F} \circ \sigma_{p,F}$$

$$\stackrel{(1)}{=} \Sigma^{-p-1} C^s(\Sigma F)^{\%} \circ \Sigma^{-p-1}\Gamma_{p+1,X} \circ \pi_{p+1,\Sigma F,X} \circ \sigma_{p,F}$$
$$- \Sigma^{-p-1}\Gamma_{p+1,Y} \circ \pi_{p+1,\Sigma F,Y} \circ \sigma_{p,F} - \Sigma^{-p-1}\widetilde{\varphi}_{\Sigma^{p+1}Y} \circ \Upsilon_{p+1,\Sigma F} \circ \sigma_{p,F}$$

$$\stackrel{(2)}{=} \Sigma^{-p-1}(C^s(\Sigma F)^{\%} \circ \Sigma^{\%}_{\Sigma^p X} \circ S_{\widetilde{C}^s(\Sigma^p X)} \circ \Sigma\Gamma_{p,X}) \circ \pi_{p,F,X}$$
$$+ \Sigma^{-p-1}(C^s(\Sigma F)^{\%} \circ \Gamma_{\Sigma^p X} \circ \Sigma\Sigma^p_X) \circ \pi_{p,F,X}$$
$$- \Sigma^{-p-1}(\Sigma^{\%}_{\Sigma^p Y} \circ S_{\widetilde{C}^s(\Sigma^p Y)} \circ \Sigma\Gamma_{p,Y}) \circ \pi_{p,F,Y}$$
$$- \Sigma^{-p-1}(\Gamma_{\Sigma^p Y} \circ \Sigma\Sigma^p_Y) \circ \pi_{p,F,Y}$$
$$- \Sigma^{-p-1}\widetilde{\varphi}_{\Sigma^{p+1}Y} \circ \Sigma^{-p-1}\Sigma_{\Sigma^p Y} \circ \Upsilon_{p,F}$$

$$\stackrel{(3)}{\simeq} \Sigma^{-p-1}(\Sigma^{\%}_{\Sigma^p Y} \circ S_{\widetilde{C}^s(\Sigma^p Y)} \circ \Sigma C^s(F)^{\%} \circ \Sigma\Gamma_{p,X}) \circ \pi_{p,F,X}$$
$$- \Sigma^{-p-1}(\Sigma^{\%}_{\Sigma^p Y} \circ S_{\widetilde{C}^s(\Sigma^p Y)} \circ \Sigma\Gamma_{p,Y}) \circ \pi_{p,F,Y}$$
$$- \Sigma^{-p-1}(\Sigma^{\%}_{\Sigma^p Y} \circ S_{\widetilde{C}^s(\Sigma^p Y)} \circ \Sigma\widetilde{\varphi}_{\Sigma^p Y}) \circ \Upsilon_{p,F} + A$$

$$\stackrel{(4)}{\simeq} \Sigma^{-p-1}(\Sigma^{\%}_{\Sigma^p Y} \circ S_{\widetilde{C}^s(\Sigma^p Y)}) \circ \beta_{p,F}$$

where in the penultimate line we abbreviate:

$$A = \Sigma^{-p-1}(C^s(\Sigma F)^{\%} \circ \Gamma_{\Sigma^p X} \circ \Sigma\Sigma^p_X) \circ \pi_{p,F,X} - \Sigma^{-p-1}(\Gamma_{\Sigma^p Y} \circ \Sigma\Sigma^p_Y) \circ \pi_{p,F,Y}.$$

Equation (1) is obtained by writing out the definition of $\beta_{p+1,\Sigma F}$ from (a). Equation (2) is obtained in the first two terms by writing out the definitions of $\Gamma_{p+1,X}$ and $\Gamma_{p+1,Y}$ from (b) and by (d), and in the third term by the equation (c). Chain homotopy (3) is induced in the first term by the equation obtained by naturality of Σ_-, functoriality of $(-)^{\%}$ and by (15.100) with $f = C^s(F)$, by the natural chain homotopy $\Sigma^{-p}\Gamma_{\Sigma^p Y}$, see Proposition 15.107 and (15.109), from the last term on the left side to the third term of the right side, and by rearranging the other terms. Chain homotopy (4) is obtained by observing that the sum of three terms on the left side yields the right hand side by the definition of $\beta_{p,F}$ from (a) and by the following calculation involving A. By naturality of $\Gamma_{\Sigma^p(-)}$ we get the second equation in the calculation

$$A = \Sigma^{-p-1}[(C^s(\Sigma F))^{\%} \circ \Gamma_{\Sigma^p X} \circ \Sigma\Sigma^p_X) \circ \pi_{p,F,X} - (\Gamma_{\Sigma^p Y} \circ \Sigma\Sigma^p_Y) \circ \pi_{p,F,Y}]$$
$$= \Sigma^{-p-1}\Gamma_{\Sigma^p Y} \circ [\Sigma^{-p}C^s(F)^{\%} \circ \Sigma^{-p}\Sigma^p_X \circ \pi_{p,F,X} - \Sigma^{-p}\Sigma^p_Y \circ \pi_{p,F,Y}] \simeq 0$$

where the last homotopy is given by the formula

$$\Sigma^{-p-1}\Gamma_{\Sigma^p Y} \circ \Upsilon_{p,F}.$$

Now define the desired chain homotopy

$$h_{p,F} = z(\beta'_{p,F}): \zeta_{p,Y} \circ \psi_{p,F} \simeq \psi_{p+1,\Sigma F} \circ \sigma_{p,F}, \qquad (15.144)$$

via Exercise 14.76, as the induced chain homotopy $z(\beta'_{p,F})$, which is preferred and natural since $\beta'_{p,F}$ is.

The calculation above can be visualised as follows. Recall that the chain homotopy $\beta_{p,F}$ is defined in terms three other chain homotopies, which correspond to three up to homotopy commuting squares in the cubical Diagram (15.132). Investigating the homotopy commutativity of the desired square involving $\beta_{-,-}$ can be viewed as studying a map between the two cubes corresponding to $C_{p,F}(X)$ and $C_{p+1,\Sigma F}(X)$, and this map can be seen as a 4-dimensional cube (alias six 3-dimensional cubes). The reader is invited to draw this large diagram or parts of it for themselves to obtain a better overview of the above calculations.

Square (3).

The horizontal maps in the square are $\zeta_{p,Y}: W_{p,Y} \to W_{p+1,Y}$ defined in (15.141) and $\zeta'_{p,Y}: W'_{p,Y} \to W'_{p+1,Y}$ defined in (15.142), and the vertical maps are $\tau_{p,Y}$ and $\tau_{p+1,Y}$ defined in (15.138) via the universal property of the homotopy fibre.

Abbreviate for better readability:

$$* := \Sigma^{-p-1} S^{p+1}_{\Sigma^{-p-1}\widetilde{C}^s(\Sigma^{p+1}Y)};$$
$$** := \Sigma^{-p} S^p_{\Sigma^{-p}\widetilde{C}^s(\Sigma^p Y)},$$

and consider the following cube

15.3 Structured Chain Complexes

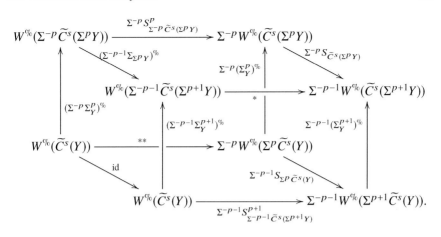

The squares other than the left and the right one commute by the previous discussion of definitions of maps $\zeta_{p,Y}$, $\zeta'_{p,Y}$ and $\tau_{p,Y}$. The left square commutes by the definition of Σ_Y^p from (15.104) and the right square commutes by Proposition 15.108. Square (3) is the induced square of the horizontal homotopy fibres and it commutes by the previous sentence.

Square (4). This square is the induced square of the homotopy fibres of maps in a cube whose arrows are the structures maps of the terms labelled 0, p and $(p + 1)$ into a homotopy colimit. The cube itself is commutative and hence also the induced square of homotopy fibres.

Square (5) obviously commutes and this concludes the proof. □

Lemma 15.145 *Let $f : X \to X'$ and $g : Y \to Y'$ be pointed maps. Consider $p \geq 0$ and maps $F : \Sigma^p X \to \Sigma^p Y$ and $F' : \Sigma^p X' \to \Sigma^p Y'$ such that the following square commutes*

$$\begin{array}{ccc} \Sigma^p X & \xrightarrow{F} & \Sigma^p Y \\ \Sigma^p f \downarrow & & \downarrow \Sigma^p g \\ \Sigma^p X' & \xrightarrow{F'} & \Sigma^p Y'. \end{array}$$

Then the analogue of Lemma 15.135 holds for entire Diagram (15.136) including the preferred chain homotopies, i.e., the Diagram (15.136) is natural.

Similarly as was the case with Lemma 15.135, there is no reasonable formulation of Lemma 15.145 where $f : X \to X'$ and $g : Y \to Y'$ are replaced by maps of the shape $f : \Sigma^p X \to \Sigma^p X'$ and $g : \Sigma^p Y \to \Sigma^p Y'$.

Proof. This amounts to studying the naturality of the vertical maps in Diagram (15.136). This has been done in Lemma 15.135 for $\psi_{p,F}$. For the other maps it follows from the naturality of constructions such as the maps Σ_-, $(-)^{\%}$, $S_{\widetilde{C}^s(-)}$.

The only harder part is that we also have to investigate the behaviour of the chain homotopy $h_{p,F}$ in the second square. Here the desired statement follows from the fact that in the construction of this chain homotopy only the natural chain homotopies $\Gamma_{\Sigma^p(-)}$ and $\Upsilon_{p,F}$ were used. □

Now we can take the homotopy colimits in each row of Diagram (15.136). Recall that there is always a natural chain homotopy equivalence from the homotopy colimit

$$\omega_C \colon \mathrm{hocolim}_{p\to\infty} \mathrm{const}_C \xrightarrow{\simeq} C$$

of the constant system const given by $C \xrightarrow{\mathrm{id}_C} C \xrightarrow{\mathrm{id}_C} \cdots$ to C itself. Using (14.106) we obtain from the Diagram (15.136) a zigzag of chain maps whose arrows pointing to the left are all chain homotopy equivalences

$$\widetilde{C}^s(X) \xleftarrow{\simeq} \mathrm{hocolim}_{p\to\infty} \mathrm{const}_{\widetilde{C}^s(X)} \xleftarrow{\simeq} \mathrm{hocolim}_{p\to\infty} C_{p,F}(X) \to$$
$$\to \mathrm{hocolim}_{p\to\infty} W_{p,Y} \xleftarrow{\simeq} \mathrm{hocolim}_{p\to\infty} W'_{p,Y} \xleftarrow{\simeq}$$
$$\xleftarrow{\simeq} \mathrm{hocolim}_{p\to\infty} \mathrm{const}_{\mathrm{hofib}(S^\infty)} \to \mathrm{hocolim}_{p\to\infty} \mathrm{const}_{W_{\%}(\widetilde{C}^s(Y))}$$
$$\xrightarrow{\simeq} W_{\%}(\widetilde{C}^s(Y)). \quad (15.146)$$

This zigzag is natural in the sense of Lemma 15.145. If we apply the functor H_n to (15.146), we get a well-defined map

$$\overline{\psi}_{p,F} \colon \widetilde{H}_n(X) := H_n(\widetilde{C}^s(X)) \to Q_n(\widetilde{C}^s(Y)). \quad (15.147)$$

Lemma 15.148

(i) *The map $\overline{\psi}_F$ of (15.147) is natural in the following sense. Let $f \colon X \to X'$ and $g \colon Y \to Y'$ be pointed maps. Consider $p \geq 0$ and pointed maps $F \colon \Sigma^p X \to \Sigma^p Y$ and $F' \colon \Sigma^p X' \to \Sigma^p Y'$ such that the following square commutes*

$$\begin{array}{ccc} \Sigma^p X & \xrightarrow{F} & \Sigma^p Y \\ \Sigma^p f \downarrow & & \downarrow \Sigma^p g \\ \Sigma^p X' & \xrightarrow{F'} & \Sigma^p Y'. \end{array}$$

Then the following diagram commutes

$$\begin{array}{ccc} \widetilde{H}_n(X) & \xrightarrow{\overline{\psi}_{p,F}} & Q_n(\widetilde{C}^s(Y)) \\ \widetilde{H}_n(f) \downarrow & & \downarrow \widetilde{C}^s(g)_{Q,n} \\ \widetilde{H}_n(X') & \xrightarrow{\overline{\psi}_{p,F'}} & Q_n(\widetilde{C}^s(Y')); \end{array}$$

(ii) *The map $\overline{\psi}_{p,F}$ depends only on the stable homotopy class $\{F\}$ of the map $F \colon \Sigma^p X \to \Sigma^p Y$.*

15.3 Structured Chain Complexes

Proof. (i) This follows from the naturality of (15.147), which is a consequence of Lemma 15.145.Lemma 15.145.

(ii) From the construction of (15.147) in terms of Diagram (15.136) we conclude that $\overline{\psi}_{p,F} = \overline{\psi}_{p+1,\Sigma F}$. Hence it suffices to show that for pointed homotopic pointed maps $F_0, F_1 \colon \Sigma^p X \to \Sigma^p Y$ we have $\overline{\psi}_{p,F_0} = \overline{\psi}_{p,F_1}$. Choose a pointed homotopy $h \colon \Sigma^p X \wedge I_+ = \Sigma^p(X \wedge I_+) \to \Sigma^p X$ between F_0 and F_1. Let $i_k \colon \Sigma^p X \to \Sigma^p(X \wedge I_+)$ be the inclusion induced by the inclusion $\{\bullet\} \to I$ with image $\{k\}$. Let $\mathrm{pr} \colon \Sigma^p(X \wedge I_+) \to \Sigma^p X$ be the projection induced by the projection $I \to \{\bullet\}$. Since this a pointed homotopy equivalence, the induced map $\widetilde{H}_n(\Sigma^p(X \wedge I_+)) \to \widetilde{H}_n(\Sigma^p X)$ is bijective. Now the claim follows from the following diagram

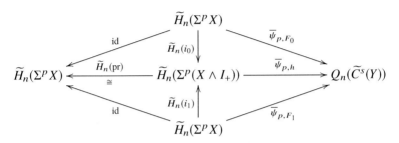

which commutes by assertion (i). □

In view of Lemma 15.148 (ii) we denote in the sequel $\overline{\psi}_{p,F}$ by

$$\overline{\psi}_{\{F\}} \colon \widetilde{H}_n(X) \to Q_n(\widetilde{C}^s(Y)), \tag{15.149}$$

and we call this map the *quadratic construction* associated to F. In view of the definition of $\psi_{p,F}$ in (15.128) via the maps $\alpha_{p,F}$ and $\beta_{p,F}$, its relation to the symmetric constructions (15.66) associated with X and Y is via the Umkehr map f from (15.122) as follows:

$$(1+T)\overline{\psi}_{\{F\}} = f^{Q,n}\overline{\varphi}_{X,\Gamma,w} - \overline{\varphi}_{Y,\Gamma,w}f_*. \tag{15.150}$$

This finishes Construction 15.124.

Remark 15.151 (The quadratic structure is an obstruction to desuspension)
Note that in the case when the pointed map $F \colon \Sigma^p X \to \Sigma^p Y$ is a p-fold suspension, i.e., $F = \Sigma^p G \colon \Sigma^p X \to \Sigma^p Y$ for some pointed map $G \colon X \to Y$, then $\psi_{\{F\}} = 0$. This is the case since by (15.121) and (15.123) the inner square of (15.132) now strictly commutes if one takes $p = 0$, and hence both the map $\alpha_{p,F}$ in (15.130) and $\beta_{p,F}$ in (15.131) are zero.

In the proof of Proposition 15.455 in Section 15.7 we will use the following additivity property of the quadratic construction.

Proposition 15.152 (Additivity of the quadratic construction) *Let* $F\colon \Sigma^p X \to \Sigma^p Y$ *and* $G\colon \Sigma^p Y \to \Sigma^p Z$ *be pointed maps of pointed spaces for some* $p > 0$, *with* $f\colon \widetilde{C}^s(X) \to \widetilde{C}^s(Y)$ *and* $g\colon \widetilde{C}^s(Y) \to \widetilde{C}^s(Z)$ *the* $\mathbb{Z}$-*chain maps associated to them as in* (15.122). *Then the quadratic constructions of* (15.149) *associated to* F, G, *and* $G \circ F$ *satisfy*

$$\overline{\psi}_{\{G\}\circ\{F\}} = g_{Q,n} \circ \overline{\psi}_{\{F\}} + \overline{\psi}_{\{G\}} \circ \widetilde{H}_n(f) \colon \widetilde{H}_n(X) \to Q_n(\widetilde{C}^s(Z)).$$

Proof. The statement is about the behaviour of the quadratic construction on homology, but the underlying construction itself was done at the level of chain complexes. Here the underlying chain maps that induce the maps appearing in the equation are, in the order from left to right, as follows:

$$\psi_{p,G\circ F}\colon C_{p,G\circ F}(X) \to W_{p,Z};$$
$$W_p(G)\colon W_{p,Y} \to W_{p,Z};$$
$$\psi_{p,F}\colon C_{p,F}(X) \to W_{p,Y};$$
$$\psi_{p,G}\colon C_{p,G}(Y) \to W_{p,Z};$$
$$f\colon \widetilde{C}^s(X) \to \widetilde{C}^s(Y).$$

The maps $\psi_{p,-}$ were defined in (15.128), the map $W_p(G)$ is defined analogously to the map $W_p(g)$ in the proof of Lemma 15.135, and the map f was defined in (15.122). Next we observe that by the definition of $C_{p,F}(X)$ and f and using (15.126) we have $f \circ \pi_{p,F,X} \simeq \pi_{p,F,Y}$, and that there is a zigzag

$$C_{p,F}(X) \xrightarrow{\pi_{p,F,Y}} \widetilde{C}^s(Y) \xleftarrow[\simeq]{\pi_{p,G,Y}} C_{p,G}(Y).$$

This allows us to replace $\widetilde{C}^s(X)$ by $C_{p,F}(X)$ and to compose the maps induced by f and $\psi_{p,G}$ in the above list of maps.

However, we see that we have a problem since by the definitions there is no obvious map between the chain complexes $C_{p,F}(X)$ (the source on the right-hand side of the desired equation) and $C_{p,G\circ F}(X)$ (the source on the left-hand side) either way. To rectify this we introduce an intermediate chain complex defined as the pullback

$$\begin{array}{ccc} C_{p,G,F}(X) & \xrightarrow{\pi_{p,G,F,Y}} & C_{p,G}(Y) \\ {\scriptstyle \pi_{p,G,F,X}} \downarrow \simeq & & \simeq \downarrow {\scriptstyle \pi_{p,G,Y}} \\ C_{p,F}(X) & \xrightarrow{\pi_{p,F,Y}} & \widetilde{C}^s(Y). \end{array}$$

The pullback is chain homotopy equivalent to the homotopy pullback in this case since the arrows $\pi_{p,F,Y}$ and $\pi_{p,G,Y}$ are fibrations. The right vertical arrow is a chain homotopy equivalence by the definition of $C_{p,G}(Y)$ and hence the left vertical arrow is also a chain homotopy equivalence.

Next recall that the complex $C_{p,G\circ F}(X)$ fits into the homotopy cartesian square

15.3 Structured Chain Complexes

$$\begin{array}{ccc}
C_{p,G\circ F}(X) & \xrightarrow{\pi_{p,G\circ F,Z}} & \widetilde{C}^s(Z) \\
{\scriptstyle \pi_{p,G\circ F,X}} \downarrow {\scriptstyle \simeq} & & \downarrow {\scriptstyle \simeq} {\scriptstyle \Sigma^{-p}\Sigma_Z^p} \\
\widetilde{C}^s(X) & \xrightarrow[\Sigma^{-p}\Sigma_X^p]{} \Sigma^{-p}\widetilde{C}^s(\Sigma^p X) \xrightarrow[\Sigma^{-p}C^s(G\circ F)]{} & \Sigma^{-p}\widetilde{C}^s(\Sigma^p Z),
\end{array}$$

and $C_{p,G\circ F}(X)$ is isomorphic to the homotopy pullback of the rest of the diagram via the isomorphism (14.74). This together with the universal property of the homotopy pullback applied to $C_{p,G\circ F}(X)$ via the diagram

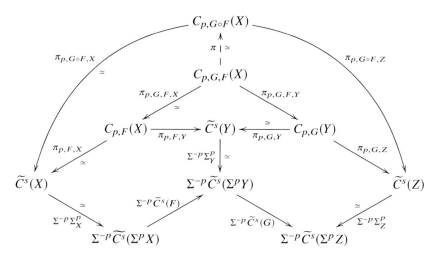

yields the arrow marked $\pi\colon C_{p,G,F}(X) \to C_{p,G\circ F}(X)$ by the following argument. In the diagram the outer part is the defining diagram of $C_{p,G\circ F}(X)$ depicted above, the lower left part is the defining Diagram (15.126) of $C_{p,F}(X)$ (and hence is commutative up to a canonical chain homotopy), the lower right part is the defining Diagram (15.126) of $C_{p,G}(Y)$ (and hence is commutative up to a canonical chain homotopy), the part just above is the defining diagram of $C_{p,G,F}(X)$ (and hence is commutative). These chain homotopies pasted together provide us with a chain homotopy between the two composites in the square

$$\begin{array}{ccc}
C_{p,G,F}(X) & \xrightarrow{\pi_{p,G,Z}\circ\pi_{p,G,F,Y}} & \widetilde{C}^s(Z) \\
{\scriptstyle \simeq} \downarrow {\scriptstyle \pi_{p,F,X}\circ\pi_{p,G,F,X}} & & \downarrow {\scriptstyle \simeq} {\scriptstyle \Sigma^{-p}\Sigma_Z^p} \\
\widetilde{C}^s(X) & \xrightarrow[\Sigma^{-p}\Sigma_X^p]{} \Sigma^{-p}\widetilde{C}^s(\Sigma^p X) \xrightarrow[\Sigma^{-p}\widetilde{C}^s(G\circ F)]{} & \Sigma^{-p}\widetilde{C}^s(\Sigma^p Z),
\end{array}$$

allowing us to use the universal property of the homotopy pullback $C_{p,G\circ F}(X)$, as we wanted.

Altogether we have the following homotopy commutative diagram

$$\begin{array}{ccc}
C_{p,G\circ F}(X) & \xleftarrow{\pi}_{\simeq} C_{p,G,F}(X) & \xrightarrow{\pi_{p,G,F,Y}} C_{p,G}(Y) \\
\pi_{p,G\circ F,X} \downarrow \simeq & \pi_{p,G,F,X} \downarrow \simeq & \downarrow \pi_{p,G,Y} \\
\widetilde{C}^s(X) & \xleftarrow[\pi_{p,F,X}]{\simeq} C_{p,F}(X) \xrightarrow{\pi_{p,F,Y}} & \widetilde{C}^s(Y),
\end{array}$$

and the map π is a chain homotopy equivalence because the other three maps in the left square are. We see that the intermediate chain complex $C_{p,G,F}(X)$ provides us with a zigzag of chain homotopy equivalences between $C_{p,F}(X)$ and $C_{p,G\circ F}(X)$, which we will use below.

There also exist explicit formulas as follows. By definition an n-chain in $C_{p,G,F}(X)$ is a 5-tuple (x, y, z, u, v) with $x \in \widetilde{C}_n^s(X)$, $y \in \widetilde{C}_n^s(Y)$, $z \in \widetilde{C}_n^s(Z)$, $u \in (\Sigma^{-p}\widetilde{C}^s(\Sigma^p Y))_{n+1}$, and $v \in (\Sigma^{-p}\widetilde{C}^s(\Sigma^p Z))_{n+1}$. It is a cycle if x, y and z are cycles and the boundary of u is $(\Sigma_Y^p(y) - C^s(F)(\Sigma_X^p(x)))$ and the boundary of v is $(\Sigma_Z^p(z) - C^s(G)(\Sigma_Y^p(y)))$.

The three maps with the source $C_{p,G,F}(X)$ are given by

$$\pi: C_{p,G,F}(X) \to C_{p,G\circ F}(X), \quad (x, y, z, u, v) \mapsto (x, z, \widetilde{C}^s(G)(u) + v);$$
$$\pi_{p,G,F,X}: C_{p,G,F}(X) \to C_{p,F}(X), \quad (x, y, z, u, v) \mapsto (x, y, u);$$
$$\pi_{p,G,F,Y}: C_{p,G,F}(X) \to C_{p,G}(Y), \quad (x, y, z, u, v) \mapsto (y, z, v).$$

The proof of the desired additivity is obtained by combining two arguments corresponding to the two following diagrams. The first argument is a calculation using formulas (15.133) and (15.134), which are applied in the context of the following diagram, where the two triangles in the left part commute up to homotopy (we do not claim that the right part of the diagram commutes, instead we prove a sum formula stated just below the diagram):

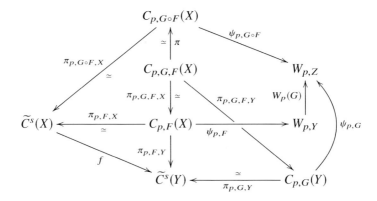

More precisely, we prove

$$\psi_{p,G\circ F} \circ \pi = W_p(G) \circ \psi_{p,F} \circ \pi_{p,G,F,X} + \psi_{p,G} \circ \pi_{p,G,F,Y}.$$

15.3 Structured Chain Complexes

To do this, using the universal property of the homotopy fibre, as we did before, we observe that a chain map $C_{p,G,F}(X) \to W_{p,Z}$ can be obtained by considering for each $n \in \mathbb{Z}$ two components

$$\alpha_{p,G,F} : C_{p,G,F}(X)_n \to W^{\%}(\Sigma^{-P}\widetilde{C}^s(\Sigma^P Z))_n;$$

$$\beta_{p,G,F} : C_{p,G,F}(X)_n \to \Sigma^{-P}W^{\%}(\widetilde{C}^s(\Sigma^P Z))_{n+1}.$$

The equality in the first component is achieved by the calculation:

$$\Sigma^{-P}\widetilde{C}^s(G)^{\%} \circ \Sigma^{-P}\widetilde{C}^s(F)^{\%} \circ (\Sigma^{-P}\Sigma^P_X)^{\%} \circ \widetilde{\varphi}_X(x) -$$
$$- \Sigma^{-P}\widetilde{C}^s(G)^{\%} \circ (\Sigma^{-P}\Sigma^P_Y)^{\%} \circ \widetilde{\varphi}_Y(y) +$$
$$+ \Sigma^{-P}\widetilde{C}^s(G)^{\%} \circ (\Sigma^{-P}\Sigma^P_Y)^{\%} \circ \widetilde{\varphi}_Y(y) - (\Sigma^{-P}\Sigma^P_Z)^{\%} \circ \widetilde{\varphi}_Z(z) =$$
$$= \Sigma^{-P}\widetilde{C}^s(G \circ F)^{\%} \circ (\Sigma^{-P}\Sigma^P_X)^{\%} \circ \widetilde{\varphi}_X(x) - (\Sigma^{-P}\Sigma^P_Z)^{\%} \circ \widetilde{\varphi}_Z(z).$$

The equality in the second component is achieved by the calculation:

$$\Sigma^{-P}\widetilde{C}^s(G)^{\%} \circ (\Sigma^{-P}C^s(F)^{\%} \circ \Sigma^{-P}\Gamma_{p,X}(x) - \Sigma^{-P}\Gamma_{p,Y}(y) + \Sigma^{-P}\widetilde{\varphi}_{\Sigma^P Y}(u)) +$$
$$+ \Sigma^{-P}\widetilde{C}^s(G)^{\%} \circ \Sigma^{-P}\Gamma_{p,Y}(y) - \Sigma^{-P}\Gamma_{p,Z}(z) + \Sigma^{-P}\widetilde{\varphi}_{\Sigma^P Z}(v) =$$
$$= \Sigma^{-P}\widetilde{C}^s(G \circ F)^{\%} \circ \Sigma^{-P}\Gamma_{p,X}(x) - \Sigma^{-P}\Gamma_{p,Z}(z) + \Sigma^{-P}\widetilde{\varphi}_{\Sigma^P Z}(\widetilde{C}^s(G)(u) + v)$$

where the following equality is used:

$$\Sigma^{-P}\widetilde{C}^s(G)^{\%} \circ \Sigma^{-P}\widetilde{\varphi}_{\Sigma^P Y}(u) = \Sigma^{-P}\widetilde{\varphi}_{\Sigma^P Z}(\widetilde{C}^s(G)(u)),$$

which is a consequence of the naturality of the symmetric construction applied to the map G.

The second argument is needed in order to relate the calculation just made to the map $g_{Q,n}$ from the statement of the proposition. It is done as follows. The left-hand side is induced by the zigzag involving $\pi_{p,G\circ F,X}$ and $\psi_{p,G\circ F}$, the first term on the right-hand side is induced by the zigzag involving $\pi_{p,F,X}, \psi_{p,F}$ and $W_p(G)$, and the second term by the zigzag involving $f, \pi_{p,G,Y}$ and $\psi_{p,G}$. We need to relate $W_p(G)$ to the map $g_{Q,n}$. Note first that the map $g \colon \widetilde{C}^s(Y) \to \widetilde{C}^s(Z)$ is obtained by choosing a chain homotopy inverse $\Sigma^{-P}_Z \colon \widetilde{C}^s(\Sigma^P Z) \to \Sigma^P \widetilde{C}^s(Z)$ and defining

$$g := \Sigma^{-P}(\Sigma^{-P}_Z \circ \widetilde{C}^s(G) \circ \Sigma^P_Y).$$

Let $h \colon \mathrm{id} \simeq \Sigma^{-P}(\Sigma^P_Z) \circ \Sigma^{-P}(\Sigma^{-P}_Z)$ be a chain homotopy. Then we have the chain homotopy

$$\overline{h} := h \circ (\Sigma^{-P}(\widetilde{C}^s(G) \circ \Sigma^P_Y)) \colon \Sigma^{-P}(\widetilde{C}^s(G) \circ \Sigma^P_Y) \simeq \Sigma^{-P}(\Sigma^P_Z) \circ g.$$

The maps $W_p(G)$ and $g_{Q,n} = H_n(g_\%)$ are compatible in the sense that we have the following diagram where the middle and the bottom square commute strictly and the top square commutes up to homotopy marked by the wobbly arrow where the required homotopy is induced by $\bar{h}$ via Exercise 15.103 and Exercise 14.77. We also indicate by subscripts of S^∞ the corresponding chain complex and the induced map.

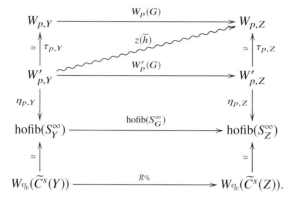

This finishes the second argument and thus also the proof of Proposition 15.152. □

For the reader's convenience we add an explanation how the calculation appearing in the first argument contained in the proof of Proposition 15.152 can be visualised using the following large diagram. For a better overview we leave it for the reader to complete the obvious indices for some of the maps of the form $\pi_{...}$:

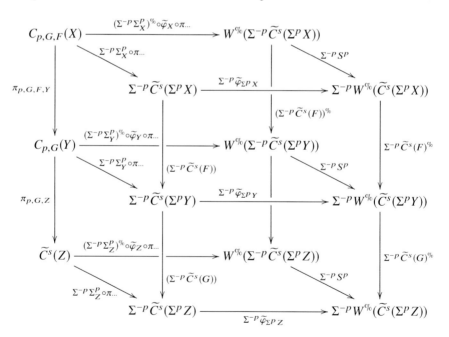

It has non-commuting squares in the back face, which yield the first component of the two summands on the right-hand side of the sum formula. The two squares in the left face commute up to homotopies $\Upsilon_{p,F}$ and $\Upsilon_{p,G}$. The three horizontal squares

15.3 Structured Chain Complexes

commute up to homotopies $\Gamma_{p,X}$, $\Gamma_{p,Y}$ and $\Gamma_{p,Z}$. The remaining squares commute strictly. Additivity corresponds in the first component to the fact that the failure of the commutativity of the whole back face is the sum of the failures of the commutativity of the two squares in that face. In the second component it corresponds to the fact that adding the homotopies of the two squares in the left face produces the homotopy for the whole face and that the sum of the four homotopies involving $\Gamma_{p,-}$ cancels the two contributions from Y, which have opposite signs, and preserves the contributions from X and Z which correspond to the top and bottom face of the whole diagram.

Construction 15.153 (The quadratic construction – equivariant version) The equivariant version of the quadratic construction goes as follows. Consider a discrete group Γ and a group homomorphism $w\colon \Gamma \to \{\pm 1\}$. Let X and Y be free pointed Γ-spaces, and let $F\colon \Sigma^p X \to \Sigma^p Y$ be a pointed Γ-map for some $p \geq 0$. Then, by similar considerations as above, but using the equivariant version of the symmetric construction from Construction 15.59 everywhere, we obtain an equivariant version of (15.146). Applying $\mathbb{Z}^w \otimes_{\mathbb{Z}\Gamma} -$ and then taking homology yields the *equivariant quadratic construction*

$$\overline{\psi}_{\{F\}_\Gamma, w}\colon \widetilde{H}_n^\Gamma(X; \mathbb{Z}^w) \to Q_n^{\mathbb{Z}\Gamma, w}(\widetilde{C}^s(Y)) \tag{15.154}$$

where

$$\widetilde{H}_n^\Gamma(X; \mathbb{Z}^w) = H_n(\mathbb{Z}^w \otimes_{\mathbb{Z}\Gamma} \widetilde{C}^s(X));$$
$$Q_n^{\mathbb{Z}\Gamma, w}(\widetilde{C}^s(Y)) = H_n(W_{\%}^{\mathbb{Z}\Gamma, w}(\widetilde{C}^s(Y))).$$

Note that $\overline{\psi}_{\{F\}_\Gamma, w}$ indeed depends only on the stable Γ-homotopy class $\{F\}_\Gamma$ of F and is natural in F in the following sense.

Lemma 15.155 (i) *Let $f\colon X \to X'$ and $g\colon Y \to Y'$ be pointed Γ-maps of free pointed Γ-spaces. Consider $p \geq 0$ and pointed Γ-maps $F\colon \Sigma^p X \to \Sigma^p Y$ and $F'\colon \Sigma^p X' \to \Sigma^p Y'$ such that the following square commutes*

$$\begin{array}{ccc} \Sigma^p X & \xrightarrow{F} & \Sigma^p Y \\ {\scriptstyle \Sigma^p f}\downarrow & & \downarrow{\scriptstyle \Sigma^p g} \\ \Sigma^p X' & \xrightarrow{F'} & \Sigma^p Y'. \end{array}$$

Then the following diagram commutes

$$\begin{array}{ccc} \widetilde{H}_n^\Gamma(X; \mathbb{Z}^w) & \xrightarrow{\overline{\psi}_{\{F\}_\Gamma,w}} & Q_n^{\mathbb{Z}\Gamma,w}(\widetilde{C}^s(Y)) \\ {\scriptstyle \widetilde{H}_n^\Gamma(f;\mathbb{Z}^w)}\downarrow & & \downarrow{\scriptstyle \widetilde{C}^s(g)_{Q,n}} \\ \widetilde{H}_n^\Gamma(X'; \mathbb{Z}^w) & \xrightarrow{\overline{\psi}_{\{F'\}_\Gamma,w}} & Q_n^{\mathbb{Z}\Gamma,w}(\widetilde{C}^s(Y')); \end{array}$$

(ii) *The map $\overline{\psi}_{\{F\}_\Gamma,w}$ depends only on the stable Γ-homotopy class $\{F\}_\Gamma$ of $F\colon \Sigma^p X \to \Sigma^p Y$.*

Proof. The proof is analogous to that of Lemma 15.148. $\square$

Proposition 15.156 (Additivity of the quadratic construction – equivariant version) *Let Γ be a group and $w\colon \Gamma \to \{\pm 1\}$ a homomorphism. Suppose that X, Y, Z are free pointed Γ-spaces and that $F\colon \Sigma^p X \to \Sigma^p Y$ and $G\colon \Sigma^p Y \to \Sigma^p Z$ are pointed Γ-maps for some $p > 0$, with the associated $\mathbb{Z}\Gamma$-chain maps $f\colon \widetilde{C}^s(X) \to \widetilde{C}^s(Y)$ and $g\colon \widetilde{C}^s(Y) \to \widetilde{C}^s(Z)$ as in (15.122). Then the equivariant quadratic constructions of (15.154) associated to F, G and $G \circ F$ satisfy*

$$\overline{\psi}_{\{G\}\circ\{F\}_\Gamma,w} = g_{Q,n} \circ \overline{\psi}_{\{F\}_\Gamma,w} + \overline{\psi}_{\{G\}_\Gamma,w} \circ \widetilde{H}_n(f; \mathbb{Z}^w):$$
$$: \widetilde{H}_n^\Gamma(X; \mathbb{Z}^w) \to Q_n^{\mathbb{Z}\Gamma,w}(\widetilde{C}^s(Z)).$$

Proof. The proof is analogous to that of Proposition 15.152. $\square$

15.3.7 Algebraic Poincaré Complexes

The following notion of an algebraic Poincaré complex corresponds to the notion of a non-singular form from Chapter 8.

Definition 15.157 (Algebraic Poincaré complexes)

An n-dimensional symmetric complex $(C, [\varphi])$ in $bR\text{-}\mathrm{CH}_{\mathrm{fgp}}$ is called an n-*dimensional symmetric Poincaré complex*, or just *Poincaré*, if the up to R-chain homotopy well-defined R-chain map $\varphi_0\colon C^{n-*} \to C$ is an R-chain homotopy equivalence.

An n-dimensional quadratic complex (C, ψ) in $bR\text{-}\mathrm{CH}_{\mathrm{fgp}}$ is called an n-*dimensional quadratic Poincaré complex*, or just *Poincaré*, if the underlying n-dimensional symmetric complex is Poincaré, or, equivalently, if $(1+T)(\psi)_0\colon C^{n-*} \to C$ is an R-chain homotopy equivalence.

We call an n-dimensional polarised symmetric complex (C, φ) in $bR\text{-}\mathrm{CH}_{\mathrm{fgp}}$ an n-*dimensional polarised symmetric Poincaré complex*, or just *Poincaré*, if the underlying n-dimensional symmetric complex $(C, [\varphi])$ is Poincaré.

We call an n-dimensional polarised quadratic complex (C, ψ) in $bR\text{-}\mathrm{CH}_{\mathrm{fgp}}$ an n-*dimensional polarised quadratic Poincaré complex*, or just *Poincaré*, if the underlying n-dimensional quadratic complex $(C, [\varphi])$ is Poincaré.

An analogous notion for hyperquadratic complexes is not defined.

Example 15.158 (Algebraic symmetric Poincaré complex associated to a finite n-dimensional Poincaré complex) Consider a connected finite n-dimensional Poincaré complex X with fundamental group π, orientation homomorphism $w\colon \pi \to \{\pm 1\}$, and fundamental class $[X]$ in $H_n(X; \mathbb{Z}^w) = H_n(\mathbb{Z}^w \otimes_{\mathbb{Z}\pi} C(\widetilde{X}))$ in the sense of Definition 5.43. It defines an n-dimensional symmetric Poincaré

15.3 Structured Chain Complexes

complex $(C(\widetilde{X}), [\varphi])$ as follows. The underlying finitely generated free $\mathbb{Z}\pi$-chain complex is the cellular $\mathbb{Z}\pi$-chain complex $C(\widetilde{X})$. The symmetric construction map $\overline{\varphi}_{\widetilde{X},\pi,w} \colon H_n(X; \mathbb{Z}^w) \to Q^n_{\mathbb{Z}\pi,w}(C^s(\widetilde{X}))$ of (15.66) yields together with the $\mathbb{Z}\pi$-chain homotopy equivalences appearing in (15.63) a homomorphism

$$H_n(X; \mathbb{Z}^w) \to Q^n_{\mathbb{Z}\pi,w}(C(\widetilde{X}))$$

and we define $[\varphi]$ to be the image of $[X]$ under this map.

We will later explain in Definition 15.312 that $(C(\widetilde{X}), [\varphi])$ defines an element in the symmetric L-group $L^n(\mathbb{Z}\pi, w)$. Moreover, we will later discuss how a normal map of degree one defines an n-dimensional quadratic Poincaré complex and thus an element in the quadratic L-group $L_n(\mathbb{Z}\pi, w)$, which turns out to be the surgery obstruction, see Example 15.162.

In the following sections it will also be convenient to consider complexes which are not Poincaré, but which satisfy some condition about the map $H_k(\varphi_0)$. In particular, we will use the following terminology.

Definition 15.159 (Connected structured complexes)

An n-dimensional symmetric complex $(C, [\varphi])$ in bR-$\mathrm{CH}_{\mathrm{fgp}}$ is called *connected* if $H_i(\varphi_0 \colon C^{n-*} \to C) = 0$ for all $i \leq 0$.

An n-dimensional quadratic complex $(C, [\psi])$ in bR-$\mathrm{CH}_{\mathrm{fgp}}$ is called *connected* if $H_i(((1+T)(\psi))_0 \colon C^{n-*} \to C) = 0$ for all $i \leq 0$.

These definitions have obvious analogues in the polarised setting.

Note that in view of Definition 14.124, this just says that the up to R-chain homotopy well-defined R-chain map φ_0 is 0-connected in the symmetric case and that the up to R-chain homotopy well-defined R-chain map $(1+T)(\psi)_0$ is 0-connected in the quadratic case.

15.3.8 Equivariant S-Duality and Umkehr Maps

In the previous section we applied the quadratic construction to a stable map between spaces. However, we would like to produce quadratic Poincaré complexes in the context of surgery theory, that is, we want to associate a quadratic Poincaré $\mathbb{Z}\Gamma$-chain complex to a normal Γ-map of degree one $(f, \overline{f}) \colon M \to X$ from a closed manifold to a finite Poincaré complex, see Definition 8.3, where Γ is a fixed (discrete) group. (The special case of the universal covering is explained in Notation 8.34.) In Chapter 6 we have already seen a close relationship between Poincaré complexes and stable homotopy via the S-duality of Section 6.9. Namely, by Theorem 6.98 the S-dual of the Thom space $\mathrm{Th}(p_X)$ of the Spivak normal fibration p_X of a Poincaré complex X is the suspension $\Sigma^p(X_+)$ of X with added base point for some $p \geq 0$.

In order to use this idea in the context of the present chapter, we need a mild generalisation of the S-duality from Chapter 6 to a certain equivariant setting, which is due to Ranicki [345]. In this subsection we summarise its definition and some of its properties.

In [345, Section 3] Ranicki works with the notions of a Γ-space, Γ-CW-complex, Γ-spectrum, and smash product $-\wedge_\Gamma -$ over Γ for a given group Γ. We will use the same setting, but will change the names slightly in order to achieve better compatibility with contemporary terminology in equivariant homotopy theory. Recall that a pointed free Γ-CW-complex is a Γ-CW-complex such that the Γ-action preserves the base point and is free outside the base point. (In [345, bottom of page 216] this is called just a Γ-*CW-complex*.) The prime example of such a pointed free Γ-CW-complex we are interested in is the p-fold reduced suspension of the universal cover of a path connected space X with added base point $\Sigma^p \widetilde{X}_+$ for $\Gamma = \pi_1(X)$. The notion of a Γ-spectrum is the obvious (naive) one, and analogously for the smash product $-\wedge_\Gamma -$.

Recall the set of stable homotopy classes of stable pointed maps between pointed spaces $\{A, B\}$ from (6.93). For pointed free Γ-spaces X and Y we denote by $\{X, Y\}_\Gamma$ the *set of equivariant stable homotopy classes of stable equivariant pointed maps from X to Y*. Note that a class $[f]_\Gamma \in \{X, Y\}_\Gamma$ need not be realised by a pointed Γ-map $f \colon X \to Y$. Following [345, page 218], given a pointed free Γ-space X and a Γ-spectrum $\mathbf{Z}$, we define

$$\{X, \mathbf{Z}\}_\Gamma := \mathrm{colim}_{p \to \infty} \{\Sigma^p X, Z_p\}_\Gamma.$$

Given pointed free Γ-spaces X, Y and a map $\alpha \colon S^N \to X \wedge_\Gamma Y$ for some $N \geq 0$, there is defined for any Γ-spectrum $\mathbf{Z}$ the slant product

$$\alpha \backslash - \colon \{X, \mathbf{Z}\}_\Gamma \to \{S^N, Y \wedge_\Gamma \mathbf{Z}\}$$

by the following formula. For a representative $f^p \colon \Sigma^p X \to Z_k$ of the class of f, the image is the class of the composite

$$S^{N+p} = \Sigma^p S^N \xrightarrow{\Sigma^p \alpha} \Sigma^p(X \wedge_\Gamma Y) = (\Sigma^p X) \wedge_\Gamma Y \to$$
$$\xrightarrow{f^p \wedge_\Gamma \mathrm{id}_Y} Z_p \wedge_\Gamma Y \xrightarrow{\mathrm{flip}} Y \wedge_\Gamma Z_p.$$

The map α is called an *S-Γ-duality* if these slant maps $\alpha \backslash -$ are bijections for every Γ-spectrum $\mathbf{Z}$. There always exists an S-Γ-dual for a Γ-covering $\widehat{X}$ of a finite CW-complex X, see [345, Proposition 3.5]. It has properties analogous to those of the S-duality from Chapter 6, see [345, Section 3] for a thorough list.

Consider a finite n-dimensional Poincaré complex X together with a Γ-covering $\widehat{X} \to X$. Let $(E, \partial E)$ be the regular neighbourhood of the embedding of X into a large euclidean space $\mathbb{R}^N$. Denote by $\widehat{E} \to E$ and $\widehat{\partial E} \to \partial E$ the Γ-coverings induced by $\widehat{X} \to X$. Then we have the S-Γ-duality

$$\alpha \colon S^N \to \widehat{X}_+ \wedge_\Gamma \mathrm{Th}_\Gamma(p_X) \tag{15.160}$$

where $\mathrm{Th}_\Gamma(p_X) = \widehat{E}/\widehat{\partial E}$ is the Thom Γ-space, see [345, Proposition 4.1].

15.3 Structured Chain Complexes   621

In the case when X is a closed n-dimensional manifold, we can proceed directly to construct a *stable Umkehr Γ-map* as follows. Given a degree one normal Γ-map $(f, \bar{f}) \colon M \to X$ between n-dimensional closed manifolds, approximate the map $f \times 0 \colon M \to X \times D^p$ for p large enough by a framed embedding, say $f' \colon M \times D^p \hookrightarrow X \times D^p$. This lifts to an embedding of Γ-coverings $\widehat{M} \times D^p \hookrightarrow \widehat{X} \times D^p$. We can use the collapse map to define

$$F \colon \Sigma^p \widehat{X}_+ = \widehat{X} \times D^p / \widehat{X} \times S^{p-1} \xrightarrow{\text{collapse}} \overline{\widehat{X} \times D^p / \widehat{X} \times D^p \smallsetminus \text{im}(f')} = \Sigma^p \widehat{M}_+.$$

If X is only a finite n-dimensional Poincaré complex, then we need to use the construction of (15.160) and its properties as in [345, Section 3] as follows.

Construction 15.161 (Stable Umkehr maps associated to normal Γ-maps of degree one) Let $(f, \bar{f}) \colon M \to X$ be a normal Γ-map of degree one from a closed n-dimensional manifold M to a finite n-dimensional Poincaré complex X as in Definition 8.3, with the generalisation that we allow M not to be connected. Then the map $\bar{f} \colon p_M \to p_X$ of the Spivak normal fibrations induces a map between the Thom Γ-spaces $\text{Th}_\Gamma(\bar{f}) \colon \text{Th}_\Gamma(p_M) \to \text{Th}_\Gamma(p_X)$. For Poincaré complexes, the S-Γ-dual of $\text{Th}_\Gamma(p_X)$ is homotopy equivalent to some suspension $\Sigma^p \widehat{X}_+$. By the functoriality of the S-Γ-duality up to stable Γ-homotopy, we obtain for some $p \geq 0$ a map, called a *stable Umkehr map*

$$F := \text{Th}_\Gamma(\widehat{f})^* \colon \Sigma^p \widehat{X}_+ \to \Sigma^p \widehat{M}_+,$$

which is well defined up to stable Γ-homotopy and represents the S-Γ-dual of $\text{Th}_\Gamma(\widehat{f})$, see [345, Proposition 4.1]. In other words, the class $\{F\}_\Gamma$ of F in $\{\widehat{X}_+, \widehat{M}_+\}_\Gamma$ is well defined. Note that $\{F\}_\Gamma$ depends on the bundle data.

Example 15.162 (Algebraic quadratic Poincaré complex associated to a normal map of degree one) Let $(f, \bar{f}) \colon M \to X$ be a normal Γ-map of degree one from a closed n-dimensional manifold M to a finite n-dimensional Poincaré complex X. Construction 15.161 produces a well-defined class $\{F\}_\Gamma$, which induces a $\mathbb{Z}\Gamma$-chain map, well defined up to $\mathbb{Z}\Gamma$-chain homotopy, on the cellular $\mathbb{Z}\Gamma$-chain complexes

$$C_*^!(\widehat{f}) \colon C(\widehat{X}) \xrightarrow{\cong} \Sigma^{-p} \widetilde{C}(\Sigma^p \widehat{X}_+) \xrightarrow{\Sigma^{-p} \widetilde{C}(F)} \Sigma^{-p} \widetilde{C}(\Sigma^p \widehat{M}_+) \xrightarrow{\cong} C(\widehat{M})$$

where the chain maps denoted by $\cong$ are the obvious suspension chain isomorphisms.

The map $C_*^!(\widehat{f})$ a priori depends on F and hence also on the bundle data $\bar{f}$. However, using an equivariant version of the relationship between the S-duality, Poincaré duality and the Thom isomorphism from Theorem 6.97, one can show that up to $\mathbb{Z}\Gamma$-chain homotopy this map coincides with the chain level Umkehr map of (8.42), which justifies the use of the same symbol.

Consider its mapping cone $\text{cone}(C_*^!(\widehat{f}))$. Recall that a group homomorphism $w \colon \Gamma \to \{\pm 1\}$ and a class $[X] \in H_n^\Gamma(X; \mathbb{Z}^w) = H_n(\mathbb{Z}^w \otimes C(\widehat{X}))$ are part of the structure of a normal Γ-map of degree one. The canonical inclusion map $e \colon C(\widehat{M}) \to \text{cone}(C_*^!(\widehat{f}))$ induces a map $e_{Q,n} \colon Q_n^{\mathbb{Z}\Gamma, w}(C(\widehat{M})) \to Q_n^{\mathbb{Z}\Gamma, w}(\text{cone}(C_*^!(\widehat{f})))$. The quadratic construction $\bar{\psi}_{\{F\}_\Gamma, w} \colon H_n^\Gamma(X; \mathbb{Z}^w) \to Q_n^{\mathbb{Z}\Gamma, w}(C^s(\widehat{M}))$ of (15.154) and the zigzag (15.63) of $\mathbb{Z}\Gamma$-chain homotopy equivalences yield a map denoted in the same way by

$$\overline{\psi}_{\{F\}_\Gamma,,w} \colon H_n^\Gamma(X;\mathbb{Z}^w) \to Q_n^{\mathbb{Z}\Gamma,w}(C(\widehat{M})).$$

Now we get an element $e_{Q,n} \circ \overline{\psi}_{\{F\}_\Gamma,w}([X])$ in $Q_n^{\mathbb{Z}\Gamma,w}(\mathrm{cone}(C_*^!(\widehat{f})))$ and thus an n-dimensional quadratic Poincaré complex

$$\left(\mathrm{cone}(C_*^!(\widehat{f})), e_{Q,n} \circ \overline{\psi}_{\{F\}_\Gamma,w}([X])\right) \tag{15.163}$$

in $b\mathbb{Z}\Gamma\text{-CH}_{\mathrm{fgf}}$. That it is Poincaré follows from Lemma 8.46 (ii).

It is well defined up to $\mathbb{Z}\Gamma$-chain isomorphism of n-dimensional finite quadratic Poincaré complexes by the following argument. Note that the element $\overline{\psi}_{\{F\}_\Gamma,w}([X])$ $\in Q_n^{\mathbb{Z}\Gamma,w}(C(\widehat{M}))$ is well defined since it depends only on $\{F\}_\Gamma$ and $[X]$. The chain map $C_*^!(\widehat{f})$ is only well defined up to $\mathbb{Z}\Gamma$-chain homotopy equivalence. If $C_*^!(\widehat{f})' \colon C(\widehat{X}) \to C(\widehat{M})$ is another representative of its $\mathbb{Z}\Gamma$-chain homotopy class, we can choose a $\mathbb{Z}\Gamma$-chain homotopy $h \colon C_*^!(\widehat{f}) \simeq C_*^!(\widehat{f})'$. By Exercise 14.56 it induces a $\mathbb{Z}\Gamma$-chain isomorphism

$$g \colon \mathrm{cone}(C_*^!(\widehat{f})) \xrightarrow{\cong_{\mathbb{Z}\Gamma}} \mathrm{cone}(C_*^!(\widehat{f})')$$

such that $g \circ e = e'$ holds for the canonical inclusions $e \colon C(\widehat{M}) \to \mathrm{cone}(C_*^!(\widehat{f}))$ and $e' \colon C(\widehat{M}) \to \mathrm{cone}(C_*^!(\widehat{f})')$. Hence $g_{Q,n} \colon Q_n^{\mathbb{Z}\Gamma,w}(C_*^!(\widehat{f})) \to Q_n^{\mathbb{Z}\Gamma,w}(C_*^!(\widehat{f}'))$ sends $e_{Q,n} \circ \overline{\psi}_{\{F\}_\Gamma,w}([X])$ to $e'_{Q,n} \circ \overline{\psi}_{\{F\}_\Gamma,w}([X])$. This shows that we obtain a $\mathbb{Z}\Gamma$-chain isomorphism of n-dimensional finite quadratic Poincaré $\mathbb{Z}\Gamma$-complexes

$$\left(\mathrm{cone}(C_*^!(\widehat{f})), e_{Q,n} \circ \overline{\psi}_{\{F\}_\Gamma,w}([X])\right)$$
$$\xrightarrow{\cong_{\mathbb{Z}\Gamma}} \left(\mathrm{cone}(C_*^!(\widehat{f})'), e'_{Q,n} \circ \overline{\psi}_{\{F\}_\Gamma,w}([X])\right).$$

Its class in the appropriate quadratic L-group, see Section 15.5, will be a well-defined element and will ultimately be identified with the surgery obstruction from Chapters 8 and 9. We emphasise that, although the underlying chain complex in (15.163) does not depend on the bundle data, the quadratic structure does depend on them. Actually the quadratic structure depends on the map of the underlying spherical fibrations only.

Remark 15.164 (Stable Umkehr maps from immersions) In Subsection 15.7.2 we will explain how an immersion $g \colon S^r \looparrowright M$ produces a stable Umkehr map whose associated quadratic structure gives an obstruction to modifying g by a regular homotopy to an embedding.

Remark 15.165 (Stable Umkehr maps from vector bundles) In Subsection 15.7.1 we will explain how a vector bundle $\alpha \colon S^r \to \mathrm{BSO}(q)$ with a stable trivialisation $\beta \colon D^{r+1} \to \mathrm{BSO}$ produces a stable map whose associated quadratic structure measures whether the bundle is itself trivial or not.

15.3 Structured Chain Complexes

15.3.9 Wu Classes

In this subsection we start by studying the behaviour of the symmetric and quadratic complexes under sums. Then we move on to the so-called Wu classes. These play a role at two stages later in this chapter. On one hand, they are obstructions to performing algebraic surgery below the dimension on a given cohomology class in the symmetric case; this will come up in Remark 15.431 in Section 15.6. On the other hand, they play a role in Section 15.7 in the identification of the surgery obstruction of a normal map of degree one, as defined in Chapters 8 and 9 with the equivalence class of a quadratic Poincaré complex obtained in Subsection 15.3.7 via the quadratic construction of Section 15.3.6, see Proposition 15.456. We omit the hyperquadratic version, which also exists, since we will not need it.

Let $C, D \in bR\text{-CH}$. Then we have the obvious inclusion

$$(C \otimes_R C) \oplus (D \otimes_R D) \subset (C \oplus D) \otimes_R (C \oplus D).$$

Definition 15.166 For $C, D \in bR\text{-CH}$ let the operations of the *direct sum*

$$W^{\%}(C) \oplus W^{\%}(D) \to W^{\%}(C \oplus D);$$
$$W_{\%}(C) \oplus W_{\%}(D) \to W_{\%}(C \oplus D),$$

be defined by the composites

$$W \xrightarrow{(1,1)} W \oplus W \to (C \otimes_R C) \oplus (D \otimes_R D) \to (C \oplus D) \otimes_R (C \oplus D);$$
$$W^{-*} \xrightarrow{(1,1)} W^{-*} \oplus W^{-*} \to (C \otimes_R C) \oplus (D \otimes_R D) \to (C \oplus D) \otimes_R (C \oplus D),$$

where in the quadratic case in the second row we use the identification $W \otimes_{\mathbb{Z}[\mathbb{Z}_2]} (C \otimes_R C) \cong \hom_{\mathbb{Z}[\mathbb{Z}_2]}(W^{-*}, C \otimes_R C)$.

By taking homology they induce maps of abelian groups

$$\oplus \colon Q^n(C) \oplus Q^n(D) \to Q^n(C \oplus D);$$
$$\oplus \colon Q_n(C) \oplus Q_n(D) \to Q_n(C \oplus D).$$

Given two n-dimensional symmetric chain complexes $(C, [\varphi])$ and $(C', [\varphi'])$, we define their *direct sum*

$$(C, [\varphi]) \oplus (C', [\varphi']) = (C \oplus C', [\varphi] \oplus [\varphi']),$$

and similarly in the quadratic case.

Before we study the properties of the direct sum operation, we need the following technical proposition.

Proposition 15.167 (Special case of Shapiro's Lemma) *Let A, B be $\mathbb{Z}$-chain complexes and let $\tau \colon A \to B$ be an isomorphism. Consider $A \oplus B$ as a $\mathbb{Z}[\mathbb{Z}/2]$-chain complex via the map*

$$A \oplus B \to A \oplus B \quad (x, y) \mapsto (\tau^{-1}(y), \tau(x)).$$

Let W be the standard $\mathbb{Z}[\mathbb{Z}/2]$-resolution of $\mathbb{Z}$ as in Definition 15.9. Then there are $\mathbb{Z}$-chain homotopy equivalences

$$\mathrm{Hom}_{\mathbb{Z}[\mathbb{Z}/2]}(W, A \oplus B) \simeq A \quad \text{and} \quad W \otimes_{\mathbb{Z}[\mathbb{Z}/2]} (A \oplus B) \simeq A.$$

Proof. We start with the proof the second statement. The $\mathbb{Z}$-chain homotopy equivalence $W \to 0[\mathbb{Z}]$ induces a $\mathbb{Z}$-chain homotopy equivalence $W \otimes_{\mathbb{Z}} A \simeq A$. On the other hand we have an obvious isomorphism of $\mathbb{Z}$-chain complexes $W \otimes_{\mathbb{Z}} A \cong W \otimes_{\mathbb{Z}[\mathbb{Z}/2]} (A \oplus B)$. Together we get the desired chain homotopy equivalence.

For the proof of the first statement, recall that for finite groups induction and coinduction of R-modules agree, see [58, Proposition III.5.9]. This extends to R-chain complexes, and so we obtain an isomorphism of $\mathbb{Z}$-chain complexes $\mathrm{hom}_{\mathbb{Z}}(W, A) \cong \mathrm{hom}_{\mathbb{Z}[\mathbb{Z}/2]}(W, A \oplus B)$. The $\mathbb{Z}$-chain homotopy equivalence $W \to 0[\mathbb{Z}]$ induces a $\mathbb{Z}$-chain homotopy equivalence $\mathrm{hom}_{\mathbb{Z}}(W, A) \simeq A$, and similarly as above we get the desired chain homotopy equivalence. □

We call the proposition a special case of Shapiro's Lemma since, when we apply H_n to the two chain homotopy equivalences, we obtain the well-known Shapiro's Lemma from cohomology of groups, see [58, Proposition 6.2]; in our case with $H = \{1\}$ and $G = \mathbb{Z}/2$.

The behaviour of the functors $W^{\%}(-)$ and $W_{\%}(-)$ with respect to the direct sum operation is as follows, see [344, Proposition 1.4 (i)].

Proposition 15.168 *Let C and D be two chain complexes in $\mathrm{b}R$-CH. Then the direct sum maps from Definition 15.166 fit into cofibration sequences that split in a suitable way so that we have*

$$W^{\%}(C \oplus D) \simeq W^{\%}(C) \oplus W^{\%}(D) \oplus (C \otimes_R D);$$
$$W_{\%}(C \oplus D) \simeq W_{\%}(C) \oplus W_{\%}(D) \oplus (C \otimes_R D).$$

Proof. Direct computation via Shapiro's Lemma from Proposition 15.167. □

For $D \in \mathrm{b}R$-CH and $n \in \mathbb{Z}$, we have the map $\alpha_D^n \colon H_n(D \otimes_R D) \to Q^n(D)$ given by $[\theta] \mapsto [\varphi]$, where $\varphi_0 = (1 + T)(\theta)$ and $\varphi_s = 0$ for $s \geq 1$ and the map $\alpha_n^D \colon H_n(D \otimes_R D) \to Q_n(D)$ given by $[\theta] \mapsto [\psi]$, where $\psi_0 = \theta$ and $\psi_s = 0$ for $s \geq 1$. The following result is from [344, Proposition 1.4 (ii)].

Proposition 15.169 *Let $f, g \colon C \to D$ be two R-chain maps between chain complexes in $\mathrm{b}R$-CH. Then we have in $Q^n(D)$, $Q_n(D)$ respectively:*

$$(f + g)^{Q,n}([\varphi]) - f^{Q,n}([\varphi]) - g^{Q,n}([\varphi]) = \alpha_D^n((1 + T)H_n(f \otimes g)([\varphi_0]));$$
$$(f + g)_{Q,n}([\psi]) - f_{Q,n}([\psi]) - g_{Q,n}([\psi]) = \alpha_n^D(H_n(f \otimes g)(1 + T)([\psi_0])).$$

Proof. Apply homology to the chain equivalences appearing in Proposition 15.168 and consider the factorisation

15.3 Structured Chain Complexes

$$(f + g)^{Q,n} = (f \oplus g)^{Q,n} \circ (\Delta_C)^{Q,n} \colon Q^n(C) \to Q^n(C \oplus C) \to Q^n(D)$$

where $\Delta_C \colon C \to C \oplus C$ is the diagonal map. Inspection shows that the composition of $(\Delta_C)^{Q,n}$ with the projection on $H_n(C \otimes_R C)$ is given by $[\varphi] \mapsto [\varphi_0]$. Similarly in the quadratic case where the composition of $(\Delta_C)_{Q,n}$ with the projection on $H_n(C \otimes_R C)$ is given by $[\psi] \mapsto [(1 + T)(\psi_0)]$. □

Now we turn our attention to Wu classes. Recall that the r-th *Wu class* of a Poincaré complex X is a characteristic cohomology class $v_r \in H^r(X; \mathbb{Z}/2)$ of X characterised by the property

$$\langle v_r \cup x, [X] \rangle = \langle \mathrm{Sq}^r(x), [X] \rangle \tag{15.170}$$

for all $x \in H^{n-r}(X; \mathbb{Z}/2)$, see for example [45, Chapter VI.17] or [308, Chapter 11]. We now introduce the Wu classes as invariants of symmetric and quadratic chain complexes generalising the classical Wu classes.

Definition 15.171 (Wu classes)

(i) Let $(C, [\varphi])$ be an n-dimensional symmetric complex over R. Define its r-th *Wu class* by
$$v_r([\varphi]) \colon H^{n-r}(C) \to Q^n((n-r)[R]), \quad [f] \mapsto [(f \otimes f)(\varphi_{n-2r})];$$

(ii) Let $(C, [\psi])$ be an n-dimensional quadratic complex over R. Define its r-th *Wu class* by
$$v^r([\psi]) \colon H^{n-r}(C) \to Q_n((n-r)[R]); \quad [f] \mapsto [(f \otimes f)(\psi_{2r-n})].$$

Here $f \in \hom_R(C_{n-r}, R)$ and $\varphi_{n-2r}, \psi_{2r-n} \in C_{n-r} \otimes C_{n-r}$ are components of polarisations of $[\varphi], [\psi]$.

It is also possible to view the Wu classes as follows. An element $[f]$ in $H^{n-r}(C)$ corresponds to a chain homotopy class of chain maps from C to $(n-r)[R]$. Given a representative $f \colon C \to (n-r)[R]$, the Wu class is $v_r([\varphi]) = f^{Q,n}([\varphi])$ in terms of the map $f^{Q,n}$ from (15.35). Similarly $v^r([\psi]) = f_{Q,n}([\psi])$.

The relationship of the maps defined in the above definition with the classical Wu class as in equation (15.170) is that the Wu class $v_r \in H^r(X, \mathbb{Z}/2)$ defines via (15.170) the map

$$\langle v_r \cup -, [X] \rangle \colon H^{n-r}(X; \mathbb{Z}/2) \to \mathbb{Z}/2. \tag{15.172}$$

For $\varphi = \varphi_X([X])$ for a Poincaré complex X and $R = \mathbb{Z}/2$, it can be shown that it is equal to the map from part (i) in the above definition. More information about the relation of the Wu classes from Definition 15.171 to classical Wu classes can be found in [345, Section 9].

Remark 15.173 (Various sorts of Q-groups) Let us consider the special case of Q-groups appearing in Definition 15.171 (ii) when $n = 2l$, $r = l$, $\varepsilon = +1$ and $R = \mathbb{Z}$ with trivial involution. When $l = 2k$, we have that as $\mathbb{Z}$-chain complexes $(2k)[\mathbb{Z}] \otimes (2k)[\mathbb{Z}] = (4k)[\mathbb{Z}]$ and the switch involution from (15.6) on the left corresponds to the trivial involution on the right and we obtain that $Q_{4k}((2k)[\mathbb{Z}]) = Q_{+1}(\mathbb{Z})$ in the sense of (8.70). When $l = 2k + 1$, we have that as $\mathbb{Z}$-chain complexes $(2k+1)[\mathbb{Z}] \otimes (2k+1)[\mathbb{Z}] = (4k+2)[\mathbb{Z}]$ and the switch involution from (15.6) on the left corresponds to the involution -1 on the right and we obtain that $Q_{4k+2}((2k+1)[\mathbb{Z}]) = Q_{-1}(\mathbb{Z})$ in the sense of (8.70).

The position of subscripts and superscripts in the notation is taken from [344, Section 1]. The logic is not clear to us, but we have decided to keep it since we will only use these Wu classes at few places in Section 15.7. The statement of the following proposition appears on page 114 in [344] and is a direct consequence of Proposition 15.169.

Proposition 15.174 *Given $C \in $ bR-CH, $[\varphi] \in Q^n(C)$, $[\psi] \in Q_n(C)$, and $[f], [g] \in H^r(C)$ with $n = 2r$, we get in $Q^n(r[R])$, $Q_n(r[R])$ respectively*

$$v_r([\varphi])([f+g]) - v_r([\varphi])([f]) - v_r([\varphi])([g]) = \alpha^n_{r[R]}((1+T)H_n(f \otimes g)([\varphi_0]));$$

$$v^r([\psi])([f+g]) - v^r([\psi])([f]) - v^r([\psi])([g]) = \alpha_n^{r[R]}(H_n(f \otimes g)(1+T)([\psi_0])).$$

15.3.10 Homotopy $\mathbb{Z}[\mathbb{Z}/2]$-Chain Maps

In Section 15.3.1 we have seen that an R-chain map $f \colon C \to D$ induces the $\mathbb{Z}[\mathbb{Z}/2]$-chain map $f \otimes f \colon C \otimes_R C \to D \otimes_R D$, which in turn induces the $\mathbb{Z}$-chain map (keeping in mind that $W^{\%}(C) = \hom_{\mathbb{Z}[\mathbb{Z}/2]}(W, C \otimes_R C)$)

$$f^{\%} = \hom_{\mathbb{Z}[\mathbb{Z}/2]}(\mathrm{id}_W, f \otimes f) \colon W^{\%}(C) \to W^{\%}(D).$$

In this section we generalise this construction using the notion of a so-called *homotopy $\mathbb{Z}[\mathbb{Z}/2]$-chain map* defined below, and we collect some properties related to this generalisation. These notions and results are interesting on their own, but we will only use them in Section 15.5.3 about the algebraic Thom Construction. We start with a preliminary general result.

Proposition 15.175 *Let $\alpha \colon X \to Y$ be a $\mathbb{Z}[\mathbb{Z}/2]$-chain map of $\mathbb{Z}[\mathbb{Z}/2]$-chain complexes, which, when considered as a $\mathbb{Z}$-chain map, is a $\mathbb{Z}$-chain equivalence. Then the induced maps*

$$\hom_{\mathbb{Z}[\mathbb{Z}/2]}(\mathrm{id}_W, \alpha) \colon \hom_{\mathbb{Z}[\mathbb{Z}/2]}(W, X) \to \hom_{\mathbb{Z}[\mathbb{Z}/2]}(W, Y);$$

$$\mathrm{id}_W \otimes_{\mathbb{Z}[\mathbb{Z}/2]} \alpha \colon W \otimes_{\mathbb{Z}[\mathbb{Z}/2]} X \to W \otimes_{\mathbb{Z}[\mathbb{Z}/2]} Y,$$

are $\mathbb{Z}$-chain equivalences.

15.3 Structured Chain Complexes

Proof. We present the proof for the first map, the proof for the second map is similar. Given integers $i \leq j$, recall the truncated $\mathbb{Z}[\mathbb{Z}/2]$-chain complexes $\widehat{W}[i,j]$ from Section 15.3.5. Namely,

$$\widehat{W}[i,j]_n = W_n \quad i \leq n \leq j \quad \text{and} \quad \widehat{W}[i,j]_n = 0 \quad \text{otherwise,}$$

and with the differential $\widehat{W}[i,j]_n \to \widehat{W}[i,j]_{n-1}$ equal to $(1 + (-1)^n T)$ when $(i+1) \leq n \leq j$ and 0 otherwise. With this notation we have for every $n \geq 0$ the degreewise split short exact sequence of $\mathbb{Z}[\mathbb{Z}/2]$-chain complexes

$$0 \to \widehat{W}[0,n] \to \widehat{W}[0, n+1] \to (n+1)[W_{n+1}] \to 0.$$

Given any $\mathbb{Z}[\mathbb{Z}/2]$-chain complex Z, this sequence induces the degreewise split exact sequence of $\mathbb{Z}$-chain complexes

$$0 \to \hom_{\mathbb{Z}[\mathbb{Z}/2]}((n+1)[W_{n+1}], Z) \to \hom_{\mathbb{Z}[\mathbb{Z}/2]}(\widehat{W}[0, n+1], Z) \to$$
$$\to \hom_{\mathbb{Z}[\mathbb{Z}/2]}(\widehat{W}[0,n], Z) \to 0.$$

Applying Proposition 14.55 to the pair of such sequences obtained by plugging X and Y for Z, we obtain by induction on n that α induces for every $n \geq 0$ a $\mathbb{Z}$-chain equivalence

$$\hom_{\mathbb{Z}[\mathbb{Z}/2]}(\mathrm{id}_{W[0,n]}, \alpha) \colon \hom_{\mathbb{Z}[\mathbb{Z}/2]}(\widehat{W}[0,n], X) \to \hom_{\mathbb{Z}[\mathbb{Z}/2]}(\widehat{W}[0,n], Y).$$

Next observe that we have

$$W = \mathrm{colim}_{n \to \infty} \widehat{W}[0,n]$$

and that the structure maps in the colimit are cofibrations, which means degreewise split injections. We have the canonical isomorphism

$$\hom_{\mathbb{Z}[\mathbb{Z}/2]}(\mathrm{colim}_{n \to \infty} \widehat{W}[0,n], X) \cong \lim_{n \to \infty} \hom_{\mathbb{Z}[\mathbb{Z}/2]}(\widehat{W}[0,n], X)$$

where the structure maps in the limit on the right-hand side are fibrations, which means degreewise split surjections. Now Proposition 14.120 yields the desired statement. □

Let X and Y be $\mathbb{Z}[\mathbb{Z}/2]$-chain complexes with the two respective involutions denoted by T_X and T_Y. By a *homotopy $\mathbb{Z}[\mathbb{Z}/2]$-chain map* from X to Y we mean a 0-dimensional cycle

$$t \in \hom_{\mathbb{Z}[\mathbb{Z}/2]}(W, \hom_{\mathbb{Z}}(X,Y))_0$$

where $\hom_{\mathbb{Z}}(X,Y)$ is viewed as a $\mathbb{Z}[\mathbb{Z}/2]$-chain complex via the involution $f \mapsto T_Y \circ f \circ T_X$. Such a map t can be described as a sequence of maps $t_i \colon X \to Y$ of degree i, for $i \geq 0$, satisfying

$$d_Y \circ t_i - (-1)^i t_i \circ d_X = t_{i-1} + (-1)^i T_Y \circ t_{i-1} \circ T_X. \tag{15.176}$$

The map t_0 is in general not a $\mathbb{Z}[\mathbb{Z}/2]$-chain map, but it is a $\mathbb{Z}$-chain map, and the map t_1 is a $\mathbb{Z}$-chain homotopy between t_0 and $T_Y \circ t_0 \circ T_X$ and so on. If $t_1 = 0$, then t_0 is a $\mathbb{Z}[\mathbb{Z}/2]$-chain map, which explains the name.

A homotopy $\mathbb{Z}[\mathbb{Z}/2]$-chain map $t \in \hom_{\mathbb{Z}[\mathbb{Z}/2]}(W, \hom_{\mathbb{Z}}(X, Y))_0$ induces a $\mathbb{Z}$-chain map

$$t^\% : \hom_{\mathbb{Z}[\mathbb{Z}/2]}(W, X) \to \hom_{\mathbb{Z}[\mathbb{Z}/2]}(W, Y)$$

as follows. Consider the composite of $\mathbb{Z}$-chain maps

$$\hom_{\mathbb{Z}[\mathbb{Z}/2]}(W, \hom_{\mathbb{Z}}(X, Y)) \otimes_{\mathbb{Z}} \hom_{\mathbb{Z}[\mathbb{Z}/2]}(W, X) \xrightarrow{-\otimes-}$$

$$\hom_{\mathbb{Z}[\mathbb{Z}/2]}(W \otimes_{\mathbb{Z}} W, \hom_{\mathbb{Z}}(X,Y) \otimes_{\mathbb{Z}} X) \xrightarrow{\hom(\Delta, c)} \hom_{\mathbb{Z}[\mathbb{Z}/2]}(W, Y)$$

where $-\otimes-$ is the tensoring map from (14.20), the $\mathbb{Z}[\mathbb{Z}/2]$-chain map $\Delta: W \to W \otimes_{\mathbb{Z}} W$ is from (15.69), and c is the evaluation map given by $f \otimes x \mapsto f(x)$, which, by using $X \cong \hom_{\mathbb{Z}}(\mathbb{Z}, X)$, can be seen as a special case of (14.19) and is a $\mathbb{Z}[\mathbb{Z}/2]$-chain map. Using the adjunction (14.16), we obtain the $\mathbb{Z}$-chain map

$$\hom_{\mathbb{Z}[\mathbb{Z}/2]}(W, \hom_{\mathbb{Z}}(X, Y)) \to \hom_{\mathbb{Z}}(\hom_{\mathbb{Z}[\mathbb{Z}/2]}(W, X), \hom_{\mathbb{Z}[\mathbb{Z}/2]}(W, Y)), \quad (15.177)$$

and the image of the cycle t is defined to be $t^\%$. Since $t^\%$ is an image of a cycle under a $\mathbb{Z}$-chain map, it is a cycle in the $\hom_{\mathbb{Z}}$-complex and as such a $\mathbb{Z}$-chain map.

Similarly, a homotopy $\mathbb{Z}[\mathbb{Z}/2]$-chain map $t \in \hom_{\mathbb{Z}[\mathbb{Z}/2]}(W, \hom_{\mathbb{Z}}(X, Y))_0$ induces for bounded X and Y a $\mathbb{Z}$-chain map

$$t_\% : W \otimes_{\mathbb{Z}[\mathbb{Z}/2]} X \to W \otimes_{\mathbb{Z}[\mathbb{Z}/2]} Y$$

by first using the canonical isomorphism $W \otimes_{\mathbb{Z}[\mathbb{Z}/2]} X \cong \hom_{\mathbb{Z}[\mathbb{Z}/2]}(W^{-*}, X)$, and then the composition of $\mathbb{Z}$-chain maps

$$\hom_{\mathbb{Z}[\mathbb{Z}/2]}(W, \hom_{\mathbb{Z}}(X, Y)) \otimes \hom_{\mathbb{Z}[\mathbb{Z}/2]}(W^{-*}, X) \xrightarrow{-\otimes-}$$

$$\hom_{\mathbb{Z}[\mathbb{Z}/2]}(W \widehat{\otimes}_{\mathbb{Z}} W^{-*}, \hom_{\mathbb{Z}}(X, Y) \otimes_{\mathbb{Z}} X) \xrightarrow{\hom(\Delta, c)} \hom_{\mathbb{Z}[\mathbb{Z}/2]}(W^{-*}, Y).$$

Here $-\otimes-$ is the tensoring map from (14.20), the $\mathbb{Z}[\mathbb{Z}/2]$-chain map $\Delta: W^{-*} \to W \widehat{\otimes}_{\mathbb{Z}} W^{-*}$ is from (15.71), and c is the evaluation map given by $f \otimes x \mapsto f(x)$ as above, which is also a $\mathbb{Z}[\mathbb{Z}/2]$-chain map. Using the canonical isomorphism $W \otimes_{\mathbb{Z}[\mathbb{Z}/2]} X \cong \hom_{\mathbb{Z}[\mathbb{Z}/2]}(W^{-*}, X)$ again and the adjunction (14.16), we obtain the $\mathbb{Z}$-chain map

$$\hom_{\mathbb{Z}[\mathbb{Z}/2]}(W, \hom_{\mathbb{Z}}(X, Y)) \to \hom_{\mathbb{Z}}(W \otimes_{\mathbb{Z}[\mathbb{Z}/2]} X, W \otimes_{\mathbb{Z}[\mathbb{Z}/2]} Y), \quad (15.178)$$

and the image of the cycle t is defined to be $t_\%$. Since $t_\%$ is an image of a cycle under a $\mathbb{Z}$-chain map it is a cycle in the $\hom_{\mathbb{Z}}$-complex and as such a $\mathbb{Z}$-chain map.

15.3 Structured Chain Complexes

Remark 15.179 (Formulas for induced maps of homotopy chain maps) Unravelling the definitions shows that the induced maps of a $\mathbb{Z}[\mathbb{Z}/2]$-chain map $t \in \hom_{\mathbb{Z}[\mathbb{Z}/2]}(W, \hom_{\mathbb{Z}}(X,Y))_0$ are given by the formulas:

$$t^{\%}(\kappa)_s = \sum_{r=0}^{s} (-1)^{r(n+s-r)} \, t_r(T^r(\kappa_{s-r})) \quad \text{for } \kappa \in \hom_{\mathbb{Z}[\mathbb{Z}/2]}(W,X)_n;$$

$$t_{\%}(\lambda)_s = \sum_{r=0}^{\infty} (-1)^{r(n-s-r)} \, t_r(T^r(\lambda_{-s-r})) \quad \text{for } \lambda \in (W \otimes_{\mathbb{Z}[\mathbb{Z}/2]} X)_n.$$

Proposition 15.180 *Let $t \in \hom_{\mathbb{Z}[\mathbb{Z}/2]}(W, \hom_{\mathbb{Z}}(X,Y))_0$ be a homotopy $\mathbb{Z}[\mathbb{Z}/2]$-chain map from X to Y with the 0-th component t_0 where X and Y are bounded chain complexes. If t_0 is a $\mathbb{Z}$-chain homotopy equivalence, then the $\mathbb{Z}$-chain maps*

$$t^{\%} \colon \hom_{\mathbb{Z}[\mathbb{Z}/2]}(W,X) \to \hom_{\mathbb{Z}[\mathbb{Z}/2]}(W,Y);$$
$$t_{\%} \colon W \otimes_{\mathbb{Z}[\mathbb{Z}/2]} X \to W \otimes_{\mathbb{Z}[\mathbb{Z}/2]} Y,$$

are $\mathbb{Z}$-chain homotopy equivalences.

Proof. Again we present the proof only for the first map, the proof for the second map is similar. The proof uses similar arguments as the proof of Proposition 15.175. Recall the truncated complexes $\widehat{W}[i,j]$ from Section 15.3.5. Given $0 \le j < \infty$, the diagonal map (15.70) can be restricted to the diagonal map $\Delta[0,j] \colon \widehat{W}[0,j] \to W \otimes \widehat{W}[0,j]$. Using this diagonal map, the homotopy $\mathbb{Z}[\mathbb{Z}/2]$-chain map t induces by the same process as described above a map

$$t^{\%}[0,j] \colon \hom_{\mathbb{Z}[\mathbb{Z}/2]}(\widehat{W}[0,j], X) \to \hom_{\mathbb{Z}[\mathbb{Z}/2]}(\widehat{W}[0,j], Y).$$

By induction on j we first prove that the map $t^{\%}[0,j]$ is a $\mathbb{Z}$-chain homotopy equivalence for every j. For the induction start with $j = 0$ and observe that the map $t^{\%}[0,0]$ is just given by t_0, hence it is a $\mathbb{Z}$-chain homotopy equivalence. For the induction step assume that the statement is proved for j. To obtain it for $(j+1)$ consider the degreewise split short exact sequence

$$0 \to \widehat{W}[0,j] \to \widehat{W}[0,j+1] \to (j+1)[W_{j+1}] \to 0.$$

We obtain the commutative diagram

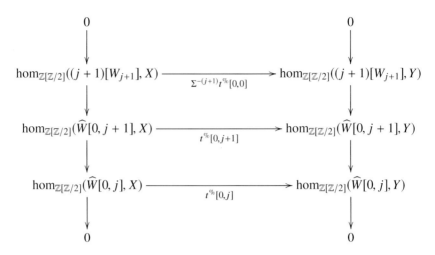

of degreewise split short exact sequences. Applying Proposition 14.55 to this map of degreewise split short exact sequences yields that the map $t^\%[0, j + 1]$ is a $\mathbb{Z}$-chain homotopy equivalence. Next, just as at the end of the proof of Proposition 15.175, an application of Proposition 14.120 yields the desired statement. □

We will also need the notion of a *homotopy* between two homotopy $\mathbb{Z}[\mathbb{Z}/2]$-chain maps. Suppose that we have two such homotopy $\mathbb{Z}[\mathbb{Z}/2]$-chain maps s, t in $\hom_{\mathbb{Z}[\mathbb{Z}/2]}(W, \hom_{\mathbb{Z}}(X, Y))_0$. By a *homotopy* from t to s we mean a 1-dimensional chain

$$k \in \hom_{\mathbb{Z}[\mathbb{Z}/2]}(W, \hom_{\mathbb{Z}}(X, Y))_1 \tag{15.181}$$

satisfying $d_{\hom}(k) = s - t$. Both maps s and t can be described as a sequence of maps $s_i, t_i \colon X \to Y$ of degree i for $i \geq 0$ satisfying the equations (15.176). A homotopy k from t to s can analogously be described as a sequence of maps $k_i \colon X \to Y$ of degree $i + 1$, for $i \geq 0$ satisfying

$$d_Y \circ k_i + (-1)^i k_i \circ d_X + k_{i-1} + (-1)^i T \circ k_{i-1} \circ T = s_i - t_i. \tag{15.182}$$

Such a homotopy k induces a $\mathbb{Z}$-chain homotopy $k^\%$ from $t^\%$ to $s^\%$ and $k_\%$ from $t_\%$ to $s_\%$ by defining $k^\%$ and $k_\%$ as the images of the 1-chain k of the $\mathbb{Z}$-chain maps (15.177) and (15.178) respectively. As 1-chains they can be viewed as $\mathbb{Z}$-chain homotopies.

Homotopy $\mathbb{Z}[\mathbb{Z}/2]$-chain maps can be composed by the following construction. Given three $\mathbb{Z}[\mathbb{Z}/2]$-chain complexes X, Y, Z consider the composition of $\mathbb{Z}$-chain maps

15.3 Structured Chain Complexes   631

$$\text{hom}_{\mathbb{Z}[\mathbb{Z}/2]}(W, \text{hom}_{\mathbb{Z}}(Y, Z)) \otimes_{\mathbb{Z}} \text{hom}_{\mathbb{Z}[\mathbb{Z}/2]}(W, \text{hom}_{\mathbb{Z}}(X, Y)) \xrightarrow{-\otimes -}$$

$$\xrightarrow{-\otimes -} \text{hom}_{\mathbb{Z}[\mathbb{Z}/2]}(W \otimes_{\mathbb{Z}} W, \text{hom}_{\mathbb{Z}}(Y, Z) \otimes_{\mathbb{Z}} \text{hom}_{\mathbb{Z}}(X, Y)) \xrightarrow{\text{hom}(\Delta, -\circ -)}$$

$$\xrightarrow{\text{hom}(\Delta, -\circ -)} \text{hom}_{\mathbb{Z}[\mathbb{Z}/2]}(W, \text{hom}_{\mathbb{Z}}(X, Z)) \quad (15.183)$$

where $-\otimes-$ is the tensoring map from (14.20), the $\mathbb{Z}[\mathbb{Z}/2]$-chain map $\Delta \colon W \to W \otimes_{\mathbb{Z}} W$ is from (15.69), and $-\circ-$ is the composition map $g \otimes f \mapsto g \circ f$ from (14.19), which is also a $\mathbb{Z}[\mathbb{Z}/2]$-chain map.

Given homotopy $\mathbb{Z}[\mathbb{Z}/2]$-chain maps s in $\text{hom}_{\mathbb{Z}[\mathbb{Z}/2]}(W, \text{hom}_{\mathbb{Z}}(X, Y))_0$ and t in $\text{hom}_{\mathbb{Z}[\mathbb{Z}/2]}(W, \text{hom}_{\mathbb{Z}}(Y, Z))_0$, their composition $t \circ s$ is defined to be the image of $t \otimes s$ under the above composition, and as such it is a homotopy $\mathbb{Z}[\mathbb{Z}/2]$-chain map. Also a composition of a homotopy $\mathbb{Z}[\mathbb{Z}/2]$-chain map (0-cycle) with a homotopy of homotopy $\mathbb{Z}[\mathbb{Z}/2]$-chain maps (suitable 1-chain) is a homotopy of homotopy $\mathbb{Z}[\mathbb{Z}/2]$-chain maps (suitable 1-chain). When we compose a homotopy $\mathbb{Z}[\mathbb{Z}/2]$-chain map t with a $\mathbb{Z}[\mathbb{Z}/2]$-chain map f to obtain a new homotopy $\mathbb{Z}[\mathbb{Z}/2]$-chain map $f \circ t$, the above procedure yields simply $(f \circ t)_i = f \circ t_i$ for all $i \geq 0$, and similarly, when we compose a homotopy $\mathbb{Z}[\mathbb{Z}/2]$-chain map t with a $\mathbb{Z}[\mathbb{Z}/2]$-chain homotopy h to obtain a new homotopy of homotopy $\mathbb{Z}[\mathbb{Z}/2]$-chain maps $h \circ t$, the above procedure yields simply $(h \circ t)_i = h \circ t_i$ for all $i \geq 0$.

Remark 15.184 (Universal property of the homotopy fibre with respect to homotopy chain maps)

The homotopy fibre $\text{hofib}(f)$ of a $\mathbb{Z}[\mathbb{Z}/2]$-chain map of $\mathbb{Z}[\mathbb{Z}/2]$-chain complexes $f \colon X \to Y$ has the following universal property with respect to homotopy $\mathbb{Z}[\mathbb{Z}/2]$-chain maps generalising Remark 14.73 about the universal property for $\mathbb{Z}[\mathbb{Z}/2]$-chain maps.

Let A be a $\mathbb{Z}[\mathbb{Z}/2]$-chain complex, let $u \in \text{hom}_{\mathbb{Z}[\mathbb{Z}/2]}(W, \text{hom}_{\mathbb{Z}}(A, X))_0$ be a homotopy $\mathbb{Z}[\mathbb{Z}/2]$-chain map, and let $h' \in \text{hom}_{\mathbb{Z}[\mathbb{Z}/2]}(W, \text{hom}_{\mathbb{Z}}(A, Y))_1$ be a homotopy of homotopy $\mathbb{Z}[\mathbb{Z}/2]$-chain maps from 0 to $f \circ u$. Then there is precisely one homotopy $\mathbb{Z}[\mathbb{Z}/2]$-chain map

$$z = z(u, h') \in \text{hom}_{\mathbb{Z}[\mathbb{Z}/2]}(W, \text{hom}_{\mathbb{Z}}(A, \text{hofib}(f)))_0$$

such that $u = \overline{e} \circ z$ and $h' = h \circ z$ holds where $\overline{e} \colon \text{hofib}(f) \to X$ and $h \colon \text{hofib}(f) \to Y$ are as in Remark 14.73.

In order to state it, we use notation $(u_i)_n \in \text{hom}_{\mathbb{Z}}(X_n, Y_{n+i})$ for the components of $u_i \in \text{hom}_{\mathbb{Z}}(X, Y)_i$ and $(h'_i)_n \in \text{hom}_{\mathbb{Z}}(X_n, Y_{n+i+1})$ for the components of $h'_i \in \text{hom}_{\mathbb{Z}}(X, Y)_{i+1}$. Then the required z is defined by

$$(z_i)_n = \begin{pmatrix} (u_i)_n \\ (-1)^{n+i} \cdot (h'_i)_n \end{pmatrix} \colon A_n \to X_{n+i} \oplus Y_{n+i+1}.$$

We observe that the $\mathbb{Z}$-chain map $z_0 \in \text{hom}_{\mathbb{Z}}(A, \text{hofib}(f))_0$ equals the map obtained by the universal property from Remark 14.73 applied to the $\mathbb{Z}$-chain map u_0 and $\mathbb{Z}$-chain homotopy h'_0.

Remark 15.185 (Terminology for homotopy chain maps) Homotopy $\mathbb{Z}[\mathbb{Z}/2]$-chain maps and some constructions with them appear in the literature for example at the following places: [344, page 100], [346, page 8], [19, page 565]. Ranicki calls these maps $\mathbb{Z}/2$-isovariant maps, but we prefer the name homotopy $\mathbb{Z}[\mathbb{Z}/2]$-chain maps since for us it better explains what they are.

15.4 Forms and Formations

Recall that an n-dimensional symmetric chain complex as introduced in Definition 15.16 is a pair $(C, [\varphi])$ such that $C \in bR\text{-CH}_{\text{fgp}}$ and $[\varphi]$ is an element in $Q^n(C)$, but there is no condition on the dimension of C. A polarisation is a choice of a cycle φ representing $[\varphi]$, see Definition 15.17. Similarly in the quadratic and hyperquadratic case.

In this section we are going to impose dimension conditions on the chain complex C itself and for this we need new terminology. In Section 15.4.1 we will define the notion of a *strictly* n-dimensional symmetric complex where we require that, in addition to the dimension condition on the symmetric structure, the underlying chain complex itself is concentrated in degrees d for $0 \leq d \leq n$. We will also consider *homotopy* n-dimensional symmetric complexes where we require that, in addition to the dimension condition on the symmetric structure, the underlying chain complex itself is chain homotopy equivalent to a chain complex concentrated between dimensions 0 and n. Both notions will be defined similarly in the quadratic case.

With the notions above we will first show that stable isomorphism classes of strictly n-dimensional structured complexes can be identified with the homotopy classes of homotopy n-dimensional structured complexes in Proposition 15.192. Then we will show that the stable isomorphism classes of strictly 0-dimensional quadratic complexes can be identified with stable isomorphism classes of ε-quadratic forms from Chapter 8. Furthermore we will show that the stable isomorphism classes of strictly 1-dimensional quadratic complexes can be identified with stable isomorphism classes of split ε-quadratic formations, which is a small modification of the ε-quadratic formations from Chapter 9. Although the modification means that the stable isomorphism classes of formations and of split formations are not the same, it will not matter after the passage to L-groups, as we will show in Section 15.6.

The technique of algebraic surgery from Section 15.5.5 will enable us in Section 15.6 to pass inside L-groups from the n-dimensional quadratic complexes in the sense of Section 15.3 to the strictly n-dimensional quadratic chain complexes. This is not quite possible in the symmetric case, see Remark 15.431, but we will nevertheless include the symmetric case in this section since it is often a good preparation for the more complicated quadratic case.

Remark 15.186 (The role of $\varepsilon = \pm 1$.) We remark again that the skew-suspension map of Section 15.6 enables us to translate between n-dimensional and $(n + 2)$-dimensional complexes. This the reason why in this section it is enough to work

15.4 Forms and Formations   633

with 0-dimensional and 1-dimensional complexes and we do not have to include 2-dimensional and 3-dimensional ones.

15.4.1 Homotopy Theory of Highly Connected Structured Chain Complexes

We start by reviewing some standard terminology and introducing some new terminology about (structured) chain complexes that we will need; mainly following [344]. Recall from Definition 14.121 that an R-chain complex C is positive n-dimensional if $C_i = 0$ unless $0 \leq i \leq n$.

Definition 15.187 (Stable isomorphism of positive n-dimensional chain complexes) Two positive n-dimensional chain complexes C, C' in bR-CH_{fgp} are called *stably isomorphic* if there exist R-contractible positive n-dimensional chain complexes D, D' in bR-CH_{fgp} and an isomorphism

$$f: C \oplus D \to C' \oplus D'.$$

Definition 15.188 (Strictly/Homotopy n-dimensional symmetric chain complex) An n-dimensional symmetric chain complex $(C, [\varphi])$ in bR-CH_{fgp} is called *strictly n-dimensional* if the underlying R-chain complex C is positive n-dimensional.

It is called *homotopy n-dimensional* if the underlying R-chain complex C is R-chain homotopy equivalent to a positive n-dimensional complex in bR-CH_{fgp}.

Definition 15.189 (Strictly/Homotopy n-dimensional quadratic chain complex) An n-dimensional quadratic chain complex $(C, [\psi])$ in bR-CH_{fgp} is called *strictly n-dimensional* if the underlying R-chain complex C is positive n-dimensional.

It is called *homotopy n-dimensional* if the underlying R-chain complex C is R-chain homotopy equivalent to a positive n-dimensional complex in bR-CH_{fgp}.

Definition 15.190 (Stable isomorphism of strictly n-dimensional symmetric complexes) A *stable isomorphism*

$$[f]: (C, [\varphi]) \xrightarrow{\cong_s} (C', [\varphi'])$$

of strictly n-dimensional symmetric chain complexes is an isomorphism of the form

$$f: (C, [\varphi]) \oplus (D, 0) \to (C', [\varphi']) \oplus (D', 0)$$

with D, D' some positive n-dimensional R-contractible chain complexes in bR-CH_{fgp}.

Definition 15.191 (Stable isomorphism of strictly n-dimensional quadratic complexes) A *stable isomorphism*

$$[f]: (C, [\psi]) \xrightarrow{\cong_s} (C', [\psi'])$$

of strictly n-dimensional quadratic chain complexes is an isomorphism of the form

$$f \colon (C, [\psi]) \oplus (D, 0) \to (C', [\psi']) \oplus (D', 0)$$

with D, D' some positive n-dimensional R-contractible chain complexes in $bR\text{-CH}_{\text{fgp}}$.

The following proposition is essentially taken from [344, Proposition 1.5].

Proposition 15.192 *There are obvious inclusion of categories*

$$\left\{ \begin{array}{c} \text{strictly } n\text{-dim. sym.} \\ \text{chain compl.} \end{array} \right\} \to \left\{ \begin{array}{c} \text{homotopy } n\text{-dim. sym.} \\ \text{chain compl.} \end{array} \right\}$$

and

$$\left\{ \begin{array}{c} \text{strictly } n\text{-dim. quadr.} \\ \text{chain compl.} \end{array} \right\} \to \left\{ \begin{array}{c} \text{homotopy } n\text{-dim. quadr.} \\ \text{chain compl.} \end{array} \right\}.$$

They induce bijections

$$\left\{ \begin{array}{c} \text{strictly } n\text{-dim. sym.} \\ \text{chain compl.} \end{array} \right\} / \cong_s \;\overset{\cong}{\to}\; \left\{ \begin{array}{c} \text{homotopy } n\text{-dim. sym.} \\ \text{chain compl.} \end{array} \right\} / \simeq$$

and

$$\left\{ \begin{array}{c} \text{strictly } n\text{-dim. quadr.} \\ \text{chain compl.} \end{array} \right\} / \cong_s \;\overset{\cong}{\to}\; \left\{ \begin{array}{c} \text{homotopy } n\text{-dim. quadr.} \\ \text{chain compl.} \end{array} \right\} / \simeq$$

where $\cong_s$ refers to stable isomorphisms classes and $\simeq$ to chain homotopy classes.

Proof. Since any strictly n-dimensional symmetric complex $(C, [\varphi])$ is also a homotopy n-dimensional symmetric complex and a stable isomorphism of strictly n-dimensional symmetric complexes

$$[f] \colon (C, [\varphi]) \to (C', [\varphi'])$$

induces a homotopy equivalence (the first and last maps are the obvious ones)

$$g \colon (C, [\varphi]) \xrightarrow{\simeq} (C, [\varphi]) \oplus (D, 0) \xrightarrow[\cong]{f} (C', [\varphi']) \oplus (D', 0) \xrightarrow{\simeq} (C', [\varphi']),$$

the inclusion of categories is well defined.

By definition, for any homotopy n-dimensional symmetric complex with underlying chain complex C, there exists a chain homotopy equivalence $g \colon C \to C'$ to a positive n-dimensional complex C' in $bR\text{-CH}_{\text{fgp}}$.

If $(C, [\varphi])$ is a homotopy n-dimensional symmetric complex, then define $[\varphi'] = g^{Q,n}([\varphi])$. We obtain the homotopy equivalence of homotopy n-dimensional complexes

15.4 Forms and Formations

$$g: (C, [\varphi]) \to (C', [\varphi'])$$

with a strictly n-dimensional target, and so the map of the equivalence classes from the left to right is onto.

To show that it is injective we have to work harder. Let

$$g: (C, [\varphi]) \to (C', [\varphi'])$$

be a homotopy equivalence of strictly n-dimensional symmetric complexes. We want to construct a stable isomorphism

$$[f]: (C, [\varphi]) \to (C', [\varphi']).$$

Since g is a homotopy equivalence of finitely generated projective R-chain complexes, we have the short exact sequence of finitely generated projective R-chain complexes

$$0 \longrightarrow C \xrightarrow{i} \text{cyl}(g) \xrightarrow{\text{pr}_0} \text{cone}(g) \longrightarrow 0$$

with $\text{cone}(g)$ contractible and

$$0 \longrightarrow C' \xrightarrow{j} \text{cyl}(g) \xrightarrow{\text{pr}_1} \text{cone}(-\text{id}_C) \longrightarrow 0$$

with $\text{cone}(-\text{id}_C)$ contractible, see Proposition 14.52 (i).

Because of Lemma 14.54 we can find R-chain maps $s: \text{cone}(g) \to \text{cyl}(g)$ with $\text{pr}_0 \circ s = \text{id}$ and $t: \text{cone}(-\text{id}_C) \to \text{cyl}(g)$ with $\text{pr}_1 \circ t = \text{id}$ such that

$$i \oplus s: C \oplus \text{cone}(g) \to \text{cyl}(g);$$
$$j \oplus t: C' \oplus \text{cone}(-\text{id}_C) \to \text{cyl}(g),$$

are isomorphisms of R-chain complexes. Then we obtain an isomorphism

$$\overline{f} = (j \oplus t)^{-1} \circ (i \oplus s): C \oplus \text{cone}(g) \xrightarrow{\cong} C' \oplus \text{cone}(-\text{id}_C),$$

which is almost what we want except that $\text{cone}(g)$ and $\text{cone}(-\text{id}_C)$ may not be n-dimensional, they may have non-trivial modules in degree $(n + 1)$. This will be corrected as follows. Since $\text{cone}(g)$ is positive and contractible, it fulfils part (ii) of Proposition 14.122 for $d = n$. Then part (i) of Proposition 14.122 tells us that we can find an n-dimensional finitely generated projective R-subchain complex $D \subseteq \text{cone}(g)$ such that the inclusion $k: D \to \text{cone}(g)$ is an R-chain homotopy equivalence.

Next, define an n-dimensional finitely generated projective quotient R-chain complex D' of $\text{cone}(-\text{id})$ by

$$D'_r = \begin{cases} C_{n-1} & \text{if } r = n; \\ \text{cone}(-\text{id})_r & \text{if } 0 \le r \le n-1; \\ 0 & \text{otherwise,} \end{cases}$$

where the n-th differential of D' is given by $\begin{pmatrix} -c_{n-1} \\ -\mathrm{id} \end{pmatrix} \colon D'_n = C_{n-1} \to D'_{n-1} = C_{n-2} \oplus C_{n-1}$. The projection $p \colon \mathrm{cone}(-\mathrm{id}_C) \to D'$ is given by

$$p_r = \begin{cases} \begin{pmatrix} \mathrm{id} & -c_n \end{pmatrix} \colon C_{n-1} \oplus C_n \to C_{n-1} & \text{if } r = n; \\ \mathrm{id}_{\mathrm{cone}(-\mathrm{id}_C)_r} & \text{if } 0 \leq r \leq n-1; \\ 0 & \text{otherwise.} \end{cases}$$

Now define the desired R-chain map f to be the composite

$$f \colon C \oplus D \xrightarrow{\mathrm{id}_C \oplus k} C \oplus \mathrm{cone}(g) \xrightarrow{\overline{f}} C' \oplus \mathrm{cone}(-\mathrm{id}_C) \xrightarrow{\mathrm{id}_{C'} \oplus p} C' \oplus D'.$$

It is an R-chain isomorphism since it agrees with $\overline{f}$ in all dimensions except in dimension n and hence f_k is an isomorphism for $k \neq n$. The source and target are n-dimensional, and f induces an isomorphism on homology, as both $i \colon D \to \mathrm{cone}(g)$ and $p \colon \mathrm{cone}(-\mathrm{id}) \to D'$ induce isomorphisms on homology.

The same argument works if one wants to stay in the realm of finitely generated free chain complexes. One has to check that then D_n is stably finitely generated free for $n \geq 1$ and actually finitely generated free for $n = 0$. Thus by adding an elementary free chain complex in the case $n \geq 1$, one can arrange additionally that D is finitely generated free. □

15.4.2 Revision of Forms and Formations

When identifying low-dimensional structured chain complexes with forms and formations, we need to recall and extend some notions from Chapters 8 and 9.

Recall the notions of an ε-symmetric form (P, φ) and an ε-quadratic form (P, ψ) from Definition 8.47 and Definition 8.77, the notions of a standard symmetric hyperbolic form from Example 8.49, and of a standard quadratic hyperbolic form from Example 8.79. Furthermore, recall the notions of a lagrangian and of a metabolic from Remark 8.96 and Lemma 8.95, which says that an ε-quadratic form that admits a lagrangian is isomorphic to the standard ε-quadratic hyperbolic form.

Remark 15.193 (Notation about elements in Q-groups) In the previous chapters we have written (P, φ) for an ϵ-symmetric form and (P, ψ) for an ε-quadratic form. We will do this for the symmetric case in this chapter as well, but in the quadratic case we would like to replace ψ by $[\psi]$. In other words, we will denote elements in $Q_\epsilon(P)$ by $[\psi]$ and call a choice of a representative $\psi \in \hom_R(P, P^*)$ of $[\psi]$ a polarisation. This makes clearer at which places we need to show that the choice of polarisation does not matter.

We also need a symmetric version of Lemma 8.95, which is stated below as Lemma 15.195. Its statement uses the following new notion.

15.4 Forms and Formations

For an ε-symmetric form (P, φ) on $P \in R\text{-MOD}_{\text{fgp}}$, define the *standard ε-symmetric hyperbolic form associated to* (P, φ) by

$$H^{\varepsilon}(P, \varphi) = (P^* \oplus P, \begin{pmatrix} 0 & 1 \\ \varepsilon & \varphi \end{pmatrix}) \in Q^{\varepsilon}(P^* \oplus P)). \qquad (15.194)$$

Lemma 15.195 *Let (P, φ) be a non-singular ε-symmetric form and $L \subset P$ be a sublagrangian. Then L is a direct summand in $L^{\perp}$ and φ induces the structure of a non-singular ε-symmetric form $(L^{\perp}/L, \varphi^{\perp}/\varphi)$ on $L^{\perp}/L$. Moreover, the inclusion $i: L \to P$ extends to an isomorphism of ε-symmetric forms*

$$H^{\varepsilon}(L, \varphi_1) \oplus (L^{\perp}/L, \varphi^{\perp}/\varphi) \xrightarrow{\cong} (P, \varphi)$$

for some ε-symmetric form $\varphi_1 \in Q^{\varepsilon}(L^*)$.

In particular, a non-singular ε-symmetric form (P, φ) is isomorphic to $H^{\varepsilon}(Q, \varphi_1)$ (for some φ_1) if and only if it contains a lagrangian $L \subset P$, which is isomorphic as an R-module to Q.

Remark 15.196 There is a significant difference between Lemma 8.95 and Lemma 15.195. In the quadratic case the standard hyperbolic form depends only on a module, and in the symmetric case the standard hyperbolic form depends on a symmetric form (L^*, φ_1), where an additional parameter φ_1 occurs. This is necessary in view of the example appearing in Remark 8.96. The reader should keep this in mind in the sequel.

Exercise 15.197 Prove Lemma 15.195.

Furthermore, recall the notion of an ε-quadratic formation $(P, [\psi]; F, G)$ from Definition 9.3, the theory around it in Subsection 9.2.1, and in particular Lemma 9.5, which says that any ε-quadratic formation is isomorphic to an ε-quadratic formation of the form $(H_{\varepsilon}(F); F, G)$. In addition we need the notion of an ε-symmetric formation and a corresponding lemma.

Definition 15.198 (ε-symmetric formation) An *ε-symmetric formation*, denoted by $(P, \varphi; F, G)$, consists of a non-singular ε-symmetric form (P, φ) together with an ordered pair of lagrangians F and G.

Here Remark 9.4 also applies, meaning that in some places, e.g., [344], formations, as we defined them, are called non-singular formations and the word "formation" is used for the case where one of F and G is a sublagrangian.

Lemma 15.199 *Every ε-symmetric formation $(P, \varphi; F, G)$ is isomorphic to the formation $(H^{\varepsilon}(F, \nu); F, G)$ for some ε-symmetric form $\nu \in Q^{\varepsilon}(F^*)$ and some lagrangian G.*

Proof. By Lemma 15.195 the symmetric form (P, φ) is isomorphic to the form $H^{\varepsilon}(F^*, \nu)$ for some ε-symmetric form $\nu \in Q^{\varepsilon}(F^*)$. □

Definition 15.200 (Trivial ε-symmetric formation) The *trivial ε-symmetric formation* associated to an ε-symmetric form (F, ν) is the non-singular formation $(H^\varepsilon(F, \nu); F, F^*)$. An ε-symmetric formation $(P, \varphi; F, G)$ is called *trivial* if it is isomorphic to the trivial ε-symmetric formation associated to some ε-symmetric form (F, ν).

Definition 15.201 (Stable isomorphism of ε-symmetric formations) A *stable isomorphism* between two ε-symmetric formations $(P, \varphi; F, G)$ and $(P', \varphi'; F', G')$ is an isomorphism

$$f : (P, \varphi; F, G) \oplus (H^\varepsilon(Q_0, 0); Q_0, Q_0^*) \cong (P', \varphi'; F', G') \oplus (H^\varepsilon(Q_1, 0); Q_1, Q_1^*)$$

for trivial formations $(H^\varepsilon(Q_0, 0); Q_0, Q_0^*)$ and $(H^\varepsilon(Q_1, 0); Q_1, Q_1^*)$.

In this case, $(P, \varphi; F, G)$ and $(P', \varphi'; F', G')$ are called *stably isomorphic* and we write $(P, \varphi; F, G) \cong_s (P', \varphi'; F', G')$.

We emphasise that in the definition above the stabilisation is made via the trivial formations associated to the ε-symmetric form with $\nu = 0$, so they are of a special shape.

Below we will also need material from Section 9.2.2 about presentations of formations and that material will later be extended to define the notion of a split formation.

15.4.3 Complexes versus Forms and Formations

In this section we establish the promised identification of the homotopy classes of low-dimensional structured chain complexes with the (stable) isomorphism classes of forms and (split) formations.

Proposition 15.202 *There are bijections*

$$\left\{ \begin{array}{c} \text{strictly 0-dim. sym.} \\ \text{chain compl.} \end{array} \right\} \Big/ \cong_s \quad \stackrel{\cong}{\longleftrightarrow} \quad \left\{ \begin{array}{c} \varepsilon\text{-symmetric} \\ \text{forms} \end{array} \right\} \Big/ \cong$$

and

$$\left\{ \begin{array}{c} \text{strictly 0-dim. quadr.} \\ \text{chain compl.} \end{array} \right\} \Big/ \cong_s \quad \stackrel{\cong}{\longleftrightarrow} \quad \left\{ \begin{array}{c} \varepsilon\text{-quadratic} \\ \text{forms} \end{array} \right\} \Big/ \cong .$$

Poincaré complexes correspond to non-singular forms in both cases.

Proof. A strictly 0-dimensional chain complex C is obviously just a module C_0. There are obvious identifications $Q^0(C) = Q^\varepsilon(C_0)$ and $Q_0(C) = Q_\varepsilon(C_0)$ between the Q-groups appearing in (8.72), (8.73), and Definition 15.14 for a strictly 0-dimensional chain complex C. For strictly 0-dimensional symmetric or quadratic complexes stable isomorphism is the same as isomorphism since a positive 0-dimensional R-contractible chain complex must be the zero complex. $\square$

15.4 Forms and Formations

Corollary 15.203 *There are bijections*

$$\left\{\begin{array}{c}\text{homotopy 0-dimensional}\\ \varepsilon\text{-symmetric chain compl.}\end{array}\right\}\Big/\simeq \quad\overset{\cong}{\longleftrightarrow}\quad \left\{\begin{array}{c}\varepsilon\text{-symmetric}\\ \text{forms}\end{array}\right\}\Big/\cong$$

and

$$\left\{\begin{array}{c}\text{homotopy 0-dimensional}\\ \varepsilon\text{-quadratic chain compl.}\end{array}\right\}\Big/\simeq \quad\overset{\cong}{\longleftrightarrow}\quad \left\{\begin{array}{c}\varepsilon\text{-quadratic}\\ \text{forms}\end{array}\right\}\Big/\cong .$$

Poincaré complexes correspond to non-singular forms in both cases.

Proof. Combine Proposition 15.192 with Proposition 15.202. □

Before turning to 1-dimensional complexes, we formulate one useful elementary exercise about the behaviour of exact sequences when changing certain signs.

Exercise 15.204 Let

$$0 \longrightarrow A \xrightarrow{\begin{pmatrix}\alpha\\ \beta\end{pmatrix}} B \oplus C \xrightarrow{(\gamma\ \delta)} D \longrightarrow 0$$

be a (split) short exact sequence of R-modules. Then the sequence

$$0 \longrightarrow A \xrightarrow{\begin{pmatrix}\alpha\\ -\varepsilon\beta\end{pmatrix}} B \oplus C \xrightarrow{(\gamma\ -\varepsilon\delta)} D \longrightarrow 0$$

is also (split) short exact.

Proposition 15.205 *There is a bijection*

$$\alpha\colon \left\{\begin{array}{c}\text{strictly 1-dim. } \varepsilon\text{-sym.}\\ \text{Poincaré complexes}\end{array}\right\}\Big/\cong_s \quad\overset{\cong}{\longleftrightarrow}\quad \left\{\begin{array}{c}\varepsilon\text{-symmetric}\\ \text{formations}\end{array}\right\}\Big/\cong_s .$$

Proof. Consider a strictly 1-dimensional ε-symmetric Poincaré complex $(C, [\varphi])$ over R. Choose a polarisation φ. It consists of the following data

$$\begin{array}{ccccc}
0 \longrightarrow & C^0 & \xrightarrow{d^*} & C^1 & \longrightarrow 0\\
& {\varphi_0}\downarrow & {\varphi_1}\swarrow & \downarrow{\widetilde{\varphi}_0} & \\
0 \longrightarrow & C_1 & \xrightarrow{d} & C_0 & \longrightarrow 0
\end{array}$$

satisfying

$$d \circ \varphi_0 = \widetilde{\varphi}_0 \circ d^*;$$
$$d \circ \varphi_1 = -\widetilde{\varphi}_0 - \varepsilon \varphi_0^*;$$
$$\varphi_1 \circ d^* = -\varphi_0 - \varepsilon \widetilde{\varphi}_0^*;$$
$$\varphi_1 = \varepsilon \varphi_1^*.$$

The condition that it is Poincaré means that the mapping cone of the chain map with components φ_0 and $\widetilde{\varphi}_0$:

$$\cdots \longrightarrow 0 \longrightarrow C^0 \xrightarrow{\begin{pmatrix} -d^* \\ \varphi_0 \end{pmatrix}} C^1 \oplus C_1 \xrightarrow{(\widetilde{\varphi}_0 \ d)} C_0 \longrightarrow 0 \longrightarrow \cdots$$

is contractible and hence a split short exact sequence.

The associated ε-symmetric formation is

$$\left(C_1 \oplus C^1, \begin{pmatrix} 0 & 1 \\ \varepsilon & \varphi_1 \end{pmatrix}; C_1, \text{im} \begin{pmatrix} \varphi_0 \\ \varepsilon d^* \end{pmatrix} : C^0 \to C_1 \oplus C^1 \right).$$

Let us verify that this is indeed an ε-symmetric formation. Firstly, observe that the equation with the left-hand side $d \circ \varphi_1$ above implies that

$$\begin{pmatrix} \varphi_0^* & \varepsilon d \end{pmatrix} \begin{pmatrix} 0 & 1 \\ \varepsilon & \varphi_1 \end{pmatrix} = \begin{pmatrix} d & -\varepsilon \widetilde{\varphi}_0 \end{pmatrix}$$

holds. Keeping this in mind, we only need to check that the following sequence is split short exact:

$$0 \longrightarrow C^0 \xrightarrow{\begin{pmatrix} \varphi_0 \\ \varepsilon d^* \end{pmatrix}} C_1 \oplus C^1 \xrightarrow{(d \ -\varepsilon \widetilde{\varphi}_0)} C_0 \longrightarrow 0.$$

This in turn follows from comparing this sequence with the above mapping cone using Exercise 15.204.

Note that the associated ε-symmetric formation depends on a choice of a polarisation φ of $[\varphi]$. The following considerations actually imply that the isomorphism class of the associated ε-symmetric formation is independent of the choice of the polarisation.

Let

$$f \colon (C, [\varphi]) \to (C', [\varphi'])$$

be an isomorphism of strictly 1-dimensional ε-symmetric complexes. Choose polarisations φ for $[\varphi]$ and φ' for $[\varphi']$. Next we show that the associated ε-symmetric formations are isomorphic. Recall that $f^{Q,1} \colon Q^1(C) \to Q^1(C')$ sends $[\varphi]$ to $[\varphi']$. Hence we have an isomorphism of chain complexes

15.4 Forms and Formations

$$\begin{array}{ccc} C_1 & \xrightarrow{d} & C_0 \\ f_1 \downarrow \cong & & \cong \downarrow f_0 \\ C'_1 & \xrightarrow{d'} & C'_0 \end{array}$$

such that

$$f^{\%}(\varphi) - \varphi' = d\chi \in \hom_{\mathbb{Z}/2}(W, (C' \otimes_R C'))_1$$

for some chain $\chi \in \hom_{\mathbb{Z}/2}(W, (C' \otimes_R C'))_2$, which has just a single component $\chi_0 \in (C'_1 \otimes_R C'_1)$ satisfying

$$f_1 \circ \varphi_0 \circ f_0^* - \varphi'_0 = \chi_0 \circ (d')^*;$$
$$f_0 \circ \widetilde{\varphi}_0 \circ f_1^* - \widetilde{\varphi}'_0 = d' \circ \chi_0;$$
$$f_1 \circ \varphi_1 \circ f_1^* - \varphi'_1 = -\chi_0 - \varepsilon\chi_0^*.$$

Using these formulas one can check that the R-module isomorphism

$$h = \begin{pmatrix} f_1 & \varepsilon\chi_0 \circ (f_1^*)^{-1} \\ 0 & (f_1^*)^{-1} \end{pmatrix} : C_1 \oplus C^1 \to C'_1 \oplus (C')^1$$

defines an isomorphism of the ε-symmetric formations associated to (C, φ) and (C', φ').

If $(C, [\varphi])$ is of the form $(C, [0])$ with contractible C, then the associated ε-symmetric formation is the trivial ε-symmetric formation $(H^\varepsilon(C_1, 0); C_1, C_1^*)$ in the sense of Definition 15.200.

To summarise, we have constructed a well-defined map α announced in the statement of Proposition 15.205.

Next we construct the inverse of α. By Lemma 15.199 any ε-symmetric formation is isomorphic to an ε-symmetric formation of the special shape

$$(F \oplus F^*, \begin{pmatrix} 0 & 1 \\ \varepsilon & \lambda \end{pmatrix}; F, G).$$

Write

$$\begin{pmatrix} \gamma \\ \delta \end{pmatrix} : G \to F \oplus F^*$$

for the inclusion of the lagrangian G. We associate to it a polarised strictly 1-dimensional ε-symmetric Poincaré complex as follows:

$$d = \varepsilon\delta^* : C_1 = F \to C_0 = G^*;$$
$$\varphi_0 = \gamma : C^0 = G \to C_1 = F;$$
$$\widetilde{\varphi}_0 = -\varepsilon\gamma^* - \varepsilon\delta^* \circ \lambda : C^1 = F^* \to C_0 = G^*;$$
$$\varphi_1 = \lambda : C^1 = F^* \to C_1 = F.$$

It is Poincaré since the mapping cone of the chain map with the components φ_0 and $\widetilde{\varphi}_0$ is a split short exact sequence by comparing it with the split short exact sequence, which verifies that G is a lagrangian, as in the previous argument.

For an isomorphism of ε-symmetric formations

$$h \colon (H(F, \lambda); F, G) \to (H(F', \lambda'); F', G')$$

define

$$\alpha = h| \colon F \to F' \quad \text{and} \quad \beta = h| \colon G \to G'.$$

Then we obtain an isomorphism of the associated 1-dimensional ε-symmetric complexes with components

$$f_1 = \alpha \quad \text{and} \quad f_0 = (\beta^*)^{-1}.$$

If h is given by the matrix $\begin{pmatrix} \alpha & \omega \\ 0 & (\alpha^*)^{-1} \end{pmatrix} \colon F \oplus F^* \to F' \oplus (F')^*$, then the required chain χ_0 is given by $\omega\alpha^*$.

To summarise, we have constructed a well-defined map, which turns out to be an inverse of α. This is seen by a direct verification. This finishes the proof of Proposition 15.205. □

Corollary 15.206 *There is a bijection*

$$\left\{ \begin{array}{c} \text{homotopy 1-dim. } \varepsilon\text{-sym.} \\ \text{Poincaré complexes} \end{array} \right\} \Big/ \simeq \quad \overset{\cong}{\longleftrightarrow} \quad \left\{ \begin{array}{c} \varepsilon\text{-symmetric} \\ \text{formations} \end{array} \right\} \Big/ \cong_s .$$

Proof. Combine Proposition 15.192 with Proposition 15.206. □

Remark 15.207 In [344, Proposition 2.3] a version of the above corollary is stated and proved where the correspondence is between 1-dimensional complexes that are only connected in the sense of Definition 15.159 and not necessarily Poincaré, and a version of formations where G is possibly only a sublagrangian, cf. Remark 9.4. As we will not need that generality, we confine ourselves to the above version, but the interested reader might either look at [344] or consider the straightforward improvement to this more general situation for himself.

The quadratic analogue of Proposition 15.206 is more complicated. To a 1-dimensional ε-quadratic complex we can associate an ε-quadratic formation, but in this process we lose information since the component ψ_1 of the quadratic structure does not appear in the resulting formation. Hence the notion of a formation has to be replaced by the notion of a *split formation*. As mentioned before, this difference vanishes when we pass to the L-groups.

Recall from Definition 9.3 that an ε-quadratic formation $(P, [\psi]; F, G)$ consists of an ε-quadratic form $(P, [\psi])$ together with an ordered pair of lagrangians F and G. (Recall also the change of notation explained in Remark 15.193.) By Lemma 9.5 any ε-quadratic formation $(P, [\psi]; F, G)$ is isomorphic to a formation of

15.4 Forms and Formations

the shape $(H_\varepsilon(F); F, G)$ for some lagrangian G. As noted in Section 9.2.2, see in particular Lemma 9.20, it is determined by the morphisms γ and δ, which make up the inclusion

$$\begin{pmatrix} \gamma \\ \delta \end{pmatrix} : G \to F \oplus F^*,$$

which is split injective by definition. There are two conditions. Firstly, the restriction of the standard ε-hyperbolic form on $F \oplus F^*$ to G has to be zero in the group $Q_\varepsilon(G)$. Secondly, for the symmetrisation we have $G = G^\perp$. This motivates the following definition, in which the first condition corresponds to the property (i) and the second condition to the property (ii).

Definition 15.208 (Split formations) A *split ε-quadratic formation* is a 5-tuple $(F, G, \gamma, \delta, [\theta])$ where

(a) $F, G \in R\text{-MOD}_{\text{fgp}}$;
(b) $\gamma \colon G \to F$, $\delta \colon G \to F^*$ are R-morphisms;
(c) $[\theta] \in Q_{-\varepsilon}(G)$,

satisfying

(i) $\gamma^* \circ \delta = \theta - \varepsilon \theta^* \colon G \to G^*$;
(ii) the sequence

$$0 \longrightarrow G \xrightarrow{\begin{pmatrix} \gamma \\ \delta \end{pmatrix}} F \oplus F^* \xrightarrow{(\delta^* \ \varepsilon \gamma^*)} G^* \longrightarrow 0$$

is exact.

The $(-\varepsilon)$-quadratic form $(G, [\theta])$ is called a *hessian* for G. So a split ε-quadratic formation is an ε-quadratic formation in the standard form together with a choice of a hessian. In other words, a split ε-quadratic formation is an ε-quadratic formation together with a *reason* why it is a formation (i.e., $(G, [\theta])$ is such a reason).

Remark 15.209 (A hessian always exists) Given any ε-quadratic formation $(P, [\psi]; F, G)$, a hessian $[\theta]$ such that $(F, G, \gamma, \delta, [\theta]))$ is a split ε-quadratic formation always exists since its existence is equivalent to the condition that G is a lagrangian. However, there may be many such choices, which is the reason why there is a difference between the notions of an ε-quadratic formation and a split ε-quadratic formation.

Definition 15.210 An *isomorphism* of split ε-quadratic formations

$$(\alpha, \beta, \xi) \colon (F, G, \gamma, \delta, [\theta]) \to (F', G', \gamma', \delta', [\theta'])$$

consists of isomorphisms $\alpha \colon F \to F'$, $\beta \colon G \to G'$ and an $(-\varepsilon)$-quadratic form (F^*, ξ) such that the diagram

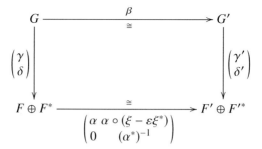

commutes and we get in $Q_{-\varepsilon}(G)$

$$Q_{-\varepsilon}(\beta)([\theta']) = [\beta^* \circ \theta' \circ \beta] = [\theta] + [\delta^* \circ \xi^* \circ \delta] = [\theta] + Q_{-\varepsilon}(\delta)([\xi^*]).$$

Definition 15.211

(i) A split ε-quadratic formation $(F, G, \gamma, \delta, [\theta])$ is *trivial* if it is isomorphic to $(F, F^*, 0, 1, [0])$;

(ii) A *stable isomorphism* of split ε-quadratic formations is an isomorphism of the type

$$(\alpha, \beta, \xi) \colon (F, G, \gamma, \delta, [\theta]) \oplus (H, H^*, 0, 1, [0])$$
$$\to (F', G', \gamma', \delta', [\theta']) \oplus (H', H'^*, 0, 1, [0])$$

for some $H, H' \in R\text{-MOD}_{\text{fgp}}$.

Remark 15.212 The condition in part (ii) implies that

$$f = \begin{pmatrix} \alpha & \alpha \circ (\xi - \varepsilon \xi^*) \\ 0 & (\alpha^*)^{-1} \end{pmatrix} \colon H_\varepsilon(F) \to H_\varepsilon(F')$$

is an isomorphism of hyperbolic ε-quadratic forms such that $f(F) = F'$ and $f(G) = G'$.

Proposition 15.213 *A stable isomorphism of split ε-quadratic formations determines a stable isomorphism of the underlying ε-quadratic formations. Conversely, given an isomorphism of ε-quadratic formations such that the source is a split ε-quadratic formation, then one can extend the structure of an ε-quadratic formation on the target to the structure of a split ε-quadratic formation for which the given isomorphism is an isomorphism of split ε-quadratic formations.*

Proof. The first statement is obvious. For the second statement let

$$f \colon (H_\varepsilon(F); F, G) \to (H_\varepsilon(F'); F', G')$$

be an isomorphism of ε-quadratic formations. We know from Lemma 9.24 that f can be expressed as

15.4 Forms and Formations

$$f = \begin{pmatrix} \alpha \, \alpha \circ (\xi - \varepsilon \xi^*) \\ 0 & (\alpha^*)^{-1} \end{pmatrix} : F \oplus F^* \to F' \oplus (F')^*$$

for some R-module morphism $\xi \colon F^* \to F$ where $\alpha = f|_F$. Let β denote $f|_G$, choose a representative $\theta' \colon G' \to (G')^*$ of a hessian $[\theta']$, and define a map $\theta = \beta^* \circ \theta' \circ \beta - \delta^* \circ \xi^* \circ \delta$. Then $\theta \colon G \to G^*$ defines a hessian $[\theta]$ for G. $\square$

Proposition 15.214 *There is a one-to-one correspondence*

$$\beta \colon \left\{ \begin{array}{c} \text{strictly 1-dim. } \varepsilon\text{-quad.} \\ \text{Poincaré complexes} \end{array} \right\} \Big/ \cong_s \;\overset{\cong}{\longleftrightarrow}\; \left\{ \begin{array}{c} \text{split } \varepsilon\text{-quadratic} \\ \text{formations} \end{array} \right\} \Big/ \cong_s .$$

Proof. Consider a strictly 1-dimensional ε-quadratic Poincaré complex $(C, [\psi])$. Choose a polarisation ψ of $[\psi]$. It consists of the following data

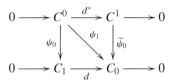

satisfying

$$d \circ \psi_0 - \widetilde{\psi}_0 \circ d^* = \varepsilon \psi_1^* - \psi_1.$$

The condition that it is Poincaré means that the mapping cone of the zeroth component of the symmetrisation of ψ is contractible. The zeroth component of the symmetrisation map itself has two components:

$$((1+T)(\psi)_0)_1 = \psi_0 - \varepsilon \widetilde{\psi}_0^* \quad \text{and} \quad ((1+T)(\psi)_0)_0 = \widetilde{\psi}_0 - \varepsilon \psi_0^*.$$

Its mapping cone is

$$\cdots \longrightarrow 0 \longrightarrow C^0 \xrightarrow{\begin{pmatrix} -d^* \\ \psi_0 - \varepsilon \widetilde{\psi}_0^* \end{pmatrix}} C^1 \oplus C_1 \xrightarrow{\left(\widetilde{\psi}_0 - \varepsilon \psi_0^* \; d \right)} C_0 \longrightarrow 0 \longrightarrow \cdots,$$

and hence this is a split short exact sequence. The associated ε-quadratic formation is

$$\left(C_1 \oplus C^1, \begin{pmatrix} 0 & 1 \\ 0 & 0 \end{pmatrix} ; C_1, \operatorname{im} \begin{pmatrix} \psi_0 - \varepsilon \widetilde{\psi}_0^* \\ \varepsilon d^* \end{pmatrix} : C^0 \to C_1 \oplus C^1 \right).$$

By setting

$$\theta = \psi_1 - d \circ \psi_0,$$

we obtain also the associated split ε-quadratic formation

$$(C_1, C^0; \psi_0 - \varepsilon \widetilde{\psi}_0^*, \varepsilon d^*, [\psi_1 - d\psi_0]).$$

Note that it depends on the choice of the polarisation and ψ_1.

To check that it is indeed a split ε-quadratic formation, one needs to verify conditions (i) and (ii) from Definition 15.208.

For condition (i) we calculate

$$\gamma^* \circ \delta = (\psi_0 - \varepsilon\widetilde{\psi_0^*})^* \circ \varepsilon d^* = \varepsilon\psi_0^* \circ d^* - \widetilde{\psi_0} \circ d^*$$
$$= \psi_1 - \varepsilon\psi_1^* + \varepsilon d \circ \widetilde{\psi_0^*} - \widetilde{\psi_0} \circ d^* = (\psi_1 - \widetilde{\psi_0} \circ d^*) - \varepsilon(\psi_1^* - d \circ \widetilde{\psi_0^*}) = \theta - \varepsilon\theta^*,$$

where for the third equation we used the formula from the beginning of the proof that says that ψ is a polarisation of an ε-quadratic structure.

For condition (ii) we need that

$$0 \longrightarrow C^0 \xrightarrow{\begin{pmatrix} \psi_0 - \varepsilon\widetilde{\psi_0^*} \\ \varepsilon d^* \end{pmatrix}} C_1 \oplus C^1 \xrightarrow{\begin{pmatrix} d\ \psi_0^* - \varepsilon\widetilde{\psi_0} \end{pmatrix}} C_0 \longrightarrow 0$$

is a split short exact sequence. This is proved by comparing it with the mapping cone of the symmetrisation map above and using Exercise 15.204, similarly as in the symmetric case.

As already stated above, the associated split ε-quadratic formation depends on a choice of a polarisation ψ of $[\psi]$. The following considerations actually imply that the isomorphism class of the associated split ε-quadratic formation is independent of the choice of the polarisation.

Let

$$f \colon (C, [\psi]) \to (C', [\psi'])$$

be an isomorphism of connected strictly 1-dimensional quadratic complexes. Choose polarisations ψ of $[\psi]$ and ψ' of $[\psi']$. Then f yields an isomorphism of chain complexes

$$\begin{array}{ccc} C_1 & \xrightarrow{d} & C_0 \\ f_1 \downarrow \cong & & \cong \downarrow f_0 \\ C'_1 & \xrightarrow{d'} & C'_0 \end{array}$$

such that

$$f_{\%}(\psi) - \psi' = d_{W_{\%}(C')}(\chi) \in (W \otimes_{\mathbb{Z}/2} (C' \otimes_R C'))_1$$

for some chain $\chi \in (W \otimes_{\mathbb{Z}/2} (C' \otimes_R C'))_2$ with components

$$\chi_0 \colon (C')^1 \to C'_1 \quad \chi_1 \colon (C')^0 \to C'_1 \quad \widetilde{\chi_1} \colon (C')^1 \to C'_0 \quad \chi_2 \colon (C')_0 \to C'_0$$

satisfying

15.4 Forms and Formations

$$f_1 \circ \psi_0 \circ f_0^* - \psi_0' = \chi_0 \circ (d')^* + \chi_1 + \varepsilon \widetilde{\chi}_1^*;$$
$$f_0 \circ \widetilde{\psi}_0 \circ f_1^* - \widetilde{\psi}_0' = d' \circ \chi_0 + \widetilde{\chi}_1 + \varepsilon \chi_1^*;$$
$$f_0 \circ \psi_1 \circ f_0^* - \psi_1' = -d' \circ \chi_1 + \widetilde{\chi}_1 \circ (d')^* + \chi_2 + \varepsilon \chi_2^*.$$

We obtain an isomorphism of the associated split ε-quadratic formations

$$(\alpha, \beta, \xi) \colon (C_1, C^0; \psi_0 - \varepsilon \widetilde{\psi}_0^*, \varepsilon d^*, [\psi_1 - d\psi_0])$$
$$\xrightarrow{\cong} (C_1', (C')^0; \psi_0' - \varepsilon \widetilde{\psi'}_0^*, \varepsilon (d')^*, [\psi_1' - d'\psi_0'])$$

by the formulas

$$\alpha = f_1; \quad \beta = (f_0^*)^{-1}; \quad \xi = f_1^{-1} \chi_0 (f_1^*)^{-1}.$$

If $(C, [\psi])$ is of the form $(C, [0])$ with contractible C, then the associated ε-quadratic formation is the trivial ε-quadratic formation $(H_\varepsilon(C_1); C_1, C_1^*)$ in the sense of Definition 9.7.

To summarise, we have constructed a well-defined map β as announced in the statement of Proposition 15.214.

To prove the bijectivity of β we construct the inverse of β next.

Let $(F, G, \gamma, \delta, [\theta])$ be a split ε-quadratic formation. Choose a polarisation θ of $[\theta]$. Our candidate for a preimage under β is the (polarised) connected ε-quadratic strictly 1-dimensional complex defined by

$$d = \varepsilon \delta^* \colon C_1 = F \to C_0 = G^*;$$
$$\psi_0 = \gamma \colon C^0 = G \to C_1 = F;$$
$$\widetilde{\psi}_0 = 0 \colon C^1 = F^* \to C_0 = G^*;$$
$$\psi_1 = \theta \colon C^0 = G \to C_0 = G^*.$$

That $(C, [\psi])$ is an ε-quadratic complex follows from directly calculating that the necessary equation connecting the above maps holds. That it is Poincaré follows from comparing the mapping cone of the symmetrisation map with the split short exact sequence of condition (ii) from Definition 15.208 and thus concluding that it is contractible.

The isomorphism class of the associated 1-dimensional ε-quadratic complex does not depend on the choice of the representative of $[\theta]$. Namely, any two choices are related by $\theta' - \theta = \tau + \tau^*$ for some $\tau \in \hom_R(G, G^*)$, and we can pick the identity as the isomorphism between the associated 1-dimensional quadratic complexes using the chain $\chi \in (W \otimes_{\mathbb{Z}/2} (C' \otimes_R C'))_2$ with components $\chi_0 = 0 = \chi_1 = \widetilde{\chi}_0$ and $\chi_2 = \tau$.

For an isomorphism of split ε-quadratic formations

$$(\alpha, \beta, \xi) \colon (F, G, \gamma, \delta, [\theta]) \to (F', G', \gamma', \delta', [\theta'])$$

we obtain an isomorphism of the associated 1-dimensional ε-quadratic complexes with components

$$f_1 = \alpha \quad \text{and} \quad f_0 = (\beta^*)^{-1}.$$

If $\tau \in \hom(G, G^*)$ satisfies

$$\beta^* \circ \theta' \circ \beta = \theta + \delta^* \circ \xi^* \circ \delta + \tau + \varepsilon\tau^* \in \hom(G, G^*),$$

then the required chain $\chi \in (W \otimes_{\mathbb{Z}/2} (C' \otimes_R C'))_2$ has components

$$\chi_0 = -\alpha \circ \xi^* \circ \alpha^*, \ \chi_1 = 0, \ \widetilde{\chi_1} = \varepsilon(\beta^*)^{-1} \circ \delta^* \circ \xi^* \circ \alpha^*, \ \chi_2 = (\beta^*)^{-1} \circ \tau \circ (\beta)^{-1}.$$

To summarise we have shown that there is a well-defined map that is a candidate for an inverse of β. To show that the two composites are the identity, we proceed as follows.

If we start with a 1-dimensional ε-quadratic complex $(C, [\psi])$ and associate to it a split ε-quadratic formation $(F, G, \gamma, \delta, [\theta])$ and to this we associate a 1-dimensional ε-quadratic complex $(C', [\psi'])$ by the above formulas, we observe that the complexes $(C, [\psi])$ and $(C', [\psi'])$ are isomorphic by taking $f_1 = \mathrm{id}$, $f_0 = \mathrm{id}$ and choosing the chain χ to have components $\chi_0 = 0$, $\chi_1 = 0$, $\widetilde{\chi_1} = \widetilde{\psi}_0$ and $\chi_2 = \psi_1$.

If we start with a split ε-quadratic formation $(F, G, \gamma, \delta, [\theta])$, associate to it a 1-dimensional ε-quadratic complex $(C, [\psi])$ and to this complex a split ε-quadratic formation, we directly see that the two split ε-quadratic formations are isomorphic. This finishes the proof of Proposition 15.214. □

Corollary 15.215 *There is a one-to-one correspondence*

$$\left\{ \begin{array}{c} \text{homotopy 1-dim. } \varepsilon\text{-quad.} \\ \text{Poincaré compl.} \end{array} \right\} \Big/ \simeq \ \stackrel{\cong}{\longleftrightarrow} \ \left\{ \begin{array}{c} \text{split } \varepsilon\text{-quadratic} \\ \text{formations} \end{array} \right\} \Big/ \cong_s .$$

Proof. Combine Proposition 15.192 with Proposition 15.214. □

The difference between formations and split formations is the choice of a hessian, which is in general not unique. Such choices occur for example in the proof of Proposition 15.213. It will be shown in Section 15.6 that these choices do not matter when we pass to the L-groups.

Remark 15.216 In [344, Proposition 2.4] a version of the above corollary is stated and proved where the correspondence is between 1-dimensional complexes that are only connected in the sense of Definition 15.159 and not necessarily Poincaré and a version of formations where G is possibly only a sublagrangian, cf. Remark 9.4. As we will not need this generality, we confine ourselves to the above version, but the interested reader might either look at [344] or consider the straightforward improvement to this more general situation for himself.

15.5 *L*-Groups in Terms of Chain Complexes

Recall that in Chapters 8 and 9 the *L*-groups were defined as the Grothendieck groups of quadratic forms or formations modulo the equivalence relation given by dividing out the hyperbolic forms or trivial and boundary formations. Now forms and formations will be replaced by structured chain complexes and these equivalence relations will be replaced by the relation of cobordism of structured chain complexes. The definition will be the same regardless of the parity of *n* and hence the problem (i) from the introduction to this chapter will be solved.

A cobordism of structured chain complexes is a notion that imitates the notion of a cobordism of manifolds just as a symmetric Poincaré complex imitates the notion of a manifold. One may ask, why should one expect the *L*-groups to be cobordism groups. An answer to this question was already provided by Chapter 13, where an interpretation of the *L*-groups was given as certain quite complicated geometric cobordism groups. A benefit of the chain complex treatment is that, once the definitions of structures on chain complexes and relative versions have been set up, the definition of the chain complex version of *L*-groups as cobordism groups is simpler than the definition in Chapter 13. In particular, we do not need to include special conditions on the boundary as in the definition of $\mathcal{L}_n^h$ and $\mathcal{L}_n^s$ in Subsection 13.4.1 (the condition on ∂f and $\partial_2 F$ in the last lines of the respective paragraphs).

Cobordism of structured chain complexes is introduced in Definition 15.281. It is proved in Proposition 15.283 that it gives an equivalence relation. This enables the construction of the chain complex *L*-groups in Definition 15.301, which is the main goal of this section. Moreover, we introduce many tools that will be needed when working with these concepts in later sections, especially the algebraic Thom Construction in Section 15.5.3, the algebraic boundary in Section 15.5.4, and the algebraic surgery in Section 15.5.5.

15.5.1 Pairs

A geometric cobordism is a special case of a pair of Poincaré complexes and in particular of topological spaces. So in order to define its algebraic analogue, we first need to present relative versions of the structured chain complexes from Section 15.3.

Recall from Definition 15.12 that, given a chain map $f \colon C \to D$, we have induced chain maps $f^{\%} \colon W^{\%}(C) \to W^{\%}(D)$ and $f_{\%} \colon W_{\%}(C) \to W_{\%}(D)$.

Definition 15.217 (Relative *Q*-groups) Let $f \colon C \to D$ be an *R*-chain map of chain complexes in h*R*-CH$_{\text{fgp}}$. Define

$$Q^{n+1}(f) = H_{n+1}(\text{cone}(f^{\%}));$$
$$Q_{n+1}(f) = H_{n+1}(\text{cone}(f_{\%})).$$

Definition 15.218 (Symmetric pair, Quadratic pair) Consider $n \in \mathbb{Z}$.

An $(n+1)$-dimensional *symmetric pair* $(f\colon C \to D, [\varphi, \delta\varphi])$ is a chain map $f\colon C \to D$ in bR-CH$_{\text{fgp}}$ together with an element $[\varphi, \delta\varphi] \in Q^{n+1}(f)$.

An $(n+1)$-dimensional *quadratic pair* $(f\colon C \to D, [\psi, \delta\psi])$ is a chain map $f\colon C \to D$ in bR-CH$_{\text{fgp}}$ together with an element $[\psi, \delta\psi] \in Q_{n+1}(f)$.

A *polarisation* is a choice of a cycle $(\varphi, \delta\varphi)$ in cone$(f^{\%})_{n+1}$ or of a cycle $(\psi, \delta\psi)$ in cone$(f_{\%})_{n+1}$ respectively, representing $[\varphi, \delta\varphi] \in Q^{n+1}(f)$ or $[\psi, \delta\psi] \in Q_{n+1}(f)$ respectively.

If $C = 0$, the notion of a pair reduces to the previous notion of a symmetric $(n+1)$-dimensional chain complex $(D, [\delta\varphi])$ using the obvious identification cone$(f^{\%}) = W^{\%}(D)$, and analogously in the quadratic case.

We warn the reader that the mapping cones cone$(f^{\%})$ and cone$(f_{\%})$ are in general not chain homotopy equivalent to the complexes $W^{\%}(\text{cone}(f))$ and $W_{\%}(\text{cone}(f))$. We already know this from Proposition 15.119 since the suspension map $S\colon \Sigma W^{\%}(C) \to W^{\%}(\Sigma C)$, which can be viewed as a map cone$(0^{\%}) \to W^{\%}(\text{cone}(0))$ for $0\colon C \to 0$, is not a chain homotopy equivalence for symmetric structures, and the analogous statement is true for quadratic structures. A full analysis of the difference will be provided in Proposition 15.340. Moreover, with the actual definition above we get obvious natural long exact sequences of Q-groups

$$\cdots \xrightarrow{\partial^{n+1}} Q^n(C) \xrightarrow{f^{Q,n}} Q^n(D) \to Q^n(f)$$
$$\xrightarrow{\partial^n} Q^{n-1}(C) \xrightarrow{f^{Q,n-1}} Q^{n-1}(D) \to \cdots \quad (15.219)$$

and

$$\cdots \xrightarrow{\partial_{n+1}} Q_n(C) \xrightarrow{f_{Q,n}} Q_n(D) \to Q_n(f)$$
$$\xrightarrow{\partial_n} Q_{n-1}(C) \xrightarrow{f_{Q,n-1}} Q_{n-1}(D) \to \cdots. \quad (15.220)$$

Definition 15.221 (Boundary of a structured pair)

For an $(n+1)$-dimensional *symmetric pair* given by $f\colon C \to D$ and $[\varphi, \delta\varphi] \in Q^{n+1}(f)$, its *boundary* is the n-dimensional symmetric chain complex $(C, \partial^{n+1}([\varphi, \delta\varphi])) = (C, [\varphi])$.

For an $(n+1)$-dimensional *quadratic pair* given by $f\colon C \to D$ and $[\psi, \delta\psi] \in Q_{n+1}(f)$, its boundary is the n-dimensional quadratic chain complex $(C, \partial_{n+1}([\psi, \delta\psi])) = (C, [\psi])$.

Consider a polarisation $(\varphi, \delta\varphi)$ of the $(n+1)$-dimensional symmetric pair $(f\colon C \to D, [\varphi, \delta\varphi])$ and a polarisation $(\psi, \delta\psi)$ of the $(n+1)$-dimensional quadratic pair $(f\colon C \to D, [\psi, \delta\psi])$. Then

$$(\varphi, \delta\varphi) \in \text{cone}(f^{\%})_{n+1} = W^{\%}(C)_n \oplus W^{\%}(D)_{n+1};$$
$$(\psi, \delta\psi) \in \text{cone}(f_{\%})_{n+1} = W_{\%}(C)_n \oplus W_{\%}(D)_{n+1}.$$

15.5 L-Groups in Terms of Chain Complexes

A polarisation for the boundary of $(f\colon C \to D, [\varphi, \delta\varphi])$ is given by (C, φ) since φ is the image of $(\varphi, \delta\varphi)$ under the canonical chain map $\mathrm{cone}(f^{\%}) \to \Sigma W^{\%}(C)$ of (14.46). A polarisation for the boundary of $(f\colon C \to D, [\psi, \delta\psi])$ is given by (C, ψ) since ψ is the image of $(\psi, \delta\psi)$ under the canonical chain map $\mathrm{cone}(f_{\%}) \to \Sigma W_{\%}(C)$ of (14.46). The cycle condition for $(\varphi, \delta\varphi)$ translates into the relation between $\delta\varphi$ and φ via the equation $d(\delta\varphi) = -f^{\%}(\varphi)$. In other words, we can think of $\delta\varphi$ as an explicit reason why the pushforward of φ with $f^{\%}$ is nullhomologous. Similarly in the quadratic case $d(\delta\psi) = -f_{\%}(\psi)$ enables us to see $\delta\psi$ as an explicit reason why the pushforward of ψ with $f_{\%}$ is nullhomologous.

Remark 15.222 We follow the notation of [344] here. We note that the symbol $\delta\varphi$ is not to be read as δ of φ since the two chains φ and $\delta\varphi$ in the chain complex $\mathrm{cone}(f^{\%})$ are a priori not related in any sense. However, if we have a symmetric pair, that means that the pair $(\varphi, \delta\varphi)$ is a cycle, then as explained above, $\delta\varphi$ can be viewed as a homology; this justifies the notation.

Remark 15.223 (Structures in terms of components relative case) When we write out the structures on $f\colon C \to D$ from Definition 15.218 in terms of their components, we obtain the following equations.

Let $(\varphi, \delta\varphi) \in \mathrm{cone}(f^{\%})_{n+1}$ be a polarisation of an $(n+1)$-dimensional symmetric pair $(f\colon C \to D, [\varphi, \delta\varphi])$. Denoting $\varphi(1_s)$ and $\delta\varphi(1_s)$ respectively by φ_s and $\delta\varphi_s$, with $1_s \in W_s$, as before, $(\varphi, \delta\varphi)$ consists of the collections $\{\varphi_s \in (C \otimes_R C)_{n+s} \mid s \in \mathbb{N}\}$ and $\{\delta\varphi_s \in (D \otimes_R D)_{n+1+s} \mid s \in \mathbb{N}\}$ of chains satisfying (15.21) and

$$d_{D \otimes_R D}(\delta\varphi_s) = (-1)^{n+1}(\delta\varphi_{s-1} + (-1)^s T(\delta\varphi_{s-1})) - (f \otimes f)(\varphi_s) \qquad (15.224)$$

in $(D \otimes_R D)_{n+s}$. These describe $\delta\varphi_0$ as an explicit reason why $(f \otimes f)(\varphi_0)$ is homologous to zero and each $\delta\varphi_s$ for $s \geq 1$ as an explicit reason why $(-1)^{n+1}(\delta\varphi_{s-1} + (-1)^s T(\delta\varphi_{s-1}))$ and $(f \otimes f)(\varphi_s)$ are homologous.

Let $(\psi, \delta\psi) \in \mathrm{cone}(f_{\%})_{n+1}$ be a polarisation of an $(n+1)$-dimensional quadratic pair $(f\colon C \to D, (\psi, \delta\psi))$. If we write out $\psi = \Sigma_{s \geq 0}(1_s \otimes \psi_s)$ and $\delta\psi = \Sigma_{s \geq 0}(1_s \otimes \delta\psi_s)$, then $(\psi, \delta\psi)$ consists of the collections of chains $\{\psi_s \in (C \otimes_R C)_{n-s} \mid s \in \mathbb{N}\}$ and $\{\delta\psi_s \in (D \otimes_R D)_{n+1+s} \mid s \in \mathbb{N}\}$ satisfying (15.22) and

$$d_{D \otimes_R D}(\delta\psi_s) = (T(\delta\psi_{s+1}) + (-1)^{s+1}\delta\psi_{s+1}) - (f \otimes f)(\psi_s) \qquad (15.225)$$

in $(D \otimes_R D)_{n+s}$. These describe each $\delta\psi_s$ for $s \geq 0$ as an explicit reason why $T(\delta\psi_{s+1}) + (-1)^{s+1}\delta\psi_{s+1}$ and $(f \otimes f)(\psi_s)$ are homologous.

Remark 15.226 (Structures in terms of components as R-module morphisms – relative case) Just as in the absolute case when working in the category $\mathrm{b}R\text{-}\mathrm{CH}_{\mathrm{fgp}}$, the chains $\varphi_s, \delta\varphi_s$, or $\psi_s, \delta\psi_s$, which constitute a symmetric or a quadratic pair, can be viewed as a collection of maps between chain complexes satisfying identities that look as follows. Let $f\colon C \to D$ be an R-chain map of chain complexes in $\mathrm{b}R\text{-}\mathrm{CH}_{\mathrm{fgp}}$.

A cycle $(\varphi, \delta\varphi) \in \mathrm{cone}(f^{\%})_{n+1}$ is a collection $\{\varphi_s\colon C^{n+s-*} \to C \mid s \in \mathbb{N}\}$ of maps satisfying (15.27) and a collection $\{\delta\varphi_s\colon D^{n+1+s-*} \to D \mid s \in \mathbb{N}\}$ of maps satisfying

$$d_D \circ \delta\varphi_s - (-1)^{n+1+s}\delta\varphi_s \circ d_{D^{-*}} =$$
$$(-1)^{n+1}(\delta\varphi_{s-1} + (-1)^s T(\delta\varphi)_{s-1}) - (f \circ (\varphi_s) \circ f^*) \colon D^{n+s-*} \to D. \quad (15.227)$$

These identities describe $\delta\varphi_0$ as a chain nullhomotopy of $f \circ (\varphi_0) \circ f^*$ and each $\delta\varphi_s$ for $s \geq 1$ as a chain homotopy between $\delta\varphi_{s-1} + (-1)^s T(\delta\varphi)_{s-1}$ and $f \circ (\varphi_s) \circ f^*$ (or their negatives).

A cycle $(\psi, \delta\psi) \in \mathrm{cone}(f_{\%})_{n+1}$ is a collection $\{\psi_s \colon C^{n-s-*} \to C \mid s \in \mathbb{N}\}$ of maps satisfying (15.29) and a collection $\{\delta\psi_s \colon D^{n+1-s-*} \to D \mid s \in \mathbb{N}\}$ of maps satisfying

$$d_D \circ \delta\psi_s - (-1)^{n+1-s}\delta\psi_s \circ d_{D^{-*}} =$$
$$(T(\delta\psi_{s+1}) + (-1)^{s+1}\delta\psi_{s+1}) - (f \circ (\psi_s) \circ f^*) \colon D^{n-s-*} \to D. \quad (15.228)$$

These identities describe $T(\delta\psi)_{s+1} + (-1)^{s+1}\delta\psi_{s+1} - (f \circ (\psi_s) \circ f^*)$ as a measure of failure for $\delta\psi_s$ to be chain map of a certain degree. Another way of interpreting them is to observe that, since C and D are bounded, there exists an $s' \geq 0$ such that $\psi_s = 0$ and $\delta\psi_s = 0$ for each $s \geq s'$. So there will be an $s'' \geq 0$ such that $\delta\psi_{s''}$ is a chain map of appropriate degree and $\delta\psi_{s''-1}$ is a chain homotopy between $(T(\delta\psi_{s''}) + (-1)^{s''}\delta\psi_{s''})$ and $(f \circ (\psi_{s''}) \circ f^*)$, and $\delta\psi_s$ for $s < s''$ are higher homotopies.

Note that since C and D are in bR-$\mathrm{CH}_{\mathrm{fgp}}$, only finitely many of the components are non-zero.

Exercise 15.229 Show that $\mathrm{cone}(f^{\%}) = \hom_{\mathbb{Z}[\mathbb{Z}/2]}(W, \mathrm{cone}(f \otimes f))$.

Definition 15.230 (Maps of structured pairs) Let C, C', D, D' be chain complexes in bR-$\mathrm{CH}_{\mathrm{fgp}}$.

(i) A *map* of $(n+1)$-dimensional symmetric pairs

$$(g, h) \colon (f \colon C \to D, [\varphi, \delta\varphi]) \to (f' \colon C' \to D', [\varphi', \delta\varphi'])$$

consists of R-chain maps $g \colon C \to C'$ and $h \colon D \to D'$ such that $f' \circ g = h \circ f$ and

$$z([\varphi, \delta\varphi]) = [\varphi', \delta\varphi'] \in Q^{n+1}(f')$$

hold, where $z \colon Q^{n+1}(f) \to Q^{n+1}(f')$ is the map induced by g and h;

(ii) A *map* of $(n+1)$-dimensional quadratic pairs

$$(g, h) \colon (f \colon C \to D, (\psi, \delta\psi)) \to (f' \colon C' \to D', (\psi', \delta\psi'))$$

consists of R-chain maps $g \colon C \to C'$ and $h \colon D \to D'$ such that $f' \circ g = h \circ f$ and

$$z([\psi, \delta\psi]) = [\psi', \delta\psi'] \in Q_{n+1}(f')$$

hold, where $z \colon Q_{n+1}(f) \to Q_{n+1}(f')$ is the map induced by g and h.

In both cases the map (g, h) is called an *isomorphism/homotopy equivalence* if the underlying R-chain maps are isomorphisms/chain homotopy equivalences.

15.5 L-Groups in Terms of Chain Complexes

The *symmetrisation map*

$$1 + T: \operatorname{cone}(f_\%) \to \operatorname{cone}(f^\%) \tag{15.231}$$

is defined as the induced map from the symmetrisations (15.42) on C and D, namely, it comes from the commutative diagram

$$\begin{array}{ccc} W_\%(C) & \xrightarrow{f_\%} & W_\%(D) \\ {\scriptstyle 1+T} \downarrow & & \downarrow {\scriptstyle 1+T} \\ W^\%(C) & \xrightarrow{f^\%} & W^\%(D). \end{array}$$

It induces a map, also called the symmetrisation map and denoted in the same way,

$$1 + T: Q_n(f) \to Q^n(f). \tag{15.232}$$

Recall from Section 15.3 that the 0-th component of a symmetric structure corresponds to the cap product. We think of the passage from φ to φ_0 as of an evaluation map.

Definition 15.233 (The evaluation maps) The *symmetric evaluation map* is

$$\operatorname{ev}: W^\%(C) \to C \otimes_R C \quad \varphi \mapsto \varphi_0$$

and the *quadratic evaluation map* is

$$\operatorname{ev}: W_\%(C) \to C \otimes_R C \quad \psi \mapsto (1+T)(\psi_0).$$

Remark 15.234 The symmetric evaluation map can and sometimes should also be thought of as a map induced by the inclusion $0[W_0] \to W$ on the functor $\hom_{\mathbb{Z}[\mathbb{Z}/2]}(-, C \otimes_R C)$. The quadratic evaluation map is the composite of the symmetrisation map and the symmetric evaluation map.

Now we proceed to the definition of the evaluation maps in the relative setting. Recall from Definition 14.61 that, given a chain map $f: C \to D$, we obtain the homotopy cofibration sequence

$$C \xrightarrow{f} D \xrightarrow{e} \operatorname{cone}(f) \tag{15.235}$$
$$h: 0 \simeq (e \circ f)$$

where $e: D \to \operatorname{cone}(f)$ is just the standard inclusion $e: D_r \to C_{r-1} \oplus D_r$ given by $e(y) = (0, y)$ and h is the chain homotopy given by

$$h: C_r \to \operatorname{cone}(f)_{r+1} = C_r \oplus D_{r+1} \quad h(x) = (x, 0). \tag{15.236}$$

The mapping cone has a universal property described in Remark 14.45.

Next we study Diagram (15.237), where all squares with full arrows are strictly commutative, all rows and columns are homotopy cofibration sequences, and the sequence of the two maps from the upper left corner in the direction down right is also a homotopy cofibration sequence. (To ease the reading the arrows with obvious maps are not labelled.) The dashed and wobbly arrows are explained in the text below.

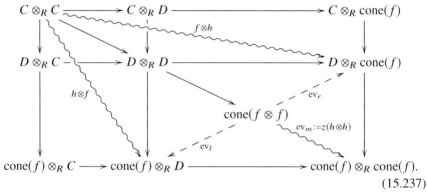

(15.237)

When h is the canonical chain homotopy as in (15.235), then $h \otimes f$ is a chain homotopy from 0 to $(e \circ f) \otimes f$ and $f \otimes h$ is a chain homotopy from 0 to $f \otimes (e \circ f)$. From these we obtain via the universal property of a mapping cone, see Remark 14.45, the induced maps from $\mathrm{cone}(f \otimes f)$ that are denoted in the diagram by ev_l and ev_r (the subscripts l and r stand for "left" and "right"). In the notation of Remark 14.45 these are

$$\mathrm{ev}_l := z(e \otimes \mathrm{id}, h \otimes f) \quad \text{and} \quad \mathrm{ev}_r := z(\mathrm{id} \otimes e, f \otimes h). \qquad (15.238)$$

Now we are interested in the bottom right part of the diagram. We ask ourselves whether the two composites $(\mathrm{id} \otimes e) \circ \mathrm{ev}_l$ and $(e \otimes \mathrm{id}) \circ \mathrm{ev}_r$ agree or not. In fact, they do not agree in general, but the construction from Exercise 14.51 provides us with a chain homotopy

$$\begin{aligned}\mathrm{ev}_m := z(h \otimes h) \colon (\mathrm{id} \otimes e) \circ \mathrm{ev}_l &= z(e \otimes e, h \otimes (e \circ f)) \\ &\simeq z(e \otimes e, (e \circ f) \otimes h) = (e \otimes \mathrm{id}) \circ \mathrm{ev}_r\end{aligned} \qquad (15.239)$$

induced by the homotopy between chain homotopies $h \otimes h \colon h \otimes (e \circ f) \simeq (e \circ f) \otimes h$.

The following lemma, which concerns the portion of Diagram (15.237) just discussed, is used in Section 15.5.3.

15.5 L-Groups in Terms of Chain Complexes

Lemma 15.240 *A chain map $f\colon C \to D$ in R-CH induces a homotopy cartesian square*

$$\begin{array}{ccc} \operatorname{cone}(f \otimes f) & \xrightarrow{\operatorname{ev}_r} & D \otimes_R \operatorname{cone}(f) \\ {\scriptstyle \operatorname{ev}_l}\downarrow & & \downarrow {\scriptstyle e \otimes \operatorname{id}_{\operatorname{cone}(f)}} \\ \operatorname{cone}(f) \otimes_R D & \xrightarrow[\operatorname{id}_{\operatorname{cone}(f)} \otimes e]{} & \operatorname{cone}(f) \otimes_R \operatorname{cone}(f) \end{array}$$

where e is from (15.235), the maps ev_l and ev_r are from (15.238), and the required chain homotopy is given by $\operatorname{ev}_m = z(h \otimes h)$ from (15.239).

Proof. Consider the strictly commutative square

$$\begin{array}{ccc} C \otimes_R C & \xrightarrow{\operatorname{id}_C \otimes f} & C \otimes_R D \\ {\scriptstyle f \otimes \operatorname{id}_C}\downarrow & & \downarrow {\scriptstyle f \otimes \operatorname{id}_D} \\ D \otimes_R C & \xrightarrow[\operatorname{id}_D \otimes f]{} & D \otimes_R D \end{array} \qquad (15.241)$$

which is the upper left square of Diagram (15.237). In the previous chapter, starting with the strictly commutative square (14.96) we constructed a certain homotopy commutative square, and in Lemma 14.98 we showed that that square was homotopy cartesian. Now we apply this technique starting with the square from (15.241) in place of (14.96).

To state the outcome of applying Lemma 14.98 in this situation, we use the following notation, compatible with the one used in Lemma 14.98. By $a\colon \operatorname{cone}(\operatorname{id}_C \otimes f) \to \operatorname{cone}(\operatorname{id}_D \otimes f)$ and $b\colon \operatorname{cone}(f \otimes \operatorname{id}_C) \to \operatorname{cone}(f \otimes \operatorname{id}_D)$ we denote the induced maps of mapping cones, by $\bar{e}_{\operatorname{id}_D \otimes f}\colon \operatorname{cone}(f \otimes f) \to \operatorname{cone}(\operatorname{id}_D \otimes f)$ the induced map associated to $(r_{\operatorname{id}_D \otimes f}, a \circ h_{\operatorname{id}_C \otimes f})$, and by $\bar{e}_{f \otimes \operatorname{id}_D}\colon \operatorname{cone}(f \otimes f) \to \operatorname{cone}(f \otimes \operatorname{id}_D)$ the induced map associated to $(r_{f \otimes \operatorname{id}_D}, b \circ h_{f \otimes \operatorname{id}_C})$. Finally $k\colon \operatorname{cone}(f \otimes f) \to \operatorname{cone}(a)$ is the map of degree one, which in degree n is of the form

$$(C \otimes_R C)_{n-1} \oplus (C \otimes_R D)_n \to$$
$$\to (C \otimes_R C)_{n-1} \oplus (C \otimes_R D)_n \oplus (D \otimes_R C)_n \oplus (D \otimes_R D)_{n+1} \quad (15.242)$$

and is given by $(x, y) \to (x, 0, 0, 0)$. It provides us with a chain homotopy $k\colon \alpha \circ e_b \circ \bar{e}_{f \otimes \operatorname{id}_D} \simeq e_a \circ \bar{e}_{\operatorname{id}_D \otimes f}$ where e_a and e_b are the canonical inclusions and $\alpha\colon \operatorname{cone}(b) \to \operatorname{cone}(a)$ is the canonical isomorphism defined just below Diagram (14.97).

Then Lemma 14.98 implies that the square

$$\begin{array}{ccc} \operatorname{cone}(f \otimes f) & \xrightarrow{\bar{e}_{\operatorname{id}_D \otimes f}} & \operatorname{cone}(\operatorname{id}_D \otimes f) \\ {\scriptstyle \bar{e}_{f \otimes \operatorname{id}_D}}\downarrow & & \downarrow {\scriptstyle e_a} \\ \operatorname{cone}(f \otimes \operatorname{id}_D) & \xrightarrow[\alpha \circ e_b]{} & \operatorname{cone}(a) \end{array} \qquad (15.243)$$

together with the chain homotopy k is homotopy cartesian.

We will construct a map from square (15.243) to the desired square, which will be an isomorphism on each entry, and this will yield the desired statement. The basic tool will be the canonical isomorphisms $\iota_{-,-}$ from (14.89) and (14.90) for various choices of maps and chain complexes.

In more detail, we consider the cube

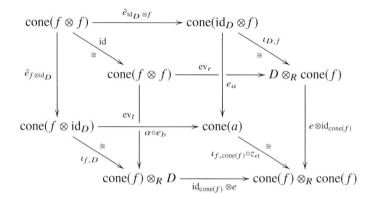

where all maps that appear have been defined before, except the isomorphism $z_a \colon \operatorname{cone}(a) \to \operatorname{cone}(f \otimes \operatorname{id}_{\operatorname{cone}(f)})$, which is described below. Moreover, we will show that all faces except the front face and the back face commute strictly and that the chain homotopy k, up to which the front face commutes, and the chain homotopy ev_m, up to which the back face commutes, are compatible in the sense that $(\iota_{f,\operatorname{cone}(f)} \circ z_a) \circ k = \operatorname{ev}_m$.

Granting this, we take the homotopy fibres of the vertical maps and thus obtain a strictly commutative square such that the left and the right map are chain isomorphisms (because they are induced by chain isomorphisms). Lemma 14.98 tells us that the induced map between the fibres corresponding to the front square is a chain homotopy equivalence. Hence the induced map between the fibres corresponding to the back square is also a chain homotopy equivalence, which proves the desired statement.

It remains to show all the claims from the paragraph around the cube above. The promised isomorphism z_a is defined as the induced map of the mapping cones of the vertical maps in the following strictly commutative square

$$\begin{CD} \operatorname{cone}(\operatorname{id}_C \otimes f) @>{\iota_{C,f}}>> C \otimes_R \operatorname{cone}(f) \\ @VaVV @VV{f \otimes \operatorname{id}_{\operatorname{cone}(f)}}V \\ \operatorname{cone}(\operatorname{id}_D \otimes f) @>>{\iota_{D,f}}> D \otimes_R \operatorname{cone}(f). \end{CD}$$

By the definition of $\iota_{f,\operatorname{cone}(f)}$ we have the commutative triangle

15.5 L-Groups in Terms of Chain Complexes

$$\begin{array}{ccc} D \otimes_R \mathrm{cone}(f) & & \\ \downarrow & \searrow^{e \otimes \mathrm{id}_{\mathrm{cone}(f)}} & \\ \mathrm{cone}(\mathrm{id}_C \otimes f) & \xrightarrow[\iota_{f,\mathrm{cone}(f)}]{} & \mathrm{cone}(f) \otimes_R \mathrm{cone}(f), \end{array}$$

which together with the definition of z_a shows that the right square in the cube under investigation strictly commutes. A calculation using formula (14.91) shows that the composition $(\iota_{f,\mathrm{cone}(f)} \circ z_a)$ in degree n is the map

$$(C \otimes_R C)_{n-2} \oplus (C \otimes_R D)_{n-1} \oplus (D \otimes_R C)_{n-1} \oplus (D \otimes_R D)_n \to (\mathrm{cone}(f) \otimes_R \mathrm{cone}(f))_n$$

given by
$$(h \otimes h)_{n-2} \oplus (h \otimes e)_{n-1} \oplus (e \otimes h)_{n-1} \oplus (e \otimes e)_n. \tag{15.244}$$

By analogous reasoning, one obtains in the bottom square that the composite $(\iota_{\mathrm{cone}(f),f} \circ z_b)$ in degree n is the map

$$(C \otimes_R C)_{n-2} \oplus (D \otimes_R C)_{n-1} \oplus (C \otimes_R D)_{n-1} \oplus (D \otimes_R D)_n \to (\mathrm{cone}(f) \otimes_R \mathrm{cone}(f))_n$$

given by
$$((-1) \cdot (h \otimes h)_{n-2}) \oplus (e \otimes h)_{n-1} \oplus (h \otimes e)_{n-1} \oplus (e \otimes e)_n.$$

Employing the isomorphism $\alpha \colon \mathrm{cone}(b) \to \mathrm{cone}(a)$ whose formula appears just below Diagram (14.97) shows that the bottom square strictly commutes.

To see the commutativity of the top square, observe first that the map $\bar{e}_{\mathrm{id}_D \otimes f}$ in degree n is the map

$$\mathrm{cone}(f \otimes f)_n = (C \otimes_R C)_{n-1} \oplus (D \otimes_R D)_n$$
$$\to \mathrm{cone}(\mathrm{id}_D \otimes f)_n = (D \otimes_R C)_{n-1} \oplus (D \otimes_R D)_n$$

given by
$$(\bar{e}_{\mathrm{id}_D \otimes f})_n = (f \otimes \mathrm{id}_D)_{n-1} \oplus (\mathrm{id}_D \otimes \mathrm{id}_D)_n.$$

Composition with $\iota_{D,f}$ in degree n yields

$$(f \otimes h)_{n-1} \oplus (\mathrm{id}_D \otimes e)_n \colon (C \otimes_R C)_{n-1} \oplus (D \otimes_R D)_n \to (D \otimes_R \mathrm{cone}(f))_n,$$

which equals $(\mathrm{ev}_r)_n$, see (15.238). The commutativity of the left square is observed analogously.

Finally, we come to the statement about the compatibility of the chain homotopies of the front and back face. Recall from the description of the chain homotopy k from (15.242) that it is given by $(x, y) \to (x, 0, 0, 0)$. Composition with $(\iota_{f,\mathrm{cone}(f)} \circ z_a)_{n+1}$, which has an analogous description as $(\iota_{f,\mathrm{cone}(f)} \circ z_a)_n$ in (15.244) above, yields $(h \otimes h)_{n-1} \oplus 0$, which equals $(\mathrm{ev}_m)_n$ from (15.239). This finishes the proof of Lemma 15.240. □

After this preparation we come to the definitions. We first observe that there is an evaluation map of the shape

$$\mathrm{ev}\colon \mathrm{cone}(f^\%) \cong \hom_{\mathbb{Z}[\mathbb{Z}/2]}(W, \mathrm{cone}(f \otimes f)) \to \mathrm{cone}(f \otimes f) \qquad (15.245)$$

given by the precomposition with the inclusion $0[W_0] \to W$. It fits into the commutative diagram

$$\begin{array}{ccccc}
W^\%(C) & \xrightarrow{f^\%} & W^\%(D) & \longrightarrow & \mathrm{cone}(f^\%) \\
\downarrow{\mathrm{ev}} & & \downarrow{\mathrm{ev}} & & \downarrow{\mathrm{ev}} \\
C \otimes_R C & \xrightarrow{f \otimes f} & D \otimes_R D & \longrightarrow & \mathrm{cone}(f \otimes f).
\end{array} \qquad (15.246)$$

Definition 15.247 (The relative evaluation maps – symmetric case) The *relative evaluation maps*

$$\mathrm{ev}_l\colon \mathrm{cone}(f^\%) \to \mathrm{cone}(f) \otimes_R D \cong \hom_R(\mathrm{cone}(f)^{-*}, D);$$
$$\mathrm{ev}_r\colon \mathrm{cone}(f^\%) \to D \otimes_R \mathrm{cone}(f) \cong \hom_R(D^{-*}, \mathrm{cone}(f)),$$

are defined as the composites of the map ev from (15.245) with the maps ev_l and ev_r from Diagram (15.237).

Taking the n-th homology yields maps, denoted the same way,

$$\mathrm{ev}_l\colon Q^{n+1}(f) \to [\mathrm{cone}(f)^{n+1-*}, D];$$
$$\mathrm{ev}_r\colon Q^{n+1}(f) \to [D^{n+1-*}, \mathrm{cone}(f)].$$

The fact that the maps in Definition 15.247 are denoted by the same symbols as the maps in Diagram (15.237) should not cause confusion since the map we have in mind is indicated by the source.

Remark 15.248 (The relative evaluation maps – formulas in the symmetric case) Unravelling the formulas using the definition of the map $\mathrm{ev}_l(\varphi_0, \delta\varphi_0)$ = $z(e \otimes \mathrm{id}, h \otimes f)(\varphi_0, \delta\varphi_0) = (e \otimes \mathrm{id})(\delta\varphi_0) + (h \otimes f)(\varphi_0)$ yields:

$$\mathrm{ev}_l\colon \mathrm{cone}(f^\%) \to \mathrm{cone}(f) \otimes_R D \cong \hom_R(\mathrm{cone}(f)^{-*}, D)$$
$$(\varphi, \delta\varphi) \mapsto \delta\varphi_0 \circ \hom(e, \mathrm{id}) + f \circ \varphi_0 \circ \hom(h, \mathrm{id})\colon \mathrm{cone}(f)^{n+1-*} \to D, \qquad (15.249)$$

and $\mathrm{ev}_r(\varphi_0, \delta\varphi_0) = z(\mathrm{id} \otimes e, f \otimes h)(\varphi_0, \delta\varphi_0) = (\mathrm{id} \otimes e)(\delta\varphi_0) + (f \otimes h)(\varphi_0)$ in the other case gives

$$\mathrm{ev}_r\colon \mathrm{cone}(f^\%) \to D \otimes_R \mathrm{cone}(f) \cong \hom_R(D^{-*}, \mathrm{cone}(f))$$
$$(\varphi, \delta\varphi) \mapsto e \circ \delta\varphi_0 + h \circ \varphi_0 \circ \hom(f, \mathrm{id})\colon D^{n+1-*} \to \mathrm{cone}(f). \qquad (15.250)$$

15.5 L-Groups in Terms of Chain Complexes

Recalling that $\hom_R(\mathrm{cone}(f)^{-*}, D)_{n+1} = \bigoplus_r \hom_R(\mathrm{cone}(f)^{n+1-r}, D_r)$ and that $\hom_R(D^{-*}, \mathrm{cone}(f))_{n+1} = \bigoplus_r \hom_R(D^{n+1-r}, \mathrm{cone}(f)_r)$, we observe that on the r-th components we obtain

$$\mathrm{ev}_l(\varphi, \delta\varphi) = ((-1)^{r+1} f \circ \varphi_{0,r}) \oplus \delta\varphi_{0,r} \colon \mathrm{cone}(f)^{n+1-r} = C^{n-r} \oplus D^{n+1-r} \to D_r, \tag{15.251}$$

and

$$\mathrm{ev}_r(\varphi, \delta\varphi) = \begin{pmatrix} \varphi_{0,r-1} \circ f^* \\ \delta\varphi_{0,r} \end{pmatrix} \colon D^{n+1-r} \to \mathrm{cone}(f)_r = C_{r-1} \oplus D_r. \tag{15.252}$$

Remark 15.253 (The homotopy commutative ladder of a symmetric pair)
Note that the discussion following Diagram (15.237) tells us that, when we pass to $\mathrm{cone}(f) \otimes \mathrm{cone}(f) \cong \hom_R(\mathrm{cone}(f)^{-*}, \mathrm{cone}(f))$, the two composites $(\mathrm{id} \otimes e) \circ \mathrm{ev}_l$ and $(e \otimes \mathrm{id}) \circ \mathrm{ev}_r$ are chain homotopic via the chain homotopy $\mathrm{ev}_m := z(h \otimes h)$ of (15.239).

As a consequence we obtain the following diagram, whose rows are extended homotopy cofibration sequences in the sense of (14.87) from Remark 14.85 and in which the left outer square commutes strictly up to signs, see (15.251), the right outer square commutes strictly, see (15.252), and the middle one only commutes up to homotopy, see Exercise 15.257:

$$\begin{array}{ccccccc}
C^{n-*} & \xrightarrow{r^{n+1-*}} & \mathrm{cone}(f)^{n+1-*} & \xrightarrow{e^{n+1-*}} & D^{n+1-*} & \xrightarrow{f^{n+1-*}} & C^{n+1-*} \\
{\scriptstyle \mathrm{ev}(\varphi)}\downarrow & & {\scriptstyle \mathrm{ev}_l(\varphi,\delta\varphi)}\downarrow & {\scriptstyle \mathrm{ev}_m}\searrow & {\scriptstyle \mathrm{ev}_r(\varphi,\delta\varphi)}\downarrow & & {\scriptstyle \mathrm{ev}(\varphi)}\downarrow \\
C & \xrightarrow{f} & D & \xrightarrow{e} & \mathrm{cone}(f) & \xrightarrow{r} & \Sigma C.
\end{array} \tag{15.254}$$

Remark 15.255 The Diagram (15.254) is a chain complex version of the following well-known diagram of (co-)homology groups as explained below in Construction 15.265.

Consider a finite $(n+1)$-dimensional CW-pair (X, A) and an element z in $H_{n+1}(X, A)$. Let $\partial_{n+1}(z)$ be the image of z under the boundary homomorphism $\partial_{n+1} \colon H_{n+1}(X, A) \to H_n(A)$. Then we get an up to signs commutative diagram for every $k \in \mathbb{Z}$

$$\begin{array}{ccccccc}
H^{k-1}(A) & \longrightarrow & H^k(X, A) & \longrightarrow & H^k(X) & \longrightarrow & H^k(A) \\
{\scriptstyle -\cap \partial_n(z)}\downarrow & & {\scriptstyle -\cap z}\downarrow & & {\scriptstyle -\cap z}\downarrow & & {\scriptstyle -\cap \partial_n(z)}\downarrow \\
H_{n-(k-1)}(A) & \longrightarrow & H_{n+1-k}(X) & \longrightarrow & H_{n+1-k}(X, A) & \longrightarrow & H_{n-k}(A),
\end{array} \tag{15.256}$$

see [45, Theorem VI.4.5 and Theorem VI.9.2] or [178, Theorem III.4.43].

Exercise 15.257 Show that the chain homotopy is given by the formula

$$(\mathrm{ev}_m)_r = (-1)^r h \circ \varphi_{0,r} \circ h^* \colon \mathrm{cone}(f)^{n+1-r} \to \mathrm{cone}(f)_{r+1}.$$

Exercise 15.258 For a given $(n+1)$-cycle $(\varphi, \delta\varphi) \in \text{cone}(f^{\%})$ consider the two $(n+1)$-cycles $\text{ev}_r(\varphi, \delta\varphi)$ and $T(\text{ev}_l(\varphi, \delta\varphi))$ in $D \otimes_R \text{cone}(f)$. Observe that the $(n+2)$-chain $(-\widetilde{\varphi}_0, \widetilde{\delta\varphi}_0)$ in $\text{cone}(f \otimes f)$ with $\widetilde{\varphi}_0$ and $\widetilde{\delta\varphi}_0$ as in Exercise 15.39 provides us with a homology from $(T(\varphi)_0, T(\delta\varphi)_0)$ to $(\varphi_0, \delta\varphi_0)$ in $\text{cone}(f \otimes f)$. Show that

$$\widetilde{\text{ev}}_{(l,r)} := \text{ev}_r(-\widetilde{\varphi}_0, \widetilde{\delta\varphi}_0)$$

yields a homology from $T(\text{ev}_l(\varphi, \delta\varphi))$ to $\text{ev}_r(\varphi, \delta\varphi)$ in $D \otimes_R \text{cone}(f)$.

Definition 15.259 (The relative evaluation maps – quadratic case) The *relative evaluation maps*

$$\text{ev}_l: \text{cone}(f_{\%}) \to \text{cone}(f) \otimes_R D \cong \hom_R(\text{cone}(f)^{-*}, D);$$
$$\text{ev}_r: \text{cone}(f_{\%}) \to D \otimes_R \text{cone}(f) \cong \hom_R(D^{-*}, \text{cone}(f)),$$

are defined as the composites of the symmetric evaluation maps ev_l and ev_r with the symmetrisation map $(1 + T)$ from (15.231).

After applying homology we obtain maps, denoted in the same way,

$$\text{ev}_l: Q_{n+1}(f) \to [\text{cone}(f)^{n+1-*}, D];$$
$$\text{ev}_r: Q_{n+1}(f) \to [D^{n+1-*}, \text{cone}(f)].$$

Remark 15.260 (The relative evaluation maps – formulas in the quadratic case) Unravelling the formulas yields:

$$\text{ev}_l: \text{cone}(f_{\%}) \to \text{cone}(f) \otimes_R D \cong \hom_R(\text{cone}(f)^{-*}, D);$$
$$(\psi, \delta\psi) \mapsto (1+T)(\delta\psi_0 \circ \hom(e, 1) + f \circ \psi_0 \circ \hom(h, \text{id})): \text{cone}(f)^{n+1-*} \to D$$
(15.261)

and

$$\text{ev}_r: \text{cone}(f_{\%}) \to D \otimes_R \text{cone}(f) \cong \hom_R(D^{-*}, \text{cone}(f));$$
$$(\psi, \delta\psi) \mapsto (1+T)(e \circ \delta\psi_0 + h \circ \psi_0 \circ \hom(f, \text{id})): D^{n+1-*} \to \text{cone}(f),$$
(15.262)

or on respective modules

$$\text{ev}_l(\psi, \delta\psi): \text{cone}(f)^{n+1-r} = C^{n-r} \oplus D^{n+1-r} \to D_r;$$
$$\text{ev}_l(\psi, \delta\psi) = ((-1)^{r+1} f \circ \psi_{0,r} + (-1)^{(r+1)n+1} \varepsilon f \circ \psi_{0,n-r}^*) \qquad (15.263)$$
$$\oplus (\delta\psi_{0,r} + (-1)^{(r+1)n+1} \varepsilon \delta\psi_{0,n+1-r}^*)$$

and

15.5 L-Groups in Terms of Chain Complexes

$$\mathrm{ev}_r(\psi, \delta\psi) \colon D^{n+1-r} \to \mathrm{cone}(f)_r = C_{r-1} \oplus D_r;$$

$$\mathrm{ev}_r(\psi, \delta\psi) = \begin{pmatrix} (\psi_{0,r-1} + (-1)^{r(n+1)+1}\varepsilon\psi^*_{0,n+1-r}) \circ f^* \\ \delta\psi_{0,r} + (-1)^{(r+1)n+1}\varepsilon\delta\psi^*_{0,n+1-r} \end{pmatrix}. \quad (15.264)$$

Construction 15.265 (The relative (pointed equivariant) symmetric construction)

Let (X, A) be a pair of topological spaces. Denote the chain map coming from the inclusion of A in X by $i \colon C^s(A) \to C^s(X)$. It induces a chain $i^{\%} \colon W_{\mathbb{Z}}^{\%}(C^s(A)) \to W_{\mathbb{Z}}^{\%}(C^s(X))$. By the naturality of the symmetric construction 15.49 we obtain a chain map

$$\varphi_{(X,A)} \colon \mathrm{cone}(i) \to \mathrm{cone}(i^{\%}), \quad (15.266)$$

which is called the *relative symmetric construction*.

Using the canonical chain homotopy equivalence $\mathrm{cone}(i) \simeq C^s(X, A)$, we see that applying H_* respectively H^* to Diagram (15.254) produces Diagram (15.256). The whole situation can be depicted by the following diagram:

$$\begin{array}{ccccc}
C^s(A) & \xrightarrow{i} & C^s(X) & \longrightarrow & \mathrm{cone}(i) \simeq C^s(X, A) \\
{\scriptstyle \varphi_A}\downarrow & & {\scriptstyle \varphi_X}\downarrow & & \downarrow{\scriptstyle \varphi_{(X,A)}} \\
W_{\mathbb{Z}}^{\%}(C^s(A)) & \xrightarrow{i^{\%}} & W_{\mathbb{Z}}^{\%}(C^s(X)) & \longrightarrow & \mathrm{cone}(i^{\%}).
\end{array}$$

If $A \hookrightarrow X$ is an inclusion of pointed spaces, we obtain the *pointed relative symmetric construction* version

$$\widetilde{\varphi}_{(X,A)} \colon \mathrm{cone}(i) \to \mathrm{cone}(i^{\%}), \quad (15.267)$$

exactly of the same shape as above except that $C^s(A)$ and $C^s(X)$ have to be replaced by $\widetilde{C}^s(A)$ and $\widetilde{C}^s(X)$.

Let Γ be a group, let (Y, B) be a pair of pointed free Γ-spaces, and let $w \colon \Gamma \to \{\pm 1\}$ be a group homomorphism. Let $\widetilde{i} \colon \widetilde{C}^s(B) \to \widetilde{C}^s(Y)$ be the $\mathbb{Z}\Gamma$-chain map coming from the inclusion of B into Y. It induces a $\mathbb{Z}$-chain map $(\widetilde{i})^{\%}_{(\mathbb{Z}\Gamma,w)} \colon W^{\%}_{(\mathbb{Z}\Gamma,w)}(\widetilde{C}^s(B)) \to W^{\%}_{(\mathbb{Z}\Gamma,w)}(\widetilde{C}^s(Y))$. Then using the naturality of the pointed equivariant symmetric construction of (15.60) we obtain the $\mathbb{Z}$-chain map

$$\widetilde{\varphi}_{(Y,B),\Gamma,w} \colon \mathbb{Z}^w \otimes_{\mathbb{Z}\Gamma} \mathrm{cone}(\widetilde{i}) \to \mathrm{cone}((\widetilde{i})^{\%}_{(\mathbb{Z}\Gamma,w)}). \quad (15.268)$$

If we take homology, we get the *pointed relative equivariant symmetric construction*,

$$\overline{\varphi}_{(Y,B),\Gamma,w} \colon H_n(Y, B; \mathbb{Z}^w) = H_n(\mathbb{Z}^w \otimes_{\mathbb{Z}\Gamma} \mathrm{cone}(\widetilde{i}))$$
$$\to Q^n_{\mathbb{Z}\Gamma,w}(\widetilde{i}) = H_n(\mathrm{cone}((\widetilde{i})^{\%}_{(\mathbb{Z}\Gamma,w)})). \quad (15.269)$$

Altogether we obtain the commutative diagram

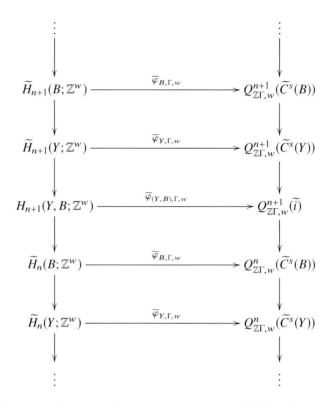

whose columns are the obvious long exact sequences and whose horizontal arrows are given by the symmetric constructions.

Construction 15.270 (The relative quadratic construction)
Let $(F, \partial F) \colon \Sigma^p(X, A) \to \Sigma^p(Y, B)$ be a map of pairs of pointed topological spaces for some $p \geq 0$. Denote by $i_X \colon A \to X$ and $i_Y \colon B \to Y$ the canonical inclusions. We have constructed in (15.146) for ∂F and F zigzags, which are compatible with the inclusions i_X and i_Y. By taking homology we obtain maps (15.147) for ∂F and F, which are again compatible with the inclusions i_X and i_Y. These constructions fit together to a map called the *relative quadratic construction*

$$\overline{\psi}_{\{F, \partial F\}} \colon H_n(X, A) \to Q_n(C^s(i_Y)). \tag{15.271}$$

The map $\overline{\psi}_{\{F, \partial F\}}$ depends only on the stable homotopy class $\{F, \partial F\}$ of $(F, \partial F)$.

Construction 15.272 (The relative equivariant quadratic construction)
Finally we describe the *relative equivariant quadratic construction* leaving the straightforward details of its construction to the reader. Consider a group Γ and a group homomorphism $w \colon \Gamma \to \{\pm 1\}$. Let (X, A) and (Y, B) be pairs of pointed free Γ-spaces. Let $\widetilde{i}_Y \colon \widetilde{C}^s(B) \to \widetilde{C}^s(Y)$ be the $\mathbb{Z}\Gamma$-chain map induced by the inclusion of B into Y. It induces a $\mathbb{Z}$-chain map $(\widetilde{i}_Y)_{\%}^{(\mathbb{Z}\Gamma, w)} \colon W_{\%}^{\mathbb{Z}\Gamma, w}(\widetilde{C}^s(B)) \to W_{\%}^{\mathbb{Z}\Gamma, w}(\widetilde{C}^s(Y))$. Let $(F, \partial F) \colon \Sigma^p(X, A) \to \Sigma^p(Y, B)$ be a pointed Γ-map for some $p \geq 0$. Then we get

15.5 L-Groups in Terms of Chain Complexes

$$\overline{\psi}_{\{F,\partial F\}_\Gamma,w} \colon H_n(X, A; \mathbb{Z}^w) = H_n(\mathbb{Z}^w \otimes_{\mathbb{Z}\Gamma} C^s(X, A))$$
$$\to Q_n^{\mathbb{Z}\Gamma,w}(\widetilde{i_Y}) = H_n(\mathrm{cone}((\widetilde{i_Y})_{\%}^{(\mathbb{Z}\Gamma,w)})). \quad (15.273)$$

The map $\overline{\psi}_{\{F,\partial F\}_\Gamma,w}$ depends only on the stable Γ-homotopy class $\{F, \partial F\}$ of $(F, \partial F)$. We obtain the commutative diagram

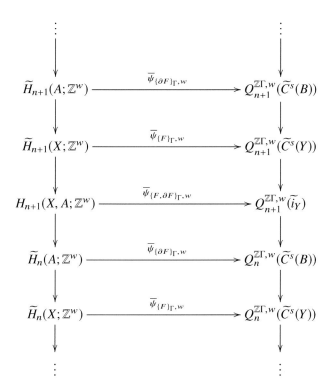

whose columns are the obvious long exact sequences and whose horizontal arrows are given by the quadratic constructions.

Similarly as in the absolute case we have the notion of a Poincaré pair.

Definition 15.274 (Symmetric Poincaré pair, Quadratic Poincaré pair)
An $(n + 1)$-dimensional *symmetric Poincaré pair* over R is a symmetric pair $(f \colon C \to D, [\varphi, \delta\varphi])$ in the sense of Definition 15.218 such that the up to R-chain homotopy unique R-chain map, see Definition 15.247,

$$\mathrm{ev}_r([\varphi, \delta\varphi]) \colon D^{n+1-*} \to \mathrm{cone}(f)$$

is an R-chain homotopy equivalence.

An $(n + 1)$-dimensional *quadratic Poincaré pair* over R is a quadratic pair $(f \colon C \to D, (\delta\psi, \psi))$ in the sense of Definition 15.218 such that the associated symmetric $(n + 1)$-dimensional pair is Poincaré, or, equivalently, the up to R-chain homotopy unique R-chain map

$$\mathrm{ev}_r \circ (1+T)([\psi, \delta\psi]) \colon D^{n+1-*} \to \mathrm{cone}(f)$$

is an R-chain homotopy equivalence where $1+T \colon Q_n(f) \to Q^n(f)$ has been defined in (15.232).

Exercise 15.275 Show that if $\mathrm{ev}_r(\varphi, \delta\varphi)$ is an R-chain homotopy equivalence, then $\mathrm{ev}_l(\varphi, \delta\varphi)$ is an R-chain equivalence and from the ladder (15.254) conclude that $\mathrm{ev}(\varphi)$ is also an R-chain equivalence. Similarly in the quadratic situation.

Exercise 15.276 Suppose (C, φ) and (C, φ') are two n-dimensional polarised symmetric complexes with the same underlying chain complex C and such that $[\varphi] = [\varphi'] \in Q^n(C)$ so that the resulting symmetric complexes satisfy $(C, [\varphi]) = (C, [\varphi'])$.

Show that there is an $(n+1)$-dimensional polarised symmetric pair of the shape $(\mathrm{id} \oplus \mathrm{id} \colon C \oplus C \to C, [-\varphi \oplus \varphi', \theta])$. If the original complex $(C, [\varphi])$ is Poincaré, then the resulting pair is Poincaré. Similarly in the quadratic situation.

The next example generalises Example 15.158 to pairs.

Example 15.277 (Symmetric Poincaré pair associated to a finite n-dimensional Poincaré pair) Let (X, A) be an $(n+1)$-dimensional geometric Poincaré pair with connected X in the sense of Definition 5.73 with fundamental group $\pi = \pi_1(X)$, orientation homomorphism $w \colon \pi \to \{\pm 1\}$, and fundamental class $[X, A]$ in $H_n(X, A; \mathbb{Z}^w) = H_n(\mathbb{Z}^w \otimes_{\mathbb{Z}\pi} C_{n+1}(\widetilde{X}, \overline{A}))$, where $\widetilde{X} \to X$ is the universal covering and $\overline{A} \subseteq \widetilde{X}$ the preimage of A under the universal covering. The equivariant symmetric construction $\overline{\varphi}_{(\widetilde{X},\overline{A}),\pi,w}$ of (15.269) together with the version of the zigzag (15.63) for pairs yields a map, denoted in the same way,

$$\overline{\varphi}_{(\widetilde{X},\overline{A}),\pi,w} \colon H_n(X, A; \mathbb{Z}^w) = H_n(\mathbb{Z}^w \otimes_{\mathbb{Z}\pi} C(\widetilde{X}, \overline{A})) \to Q^n_{\mathbb{Z}\pi,w}(\overline{i}), \qquad (15.278)$$

where here the singular $\mathbb{Z}\pi$-chain complexes are replaced by the cellular $\mathbb{Z}\pi$-chain complexes everywhere. Then the inclusion of the cellular $\mathbb{Z}\pi$-chain complexes $\overline{i} \colon C(\overline{A}) \to C(\widetilde{X})$ together with the image $[\varphi, \delta\varphi]$ of $[X, A]$ under the map (15.278) is an $(n+1)$-dimensional symmetric Poincaré pair $(\overline{i} \colon C(\overline{A}) \to C(\widetilde{X}), [\varphi, \delta\varphi])$ in the sense of Definition 15.274. The up to $\mathbb{Z}\pi$-chain homotopy unique $\mathbb{Z}\pi$-chain map

$$\mathrm{ev}_r \circ \varphi_{(X,A),\pi,w}([X, A]) \colon C^{n+1-*}(\widetilde{X}) \to C(\widetilde{X}, \overline{A})$$

is the same as the $\mathbb{Z}\pi$-chain map $- \cap [X, A] \colon C^{n+1-*}(\widetilde{X}) \to C(\widetilde{X}, \overline{A})$ appearing in Definition 5.73 and is a $\mathbb{Z}\pi$-chain homotopy equivalence.

The next construction generalises Construction 15.161 and Example 15.162 to pairs.

Example 15.279 (Stable Umkehr maps and the algebraic quadratic Poincaré complex associated to normal Γ-maps of degree one of pairs) We indicated at the beginning of Subsection 8.8.1 how to extend the Definition 8.3 of a normal

15.5 L-Groups in Terms of Chain Complexes

Γ-map of degree one to Γ-pairs. So let $((f, \partial f), (\overline{f}, \overline{\partial f})) : (M, \partial M) \to (X, \partial X)$ be normal Γ-map of degree one of pairs from a compact $(n + 1)$-dimensional manifold with boundary to a finite $(n + 1)$-dimensional geometric Poincaré pair, where we do not assume that $\partial f : \partial M \to \partial X$ is a homotopy equivalence, only that it is a normal map of degree one. Similarly as in Construction 15.161 we allow that M and ∂M are not connected.

Recall that we have a Γ-covering $\widehat{X} \to X$, which induces Γ-coverings $\widehat{M} \to M$, $\widehat{\partial M} \to \partial M$, and $\widehat{\partial X} \to \partial X$. Then the relative version of the equivariant S-duality from Section 15.3.8 provides us with a map of pairs of pointed free Γ-pairs of the shape $(F, \partial F) : \Sigma^p(\widehat{X}_+, \widehat{\partial X}_+) \to \Sigma^p(\widehat{M}_+, \widehat{\partial M}_+)$ for some $p \geq 0$.

We get from $\{F, \partial F\}_\Gamma$ by the Construction 15.161 two $\mathbb{Z}\Gamma$-chain maps on the cellular $\mathbb{Z}\Gamma$-chain complexes

$$C^!_*(\widehat{f}) : \widetilde{C}(\widehat{X}) \xrightarrow{\cong} \Sigma^{-p}\widetilde{C}(\Sigma^p\widehat{X}) \xrightarrow{\Sigma^{-p}\widetilde{C}(F)} \Sigma^{-p}\widetilde{C}(\Sigma^p\widehat{M}_+) \xrightarrow{\cong} C(\widehat{M})$$

and

$$C^!_*(\widehat{\partial f}) : C(\widehat{\partial X}) \xrightarrow{\cong} \Sigma^{-p}\widetilde{C}(\Sigma^p\widehat{\partial X}_+) \xrightarrow{\Sigma^{-p}C(\partial F)} \Sigma^{-p}\widetilde{C}(\Sigma^p\widehat{\partial M}_+) \xrightarrow{\cong} C(\widehat{\partial M})$$

where the chain maps denoted by $\cong$ are the obvious suspension chain isomorphisms. They fit together into a commutative diagram

$$\begin{array}{ccc} C(\widehat{\partial X}) & \xrightarrow{C^!_*(\widehat{\partial f})} & C(\widehat{\partial M}) \\ {\scriptstyle i_X} \downarrow & & \downarrow {\scriptstyle i_M} \\ C(\widehat{X}) & \xrightarrow{C^!_*(\widehat{f})} & C(\widehat{M}) \end{array}$$

where the vertical arrows are given by the inclusions. We obtain a chain map $j : \operatorname{cone}(C^!_*(\widehat{\partial f})) \to \operatorname{cone}(C^!_*(\widehat{f}))$ and a commutative diagram

$$\begin{array}{ccc} C(\widehat{\partial M}) & \xrightarrow{\partial e} & \operatorname{cone}(C^!_*(\widehat{\partial f})) \\ {\scriptstyle i_M} \downarrow & & \downarrow {\scriptstyle j} \\ C(\widehat{M}) & \xrightarrow{e} & \operatorname{cone}(C^!_*(\widehat{f})). \end{array}$$

The pair $(e, \partial e)$ induces a map on Q-groups

$$(e, \partial e)_{Q, n+1} : Q^{\mathbb{Z}\Gamma, w}_{n+1}(i_M) \to Q^{\mathbb{Z}\Gamma, w}_{n+1}(j).$$

Recall that a group homomorphism $w : \Gamma \to \{\pm 1\}$ and a class $[X, \partial X]$ in $H_{n+1}(X, \partial X; \mathbb{Z}^w)$ are part of the structure of a normal Γ-map of degree one. The relative quadratic construction of (15.273) together with the version of the zigzag (15.63) for pairs yields a map, denoted in the same way,

$$\overline{\psi}_{\{F,\partial F\}_{\Gamma},w}\colon H_{n+1}(X,\partial X;\mathbb{Z}^w)\to Q^{\mathbb{Z}\Gamma,w}_{n+1}(i_M),$$

where here the singular $\mathbb{Z}\Gamma$-chain complexes are replaced by the cellular $\mathbb{Z}\Gamma$-chain complexes everywhere. Now we get an element $(e,\partial e)_{Q,n+1}\circ\overline{\psi}_{\{F,\partial F\}_{\Gamma},w}([X,\partial X])$ in $Q^{\mathbb{Z}\Gamma,w}_{n+1}(j)$ and thus an $(n+1)$-dimensional finite quadratic Poincaré pair over $\mathbb{Z}\Gamma$

$$\left(j\colon \operatorname{cone}(C^!_*(\widehat{\partial f}))\to \operatorname{cone}(C^!_*(\widehat{f})), (e,\partial e)_{Q,n+1}\circ \overline{\psi}_{\{F,\partial F\}_{\Gamma},w}([X,\partial X])\right). \quad (15.280)$$

The fact that it is Poincaré follows from Lemma 8.128 (ii). It is well defined up to $\mathbb{Z}\Gamma$-chain homotopy equivalence of $(n+1)$-dimensional finite quadratic Poincaré complexes.

15.5.2 Cobordisms

Now we pass to the definition of cobordism. In accordance with geometric intuition, this is a Poincaré pair of chain complexes where the boundary chain complex has two components. The aim of this subsection is to show that cobordism gives an equivalence relation, which enables us to give a new Definition 15.301 of the L-groups.

Definition 15.281 (Algebraic cobordism) A *cobordism* between n-dimensional symmetric Poincaré complexes $(C,[\varphi])$ and $(C',[\varphi'])$ over R is an $(n+1)$-dimensional symmetric Poincaré pair over R of the special shape

$$(f\oplus f'\colon C\oplus C'\to D,[\Phi,\delta\Phi])$$

such that the image of the element $[\Phi,\delta\Phi]$ under the boundary homomorphism $\partial^{n+1}\colon Q^{n+1}(f\oplus f')\to Q^n(C\oplus C')$, see (15.219), and the image of the element $(-[\varphi],[\varphi'])$ under the map $\oplus\colon Q^n(C)\oplus Q^n(C')\to Q^n(C\oplus C')$ appearing in Definition 15.166 agree.

Two symmetric Poincaré complexes $(C,[\varphi])$ and $(C',[\varphi'])$ such that there exists a cobordism between them are said to be *cobordant*, and this fact is denoted by $(C,[\varphi])\sim (C',[\varphi'])$.

A *cobordism* between n-dimensional quadratic Poincaré complexes $(C,[\psi])$ and $(C',[\psi'])$ over R is an $(n+1)$-dimensional quadratic Poincaré pair over R of the special shape

$$(f\oplus f'\colon C\oplus C'\to E,[\Psi,\delta\Psi])$$

such that the image of the element $[\Psi,\delta\Psi]$ under the boundary homomorphism $\partial_{n+1}\colon Q_{n+1}(f\oplus f')\to Q_n(C\oplus C')$, see (15.220), and the image of the element $(-[\psi],[\psi'])$ under the map $\oplus\colon Q_n(C)\oplus Q_n(C')\to Q_n(C\oplus C')$ appearing in Definition 15.166 agree.

15.5 L-Groups in Terms of Chain Complexes

Two quadratic Poincaré complexes $(C, [\psi])$ and $(C', [\psi'])$ such that there exists a cobordism between them are said to be *cobordant*, and this fact is denoted by $(C, [\psi]) \sim (C', [\psi'])$.

Remark 15.282 (Cobordism and polarisations) In the literature a cobordism is sometimes denoted by $(f \oplus f' : C \oplus C' \to D, (-\varphi \oplus \varphi', \delta\varphi))$ for $\varphi \in W^{\%}(C)_n$, $\varphi' \in W^{\%}(C')_n$, and $\delta\varphi \in W^{\%}(D)_{n+1}$ in the symmetric case, or $(f \oplus f' : C \oplus C' \to D, (-\psi \oplus \psi', \delta\psi))$ for $\psi \in W_{\%}(C)_n$, $\psi' \in W_{\%}(C')_n$, and $\delta\psi \in W_{\%}(D)_{n+1}$ in the quadratic case. This sloppy but also intuitive notation, which we will use in the sequel as well, reflects the following two observations. We only explain the symmetric case, the quadratic case is completely analogous.

Consider n-dimensional symmetric Poincaré complexes $(C, [\varphi])$ and $(C', [\varphi'])$ over R and a cobordism $(f \oplus f' : C \oplus C' \to D, [\Phi, \delta\Phi])$ between them. Suppose that we have chosen polarisations φ and φ' for $[\varphi]$ and $[\varphi']$. Then we can find a polarisation of the shape $(-\varphi \oplus \varphi', \delta\varphi)$ for $[\Phi, \delta\Phi]$. On the other hand, suppose that we have chosen a polarisation $(\Phi, \delta\Phi)$ of $[\Phi, \delta\Phi]$. We can write $\Phi = -\varphi \oplus \varphi'$. Then φ and φ' are polarisations for $[\varphi]$ and $[\varphi']$.

Proposition 15.283 (Algebraic cobordism is an equivalence relation)

(i) *Cobordism is an equivalence relation on n-dimensional symmetric Poincaré chain complexes over R;*
(ii) *Cobordism is an equivalence relation on n-dimensional quadratic Poincaré chain complexes over R.*

Proof. We confine ourselves to the symmetric case, the quadratic case being entirely analogous.

1. Reflexivity. Let $(C, [\varphi])$ be an n-dimensional symmetric Poincaré complex. Choose a polarisation φ for $[\varphi]$. Consider the 1-dimensional chain $\omega \in W^{\%}(I)$ from Subsection 15.3.4 where I was the cellular chain complex of the unit interval $[0, 1]$. (Remark 15.81 contains an explicit choice for the chain ω.) Recall that we have the formula $d_{W^{\%}(I)}(\omega) = i_1^{\%}(\nu) - i_0^{\%}(\nu)$, which means that the chain $-\omega$ may be viewed as a homology from $i_1^{\%}(\nu) + i_0^{\%}(-\nu)$ to 0 and therefore we obtain a 1-dimensional symmetric pair $((i_0 \oplus i_1) : 0[\mathbb{Z}] \oplus 0[\mathbb{Z}] \to I, ((-\nu \oplus \nu), -\omega))$. Obviously it is a Poincaré pair and hence a cobordism. The product construction (15.73) provides us with a polarised $(n+1)$-dimensional symmetric cobordism $(i_0 \oplus i_1 : C \oplus C \to I \otimes C, (-\varphi \oplus \varphi, -\omega \otimes \varphi))$ between $(C, [\varphi])$ and $(C, [\varphi])$.

2. Symmetry. Starting with a polarised cobordism

$$(f_C \oplus f_{C'} : C \oplus C' \to D, (-\varphi \oplus \varphi', \delta\varphi))$$

between $(C, [\varphi])$ and $(C', [\varphi'])$, the pair

$$(f_{C'} \oplus f_C : C' \oplus C \to D, (-\varphi' \oplus \varphi, -\delta\varphi))$$

defines a polarised cobordism between $(C', [\varphi'])$ and $(C, [\varphi])$.

3. Transitivity. To see the transitivity a new concept is need, namely the union of cobordisms along a common boundary component, which is presented in Construction 15.285 below. Note that in Construction 15.285 we have the condition that the φ' appearing in c and c' agree. This condition can always be achieved by an appropriate choice of polarisations, which is possible by Exercise 15.276, see also Remark 15.282. □

Remark 15.284 (Gluing chain complexes versus gluing manifolds)
Construction 15.285 below mimics the gluing of a bordism W between two closed manifolds M and M' and a bordism W' between two closed manifolds M' and M'' along M' to obtain a bordism $W \cup_{M'} W'$ between M and M''. Note that for manifolds the inclusion of the boundary to the cobordism is a cofibration, which is not demanded on the chain complex level. This makes the exposition a little bit more complicated and is the reason for the appearance of all the mapping cones in the constructions, but gives very useful flexibility in the algebraic setting.

This gluing construction is the main reason why one needs all higher coherences in the setting of Poincaré chain complexes. If one has two quadratic pairs with a common boundary such that the Poincaré chain homotopy equivalences fit together on the boundary, one can glue these together to a chain map on the union, but it is no longer clear that the result is again a chain homotopy equivalence that is up to chain homotopy the same as its dual. The higher coherences guarantee this. For an explicit setting where this can be seen, observe the occurrence of ψ_1' in the formula for $\delta\psi_0''$ in Remark 15.296.

Construction 15.285 (The union of polarised symmetric pairs along a common boundary) Next we introduce the operation of the *union* of polarised cobordisms along a common boundary (component). Let

$$c = (f_C \oplus f_{C'} : C \oplus C' \to D, (-\varphi \oplus \varphi', \delta\varphi));$$
$$c' = (f'_{C'} \oplus f'_{C''} : C' \oplus C'' \to D', (-\varphi' \oplus \varphi'', \delta\varphi')),$$

be two polarised $(n+1)$-dimensional symmetric cobordisms. Note that the same φ' appears in c and c'. Their union is the $(n+1)$-dimensional polarised cobordism

$$c'' = c \cup c' = (f''_C \oplus f''_{C''} : C \oplus C'' \to D'', (-\varphi \oplus \varphi'', \delta\varphi'')),$$

whose definition is presented below. For simplicity we restrict ourselves to the case when $C = C'' = 0$, which contains the essential ideas. We will frequently use notions from the homotopy theory of chain complexes such as homotopy pushouts and homotopy pullbacks of chain complexes and relations among them, as described in Chapter 14.

The underlying chain complex D'' is defined to be the homotopy pushout

$$D'' = D \tilde{\cup}_{C'} D' = \mathrm{cone}\left(\begin{pmatrix} f_{C'} \\ -f'_{C'} \end{pmatrix} : C' \to D \oplus D' \right). \quad (15.286)$$

15.5 L-Groups in Terms of Chain Complexes

Recall that, given a chain map $f: C \to D$, we have the canonical homotopy cofibration sequence $W^{\%}(C) \xrightarrow{f^{\%}} W^{\%}(D) \xrightarrow{j} \mathrm{cone}(f^{\%})$ and the canonical cofibration sequence $W^{\%}(D) \xrightarrow{j} \mathrm{cone}(f^{\%}) \xrightarrow{q} \Sigma W^{\%}(C)$, where the R-chain maps j and q are the obvious ones. Define HPO by the homotopy pushout

$$\begin{array}{ccc} W^{\%}(C') \oplus W^{\%}(C') & \xrightarrow{\mathrm{id} \oplus \mathrm{id}} & W^{\%}(C') \\ {\scriptstyle f^{\%} \oplus f'^{\%}} \downarrow & & \downarrow {\scriptstyle \overline{f^{\%} \oplus f'^{\%}}} \\ W^{\%}(D) \oplus W^{\%}(D') & \xrightarrow[\mathrm{id} \oplus \mathrm{id}]{} & \mathrm{HPO} \end{array}$$

where we have abbreviated $f = f_{C'}$ and $f' = f'_{C'}$. Let $k: D \to D''$ and $k': D' \to D''$ be the canonical inclusions. Then we obtain R-chain maps

$$k^{\%} \oplus k'^{\%}: W^{\%}(D) \oplus W^{\%}(D') \to W^{\%}(D'')$$

and

$$(k \circ f)^{\%}: W^{\%}(C') \to W^{\%}(D'').$$

There is a canonical chain homotopy $h: k \circ f \simeq k' \circ f'$ of R-chain maps $C' \to D''$ since D'' is defined as a homotopy pushout. It induces a preferred chain homotopy $h^{\%}: (k \circ f)^{\%} \simeq (k' \circ f')^{\%}$ of R-chain maps $W^{\%}(C') \to W^{\%}(D'')$. Thus we obtain by $0 \oplus h^{\%}$ a preferred chain homotopy from the chain map

$$(k \circ f)^{\%} \circ (\mathrm{id} \oplus \mathrm{id}) = (k \circ f)^{\%} \oplus (k \circ f)^{\%}: W^{\%}(C') \oplus W^{\%}(C') \to W^{\%}(D'')$$

to the chain map

$$(k^{\%} \oplus k'^{\%}) \circ (f^{\%} \oplus f'^{\%}) = (k \circ f)^{\%} \oplus (k' \circ f')^{\%}: W^{\%}(C') \oplus W^{\%}(C') \to W^{\%}(D'').$$

From these data the universal property of the homotopy pushout yields a chain map

$$\Phi: \mathrm{HPO} \to W^{\%}(D'').$$

We get a diagram whose uppermost square commutes up to preferred chain homotopy and whose other two squares commute

$$\begin{array}{ccc}
W^{\%}(C') \oplus W^{\%}(C') & \xrightarrow{\mathrm{id} \oplus \mathrm{id}} & W^{\%}(C') \\
{\scriptstyle f^{\%} \oplus f'^{\%}} \downarrow & & \downarrow {\scriptstyle \overline{f^{\%} \oplus f'^{\%}}} \\
W^{\%}(D) \oplus W^{\%}(D') & \xrightarrow{\overline{\mathrm{id} \oplus \mathrm{id}}} & \mathrm{HPO} \\
{\scriptstyle j \oplus j'} \downarrow & & \downarrow {\scriptstyle j''} \\
\mathrm{cone}(f^{\%}) \oplus \mathrm{cone}(f'^{\%}) & \xrightarrow{z} & \mathrm{cone}(\overline{f^{\%} \oplus f'^{\%}}) \\
{\scriptstyle p \oplus p'} \downarrow & & \downarrow {\scriptstyle p''} \\
\Sigma(W^{\%}(C') \oplus W^{\%}(C')) & \xrightarrow{\Sigma(\mathrm{id} \oplus \mathrm{id})} & \Sigma W^{\%}(C).
\end{array} \qquad (15.287)$$

Here j'' is the obvious inclusion, p'' is the obvious projection, and the chain map z comes from the universal property of the mapping cone and the canonical chain homotopy associated to the homotopy pushout above.

We observe that the element

$$(-(\varphi', \delta\varphi), (\varphi', \delta\varphi')) \in \mathrm{cone}(f^{\%})_{n+1} \oplus \mathrm{cone}(f'^{\%})_{n+1}$$

is sent under the map $\Sigma(\mathrm{id} \oplus \mathrm{id}) \circ p \oplus p' = p'' \circ z$ to zero in $\Sigma W^{\%}(C)$. Hence there is precisely one element $\widehat{\delta\varphi}$ in HPO_{n+1} satisfying

$$j''_{n+1}(\widehat{\delta\varphi}) = z(-(\varphi', \delta\varphi), (\varphi', \delta\varphi')).$$

Now we define

$$\delta\varphi'' := \Phi(\widehat{\delta\varphi}) \in W^{\%}(D'')_{n+1}. \qquad (15.288)$$

Define $\overline{\mathrm{HPO}}$ by the homotopy pushout

$$\begin{array}{ccc}
(C' \otimes_R C') \oplus (C' \otimes_R C') & \xrightarrow{\mathrm{id} \oplus \mathrm{id}} & C' \otimes_R C' \\
{\scriptstyle (f \otimes f) \oplus (f' \otimes f')} \downarrow & & \downarrow {\scriptstyle \overline{(f \otimes f) \oplus (f' \otimes f')}} \\
(D \otimes_R D) \oplus (D' \otimes_R D') & \xrightarrow[\overline{\mathrm{id} \oplus \mathrm{id}}]{} & \overline{\mathrm{HPO}}.
\end{array}$$

Then we obtain R-chain maps

$$(k \oplus k) \oplus (k' \otimes k') \colon (D \otimes_R D) \oplus (D' \otimes_R D') \to D'' \otimes_R D''$$

and

$$(k \circ f) \otimes (k \circ f) \colon C' \otimes_R C' \to D'' \otimes_R D''.$$

The canonical chain homotopy $h \colon k \circ f \simeq k' \circ f'$ yields a preferred chain homotopy from the chain map

$$(k \circ f) \otimes (k \circ f) \circ (\mathrm{id} \oplus \mathrm{id}) \colon (C' \otimes_R C') \oplus (C' \otimes_R C') \to D'' \otimes_R D''$$

15.5 L-Groups in Terms of Chain Complexes

to the chain map

$$(k \otimes k) \oplus (k' \otimes k') \circ (f \otimes f) \oplus (f' \otimes f') \colon (C' \otimes_R C') \oplus (C' \otimes_R C') \to D'' \otimes_R D''.$$

Using these data the universal property of the homotopy pushout yields a chain map

$$\overline{\Phi} \colon \overline{\mathrm{HPO}} \to D'' \otimes_R D''.$$

Recall that the evaluation maps ev defined in (15.245) were essentially given by the inclusion $0[W_0] \to W$. One easily checks that everything is compatible with these evaluation maps. Hence we get a chain map

$$w \colon \mathrm{HPO} \to \overline{\mathrm{HPO}} \tag{15.289}$$

for which the following diagram commutes

$$\begin{array}{ccc} \mathrm{HPO} & \xrightarrow{\Phi} & W^{\%}(D'') \\ w \downarrow & & \downarrow \overline{\mathrm{ev}} \\ \overline{\mathrm{HPO}} & \xrightarrow{\overline{\Phi}} & D'' \otimes_R D''. \end{array}$$

Analogously to (15.287) we construct using (15.246) a commutative diagram

$$\begin{array}{ccc}
(C' \otimes_R C') \oplus (C' \otimes_R C') & \xrightarrow{\mathrm{id} \oplus \mathrm{id}} & C' \otimes_R C' \\
{\scriptstyle (f \otimes f) \oplus (f' \otimes f')} \downarrow & & \downarrow {\scriptstyle \overline{(f \otimes f) \oplus (f' \otimes f')}} \\
(D \otimes_R D) \oplus (D' \otimes_R D') & \longrightarrow & \overline{\mathrm{HPO}} \\
\downarrow & & \downarrow \\
(\mathrm{cone}(f) \otimes_R \mathrm{cone}(f)) \oplus (\mathrm{cone}(f') \otimes_R \mathrm{cone}(f')) & \to & \mathrm{cone}\bigl(\overline{(f \otimes f) \oplus (f' \otimes f')}\bigr) \\
\downarrow & & \downarrow \\
\Sigma((C \otimes_R C) \oplus (C' \otimes_R C')) & \xrightarrow{\Sigma(\mathrm{id} \oplus \mathrm{id})} & \Sigma(C \otimes_R C).
\end{array} \tag{15.290}$$

Starting with the element

$$(-\mathrm{ev}(\varphi', \delta\varphi), \mathrm{ev}(\varphi', \delta\varphi')) \in (\mathrm{cone}(f) \otimes \mathrm{cone}(f)) \oplus (\mathrm{cone}(f') \otimes \mathrm{cone}(f')),$$

one constructs a unique element $\overline{\delta\varphi}$ in $\overline{\mathrm{HPO}}_{n+1}$ as done above for $\widehat{\delta\varphi}$ in HPO_{n+1}. Now define

$$\overline{\delta\varphi''} := \overline{\Phi}(\overline{\delta\varphi}) \in (D'' \otimes_R D'')_{n+1}. \tag{15.291}$$

Then the image of $\delta\varphi$ under ev: $W^{\%}(D'')_{n+1} \to D'' \otimes_R D''$ is $\overline{\delta\varphi''}$.

This implies by inspecting the definition of the map ev_r from Definition 15.247 that we obtain a homotopy commutative diagram

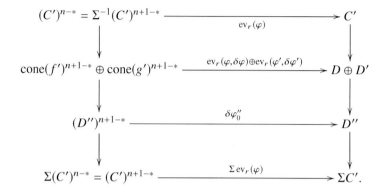

Since the two lower squares in the diagram above have as vertical arrows cofibration sequences, and the chain maps $\mathrm{ev}_r(\varphi, \delta\varphi) \oplus \mathrm{ev}_r(\varphi', \delta\varphi')$ and $\Sigma^{-1}\mathrm{ev}(\varphi)$ are chain homotopy equivalences, the chain map in the third row $\delta\varphi_0'' \colon (D'')^{n+1-*} \to D''$ is an R-chain homotopy equivalence. This finishes Construction 15.285.

Remark 15.292 (Gluing depends on a choice of polarisations) Starting with the two cobordisms with one common end, we were able to glue them along the common end to obtain another cobordism, thus proving the transitivity of the cobordism relation. The starting data, meaning the symmetric structures of the original cobordisms, were by definition cohomology classes. However, in the process we used polarisations to obtain the glued cobordism. In particular, we used Exercise 15.276 where, given two polarisations φ, φ' of a symmetric structure $[\varphi]$, a choice of a chain θ such that $d_{W^{\%}(C)}(\theta) = \varphi - \varphi'$ was made. The symmetric structure on the glued cobordism depends on this choice. This is good enough for our purposes since we were only interested in obtaining *some* cobordism, which manifests transitivity. But for the interested reader we mention the above dependence on the choice since for example it means that our construction does not give a map from the Q-groups of the underlying pairs of the two original cobordisms to the Q-group of the underlying pair of the resulting cobordism.

Exercise 15.293 Show that the polarisation $\delta\varphi''$ of the resulting glued symmetric structure on D'' constructed in Construction 15.285 has the property that it fits into the following commutative diagram whose rows are short exact sequences:

$$\begin{array}{ccccccccc}
0 & \to & \mathrm{cone}(f)^{n+1-*} & \to & (D'')^{n+1-*} & \to & (D')^{n+1-*} & \to & 0 \\
& & \downarrow \mathrm{ev}_l(\varphi, \delta\varphi) & & \downarrow \delta\varphi_0'' & & \downarrow \mathrm{ev}_r(\varphi', \delta\varphi') & & \\
0 & \to & D & \to & D'' & \to & \mathrm{cone}(f') & \to & 0.
\end{array}$$

15.5 L-Groups in Terms of Chain Complexes

Remark 15.294 (The union of polarised symmetric pairs – formulas) Unravelling the definitions yields the following explicit formulas for $c'' = c \cup c'$:

$$D''_r = C'_{r-1} \oplus D_r \oplus D'_r,$$
$$d_{D''} = \begin{pmatrix} -d_{C'} & 0 & 0 \\ -f_{C'} & d_D & 0 \\ f'_{C'} & 0 & d_{D'} \end{pmatrix} : C'_r \oplus D_{r+1} \oplus D'_{r+1} \to C'_{r-1} \oplus D_r \oplus D'_r, \tag{15.295}$$

and

$$\delta\varphi''_{0,r} : (C')^{n+r} \oplus D^{n+1-r} \oplus (D')^{n+1-r} \to C'_{r-1} \oplus D_r \oplus D'_r,$$
$$\delta\varphi''_{0,r} = \begin{pmatrix} 0 & -\varphi'_{0,r-1} \circ f^* & 0 \\ 0 & \delta\varphi_{0,r} & 0 \\ (-1)^r f' \circ \varphi'_{0,r} & 0 & \delta\varphi'_{0,r} \end{pmatrix};$$

$$\delta\varphi''_{s,r} : (C')^{n+s+r} \oplus D^{n+1+s-r} \oplus (D')^{n+1+s-r} \to C'_{r-1} \oplus D_r \oplus D'_r,$$
$$\delta\varphi''_{s,r} = \begin{pmatrix} (-1)^{\gamma(n,s,r)}\varepsilon e^{-1}_{r-1} \circ (\varphi'_{s-1,n+s-r})^* & -\varphi'_{s,r-1} \circ f^* & 0 \\ 0 & \delta\varphi_{s,r} & 0 \\ (-1)^r f' \circ \varphi'_{s,r} & 0 & \delta\varphi'_{s,r} \end{pmatrix},$$

with $\gamma(n,s,r) = (r+1)(n+s) + r$, $e_{r-1} = e(C'_{r-1})$.

Remark 15.296 (The union of polarised quadratic pairs – formulas) Unravelling the definitions yields the following explicit formulas for $c'' = c \cup c'$:

$$D''_r = C'_{r-1} \oplus D_r \oplus D'_r,$$
$$d_{D''} = \begin{pmatrix} -d_{C'} & 0 & 0 \\ -f_{C'} & d_D & 0 \\ f'_{C'} & 0 & d_{D'} \end{pmatrix} : C'_r \oplus D_{r+1} \oplus D'_{r+1} \to C'_{r-1} \oplus D_r \oplus D'_r, \tag{15.297}$$

and

$$\delta\psi''_{0,r} : (C')^{n+r} \oplus D^{n+1-r} \oplus (D')^{n+1-r} \to C'_{r-1} \oplus D_r \oplus D'_r,$$
$$\delta\psi''_{0,r} = \begin{pmatrix} (-1)^{\gamma(n,0,r)}\varepsilon e^{-1}_{r-1} \circ (\psi'_{1,n-r})^* & \psi'_{0,r-1} \circ f^* & 0 \\ 0 & \delta\psi_{0,r} & 0 \\ (-1)^r f' \circ \psi'_{0,r} & 0 & \delta\psi'_{0,r} \end{pmatrix};$$
$$\delta\psi''_{s,r} : (C')^{n-s+r} \oplus D^{n+1-s-r} \oplus (D')^{n+1-s-r} \to C'_{r-1} \oplus D_r \oplus D'_r,$$
$$\delta\psi''_{s,r} = \begin{pmatrix} (-1)^{\gamma(n,s,r)}\varepsilon e^{-1}_{r-1} \circ (\psi'_{s+1,n-s-r})^* & \psi'_{s,r-1} \circ f^* & 0 \\ 0 & \delta\psi_{s,r} & 0 \\ (-1)^r f' \circ \psi'_{s,r} & 0 & \delta\psi'_{s,r} \end{pmatrix},$$

with $\gamma(n,s,r) = (r+1)(n+s) + r$, $e_{r-1} = e(C'_{r-1})$.

Remark 15.298 (Universal constructions versus formulas) A difference between our treatment in Construction 15.285 and the definitions in the original source, Ranicki's paper [344], is that we provide the reader with constructions based on universal properties of homotopy pushouts whereas in [344, p. 135-136] an explicit formula is given without an explanation of where it comes from. Our presentation provides an abstract explanation. Moreover, observe that the shape of the formula depends on various choices made in Chapter 14. Working with a different set of choices, the reader will be able to feed them into Construction 15.285 and obtain the formula in such an alternative setting.

Recall from Definition 15.38 that a map $f : (C, [\varphi]) \to (C', [\varphi'])$ of n-dimensional symmetric chain complexes is called a *homotopy equivalence* if f is an R-chain homotopy equivalence and similarly in the quadratic case. Cobordism of structured chain complexes has the following property with respect to this notion, which is taken from [344, Proposition 3.2].

Proposition 15.299 (Homotopy equivalent structured chain complexes are cobordant)

(i) Let $f : (C, [\varphi]) \to (C', [\varphi'])$ be a homotopy equivalence of n-dimensional symmetric chain complexes. Then $(C, [\varphi])$ and $(C', [\varphi'])$ are cobordant;
(ii) Let $f : (C, [\psi]) \to (C', [\psi'])$ be a homotopy equivalence of n-dimensional quadratic chain complexes. Then $(C, [\psi])$ and $(C', [\psi'])$ are cobordant.

Proof. We only discuss the symmetric case, the quadratic being entirely analogous. Note that a special case, when f is the identity, has already been proved in the proof of the reflexivity of the cobordism relation. In the general case the desired cobordism is provided by the mapping cylinder of f as follows.

Choose a polarisation φ of $[\varphi]$. Then $\varphi' = f^{\%}(\varphi)$ is a polarisation of $[\varphi']$. We use these polarisations in the sequel.

Recall again the 1-dimensional chain $\omega \in W^{\%}(I)$ from Subsection 15.3.4 whose one choice is explicitly described in Remark 15.81. As explained in the proof of Proposition 15.283, it provides us with an $(n+1)$-dimensional symmetric cobordism $(i_0 \oplus i_1 : C \oplus C \to I \otimes C, (-\varphi \oplus \varphi, -\omega \otimes \varphi))$.

Recall that, as in (14.39), the chain homotopy equivalence f provides us with a pushout diagram

$$\begin{array}{ccc} 0[\mathbb{Z}] \otimes_{\mathbb{Z}} C = C & \xrightarrow{i_1 \otimes_{\mathbb{Z}} \mathrm{id}_C} & I \otimes_{\mathbb{Z}} C \\ {\scriptstyle f} \downarrow & & \downarrow {\scriptstyle \overline{f}} \\ C' & \xrightarrow{j} & \mathrm{cyl}(f). \end{array}$$

15.5 L-Groups in Terms of Chain Complexes

This diagram induces a homotopy pushout diagram

$$\begin{array}{ccc} W^{\%}(C) & \xrightarrow{(i_1)^{\%}} & W^{\%}(I \otimes_{\mathbb{Z}} C) \\ {\scriptstyle f^{\%}}\downarrow & & \downarrow \\ W^{\%}(C') & \longrightarrow & \mathrm{HPO}. \end{array}$$

The triple $(\varphi, \varphi', -\omega \otimes \varphi)$ gives a cycle in HPO. By the universal property of a homotopy pushout we have a chain map $u \colon \mathrm{HPO} \to W^{\%}(\mathrm{cyl}(f))$. Denote by $\varphi_{\mathrm{cyl}(f)}$ the image of this triple under the map u:

$$\varphi_{\mathrm{cyl}(f)} := u(\varphi, \varphi', -\omega \otimes \varphi).$$

This produces an $(n+1)$-dimensional symmetric cobordism which is of the shape $(i \oplus j \colon C \oplus C' \to \mathrm{cyl}(f), (-\varphi \oplus \varphi', \varphi_{\mathrm{cyl}(f)}))$. □

Remark 15.300 (Homotopy invariance of the cobordism of chain complexes)
Note that there is a difference between the behaviour of the notion of cobordism between the category of manifolds and the categories of structured chain complexes. Homotopy equivalent manifolds are not necessarily cobordant, as we have already seen in Example 12.66. Hence the analogy between the two worlds that we have been appealing to so far does not fully extend.

On the other hand, if we imitate the construction from the proof of Proposition 15.299 starting with a homotopy equivalence between manifolds, we produce its mapping cylinder, which is a Poincaré pair in the sense of Definition 5.73 with boundary consisting of two parts. It is indeed the case that there is a closer relationship between the cobordisms of structured chain complexes and such "cobordisms" of Poincaré complexes. We give a few more details about this in Remark 15.317 below.

The following definitions are adaptations to our setup of the definitions from [344, Proposition. 3.2] and [348, Definition 1.8]. The precise relation between these definitions is described in Remark 15.310.

Definition 15.301 (Projective L-groups in terms of chain complexes) Let $n \in \mathbb{Z}$ and let R be a ring with involution.

The *projective symmetric L-groups in terms of chain complexes* of R are defined by

$$L^n_p(R)_{\mathrm{chain}} := \left\{ \begin{array}{c} \text{cobordism classes of } n\text{-dimensional symmetric} \\ \text{algebraic Poincaré complexes over } R \end{array} \right\}.$$

The *projective quadratic L-groups in terms of chain complexes* of R are defined by

$$L^p_n(R)_{\mathrm{chain}} := \left\{ \begin{array}{c} \text{cobordism classes of } n\text{-dimensional quadratic} \\ \text{algebraic Poincaré complexes over } R \end{array} \right\}.$$

The group operation is the direct sum of the structured chain complexes in both cases. The inverse of a symmetric Poincaré complex $(C, [\varphi])$ is given by $(C, -[\varphi])$, and the inverse of a quadratic Poincaré complex $(C, [\psi])$ is given by $(C, -[\psi])$.

Note that so far all chain complexes have lived over $bR\text{-}CH_{\text{fgp}}$. There is also a version over $bR\text{-}CH_{\text{fgf}}$.

Definition 15.302 ((Free)L-groups in terms of chain complexes) Let $n \in \mathbb{Z}$ and let R be a ring with involution.

If we demand that all chain complexes live over $bR\text{-}CH_{\text{fgf}}$, then we get the *(free) symmetric L-groups in terms of chain complexes* of R

$$L_h^n(R)_{\text{chain}} := \left\{ \begin{array}{c} \text{cobordism classes of } n\text{-dimensional symmetric} \\ \text{algebraic Poincaré complexes over } R \text{ in } bR\text{-}CH_{\text{fgf}} \end{array} \right\}$$

and the *(free) quadratic L-groups in terms of chain complexes* of R

$$L_n^h(R)_{\text{chain}} := \left\{ \begin{array}{c} \text{cobordism classes of quadratic } n\text{-dimensional} \\ \text{algebraic Poincaré complexes in } R \text{ over } bR\text{-}CH_{\text{fgf}} \end{array} \right\}.$$

We often abbreviate

$$L^n(R)_{\text{chain}} = L_h^n(R)_{\text{chain}} \quad \text{and} \quad L_n(R)_{\text{chain}} = L_n^h(R)_{\text{chain}}.$$

The symmetrisation maps (15.42) and (15.231) induce the *symmetrisation map* on the L-groups

$$1 + T \colon L_n(R)_{\text{chain}} \to L^n(R)_{\text{chain}}. \tag{15.303}$$

As expected this map is not in general an isomorphism, see Sections 7 and 10 of [344]. However, it is an isomorphism modulo 8-torsion and it is an isomorphism if 2 is invertible in R, see Propositions 8.2 and 3.3 in [344] and the following exercise.

Exercise 15.304 Show that, if $1/2 \in R$, then the symmetrisation map from (15.303) is an isomorphism.

Remark 15.305 (Normal L-groups) The symmetrisation map can be studied significantly further. In [346, page 136] and in [348] a third version of L-groups is defined that fits into a long exact sequence with the symmetrisation map. These groups are called *hyperquadratic* in [346] and *normal* in [348], the latter name being now standard. We will not need these groups in this book, but refer the interested reader to the above references.

Remark 15.306 (The identification of the L-groups) It will be proven in Section 15.6 that the groups $L_n(R)_{\text{chain}}$ are isomorphic to Wall's surgery obstruction groups $L_n(R)$ from Chapters 8 and 9.

15.5 L-Groups in Terms of Chain Complexes

Remark 15.307 (Chain complex L-groups of $\mathbb{Z}$) We have not touched the topic of calculation of the chain complex L-groups yet. At this stage we at least mention what the situation is for the ring $\mathbb{Z}$ with the trivial involution. Given $n \in \mathbb{Z}$ we have

$$L^n(\mathbb{Z})_{\text{chain}} \cong \mathbb{Z}, \mathbb{Z}/2, 0, 0 \quad n \equiv 0, 1, 2, 3 \mod 4, \qquad (15.308)$$

and

$$L_n(\mathbb{Z})_{\text{chain}} \cong \mathbb{Z}, 0, \mathbb{Z}/2, 0 \quad n \equiv 0, 1, 2, 3 \mod 4. \qquad (15.309)$$

The symmetrisation map $(1 + T)\colon L_0(\mathbb{Z}) \to L^0(\mathbb{Z})$ is injective, in fact it is given by multiplication by 8. For other n the map is zero for obvious reasons. Of course, the calculation in (15.309) coincides with the L-groups of $\mathbb{Z}$ from Chapters 8 and 9 as indicated more generally in Remark 15.306. References for both calculations are discussed in Remark 15.310 below.

Remark 15.310 (Versions of L-groups depending on connectivity) Both symmetric and quadratic L-groups are 4-periodic for any R, see Proposition 15.427 in Section 15.6 below or [348, Proposition 1.10]. This is due to the fact that we put no dimension conditions on the underlying chain complexes, we only require that they are bounded.

Our definition differs from the definition in [348, Definition 1.8] only in that for us symmetric and quadratic structures are homology classes rather than cycles. This does not change the L-groups and our L-groups agree with those from [348, Definition 1.8].

When working with homotopy (or strictly) n-dimensional complexes in the sense of Definition 15.187, as Ranicki does in [344], the situation is as follows. In the quadratic case the forgetful map from the L-groups (where the chain complexes are assumed to be homotopy n-dimensional) to the quadratic L-groups (where there are no dimension assumptions) is an isomorphism, see [348, Example 1.11 (i)]. This is not true in the symmetric case since if one imposes the condition that the chain complexes are homotopy n-dimensional, then the symmetric L-groups are not 4-periodic anymore. A counterexample can be found in [344, Proposition 10.2]. The precise relation between these two sorts of symmetric L-groups is explained in [348, Example 1.11 (ii)]. It says that our groups $L^n(R)_{\text{chain}}$, which use bounded chain complexes, are isomorphic to the colimit as $k \to \infty$ of the sequence of the version of L^{n+4k}-groups of R which use homotopy $(n + 4k)$-dimensional chain complexes, and where the structure maps in the colimit are compositions of two skew suspension maps, which we discuss later in Proposition 15.427.

However, for Dedekind rings, such as $\mathbb{Z}$, the composition of two skew suspensions on the version of the L-groups which uses homotopy n-dimensional complexes is an isomorphism for $n \geq 0$, see [344, page 169]. Hence we get, for any $n \in \mathbb{Z}$ and any $k \geq 0$ satisfying $n + 4k \geq 0$, that the group $L^n(\mathbb{Z})_{\text{chain}}$ from Remark 15.307 is isomorphic to the group L^{n+4k} of $\mathbb{Z}$ defined in terms of homotopy n-dimensional complexes from [344, Section 7] and the calculation in Remark 15.307 then follows from Proposition 7.2 in [344].

The calculation of the quadratic L-groups can either be obtained from Proposition 7.2 in [344] and Example 1.11 (i) of [348], or using our Theorem 15.3 and calculations from Theorem 8.99 and Theorem 9.18.

Remark 15.311 (Invariants of chain complex L-groups of $\mathbb{Z}$) The invariants which realise the isomorphisms from Remark 15.307 are obtained by passing to cohomology as follows. Given $C \in b\mathbb{Z}\text{-CH}_{\text{fgp}}$ consider its homology groups $H_k(C)$ and cohomology groups $H^k(C)$ and denote by $T_k(C)$ the torsion subgroup of $H_k(C)$ and by $T^k(C)$ the torsion subgroup of $H^k(C)$. Then the free part is $F_k(C) = H_k(C)/T_k(C)$ and $F^k(C) = H^k(C)/T^k(C)$.

Given a $(4k)$-dimensional symmetric Poincaré complex $(C, [\varphi])$, any choice of polarisation induces a non-singular symmetric bilinear form $(F^{2k}(C), \lambda)$ in the sense of (8.48) over $\mathbb{Z}$ by $\lambda \colon ([f], [g]) \mapsto (f \otimes g)(\varphi_0) \in \mathbb{Z}$ and we can consider its signature, as explained just before Theorem 8.99.

Given a $(4k+1)$-dimensional symmetric Poincaré complex $(C, [\varphi])$, it is explained in [344, Section 7] how to obtain a non-singular (-1)-symmetric linking form over $\mathbb{Z}$ (it is also explained there what this notion is) with the underlying $\mathbb{Z}$-module $T^{2k+1}(C)$. Such forms have the so-called de Rham invariant, which is an element in $\mathbb{Z}/2$, see [312, Section 6], and which in our case turns out to be just the rank of the 2-torsion in $T^{2k+1}(C)$, see [312, middle of page 510].

Given a $(4k)$-dimensional quadratic Poincaré complex $(C, [\psi])$, any choice of polarisation induces a non-singular quadratic form $(F^{2k}(C), \lambda, \mu)$ over $\mathbb{Z}$ (in the sense Remark 8.78) by $\lambda \colon ([f], [g]) \mapsto (f \otimes g)((1+T)(\psi_0))$ and $\mu = v^{2k}([\psi])$, where v^{2k} is the Wu class from Definition 15.171 and the identification $Q_{4k}((2k)[\mathbb{Z}]) \cong Q_{+1}(\mathbb{Z})$ (in the sense of (8.69), see Remark 15.173) is used. By the proof of Theorem 8.99 its signature is divisible by 8 and the desired invariant is $(1/8)$ times the signature.

Given a $(4k+2)$-dimensional quadratic Poincaré complex $(C, [\psi])$, any choice of polarisation induces a non-singular (-1)-quadratic form of the form $(F^{2k+1}(C), \lambda, \mu)$ over $\mathbb{Z}$ (in the sense Remark 8.78) by $\lambda \colon ([f], [g]) \mapsto (f \otimes g)((1+T)(\psi_0))$ and $\mu = v^{2k+1}([\psi])$ where v^{2k+1} is the Wu class from Definition 15.171 and the identification $Q_{4k+2}((2k+1)[\mathbb{Z}]) \cong Q_{-1}(\mathbb{Z})$ (in the sense of (8.69), see Remark 15.173) is used. Taking its Arf invariant in the sense of Definition 8.110 yields the desired invariant.

Summarising, the invariants realising the isomorphisms in Remark 15.307 are:

$$\text{signature} \colon L^{4k}(\mathbb{Z}) \to \mathbb{Z};$$
$$\text{de Rham invariant} \colon L^{4k+1}(\mathbb{Z}) \to \mathbb{Z}/2;$$
$$(1/8) \cdot \text{signature} \colon L_{4k}(\mathbb{Z}) \to \mathbb{Z};$$
$$\text{Arf invariant} \colon L_{4k+2}(\mathbb{Z}) \to \mathbb{Z}.$$

This statement is presented in [344, Proposition 7.2] for a version of L-groups in terms of homotopy n-dimensional complexes for $n \geq 0$. That it also holds for our version of L-groups for all $n \in \mathbb{Z}$ follows from the discussion in Remark 15.310.

15.5 L-Groups in Terms of Chain Complexes

Definition 15.312 (Symmetric signature of a finite Poincaré complex) Let X be a connected finite n-dimensional Poincaré complex with fundamental group $\pi_1(X) = \pi$, orientation homomorphism $w \colon \pi \to \{\pm 1\}$, and fundamental class $[X] \in H_n(X; \mathbb{Z}^w)$ in the sense of Definition 5.43. We have associated to it an n-dimensional symmetric Poincaré complex $(C(\widetilde{X}), [\varphi])$ in Example 15.158. In view of Definition 15.301 we obtain an element called *the symmetric signature* of X

$$\mathrm{ssign}_\pi(X, [X]) \in L^n(\mathbb{Z}\pi, w)_{\mathrm{chain}}. \tag{15.313}$$

There is an intrinsic version for a w-oriented finite Poincaré complex $(X, O_X, [[X]])$, see Remark 5.45,

$$\mathrm{ssign}_\Pi(X, [[X]]) \in L^n(\mathbb{Z}\Pi(X), O_X)_{\mathrm{chain}} \tag{15.314}$$

taking values in the intrinsic version of the symmetric L-groups from Definition 15.301, cf., Subsections 8.7.3 and 9.4.1.

Remark 15.315 (Bordism invariance of the symmetric signature) Note that $L^n(\mathbb{Z}\Pi(X), O_X)_{\mathrm{chain}}$ is a functor $\mathcal{R} \to \mathbb{Z}\text{-MOD}$ for the category $\mathcal{R}$ defined before in (8.154). One obtains a functor $\Omega_n \colon \mathcal{R} \to \mathbb{Z}\text{-MOD}$ by sending an object (X, O_X) to the bordism groups of w-oriented n-dimensional closed manifolds M with reference to (X, O_X), where such a reference is a map $f \colon M \to X$ covered by a map of infinite cyclic coefficient systems from O_M to O_X. Since the L-group $L^n(\mathbb{Z}\Pi(X), O_X)_{\mathrm{chain}}$ is defined in terms of cobordism classes and we have the relative pointed equivariant symmetric construction (15.265), we obtain a natural transformation of functors of the form $\mathcal{R} \to \mathbb{Z}\text{-MOD}$

$$\mathrm{ssign}_\Pi \colon \Omega_n(X, O_X) \to L^n(\mathbb{Z}\Pi(X), O_X)_{\mathrm{chain}}. \tag{15.316}$$

Remark 15.317 (Poincaré cobordism) As indicated in Remark 15.300, it is possible to consider $(n + 1)$-dimensional Poincaré pairs $(Y, X \sqcup X')$ in the sense of Definition 5.73, where the "smaller" complex is a disjoint union of two n-dimensional complexes X and X'. Such pairs are called *Poincaré cobordisms* and one can study the group $\Omega_n^P(X, O_X)$ defined analogously as $\Omega_n(X, O_X)$ only with Poincaré complexes instead of manifolds and Poincaré cobordisms instead of manifold cobordisms. When this is done, one can see that the symmetric signature map from (15.316) factors through $\Omega_n^P(X, O_X)$. For more about this and related topics we refer the reader to [179, 221, 343].

Definition 15.318 (Quadratic signature of a normal map degree one) Let $(f, \overline{f}) \colon M \to X$ be a normal map of degree one from a closed n-dimensional manifold M to a connected finite n-dimensional Poincaré complex X with fundamental group $\pi_1(X) = \pi$, orientation homomorphism $w \colon \pi \to \{\pm 1\}$, and fundamental class $[X] \in H_n(X; \mathbb{Z}^w)$ in the sense of Definition 5.43. We have assigned to it an n-dimensional quadratic Poincaré complex in (15.163), where we here consider only the universal covering case, see Notation 8.34, but as in Construction 15.161 we allow M not to be connected. In view of Definition 15.301 and the equivariant quadratic construction appearing in Example 15.162 we obtain an element called *the quadratic signature* of $(f, \overline{f})$

$$\text{qsign}_\pi(f, \overline{f}) \in L_n(\mathbb{Z}\pi, w)_{\text{chain}}, \tag{15.319}$$

which depends only on the normal bordism class of $(f, \overline{f})$ thanks to the relative equivariant quadratic construction from Construction 15.272.

There is an intrinsic version for a w-oriented finite Poincaré complex $(X, O_X, [[X]])$, see Remark 5.45

$$\text{qsign}_\Pi(f, \overline{f}) \in L_n(\mathbb{Z}\Pi(X), O_X)_{\text{chain}} \tag{15.320}$$

taking values in the intrinsic version of the quadratic L-groups from Definition 15.301, cf., Subsections 8.7.3 and 9.4.1. It depends only on the normal bordism class of $(f, \overline{f})$.

Proposition 15.321 *Let $(f, \overline{f}) \colon M \to X$ be a normal map of degree one from a closed n-dimensional manifold M to a w-oriented finite Poincaré complex $(X, O_X, [[X]])$ in the sense of Remark 5.45. The symmetric and quadratic signatures arising from $(f, \overline{f})$, M and X are related by*

$$(1 + T)(\text{qsign}_\Pi(f, \overline{f})) = (f, \overline{f})_*(\text{ssign}_\Pi(M, [[M]])) - \text{ssign}_\Pi(X, [[X]])$$
$$\in L^n(\mathbb{Z}\Pi(X); O_X)_{\text{chain}}$$

where $(f, \overline{f})_* \colon L^n(\mathbb{Z}\Pi(M); O_M)_{\text{chain}} \to L^n(\mathbb{Z}\Pi(X); O_X)_{\text{chain}}$ is the map induced by $(f, \overline{f})$ and $(1 + T) \colon L_n(\mathbb{Z}\Pi(X); O_X)_{\text{chain}} \to L^n(\mathbb{Z}\Pi(X); O_X)_{\text{chain}}$ is the intrinsic version of the symmetrisation map (15.303).

Proof. We explain the case when M and X are connected. The general case follows by arguments analogous to those used in similar situations in Chapters 8 and 9.

Let $(f, \overline{f}) \colon M \to X$ be a normal π-map of degree one from a closed connected n-dimensional manifold M to a finite connected n-dimensional Poincaré complex X with fundamental group $\pi_1(X) = \pi$ and orientation homomorphism $w \colon \pi \to \{\pm 1\}$, so that in the notation of Example 15.162 we are presently in the universal covering case.

Observe first that on the level of chain complexes we have a chain homotopy equivalence

$$e \oplus C_*(f) \colon C(\widehat{M}) \xrightarrow{\simeq} \text{cone}(C_*^!(\widehat{f})) \oplus C(\widetilde{X}), \tag{15.322}$$

where $e \colon C(\widehat{M}) \to \text{cone}(C_*^!(\widehat{f}))$ is the canonical map, by arguments as in the proof of Lemma 8.46.

Let $F \colon \Sigma^p \widetilde{X}_+ \to \Sigma^p \widehat{M}_+$ for some $p \geq 0$ be the Umkehr map from Construction 15.161 associated to $(f, \overline{f})$. Recall that by definition of $\overline{\psi}_{\{F\}_{\pi,w}}$ from (15.154) we have on M

$$(1 + T)\overline{\psi}_{\{F\}_{\pi,w}}([X]) = \overline{\varphi}_{\widehat{M},\pi,w}([M]) - (C_*^!(\widehat{f}))^{Q,n} \overline{\varphi}_{\widetilde{X},\pi,w}([X]).$$

15.5 L-Groups in Terms of Chain Complexes

The symmetrisation map commutes with the maps $e^{Q,n}$ and $e_{Q,n}$ and the composite $e \circ C_*^!(\widehat{f}) \simeq 0$, so the symmetrisation of the quadratic structure on $\mathrm{cone}(C_*^!(\widehat{f}))$ is $e^{Q,n}(\overline{\varphi}_{\widehat{M},\pi,w}([M]))$. The result now follows from (15.322) and from the fact that addition on L-groups is given by direct sum. □

Remark 15.323 The importance of Proposition 15.321 lies in the fact that $\mathrm{qsign}_\pi(f, \overline{f})$ depends on the bundle data whereas $(1+T)(\mathrm{qsign}_\pi(f, \overline{f}))$ is independent of the bundle data, and moreover it is the difference between two absolute invariants associated to M and X.

If f is 2-connected and we identify $\Pi(M)$ and $\Pi(X)$ via $\Pi(f)$, it says that the symmetrisation of the quadratic signature of $(f, \overline{f})$ is just the difference of the symmetric signature of M and X and hence independent of f and $\overline{f}$. Recall that the kernel and cokernel of the map $(1+T)$ are 8-torsion.

Remark 15.324 (Bordism invariance with not necessarily cylindrical target) In Subsection 8.7.5 we stated Theorem 8.168 and in Subsection 9.4.3 we stated its odd-dimensional analogue Theorem 9.109 about the bordism invariance of the surgery obstruction with not necessarily cylindrical target, but we relegated the proofs in both cases to this chapter. The relative quadratic construction from Construction 15.270 provides us with a proof of such a theorem for the quadratic signature $\mathrm{qsign}_\pi(f, \overline{f})$. Thus, when we identify the L-groups in Section 15.6 and when we identify the surgery obstructions with quadratic signatures in Section 15.7, we will automatically obtain the proofs of Theorem 8.168 and Theorem 9.109.

Remark 15.325 (Gluing complexes along homotopy equivalences) Let $c = (f : C \to D, [\varphi, \delta\varphi])$ and $c' = (f' : C' \to D', [\varphi', \delta\varphi'])$ be two $(n+1)$-dimensional symmetric Poincaré pairs and let $h : (C, [\varphi]) \to (C', [\varphi'])$ be a homotopy equivalence of n-dimensional symmetric Poincaré complexes. Choose polarisations $(\varphi, \delta\varphi)$ of $[\varphi, \delta\varphi]$ and $(\varphi', \delta\varphi')$ of $[\varphi', \delta\varphi']$ such that $h^\%(\varphi) = \varphi'$ holds. In this situation we might wish to "glue" the two pairs c and c' along h. This can be done in two ways. By Proposition 15.299 the homotopy equivalence h produces the cobordism $c(h) := (C \oplus C', \mathrm{cyl}(h), (-\varphi \oplus \varphi', \varphi_{\mathrm{cyl}(h)}))$ whose ends are the polarised complexes (C, φ) and (C', φ'). Using Construction 15.285 we can then glue together the three cobordisms to get

$$(D \cup_h D', \delta\varphi \cup_h \delta\varphi') := c \cup c(h) \cup c'. \qquad (15.326)$$

Alternatively, one can observe that the homotopy equivalence h can be used to obtain from c' a new $(n+1)$-dimensional polarised symmetric Poincaré pair $c'' = (f' \circ h : C \to D', (\varphi, \delta\varphi'))$. Then we can glue c and c'' using Construction 15.285, since they have a common boundary, to get the symmetric Poincaré complex $c \cup c''$ homotopy equivalent to (15.326).

If one uses the second approach, then one easily obtains a modification of the formulas from Remark 15.294 for this situation. Moreover, it is obvious that there is a homotopy equivalence $c \cup c(h) \cup c' \to c \cup c''$, which corresponds to the canonical homotopy equivalence $\mathrm{cyl}(h) \to C'$.

Remark 15.327 (Three pairs with the same source) Consider three polarised $(n+1)$-dimensional symmetric Poincaré pairs $c = (f \colon C \to D, (\varphi, \delta\varphi))$, $c' = (f' \colon C \to D', (\varphi, \delta\varphi'))$, and $c'' = (f'' \colon C \to D'', (\varphi, \delta\varphi''))$ with the common source (C, φ). We claim that the following formula holds:

$$[c \cup c'] + [c' \cup c''] = [c \cup c''] \in L^{n+1}(R).$$

For the proof some preparation is needed. First, to save space, let us denote by $\iota = ((i_0 \oplus i_1) \colon \mathbb{Z} \oplus \mathbb{Z} \to I, (-\nu \oplus \nu, -\omega))$ the 1-dimensional symmetric Poincaré pair already considered in the proof of Proposition 15.283. Given an $(n+1)$-dimensional Poincaré pair c, the tensor product $c \otimes \iota$ becomes an $(n+2)$-dimensional symmetric Poincaré triad in the sense of [346, Chapter 2]. Such a triad of structured chain complexes is an algebraic analogue of a triad of manifolds, which we considered already in Chapter 8 (more specifically Section 8.8.4). Intuitively, cobordisms between manifolds with boundary are triads and the same holds for triads of chain complexes. Informally, such an $(n+2)$-dimensional triad consists of a square diagram of R-chain maps

$$\begin{array}{ccc} X & \longrightarrow & Y \\ \downarrow & & \downarrow \\ U & \longrightarrow & V \end{array}$$

together with chains $\varphi_X \in W^{\%}(X)_n$, $\varphi_Y \in W^{\%}(Y)_{n+1}$, $\varphi_U \in W^{\%}(U)_{n+1}$, and $\varphi_V \in W^{\%}(V)_{n+2}$, so that $(X \to Y, (\varphi_X, \varphi_Y))$ and $(X \to U, (-\varphi_X, \varphi_U))$ are $(n+1)$-dimensional symmetric Poincaré pairs and in addition we have that $(U \cup_X Y \to V, ((\varphi_U \cup_{\varphi_Y} \varphi_Y), \varphi_V))$ is an $(n+2)$-dimensional symmetric Poincaré pair.

For example, the triad $c \otimes \iota$ will have the following underlying square of chain complexes

$$\begin{array}{ccc} C \otimes \partial I & \longrightarrow & C \otimes I \\ \downarrow & & \downarrow \\ D \otimes \partial I & \longrightarrow & D \otimes I. \end{array}$$

We leave it for the reader to obtain the necessary structures using products from Section 15.3.3.

Such triads can be glued along common $(n+1)$-dimensional parts and we denote such gluing by the symbol $- \cup -$. The desired cobordism between $(c \cup c') \oplus (c' \cup c'')$ and $(c \cup c'')$ is obtained in this way by

$$\big((c \otimes \iota) \cup (c(\mathrm{id}_C) \otimes \iota) \cup (c'' \otimes \iota)\big) \cup_\alpha (c' \otimes \iota).$$

Here α is the identification of $(C, \varphi) \otimes \iota$ considered as a part of $c' \otimes \iota$ with the part $c(\mathrm{id}_C) \otimes \partial_1 \iota$ of $c(\mathrm{id}_C) \otimes \iota$, where $\partial_1 \iota = (0[\mathbb{Z}], \nu)$ corresponds to $i_1 \colon 0[\mathbb{Z}] \to I$. An analogous formula holds in the quadratic case.

15.5.3 The Algebraic Thom Construction

The aim of this section is to introduce the *algebraic Thom Construction* that produces from an $(n+1)$-dimensional symmetric pair with underlying chain map $f: C \to D$ an $(n+1)$-dimensional symmetric complex with underlying chain complex the mapping cone $\text{cone}(f)$. It corresponds to the passage from a (nice) pair of spaces (X, A) equipped with an element in $H_{n+1}(X, A)$ to the quotient X/A and the corresponding element in $H_{n+1}(X, A)$. In other words, it is a passage from the relative cap products on the pair (X, A) to the absolute cap products on the space X/A, both of which come from an element in $H_{n+1}(X, A)$.

Such a construction and its properties are interesting on their own. For us its importance comes from the homotopy-theoretic point of view on geometric surgery and cobordism. As explained in more detail in Remark 15.390, from the homotopy-theoretic point of view, a geometric surgery is a process which involves attaching a cell and detaching a cell. In the chain complex world attaching a cell is viewed as taking a mapping cone of some chain map and therefore it is desirable to understand the behaviour of the symmetric and quadratic structures when passing to mapping cones. The reverse process, corresponding to detaching a cell, is discussed in Subsections 15.5.4 and 15.5.5.

Besides the original sources of Ranicki, the ideas in this and the following sections are taken from the unpublished lecture notes of Lurie [271]. Lurie works in the setting of ∞-categories, so the statements and proofs here are translated into the world of chain complexes.

Construction 15.328 (The algebraic Thom Construction – symmetric case)
The *algebraic Thom Construction* in the symmetric case associates to an $(n + 1)$-dimensional symmetric pair $(f: C \to D, [\varphi, \delta\varphi])$ an $(n+1)$-dimensional symmetric algebraic complex $(\text{cone}(f), [\delta\varphi/\varphi])$ with the underlying chain complex the mapping cone of f as follows.

Choose a polarisation $(\varphi, \delta\varphi)$ of $[\varphi, \delta\varphi]$. The homotopy cofibration sequence of chain complexes

$$C \xrightarrow{f} D \xrightarrow{e} \text{cone}(f)$$
$$h: 0 \simeq (e \circ f)$$

induces a sequence of chain maps

$$W^\%(C) \xrightarrow{f^\%} W^\%(D) \xrightarrow{e^\%} W^\%(\text{cone}(f)).$$

Although, as we know, this is in general not a cofibration sequence anymore, the composition $e^\% \circ f^\%$ is nullhomotopic, for example via the nullhomotopy $h^\%$ as in (15.78). This choice of a nullhomotopy induces a chain map

$$z(e^\%, h^\%): \text{cone}(f^\%) \to W^\%(\text{cone}(f)), \quad (\varphi, \delta\varphi) \mapsto e^\%(\delta\varphi) + h^\%(\varphi). \quad (15.329)$$

Now we define
$$\delta\varphi/\varphi := z(e^{\%}, h^{\%})(\varphi, \delta\varphi). \tag{15.330}$$

The class of $[\delta\varphi/\varphi]$ in $Q^{n+1}(\operatorname{cone}(f))$ depends only on the class $[\varphi, \delta\varphi]$ in $Q^{n+1}(f: C \to D)$ since the map on the Q-groups is induced by the chain map from (15.329).

Remark 15.331 (The algebraic Thom Construction – formulas) We also need to understand the behaviour of the algebraic Thom Construction with respect to the evaluation maps. Unravelling the definitions via (15.83) yields that $(\delta\varphi/\varphi)_0$ is the $(n+1)$-dimensional cycle in $\operatorname{cone}(f) \otimes_R \operatorname{cone}(f)$ given by

$$(\delta\varphi/\varphi)_0 = (\operatorname{id} \otimes e) \circ \operatorname{ev}_r(\varphi, \delta\varphi) = (e \otimes e)(\delta\varphi_0) + ((e \circ f) \otimes h)(\varphi_0). \tag{15.332}$$

Alternatively, viewed as an R-chain map $(\delta\varphi/\varphi)_0 \colon \operatorname{cone}(f)^{n+1-*} \to \operatorname{cone}(f)$, its component for each $r \in \mathbb{Z}$ is

$$(\delta\varphi/\varphi)_{0,r} = \begin{pmatrix} 0 & \varphi_{0,r-1} \circ f^* \\ 0 & \delta\varphi_{0,r} \end{pmatrix} \colon C^{n-r} \oplus D^{n+1-r} \to C_{r-1} \oplus D_r. \tag{15.333}$$

This equation is important in view of the homotopy commutative ladder (15.254) since it equals the composition first right then down in the middle square.

Similarly we can write out the remaining components, again via (15.83), viewed as chains $(\delta\varphi/\varphi)_s \in (\operatorname{cone}(f) \otimes_R \operatorname{cone}(f))_{n+1+s}$ as

$$(\delta\varphi/\varphi)_s = (e \otimes e)(\delta\varphi_s) + ((e \circ f) \otimes h)(\varphi_s) + (-1)^{n+s}(h \otimes h)(T(\varphi_{s-1})).$$

Alternatively, viewed as chains in $\hom_R(\operatorname{cone}(f)^{n+1-*}, \operatorname{cone}(f))$, the component $(\delta\varphi/\varphi)_{s,r}$ for each $r \in \mathbb{Z}$ is given by

$$(\delta\varphi/\varphi)_{s,r} \colon C^{n+s-r} \oplus D^{n+1+s-r} \to C_{r-1} \oplus D_r;$$
$$(\delta\varphi/\varphi)_{s,r} = \begin{pmatrix} (-1)^{\delta(n,s,r)}\varepsilon e_{r-1}^{-1} \circ \varphi^*_{s-1,n+s-r} & \varphi_{s,r-1} \circ f^* \\ 0 & \delta\varphi_{s,r} \end{pmatrix},$$

where $\delta(n,s,r) = (r+1)(n+s) + r$ and $e_{r-1} = e(C_{r-1})$. This is obtained using the formulas for $\hom(I(\alpha), \beta)$ from (14.21) and (14.32) on the respective components.

Exercise 15.334 Find the formulas for the algebraic Thom construction using the other choice of ω from Remark 15.81, which gives a different formula for $h^{\%}$.

Exercise 15.335 Check that the formulas for the element $(\delta\varphi/\varphi)_s$ in $(\operatorname{cone}(f)\otimes_R\operatorname{cone}(f))_{n+1+s}$ upon applying $\hom(I(\alpha), \beta)$ of (14.21) and (14.32) indeed provide us with the formulas for $(\delta\varphi/\varphi)_{s,r}$.

Exercise 15.336 Prove the equation:

$$\widetilde{(\delta\varphi/\varphi)}_0 = \widetilde{\operatorname{ev}}_{(l,r)} \circ e - T(\operatorname{ev}_m),$$

with the notation $\widetilde{\delta\varphi/\varphi}$ from Exercise 15.39 and $\widetilde{\operatorname{ev}}_{(l,r)}$ from Exercise 15.258.

15.5 L-Groups in Terms of Chain Complexes

Remark 15.337 (Poincaré pairs in general collapse to non-Poincaré complexes)
We note that in general, if we start with a symmetric Poincaré pair, we end up with a symmetric complex that is not Poincaré, see formula (15.333). This is in accordance with the geometric example in which we start with a manifold with boundary $(W, \partial W)$ and pass to the quotient $W/\partial W$ that is in general not a manifold.

As already noted, the algebraic Thom Construction map $z(e^{\%}, h^{\%})$ from (15.329) is in general not a chain equivalence, we think of it as losing some information. In Proposition 15.340 below we determine its failure to be a chain equivalence. This is very useful since it will give us a way to reverse the algebraic Thom Construction in some sense. It will tell us precisely what additional information we need besides the symmetric structure on the quotient to recover a pair. We found a more general version of this proposition stated as Proposition 5 in Lecture 9 of the unpublished notes of Lurie [271]. Its slightly modified homological version can also be found in the paper [19, Proposition 18] of Banagl and Ranicki.

For the desired proposition we need some preparation that reformulates the algebraic Thom construction $z(e^{\%}, h^{\%})$ from (15.329) in terms of homotopy $\mathbb{Z}[\mathbb{Z}/2]$-chain maps from Section 15.3.10. Given an R-chain map $f\colon C \to D$, we see using Exercise 15.229 that the source of $z(e^{\%}, h^{\%})$ is the $\mathbb{Z}$-chain complex with an isomorphism

$$\mathrm{cone}(f^{\%}) \cong \hom_{\mathbb{Z}[\mathbb{Z}/2]}(W, \mathrm{cone}(f \otimes f))$$

and its target is the $\mathbb{Z}$-chain complex

$$W^{\%}(\mathrm{cone}(f)) = \hom_{\mathbb{Z}[\mathbb{Z}/2]}(W, \mathrm{cone}(f) \otimes_R \mathrm{cone}(f)).$$

So both of these $\mathbb{Z}$-chain complexes are of the shape $\hom_{\mathbb{Z}[\mathbb{Z}/2]}(W, -)$, but the map $z(e^{\%}, h^{\%})$ is not induced by applying $\hom_{\mathbb{Z}[\mathbb{Z}/2]}(\mathrm{id}_W, -)$ to a $\mathbb{Z}[\mathbb{Z}/2]$-chain map of the shape $\mathrm{cone}(f \otimes f) \to \mathrm{cone}(f) \otimes_R \mathrm{cone}(f)$. However, it can be described as induced by a homotopy $\mathbb{Z}[\mathbb{Z}/2]$-chain map from Section 15.3.10 as follows.

Consider the canonical homotopy cofibration sequence $C \xrightarrow{f} D \xrightarrow{e} \mathrm{cone}(f)$ with the nullhomotopy $h\colon 0 \simeq (e \circ f)$. In this situation we construct a homotopy $\mathbb{Z}[\mathbb{Z}/2]$-chain map

$$t(f, h) \in \hom_{\mathbb{Z}[\mathbb{Z}/2]}(W, \hom_{\mathbb{Z}}(\mathrm{cone}(f \otimes f), \mathrm{cone}(f) \otimes_R \mathrm{cone}(f))) \qquad (15.338)$$

by the assignments

$$t(f, h)_0\colon (\varphi, \delta\varphi) \mapsto (e \otimes e)(\delta\varphi) + ((e \circ f) \otimes h)(\varphi) = (e \otimes \mathrm{id}) \circ \mathrm{ev}_r(\varphi, \delta\varphi);$$
$$t(f, h)_1\colon (\varphi, \delta\varphi) \mapsto (h \otimes h)(\varphi) = \mathrm{ev}_m(\varphi, \delta\varphi),$$

where ev_r comes from (15.238) and ev_m comes from (15.239).

Proposition 15.339 (Algebraic Thom Construction is induced by a homotopy $\mathbb{Z}/2$-equivariant map) *Consider a chain map $f\colon C \to D$ in $\mathrm{b}R\text{-}\mathrm{CH}_{\mathrm{fgp}}$. Then we have*

$$z(e^{\%}, h^{\%}) = t(f, h)^{\%}\colon \mathrm{cone}(f^{\%}) \to W^{\%}(\mathrm{cone}(f)).$$

Proof. The statement results from the following calculation. Remark 15.331 reveals that given an $(n + 1)$-chain $(\varphi, \delta\varphi) \in \text{cone}(f^{\%})$ the s-th component of the image under $z(e^{\%}, h^{\%})$ is given by the formula:

$$z(e^{\%}, h^{\%})(\varphi, \delta\varphi)_s = (e \otimes e)(\delta\varphi_s) + ((e \circ f) \otimes h)(\varphi_s) + (-1)^{n+s}(h \otimes h)(T(\varphi_{s-1})).$$

The formula for $t(f, h)^{\%}$ in Remark 15.179 tells us that the s-th component of the image under this map equals the right-hand side. For the sign in front of the last summand, we need to keep in mind that the formula is being applied to an $(n + 1)$-dimensional structure $\kappa = (\varphi, \delta\varphi)$. □

Proposition 15.340 (Symmetric pairs versus quotients) *Consider a chain map $f \colon C \to D$ in $\text{bR-CH}_{\text{fgp}}$. Then we obtain a homotopy cartesian square*

$$\begin{array}{ccc} \text{cone}(f^{\%}) & \xrightarrow{\text{ev}_r} & D \otimes_R \text{cone}(f) \\ {\scriptstyle z(e^{\%}, h^{\%})} \downarrow & & \downarrow {\scriptstyle e \otimes \text{id}} \\ W^{\%}(\text{cone}(f)) & \xrightarrow{\text{ev}} & \text{cone}(f) \otimes_R \text{cone}(f). \end{array}$$

Proof. We derive the proof from the diagram in Lemma 15.240, but we look at it from a slightly different point of view.

Namely, we study the following diagram where we use the abbreviated notation $M^{\otimes 2} := M \otimes_R M$, the symbol Δ is the diagonal map, and $t(f, h)_0$ is the 0-th component of the homotopy $\mathbb{Z}[\mathbb{Z}/2]$-chain map $t(f, h)$ from (15.338):

$$\begin{array}{ccc} \text{cone}(f \otimes f) & \xrightarrow{\text{ev}_l \oplus \text{ev}_r} & \text{cone}(f) \otimes_R D \oplus D \otimes_R \text{cone}(f) \\ {\scriptstyle t(f,h)_0} \downarrow & & \downarrow {\scriptstyle \text{id} \otimes e \oplus e \otimes \text{id}} \\ \text{cone}(f)^{\otimes 2} & \xrightarrow{\Delta} & \text{cone}(f)^{\otimes 2} \oplus \text{cone}(f)^{\otimes 2}. \end{array} \quad (15.341)$$

In this diagram all the maps except $t(f, h)_0$ in the left vertical arrow are $\mathbb{Z}[\mathbb{Z}/2]$-chain maps. The map $t(f, h)_0$ is only a $\mathbb{Z}$-chain map, but later we also use the fact that it is the 0-th component of the homotopy $\mathbb{Z}[\mathbb{Z}/2]$-chain map $t(f, h)$. The diagram commutes up to $\mathbb{Z}$-chain homotopy given by the formula

$$(\varphi, \delta\varphi) \mapsto (-\text{ev}_m(\varphi, \delta\varphi), 0).$$

When viewed as a diagram of $\mathbb{Z}$-chain maps, it is a homotopy cartesian square by the following argument, which involves comparing the mapping cones of the horizontal maps. Using Lemma 14.68 and Lemma 15.240, we observe that the induced map

$$z((-\text{id} \otimes e) \oplus (e \otimes \text{id}), \text{ev}_m) \colon \text{cone}(\text{ev}_l \oplus \text{ev}_r) \to \text{cone}(f) \otimes_R \text{cone}(f)$$

is a $\mathbb{Z}$-chain homotopy equivalence. Similarly the induced map

15.5 L-Groups in Terms of Chain Complexes

$$z((-\pi_1) \oplus \pi_2, 0) \colon \mathrm{cone}(\Delta) \to \mathrm{cone}(f) \otimes_R \mathrm{cone}(f)$$

for π_j the projection on the j-th summand is a $\mathbb{Z}$-chain homotopy equivalence since Δ is obviously a cofibration and so by Lemma 14.59 the induced map from the mapping cone to the cokernel is a $\mathbb{Z}$-chain homotopy equivalence. Moreover, by the discussion below Diagram (14.67), the induced map $a \colon \mathrm{cone}(\mathrm{ev}_l \oplus \mathrm{ev}_r) \to \mathrm{cone}(\Delta)$ of the cones of the horizontal maps is given by the formula

$$a = \begin{pmatrix} (e \otimes \mathrm{id}) \circ \mathrm{ev}_r & 0 & 0 \\ -\mathrm{ev}_m & \mathrm{id} \otimes e & 0 \\ 0 & 0 & e \otimes \mathrm{id} \end{pmatrix} \colon$$

$$\mathrm{cone}(f^{\otimes 2})_p \oplus ((\mathrm{cone}(f) \otimes_R D) \oplus (D \otimes_R \mathrm{cone}(f)))_{p+1} \to$$
$$\to \mathrm{cone}(f^{\otimes 2})_p \oplus (\mathrm{cone}(f^{\otimes 2}) \oplus \mathrm{cone}(f^{\otimes 2}))_{p+1}.$$

An inspection shows that $z((-\mathrm{id} \otimes e) \oplus (e \otimes \mathrm{id}), \mathrm{ev}_m) = z((-\pi_1) \oplus \pi_2, 0) \circ a$, and it follows that the induced map a is a $\mathbb{Z}$-chain equivalence.

Let $\mathcal{P}$ denote the homotopy pullback of the lower right part of the diagram alias

$$\mathcal{P} := \mathrm{hofib}\bigl(-\Delta \oplus (\mathrm{id} \otimes e \oplus e \otimes \mathrm{id})\bigr).$$

By the universal property of the homotopy pullback, we obtain a $\mathbb{Z}$-chain map $u_0 \colon \mathrm{cone}(f \otimes f) \to \mathcal{P}$, which is by Lemma 14.68, Theorem 14.95 and by the above arguments a $\mathbb{Z}$-chain equivalence.

Next, a direct calculation reveals that the diagram of homotopy $\mathbb{Z}[\mathbb{Z}/2]$-chain maps

$$\begin{array}{ccc} \mathrm{cone}(f \otimes f) & \xrightarrow{\mathrm{ev}_l \oplus \mathrm{ev}_r} & \mathrm{cone}(f) \otimes_R D \oplus D \otimes_R \mathrm{cone}(f) \\ {\scriptstyle t(f,h)}\downarrow & & \downarrow{\scriptstyle \mathrm{id} \otimes e \oplus e \otimes \mathrm{id}} \\ \mathrm{cone}(f)^{\otimes 2} & \xrightarrow{\Delta} & \mathrm{cone}(f)^{\otimes 2} \oplus \mathrm{cone}(f)^{\otimes 2} \end{array}$$

commutes up to the homotopy of homotopy $\mathbb{Z}[\mathbb{Z}/2]$-chain maps, given by

$$k_0(\varphi, \delta\varphi) = (-\mathrm{ev}_m(\varphi, \delta\varphi), 0), \quad k_i(\varphi, \delta\varphi) = 0 \quad \text{for } i \geq 1,$$

see (15.181) and (15.182) for the notion of the homotopy of homotopy $\mathbb{Z}[\mathbb{Z}/2]$-chain maps, as well as the definition of composition of homotopy $\mathbb{Z}[\mathbb{Z}/2]$-chain maps in (15.183).

Looking again at the homotopy pullback $\mathcal{P}$ as a homotopy fibre of a sum of two $\mathbb{Z}[\mathbb{Z}/2]$-chain maps, we observe that $\mathcal{P}$ is in fact a $\mathbb{Z}[\mathbb{Z}/2]$-chain complex as well. By the above arguments and by the universal property of the homotopy pullback with respect to homotopy $\mathbb{Z}[\mathbb{Z}/2]$-chain maps from Remark 15.184, we obtain a homotopy $\mathbb{Z}[\mathbb{Z}/2]$-chain map $u \colon \mathrm{cone}(f \otimes f) \to \mathcal{P}$ whose 0-th component u_0 is a $\mathbb{Z}$-chain equivalence. By Proposition 15.180, the induced map

$$u^{\%} \colon \operatorname{cone}(f^{\%}) \cong \hom_{\mathbb{Z}[\mathbb{Z}/2]}(W, \operatorname{cone}(f \otimes f)) \to \hom_{\mathbb{Z}[\mathbb{Z}/2]}(W, \mathcal{P})$$

is a $\mathbb{Z}$-chain homotopy equivalence.

Next we can apply Proposition 14.93, which tells us that the functor $\hom_{\mathbb{Z}[\mathbb{Z}/2]}(W, -)$ commutes with homotopy pullbacks. Moreover, we can apply Shapiro's Lemma 15.167 to $\hom_{\mathbb{Z}[\mathbb{Z}/2]}(W, -)$ if the argument is a coinduced chain complex, which is the case for both chain complexes in the right column. From this we obtain a cartesian square of the desired shape. □

Proposition 15.340 has the following interpretation in terms of Q-groups.

Corollary 15.342 (Symmetric pairs versus quotients – homology) *Consider a chain map $f \colon C \to D$ in $\mathrm{b}R\text{-}\mathrm{CH}_{\mathrm{fgp}}$. Then the homotopy cartesian square from Proposition 15.340 induces the long exact sequence of abelian groups*

$$\cdots \to Q^n(f) \to H_n(D \otimes_R \operatorname{cone}(f)) \oplus Q^n(\operatorname{cone}(f))$$
$$\to H_n(\operatorname{cone}(f)^{\otimes 2}) \to Q^{n-1}(f) \to \cdots.$$

The left-hand vertical arrow of the homotopy cartesian square from Proposition 15.340 induces the long exact sequence of abelian groups

$$\cdots \to H_{n+1}(C \otimes_R \operatorname{cone}(f)) \to Q^n(f) \to Q^n(\operatorname{cone}(f))$$
$$\to H_n(C \otimes_R \operatorname{cone}(f)) \to Q^{n-1}(f) \to \cdots.$$

Another tool that helps us to understand $\operatorname{cone}(f^{\%})$ is the following proposition.

Proposition 15.343 (The algebraic Thom Construction versus the suspension – symmetric case) *Let $f \colon C \to D$ be a chain map in $\mathrm{b}R\text{-}\mathrm{CH}_{\mathrm{fgp}}$ so that we have a homotopy cofibration sequence*

$$C \xrightarrow{f} D \xrightarrow{e} \operatorname{cone}(f).$$
$$h \colon 0 \simeq (e \circ f)$$

Then we have the following commutative diagram, in which the top row is a part of the extended homotopy cofibration sequence (14.87)

$$W^{\%}(C) \xrightarrow{f^{\%}} W^{\%}(D) \xrightarrow{\bar{e}} \operatorname{cone}(f^{\%}) \xrightarrow{\bar{r}} \Sigma W^{\%}(C) \xrightarrow{\Sigma(f^{\%})} \Sigma W^{\%}(D)$$

with diagonal $e^{\%}$ from $W^{\%}(D)$, vertical $z(e^{\%}, h^{\%})$ to $W^{\%}(\operatorname{cone}(f))$, and vertical s maps,

$$W^{\%}(\operatorname{cone}(f)) \xrightarrow{r^{\%}} W^{\%}(\Sigma C) \xrightarrow{(\Sigma f)^{\%}} W^{\%}(\Sigma D).$$

15.5 *L*-Groups in Terms of Chain Complexes

Proof. Consider the commutative square

$$\begin{array}{ccc} C & \xrightarrow{f} & D \\ \downarrow{\scriptstyle\mathrm{id}} & & \downarrow{\scriptstyle 0} \\ C & \xrightarrow{0} & 0 \end{array}$$

and apply the algebraic Thom Construction to its rows. Observe that the algebraic Thom construction applied to $0\colon C \to 0$ can be identified with the suspension map $S\colon \Sigma W^{\%}(C) \to W^{\%}(\Sigma C)$. The naturality of the algebraic Thom construction means that the square involving the maps z and S in the desired diagram is commutative. □

Next we collect several ideas about the intuitive and practical meaning of Propositions 15.340 and 15.343.

Remark 15.344 (Reversing the algebraic Thom Construction – the symmetric case) We will use Proposition 15.340 to produce symmetric pairs in Section 15.5.4 about the algebraic boundary and in Section 15.5.5 about algebraic surgery. Informally, the mechanism for such a production is based on the following observations.

By definition a polarised $(n+1)$-dimensional symmetric pair is given by an $(n+1)$-dimensional cycle in $\mathrm{cone}(f^{\%})$. Proposition 15.340 tells us that we have a chain homotopy equivalence from the upper left corner to the homotopy pullback of the rest of the diagram

$$u(f)\colon \mathrm{cone}(f^{\%}) \xrightarrow{\simeq} W^{\%}(\mathrm{cone}(f)) \tilde{\times}_{\mathrm{cone}(f)^{\otimes 2}} D \otimes \mathrm{cone}(f). \tag{15.345}$$

This means that up to homotopy any $(n+1)$-dimensional cycle in the target of the chain homotopy equivalence $u(f)$ from (15.345) also gives us an $(n+1)$-dimensional symmetric pair. The point here is, of course, that in certain situations in the above mentioned sections we will be able to construct such a cycle in the target of $u(f)$.

In more detail, suppose that $(f\colon C \to D, (\varphi, \delta\varphi))$ is a polarised $(n+1)$-dimensional symmetric pair and let us look at the image of $(\varphi, \delta\varphi)$ under $u(f)$. Note that we actually land in the honest pullback

$$W^{\%}(\mathrm{cone}(f)) \times_{\mathrm{cone}(f)^{\otimes 2}} D \otimes_R \mathrm{cone}(f)$$

and the image is given by the pair $(\delta\varphi/\varphi, \mathrm{ev}_r(\varphi, \delta\varphi))$.

We can look at this situation as follows. Suppose we start with a symmetric pair $(\varphi, \delta\varphi)$ and we pass to the quotient $\delta\varphi/\varphi$ and forget that this symmetric structure came from a pair. To emphasise the forgetting, denote in this situation $\mathrm{cone}(f)$ by E and $\delta\varphi/\varphi$ by φ_E. Now we ask ourselves, given (E, φ), what additional information do we need to remember to recover the pair $(\varphi, \delta\varphi)$? Proposition 15.340 tells us that we need the underlying chain complexes, that means, we need to know that E is the mapping cone of $f\colon C \to D$, or equivalently that C is the homotopy fibre of $e\colon D \to E$ and we also need the evaluation map $\mathrm{ev}_r(\varphi, \delta\varphi) \in D \otimes E$, but that is all.

In other words, the information contained in φ_s for $s \geq 0$, and in the maps $e \colon D \to E$ and $\mathrm{ev}_r(\varphi, \delta\varphi)$, enables us to recover $\delta\varphi_s$ and φ_s for all $s \geq 0$ up to homotopy.

In Sections 15.5.4 and 15.5.5 variations on this idea will be used and additional intuition will be provided in the respective situations.

The algebraic Thom Construction in the quadratic case and its properties are as follows.

Construction 15.346 (The algebraic Thom Construction – the quadratic case)
The *algebraic Thom Construction* associates to an $(n+1)$-dimensional quadratic pair $(f \colon C \to D, [\psi, \delta\psi])$ an $(n+1)$-dimensional quadratic complex $(\mathrm{cone}(f), [\delta\psi/\psi])$ with underlying chain complex the mapping cone of f as follows.

Choose a polarisation $(\psi, \delta\psi)$ of $[\psi, \delta\psi]$. As in the symmetric case the sequence of chain maps

$$W_\%(C) \xrightarrow{f_\%} W_\%(D) \xrightarrow{e_\%} W_\%(\mathrm{cone}(f))$$

is in general not a cofibration sequence, but the composite $e_\% \circ f_\%$ is nullhomotopic, for example via the nullhomotopy $h_\%$ as in (15.79). This choice of nullhomotopy induces a chain map

$$z(e_\%, h_\%) \colon \mathrm{cone}(f_\%) \to W_\%(\mathrm{cone}(f)), \ (\psi, \delta\psi) \mapsto e_\%(\delta\psi) + h_\%(\psi). \qquad (15.347)$$

Now we define

$$\delta\psi/\psi := z(e_\%, h_\%)(\psi, \delta\psi). \qquad (15.348)$$

The class of $[\delta\psi/\psi]$ in $Q_{n+1}(\mathrm{cone}(f))$ depends only on the class $[(\psi, \delta\psi)]$ in $Q_{n+1}(f \colon C \to D)$ since $[\delta\psi/\psi]$ is the image of $[(\psi, \delta\psi)]$ under the map of Q-groups induced by the chain map (15.347).

Remark 15.349 (The algebraic Thom Construction in the quadratic case – formulas) We also need to understand the behaviour of the algebraic Thom Construction with respect to the evaluation maps. Unravelling the definitions yields that

$$(\delta\psi/\psi)_0 = (e \otimes e)(\delta\psi_0) + ((e \circ f) \otimes h)(\psi_0) + (-1)^{n+1}(h \otimes h)(T(\psi_1)). \qquad (15.350)$$

Alternatively, viewed as a map $(\delta\psi/\psi)_0 \colon \mathrm{cone}(f)^{n+1-*} \to \mathrm{cone}(f)$, its component for each $r \in \mathbb{Z}$ is

$$(\delta\psi/\psi)_{0,r} \colon C^{n-r} \oplus D^{n-r+1} \to C_{r-1} \oplus D_r,$$
$$(\delta\psi/\psi)_{0,r} = \begin{pmatrix} (-1)^{(r+1)n+1} \varepsilon e_{r-1}^{-1} \circ \psi^*_{1,n-1-r} & \psi_{0,r-1} \circ f^* \\ 0 & \delta\psi_{0,r} \end{pmatrix}. \qquad (15.351)$$

Similarly we can write out the remaining components, again via (15.84), viewed as chains $(\delta\psi/\psi)_s \in (\mathrm{cone}(f) \otimes_R \mathrm{cone}(f))_{n+1-s}$ as

$$(\delta\psi/\psi)_s = (e \otimes e)(\delta\psi_s) + ((e \circ f) \otimes h)(\psi_s) + (-1)^{n+s}(h \otimes h)(T(\varphi_{s+1})).$$

15.5 L-Groups in Terms of Chain Complexes

Alternatively, viewed as chains in $\hom_R(\text{cone}(f)^{n+1-*}, \text{cone}(f))$, the component $(\delta\psi/\psi)_{s,r}$ for each $r \in \mathbb{Z}$ is given by

$$(\delta\psi/\psi)_{s,r} \colon C^{n-s-r} \oplus D^{n+1-s-r} \to C_{r-1} \oplus D_r,$$

$$(\delta\psi/\psi)_{s,r} = \begin{pmatrix} (-1)^{\delta(n,s,r)} \varepsilon e_{r-1}^{-1} \circ \psi^*_{s+1,n-s-r} & \psi_{s,r-1} \circ f^* \\ 0 & \delta\psi_{s,r} \end{pmatrix},$$

where $\delta(n, s, r) = (r + 1)(n + s) + r$ and $e_{r-1} = e(C_{r-1})$. This is obtained using formulas for $\hom(I(\alpha), \beta)$ from (14.21) and (14.32) on the respective components.

Exercise 15.352 Find the formulas for the algebraic Thom construction using the other choice of ω from Remark 15.81, which gives a different formula for $h_{\%}$.

For the further analysis of the quadratic algebraic Thom Construction we will use, similarly as in the symmetric case, the homotopy $\mathbb{Z}[\mathbb{Z}/2]$-chain maps. Firstly, given the canonical homotopy cofibration sequence $C \xrightarrow{f} D \xrightarrow{e} \text{cone}(f)$ with the nullhomotopy $h \colon 0 \simeq (e \circ f)$, we consider the homotopy $\mathbb{Z}[\mathbb{Z}/2]$-chain map $t(f, h)$ from (15.338), for which we obtain the following analogue of Proposition 15.339.

Proposition 15.353 (Algebraic Thom Construction is induced by a homotopy $\mathbb{Z}/2$-equivariant map – quadratic case) *Consider a chain map $f \colon C \to D$ in $bR\text{-CH}_{\text{fgp}}$. Then we have*

$$z(e_\%, h_\%) = t(f, h)_\% \colon \text{cone}(f_\%) \to W_\%(\text{cone}(f)).$$

Proof. The statement results from the following calculation. Remark 15.349 reveals that given an $(n + 1)$-chain $(\psi, \delta\psi) \in \text{cone}(f_\%)$ the s-th component of the image under $z(e_\%, h_\%)$ is given by the formula:

$$z(e_\%, h_\%)(\psi, \delta\psi)_s = (e \otimes e)(\delta\psi_s) + ((e \circ f) \otimes h)(\psi_s) + (-1)^{n-s}(h \otimes h)(T(\psi_{s+1})).$$

The formula for $t(f, h)_\%$ in Remark 15.179 tells us that the s-th component of the image under this map equals the right-hand side. For the sign in front of the last summand, we need to keep in mind that the formula is being applied to an $(n + 1)$-dimensional structure $\lambda = (\psi, \delta\psi)$. □

Proposition 15.354 (Quadratic pairs versus quotients) *Consider a chain map $f \colon C \to D$ in $hR\text{-CH}_{\text{fgp}}$. Then we obtain a homotopy cartesian square*

$$\begin{array}{ccc} \text{cone}(f_\%) & \xrightarrow{\text{ev}_r} & D \otimes_R \text{cone}(f) \\ {\scriptstyle z(e_\%,h_\%)}\downarrow & & \downarrow{\scriptstyle e \otimes \text{id}} \\ W_\%(\text{cone}(f)) & \xrightarrow{\text{ev}} & \text{cone}(f) \otimes_R \text{cone}(f). \end{array}$$

Proof. The proof is analogous to the proof of Proposition 15.340 except we use the functor $W \otimes - = \hom(W^{-*}, -)$ instead of $\hom(W, -)$ and apply the quadratic part of Proposition 15.180 and Shapiro's Lemma 15.167 in homology rather than cohomology. □

Proposition 15.355 (The algebraic Thom Construction versus the suspension – quadratic case) *Let $f\colon C \to D$ be a chain map in hR-CH_{fgp} so that we have a homotopy cofibration sequence*

$$C \xrightarrow{f} D \xrightarrow{e} \text{cone}(f).$$
$$h\colon 0 \simeq (e \circ f)$$

Then we have the following commutative diagram in which the top row is a part of the extended homotopy cofibration sequence (14.87)

$$\begin{array}{ccccccccc}
W_{\%}(C) & \xrightarrow{f_{\%}} & W_{\%}(D) & \xrightarrow{\bar{e}} & \text{cone}(f_{\%}) & \xrightarrow{\bar{r}} & \Sigma W_{\%}(C) & \xrightarrow{\Sigma(f_{\%})} & W_{\%}(D) \\
 & & {\scriptstyle e_{\%}} \searrow & & \downarrow {\scriptstyle z(e_{\%}, h_{\%})} & & \downarrow {\scriptstyle S} & & \downarrow {\scriptstyle S} \\
 & & & & W_{\%}(\text{cone}(f)) & \xrightarrow{r_{\%}} & W_{\%}(\Sigma C) & \xrightarrow{(\Sigma f)_{\%}} & W_{\%}(\Sigma D).
\end{array}$$

Proof. Consider the same square containing $f\colon C \to D$ as in the symmetric case and analogously use the naturality of the quadratic version of the algebraic Thom construction. □

Remark 15.356 (Reversing the algebraic Thom Construction – the quadratic case) Proposition 15.354 has similar consequences as Proposition 15.340 only adapted to the quadratic case. In the symmetric case they were described in Remark 15.344 in detail. The changes that appear in the quadratic case are the following.

A chain map $f\colon C \to D$ in hR-CH_{fgp} induces a chain homotopy equivalence

$$u(f)\colon \text{cone}(f_{\%}) \xrightarrow{\simeq} W_{\%}(\text{cone}(f)) \tilde{\times}_{\text{cone}(f)^{\otimes 2}} D \otimes_R \text{cone}(f). \tag{15.357}$$

The consequence is that the information contained in $(\delta\psi/\psi)_s$, for $s \geq 0$, in the map $e\colon D \to \text{cone}(f)$ and in the evaluation map $\text{ev}_r(1+T)(\psi, \delta\psi)$ enables us to recover $(\delta\psi)_s$ and ψ_s for all $s \geq 0$ up to homotopy.

Proposition 15.354 has the following interpretation in terms of Q-groups.

Corollary 15.358 (Quadratic pairs versus quotients – homology) *Consider a chain map $f\colon C \to D$ in bR-CH_{fgp}. Then the homotopy cartesian square from Proposition 15.354 induces the long exact sequence of abelian groups*

$$\cdots \to Q_n(f) \to H_n(D \otimes_R \text{cone}(f)) \oplus Q_n(\text{cone}(f))$$
$$\to H_n(\text{cone}(f)^{\otimes 2}) \to Q_{n-1}(f) \to \cdots.$$

The left-hand vertical arrow of the homotopy cartesian square from Proposition 15.354 induces the long exact sequence of abelian groups

15.5 L-Groups in Terms of Chain Complexes

$$\cdots \to H_{n+1}(C \otimes_R \operatorname{cone}(f)) \to Q_n(f) \to Q_n(\operatorname{cone}(f))$$
$$\to H_n(C \otimes_R \operatorname{cone}(f)) \to Q_{n-1}(f) \to \cdots.$$

We close this section with the following exercise, which contains a formula that we will need in the following section. The formula is an elementary consequence of our sign conventions.

Exercise 15.359 Prove that we obtain a well-defined isomorphism of chain complexes for any $m \in \mathbb{Z}$ and $E, D \in R\text{-CH}_{\mathrm{fgp}}$:

$$\Sigma^m \hom_R(E, D) \xrightarrow{\cong} \hom_R((E^{m-*})^{-*}, D);$$
$$(f_r : E_{r-(a-m)} \to D_r) \mapsto ((-1)^{r(m+1)+a} f_r \circ e^{-1} : E^{**}_{r-(a-m)} \to D_r)$$

where the bottom row is the description in degree $a \in \mathbb{Z}$ and $e = e(E_{r-(a-m)})$. Show that as a consequence in the case $a = m$ we have that

$$((\mathrm{id}_E)_r : E_r \to E_r) \mapsto ((-1)^{(r+1)(m+1)+1} e(E_r)^{-1} : E^{**}_r \to E_r).$$

Taking $m = n + 1$, this yields

$$(\varepsilon \cdot (\mathrm{id}_E)_r : E_r \to E_r) \mapsto ((-1)^{(r+1)n+1} \varepsilon \cdot e(E_r)^{-1} : E^{**}_r \to E_r).$$

15.5.4 The Algebraic Boundary

Suppose we start with an $(n + 1)$-dimensional symmetric complex $(E, [\varphi])$ and we would like to associate to it some $(n+1)$-dimensional symmetric pair whose algebraic Thom construction is $(E, [\varphi])$. This can be done in many different ways. However, if we demand that the symmetric pair is Poincaré, then it turns out that there is up to homotopy equivalence only one such symmetric pair. The main aim of this section is to present Construction 15.360, which yields this pair in the symmetric case, and Construction 15.380, which deals with the quadratic version. The source in the pair is an n-dimensional symmetric Poincaré complex called the algebraic boundary of $(E, [\varphi])$ and it is denoted by $(\partial E, [\partial \varphi])$ and analogously in the quadratic case.

Some comments about the usefulness of the algebraic boundary in the following sections are provided in Remark 15.379, while some intuition can be obtained from the geometric motivation recalled in Remark 15.374.

Construction 15.360 (The algebraic boundary of a symmetric complex) Let $(E, [\varphi])$ be an $(n + 1)$-dimensional symmetric chain complex (not necessarily Poincaré). We are going to associate to it an $(n+1)$-dimensional symmetric Poincaré pair

$$(\partial E \xrightarrow{i} E^{n+1-*}, [\partial \varphi, \overline{\varphi}]),$$

which will be unique up to chain isomorphism of $(n + 1)$-dimensional symmetric Poincaré pairs, as follows.

First of all we choose a polarisation φ of $[\varphi]$. We define

$$\partial E := \Sigma^{-1} \operatorname{cone}(\varphi_0 \colon E^{n+1-*} \to E). \tag{15.361}$$

Recall that for the chain map φ_0 we have the canonical homotopy cofibration sequence $E^{n+1-*} \xrightarrow{\varphi_0} E \xrightarrow{e} \operatorname{cone}(\varphi_0)$ and in fact, as explained in (14.87) in Remark 14.85, also the part of the associated extended homotopy cofibration sequence $\Sigma^{-1}\operatorname{cone}(\varphi_0) \xrightarrow{i} E^{n+1-*} \xrightarrow{\varphi_0} E \xrightarrow{e} \operatorname{cone}(\varphi_0)$. Using (15.361) we get the extended homotopy cofibration sequence

$$\partial E \xrightarrow{i} E^{n+1-*} \xrightarrow{\varphi_0} E \xrightarrow{e} \Sigma \partial E.$$

Applying Proposition 15.340 to $i \colon \partial E \to E^{n+1-*}$, we obtain the homotopy cartesian square

$$\begin{array}{ccc} \operatorname{cone}(i^\%) & \xrightarrow{\operatorname{ev}_r} & E^{n+1-*} \otimes_R \operatorname{cone}(i) \\ {\scriptstyle z(e^\%, j^\%)} \downarrow & & \downarrow {\scriptstyle \bar{e} \otimes \operatorname{id}} \\ W^\%(\operatorname{cone}(i)) & \xrightarrow{\operatorname{ev}} & \operatorname{cone}(i) \otimes_R \operatorname{cone}(i) \end{array}$$

where $\bar{e} \colon E^{n+1-*} \to \operatorname{cone}(i)$ is the canonical map and $j \colon 0 \simeq \bar{e} \circ i$ the canonical chain homotopy. Using the canonical R-chain homotopy equivalence $v \colon \operatorname{cone}(i) \xrightarrow{\simeq} E$, which by definition satisfies $v \circ \bar{e} = \varphi_0$, we get a commutative diagram whose vertical arrows are R-chain homotopy equivalences

$$\begin{array}{ccccc} W^\%(\operatorname{cone}(i)) & \xrightarrow{\operatorname{ev}} & \operatorname{cone}(i) \otimes_R \operatorname{cone}(i) & \xleftarrow{\bar{e} \otimes \operatorname{id}} & E^{n+1-*} \otimes_R \operatorname{cone}(i) \\ {\scriptstyle W^\%(v)} \downarrow & & \downarrow {\scriptstyle v \otimes v} & & \downarrow {\scriptstyle \operatorname{id} \otimes v} \\ W^\%(E) & \xrightarrow{\operatorname{ev}_E} & E \otimes_R E & \xleftarrow{\varphi_0 \otimes \operatorname{id}} & E^{n+1-*} \otimes_R E. \end{array}$$

Combining these two diagrams, we obtain a new homotopy cartesian square

$$\begin{array}{ccc} \operatorname{cone}(i^\%) & \xrightarrow{(\operatorname{id} \otimes v) \circ \operatorname{ev}_r} & E^{n+1-*} \otimes_R E \\ {\scriptstyle W^\%(v) \circ z(e^\%, j^\%)} \downarrow & & \downarrow {\scriptstyle \varphi_0 \otimes \operatorname{id}} \\ W^\%(E) & \xrightarrow{\operatorname{ev}_E} & E \otimes_R E. \end{array}$$

Define HPB to be the homotopy pullback

$$\begin{array}{ccc} \operatorname{HPB} & \longrightarrow & E^{n+1-*} \otimes_R E \\ \downarrow & & \downarrow {\scriptstyle \varphi_0 \otimes \operatorname{id}} \\ W^\%(E) & \xrightarrow{\operatorname{ev}_E} & E \otimes_R E. \end{array}$$

15.5 L-Groups in Terms of Chain Complexes

The universal property of the homotopy pullback yields an R-chain homotopy equivalence

$$u \colon \mathrm{cone}(i^{\%}) \xrightarrow{\simeq} \mathrm{HPB} . \qquad (15.362)$$

For any $E, D \in bR\text{-}\mathrm{CH}_{\mathrm{fgp}}$, we can combine the adjunctions from (14.31) with the isomorphisms from Exercise 15.359 to obtain isomorphisms

$$(E^{n+1-*} \otimes_R D) \cong \mathrm{hom}_R((E^{n+1-*})^{-*}, D) \cong \Sigma^{n+1} \mathrm{hom}_R(E, D);$$
$$(E \otimes_R D) \cong \mathrm{hom}_R(E^{-*}, D).$$

Inserting $D = E$ and looking at the degree $n+1$, they yield in particular identifications

$$(E^{n+1-*} \otimes_R E)_{n+1} \cong \mathrm{hom}_R(E, E)_0;$$
$$(E \otimes_R E)_{n+1} \cong \mathrm{hom}_R(E^{n+1-*}, E)_0.$$

Exercise 15.359 together with Diagram (14.32) tell us that the cycle in $\mathrm{hom}_R(E, E)_0$ given by the chain map $\varepsilon \cdot \mathrm{id}_E$ maps under $\varphi_0 \otimes \mathrm{id}_E$ to a cycle in $(E \otimes_R E)_{n+1} \cong \mathrm{hom}_R(E^{n+1-*}, E)_0$, which equals $\mathrm{ev}(T(\varphi))$, with $T(\varphi) = T \circ \varphi$. Using the above identifications we obtain that the triple

$$(T(\varphi), \varepsilon \cdot \mathrm{id}_E, 0) \in \mathrm{HPB}_{n+1} = W^{\%}(E)_{n+1} \oplus (E^{n+1-*} \otimes_R E)_{n+1} \oplus (E \otimes_R E)_{n+2}$$

is a cycle in HPB. Let $(\partial\varphi, \overline{\varphi})$ be an $(n+1)$-dimensional cycle in $\mathrm{cone}(i^{\%})$ given as the image of $(T(\varphi), \varepsilon \cdot \mathrm{id}_E, 0)$ under some chain homotopy inverse of the R-chain map u from (15.362). It is well defined only up to chain homotopy, but this will be good enough for our purposes since its class $[\partial\varphi, \overline{\varphi}]$ in $Q^{n+1}(i) = H_{n+1}(\mathrm{cone}(i^{\%}))$ is well defined.

We have to show that with this definition $(\partial E \xrightarrow{i} E^{n+1-*}, [\partial\varphi, \overline{\varphi}])$ is a Poincaré pair. Consider the homotopy commutative ladder associated to a symmetric pair as described in Diagram (15.254) applied to our situation and remember that we have the canonical R-chain homotopy equivalence $v \colon \mathrm{cone}(i) \to E$. Substituting E into the ladder using v it looks as follows:

$$\begin{array}{ccccccc}
(\Sigma\partial E)^{n+1-*} & \xrightarrow{e^*} & E^{n+1-*} & \xrightarrow{\varphi_0^*} & E & \xrightarrow{i^*} & (\partial E)^{n+1-*} \\
{\scriptstyle \mathrm{ev}(\partial\varphi)}\downarrow & & {\scriptstyle \mathrm{ev}_l(\partial\varphi,\overline{\varphi})\circ v^*}\downarrow & {\scriptstyle \mathrm{ev}_m}\searrow & {\scriptstyle \varepsilon\cdot\mathrm{id}_E}\downarrow & & \downarrow{\scriptstyle \mathrm{ev}(\partial\varphi)} \\
\partial E & \xrightarrow{i} & E^{n+1-*} & \xrightarrow{\varphi_0} & E & \xrightarrow{e} & \Sigma\partial E.
\end{array} \qquad (15.363)$$

By definition we have

$$v \circ \mathrm{ev}_r(\partial\varphi, \overline{\varphi}) = \varepsilon \cdot \mathrm{id}_E,$$

so $\mathrm{ev}_r(\partial\varphi, \overline{\varphi})$ is an R-chain equivalence and therefore Exercise 15.275 implies that $\mathrm{ev}_l(\partial\varphi, \overline{\varphi})$ is an R-chain homotopy equivalence. The rows of the ladder are extended homotopy cofibration sequences. Hence Lemma 14.60 implies the map $ev(\partial \varphi_E)$ is an R-chain homotopy equivalence and the n-dimensional symmetric complex $(\partial E, [\partial \varphi])$ is Poincaré.

Finally we define the $(n + 1)$-dimensional symmetric Poincaré pair

$$(\partial E \xrightarrow{i} E^{n+1-*}, [\partial \varphi, \overline{\varphi}]) \qquad (15.364)$$

and in particular we get an n-dimensional symmetric Poincaré complex

$$(\partial E, [\partial \varphi]). \qquad (15.365)$$

The dependence of our construction on the various choices, mainly the polarisation φ of $[\varphi]$, is discussed in Proposition 15.367.

Definition 15.366 (The algebraic boundary of a symmetric complex) Given an $(n + 1)$-dimensional symmetric chain complex $(E, [\varphi])$, its *boundary* is the n-dimensional symmetric Poincaré complex $(\partial E, [\partial \varphi])$ of (15.365), which is well defined up to chain isomorphism of n-dimensional symmetric Poincaré complexes.

Proposition 15.367 (Dependence of algebraic boundary on the homotopy type of the input) *Let $(E, [\varphi])$ be an $(n + 1)$-dimensional symmetric chain complex. Then the following statements hold:*

(i) *The $(n + 1)$-dimensional symmetric Poincaré pair $(\partial E \xrightarrow{i} E^{n+1-*}, [\partial \varphi, \overline{\varphi}])$ of (15.364) is well defined up to chain isomorphism of $(n + 1)$-dimensional symmetric Poincaré pairs.*

In particular, the n-dimensional symmetric Poincaré complex $(\partial E, [\partial \varphi])$ of Definition 15.366 is well defined up to chain isomorphism of n-dimensional symmetric Poincaré complexes;

(ii) *A homotopy equivalence $f \colon (E, [\varphi]) \to (E^\flat, [\varphi^\flat])$ of $(n + 1)$-dimensional symmetric complexes induces a zigzag of homotopy equivalences between the associated boundaries $(\partial E, [\partial \varphi])$ and $(\partial E^\flat, [\partial \varphi^\flat])$.*

Proof. Throughout the proof we use notation from Construction 15.360 and we also add some temporary notation at suitable places to reflect various choices made.

(i) Consider two polarisations φ and φ' of $[\varphi]$ and let $\partial_\varphi E := \Sigma^{-1} \operatorname{cone}(\varphi_0)$ and $\partial_{\varphi'} E := \Sigma^{-1} \operatorname{cone}(\varphi'_0)$ to distinguish the two underlying chain complexes occurring in Construction 15.360 depending on the choice of the polarisation. Moreover let $i_\varphi \colon \partial_\varphi E \to E^{n+1-*}$ and $i_{\varphi'} \colon \partial_{\varphi'} E \to E^{n+1-*}$ be the two canonical maps. Since $[\varphi] = [\varphi'] \in Q^{n+1}(E)$, there is a chain homotopy $\widetilde{\varphi}_0 \colon \varphi_0 \simeq \varphi'_0$, which yields a chain isomorphism $a(\widetilde{\varphi}) \colon \partial_\varphi E \xrightarrow{\cong} \partial_{\varphi'} E$ by Exercise 14.56, and one easily checks that $i_{\varphi'} \circ a(\widetilde{\varphi}) = i_\varphi$.

Denote by $\operatorname{HPB}_\varphi$ the homotopy pullback of the diagram

$$W^{\%}(E) \xrightarrow{ev_E} E \otimes_R E \xleftarrow{\varphi_0 \otimes \operatorname{id}} E^{n+1-*} \otimes E$$

and analogously $\operatorname{HPB}_{\varphi'}$. Again by Exercise 14.56 we obtain a chain isomorphism $b(\widetilde{\varphi}) \colon \operatorname{HPB}_\varphi \to \operatorname{HPB}_{\varphi'}$.

15.5 L-Groups in Terms of Chain Complexes

Let u_φ and $u_{\varphi'}$ be the two versions of the map (15.362) obtained from the two choices of the polarisation of $[\varphi]$. A straightforward verification shows that the following diagram is commutative

$$\begin{array}{ccc} \mathrm{cone}(i_\varphi^{\%}) & \xrightarrow{u_\varphi} & \mathrm{HPB}_\varphi \\ {\scriptstyle z(a(\widetilde{\varphi}),\mathrm{id})}\downarrow \cong & & \cong \downarrow {\scriptstyle b(\widetilde{\varphi})} \\ \mathrm{cone}(i_{\varphi'}^{\%}) & \xrightarrow[u_{\varphi'}]{\simeq} & \mathrm{HPB}_{\varphi'} \end{array}$$

where $z(a(\widetilde{\varphi}), \mathrm{id})$ is an isomorphism induced by $a(\widetilde{\varphi})$ and $\mathrm{id}\colon E^{n+1-*} \to E^{n+1-*}$. Applying homology yields a commutative diagram of abelian groups

$$\begin{array}{ccccc} Q^n(\partial_\varphi E) & \xleftarrow{\partial_\varphi^{n+1}} & Q^{n+1}(i_\varphi) & \xrightarrow{H_{n+1}(u_\varphi)}_{\cong} & H_{n+1}(\mathrm{HPB}_\varphi) \\ {\scriptstyle a(\widetilde{\varphi})^{Q,n}}\downarrow \cong & & {\scriptstyle z(a(\widetilde{\varphi}),\mathrm{id})^{Q,n+1}}\downarrow \cong & & {\scriptstyle H_{n+1}(b(\widetilde{\varphi}))}\downarrow \cong \\ Q^n(\partial_{\varphi'} E) & \xleftarrow[\partial_{\varphi'}^{n+1}]{} & Q^{n+1}(i_{\varphi'}) & \xrightarrow[H_{n+1}(u_{\varphi'})]{\cong} & H_{n+1}(\mathrm{HPB}_{\varphi'}). \end{array}$$

In Construction 15.360 we considered cycles $(T(\varphi), \varepsilon \cdot \mathrm{id}_E, 0) \in (\mathrm{HPB}_\varphi)_{n+1}$ and $(T(\varphi'), \varepsilon \cdot \mathrm{id}_E, 0) \in (\mathrm{HPB}_{\varphi'})_{n+1}$. They define elements $[T(\varphi), \varepsilon \cdot \mathrm{id}_E, 0]$ in $H_{n+1}(\mathrm{HPB}_\varphi)$ and $[T(\varphi'), \varepsilon \cdot \mathrm{id}_E, 0]$ in $H_{n+1}(\mathrm{HPB}_{\varphi'})$ which are mapped to one another by $H_{n+1}(b(\widetilde{\varphi}))$ by the following argument. The chain isomorphism $b(\widetilde{\varphi})$ is given by the formula

$$b(\widetilde{\varphi})\colon (\kappa, \lambda, \mu) \mapsto (\kappa, \lambda, \mu + ((-1)^{n+1}\widetilde{\varphi}_0 \otimes \mathrm{id})(\lambda)).$$

Hence we get that $b(\widetilde{\varphi})(T(\varphi), \varepsilon \cdot \mathrm{id}_E, 0) = (T(\varphi), \varepsilon \cdot \mathrm{id}_E, T(\widetilde{\varphi})_0)$. Now the differential of $\mathrm{HPB}_{\varphi'}$ is given by the matrix

$$d_{\mathrm{HPB}_{\varphi'}} = \begin{pmatrix} d_{W^{\%}(E)} & 0 & 0 \\ 0 & d_{E^{n+1-*} \otimes_R E} & 0 \\ (-1)^{n+1}\mathrm{ev}_E & (-1)^n \varphi'_0 \otimes \mathrm{id} & d_{E \otimes_R E} \end{pmatrix} \colon (\mathrm{HPB}_{\varphi'})_{n+2} \to (\mathrm{HPB}_{\varphi'})_{n+1}.$$

Considering the $(n+1)$-chain

$$(T(\widetilde{\varphi}), 0, 0) \in (\mathrm{HPB}_{\varphi'})_{n+2}$$

and applying $d_{\mathrm{HPB}_{\varphi'}}$ we conclude

$$\begin{aligned} d_{\mathrm{HPB}_{\varphi'}}(T(\widetilde{\varphi}), 0, 0) &= (T(\varphi) - T(\varphi'), 0, (-1)^{n+1} T(\widetilde{\varphi})_0) \\ &= (T(\varphi), \varepsilon \cdot \mathrm{id}_E, (-1)^{n+1} T(\widetilde{\varphi})_0) - (T(\varphi'), \varepsilon \cdot \mathrm{id}_E, 0), \end{aligned}$$

which proves $[b(\widetilde{\varphi})(T(\varphi), \varepsilon \cdot \mathrm{id}_E, 0)] = [T(\varphi'), \varepsilon \cdot \mathrm{id}_E, 0]$.

We defined $[\partial\varphi, \overline{\varphi}]$ in $Q^{n+1}(i_\varphi)$ and $[\partial\varphi', \overline{\varphi'}]$ in $Q^{n+1}(i_{\varphi'})$ as the preimages of these triples under the horizontal isomorphisms $H_{n+1}(u_\varphi)$ and $H_{n+1}(u_{\varphi'})$ in the above diagram, and $[\partial\varphi]$ in $Q^n(\partial_\varphi E)$, and $[\partial\varphi']$ in $Q^n(\partial_{\varphi'} E)$ as their images under the horizontal maps ∂_φ^{n+1} and $\partial_{\varphi'}^{n+1}$. By the above discussion we obtain that the middle vertical arrow $a(\widetilde{\varphi})^{Q,n}$ sends $[\partial\varphi, \overline{\varphi}]$ to $[\partial\varphi', \overline{\varphi'}]$. Thus we obtain a chain isomorphism of $(n+1)$-dimensional symmetric Poincaré pairs

$$(\partial_\varphi E \xrightarrow{i_\varphi} E^{n+1-*}, [\partial\varphi, \overline{\varphi}]) \to (\partial_{\varphi'} E \xrightarrow{i_{\varphi'}} E^{n+1-*}, [\partial\varphi', \overline{\varphi'}])$$

and a chain isomorphism of n-dimensional symmetric Poincaré complexes

$$a(\widetilde{\varphi}) \colon (\partial_\varphi E, [\partial\varphi]) \xrightarrow{\cong} (\partial_{\varphi'} E, [\partial\varphi']).$$

(ii) Choose a polarisation φ of $[\varphi]$. Via $\varphi^\flat := f^{\%}(\varphi) = f \circ \varphi \circ f^*$, we obtain a polarisation of $[\varphi^\flat]$. The homotopy equivalence $f \colon (E, [\varphi]) \to (E^\flat, [\varphi^\flat])$ induces the following commutative diagram

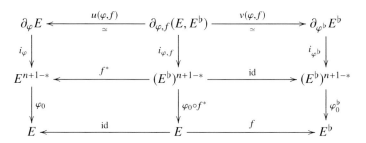

where

$$\partial_{\varphi, f}(E, E^\flat) := \Sigma^{-1} \operatorname{cone}(\varphi_0 \circ f^* \colon (E^\flat)^{n+1-*} \to E)$$

and the maps $u(\varphi, f)$ and $v(\varphi, f)$ are induced by the strictly commutative bottom squares and yield R-chain homotopy equivalences of pairs.

Denote by $\operatorname{HPB}_{\varphi, f}$ the homotopy pullback of the diagram

$$W^{\%}(E) \xrightarrow{\operatorname{ev}_E} E \otimes_R E \xleftarrow{(\varphi_0 \circ f^*) \otimes \operatorname{id}} (E^\flat)^{n+1-*} \otimes E.$$

The above diagram induces a corresponding zigzag of homotopy pullbacks, which appears after applying homology as the bottom row in the following diagram, where the bottom vertical maps are the respective versions of (15.362):

15.5 L-Groups in Terms of Chain Complexes

$$\begin{CD}
Q^n(\partial_\varphi E) @<{u(\varphi,f)^{Q,n}}<{\cong}< Q^n(\partial_{\varphi,f}(E, E^\flat)) @>{v(\varphi,f)^{Q,n}}>{\cong}> Q^n(\partial_{\varphi^\flat} E^\flat) \\
@A{\partial_\varphi^{n+1}}AA @A{\partial_{\varphi,f}^{n+1}}AA @A{\partial_{\varphi^\flat}^{n+1}}AA \\
Q^{n+1}(i_\varphi) @<{(u(\varphi,f),\mathrm{id})^{Q,n+1}}<{\cong}< Q^{n+1}(i_{\varphi,f}) @>{(v(\varphi,f),\mathrm{id})^{Q,n+1}}>{\cong}> Q^{n+1}(i_{\varphi^\flat}) \\
@V{H_{n+1}(u_\varphi)}V{\cong}V @V{H_{n+1}(u_{\varphi,f})}V{\cong}V @V{H_{n+1}(u_{\varphi^\flat})}V{\cong}V \\
H_{n+1}(\mathrm{HPB}_\varphi) @<{u_{\mathrm{HPB}}(\varphi,f)}<{\cong}< H_{n+1}(\mathrm{HPB}_{\varphi,f}) @>{v_{\mathrm{HPB}}(\varphi,f)}>{\cong}> H_{n+1}(\mathrm{HPB}_{\varphi^\flat}).
\end{CD}$$

Choose an R-chain homotopy inverse $g \colon E^\flat \to E$ and an R-chain homotopy $h \colon \mathrm{id}_E \simeq g \circ f$. Then the triple $(T(\varphi), \varepsilon \cdot g, \varepsilon \cdot h \circ \varphi_0^*)$ defines an $(n+1)$-dimensional cycle in $\mathrm{HPB}_{\varphi,f}$, and one can check

$$u_{\mathrm{HPB}}(\varphi, f) \colon [(T(\varphi), \varepsilon \cdot g, \varepsilon \cdot h \circ \varphi_0^*)] \mapsto [(T(\varphi), \varepsilon \cdot \mathrm{id}_E, 0)];$$
$$v_{\mathrm{HPB}}(\varphi, f) \colon [(T(\varphi), \varepsilon \cdot g, \varepsilon \cdot h \circ \varphi_0^*)] \mapsto [(T(\varphi^\flat), \varepsilon \cdot \mathrm{id}_{E^\flat}, 0)].$$

In the proof of the right equality the chain complex version of Vogt's lemma from [401, Proposition on page 627] is used. It tells us that there exists $k \colon \mathrm{id}_{E^\flat} \simeq f \circ g$ and $l \colon E \to E^\flat$ of degree 2 satisfying $dl - ld = f \circ h \simeq k \circ f$. Using these maps, one shows

$$d_{\mathrm{HPB}_{\varphi^\flat}}(0, \varepsilon \cdot k, \varepsilon \cdot l \circ \varphi_0^*) = (T(\varphi^\flat), \varepsilon \cdot f \circ g, \varepsilon \cdot f \circ h \circ \varphi_0^* \circ f^*) - (T(\varphi^\flat), \varepsilon \cdot \mathrm{id}_{E^\flat}, 0),$$

which proves the desired equation.

Define $[\partial(\varphi, f)] := \partial_{\varphi,f}^{n+1} H_{n+1}(u_{\varphi,f})^{-1}[(T(\varphi), \varepsilon \cdot g, \varepsilon \cdot h \circ \varphi_0^*)]$. By the above calculation and by the commutativity of the above diagram, we obtain the necessary zigzag. Note that the class $[(T(\varphi), \varepsilon \cdot g, \varepsilon \cdot h \circ \varphi_0^*)]$ is independent of the choice of the pair (g, h), which justifies the notation. $\square$

Exercise 15.368 Let $f \colon C \to D$ be an R-chain homotopy equivalence. Consider the induced map of chain complexes $\hom(\mathrm{id}_D, f) \colon \hom_R(D, C) \to \hom_R(D, D)$ that is given by $\hom(\mathrm{id}, f)(g) = f \circ g$.

Show that the homotopy fibre $\mathrm{hofib}(\hom(\mathrm{id}, f))$ is a contractible chain complex.

Remark 15.369 (Naturality properties of the zigzag from Proposition 15.367 (ii))

Let φ and φ' be two choices of a polarisation of $[\varphi]$ and let $\widetilde{\varphi} \in W^{\%}(E)_{n+1}$ be such that $d_{W^{\%}(E)}(\widetilde{\varphi}) = \varphi' - \varphi$. These induce two choices $\varphi^\flat$ and $(\varphi')^\flat$ of a polarisation of $[\varphi^\flat]$ by $\varphi^\flat := f^{\%}(\varphi)$ and $(\varphi')^\flat := f^{\%}(\varphi')$ and $\widetilde{\varphi}^\flat := f^{\%}(\widetilde{\varphi})$ satisfying $d_{W^{\%}(E')}(\widetilde{\varphi}^\flat) = (\varphi')^\flat - \varphi^\flat$.

Let $(\partial_\varphi E, [\partial \varphi])$ be the boundary obtained from the choice of the polarisation φ, and let $a(\widetilde{\varphi}) \colon (\partial_\varphi E, [\partial \varphi]) \to (\partial_{\varphi'} E, [\partial \varphi'])$ be the isomorphism of n-dimensional Poincaré complexes obtained in part (i) of the proof of Proposition 15.367. Analogous reasoning produces an isomorphism

$$a(\widetilde{\varphi}, f) \colon (\partial_{\varphi, f}(E, E^\flat), [\partial(\varphi, f)]) \to (\partial_{\varphi', f}(E, E^\flat), [\partial(\varphi', f)])$$

and a straightforward inspection shows that we have a strictly commutative diagram

$$\begin{array}{ccccc}
(\partial_\varphi E, [\partial\varphi]) & \xleftarrow{u(f,\varphi)}_{\simeq} & (\partial_{\varphi,f}(E, E^\flat), [\partial(\varphi, f)]) & \xrightarrow{v(f,\varphi)}_{\simeq} & (\partial_{\varphi^\flat} E^\flat, [\partial\varphi^\flat]) \\
a(\widetilde{\varphi}) \downarrow \cong & & a(\widetilde{\varphi},f) \downarrow \cong & & a(\widetilde{\varphi}^\flat) \downarrow \cong \\
(\partial_{\varphi'} E, [\partial\varphi']) & \xleftarrow{u(f,\varphi')}_{\simeq} & (\partial_{\varphi',f}(E, E^\flat), [\partial(\varphi', f)]) & \xrightarrow{v(f,\varphi')}_{\simeq} & (\partial_{(\varphi')^\flat} E^\flat, [\partial(\varphi')^\flat])
\end{array}$$

of n-dimensional symmetric complexes.

Next we discuss dependence of the zigzag when we vary f inside its chain homotopy class. Let $\widetilde{f} \colon f \simeq f'$ and denote by

$$a(\varphi, \widetilde{f}) \colon (\partial_{\varphi, f}(E, E^\flat), [\partial(\varphi, f)]) \to (\partial_{\varphi, f'}(E, E^\flat), [\partial(\varphi, f')])$$

the isomorphism induced via Exercise 14.56. We claim that there is a diagram

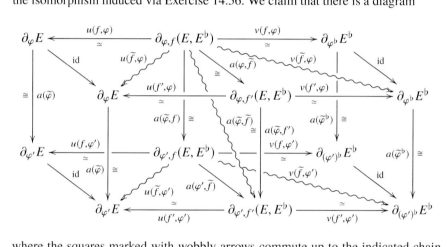

where the squares marked with wobbly arrows commute up to the indicated chain homotopy and the other squares commute strictly. Here the chain homotopy $u(\widetilde{f}, \varphi)$ is defined analogously to the map $u(f, \varphi)$, only replacing f by $\widetilde{f}$, and similarly $v(\widetilde{f}, \varphi)$ and $u(\widetilde{f}, \varphi')$, $v(\widetilde{f}, \varphi')$. The chain homotopy $a(\widetilde{\varphi}, \widetilde{f})$ is defined analogously to the map $a(\varphi, \widetilde{f})$, only replacing φ by $\widetilde{\varphi}$. Moreover, the three chain homotopies in each of the cubes are compatible.

The above diagram induces a strictly commutative diagram of Q-groups with all maps isomorphisms. We have already proved that the four left-right zigzags in this diagram are zigzags of homotopy equivalences of symmetric complexes. Moreover, above we have shown that the front and the back of the diagram are diagrams of the symmetric complexes. An obvious diagram chase tells us that then also the top and the bottom of the diagram are diagrams of symmetric complexes. An inspection shows that the structures are mapped to each other.

15.5 *L*-Groups in Terms of Chain Complexes

Remark 15.370 (The algebraic boundary of a symmetric complex – formulas)
Let $(E, [\varphi])$ be an $(n + 1)$-dimensional symmetric complex over R with a choice φ of a polarisation and let $d_r \colon E_r \to E_{r-1}$ denote the differential in the underlying chain complex.

Unravelling the definitions yields the following explicit formulas for $(\partial E, \partial \varphi)$ after we have chosen a polarisation φ of $[\varphi]$

$$\partial E_r = E^{n+1-r} \oplus E_{r+1};$$
$$(d_{\partial E})_r \colon \partial E_r = E^{n+1-r} \oplus E_{r+1} \to \partial E_{r-1} = E^{n+2-r} \oplus E_r, \quad (15.371)$$
$$(d_{\partial E})_r = \begin{pmatrix} (-1)^{r+1} d^*_{n+2-r} & 0 \\ -\varphi_{0,r} & -d_{r+1} \end{pmatrix},$$

and

$$\partial \varphi_{0,r} \colon \partial E^{n-r} = E_{r+1} \oplus E^{n+1-r} \to \partial E_r = E^{n+1-r} \oplus E_{r+1},$$
$$\partial \varphi_{0,r} = \begin{pmatrix} 0 & (-1)^{r+1} \\ (-1)^{rn+1} \varepsilon & (-1)^{n+r+1} \varphi_{1,r+1} \end{pmatrix};$$
$$\partial \varphi_{s,r} \colon \partial E^{n+s-r} = E_{r-s+1} \oplus E^{n+s-r} \to \partial E_r = E^{n+1-r} \oplus E_{r+1},$$
$$\partial \varphi_{s,r} = \begin{pmatrix} 0 & 0 \\ 0 & (-1)^{n+s+r+1} \varphi_{s+1,r+1} \end{pmatrix}.$$

For the nullhomology we can take $\overline{\varphi} = 0$.

These formulas hold since the chain map $u \colon \mathrm{cone}(i^{\%}) \xrightarrow{\simeq}$ HPB of (15.362) sends the specific cycle in $\mathrm{cone}(i^{\%})_{n+1}$ described by these formulas to the cycle $(T(\varphi), \varepsilon \cdot \mathrm{id}_E, 0)$ in HPB_{n+1}. The verification of this observation, whose details we leave to the reader, involves checking the requirements that $\nu \circ \mathrm{ev}_r(\partial \varphi, \overline{\varphi}) = \varepsilon \cdot \mathrm{id}_E$, that $\partial \varphi$ is a symmetric structure (i.e. it fulfils formulas (15.21)) and that $\overline{\varphi}/\partial \varphi$ maps to $T(\varphi)$ via $\nu^{\%}$.

Exercise 15.372 Prove that the formulas from Remark 15.370 yield an n-dimensional symmetric structure on ∂E.

Remark 15.373 (The algebraic boundary measures the failure of Poincaré duality) Note that an $(n + 1)$-dimensional symmetric complex (E, φ_E) is Poincaré if and only if the underlying chain complex ∂E of its boundary is contractible. More generally, an $(n + 1)$-dimensional symmetric complex (E, φ_E) is connected in the sense of Definition 15.159 if and only if the underlying chain complex ∂E of its boundary is (-1)-connected in the sense of Definition 14.121.

Remark 15.374 (Geometric interpretation of the algebraic boundary) Construction 15.360 also has a geometric analogue that arises from considering an n-dimensional manifold with boundary, say $(N, \partial N)$. Consider the chain complex $C(N, \partial N)$ and its suspended dual $C^{n-*}(N, \partial N)$. There is a symmetric structure on $C(N, \partial N)$, which is not Poincaré. However, there is the Poincaré duality $C^{n-*}(N, \partial N) \simeq C(N)$. Thus the mapping cone of the composition

$C^{n-*}(N, \partial N) \simeq C(N) \to C(N, \partial N)$ becomes homotopy equivalent to the mapping cone of the map $C(N) \to C(N, \partial N)$, which is $\Sigma C(\partial N)$.

The following proposition describes in which sense the algebraic boundary reverses the algebraic Thom Construction.

Proposition 15.375 (Algebraic boundary reverses algebraic Thom Construction – symmetric case) *Let $(f: C \to D, (\varphi, \delta\varphi))$ be an $(n+1)$-dimensional symmetric Poincaré pair. Consider the associated algebraic Thom Construction complex $(E, \varphi_E) = (\text{cone}(f), \delta\varphi/\varphi)$, which is an $(n+1)$-dimensional symmetric complex, see Construction 15.328 and Remark 15.331. Let $(i: \partial E \to E^{n+1-*}, [\partial \varphi_E, \overline{\varphi_E}])$ be its algebraic boundary, see Definition 15.366 and Remark 15.370.*

Then there exists a homotopy equivalence of $(n+1)$-dimensional symmetric Poincaré pairs

$$(g, k): (f: C \to D) \to (i: \partial E \to E^{n+1-*});$$
$$[z(g, k)((\varphi, \delta\varphi))] = [(\partial \varphi_E, \overline{\varphi_E})] \in Q^{n+1}(i).$$

Proof. The definitions of the algebraic Thom Construction and of the algebraic boundary lead to the following homotopy commutative diagram in which both rows are homotopy cofibration sequences and all vertical maps are R-chain homotopy equivalences:

$$\begin{array}{ccccc}
\partial E & \xrightarrow{i} & E^{n+1-*} & \xrightarrow{(\varphi_E)_0} & E \\
{\scriptstyle z} \downarrow {\scriptstyle \simeq} & & {\scriptstyle \text{ev}_l} \downarrow {\scriptstyle \simeq} & & \downarrow = \\
C & \xrightarrow{f} & D & \xrightarrow{e} & \text{cone}(f).
\end{array}$$

The right-hand square is homotopy commutative by the R-chain homotopy $\text{ev}_m : e \circ \text{ev}_l \simeq \text{ev}_r \circ e^* = (\varphi_E)_0$ obtained in Exercise 15.257, see also Diagram (15.254). Then by Lemma 14.60 we obtain the induced map z, which is also an R-chain homotopy equivalence. Setting $g = z$ and $k = \text{ev}_l$ provides us with the desired R-chain map (g, k).

To obtain the statement about the symmetric structures, consider the homotopy cartesian square of Proposition 15.340 applied to the chain map $f: C \to D$

$$\begin{array}{ccc}
\text{cone}(f^\%) & \xrightarrow{\text{ev}_r} & D \otimes_R \text{cone}(f) \\
{\scriptstyle z(e^\%, h^\%)} \downarrow & & \downarrow {\scriptstyle e \otimes \text{id}} \\
W^\%(\text{cone}(f)) & \xrightarrow{\text{ev}} & \text{cone}(f) \otimes_R \text{cone}(f)
\end{array}$$

and the same square applied to the map $i: \partial E \to E^{n+1-*}$, modified by substituting via $v: \text{cone}(i) \simeq E$

15.5 L-Groups in Terms of Chain Complexes

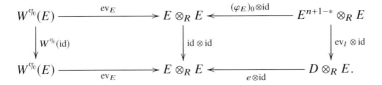

The R-chain homotopy equivalence of pairs (g, k) induces a map from the second square to the first. Removing the upper left entry from both diagrams yields the following diagram whose left square is strictly commutative and whose right square is homotopy commutative

$$\begin{array}{ccccc}
W^{\%}(E) & \xrightarrow{\mathrm{ev}_E} & E \otimes_R E & \xleftarrow{(\varphi_E)_0 \otimes \mathrm{id}} & E^{n+1-*} \otimes_R E \\
\downarrow W^{\%}(\mathrm{id}) & & \downarrow \mathrm{id} \otimes \mathrm{id} & & \downarrow \mathrm{ev}_l \otimes \mathrm{id} \\
W^{\%}(E) & \xrightarrow{\mathrm{ev}_E} & E \otimes_R E & \xleftarrow{e \otimes \mathrm{id}} & D \otimes_R E.
\end{array}$$

Recall that the pair $(\partial \varphi_E, \overline{\varphi_E})$ is obtained from the cycle $(T(\varphi_E), \varepsilon \cdot \mathrm{id}_E, 0)$ in the homotopy pullback associated to the first row. The pair $(\varphi, \delta\varphi)$ corresponds to the cycle $(\varphi, \mathrm{ev}_r, 0)$ in the homotopy pullback associated to the second row. The induced map of the homotopy pullbacks sends $(T(\varphi_E), \varepsilon \cdot \mathrm{id}_E, 0)$ to $(T(\varphi_E), T(\mathrm{ev}_l), (-1)^{n+1} T(\mathrm{ev}_m))$, which is homologous to $(\varphi_E, \mathrm{ev}_r, 0)$ via the $(n+2)$-chain $(\widetilde{\varphi}_E, \widetilde{\mathrm{ev}}_{(l,r)}, 0)$, where $\widetilde{\varphi}_E$ is the chain associated to φ_E as in Exercise 15.39 and $\widetilde{\mathrm{ev}}_{(l,r)}$ is taken from Exercise 15.258. This is seen by direct calculation using the equation $\widetilde{\varphi}_{E0} = \widetilde{\mathrm{ev}}_{(l,r)} \circ e - T(\mathrm{ev}_m)$ from Exercise 15.336. □

Remark 15.376 In Proposition 15.375 the assumption that the pair we start with is Poincaré is crucial. If we start with a pair that is not Poincaré and pass to the algebraic Thom Construction and then construct the boundary we end up with a different pair since the boundary always produces a Poincaré pair.

Remark 15.377 (Relation to Remark 15.344) At first glance Proposition 15.375 might seem to be in contradiction with Remark 15.344. There we identified additional data that we need to recover a symmetric pair from its algebraic Thom Construction, and now the construction of the algebraic boundary recovers the pair seemingly without additional data. But that is not quite right. In fact, we use the duality map φ_0 as the additional data. So in fact the map φ_0 serves two purposes in this construction. In the notation of Remark 15.344 it is both the map e and the map $\mathrm{ev}_r(\varphi, \delta\varphi)$. Hence the additional data are secretly present in Proposition 15.375 and there is no contradiction.

Remark 15.378 (Terminology regarding boundary) Note that, starting with an $(n+1)$-dimensional symmetric complex $(E, [\varphi])$, the boundary construction of this section produces an n-dimensional symmetric Poincaré complex $(\partial E, [\partial \varphi])$. This n-dimensional Poincaré complex is nullcobordant, but note that the nullcobordism is given by the chain complex E^{n+1-*}, it is not E itself. Similarly in the quadratic case below. This is in contrast with the terminology in the world of manifolds where an $(n+1)$-dimensional manifold with boundary $(W, \partial W)$ contains an n-dimensional closed manifold ∂W as its boundary and this manifold is nullcobordant with the nullcobordism being the manifold W itself. The terminology in algebraic surgery comes from the foundational paper [344] and is common in the area, so we do not want to change it. It also should not be confused with the notion of a boundary appearing Definition 15.221.

Remark 15.379 (Applications of the algebraic boundary) The algebraic boundary is important for several reasons. Since it is the source of a Poincaré pair, it is nullcobordant and thus its construction can be viewed as a device for producing nullcobordant complexes. Moreover, as we have seen in Proposition 15.375, any nullcobordant Poincaré complex is homotopy equivalent to a boundary and hence we not only have a construction, but also a characterisation of nullcobordant Poincaré complexes. This will be used in Section 15.6 in proving that the cobordism definition of the L-groups coincides with the form and formation definition.

The second useful feature recalled in Remark 15.373 is that the underlying chain complex of the boundary $(\partial E, [\partial \varphi])$ is contractible if and only if the starting complex $(E, [\varphi])$ is Poincaré and thus the boundary measures the failure of Poincaré duality. This will be used in Section 15.5.5 to show that algebraic surgery on Poincaré complexes produces Poincaré complexes.

The algebraic boundary has many other applications, for example it is essential in constructing the algebraic surgery exact sequence, see Section 15.9.5, but for us only the above two properties will be of mathematical concern. A pedagogically useful fact is that, as we will see, this construction is also an (easy) special case of algebraic surgery, so it makes sense to study it, before we proceed to the general algebraic surgery in the following section.

Construction 15.380 (The algebraic boundary of a quadratic complex) Let $(E, [\psi])$ be an $(n+1)$-dimensional quadratic complex (not necessarily Poincaré). We are going to associate to it an $(n+1)$-dimensional quadratic Poincaré pair

$$(\partial E \xrightarrow{i} E^{n+1-*}, [\partial \psi, \overline{\psi}]),$$

which will be unique up to chain isomorphism of $(n+1)$-dimensional quadratic Poincaré pairs. Note this in particular yields an n-dimensional quadratic Poincaré complex $(\partial E, [\partial \psi])$, which is unique up to chain isomorphism of n-dimensional quadratic Poincaré complexes.

We proceed analogously to the symmetric case. Choose a polarisation ψ or $[\psi]$. Define

$$\partial E := \Sigma^{-1} \operatorname{cone}((1+T)(\psi_0) \colon E^{n+1-*} \to E)$$

so that we have an extended homotopy cofibration sequence

$$\partial E \xrightarrow{i} E^{n+1-*} \xrightarrow{(1+T)(\psi_0)} E \xrightarrow{e} \Sigma \partial E$$

15.5 L-Groups in Terms of Chain Complexes

just as in the symmetric case. Applying Proposition 15.354 to the R-chain map $\partial E \xrightarrow{i} E^{n+1-*}$, we obtain the homotopy cartesian square

$$\begin{array}{ccc} \operatorname{cone}(i_\%) & \xrightarrow{\operatorname{ev}_r} & E^{n+1-*} \otimes_R \operatorname{cone}(i) \\ {\scriptstyle z(e_\%, j_\%)} \downarrow & & \downarrow {\scriptstyle \bar{e} \otimes \operatorname{id}} \\ W_\%(\operatorname{cone}(i)) & \xrightarrow{\operatorname{ev}} & \operatorname{cone}(i) \otimes_R \operatorname{cone}(i) \end{array}$$

where $\bar{e} \colon E^{n+1-*} \to \operatorname{cone}(i)$ is the canonical map and $j \colon 0 \simeq \bar{e}' \circ i$ the canonical chain homotopy. Using the canonical R-chain homotopy equivalence $v \colon \operatorname{cone}(i) \xrightarrow{\simeq} E$, which by definition satisfies $v \circ \bar{e} = (1+T)(\psi_0)$, we get a commutative diagram whose vertical arrows are $\mathbb{Z}$-chain homotopy equivalences

$$\begin{array}{ccccc} W_\%(\operatorname{cone}(i)) & \xrightarrow{\operatorname{ev}} & \operatorname{cone}(i) \otimes_R \operatorname{cone}(i) & \xleftarrow{\bar{e} \otimes \operatorname{id}} & E^{n+1-*} \otimes_R \operatorname{cone}(i) \\ {\scriptstyle W_\%(v)} \downarrow & & \downarrow {\scriptstyle v \otimes v} & & \downarrow {\scriptstyle \operatorname{id} \otimes v} \\ W_\%(E) & \xrightarrow{\operatorname{ev}_E} & E \otimes_R E & \xleftarrow{(1+T)(\psi_0) \otimes \operatorname{id}} & E^{n+1-*} \otimes_R E. \end{array}$$

Combining these two diagrams, we obtain a homotopy cartesian square

$$\begin{array}{ccc} \operatorname{cone}(i_\%) & \xrightarrow{(\operatorname{id} \otimes v) \circ \operatorname{ev}_r} & E^{n+1-*} \otimes_R E \\ {\scriptstyle W^\%(v) \circ z(e_\%, j_\%)} \downarrow & & \downarrow {\scriptstyle (1+T)(\psi_E)_0 \otimes \operatorname{id}} \\ W_\%(E) & \xrightarrow{\operatorname{ev}_E} & E \otimes_R E. \end{array}$$

Define HPB to be the homotopy pullback

$$\begin{array}{ccc} \operatorname{HPB} & \longrightarrow & E^{n+1-*} \otimes_R E \\ \downarrow & & \downarrow {\scriptstyle (1+T)(\psi_E)_0 \otimes \operatorname{id}} \\ W_\%(E) & \xrightarrow{\operatorname{ev}_E} & E \otimes_R E. \end{array}$$

Then the universal property of the homotopy pullback yields a $\mathbb{Z}$-chain homotopy equivalence

$$u \colon \operatorname{cone}(i_\%) \xrightarrow{\simeq} \operatorname{HPB}. \tag{15.381}$$

Now we use the same adjunctions as in the symmetric case to observe that the chain map $\varepsilon \cdot \operatorname{id}_E$ defines a cycle in $(E^{n+1-*} \otimes_R E)_{n+1} \cong \hom_R(E, E)_0$ whose image under $(1+T)(\psi_E) \otimes \operatorname{id}_E$ is a cycle in $(E \otimes_R E)_{n+1} \cong \hom_R(E^{n+1-*}, E)_0$, which equals $\operatorname{ev}(T((1+T)(\psi))) = \operatorname{ev}(1+T)(\psi)$. Therefore the triple

$$(T(\psi), \varepsilon \cdot \operatorname{id}_E, 0) \in \operatorname{HPB}_{n+1} = W_\%(E)_{n+1} \oplus (E^{n+1-*} \otimes_R E)_{n+1} \oplus (E \otimes_R E)_{n+2}$$

is a cycle in HPB. Let $(\partial\psi, \overline{\psi})$ be an $(n + 1)$-cycle in $\mathrm{cone}(i_\%)$ given as the image of $(T(\psi), \varepsilon \cdot \mathrm{id}_E, 0)$ under some chain homotopy inverse of the R-chain map u from (15.381). It is well defined only up to chain homotopy, but this will be good enough for our purposes since its class $[\partial\psi, \overline{\psi}]$ in $Q_{n+1}(i) = H_{n+1}(\mathrm{cone}(i_\%))$ is well defined.

Now we have to show that $(i\colon \partial E \to E^{n+1-*}, [\partial\psi, \overline{\psi}])$ is a Poincaré pair. This can be quickly seen using the observation that its symmetrisation is the symmetric Poincaré pair $(i\colon \partial E \to E^{n+1-*}, [\partial\varphi, \overline{\varphi}])$ obtained via Construction 15.360 from the symmetrisation (E, φ) of the original quadratic complex (E, ψ).

Finally we define the $(n + 1)$-dimensional quadratic Poincaré pair

$$(\partial E \xrightarrow{i} E^{n+1-*}, [\partial\psi, \overline{\psi}]) \qquad (15.382)$$

and in particular we get an n-dimensional quadratic Poincaré complex

$$(\partial E, [\partial\psi]). \qquad (15.383)$$

Proposition 15.384 (Dependence of algebraic boundary on the homotopy type of the input – quadratic case) *Let $(E, [\psi])$ be an $(n + 1)$-dimensional quadratic chain complex. Then the following statements hold:*

(i) *The $(n + 1)$-dimensional quadratic Poincaré pair $(\partial E \xrightarrow{i} E^{n+1-*}, [\partial\psi, \overline{\psi}])$ of (15.382) is well defined up to chain isomorphism of $(n + 1)$-dimensional quadratic Poincaré pairs.*
In particular, the n-dimensional quadratic Poincaré complex $(\partial E, [\partial\psi])$ of Definition 15.385 is well defined up to chain isomorphism of n-dimensional quadratic Poincaré complexes;
(ii) *A homotopy equivalence $f\colon (E, [\psi]) \to (E^\flat, [\psi^\flat])$ of $(n + 1)$-dimensional quadratic complexes induces a zigzag of homotopy equivalences between the associated boundaries $(\partial E, [\partial\psi])$ and $(\partial E^\flat, [\partial\psi^\flat])$.*

Proof. The proof is analogous to the proof of Proposition 15.367. □

Naturality properties of the zigzag from Proposition 15.384 (ii) are straightforward analogues of their counterparts in the symmetric case, as described in Remark 15.369. We leave the exact formulation and proof to the reader.

Definition 15.385 (The algebraic boundary of a quadratic complex) Let (E, ψ) be an $(n + 1)$-dimensional quadratic chain complex. The *boundary* of (E, ψ) is the n-dimensional quadratic Poincaré complex $[\partial E, \partial\psi]$ obtained via Construction 15.360, which is well defined up to chain isomorphism of n-dimensional quadratic Poincaré complexes.

Remark 15.386 Note that an n-dimensional quadratic complex (E, ψ) is Poincaré if and only if its boundary is contractible.

15.5 L-Groups in Terms of Chain Complexes

Remark 15.387 (Boundary of a quadratic complex – formulas) Unravelling the definitions yields the following explicit formulas for $(\partial E, \partial \psi)$:

$$\partial E_r = E^{n+1-r} \oplus E_{r+1},$$
$$(d_{\partial E})_r \colon \partial E_r = E^{n+1-r} \oplus E_{r+1} \to \partial E_{r-1} = E^{n-r} \oplus E_r, \qquad (15.388)$$
$$(d_{\partial E})_r = \begin{pmatrix} (-1)^{r+1} d^*_{n+2-r} & 0 \\ -(1+T)(\psi)_{0,r} & -d_{r+1} \end{pmatrix},$$

and

$$\partial \psi_{0,r} \colon \partial E^{n-r} = E_{r+1} \oplus E^{n-r} \to \partial E_r = E^{n-r} \oplus E_{r+1},$$
$$\partial \psi_{0,r} = \begin{pmatrix} 0 & (-1)^{r+1} \\ 0 & 0 \end{pmatrix};$$
$$\partial \psi_{s,r} \colon \partial E^{n-s-r} = E_{r+s+1} \oplus E^{n-s-r} \to \partial E_r = E^{n-r} \oplus E_{r+1},$$
$$\partial \psi_{s,r} = \begin{pmatrix} 0 & 0 \\ 0 & (-1)^{n+s+r+1} \psi_{s-1,r+1} \end{pmatrix}.$$

Again, for the nullhomotopy we can take $\overline{\psi} = 0$.

These formulas hold since the chain map $u \colon \mathrm{cone}(i_{\%}) \xrightarrow{\simeq} \mathrm{HPB}$ of (15.381) sends the specific cycle in $\mathrm{cone}(i_{\%})_{n+1}$ described by these formulas to the cycle $(T(\psi), \varepsilon \cdot \mathrm{id}_E, 0)$ in HPB_{n+1}. The verification of this observation, whose details we leave to the reader, involves checking the requirements that $\nu \circ \mathrm{ev}_r(\partial \psi, \overline{\psi}) = \varepsilon \cdot \mathrm{id}_E$, that $\partial \psi$ is a quadratic structure (i.e. it fulfils formulas (15.22)) and that $\overline{\psi}/\partial \psi$ maps to $T(\psi)$ via $\nu_{\%}$.

Proposition 15.389 (Algebraic boundary reverses algebraic Thom – quadratic case) Let $(f \colon C \to D, (\psi, \delta\psi))$ be an $(n+1)$-dimensional quadratic Poincaré pair. Consider its algebraic Thom Construction $(E, \psi) = (\mathrm{cone}(f), \delta\psi/\psi)$, which is an $(n+1)$-dimensional quadratic complex, see Construction 15.346 and Remark 15.349. Let $(i \colon \partial E \to E^{n+1-*}, (\partial \psi, \overline{\psi}))$ be its algebraic boundary, see Definition 15.385 and Remark 15.387.

Then there exists a homotopy equivalence of $(n+1)$-dimensional quadratic Poincaré pairs

$$(g, k) \colon (f \colon C \to D) \to (i \colon \partial E \to E^{n+1-*});$$
$$[z(g, k)((\psi, \delta\psi))] = [(\partial \psi, \overline{\psi})] \in Q_{n+1}(i).$$

Proof. The proof is analogous to the proof of Proposition 15.375. □

15.5.5 Algebraic Surgery

Algebraic surgery is a technical tool that is needed in the following Section 15.6, where it is shown that the L-groups defined in the previous Section 15.5 coincide with the L-groups defined in Chapters 8 and 9.

Recall the well-known fact that, using Morse theory, any geometric cobordism can be decomposed into elementary cobordisms, which are in turn obtained via surgery, see [189, Chapter 6], [298, part I]. The notion of algebraic surgery, which we describe next, is an algebraic analogue of geometric surgery. We start with the symmetric case and later proceed to quadratic complexes.

The basic idea, how this is used in the identification of the L-groups, is that in the quadratic case algebraic surgery can make Poincaré complexes highly connected and highly connected Poincaré complexes correspond to forms and formations, as already noted in Section 15.4. An interesting feature of algebraic surgery is that the process of obtaining highly connected cobordant complexes can be made in one step, see the proof of Proposition 15.429 whereas the process of making a normal map of degree one highly connected via the geometric surgery in Chapters 4, 8 and 9 had to be made step-by-step. Full details are to be found in Section 15.6.

Remark 15.390 (Review of geometric surgery) Recall the basic setup for a surgery on an n-dimensional closed manifold M. For simplicity we assume that we are in a simply connected situation. The starting *data* is an embedding $q \colon S^k \times D^{n-k} \hookrightarrow M$. Then we define the *effect* of the surgery on M using x to be

$$M' = (M \smallsetminus \mathrm{int}(q)) \cup_{S^k \times S^{n-k-1}} D^{k+1} \times S^{n-k-1},$$

and the *trace* of the surgery is the cobordism

$$W = M \times [0, 1] \cup_{S^k \times D^{n-k}} D^{k+1} \times D^{n-k}$$

between M and M'. From the homotopy-theoretic point of view, we have

$$M \cup_q D^{k+1} \simeq W \simeq M' \cup_{q'} D^{n-k}$$

for some embedding $q' \colon D^{k+1} \times S^{n-k-1} \hookrightarrow M'$. Hence one may view surgery on M as a process in which we attach one $(k+1)$-dimensional cell and detach another $(n-k)$-dimensional cell, see also Remark 4.17. The whole situation can be conveniently described in a commutative braid as in Diagram (15.391) where q^* and $(q')^*$ stand for chain maps that represent Poincaré duals of cycles defined by q and q'.

15.5 *L*-Groups in Terms of Chain Complexes

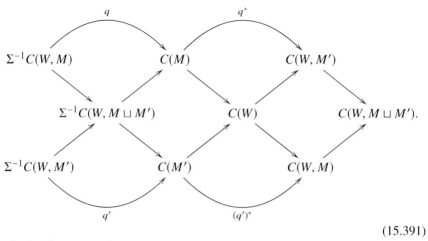

(15.391)

In the diagram we have

$$C(W, M) = (k+1)[\mathbb{Z}];$$
$$C(W, M') = (n-k)[\mathbb{Z}];$$
$$C^{n+1-*}(W, M) \simeq C(W, M').$$

In the sequel we fix a ring R with involution and all chain complexes and forthcoming constructions are to be understood over R.

The following construction is due to Ranicki [348, Definition 1.12].

Construction 15.392 (Algebraic surgery – symmetric case)

Let $(C, [\varphi])$ be an n-dimensional symmetric chain complex.

The *data* for an algebraic surgery on $(C, [\varphi])$ is an $(n+1)$-dimensional symmetric pair $(f \colon C \to D, [\varphi, \delta\varphi])$ whose boundary in the sense of Definition 15.221 is $(C, [\varphi])$.

The *effect* of the algebraic surgery on $(C, [\varphi])$ using $(f \colon C \to D, [\varphi, \delta\varphi])$ is the n-dimensional symmetric chain complex $(C', [\varphi'])$, which will be defined below and is unique up to chain isomorphism of n-dimensional symmetric chain complexes.

Choose a polarisation $(\varphi, \partial\varphi)$ of $[\varphi, \partial\varphi]$. Note that it induces a polarisation φ of $[\varphi]$.

The underlying chain complex is defined by

$$C' := \Sigma^{-1} \operatorname{cone}(\operatorname{ev}_r(\varphi, \delta\varphi)).$$

We have the following diagram

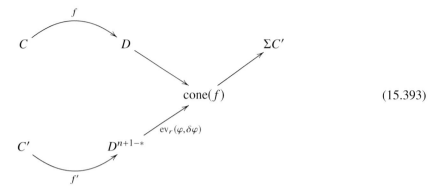

(15.393)

where the sequence involving the map $\mathrm{ev}_r(\varphi, \delta\varphi)$ is a part of an extended homotopy cofibration sequence induced by this map, as explained in (14.87) in Remark 14.85. The map f' is defined via desuspending the corresponding map in that sequence.

The following part of Diagram (15.391)

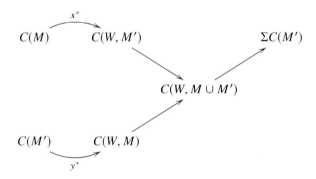

corresponds to Diagram (15.393).

Next we define the symmetric structure on C'. Applying Proposition 15.340 to $f' \colon C' \to D^{n+1-*}$ yields the homotopy cartesian square

$$\begin{array}{ccc} \mathrm{cone}((f')^{\%}) & \xrightarrow{\mathrm{ev}_r} & D^{n+1-*} \otimes_R \mathrm{cone}(f') \\ {\scriptstyle z((e')^{\%},(j')^{\%})} \Big\downarrow & & \Big\downarrow {\scriptstyle e' \otimes \mathrm{id}} \\ W^{\%}(\mathrm{cone}(f')) & \xrightarrow[\mathrm{ev}]{} & \mathrm{cone}(f') \otimes_R \mathrm{cone}(f') \end{array} \qquad (15.394)$$

where $e' \colon D^{n+1-*} \to \mathrm{cone}(f')$ is the canonical map and $j' \colon 0 \simeq e' \circ f'$ the canonical chain homotopy. We have the canonical chain homotopy equivalence of Definition 14.61 associated to the homotopy cofibration sequence $C' \xrightarrow{f'} D^{n+1-*} \xrightarrow{\mathrm{ev}_r(\varphi, \delta\varphi)} \mathrm{cone}(f)$ appearing in Diagram (15.393).

15.5 L-Groups in Terms of Chain Complexes

$$z \colon \operatorname{cone}(f') \xrightarrow{\simeq} \operatorname{cone}(f). \tag{15.395}$$

It provides us with a commutative diagram whose vertical arrows are chain homotopy equivalences

$$\begin{array}{ccccc}
W^{\%}(\operatorname{cone}(f')) & \xrightarrow{\operatorname{ev}'} & \operatorname{cone}(f') \otimes_R \operatorname{cone}(f') & \xleftarrow{e' \otimes \operatorname{id}} & D^{n+1-*} \otimes_R \operatorname{cone}(f') \\
{\scriptstyle W^{\%}(z)} \downarrow \simeq & & {\scriptstyle z \otimes z} \downarrow \simeq & & {\scriptstyle \operatorname{id} \otimes z} \downarrow \simeq \\
W^{\%}(\operatorname{cone}(f)) & \xrightarrow{\operatorname{ev}} & \operatorname{cone}(f) \otimes_R \operatorname{cone}(f) & \xleftarrow[\operatorname{ev}_r(\varphi,\delta\varphi) \otimes \operatorname{id}]{} & D^{n+1-*} \otimes_R \operatorname{cone}(f).
\end{array} \tag{15.396}$$

Define HPB to be the homotopy pullback

$$\begin{array}{ccc}
\operatorname{HPB} & \longrightarrow & D^{n+1-*} \otimes_R \operatorname{cone}(f) \\
\downarrow & & \downarrow {\scriptstyle \operatorname{ev}_r(\varphi,\delta\varphi) \otimes \operatorname{id}} \\
W^{\%}(\operatorname{cone}(f)) & \xrightarrow{\operatorname{ev}} & \operatorname{cone}(f) \otimes_R \operatorname{cone}(f).
\end{array}$$

The universal property of the homotopy pullback together with the homotopy cartesian square (15.394) and the Diagram (15.396) provides us with a chain homotopy equivalence

$$u \colon \operatorname{cone}((f')^{\%}) \xrightarrow{\simeq} \operatorname{HPB}. \tag{15.397}$$

Now we use adjunctions in a similar way as in the Construction 15.360 of algebraic boundary to observe that the chain map $e \colon D \to \operatorname{cone}(f)$ defines a cycle in

$$(D^{n+1-*} \otimes_R \operatorname{cone}(f))_{n+1} \cong \hom_R(D, \operatorname{cone}(f))_0$$

and the chain map $((\delta\varphi/\varphi)_0)$ defines a cycle in

$$(\operatorname{cone}(f)^{\otimes 2})_{n+1} = \hom_R(\operatorname{cone}(f)^{n+1-*}, \operatorname{cone}(f))_0.$$

By Diagram (15.254) and abbreviating $\operatorname{ev}_r = \operatorname{ev}_r(\varphi, \delta\varphi)$ we have

$$(\operatorname{ev}_r \otimes \operatorname{id})(\varepsilon \cdot e) = \operatorname{id} \circ e \circ \operatorname{ev}_r^* = T(\operatorname{ev}_r \circ e^*) = T(\delta\varphi/\varphi)_0.$$

Hence we obtain a preferred cycle in the target of the chain map u of (15.397), namely the triple

$$(T(\delta\varphi/\varphi), \varepsilon \cdot e, 0). \tag{15.398}$$

Let $(-\varphi', \delta\varphi')$ be the image of $(T(\delta\varphi/\varphi), \varepsilon \cdot e, 0)$ under some chain homotopy inverse of the chain homotopy equivalence u of (15.397). This is well defined only up to chain homotopy, but that will be enough for our purposes since only the class of $(-\varphi', \delta\varphi')$ in the corresponding Q-group will matter.

Altogether we have a polarised $(n+1)$-dimensional symmetric pair

$$(f' \colon C' \to D^{n+1-*}, (\varphi', -\delta\varphi')),$$

which contains the polarised n-dimensional symmetric chain complex (C', φ').

We claim that the $(n+1)$-dimensional symmetric pair

$$(f' \colon C' \to D^{n+1-*}, [\varphi', -\delta\varphi']) \qquad (15.399)$$

and the n-dimensional symmetric chain complex

$$(C', [\varphi']) \qquad (15.400)$$

are well defined, i.e., are independent of the choice of polarisation $(\varphi, \partial\varphi)$ of $[\varphi, \partial\varphi]$, up to chain isomorphism of $(n+1)$-dimensional symmetric pairs and of n-dimensional symmetric chain complexes. To see that the underlying chain complexes obtained from two different polarisations are isomorphic, it is enough to observe that the two evaluation maps obtained from these polarisations are chain homotopic, see Definition 15.247 and Exercise 14.56. We leave the details of the rest of the proof of the independence claim to the reader since it is analogous to the proof of Proposition 15.367 (i). This finishes the Construction 15.392.

Definition 15.401 (The effect of algebraic surgery – symmetric case) We will call the n-dimensional symmetric chain complex $(C', [\varphi'])$ of (15.400), which is well defined up to chain isomorphism of symmetric n-dimensional Poincaré complexes, the *effect of the algebraic surgery on* $(C, [\varphi])$ *using the data* $(f \colon C \to D, [\varphi, \delta\varphi])$.

Remark 15.402 (Summary diagram for symmetric algebraic surgery) The two polarised $(n+1)$-dimensional symmetric pairs in Construction 15.392

$$(C \xrightarrow{f} D, (\varphi, \delta\varphi)) \quad \text{and} \quad (C' \xrightarrow{f'} D^{n+1-*}, (\varphi', -\delta\varphi'))$$

have associated homotopy commutative ladders, see 15.254, which involve the extended homotopy cofibration sequences

$$C \xrightarrow{f} D \xrightarrow{e} \operatorname{cone}(f) \xrightarrow{r} \Sigma C$$

and

$$C' \xrightarrow{f'} D^{n+1-*} \xrightarrow{e'} \operatorname{cone}(f') \xrightarrow{r'} \Sigma C'.$$

Using the canonical chain homotopy equivalence $z \colon \operatorname{cone}(f') \xrightarrow{\simeq} \operatorname{cone}(f)$ of (15.395), which satisfies $z \circ e' = \operatorname{ev}_r(\varphi, \delta\varphi)$, these fit together into the following homotopy commutative diagram

15.5 L-Groups in Terms of Chain Complexes

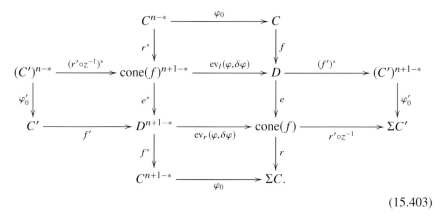

(15.403)

Remark 15.404 (Geometric versus algebraic surgery setup) One formal difference between the situation described in Remark 15.390 and Construction 15.392 is that the starting data in 15.392 are "cohomological", rather than "homological", meaning, in the notation of Diagram (15.391), we start with x^* rather than x. But this is really just a formal difference, and it is possible to translate between these data via the duality.

The philosophy of the approach to algebraic surgery is to express the process of surgery in terms of the cobordism given by the trace. The advantage of this approach is that, when we start with the cohomological data, these give us an $(n+1)$-dimensional symmetric pair since they correspond to the inclusion $M \to W/M'$ and the pair $(W, M \sqcup M')$ has the fundamental class $[W]$. The algebraic Thom Construction provides us with an $(n+1)$-dimensional symmetric structure on the quotient $W/M \sqcup M'$, which can be pushed further to an $(n+1)$-dimensional symmetric structure on $\Sigma M'$. This turns out to have a desuspension since it in fact comes from the suspension of the fundamental class $[M']$. The point here is that the cohomological setup enables us to use the technology of constructing symmetric complexes and pairs by the reverse of the algebraic Thom Construction given by Proposition 15.340 and Proposition 15.343.

Remark 15.405 (Motivation for the triple from display (15.398)) Algebraic surgery can also be viewed as a variation on the ideas described in Remark 15.344. Indeed, suppose we start with a symmetric pair $(\varphi, \delta\varphi)$ and we pass to the quotient $\delta\varphi/\varphi$ and we forget that this symmetric structure came from a pair. We can also ask ourselves whether the quotient $\delta\varphi/\varphi$ is the result of the algebraic Thom Construction on some other symmetric pair $(f' \colon C' \to D', (\varphi', \delta\varphi'))$ such that $\mathrm{cone}(f) \simeq \mathrm{cone}(f')$. And, if yes, then what information do we need to recover the pair $(\varphi', \delta\varphi')$? The answer is that, if we know the map of underlying chain complexes $e' \colon D' \to \mathrm{cone}(f')$ and hence we have that C' is the homotopy fibre of e', then we also need a cycle, say $\bar{e} \in D' \otimes \mathrm{cone}(f') \cong \hom((D')^{-*}, \mathrm{cone}(f'))$, which means a hypothetical evaluation map $\bar{e} = \mathrm{ev}_r(\varphi', \delta\varphi')$, but up to homotopy that is all we need. In Construction 15.392 such a map e' is up to homotopy provided by $\mathrm{ev}_r(\varphi, \delta\varphi)$ and $\bar{e}$ is provided by $e \colon D \to \mathrm{cone}(f)$ using $\mathrm{cone}(f) \simeq \mathrm{cone}(f')$.

Remark 15.406 Let $(C, [\varphi])$ be an n-dimensional symmetric complex. Then the inverse of the algebraic boundary $(\partial C, -[\partial \varphi])$, which is by Definition 15.366 an $(n-1)$-dimensional symmetric Poincaré complex, is isomorphic to the effect of the surgery on $(0,0)$ with data the symmetric pair $(0 \to C, [0, \varphi])$. To see this, observe that in this case, given a polarisation φ, the relative evaluation map $\mathrm{ev}_r(0, \varphi)$ used in Construction 15.392 is just the 0-th component φ_0.

Remark 15.407 (Algebraic surgery on a symmetric complex – formulas) Let (C, φ) be a polarised n-dimensional symmetric complex and let the polarised data for algebraic surgery on (C, φ) be $(f \colon C \to D, (\varphi, \delta\varphi))$. The polarised effect (C', φ') is given by the following explicit formulas:

$$C'_r = D^{n+1-r} \oplus C_r \oplus D_{r+1};$$

$$(d_{C'})_r \colon C'_r = D^{n+1-r} \oplus C_r \oplus D_{r+1} \to C'_{r-1} = D^{n+2-r} \oplus C_{r-1} \oplus D_r;$$

$$(d_{C'})_r = \begin{pmatrix} (-1)^{r+1}(d_D)^*_{n+2-r} & 0 & 0 \\ -\varphi_{0,r-1} \circ f^* & (d_C)_r & 0 \\ -\delta\varphi_{0,r} & -f & -(d_D)_{r+1} \end{pmatrix}, \text{ and}$$

$$\varphi'_{0,r} \colon (C')^{n-r} = D_{r+1} \oplus C^{n-r} \oplus D^{n-r+1} \to C'_r = D^{n+1-r} \oplus C_r \oplus D_{r+1};$$

$$\varphi'_{0,r} = \begin{pmatrix} 0 & 0 & (-1)^{r+1} \\ 0 & (-1)^{\overline{\delta}(n,0,r)} \varepsilon e_r^{-1} \circ \varphi^*_{0,n-r} & (-1)^{n+r+1} \varphi_{1,r} \circ f^* \\ (-1)^{rn+1} \varepsilon & 0 & (-1)^{n+r+1} \delta\varphi_{1,r+1} \end{pmatrix};$$

$$\varphi'_{s,r} \colon (C')^{n+s-r} = D_{r-s+1} \oplus C^{n+s-r} \oplus D^{n+s-r+1} \to$$

$$C'_r = D^{n+1-r} \oplus C_r \oplus D_{r+1};$$

$$\varphi'_{s,r} = \begin{pmatrix} 0 & 0 & 0 \\ 0 & (-1)^{\overline{\delta}(n,s,r)} \varepsilon e_r^{-1} \circ \varphi^*_{s,n+s-r} & (-1)^{n+s+r+1} \varphi_{s+1,r} \circ f^* \\ 0 & 0 & (-1)^{n+s+r+1} \delta\varphi_{s+1,r+1} \end{pmatrix},$$

where $\overline{\delta}(n,s,r) = r(n+s) + 1 = \delta(n,s,r) + n + s + 1 + r$ and $e_r = e(C_r)$. The component $\delta\varphi' \in W^{\%}(D^{n+1-*})$ is defined to be zero.

These formulas hold since the chain map $u \colon \mathrm{cone}((f')^{\%}) \xrightarrow{\simeq} \mathrm{HPB}$ from (15.397) sends the specific cycle in $\mathrm{cone}((f')^{\%})_{n+1}$ described by these formulas to the cycle $(T(\delta\varphi/\varphi), \varepsilon \cdot e, 0)$ in HPB_{n+1}. The verification of this observation, whose details we leave to the reader, includes checking the requirements that the element $\delta\varphi'/\varphi'$ is mapped to $T(\delta\varphi/\varphi)$ under the chain homotopy equivalence $W^{\%}(z) \colon W^{\%}(\mathrm{cone}(f')) \xrightarrow{\simeq} W^{\%}(\mathrm{cone}(f))$ induced by the chain homotopy equivalence z of (15.395).

This proves that the above formulas give explicit representatives of the symmetric structures, which are a priori only well defined as a class in the appropriate Q-group.

15.5 L-Groups in Terms of Chain Complexes

The following construction is due to [348, Definition 1.12].

Construction 15.408 (Algebraic surgery – quadratic case)
Let $(C, [\psi])$ be an n-dimensional quadratic chain complex. We proceed analogously to Construction 15.392 making necessary amendments.

The *data for an algebraic surgery* on $(C, [\psi])$ is an $(n + 1)$-dimensional quadratic pair $(f\colon C \to D, [\psi, \delta\psi])$ whose boundary in the sense of Definition 15.221 is $(C, [\psi])$.

The *effect of the algebraic surgery on* (C, ψ) using $(f\colon C \to D, [\psi, \delta\psi])$ is the n-dimensional quadratic chain complex $(C', [\psi'])$, which will be defined below and is unique up to chain isomorphism of n-dimensional quadratic chain complexes.

Choose a polarisation $(\psi, \partial\psi)$ of $[\psi, \delta\psi]$. Note that it induces a polarisation ψ of $[\psi]$. The underlying chain complex is defined by

$$C' = \Sigma^{-1} \operatorname{cone}(\operatorname{ev}_r(\psi, \delta\psi)).$$

We have an extended homotopy cofibration sequence

$$C' \xrightarrow{f'} D^{n+1-*} \xrightarrow{\operatorname{ev}_r(\psi, \delta\psi)} \operatorname{cone}(f) \xrightarrow{r'} \Sigma C'$$

where f' is defined by desuspending the chain map $\Sigma C' \to \Sigma D^{n+1-*}$ appearing in the extended homotopy cofibration sequence associated to $\operatorname{ev}_r(\psi, \delta\psi)$ via the rotation trick from the proof of Theorem 14.79.

Next we define the quadratic structure on C'. Applying Proposition 15.354 to $f'\colon C' \to D^{n+1-*}$ yields the homotopy cartesian square

$$\begin{array}{ccc}
\operatorname{cone}((f')_\%) & \xrightarrow{\operatorname{ev}_r} & D^{n+1-*} \otimes_R \operatorname{cone}(f') \\
{\scriptstyle z(e_\%, j_\%)}\downarrow & & \downarrow{\scriptstyle \bar{e} \otimes \operatorname{id}} \\
W_\%(\operatorname{cone}(f')) & \xrightarrow{\operatorname{ev}} & \operatorname{cone}(f') \otimes_R \operatorname{cone}(f').
\end{array}$$

The canonical chain homotopy equivalence $z\colon \operatorname{cone}(f') \xrightarrow{\simeq} \operatorname{cone}(f)$ of (15.395) provides us with a commutative diagram whose vertical arrows are chain homotopy equivalences

$$\begin{array}{ccccc}
W_\%(\operatorname{cone}(f')) & \xrightarrow{\operatorname{ev}} & \operatorname{cone}(f') \otimes_R \operatorname{cone}(f') & \xleftarrow{e' \otimes \operatorname{id}} & D^{n+1-*} \otimes_R \operatorname{cone}(f') \\
{\scriptstyle W_\%(z)}\downarrow{\scriptstyle \simeq} & & {\scriptstyle z \otimes z}\downarrow{\scriptstyle \simeq} & & {\scriptstyle \operatorname{id} \otimes z}\downarrow{\scriptstyle \simeq} \\
W_\%(\operatorname{cone}(f)) & \xrightarrow{\operatorname{ev}} & \operatorname{cone}(f) \otimes_R \operatorname{cone}(f) & \xleftarrow{\operatorname{ev}_r(\psi, \delta\psi) \otimes \operatorname{id}} & D^{n+1-*} \otimes_R \operatorname{cone}(f).
\end{array}$$

Define HPB to be the homotopy pullback

$$\begin{array}{ccc} \text{HPB} & \longrightarrow & D^{n+1-*} \otimes_R \text{cone}(f) \\ \downarrow & & \downarrow {\scriptstyle \text{ev}_r(\psi,\delta\psi)\otimes\text{id}} \\ W_{\%}(\text{cone}(f)) & \xrightarrow{\text{ev}} & \text{cone}(f) \otimes_R \text{cone}(f). \end{array}$$

The universal property of the homotopy pullback provides us with a chain homotopy equivalence

$$u \colon \text{cone}((f')_{\%}) \xrightarrow{\simeq} \text{HPB} . \tag{15.409}$$

Now we use adjunctions in a similar way as in the symmetric case to observe that we have a preferred element in the target of this map, namely, the triple

$$(T(\delta\psi/\psi), \varepsilon \cdot e, 0).$$

Let $(-\psi', \delta\psi')$ be the image of $(T(\delta\psi/\psi), \varepsilon \cdot e, 0)$ under some chain homotopy inverse of u. This is well defined only up to chain homotopy, but that will be enough for our purposes since only its class in the Q-group will be relevant.

Altogether we have a polarised $(n + 1)$-dimensional quadratic pair

$$(f' \colon C' \to D^{n+1-*}, (\psi', -\delta\psi')).$$

It contains the polarised n-dimensional quadratic chain complex (C', ψ').

We claim that the $(n + 1)$-dimensional quadratic pair

$$(f' \colon C' \to D^{n+1-*}, [\psi', -\delta\psi']) \tag{15.410}$$

and the n-dimensional quadratic chain complex

$$(C', [\psi']) \tag{15.411}$$

are well defined, i.e., are independent of the choice of polarisation $(\psi, \partial\psi)$ of $[\psi, \partial\psi]$ up to chain isomorphism of $(n+1)$-dimensional quadratic pairs and of n-dimensional quadratic chain complexes. For the underlying chain complexes the argument is the same as in the symmetric case. We leave the details of the rest of the proof of the independence claim to the reader since it is analogous to the proof of Proposition 15.384 (i). This finishes the Construction 15.392.

Definition 15.412 (The effect of algebraic surgery – quadratic case) We will call the n-dimensional quadratic chain complex $(C', [\psi'])$ of (15.411), which is well defined up to chain isomorphism of quadratic n-dimensional Poincaré complexes, the *effect of the algebraic surgery on* $(C, [\psi])$ *using the data* $(f \colon C \to D, [\psi, \delta\psi])$.

Remark 15.413 Let $(C, [\psi])$ be an n-dimensional quadratic complex. Then inverse of its boundary $(\partial C, [-\partial \psi])$, which is by Definition 15.385 an $(n-1)$-dimensional quadratic Poincaré complex, is isomorphic to the effect of the surgery on $(0, 0)$ with

15.5 *L-Groups in Terms of Chain Complexes* 717

data the quadratic pair $(0 \to C, [0, \psi])$ by an argument analogous to the symmetric case.

Remark 15.414 (Algebraic surgery on a quadratic complex – formulas) Let (C, ψ) be a polarised n-dimensional quadratic complex, and let the polarised data for algebraic surgery on (C, ψ) be $(f \colon C \to D, (\psi, \delta\psi))$. The polarised effect (C', ψ') is given by the following explicit formulas:

$$C'_r = D^{n+1-r} \oplus C_r \oplus D_{r+1};$$

$$d_{C'} \colon C'_r = D^{n+1-r} \oplus C_r \oplus D_{r+1} \to C'_{r-1} = D^{n+2-r} \oplus C_{r-1} \oplus D_r;$$

$$(d_{C'})_r = \begin{pmatrix} (-1)^{r+1}(d_D)^*_{n+2-r} & 0 & 0 \\ -(1+T)(\psi)_{0,r-1} \circ f^* & (d_C)_r & 0 \\ -(1+T)(\delta\psi)_{0,r} & -f & -(d_D)_{r+1} \end{pmatrix},$$

and

$$\psi'_{0,r} \colon (C')^{n-r} = D_{r+1} \oplus C^{n-r} \oplus D^{n-r+1} \to C'_r = D^{n+1-r} \oplus C_r \oplus D_{r+1},$$

$$\psi'_{0,r} = \begin{pmatrix} 0 & 0 & (-1)^{r+1} \\ 0 & (-1)^{\overline{\delta}(n,0,r)} \varepsilon e_r^{-1} \circ \psi^*_{0,n-r} & 0 \\ 0 & 0 & 0 \end{pmatrix};$$

$$\psi'_{s,r} \colon (C')^{n-s-r} = D_{r+s+1} \oplus C^{n-s-r} \oplus D^{n-s-r+1} \to$$
$$C'_r = D^{n+1-r} \oplus C_r \oplus D_{r+1},$$

$$\psi'_{s,r} = \begin{pmatrix} 0 & 0 & 0 \\ 0 & (-1)^{\overline{\delta}(n,s,r)} \varepsilon e_r^{-1} \circ \psi^*_{s,n-s-r} & (-1)^{n+s+r+1}\psi_{s-1,r} \circ f^* \\ 0 & 0 & (-1)^{n+s+r+1}\delta\psi_{s-1,r+1} \end{pmatrix},$$

where $\overline{\delta}(n, s, r) = r(n+s) + 1 = \delta(n, s, r) + n + s + 1 + r$ and $e_r = e(C_r)$. The component $\delta\psi' \in W_{\%}(D^{n+1-*})$ is defined to be zero.

These formulas hold as the chain map $u \colon \mathrm{cone}((f')_{\%}) \xrightarrow{\simeq} \mathrm{HPB}$ of (15.409) sends the specific cycle in $\mathrm{cone}((f')_{\%})_{n+1}$ described by these formulas to the cycle $(T(\delta\psi/\psi), \varepsilon \cdot e, 0)$ in HPB_{n+1}. The verification of this observation, whose details we leave to the reader, includes checking the requirements that the element $\delta\psi'/\psi'$ is mapped to $T(\delta\psi/\psi)$ under the chain homotopy equivalence $W_{\%}(z) \colon W_{\%}(\mathrm{cone}(f')) \xrightarrow{\simeq} W_{\%}(\mathrm{cone}(f))$ induced by the chain homotopy equivalence z of (15.395).

This proves that the above formulas give explicit representatives of the (a priori only up to homotopy) well-defined quadratic structures.

The next proposition is taken from [344, Proposition 4.1].

Proposition 15.415 *Algebraic surgery preserves the homotopy type of the boundary of $(C, [\varphi])$ in the symmetric case and of $(C, [\psi])$ in the quadratic case. In particular, we have that*

$$(C, [\varphi]) \text{ is Poincaré} \Leftrightarrow (C', [\varphi']) \text{ is Poincaré},$$

and
$$(C, [\psi]) \text{ is Poincaré} \Leftrightarrow (C', [\psi']) \text{ is Poincaré.}$$

Proof. The symmetric case follows from Diagram (15.403). The boundary ∂C is isomorphic to the homotopy fibre of the top row and the boundary $\partial C'$ is isomorphic to the homotopy fibre of the left vertical map. Hence both of them are iterated homotopy fibres of the middle square and as such are homotopy equivalent by Lemma 14.68.

The quadratic case follows from the same diagram taking into account that the underlying chain complex of the boundary of a quadratic complex is obtained via its symmetrisation and the symmetrisation of $[\partial \psi']$ is $[\partial \varphi']$. □

As indicated in the introduction to this subsection, algebraic surgery is analogous to geometric surgery in many aspects. Now we would like to concentrate on the relation between algebraic surgery and cobordisms. Therefore we assume that the n-dimensional symmetric complex $(C, [\varphi])$ in Construction 15.392 is Poincaré and hence by Proposition 15.415 also the effect $(C', [\varphi'])$ is Poincaré.

Observe that in this case it is more convenient to rewrite Diagram (15.403) in the shape of a homotopy commutative braid as in Diagram (15.416). This braid is completely analogous to the braid (15.391), but with a shift in degrees to the right since that is the part we are focusing on now.

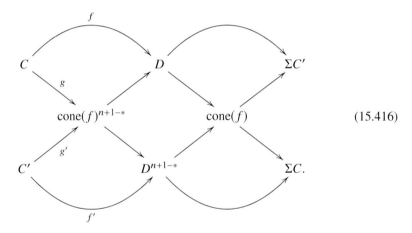

(15.416)

In more detail, in the notation from Diagram (15.403) the maps g and g' are defined as
$$\begin{aligned} g &:= r^* \circ \varphi_0^{-1}; \\ g &:= (r' \circ z^{-1})^* \circ (\varphi_0')^{-1}, \end{aligned}$$
(15.417)

where φ_0^{-1} and $(\varphi_0')^{-1}$ are some chain homotopy inverses of φ_0 and φ_0' respectively. These are well defined only up to chain homotopy, but that will be good enough for our purposes in the following proposition.

15.5 L-Groups in Terms of Chain Complexes

Proposition 15.418 *Let $(C, [\varphi])$ be an n-dimensional symmetric Poincaré complex, and let $(f: C \to D, [\varphi, \delta\varphi])$ be data for algebraic surgery with effect the n-dimensional symmetric Poincaré complex $(C', [\varphi'])$.*

Then there exists a cobordism

$$(C \oplus C' \xrightarrow{g \oplus g'} F, [(-\varphi \oplus \varphi'), \delta\varphi_F])$$

between $(C, [\varphi])$ and $(C', [\varphi'])$.

Proof. Choose a polarisation $(\varphi, \delta\varphi)$ of $[\varphi, \delta\varphi]$. The underlying chain complex is defined as

$$F := \operatorname{cone}(f)^{n+1-*},$$

and the maps g and g' are those from (15.417). These depend on the choices of φ_0^{-1} and $(\varphi_0')^{-1}$, but, since we are only interested in producing some cobordism, any choices are good for us here. We see that we have the following homotopy commutative diagram where all rows are homotopy cofibration sequences and s and s' are the corresponding induced maps

$$\begin{array}{ccccc}
C & \xrightarrow{f} & D & \xrightarrow{e} & \operatorname{cone}(f) \\
\uparrow{\scriptstyle \operatorname{pr}_1} & & \uparrow{\scriptstyle \operatorname{ev}_l} & & \uparrow{\scriptstyle \simeq}\ {\scriptstyle s} \\
C \oplus C' & \xrightarrow{g \oplus g'} & F & \xrightarrow{e''} & \operatorname{cone}(g \oplus g') \\
\downarrow{\scriptstyle \operatorname{pr}_2} & & \downarrow{\scriptstyle e^*} & & \downarrow{\scriptstyle \simeq}\ {\scriptstyle s'} \\
C' & \xrightarrow{f'} & D^{n+1-*} & \xrightarrow{e'} & \operatorname{cone}(f').
\end{array}$$

The fact that the induced maps s and s' are chain homotopy equivalences is seen as follows. For s look at the upper left square. Diagram (15.403) shows that the induced map of the homotopy fibres of the vertical maps is a chain homotopy inverse of the chain homotopy equivalence $\varphi_0': (C')^{n-*} \to C'$. Hence the induced map of the mapping cones of the horizontal maps is also a chain homotopy equivalence. For s' look at the lower left square and apply analogous reasoning using $\varphi_0: C^{n-*} \to C$.

Applying Proposition 15.340 to $g \oplus g': C \oplus C' \to F$ yields the homotopy cartesian square

$$\begin{array}{ccc}
\operatorname{cone}((g \oplus g')^{\%}) & \xrightarrow{\operatorname{ev}_r''} & F \otimes \operatorname{cone}(g \oplus g') \\
\downarrow{\scriptstyle z((e'')^{\%}, (j'')^{\%})} & & \downarrow{\scriptstyle e'' \otimes \operatorname{id}} \\
W^{\%}(\operatorname{cone}(g \oplus g')) & \xrightarrow{\operatorname{ev}''} & \operatorname{cone}(g \oplus g') \otimes \operatorname{cone}(g \oplus g')
\end{array}$$

where $e'' \colon F \to \operatorname{cone}(g \oplus g')$ is the canonical map and $j'' \colon 0 \simeq e'' \circ (g \oplus g')$ the canonical chain homotopy. The chain homotopy equivalence z provides us with a homotopy commutative diagram whose vertical arrows are chain homotopy equivalences

$$\begin{array}{ccccc}
W^{\%}(\operatorname{cone}(g \oplus g')) & \xrightarrow{\operatorname{ev}''} & \operatorname{cone}(g \oplus g')^{\otimes 2} & \xleftarrow{e'' \otimes \operatorname{id}} & F \otimes_R \operatorname{cone}(g \oplus g') \\
{\scriptstyle W^{\%}(s)} \downarrow \simeq & & {\scriptstyle s^{\otimes 2}} \downarrow \simeq & & \downarrow {\scriptstyle \operatorname{id} \otimes s} \simeq \\
W^{\%}(\operatorname{cone}(f)) & \xrightarrow{\operatorname{ev}} & \operatorname{cone}(f)^{\otimes 2} & \xleftarrow{e^* \otimes \operatorname{id}} & F \otimes_R \operatorname{cone}(f).
\end{array}$$

Define HPB to be the homotopy pullback

$$\begin{array}{ccc}
\operatorname{HPB} & \longrightarrow & F \otimes_R \operatorname{cone}(f) \\
\downarrow & & \downarrow {\scriptstyle e^* \otimes \operatorname{id}} \\
W^{\%}(\operatorname{cone}(f)) & \xrightarrow{\operatorname{ev}} & \operatorname{cone}(f) \otimes_R \operatorname{cone}(f).
\end{array}$$

The universal property of the homotopy pullback provides us with a chain homotopy equivalence

$$u'' \colon \operatorname{cone}((g \oplus g')^{\%}) \xrightarrow{\simeq} \operatorname{HPB}.$$

Moreover, extending the homotopy cofibration sequence induced by the map $(g \oplus g')^{\%}$ to the right, we obtain a map

$$\bar{r}'' \colon \operatorname{cone}((g \oplus g')^{\%}) \to \Sigma W^{\%}(C \oplus C'),$$

and the homotopy cofibration sequence, see Proposition 15.343,

$$\operatorname{cone}((g \oplus g')^{\%}) \xrightarrow{\bar{r}''} \Sigma W^{\%}(C \oplus C') \xrightarrow{\Sigma(g \oplus g')^{\%}} \Sigma W^{\%}(F). \tag{15.419}$$

The polarised symmetric structure $-\varphi \oplus \varphi'$ is a cycle in the middle term. To prove the desired statement, it suffices to show that its pushforward to the right is nullhomologous, or equivalently, that it is the image under $\bar{r}''$ of a cycle from the left term, and that the resulting symmetric pair is Poincaré.

To see this, let us start by considering the following commutative diagram

$$\begin{array}{ccc}
\operatorname{cone}(f^{\%}) \oplus \operatorname{cone}((f')^{\%}) & \xrightarrow{\bar{r} \oplus \bar{r}'} & \Sigma W^{\%}(C) \oplus \Sigma W^{\%}(C') \\
{\scriptstyle W^{\%}(s) \oplus W^{\%}(s')} \uparrow & & \downarrow {\scriptstyle \oplus} \\
\operatorname{cone}((g \oplus g')^{\%}) & \xrightarrow{\bar{r}''} & \Sigma W^{\%}(C \oplus C').
\end{array}$$

15.5 L-Groups in Terms of Chain Complexes

We also have the commutative diagram

$$\begin{array}{ccc} \mathrm{HPB}_f \oplus \mathrm{HPB}_{f'} & \xrightarrow{r \oplus r'} & \Sigma W^{\%}(C) \oplus \Sigma W^{\%}(C') \\ \alpha \oplus \alpha' \uparrow & & \downarrow \oplus \\ \mathrm{HPB} & \xrightarrow{r''} & \Sigma W^{\%}(C \oplus C') \end{array}$$

where

$$\mathrm{HPB}_f := W^{\%}(\mathrm{cone}(f)) \widetilde{\times}_{\mathrm{cone}(f)^{\otimes 2}} D \otimes_R \mathrm{cone}(f);$$
$$\mathrm{HPB}_{f'} := W^{\%}(\mathrm{cone}(f)) \widetilde{\times}_{\mathrm{cone}(f)^{\otimes 2}} D^{n+1-*} \otimes_R \mathrm{cone}(f),$$

and

$$\alpha = (\mathrm{id} \,\widetilde{\times}\, (\mathrm{ev}_l \otimes \mathrm{id})) \colon \mathrm{HPB} \to \mathrm{HPB}_f;$$
$$\alpha' = (\mathrm{id} \,\widetilde{\times}\, (e^* \otimes \mathrm{id})) \colon \mathrm{HPB} \to \mathrm{HPB}_{f'}$$

are chain maps. Moreover, recall that we have chain homotopy equivalences

$$u \colon \mathrm{cone}(f^{\%}) \xrightarrow{\simeq} \mathrm{HPB}_f \quad \text{and} \quad u' \colon \mathrm{cone}((f')^{\%}) \xrightarrow{\simeq} \mathrm{HPB}_{f'},$$

which fit into the commutative diagram

$$\begin{array}{ccc} \mathrm{cone}(f^{\%}) \oplus \mathrm{cone}((f')^{\%}) & \xrightarrow{u \oplus u'} & \mathrm{HPB}_f \oplus \mathrm{HPB}_{f'} \\ W^{\%}(s) \oplus W^{\%}(s') \uparrow & & \alpha \oplus \alpha' \uparrow \\ \mathrm{cone}((g \oplus g')^{\%}) & \xrightarrow{u''} & \mathrm{HPB} \,. \end{array}$$

Using the adjunction

$$F \otimes_R \mathrm{cone}(f) = \mathrm{cone}(f)^{n+1-*} \otimes_R \mathrm{cone}(f) \cong \hom_R(\mathrm{cone}(f), \mathrm{cone}(f)),$$

we see that in HPB we have a preferred cycle

$$(T(\delta\varphi/\varphi), \varepsilon \cdot \mathrm{id}_{\mathrm{cone}(f)}, 0),$$

which under the maps α and α' goes to the two cycles

$$(T(\delta\varphi/\varphi), T(\mathrm{ev}_l), 0) \quad \text{and} \quad (T(\delta\varphi/\varphi), \varepsilon \cdot e, 0).$$

The second cycle induces the symmetric structure $-\varphi'$. The first cycle is homologous to the cycle $(\delta\varphi/\varphi, \mathrm{ev}_r, 0)$, which induces the symmetric structure φ. Hence, using the chain homotopy equivalence u'', we conclude that the symmetric structure represented by $-\varphi \oplus \varphi'$, which sits in the middle entry of the sequence (15.419),

comes from the left and hence we have a preferred homology to 0 on the right. This homology gives
$$\delta \varphi_F : (g \oplus g')^{\%}(-\varphi \oplus \varphi') \sim 0.$$
To see that the resulting pair is Poincaré, one observes that the "middle" evaluation map
$$\text{ev}_m : \text{cone}(g')^{n+1-*} \simeq D^{n+1-*} \to \text{cone}(g) \simeq D^{n+1-*}$$
is homotopic to the identity. □

Proposition 15.420 *Let $(C, [\psi])$ be an n-dimensional quadratic Poincaré complex, and let $(f : C \to D, [\psi, \delta\psi])$ be data for algebraic surgery with effect the n-dimensional quadratic Poincaré complex $(C', [\psi'])$.*
Then there exists a cobordism
$$(C \oplus C' \xrightarrow{g \oplus g'} F, [(-\psi \oplus \psi'), \delta\psi_F])$$
between $(C, [\psi])$ and $(C', [\psi'])$.

The proof of Proposition 15.420 is analogous to that of Proposition 15.418, and we leave it to the reader.

The relationship between the algebraic cobordism and algebraic surgery is further clarified by the following result taken from [344, Proposition 4.1].

Proposition 15.421 *The equivalence relation generated by algebraic surgery and homotopy equivalence is the same as the equivalence relation given by cobordism in both the symmetric and quadratic case.*

Proof. By Propositions 15.299, 15.418, and 15.420 it remains to show that, if we have a cobordism, then up to chain homotopy we can obtain the second end of the cobordism from the first one via some algebraic surgery. Let us concentrate on the symmetric case, the quadratic being completely analogous.
Suppose we have a cobordism
$$(C \oplus C' \xrightarrow{g \oplus g'} F, [(-\varphi \oplus \varphi'), \delta\varphi_F])$$
between n-dimensional symmetric Poincaré complexes $(C, [\varphi])$ and $(C', [\varphi'])$. Choose a polarisation $((-\varphi \oplus \varphi'), \delta\varphi_F)$ of $[(-\varphi \oplus \varphi'), \delta\varphi_F]$. Consider the commutative diagram

$$\begin{array}{ccccc}
C \oplus C' & \xrightarrow{g \oplus g'} & F & \longrightarrow & \text{cone}((g \oplus g')^{\%}) \\
{\scriptstyle \text{pr}_{C'}} \downarrow & & {\scriptstyle \text{id}} \downarrow & & \downarrow {\scriptstyle z_2} \\
C' & \xrightarrow{g'} & F & \longrightarrow & \text{cone}((g')^{\%})
\end{array}$$

where the chain map z_2 is induced by the universal property of the mapping cone. Let $\delta\varphi_F/\varphi' := z_2((-\varphi \oplus \varphi'), \delta\varphi_F)$, $D := \text{cone}(g')$, and denote by $\overline{e}' : F \to D$ the

15.5 *L*-Groups in Terms of Chain Complexes

canonical map. The universal property of the mapping cone provides us with a chain map
$$z' \colon \mathrm{cone}((g')^{\%}) \to W^{\%}(D).$$
Define $\delta\varphi_D := z'(\delta\varphi_F/\varphi')$. Then
$$(f \colon C \to D, [\varphi, \delta\varphi_D])$$
is an $(n+1)$-dimensional symmetric pair. If we use it as data for algebraic surgery according to Construction 15.392, then we claim that the result is homotopy equivalent to $(C', [\varphi'])$. To see this, consider the two homotopy commutative braids of the same shape as in Diagram (15.416), one coming from the cobordism we start with and the other one coming from the cobordism provided by the algebraic surgery. The desired homotopy equivalence is obtained by comparing these two braids. □

The next proposition is taken from [344, Proposition 4.6]. In the statement we use the notion of a connected structured complex from Definition 15.159 and of a k-connected chain complex from Definition 14.121.

Proposition 15.422 (i) *Let $(C, [\varphi_C])$ and $(C', [\varphi_{C'}])$ be two n-dimensional symmetric Poincaré complexes. Then they are cobordant, i.e.,*
$$(C, [\varphi_C]) \sim (C', [\varphi_{C'}])$$
if and only if there exist two $(n+1)$-dimensional symmetric complexes $(D, [\varphi_D])$ and $(D', [\varphi_{D'}])$ and a homotopy equivalence of n-dimensional symmetric Poincaré complexes
$$(C, [\varphi_C]) \oplus \partial(D, [\varphi_D]) \simeq (C', [\varphi_{C'}]) \oplus \partial(D', [\varphi_{D'}]).$$
Both $(D, [\varphi_D])$ and $(D', [\varphi_{D'}])$ can be chosen to be connected if C and C' are (-1)-connected;

(ii) *Let $(C, [\psi_C])$ and $(C', [\psi_{C'}])$ be two n-dimensional quadratic Poincaré complexes. Then they are cobordant, i.e.,*
$$(C, [\psi_C]) \sim (C', [\psi_{C'}])$$
if and only if there exist two $(n+1)$-dimensional quadratic complexes $(D, [\psi_D])$ and $(D', [\psi_{D'}])$ and a homotopy equivalence of n-dimensional quadratic Poincaré complexes
$$(C, [\psi_C]) \oplus \partial(D, [\psi_D]) \simeq (C', [\psi_{C'}]) \oplus \partial(D', [\psi_{D'}]).$$
Both $(D, [\psi_D])$ and $(D', [\psi_{D'}])$ can be chosen to be connected if C and C' are (-1)-connected.

Proof. (i) Consider the identity isomorphism $C \oplus C' \oplus C' \to C \oplus C' \oplus C'$. If $(C, [\varphi_C]) \sim (C', [\varphi_{C'}])$, then both the sum $(C, [\varphi_C]) \oplus (C', [-\varphi_{C'}])$ and the sum

$(C', [\varphi_{C'}]) \oplus (C', [-\varphi_{C'}])$ are boundaries up to homotopy equivalence by Propositions 15.299 (i) and 15.375. The connectivity statement follows from Remark 15.373.

Conversely, assume there are symmetric complexes $(D, [\varphi_D])$ and $(D', [\varphi_{D'}])$ and a homotopy equivalence $(C, [\varphi_C]) \oplus \partial(D, [\varphi_D]) \xrightarrow{\simeq} (C', [\varphi_{C'}]) \oplus \partial(D', [\varphi_{D'}])$. Then $(C, [\varphi_C]) \oplus \partial(D, [\varphi_D])$ and $(C', [\varphi_{C'}]) \oplus \partial(D', [\varphi_{D'}])$ are cobordant by Proposition 15.299 (i). Since $\partial(D, [\varphi_D])$ and $\partial(D', [\varphi_{D'}])$ are nullbordant by construction, $(C, [\psi_C])$ and $(C', [\psi_{C'}])$ are cobordant.

(ii) The quadratic case is proved analogously. □

15.6 The Identification of the L-Groups

This section is devoted to the proof of Theorem 15.3, which states that the L-groups from Chapters 8 and 9 defined via quadratic forms and quadratic formations and the L-groups from Section 15.5 defined as cobordism groups of quadratic chain complexes are isomorphic. The identification will be via a sequence of propositions where we will step-by-step modify the chain complex L-groups to look more and more like the forms/formations L-groups. For the convenience of the reader the summary of all these steps is provided in Diagram (15.5) from the Overview Section 15.2.3 together with some basic ideas, which will be presented in detail in this section.

Throughout the process various ideas are involved, but in general the main tool will be algebraic surgery below middle dimension from the previous section, which is of course inspired by the geometric surgery below middle dimension. We also note that the corresponding statement about L-groups in the symmetric case is not true in general, see Remark 15.428 for more information.

A useful way to start is to establish 4-periodicity of the chain complex L-groups in Proposition 15.427, so that we only have to deal with four cases. In fact, using the unit $\varepsilon = \pm 1$ appropriately, we only have to deal with two cases. The periodicity is obtained using the following isomorphism.

Lemma 15.423 *Let C be a chain complex in $R\text{-CH}_{\text{fgp}}$ and let $\varepsilon = \pm 1$. Then the map*

$$\overline{\mu_1} \colon (C \otimes_R C) \to (\Sigma C) \otimes_R (\Sigma C)$$
$$x \otimes y \mapsto (-1)^{|x|} x \otimes y$$

is a $\mathbb{Z}/2$-equivariant chain isomorphism of degree 2 with respect to the involution T_ε in the source and the involution $T_{-\varepsilon}$ in the target.

We have already encountered a map called μ_1 defined by the same formula in (15.95). However, there this map was viewed as a self map of $C \otimes_R C$ of degree 0 that was only a degreewise isomorphism whereas here the same formula defines a chain map of degree 2 with a different source and target. Hence we distinguish the two maps via the bar in $\overline{\mu_1}$.

15.6 The Identification of the L-Groups

Recall that the involution $\varepsilon = \pm 1$ plays a role in the definition of the chain complexes $W^\%(C)$, $W_\%(C)$, and $\widehat{W}^\%(C)$, although it does not appear in the notation. Namely, it is used in the definition of the involution on $C \otimes_R C$. Since now we need to use different choices of ε, we temporarily introduce the symbols $W^\%_\varepsilon(C)$, $W^\varepsilon_\%(C)$, and $\widehat{W}^\%_\varepsilon(C)$ to keep track of these choices; their meaning should be obvious.

The map $\overline{\mu_1}$ from Lemma 15.423 induces chain isomorphisms between W-complexes, which we call *skew suspensions*, as follows:

$$\overline{S} \colon \Sigma^2 W^\%_\varepsilon(C) \xrightarrow{\cong} W^\%_{-\varepsilon}(\Sigma C), \quad \overline{S}(\varphi) := \overline{\mu_1} \circ \varphi;$$
$$\overline{S} \colon \Sigma^2 W^\varepsilon_\%(C) \xrightarrow{\cong} W^{-\varepsilon}_\%(\Sigma C), \quad \overline{S}(\psi) := \overline{\mu_1} \circ \psi; \qquad (15.424)$$
$$\overline{S} \colon \Sigma^2 \widehat{W}^\%_\varepsilon(C) \xrightarrow{\cong} \widehat{W}^\%_{-\varepsilon}(\Sigma C), \quad \overline{S}(\theta) := \overline{\mu_1} \circ \theta.$$

Note that, using these maps, we obtain from an n-dimensional symmetric Poincaré complex $(C, [\varphi])$ an $(n+2)$-dimensional symmetric Poincaré complex $(\Sigma C, [\overline{S}(\varphi)])$. We emphasise again that the dimension aims at $[\varphi]$ in the sense that the dimension of $[\overline{S}(\varphi)]$ is two plus the dimension of $[\varphi]$, and not at the dimension of C. Analogous statements hold for the quadratic case and hyperquadratic case. Further, the maps $\overline{S}$ are isomorphisms since $\overline{\mu_1}$ from Lemma 15.423 is an isomorphism of chain complexes. Hence it can be used to get an inverse of $\overline{S}$.

Remark 15.425 (The difference between the suspension and the skew suspension) We warn the reader that the notions of suspension from Section 15.3.5 and the above skew suspension are quite different constructions on a structured chain complex and are not directly related. For example, starting with an n-dimensional symmetric complex (C, φ) the suspension is an $(n+1)$-dimensional symmetric complex with the underlying chain complex ΣC whereas the skew suspension is an $(n+2)$-dimensional symmetric complex with (the same) underlying chain complex ΣC.

Next we turn to the relative situation. Given an R-chain map $f \colon C \to D$, recall the isomorphism $\mathrm{cone}(f^\%) = \mathrm{hom}_{\mathbb{Z}[\mathbb{Z}/2]}(W, \mathrm{cone}(f \otimes f))$ from Exercise 15.229. The isomorphisms $\overline{\mu_1}$ for C and D are compatible with $f \otimes f$ and $\Sigma f \otimes \Sigma f$. Using the universal property of the mapping cone, we obtain a chain isomorphism of degree 2, which is equivariant with respect to appropriate involutions indicated by subscripts

$$\overline{\mu_1} \colon \mathrm{cone}(f \otimes f)_\varepsilon \xrightarrow{\cong} \mathrm{cone}(\Sigma f \otimes \Sigma f)_{-\varepsilon}.$$

It also induces *skew suspension* isomorphisms (with subscripts and superscripts $\pm \varepsilon$ again indicating the involution used) as follows:

$$\overline{S}\colon \Sigma^2 \operatorname{cone}(f_\varepsilon^{\%}) \xrightarrow{\cong} \operatorname{cone}((\Sigma f)_{-\varepsilon}^{\%});$$
$$\overline{S}\colon \Sigma^2 \operatorname{cone}(f_{\%}^{\varepsilon}) \xrightarrow{\cong} \operatorname{cone}((\Sigma f)_{\%}^{-\varepsilon}); \qquad (15.426)$$
$$\overline{S}\colon \Sigma^2 \operatorname{cone}(\widehat{f}_\varepsilon^{\%}) \xrightarrow{\cong} \operatorname{cone}(\widehat{(\Sigma f)}_{-\varepsilon}^{\%}).$$

Again $(n+1)$-dimensional Poincaré pairs go to $(n+3)$-dimensional Poincaré pairs in both the symmetric and quadratic case.

At the level of L-groups we need to keep track of ε as well, which we do by using the notation $L^n(R, \varepsilon)_{\text{chain}}$ and $L_n(R, \varepsilon)_{\text{chain}}$; again the meaning should be clear. With this notation we obtain the following proposition.

Proposition 15.427 (Skew suspension isomorphism on L-groups) *Let $n \in \mathbb{Z}$. The skew suspension maps from* (15.424) *induce isomorphisms of L-groups*

$$L^n(\overline{S})\colon L^n(R, \varepsilon)_{\text{chain}} \to L^{n+2}(R, -\varepsilon)_{\text{chain}};$$
$$L_n(\overline{S})\colon L_n(R, \varepsilon)_{\text{chain}} \to L_{n+2}(R, -\varepsilon)_{\text{chain}};$$
$$L_p^n(\overline{S})\colon L_p^n(R, \varepsilon)_{\text{chain}} \to L_p^{n+2}(R, -\varepsilon)_{\text{chain}};$$
$$L_n^p(\overline{S})\colon L_n^p(R, \varepsilon)_{\text{chain}} \to L_{n+2}^p(R, -\varepsilon)_{\text{chain}}.$$

Proof. Via (15.424) and (15.426) we obtain well-defined maps. Since the process of skew suspension is reversible, we obtain that the induced maps on the L-groups have inverses. □

We emphasise again that in the above proposition $n \in \mathbb{Z}$, meaning that negative values of n are allowed, and also even if $n = 0$, the chain complexes and cobordisms are allowed to have non-zero modules in negative dimensions and also non-zero homology in negative dimensions, they are only required to be bounded.

Remark 15.428 (The effect of the connectivity assumptions of the category of chain complexes on the periodicity) We also warn the reader that here we use a different definition of an n-dimensional structured complex from the one in [344, 345]. We use the definition from [348]. With our definition the skew suspension maps give isomorphisms in both the quadratic and symmetric case. In [344, 345] this was not the case. See Example 1.11 in [348] for more explanation.

At the same time we warn the reader that, although we do have 4-periodicity in both the symmetric case and the quadratic case here, the identification of the chain complex L-groups with the L-groups from Chapters 8 and 9 will only work in the quadratic case. Of course, in Chapters 8 and 9 we only defined quadratic L-groups in terms of quadratic forms and formations. Nevertheless, with some modifications, it is also possible to define analogous groups using symmetric forms and formations, for example as in [344, Section 5]. But even if we did so, the identification with the L-groups via chain complexes would still only work in the quadratic case. This is so because surgery below middle dimension, which will be used below, is obstructed in the symmetric case in general, see Remark 15.431.

15.6 The Identification of the L-Groups

As promised the main ingredient will be algebraic surgery below middle dimension. We need two cases that are presented in the following propositions. We remind the reader that the notion of the cohomology of a chain complex that is used in the proofs satisfies $H^k(C) = H_{-k}(C^{-*})$, see (14.24).

Proposition 15.429 (Algebraic surgery below middle dimension) *Let $n \geq 0$ and let $(C, [\psi])$ be an n-dimensional quadratic Poincaré complex. Then there exists another n-dimensional quadratic Poincaré complex $(C', [\psi'])$ such that*

$$(C, [\psi]) \sim (C', [\psi']) \quad \text{and} \quad H_k(C') = 0 \quad \text{unless} \quad 0 \leq k \leq n.$$

The same statement holds in the category $bR\text{-CH}_{\text{fgf}}$.

Proof. Let D be the chain complex defined as follows

$$D_r = \begin{cases} C_r & r \geq n+1; \\ 0 & r \leq n. \end{cases}$$

Let $j \colon C \to D$ be the obvious projection map. Choosing a polarisation ψ of $[\psi]$ and observing that $j_{\%}(\psi) = 0$, we obtain an $(n+1)$-dimensional quadratic pair

$$(j \colon C \to D, [\psi, 0]).$$

Let $(C', [\psi'])$ be the n-dimensional quadratic Poincaré complex obtained as the effect of algebraic surgery on $(C, [\psi])$ with data $(j \colon C \to D, [\psi, 0])$, as described in Construction 15.408. We claim that $(C', [\psi'])$ has the desired properties. This follows from the following computations, which are based on studying the long exact sequences of homology groups applied to homotopy cofibration sequences in Diagram (15.416).

The homotopy cofibration sequence $C \to D \to \text{cone}(j)$ yields

$$H_k(j) \colon H_k(C) \xrightarrow{\cong} H_k(D) \text{ for } k \geq n+2;$$

$$H_k(\text{cone}(j)) = 0 \text{ for } k \geq n+2;$$

$$0 \to H_{n+1}(C) \to H_{n+1}(D) \to H_{n+1}(\text{cone}(j)) \to H_n(C) \to 0;$$

$$H_k(D) = 0 \text{ for } k \leq n;$$

$$H_k(\text{cone}(j)) \xrightarrow{\cong} H_{k-1}(C) \text{ for } k \leq n,$$

and

$$H^k(j) \colon H^k(C) \xleftarrow{\cong} H^k(D) \text{ for } k \geq n+2;$$

$$H^k(\text{cone}(j)) = 0 \text{ for } k \geq n+2;$$

$$0 \leftarrow H^{n+1}(C) \leftarrow H^{n+1}(D) \leftarrow H^{n+1}(\text{cone}(j)) \leftarrow H^n(C) \leftarrow 0;$$

$$H^k(D) = 0 \text{ for } k \leq n;$$

$$H^k(\text{cone}(j)) \xleftarrow{\cong} H^{k-1}(C) \text{ for } k \leq n.$$

The homotopy cofibration sequence $C' \to D^{n+1-*} \to \text{cone}(j)$ yields the following results. For $k \geq n + 1$ we have the exact sequence

$$0 \cong H_{k+1}(\text{cone}(j)) \to H_k(C') \to H^{n+1-k}(D) \cong 0,$$

and hence $H_k(C') = 0$. Analogously, still for $k \geq n + 1$, we have the exact sequence

$$0 \cong H_{n+1-k}(D) \to H^k(C') \to H^{k+1}(\text{cone}(j)) \cong 0,$$

and hence $H^k(C') = 0$. We also have Poincaré duality $H^{n-k}(C') \cong H_k(C')$ for all $k \in \mathbb{Z}$ and this gives us $H_k(C') = 0$ unless $0 \leq k \leq n$. □

Proposition 15.430 (Algebraic surgery below middle dimension on a cobordism) *Consider $n \geq 0$. Let $(C, [\psi_C])$ and $(C', [\psi_{C'}])$ be two n-dimensional cobordant quadratic Poincaré complexes with $H_k(C) = 0$ and $H_k(C') = 0$ unless $0 \leq k \leq n$ such that*

$$(C, [\psi_C]) \oplus (C', [-\psi_{C'}]) \simeq \partial(D, [\psi_D])$$

for some $(n + 1)$-dimensional quadratic complex $(D, [\psi_D])$.

In this situation there exists an $(n+1)$-dimensional quadratic complex $(D', [\psi_{D'}])$ such that $H_k(D') = 0$ unless $0 \leq k \leq n + 1$ and a homotopy equivalence of n-dimensional quadratic Poincaré complexes

$$(C, [\psi_C]) \oplus (C', [-\psi_{C'}]) \simeq \partial(D', [\psi_{D'}]).$$

The same statement holds in the category $\text{b}R\text{-CH}_{\text{fgf}}$.

Proof. Let E be a chain complex defined as follows

$$E_r = \begin{cases} D_r & r \geq n + 2; \\ 0 & r \leq n + 1, \end{cases}$$

and let $j: D \to E$ be the obvious projection map. Choose a polarisation ψ_D of $[\psi_D]$. Since $j_{\%}(\psi_D) = 0$, we get a $(n + 2)$-dimensional quadratic pair

$$(j: D \to E, [\psi_D, 0]).$$

Let $(D', [\psi_{D'}])$ be the $(n + 1)$-dimensional quadratic complex obtained as the effect of algebraic surgery on $(D, [\psi_D])$ with data $(j: D \to E, [\psi_D, 0])$ as described in Construction 15.408. We claim that $(D', [\psi_{D'}])$ has the desired properties. This follows from computations that are based on studying the long exact sequences of homology groups applied to homotopy cofibration sequences in Diagram (15.403), which depicts complexes and maps arising when doing algebraic surgery. This time we need to use this diagram rather than Diagram (15.416) since $(D, [\psi_D])$ is not Poincaré to begin with. On the other hand since algebraic surgery preserves the homotopy type of the boundary, see Proposition 15.415, we automatically have the desired homotopy equivalence, only the statements about the homology groups need to be verified.

15.6 The Identification of the L-Groups

The homotopy cofibration sequence $D \to E \to \mathrm{cone}(j)$ yields

$$H_k(j) \colon H_k(D) \xrightarrow{\cong} H_k(E) \text{ for } k \geq n+3;$$
$$H_k(\mathrm{cone}(j)) = 0 \text{ for } k \geq n+3;$$
$$0 \to H_{n+2}(D) \to H_{n+2}(E) \to H_{n+2}(\mathrm{cone}(j)) \to H_{n+1}(D) \to 0;$$
$$H_k(E) = 0 \text{ for } k \leq n+1;$$
$$H_k(\mathrm{cone}(j)) \xrightarrow{\cong} H_{k-1}(D) \text{ for } k \leq n+1,$$

and

$$H^k(j) \colon H^k(E) \xleftarrow{\cong} H^k(D) \text{ for } k \geq n+3;$$
$$H^k(\mathrm{cone}(j)) = 0 \text{ for } k \geq n+3;$$
$$0 \leftarrow H^{n+2}(D) \leftarrow H^{n+2}(E) \leftarrow H^{n+2}(\mathrm{cone}(j)) \leftarrow H^{n+1}(D) \leftarrow 0;$$
$$H^k(E) = 0 \text{ for } k \leq n+1;$$
$$H^k(\mathrm{cone}(j)) \xleftarrow{\cong} H^{k-1}(D) \text{ for } k \leq n+1.$$

The homotopy cofibration sequence $D' \to E^{n+2-*} \to \mathrm{cone}(j)$ yields the following results. For $k \geq n+2$ we have the exact sequence

$$0 \cong H_{k+1}(\mathrm{cone}(j)) \to H_k(D') \to H^{n+2-k}(E) \cong 0,$$

and hence $H_k(D') = 0$. Now we do not have the full duality for D since $(D, [\psi_D])$ was not assumed to be Poincaré, but we do have duality in a range since $(D, [\psi_D])$ is connected in the sense of Definition 15.159, which follows from the assumption that C and C' are (-1)-connected in the sense of Definition 14.121 via Remark 15.373. In other words, the map

$$H_k((1+T)(\psi_D)_0) \colon H^{n+1-k}(D) \to H_k(D)$$

is an isomorphism for $k \leq -1$ and onto for $k = 0$. To use this knowledge recall again in more detail Diagram (15.403) that we are using. When we substitute into the diagram chain complexes corresponding to our situation, we see that the homotopy cofibration sequence $D' \to E^{n+2-*} \to \mathrm{cone}(j)$ is in the third row of the diagram and we can consider its associated long exact homology sequence. This is the top row of the diagram below. The diagram below also contains two squares on the left and on the right which are induced by the lower square of Diagram (15.403). The calculations above imply that for $k \leq -1$ the maps are as follows with the upper row an exact sequence

$$\begin{array}{ccccccccc}
H^{n+1-k}(E) & \to & H_{k+1}(\mathrm{cone}(j)) & \longrightarrow & H_k(D') & \longrightarrow & H^{n+2-k}(E) & \longrightarrow & H_k(\mathrm{cone}(j)) \\
{\scriptstyle \mathrm{onto}}\downarrow & & \downarrow{\scriptstyle \cong} & & & & \downarrow{\scriptstyle \cong} & & \downarrow{\scriptstyle \cong} \\
H^{n+1-k}(D) & \underset{\cong}{\longrightarrow} & H_k(D) & & & & H^{n+2-k}(D) & \underset{\cong}{\longrightarrow} & H_{k-1}(D)
\end{array}$$

and hence we get $H_k(D') = 0$ in this case too. $\square$

Remark 15.431 (Symmetric surgery below middle dimension is obstructed)
As noted above the analogous propositions do not hold in the symmetric case. The reason for this is the following. If $(C, [\varphi])$ is a 0-dimensional symmetric complex with a polarisation φ of $[\varphi]$ and we define the chain complex D as in Propositions 15.429 or 15.430, then the pushforward $j^{\%}(\varphi)$ might not be nullhomologous in $W^{\%}(D)$ and so we do not obtain a symmetric pair, which we could use as data for surgery. Obstructions for the nullhomology can be formulated in terms of the Wu classes from Section 15.3.9 and conditions for their vanishing can be studied. This is done in Sections 4 and 7 of [344], but we will not go into that here.

At this stage it is convenient to first introduce the notion of a (strictly) well-connected structured complex. Recall from Definition 15.159 that an n-dimensional symmetric complex is connected if $H_i(\varphi_0 \colon C^{n-*} \to C) = 0$ for $i \leq 0$, and analogously in the quadratic case. Also recall from Definition 14.121 that an R-chain complex C is called k-connected if $H_i(C) = 0$ for $i \leq k$.

Definition 15.432 (Well-connected structured complexes)

(i) An n-dimensional symmetric complex $(C, [\varphi])$ is called *well-connected* if it is connected and the R-chain complex C is 0-connected.
(ii) An n-dimensional quadratic complex $(C, [\psi])$ is called *well-connected* if it is connected and the R-chain complex C is 0-connected.
(iii) An n-dimensional symmetric complex $(C, [\varphi])$ is called *strictly well-connected* if it is connected and the R-chain complex C satisfies $C_i = 0$ for $i \leq 0$.
(iv) An n-dimensional quadratic complex $(C, [\psi])$ is called *strictly well-connected* if it is connected and the R-chain complex C satisfies $C_i = 0$ for $i \leq 0$.

This allows us to obtain the following improvement on Proposition 15.422 in the quadratic case. It is taken from [344, Proposition 4.6], which is formulated in both the symmetric and quadratic case, but in a different connectivity setting. In our setting a symmetric version would need an additional assumption, making it more complicated, so we refrain from spelling it out, as we do not need it.

Proposition 15.433 (Cobordisms as boundaries of well-connected complexes)
Fix $n \geq 0$. Let $(C, [\psi_C])$ and $(C', [\psi_{C'}])$ be two n-dimensional quadratic Poincaré complexes such that $H_k(C) \cong H_k(C') = 0$ unless $0 \leq i \leq n$. Then

$$(C, [\psi_C]) \sim (C', [\psi_{C'}])$$

if and only if there exist two well-connected $(n+1)$-dimensional quadratic complexes $(D, [\psi_D])$ and $(D', [\psi_{D'}])$ and a homotopy equivalence of n-dimensional quadratic

15.6 The Identification of the L-Groups

Poincaré complexes

$$(C, [\psi_C]) \oplus \partial(D, [\psi_D]) \simeq (C', [\psi_{C'}]) \oplus \partial(D', [\psi_{D'}]).$$

The analogous statement holds in the category $\mathbf{b}R\text{-}\mathbf{CH}_{\mathrm{fgf}}$.

Proof. The if part is obvious by Proposition 15.422. For the only if part, we first employ Proposition 15.389 and Proposition 15.430 to obtain a quadratic complex $(D, [\psi_D])$ such that D has $H_k(D) = 0$ unless $0 \leq k \leq n+1$ and

$$(C, [\psi_C]) \oplus (C', [-\psi_{C'}]) \simeq \partial(D, [\psi_D]).$$

By the same reasoning as in the proof of Proposition 15.422, it is enough to show that, for any connected $(D, [\psi_D])$ with D as above, we can find some well-connected $(D', [\psi_{D'}])$ and $(D'', [\psi_{D''}])$ together with a homotopy equivalence of n-dimensional quadratic Poincaré complexes

$$h \colon \partial(D, [\psi_D]) \oplus \partial(D', [\psi_{D'}]) \xrightarrow{\simeq} \partial(D'', [\psi_{D''}]).$$

In that case we can glue the mapping cylinder of h to $(D', [\psi_{D'}])$ and $(D'', [\psi_{D''}])$ to get a cobordism between $(C, [\psi_C])$ and $(C', [\psi_{C'}])$, which is obtained as a boundary of a well-connected connected quadratic complex and, again by the method of the proof of Proposition 15.422, gives the desired homotopy equivalence. The assumption that $(D, [\psi_D])$ is connected means for any choice of the polarisation ψ_D of $[\psi_D]$ that $H_k((1+T)(\psi_D)_0) = 0$ for $k \leq 0$. It is fulfilled since $H_k((1+T)(\psi_D)_0) \cong H_{k-1}(C \oplus C') = 0$ for $k \leq 0$.

Starting with $(D, [\psi_D])$ we define

$$D'_r = \begin{cases} D_r & r \geq n+1; \\ 0 & r \leq n, \end{cases}$$

and let $g \colon D \to D'$ be the obvious projection. Next let $\psi' := -g_{\%}(\psi_D)$ and consider the map $j = g \oplus \mathrm{id}_{D'} \colon D \oplus D' \to D'$. Observe $j_{\%}(\psi_D \oplus \psi') = 0$. Therefore we have an $(n+2)$-dimensional quadratic pair of the form $(j \colon D \oplus D' \to D', [\psi_D \oplus \psi', 0])$. We perform algebraic surgery on the quadratic complex $(D \oplus D', [\psi_D \oplus \psi'])$ using this pair as data, as described in Construction 15.392, and call the result $(D'', [\psi_{D''}])$.

Thanks to Proposition 15.415, we obtain the desired homotopy equivalence h. It remains to check that $(D', [\psi_{D'}])$ and $(D'', [\psi_{D''}])$ have the property that they are well-connected. For $(D', [\psi_{D'}])$ it is obvious. For $(D'', [\psi_{D''}])$, arguing as in the proof of Proposition 15.430, we obtain the following diagram with the upper row an exact sequence

$$H^{n+1}(D') \longrightarrow H_1(\text{cone}(j)) \longrightarrow H_0(D'') \longrightarrow H^{n+2}(D') \longrightarrow H_0(\text{cone}(j))$$
$$\downarrow \text{onto} \qquad \downarrow \cong \qquad \qquad \downarrow \cong \qquad \qquad \downarrow \cong$$
$$H^{n+1}(D) \xrightarrow[\text{onto}]{} H_0(D) \qquad\qquad\qquad H^{n+2}(D) \xrightarrow[\text{one-one}]{} H_{-1}(D)$$

where the left lower horizontal map is onto and the right lower horizontal arrow is one-to-one since $(D, [\psi_D])$ is connected, which in particular means $H_0((1+T)(\psi_D)_0) = 0$. The other maps are as indicated since by the definition of D' we have

$$H_k(\text{cone}(j)) \xrightarrow{\cong} H_{k-1}(D) \text{ for } k \leq n;$$
$$H^k(D') \xrightarrow{\cong} H^k(D) \text{ for } k \geq n+2;$$
$$H^k(D') \xrightarrow{\text{onto}} H^k(D) \text{ for } k = n+1.$$

This finishes the proof of Proposition 15.433. □

The propositions above tell us that we can modify n-dimensional quadratic chain complexes $(C, [\psi])$ using algebraic surgery to n-dimensional quadratic chain complexes whose underlying chain complexes are highly connected. Similarly we can modify cobordisms to cobordisms that are obtained via the algebraic boundary construction from quadratic complexes with highly connected underlying chain complexes. However, in order to achieve the desired identification, we need to improve on this by replacing these quadratic complexes with highly connected underlying chain complexes by quadratic complexes whose underlying chain complexes are actually concentrated in dimensions between 0 and n, and analogously for cobordisms. In other words, in the terminology of Section 15.4 we need to pass to strictly n-dimensional quadratic chain complexes. By 4-periodicity it enough to consider the cases $n = 0, 1$.

Proposition 15.434 (i) Let $(C, [\psi])$ be a 0-dimensional quadratic Poincaré complex such that $H_k(C) = 0$ for $k \neq 0$. Then C is chain homotopy equivalent to a chain complex C' in $\text{b}R\text{-CH}_{\text{fgp}}$ concentrated in dimension 0, namely $C' = 0[H_0(C)]$;

(ii) Let $(C, [\psi])$ be a 1-dimensional quadratic Poincaré complex such that $H_k(C) = 0$ for $k \neq 0, 1$. Then C is chain homotopy equivalent to a chain complex C' in $\text{b}R\text{-CH}_{\text{fgp}}$ concentrated in dimensions 0 and 1, namely

$$C' = \cdots \to 0 \to C_1^\perp \to \ker(c_0) \to 0 \to \cdots$$

where $C_1^\perp$ is the orthogonal complement of $\text{im}(c_2)$ in C_1, which exists;

(iii) For chain complexes in $\text{b}R\text{-CH}_{\text{fgf}}$, in the 0-dimensional case the resulting chain complex can be made to consist of stably finitely generated free modules and in the 1-dimensional case it can be made to be in $\text{b}R\text{-CH}_{\text{fgf}}$.

15.6 The Identification of the L-Groups 733

Proof. The proof uses Corollary 14.123, which says that, for a chain complex C in $bR\text{-CH}_{\text{fgp}}$, under certain conditions on its homology and cohomology C is chain homotopy equivalent to a chain complex in $bR\text{-CH}_{\text{fgp}}$ concentrated in a specific range of dimensions. The assumptions appearing in Proposition 15.434 correspond precisely to the assumptions on homology in Corollary 14.123. The assumption on the cohomology is shown to hold using the quadratic structure and the associated duality of its symmetrisation as follows. The reasoning is similar to that in the proof of Lemma 8.55.

(i) Corollary 14.123 (i) implies that C is chain homotopy equivalent to D such that $D_i = 0$ for $i < 0$. Since $(C, [\psi])$ is a 0-dimensional quadratic Poincaré complex, we have the symmetrisation R-chain homotopy equivalence $C^{-*} \simeq C$. We therefore get for any R-module V the chain homotopy equivalence $\hom_R(C, V) \xrightarrow{\simeq} \hom_R(C^{-*}, V)$. Altogether we have

$$H^1(C, V) \cong H_{-1}(\hom_R(C, V)) \cong H_{-1}(\hom_R(C^{-*}, V))$$
$$\cong H_{-1}(\hom_R(D^{-*}, V)) \cong 0,$$

where the last isomorphism holds since $D^{-1} = 0$ and the result follows from Corollary 14.123 (ii).

(ii) Corollary 14.123 (i) implies that $(C, [\psi])$ is chain homotopy equivalent to D such that $D_i = 0$ for $i < 0$. Since $(C, [\psi])$ is a 1-dimensional quadratic Poincaré complex, we have the symmetrisation chain homotopy equivalence $C^{1-*} \simeq C$. Therefore we get for any R-module V the chain homotopy equivalence $\hom_R(C, V) \xrightarrow{\simeq} \hom_R(C^{1-*}, V)$. Altogether we have

$$H^2(C, V) \cong H_{-2}(\hom_R(C, V)) \cong H_{-2}(\hom_R(C^{1-*}, V))$$
$$\cong H_{-2}(\hom_R(D^{1-*}, V)) \cong 0,$$

where the last isomorphism holds since $D^{-1} = 0$, and the result follows from Corollary 14.123 (ii).

(iii) The 0-dimensional statement follows by the same reasoning and applying Corollary 14.123 (iv) at the end. The 1-dimensional statement follows by the same reasoning and applying Corollary 14.123 (iii) at the end. □

Now we need an analogous proposition for cobordisms.

Proposition 15.435 (i) *For $n \geq 0$ let $(C, [\psi_C])$ and $(C', [\psi_{C'}])$ be two n-dimensional cobordant quadratic Poincaré complexes with $C_k = 0$ and $C'_k = 0$ unless $0 \leq k \leq n$, such that*

$$(C, [\psi_C]) \oplus (C', [-\psi_{C'}]) \simeq \partial(D, [\psi_D])$$

for some $(n+1)$-dimensional quadratic complex $(D, [\psi_D])$ for which $H_k(D) = 0$ holds unless $1 \leq k \leq n+1$.

In this situation there exists an $(n + 1)$-dimensional quadratic complex $(D', [\psi_{D'}])$ such that $D'_k = 0$ unless $1 \leq k \leq n + 1$ and a homotopy equivalence of n-dimensional quadratic Poincaré complexes

$$(C, [\psi_C]) \oplus (C', [-\psi_{C'}]) \simeq \partial(D', [\psi_{D'}]);$$

(ii) *For chain complexes in $bR\text{-}CH_{fgf}$, in the case $n = 0$ the resulting cobordism can be made to consist of stably finitely generated free modules and in the case when $n \geq 1$ we it can be made to be in $bR\text{-}CH_{fgf}$.*

Proof. (i) Corollary 14.123 (i) implies that D is chain homotopy equivalent to E such that $E_i = 0$ for $i < 1$. By our assumptions for any choice of a polarisation φ_D of $[\varphi_D]$ the symmetrisation map $(1 + T)(\psi_D)_0 \colon D^{n+1-*} \to D$ is not a chain homotopy equivalence, but its homotopy fibre is chain homotopy equivalent to $C'' := C \oplus C'$. This leads for any R-module V to the long exact sequence

$$\cdots \to H^{n+1}(C''; V) \to H^{n+2}(D; V) \to H_{-1}(D; V) \to H^{n+2}(C'', V) \to \cdots.$$

By the proof of the previous proposition, $H^k(C''; V) = 0$ for $k = n + 1, n + 2$, and hence the middle map in the above sequence is an isomorphism. Altogether we have

$$H^{n+2}(D; V) \cong H_{-1}(D; V) \cong H_{-1}(E; V) \cong 0,$$

where the last isomorphism holds because of $E_{-1} = 0$ and the result follows from Corollary 14.123 (ii).

(ii) The 0-dimensional statement follows by the same reasoning and applying Corollary 14.123 (iv) at the end. The 1-dimensional statement follows by the same reasoning and applying Corollary 14.123 (iii) at the end. □

The above propositions allow us to replace in the following corollary the chain complex L-groups by cobordism groups of chain complexes, where we require that the complexes are strictly n-dimensional. In view of Section 15.4 this brings us closer to the desired forms and formations groups.

Corollary 15.436 (*L-groups via strictly n-dimensional complexes*) *Suppose $n = 0, 1$. Let $(C, [\psi_C])$ and $(C', [\psi_{C'}])$ be two strictly n-dimensional quadratic Poincaré complexes. Define*

$$(C, [\psi_C]) \sim' (C', [\psi_{C'}])$$

if there exist two strictly $(n + 1)$-dimensional quadratic chain complexes $(D, [\psi_D])$ and $(D', [\psi_{D'}])$ that are strictly well connected, and a chain homotopy equivalence of n-dimensional quadratic Poincaré complexes

$$(C, [\psi_C]) \oplus \partial(D, [\psi_D]) \simeq (C', [\psi_{C'}]) \oplus \partial(D', [\psi_{D'}]).$$

Then the version of $L_n^p(R)_{\text{chain}}$ and $L_n(R)_{\text{chain}}$ with respect to the relation $\sim'$ is isomorphic to $L_n^p(R)_{\text{chain}}$ and $L_n(R)_{\text{chain}}$ with respect to the relation $\sim$ as they appear in Definitions 15.301 and 15.302.

15.6 The Identification of the L-Groups

Suppose $n = 0$ and the same assumptions as above except that we require C, C' to be in bR-CH_{fgf}. Then the same conclusion holds, but with D, D' the chain complexes of stably free finitely generated R-modules.

Suppose $n = 1$ and the same assumptions except $C, C' \in bR$-CH_{fgf}. Then the same conclusion holds, but with $D, D' \in bR$-CH_{fgf}.

Proof. The relation $\sim'$ is an equivalence relation since the previous propositions show that $(C, \psi_C) \sim' (C', \psi_{C'})$ if and only if $(C, \psi_C) \sim (C', \psi_{C'})$, and $\sim$ was shown to be an equivalence relation in Proposition 15.283. □

The desired identification of the groups $L_n(R)_{\text{chain}}$ with the L-groups from Chapters 8 and 9 will be established by defining maps from the chain complex versions obtained in Corollary 15.436 to the module versions of the L-groups, which will then be shown to be isomorphisms. To see that the maps are well defined, we will need two propositions, which correspond to the cases $n = 0$ and $n = 1$ respectively. We first state their symmetric versions, which are not used, but their proofs shall illuminate the harder quadratic case. We will use the correspondences coming from Section 15.4, which link low-dimensional structured complexes to forms and formations.

Proposition 15.437 *The boundary of a strictly well-connected 1-dimensional ε-symmetric complex $(D, [\varphi])$ is a strictly 0-dimensional ε-symmetric Poincaré complex $(\partial D, [\partial \varphi])$, which via Proposition 15.202 corresponds to the standard symmetric ε-hyperbolic form associated to some symmetric form.*

If in addition D is a chain complex of stably free finitely generated R-modules, then ∂D is also a chain complex of stably free finitely generated R-modules.

Proof. A polarised strictly well-connected 1-dimensional ε-symmetric complex (D, φ) consists of the following data

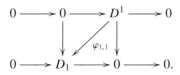

Its polarised algebraic boundary is the 0-dimensional ε-symmetric Poincaré complex $(\partial D, \partial \varphi)$ with

$$\partial D_0 = D^1 \oplus D_1, \quad \partial D_i = 0, \ i \geq 1, \quad \partial \varphi_{0,0} = \begin{pmatrix} 0 & -1 \\ -\varepsilon & -\varphi_{1,1} \end{pmatrix} \quad \partial \varphi_i = 0, \ i \geq 1.$$

Via Proposition 15.202 it corresponds to the standard ε-symmetric hyperbolic form associated to the ε-symmetric form $(D_1, \varphi_{1,1})$ in the sense of (15.194).

If $D \in bR$-CH_{fgp}, then clearly $\partial D \in bR$-CH_{fgp} and the same holds in the stably free case. □

Proposition 15.438 *The boundary of a strictly well-connected 2-dimensional ε-symmetric complex $(D, [\varphi])$ is a strictly 1-dimensional ε-symmetric Poincaré complex $(\partial D, [\partial \varphi])$ which via Proposition 15.205 corresponds to an ε-symmetric formation with the property that the dual of one of the lagrangians is complementary to the other lagrangian.*

The same holds in the category $bR\text{-}CH_{fgf}$.

In the quadratic version the above property of the lagrangians that the dual of one of them is complementary to the other means that the formation is isomorphic to a boundary formation. In the symmetric case we do not use this terminology since we do not have a good notion of a symmetric boundary formation.

Proof. A polarised strictly well-connected 2-dimensional ε-symmetric complex (D, φ) consists of the following data:

$$\begin{array}{ccccccccc} 0 & \longrightarrow & 0 & \longrightarrow & D^1 & \xrightarrow{d^*} & D^2 & \longrightarrow & 0 \\ & & \downarrow & \swarrow^{\varphi_{1,2}} & \downarrow^{\varphi_{0,1}} & \swarrow^{\varphi_{1,1}} & \downarrow & & \\ 0 & \longrightarrow & D_2 & \xrightarrow{d} & D_1 & \longrightarrow & 0 & \longrightarrow & 0 \end{array}$$

plus $\varphi_{2,2}\colon D^2 \to D_2$ satisfying certain relations. Its polarised boundary is the strictly 1-dimensional symmetric Poincaré complex

$$\begin{array}{ccccccc} 0 & \longrightarrow & D_2 \oplus D^1 & \xrightarrow{\begin{pmatrix} d & -\varphi_{0,1}^* \\ 0 & -d^* \end{pmatrix}} & D_1 \oplus D^2 & \longrightarrow & 0 \\ & & \downarrow{\begin{pmatrix} 0 & 1 \\ \varepsilon & -\varphi_{1,2} \end{pmatrix}} & \swarrow{\begin{pmatrix} 0 & 0 \\ 0 & \varphi_{2,2} \end{pmatrix}} & \downarrow{\begin{pmatrix} 0 & -1 \\ -\varepsilon & -\varphi_{1,1} \end{pmatrix}} & & \\ 0 & \longrightarrow & D^1 \oplus D_2 & \xrightarrow{\begin{pmatrix} d^* & 0 \\ -\varphi_{0,1} & -d \end{pmatrix}} & D^2 \oplus D_1 & \longrightarrow & 0. \end{array}$$

Via Proposition 15.205 it corresponds to the following ε-symmetric formation:

$$(H^\varepsilon(F, \theta); F, G) \quad \text{with} \quad F = D^1 \oplus D_2 \quad \text{and} \quad G = \mathrm{im}\begin{pmatrix} \gamma \\ \delta \end{pmatrix}$$

where

$$\gamma = \begin{pmatrix} 0 & 1 \\ \varepsilon & -\varphi_{1,2} \end{pmatrix} \quad \text{and} \quad \delta = \begin{pmatrix} \varepsilon d & -\varepsilon \varphi_{0,1}^* \\ 0 & -\varepsilon d^* \end{pmatrix} \quad \text{and} \quad \theta = \begin{pmatrix} 0 & 0 \\ 0 & \varphi_{2,2} \end{pmatrix}.$$

The fact that the map γ is an isomorphism implies that the lagrangians G and F^* are complementary.

15.6 The Identification of the L-Groups

If $D \in \mathrm{b}R\text{-}\mathrm{CH}_{\mathrm{fgp}}$, then clearly $\partial D \in \mathrm{b}R\text{-}\mathrm{CH}_{\mathrm{fgp}}$ and the same holds in the category $\mathrm{b}R\text{-}\mathrm{CH}_{\mathrm{fgf}}$. □

And now, as promised, we proceed with the quadratic versions, which is the case we really need.

Proposition 15.439 *The boundary of a strictly well-connected 1-dimensional quadratic complex $(D, [\psi])$ is a strictly 0-dimensional quadratic Poincaré complex $(\partial D, [\partial \psi])$ which via Proposition 15.202 corresponds to a hyperbolic form.*

If in addition D is a chain complex of stably free finitely generated R-modules, then ∂D is also a chain complex of stably free finitely generated R-modules.

Proof. A polarised strictly well-connected 1-dimensional quadratic complex (D, ψ) consists of the following data

$$\begin{array}{ccccccc} 0 & \longrightarrow & 0 & \longrightarrow & D^1 & \longrightarrow & 0 \\ & & \downarrow & & \downarrow & & \\ 0 & \longrightarrow & D_1 & \longrightarrow & 0 & \longrightarrow & 0. \end{array}$$

Its boundary is given by the polarised strictly 0-dimensional quadratic Poincaré complex $(\partial D, \partial \psi)$ with

$$\partial D_0 = D_1 \oplus D^1, \quad \partial D_i = 0, \ i \geq 1, \quad \partial \psi_{0,0} = \begin{pmatrix} 0 & -1 \\ 0 & 0 \end{pmatrix} \quad \partial \psi_i = 0, \ i \geq 1.$$

Via Proposition 15.202 this corresponds to the standard quadratic hyperbolic form.

If $D \in \mathrm{b}R\text{-}\mathrm{CH}_{\mathrm{fgp}}$, then clearly $\partial D \in \mathrm{b}R\text{-}\mathrm{CH}_{\mathrm{fgp}}$ and the same holds in the stably free case. □

Proposition 15.440 *The boundary of a strictly well-connected 2-dimensional quadratic complex $(D, [\psi])$ is a strictly 1-dimensional quadratic Poincaré complex $(\partial D, [\partial \psi])$ which via Proposition 15.214 corresponds to a boundary formation.*

The same holds in the category $\mathrm{b}R\text{-}\mathrm{CH}_{\mathrm{fgf}}$.

Proof. A polarised strictly well-connected 2-dimensional quadratic complex (D, ψ) consists of the following data:

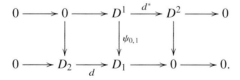

Its algebraic boundary is given by the polarised strictly 1-dimensional quadratic Poincaré complex

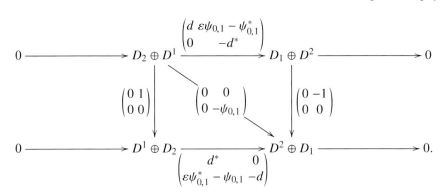

Via Proposition 15.214 it corresponds to a split ε-quadratic formation whose underlying quadratic formation is given as follows:

$$(H_\varepsilon(F); F, G) \quad \text{with} \quad F = D^1 \oplus D_2 \quad \text{and} \quad G = \text{im}\begin{pmatrix} \gamma \\ \delta \end{pmatrix}$$

with

$$\gamma = \begin{pmatrix} 0 & -1 \\ -\varepsilon & 0 \end{pmatrix} \quad \text{and} \quad \delta = \begin{pmatrix} \varepsilon d \ \psi_{0,1} - \varepsilon \psi_{0,1}^* \\ 0 & -\varepsilon d^* \end{pmatrix}.$$

The fact that the map γ is an isomorphism implies that the lagrangians G and F^* are complementary, which shows via Lemma 9.13(ii) that it is isomorphic to a boundary in the sense of Definition 9.12.

If $D \in bR\text{-CH}_{\text{fgp}}$ then clearly $\partial D \in bR\text{-CH}_{\text{fgp}}$ and the same holds in the category $bR\text{-CH}_{\text{fgf}}$. $\square$

For the proof of the bijectivity in the desired identification, we will also need another result, which is simultaneously an improvement on Proposition 15.214, Corollary 15.215 and Proposition 15.440. To prepare for it, consider a $(-\varepsilon)$-quadratic form $(Q, [\mu])$. Let $\partial(Q, [\mu]) = (H_\varepsilon(Q); Q, \Gamma_{[\mu]})$ be the associated boundary ε-quadratic formation in the sense of Definition 9.9. Then the class $[\mu]$ represented by $\mu \colon Q \to Q^*$ provides us with a choice of a hessian. Hence $\partial(Q, [\mu])$ is in a preferred way a split ε-quadratic formation. We will call a split ε-quadratic formation that is isomorphic to $\partial(Q, [\mu])$ with hessian $[\mu]$, for some $(-\varepsilon)$-quadratic form $(Q, [\mu])$, a *split ε-quadratic boundary formation*.

Proposition 15.441 (Split quadratic boundary formations correspond to boundary quadratic complexes) *Let $(C, [\psi])$ be a strictly 1-dimensional quadratic Poincaré complex, which corresponds via Proposition 15.214 to a split ε-quadratic formation $(F, G; \gamma, \delta, [\theta])$ that is stably isomorphic to a split ε-quadratic boundary formation $\partial(Q, [\mu])$ of a $(-\varepsilon)$-quadratic form $(Q, [\mu])$.*

Then $(C, [\psi])$ is homotopy equivalent to a boundary of a strictly well-connected 2-dimensional quadratic complex.

15.6 The Identification of the L-Groups

Proof. Consider the polarised strictly well-connected 2-dimensional quadratic chain complex $(D, [\psi_D])$ with $D_1 = Q$, $D_i = 0$ for $i \neq 1$ and $(\psi_D)_0 = \mu$. Then its polarised boundary $\partial(D, [\psi_D])$ is the polarised strictly 1-dimensional Poincaré quadratic complex

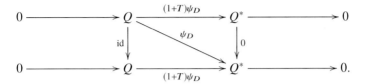

Via the correspondence of Proposition 15.214, we see that its associated split quadratic formation is $\partial(Q, [\mu])$. Since by Corollary 15.215 stable isomorphism classes of split formations correspond to the homotopy classes of homotopy 1-dimensional complexes, we obtain the desired statement. □

Closely related is the following proposition, which describes the effect of the choice of a hessian for boundary formations.

Proposition 15.442 (Choice of a hessian for a boundary formation) *Let $(Q, [\mu])$ be an $(-\varepsilon)$-quadratic form. Consider its boundary ε-quadratic split formation $\partial(Q, [\mu])$. Then any hessian $[\theta] \in Q_{-\varepsilon}(Q)$ for $\partial(Q, [\mu])$ has the property that there is an isomorphism of the underlying quadratic formations*

$$\partial(Q, [\mu]) \cong \partial(Q, [\theta]).$$

Proof. Any hessian $[\theta] \in Q_{-\varepsilon}(Q)$ for the boundary formation $\partial(Q, [\mu])$ has the property that $\theta - \varepsilon T(\theta) = \mu - \varepsilon T(\mu)$. □

Finally we come to the proof of Theorem 15.3 where the desired identification of the L-groups is achieved, which is one of the main results of this chapter. Recall that $L_n(R)$ denotes the group from Definition 8.90, when n is even, and the group from Definition 9.15, when n is odd. The groups denoted by $L_n(R)_{\text{chain}}$ were introduced in Definition 15.301 in both cases.

Proof of Theorem 15.3. We only treat isomorphism $L_n^p(R)_{\text{chain}} \xrightarrow{\cong} L_n^p(R)$. The proof for $L_n^h(R)_{\text{chain}} \xrightarrow{\cong} L_n^h(R)$ is an obvious variation if one takes Remark 8.92 and Exercise 9.16 into account.

By the skew-suspension isomorphisms from Proposition 15.427, it is enough to show the result for $n = 0, 1$. Also, by Corollary 15.436 we can assume that $L_n^p(R)_{\text{chain}}$ is defined using strictly n-dimensional complexes and strictly well-connected cobordisms.

For $n = 0$ the desired map is given by associating to a strictly 0-dimensional ε-quadratic Poincaré complex $(C, [\psi])$ the non-singular ε-quadratic form $(C^0, [\psi_0])$, which can be done as explained in Proposition 15.202. This map is well defined since by Corollary 15.203 a homotopy equivalence of strictly 0-dimensional ε-quadratic complexes induces a stable isomorphism of associated ε-quadratic forms and by

Proposition 15.439 the boundary of a strictly 1-dimensional ε-quadratic complex is sent to an ε-hyperbolic form. The map is onto since by Proposition 15.202 any non-singular ε-quadratic form comes from a 0-dimensional ε-quadratic Poincaré complex. The map is injective by Proposition 15.192 and Proposition 15.439.

If a strictly 0-dimensional ε-quadratic Poincaré complex maps to a form stably isomorphic to the standard hyperbolic form $H_\varepsilon(F)$, then it is homotopy equivalent to the boundary of the strictly well-connected 1-dimensional ε-quadratic complex $(D, [\psi])$ with $D_1 = F$ and $\psi = 0$.

For $n = 1$ the map is given by associating to a strictly 1-dimensional ε-quadratic Poincaré complex $(C, [\psi])$ the underlying non-singular ε-quadratic formation $(F, G; \gamma, \delta)$ of the split ε-quadratic formation obtained in the proof of Proposition 15.214. The map is well defined since by Corollary 15.215 a homotopy equivalence of strictly 1-dimensional ε-quadratic complexes induces a stable isomorphism of the corresponding ε-quadratic split formations (and hence also of ε-quadratic formations) and by Proposition 15.440 the boundary of a strictly well-connected 2-dimensional ε-quadratic complex is sent to a boundary ε-quadratic formation. The map is onto since any ε-quadratic formation comes from a split ε-quadratic formation by a choice of a hessian, which always exists by Remark 15.209, and by Proposition 15.214 any non-singular split ε-quadratic formation comes from a strictly 1-dimensional ε-quadratic Poincaré complex.

The map is injective by the following argument. Let $(C, [\psi])$ be a strictly 1-dimensional ε-quadratic Poincaré complex which maps to the underlying ε-quadratic formation of a split ε-quadratic formation $(F, G; \gamma, \delta, [\theta])$ with the property that the underlying ε-quadratic formation $(F, G; \gamma, \delta)$ is stably isomorphic to a boundary ε-quadratic formation $\partial(Q, [\mu])$ of a $(-\varepsilon)$-quadratic form $(Q, [\mu])$ in the sense of Definition 9.12. Then by Proposition 15.213 we can extend the structure of an ε-quadratic formation on $\partial(Q, [\mu])$ to a split ε-quadratic formation by some choice of a hessian $[\theta']$ for some $\theta' : Q \to Q^*$. Hence there is a stable isomorphism of split ε-quadratic formations between $(F, G; \gamma, \delta, [\theta])$ and $\partial(Q, [\mu])$ with the new hessian $[\theta']$. By Proposition 15.442 the formation $\partial(Q, [\mu])$ with the new hessian $[\theta']$ is isomorphic to the split ε-quadratic boundary formation $\partial(Q, [\theta'])$. By Proposition 15.441 we obtain that $(C, [\psi])$ itself is homotopy equivalent to a boundary of a 2-dimensional ε-quadratic complex, which is what we wanted. □

The simple case will be treated in Theorem 15.464.

Thanks to Theorem 15.3 we can leave out the subscript "chain" from the notation for L-groups defined in this chapter in the sequel.

15.7 The Identification of the Surgery Obstructions

In this section we identify the quadratic signature $\text{qsign}_\pi(f, \overline{f}) \in L_n(\mathbb{Z}\pi, w)$ of a normal map of degree one $(f, \overline{f}): M \to X$ from Definition 15.318 with the surgery obstruction defined in Chapters 8 and 9 by proving Theorem 15.4, where we use the

15.7 The Identification of the Surgery Obstructions

decoration h, so that the L-groups $L_n(R)$ are cobordism groups of quadratic chain complexes in $bR\text{-CH}_{\text{fgf}}$. The simple version will be treated in Theorem 15.464. Just as in Chapters 8 and 9, there are also versions of Theorem 15.4 where one starts with a normal map of degree one of manifolds with boundary assuming a homotopy equivalence on the boundary. These are obtained by a straightforward adaptation, which we leave for the reader.

We start by recalling (and slightly adapting) the notation in the case $n = 2k$. Let $(f, \overline{f}): M \to X$ be a normal map of degree one from an n-dimensional manifold M to an n-dimensional Poincaré complex X with $\pi = \pi_1(X)$ and orientation homomorphism $w: \pi \to \{\pm 1\}$. The surgery obstruction

$$\sigma(f, \overline{f}) = [(K_k(\widetilde{M}), s, t)] \in L_{2k}(\mathbb{Z}\pi, w)$$

was obtained in Theorem 8.112 by first arranging the map $(f, \overline{f})$ to be k-connected by surgery below the middle dimension and then taking the surgery kernel $(K_k(\widetilde{M}), s, t)$ as explained in Definition 8.36, Lemma 8.59, and Example 8.86.

On the other side recall from Definition 15.318 and (15.163) the *quadratic signature* of the normal map of degree one $(f, \overline{f}): M \to X$

$$\mathrm{qsign}_\pi((f, \overline{f}): M \to X) = \left(\mathrm{cone}(C_*^!(\widehat{f})), e_{Q,n} \circ \overline{\psi}_{\{F\}_{\Gamma, w}}([X])\right) \in L_n(\mathbb{Z}\pi, w).$$

The obstruction groups are identified in Section 15.6. Recall that the iteration of the skew-suspension isomorphisms from Proposition 15.427 produced the isomorphism

$$L_{n-2k}(\overline{S})^k := L_{n-2}(\overline{S}) \circ \cdots \circ L_{n-2k}(\overline{S}) \colon L_{n-2k}(\mathbb{Z}\pi, w) \to L_n(\mathbb{Z}\pi, w).$$

Further recall that in the process of this identification, an ε-quadratic form is thought of as an n-dimensional quadratic complex simply by shifting the underlying module from dimension 0 to the middle dimension k and applying Proposition 15.202. In the reverse direction one passes from an n-dimensional quadratic complex to a complex concentrated in a single dimension via algebraic surgery and homotopy invariance and the other direction of Proposition 15.202 produces a quadratic form.

Also note that we have proved normal bordism invariance for the surgery obstruction $\sigma(f, \overline{f})$ with the cylindrical target, see Theorems 8.112 and 9.60, and for the quadratic signature $\mathrm{qsign}_\pi(f, \overline{f})$ even for the not necessarily cylindrical target, see Definition 15.318 and Construction 15.270. Hence, thanks to the surgery below middle dimension, in order to identify them, it is enough to show the identification for k-connected maps.

The identification for highly connected maps boils down to comparing the Wu classes v^k of the quadratic signature applied to the Poincaré dual of the kernel homology class of a pointed immersion $\alpha = [g, w]$ with $g: S^k \hookrightarrow M$ with the self-intersection number $\mu(\alpha)$ from (8.23) in Chapter 8, see also Example 8.86. We will show in Proposition 15.456 that the Wu class v^k of the quadratic signature at the Poincaré dual of g equals the sum of the self-intersection number $\mu(\alpha)$ and a certain element $j_2(v_g, v_h)$, which is an obstruction to framing the normal bundle v_g of g.

Because g represents a class in the surgery kernel, we can always choose g so that ν_g can be framed, and so we obtain the desired identification in Proposition 15.460.

The ingredients needed for the proof of the formula in Proposition 15.456 are the following. Besides the definition of the quadratic construction from Construction 15.153, we will need its additivity upon composition of maps from Proposition 15.156. Furthermore we will need to understand how quadratic structures measure the failure of a stably trivial bundle to be trivial. This is done in Section 15.7.1, in Lemma 15.447, and the translation goes via associating to the geometric problem an appropriate stable map, which becomes unstable in the desired case. Next we will need to investigate how the quadratic structure can measure the failure of an immersion to be regularly homotopic to an embedding. Again this is done by producing out of an immersion a stable (Umkehr) map, which becomes unstable in the desired case. This is done in Proposition 15.455 after some preparation in Section 15.7.2. In both cases the equivariant S-duality from Section 15.3.8 will also be used.

In the odd-dimensional case $n = 2k + 1$ we can similarly pass to k-connected maps and we have that the skew suspension isomorphisms from Proposition 15.427 induces the isomorphism

$$L_{n-2k}(\overline{S})^k := L_{n-2}(\overline{S}) \circ \cdots \circ L_{n-2k}(\overline{S}) \colon L_{n-2k}(\mathbb{Z}\pi, w) \to L_n(\mathbb{Z}\pi, w).$$

This isomorphism together with Proposition 15.214 yields the correspondence between quadratic formations and 1-dimensional quadratic complexes. The correspondence between the underlying modules of the formations and 1-dimensional complexes is just a standard verification via homological algebra. The next key idea is that the surgery obstruction in the odd-dimensional case is in fact defined via surgery kernels of even-dimensional highly connected maps. Hence we can use what we have already shown in the even-dimensional case and obtain finally the identification also in the odd-dimensional case in Proposition 15.461.

In this section we follow very closely the original source, which is in Ranicki's paper [345, Section 5].

15.7.1 Euler Classes

As promised above, in this section we aim to understand how the quadratic construction measures the failure of a stably trivial normal bundle to be actually trivial.

The homomorphism

$$j_1 \colon \pi_m(\mathrm{BSO}(q)) = \pi_{m-1}(\mathrm{SO}(q)) \to Q^{m+q}(q[\mathbb{Z}]) \qquad (15.443)$$

is defined as follows. Given an oriented q-plane bundle $\alpha \colon S^m \to \mathrm{BSO}(q)$, the Thom space $\mathrm{Th}(\alpha)$ can be obtained by attaching an $(m + q)$-cell to S^q

$$\mathrm{Th}(\alpha) = S^q \cup_{J(\alpha)} e^{m+q}$$

15.7 The Identification of the Surgery Obstructions

where $J\colon \pi_{m-1}(\mathrm{SO}(q)) \to \pi_{m+q-1}(S^q)$ is the J-homomorphism as in Definition 12.29. Recall from (15.66) the symmetric construction $\overline{\varphi}_{\mathrm{Th}(\alpha)}$ and from Definition 15.171 (i) the Wu class $v_m([\varphi])$, which we view as a function of $[\varphi] \in Q^{m+q}(\widetilde{C}(\mathrm{Th}(\alpha)))$. Consider the composite

$$H_{m+q}(\mathrm{Th}(\alpha)) \xrightarrow{\overline{\varphi}_{\mathrm{Th}(\alpha)}} Q^{m+q}(\widetilde{C}(\mathrm{Th}(\alpha))) \xrightarrow{v_m} \hom_{\mathbb{Z}}(H^q(\mathrm{Th}(\alpha)), Q^{m+q}(q[\mathbb{Z}])).$$

Denote by $[S^m] \in H_m(S^m)$ the fundamental class obtained from the standard orientation of S^m and by $u_\alpha \in H^q(\mathrm{Th}(\alpha))$ the Thom class associated to the given orientation of α. The Thom isomorphism Theorem 6.54 tells us that the class $t_\alpha := (- \cap u_\alpha)^{-1}[S^m] \in H_{m+q}(\mathrm{Th}(\alpha))$ is a generator of this infinite cyclic group. Define

$$j_1(\alpha) := v_m(\overline{\varphi}_{\mathrm{Th}(\alpha)}(t_\alpha))(u_\alpha) \in Q^{m+q}(q[\mathbb{Z}]). \tag{15.444}$$

Next we define the homomorphism

$$j_2 \colon \pi_{m+1}(\mathrm{BSO}(p+q), \mathrm{BSO}(q)) \to Q_{m+q}(q[\mathbb{Z}]). \tag{15.445}$$

Suppose in addition to the above data that there is given a nullhomotopy $\beta \colon D^{m+1} \to \mathrm{BSO}(p+q)$ of $\alpha \oplus \underline{\mathbb{R}}^p \colon S^m \to \mathrm{BSO}(p+q)$. The induced isomorphism of bundles $\beta \colon \underline{\mathbb{R}}^{p+q} \to \alpha \oplus \underline{\mathbb{R}}^p$ induces a homotopy equivalence of Thom spaces

$$\mathrm{Th}(\beta) \colon \mathrm{Th}(\underline{\mathbb{R}}^{p+q}) \simeq S^{p+q} \vee S^{m+p+q} \to \mathrm{Th}(\alpha \oplus \underline{\mathbb{R}}^p) \simeq \Sigma^p \mathrm{Th}(\alpha).$$

For the trivial bundle $\underline{\mathbb{R}}^{p+q}$ the generator $t_{\underline{\mathbb{R}}^{p+q}} \in H_{m+p+q}(\mathrm{Th}(\underline{\mathbb{R}}^{p+q})) = \mathbb{Z}$ is represented by the map $S^{m+p+q} \xrightarrow{\mathrm{incl}} S^{p+q} \vee S^{m+p+q} \simeq \mathrm{Th}(\underline{\mathbb{R}}^{p+q})$. Therefore, the composite

$$I(\beta) \colon \Sigma^p S^{m+q} \cong S^{m+p+q} \xrightarrow{\mathrm{incl}} S^{p+q} \vee S^{m+p+q} \xrightarrow{\mathrm{Th}(\beta)} \Sigma^p \mathrm{Th}(\alpha)$$

is a map which represents the generator $t_{\alpha \oplus \underline{\mathbb{R}}^p} \in H_{m+p+q}(\Sigma^p \mathrm{Th}(\alpha)) = \mathbb{Z}$. Now we can apply the quadratic construction (15.149) and the Wu class from Definition 15.171 (ii)

$$H_{m+q}(S^{m+q}) \xrightarrow{\overline{\psi}_{\{I(\beta)\}}} Q_{m+q}(\widetilde{C}(\mathrm{Th}(\alpha))) \xrightarrow{v^m} \hom_{\mathbb{Z}}(H^q(\mathrm{Th}(\alpha)), Q_{m+q}(q[\mathbb{Z}])),$$

and define

$$j_2(\alpha, \beta) = v^m(\overline{\psi}_{\{I(\beta)\}}([S^{m+q}]))(u_\alpha) \in Q_{m+q}(q[\mathbb{Z}]) \tag{15.446}$$

where $[S^{m+q}] \in H_{m+q}(S^{m+q}) \cong \mathbb{Z}$ is the fundamental class obtained from the standard orientation, and $u_\alpha \in H^q(\mathrm{Th}(\alpha))$ is the Thom class associated to the given orientation of α.

Lemma 15.447 Let $\alpha\colon S^m \to \mathrm{BSO}(q)$ be a q-dimensional bundle and let $\beta\colon D^{m+1} \to \mathrm{BSO}(p+q)$ be a nullhomotopy of $\alpha \oplus \mathbb{R}^p\colon S^m \to \mathrm{BSO}(p+q)$. If β can be factored through $\mathrm{BSO}(q)$, then

$$j_2(\alpha, \beta) = 0 \in Q_{m+q}(q[\mathbb{Z}]).$$

Proof. This follows from the fact that the quadratic construction is an obstruction to desuspension of a stable map, see Remark 15.151. □

Remark 15.448 (Euler class terminology) Recall that the *Euler class* of an oriented q-dimensional bundle $\alpha\colon X \to \mathrm{BSO}(q)$ is a characteristic class $e(\alpha) \in H^q(X)$ with the property that, if α has a section, then $e(\alpha) = 0$, see [308, Chapter 9]. If X is an oriented q-dimensional Poincaré complex with the fundamental class $[X] \in H_q(X)$, then the *Euler number* of α is defined to be $\chi(\alpha) = e(\alpha)([X]) \in \mathbb{Z}$. The function j_1 above is related to these concepts. Namely, in the case $q = m = 2k$, using the identification $Q^{m+q}(q[\mathbb{Z}]) = \mathbb{Z}$, it is in fact the function that assigns to a class $\alpha \in \pi_m(\mathrm{BSO}(q))$ viewed as a q-dimensional bundle over S^m its Euler number $\chi(\alpha)$.

15.7.2 Self-Intersections

In this subsection we define certain invariants λ^{alg} and μ^{alg} of immersions.

Let M be a connected n-dimensional manifold with a base point $x_0 \in M$, fundamental group $\pi = \pi_1(M)$, and orientation homomorphism $w\colon \pi \to \{\pm 1\}$. Choose a base point $\widetilde{x}_0 \in \widetilde{M}$. Recall from (8.9) the $\mathbb{Z}\pi$-module $I_r(M)$ of pointed regular homotopy classes of oriented immersions $g\colon S^r \looparrowright M$ equipped with a path v from the previously fixed base point of S^r to x_0. With these choices the path v can be used to obtain a preferred lift $\widehat{g}\colon \widehat{S^r} = \pi \times S^r \looparrowright \widetilde{M}$ to the universal cover.

For an immersion g representing an element $\alpha = [(g, v)] \in I_r(M)$, let $\nu_g\colon S^r \to \mathrm{BSO}(n-r)$ be its normal bundle, let $\mathrm{Th}_\pi(\nu_g)$ be the associated Thom π-space and let $u_{\nu_g} \in H^{n-r}(\mathrm{Th}_\pi(\nu_g))$ be the Thom class.

Define the Umkehr map $C^!(\widehat{g})\colon C(\widetilde{M}) \to \widetilde{C}(\mathrm{Th}_\pi(\nu_g))$ as the composite

$$C^!(\widehat{g})\colon C(\widetilde{M}) \xrightarrow{(-\cap [M])^{-1}} C^{n-*}(\widetilde{M}) \xrightarrow{\widehat{g}^*} C^{n-*}(\widehat{S^r})$$

$$\xrightarrow{-\cap [S^r]} \Sigma^{n-r} C(\widehat{S^r}) \xrightarrow{(-\cap [u_{\nu_g}])^{-1}} \widetilde{C}(\mathrm{Th}_\pi(\nu_g)) \quad (15.449)$$

where we remind the reader that $\widehat{S^r}$ is the cover of S^r induced from $\widetilde{M}$. Moreover, note that the map $C^!(\widehat{g})$ is only well defined up to π-homotopy, but that is good enough for our purposes.

Denote $(C^!(\widehat{g}))^*(u_{\nu_g}) \in H^{n-r}(\widetilde{M})$ by x and, using the symmetric construction $\overline{\varphi}_{M,\pi,w}$ from (15.66), define

$$\lambda^{\mathrm{alg}}(\alpha) = \nu_r(\overline{\varphi}_{M,\pi,w}([M]))(x) \in Q^n((n-r)[\mathbb{Z}\pi]). \quad (15.450)$$

15.7 The Identification of the Surgery Obstructions

The following equality is an elementary consequence of the definition of v_r

$$\lambda^{\text{alg}}(\alpha) = v_r(\overline{\varphi}_{M,\pi,w}([M]))(x) = v_r(C^!(\widehat{g})^{Q,n}\overline{\varphi}_{M,\pi,w}([M]))(u_{v_g}). \quad (15.451)$$

For future purposes we also record that the Umkehr map $C^!(\widehat{g})$ from (15.449) satisfies the following two equations:

$$((C^!(\widehat{g}))^*(u_{v_g})) \cap [M] = g_*([S^r]) \quad \text{and} \quad C^!(\widehat{g})_*([M]) = t_{v_g}. \quad (15.452)$$

We now turn to μ^{alg}. Given $\alpha = [(g,v)] \in I_r(M)$, let $g\colon S^r \looparrowright M$, v_g, and $\text{Th}_\pi(v_g)$ be as above, and assume in addition that $p > 2r - n + 1$. Then it is possible to deform the immersion $g \times 0\colon S^r \looparrowright M \times D^p$ into an embedding $g'\colon S^r \hookrightarrow M \times D^p$ by a small regular homotopy so that $v_{g'} = v_g \oplus \mathbb{R}^p$. Let E be the tubular neighbourhood of $g'(S^r)$. Then the Umkehr map

$$G\colon \Sigma^p(\widetilde{M}_+) = \widetilde{M} \times D^p/\widetilde{M} \times S^{p-1} \xrightarrow{\text{collapse}}$$

$$\widetilde{E}/\partial\widetilde{E} = \text{Th}_\pi(v_{g'}) \cong \Sigma^p \text{Th}_\pi(v_g) \quad (15.453)$$

induces the map $C^!(\widehat{g})\colon C(\widetilde{M}) \to \widetilde{C}(\text{Th}_\pi(v_g))$ above.

Using the quadratic construction $\overline{\psi}_{\{G\}_{\pi,w}}$ from (15.154), define for such an α in $I_r(M)$ the invariant

$$\mu^{\text{alg}}(\alpha) = -v^r(\overline{\psi}_{\{G\}_{\pi,w}}[M])(u_{v_g}) \in Q_n((n-r)[\mathbb{Z}\pi]). \quad (15.454)$$

15.7.3 Putting it Together

In this section we prove Theorem 15.4. The main technical result will be a relation between the invariants λ^{alg} and μ^{alg} from the previous section and the geometric intersection numbers and the self-intersection numbers from Chapter 8, which for the purposes of the present section we denote by λ^{geo} for (8.11) and μ^{geo} for (8.23).

We start with a proposition which establishes that the invariants λ^{alg} and μ^{alg} have analogous properties to those described in Lemma 8.26 for λ^{geo} and μ^{geo}. The proposition is taken from [345, Proposition 5.2], but the formulation is slightly modified to accommodate the operation ρ from (8.25) in item (i) and the cup product pairing $s\colon H_k(\widetilde{M}) \otimes H_k(\widetilde{M}) \to \mathbb{Z}\pi$ from (8.15).

We also remind the reader of the equation (8.18), which established the relation between the geometric intersection pairing λ^{geo} and the cup product pairing s. Hence as a consequence of the first equation from (15.452) we obtain

$$\lambda^{\text{alg}}(\alpha) = s([\widehat{g}],[\widehat{g}]) = \lambda^{\text{geo}}(\alpha,\alpha).$$

Just as in the previous section, we assume that M is a connected n-dimensional manifold with a base point $x_0 \in M$, fundamental group $\pi = \pi_1(M)$, and orientation homomorphism $w \colon \pi \to \{\pm 1\}$, and that we made a choice of a base point $\widetilde{x}_0 \in \widetilde{M}$.

Proposition 15.455 (The symmetric and quadratic self-intersections) *The functions*

$$\lambda^{\mathrm{alg}} \colon I_r(M) \to Q^n((n-r)[\mathbb{Z}\pi]) \quad (\alpha = [(g \colon S^r \looparrowright M, v)]) \mapsto \lambda^{\mathrm{alg}}(\alpha);$$

$$\mu^{\mathrm{alg}} \colon I_r(M) \to Q_n((n-r)[\mathbb{Z}\pi]) \quad (\alpha = [(g \colon S^r \looparrowright M, v)]) \mapsto \mu^{\mathrm{alg}}(\alpha),$$

from (15.450) *and* (15.454) *satisfy the following statements:*

(i) $\lambda^{\mathrm{alg}}(a\alpha) = a\lambda^{\mathrm{alg}}(\alpha)\overline{a}$, $\mu^{\mathrm{alg}}(a\alpha) = \rho(a, \mu^{\mathrm{alg}}(\alpha))$;

(ii) $\lambda^{\mathrm{alg}}(\alpha) = (j_1(v_g), 0) + (1+T)\mu^{\mathrm{alg}}(\alpha)$, *where* j_1 *is from* (15.443);

(iii) $\mu^{\mathrm{alg}}(\alpha_1 + \alpha_2) - \mu^{\mathrm{alg}}(\alpha_1) - \mu^{\mathrm{alg}}(\alpha_2) = \begin{cases} [s([\widehat{g}_1],[\widehat{g}_2])] & n = 2r; \\ 0 & n \neq 2r; \end{cases}$

(iv) *If* $\alpha \in I_r(M)$ *contains an embedding, then* $\mu^{\mathrm{alg}}(\alpha) = 0$, *and, if it contains a framed embedding, then also* $\lambda^{\mathrm{alg}}(\alpha) = 0$;

(v) *If* $n = 2r \geq 6$ *and* $\mu^{\mathrm{alg}}(\alpha) = 0$, *then* $\alpha \in I_r(M)$ *contains an embedding.*

Proof. (i) This is an elementary consequence of Definition 15.171.

(ii) By the property (15.150) about the symmetrisation of the quadratic construction we have the relation

$$(C^!(\widehat{g}))^{Q,n} \overline{\varphi}_{M,\pi,w}([M]) = \overline{\varphi}_{\mathrm{Th}_\pi(v_g),\pi,w} C^!(\widehat{g})_*([M]) - (1+T)\overline{\psi}_{\{G\}_{\pi,w}}([M]).$$

Now apply $v_r(-)(u_{v_g})$ and use (15.451) on the left-hand side, and (15.444) with the second equation in (15.452) in the first summand, and (15.454) in the second summand on the right-hand side.

(iii) Let $\alpha_1, \alpha_2 \in I_r(M)$ be two elements represented by immersions g_1, g_2, and let $\alpha_1 + \alpha_2 \in I_r(M)$ be their sum represented by the connected sum $g_1 \# g_2$. Then we have for the Thom spaces that

$$\mathrm{Th}_\pi(v_{g_1 \# g_2}) \simeq \mathrm{Th}_\pi(v_{g_1}) \vee \mathrm{Th}_\pi(v_{g_2}),$$

which gives us the identification of the chain complexes

$$\widetilde{C}(\mathrm{Th}_\pi(v_{g_1+g_2})) \simeq \widetilde{C}(\mathrm{Th}_\pi(v_{g_1})) \oplus \widetilde{C}(\mathrm{Th}_\pi(v_{g_2})).$$

For the quadratic self-intersection we have

$$\mu^{\mathrm{alg}}(\alpha_1 + \alpha_2) = -v^r(\overline{\psi}_{\{G\}_{\pi,w}}[M])(u_{v_{g_1}}, u_{v_{g_2}}) \in Q_n((n-r)[\mathbb{Z}\pi])$$

with

$$G = G_1 \vee G_2 \colon \Sigma^p \widetilde{M}_+ \to \Sigma^p \mathrm{Th}_\pi(v_{g_1}) \vee \Sigma^p \mathrm{Th}_\pi(v_{g_2}).$$

By the additivity properties of the Q-groups from Proposition 15.168 the quadratic construction of G is of the form

15.7 The Identification of the Surgery Obstructions   747

$$\overline{\psi}_{\{G\}_{\pi,w}} = \begin{pmatrix} \overline{\psi}_{\{G_1\}_{\pi,w}} \\ \overline{\psi}_{\{G_2\}_{\pi,w}} \\ (C^!(\widehat{g}_1) \otimes C^!(\widehat{g}_2))\Delta_0 \end{pmatrix} : H_n(\mathbb{Z}^w \otimes_{\mathbb{Z}\pi} C(\widetilde{M})) \to$$

$$\to Q_n(\widetilde{C}(\mathrm{Th}_\pi(\nu_{g_1}))) \oplus Q_n(\widetilde{C}(\mathrm{Th}_\pi(\nu_{g_2}))) \oplus$$
$$\oplus H_n(\widetilde{C}(\mathrm{Th}_\pi(\nu_{g_1})) \otimes_{\mathbb{Z}\pi} \widetilde{C}(\mathrm{Th}_\pi(\nu_{g_2}))),$$

where $\Delta_0 \colon H_n(\mathbb{Z}^w \otimes_{\mathbb{Z}\pi} C(\widetilde{M})) \to H_n(C(\widetilde{M}) \otimes_{\mathbb{Z}\pi} C(\widetilde{M}))$ is the map induced by the diagonal approximation from Construction 15.48. Applying the additivity property of the Wu class v^r from Proposition 15.174 and equation (15.452) yields the required identity upon observing that

$$H_n(C^!(\widehat{g}_1)^*(u_{\nu_{g_1}}) \otimes C^!(\widehat{g}_2)^*(u_{\nu_{g_2}}))(\Delta_0[M]) =$$
$$= ((C^!(\widehat{g}_1)^*(u_{\nu_{g_1}})) \cup (C^!(\widehat{g}_2)^*(u_{\nu_{g_2}}))) \cap [M] = s([\widehat{g}_1], [\widehat{g}_2]).$$

(iv) If g is regularly homotopic to an embedding, then the map G can be desuspended and $\psi_{\{G\}_{\pi,w}} = 0$ by Remark 15.151. If it is regularly homotopic to a framed embedding, then also the normal bundle ν_g is trivial and so the term $j_1(\nu_g)$ in the property (ii) of Proposition 15.455 vanishes, and we obtain the equation $\lambda^{\mathrm{alg}}(\alpha) = (1+T)\mu^{\mathrm{alg}}(\alpha) = 0$.

(v) Recall that the geometric self-intersection $\mu^{\mathrm{geo}} \colon I_r(M) \to Q_0(0[\mathbb{Z}\pi])$ from (8.24) satisfies by Lemma 8.26 (ii) and (8.18) the relation

$$\mu^{\mathrm{geo}}(\alpha_1 + \alpha_2) - \mu^{\mathrm{geo}}(\alpha_1) - \mu^{\mathrm{geo}}(\alpha_2) = [s([\widehat{g}_1], [\widehat{g}_2])]$$

and has the property that $\mu^{\mathrm{geo}}(\alpha) = 0$ with $\alpha = [(g, \nu)]$ if and only if g is regularly homotopic to an embedding, see Theorem 8.28.

Next we show that $\mu^{\mathrm{alg}}(\alpha) = \mu^{\mathrm{geo}}(\alpha)$ by a generalisation of a trick by Browder [55, Lemma IV.7]. Let $a \in \mathbb{Z}\pi$ be a lift of $\mu^{\mathrm{geo}}(\alpha)$ and consider the immersion

$$g' = S^r \hookrightarrow M' = M \# (S^r \times S^r)$$

from Example 8.32 representing the homology class

$$C(g')_*[S^r] = (0, a, 1) \in H_r(M') = H_r(M) \oplus \mathbb{Z}\pi \oplus \mathbb{Z}\pi.$$

Define the immersion

$$g'' = g\#0 + g' \colon S^r \hookrightarrow M',$$

and apply to $\alpha'' := [(g'', \nu)]$ the sum formulas, which are available for μ^{geo} by Lemma 8.26, and for μ^{alg} by item (iii) of this Proposition 15.455

$$\mu^{\mathrm{alg}}(\alpha'') = \mu^{\mathrm{alg}}(\alpha\#0) + \mu^{\mathrm{alg}}(\alpha') = \mu^{\mathrm{alg}}(\alpha) - [a] = \mu^{\mathrm{alg}}(\alpha) - \mu^{\mathrm{geo}}(\alpha);$$
$$\mu^{\mathrm{geo}}(\alpha'') = \mu^{\mathrm{geo}}(\alpha\#0) + \mu^{\mathrm{geo}}(\alpha') = \mu^{\mathrm{geo}}(\alpha) - [a] = 0$$

for $\alpha' := [(g', v)]$. From item (iv) of this Proposition 15.455 we conclude that $\mu^{\mathrm{geo}}(\alpha'') = 0$ implies $\mu^{\mathrm{alg}}(\alpha'') = 0$, and hence $\mu^{\mathrm{alg}}(\alpha) = \mu^{\mathrm{geo}}(\alpha)$, which yields the desired statement. This finishes the proof of Proposition 15.455. □

To a degree one normal map $(f, \overline{f}) \colon M \to X$ we associated in (15.163) a quadratic Poincaré complex which represents the quadratic signature $\mathrm{qsign}_\pi(f, \overline{f})$ in the quadratic L-group of the group ring of X. In the identification of the quadratic signature with the surgery obstructions from Chapters 8 and 9 it will be important to relate the Wu classes of the quadratic Poincaré complex representing $\mathrm{qsign}_\pi(f, \overline{f})$ with the invariant μ^{alg} from (15.454).

This is done in the next proposition. It can be viewed as an improvement of the property (ii) from the previous Proposition 15.455, which only describes the symmetrisation of μ^{alg}. Below we describe μ^{alg} itself in the situation where we only consider immersions representing elements in the surgery kernel of a degree one normal map. Since we will ultimately apply it only in the case where the degree one normal map is highly connected, we can assume that we are in the universal covering case as in Notation 8.34 and we made the choices stated there. Since we have shown $\mu^{\mathrm{alg}} = \mu^{\mathrm{geo}}$ in the above proof, we drop the superscripts and just use μ. In [345] this proposition appears as Proposition 5.3.

Proposition 15.456 (Wu classes of the quadratic kernel)
Let $(f, \overline{f}) \colon M \to X$ be a normal π-map of degree one from an n-dimensional closed manifold to an n-dimensional Poincaré complex X with $\pi_1(X) = \pi$ and assume that we are in the universal covering case as in Notation 8.34.

Let $\alpha \in I_r(M)$ be represented by an immersion $g \colon S^r \looparrowright M$ with the normal bundle $v_g \colon S^r \to \mathrm{BSO}(n-r)$, and let there be a nullhomotopy $h \colon fg \simeq 0 \colon S^r \to X$ inducing the stable trivialisation $v_h \colon D^{r+1} \to \mathrm{BSO}$ of v_g determined by the bundle map $\overline{f} \colon TM \oplus \mathbb{R}^a \to \xi$.

Let $C_*^!(\widehat{f})$ be the Umkehr map and let e, F, and $\psi_{\{F\}_{\pi, w}}$ be the maps associated to $(f, \overline{f})$ as in Construction 15.161 and Example 15.162. Consider the n-dimensional quadratic complex

$$\left(C = \mathrm{cone}(C_*^!(\widehat{f})), [\psi] = e_{Q,n} \circ \overline{\psi}_{\{F\}_{\pi, w}}([X])\right)$$

from (15.163) and the Wu class

$$v^r([\psi]) \colon H^{n-r}(C) \to Q_n((n-r)[\mathbb{Z}\pi]) \cong Q_{n-2r}((n-2r)[\mathbb{Z}\pi])$$

from Definition 15.171.

Then for the Poincaré dual $x \in K^{n-r}(\widetilde{M}) \cong H^{n-r}(C)$ of the image of the element $(g, h) \in \pi_{r+1}(f) = \pi_{r+1}(\widehat{f})$ in $H_{r+1}(\widehat{f}) \cong K_r(\widetilde{M})$ we obtain

$$v^r([\psi])(x) = (j_2(v_g, v_h), 0) + \mu(\alpha) \in Q_n((n-r)[\mathbb{Z}\pi])$$

where j_2 is the map from (15.445).

15.7 The Identification of the Surgery Obstructions

Proof. The maps f, g and h fit into the commutative diagram

$$\begin{array}{ccc} S^r & \xrightarrow{g} & M \\ \downarrow & & \downarrow f \\ D^{r+1} & \xrightarrow{h} & X \end{array} \qquad (15.457)$$

with the corresponding normal bundle data diagram

$$\begin{array}{ccc} \mathrm{Th}_\pi(g^*\nu_M) & \longrightarrow & \mathrm{Th}_\pi(\nu_M) \\ \downarrow & & \downarrow \overline{f} \\ \mathrm{Th}_\pi(h^*\nu_X) & \longrightarrow & \mathrm{Th}_\pi(\nu_X) \end{array} \qquad (15.458)$$

and its S-dual

$$\begin{array}{ccc} \Sigma^p \, \mathrm{Th}_\pi(\nu_g) & \xleftarrow{G} & \Sigma^p \widetilde{M}_+ \\ I_h \uparrow & & \uparrow F \\ \Sigma^p \vee_\pi S^n & \xleftarrow{H} & \Sigma^p \widetilde{X}_+ \end{array} \qquad (15.459)$$

From the commutativity of this diagram and from the additivity formula of the equivariant quadratic construction for compositions in Proposition 15.156 we obtain

$$C^!(\widehat{g})_{Q,n} \circ \overline{\psi}_{\{F\}_{\pi,w}}[X] + \overline{\psi}_{\{G\}_{\pi,w}}[M] = \overline{\psi}_{\{I(\nu_h)\}_{\pi,w}}[S^n] + I(\nu_h)_{Q,n} \circ \overline{\psi}_{\{H\}_{\pi,w}}[X].$$

Recall from [408, Theorem 2.4] that X is homotopy equivalent to a complex obtained from its $(n-1)$-skeleton by attaching a single n-cell, so we can assume $X = X^{(n-1)} \cup e^n$. After passing to the universal covers and adding the base points, we can collapse $X^{(n-1)}$ and obtain a map $\widetilde{X}_+ \to \vee_\pi S^n$, which is an S-dual to $\mathrm{Th}_\pi(h^*\nu_X) \to \mathrm{Th}_\pi(\nu_X)$, and so we can assume that H is its p-th suspension. From Remark 15.151 we obtain $\overline{\psi}_{\{H\}_{\pi,w}} = 0$. Next we apply the Wu class

$$v^r : Q_n(\widetilde{C}(\mathrm{Th}_\pi(\nu_g))) \to \hom_{\mathbb{Z}\pi}(\widetilde{H}^{n-r}(\mathrm{Th}_\pi(\nu_g)), Q_n((n-r)[\mathbb{Z}\pi]))$$

to the three remaining terms and evaluate what we obtain at the Thom class $u_{\nu_g} \in \widetilde{H}^{n-r}(\mathrm{Th}_\pi(\nu_g))$. Unravelling the definitions shows the following equations. The definition of $[\psi]$ from Construction 15.161 and Example 15.162 and an elementary naturality property of v^r give

$$v^r([\psi])(x) = v^r(\overline{\psi}_{\{F\}_{\pi,w}}[X])((C^!(\widehat{g}))^*(u_{\nu_g})) = v^r(C^!(\widehat{g})_{Q,n}\overline{\psi}_{\{F\}_{\pi,w}}[X])(u_{\nu_g}).$$

The definition of μ from (15.454) gives
$$\mu(\alpha) = -v^r(\overline{\psi}_{\{G\}_{\pi,w}}[M])(u_{v_g}).$$

And finally the definition of j_2 from (15.446) gives
$$j_2(v_g, v_h) = v^r(\overline{\psi}_{\{I(v_h)\}_{\pi,w}}[S^n])(u_{v_g}).$$

Putting these equations together finishes the proof. □

The proof of the identification in Theorem 15.4 is finalised in the following two propositions corresponding to the two cases depending on the parity of n. They appear as Proposition 5.4 in [345]. In both cases we assume that we are in the universal covering case as in Notation 8.34.

Proposition 15.460 *Let $n = 2k \geq 5$ and let $(f, \overline{f}) \colon M \to X$ be a k-connected normal map of degree one. Then we have*
$$\sigma(f, \overline{f}) = \mathrm{qsign}_\pi(f, \overline{f}) \in L_n(\mathbb{Z}\pi, w).$$

Proof. Since f is k-connected, we get
$$\mathrm{cone}(C_*^!(\widehat{f})) \simeq k[K_k(\widetilde{M})].$$

Since any class $y \in K_k(\widetilde{M})$ can be represented by a framed immersion $\alpha = [g, v]$ with $g \colon S^k \looparrowright M$, we get from Lemma 15.447 that $j_2(v_g, v_h) = 0$ for the nullhomotopy $h \colon fg \simeq 0$ as in Proposition 15.456. Thus the formula from Proposition 15.456 implies for the Poincaré dual $x \in K^k(\widetilde{M})$ of y
$$v^k([\psi])(x) = \mu(\alpha).$$

This identifies the quadratic form associated with the quadratic structure on a chain complex concentrated in a single dimension k with the self-intersection quadratic form of Chapter 8. □

Proposition 15.461 *Consider $n = 2k+1 \geq 5$. Let $(f, \overline{f}) \colon M \to X$ be a k-connected normal map of degree one. Then we have*
$$\sigma(f, \overline{f}) = \mathrm{qsign}_\pi(f, \overline{f}) \in L_n(\mathbb{Z}\pi, w).$$

Proof. The chain complex $\mathrm{cone}(f^!)$ is chain homotopy equivalent to the 1-dimensional chain complex
$$\cdots \to 0 \to K_{k+1}(\widetilde{M}, \widetilde{U}) \to K_k(\widetilde{U}) \to 0 \to \cdots$$

whose homology groups are $K_k(\widetilde{M})$ and $K_{k+1}(\widetilde{M})$. The associated quadratic formation has as its underlying quadratic form the self-intersection form of the map $\partial U \to S^{n-1}$, which is the quadratic form associated with a k-connected $2k$-dimensional normal map of degree one by the previous proposition. The lagrangians of the quadratic formation associated with the above 1-dimensional complex are precisely the lagrangians of the kernel formation of $(f, \overline{f})$ in the sense of Chapter 9. □

Combining Propositions 15.460 and 15.461 with the bordism invariance with cylindrical ends of the quadratic signature coming from Construction 15.270 finishes the proof of Theorem 15.4.

As we already indicated in Remark 15.324, we also finally obtain the proofs of both Theorem 8.168 and Theorem 9.109 from Theorem 15.4 and Construction 15.270.

15.8 Simple L-groups and Simple Surgery Obstruction in Terms of Chain Complexes

In this section we briefly state what happens in the decorated and in particular in the simple setting.

So far we have only considered the L-groups defined in terms of chain complexes which correspond to the decorations p and h, see Definitions 15.301 and 15.302. Now we introduce further decorations to the L-theory of chain complexes analogously to the situation of Definitions 10.11 and 10.12.

Suppose that we have specified a subgroup $U \subset K_1(R)$ such that U is closed under the involution (3.25) on $K_1(R)$ coming from the involution of R and contains the image of the change of rings homomorphism $K_1(\mathbb{Z}) \to K_1(R)$. We call a chain homotopy equivalence of stably U-based finite free chain complexes U-*simple* if its U-torsion taking values in $K_1(R)/U$ vanishes, see (10.9).

Definition 15.462 (Decorated L-groups via chain complexes) Let R be an associative ring with unit and involution, and let n be an integer.

Define the *n-th U-decorated symmetric L-group in terms of chain complexes*

$$L^n_U(R)_{\text{chain}}$$

to be the cobordism group of n-dimensional symmetric Poincaré complexes $(C, [\varphi])$ over R whose underlying chain complexes belong to $bR\text{-}CH_{\text{fgf}}$ and are stably U-based such that the chain homotopy equivalence $\varphi_0 \colon C^{n-*} \to C$ is U-simple, i.e.,

$$\tau^U(\varphi_0 \colon C^{n-*} \to C) = 0 \text{ in } K_1(R)/U.$$

Define the *n-th U-decorated quadratic L-group in terms of chain complexes*

$$L^U_n(R)_{\text{chain}}$$

to be the cobordism group of n-dimensional quadratic Poincaré complexes $(C, [\psi])$ whose underlying chain complexes belong to $bR\text{-}CH_{\text{fgf}}$ and are stably U-based such that the chain homotopy equivalence $(1+T)\psi_0 \colon C^{n-*} \to C$ is U-simple, i.e.,

$$\tau^U((1+T)\psi_0 \colon C^{n-*} \to C) = 0 \text{ in } K_1(R)/U.$$

Note that in the definition of cobordism in this setting stable U-bases have to appear and any chain homotopy equivalence has to be U-simple.

Also note that $L_n^h(R) = L_n(R)$ agrees with $L_n^U(R)$ if $U = K_1(R)$. If $R = \mathbb{Z}\pi$ and $w\colon \pi \to \{\pm 1\}$ is a group homomorphism, then $L_n^s(\mathbb{Z}\pi, w)$ is by definition $L_n^T(\mathbb{Z}\pi)$ for $\mathbb{Z}\pi$ equipped with the w-twisted involution and T the subgroup of $K_1(\mathbb{Z}\pi)$ given by the classes represented by trivial units $\pm x$ for $x \in \pi$. This motivates the following definition, which is a special case of Definition 15.462.

Definition 15.463 (Simple L-groups in terms of chain complexes) Consider $n \in \mathbb{Z}$. Let π be a group and $w\colon \pi \to \{\pm 1\}$ be a group homomorphisms. Define the *simple symmetric L-groups in terms of chain complexes* of $\mathbb{Z}\pi$ with the w-twisted involution

$$L_s^n(\mathbb{Z}\pi, w)_{\text{chain}} := L_T^n(\mathbb{Z}\pi)_{\text{chain}}$$

and the *simple quadratic L-groups in terms of chain complexes* of $\mathbb{Z}\pi$ with the w-twisted involution

$$L_n^s(\mathbb{Z}\pi, w)_{\text{chain}} := L_n^T(\mathbb{Z}\pi)_{\text{chain}}.$$

Theorem 15.464 (Identification of the form/formation and the chain complex version of quadratic L-groups in the simple case)

(i) *Let R be an associative ring with unit and involution and let n be an integer. Consider a subgroup $U \subset K_1(R)$ such that U is closed under the involution (3.25) on $K_1(R)$ coming from the involution of R and contains the image of the change of rings homomorphism $K_1(\mathbb{Z}) \to K_1(R)$.*
Then there is for all $n \in \mathbb{Z}$ a natural isomorphism

$$L_n^U(R)_{\text{chain}} \xrightarrow{\cong} L_n^U(R)$$

between the decorated L-groups in terms of chain complexes, see Definition 15.462, and the decorated L-groups in terms of forms and formations, see Definitions 10.11 and 10.12;

(ii) *Let π be a group and $w\colon \pi \to \{\pm 1\}$ be a group homomorphism.*
Then there is for all $n \in \mathbb{Z}$ a natural isomorphism

$$L_n^s(\mathbb{Z}\pi, w)_{\text{chain}} \xrightarrow{\cong} L_n^s(\mathbb{Z}\pi, w)$$

between the simple L-groups defined in term of chain complexes, see Definition 15.463, and the simple L-groups defined in terms of forms and formations, see Notation 10.13.

Proof. (i) This is an obvious variation of the proof of Theorem 15.3.

(ii) This is a special case of assertion (i). □

15.9 Applications of Algebraic Surgery

Theorem 15.465 (Identification of the simple surgery obstruction with the simple quadratic signature) *Let* $(f, \overline{f}) \colon M \to X$ *be a normal map of degree one of dimension* $n \geq 5$ *in the simple setting. Then we get*

$$\sigma^s(f, \overline{f}) = \mathrm{qsign}_\pi^s(f, \overline{f}) \in L_n^s(\mathbb{Z}\pi, w)$$

under the identification of simple L-groups of Theorem 15.464 (ii) where $\sigma^s(f, \overline{f})$ *has been defined in (10.26) and* $\mathrm{qsign}_\pi^s(f, \overline{f})$ *is the simple version of the quadratic signature of* $\mathrm{qsign}_\pi(f, \overline{f})$ *introduced in Definition 15.318.*

15.9 Applications of Algebraic Surgery

We have seen that algebraic surgery rectifies the two deficiencies of the geometric surgery theory mentioned in the introduction to this chapter. Moreover, we have discussed one more application, namely, the bordism invariance with varying target, see Remark 15.324. There are many more applications, many of which the reader may find in the numerous papers and books by Andrew Ranicki, for example in [346] and [348]. Here we would like to highlight just some of them, namely, those that fit well with the other topics in this book.

15.9.1 Product formulas

We discussed products already in Subsection 15.3.3 on the level of structured chain complexes. These products descend via standard observations to the *L*-groups and further analysis of the geometric constructions also yields product formulas for surgery obstructions. There are many versions, but for the sake of clarity we confine ourselves to one case, which we need the most. For the proofs see [344, Section 8] and [345, Section 8].

Theorem 15.466 *The products from Definition 15.72 descend to products*

$$- \otimes - \colon L^m(R) \otimes L_n(S) \to L_{m+n}(R \otimes S).$$

Theorem 15.467 *Consider an m-dimensional closed connected manifold N with* $w_N \colon \pi = \pi_1(N) \to \{\pm 1\}$, *a choice of* $[N] \in H_m(N; \mathbb{Z}^{w_N})$ *and an n-dimensional normal map of degree one* $(f, \overline{f}) \colon M \to X$ *with X connected,* $w_X \colon \pi' = \pi_1(X) \to \{\pm 1\}$, *and a choice* $[X] \in H_n(X; \mathbb{Z}^{w_X})$. *There is the product formula*

$$\mathrm{qsign}_{\pi \times \pi'}((f, \overline{f}) \times \mathrm{id}_N) = \mathrm{ssign}_\pi(N, [N]) \otimes \mathrm{qsign}_{\pi'}(f, \overline{f})$$

$$\in L_{m+n}(\mathbb{Z}[\pi \times \pi'], w_{N \times X})$$

relating the signatures from Definitions 15.312 and 15.318.

Exercise 15.468 Let $(f, \bar{f}) \colon M \to X$ be an n-dimensional normal map of degree one. Show that for every natural number m with $2 \leq m$ and $m + n \geq 5$, one can turn the normal map of degree one obtained by taking the product of $(f, \bar{f}) \colon M \to X$ with S^m by surgery into a homotopy equivalence.

15.9.2 The Algebraic Surgery Transfers for Fibrations

The product formulas can be generalised to a more general setting. Suppose we have a fibration $F \to E \to B$ of (for simplicity) connected CW-complexes with a finite d-dimensional connected Poincaré complex F as fibre. Then using the chain complex version of the L-groups and the theory of chain homotopy representations as they appear in the K-theoretic transfer of [248, 249] one can define an *algebraic surgery transfer map*

$$p^* \colon L_n(\mathbb{Z}\pi_1(B), w_B) \to L_{n+d}(\mathbb{Z}\pi_1(E), w_E) \tag{15.469}$$

for appropriate homomorphisms $w_B \colon \pi_1(B) \to \{\pm 1\}$ and $w_E \colon \pi_1(E) \to \{\pm 1\}$, see [264, 265] and also [348, Section § 21].

This has the following geometric meaning. Suppose that B is a finite n-dimensional Poincaré complex with orientation homomorphism w_B, F is a d-dimensional $w_1(F)$-oriented closed manifold, and p is a locally trivial bundle with typical fibre F. Then E is a finite $(n + d)$-dimensional Poincaré complex with w_E as orientation homomorphism. Let $(f, \bar{f})$ be a normal map of degree one with B as target. Then the pullback of p with f defines a normal map of degree one $(g, \bar{g})$ with target E and we get for the surgery obstructions by [264, Theorem 6.2]

$$p^*(\sigma(f, \bar{f})) = \sigma(g, \bar{g}). \tag{15.470}$$

So in particular vanishing results of p^* are interesting since they imply the vanishing of $\sigma(g, \bar{g})$. In this context the following two results are useful, see [265, Corollary 2.2 and Corollary 6.1].

Theorem 15.471 (up-down formula for the surgery transfer) *Suppose that the fibration $F \to E \xrightarrow{p} B$ is orientable, i.e., the homotopy action of $\pi_1(B)$ on F given by the fibre transport is trivial, $w_1(F)$ is trivial, and F is oriented. Let $w_B \colon \pi_1(B) \to \{\pm 1\}$ be a group homomorphism. Then:*

(i) *The composite*

$$L_n(\mathbb{Z}\pi_1(B), w_B) \xrightarrow{p^*} L_{n+d}(\mathbb{Z}\pi_1(E), w_B \circ \pi_1(p)) \xrightarrow{p_*} L_{n+d}(\mathbb{Z}\pi_1(B), w_B)$$

of the surgery transfer of (15.469) *with the homomorphism p_* induced by $\pi_1(p) \colon \pi_1(E) \to \pi_1(B)$ is given by multiplication from Theorem 15.466 with the simply connected symmetric signature* $\mathrm{ssign}_{\{1\}}(F) \in L^d(\mathbb{Z})$;

(ii) *Suppose that $\pi_1(p)\colon \pi_1(E) \to \pi_1(B)$ is an isomorphism. Then p^* is trivial if $d \equiv 0 \mod 4$ and $\mathrm{ssign}_{\{1\}}(F)$ in $L^d(\mathbb{Z}) \cong \mathbb{Z}$, which is the classical signature $\mathrm{sign}(F)$ of F, vanishes, or if $d \equiv 1 \mod 4$ and $\mathrm{ssign}_{\{1\}}(F)$ in $L^d(\mathbb{Z}) \cong \mathbb{Z}/2$ vanishes, or if $d \equiv 2, 3 \mod 4$.*

Theorem 15.472 (Vanishing of the surgery transfer) *Let G be a compact connected d-dimensional Lie group that is not a torus. Let $G \to E \xrightarrow{p} B$ be a principal G-bundle. Then the surgery transfer of (15.469) is trivial.*

Exercise 15.473 Let G be a compact connected Lie group that is not a torus. Let X be a free G-CW-complex such that X/G is a finite Poincaré complex of dimension ≥ 5 such that there exists a vector bundle reduction of its Spivak normal fibration. Then X is homotopy equivalent to a closed manifold.

15.9.3 The Algebraic Rothenberg Sequence

In this section we establish the algebraic Rothenberg sequence, which is isomorphic to the geometric Rothenberg sequence from Section 13.5 in the simple case. The primary reference is [344, Proposition 9.1]. Our more conceptual proof will follow the ideas appearing already in the proof of the geometric Rothenberg sequence, see Theorem 13.52.

The next theorem deals with the dependency of $L_n^U(R)_{\mathrm{chain}} = L_n^U(R)$, see Definition 15.462 and Theorem 15.464, in terms of the decoration U.

Theorem 15.474 (The Algebraic Rothenberg sequence) *Let R be a ring with involution. Consider two subgroups $U, V \subseteq K_1(R)$ such that both U and V are closed under the involution (3.25) on $K_1(R)$ coming from the involution of R, contain the image of the change of rings homomorphism $K_1(\mathbb{Z}) \to K_1(R)$, and satisfy $U \subseteq V$. There is a long exact sequence of abelian groups*

$$\cdots \xrightarrow{\partial_{n+2}} \widehat{H}^{n+2}(\mathbb{Z}/2, V/U) \xrightarrow{i_{n+2}} L_{n+1}^U(R) \xrightarrow{j_{n+1}} L_{n+1}^V(R)$$
$$\xrightarrow{\partial_{n+1}} \widehat{H}^{n+1}(\mathbb{Z}/2, V/U) \xrightarrow{i_n} L_n^U(R) \xrightarrow{j_n} L_n^V(R) \xrightarrow{\partial_{n-1}} \cdots,$$

which is infinite in both directions.

Proof. The proof follows a similar pattern to the proof of Theorem 13.52. First let us consider for each n an intermediate L-group

$$L_n^{U \subseteq V}(R).$$

It is constructed analogously to the group $\mathcal{L}_n^{s,h}(K, \mathcal{O}_K)$ from Section 13.5. An element of $L_n^{U \subseteq V}(R)$ is represented by an $(n+1)$-dimensional quadratic Poincaré pair $(f\colon C \to D, [\psi, \delta\psi])$ such that the underlying chain complexes belong to $bR\text{-}\mathrm{CH}_{\mathrm{fgf}}$, both C and D are stably U-based, and the $(n+1)$-dimensional quadratic

Poincaré pair $(f\colon C \to D, [\psi, \delta\psi])$ is V-simple, and the n-dimensional quadratic Poincaré complex $(C, [\psi])$ is U-simple. Two such pairs $(f\colon C \to D, [\psi, \delta\psi])$ and $(f'\colon C' \to D', [\psi', \delta\psi'])$ are called cobordant, if there exists a stably V-based $(n + 2)$-dimensional quadratic Poincaré triad of chain complexes in bR-CH$_{\text{fgf}}$ over $U \subseteq V$ in the following sense, cf. Remark 15.327. Firstly we require a U-simple $(n + 1)$-dimensional quadratic cobordism between $(C, [\psi])$ and $(C', [\psi'])$. Gluing $(f\colon C \to D, [\psi, \delta\psi])$ and $(f'\colon C' \to D', [\psi', \delta\psi'])$ along this U-simple $(n + 1)$-dimensional quadratic cobordism, we obtain an $(n + 1)$-dimensional V-simple Poincaré complex for which we secondly require the existence of a V-simple $(n + 2)$-dimensional quadratic nullbordism.

One easily checks that we obtain a long exact sequence

$$\cdots \xrightarrow{j_{n+2}} L^{U \subseteq V}_{n+2}(R) \xrightarrow{\partial_{n+2}} L^U_{n+1}(R) \xrightarrow{i_{n+1}} L^V_{n+1}(R)$$
$$\xrightarrow{j_{n+1}} L^{U \subseteq V}_{n+1}(R) \xrightarrow{\partial_{n+1}} L^U_n(R) \xrightarrow{i_n} L^V_n(R) \xrightarrow{j_n} \cdots$$

where ∂_{n+1} assigns to an element represented by an algebraic $(n + 1)$-dimensional Poincaré pair its boundary in the sense of Definition 15.221, i_n is the obvious map given by viewing a stable U-basis as a stable V-basis, and j_n is given by considering an algebraic n-dimensional Poincaré complex as an algebraic n-dimensional Poincaré pair with boundary the trivial chain complex.

Next, arguments analogous to those used in the proof of Lemma 13.51 and the obvious analogue of Lemma 3.18 for U-torsion show that the map given by taking the U-torsion

$$\tau\colon L^{U \subseteq V}_{n+1}(R) \to \widehat{H}^{n+1}(\mathbb{Z}/2, V/U)$$
$$(f\colon C \to D, (\psi, \delta\psi)) \mapsto \tau^U((1 + T)\,\text{ev}_r(\psi, \delta\psi))$$

is a well-defined homomorphism of abelian groups.

Next we show that it is bijective. We begin with injectivity. Consider an $(n + 1)$-dimensional quadratic Poincaré pair $(f\colon C \to D, [\psi, \delta\psi])$ as it appears in the definition of $L^{U \subseteq V}_n(R)$ satisfying $\tau^U((1 + T)\,\text{ev}_r(\psi, \delta\psi)) = 0$ in $\widehat{H}^{n+1}(\mathbb{Z}/2, V/U)$. Choose an element $x \in V/U$ such that we get in V/U

$$\tau^U((1 + T)\,\text{ev}_r(\psi, \delta\psi)) = x + (-1)^{n+1} \cdot *(x).$$

Recall that both C and D come with a stable U-basis. Now equip D with a new U-stable basis and call this D' such that the chain map $D \to D'$ given by the identity has as U-torsion the given element $x \in V/U$. Then the new pair $C \to D'$ has trivial U-torsion. Moreover, recalling the cellular chain complex I of the unit interval from (14.33), a suitable V-simple nullbordism for the union of $f\colon C \to D$, $C \otimes (0[\mathbb{Z}] \oplus 0[\mathbb{Z}]) \to C \otimes I$ and $f\colon C \to D'$ is given by $D \times [0, 1]$. These data imply that the class of $(f\colon C \to D, [\psi, \delta\psi])$ in $L^{U \subseteq V}_n(R)$ is trivial.

Finally we prove surjectivity. Choose an n-dimensional quadratic Poincaré complex $(C, [\psi])$ such that C comes with a stable U-basis for which it is U-simple, i.e., $\tau^U(((1+T)\psi)_0: C^{n-*} \to C) = 0$ holds in V/U. Consider any element $[x] \in \widehat{H}^{n+1}(\mathbb{Z}/2, V/U)$ represented by $x \in V/U$ satisfying $x = (-1)^{n+1} \cdot *(x)$. Now equip C with a new U-stable basis and call this C' such that the chain map $C \to C'$ given by the identity has U-torsion the element $x \in V/U$. Then $(C', [\psi])$ is a U-simple n-dimensional quadratic Poincaré complex and the obvious inclusion $C \oplus C' \to C \otimes I$, which corresponds after forgetting the stable U-basis to the inclusion $C \otimes (0[\mathbb{Z}] \oplus 0[\mathbb{Z}]) \to C \otimes I$, defines an element in $L_{n+1}^{U \subseteq V}(R)$ whose image under $\tau: L_{n+1}^{U \subseteq V}(R) \to \widehat{H}^{n+1}(\mathbb{Z}/2, V/U)$ is $[x]$. □

As a special case of Theorem 15.474 we obtain the following version relating h and s

$$\cdots \xrightarrow{j_{n+2}} \widehat{H}^{n+2}(\mathbb{Z}/2, \mathrm{Wh}(\pi)) \xrightarrow{\partial_{n+2}} L_{n+1}^s(\mathbb{Z}\pi, w) \xrightarrow{i_{n+1}} L_{n+1}^h(\mathbb{Z}\pi, w)$$
$$\xrightarrow{j_{n+1}} \widehat{H}^{n+1}(\mathbb{Z}/2, \mathrm{Wh}(\pi)) \xrightarrow{\partial_{n+1}} L_n^s(\mathbb{Z}\pi, w) \xrightarrow{i_n} L_n^h(\mathbb{Z}\pi, w) \xrightarrow{j_n} \cdots . \quad (15.475)$$

Exercise 15.476 Let G be a group. Let Z be the image of $K_1(\mathbb{Z}) \to K_1(\mathbb{Z}G)$. Show that the change of decoration map $L_n^Z(\mathbb{Z}G) \to L_n^s(\mathbb{Z}G)$ is bijective for all $n \in \mathbb{Z}$ if and only if $2 \cdot \mathrm{id}_{G_{\mathrm{ab}}}: G_{\mathrm{ab}} \to G_{\mathrm{ab}}$ is bijective for the abelianisation G_{ab} of G.

There is also a version of the Rothenberg sequences comparing decorations h and p. We refer the reader to [349, Proposition 17.2] for more details.

Since for the chain complex L-groups we also have a symmetric version, we may wonder about the corresponding version of the Rothenberg sequence. It does exist and we refer the reader to [344, Section 9].

15.9.4 The Algebraic Shaneson Splitting

In this section we briefly mention the algebraic analogue of the geometric Shaneson splitting from Section 13.6. We give two statements that correspond to two choices of decorations.

Theorem 15.477 (Algebraic Shaneson splitting – decorations h, p)
For any $n \in \mathbb{Z}$ and any ring with involution R there is an isomorphism

$$i_n \oplus s_{n-1}: L_n^h(R) \oplus L_{n-1}^p(R) \xrightarrow{\cong} L_n^h(R[z, z^{-1}]).$$

The map i_n is induced by the inclusion of rings $R \hookrightarrow R[z, z^{-1}]$, but the map s_{n-1} has a more complicated definition, which is beyond our scope.

Theorem 15.478 (Algebraic Shaneson splitting – decorations s, h)
Let π be a group and let $w: \pi \to \{\pm 1\}$ be a homomorphism. For any $n \in \mathbb{Z}$ there is an isomorphism

$$i_n \oplus s_{n-1}: L_n^s(\mathbb{Z}\pi, w) \oplus L_{n-1}^h(\mathbb{Z}\pi, w) \xrightarrow{\cong} L_n^s(\mathbb{Z}[\pi \times \mathbb{Z}], w \circ \mathrm{pr})$$

where pr: $\pi \times \mathbb{Z} \to \pi$ *is the projection.*

The map s_{n-1} is given via the product from Theorem 15.466 with the symmetric signature $\mathrm{ssign}_{\mathbb{Z}}(S^1) \in L^1(\mathbb{Z})$ of the circle. That such a map is well defined in terms of decorations follows from an algebraic version of Theorem 3.1 (iv).

Unfortunately, in the literature we could not find a fully satisfactory treatment of these theorems via the theory of structured chain complexes that would give an exact analogue of the geometric result. In Chapter 16 of the book [349] Ranicki states and proves the version that relates decorations h and p, see Theorem 16.2 and the discussion after it. He then states a version that relates decorations s and h at the end of that chapter at the bottom of page 144, but the statement is in the more general setting of additive categories with chain duality, and his definitions mean that in the special case of the category of $\mathbb{Z}\pi$-modules for a group π the s decoration differs from our s decoration, see the discussion between Definition 8.14 and Example 8.15 on page 78. On the other hand in the paper [347] he explains how to implement the decoration s in the sense as we use it also in the setting of additive categories with chain duality, see Example 7.23 in [347]. So perhaps using this implementation it is possible to follow and adapt the proofs in the book [349] to obtain the exact analogue of the Shaneson splitting from Section 13.6, but since the proofs are rather involved, we have to leave it for the reader to check by themselves if she or he is interested in such a version.

On the other hand, we note that there is another proof of Theorem 15.478 since we can use the geometric Shaneson splitting of Section 13.6, the identification of the geometric L-groups from Chapter 13 with the forms and formation groups from Chapters 8 and 9, and the identification in Theorem 15.3 of these with the chain complex L-groups of this chapter, but such a proof is not entirely in the setting of chain complexes.

Yet another viewpoint, which we were informed of by Wolfgang Steimle, is that the L-theory for Poincaré ∞-categories, as developed to great extent in [69, 70, 71], allows for general formulations of Shaneson splittings from which one can also deduce the two Shaneson splittings stated above.

15.9.5 The Algebraic Surgery Exact Sequence

The most important application of algebraic surgery is the construction of the algebraic surgery exact sequence, which can be identified with the geometric surgery exact sequence of Chapter 11 in the topological category. Via this identification the topological geometric surgery exact sequence becomes an exact sequence of abelian groups, which is in itself an interesting result. Moreover, the algebraic surgery exact sequence has at the place corresponding to the surgery obstruction map the so-called *assembly map*, which has a certain universal property, and about which a great deal has been shown in recent decades. We will say more about these developments in Chapter 19. Here we just want to state it and give a few hints about ideas that go

15.9 Applications of Algebraic Surgery

into the construction and into the identification. The ultimate source is [348], see also [234] and [353].

Firstly, algebraic surgery can be used to define L-spaces and L-spectra whose homotopy groups are the L-groups. Intuitively, the idea is to incorporate into the setup the notion of n-ads as in [414, Chapter 0]. This is quite involved technically, in particular the notion of an additive category with chain duality is needed, which is presented in Chapter 1 of [348]. The result is that one can think of a collection of chain complexes indexed by faces of a simplex and a collection of symmetric or quadratic structures on these complexes, which satisfy a lot of compatibility conditions, as a (single) symmetric or quadratic complex in such a category. The usual notation is as follows. For a ring with involution R and $n \in \mathbb{Z}$, there are defined L-spaces $\mathbf{L}_n(R)$ such that $\pi_k(\mathbf{L}_n(R)) \cong L_{n+k}(R)$ for all $k \geq 0$. These spaces form an Ω-spectrum $\mathbf{L}(R)$ such that $\pi_k(\mathbf{L}(R)) \cong L_k(R)$ for all $k \in \mathbb{Z}$. For $l \in \mathbb{Z}$ there are connective versions $\mathbf{L}(R)\langle l \rangle$ such that

$$\pi_k(\mathbf{L}(R)\langle l \rangle) \cong \begin{cases} L_k(R) & k \geq l; \\ 0 & k < l. \end{cases} \quad (15.479)$$

When $R = \mathbb{Z}$, with the trivial involution, then one sometimes just writes $\mathbf{L}$ (or $\mathbf{L}_\bullet$). In particular, $\mathbf{L}\langle 1 \rangle = \mathbf{L}(\mathbb{Z})\langle 1 \rangle$.

Having done this, one can define (co-)homology theories on simplicial complexes with coefficients in these spectra. Moreover, one can vary the additive category with chain duality suitably (via the notion of an algebraic bordism category) so that one obtains exact sequences of various types of L-groups. Appropriate choices produce the *algebraic surgery exact sequence* of abelian groups for a locally finite simplicial complex K and the decoration $\langle i \rangle = h, s$

$$\cdots \to H_{n+1}(K; \mathbf{L}(\mathbb{Z})\langle 1 \rangle) \to L_{n+1}^{\langle i \rangle}(\mathbb{Z}\pi_1(K)) \to \mathbb{S}_{n+1}^{\langle i \rangle}(K)$$
$$\to H_n(K; \mathbf{L}(\mathbb{Z})\langle 1 \rangle) \to L_n^{\langle i \rangle}(\mathbb{Z}\pi_1(K)) \to \cdots, \quad (15.480)$$

which is natural with respect to simplicial maps and also homotopy invariant, see [348, Definition 15.19 on page 169].

Informally we can describe the terms and maps in the algebraic surgery exact sequence from (15.480) as follows. An element in each of these groups is represented by a collection of quadratic chain complexes indexed by simplices of K and these complexes posses various dualities, which are in some sense compatible. In different groups different additional conditions on these collections are imposed. In the L-groups $L_n^{\langle i \rangle}(\mathbb{Z}\pi)$ one requires global duality, in the source of the assembly maps one requires local duality, which leads to the term being a homology theory, and in the term $\mathbb{S}_n^{\langle i \rangle}(K)$ one requires local duality and simultaneously that the underlying chain complex is globally contractible. Hence the assembly map is seen as a passage from local to global duality and the structure set term is seen as measuring the failure of this passage to be an isomorphism. This is an important philosophy, which is applied for example in questions such as those discussed in Chapter 19.

If M is a (not necessarily triangulable) n-dimensional closed topological manifold, one can define its algebraic surgery exact sequence as follows. Consider the following category $C(M)$. Objects are (simple) homotopy equivalences $f\colon X \to M$ with X a finite simplicial complex. Note that the notion of a simple homotopy equivalence does make sense since a topological manifold of dimension $\neq 4$ has a CW-structure, see [159, Section 9.2] or [219, Theorem 2.2 in III.2 on page 107], and in dimension 4 there is at least a preferred simple structure on it, see [219, Theorem 4.1 in III.4 on page 118]. A morphism from $f_0\colon X_0 \to M$ to $f_1\colon X_1 \to M$ is a (simple) homotopy equivalence $u\colon X_0 \to X_1$ satisfying $f_1 \circ u \simeq f_0$. Then we obtain a functor from $C(M)$ to the category of exact sequences of abelian groups by sending $f\colon X \to M$ to the algebraic surgery exact sequence of X in the sense of (15.480). Now taking the colimit of this functor yields the *algebraic surgery exact sequence* of abelian groups for a (not necessarily triangulable) n-dimensional closed topological manifold M and the decoration $\langle i \rangle = h, s$

$$\cdots \to H_{n+1}(M; \mathbf{L}(\mathbb{Z})\langle 1\rangle) \to L_{n+1}^{\langle i \rangle}(\mathbb{Z}\pi_1(M)) \to \mathbb{S}_{n+1}^{\langle i \rangle}(M)$$
$$\to H_n(M; \mathbf{L}(\mathbb{Z})\langle 1\rangle) \to L_n^{\langle i \rangle}(\mathbb{Z}\pi_1(M)) \to \cdots . \quad (15.481)$$

Theorem 15.482 (Identification of the geometric and algebraic surgery exact sequence) *Consider a connected n-dimensional closed topological manifold M with $\pi = \pi_1(M)$ such that $w_1(M) = 0$ and $n \geq 5$ hold. Make a choice (ABP), see Notation 8.13, which essentially is a choice of a fundamental class $[M] \in H_n(M; \mathbb{Z})$.*

Then there is a bijection between the (simple) geometric surgery exact sequence in the topological category of Theorem 11.25 and the algebraic surgery exact sequence of (15.481) for the decorations $\langle i \rangle = s, h$.

15.9 Applications of Algebraic Surgery

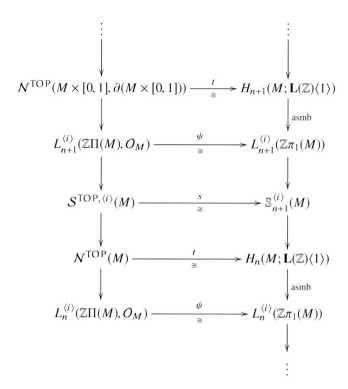

Via this identification all terms in the geometric surgery exact sequence become L-groups of some category since all terms in the right-hand column are. The horizontal arrows in the ladder in Theorem 15.482 are constructed by refining the quadratic construction of Section 15.3.6 and the map denoted asmb in the right column is the assembly map mentioned above. The theorem is proved in [348, Theorem 18.5] for the h decoration, see Introduction of [348] and [343, Corollary 2] for the s decoration. We emphasise that no triangulation on M is required in Theorem 15.482.

There exists a version for manifolds with boundary $(M, \partial M)$, for which we refer the reader to [346, pages 558-561] and [348, pages 207-208].

By Remark 11.39 the smooth geometric surgery exact sequence cannot be an exact sequence of abelian groups.

Remark 15.483 (Intrinsic version of the algebraic surgery exact sequence)
There is also a version in the non-orientable case, see [352, Appendix A]. We have not checked the details but the following should be true. Consider an n-dimensional closed topological manifold M, which is not necessarily triangulable and may have non-trivial first Stiefel–Whitney class $w_1(M)$. Then there is an intrinsic version of the algebraic surgery exact sequence of abelian groups for the decorations $\langle i \rangle = s, h$

$$\cdots \to L_{n+1}^{\langle i \rangle}(\mathbb{Z}\Pi(M); \mathcal{O}_M) \to \mathbb{S}_{n+1}^{\langle i \rangle}(M)$$
$$\to H_n^{\Pi(M)}(\widetilde{M}; \underline{\mathbf{L}(\mathbb{Z})}\langle 1 \rangle) \to L_n^{\langle i \rangle}(\mathbb{Z}\Pi(M), \mathcal{O}_M) \to \cdots$$

where $H_n^{\Pi(M)}(\widetilde{M}; \mathbf{L}(\mathbb{Z})\langle 1 \rangle)$ is $\pi_n(\widetilde{M} \wedge_{\Pi(M)} \mathbf{L}(\mathbb{Z})\langle 1 \rangle)$ for the universal covering functor $\widetilde{M} \colon \Pi(M) \to \mathrm{SPACES}$, see [250, Definition 8.22 on page 146] and an appropriate covariant functor $\mathbf{L}(\mathbb{Z})\langle 1 \rangle \colon \Pi(M) \to \mathrm{SPECTRA}$. This sequence can be viewed as a functor from the category $\mathcal{M}$ defined in Subsection 11.10. Moreover there is a natural transformation of functors on $\mathcal{M}$ of the shape

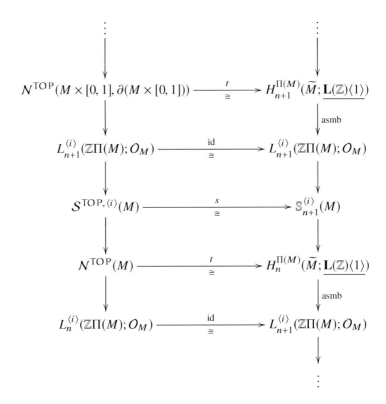

The algebraic surgery exact sequence can be used to obtain an algebraic analogue of the Rothenberg sequence for structure sets from Theorem 13.54. Namely, we get for a n-dimensional closed connected topological manifold M

$$\cdots \to \widehat{H}^{n+2}(\mathbb{Z}/2, \mathrm{Wh}(\pi_1(M))) \to \mathbb{S}_{n+1}^s(M) \to \mathbb{S}_{n+1}^h(M) \to$$
$$\to \widehat{H}^{n+1}(\mathbb{Z}/2, \mathrm{Wh}(\pi_1(M))) \to \cdots \quad (15.484)$$

for all $n \in \mathbb{Z}$, see [348, page 11]. Also there is an algebraic analogue of the Shaneson splitting for structure sets from Theorem 13.57 with decorations h and p for all $n \in \mathbb{Z}$, see [348, Example 23.5 B]

$$\mathbb{S}_{n+1}^h(M \times S^1) \cong \mathbb{S}_{n+1}^h(M) \oplus \mathbb{S}_n^p(M). \quad (15.485)$$

15.9.6 The Total Surgery Obstruction

Another aspect of algebraic surgery that we would like to highlight is the so-called *total surgery obstruction* (TSO). Recall that one version of the surgery program (Problem 4.1) asks when a topological space X is homotopy equivalent to a closed (topological) manifold, see also Remarks 7.46 and 7.47. We have seen that X must satisfy Poincaré duality. This leads to the modified question: when is a finite n-dimensional Poincaré complex X homotopy equivalent to a closed (topological) manifold? So far the material of this book provides us with the answer that the spherical Spivak normal fibration of X has to have a topological reduction and, if that is the case, then the surgery obstruction of at least one of the surgery problems with target X has to vanish. Altogether we have a two-stage obstruction answer. Ranicki came up with an idea to define a single obstruction $s(X)$, called the *total surgery obstruction*,

$$s(X) \in \mathbb{S}_n(X)$$

with the property that, provided $n \geq 5$, then $s(X) = 0$ if and only if there exists an n-dimensional topological manifold M homotopy equivalent to X. For more details, see [234], [343], [348], and [353]. This theorem is closely related to Theorem 15.482 about the identification of the geometric surgery exact sequence and the algebraic surgery exact sequence. In fact, Theorem 15.482 can be viewed as the relative version of the main theorem about TSO.

15.10 Notes

There are many more aspects of the algebraic theory of surgery that we have not covered in this chapter. In addition to the topics we have hinted at in Section 15.9, these include for example all the material contained in the books [346], [348], [349] and the paper [429]. To name just a few topics let us mention localisation, calculations for specific rings, transfer (here see also [264] and [265]), splitting theorems, and a more comprehensive study of decorations of L-groups.

There is also a recent development that originated with Lurie's generalisation of algebraic surgery to the setting of ∞-categories (in unpublished notes available on the internet) and found spectacular applications in the joint work of Calmes, Dotto, Harpaz, Hebestreit, Land, Moi, Nardin, Nikolaus, and Steimle that relates algebraic surgery to hermitian K-theory and is at the time of writing available as a series of preprints and papers [69, 70, 71].

Chapter 16
Brief Survey of Computations of L-Groups

16.1 Introduction

In this chapter we give a brief survey of computations of L-groups.

In Section 16.2 we introduce decorated L-groups $L_n^\epsilon(\mathbb{Z}G, w)$ for the decoration ϵ given by s, h, p, as before, and also for $\langle j \rangle$ for $j \leq 2$, or $\langle -\infty \rangle$. They are related by Rothenberg sequences. The decoration $\langle -\infty \rangle$ will be needed for the Farrell–Jones Conjecture, see Remark 16.39.

In Section 16.3 we state the main general properties about L-groups of integral group rings $\mathbb{Z}G$ for finite groups. For instance, they are always finitely generated and contain no p-torsion for odd primes p.

In Section 16.4 we state what is for us the most relevant version of the Farrell–Jones Conjecture 16.19, namely the one for torsionfree groups and integral coefficients, and explain how it can be used for computations. It will be one of the key ingredients when we will deal with the question whether an aspherical closed manifold is topologically rigid in Section 19.4. It says, for a torsionfree group G, that $\text{Wh}(G)$ and $\widetilde{K}_n(\mathbb{Z}G)$ for $n \leq 0$ vanish, see Remark 16.8, and the L-groups $L_n^\epsilon(\mathbb{Z}G, w)$ are independent of the decoration ϵ. Moreover, if w is trivial, $L_n^\epsilon(\mathbb{Z}G, w)$ can be computed by $H_n(BG; \mathbf{L}(\mathbb{Z})) = \pi_n(BG_+ \wedge \mathbf{L}(\mathbb{Z}))$ for the L theory spectrum $\mathbf{L}(\mathbb{Z})$ of $\mathbb{Z}$, which satisfies $\pi_n(\mathbf{L}(\mathbb{Z})) = L_n(\mathbb{Z})$ for $n \in \mathbb{Z}$.

In Section 16.5 we present the Farrell–Jones Conjecture in its most general form, namely the Full Farrell–Jones Conjecture 16.42. We explain the rather large class $\mathcal{FJ}$ of Farrell–Jones groups, i.e., groups for which the Full Farrell Jones Conjecture 16.42 is known to be true, see Theorem 16.44. It contains for instance all hyperbolic groups, all finite-dimensional CAT(0)-groups, fundamental groups of 3-manifolds, and lattices in almost connected Lie groups, We also explain why it implies the version for torsionfree groups.

Guide 16.1 It is worthwhile to browse through the results of Section 16.2 and Section 16.4, and one may skip the other sections for the first reading since the torsionfree version of the Farrell–Jones Conjecture 16.19 is independent of the decorations, is the key ingredient in further geometric applications, and is much easier to understand than the Full Farrell–Jones Conjecture 16.42.

There are two reasons why the Full Farrell–Jones Conjecture 16.42 is important, namely, it is formulated for all groups and in particular applies to group rings RG for any coefficient ring R and has much better inheritance properties than the Farrell–Jones Conjecture 16.19.

16.2 More Decorated L-Groups

We have introduced the L-groups $L_n^s(\mathbb{Z}G, w)$ and $L_n^h(\mathbb{Z}G, w) = L_n(\mathbb{Z}G, w)$ in Notation 10.13. If we replace 'finitely generated free' by 'finitely generated projective' everywhere in the definition of $L_n^h(\mathbb{Z}G, w)$, then we get the *projective quadratic L-group* $L_n^p(\mathbb{Z}G, w)$, see also Definition 15.2. If in the definition of $L_n^s(\mathbb{Z}G, w)$ we take the subgroup $U \subseteq K_1(\mathbb{Z}G)$ to be the image of $K_1(\mathbb{Z}) \to K_1(\mathbb{Z}G)$, or, equivalently, $U = \{\pm 1\}$, instead of $\{\pm g \mid g \in g\}$, then we obtain $L_n^{\langle 2 \rangle}(\mathbb{Z}G, w)$. Often we define

$$L_n^{\langle 1 \rangle}(\mathbb{Z}G, w) = L_n^h(\mathbb{Z}G, w);$$
$$L_n^{\langle 0 \rangle}(\mathbb{Z}G, w) = L_n^p(\mathbb{Z}G, w).$$

There are also L-groups $L_n^{\langle j \rangle}(\mathbb{Z}G, w)$ for $j \in \mathbb{Z}$, $j \leq -1$. They come with canonical maps $L_n^{\langle j+1 \rangle}(\mathbb{Z}G, w) \to L_n^{\langle j \rangle}(\mathbb{Z}G, w)$ for $j \in \mathbb{Z}, j \leq 1$, which correspond for $j = 1, 2$ to the obvious forgetful maps. The group $L_n^{\langle -\infty \rangle}(\mathbb{Z}G, w)$ is defined as the colimit

$$L_n^{\langle -\infty \rangle}(\mathbb{Z}G, w) = \operatorname{colim}_{j \to -\infty} L_n^{\langle j \rangle}(\mathbb{Z}G, w).$$

Hence we have L-groups $L_n^\epsilon(\mathbb{Z}G, w)$ for the decoration ϵ given by s, h, p, $\langle j \rangle$ for $j \leq 2$, or $\langle -\infty \rangle$.

There are Rothenberg sequences, infinite in both directions, for $j \in \mathbb{Z}, j \leq 0$ of the shape

$$\cdots \to L_n^{\langle j+1 \rangle}(\mathbb{Z}G, w) \to L_n^{\langle j \rangle}(\mathbb{Z}G, w) \to \widehat{H}^n(\mathbb{Z}/2, \widetilde{K}_j(\mathbb{Z}G))$$
$$\to L_{n-1}^{\langle j+1 \rangle}(\mathbb{Z}G, w) \to L_{n-1}^{\langle j \rangle}(\mathbb{Z}G, w) \to \cdots \quad (16.2)$$

with the groups $\widetilde{K}_j(\mathbb{Z}G)$ discussed in Remark 16.8 below, and of the shape

$$\cdots \to L_n^s(\mathbb{Z}G, w) \to L_n^h(\mathbb{Z}G, w) \to \widehat{H}^n(\mathbb{Z}/2, \mathrm{Wh}(G))$$
$$\to L_{n-1}^s(\mathbb{Z}G, w) \to L_{n-1}^h(\mathbb{Z}G, w) \to \cdots \quad (16.3)$$

where the last one has already been explained in Theorem 13.52 and Theorem 15.474, see also (15.475). The Shaneson splitting yields isomorphism for $j \leq 1$ and $w \colon G \to \{\pm 1\}$, if $\widehat{w} \colon G \times \mathbb{Z} \to \{\pm 1\}$ sends (g, m) to $w(g)$:

$$L_n^s(\mathbb{Z}[G \times \mathbb{Z}], \widehat{w}) \cong L_n^s(\mathbb{Z}G, w) \oplus L_{n-1}^h(\mathbb{Z}G, w); \tag{16.4}$$

$$L_n^{\langle j \rangle}(\mathbb{Z}[G \times \mathbb{Z}], \widehat{w}) \cong L_n^{\langle j \rangle}(\mathbb{Z}G, w) \oplus L_{n-1}^{\langle j-1 \rangle}(\mathbb{Z}G, w); \tag{16.5}$$

$$L_n^{\langle -\infty \rangle}(\mathbb{Z}[G \times \mathbb{Z}], \widehat{w}) \cong L_n^{\langle -\infty \rangle}(\mathbb{Z}G, w) \oplus L_{n-1}^{\langle -\infty \rangle}(\mathbb{Z}G, w), \tag{16.6}$$

where the first one has already been explained in Theorem 13.55 and Theorem 15.478 and the second one for $j = 1$ in Theorem 15.477.

For more details about the decorated L-groups $L_n^\epsilon(\mathbb{Z}G, w)$ the reader should consult [342, 349].

Exercise 16.7 Show that for any decoration ϵ the canonical map from $L_n^\epsilon(\mathbb{Z}G, w)$ to $L_n(\mathbb{Z}G, w)$ or the canonical map from $L_n(\mathbb{Z}G, w)$ to $L_n^\epsilon(\mathbb{Z}G, w)$ has the property that the kernel and the cokernel consists of elements of order two and in particular becomes an isomorphism after inverting 2.

Remark 16.8 (Algebraic K-theory) The exact sequence (16.2) contains certain abelian groups $\widetilde{K}_j(\mathbb{Z}G)$ for $j \le 0$. Recall first that given $j \in \mathbb{Z}$ and a ring R there exists abelian groups $K_j(R)$, called the *algebraic K-theory groups* of R, which have many applications to geometric topology and about which a great deal is known. We have already encountered the case $j = 1$ in Chapters 2 and 3, where $K_1(\mathbb{Z}G)$ played a prominent role in the s-cobordism Theorem 2.1 via its quotient, the Whitehead group $\mathrm{Wh}(G)$, see Definition 3.4. Since we do not have room for more details about $K_j(R)$ for $j \ne 1$, we refer the reader to the book by Rosenberg [361]. At least we mention that the case $j = 0$ is the well-studied projective class group, defined in terms of finitely generated projective R-modules. Chapter 1 of [361] is devoted to its study, while $K_j(R)$ for $j \le -1$ are treated in Section 3.3 and $K_j(R)$ for $j \ge 2$ in Chapter 5 of the same book. The group $\widetilde{K}_j(\mathbb{Z}G)$ appearing in (16.2) is defined as the cokernel of the map $K_j(\mathbb{Z}) \to K_j(\mathbb{Z}G)$ induced by the inclusion $\mathbb{Z} \to \mathbb{Z}G$ and we have $\widetilde{K}_j(\mathbb{Z}G) = K_j(\mathbb{Z}G)$ for $j \le -1$ since then $K_j(\mathbb{Z}) = 0$.

Other textbooks on algebraic K-groups K_n for $n \in \mathbb{Z}$ are [31, 261, 384, 422].

16.3 Finite Groups

The algebraic L-groups $L_n^\epsilon(\mathbb{Z}G, w)$ for $G = \{1\}$ and $\mathbb{Z}/2$ are given and we state some general properties about $L_n^\epsilon(\mathbb{Z}G)$ for a finite group G. Many of the results below have analogues when one allows a non-trivial orientation homomorphism w.

Notation 16.9 Define the reduced simple L-group $\widetilde{L}_n^s(\mathbb{Z}G)$ to be the kernel of the homomorphism $L_n^s(\mathbb{Z}G) \to L_n^s(\mathbb{Z})$ induced by the projection $G \to \{1\}$.

Since the composite of the inclusion $\{1\} \to G$ with the projection $G \to \{1\}$ is the identity, we obtain a splitting

$$L_n^s(\mathbb{Z}G) = L_n(\mathbb{Z}) \oplus \widetilde{L}_n^s(\mathbb{Z}G). \tag{16.10}$$

Theorem 16.11 (Algebraic L-theory of $\mathbb{Z}G$ for finite groups)
Let G be a finite group. Then

(i) *The groups $L_n^\epsilon(\mathbb{Z})$ are independent of the decoration ϵ and are given by $\mathbb{Z}$, 0, $\mathbb{Z}/2$, 0 for $n \equiv 0, 1, 2, 3 \bmod (4)$;*

(ii) *The groups $L_n^\epsilon(\mathbb{Z}[\mathbb{Z}/2], w)$ are independent of the decoration ϵ. For trivial w they are given by $\mathbb{Z} \oplus \mathbb{Z}$, 0, $\mathbb{Z}/2$, $\mathbb{Z}/2$ for $n \equiv 0, 1, 2, 3 \bmod (4)$. For non-trivial w they are given by $\mathbb{Z}/2$, 0, $\mathbb{Z}/2$, 0 for $n \equiv 0, 1, 2, 3 \bmod (4)$;*

(iii) *For each decoration ϵ and integer n, the algebraic L-groups $L_n^\epsilon(\mathbb{Z}G)$ are finitely generated as abelian groups and contain no p-torsion for odd primes p. Moreover, they are finite for odd n;*

(iv) *Let $r(G)$ be the number of isomorphisms classes of irreducible real G-representations. Let $r_\mathbb{C}(G)$ be the number of isomorphisms classes of irreducible real G-representations that are of complex type. For every decoration ϵ we have*

$$L_n^\epsilon(\mathbb{Z}G)[1/2] \cong L_n^\epsilon(\mathbb{Q}G)[1/2] \cong L_n^\epsilon(\mathbb{R}G)[1/2]$$

$$\cong \begin{cases} \mathbb{Z}[1/2]^{r(G)} & n \equiv 0 \ (4); \\ \mathbb{Z}[1/2]^{r_\mathbb{C}(G)} & n \equiv 2 \ (4); \\ 0 & n \equiv 1, 3 \ (4); \end{cases}$$

(v) *Suppose that G has odd order and n is odd. Then $L_n^\epsilon(\mathbb{Z}G) = 0$ for any decoration $\epsilon \in \{s, h, p\}$. For the decoration $j \in \{-1, -2, \ldots\} \sqcup \{-\infty\}$ we get $L_n^{\langle j \rangle}(\mathbb{Z}G) = (\mathbb{Z}/2)^r$ where r is the rank of the finitely generated free abelian group $K_{-1}(\mathbb{Z}G)$;*

(vi) *If G is a group of odd order, then $\widetilde{L}_n^s(\mathbb{Z}G)$ is torsionfree. In particular, $\mathrm{tors}(L_n^s(\mathbb{Z}G))$ is $\mathbb{Z}/2$ if $n \equiv 2 \bmod 4$ and trivial otherwise;*

(vii) *If G is a non-trivial cyclic group of even order, then*

$$\mathrm{tors}(\widetilde{L}_n^s(\mathbb{Z}G)) \cong \begin{cases} 0 & \text{if } n \equiv 0, 1, 2 \bmod 4; \\ \mathbb{Z}/2 & \text{if } n \equiv 3 \bmod 4. \end{cases}$$

In particular, $\mathrm{tors}(L_n^s(\mathbb{Z}G))$ *is $\mathbb{Z}/2$ if $n \equiv 2, 3 \bmod 4$ and is trivial otherwise;*

Proof. (i) See Theorem 8.99, Theorem 8.111 and Theorem 9.18.

(ii) See for instance [414, Theorem 13A.1 on page 172].

(iii) See [414, Theorem 13.A.4 (i) on page 177], [177] for the decoration s. Now the claim follows for all decorations from the Rothenberg sequences (16.2) and (16.3) since the relevant K-groups of $\mathbb{Z}G$ are all finitely generated abelian groups.

(iv) See [348, Proposition 22.34 on page 252].

(v) See [16], [177, Theorem 10.1] for $\epsilon \in \{s, h, p\}$. We could not find a good reference in the literature for the other decorations so we at least sketch a proof below for some claims which are just stated and whose proof seem to follow by inspecting [86], [175, Section 3], and the proof of [177, Theorem 10.1]. Note that $K_n(\mathbb{Z}G) = 0$ for $n \leq -2$ and $K_{-1}(\mathbb{Z}G) = \mathbb{Z}^r$, see [86], and also [31, Theo-

16.3 Finite Groups

rem 10.6 on page 695]. The involution on $K_{-1}(\mathbb{Z}G) = \mathbb{Z}^r$ is given by $-\mathrm{id}$. Hence $\widehat{H}^0(\mathbb{Z}/2, K_{-1}(\mathbb{Z}G)) = 0$ and $H^1(\mathbb{Z}/2, K_{-1}(\mathbb{Z}G)) = (\mathbb{Z}/2)^r$. Since $L_n^p(\mathbb{Z}G) = 0$ for odd n and $\widetilde{L}_n^p(\mathbb{Z}G) = \mathrm{coker}(L_n^p(\mathbb{Z}) \to L_n^p(\mathbb{Z}G))$ is known to be torsionfree for even n, the claim follows from the Rothenberg sequences (16.2), using the fact that the Arf invariant in $L_2^p(\mathbb{Z})$ goes forward to $L_2^{\langle -1 \rangle}(\mathbb{Z}G)$.

(vi) See [414, Theorem 13.A.4 (ii) on page 177], [177, page 227 and Section 10].

(vii) See [177, page 227 and Sections 10, 11, and 12]. □

For some concrete computations and for the general techniques to carry them out, we refer for instance to the survey article [177].

Part of the proof of the theorem above is obtained using representation theory. Since this will be needed in Chapter 18, we briefly recall the idea. Given a finite-dimensional complex vector space V, define the dual complex vector space $V' := \hom_{\mathbb{C}}(V, \mathbb{C})$ where the addition is given by $(f + f')(v) = f(v) + f'(v)$ and the $\mathbb{C}$-scalar multiplication by $(\lambda \cdot f)(v) = f(v) \cdot \lambda$. If V is a G-representation, i.e., comes with a G-action by linear automorphisms, then V' becomes a G-representation by $(g \cdot f)(v) = f(g^{-1}v)$ for $f \in V'$, $g \in G$, and $v \in V$. It is sometimes called the *contragredient representation*. The character $\chi_{V'}$ satisfies $\chi_{V'}(g) = \overline{\chi_V(g)}$. Note that the canonical map $V \to V''$ that sends v to the element in V'' that assigns to $f \in V'$ the complex number $f(v)$ is an isomorphism of complex G-representations. Hence we obtain an involution $*: R_{\mathbb{C}}(G) \xrightarrow{\cong} R_{\mathbb{C}}(G)$ on the complex representation ring $R_{\mathbb{C}}(G)$ by sending the class $[V]$ of V to the class $[V']$ of V'. One can define (± 1)-eigenspaces. In terms of characters the $(+1)$-eigenspace corresponds to real characters whereas the (-1)-eigenspace corresponds to purely imaginary characters. We define

$$R_{\mathbb{C}}^{\pm}(G) = \{[V] = \pm[V'] \mid [V] \in R_{\mathbb{C}}(G)\}. \tag{16.12}$$

Exercise 16.13 Let G be a finite group. Let V be a finite-dimensional complex G-representation. Show that the contragredient representation V' is isomorphic to $V^* = \hom_{\mathbb{C}G}(V, \mathbb{C}G)$ in the sense of Subsection 5.6.1, where we equip $\mathbb{C}G$ with the involution sending $\sum_{g \in G} \lambda_g \cdot g$ to $\sum_{g \in G} \lambda_g \cdot g^{-1}$.

The following definition of the *G-signature* of a non-singular G-invariant $(-1)^k$-symmetric bilinear form $B: H \times H \to \mathbb{R}$ over $\mathbb{R}$

$$G\text{-sign}(B) \in R_{\mathbb{C}}^{(-1)^k}(G) \tag{16.14}$$

is taken from [15, (6.7) and (6.9)]. Choose a G-invariant scalar product $\langle -, - \rangle$ on the finite-dimensional real vector space H. Define the $\mathbb{R}$-linear map $A: H \to H$ by requiring $\langle x, A(y) \rangle = B(x, y)$ for all $x, y \in H$. Then A commutes with the G-action and its adjoint A^* with respect to $\langle -, - \rangle$ satisfies $A^* = (-1)^k \cdot A$.

Next we consider the case where k is even. Then A is self-adjoint and the eigenspaces associated to the positive and negative eigenvalues of A give a decomposition of real vector spaces $H = H^+ \oplus H^-$, which is invariant under the G-action. Hence H^+ and H^- are real G-representations, and we can define the element $[H^+] - [H^-]$ in the real representation ring $R_{\mathbb{R}}(G)$. The map $R_{\mathbb{R}}(G) \to R_{\mathbb{C}}(G)$ given by extending scalars has image in $R_{\mathbb{C}}^+(G)$, and we define G-sign(B) to be the image of $[H^+] - [H^-]$ under this map.

Finally we consider the case where k is odd. Then A is skew-adjoint. Let $(AA^*)^{1/2} \colon H \to H$ be the positive square root of AA^*. The operator $J \colon H \to H$ uniquely determined by $A = J \circ (AA^*)^{1/2}$ satisfies $J^2 = -\mathrm{id}$. Hence it defines the structure of a complex vector space on H. Since J commutes with the G-action, we get a complex G-representation H_c whose underlying real G-representation is H. Now define G-sign(B) to be $[H_c] - [H_c']$. All these definitions are independent of the choice of the inner G-invariant scalar product $\langle -, - \rangle$ since two such choices are homotopic.

Next we briefly explain the *G-signature homomorphism*, see also [414, Chapter 13] or [348, Chapter 22]

$$G\text{-sign} \colon L_{2k}^s(\mathbb{Z}G) \to R_{\mathbb{C}}^{(-1)^k}(G). \tag{16.15}$$

Consider an element $x \in L_{2k}^s(\mathbb{Z}G)$. It is represented by a non-singular $(-1)^k$-quadratic form. Its symmetrisation is a non-singular $(-1)^k$-symmetric form $s \colon P \to P^*$, see Definition 8.77, which yields a $(-1)^k$-symmetric bilinear pairing

$$\lambda \colon P \times P \to \mathbb{Z}G, \quad (p, q) \mapsto s(p)(q)$$

as explained in (8.48). It induces a non-singular $(-1)^k$-symmetric bilinear form over $\mathbb{R}$ by induction with the inclusion $\mathbb{Z} \to \mathbb{R}$ and taking the value at $e \in G$

$$B \colon (\mathbb{R} \otimes_{\mathbb{Z}} P) \times (\mathbb{R} \otimes_{\mathbb{Z}} P) \to \mathbb{R}, \quad (a \otimes p, b \otimes q) \mapsto a \cdot \lambda(p, q)_e \cdot b.$$

Define G-sign(x) to be G-sign(B) as defined in (16.14).

Theorem 16.16 (G-signature homomorphism) *Let G be a finite group.*

(i) *The kernel and the cokernel of the G-signature homomorphism G-sign of (16.15) are finite 2-groups;*
(ii) *Suppose that G is finite cyclic. Then the G-signature homomorphism G-sign of (16.15) induces a homomorphism of finitely generated free abelian groups*

$$L_{2k}^s(G) \to R_{\mathbb{C}}^{(-1)^k}(G)$$

whose image is $\{4 \cdot ([V] + (-1)^k \cdot [V']) \mid [V] \in R_{\mathbb{C}}(G)\} \subseteq R_{\mathbb{C}}^{(-1)^k}(G)$.
Its restriction to $\widetilde{L}_{2k}^s(G)$ is injective and induces an isomorphism of finitely generated free abelian groups

16.4 Torsionfree Groups   771

$$\widetilde{L}_{2k}^s(G) \xrightarrow{\cong} \{4 \cdot ([V] + (-1)^k \cdot [V']) \mid [V] \in R_{\mathbb{C}}(G)\} / \{8l \cdot [\mathbb{C}G] \mid l \in \mathbb{Z}\}$$

if k is even, and an isomorphism of finitely generated free abelian groups

$$\widetilde{L}_{2k}^s(G) \xrightarrow{\cong} \{4 \cdot ([V] + (-1)^k \cdot [V']) \mid [V] \in R_{\mathbb{C}}(G)\}$$

if k is odd.

Proof. (i) This follows from [414, Theorem 13.A3 on page 176 and Theorem 13.A4 (i) on page 177].

(ii) For G a cyclic group of odd order, this follows from [414, Theorem 13.A4 (ii) on page 177]. If G is cyclic, see [177, page 227, Section 11 and 12]. □

The isomorphisms appearing in Theorem 16.11 (iv) come from Theorem 16.16 (i).

Exercise 16.17 Since a G-representation is the same as a finitely generated projective $\mathbb{C}G$-module, we get an obvious map $\alpha \colon R_{\mathbb{C}}(G) \to K_0(\mathbb{C}G)$. We have defined an involution on $R_{\mathbb{C}}(G)$ by $[V] \mapsto [V']$. Using the involution on $\mathbb{C}G$ sending $\sum_{g \in G} \lambda_g \cdot g$ to $\sum_{g \in G} \lambda_g \cdot g^{-1}$, we obtain an involution on $K_0(\mathbb{C}G)$ by sending $[P]$ to $[P^*]$ for P^* defined in Subsection 5.6.1.

Show that α is an isomorphism compatible with the involutions.

Exercise 16.18 Let G be a finite cyclic group G. Show

$$L_0^s(\mathbb{Z}G) \cong \begin{cases} \mathbb{Z}^{\frac{|G|+1}{2}} & |G| \text{ odd}; \\ \mathbb{Z}^{\frac{|G|}{2}+1} & |G| \text{ even}; \end{cases}$$

$$L_1^s(\mathbb{Z}G) \cong \{0\};$$

$$L_2^s(\mathbb{Z}G) \cong \begin{cases} \mathbb{Z}^{\frac{|G|-1}{2}} \oplus \mathbb{Z}/2 & |G| \text{ odd}; \\ \mathbb{Z}^{\frac{|G|}{2}-1} \oplus \mathbb{Z}/2 & |G| \text{ even}; \end{cases}$$

$$L_3^s(\mathbb{Z}G) \cong \begin{cases} \{0\} & |G| \text{ odd}; \\ \mathbb{Z}/2 & |G| \text{ even}. \end{cases}$$

16.4 Torsionfree Groups

For a torsionfree group we have the following conjecture that will be a special case of the more general Full Farrell–Jones Conjecture, which we will discuss in Section 16.5. In particular, the next conjecture is known to be true for the family $\mathcal{FJ}$ of Farrell–Jones groups introduced in Definition 16.43, which for instance contains all hyperbolic groups, all finite-dimensional CAT(0)-groups, fundamental groups of 3-manifolds and lattices in almost connected Lie groups, see Theorem 16.44.

Conjecture 16.19 (Farrell–Jones Conjecture for torsionfree groups and integer coefficients) Let G be a torsionfree group and $w \colon G \to \{\pm 1\}$ a group homomorphism. Then

(i) The groups $\mathrm{Wh}(G)$ and $\widetilde{K}_n(\mathbb{Z}G)$ for $n \leq 0$ vanish;
(ii) For $j \leq 0$ the natural maps

$$L_n^s(\mathbb{Z}G, w) \xrightarrow{\cong} L_n^h(\mathbb{Z}G, w) \xrightarrow{\cong} L_n^p(\mathbb{Z}G, w) \xrightarrow{\cong} L_n^{\langle j \rangle}(\mathbb{Z}G, w) \xrightarrow{\cong} L_n^{\langle -\infty \rangle}(\mathbb{Z}G, w)$$

are isomorphisms;
(iii) There is an isomorphism, the so-called assembly map,

$$\mathrm{asmb}_n^{-\infty} \colon H_n^G(EG; \mathbf{L}(\mathbb{Z})^w) = \pi_n(EG_+ \wedge_G \mathbf{L}(\mathbb{Z})^w) \xrightarrow{\cong} L_n^{\langle -\infty \rangle}(\mathbb{Z}G, w),$$

where EG is the total space of the universal G-covering $EG \to BG$ and $\mathbf{L}(\mathbb{Z})^w$ is a spectrum with G-action such that $g \in G$ acts on $\pi_n(\mathbf{L}(\mathbb{Z}))$ by multiplication with $w(g)$ and $\pi_n(\mathbf{L}(\mathbb{Z})) \cong L_n(\mathbb{Z})$ holds for all $n \in \mathbb{Z}$.

Remark 16.20 (On the source of the assembly map) There is a twisted Atiyah–Hirzebruch spectral sequence converging to $H_n^G(EG; \mathbf{L}(\mathbb{Z})^w)$, see for instance [292, 11.16 on page 496] or [393, Theorem 15.7 on page 341]. Its E^2-term $E_{p,q}^2$ is the twisted homology of EG with coefficients in $L_q(\mathbb{Z})^w$, i.e., $H_p^G(EG; L_q(\mathbb{Z})^w) := H_p(C_*(EG) \otimes_{\mathbb{Z}G} L_q(\mathbb{Z})^w)$, where $L_q(\mathbb{Z})^w$ is the $\mathbb{Z}G$ module whose underlying abelian group is $L_q(\mathbb{Z})$ and on which $g \in G$ acts by multiplication with $w(g)$.

If w is trivial, then $H_n^G(EG; \mathbf{L}(\mathbb{Z})^w) = \pi_n(EG_+ \wedge_G \mathbf{L}(\mathbb{Z})^w)$ boils down to the value of the homology theory associated to the spectrum $\mathbf{L}(\mathbb{Z})$ on BG, i.e., $H_n(BG; \mathbf{L}(\mathbb{Z})) = \pi_n(BG_+ \wedge \mathbf{L}(\mathbb{Z}))$ and the E^2-term in the Atiyah–Hirzebruch spectral sequence converging to $H_n(BG; \mathbf{L}(\mathbb{Z}))$ is given by

$$E_{p,q}^2 = \begin{cases} H_p(BG; \mathbb{Z}) & \text{for } q \equiv 0 \bmod (4); \\ H_p(BG; \mathbb{Z}/2) & \text{for } q \equiv 2 \bmod (4); \\ 0 & q \text{ odd}. \end{cases}$$

The latter claim follows from Theorem 16.11 (i).

Exercise 16.21 Let F_g be the closed orientable surface of genus g. Compute $L_n^\epsilon(\mathbb{Z}[\pi_1(F_g)])$ for all decorations ϵ and $n \in \mathbb{Z}$.

Example 16.22 Consider an odd natural number $m \geq 3$. Choose a self-diffeomorphism $f \colon T^2 \to T^2$ that induces on $H_1(T^2) \cong \mathbb{Z}^2$ the automorphism given by the invertible $(2,2)$-matrix $A = \begin{pmatrix} m & 2m-1 \\ 1 & 2 \end{pmatrix}$. Note that A has determinant 1. Let M be the mapping torus of f. It is an aspherical closed orientable 3-manifold since it is a locally trivial T^2-bundle over S^1. In particular, $M = B\pi_1(M)$. By the Wang sequence one computes

$$H_p(M;\mathbb{Z}) = \begin{cases} \mathbb{Z} & \text{for } p = 0, 2, 3; \\ \mathbb{Z} \oplus \mathbb{Z}/m & \text{for } p = 1; \\ 0 & \text{otherwise}; \end{cases}$$

$$H_p(M;\mathbb{Z}/2) = \begin{cases} \mathbb{Z}/2 & \text{for } p = 0, 1, 2, 3; \\ 0 & \text{otherwise}. \end{cases}$$

Consider the Atiyah–Hirzebruch spectral sequence converging to $H_n(M;\mathbf{L}(\mathbb{Z}))$. Using the edge homomorphism and the fact that the composite $H_n(\{\bullet\}) \to H_n(M;\mathbf{L}(\mathbb{Z})) \to H_n(\{\bullet\};\mathbf{L}(\mathbb{Z}))$ of the maps induced by the inclusion $\{\bullet\} \to M$ and the projection $M \to \{\bullet\}$ is the identity, one computes

$$H_n(M;\mathbf{L}(\mathbb{Z})) \cong \begin{cases} \mathbb{Z} \oplus \mathbb{Z}/2 & \text{for } n \equiv 0, 2, 3 \bmod 4; \\ G & \text{for } n \equiv 1 \bmod 4, \end{cases}$$

where the abelian group G can be written as an extension of the shape $0 \to \mathbb{Z} \oplus \mathbb{Z}/m \to G \to \mathbb{Z}/2 \to 0$. Since Conjecture 16.19 holds for fundamental groups of 3-manifolds, we have $L_n^\epsilon(\mathbb{Z}[\pi_1(M)]) \cong H_n(B\pi_1(M);\mathbf{L}(\mathbb{Z}))$. Hence $L_1^\epsilon(\mathbb{Z}[\pi_1(M)])$ contains $\mathbb{Z}/m$ as a subgroup for all decorations ϵ. This is in sharp contrast to Theorem 16.11 (iii).

Exercise 16.23 Let p be any odd prime, n be any integer and d any natural number with $d \geq 6$. Construct an aspherical orientable closed manifold N of dimension d such that $L_n^\epsilon(\mathbb{Z}[\pi_1(N)])$ contains non-trivial p-torsion.

Exercise 16.24 Let G be a torsionfree group that is a Farrell–Jones group. Then we get for any decoration ϵ and any integer n

$$L_n^\epsilon(\mathbb{Z}G) \otimes_\mathbb{Z} \mathbb{Q} \cong \bigoplus_{k \in \mathbb{Z}} H_{4k+n}(BG;\mathbb{Q}).$$

Combining (16.2), (16.3), (16.4), (16.5) and (16.6), we obtain for $G = \mathbb{Z}^d$ with $d \geq 1$ and all $j \leq 1$

$$L_n^s(\mathbb{Z}[\mathbb{Z}^d]) \cong L_n^{\langle j \rangle}(\mathbb{Z}[\mathbb{Z}^d]) \cong \bigoplus_{i=0}^d L_{n-i}(\mathbb{Z})^{\binom{d}{i}}. \tag{16.25}$$

16.5 The Farrell–Jones Conjecture

In this section we briefly introduce the Full Farrell–Jones Conjecture 16.42, which identifies for any group the algebraic K- and L-groups of a group ring RG with a certain G-homology theory evaluated on the classifying spaces for the family of virtually cyclic subgroups of G and actually makes sense for additive G-categories (with involution) as coefficients. We explain why it implies Conjecture 16.19 and give a status report for which groups it is known, see Theorem 16.44.

16.5.1 The K-theoretic Farrell–Jones Conjecture with Coefficients in Additive G-Categories

Fix a group G. Let $\mathcal{A}$ be an additive G-category in the sense of [26, Definition 2.1], i.e., an additive category with right G-action by functors of additive categories. For a group G we denote by $I(G)$ the groupoid with one object and G as its automorphism group. Let GROUPOIDS $\downarrow G$ be the category of connected groupoids over $I(G)$.

We obtain from [26, Section 5] a contravariant functor to the category ADD-CAT of small additive categories

$$\text{GROUPOIDS} \downarrow G \to \text{ADD-CAT}, \quad \text{pr}\colon \mathcal{G} \to I(G) \mapsto \int_G \mathcal{A} \circ \text{pr}. \quad (16.26)$$

We have the functor sending an additive category to its non-connective K-theory spectrum

$$\mathbf{K}\colon \text{ADD-CAT} \to \text{SPECTRA}, \quad (16.27)$$

see for instance [84, 269, 329]. For a G-set S we denote by $\mathcal{G}^G(S)$ its associated *transport groupoid*. Its objects are the elements of S. The set of morphisms from s_0 to s_1 consists of those elements $g \in G$ that satisfy $gs_0 = s_1$. Composition in $\mathcal{G}^G(S)$ comes from the multiplication in G. Let $\text{Or}(G)$ be the *orbit category* whose objects are homogeneous G-spaces G/H and whose morphisms are G-maps between them. Thus we obtain for a group G a covariant functor

$$\mathcal{G}^G\colon \text{Or}(G) \to \text{GROUPOIDS} \downarrow G, \quad G/H \mapsto \mathcal{G}^G(G/H) \quad (16.28)$$

where we view $\mathcal{G}^G(G/H)$ as category over $I(G) = \mathcal{G}^G(G/G)$ by the functor induced by the projection $G/H \to G/G$.

Composing the three functors (16.26), (16.27), and (16.28) yields a covariant functor

$$\mathbf{K}_{\mathcal{A}}\colon \text{Or}(G) \to \text{SPECTRA}. \quad (16.29)$$

It determines by [266, Proposition 156 on page 795] a G-homology theory $H_*^G(-, \mathbf{K}_{\mathcal{A}})$ in the sense of [266, page 739-740 in Subsection 2.3.1]. Recall that we get for every G-CW-complex X and integer $n \in \mathbb{Z}$ an abelian group $H_n^G(X; \mathbf{K}_{\mathcal{A}})$, functorial in X, such that $H_n^G(G/H; \mathbf{K}_{\mathcal{A}}) \cong \pi_n(\mathbf{K}_{\mathcal{A}}(G/H))$ holds for all $n \in \mathbb{Z}$ and G-homotopy invariance, excision and the disjoint union axiom hold.

Let $\mathcal{F}$ be a family of subgroups of G, i.e., a collection of subgroups closed under conjugation and passing to subgroups. Examples are the trivial family $\mathcal{TR}$ consisting of one element, namely, the trivial subgroup, the family $\mathcal{FIN}$ of finite subgroups

16.5 The Farrell–Jones Conjecture

and the family $\mathcal{VCY}$ of virtually cyclic subgroups. Recall that *virtually cyclic* means that the group is finite or contains $\mathbb{Z}$ as a subgroup of finite index.

We denote by $E_\mathcal{F}(G)$ a model for the *classifying space for the family* $\mathcal{F}$, i.e., a G-CW-complex such that for a subgroup $H \subseteq G$ its H-fixed point set is contractible if H is in $\mathcal{F}$, and empty otherwise. If X is a G-CW-complex whose isotropy groups belong to $\mathcal{F}$, then there is up to G-homotopy precisely one G-map from X to $E_\mathcal{F}(G)$. This implies that two models for $E_\mathcal{F}(G)$ are G-homotopy equivalent. Note that $E_{\mathcal{TR}}(G)$ is the same as EG, the total space of the universal principle G-bundle $G \to EG \to BG$.

Exercise 16.30 Let D be the infinite dihedral group $D = \mathbb{Z} \rtimes \mathbb{Z}/2 \cong \mathbb{Z}/2 * \mathbb{Z}/2$. Show that any model for ED is infinite-dimensional whereas $E_{\mathcal{FIN}}(D)$ has a 1-dimensional model.

For more information about classifying spaces for families of subgroups we refer for instance to the survey article [256].

Conjecture 16.31 (K-theoretic Farrell–Jones Conjecture with coefficients in additive G-categories) We say that G satisfies the *K-theoretic Farrell–Jones Conjecture with coefficients in additive G-categories* if for any additive G-category $\mathcal{A}$ and any $n \in \mathbb{Z}$ the assembly map induced by the projection $\mathrm{pr} \colon E_{\mathcal{VCY}}(G) \to G/G$

$$H_n(\mathrm{pr}) \colon H_n^G(E_{\mathcal{VCY}}(G); \mathbf{K}_\mathcal{A}) \to H_n^G(G/G; \mathbf{K}_\mathcal{A}) = \pi_n\bigl(\mathbf{K}_\mathcal{A}(I(G))\bigr)$$

is bijective.

Remark 16.32 (**The setting of additive G-categories encompasses the setting with rings as coefficients**) Let R be a ring with unit. Define a category $\mathcal{A}(R)$ as follows. For each integer $m \in \mathbb{Z}$ with $m \geq 0$ we have one object $[m]$. For $m, n \geq 1$ the set of morphisms from $[m]$ to $[n]$ is the set $\mathrm{M}(m, n; R)$ of (m, n)-matrices with entries in R. The set of morphisms from $[0]$ to $[m]$ and from $[m]$ to $[0]$ consist of precisely one element. Composition is given by matrix multiplication. Obviously $\mathcal{A}(R)$ can be equipped with the structure of a small additive category and is equivalent as an additive category to the category of finitely generated free R-modules. Hence $\mathbf{K}(\mathcal{A}(R))$ is the non-connective K-theory spectrum $\mathbf{K}(R)$ of the ring R.

Equip $\mathcal{A}(R)$ with the trivial G-action. Then $\mathbf{K}_{\mathcal{A}(R)}(G/H)$ is weakly homotopy equivalent to the non-connective K-theory spectrum $\mathbf{K}^\infty(RH)$ of the group ring RH. In particular, we get $\pi_n(\mathbf{K}_{\mathcal{A}(R)}(G/H)) = K_n(RH)$ for $n \in \mathbb{Z}$. Hence the assembly map appearing in Conjecture 16.31 reduces to

$$H_n(\mathrm{pr}) \colon H_n^G(E_{\mathcal{VCY}}(G); \mathbf{K}_{\mathcal{A}(R)}) \to H_n^G(G/G; \mathbf{K}_{\mathcal{A}(R)}) = K_n(RG).$$

One can also consider twisted group rings $R_\rho G$ if $\rho \colon G \to \mathrm{aut}(R)$ is a group homomorphism with the group of unital ring automorphisms of R as target. All this is explained in [26].

Remark 16.33 (Involutions and K-theory) If $\mathcal{A}$ is an additive G-category with involution in the sense of [26, Definition 4.22], then the involution induces involutions on the source and target of the K-theoretic assembly map

$$H_n(\mathrm{pr}) \colon H_n^G(E_{\mathcal{VCY}}(G); \mathbf{K}_\mathcal{A}) \to H_n^G(G/G; \mathbf{K}_\mathcal{A}) = \pi_n\big(\mathbf{K}_\mathcal{A}(I(G))\big)$$

of Conjecture 16.31 and the assembly map is compatible with them.

16.5.2 The L-theoretic Farrell–Jones Conjecture with Coefficients in Additive G-Categories with Involution

Let $\mathcal{A}$ be an additive G-category with involution in the sense of [26, Definition 4.22]. We obtain from [26, Section 7] a contravariant functor to the category $\mathrm{ADD\text{-}CAT}_{\mathrm{inv}}$ of small additive categories with involution

$$\mathrm{GROUPOIDS} \downarrow G \to \mathrm{ADD\text{-}CAT}_{\mathrm{inv}}, \quad \mathrm{pr} \colon \mathcal{G} \to I(G) \mapsto \int_{\mathcal{G}} \mathcal{A} \circ \mathrm{pr}.$$

We have the functor sending an additive category with involution to its L-theory spectrum $\mathbf{L}^{\langle -\infty \rangle}(\mathcal{A})$ with decoration $\langle -\infty \rangle$

$$\mathbf{L}_\mathcal{A}^{\langle -\infty \rangle} \colon \mathrm{ADD\text{-}CAT}_{\mathrm{inv}} \to \mathrm{SPECTRA}, \qquad (16.34)$$

see Ranicki [348, Chapter 13]. Now we obtain analogously to the construction in Subsection 16.5.1 a G-homology theory $H_*^G(-, \mathbf{L}_\mathcal{A}^{\langle -\infty \rangle})$ such that $H_n^G(G/H; \mathbf{L}_\mathcal{A}^{\langle -\infty \rangle}) \cong \pi_n(\mathbf{L}_\mathcal{A}^{\langle -\infty \rangle}(G/H))$ holds for all $n \in \mathbb{Z}$.

Conjecture 16.35 (L-theoretic Farrell–Jones Conjecture with coefficients in additive G-categories with involution) We say that G satisfies the L-theoretic Farrell–Jones Conjecture with coefficients in additive G-categories with involution if for any additive G-category with involution $\mathcal{A}$ the assembly map induced by the projection $\mathrm{pr} \colon E_{\mathcal{VCY}}(G) \to G/G$

$$H_n(\mathrm{pr}) \colon H_n^G(E_{\mathcal{VCY}}(G); \mathbf{L}_\mathcal{A}^{\langle -\infty \rangle}) \to H_n^G(G/G; \mathbf{L}_\mathcal{A}^{\langle -\infty \rangle}) = \pi_n\big(\mathbf{L}_\mathcal{A}^{\langle -\infty \rangle}(I(G))\big)$$

is bijective.

Remark 16.36 (The setting of additive G-categories with involution encompasses the setting with rings with involution as coefficients) Let R be a ring with involution. Fix a group homomorphism $w \colon G \to \{\pm 1\}$. Then one can construct a specific functor

$$\mathbf{L}_{R,w}^{\langle -\infty \rangle} \mathrm{GROUPOIDS} \downarrow G \to \mathrm{ADD\text{-}CAT}_{\mathrm{inv}}$$

such that $H_n^G(G/H; \mathbf{L}_{R,w}^{\langle -\infty \rangle})$ is isomorphic to $L_n^{\langle -\infty \rangle}(RH, w|_H)$ for every subgroup $H \subseteq G$. The assembly map appearing in Conjecture 16.35 reduces to the assembly map

$$H_n^G(\mathrm{pr}) \colon H_n^G(E_{\mathcal{VCY}}(G); \mathbf{L}_{R,w}^{\langle -\infty \rangle}) \to H_n^G(G/G; \mathbf{L}_{R,w}^{\langle -\infty \rangle}) = L_n^{\langle -\infty \rangle}(RG, w).$$

One can also consider twisted group rings $R_\rho G$ if $\rho \colon G \to \mathrm{aut}(R)$ is a group homomorphism with the group of unital ring automorphisms of R as target. All this is explained in [26].

Remark 16.37 (Assembly as homotopy colimit) The quickest and probably for a homotopy theorist most convenient approach to assembly maps is via homotopy colimits. Let $\mathcal{F}$ be a family of subgroups of G, i.e., a collection of subgroups closed under conjugation and passing to subgroups. Let $\mathrm{Or}_\mathcal{F}(G)$ be the full subcategory of the orbit category $\mathrm{Or}(G)$ consisting of objects G/H satisfying $H \in \mathcal{F}$. Consider a covariant functor $\mathbf{E} \colon \mathrm{Or}(G) \to \mathrm{SPECTRA}$ to the category of spectra. We get from the inclusion $\mathrm{Or}_\mathcal{F}(G) \to \mathrm{Or}(G)$ and the fact that G/G is a terminal object in $\mathrm{Or}(G)$ a map

$$\mathrm{hocolim}_{\mathrm{Or}_\mathcal{F}(G)} \mathbf{E}|_{\mathrm{Or}_\mathcal{F}(G)} \to \mathrm{hocolim}_{\mathrm{Or}(G)} \mathbf{E} = \mathbf{E}(G/G). \quad (16.38)$$

Taking homotopy groups yields the assembly map appearing in Conjecture 16.31 or Conjecture 16.35 respectively if we take $\mathbf{E} = \mathbf{K}_\mathcal{A}$ or $\mathbf{E} = \mathbf{L}_\mathcal{A}^{\langle -\infty \rangle}$ respectively.

Remark 16.39 (Decoration $\langle -\infty \rangle$ is necessary) We had to introduce the decoration $\langle -\infty \rangle$ as the Farrell–Jones Conjecture is not true in general if one replaces $\langle -\infty \rangle$ by s, h or p. Counterexamples are constructed in [151] for $G = \mathbb{Z}^2 \times F$ for appropriate finite groups F and its integral group ring $\mathbb{Z}G$.

16.5.3 Reducing the Family

Recall that in the general version of the Farrell Conjectures 16.31 and 16.35 the family $\mathcal{VCY}$ occurs. In some favourite situations one can choose smaller families thanks to the so-called Transitivity Principle. We mention the following examples. A virtually cyclic subgroup is called of type I if it is finite or admits a surjection with finite kernel onto $\mathbb{Z}$ and of type II if it is finite or admits a surjection with finite kernel on the infinite dihedral group $\mathbb{Z} \rtimes \mathbb{Z}/2 \cong \mathbb{Z}/2 * \mathbb{Z}/2$. Every virtually cyclic subgroup is of type I or type II and there is no infinite virtually cyclic group that is both of type I and type II. We denote by $\mathcal{VCY}_I$ or $\mathcal{VCY}_{II}$ respectively the family of virtually cyclic subgroups of type I or type II respectively.

If $\mathcal{F} \subseteq \mathcal{G}$ is an inclusion of families of subgroups of the group G, then there is up to G-homotopy precisely one G-map $i_{\mathcal{F} \subseteq \mathcal{G}} \colon E_\mathcal{F}(G) \to E_\mathcal{G}(G)$.

Theorem 16.40 *Let G be a group and $n \in \mathbb{Z}$ any integer. Then the following maps are isomorphisms*

(i) *The map*

$$H_n(i_{\mathcal{VCY}_I,\mathcal{VCY}};\mathbf{K}_{\mathcal{A}})\colon H_n(E_{\mathcal{VCY}_I}(G);\mathbf{K}_{\mathcal{A}}) \to H_n(E_{\mathcal{VCY}}(G);\mathbf{K}_{\mathcal{A}})$$

is bijective where $\mathcal{A}$ is an additive G-category;

(ii) *The map*

$$H_n(i_{\mathcal{FIN},\mathcal{VCY}_I};\mathbf{L}_{\mathcal{A}}^{\langle-\infty\rangle})\colon H_n(E_{\mathcal{FIN}}(G);\mathbf{L}^{\langle-\infty\rangle}) \to H_n(E_{\mathcal{VCY}_I}(G);\mathbf{L}^{\langle-\infty\rangle})$$

is bijective where $\mathcal{A}$ is an additive G-category with involution;

(iii) *If G is torsionfree and R is a regular, e.g., $R = \mathbb{Z}$, then the map*

$$H_n(i_{\mathcal{TR},\mathcal{VCY}};\mathbf{K}_{\mathcal{A}(R)})\colon H_n(E_{\mathcal{TR}}(G);\mathbf{K}_{\mathcal{A}(R)}) \to H_n(E_{\mathcal{VCY}}(G);\mathbf{K}_{\mathcal{A}(R)})$$

is bijective;

(iv) *If G is torsionfree, then the map*

$$H_n(i_{\mathcal{TR},\mathcal{VCY}};\mathbf{L}_{\mathcal{A}}^{\langle-\infty\rangle})\colon H_n(E_{\mathcal{TR}}(G);\mathbf{L}_{\mathcal{A}}^{\langle-\infty\rangle}) \to H_n(E_{\mathcal{VCY}}(G);\mathbf{L}_{\mathcal{A}}^{\langle-\infty\rangle})$$

is bijective where $\mathcal{A}$ is an additive G-category with involution.

Proof. (i) See [121, Remark 1.6].

(ii) The argument given in [255, Lemma 4.2] goes through since it is based on the Wang sequence for a semidirect product $G \rtimes \mathbb{Z}$, which can be generalised for additive G-categories with involutions as coefficients.

(iii) and (iv) Every torsionfree virtually cyclic group is isomorphic to $\mathbb{Z}$. By the Transitivity Principle, see for instance [25, Theorem 1.7], applied to $\mathcal{TR} \subseteq \mathcal{VCY}$, it suffices to show that the assembly maps

$$H_n^{\mathbb{Z}}(E\mathbb{Z};\mathbf{K}_{\mathcal{A}(R)}) \to H_n^{\mathbb{Z}}(\mathbb{Z}/\mathbb{Z};\mathbf{K}_{\mathcal{A}(R)});$$
$$H_n^{\mathbb{Z}}(E\mathbb{Z};\mathbf{L}_{\mathcal{A}}^{\langle-\infty\rangle}) \to H_n^{\mathbb{Z}}(\mathbb{Z}/\mathbb{Z};\mathbf{L}_{\mathcal{A}}^{\langle-\infty\rangle}),$$

are bijective for $n \in \mathbb{Z}$. This follows for K-theory from the Bass–Heller–Swan decomposition for regular rings R, see for instance [384, Theorem 9.8 on page 207], [360, Theorem 5.3.30 on page 295], or [270], and for L-theory from the Shaneson splitting for additive category with involution, see [349, 17.2]. □

Corollary 16.41 *If the torsionfree group G satisfies both Farrell–Jones Conjectures 16.31 and 16.35, then it satisfies Conjecture 16.19.*

Proof. We have to check the various assertions appearing in Conjecture 16.19.

(i) This follows from the canonical identification

$$H_n(E_{\mathcal{TR}}(G);\mathbf{K}_{\mathcal{A}(\mathbb{Z})}) = H_n(BG;\mathbf{K}_{\mathbb{Z}})$$

16.5 The Farrell–Jones Conjecture

and the Atiyah–Hirzebruch spectral sequence applied to $H_n(BG; \mathbf{K}_{\mathbb{Z}})$ using the facts that $\widetilde{K}_n(\mathbb{Z}) = 0$ for $n = 0, 1$ and $K_n(\mathbb{Z}) = 0$ for $n \leq -1$ hold.

(ii) This follows now from the Rothenberg sequences (16.2) and (16.3).

(iii) We define the desired map $\mathrm{asmb}_n^{\langle-\infty\rangle}$ to be the map induced by the projection $EG \to \{\bullet\}$ on $H_n^G(-; \mathbf{L}_{\mathcal{A}(\mathbb{Z})}^{\langle-\infty\rangle})$ using $EG = E_{\mathcal{TR}}(G)$ and the fact that $L_n^\epsilon(\mathbb{Z})$ does not depend on ϵ. Now apply Theorem 16.40 (iv).

This finishes the proof of Corollary 16.41, which can also be found in [266, Proposition 66 on page 743]. □

16.5.4 The Full Farrell–Jones Conjecture

The versions of the Farrell–Jones conjecture 16.31 and 16.35 discussed above do not carry over to overgroups of finite index. To handle this difficulty, we consider finite wreath products.

Let G and F be groups. Their *wreath product* $G \wr F$ is defined as the semidirect product $(\prod_F G) \rtimes F$ where F acts on $\prod_F G$ by permuting the factors. The main feature for us is the following. Let G be an overgroup of H of finite index. Then there is subgroup $N \subseteq H$ of H that satisfies $[G : H] < \infty$ and is normal in G, and a finite group F such that G embeds into $N \wr F$.

Conjecture 16.42 (Full Farrell–Jones Conjecture) We say that a group G satisfies the *Full Farrell–Jones Conjecture* if for any finite group F the group $G \wr F$ satisfies both the K-theoretic Farrell–Jones Conjecture 16.31 and the L-theoretic Farrell–Jones Conjecture 16.35 with coefficients in additive G-categories (with involution).

Definition 16.43 (Farrell–Jones group) We call a group G a *Farrell–Jones group* if it satisfies the Full Farrell–Jones Conjecture 16.42. We denote by $\mathcal{FJ}$ the class of Farrell–Jones groups.

Theorem 16.44 (The class $\mathcal{FJ}$) *The class $\mathcal{FJ}$ of Farrell–Jones groups has the following properties:*

(i) *The following classes of groups belong to $\mathcal{FJ}$:*

   (a) *Hyperbolic groups;*
   (b) *Finite-dimensional CAT(0)-groups;*
   (c) *Virtually solvable groups;*
   (d) *(Not necessarily cocompact) lattices in second countable locally compact Hausdorff groups with finitely many path components;*
   (e) *Fundamental groups of (not necessarily compact) connected manifolds (possibly with boundary) of dimension ≤ 3;*
   (f) *The groups $\mathrm{GL}_n(\mathbb{Q})$ and $\mathrm{GL}_n(F(t))$ for $F(t)$ the function field over a finite field F;*
   (g) *S-arithmetic groups;*
   (h) *Mapping class groups;*

(i) *Normally poly-free groups G, i.e., there exists a finite filtration* $\{1\} = G_0 \subseteq G_1 \subseteq \cdots \subseteq G_d = G$ *such that* $G_{i-1} \subseteq G$ *is normal and* G_i/G_{i-1} *is free (of possibly infinite rank) for* $i = 1, 2, \ldots, d$;

(ii) *The class* $\mathcal{FJ}$ *has the following inheritance properties:*

   (a) Passing to subgroups

   Let $H \subseteq G$ be an inclusion of groups. If G belongs to $\mathcal{FJ}$, then H belongs to $\mathcal{FJ}$;

   (b) Passing to finite direct products

   If the groups G_0 and G_1 belong to $\mathcal{FJ}$, then also $G_0 \times G_1$ belongs to $\mathcal{FJ}$;

   (c) Group extensions

   Let $1 \to K \to G \to Q \to 1$ be an extension of groups. Suppose that for any infinite cyclic subgroup $C \subseteq Q$ the group $p^{-1}(C)$ belongs to $\mathcal{FJ}$ and that the groups K and Q belong to $\mathcal{FJ}$.
   Then G belongs to $\mathcal{FJ}$;

   (d) Group extensions with virtually torsionfree hyperbolic groups as kernel

   Let $1 \to K \to G \to Q \to 1$ be an extension of groups such that K is virtually torsionfree hyperbolic and Q belongs to $\mathcal{FJ}$. Then G belongs to $\mathcal{FJ}$;

   (e) Directed colimits

   Let $\{G_i \mid i \in I\}$ be a direct system of groups indexed by the directed set I (with arbitrary structure maps). Suppose that for each $i \in I$ the group G_i belongs to $\mathcal{FJ}$.
   Then the colimit $\mathrm{colim}_{i \in I} G_i$ belongs to $\mathcal{FJ}$;

   (f) Passing to finite free products

   If the groups G_0 and G_1 belong to $\mathcal{FJ}$, then $G_0 * G_1$ belongs to $\mathcal{FJ}$;

   (g) Passing to overgroups of finite index

   Let G be an overgroup of H with finite index $[G : H]$. If H belongs to $\mathcal{FJ}$, then G belongs to $\mathcal{FJ}$.

Proof. See [22, 23, 24, 27, 28, 30, 39, 61, 143, 204, 210, 370, 419, 420]. □

It is not known whether all amenable groups belong to $\mathcal{FJ}$.

Exercise 16.45 Let $w \colon \mathbb{Z} \to \{\pm 1\}$ be the non-trivial group homomorphism. Compute $L_n^\epsilon(\mathbb{Z}[\mathbb{Z}], w)$ for all possible decorations ϵ and $n \in \mathbb{Z}$.

Exercise 16.46 Let $1 \to \pi_1(S) \to G \xrightarrow{p} \pi_1(M) \to 1$ be an extension of groups where S is a connected closed surface and M a connected manifold of dimension ≤ 3. Show that G is a Farrell–Jones group.

Nearly all computations of L-groups for integral group rings of infinite groups are carried out using the Farrell Conjecture by computing the left-hand side of the assembly map. In particular, if G contains non-trivial torsion, this is not at all easy and not even finished for some groups that at first glance seem to be accessible, e.g., crystallographic groups. Some special cases and tools are described for instance in [117, 255, 267, 268].

The situation is better when one rationalises. The change of rings map $L_n^\epsilon(\mathbb{Z}G) \to L_n^\epsilon(\mathbb{Q}G)$ is an isomorphism after inverting 2 for every group G, see [346, page 376]. One can compute $\mathbb{Q} \otimes_\mathbb{Z} L_n^\epsilon(\mathbb{Q}G)$ and hence $\mathbb{Q} \otimes_\mathbb{Z} L_n^\epsilon(\mathbb{Z}G)$ explicitly in terms of the group homology $H_n(C_G C; \mathbb{Q})$ of the centralisers of the finite cyclic subgroups $C \subseteq G$ using equivariant Chern characters, provided that G is a Farrell–Jones group, see [253, Theorem 0.4].

16.6 Notes

A systematic presentation of the statements and proofs about the Farrell–Jones Conjecture of this chapter, which are stated or cited in this chapter and scattered over the literature, will be given in the book [261]. For survey articles about the Farrell–Jones Conjecture we refer, for instance, to [150, 257, 259, 266].

Chapter 17
The Homotopy Type of G/TOP, G/PL and G/O

17.1 Introduction

In this chapter we review how to determine the homotopy types of the spaces G/PL and G/TOP and discuss some related results. We have already seen the statements in Section 11.9, here we deal with the proofs. Both the statements and the proofs are due to Sullivan in the PL case. In the topological case they are also due to Sullivan, but they are based on the foundational work of Kirby and Siebenmann [219]. The PL case is an input in the work of Kirby and Siebenmann, hence the topological case is a corollary of the PL case in this sense. However, we will concentrate on the topological case since there the statement is simpler, and we will rely the work of Kirby and Siebenmann without further explanations.

The starting point is that, by the work of Kirby and Siebenmann, we have the geometric surgery exact sequence in the topological category, see Theorem 11.25, for any compact topological manifold X with $\dim(X) = n \geq 5$. The Poincaré conjecture tells us that the topological structure set of the disk D^n is trivial $\mathcal{S}^{\mathrm{TOP}}(D^n, S^{n-1}) = \{\mathrm{id}_{D^n}\}$. The surgery sequence is of the form

$$\mathcal{S}^{\mathrm{TOP}}(D^n, S^{n-1}) \to [D^n, S^{n-1}; \mathrm{G/TOP}, *] \to L_n(\mathbb{Z}) \to \mathcal{S}^{\mathrm{TOP}}(D^{n-1}, S^{n-2}),$$

which gives for $n \geq 6$ the calculation

$$\pi_n(\mathrm{G/TOP}) \cong L_n(\mathbb{Z}), \tag{17.1}$$

as already indicated in Exercise 11.37. Special arguments can be employed to obtain the isomorphism for all $n \geq 1$, see Theorem C.1 on page 274 of [219, Essay V], and for $n = 0$ we have that G/TOP is connected.

Now we also already know that the space G/TOP is an infinite loop space, either via the Whitney sum operation $\oplus$, or via the operation obtained by Quinn–Ranicki–Siebenmann, see Remark 11.38 and also Remark 17.19 below, and so the normal invariants in the topological category form a generalised cohomology theory. By the above result we know the homotopy groups of the coefficient spectrum. This may be enough to calculate the normal invariants $\mathcal{N}^{\mathrm{TOP}}(X) \cong [X, \mathrm{G/TOP}]$ for some manifolds X via the Atiyah–Hirzebruch spectral sequence since for example the sequence may collapse on the second page for dimensional reasons, see Section 18.4 for the case $X = \mathbb{C}P^n$. Note that in general it is not enough to know the homotopy groups of G/TOP to calculate $[X, \mathrm{G/TOP}]$, we need the homotopy type. The calculation of the homotopy type of G/TOP is an illuminating example which highlights the difference between knowing the homotopy groups and the homotopy type. In addition, the technique of localisation of spaces is used, and hence it provides us with a good opportunity to illustrate the use of this technique in topology.

Sullivan described his work in his 1966 PhD thesis [391], which was never published. Notes based on the thesis circulated, but only appeared in print as [392] in 1996. In the meantime the main reference became the book of Madsen and Milgram [279]. Another useful reference, that is relevant when discussing the H-space structures on G/TOP, is the paper of Taylor and Williams [394].

17.2 Localisation

Recall the localisation in the theory of rings, as explained for example in Chapter 3 of the standard book [13]. Here we will only need the case of the ground ring $\mathbb{Z}$. Given any prime p, we can invert this prime and obtain the ring $\mathbb{Z}[1/p]$ of all fractions with denominator a power of p. Given an abelian group M, we obtain the $\mathbb{Z}[1/p]$-module $M[1/p] = M \otimes_{\mathbb{Z}} \mathbb{Z}[1/p]$. We can also invert a set of primes. When we choose the set of all primes except a given p we obtain the ring $\mathbb{Z}_{(p)}$ of all fractions with denominator coprime to p and, given an abelian group M, putting $M_{(p)} = M \otimes_{\mathbb{Z}} \mathbb{Z}_{(p)}$, we say that we *localise M at p*. The terminology comes from the fact that $\mathbb{Z}_{(p)}$ is a local ring, i.e., has a unique maximal ideal. In this spirit we also say that, when we form $M[1/p]$, we *localise M away from p*. We can also invert all primes, in which case we obviously obtain the field of rational numbers $\mathbb{Q}$, and we can also *rationalise* an abelian group M to obtain $M \otimes_{\mathbb{Z}} \mathbb{Q}$. In general, given a multiplicatively closed subset S of $\mathbb{Z}$ containing 1, we obtain the ring of all fractions $\mathbb{Z}_S$ with denominators in S and for an abelian group M we get the $\mathbb{Z}_S$-module $M_S = M \otimes_{\mathbb{Z}} \mathbb{Z}_S$.

The technique of localisation can be used to extract from a finitely generated abelian group M the free part and the p-primary torsion. On the other hand it is possible to reconstruct the abelian group if we know the appropriate localisation, in our case if we know for a given prime the localisation at p, away from p, and the rationalisation. Namely, for any finitely generated abelian group M and any prime p, we have the following diagram of abelian groups

$$\begin{array}{ccc} M & \longrightarrow & M[1/p] \\ \downarrow & & \downarrow \\ M_{(p)} & \longrightarrow & M_{\mathbb{Q}} \end{array} \qquad (17.2)$$

which is both a pullback and a pushout diagram.

Sullivan introduced the technique of localisation at the level of (nice) topological spaces. For simplicity we concentrate on the simply connected case, which can be found in [390, Corollary on page 19] or [290, Chapter 5].

Theorem 17.3 (Localisation of simply connected spaces) *Let X be a simply connected space and suppose that S is a multiplicatively closed subset of $\mathbb{Z}$ containing 1. Then there exists a space X_S such that for the homotopy and homology groups we have $\pi_n(X_S) \cong (\pi_n(X))_S$ and $H_n(X_S) \cong (H_n(X))_S$.*

The construction of X_S if X is an H-space is relatively simple, it uses the mapping telescope. It is also sufficiently functorial, so that for any prime p we obtain the square

$$\begin{array}{ccc} X & \longrightarrow & X[1/p] \\ \downarrow & & \downarrow \\ X_{(p)} & \longrightarrow & X_\mathbb{Q} \end{array} \tag{17.4}$$

which is both a homotopy fibre square and a homotopy cofibre square, again by [390]. Hence we are able to reconstruct the homotopy type of X from the homotopy types of the localisation and the rationalisation of X. The point is that in some cases, such as $X = $ G/TOP, it is easier to determine the homotopy types of the localisation and the rationalisation separately. For this purpose one uses the universal property of localisation, see the diagram on page 18 of [390] or [290, Chapter 5.2].

Exercise 17.5 Show that $(S^n)_\mathbb{Q}$ and $K(\mathbb{Z}, n)_\mathbb{Q}$ are homotopy equivalent for $n \geq 1$ if and only if n is odd. (Use the fact that $\pi_m(S^n)$ is finite for $m \geq n + 1$ and n odd and $\pi_{2n-1}(S^n)$ contains $\mathbb{Z}$ for n even.)

17.3 G/TOP

Our aim in this section is to sketch the main ideas in the proof of the following theorem proved by Sullivan in [392].

Theorem 17.6 (Homotopy type of G/TOP) *We have*

$$\text{G/TOP}_{(2)} \simeq \prod_{i \geq 1} K(\mathbb{Z}_{(2)}, 4i) \times \prod_{i \geq 1} K(\mathbb{Z}/2, 4i - 2);$$
$$\text{G/TOP}[1/2] \simeq \text{BO}[1/2];$$
$$\text{G/TOP}_\mathbb{Q} \simeq \text{BO}_\mathbb{Q} \simeq \prod_{i \geq 1} K(\mathbb{Q}, 4i).$$

This theorem also has a geometric version, which goes by the name of *the characteristic variety theorem*, see Remark 17.28 below.

The main ingredients that come into the proof are the already presented knowledge of the homotopy groups of G/TOP, bordism invariance of surgery obstructions with varying target from Remark 15.324, the calculation of the cobordism groups localised at 2 by Thom, see Theorem 17.7, and away from 2 by Conner–Floyd, see Theorem 17.8, and the product formula for the surgery obstruction from Theorem 15.467.

We start with the 2-local homotopy type. The desired homotopy equivalence from the left-hand side to the right-hand side is obtained by the universal property of the localisation from the maps to the respective factors, which we will denote

$$\kappa_{4i} \colon \mathrm{G/TOP} \to K(\mathbb{Z}_{(2)}, 4i) \quad \text{and} \quad \kappa_{4i-2} \colon \mathrm{G/TOP} \to K(\mathbb{Z}/2, 4i-2),$$

such that they induce isomorphisms on $\pi_{4i}(-)_{(2)}$ and $\pi_{4i-2}(-)_{(2)}$. Since homotopy classes of maps to Eilenberg–MacLane spaces are the same as cohomology classes, it is enough to construct elements $\kappa_{4i} \in H^{4i}(\mathrm{G/TOP}; \mathbb{Z}_{(2)})$ and $\kappa_{4i-2} \in H^{4i-2}(\mathrm{G/TOP}; \mathbb{Z}/2)$ with the property that $\bar{\kappa}_{4i}(h(\gamma_{4i})) = 1$ and $\bar{\kappa}_{4i-2}(h(\gamma_{4i-2})) = 1$, where $\gamma_{2j} \in \pi_{2j}(\mathrm{G/TOP})$ is a generator of the cyclic group $\pi_{2j}(\mathrm{G/TOP}) \cong L_{2j}(\mathbb{Z})$, the map $h \colon \pi_{2j}(\mathrm{G/TOP}) \to H_{2j}(\mathrm{G/TOP})$ is the Hurewicz homomorphism, and

$$\bar{\kappa}_{4i} \colon H_{4i}(\mathrm{G/TOP}) \to \mathbb{Z}_{(2)};$$
$$\bar{\kappa}_{4i-2} \colon H_{4i-2}(\mathrm{G/TOP}) \to \mathbb{Z}/2$$

are the homomorphisms given by applying to κ_{4i} and κ_{4i-2} the evaluation maps

$$H^{4i}(\mathrm{G/TOP}; \mathbb{Z}_{(2)}) \to \mathrm{Hom}(H_{4i}(\mathrm{G/TOP}), \mathbb{Z}_{(2)});$$
$$H^{4i-2}(\mathrm{G/TOP}; \mathbb{Z}/2) \to \mathrm{Hom}(H_{4i-2}(\mathrm{G/TOP}), \mathbb{Z}/2).$$

In fact, it is enough to construct the homomorphisms $\bar{\kappa}_{2j}$ since by the universal coefficient theorems the above evaluation maps are surjective, and so the homomorphisms $\bar{\kappa}_{2j}$ can be lifted to the classes κ_{2j}.

Next we recall the basic notation of bordism spectra. In Section 4.7 we introduced for any space X the symbol $\Omega_n(X)$ to mean the oriented cobordism group of n-dimensional closed oriented smooth manifolds equipped with a map to X. Now, by $\Omega_n^{\mathrm{unor}}(X)$ we denote the unoriented cobordism group of n-dimensional closed smooth manifolds equipped with a map to X. Theorem 4.57 and Remark 4.58 explain that we have the Pontrjagin–Thom isomorphism $\Omega_n(X) \cong \pi_n(\mathbf{MSO})$ for all $n \geq 0$, where **MSO** is the Thom spectrum obtained from the canonical bundles over the spaces $\mathrm{BSO}(k)$ for varying k. Similarly we have $\Omega_n^{\mathrm{unor}}(X) \cong \pi_n(\mathbf{MO})$ for all $n \geq 0$, where **MO** is the Thom spectrum obtained from the canonical bundles over the spaces $\mathrm{BO}(k)$ for varying k. Let $\mathbf{K}_{\mathrm{EM}}(G)$ be the Eilenberg–MacLane spectrum associated to an abelian group G. The homotopy type of spectra **MO** and **MSO** was analysed by Thom, who proved the following theorem, see [388].

17.3 G/TOP

Theorem 17.7 (**The 2-local homotopy type of bordism spectra**) *There exist natural numbers a_i, b_i, c_i such that*

$$\alpha_{\mathbf{MO}} \colon \mathbf{MO} \simeq \prod_{i=0}^{\infty} \Sigma^{a_i} \mathbf{K}_{\mathrm{EM}}(\mathbb{Z}/2);$$

$$\alpha_{\mathbf{MSO}} \colon \mathbf{MSO}_{(2)} \simeq \prod_{i=0}^{\infty} \Sigma^{b_i} \mathbf{K}_{\mathrm{EM}}(\mathbb{Z}_{(2)}) \times \prod_{i=0}^{\infty} \Sigma^{c_i} \mathbf{K}_{\mathrm{EM}}(\mathbb{Z}/2).$$

Sketch of proof of Theorem 17.6, 2-local part. The proof is organised around the following commutative diagrams (explanations follow):

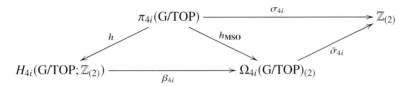

and

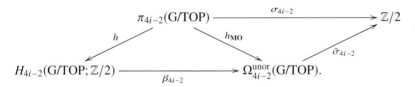

We start with the right triangles in both diagrams. Consider the surgery obstruction maps $\sigma_{2j} \colon \pi_{2j}(\mathrm{G/TOP}) \to L_{2j}(\mathbb{Z})$ of Theorem 8.167, see also Theorem 11.24. Let $h_{\mathbf{MSO}}$ and $h_{\mathbf{MO}}$ be the Hurewicz maps for the (oriented) cobordism, which take a representative $\lambda \colon S^{2j} \to \mathrm{G/TOP}$ of an element in $\pi_{2j}(\mathrm{G/TOP})$, and think of it as a representative of a bordism class in $\Omega_{4i}(\mathrm{G/TOP})$ or $\Omega_{4i-2}^{\mathrm{unor}}(\mathrm{G/TOP})$. Via Theorem 11.24 any map of the form $M \to \mathrm{G/TOP}$ represents some surgery problem $(f, \overline{f}) \colon N \to M$. By the bordism invariance of surgery obstructions with varying target, see Remark 15.324, the maps σ_{2j} factor through the term $\Omega_{4i}(\mathrm{G/TOP})$ or through the term $\Omega_{4i-2}^{\mathrm{unor}}(\mathrm{G/TOP})$; call the factoring maps after localisation $\bar{\sigma}_{2j}$.

The left maps h are just the classical Hurewicz homomorphisms. The construction of the maps β_{2j} so that the left triangles commute is explained on page 77 in [279] with a further reference to [278]. It is at this place that Theorem 17.7 is used, but more details are beyond our scope.

The desired classes $\bar{\kappa}_{4i}$ and $\bar{\kappa}_{4i-2}$ then come up as the compositions from the lower left entry to the upper right entry in the respective diagrams. □

We proceed to the odd-local homotopy type. The starting point is that $\pi_n(\mathrm{G/TOP})[1/2]$ are concentrated in dimensions $n = 4i$ where we have $\pi_{4i}(\mathrm{G/TOP})[1/2] \cong \mathbb{Z}[1/2]$ for $i > 0$. By the calculation of the homotopy groups of

788   17 The Homotopy Type of G/TOP, G/PL and G/O

BO by Bott, see the table below Lemma 12.30, we know that the $\pi_n(\mathrm{BO}[1/2])$-s are also concentrated in dimensions $n = 4i$ and here we also have $\pi_{4i}(\mathrm{BO}[1/2]) \cong \mathbb{Z}[1/2]$ for $i > 0$. It remains to find a map $\nu \colon \mathrm{G/TOP} \to \mathrm{BO}[1/2]$ that induces isomorphisms on $\pi_{4i}(-)[1/2]$. It turns out that the calculation of the oriented bordism ring at odd primes differs from the calculation in Theorem 17.7. Instead of the relation to singular homology there is a relation with the topological K-theory of real vector bundles, denoted by KO. Recall that, after inverting 2, this is a generalised cohomology theory based on the 4-periodic spectrum whose infinite loop space is $\mathbb{Z}[1/2] \times \mathrm{BO}[1/2]$, and we consider $\mathrm{KO}_n(X; \mathbb{Z}[1/2])$ as $\mathbb{Z}/4$-graded.

Following [279, page 84] we summarise. Firstly, in this part the $\mathbb{Z}/4$-grading is also used for the oriented cobordism, that means, we define $\Omega_n^{\mathrm{per}}(X) = \oplus_{l \geq 0} \Omega_{n+4l}(X)$ for $n \in \mathbb{Z}/4$. Cobordism theory from [388, Chapter IX] yields a natural transformation $\delta \colon \Omega_*^{\mathrm{per}}(X) \to \mathrm{KO}_*(X; \mathbb{Z}[1/2])$, which in degree 0 and for $X = \bullet$ is given by the signature. Moreover, $\Omega_*^{\mathrm{per}}(X)$ is a graded module over $\Omega_*^{\mathrm{per}} = \Omega_*^{\mathrm{per}}(\bullet)$. Denoting by $\mathbb{Z}[1/2]_*$ the $\mathbb{Z}/4$-graded module, which has $\mathbb{Z}[1/2]$ in degree 0 and zero in all other degrees, the signature map gives $\mathbb{Z}[1/2]_*$ the structure of a module over Ω_*^{per}. Thus the map δ factors through the tensor product over Ω_*^{per}, yielding a map denoted $\bar{\delta}$, and the following theorem of Conner and Floyd identifies the tensor product, see [100].

Theorem 17.8 (The odd-local part of the oriented bordism groups) *The map*

$$\bar{\delta} \colon (\Omega_*^{\mathrm{per}}(X) \otimes_{\Omega_*^{\mathrm{per}}} \mathbb{Z}[1/2]_*)_n \xrightarrow{\cong} \mathrm{KO}_n(X; \mathbb{Z}[1/2])$$

is an isomorphism, for all $n \in \mathbb{Z}/4$.

Sketch of proof of Theorem 17.6, odd-local part. We study the commutative diagram (explanations follow):

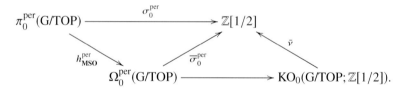

Here we need to consider the surgery obstruction maps $\sigma_{4i} \colon \pi_{4i}(\mathrm{G/TOP}) \to \mathbb{Z}$ constructed in Theorem 8.167, see also Theorem 11.24, for all i at once, so we use the notation $\pi_0^{\mathrm{per}} = \oplus_i \pi_{4i}$ and σ_0^{per}. Similarly, the map $h_{\mathrm{MSO}}^{\mathrm{per}}$ is the sum of Hurewicz maps for the oriented cobordism, and the bordism invariance of surgery obstructions with varying target, see Remark 15.324, guarantees that the map σ_0^{per} factors through the term $\Omega_0^{\mathrm{per}}(\mathrm{G/TOP})$; called the factoring map $\bar{\sigma}_0^{\mathrm{per}}$. The map $\bar{\sigma}_0^{\mathrm{per}}$ itself factors as

$$\bar{\sigma}_0^{\mathrm{per}} \colon \Omega_0^{\mathrm{per}}(\mathrm{G/TOP}) \xrightarrow{\iota} (\Omega_*^{\mathrm{per}}(\mathrm{G/TOP}) \otimes_{\Omega_*^{\mathrm{per}}} \mathbb{Z}[1/2]_*)_0 \xrightarrow{\bar{\bar{\sigma}}_0^{\mathrm{per}}} \mathbb{Z}[1/2]$$

where ι is the canonical map. This follows from the product formula of Theorem 15.467 (for the case when $\pi = \pi'$ is trivial) by the following reasoning. Any map $g \colon X \to \mathrm{G/TOP}$ corresponds to a surgery problem $(f, \bar{f}) \colon M \to X$. Given an oriented manifold N, the composite of g with the canonical projection $X \times N \to X \to$ G/TOP corresponds to the surgery problem $(f, \bar{f}) \times \mathrm{id}_N : M \times N \to X \times N$. The product formula applied to this surgery problem now implies the factorisation.

Finally, we use the case of Theorem 17.8 when $k = 0$, which identifies $(\Omega_*^{\mathrm{per}}(\text{G/TOP}) \otimes_{\Omega_*^{\mathrm{per}}} \mathbb{Z}[1/2]_*)_0$ with $\mathrm{KO}_0(\text{G/TOP}; \mathbb{Z}[1/2])$. Under this identification the horizontal map in the above diagram corresponds to ι and $\bar{\nu}$ corresponds to $\overline{\overline{\sigma}}_0^{\mathrm{per}}$.

Homotopy classes of maps of the form G/TOP $\to$ BO[1/2] are elements in the localised reduced real KO-theory of G/TOP, so it is enough to construct such a cohomology element $\nu \in \widetilde{\mathrm{KO}}^0(\text{G/TOP}; \mathbb{Z}[1/2])$. Again, we have the evaluation map

$$\mathrm{KO}^0(\text{G/TOP}; \mathbb{Z}[1/2]) \to \mathrm{Hom}(\mathrm{KO}_0(\text{G/TOP}; [1/2]); \mathbb{Z}[1/2]),$$

which is surjective by the universal coefficient theorem of KO-theory from [438]. The homomorphism $\bar{\nu}$ yields a map $\overline{\bar{\nu}} \colon \widetilde{\mathrm{KO}}_0(\text{G/TOP}; \mathbb{Z}[1/2]) \to \mathbb{Z}[1/2]$, which via the above surjection lifts to $\nu \in \widetilde{\mathrm{KO}}^0(\text{G/TOP}; \mathbb{Z}[1/2])$.

That the map G/TOP[1/2] $\to$ BO[1/2] induces an isomorphism on homotopy groups is shown in [279, Chapter 4.D on page 89] using additional arguments about the behaviour of ν with respect to the Pontrjagin character, which relates KO-theory to singular cohomology. These are beyond our scope so we refer the reader to this source for the rest of the proof. □

For the rationalisation it turns out that it is enough to know the homotopy groups and we instantly obtain

$$\text{G/TOP}_\mathbb{Q} \simeq \text{BO}_\mathbb{Q} \simeq \prod_{j \geq 1} K(\mathbb{Q}, 4j).$$

The integral homotopy type is obtained from the homotopy pullback (17.4).

Remark 17.9 We note that the homotopy classes of the maps κ_{2j} and $\bar{\nu}$ sketched above are not unique. Hence also the homotopy equivalences appearing in Theorem 17.6 are not unique. Discussion of various choices for them as well as further details can be found in [279, Chapter 4].

17.4 G/PL

The statement corresponding to Theorem 17.6 in the PL case is similar, the only difference is that, when localised at 2, the first two factors on the right-hand side are replaced by a two-stage Postnikov system, we have

$$\text{G/PL}_{(2)} \simeq E \times \prod_{j \geq 2} K(\mathbb{Z}_{(2)}, 4j) \times \prod_{j \geq 2} K(\mathbb{Z}/2, 4j-2),$$

where E is the total space of the fibration $K(\mathbb{Z}, 4) \to E \to K(\mathbb{Z}/2, 2)$ with k-invariant the element of order 2 in $H^5(K(\mathbb{Z}/2, 2); \mathbb{Z}) \cong \mathbb{Z}/4$.

Informally, the difference comes from the fact that, unlike in the topological case, the surgery obstruction map $\sigma_4 \colon \pi_4(G/PL) \to \mathbb{Z}$ is not an isomorphism, but is an inclusion of infinite cyclic groups of index 2. This is a consequence of the famous Rochlin Theorem, which says that the signature of a closed 4-dimensional spin PL manifold is divisible by 16, see [218, Chapter XI]. (In the topological case Freedman constructed a closed 4-dimensional spin topological manifold with signature 8, see [157].) For how this is used in the calculation of the homotopy type of G/PL, we refer the reader to [279, Theorem 4.8, page 79] or [93, Theorem 3.37, page 110].

We have also already stated Theorem 11.40 of Kirby and Siebenmann, which describes the space TOP/PL as an Eilenberg–MacLane space. In fact, they used this description to obtain an obstruction that describes the difference between PL and topological structures on manifolds:

Theorem 17.10 (Kirby–Siebenmann invariant) *Let M^n be a closed n-dimensional topological manifold with $n \geq 5$. There exists an invariant* $\mathrm{ks}(M) \in H^4(M, \mathbb{Z}/2)$ *such that* $\mathrm{ks}(M) = 0$ *if and only if M admits a PL structure.*

If M admits a PL structure, then the concordance classes of the PL structures on M are in bijection with $H^3(M, \mathbb{Z}/2)$.

For the definition of the notion of concordance of PL structures and for a more thorough discussion we refer the reader to [219], Classification Theorem 5.4 on page 318, and Section 3.7 in [93]. Another useful reference is "The Hauptvermutung book" [355]. The title of this book refers to the "Hauptvermutung" ("Main Conjecture" in German) of the combinatorial topology, one version of which, see [350, page 4], is as follows.

Conjecture 17.11 (Manifold Hauptvermutung) Every homeomorphism between PL manifolds is homotopic to a PL homeomorphism.

This conjecture is known to be false. Counterexamples can be found by studying the forgetful map $\mathcal{S}^{\mathrm{PL}}(M) \to \mathcal{S}^{\mathrm{TOP}}(M)$ for suitable PL manifolds M. We will encounter such examples in Chapter 18 for M a real projective space, see Section 18.8, or an aspherical manifold such as the tori in Section 18.10.

Exercise 17.12 Let M be a closed n-dimensional topological manifold for $n \geq 5$ satisfying $H_k(M; \mathbb{Z}/2) = 0$ for $k = 3, 4$. Show that it carries a PL structure unique up to concordance.

17.5 G/O

The computation of the homotopy type of G/O is much harder than that of G/TOP. This is the main reason why classification problems in the smooth category are often much harder than in the topological category. An example of this is the classification of homotopy spheres in the smooth category of Chapter 12, which in the topological category boils down to saying that the standard sphere is topologically rigid.

We have already given some information about G/O in Theorem 11.42 and Remark 11.43. Another approach, given the fact that we know G/TOP, is to analyse the difference between G/TOP and G/O, which means that we have to understand TOP/O. The space TOP/O plays a key role in smoothing theory, see [219, Essay V].

The homotopy groups of $\pi_k(\text{TOP/O})$ are $\mathbb{Z}/2$ if $k = 3$, are trivial for $k = 0, 1, 2, 4, 5, 6$ and are given by the groups Θ_k of Chapter 12 for $k \neq 3$, see (12.61) and Theorem 11.40. The value $\pi_3(\text{TOP/O}) \cong \mathbb{Z}/2$, which is different from $\Theta_3 = \{0\}$, is due to the Kirby–Siebenmann invariant.

The main problem in understanding the homotopy type is that the k-invariants of TOP/O are not known. The best result so far is due to Kupers [235]. He shows that there is a 9-connected map

$$\text{TOP/O} \to K(\mathbb{Z}/2, 3) \times K(\mathbb{Z}/28, 7) \times K(\mathbb{Z}/2, 8).$$

Exercise 17.13 Show that $\pi_{13}(\text{TOP/O})$ is isomorphic to $\mathbb{Z}/3$.

17.6 H-Space Structures on G/TOP and G/PL

The existence of two different H-space structures on G/TOP has already been mentioned in Remark 11.38. We recall that one of them comes from the Whitney sum, and is denoted by $\oplus$, and the other one comes from either Siebenmann's work using Quinn's surgery spaces or from Ranicki's L-spectra, and it will be denoted by $+$.

More specifically, using the theory of n-ads Ranicki constructs a homotopy equivalence

$$\sigma \colon \text{G/TOP} \to \mathbf{L}_0(\mathbb{Z})\langle 1 \rangle \tag{17.14}$$

where the target is the zeroth space in the connective version of his quadratic L-spectrum discussed in Section 15.9.5 and which induces the surgery obstruction isomorphism $\pi_n(\text{G/TOP}) \cong L_n(\mathbb{Z})$ on homotopy groups for $n \geq 1$. Since $\mathbf{L}_0(\mathbb{Z})\langle 1 \rangle$ is the zeroth space in an Ω-spectrum, it has an H-space structure. It turns out that the map σ does not map the H-space structure on the source given by the Whitney sum to this H-spaces structure. On the other hand, the map σ can be used to introduce a new H-space structure on G/TOP, which, following Ranicki, we denote by $+$. Ranicki also calls it the "disjoint union" structure, for reasons we explain later in Remark 17.19. The relation between the two H-space structures is described using products on L-spectra, which are discussed in Appendix B of [348]. Here we only mention that one of these products induces on L-groups the product from Theorem 15.466. We also note that the symmetrisation map $(1+T)\colon L_m(\mathbb{Z}) \to L^m(\mathbb{Z})$ can also be used to obtain the product of the shape $-\otimes_q-\colon L_m(\mathbb{Z})\otimes L_n(\mathbb{Z}) \xrightarrow{(1+T)\otimes \mathrm{id}} L^m(\mathbb{Z})\otimes L_n(\mathbb{Z}) \xrightarrow{\otimes} L_{m+n}(\mathbb{Z})$ where the second map comes from Theorem 15.466. The desired formula relating the two H-space structures is

$$a \oplus b = a + b + a \otimes_q b,$$

where $\otimes_q$ is the above described product at the level of quadratic L-spectra. See [343, page 292], and [340, Section 3] for an extensive discussion.

Taylor and Williams study the homotopy type of Ranicki's L-spectra and hence by the above discussion of the space G/TOP with the H-space structure given by $+$. Using the work from [312, 343, 368], they prove the following theorem, see [394].

Theorem 17.15 (Homotopy type of G/TOP as infinite loop space) *There are homotopy equivalences of infinite loop spaces*

$$l \times k\colon \mathrm{G/TOP}_{(2)} \simeq \prod_{i\geq 1} K(\mathbb{Z}_{(2)}, 4i) \times \prod_{i\geq 1} K(\mathbb{Z}/2, 4i-2);$$

$$\sigma\colon \mathrm{G/TOP}[1/2] \simeq \mathrm{BO}[1/2];$$

$$\sigma_\mathbb{Q}\colon \mathrm{G/TOP}_\mathbb{Q} \simeq \mathrm{BO}_\mathbb{Q} \simeq \prod_{i\geq 1} K(\mathbb{Q}, 4i),$$

for the infinite loop space structure on the left-hand side that induces the H-space structure $+$ on G/TOP.

As a consequence we obtain a Mayer–Vietoris type long exact sequence that, for a w-oriented finite Poincaré CW-complex X, reduces to the following short exact sequence, which calculates $\mathcal{N}^{\mathrm{TOP}}(X) \cong [X; \mathrm{G/TOP}]$ with the abelian group structure induced by the H-space structure $+$ on G/TOP:

$$0 \to \mathcal{N}^{\mathrm{TOP}}(X) \to$$
$$\to \bigoplus_{i=1}^{\infty} H^{4i}(X;\mathbb{Z}_{(2)}) \oplus \bigoplus_{i=1}^{\infty} H^{4i-2}(X;\mathbb{Z}/2) \oplus \widetilde{\mathrm{KO}}^0(X;\mathbb{Z}[1/2]) \to$$
$$\to \bigoplus_{i=1}^{\infty} H^{4i}(X;\mathbb{Q}) \to 0. \quad (17.16)$$

This is so since by the Atiyah–Hirzebruch spectral sequence

$$\mathcal{N}^{\mathrm{TOP}}(X) \cong [X; \mathrm{G/TOP}]$$

is a finitely generated abelian group and the term to the left is a $\mathbb{Q}$-vector space, so the map to $\mathcal{N}^{\mathrm{TOP}}(X)$ must be zero.

17.6 H-Space Structures on G/TOP and G/PL

At the level of spaces Theorem 17.15 looks very similar to Theorem 17.6, but we warn the reader that the maps inducing homotopy equivalences are different. Maps from both theorems have an interpretation in terms of characteristic classes and the fact that they are different means that the resulting characteristic classes are different.

We at least mention that for the rationalisation we have the following theorem, see Example 18.4 in [348] and also Remark 19.8 below.

Theorem 17.17 (Normal invariants in the topological category after rationalisation) *Let X be a closed connected n-dimensional topological manifold with $w_1(X)$ trivial. Make a choice (ABP), see Notation 8.13, which in particular yields a choice of a fundamental class $[X] \in H_n(X; \mathbb{Z})$. Consider the normal invariants in the topological category $\mathcal{N}^{TOP}(X) \cong [X, G/TOP]$, see Theorem 11.24, with the structure of an abelian group given by the H-space structure + on G/TOP. Then the map*

$$t \colon \mathcal{N}^{TOP}(X) \otimes \mathbb{Q} \xrightarrow{\cong} \oplus_{i \geq 1} H_{n-4i}(X; \mathbb{Q});$$
$$(f, \overline{f}) \colon M \to X \mapsto f_*(\mathcal{L}(M) \cap [N]) - \mathcal{L}(X) \cap [X],$$

with $\mathcal{L}(-)$ the Hirzebruch polynomial, is a $\mathbb{Q}$-isomorphism.

The Hirzebruch polynomial $\mathcal{L}(-)$ was discussed in Section 8.7.6 for smooth manifolds. For the generalisation to the categories PL and TOP, see Chapter 20 and Epilogue in [308], Chapter 4 in [279], and Siebenmann's article with the title "Topological manifolds" reprinted in [219]

Exercise 17.18 Show that the rank of the abelian group $\mathcal{N}^{TOP}(\mathbb{CP}^3 \times S^2)$ is 3.

Remark 17.19 (Quinn–Siebenmann versus Ranicki H-space structure on G/TOP) The Quinn–Siebenmann H-space structure on G/TOP is obtained from Quinn spaces $\widetilde{\mathcal{L}}_n(\bullet)$, discussed in the Notes section of Chapter 11, which also yield an Ω-spectrum, see the Notes section of Chapter 13. In fact, Quinn's theory gives a map G/TOP $\to \widetilde{\mathcal{L}}_0(\bullet)$ and Ranicki's theory yields a map $\widetilde{\mathcal{L}}_0(\bullet) \to \mathbf{L}_0(\mathbb{Z})\langle 1 \rangle$, induced by a map of spectra, such that their composite is the map σ from (17.14). Both factoring maps are homotopy equivalences, and this gives the identification of the H-space structures of Quinn–Siebenmann and Ranicki. (Strictly speaking there are some low-dimensional issues in the above arguments, but they can be overcome.)

Sometimes the name "disjoint union" is used for the H-space structure +. This refers to the fact that the spaces $\widetilde{\mathcal{L}}_n(K)$ are constructed out of surgery problems that do not necessarily have a connected target, and in fact the loop space "addition" corresponds to taking the disjoint union of such problems. Note that in Chapters 8 and 9 we carefully explained the surgery obstruction theory of such problems. This "disjoint union" operation corresponds simply to the direct sum of quadratic chain complexes when we pass to Ranicki's algebraic setting of $\mathbf{L}$-spectra.

Remark 17.20 (Normal invariants as a homology theory) By the above discussion we have for an n-dimensional topological manifold X that the normal invariants of X in the topological category satisfy

$$\mathcal{N}^{\mathrm{TOP}}(X) \cong [X, \mathrm{G/TOP}] \cong [X, \mathbf{L}_0(\mathbb{Z})\langle 1\rangle] \cong H^0(X; \mathbf{L}(\mathbb{Z})\langle 1\rangle),$$

where all the maps are isomorphisms of abelian groups with the H-space structure $+$ on G/TOP. Moreover, in view of this, Ranicki's identification from Theorem 15.482 of the normal invariants with the homology theory, which is the source of the assembly maps, can in fact be identified with the Poincaré duality isomorphism from the generalised cohomology theory to the generalised homology theory with respect to the spectrum $\mathbf{L}(\mathbb{Z})\langle 1\rangle$:

$$t\colon \mathcal{N}^{\mathrm{TOP}}(X) \cong H^0(X; \mathbf{L}(\mathbb{Z})\langle 1\rangle) \xrightarrow{\cong} H_n(X; \mathbf{L}(\mathbb{Z})\langle 1\rangle).$$

For more details, see [348, Proposition 18.3] and [234].

Remark 17.21 (Surgery obstruction map is a homomorphism with respect to the H-space structure $+$ on G/TOP) For an n-dimensional topological manifold X with $\pi = \pi_1(X)$, the surgery obstruction map $\sigma\colon \mathcal{N}^{\mathrm{TOP}}(X) \to L_n(\mathbb{Z}\pi)$ is a homomorphism with respect to the H-space structure $+$ on G/TOP. Namely, by the identification of the geometric and algebraic surgery exact sequences from Theorem 15.482 and Remark 17.20, it is the composition of an isomorphism of abelian groups with the assembly map, which is a homomorphism of abelian groups by its construction.

Remark 17.22 ((Near) 4-periodicity of G/TOP) It turns out that the 4-periodicity of the homotopy groups of G/TOP obtained via the surgery obstruction map and the periodicity of L-groups can be realised at the level of spaces. Namely, there is a homotopy equivalence

$$\mathbb{Z} \times \mathrm{G/TOP} \simeq \Omega^4(\mathrm{G/TOP}),$$

which we refer to as the (near) 4-periodicity of the space G/TOP.

The infinite loop space structure obtained in this way on G/TOP is compatible with the one obtained from the Quinn spectra $\widetilde{\mathcal{L}}_\bullet(\bullet)$ and the Ranicki spectrum $\mathbf{L}(\mathbb{Z})\langle 1\rangle$ via the map σ from (17.14).

The PL version, which predates the above discussion, is attributed to Casson and Sullivan, see [87, page 55] and the topological version comes from Siebenmann in Section 14 on page 327 of his article "Topological manifolds" reprinted in [219]. It also has a geometric interpretation, see [74] and [110].

For a discussion of H-space structures on G/PL and applications we refer the reader to [320, pages 81-82].

Remark 17.23 (Comparing structure sets in smooth and topological category) Sometimes information about the structure sets in the smooth category can be obtained by comparing the surgery exact sequence in the smooth category and topological category. For a smooth manifold M and the obvious forgetful map, $F^s\colon \mathcal{S}^s(M) \to \mathcal{S}^{\mathrm{TOP},s}(M)$ is in general neither injective nor surjective. However, there is an exact sequence of sets, see [219, Essay V, Appendix C, page 289]:

$$\mathcal{S}^{\mathrm{TOP},s}(M \times I, \partial(M \times I)) \to [M, \mathrm{TOP/O}] \to \mathcal{S}^s(M) \xrightarrow{F^s} \mathcal{S}^{\mathrm{TOP},s}(M).$$

17.7 Splitting Invariants   795

Using the fact that the homotopy groups of TOP/O are finite, see Section 17.5, Weinberger showed in [423] that the preimage with respect to F^s of each element in $\mathcal{S}^{\mathrm{TOP},s}(M)$ is finite and that the image of F^s contains a subgroup of a finite index when $\mathcal{S}^{\mathrm{TOP},s}(M)$ is considered as an abelian group with the group structure from Remark 11.38, see also Section 17.6 below. In the case $M = S^3 \times S^4$ Crowley found elements in $\mathcal{S}^{\mathrm{TOP}}(S^3 \times S^4)$ whose preimages have different cardinality in [108].

17.7 Splitting Invariants

We start with a general discussion about splitting invariants. These provide us in some cases with a tool to determine whether a given degree one normal map $(f, \overline{f}): M \to X$ represents a non-trivial element in the set of normal invariants $\mathcal{N}(X)$. Using the map $\mathcal{S}(X) \to \mathcal{N}(X)$ this method can also be used to detect non-trivial elements in the structure set of X. Sources where it is used include [354, page 73], [414, § 14C,D,E], [247, III.1.3]. We will also use it in Chapter 18, see Theorems 18.11, 18.15, 18.20, and (18.25), (18.43). The splitting invariants also fit into a more general context, see Remarks 17.28 and 17.29 below. The following theorem is based on [247, III.1.3 on page 30].

We remind the reader that, if $(X, \mathcal{O}_X)$ is a finite Poincaré complex, then according to Definition 7.6 the normal invariants $\mathcal{N}(X)$ are defined after we made a choice of $[[X]] \in H_n(X, \mathcal{O}_X)$, while according to Definition 11.7 when X is a manifold we choose $[[X]]$ to be the intrinsic fundamental class $[[X]] \in H_n(X, \mathcal{O}_X)$ from Definition 5.29.

Theorem 17.24 (Normal splitting along a submanifold) *Let X be a closed n-dimensional smooth manifold and let $i: Y \hookrightarrow X$ be an embedding of a closed m-dimensional smooth submanifold with $n \geq m \geq 5$.*

(i) *There is a well-defined map of sets*

$$s^{\mathcal{N}}_{Y \subseteq X}: \mathcal{N}(X) \to \mathcal{N}(Y), \quad [(f, \overline{f}): M \to X] \mapsto [(g, \overline{g}): N \to Y]$$

called the normal splitting along Y. It associates to a degree one normal map $(f, \overline{f}): M \to X$ representing an element in $\mathcal{N}(X)$ the degree one normal map $(g, \overline{g}): N \to Y$ obtained from $(f, \overline{f})$ by transversality;

(ii) *The following diagram commutes*

$$\begin{array}{ccc} [X, G/O] & \xrightarrow{i^*} & [Y, G/O] \\ \cong \downarrow & & \downarrow \cong \\ \mathcal{N}(X) & \xrightarrow{s^{\mathcal{N}}_{Y \subseteq X}} & \mathcal{N}(Y) \end{array}$$

where the upper vertical arrow is induced by the embedding $i: Y \hookrightarrow X$ and the vertical bijections stem from Theorem 7.34 (ii);

(iii) *The analogues of assertions* (i) *and* (ii) *hold also in the* PL *category and in the topological category.*

Proof. (i) Consider a degree one normal map $(f, \overline{f}) \colon M \to X$ representing an element in $N(X)$. We remind the reader that according to Definition 7.13 such an element is specified by the tuple $(M, i_M, f, \xi, \overline{f}, o_\xi)$, where $i_M \colon M \hookrightarrow \mathbb{R}^{n+k}$ is an embedding, ξ is a vector bundle over X, $\overline{f} \colon \nu(i_M) \to \xi$ is a bundle map, $o_\xi \colon O_\xi \to O_X$ is an isomorphism of local coefficient systems, and we require $H_n(f, \alpha)([[M]]) = [[X]]$ where $[[M]] \in H_n(M; O_M)$ and $[[X]] \in H_n(X; O_X)$ are the intrinsic fundamental classes and $\alpha \colon O_M \to f^*O_X$ is a certain isomorphism obtained from o_ξ.

From these data we construct a normal map of degree one $(g, \overline{g}) \colon N \to Y$, representing an element in $N(Y)$, specified by a tuple $(N, i_N, g, \eta, \overline{g}, o_\eta)$, as follows. Make f transverse to Y and define $N := f^{-1}(Y)$, $g := f|_N \colon N \to Y$, and $\eta = i^*(\xi) \oplus \nu(i)$. Denoting by $j \colon N \hookrightarrow M$ the canonical embedding, consider $i_N := i_M \circ j$ so that we get an isomorphism $\nu(i_N) \cong j^*\nu(i_M) \oplus \nu(j)$ of bundles over N. Define the bundle map $\overline{g}$ covering g by

$$\overline{g} := \nu(g) \oplus \overline{f}|_{j^*(\nu(i_M))} \colon \nu(j) \oplus j^*(\nu(i_M)) \to \nu(i) \oplus i^*(\xi),$$

where $\nu(g)$ is the bundle map obtained by transversality and $\overline{f}|_{j^*(\nu_M)}$ is the restriction of the bundle map $\overline{f} \colon \nu_M \to \xi$ to $j^*(\nu(i_M)) \subset \nu(i_M)$.

The embedding $i \colon Y \hookrightarrow X$ induces an isomorphism of vector bundles $i^*TX \cong TY \oplus \nu(i)$ over Y. A choice of an embedding $i_X \colon X \hookrightarrow \mathbb{R}^{n+k}$ can be used to obtain an embedding $i_Y \colon Y \hookrightarrow \mathbb{R}^{n+k}$ by setting $i_Y := i_X \circ i$ and for the normal bundles we obtain $\nu(i_Y) = i^*(\nu(i_X)) \oplus \nu(i)$. Recalling Lemma 5.23 (i) we obtain isomorphisms of local coefficients systems

$$i^*O_X \cong O_Y \otimes O_{\nu(i)}, \quad O_{\nu(i_Y)} \cong i^*O_{\nu(i_X)} \otimes O_{\nu(i)}, \quad O_\eta \cong i^*O_\xi \otimes O_{\nu(i)},$$

and recalling Lemma 5.23 (iv) also $O_X \cong O_{\nu(i_X)}$ and $O_Y \cong O_{\nu(i_Y)}$. Using these and the given $o_\xi \colon O_\xi \to O_X$, the desired isomorphism $o_\eta \colon O_\eta \to O_Y$ is obtained as the composition

$$o_\eta \colon O_\eta \cong i^*O_\xi \otimes O_{\nu(i)} \cong i^*O_X \otimes O_{\nu(i)} \cong i^*O_{\nu(i_X)} \otimes O_{\nu(i)} \cong O_{\nu(i_Y)} \cong O_Y.$$

Similarly, the canonical embedding $j \colon N \hookrightarrow M$ yields $j^*TM \cong TN \oplus \nu(j)$ and analogous isomorphisms of various local coefficients systems.

To verify the statement about the compatibility of the intrinsic fundamental classes, we need to recall the Thom isomorphism for the normal bundles from Chapter 6. Firstly, consider the embedding $i_{D(\nu(i))} \colon D(\nu(i)) \hookrightarrow X$ where $D(\nu(i))$ is the normal disk bundle of i embedded as a tubular neighbourhood in X. Then a zigzag of isomorphisms induced by inclusion and excision, and the canonical isomorphism $(i_{D(\nu(i))})^*O_X \cong O_{D(\nu(i))}$, tells us that $[[X]] \in H_n(X; O_X)$ and $[[D(\nu(i)), S(\nu(i))]] \in H_n(D(\nu(i)), S(\nu(i)); O_{D(\nu(i))})$ correspond to each other. Next arguing as in the proof of Lemma 6.67, we see that for the Thom class

17.7 Splitting Invariants

$\overline{U_{\nu(i)}} \in H^{n-m}(D(\nu(i)), S(\nu(i)); \mathcal{O}_{D(\nu(i))})$, see also Theorem 6.54, we obtain $\overline{U_{\nu(i)}} \cap [[D(\nu(i)), S(\nu(i))]] = [[Y]]$ in $H_m(Y; \mathcal{O}_Y)$. Similarly we get a correspondence between $[[M]] \in H_n(M; \mathcal{O}_M)$ and $[[D(\nu(j)), S(\nu(j))]] \in H_n(D(\nu(j)), S(\nu(j)); \mathcal{O}_{D(\nu(j))})$ and hence the equation

$$\overline{U_{\nu(j)}} \cap [[D(\nu(j)), S(\nu(j))]] = [[N]]$$

in $H_m(N; \mathcal{O}_N)$. Using this and the fact that the pullback of $\overline{U_{\nu(i)}}$ is $\overline{U_{\nu(j)}}$ (because of the bundle map $\nu(g)$), we obtain

$$H_m(g, \alpha_N)([[N]]) = [[Y]] \in H_m(Y; \mathcal{O}_Y),$$

where $\alpha_N \colon \mathcal{O}_N \to g^*\mathcal{O}_Y$ is the isomorphism induced by o_η.

The assignment just described produces a well-defined function of the form $s^\mathcal{N}_{Y \subseteq X} \colon \mathcal{N}(X) \to \mathcal{N}(Y)$. To see this assume that we have a normal bordism with cylindrical end $(F, \overline{F})$ between two degree one normal maps $(f, \overline{f})$ and $(f', \overline{f'})$ with the target X. Then we can make an analogous transversality construction with $(F, \overline{F})$ and obtain a normal bordism with cylindrical end between the associated degree one normal maps $(g, \overline{g})$ and $(g', \overline{g'})$.

(ii) Note that we assume that X is a manifold, so we have the stable normal bundle $\nu_X \colon X \to BO$. Given $\zeta \colon X \to G/O$ representing an element in $[X, G/O]$, its image under the bijection $[X, G/O] \xrightarrow{\cong} \mathcal{N}(X)$ is represented by a degree one normal map $(f, \overline{f})$ with bundle data $\overline{f} \colon \nu_M \to \nu_X \oplus \zeta$. Here we interpret ζ as a vector bundle with a preferred trivialisation of the associated spherical fibration, which is possible since G/O is the homotopy fibre of the map $J \colon BO \to BG$ from (6.70). We get

$$(\zeta \circ i) \oplus \nu_Y = i^*(\zeta) \oplus \nu_Y \cong i^*(\zeta) \oplus i^*(\nu_X) \oplus \nu(i) \cong i^*(\zeta \oplus \nu_X) \oplus \nu(i).$$

This finishes the proof of Theorem 17.24 in the smooth category.

(iii) The proofs for the categories PL and TOP are analogous. □

Definition 17.25 (Splitting invariant along a submanifold) Let X be a closed n-dimensional manifold and let $i \colon Y \hookrightarrow X$ be an embedding of a closed m-dimensional submanifold, with $n \geq m \geq 5$. Then the function

$$s_{Y \subseteq X} \colon \mathcal{N}(X) \to L_m(\mathbb{Z}\Pi(Y), \mathcal{O}_Y);$$
$$[(f, \overline{f})] \mapsto (\sigma_m \circ s^\mathcal{N}_{Y \subseteq X})([(f, \overline{f})]),$$

where $s^\mathcal{N}_{Y \subseteq X}$ is the normal splitting along Y from Theorem 17.24 and $\sigma_m \colon \mathcal{N}(Y) \to L_m(\mathbb{Z}\Pi(Y), \mathcal{O}_Y)$ is the surgery obstruction map from Theorem 8.167 or 9.108, is called the *splitting invariant along* Y.

In the topological category if we view $\mathcal{N}^{TOP}(X)$ as an abelian group via $+$ on G/TOP, this function is a homomorphism of abelian groups because of Theorem 17.24 (ii) and (iii).

Next we discuss the naturality properties of the splitting map appearing in Definition 17.25 under diffeomorphisms. Let $a\colon X \to X'$ be a diffeomorphism. The differential of a induces an isomorphism of infinite cyclic local coefficient systems $O_a\colon O_X \to a^*O_{X'}$. There is precisely one automorphism $u\colon O_{X'} \to O_{X'}$ of infinite cyclic local coefficient systems, namely id or $-$ id, such that the map $H_n(X, O_X) \to H_n(X'; O_{X'})$ induced by $a\colon X \to X'$ and the isomorphism of infinite local coefficient systems $a^*u \circ O_a\colon O_X \to a^*O_{X'}$ sends $[[X]]$ to $[[X']]$. We call a *orientation preserving* if u can be chosen to be the identity and *orientation reversing* otherwise.

Recall that we have already studied functoriality of the geometric surgery exact sequence in Section 11.10. In the sequel we will use the notation appearing there. The following diagram is part of Diagram (11.49), where the rows are part of the geometric surgery exact sequence and I is from (11.48):

$$\begin{array}{ccccc}
\mathcal{S}(X) & \xrightarrow{\eta_X} & \mathcal{N}(X) & \xrightarrow{\sigma_X} & L_n(\mathbb{Z}\Pi(X), O_X) \\
\mathcal{S}(I(a)) \downarrow \cong & & \mathcal{N}(I(a)) \downarrow \cong & & L_n(I(a)) \downarrow \cong \\
\mathcal{S}(X') & \xrightarrow{\eta_{X'}} & \mathcal{N}(X') & \xrightarrow{\sigma_{X'}} & L_n(\mathbb{Z}\Pi(X'), O_{X'}).
\end{array}$$

Now suppose that $Y \subseteq X$ is a closed submanifold and put $Y' = a(Y)$. Then a induces a diffeomorphism $b\colon Y \to Y'$. Note that we do *not* require compatibility between the orientation behaviour of a and b. One checks by unravelling the definitions that the following diagram commutes

$$\begin{array}{ccc}
\mathcal{N}(X) & \xrightarrow{s^N_{Y \subseteq X}} & \mathcal{N}(Y) \\
\mathcal{N}(I(a)) \downarrow \cong & & \cong \downarrow \mathcal{N}(I(b)) \\
\mathcal{N}(X') & \xrightarrow{s^N_{Y' \subseteq X'}} & \mathcal{N}(Y').
\end{array}$$

This implies that the following diagram commutes

$$\begin{array}{ccc}
\mathcal{S}(X) & \xrightarrow{s_{Y \subseteq X}} & L_n(\mathbb{Z}\Pi(Y), O_Y) \\
\mathcal{S}(I(a)) \downarrow \cong & & \cong \downarrow L_n(I(b)) \\
\mathcal{S}(X') & \xrightarrow{s_{Y' \subseteq X'}} & L_n(\mathbb{Z}\Pi(Y'), O_{Y'})
\end{array}$$

where $s_{Y \subseteq X}$ and $s_{Y' \subseteq X'}$ are introduced in Definition 17.25 and $\mathcal{S}(I(a))$ is given by composition with a.

Now suppose that X and Y are connected and $X = X'$ and $Y = Y'$ holds, in other words, a is a self-diffeomorphism of X satisfying $a(Y) = (Y)$. Then $L_n(I(b))\colon L_n(\mathbb{Z}\Pi(Y), O_Y) \xrightarrow{\cong} L_n(\mathbb{Z}\Pi(Y), O_Y)$ is $\pm$ id, where the sign is 1 if b is orientation preserving, and -1 if b is orientation reversing. Note that the sign $\pm$ is

17.7 Splitting Invariants

independent of the orientation behaviour of a. Then we get

$$\mathbf{s}_{Y \subseteq X} \circ \mathcal{S}(I(a)) = \pm \mathbf{s}_{Y \subseteq X}. \qquad (17.26)$$

Exercise 17.27 Let $p, q \geq 5$ and consider $X = S^p \times S^q$ and $Y = S^p \times \{s_q\}$ for some base point $s_q \in S^q$. In the situation as in (17.26) find a self-diffeomorphism a of $X = S^p \times S^q$ such that $\mathbf{s}_{Y \subseteq X} \circ \mathcal{S}(I(a)) = -\mathbf{s}_{Y \subseteq X}$.

An obvious immediate application of the splitting invariants is that, if the original degree one normal map $(f, \bar{f}) \colon M \to X$ represents the zero element in $\mathcal{N}(X)$, then we also must have that $0 = \mathbf{s}_{Y \subseteq X}(f, \bar{f}) = \sigma_m(g, \bar{g})$ in $L_m(\mathbb{Z}\Pi(Y), \mathcal{O}_Y)$. This is so since the assumption means that $(f, \bar{f})$ is normally cobordant to a diffeomorphism. By transversality we obtain a normal cobordism of $(g, \bar{g})$ to a diffeomorphism, and hence the surgery obstruction of $(g, \bar{g})$ is zero, and similarly for PL and TOP. In this way we obtain a useful method for detecting non-trivial elements in the set (abelian group) of normal invariants and structure sets.

Remark 17.28 (The characteristic variety theorem) The ideas about splitting invariants can be generalised to include the notion of $\mathbb{Z}/r$-manifolds. A $\mathbb{Z}/r$-manifold is a space that is obtained from a compact manifold with boundary that consists of the disjoint union of r copies of the same closed manifold, by gluing these copies together. As such, it is a singular space, but it turns out that the notion of the signature modulo $\mathbb{Z}/r$ is still well defined.

Sullivan gives a geometric flavour to Theorem 17.6 by asserting that for any PL or topological closed manifold X one can find a collection of $\mathbb{Z}/r$-manifolds (for different r's) embedded in X, called *the characteristic variety of X*, such that the splitting invariants along these $\mathbb{Z}/r$-manifolds detect the normal invariants $\mathcal{N}^{\mathrm{PL}}(X)$ or $\mathcal{N}^{\mathrm{TOP}}(X)$. We refer to [392] for more details.

Although such statements certainly hold in some cases, for example for $X = \mathbb{CP}^n$ such a characteristic variety is provided by the collection of $\mathbb{CP}^m$'s with $m \leq n$, we warn the reader that the details in the general case are somewhat uncertain. The participants of the Oberwolfach Mini-Workshop: The Hauptvermutung for High-Dimensional Manifolds in 2006, who were studying the notes of Sullivan, state in [220] that "The rigorous proof... would require the characteristic classes formulae for surgery obstructions [394], as well as the addressing of certain issues arising from the different shape of the universal coefficient theorem for real K-theory. The latter seems to be the most controversial part of the argument."

Remark 17.29 (More general splitting problems) In the setting of Proposition 17.25, the question whether for a given element represented by $(f, \bar{f})$ we have $s_Y(f, \bar{f}) = 0$ is often referred to as the question "whether we can split $(f, \bar{f})$ along Y". However, there are more general questions that can be asked in this situation, which we now informally discuss. In our setup let us denote the tubular neighbourhood of Y in X by E_Y and the closure of its complement by C_Y. Then the intersection $S_Y = E_Y \cap C_Y$ gives an $(n - m - 1)$-dimensional sphere bundle over Y and we have a decomposition $X = E_Y \cup_{S_Y} C_Y$. Using transversality one obtains via $(f, \bar{f})$ a similar decomposition $M = E_N \cup_{S_N} C_N$ in the source, so that the map f respects the decomposition. One may ask whether, given a homotopy equivalence f, is it possible to change it by a homotopy to a map that is a homotopy equivalence on all pieces? (Formally this can be phrased in terms of triads.) There is a general obstruction theory of such problems, the obstruction groups are denoted $LS_n(\Phi)$, where Φ stands for the diagram

$$\begin{array}{ccc} \pi_1(S) & \longrightarrow & \pi_1(C) \\ \downarrow & & \downarrow \\ \pi_1(Y) & \longrightarrow & \pi_1(X) \end{array}$$

and they depend only on these fundamental groups and the dimensions n and m. This theory is developed extensively in Chapters 11 and 12 of Wall's book [414], but we will not go into it.

It has many applications, some of which are contained in Chapter 14 of the same book. Here we would at least like to mention that there are interesting special cases. For example, when $m = n-1$ and $X = X_1 \cup_Y X_2$, which means we have a codimension 1 submanifold Y of X and it splits X into two codimension 0 submanifolds with boundary X_1 and X_2. A special case of this special case is when we have the connected sum decomposition $X = X_1 \sharp X_2$, and then the question of splitting is closely related to the interesting question whether any manifold homotopy equivalent to X is also a connected sum. The answer is in general no, and in this setting the obstruction groups were related to the so-called UNil-groups $\text{UNil}_*(\Phi)$ studied by Cappell [77, 78, 79]. For information about them, see e.g., the comments at the end of Chapter 11A in [414] and Chapters 6.3 and 6.4 in [93].

17.8 Milnor Manifolds and Kervaire Manifolds

Part (i) of the following theorem was promised in Theorem 11.26.

Theorem 17.30 (Simply connected surgery obstruction maps in the PL and topological category are surjective)

(i) *Let X be a simply connected finite Poincaré complex of dimension $n \geq 5$ such that the Spivak normal fibration of X admits a topological reduction. Then the surgery obstruction map*

$$\sigma_n^{\text{TOP}} : \mathcal{N}^{\text{TOP}}(X) \to L_n(\mathbb{Z})$$

is surjective;

(ii) *Let $(X, \partial X)$ be a simply connected compact topological manifold with boundary of dimension $n \geq 5$. Then the surgery obstruction map*

$$\sigma_{n+1}^{\text{TOP}} : \mathcal{N}^{\text{TOP}}(X, \partial X) \to L_{n+1}(\mathbb{Z})$$

is surjective;

17.8 Milnor Manifolds and Kervaire Manifolds

(iii) *Both statements hold also in the* PL *category.*

Proof. (i) Since $L_{2k+1}(\mathbb{Z}) = 0$ it is enough to consider the case $n = 2k$. By Theorem 8.195 we can realise any element $x \in L_{2k}(\mathbb{Z})$ as the surgery obstruction of a problem of the form

$$(f, \overline{f}) \colon (W; M_0, M_1) \to (S^{n-1} \times [0,1]; S^{n-1} \times \{0\}, S^{n-1} \times \{1\})$$

where the restriction $(f_0, \overline{f}_0) \colon M_0 \to S^{n-1} \times \{0\}$ is a homeomorphism and the restriction $(f_1, \overline{f}_1) \colon M_1 \to S^{n-1} \times \{1\}$ is a homotopy equivalence. By the generalised Poincaré conjecture from Theorem 2.5 we know that f_1 can be homotoped to a homeomorphism. So we may assume that f_1 is also a homeomorphism. Therefore we can "close" the surgery problem by adding disks to both parts of the boundary both in the source and in the target and extend the map f via a homeomorphism to these disks. Since this does not change the surgery obstruction, we realise x by a surgery problem of closed manifolds with the target S^n and get the desired statement for $X = S^n$.

For general X recall that by [408, Theorem 2.4] it is homotopy equivalent to a CW complex with precisely one cell in the top dimension n, so we may assume that $X = X_0 \cup D^n$. (This idea was already used in Section 9.3.1.) Since X has a topological reduction of the Spivak normal fibration, there exists a normal map of degree one $(g, \overline{g}) \colon N \to X$ with some surgery obstruction $y = \sigma(g, \overline{g}) \in L_n(\mathbb{Z})$. Given $x \in L_{2k}(\mathbb{Z})$, the connected sum $(f, \overline{f}) \sharp (g, \overline{g})$, where in the target we take the connected sum in the interior of the top cell $D^n \subset X$, will have surgery obstruction $x + y \in L_{2k}(\mathbb{Z})$. Since x runs through all possible values, we can realise all elements this way. When X is a simply connected closed manifold X, then $(g, \overline{g})$ can be taken to be $(\mathrm{id}_X, \mathrm{id}_{\nu_X}) \colon X \to X$.

(ii) Here we start with $(\mathrm{id}_X, \mathrm{id}_{\nu_X}) \colon (X, \partial X) \to (X, \partial X)$ and make the connected sum in the interior of X.

(iii) The proof is analogous. □

Definition 17.31 (Milnor manifolds and Kervaire manifolds) The source manifolds of the degree one normal maps constructed in the proof of Theorem 17.30 for the case $X = S^n$ are called the *Milnor manifolds* in the case $n = 4l$ and the *Kervaire manifolds* in the case $n = 4l + 2$.

An even further generalisation can be achieved. Recall from (16.10) that $L_n(\mathbb{Z})$ is a direct summand of $L_n^s(\mathbb{Z}\pi)$ for any π. Then, in the same setting as in the above theorem, except with $\pi = \pi_1(X)$, any element $x \in L_n^s(\mathbb{Z}\pi)$ in the summand $L_n(\mathbb{Z})$ can be realised as the surgery obstruction of a surgery problem with the target X or $(X, \partial X)$ respectively. The proof is by the same argument via the connected sum.

It should be stressed that, of course, the above constructions only work in the categories TOP and PL since we have the generalised Poincaré conjecture only in those categories. In the smooth category one can see that these constructions cannot work, for example by the results in Chapter 12 about the homotopy spheres.

Exercise 17.32 Let X be an n-dimensional closed TOP manifold with a collapse map $c\colon X \to S^n$. Show that under $[Y, \text{G/TOP}] \cong \mathcal{N}^{TOP}(Y)$ from Theorem 11.24 the induced map $c^*\colon [S^n, \text{G/TOP}] \to [X, \text{G/TOP}]$ corresponds to the connected sum of a surgery problem with target S^n with the identity on X.

17.9 Notes

There exist formulas for surgery obstructions in terms of the characteristic classes in the cohomology of G/TOP obtained via the maps from Theorems 17.6 and 17.15. We refer the reader to [279, Chapter 4], [312], [368], [394], [412], [413], [414, Chapter 13B] and the recent preprint [196].

Besides the space TOP/O mentioned in Section 17.5 one can also study the closely related space PL/O. By [91] it is known that it is 6-connected. See also Remark 12.72.

Chapter 18
Computations of Topological Structure Sets of some Prominent Closed Manifolds

18.1 Introduction

In this chapter we present some calculations of the (simple) structure sets of connected closed manifolds M. We concentrate on the topological category where most known result are located. In the PL category the results are quite similar due to the discussion from Section 17.4. In turn, the results in the smooth category are dramatically different. For $M = S^n$ we have explained the state of the art in Chapter 12, see in particular in Remark 12.36 where it is conjectured that the smooth structure set $\mathcal{S}(S^n)$ of S^n is trivial only for finitely many values of n whereas in the topological case we have $\mathcal{S}^{\mathrm{TOP}}(S^n) = \{[\mathrm{id}_{S^n}]\}$ for all n. For M other than S^n only very few calculations are known in the smooth category. We touch on this at appropriate places throughout this chapter. We also study the action of various groups of (simple) self-homotopy equivalences on the structure sets and thus achieve in some cases a full classification of manifolds inside a given (simple) homotopy type, see Problem 1.5.

The main tool for calculating the structure sets is the surgery exact sequence, which was presented in Theorem 11.22 in the smooth case and in Theorem 11.25 in the PL and topological cases. To effectively use the surgery exact sequence, we need to understand the other terms in it. The computation of the L-groups was discussed in Chapter 16 and the computation of the set of normal invariants in the topological category in Chapter 17. Hence we may put together these ingredients and finalise the calculation of the structure sets of some manifolds. We start with the easiest case, namely simply connected manifolds in the topological category. Then we move on to the non-simply connected cases in the topological category, which are quite different depending on whether the fundamental group is finite or not. In particular, for aspherical closed manifolds, which always have infinite and torsionfree fundamental groups, the situation turns out to be so strikingly different that a separate Chapter 19 about topological rigidity is devoted to it.

In more detail, given an n-dimensional closed topological manifold M with $n \geq 5$, we start investigating the geometric surgery exact sequence of Theorem 11.25 in the topological category:

$$\mathcal{N}^{\mathrm{TOP}}(M \times I, \partial(M \times I)) \xrightarrow{\sigma^s_{n+1}} L^s_{n+1}(\mathbb{Z}\Pi(M), \mathcal{O}_M) \xrightarrow{\rho^s_{n+1}}$$

$$\xrightarrow{\rho^s_{n+1}} \mathcal{S}^{\mathrm{TOP},s}(M) \xrightarrow{\eta^s_n} \mathcal{N}^{\mathrm{TOP}}(M) \xrightarrow{\sigma^s_n} L^s_n(\mathbb{Z}\Pi(M), \mathcal{O}_M). \quad (18.1)$$

Let us summarise what we already know about the terms and the maps:

- The sequence (18.1) is an exact sequence of abelian groups by Theorem 15.482;
- $\mathcal{N}^{\mathrm{TOP}}(M) \cong [M; \mathrm{G/TOP}]$, see Theorem 11.24 and Theorem 17.17. Moreover, the homotopy type of G/TOP is determined in Theorem 17.6 and the group itself can be calculated via (17.16);
- $\mathcal{N}^{\mathrm{TOP}}(M \times I, \partial(M \times I)) \cong [M \times I/M \times \{0,1\}, \mathrm{G/TOP}]$;
- $L_n(\mathbb{Z}) \cong \mathbb{Z}, 0, \mathbb{Z}/2, 0$ for $n \equiv 0, 1, 2, 3$, see Sections 8.5.2, 8.5.3, and 9.2.4;
- We have a lot of information about $L^s_n(\mathbb{Z}\pi, w)$ via results in Chapter 16, in particular for finite π and for torsionfree π;
- The surgery obstruction maps $\sigma_n \colon \mathcal{N}^{\mathrm{TOP}}(M) \to L_n(\mathbb{Z})$ for all $n \geq 5$ and $\sigma_{n+1} \colon \mathcal{N}^{\mathrm{TOP}}(M \times I, \partial(M \times I)) \to L_{n+1}(\mathbb{Z})$ for all $n \geq 4$ are surjective by Theorem 17.30 if M is simply connected.

We will take various concrete manifolds M and use these results to calculate $\mathcal{S}^{\mathrm{TOP},s}(M)$. In the simply connected case, or more generally whenever $\mathrm{Wh}(\pi_1(M))$ is trivial, we leave out the decoration s from the notation since in those cases there is no difference between the decorations s and h. Before we delve into this, we devote one section to a general discussion of the action of various groups of (simple) self-homotopy equivalences on the (simple) structures sets.

Guide 18.2 On the first reading it is possible to skip Sections 18.6 and 18.7 about the join construction, the transfer and the ρ-invariant and proceed to Sections 18.8 and 18.9 about calculations, returning to the two sections when these tools are needed.

18.2 Fake Spaces

Definition 18.3 (Fake spaces) Let N be a connected closed topological manifold. A *fake N* is a closed topological manifold M that is homotopy equivalent to N. A *simple fake N* is a closed topological manifold M that is simple homotopy equivalent to N.

One may also consider a closed PL manifold or a closed smooth manifold N and define a *(simple) PL fake N* or *smooth fake N* to be a closed PL manifold or closed smooth manifold that is (simple) homotopy equivalent to N.

18.2 Fake Spaces

If N is simply connected, or more generally, if the Whitehead group $\text{Wh}(\pi_1(N))$ vanishes, then fake N and simple fake N are the same.

Sometimes in the literature smooth fake N is called *homotopy N* or *exotic N*. So a smooth fake S^n in the sense of Definition 18.3 is the same as a homotopy sphere in the sense of the Introduction 12.1 of Chapter 12.

Given a connected closed manifold N in TOP, PL or DIFF one may ask for a classification of (simple) fake N-s up to homeomorphism, PL homeomorphism, or diffeomorphism, in other words to determine the (simple) manifold set $\mathcal{M}^s(N)$ in the sense of Definition 11.3, or its TOP or PL version. This boils down to investigating the quotient of the relevant structure sets of N under the operation of the relevant groups of (simple) homotopy classes of self-homotopy equivalences of N, as explained next.

Let ho-aut(N) be the group of homotopy classes of self-homotopy equivalences of N and ho-aut$^s(N)$ be the group of homotopy classes of simple self-homotopy equivalences of N. Composition yields an action of ho-aut(N) on the structure sets $\mathcal{S}^{\text{TOP},h}(N)$, $\mathcal{S}^{\text{PL},h}(N)$, and $\mathcal{S}^h(N) = \mathcal{S}^{\text{DIFF},h}(N)$ and an action of ho-aut$^s(N)$ on the structure sets $\mathcal{S}^{\text{TOP},s}(N)$, $\mathcal{S}^{\text{PL},s}(N)$, and $\mathcal{S}^s(N) = \mathcal{S}^{\text{DIFF},s}(N)$. By Exercise 11.4 we get the following identifications

$$\text{ho-aut}^s(N) \backslash \mathcal{S}^{\text{TOP},s}(N) = \{\text{homeomorphism classes of simple fake } N\text{-s}\};$$
$$\text{ho-aut}^s(N) \backslash \mathcal{S}^{\text{PL},s}(N) = \{\text{PL homeomorphism classes of simple fake } N\text{-s}\};$$
$$\text{ho-aut}^s(N) \backslash \mathcal{S}^{\text{DIFF},s}(N) = \{\text{diffeomorphism classes of simple fake } N\text{-s}\}.$$

In general it is hard to compute the (simple) structure sets and the (simple) homotopy automorphisms, and it is even harder to compute the quotients. It is already interesting to compute the (simple) structure sets in some prominent examples, which we will do below in some cases.

Example 18.4 (Fake S^n) A fake S^n, or, equivalently, a simple fake S^n, is any closed topological manifold M that is homotopy equivalent to S^n. By Theorem 2.5 any fake S^n is homeomorphic to S^n.

A smooth fake S^n is a homotopy sphere in the sense of Chapter 12. By the s-cobordism Theorem 2.1 and the definition of Θ_n from Section 12.2 we get for $n \neq 4$ a bijection

$$\{\text{oriented diffeomorphism classes of oriented smooth fake } S^n\text{-s}\} \xrightarrow{\cong} \Theta_n.$$

Note that ho-aut(S^n) is isomorphic to $\mathbb{Z}/2$ and a generator can be represented by an orientation reversing self-diffeomorphism of order two. Lemma 12.5, Lemma 12.6, and the fact that the inverse of $[\Sigma] \in \Theta_n$ is given by $[\Sigma^-]$ imply that when $n \neq 4$ there is an isomorphism of abelian groups

$$\mathcal{S}(S^n) \xrightarrow{\cong} \Theta_n.$$

compatible with the $\mathbb{Z}/2$-actions where the generator of $\mathbb{Z}/2$ acts on $\mathcal{S}(S^n)$ by the action of ho-aut$(S^n) = \mathbb{Z}/2$ and on Θ_n by $-$ id. Hence we get for $n \neq 4$ a bijection

$$\{\text{diffeomorphism classes of smooth fake } S^n\text{-s}\} \xrightarrow{\cong} \Theta_n/(\mathbb{Z}/2).$$

Exercise 18.5 How many diffeomorphism classes of smooth fake spheres are there in dimension seven?

Example 18.6 The operation of ho-aut$^s(N)$ on the various structure sets is in general complicated. We will for instance see that $\mathcal{S}^{\text{TOP}}(S^2 \times S^2)$ consists of four elements, actually is isomorphic as an abelian group to $\mathbb{Z}/2 \times \mathbb{Z}/2$, see Theorem 18.11, but ho-aut$(S^2 \times S^2) \backslash \mathcal{S}^{\text{TOP}}(S^2 \times S^2)$ is trivial, see [226, Theorem 0.13]. Its proof is based on a certain construction of self-homotopy equivalences, which can be used to show in some cases that the group of self-homotopy equivalences acts transitively on the structure set, see [226, Section 2]. So every fake $S^2 \times S^2$ is homeomorphic to $S^2 \times S^2$, but there are self-homotopy equivalences of $S^2 \times S^2$ that are not homotopy equivalent to a homeomorphism, as we have seen already in Remark 6.86.

In Chapter 19 we will study those connected closed manifolds N for which the structure set $\mathcal{S}^{\text{TOP},h}(N)$ is trivial. Provided that $\text{Wh}(\pi_1(N))$ vanishes, we have $\mathcal{S}^{\text{TOP},s}(N) = \mathcal{S}^{\text{TOP},h}(N)$ and $\mathcal{S}^{\text{TOP},h}(N)$ is trivial if and only if N is topologically rigid, i.e., if any homotopy equivalence $M \to N$ for some closed manifold M as source and N a target is homotopic to a homeomorphism, see Lemma 11.32.

The next result is due to Ranicki [340].

Theorem 18.7 (Composition formula in the topological structure set) *Consider simple homotopy equivalences $f \colon M \to N$ and $g \colon N \to N'$ of closed topological manifolds of dimension ≥ 5. Let $[f]$ and $[g \circ f]$ be the elements f and $g \circ f$ represent in the topological structure sets $\mathcal{S}^{\text{TOP},s}(N)$ and $\mathcal{S}^{\text{TOP},s}(N')$, and let $g_* \colon \mathcal{S}^{\text{TOP},s}(N) \xrightarrow{\cong} \mathcal{S}^{\text{TOP},s}(N')$ be the group isomorphism induced by g, see (15.481) and Theorem 15.482.*

Then we get in $\mathcal{S}^{\text{TOP},s}(N')$

$$[g \circ f] = [g] + g_*([f]).$$

The analogous statement is also true for $\mathcal{S}^{\text{TOP},h}$.

Exercise 18.8 Let $g \colon N \xrightarrow{\cong} N$ be a self-homeomorphism of a closed topological manifold of dimension ≥ 5, and let $f \colon M \to N$ be a simple homotopy equivalence. Show that we get in $\mathcal{S}^{\text{TOP},s}(N)$

$$[g \circ f] = g_*([f]).$$

18.3 Products of Spheres

Exercise 18.9 Let $f: M \xrightarrow{\cong} M'$ be a homeomorphism of a closed topological manifold of dimension ≥ 5, and let $g: M' \to N$ be a simple homotopy equivalence. Show that we get in $\mathcal{S}^{\text{TOP},s}(N)$

$$[g \circ f] = [g].$$

Some general results about possible cardinalities of the quotients are obtained in [111].

18.3 Products of Spheres

For a simply connected closed manifold M we will give a general expression for $\mathcal{S}^{\text{TOP}}(M)$ in Theorem 19.15. Here we want to make it more explicit for the products of two spheres and for projective spaces.

We start with the example $M = S^p \times S^q$ with $p, q \geq 2$. Obviously we have $\pi_1(S^p \times S^q) = \{1\}$ and so the surgery exact sequence for $S^p \times S^q$ becomes the short exact sequence

$$0 \to \mathcal{S}^{\text{TOP}}(S^p \times S^q) \xrightarrow{\eta_{p+q}} \mathcal{N}^{\text{TOP}}(S^p \times S^q) \xrightarrow{\sigma_{p+q}} L_{p+q}(\mathbb{Z}) \to 0. \quad (18.10)$$

The normal invariants can be calculated even without the knowledge of the homotopy type of G/TOP, only the knowledge of the homotopy groups is needed and the fact that G/TOP is a loop space. Recall that given two pointed CW-complexes X and Y we have $\Sigma(X \times Y) \simeq \Sigma X \vee \Sigma Y \vee \Sigma(X \wedge Y)$, where Σ is the reduced suspension, see [178, Proposition 4I.1]. Combining with the adjunction $[\Sigma X, Y] \cong [X, \Omega Y]$ we obtain that

$$[S^p \times S^q; \text{G/TOP}] \cong [S^p; \text{G/TOP}] \oplus [S^q; \text{G/TOP}] \oplus [S^{p+q}; \text{G/TOP}].$$

The projections on the first two summands are induced by the inclusion maps of the two factors and hence are detected by the splitting invariants along S^p and S^q, see Definition 17.25. The inclusion of the last summand into the left-hand side can be interpreted as the map that is given on a surgery problem with the target a sphere by connected sum with the identity map on the product of spheres producing a surgery problem with the target a product of spheres. Hence the last summand maps isomorphically onto $L_n(\mathbb{Z})$ by Theorem 17.30. Altogether the desired calculation is contained in the following result.

Theorem 18.11 (Topological structure sets of products of spheres) *Let $p, q \geq 2$. Then we have the isomorphism of abelian groups*

$$(\mathbf{s}_p, \mathbf{s}_q) \colon \mathcal{S}^{\text{TOP}}(S^p \times S^q) \xrightarrow{\cong} L_p(\mathbb{Z}) \oplus L_q(\mathbb{Z})$$

where $\mathbf{s}_p$ and $\mathbf{s}_q$ denote the splitting invariants along S^p and S^q respectively, in the sense of Definition 17.25.

In particular, if both p and q are odd, the structure set is zero, but if at least one of them is even, then we have a non-trivial structure set. Explicitly, for $p, q \geq 2$ we have

$$\mathcal{S}^{\mathrm{TOP}}(S^p \times S^q) \cong \begin{cases} 0 & \text{if } p, q \text{ odd}; \\ \mathbb{Z} \oplus \mathbb{Z} & \text{if } p, q \equiv 0 \bmod 4; \\ \mathbb{Z} & \text{if } p \equiv 0 \bmod 4 \text{ and } q \text{ odd}; \\ \mathbb{Z} & \text{if } p \text{ odd and } q \equiv 0 \bmod 4; \\ \mathbb{Z}/2 \oplus \mathbb{Z}/2 & \text{if } p, q \equiv 2 \bmod 4; \\ \mathbb{Z}/2 & \text{if } p \equiv 2 \bmod 4 \text{ and } q \text{ odd}; \\ \mathbb{Z}/2 & \text{if } p \text{ odd and } q \equiv 2 \bmod 4; \\ \mathbb{Z} \oplus \mathbb{Z}/2 & \text{if } p \equiv 0 \bmod 4 \text{ and } q \equiv 2 \bmod 4; \\ \mathbb{Z}/2 \oplus \mathbb{Z} & \text{if } p \equiv 2 \bmod 4 \text{ and } q \equiv 0 \bmod 4. \end{cases}$$

We may also ask about the realisation problem: how do we construct non-trivial elements in $\mathcal{S}^{\mathrm{TOP}}(S^p \times S^q)$? We offer two examples.

Given $p \geq 5$ and an element $\omega \in L_p(\mathbb{Z})$, there exists by Theorem 17.30 a surgery problem $(f_\omega, \overline{f}_\omega) \colon M(\omega) \to S^p$ with the surgery obstruction $\sigma_p(f_\omega, \overline{f}_\omega) = \omega$. By the product formula from Theorem 15.466 the surgery obstruction of the degree one normal map $(f_\omega, \overline{f}_\omega) \times \mathrm{id}_{S^q} \colon M(\omega) \times S^q \to S^p \times S^q$ vanishes. Hence it is normally cobordant to a homotopy equivalence and this homotopy equivalence represents an element in $\mathcal{S}^{\mathrm{TOP}}(S^p \times S^q)$. By the transversality arguments as in the proof of Definition 17.25 its splitting invariant along S^p is ω.

Alternatively, in [348, Example 20.4] it is explained how under certain assumptions on p and q one may realise an element in $\mathcal{S}^{\mathrm{TOP}}(S^p \times S^q)$ with a prescribed splitting invariant along S^p by a homotopy equivalence from the sphere bundle $S(\eta)$ of a topological bundle $\eta \colon S^p \to B\mathrm{TOP}$ whose associated spherical fibration is trivial.

Note that in Theorem 18.11 we can allow the case $p = 2$ and $q = 2$ because of Remark 8.30.

The calculation of $\mathcal{S}^{\mathrm{PL}}(S^p \times S^q)$ is similar for $p + q \geq 5$, see [350, page 21]. The forgetful map $\mathcal{S}^{\mathrm{PL}}(S^p \times S^q) \to \mathcal{S}^{\mathrm{TOP}}(S^p \times S^q)$ is injective, but when $p = 4$ and $q \geq 2$ is arbitrary (or vice versa) it is not surjective and one obtains topological manifolds that do not admit PL structure, see [350, page 22].

The calculation of $\mathcal{S}^{\mathrm{DIFF}}(S^p \times S^q)$ for $p, q \geq 2$ and $p + q \geq 5$ is described in [108, Theorem 1.5, 1.6]. Its input consists of information about bP_n, Θ_n from Chapter 12 and about the classifying spaces G/O and G/TOP. Example 13.26 of [352] contains an incorrect result, see Remark 1.8 of [108].

Exercise 18.12 Compute $\mathcal{S}^{\mathrm{TOP}}(S^p \times D^q, S^p \times S^{q-1})$ for $p, q \geq 2$, $p + q \geq 5$.

18.4 Complex Projective Spaces

The complex projective space $\mathbb{CP}^d$ is defined to be the quotient of the diagonal S^1-action on $S^{2d+1} = S(\mathbb{C}^{d+1})$. As a real manifold it has dimension $2d$ and $\pi_1(\mathbb{CP}^d) = \{1\}$. Hence the surgery exact sequence for $\mathbb{CP}^d$ becomes the short exact sequence

$$0 \to \mathcal{S}^{\mathrm{TOP}}(\mathbb{CP}^d) \xrightarrow{\eta_{2d}} \mathcal{N}^{\mathrm{TOP}}(\mathbb{CP}^d) \xrightarrow{\sigma_{2d}} L_{2d}(\mathbb{Z}) \to 0. \tag{18.13}$$

To identify the normal invariants, observe, following [414, page 202], that the cofibration sequence $\mathbb{CP}^{d-1} \to \mathbb{CP}^d \to S^{2d}$ induces the short exact sequence of abelian groups

$$\pi_{2d}(G/\mathrm{TOP}) \to [\mathbb{CP}^d, G/\mathrm{TOP}] \to [\mathbb{CP}^{d-1}, G/\mathrm{TOP}] \to \pi_{2d-1}(G/\mathrm{TOP}) = 0.$$

This sequence splits since the surgery obstruction map $\sigma_{2d}^{\mathrm{TOP}} : \mathcal{N}^{\mathrm{TOP}}(S^{2d}) = \pi_{2d}(G/\mathrm{TOP}) \to L_{2d}(\mathbb{Z})$ is bijective by (17.1) and is the composite of the surgery obstruction map $\sigma_{2d}^{\mathrm{TOP}} : \mathcal{N}^{\mathrm{TOP}}(\mathbb{CP}^d) = [\mathbb{CP}^{d-1}, G/\mathrm{TOP}] \to L_{2d}(\mathbb{Z})$ with $\pi_{2d}(G/\mathrm{TOP}) \to [\mathbb{CP}^d, G/\mathrm{TOP}]$ by Exercise 17.32. By induction we obtain a calculation in terms of the splitting invariants from Definition 17.25:

$$\bigoplus_{0 < i \leq d} \mathbf{s}_{2i} : \mathcal{N}^{\mathrm{TOP}}(\mathbb{CP}^d) \xrightarrow{\cong} \bigoplus_{0 < i \leq d} L_{2i}(\mathbb{Z}) \cong \bigoplus_{i=1}^{\overline{c}_4(d)} \mathbb{Z} \oplus \bigoplus_{i=1}^{\overline{c}_2(d)} \mathbb{Z}/2 \tag{18.14}$$

where we are abbreviating $\mathbf{s}_{2j} = \mathbf{s}_{\mathbb{CP}^j \subseteq \mathbb{CP}^d}$ and where $\overline{c}_4(d) = e - 1$ and $\overline{c}_2(d) = e$ if $d = 2e - 1$, and $\overline{c}_4(d) = e$ and $\overline{c}_2(d) = e$ if $d = 2e$. The surgery obstruction map $\sigma_{2d}^{\mathrm{TOP}}$ takes the summand of $\mathcal{N}^{\mathrm{TOP}}(\mathbb{CP}^d)$ in the top dimension isomorphically onto $L_{2d}(\mathbb{Z})$ by Theorem 17.30. Hence the short exact sequence (18.13) splits and we obtain the calculation $\mathcal{S}^{\mathrm{TOP}}(\mathbb{CP}^d)$ in terms of the splitting invariants $\mathbf{s}_{2i}$ as follows.

Theorem 18.15 (Topological structure sets of complex projective spaces) *Let $d \geq 2$. Then we have the isomorphism of abelian groups*

$$\bigoplus_{0 < i < d} \mathbf{s}_{2i} : \mathcal{S}^{\mathrm{TOP}}(\mathbb{CP}^d) \xrightarrow{\cong} \bigoplus_{0 < i < d} L_{2i}(\mathbb{Z}) \cong \bigoplus_{i=1}^{c_4(d)} \mathbb{Z} \oplus \bigoplus_{i=1}^{c_2(d)} \mathbb{Z}/2$$

where $c_4(d) = e - 1$ and $c_2(d) = e - 1$ if $d = 2e - 1$, and $c_4(d) = e - 1$ and $c_2(d) = e$ if $d = 2e$.

Realisation of elements in the structure set of $\mathcal{S}^{\mathrm{TOP}}(\mathbb{CP}^d)$ is explained in detail in [279, Chapter 8.C] where it is also discussed in Theorem 8.31 on page 173 that they provide some generators of the torsionfree part of the topological cobordism ring Ω_*^{TOP}.

Exercise 18.16 Consider the map ho-aut($\mathbb{CP}^d$) → {±1} sending the class of a self-homotopy equivalence $f \colon \mathbb{CP}^d \to \mathbb{CP}^d$ to ±1 if the automorphism

$$H^2(f) \colon H^2(\mathbb{CP}^d) \xrightarrow{\cong} H^2(\mathbb{CP}^d)$$

of the infinite cyclic group $H^2(\mathbb{CP}^d)$ is ± id. Show that it is an isomorphism of groups.

Remark 18.17 The action of ho-aut($\mathbb{CP}^d$) on $\mathcal{S}^{\mathrm{TOP}}(\mathbb{CP}^d)$ is trivial by the following reasoning. Observe first that the non-trivial element $a \colon \mathbb{CP}^d \to \mathbb{CP}^d$ in ho-aut($\mathbb{CP}^d$) is given by a homeomorphism induced by the complex conjugation. It leaves each $\mathbb{CP}^i$ invariant and its restriction $a|_{\mathbb{CP}^i}$ is orientation reversing if i is odd, and is orientation preserving if i is even.

By Theorem 18.15 we need to investigate the effect of the composition of a homotopy equivalence $h \colon Q \to \mathbb{CP}^d$ with the homeomorphism $a \colon \mathbb{CP}^d \to \mathbb{CP}^d$ on the splitting invariants $\mathbf{s}_{2i}$. We have discussed the effect of post-composition by a self-homeomorphism on splitting invariants in general in Section 17.7 in the text following Definition 17.25 culminating in formula (17.26). This formula implies that $\mathbf{s}_{4i}([h]) = \mathbf{s}_{4i}([a \circ h]) \in L_{4i}(\mathbb{Z}) \cong \mathbb{Z}$ and that $\mathbf{s}_{4i-2}([h]) = -\mathbf{s}_{4i-2}([a \circ h]) = \mathbf{s}_{4i-2}([a \circ h]) \in L_{4i-2}(\mathbb{Z}) \cong \mathbb{Z}/2$.

An incorrect statement about this action was given in [111, Example 3.8]. The present remark repairs this mistake.

The calculation of $\mathcal{S}^{\mathrm{PL}}(\mathbb{CP}^d)$ is similar, see [414, 14C]. The forgetful map $\mathcal{S}^{\mathrm{PL}}(\mathbb{CP}^d) \to \mathcal{S}^{\mathrm{TOP}}(\mathbb{CP}^d)$ is injective, but not surjective, see Example 15.7 in [219, page 331]. This yields topological manifolds that do not admit a PL structure.

18.5 Quaternionic Projective Spaces

The quaternionic projective space $\mathbb{HP}^d$ is defined to be the quotient of the diagonal S^3-action on $S^{4d+3} = S(\mathbb{H}^{d+1})$. As a real manifold it has dimension $4d$ and $\pi_1(\mathbb{HP}^d) = \{1\}$. Hence we have that the surgery exact sequence for $\mathbb{HP}^d$ becomes the short exact sequence

$$0 \to \mathcal{S}^{\mathrm{TOP}}(\mathbb{HP}^d) \xrightarrow{\eta_{4d}} \mathcal{N}^{\mathrm{TOP}}(\mathbb{HP}^d) \xrightarrow{\sigma_{4d}} L_{4d}(\mathbb{Z}) \to 0. \tag{18.18}$$

Similarly, as in the previous section, there exists the cofibration sequence $\mathbb{HP}^{d-1} \to \mathbb{HP}^d \to S^{4d}$ inducing a split exact sequence of abelian groups, which by induction identifies normal invariants as

$$\bigoplus_{0 < i \le d} \mathbf{s}_{4i} \colon \mathcal{N}^{\mathrm{TOP}}(\mathbb{HP}^d) \xrightarrow{\cong} \bigoplus_{0 < i \le d} L_{4i}(\mathbb{Z}) \cong \bigoplus_{0 < i \le d} \mathbb{Z} \tag{18.19}$$

where $\mathbf{s}_{4i} = \mathbf{s}_{\mathbb{HP}^i \subseteq \mathbb{HP}^d}$ in the sense of Definition 17.25.

18.6 The Join and the Transfer

The surgery obstruction map σ_{4d}^{TOP} takes the summand of $\mathcal{N}^{TOP}(\mathbb{HP}^d)$ in the top dimension isomorphically onto $L_{4d}(\mathbb{Z})$ by the same argument as in the previous section. Hence the short exact sequence (18.13) splits, and we obtain the bijection of $\mathcal{S}^{TOP}(\mathbb{HP}^d)$ given by the splitting invariants s_{4i} as follows.

Theorem 18.20 (Topological structure sets of quaternionic projective spaces) *Let $d \geq 1$. Then we have the isomorphism of abelian groups*

$$\bigoplus_{0<i<d} s_{4i} : \mathcal{S}^{TOP}(\mathbb{HP}^d) \xrightarrow{\cong} \bigoplus_{0<i<d} L_{4i}(\mathbb{Z}) \cong \bigoplus_{0<i<d} \mathbb{Z}.$$

The set ho-aut($\mathbb{HP}^d$) seems to be unknown for $d \geq 4$, see [166, page 90]. If $d \geq 2$, then the induced map ho-aut($\mathbb{HP}^d$) $\to$ aut($H^*(\mathbb{HP}^d)$) is trivial by [203, Proposition on page 32], but it is not clear to us what this implies about the action of ho-aut($\mathbb{HP}^d$) on $\mathcal{S}^{TOP}(\mathbb{HP}^d)$.

18.6 The Join and the Transfer

Before proceeding to the next examples in Sections 18.8 and 18.9, we recall two structural properties of the surgery setup, which are useful when studying the structure sets of manifolds whose universal covers are spheres. The first is the join construction, which provides us with maps between structure sets of such manifolds of different dimensions. The second, the transfer maps, can in fact be defined for any fibre bundle.

We have introduced generalised lens spaces in Subsection 3.5.2 and classified them in Theorem 3.49 up to homotopy equivalence and in Theorem 3.61 up to simple homotopy equivalence using Reidemeister torsion. Note that Lemma 3.53 implies that a generalised lens space is the same as a fake lens space.

Notation 18.21 In Section 3.5 we denoted lens spaces by $L(V)$ and generalised lens spaces by $L(\alpha)$. We sometime abbreviate this to L or add decorations $L_G^n(\alpha)$ where n refers to the dimension of the (generalised) lens space and G to the underlying finite cyclic group. We may also denote by L a (fake) complex projective space.

We recall and extend the *join construction* from (3.48), see also [414, Chapter 14A]. Let G be a group, which will be in our case a finite cyclic group or the Lie group S^1, acting freely on the spheres S^m and S^n. Then the two actions extend to the join $S^{m+n+1} \cong S^m * S^n$ and the resulting action remains free. When we are given two lens spaces, generalised lens spaces, complex projective spaces, or fake complex projective spaces L and L', we can pass to the universal coverings, form the join, and then pass to the quotient again. The resulting space is again a lens space, generalised lens space, complex projective space, or fake complex projective space respectively. It will be denoted by $L * L'$ and called the *join*.

Let G be a finite cyclic group. Fix a 1-dimensional lens space $L_0 = L(V_0)$ for some 1-dimensional complex G-representation V_0. In the special case $L' = L_0$, the join operation $L \mapsto L * L_0 =: \Sigma L$ is also called the *suspension*. It induces for any lens space L a map

$$\Sigma \colon \mathcal{S}^{\mathrm{TOP},s}(L) \to \mathcal{S}^{\mathrm{TOP},s}(\Sigma L) \tag{18.22}$$

as follows. Consider a simple homotopy equivalence $f \colon M \to L$ representing a class $[f]$ in $\mathcal{S}^{\mathrm{TOP},s}(L)$. Then we can choose a homeomorphism $h \colon L(\alpha) \xrightarrow{\cong} M$ for some generalised lens space $L(\alpha)$ by Lemma 3.53 (ii). We obtain a simple homotopy equivalence $(f \circ h) * \mathrm{id}_{L_0} \colon L(\alpha) * L_0 \to L * L_0 = \Sigma L$. Its class $[(f \circ h) * \mathrm{id}_{L_0}]$ in $\mathcal{S}^{\mathrm{TOP},s}(\Sigma L)$, which is independent of the choice of h and $f \in [f]$, will be defined to be the image of $[f]$ under the map (18.22). A similar map can be obtained for the decoration h.

There is a version of the suspension map for real projective spaces, which only raises the dimension by one. Namely, starting with a fake real projective space Q^d, pass to S^d with the free $\mathbb{Z}/2$-action and consider $S^{d+1} = S^d * S^0$ with the free antipodal action on S^0. These actions extend to a free action on S^{d+1} with the quotient denoted by ΣQ. For $Q = \mathbb{R}\mathbb{P}^d$ we get $\mathbb{R}\mathbb{P}^{d+1} = \Sigma \mathbb{R}\mathbb{P}^d$. Analogously to the construction of the map (18.22), we can define a map

$$\Sigma \colon \mathcal{S}^{\mathrm{TOP}}(\mathbb{R}\mathbb{P}^d) \to \mathcal{S}^{\mathrm{TOP}}(\mathbb{R}\mathbb{P}^{d+1}). \tag{18.23}$$

Proposition 18.24 *The suspension maps* (18.23) *and* (18.22) *are homomorphisms of abelian groups.*

The proof is beyond our scope. It is discussed in [274, Section 3.3] using the fact that certain transfer maps, see (18.27) below, are homomorphisms, which is a result thoroughly investigated in [320].

The suspension maps are related to the splitting invariants from Section 17.7 as follows. The normal splitting along a submanifold from Theorem 17.24, applied to the inclusion $L \subset \Sigma L$, induces a homomorphism of abelian groups on the normal invariants

$$\mathbf{s}^N_{L \subset \Sigma L} \colon \mathcal{N}^{\mathrm{TOP}}(\Sigma L) \to \mathcal{N}^{\mathrm{TOP}}(L). \tag{18.25}$$

It follows from the constructions that we have a commutative diagram of abelian groups

$$\begin{array}{ccc} \mathcal{S}^{\mathrm{TOP},s}(L) & \xrightarrow{\eta^s_L} & \mathcal{N}^{\mathrm{TOP}}(L) \\ \Sigma \downarrow & & \uparrow \mathbf{s}^N_{L \subset \Sigma L} \\ \mathcal{S}^{\mathrm{TOP},s}(\Sigma L) & \xrightarrow{\eta^s_{\Sigma L}} & \mathcal{N}^{\mathrm{TOP}}(\Sigma L). \end{array} \tag{18.26}$$

See [414, Lemma 14A.3 on page 197] for details where Wall works in the PL category, but later states on pages 209–210 that everything works also in the topological category. There is a similar diagram of abelian groups associated to the inclusion $\mathbb{R}\mathbb{P}^d \subset \mathbb{R}\mathbb{P}^{d+1}$, see also [247, page 44].

We briefly discuss transfer maps. Let B be a closed manifold and let $F \to E \xrightarrow{p} B$ be a locally trivial fibre bundle with fibre a (possibly disconnected) manifold F. (The case when F is a discrete set, or, equivalently, when $p\colon E \to B$ is a covering space, is included.) Utilising pullback, we get transfer maps

$$p^!\colon \mathcal{S}^{\mathrm{TOP},s}(B) \to \mathcal{S}^{\mathrm{TOP},s}(E) \tag{18.27}$$

and

$$p^!\colon \mathcal{N}^{\mathrm{TOP}}(B) \to \mathcal{N}^{\mathrm{TOP}}(E) \tag{18.28}$$

by the following constructions.

Given a (simple) homotopy equivalence $h\colon M \xrightarrow{\simeq_s} X$ the pullback yields $h^*p\colon N := h^*E \to E$, which is a simple homotopy equivalence, see for instance [7].

Given a degree one normal map $(f, \overline{f})\colon M \to B$ the pullback construction yields $f^*p\colon N := f^*E \to E$, which is a degree one map by a spectral sequence argument. For the bundle data we need some preparation. Recall that the vertical tangent bundle of the total space E is defined as $T_v E = \ker(dp\colon TE \to TB)$, where dp is the differential of p. Then we have $TE \cong p^*TB \oplus T_v E$ and a bundle map $f_v\colon T_v N \to T_v E$. The required bundle map is $p^*(\overline{f}) \oplus f_v\colon TN \cong p_N^*TM \oplus T_v N \to p^*\xi \oplus T_v E$. Moreover, recall that by Definition 7.13 the normal invariants also include certain local coefficients systems and compatibility of intrinsic fundamental classes. These are obtained by the pullback construction as well, we leave out the details.

Both transfer maps (18.27) and (18.28) are homomorphisms of abelian groups by Theorem 5.1.21 and the following discussion on page 81 of [320] and the maps from (18.27) and (18.28) are compatible with the maps $\eta\colon \mathcal{S}^{\mathrm{TOP}} \to \mathcal{N}^{\mathrm{TOP}}$. More information on transfer maps and their properties can be obtained from [248, 249, 264, 265, 320], and [348, § 21].

18.7 The ρ-Invariant

We review the definition of the ρ-invariant for odd-dimensional manifolds and some of its properties from [15, § 7] and [414, Theorem 13B.22 on page 182]. Later, when studying the surgery exact sequence for real projective spaces and lens spaces, we will be able to calculate the normal invariants and L-groups, but we will be left with an extension problem. The ρ-invariant will provide us with a map from the structure set to a certain abelian group coming from the representation theory of the fundamental group. Information about this map is used to solve that extension problem.

Notation 18.29 The complex representation ring is denoted by $R_{\mathbb{C}}(G)$. Let $\mathbb{Q}R_{\mathbb{C}}(G)$ be the $\mathbb{Q}$-algebra $\mathbb{Q} \otimes_{\mathbb{Z}} R_{\mathbb{C}}(G)$. Let reg in $R_{\mathbb{C}}(G)$ and reg in $\mathbb{Q}R_{\mathbb{C}}(G)$ be the classes given by the regular representation $\mathbb{C}G$. Let $\overline{R_{\mathbb{C}}(G)} = R_{\mathbb{C}}(G)/(\mathrm{reg})$ and $\overline{\mathbb{Q}R_{\mathbb{C}}(G)} = (\mathbb{Q} \otimes_{\mathbb{Z}} R_{\mathbb{C}}(G))/(\mathrm{reg})$ be the rings obtained from $R_{\mathbb{C}}(G)$ and $\mathbb{Q}R_{\mathbb{C}}(G)$ by dividing out the ideal (reg) generated by reg.

We have the involution $*\colon R_{\mathbb{C}}(G) \to R_{\mathbb{C}}(G)$ sending $[V]$ to $[V']$ for the class of contragredient representation V' associated to V. It induces involutions $*\colon \mathbb{Q}R_{\mathbb{C}}(G) \to \mathbb{Q}R_{\mathbb{C}}(G)$ and $*\colon \overline{\mathbb{Q}R_{\mathbb{C}}(G)} \to \overline{\mathbb{Q}R_{\mathbb{C}}(G)}$. Let $R_{\mathbb{C}}(G)^{\pm 1}$, $\mathbb{Q}R_{\mathbb{C}}(G)^{\pm 1}$, and $\overline{\mathbb{Q}R_{\mathbb{C}}(G)}^{\pm 1}$ be the eigenspaces of ± 1 for these involutions.

Let G be a compact Lie group acting orientation preserving and smoothly on a smooth compact oriented manifold Y^{2d}, possibly with non-empty boundary ∂Y. Let $[Y, \partial Y] \in H_{2d}(Y, \partial Y)$ be its fundamental class. The $(-1)^d$-symmetric $\mathbb{R}$-bilinear intersection pairing is defined by

$$I^{2d}\colon H^d(Y, \partial Y; \mathbb{R}) \times H^d(Y, \partial Y; \mathbb{R}) \to \mathbb{R}, \quad (x, y) \mapsto \langle x \cup y, [Y, \partial Y]_{\mathbb{R}} \rangle,$$

where $[Y, \partial Y]_{\mathbb{R}} \in H_{2d}(Y, \partial Y; \mathbb{R})$ is the image of $[Y, \partial Y]$ under the obvious change of coefficients map $H_{2d}(Y, \partial Y; \mathbb{Z}) \to H_{2d}(Y, \partial Y; \mathbb{R})$. The pairing I^{2d} is not in general non-degenerate since its null space is the kernel of the map $j_Y\colon H^d(Y, \partial Y; \mathbb{R}) \to H^d(Y; \mathbb{R})$ induced by the inclusion $Y \to (Y, \partial Y)$. It follows that I^{2d} induces a non-degenerate $(-1)^d$-symmetric bilinear pairing $\overline{I^{2d}}$ on the image of j_Y, which is compatible with the G-actions. As explained in Section 16.3, such a form $\overline{I^{2d}}$ yields an element denoted by G-sign$(Y) \in R_{\mathbb{C}}^{(-1)^d}(G)$. In terms of characters we obtain a real character if d is even and a purely imaginary character if d is odd. We will denote the character as the function G-sign$(-, Y)\colon g \in G \mapsto G\text{-sign}(g, Y) \in \mathbb{C}$.

The (cohomological version of the) G-Signature Theorem due to Atiyah–Singer [15, Theorem (6.12)] tells us that, if Y is closed, then for all $g \in G$ there is an equation

$$G\text{-sign}(g, Y) = L(g, Y) \in \mathbb{C} \tag{18.30}$$

where $L(g, Y)$ is a certain invariant defined in [15, Theorem (6.12)], which vanishes if $Y^{\langle g \rangle}$ is empty and is the Hirzebruch L-genus for $g = e$.

Equation (18.30) was generalised by Wall [414, Chapter 14B] to topological semifree tame actions on topological manifolds, which is the case we will need here. An action is *semifree* if there is a fixed point set $F = Y^{\langle g \rangle}$ for all $g \in G$, so the action on $Y \setminus F$ is free. A semifree action is called *tame* if F is a manifold, there is a G-vector bundle N over F equivariantly homeomorphic to a neighbourhood of F in Y and if, in addition, $(Y \setminus F)/G$ is a manifold. Of course, any free action is semifree tame.

Note that $L(g, Y)$ is still defined if ∂Y is not empty, but (18.30) holds only for $\partial Y = \emptyset$. The assumption that Y is closed is essential in the validity of (18.30) and motivates the definition of the ρ-invariant. In fact, Atiyah and Singer provide two definitions. For the first one, which is in the setting of finite groups G, the result of Conner and Floyd from [99] (and from [436, 279] in the PL category and in the topological category) is also needed. It states that for an odd-dimensional oriented closed G-manifold X there always exists a $k \in \mathbb{N}$ and an oriented compact G-manifold with boundary $(Y, \partial Y)$ such that $\partial Y = k \cdot X$ holds, where $k \cdot X$ denotes the disjoint union of k copies of X. Let

$$\overline{G\text{-sign}(Y)} \in \overline{\mathbb{Q}R_{\mathbb{C}}(G)}^{(-1)^d}$$

18.7 The ρ-Invariant

be the image of G-sign(Y) under the obvious map $R_{\mathbb{C}}(G)^{(-1)^d} \to \overline{\mathbb{Q}R_{\mathbb{C}}(G)}^{(-1)^d}$. The following definition is taken from [15, Remark, page 590].

Definition 18.31 (The ρ-invariant for finite groups) Let G be a finite group and let X be a $(2d-1)$-dimensional closed oriented G-manifold. Define

$$\rho_G(X) = \frac{1}{k} \cdot \overline{G\text{-sign}(Y)} \in \overline{\mathbb{Q}R_{\mathbb{C}}(G)}^{(-1)^d}$$

for some $k \in \mathbb{N}$ and $(Y, \partial Y)$ a $2d$-dimensional oriented compact G-manifold with boundary and $\partial Y = k \cdot X$.

The ρ-invariant from Definition 18.31 is well defined by the following arguments. Given any two choices of $(Y, \partial Y)$ and $(Y', \partial Y')$ with $\partial Y = k \cdot X$ and $\partial Y' = k' \cdot X$, one can consider the closed manifold $(k' \cdot Y) \cup_{(kk') \cdot X} (k \cdot Y')^-$, where the superscript $^-$ denotes the opposite orientation. By the additivity property for the G-signature from [15, Proposition (7.1)], the G-signature of this closed manifold is the difference of the G-signatures of $k' \cdot Y$ and $k \cdot Y'$ and by the Atiyah–Singer G-index theorem [15, Theorem (6.12)] this is a multiple of the regular representation.

The ρ-invariant can be defined in more general settings. Namely there is a definition for compact Lie groups, see [15, Theorem (7.4)], and other versions appear in [110, 118, 426], but we will not go into these here.

The ρ-invariant is an h-cobordism invariant by [15, Corollary 7.5] and therefore also an s-cobordism and homeomorphism invariant. For an oriented connected closed $(2d-1)$-dimensional manifold X with finite fundamental group $\pi = \pi_1(X)$ it defines a function

$$\widetilde{\rho}_\pi \colon \mathcal{S}^{\text{TOP},s}(X) \to \overline{\mathbb{Q}R_{\mathbb{C}}(\pi)}^{(-1)^d}; \qquad (18.32)$$
$$[h \colon M \xrightarrow{\simeq_s} X] \mapsto \rho_\pi(\widetilde{M}) - \rho_\pi(\widetilde{X}),$$

where we use the simple homotopy equivalence h to identify the fundamental group of M with π and to orient M compatibly with the orientation of X and $\widetilde{M}$ and $\widetilde{X}$ are the universal coverings, which are oriented closed π-manifolds.

Although it is not immediately clear from the definitions, it is true that if we put on $\mathcal{S}^{\text{TOP},s}(X)$ the abelian group structure obtained via Theorem 15.482, the map $\widetilde{\rho}_\pi$ is a homomorphism by [110]. We also have the following proposition.

Proposition 18.33 *For a $(2d-1)$-dimensional oriented connected closed manifold X with finite fundamental group $\pi = \pi_1(X)$ and $d \geq 3$, there is a commutative diagram of abelian groups*

$$\begin{array}{ccc} L_{2d}^s(\mathbb{Z}\pi) & \xrightarrow{\rho_{2d}^s} & \mathcal{S}^{\text{TOP},s}(X) \\ \downarrow{\pi\text{-sign}} & & \downarrow{\widetilde{\rho}_\pi} \\ R_{\mathbb{C}}(\pi)^{(-1)^d} & \longrightarrow & \overline{\mathbb{Q}R_{\mathbb{C}}(\pi)}^{(-1)^d} \end{array}$$

Proof. See [334, Theorem 2.3]. Informally, the main idea is to use the Realisation Theorem 8.195 to represent an element of the L-group by a surgery problem of manifolds with boundary with the source some cobordism W and with the target $X \times I$ where on one end we have a homeomorphism and on the other end a simple homotopy equivalence with a source some M^{2d-1} simple homotopy equivalent to X. The ρ-invariant of M is obtained from some $(2d)$-dimensional compact manifold N with boundary several copies of M, while the ρ-invariant of X is obtained from some $(2d)$-dimensional compact manifold Y with boundary several copies of X. The cobordism W has ends M and X, so we may glue to N several copies of W to obtain a manifold with boundary several copies of X, and hence the glued manifold is a possible choice for Y. The desired commutativity is then obtained by the additivity of the G-signature upon gluing along components of boundary. □

Corollary 18.34 *For a $(2d-1)$-dimensional oriented connected closed manifold X with finite cyclic fundamental group $\pi = \pi_1(X)$ and $d \geq 3$, the restriction*

$$\widetilde{L}_{2d}^s(\mathbb{Z}\pi) \xrightarrow{\rho_{2d}^s} \mathcal{S}^{\mathrm{TOP},s}(X)$$

of the map ρ_{2d}^s from the surgery exact sequence of X in the situation of Proposition 18.33 is injective. Here $\widetilde{L}_{2d}^s(\mathbb{Z}\pi)$ denotes the reduced L-group from (16.9).

Proof. We know from Theorem 17.30 (ii) that the summand $L_{2d}(\mathbb{Z})$ of $L_{2d}^s(\mathbb{Z}\pi)$ is hit by the surgery obstruction from the left. We get from Theorem 16.16 (ii) isomorphisms of finitely generated free abelian groups

$$\widetilde{L}_{2d}^s(\mathbb{Z}\pi) \xrightarrow{\cong} \begin{cases} \{4 \cdot ([V]+[V']) \mid [V] \in R_{\mathbb{C}}(\pi)\}/\{8l \cdot \mathrm{reg} \mid l \in \mathbb{Z}\}, & \text{if } d \text{ is even;} \\ \{4 \cdot ([V]-[V']) \mid [V] \in R_{\mathbb{C}}(\pi)\}, & \text{if } d \text{ odd.} \end{cases}$$
(18.35)

Thus the commutative diagram in Proposition 18.33 yields the commutative diagrams with an isomorphism as left vertical arrow

$$\begin{array}{ccc} L_{2d}^s(\mathbb{Z}\pi) & \xrightarrow{\rho_{2d}^s} & \mathcal{S}^{\mathrm{TOP},s}(X) \\ {\scriptstyle \cong}\Big\downarrow{\scriptstyle \pi\text{-sign}} & & \Big\downarrow{\widetilde{\rho}_\pi} \\ \{4 \cdot ([V]+[V']) \mid [V] \in R_{\mathbb{C}}(\pi)\}/\{8l \cdot \mathrm{reg} \mid l \in \mathbb{Z}\} & \longrightarrow & \overline{\mathbb{Q}R_{\mathbb{C}}(\pi)}^{(+1)} \end{array}$$

if d is even, and

$$\begin{array}{ccc} L_{2d}^s(\mathbb{Z}\pi) & \xrightarrow{\rho_{2d}^s} & \mathcal{S}^{\mathrm{TOP},s}(X) \\ {\scriptstyle \cong}\Big\downarrow{\scriptstyle \pi\text{-sign}} & & \Big\downarrow{\widetilde{\rho}_\pi} \\ \{4 \cdot ([V]-[V']) \mid [V] \in R_{\mathbb{C}}(\pi)\} & \longrightarrow & \overline{\mathbb{Q}R_{\mathbb{C}}(\pi)}^{(-1)} \end{array}$$

18.7 The ρ-Invariant

if d is odd. Since $*(\text{reg}) = \text{reg}$, the lower horizontal arrow is in both cases injective. Hence the upper horizontal arrow is injective in both cases. □

Remark 18.36 (The ρ-invariant is natural in G) It follows from the definitions that the ρ-invariant is natural with respect to inclusions $H < G$.

We are also interested in the ρ-invariant of the join $L * L'$ of (generalised) lens spaces from the previous section. For this observe first that the multiplication in $R_{\mathbb{C}}(G)$ defines a pairing

$$\overline{\mathbb{Q}R_{\mathbb{C}}(G)}^i \otimes \overline{\mathbb{Q}R_{\mathbb{C}}(G)}^j \to \overline{\mathbb{Q}R_{\mathbb{C}}(G)}^{ij} \tag{18.37}$$

for $i, j \in \{\pm 1\}$. We have [414, Theorem 14A.1 on page 196]

$$\rho_G(L * L') = \rho_G(L) \cdot \rho_G(L') \tag{18.38}$$

where the dot $\cdot$ on the right-hand side is the paring from (18.37).

Let G be a finite cyclic group. We denote and define the *Pontrjagin dual* of G by $\widehat{G} = \text{Hom}_{\mathbb{Z}}(G, S^1)$. There is a ring isomorphism $\mathbb{Z}\widehat{G} \xrightarrow{\cong} R_{\mathbb{C}}(G)$ that sends the element $u \colon G \to S^1$ in $\widehat{G}$ to the class of the 1-dimensional complex G-representation having u as character. It induces a ring isomorphism

$$\mathbb{Q}\widehat{G} \xrightarrow{\cong} \mathbb{Q}R_{\mathbb{C}}(G). \tag{18.39}$$

Note that it sends the norm element $N = \sum_{\widehat{g} \in \widehat{G}} \widehat{g}$ to the class reg of the regular representation. Recall that we have chosen a 1-dimensional lens space $L_0 = L(V_0)$. Let χ be the character of V_0. It is a generator of $\widehat{G}$. Note that $1 - \chi \in \mathbb{Q}\widehat{G}$ is invertible when considered as an element in $\mathbb{Q}\widehat{G}/(N)$, because of

$$(1-\chi) \cdot \left(-\frac{1}{|G|}(1 + 2 \cdot \chi + 3 \cdot \chi^2 + \cdots + |G| \cdot \chi^{(|G|-1)})\right) = 1 - \frac{N}{|G|}. \tag{18.40}$$

Hence we can consider the element $\frac{1+\chi}{1-\chi}$ in $\mathbb{Q}\widehat{G}/(N)$. Let f be its image under the isomorphism

$$\mathbb{Q}\widehat{G}/(N) \xrightarrow{\cong} \overline{\mathbb{Q}R_{\mathbb{C}}(G)}$$

induced by the isomorphism (18.39). Then f lies in $\overline{\mathbb{Q}R_{\mathbb{C}}(G)}^{(-1)}$ and we get following formula [414, Proof of Theorem 14C.4 on page 224]

$$\rho_G(L_0) = f \quad \in \overline{\mathbb{Q}R_{\mathbb{C}}(G)}^{(-1)}. \tag{18.41}$$

18.8 Real Projective Spaces

The real projective space $\mathbb{RP}^d$ is a d-dimensional closed manifold with $\pi = \pi_1(\mathbb{RP}^d) = \mathbb{Z}/2$ for $d \geq 2$. Recall that $\text{Wh}(\mathbb{Z}/2) = \{1\}$, see e.g. [361, Theorem 2.4.3]. Hence there is no difference between the s and h decoration and so we leave it out from the notation. Moreover, $\mathbb{RP}^d$ is orientable if d is odd, which gives the trivial orientation homomorphism $+\colon \pi \to \{\pm 1\}$, and not orientable if d is even, which gives the non-trivial orientation homomorphism $-\colon \pi \to \{\pm 1\}$. From Theorem 16.11 we have for the L-groups:

$$L_d(\mathbb{Z}[\mathbb{Z}/2], +) = \mathbb{Z} \oplus \mathbb{Z}, 0, \mathbb{Z}/2, \mathbb{Z}/2 \quad \text{for } d \equiv 0, 1, 2, 3 \pmod 4;$$
$$L_d(\mathbb{Z}[\mathbb{Z}/2], -) = \mathbb{Z}/2, 0, \mathbb{Z}/2, 0 \quad \text{for } d \equiv 0, 1, 2, 3 \pmod 4.$$

In the case $d \equiv 0 \mod 4$ with the trivial orientation homomorphism, the isomorphism is given by the $\mathbb{Z}/2$-signature. The three cases in even dimensions when we get $\mathbb{Z}/2$ are all given by the Arf invariant. The $\mathbb{Z}/2$ in dimension $d \equiv 3 \mod 4$ in the orientable case is more complicated; full information can be found in [414, Theorem 13A.1 on page 183].

The surgery exact sequence for $\mathbb{RP}^d$ does not obviously split into short exact sequences, as was the case in Section 18.8, since we cannot apply Theorem 17.30. Hence we proceed with the analysis of normal invariants and then we see what information can we extract about the whole sequence in respective cases. For the normal invariants we obtain from (17.16) the isomorphism of abelian groups

$$\mathcal{N}^{\text{TOP}}(\mathbb{RP}^d) \cong \bigoplus_{i=1}^{\infty} H^{4i}(\mathbb{RP}^d; \mathbb{Z}) \oplus \bigoplus_{i=1}^{\infty} H^{4i-2}(\mathbb{RP}^d; \mathbb{Z}/2). \tag{18.42}$$

Using appropriate choices from Remark 17.9 and denoting the factors

$$\mathbf{r}_{4i} \colon \mathcal{N}^{\text{TOP}}(\mathbb{RP}^d) \to H^{4i}(\mathbb{RP}^d; \mathbb{Z}) \cong \mathbb{Z}/2;$$
$$\mathbf{r}_{4i-2} \colon \mathcal{N}^{\text{TOP}}(\mathbb{RP}^d) \to H^{4i-2}(\mathbb{RP}^d; \mathbb{Z}/2) \cong \mathbb{Z}/2,$$

it is possible to identify $\mathbf{r}_{4i-2}$ as splitting invariants along $\mathbb{RP}^{4i-2}$, in the sense of Definition 17.25, where we abbreviate $\mathbf{r}_{4i-2} = \mathbf{s}_{\mathbb{RP}^{4i-2} \subset \mathbb{RP}^d}$, while $\mathbf{r}_{4i}$ have a more complicated interpretation in terms of splitting invariants along $\mathbb{Z}/2$-submanifolds (in the sense of Remark 17.28), as explained in [247, Remarks on page 43], see also [414, page 208]. Altogether we get an isomorphism of abelian groups

$$\mathcal{N}^{\text{TOP}}(\mathbb{RP}^d) \cong \bigoplus_{i=1}^{\lfloor d/2 \rfloor} \mathbb{Z}/2. \tag{18.43}$$

Next recall that we have the suspension maps Σ from (18.23), that the inclusions $\mathbb{RP}^{d-1} \subset \mathbb{RP}^d$ induce the normal splitting maps, see (18.25), of the shape $s^{\mathcal{N}}_{\mathbb{RP}^{d-1} \subset \mathbb{RP}^d} \colon \mathcal{N}^{\text{TOP}}(\mathbb{RP}^d) \to \mathcal{N}^{\text{TOP}}(\mathbb{RP}^{d-1})$, and that an obvious modification

18.8 Real Projective Spaces   819

of Diagram (18.26) gives compatibility between Σ and $s^{\mathcal{N}}_{\mathbb{RP}^{d-1} \subset \mathbb{RP}^d}$. Moreover, recall that in the case $d = 4k + 3$, we have the L-group $L_{d+1}(\mathbb{Z}[\mathbb{Z}/2], +) \cong L_{d+1}(\mathbb{Z}) \oplus \widetilde{L}_{d+1}(\mathbb{Z}[\mathbb{Z}/2], +) \cong \mathbb{Z} \oplus \mathbb{Z}$ to the left of the structure set $\mathcal{S}^{\text{TOP}}(\mathbb{RP}^d)$. The summand $L_d(\mathbb{Z})$ is hit from the left by the discussion following Definition 17.31 and Corollary 18.34 implies that the summand $\widetilde{L}_{d+1}(\mathbb{Z}[\mathbb{Z}/2], +) \cong \mathbb{Z}$ maps injectively to the structure set. All the above information tells us that we can summarise the situation in the respective cases in the following commutative diagram of abelian groups defined for any $k \geq 1$, where the rows are exact sequences,

$$
\begin{array}{ccccccccc}
& & 0 & \longrightarrow & \mathcal{S}^{\text{TOP}}(\mathbb{RP}^{4k+1}) & \xrightarrow[\cong]{\eta_{4k+1}} & \mathcal{N}^{\text{TOP}}(\mathbb{RP}^{4k+1}) & \longrightarrow & 0 \\
& & & & \Sigma \downarrow & & \uparrow s^{\mathcal{N}}_{\mathbb{RP}^{4k+1} \subset \mathbb{RP}^{4k+2}} & & \\
& & 0 & \longrightarrow & \mathcal{S}^{\text{TOP}}(\mathbb{RP}^{4k+2}) & \xrightarrow{\eta_{4k+2}} & \mathcal{N}^{\text{TOP}}(\mathbb{RP}^{4k+2}) & \xrightarrow{\sigma_{4k+2}} & \mathbb{Z}/2 \\
& & & & \Sigma \downarrow & & \cong \uparrow s^{\mathcal{N}}_{\mathbb{RP}^{4k+2} \subset \mathbb{RP}^{4k+3}} & & \\
0 & \longrightarrow & \mathbb{Z} & \xrightarrow{\rho_{4k+4}} & \mathcal{S}^{\text{TOP}}(\mathbb{RP}^{4k+3}) & \xrightarrow{\eta_{4k+3}} & \mathcal{N}^{\text{TOP}}(\mathbb{RP}^{4k+3}) & \xrightarrow{\sigma_{4k+3}} & \mathbb{Z}/2 \\
& & & & \Sigma \downarrow & & \uparrow s^{\mathcal{N}}_{\mathbb{RP}^{4k+3} \subset \mathbb{RP}^{4k+4}} & & \\
& & 0 & \longrightarrow & \mathcal{S}^{\text{TOP}}(\mathbb{RP}^{4k+4}) & \xrightarrow{\eta_{4k+4}} & \mathcal{N}^{\text{TOP}}(\mathbb{RP}^{4k+4}) & \xrightarrow{\sigma_{4k+4}} & \mathbb{Z}/2 \\
& & & & \Sigma \downarrow & & \cong \uparrow s^{\mathcal{N}}_{\mathbb{RP}^{4k+4} \subset \mathbb{RP}^{4k+5}} & & \\
& & 0 & \longrightarrow & \mathcal{S}^{\text{TOP}}(\mathbb{RP}^{4k+5}) & \xrightarrow[\cong]{\eta_{4k+5}} & \mathcal{N}^{\text{TOP}}(\mathbb{RP}^{4k+5}) & \longrightarrow & 0.
\end{array}
$$

We conclude from Theorem 17.24 (ii) and (iii) that the normal splitting maps coincide with the maps induced by inclusions on $[-, G/\text{TOP}]$. This implies that all vertical maps in the right columns are surjective and the two labelled maps are isomorphisms. We get for $k \geq 1$ from (18.43) isomorphisms of abelian groups

$$\mathcal{N}^{\text{TOP}}(\mathbb{RP}^5) \xrightarrow{\cong} (\mathbb{Z}/2)^2;$$
$$\mathcal{N}^{\text{TOP}}(\mathbb{RP}^{4k+2}) \xrightarrow{\cong} \mathcal{N}^{\text{TOP}}(\mathbb{RP}^{4k+1}) \oplus \mathbb{Z}/2;$$
$$\mathcal{N}^{\text{TOP}}(\mathbb{RP}^{4k+3}) \xrightarrow{\cong} \mathcal{N}^{\text{TOP}}(\mathbb{RP}^{4k+1}) \oplus \mathbb{Z}/2;$$
$$\mathcal{N}^{\text{TOP}}(\mathbb{RP}^{4k+4}) \xrightarrow{\cong} \mathcal{N}^{\text{TOP}}(\mathbb{RP}^{4k+1}) \oplus (\mathbb{Z}/2)^2;$$
$$\mathcal{N}^{\text{TOP}}(\mathbb{RP}^{4k+5}) \xrightarrow{\cong} \mathcal{N}^{\text{TOP}}(\mathbb{RP}^{4k+1}) \oplus (\mathbb{Z}/2)^2.$$

We see that all that is left is to determine the three surgery obstruction maps σ_{4k+i} with the target $\mathbb{Z}/2$ for $i = 2, 3, 4$ and to solve the extension problem in the middle row. The first problem is addressed in the following proposition. One proof of it can be found in [414, Theorem 14D.2 on page 208] and another in [247, IV.3.2 and IV.3.3 on pages 44-46].

Proposition 18.44 *The map*

$$\sigma_d \colon \mathcal{N}^{\mathrm{TOP}}(\mathbb{RP}^d) \to \mathbb{Z}/2$$

is surjective in the above three cases where $d = 4k + i$ with $i = 2, 3, 4$ and it is given by the splitting invariant $\mathbf{r}_{4k+2}$.

This also means that if we put the equal signs between the three copies of $\mathbb{Z}/2$ in the last column of the above diagram, then the two resulting squares commute, which yields an interesting relation between surgery obstructions in different dimensions in this case, but we will not pursue this further.

Proposition 18.44 completely takes care of all the cases except $d = 4k + 3$, which is analysed as follows. In this case Proposition 18.44 together with Corollary 18.34 identify for us a short exact sequence with the middle term the desired structure set. Moreover, we have the ρ-invariant map from (18.32) with the source the structure set and the commutative diagram from Proposition 18.33 and Corollary 18.34. Using all this we obtain a map of short exact sequences as follows:

$$\begin{array}{ccccccccc}
0 & \to & \mathbb{Z} & \xrightarrow{\rho_{4k+4}} & \mathcal{S}^{\mathrm{TOP}}(\mathbb{RP}^{4k+3}) & \xrightarrow{\eta_{4k+3}} & \mathrm{im}(\eta_{4k+3}) & \to & 0 \\
& & \cong \downarrow \mathbb{Z}/2\text{-sign} & & \downarrow \widetilde{\rho}_{\mathbb{Z}/2} & & \downarrow [\widetilde{\rho}_{\mathbb{Z}/2}] & & \\
0 & \to & 8\mathbb{Z} & \to & \mathbb{Q} & \to & \mathbb{Q}/8\mathbb{Z} & \to & 0
\end{array} \qquad (18.45)$$

where $\mathrm{im}(\eta_{4k+3})$ can be identified with $\mathcal{N}^{\mathrm{TOP}}(\mathbb{RP}^{4k+1})$ and $[\widetilde{\rho}_{\mathbb{Z}/2}]$ is the homomorphism induced by $\widetilde{\rho}_{\mathbb{Z}/2}$. The extension problem given by the upper exact row in (18.45) is resolved by showing that $[\widetilde{\rho}_{\mathbb{Z}/2}] = 0$, which implies that the $\widetilde{\rho}_{\mathbb{Z}/2}$-invariant map in fact splits the map ρ_{4k+4} and the extension problem is solved.

We sketch the argument for the statement $[\widetilde{\rho}_{\mathbb{Z}/2}] = 0$. Proposition 18.44 together with the large diagram above yield that any element in $\mathrm{im}(\eta_{4k+3})$ has a representative that is in the image of $\eta_{4k+3} \circ \Sigma \colon \mathcal{S}^{\mathrm{TOP}}(\mathbb{RP}^{4k+2}) \to \mathrm{im}(\eta_{4k+3})$. But any element in $\mathcal{S}^{\mathrm{TOP}}(\mathbb{RP}^{4k+3})$ that is in the image of the suspension $\Sigma \colon \mathcal{S}^{\mathrm{TOP}}(\mathbb{RP}^{4k+2}) \to \mathcal{S}^{\mathrm{TOP}}(\mathbb{RP}^{4k+3})$ possess an orientation reversing homeomorphism, given by interchanging the two suspension points on the universal covers. This implies that the $\widetilde{\rho}_{\mathbb{Z}/2}$-invariant of such an element vanishes and hence also the $[\widetilde{\rho}_{\mathbb{Z}/2}]$-invariant of its image under η_{4k+3}.

Summarising we obtain the final result.

Theorem 18.46 (Topological structure sets of real projective spaces) *Let $d \geq 4$. Then we get isomorphisms of abelian groups*

$$\mathcal{S}^{\mathrm{TOP}}(\mathbb{RP}^d) \cong \begin{cases} (2k) \cdot \mathbb{Z}/2 & \text{for } d = 4k+1; \\ (2k) \cdot \mathbb{Z}/2 & \text{for } d = 4k+2; \\ \mathbb{Z} \oplus (2k) \cdot \mathbb{Z}/2 & \text{for } d = 4k+3; \\ (2k+1) \cdot \mathbb{Z}/2 & \text{for } d = 4k+4. \end{cases}$$

18.8 Real Projective Spaces

A full proof of this theorem is given in [414, Chapter 14D] and another account can be found in [247]. It is interesting to compare the calculations from these two sources. We have followed quite closely the calculation by Wall, while the calculation of López de Medrano is different. Instead of the ρ-invariant he uses the so-called Browder–Livesay invariants. These are equivalent to the ρ-invariant in this setting, but their definition is quite different, see [247, Section IV.4]. They are designed to be obstructions to desuspension, that means, they measure the deviation of the suspension map Σ from (18.23) from being surjective. As such, they are closely related to the splitting invariants along $\mathbb{R}\mathrm{P}^{d-1}$ in the sense of Remark 17.29. The theory can be developed more generally into what is called *surgery on submanifolds*, but it transcends the goals of this book. We refer the reader to [247, Chapters I and II], [414, Chapter 12C] and [346, Chapter 7] and we note that the maps Σ in the large diagram above are also completely determined in [247].

The realisation question for fake real projective spaces can be addressed as follows. Any $2d$-dimensional fake complex projective space is obtained by a free action of S^1 on S^{2d+1} and restricting the action to $\mathbb{Z}/2 < S^1$ produces a fake real projective space in dimension $(2d + 1)$. In dimension $2d + 1 = 4k + 3$ we can apply the realisation Theorem 8.195 and map ρ_{4k+4} to produce fake real projective spaces with the prescribed ρ-invariant. In the even dimensions one can realise fake real projective spaces by taking odd-dimensional examples and applying the suspension map.

Remark 18.47 Finally we report on the quotient $\mathcal{S}^{\mathrm{TOP}}(\mathbb{R}\mathrm{P}^d)/\mathrm{ho\text{-}aut}(\mathbb{R}\mathrm{P}^d)$, see also [51, Theorem 4]. Since $\mathrm{ho\text{-}aut}(\mathbb{R}\mathrm{P}^d)$ is known to be trivial for d even, we only have to deal with the case where d is odd.

Then $\mathbb{R}\mathrm{P}^d$ is orientable and one obtains an isomorphism of abelian groups $\mathrm{ho\text{-}aut}(\mathbb{R}\mathrm{P}^d) \xrightarrow{\cong} \{\pm 1\}$ by the degree of a self-homotopy equivalence. Moreover, the non-trivial element is represented by a self-homeomorphism obtained from an equivariant self-homeomorphism of degree -1 of S^d.

Then the corresponding action on the right-hand side of the isomorphism appearing in Theorem 18.46 is compatible with the direct sum decomposition. It is trivial on each summand of the shape $\mathbb{Z}/2$, by the same reasoning as for $\mathbb{C}\mathrm{P}^d$ in Remark 18.17, and by $-\mathrm{id}$ on each summand of the shape $\mathbb{Z}$ since the ρ-invariant changes sign upon reversing orientation. Hence we get a bijection of sets

$$\mathcal{S}^{\mathrm{TOP}}(\mathbb{R}\mathrm{P}^d)/\mathrm{ho\text{-}aut}(\mathbb{R}\mathrm{P}^d) \cong \begin{cases} (2k) \cdot \mathbb{Z}/2 & \text{for } d = 4k+1; \\ (2k) \cdot \mathbb{Z}/2 & \text{for } d = 4k+2; \\ \mathbb{Z}^{\geq 0} \oplus (2k) \cdot \mathbb{Z}/2 & \text{for } d = 4k+3; \\ (2k+1) \cdot \mathbb{Z}/2 & \text{for } d = 4k+4. \end{cases}$$

The calculation of $\mathcal{S}^{\mathrm{PL}}(\mathbb{R}\mathrm{P}^d)$ is similar, but slightly different. Namely, the sum $\mathbb{Z}/2 \oplus \mathbb{Z}/2$ in the lowest dimensions is replaced by $\mathbb{Z}/4$ and on this $\mathbb{Z}/4$ the forgetful map $\mathcal{S}^{\mathrm{PL}}(\mathbb{R}\mathrm{P}^d) \to \mathcal{S}^{\mathrm{TOP}}(\mathbb{R}\mathrm{P}^d)$ is the projection on the first summand of $\mathbb{Z}/2 \oplus \mathbb{Z}/2$. So it is neither injective nor surjective, which yields counterexamples to Manifold

Hauptvermutung 17.11 as well as examples of topological manifolds that do not admit a PL structure. See [93, Theorem 6.20], [247, IV.5], [219, § 16 on page 331].

18.9 Lens Spaces

In this section we will continue to use Notation 18.21 for generalised lens spaces. Recall from Definition 3.41 that a generalised lens space is the orbit space $L_G^n(\alpha)$ of a free topological action $\alpha \colon G \to \mathrm{aut}(S^n)$ of a finite cyclic group G of order $|G|$. Note that it comes with a preferred orientation and an identification of G and $\pi_1(L_G^n(\alpha))$, as explained in (3.43). Recall from Lemma 3.53 (ii) that a generalised lens space is the same as a fake lens space in the sense of Definition 18.3.

When n is even, the only group admitting such an action is $\mathbb{Z}/2$. So it makes sense to treat that case separately, which we did in Section 18.8. Here we present the case $|G| \geq 3$ and n odd with $n \geq 5$. In this section we follow the survey [273], and the original sources [414, Chapter 14E], [274] and [275], but note that we have changed the notation slightly.

First we state the computation of the simple structure set of a fake lens space. The case of a fundamental group G of odd order has been treated by Wall [414, Theorem 14E.7 on page 224], see also [57, Theorem 4] and Remark 18.68 below.

Theorem 18.48 (The simple topological structure set of a fake lens space with fundamental group of odd order) *Let L^{2d-1} be a generalised lens space of dimension $2d - 1$ for $d \geq 3$ whose fundamental group G is finite cyclic of odd order. Then the homomorphism of abelian groups*

$$\widetilde{\rho}_G \colon \mathcal{S}^{\mathrm{TOP},s}(L^{2d-1}) \to \overline{\mathbb{Q}R_{\mathbb{C}}(G)}^{(-1)^d}$$

defined in (18.32) *is injective. Its image is a finitely generated free abelian group of rank* $(|G| - 1)/2$.

Remark 18.49 (The simple topological structure set of a fake lens space with fundamental group of even order) The more complicated case of the fundamental group of even order has been treated in [275, Theorem 2.2] using an earlier calculation for the fundamental group of order a power of 2 in [274, Theorem 1.2]. However, during the late stages of writing this chapter we found that in both papers there is the same mistake, in [274, Proposition 4.5] and [275, Proposition 4.5], described in more detail in Remark 18.67. When corrected, it makes the calculation of the simple structure set potentially smaller by index 2 when d is even. There is some indication that the calculations remain the same, but a more thorough examination of the papers is needed to confirm this.

18.9 Lens Spaces

We have introduced the Reidemeister torsion of a generalised lens space $L(\alpha) = L_G^{2d-1}(\alpha)$

$$\Delta(L(\alpha)) \in D(G)$$

in Definition 3.60, with $D(G)$ from (3.55). We have defined the ρ-invariant

$$\rho_G(L(\alpha)) \in \overline{\mathbb{Q}R_{\mathbb{C}}(G)}^{(-1)^d}$$

in Definition 18.31. Moreover, we have introduced the element

$$c(L(\alpha)) \in H_{2d-1}(BG)$$

in (3.44).

Next we want to classify the fake lens spaces in terms of these invariants using the calculation of the structure sets above. Given arbitrary fake lens spaces L and L', we first need to determine their simple homotopy type to be able to see if L' determines an element in the simple structure set of L. This we have already done in Theorem 3.61.

Exercise 18.50 Let $L(\alpha)$ be a generalised lens space. Then the forgetful map

$$\{f \in \text{ho-aut}^s(L(\alpha)) \mid \pi_1(f) = \text{id}\} \to \{f \in \text{ho-aut}(L(\alpha)) \mid \pi_1(f) = \text{id}\}$$

is an isomorphism of abelian groups.

The structure of the group ho-aut($L(\alpha)$) can easily be derived from Theorem 3.49. We get

$$\{[f] \in \text{ho-aut}(L(\alpha)) \mid \pi_1(f) = \text{id}_G, \deg(f) = 1\} \cong \{[\text{id}_{L(\alpha)}]\}, \quad (18.51)$$

and isomorphisms

$$\{[f] \in \text{ho-aut}(L^{2d-1}(\alpha)) \mid \deg(f) = 1\} \xrightarrow{\cong} \text{aut}_d^+(G), \quad [f] \mapsto \pi_1(f) \quad (18.52)$$

with

$$\text{aut}_d^+(G) = \{\gamma \in \text{aut}(G) \mid \gamma^d = \text{id}\} \subseteq \text{aut}(G),$$

and

$$\text{ho-aut}(L(\alpha)) \xrightarrow{\cong} \text{aut}_d^{\pm}(G), \quad [f] \mapsto (\pi_1(f)). \quad (18.53)$$

with

$$\text{aut}_d^{\pm}(G) = \{\gamma \in \text{aut}(G) \mid \gamma^d = \pm\text{id}\} \subseteq \text{aut}(G).$$

In Theorem 3.68 we classified lens spaces up to homeomorphism. The following theorem extends this to fake lens spaces in a special case. Given an automorphism $a: G \xrightarrow{\cong} G$, we denote by a_* the obvious automorphisms of $D(G), \overline{\mathbb{Q}R_{\mathbb{C}}(G)}^{(-1)^d}$, and $H_{2d-1}(BG)$ induced by a.

Theorem 18.54 (The topological classification of fake lens spaces with fundamental groups of odd order) Let $L = L_G^{2d-1}(\alpha)$ and $L' = L_G^{2d-1}(\alpha')$ be two $(2d-1)$-dimensional generalised lens spaces with finite cyclic fundamental groups $G = \pi_1(L) = \pi_1(L')$ of odd order. Suppose $d \geq 3$.

(i) *There is an orientation preserving homeomorphism $L \to L'$ inducing the identity on $G = \pi_1(L) = \pi_1(L')$ if and only if $c(L) = c(L')$, $\Delta(L) = \Delta(L')$, and $\rho_G(L) = \rho_G(L')$ holds;*
(ii) *There is an orientation preserving homeomorphism $L \to L'$ if and only if there exists an automorphism $a \colon G \xrightarrow{\cong} G$ such that $a_*(c(L)) = c(L')$, $a_*(\Delta(L)) = \Delta(L')$, and $a_*(\rho_G(L)) = \rho_G(L')$ holds;*
(iii) *There is a homeomorphism $L \to L'$ if and only if there exists an automorphism $a \colon G \xrightarrow{\cong} G$ and $\epsilon \in \{\pm 1\}$ such that $a_*(c(L)) = \epsilon \cdot c(L')$, $a_*(\Delta(L)) = \Delta(L')$, and $a_*(\rho_G(L)) = \epsilon \cdot \rho_G(L')$ hold.*

Proof. We have $c(L) = -c(L^-)$, $\Delta(L) = \Delta(L^-)$, and $\rho_G(L) = -\rho_G(L^-)$ for a generalised lens space L, where L^- is obtained from L by reversing the orientation. If $L(\alpha)$ is a generalised lens space and $a \colon G \xrightarrow{\cong} G$ is an automorphism, then $c(L(\alpha \circ a)) = a_* c(L(\alpha))$, $\Delta(L(\alpha \circ a)) = a_*(\Delta(L(\alpha)))$, and $\rho_G(L(\alpha \circ a)) = a_*(\rho_G(L(\alpha)))$ hold. Now the claim follows from Lemma 3.57, Theorems 3.61 and 18.48, together with (18.51), (18.52), and (18.53). □

Remark 18.55 (Comparing our setup with the one of Wall) In Wall [414, Theorem 14E.7 on page 224] the Reidemeister torsion $\Delta_W(L(\alpha))$ takes values in $D_W(G)$, which is defined to be

$$D_W(G) := \mathbb{Q}G/(N)^\times / T_W(G), \tag{18.56}$$

where $T_W(G)$ is the image of the subgroup $\{g \mid g \in G\}$ under the composite $\mathbb{Z}G^\times \xrightarrow{i} \mathbb{Q}G^\times \xrightarrow{p} \mathbb{Q}G/(N)^\times$ for i the obvious inclusion. Note that $D(G)$ is the quotient of $D_W(G)$ by the cyclic subgroup of order two given by $\{\pm 1\}$ and $\Delta(L(\alpha))$ is the image of $\Delta_W(L(\alpha))$ under the canonical projection $D_W(G) \to D(G)$. However, note that Wall's definition depends on a choice of a generator $t \in G$, the fact that any fake lens space $L_G^{2d-1}(\alpha)$ can be obtained up to simple homotopy equivalence from the $(2d-2)$-skeleton $L_0^{(2d-2)}$ of the standard lens space $L_0 = L(|G|; 1, 1, \ldots, 1)$ by attaching one $(2d-1)$-cell, and a choice of orientations for all the cells in $L_0^{(2d-2)}$. Then the orientation of $L(\alpha)$ induces an orientation of the additional $(2d-1)$-cell and hence all cells are oriented. This allows us to lift $\Delta(L(\alpha)) \in D(G)$ to an element $\Delta_W(L(\alpha)) \in D_W(G)$. Different choices may of course yield different lifts.

Given two generalised lens spaces L and L', the quotient $\Delta_W(L')/\Delta_W(L)$ lies in $\mathbb{Z}G/(N)^\times / T_W(G)$ and is mapped under the canonical homomorphism $\mathbb{Z}G/(N)^\times / T_W(G) \to \mathbb{Z}/|G|^\times$ to $c(L')/c(L)$, where $c(L')/c(L)$ denotes the element $u \in \mathbb{Z}/|G|^\times$ uniquely determined by the equation $c(L') = u \cdot c(L)$ in $H_{2d-1}(BG)$. Hence $\Delta_W(L) = \Delta_W(L')$ implies $c(L) = c(L')$ whereas $\Delta(L) = \Delta(L')$ only implies $c(L) = \pm c(L')$. Note that $\Delta_W(L)$ contains the same information as the pair

18.9 Lens Spaces

$(\Delta(L)), c(L))$, but the definition of $\Delta_W(L)$ depends on extra choices, which do not occur in the definition of $(\Delta(L(\alpha)), c(L(\alpha)))$.

There is a more intrinsic lift of the Reidemeister torsion presented in [250, Definition 18.10 on page 362], which actually works also for non-free actions. We at least describe briefly what it yields for generalised lens spaces. We put

$$\widetilde{D}(G) := \mathbb{Q}G^\times / T(G), \tag{18.57}$$

where $T(G)$ is the subgroup $\{\pm g \mid g \in G\}$. Note that in the definition of $\widetilde{D}(G)$ we do not divide out the ideal (N) of the norm element, but do divide out the subgroup $\{\pm 1\}$. Then one can assign to an oriented generalised lens space $L(\alpha)$ an element

$$\widetilde{\Delta}(L(\alpha)) \in \widetilde{D}(G) \tag{18.58}$$

whose image under the obvious projection $\widetilde{D}(G) \to D(G)$ is $\Delta(L(\alpha))$. The construction of the element $\widetilde{\Delta}(L(\alpha))$ is a special case of the element $\overline{\rho}^G(L(\alpha))$ defined in [250, Definition 18.10 on page 362]. The basic idea is to consider the $\mathbb{Q}G$-chain complex $H_*(L(\alpha); \mathbb{Q}) := H_*(\mathbb{Q} \otimes_\mathbb{Z} C_*(L(\alpha)))$ that is given by the homology with rational coefficients of the cellular $\mathbb{Z}G$-chain complex $C_*(L(\alpha))$ and concentrated in dimensions 0 and $2d - 1$, and the $\mathbb{Q}G$-chain homotopy equivalence $i_* \colon H_*(L(\alpha); \mathbb{Q}) \to \mathbb{Q} \otimes_\mathbb{Z} C_*(L(\alpha))$ uniquely determined up to $\mathbb{Q}G$-chain homotopy by the property that it induces the identity on the homology, and to use the fact that the orientation of $L(\alpha)$ yields an isomorphism of $\mathbb{Q}G$-modules $H_{2d-1}(L(\alpha); \mathbb{Q}) \xrightarrow{\cong} H_0(L(\alpha); \mathbb{Q})$. Let $\partial \colon \widetilde{D}(G) \to \mathbb{Q}G^\times / \mathbb{Z}_{(|G|)} G^\times$ be the canonical projection and let $\overline{sw} \colon \mathbb{Z}/|G|^\times \to \mathbb{Q}G^\times / \mathbb{Z}_{(|G|)} G^\times$ be the injective map coming from the generalised Swan homomorphism of [250, 19.2 and Theorem 19.4 on page 381], where $\mathbb{Z}_{(-)}$ denotes localisation. Then

$$\overline{sw}(c(L(\alpha))) = \partial(\widetilde{\Delta}(L(\alpha))). \tag{18.59}$$

Hence $\widetilde{\Delta}(L(\alpha))$ determines both $\Delta(L(\alpha))$ and $c(L(\alpha))$.

Another difference to the setup of Wall is that $c(L(\alpha)) \in H_{2d-1}(BG)$ from (3.44) replaces the k-invariant $k_{2d-1}(L_G^{2d-1}(\alpha)) \in H^{2d}(BG; \mathbb{Z})$ (in the sense of homotopy theory, as in [178, Section 4.3]) that he uses for the homotopy classification in [414, 14E.1, 14E.3].

Remark 18.60 (Realisation of invariants for π of odd order) For a finite cyclic group G of odd order and an integer $d \geq 3$, Wall explicitly describes in [414, Theorem 14E.7 on page 224] which pairs (Δ, ρ) in $D_W(G) \times \overline{\mathbb{Q}R_\mathbb{C}(G)}^{(-1)^d}$ can be realised as $\left(\Delta_W(L_G^{2d-1}(\alpha)), \rho_G(L_G^{2d-1}(\alpha))\right)$ for some generalised lens space $L_G^{2d-1}(\alpha)$.

Next we offer some ideas that go into investigating the terms in the simple geometric surgery exact sequence of a fake lens space $L_G^n(\alpha)$. When $|G|$ is odd, these go into the proof of Theorem 18.48. For simplicity we concentrate on the case of the standard lens space $L = L(\oplus_{i=1}^d V_1)$.

The simple L-groups were described in Theorem 16.11, so the next step is to consider the normal invariants. Since $H^j(L, \mathbb{Q}) = \mathbb{Q}$ for $j \in \{0, (2d-1)\}$ and zero otherwise, and keeping in mind that $H^{2d-1}(L, \mathbb{Q}) \cong \mathbb{Q}$ comes from $H^{2d-1}(L, \mathbb{Z}) \cong \mathbb{Z}$, we see from the long exact sequence (17.16) that

$$\mathcal{N}^{\mathrm{TOP}}(L) \cong \bigoplus_{i=1}^{\lfloor (d-1)/2 \rfloor} H^{4i}(L; \mathbb{Z}_{(2)}) \oplus \bigoplus_{i=1}^{\lfloor d/2 \rfloor} H^{4i-2}(L; \mathbb{Z}/2);$$
$$\oplus \widetilde{\mathrm{KO}}^0(L, \mathbb{Z}[1/2]).$$

Setting $|G| = 2^k \cdot m$ for odd m, the first two summands can be calculated explicitly and as explained in [414, page 218], appropriate choices from Remark 17.9 allow us to view the factors as splitting invariants, which are denoted by

$$\mathbf{t}_{4i} \colon \mathcal{N}^{\mathrm{TOP}}(L) \to H^{4i}(L^{2d-1}; \mathbb{Z}_{(2)}) \cong \mathbb{Z}/2^k; \qquad (18.61)$$

$$\mathbf{t}_{4i-2} \colon \mathcal{N}^{\mathrm{TOP}}(L) \to H^{4i-2}(L^{2d-1}; \mathbb{Z}/2) \cong \mathbb{Z}/2. \qquad (18.62)$$

The $\widetilde{\mathrm{KO}}^0(-, \mathbb{Z}[1/2])$-summand is more difficult to calculate explicitly, but we will not need the full calculation. Note that, when Wall analysed the case $|G| = m$ in [414, Chapter 14E] he only needed that the order of the group $\widetilde{\mathrm{KO}}^0(L, \mathbb{Z}[1/2])$ is m^c with $c = \lfloor (d-1)/2 \rfloor$, which follows easily from the corresponding Atiyah–Hirzebruch spectral sequence.

Next we have some information about the surgery obstruction maps.

Theorem 18.63 (Surgery obstruction maps for lens spaces)
Suppose $|G| = 2^k \cdot m$ with m odd and $k \geq 1$.

(i) *If $d = 2e$, then the surgery obstruction map*

$$\sigma_{2d-1}^s \colon \mathcal{N}^{\mathrm{TOP}}(L) \to L_{2d-1}^s(\mathbb{Z}G) = L_{4e-1}^s(\mathbb{Z}G) = \mathbb{Z}/2$$

is given by $\sigma(x) = \mathbf{t}_{4e-2}(x) \in \mathbb{Z}/2$;

(ii) *The map*

$$\sigma_{2d}^s \colon \mathcal{N}^{\mathrm{TOP}}(L \times I, \partial(L \times I)) \to L_{2d}^s(\mathbb{Z}G)$$

maps onto the summand $L_{2d}(\mathbb{Z})$.

Part (i) is Theorem 14E.4 in [414], part (ii) follows from Proposition 18.33 and Section 17.8.

Putting all the knowledge together, we obtain the short exact sequence

$$0 \to \widetilde{L}_{2d}^s(\mathbb{Z}G) \xrightarrow{\partial} \mathcal{S}^{\mathrm{TOP},s}(L^{2d-1}) \xrightarrow{\eta} \widetilde{\mathcal{N}}^{\mathrm{TOP}}(L^{2d-1}) \to 0, \qquad (18.64)$$

the isomorphism $\widetilde{\mathcal{N}}^{\mathrm{TOP}}(L^{4e-1}) \cong \ker(\mathbf{t}_{4e-2})$, the equality $\widetilde{\mathcal{N}}^{\mathrm{TOP}}(L^{4e+1})) = \mathcal{N}(L^{4e+1})$, and hence the isomorphism

$$\widetilde{\mathcal{N}}^{\mathrm{TOP}}(L^{2d-1}) \cong \bigoplus_{i=1}^{c} \mathbb{Z}/2^k \oplus \bigoplus_{i=1}^{c} \mathbb{Z}/2 \oplus \widetilde{\mathrm{KO}}^0(L^{2d-1}, \mathbb{Z}[1/2]), \qquad (18.65)$$

18.9 Lens Spaces

where $c = \lfloor (d-1)/2 \rfloor$ and where the order of the last summand is m^c. The first term in the sequence (18.64) is understood by Theorem 16.11, and the third term is understood by (18.65). Hence we are left with an extension problem, which is addressed via the following proposition, which summarises what we know so far. (The general method is described in [57].)

Proposition 18.66 *There is the following commutative diagram of abelian groups and homomorphisms with exact rows*

$$
\begin{array}{ccccccccc}
0 & \longrightarrow & \widetilde{L}^s_{2d}(\mathbb{Z}G) & \xrightarrow{\rho_{2d}} & \mathcal{S}^{\mathrm{TOP},s}(L^{2d-1}) & \xrightarrow{\eta_{2d-1}} & \widetilde{\mathcal{N}}^{\mathrm{TOP}}(L^{2d-1}) & \longrightarrow & 0 \\
 & & \cong \Big\downarrow \text{G-sign} & & \Big\downarrow \widetilde{\rho}_G & & \Big\downarrow [\widetilde{\rho}_G] & & \\
0 & \longrightarrow & A(d,G) & \longrightarrow & \overline{\mathbb{Q}R_{\mathbb{C}}(G)}^{(-1)^d} & \longrightarrow & \overline{\mathbb{Q}R_{\mathbb{C}}(G)}^{(-1)^d}/A(d,G) & \longrightarrow & 0
\end{array}
$$

where

$$
A(d,G) = \begin{cases} \{4 \cdot ([V] + [V']) \mid [V] \in R_{\mathbb{C}}(G)\}/\{8l \cdot \text{reg} \mid l \in \mathbb{Z}\} & \text{for } d \text{ even;} \\ \{4 \cdot ([V] - [V']) \mid [V] \in R_{\mathbb{C}}(G)\} & \text{for } d \text{ odd,} \end{cases}
$$

and where $[\widetilde{\rho}_G]$ is the homomorphism induced by $\widetilde{\rho}_G$.

Hence, one faces a similar extension problem as in (18.45), but in this case the results are different. Namely, for $|G|$ odd Wall shows that the map $[\widetilde{\rho}_G]$ is injective, which leads to Theorem 18.48. If $|G| = 2^k \cdot m$ with $k \geq 2$ and m odd it follows from ad hoc calculations that the map $[\widetilde{\rho}_G]$ is neither 0 nor injective in general. The Snake lemma says that $\ker[\widetilde{\rho}_G]$ splits off the simple structure set and so the quest becomes to find $\ker[\widetilde{\rho}_G]$. This is investigated in [274, 275].

Remark 18.67 The mistake in [274] and [275] indicated in Remark 18.49 is that in both cases Proposition 4.5 should be replaced by Proposition 18.66 above. They differ in the case when d is even in the left vertical arrow. When compared one sees that these arrows in [274] and [275] are not isomorphisms if d is even, they are only injective with the cokernel of order 2.

We close with a remark describing the passage to generalised lens spaces.

Remark 18.68 (Theorem 18.48 for generalised lens spaces) Theorem 18.48 for generalised lens spaces is obtained from the same theorem for lens spaces by the following arguments.

Given a generalised lens space $L(\alpha)$, we know that by Lemma 3.53 (i) we can find a lens space $L(V)$ and a homotopy equivalence $f: L(\alpha) \to L(V)$. Then the following diagram commutes

$$\begin{array}{ccccc}
\mathcal{S}^{\mathrm{TOP},s}(L(\alpha)) & \longrightarrow & \mathcal{S}^{\mathrm{TOP},h}(L(\alpha)) & \xrightarrow{\widetilde{\rho}_G} & \overline{\mathbb{Q}R_{\mathbb{C}}(G)}^{(-1)^d} \\
{\scriptstyle f_*}\downarrow {\scriptstyle \cong} & & {\scriptstyle f_*}\downarrow {\scriptstyle \cong} & & \downarrow = \\
\mathcal{S}^{\mathrm{TOP},s}(L(V)) & \longrightarrow & \mathcal{S}^{\mathrm{TOP},h}(L(V)) & \xrightarrow{\widetilde{\rho}_G} & \overline{\mathbb{Q}R_{\mathbb{C}}(G)}(-1)^d
\end{array}$$

where the vertical arrows f_* are the group isomorphism induced by f in the sense of Ranicki's algebraic theory of surgery as in Theorem 18.7. (Note that this is an isomorphism for both s and h decorations despite the fact that f is not required to be simple.) The left horizontal maps are forgetful maps and the left-hand square commutes by the Ranicki's theory. The horizontal compositions are the factorisations of the ρ-invariant maps, which exists since the ρ-invariant is an h-cobordism invariant. The right square commutes by the following calculation for $[h\colon L \to L(\alpha)] \in \mathcal{S}^{\mathrm{TOP},h}(L(\alpha))$, where we use the composition formula from Theorem 18.7 and the fact that $\widetilde{\rho}_G$ is a homomorphism:

$$\begin{aligned}
\widetilde{\rho}_G(f_*([h])) &= \widetilde{\rho}_G([f \circ h] - [f]) = \widetilde{\rho}_G([f \circ h]) - \widetilde{\rho}_G([f]) \\
&= \rho_G(L) - \rho_G(L(V)) - (\rho_G(L(\alpha)) - \rho_G(L(V))) = \widetilde{\rho}_G([h]).
\end{aligned}$$

Hence $\widetilde{\rho}_G\colon \mathcal{S}^{\mathrm{TOP},s}(L(\alpha)) \to \overline{\mathbb{Q}R_{\mathbb{C}}(G)}^{(-1)^d}$ is an isomorphism onto its image since $\widetilde{\rho}_G \circ f_*$ is.

18.10 Aspherical Manifolds

Theorem 18.69 (Topological Structure sets of closed aspherical manifolds) *Let M be a closed aspherical topological manifold of dimension ≥ 5 whose fundamental group is a Farrell–Jones group in the sense of Definition 16.43. Then*

$$\mathcal{S}^{\mathrm{TOP},s}(M) = \mathcal{S}^{\mathrm{TOP},h}(M) = \{[\mathrm{id}_M]\},$$

and for every $k \geq 1$

$$\mathcal{S}^{\mathrm{TOP},s}(M \times D^k, \partial(M \times D^k)) = \mathcal{S}^{\mathrm{TOP},h}(M \times D^k, \partial M \times D^k) = \{[\mathrm{id}_{M \times D^k}]\}.$$

If additionally M admits a PL structure, then we get an isomorphism of abelian groups

$$\mathcal{S}^{\mathrm{PL},s}(M) \cong \mathcal{S}^{\mathrm{PL},s}(M) \cong H^3(M, \mathbb{Z}/2),$$

and for every $k \geq 1$

$$\mathcal{S}^{\mathrm{PL},s}(M \times D^k, \partial(M \times D^k)) = \mathcal{S}^{\mathrm{PL},h}(M \times D^k, \partial M \times D^k) \cong H^{3-k}(M, \mathbb{Z}/2).$$

18.11 Some Torus Bundles over Lens Spaces

Proof. We will get the assertions for $\mathcal{S}^{\mathrm{TOP}}$ as special cases from Theorem 19.32 and Remark 19.35. The computation for $\mathcal{S}^{\mathrm{PL}}$ follows from the comparison between the surgery exact sequences in the PL category and on the topological category analogous to the comparison between the smooth and topological category in Remark 17.23 and from the fact that TOP/PL $\simeq K(\mathbb{Z}/2, 3)$, see Theorem 11.40. □

Hence in the above situation the forgetful map $\mathcal{S}^{\mathrm{PL}}(M) \to \mathcal{S}^{\mathrm{TOP}}(M)$ is not injective and we obtain counterexamples to the Manifold Hauptvermutung 17.11. Using Remark 17.23 we obtain in the situation from Theorem 18.69 under the additional assumption that M is smooth

$$\mathcal{S}^{\mathrm{DIFF},h}(M) \cong \mathcal{S}^{\mathrm{DIFF},s}(M) = [M, \mathrm{TOP}/O].$$

For the torus T^n with $n \geq 4$ this calculation has the following nice expression

$$\mathcal{S}^{\mathrm{DIFF},h}(T^n) \cong \mathcal{S}^{\mathrm{DIFF},s}(T^n) = [T^n, \mathrm{TOP}/O] \cong \bigoplus_{i=1}^{n} \pi_i(\mathrm{TOP}/O)^{\binom{n}{i}}, \quad (18.70)$$

where we remind the reader that $\pi_i(\mathrm{TOP}/O) \cong \Theta_i$ for $i \neq 3$, see Section 17.5 and [144, Lecture 16] for more information.

Remark 18.71 (Fake PL tori in the work of Kirby–Siebenmann) Assuming the part of Theorem 18.69 for $\mathcal{S}^{\mathrm{TOP},s}(T^n)$, and Theorem 17.10, the result of Theorem 18.69 for $\mathcal{S}^{\mathrm{PL},s}(T^n)$ is an immediate corollary. However, it should be pointed out that the proof of $\mathcal{S}^{\mathrm{PL},s}(T^n) \cong H^3(T^n, \mathbb{Z}/2)$ by Wall [414, Chapter 15A] is independent of the work of Kirby and Siebenmann in the topological category. More importantly, the PL case is actually used in the work of Kirby and Siebenmann in analysing the space TOP (and hence G/TOP) where in addition they study the behaviour of the fake PL tori under covering projections. They utilise a so-called "torus trick", see the paper reprinted as Annex 1 in [219].

18.11 Some Torus Bundles over Lens Spaces

So far in this Chapter 18 the fundamental groups were finite or torsionfree. Here is an example where the fundamental group is infinite and contains torsion. Consider the following setup and notation:

- Let p be an odd prime;
- Let $\rho \colon \mathbb{Z}/p \to \mathrm{aut}(\mathbb{Z}^n) = GL(n, \mathbb{Z})$ be a group homomorphism such that the induced action of $\mathbb{Z}/p$ on $\mathbb{Z}^n - \{0\}$ is free;
- The homomorphism ρ defines an action of $\mathbb{Z}/p$ on the torus $T^n = \mathbb{R}^n/\mathbb{Z}^n$;
- Fix a free action of $\mathbb{Z}/p$ on a sphere S^l for an odd integer $l \geq 3$. The orbit space $L^l := S^l/(\mathbb{Z}/p)$ is a generalised lens space;

- Define a closed $(m+l)$-manifold

$$M := T_\rho^n \times_{\mathbb{Z}/p} S^l;$$

- The fundamental group of M^{n+l} is the semidirect product denoted by

$$\Gamma := \mathbb{Z}^n \times_\rho \mathbb{Z}/p.$$

An example of such an action ρ is given by the regular representation $\mathbb{Z}[\mathbb{Z}/p]$ modulo the ideal generated by the norm element, in which case we have $\rho \colon \mathbb{Z}/p \to \mathrm{aut}(\mathbb{Z}^{p-1})$.

Equip the torus and the sphere with the standard orientations. This determines an orientation on M. Since the $\mathbb{Z}/p$-action on S^l is free, there is a fibre bundle $T_\rho^n \to M^{n+l} \to L^l$. Note that our assumptions imply $\dim(M) = n + l \geq 5$.

In the situation described above, there exists a natural number k uniquely determined by $n = k(p-1)$. Define

$$r_j := \dim_{\mathbb{Z}}\left((\Lambda^j(\mathbb{Z}[\zeta_p]^k)^{\mathbb{Z}/p})\right), \tag{18.72}$$

where Λ^j means the j-th exterior power of a $\mathbb{Z}$-module and ζ_p is a primitive p-th root of unity. Then one has $r_j = \dim_{\mathbb{Z}}\left(H^j(T^n)^{\mathbb{Z}/p}\right)$. If $k = 1$, then $r_j = 0$ for $j \geq p$, and we get for $0 \leq j \leq (p-1)$

$$r_j = \frac{1}{p}\left(\binom{p-1}{j} + (-1)^j(p-1)\right).$$

The next result is [118, Theorem 0.1].

Theorem 18.73 (Topological structure sets of certain torus bundles over lens spaces) *There is an isomorphism of abelian groups*

$$\mathcal{S}^{\mathrm{TOP},s}(M) \cong \mathbb{Z}^{p^k(p-1)/2} \oplus \bigoplus_{i=0}^{n-1} L_{n-i}(\mathbb{Z})^{r_j}$$

where the natural number k is determined by the equality $n = k(p-1)$ and the numbers r_i are defined in (18.72).

One can describe a set of invariants that decide whether a simple homotopy equivalence of closed manifolds with M as target is homotopic to a homeomorphism, see [118, Theorem 0.2]. They consist of splitting invariants and ρ-invariants.

Wang [418] treats the more general case where it is not required that the action $\rho \colon \mathbb{Z}/p \to \mathrm{aut}(\mathbb{Z}^n) = \mathrm{GL}(n, \mathbb{Z})$ induces a free action of $\mathbb{Z}/p$ on $\mathbb{Z}^n - \{0\}$.

18.12 Notes

We have mentioned some results on $S^{\text{DIFF}}(X)$ for X a product of spheres and aspherical manifolds in the respective sections. Further examples of calculations for X a complex and real projective space, mainly in low dimensions, can be found for instance in [62, 247, 334, 392]. Examples of non-trivial elements in $S^{\text{DIFF}}(L)$, where L is a lens space, can be found in [325]. See also the discussions in Section 19.6.

In the smooth case the problem of determining the so-called inertia subgroup is interesting. Given an oriented smooth manifold M of dimension n, the *inertia group* $I(M) \subset \Theta_n$, where Θ_n is the group of oriented homotopy n-spheres from Chapter 12, is defined as the set of $[\Sigma] \in \Theta_n$ such there exists an orientation preserving diffeomorphism $\varphi \colon M \to M\#\Sigma$. This group is often non-trivial, but there are also limitations on its size. Some information can be found e.g., in [32, 33, 53, 148, 149, 205, 211, 212, 223, 244, 402]. See also Conjecture 19.45.

There are also other calculations of (simple) structure sets in the categories TOP and PL. Without attempting to be exhaustive we list some that are related to the calculations presented in this chapter: [17, 18, 51, 200, 227, 217, 280, 313, 314, 315] and also Theorem 19.14 below.

The action of ho-aut on the structure set has also been studied for connected sums of real projective spaces in [51], see also in [93, Theorem 6.49]. The following general result of Chang–Weinberger [92, Theorem 1] is obtained using an L^2-version of the ρ-invariant from Section 18.7.

Theorem 18.74 *Let M have dimension $n = 4k-1 \geq 7$ and suppose $\pi = \pi_1(M)$ contains torsion. Then there are infinitely many pairwise non-homeomorphic manifolds that are simple homotopy equivalent to M, but not homeomorphic to M.*

Many examples and applications of surgery theory can be found in the recent book by Chang and Weinberger [93]. We further recommend the following collections of the state of the art at the times of publication: [72, 73, 153, 154].

Chapter 19
Topological Rigidity

19.1 Introduction

Recall (Definition 11.31) that a closed topological manifold M is called *topologically rigid* if any homotopy equivalence $N \to M$ with a closed topological manifold N as source and M as target is homotopic to a homeomorphism. In this chapter we want to study the question of which closed topological manifolds are topologically rigid.

In Section 19.2 we give a necessary and sufficient condition for topological rigidity in terms of the surgery exact sequence.

After we have given some information about aspherical closed manifolds in Section 19.3, we discuss in Section 19.4 the Borel Conjecture predicting that any aspherical closed topological manifold is topologically rigid.

Examples of non-aspherical closed manifolds that are topologically rigid are presented in Section 19.5.

In Section 19.6 we briefly discuss that smooth rigidity is a very rare phenomenon in high dimensions.

Guide 19.1 One may browse through this chapter even before one has absorbed the basics about surgery theory since the results can be understood without much previous knowledge and illustrate the potential of surgery theory.

19.2 Topological Rigidity and the Surgery Exact Sequence

Lemma 19.2 *Let M be a connected closed topological manifold of dimension n. Suppose that $n \geq 5$ holds or that $n = 4$ and π is good in the sense of Freedman, see Remark 8.30. Consider the following two statements:*

(i) *M is topologically rigid.*
(ii) *The surgery obstruction map*

$$\sigma_{n+1}^{\text{TOP},s} : \mathcal{N}^{\text{TOP}}(M \times D^1, \partial(M \times D^1)) \to L_{n+1}^s(\mathbb{Z}\Pi(M), O_M)$$

is surjective and the surgery obstruction map

$$\sigma_n^{\mathrm{TOP},s} \colon \mathcal{N}^{\mathrm{TOP}}(M) \to L_n^s(\mathbb{Z}\Pi(M), O_M)$$

is injective.

Then (i) $\implies$ (ii). *If* $\mathrm{Wh}(\Pi(M)) = 0$, *then* (i) $\iff$ (ii)

Proof. This follows from Lemma 11.32 and Theorem 11.25. $\square$

Let $\mathbf{E}$ be a spectrum. One can assign to $\mathbf{E}$ its *1-connective cover* $\mathbf{E}\langle 1\rangle$. It comes with a map of spectra $\mathbf{i}(\mathbf{E}) \colon \mathbf{E}\langle 1\rangle \to \mathbf{E}$. The main properties are that $\pi_m(\mathbf{i}(\mathbf{E})) \colon \pi_m(\mathbf{E}\langle 1\rangle) \to \pi_m(\mathbf{E})$ is bijective for $m \geq 1$ and $\pi_m(\mathbf{E}\langle 1\rangle) = 0$ for $m \leq 0$. This construction can be arranged to be functorial in $\mathbf{E}$.

Let M be a connected closed topological manifold of dimension $n \geq 5$ with fundamental group $\pi = \pi_1(M)$ and first Stiefel–Whitney class $w \colon \pi \to \{\pm 1\}$. Let $\widetilde{M} \to M$ be its universal covering. Let ϵ be a decoration.

We get a homomorphism

$$H_n^\pi(\mathrm{id}_{\widetilde{M}}; \mathbf{i}(\mathbf{L}(\mathbb{Z}))^w) \colon H_n^\pi(\widetilde{M}; \mathbf{L}(\mathbb{Z})\langle 1\rangle^w) \to H_n^\pi(\widetilde{M}; \mathbf{L}(\mathbb{Z})^w), \qquad (19.3)$$

where $\mathbf{L}(\mathbb{Z})$ is the spectrum from (15.479), see also (16.34) and Subsection 16.5.1.

Let $f_{\widetilde{M}} \colon \widetilde{M} \to E\pi$ be the up to π-homotopy unique π-map. It induces for $m \in \mathbb{Z}$ homomorphisms

$$H_m^\pi(f_{\widetilde{M}}; \mathbf{L}(\mathbb{Z})^w) \colon H_m^\pi(\widetilde{M}; \mathbf{L}(\mathbb{Z})^w) \to H_m^\pi(E\pi; \mathbf{L}(\mathbb{Z})^w); \qquad (19.4)$$

$$H_m^\pi(f_{\widetilde{M}}; \mathbf{L}(\mathbb{Z})\langle 1\rangle^w) \colon H_m^\pi(\widetilde{M}; \mathbf{L}(\mathbb{Z})\langle 1\rangle^w) \to H_m^\pi(E\pi; \mathbf{L}(\mathbb{Z})\langle 1\rangle^w). \qquad (19.5)$$

The functor appearing in (16.34), which was defined for $\epsilon = \langle -\infty\rangle$, also makes sense for other decorations. Since the decoration does not matter for $L_n^\epsilon(\mathbb{Z})$, if we take $G = \pi$, the map appearing in Conjecture 16.19 (iii) has for $m \in \mathbb{Z}$ also a version of the shape

$$\mathrm{asmb}_m^\epsilon \colon H_m^\pi(E\pi; \mathbf{L}(\mathbb{Z})^w) \to L_m^\epsilon(\mathbb{Z}\pi). \qquad (19.6)$$

Proposition 19.7 *Let M be a connected closed topological manifold of dimension $n \geq 4$ with fundamental group $\pi = \pi_1(M)$ and first Stiefel–Whitney class $w \colon \pi \to \{\pm 1\}$, and make the choice (ABP) of (8.13).*

Then there are bijections

$$\mu_{n+1} \colon \mathcal{N}^{\mathrm{TOP}}(M \times D^1, \partial(M \times D^1)) \xrightarrow{\cong} H_{n+1}^\pi(\widetilde{M}; \mathbf{L}(\mathbb{Z})\langle 1\rangle^w);$$

$$\mu_n \colon \mathcal{N}^{\mathrm{TOP}}(M) \xrightarrow{\cong} H_n^\pi(\widetilde{M}; \mathbf{L}(\mathbb{Z})\langle 1\rangle^w),$$

such that the composite

19.2 Topological Rigidity and the Surgery Exact Sequence   835

$$\mathcal{N}^{\mathrm{TOP}}(M \times D^1, \partial(M \times D^1)) \xrightarrow{\mu_{n+1}} H^\pi_{n+1}(\widetilde{M}; \mathbf{L}(\mathbb{Z})\langle 1 \rangle^w)$$

$$\xrightarrow{H^\pi_{n+1}(\mathrm{id}_{\widetilde{M}}; \mathbf{i}(\mathbf{L}(\mathbb{Z}))^w)} H^\pi_{n+1}(\widetilde{M}; \mathbf{L}(\mathbb{Z})^w)$$

$$\xrightarrow{H^\pi_{n+1}(f_{\widetilde{M}}; \mathbf{L}(\mathbb{Z})^w)} H^\pi_{n+1}(E\pi; \mathbf{L}(\mathbb{Z})^w) \xrightarrow{\mathrm{asmb}^s_{n+1}} L^s_{n+1}(\mathbb{Z}\pi)$$

is $\sigma^{\mathrm{TOP},s}_{n+1}$ and the composite

$$\mathcal{N}^{\mathrm{TOP}}(M) \xrightarrow{\mu_n} H^\pi_n(\widetilde{M}; \mathbf{L}(\mathbb{Z})\langle 1 \rangle^w) \xrightarrow{H^\pi_n(\mathrm{id}_{\widetilde{M}}; \mathbf{L}(\mathbb{Z})\langle 1 \rangle^w)} H^\pi_n(\widetilde{M}; \mathbf{L}(\mathbb{Z})^w)$$

$$\xrightarrow{H^\pi_n(f_{\widetilde{M}}; \mathbf{L}(\mathbb{Z})^w)} H^\pi_n(E\pi; \mathbf{L}(\mathbb{Z})^w) \xrightarrow{\mathrm{asmb}^s_n} L^s_n(\mathbb{Z}\pi)$$

is $\sigma^{\mathrm{TOP},s}_n$

Proof. See [348, Theorem 18.5 on page 198], [343], [234], and Remark 15.483. □

Remark 19.8 (The surgery exact sequence and L-classes) For every homology theory satisfying the disjoint union axiom and hence in particular for $H_*(-; \mathbf{L}(\mathbb{Z})\langle 1 \rangle)$, there is a natural Chern character, see Dold [131]. Thus we get for every CW-complex natural isomorphisms

$$\mathrm{ch}_n(X) \colon \bigoplus_{i \in \mathbb{Z}, i \geq 1} H_{n-4i}(X; \mathbb{Q}) \xrightarrow{\cong} H_n(X; \mathbf{L}(\mathbb{Z})\langle 1 \rangle) \otimes_\mathbb{Z} \mathbb{Q}. \quad (19.9)$$

Consider an n-dimensional connected closed topological manifold M with fundamental group $\pi = \pi_1(M)$ such that its first Stiefel–Whitney class $w_1(M)$ vanishes. Make the choice (ABP) of (8.13), which includes fixing a fundamental class $[M] \in H_n(M; \mathbb{Z})$. Recall the isomorphism

$$\mu_n \colon \mathcal{N}^{\mathrm{TOP}}(M) \xrightarrow{\cong} H^\pi_n(\widetilde{M}; \mathbf{L}(\mathbb{Z})\langle 1 \rangle)$$

from Proposition 19.7. We get a rational isomorphism

$$\omega_n \colon \mathcal{N}^{\mathrm{TOP}}(M) \otimes_\mathbb{Z} \mathbb{Q} \xrightarrow{\mu_n \otimes \mathrm{id}_\mathbb{Q}} H_n(M; \mathbf{L}(\mathbb{Z})\langle 1 \rangle) \otimes_\mathbb{Z} \mathbb{Q}$$

$$\xrightarrow{\mathrm{ch}_n(M)^{-1}} \bigoplus_{i \in \mathbb{Z}, i \geq 1} H_{n-4i}(M; \mathbb{Q}). \quad (19.10)$$

Then the composite of

$$\eta^s_n \colon \mathcal{S}^{\mathrm{TOP},s}(M) \to \mathcal{N}^{\mathrm{TOP}}(M) \to \mathcal{N}^{\mathrm{TOP}}(M) \otimes_\mathbb{Z} \mathbb{Q}$$

with ω_n sends the class of a homotopy equivalence $f \colon N \to M$ to the element $\{(f_*(\mathcal{L}(N) \cap [N]) - \mathcal{L}(M) \cap [M])_i \mid i \in \mathbb{Z}, i \geq 1\}$, where $[N] \in H_n(N; \mathbb{Z})$ is determined by the property that it is sent under $H_n(f; \mathbb{Z})$ to $[M]$, see [348, Example 18.4 on page 198].

The contents of this remark have already been presented from another point of view in Theorem 17.17.

Proposition 19.11 *Let M be a connected closed topological manifold of dimension n with fundamental group $\pi = \pi_1(M)$ and first Stiefel–Whitney class $w\colon \pi \to \{\pm 1\}$. Suppose that $n \geq 5$ holds or that $n = 4$ and π is good in the sense of Freedman, see Remark 8.30.*

(i) *Suppose that M is topologically rigid. Then the map of (19.5)*

$$H_n^\pi(f_{\widetilde{M}};\mathbf{L}(\mathbb{Z})\langle 1\rangle^w)\colon H_n^\pi(\widetilde{M};\mathbf{L}(\mathbb{Z})\langle 1\rangle^w) \to H_n^\pi(E\pi;\mathbf{L}(\mathbb{Z})\langle 1\rangle^w)$$

is injective;

(ii) *Suppose that π is torsionfree and a Farrell–Jones group in the sense of Definition 16.43 and that M is topological rigid. Then the map of (19.4)*

$$H_{n+1}^\pi(f_{\widetilde{M}};\mathbf{L}(\mathbb{Z})^w)\colon H_{n+1}^\pi(\widetilde{M};\mathbf{L}(\mathbb{Z})^w) \to H_{n+1}^\pi(E\pi;\mathbf{L}(\mathbb{Z})^w)$$

is surjective;

(iii) *Suppose that π is torsionfree and a Farrell–Jones group in the sense of Definition 16.43. Then M is topologically rigid if the map of (19.4)*

$$H_{n+1}^\pi(f_{\widetilde{M}};\mathbf{L}(\mathbb{Z})^w)\colon H_{n+1}^\pi(\widetilde{M};\mathbf{L}(\mathbb{Z})^w) \to H_{n+1}^\pi(E\pi;\mathbf{L}(\mathbb{Z})^w)$$

is surjective and the map of (19.4)

$$H_n^\pi(f_{\widetilde{M}};\mathbf{L}(\mathbb{Z})^w)\colon H_n^\pi(\widetilde{M};\mathbf{L}(\mathbb{Z})^w) \to H_n^\pi(E\pi;\mathbf{L}(\mathbb{Z})^w)$$

is injective.

Proof. The composite of $\mathrm{asmb}_m^s\colon H_m^\pi(E\pi;\mathbf{L}(\mathbb{Z})^w) \to L_m^s(\mathbb{Z}\pi, w)$ of (19.6) and $L_m^s(\mathbb{Z}\pi) \to L_m^{\langle-\infty\rangle}(\mathbb{Z}\pi, w)$ is $\mathrm{asmb}_m^{\langle-\infty\rangle}\colon H_m^\pi(E\pi;\mathbf{L}(\mathbb{Z})^w) \to L_m^{\langle-\infty\rangle}(\mathbb{Z}\pi, w)$. Hence Corollary 16.41 implies that $\mathrm{Wh}(\Pi(M)) = 0$ holds and that the map

$$\mathrm{asmb}_m^s\colon H_m^\pi(E\pi;\mathbf{L}(\mathbb{Z})^w) \to L_m^s(\mathbb{Z}\pi, w)$$

is bijective for all $m \in \mathbb{Z}$ provided π is a torsionfree Farrell–Jones group.

An easy argument using the Atiyah–Hirzebruch spectral sequence shows that the map $H_m^\pi(\mathrm{id}_{\widetilde{M}};\mathbf{i}(\mathbf{L}(\mathbb{Z}))^w)\colon H_m^\pi(\widetilde{M};\mathbf{L}(\mathbb{Z})\langle 1\rangle^w) \to H_m^\pi(\widetilde{M};\mathbf{L}(\mathbb{Z})^w)$ of (19.3) is bijective for $m = n+1$ and injective for $m = n$.

Note that the following diagram commutes

$$\begin{array}{ccc}
H_m^\pi(\widetilde{M};\mathbf{L}(\mathbb{Z})\langle 1\rangle^w) & \xrightarrow{H_m^\pi(\mathrm{id}_{\widetilde{M}};\mathbf{i}(\mathbf{L}(\mathbb{Z}))^w)} & H_m^\pi(\widetilde{M};\mathbf{L}(\mathbb{Z})^w) \\
{\scriptstyle H_n^\pi(f_{\widetilde{M}};\mathbf{L}(\mathbb{Z})\langle 1\rangle^w)}\downarrow & & \downarrow{\scriptstyle H_n^\pi(f_{\widetilde{M}};\mathbf{L}(\mathbb{Z})^w)} \\
H_m^\pi(E\pi;\mathbf{L}(\mathbb{Z})\langle 1\rangle^w) & \xrightarrow{H_m^\pi(E\pi;\mathbf{i}(\mathbf{L}(\mathbb{Z}))^w)} & H_m^\pi(E\pi;\mathbf{L}(\mathbb{Z})^w).
\end{array}$$

19.2 Topological Rigidity and the Surgery Exact Sequence   837

Now Proposition 19.11 follows from Lemma 19.2 and Proposition 19.7. □

Example 19.12 (A necessary condition for topological rigidity) Consider an n-dimensional connected closed topological manifold M with trivial first Stiefel–Whitney class $w_1(M)$ and fundamental group $\pi = \pi_1(M)$. Let $f_M : M \to B\pi$ be the classifying map. Now we have the commutative square

$$\begin{array}{ccc} \bigoplus_{i \in \mathbb{Z}, i \geq 1} H_{n-4i}(M; \mathbb{Q}) & \xrightarrow{\bigoplus_{i \in \mathbb{Z}, i \geq 1} H_{n-4i}(f_M; \mathbb{Q})} & \bigoplus_{i \in \mathbb{Z}, i \geq 1} H_{n-4i}(B\pi; \mathbb{Q}) \\ \mathrm{ch}_n(M) \downarrow \cong & & \cong \downarrow \mathrm{ch}_n(B\pi) \\ H_n(M; \mathbf{L}(\mathbb{Z})\langle 1\rangle) \otimes_{\mathbb{Z}} \mathbb{Q} & \xrightarrow{H_n(f_M; \mathbf{L}(\mathbb{Z})\langle 1\rangle) \otimes_{\mathbb{Z}} \mathrm{id}_{\mathbb{Q}}} & H_n(B\pi; \mathbf{L}(\mathbb{Z})\langle 1\rangle) \otimes_{\mathbb{Z}} \mathbb{Q} \end{array}$$

where the vertical isomorphisms have been introduced in (19.9). If M is topologically rigid, the lower horizontal map is injective by Proposition 19.11 (i).

Hence for all $i \in \mathbb{Z}$ with $i \geq 1$ the map

$$H_{n-4i}(f_M; \mathbb{Q}) \colon H_{n-4i}(M; \mathbb{Q}) \to H_{n-4i}(B\pi; \mathbb{Q})$$

is injective if M is topologically rigid.

Analogously one shows for all $i \in \mathbb{Z}$ that the map

$$H_{n+1-4i}(f_M; \mathbb{Q}) \colon H_{n+1-4i}(M; \mathbb{Q}) \to H_{n+1-4i}(B\pi; \mathbb{Q})$$

is surjective if M is topologically rigid and π is a torsionfree Farrell–Jones group.

Exercise 19.13 Let M be a simply connected n-dimensional closed topological manifold, which is topologically rigid. Then $H_{n-4i}(M; \mathbb{Q})$ is zero for $i \in \mathbb{Z}$ with $i \geq 1$ and $4i \neq n$.

Theorem 19.14 (The structure set for simply connected manifold times aspherical manifold) *Let M be a simply connected closed topological manifold of dimension $m \geq 1$ and N be an aspherical closed topological manifold of dimension $n \geq 0$ with fundamental group π and first Stiefel–Whitney class $w \colon \pi \to \{\pm 1\}$. Assume that π is a Farrell–Jones group in the sense of Definition 16.43. Suppose that $m + n \geq 5$ or that $m + n \geq 4$ and π is good in the sense of Freedman, see Remark 8.30. Fix an element $x \in M$.*

Then we obtain an isomorphism of abelian groups.

$$\mathcal{S}^{\mathrm{TOP},s}(M \times N) \xrightarrow{\cong} H_m^\pi((M, \{x\}) \times E\pi; \mathbf{L}(\mathbb{Z})\langle 1\rangle^w).$$

Proof. We begin with the case $m + n \geq 5$. Note that $\widetilde{N}$ is a model for $E\pi$ and $M \times \widetilde{N} \to M \times N$ is the universal covering of $M \times N$. The canonical map

$$H^\pi_{m+n+l}(\widetilde{N}; \mathbf{L}(\mathbb{Z})\langle 1\rangle^w) \to H^\pi_{m+n+l}(\widetilde{N}; \mathbf{L}(\mathbb{Z})^w)$$

is bijective for $l = 0, 1$ by an easy spectral sequence argument since $m \geq 1$. Hence the geometric surgery exact sequence in the topological category, see Theorem 11.25

and Remark 11.38, reduces by Proposition 19.7 and Corollary 16.41 to the exact sequence of abelian groups

$$H^\pi_{m+n+1}(M \times \widetilde{N}; \mathbf{L}(\mathbb{Z})\langle 1\rangle^w) \xrightarrow{H^\pi_{m+n+1}(\mathrm{pr};\mathbf{L}(\mathbb{Z})\langle 1\rangle^w)} H^\pi_{m+n+1}(\widetilde{N}; \mathbf{L}(\mathbb{Z})\langle 1\rangle^w)$$
$$\to \mathcal{S}^{\mathrm{TOP},s}(M \times N) \to H^\pi_{m+n}(M \times \widetilde{N}; \mathbf{L}(\mathbb{Z})\langle 1\rangle^w)$$
$$\xrightarrow{H^\pi_{m+n}(\mathrm{pr};\mathbf{L}(\mathbb{Z})\langle 1\rangle^w)} H^\pi_{m+n}(\widetilde{N}; \mathbf{L}(\mathbb{Z})\langle 1\rangle^w)$$

where $\mathrm{pr}\colon M \times \widetilde{N} \to \widetilde{N}$ is the projection. Now use the fact that the inclusion $\{x\} \times \widetilde{N} \to M \times \widetilde{N}$ composed with pr is a homeomorphism and the long exact homology sequence associated to the pair $(M, x) \times \widetilde{N}$.

For the case $m + n = 4$ we refer to Remark 8.30. □

If in Theorem 19.14 we take the manifold N to be $\{\bullet\}$, then taking Remark 8.30 into account, we get the next result.

Theorem 19.15 (The structure set for simply connected closed manifolds) *Let M be a simply connected closed topological manifold of dimension $n \geq 4$. Fix an element $x \in M$.*

Then we obtain an isomorphism of abelian groups

$$\mathcal{S}^{\mathrm{TOP},s}(M) \xrightarrow{\cong} H_n(M, \{x\}; \mathbf{L}(\mathbb{Z})\langle 1\rangle).$$

Exercise 19.16 Let M be a simply connected closed topological manifold of dimension 5. Show that $\mathcal{S}^{\mathrm{TOP},s}(M)$ is isomorphic to $H_3(M; \mathbb{Z}/2)$.

19.3 Aspherical Closed Manifolds

Definition 19.17 (Aspherical) A space X is called *aspherical* if it is path connected and all its higher homotopy groups vanish, i.e., $\pi_n(X)$ is trivial for $n \geq 2$.

Exercise 19.18 Let $F \to E \to B$ be a fibration of path connected spaces. Show that E is aspherical if F and B are aspherical, and that F is aspherical if E and B are aspherical. Give an example such that B is not aspherical but F and E are aspherical.

19.3.1 Homotopy Classification of Aspherical CW-Complexes

A CW-complex is aspherical if and only if it is connected and its universal covering is contractible. Given two aspherical CW-complexes X and Y, the map from the set of homotopy classes of maps $X \to Y$ to the set of group homomorphisms $\pi_1(X) \to \pi_1(Y)$ modulo inner automorphisms of $\pi_1(Y)$ given by the map induced on the fundamental groups is a bijection. In particular, two aspherical CW-complexes are homotopy equivalent if and only if they have isomorphic fundamental groups and every isomorphism between their fundamental groups comes from a homotopy equivalence.

19.3.2 Low Dimensions

A connected closed 1-dimensional manifold is homeomorphic to S^1 and hence aspherical.

Let M be a connected closed 2-dimensional manifold. Then M is either aspherical or homeomorphic to S^2 or $\mathbb{RP}^2$. The following statements are equivalent:

(i) M is aspherical;
(ii) M admits a Riemannian metric that is *flat*, i.e., with sectional curvature constant 0, or that is *hyperbolic*, i.e., with sectional curvature constant -1;
(iii) The universal covering of M is homeomorphic to $\mathbb{R}^2$.

A connected closed 3-manifold M is called *prime* if for any decomposition as a connected sum $M \cong M_0 \sharp M_1$ one of the summands M_0 or M_1 is homeomorphic to S^3. It is called *irreducible* if any embedded sphere S^2 bounds a disk D^3. Every irreducible closed 3-manifold is prime. A prime closed 3-manifold is either irreducible or an S^2-bundle over S^1, see [182, Lemma 3.13 on page 28]. An orientable closed 3-manifold is aspherical if and only if it is irreducible and has infinite fundamental group. A closed 3-manifold is aspherical if and only if it is irreducible and its fundamental group is infinite and contains no element of order 2. This follows from the Sphere Theorem [182, Theorem 4.3 on page 40].

Thurston's Geometrisation Conjecture implies that a closed 3-manifold is aspherical if and only if its universal covering is homeomorphic to $\mathbb{R}^3$. This follows from [182, Theorem 13.4 on page 142] and the fact that the 3-dimensional geometries which have compact quotients and whose underlying topological spaces are contractible have as underlying smooth manifold $\mathbb{R}^3$, see [374]. A proof of Thurston's Geometrisation Conjecture is given in [222, 311] following ideas of Perelman.

There are examples of orientable closed 3-manifolds that are aspherical but do not support a Riemannian metric with non-positive sectional curvature, see [242].

For more information about 3-manifolds we refer, for instance, to [10, 182, 374].

19.3.3 Non-Positive Curvature

Let M be a closed smooth manifold. Suppose that it possesses a Riemannian metric whose sectional curvature is non-positive, i.e., is ≤ 0 everywhere. Then the universal covering $\widetilde{M}$ inherits a complete Riemannian metric whose sectional curvature is non-positive. Since $\widetilde{M}$ is simply connected and has non-positive sectional curvature, the Hadamard–Cartan Theorem, see [164, 3.87 on page 134], implies that $\widetilde{M}$ is diffeomorphic to $\mathbb{R}^n$ and hence contractible. We conclude that M is aspherical.

19.3.4 Torsionfree Discrete Subgroups of Almost Connected Lie Groups

Let L be a Lie group with finitely many path components. Let $K \subseteq L$ be a maximal compact subgroup. Let $G \subseteq L$ be a discrete torsionfree subgroup. Then $M = G\backslash L/K$ is an aspherical closed smooth manifold with fundamental group G since its universal covering L/K is diffeomorphic to $\mathbb{R}^n$ for appropriate n, see [181, Theorem 1. in Chapter VI].

19.3.5 Hyperbolisation

A very important construction of aspherical manifolds comes from the *hyperbolisation technique* due to Gromov [169]. It turns a cell complex into a non-positively curved (and hence aspherical) polyhedron. The rough idea is to define this procedure for simplices such that it is natural under inclusions of simplices and then define the hyperbolisation of a simplicial complex by gluing the results for the simplices together as described by the combinatorics of the simplicial complex. The goal is to achieve that the result shares some of the properties of the simplicial complexes one has started with, but additionally to produce a non-positively curved and hence aspherical polyhedron. Since this construction preserves local structures, it turns topological manifolds into topological manifolds.

We briefly explain what the *orientable hyperbolisation procedure* gives. Further expositions of this construction can be found in [96, 122, 124, 126]. We start with a finite-dimensional simplicial complex Σ and assign to it a cubical cell complex $h(\Sigma)$ and a natural map $c \colon h(\Sigma) \to \Sigma$ with the following properties:

(i) $h(\Sigma)$ is non-positively curved and in particular aspherical;
(ii) The natural map $c \colon h(\Sigma) \to \Sigma$ induces a surjection on the integral homology;
(iii) $\pi_1(f) \colon \pi_1(h(\Sigma)) \to \pi_1(\Sigma)$ is surjective;
(iv) If Σ is an oriented topological manifold, then

   (a) $h(\Sigma)$ is an oriented topological manifold;
   (b) The natural map $c \colon h(\Sigma) \to \Sigma$ has degree one;
   (c) There is a stable isomorphism between the tangent bundle $Th(\Sigma)$ and the pullback $c^*T\Sigma$.

There are also criteria when the output of hyperbolisation is a PL manifold or smooth manifold.

19.3 Aspherical Closed Manifolds

Remark 19.19 (Characteristic numbers and aspherical manifolds) Suppose that M is a closed manifold. Then the pullbacks of the characteristic classes of M under the natural map $c \colon h(M) \to M$ yield the characteristic classes of $h(M)$, and M and $h(M)$ have the same characteristic numbers. This shows that the aspherical condition does not impose any restrictions on the characteristic numbers of a manifold.

Remark 19.20 (Bordism and aspherical manifolds) The conditions above say that c is a normal map in the sense of surgery. One can show that c is normally bordant to the identity map on M. In particular, M and $h(M)$ are oriented bordant.

Consider a bordism theory Ω_* for PL manifolds or smooth manifolds that is given by imposing conditions on the stable tangent bundle. Examples are unoriented bordism, oriented bordism, framed bordism. Then any bordism class can be represented by an aspherical closed manifold. If two aspherical closed manifolds represent the same bordism class, then one can find an aspherical bordism between them. See [122, Remarks 15.1] and [126, Theorem B].

For vanishing results about tautological classes for aspherical closed manifolds we refer to [180].

19.3.6 Exotic Aspherical Closed Manifolds

The following result is taken from Davis–Januszkiewicz [126, Theorem 5a.1].

Theorem 19.21 (Exotic aspherical closed manifolds in dimension 4) *There is an aspherical closed topological 4-manifold N with the following properties:*

(i) *N is not homotopy equivalent to a PL manifold;*
(ii) *N is not triangulable, i.e., not homeomorphic to a simplicial complex;*
(iii) *The universal covering $\widetilde{N}$ is not homeomorphic to $\mathbb{R}^4$;*
(iv) *N is homotopy equivalent to a piecewise flat, non-positively curved polyhedron.*

The next result is due to Davis–Januszkiewicz [126, Theorem 5a.4].

Theorem 19.22 (Aspherical closed non-PL manifolds)
For every $n \geq 4$, there exists an aspherical closed topological n-manifold that is not homotopy equivalent to a PL manifold.

The following result is proved by Davis–Fowler–Lafont [125] using the work of Manolescu [287, 286].

Theorem 19.23 (Non-triangulable aspherical closed manifolds)
For every $n \geq 6$, there exists an n-dimensional aspherical closed topological manifold that cannot be triangulated.

The proof of the following theorem can be found in [123], [126, Theorem 5b.1].

Theorem 19.24 (Exotic universal covering) *For each $n \geq 4$, there exists an aspherical closed topological n-manifold such that its universal covering is not homeomorphic to $\mathbb{R}^n$.*

By the Hadamard–Cartan Theorem, see [164, 3.87 on page 134], the manifold appearing in Theorem 19.24 above cannot be homeomorphic to a smooth manifold with Riemannian metric with non-positive sectional curvature.

The following theorem is proved in [126, Theorem 5c.1 and Remark on page 386] by considering the ideal boundary, which is a quasiisometry invariant in the negatively curved case.

Theorem 19.25 (Exotic aspherical closed manifolds with hyperbolic fundamental group) *For every $n \geq 5$, there exists an aspherical closed smooth n-dimensional manifold N that is homeomorphic to a strictly negatively curved polyhedron and which has in particular a hyperbolic fundamental group such that the universal covering is homeomorphic to $\mathbb{R}^n$, but N is not homeomorphic to a smooth manifold with Riemannian metric with negative sectional curvature.*

The next results are due to Belegradek [38, Corollary 5.1], Mess [293] and Weinberger, see [122, Section 13].

Theorem 19.26 (Aspherical closed manifolds with exotic fundamental groups)

(i) *For every $n \geq 4$, there is an aspherical closed topological manifold of dimension n whose fundamental group contains an infinite divisible abelian group;*

(ii) *For every $n \geq 4$, there is an aspherical closed PL manifold of dimension n whose fundamental group has an unsolvable word problem and whose simplicial volume in the sense of Gromov [168] is non-zero.*

Note that a finitely presented group with unsolvable word problem is not a CAT(0)-group, not hyperbolic, not automatic, not asynchronously automatic, not residually finite, and not linear over any commutative ring, see [38, Remark 5.2].

The proof of Theorem 19.26 is based on the *reflection group trick* as it appears for instance in [122, Sections 8,10 and 13]. It can be summarised as follows.

Theorem 19.27 (Reflection group trick) *Let G be a group that possesses a finite model for BG. Then there is an aspherical closed topological manifold M and a map $i: BG \to M$ and $r: M \to BG$ such that $r \circ i = \mathrm{id}_{BG}$.*

Remark 19.28 (Reflection group trick and various conjectures) Another interesting immediate consequence of the reflection group trick is that many well-known conjectures about groups hold for every group that possesses a finite model for BG if and only if it holds for the fundamental group of every aspherical closed topological manifold, see [122, Sections 11]. This applies for instance to the Kaplansky Conjecture, the Unit Conjecture, the Zero-divisor-conjecture, the Baum–Connes Conjecture, the Farrell–Jones Conjecture for algebraic K-theory for regular R, the Farrell–Jones Conjecture for algebraic L-theory, and to the vanishing of $\widetilde{K}_0(\mathbb{Z}G)$ and of $\mathrm{Wh}(G) = 0$. For information about these conjectures and their links, we refer for instance to [29, 254, 266]. Further similar consequences of the reflection group trick can be found in Belegradek [38].

For more information about fundamental groups of closed aspherical manifolds with unusual properties we refer, for instance, to [371]. Another survey article on aspherical manifolds is [258].

19.3.7 Non-Aspherical Closed Manifolds

A closed manifold of dimension ≥ 1 with finite fundamental group is never aspherical. So prominent non-aspherical manifolds are spheres, lens spaces, real projective spaces, and complex projective spaces.

Lemma 19.29 *The fundamental group of an aspherical finite-dimensional CW-complex X is torsionfree.*

Proof. Let $C \subseteq \pi_1(X)$ be a finite cyclic subgroup of $\pi_1(X)$. We have to show that C is trivial. Since X is aspherical, $C \backslash \widetilde{X}$ is a finite-dimensional model for BC. Hence $H_k(BC) = 0$ for large k. This implies that C is trivial. □

Exercise 19.30 If M is a connected sum $M_1 \sharp M_2$ of two closed manifolds M_1 and M_2 of dimension $n \geq 3$ that are not homotopy equivalent to a sphere, then M is not aspherical.

19.4 The Borel Conjecture

Conjecture 19.31 (Borel Conjecture) Every aspherical closed topological manifold is topologically rigid.

Theorem 19.32 (Status of the Borel Conjecture) *Let M be an aspherical closed topological manifold with fundamental group π and dimension $n \geq 5$. Suppose that π is a Farrell–Jones group in the sense of Definition 16.43.*
Then M satisfies the Borel Conjecture 19.31.

Proof. This follows from Proposition 19.11 (iii) since the map $f_{\widetilde{M}} \colon \widetilde{M} \to E\pi$ is a homotopy equivalence as M is by assumption aspherical. □

Remark 19.33 (Dimension 4) The conclusion of Theorem 19.32 also holds in dimension 4, provided that the fundamental group π is good in the sense of Freedman, see Remark 8.30.

Remark 19.34 (The Borel Conjecture in low dimensions) The Borel Conjecture is true in dimension ≤ 2 by the classification of closed manifolds of dimension 2. It is true in dimension 3 if Thurston's Geometrisation Conjecture is true. This follows from results of Waldhausen, see Hempel [182, Lemma 10.1 and Corollary 13.7] and Turaev [400], as explained for instance in [227, Section 5]. A proof of Thurston's Geometrisation Conjecture is given in [222, 311] following ideas of Perelman.

Remark 19.35 (The Borel Conjecture and the higher structure sets) In the situation as in Theorem 19.32 we also get for $k \geq 1$ that the simple structure set $\mathcal{S}^{\mathrm{TOP},s}(M \times D^k, \partial(M \times D^k)) \cong \mathcal{S}^{\mathrm{TOP},h}(M \times D^k, \partial(M \times D^k))$ consists precisely of one element. The reasoning is analogous to the proof of Theorem 19.32 since there are analogues of Lemma 19.2, Proposition 19.7 and Proposition 19.11 where M is replaced by $M \times D^k$, $M \times D^1$ is replaced by $M \times D^{k+1}$ and the dimensions of L-groups are adjusted appropriately.

Remark 19.36 (The Borel Conjecture does not hold in the smooth category) The Borel Conjecture 19.31 is false in the smooth category, i.e., if one replaces 'topological manifold' by 'smooth manifold' and 'homeomorphism' by 'diffeomorphism'. The torus T^n for $n \geq 5$ is an example, see (18.70) and [414, 15A]. Other counterexamples involving negatively curved manifolds are constructed by Farrell–Jones [149, Theorem 0.1].

Remark 19.37 (The Borel Conjecture versus Mostow rigidity) The examples of Farrell–Jones [149, Theorem 0.1] actually give more. Namely, they yield for given $\epsilon > 0$ a closed Riemannian manifold M_0 whose sectional curvature takes values in the interval $[-1+\epsilon, -1]$ and a hyperbolic closed manifold M_1 such that M_0 and M_1 are homeomorphic but not diffeomorphic. The idea of the construction is essentially to take the connected sum of M_1 with exotic spheres. Note that by definition M_0 would be hyperbolic if we were to take $\epsilon = 0$. Hence this example is remarkable in view of *Mostow rigidity*, which predicts for two hyperbolic closed manifolds N_0 and N_1 that they are isometrically diffeomorphic if and only if $\pi_1(N_0) \cong \pi_1(N_1)$ and that every homotopy equivalence $N_0 \to N_1$ is homotopic to an isometric diffeomorphism.

One may view the Borel Conjecture as the topological version of Mostow rigidity. The conclusion in the Borel Conjecture is weaker, one gets only homeomorphisms and not isometric diffeomorphisms, but the assumption is also weaker since there are many more aspherical closed topological manifolds than hyperbolic closed manifolds.

Exercise 19.38 Let $F \to E \to B$ be a fibration of closed manifolds such that F and B are aspherical and $\dim(F) \leq 2$ and $\dim(B) \leq 3$. Show that then E is topologically rigid.

19.5 Non-Aspherical Topologically Rigid Closed Manifolds

The Borel Conjecture 19.31 predicts that an aspherical closed topological manifold is topologically rigid. There are also non-aspherical closed manifolds that are topologically rigid, as we will illustrate next.

19.5 Non-Aspherical Topologically Rigid Closed Manifolds

19.5.1 Spheres

Since any self-homotopy equivalence of a sphere is homotopic to a homeomorphism, the Poincaré Conjecture, which is known to be true in all dimensions, see Theorem 2.5 and Section 2.6, is equivalent to the statement that every sphere is topologically rigid.

19.5.2 3-Manifolds

Theorem 19.39 (Topologically rigid 3-manifolds) *A closed 3-manifold with torsionfree fundamental group is topological rigid.*

Proof. Turaev [400] showed that a simple homotopy equivalence between closed 3-manifolds with torsionfree fundamental group is homotopic to a homeomorphism, provided that Thurston's Geometrisation Conjecture for irreducible 3-manifolds with infinite fundamental group and the 3-dimensional Poincaré Conjecture are true. Nowadays we know that the latter assumptions are satisfied. Since the Full Farrell–Jones Conjecture 16.42 holds for the fundamental group of any 3-manifold, see Theorem 16.44 (ie), we conclude from Corollary 16.41 that every homotopy equivalence of closed 3-manifolds with torsionfree fundamental groups is simple and hence homotopic to a homeomorphism. □

19.5.3 Products of Two Spheres

Theorem 19.40 (Topological rigidity and products of two spheres)
(i) *Consider $k, l \in \mathbb{Z}$ with $k, l \geq 2$. Then $S^k \times S^l$ is topologically rigid if and only if both k and l are odd;*
(ii) *Consider $k \in \mathbb{Z}$ with $k \geq 1$. Then $S^k \times S^1$ is topologically rigid.*

Proof. (i) This follows for $k + l \geq 4$ and $k, l \geq 2$ from Theorem 19.15 since an easy computation shows for a base point $z \in S^k \times S^l$

$$H_{k+l}(S^k \times S^l, \{z\}; \mathbf{L}(\mathbb{Z})\langle 1 \rangle) \cong L_k(\mathbb{Z}) \oplus L_l(\mathbb{Z}),$$

and $L_k(\mathbb{Z})$ vanishes if and only if k is odd, see Theorem 16.11 (i).
(ii) Suppose that $k \geq 3$. An easy calculation shows for a base point $x \in S^k$

$$H^{\mathbb{Z}}_{k+1}((S^k, \{x\}) \times E\mathbb{Z}; \mathbf{L}(\mathbb{Z})\langle 1 \rangle) \cong L_1(\mathbb{Z}) = \{0\}.$$

Now apply Theorem 19.14 using the fact that $\mathbb{Z}$ is good.
 The case $k = 2$ follows from Theorem 19.39.
 The case $S^1 \times S^1$ follows from the classification of connected closed surfaces and of the self-homotopy equivalences of $S^1 \times S^1$. □

19.5.4 Homology Spheres

The next result follows from [227, Theorem 0.23].

Theorem 19.41 (Topological rigidity and homology spheres) *Let M be an n-dimensional closed topological manifold of dimension $n \geq 5$ with fundamental group $\pi = \pi_1(M)$. Suppose that M is an integral homology sphere and $\mathrm{Wh}(\pi)$ vanishes.*

Then M is topologically rigid if and only if the inclusion $j \colon \mathbb{Z} \to \mathbb{Z}\pi$ induces an isomorphism

$$L_{n+1}^s(j) \colon L_{n+1}^s(\mathbb{Z}) \xrightarrow{\cong} L_{n+1}^s(\mathbb{Z}\pi)$$

on the simple L-groups.

19.5.5 Connected Sums

The next result is taken from [227, Theorem 0.9].

Theorem 19.42 (Topological rigidity and connected sums) *Let M and N be orientable closed topological manifolds of the same dimension $n \geq 5$ such that neither $\pi_1(M)$ nor $\pi_1(N)$ contains elements of order 2 or that $n = 0, 3 \mod 4$. If both M and N are topologically rigid, then the same is true for their connected sum $M\#N$.*

19.5.6 Homology Isomorphisms and Topological Rigidity

The next result follows from [227, Theorem 0.24] or from Lemma 19.2, Proposition 19.7, Proposition 19.11, and the Atiyah–Hirzebruch spectral sequence. Note that a map of CW-complexes which induces an isomorphism on π_1 and on the homology with integral coefficients is not necessarily a homotopy equivalence.

Theorem 19.43 (Homology isomorphisms and topologically rigidity)

(i) *Let $f \colon M \to N$ be a map of orientable connected closed topological manifolds of dimensions ≥ 5 that induces an isomorphism on π_1 and on the homology with integral coefficients. Suppose $\mathrm{Wh}(\pi_1(M)) = 0$. Then M is topologically rigid if and only if N is topologically rigid.*
(ii) *Let M be an orientable connected closed topological manifold of dimension ≥ 5 that is homological aspherical, i.e., its classifying map $M \to B\pi_1(M)$ induces an isomorphism on homology with integral coefficients. Suppose that $\pi_1(M)$ is a torsionfree Farrell–Jones group in the sense of Definition 16.43. Then M is topologically rigid.*

19.6 Smooth Rigidity

One may also study rigidity in the smooth category. Recall that a closed smooth manifold M is smoothly rigid if any homotopy equivalence $N \to M$ with a closed smooth manifold N as source and M as target is homotopic to a diffeomorphism. There is an analogous notion of PL rigidity.

In dimensions ≤ 3 there is no difference between topologically rigidity and smoothly rigidity because there is no difference between the topological and smooth category in these dimensions.

The four-sphere S^4 is known to be topologically rigid. It is smoothly rigid if and only if the smooth version of the four-dimensional Poincaré Conjecture is true, which is an open problem.

The question for which n the standard sphere S^n is smoothly rigid has already been discussed in Remark 12.36.

The following conjectures are based on Theorem 18.69 and a few computations of inertia groups, see for instance [148, 149]. One has to be very optimistic if one believes that they hold in general.

Conjecture 19.44 (Rigidity for closed aspherical manifolds) Let M be an aspherical closed topological manifold.

(i) Then M is topologically rigid;
(ii) Suppose additionally that M carries a PL structure. Then M is PL rigid if and only if $H^3(M; \mathbb{Z}/2)$ vanishes;
(iii) Suppose additionally that M carries a smooth structure. Then M is smoothly rigid if $\dim(M) \leq 3$. It is not smoothly rigid if $\dim(M) \geq 4$.

Conjecture 19.45 (Inertia Conjecture for closed aspherical manifolds) The inertia group of an aspherical closed manifold is trivial.

Let M is a smooth manifold of dimension ≥ 5 that is topologically rigid. Then the smooth structure set $\mathcal{S}^s(M)$ is finite by Remark 17.23.

Conjecture 19.45 has been proved in special cases (for example for a torus), see [144, Corollary 16.4].

19.7 Notes

We do not know a closed manifold that is topologically rigid and whose fundamental group contains torsion. Theorem 18.74 shows that in dimension $4k + 3$ torsion in the fundamental group indeed excludes topological rigidity.

One can also study a weaker form of topological rigidity where one wants to find for a homotopy equivalence $f \colon M \to N$ a homeomorphism $f' \colon M \to N$ such that f and f' induce the same map on the fundamental groups up to conjugation and does not require that f and f' are homotopic. Of course this is the same as topological rigidity if N is aspherical, but not in general. This notion is studied in [227].

Further information about the Borel Conjecture can be found in [425].

There is an equivariant version of the Borel Conjecture where one replaces EG with the classifying space for proper G-actions $\underline{E}G = E_{\mathcal{FIN}}(G)$, see [256]. One may ask whether for two compact proper topological G-manifolds M and N, both of which are models for $\underline{E}G$, any G-homotopy equivalence between them is G-homotopic to a G-homeomorphism. This version is not true in general and is investigated for instance in [101, 102, 103, 106, 119, 260], see also [425].

Chapter 20
Modified Surgery

20.1 Introduction

In this chapter we digress from the main line of the book, which treats the classification of manifolds with a given homotopy type via classical surgery, and discuss aspects of the use of surgery theory to classify manifolds with less homotopy-theoretic input. Specifically we discuss variations of the Surgery Program for classifying manifolds, which were pioneered by Kreck [225] and are often called *modified surgery*. Modified surgery might not have the general structural impact as surgery has on the classification of manifolds, on prominent conjectures such as those of Borel or Novikov, or on index theory and the theory of C^*-algebras, but it leads in some special but very interesting cases to better and beautiful results.

We will work in the smooth category unless explicitly stated otherwise.

The simple surgery exact sequence aims at the computation of the simple geometric structure set, see Definition 11.2 and Theorem 11.22. The computation of the simple structure set leads to a solution of the problem when a simple homotopy equivalence between two closed manifolds is homotopic to a homeomorphism or diffeomorphism. Therefore it is, for instance, a crucial ingredient for deciding whether a manifold is topologically rigid, see Definition 11.31 and Lemma 11.32. The uniqueness part of the Surgery Program, see Remark 2.9, aims at the question whether two closed manifolds M and N are homeomorphic or diffeomorphic. Before it can be applied, one has to find a simple homotopy equivalence $f \colon M \to N$, which can be fed into the structure set. It is possible that some simple homotopy equivalence $f \colon M \to N$ is homotopic to homeomorphism or diffeomorphism whereas another one $f' \colon M \to N$ is not. Therefore one has to consider the action of the group ho-aut$^s(N)$ of homotopy equivalence classes of simple self-homotopy equivalences of N on the simple structure set $\mathcal{S}^s(N)$ in order to decide whether two simply homotopy equivalent manifolds are homeomorphic or diffeomorphic. Namely, one chooses any simple homotopy equivalence $f \colon M \to N$ and then decides whether its class in the structure set lies in the ho-aut$^s(N)$-orbit of the class of id_N. It can happen that via surgery theory one can compute $\mathcal{S}^s(N)$, but it is hard to figure out ho-aut$^s(N)$ and its action on $\mathcal{S}^s(N)$. In most cases where this can be done, the closed manifolds are spheres or skeletons of Eilenberg–MacLane spaces.

The basic idea of modified surgery is that one a priori does not have to determine the homotopy type of N, but only its *normal k-type*, see Definition 20.36. Roughly speaking, this is the homotopy type of the k-skeleton of M together with the isomorphism type of the normal bundle over that skeleton. Then one can already start modified surgery theory. The drawback is that then the L-groups have to be replaced by certain l-monoids, which are very hard to handle. One knows much more about the L-groups than about these l-monoids. However, one can apply modified surgery in some special cases where one has some information about the normal k-type from the very beginning, and get beautiful results, see for instance Theorem 20.50 and Theorem 20.51. Another instance where modified surgery is useful appears if one is willing to allow stabilisation by connected sums with $S^k \times S^k$, see Definition 20.2. Then the L-groups disappear from the picture and one has to deal with bordism groups instead, see Theorem 20.44. This is sometimes easier than calculating the simple surgery exact sequence, for instance for 4-dimensional manifolds. In dimension 4 this is due to the facts that the Atiyah–Hirzebruch spectral sequence becomes tractable because its E^2-page $E^2_{p,q}$ has input only for $0 \leq q \leq 4$, and that the 1-type can be easily computed, see Example 20.38.

In this overview chapter we only deal with the basics of modified surgery. In particular, we are careful about setting up the bundle information in detail, just as we did in the classical surgery setting. However, we do not go into the definition of the above mentioned l-monoids as this would be too technical and long. The interested reader can consult the original source [225]. The process of surgery is used in Lemma 20.24 and Theorem 20.25, which are in turn used in the main theorem of this chapter, namely, Theorem 20.44.

Guide 20.1 In order to get an intuition for which cases modified surgery yields interesting results, one may browse through Section 20.8.

20.2 Stable Diffeomorphisms

The classification of closed manifolds up to diffeomorphism is a very difficult problem. Sometimes one may not achieve it but it may be possible to obtain a weaker classification, namely, up to stable diffeomorphism.

For a natural number g, define the manifold $W_g := \natural_g (S^k \times S^k)$ to be the g-fold connected sum of $S^k \times S^k$ with itself where we put $W_0 = S^{2k}$.

Definition 20.2 (Stable diffeomorphism) Let M_0 and M_1 be connected closed $2k$-manifolds. A *stable diffeomorphism* between closed connected $2k$-manifolds M_0 and M_1 is a diffeomorphism

$$h: M_0 \natural W_{g_0} \to M_1 \natural W_{g_1}$$

for some natural numbers g_0 and g_1. We call M_0 and M_1 *stably diffeomorphic* if there exists a stable diffeomorphism between M_0 and M_1.

The advantage of classification up to stable diffeomorphism is that by taking the iterated connected sum with $S^k \times S^k$ one can avoid the problem occurring when turning a k-connected normal map with a closed $2k$-dimensional manifold as target into a $(k+1)$-connected map, which is then automatically a homotopy equivalence. Remember that the problem is that we may have two immersions $i_0, i_1 \colon S^k \to M$ in general position, which may have intersection points or self-intersection points, which we want to get rid of. If we allow taking the connected sum with $S^k \times S^k$, we can get rid of an intersection point by redirecting the two branches along the attached $S^k \times S^k$, so that the intersection point under consideration disappears.

Exercise 20.3 Let M_0 and M_1 be connected closed $2k$-dimensional manifolds. Let $h \colon M_0 \sharp W_{g_0} \to M_1 \sharp W_{g_1}$ be a homotopy equivalence. Show

$$g_1 - g_0 = (-1)^k \cdot \frac{\chi(M_0) - \chi(M_1)}{2}.$$

Remark 20.4 Let M and N be two smooth simply connected oriented 4-manifolds. Milnor [295] showed that they are oriented homotopy equivalent if and only if their intersection forms are isomorphic. Wall [404] showed that they are stably oriented diffeomorphic if their intersection forms are isomorphic. This implies that they are stably oriented diffeomorphic if and only if their signatures agree. We will show how modified surgery can be used to prove Wall's result in Example 20.46. Wall used the connected sum with $S^2 \times S^2$ to circumvent the problem that surgery a priori does not work in dimension 4.

The much harder problem about the classification without stabilisation was solved by Freedman [157, Theorem 1.5]. He showed that two closed topological oriented 4-manifolds are oriented homeomorphic if and only if their intersection forms agree and if they have the same Kirby–Siebenmann invariant. Note that the Kirby–Siebenmann invariant vanishes in the smooth case. More information about surgery in the topological category in dimension 4 can be found in [159].

For more information about 4-manifolds in the smooth category and dimension 4, we refer for instance to Donaldson [135].

20.3 Bordism Groups Associated to Spaces over BO

Let ξ_l be the universal l-dimensional vector bundle over $BO(l)$. Fix a bundle map $(i_{l,l+1}, \overline{i_{l,l+1}}) \colon \xi_l \oplus \mathbb{R} \to \xi_{l+1}$ covering a map $i_{l,l+1} \colon BO(l) \to BO(l+1)$. Up to homotopy of bundle maps this map $(i_{l,l+1}, \overline{i_{l,l+1}})$ is unique. For convenience we will assume that each $i_{l,l+1}$ is a cofibration. Define BO to be the colimit $\mathrm{colim}_{l \to \infty} BO(l)$. This definition agrees with the one from Section 6.2 up to homotopy. Let $i_l \colon BO(l) \to BO$ be the canonical inclusion.

Consider a space B together with a fibration $b\colon B \to \mathrm{BO}$. Define B_l by the pullback

$$\begin{array}{ccc} B_l & \xrightarrow{j_l} & B \\ {\scriptstyle b_l}\downarrow & & \downarrow{\scriptstyle b} \\ \mathrm{BO}(l) & \xrightarrow{i_l} & \mathrm{BO}\,. \end{array} \qquad (20.5)$$

Then we get for every l a pullback

$$\begin{array}{ccc} B_l & \xrightarrow{j_{l,l+1}} & B_{l+1} \\ {\scriptstyle b_l}\downarrow & & \downarrow{\scriptstyle b_{l+1}} \\ \mathrm{BO}(l) & \xrightarrow{i_{l,l+1}} & \mathrm{BO}(l+1)\,. \end{array} \qquad (20.6)$$

Let $\mathrm{pr}_l^* \xi_l$ be the l-dimensional vector bundle over B_l given by the pullback of the l-dimensional vector bundle ξ_l over $\mathrm{BO}(l)$. Then, by the universal property of the pullback construction, we obtain from the bundle map $(i_{l,l+1}, \overline{i_{l,l+1}})\colon \xi_l \oplus \mathbb{R} \to \xi_{l+1}$ a bundle map

$$\begin{array}{ccc} E_{\mathrm{pr}_l^* \xi_l \oplus \mathbb{R}} & \xrightarrow{\overline{j_{l,l+1}}} & E_{\mathrm{pr}_{l+1}^* \xi_{l+1}} \\ {\scriptstyle p_{\mathrm{pr}_l^* \xi_l}}\downarrow & & \downarrow{\scriptstyle p_{\mathrm{pr}_{l+1}^* \xi_{l+1}}} \\ B_l & \xrightarrow{j_{l,l+1}} & B_{l+1}\,. \end{array} \qquad (20.7)$$

Now we get a directed system of sets

$$\Omega_n(\mathrm{pr}_l^* \xi_l) \xrightarrow{\Omega_n(\overline{j_{l,l+1}})} \Omega_n(\mathrm{pr}_{l+1}^* \xi_{l+1}) \xrightarrow{\Omega_n(\overline{j_{l+1,l+2}})}$$
$$\Omega_n(\mathrm{pr}_{l+2}^* \xi_{l+2}) \xrightarrow{\Omega_n(\overline{j_{l+2,l+3}})} \Omega_n(\mathrm{pr}_{l+3}^* \xi_{l+3}) \xrightarrow{\Omega_n(\overline{j_{l+3,l+4}})} \cdots$$

where the set $\Omega_n(\mathrm{pr}_l^* \xi_l)$ has been defined in (4.51).

Remark 20.8 Note that each $\Omega_n(\mathrm{pr}_l^* \xi_l)$ is an abelian semigroup under taking disjoint union and the system above is a directed system of abelian semigroups. Note that $\Omega_n(\mathrm{pr}_l^* \xi_l)$ is not necessarily an abelian group. However, for every element $x \in \Omega_n(\mathrm{pr}_l^* \xi_l)$ there exists an element $y \in \Omega_n(\mathrm{pr}_{l+1}^* \xi_{l+1})$ such that $\Omega_n(\overline{j_{l,l+1}})(x) + y = 0$ holds.

This complication is due to the identification appearing in Remark 4.54. The point is essentially that for a vector bundle ξ over X the automorphism $\mathrm{id}_\xi \oplus -\mathrm{id}_\mathbb{R}\colon \xi \oplus \mathbb{R} \to \xi \oplus \mathbb{R}$ is not homotopic through bundle automorphisms covering id_X to $\mathrm{id}_{\xi \oplus \mathbb{R}}$

20.3 Bordism Groups Associated to Spaces over BO

whereas $\mathrm{id}_\xi \oplus -\mathrm{id}_{\underline{\mathbb{R}}} \oplus -\mathrm{id}_{\underline{\mathbb{R}}} \colon \xi \oplus \underline{\mathbb{R}} \oplus \underline{\mathbb{R}} \to \xi \oplus \underline{\mathbb{R}} \oplus \underline{\mathbb{R}}$ is homotopic through bundle automorphisms covering id_X to $\mathrm{id}_{\xi \oplus \underline{\mathbb{R}} \oplus \underline{\mathbb{R}}}$.

Suppose that the element x is given by the closed manifold M with embedding $i\colon M \hookrightarrow \mathbb{R}^{n+l}$ and the diagram

$$\begin{array}{ccc} E_{\nu(i)} & \xrightarrow{\overline{f}} & E_{\mathrm{pr}_l^* \xi_l} \\ {\scriptstyle p_{\nu(i)}}\downarrow & & \downarrow{\scriptstyle p_{\mathrm{pr}_l^* \xi_l}} \\ M & \xrightarrow{f} & B_l. \end{array} \tag{20.9}$$

Let $j\colon M \hookrightarrow \mathbb{R}^{n+l+1}$ be the composite of i with the standard embedding $\mathbb{R}^{n+l} = \mathbb{R}^{n+l} \times \{0\} \subseteq \mathbb{R}^{n+l+1}$. Then there is a canonical identification $\nu(j) = \nu(i) \oplus \underline{\mathbb{R}}$ and the desired element y is given by

$$\begin{array}{ccccc} E_{\nu(j)} = E_{\nu(i) \oplus \underline{\mathbb{R}}} & \xrightarrow{\overline{f} \oplus -\mathrm{id}_{\underline{\mathbb{R}}}} & E_{\mathrm{pr}_l^* \xi_l \oplus \underline{\mathbb{R}}} & \xrightarrow{\overline{j_{l,l+1}}} & E_{\mathrm{pr}_{l+1}^* \xi_{l+1}} \\ {\scriptstyle p_{\nu(j)}}\downarrow & & \downarrow{\scriptstyle p_{\mathrm{pr}_l^* \xi_l \oplus \underline{\mathbb{R}}}} & & \downarrow{\scriptstyle p_{\mathrm{pr}_{l+1}^* \xi_{l+1}}} \\ M & \xrightarrow{f} & B_l & \xrightarrow{j_{l,l+1}} & B_{l+1}. \end{array}$$

Remark 20.8 implies that the colimit of abelian semigroups above is actually an abelian group and hence we can define the abelian group

$$\Omega_n(b) = \Omega_n(b\colon B \to \mathrm{BO}) := \mathrm{colim}_{l \to \infty} \Omega_n(\mathrm{pr}_l^* \xi_l). \tag{20.10}$$

Lemma 20.11 *Let l be a natural number satisfying $l \geq (n+1)$ if $n \geq 2$, or $l \geq 4$ if $n = 1$, or $l \geq 2$ if $n = 0$.*

(i) *Consider an element y in $\Omega_n(\mathrm{pr}_{l+1}^* \xi_{l+1})$ given by a diagram*

$$\begin{array}{ccc} E_{\nu(i)} & \xrightarrow{\overline{f}} & E_{\mathrm{pr}_{l+1}^* \xi_{l+1}} \\ {\scriptstyle p_{\nu(i)}}\downarrow & & \downarrow{\scriptstyle p_{\mathrm{pr}_{l+1}^* \xi_{l+1}}} \\ M & \xrightarrow{f} & B_{l+1}. \end{array} \tag{20.12}$$

Then there is an element x in $\Omega_n(\mathrm{pr}_l^* \xi_l)$ represented by a diagram

$$\begin{array}{ccc} E_{\nu(i')} & \xrightarrow{\overline{f'}} & E_{\mathrm{pr}_l^* \xi_l} \\ {\scriptstyle p_{\nu(i')}}\downarrow & & \downarrow{\scriptstyle p_{\mathrm{pr}_l^* \xi_l}} \\ M & \xrightarrow{f'} & B_l. \end{array} \tag{20.13}$$

such that the map $\Omega_n(\overline{j_{l,l+1}})\colon \Omega_n(\mathrm{pr}_l^*\,\xi_l) \to \Omega_n(\mathrm{pr}_{l+1}^*\,\xi_{l+1})$ sends x to y and $j_{l,l+1} \circ f' \simeq f$ holds;

(ii) *The map*

$$\Omega_n(\overline{j_{l,l+1}})\colon \Omega_n(\mathrm{pr}_l^*\,\xi_l) \xrightarrow{\cong} \Omega_n(\mathrm{pr}_{l+1}^*\,\xi_{l+1})$$

is bijective;

(iii) *The structure map*

$$\psi_l\colon \Omega_n(\mathrm{pr}_l^*\,\xi_l) \xrightarrow{\cong} \Omega_n(b\colon B \to \mathrm{BO})$$

is bijective. In particular, the abelian semigroup $\Omega_n(\mathrm{pr}_l^*\,\xi_l)$ *is an abelian group.*

Proof. (i) Let $i\colon M \hookrightarrow \mathbb{R}^{n+l+1}$ be the embedding appearing in the diagram (20.12). Our assumptions on l guarantee that we can find an embedding $i'\colon M \hookrightarrow \mathbb{R}^{n+l}$ and an isotopy $h\colon M \times [0,1] \to \mathbb{R}^{n+l+1}$ between the embeddings i and $\iota_{n+l} \circ i'$ for the standard embedding $\iota_{n+l}\colon \mathbb{R}^{n+l} = \mathbb{R}^{n+l} \times \{0\} \hookrightarrow \mathbb{R}^{n+l+1}$. This isotopy can be embedded into a diffeotopy $\mathbb{R}^{n+l+1} \times [0,1] \to \mathbb{R}^{n+l+1}$. Now we can consider the embedding $M \times [0,1] \hookrightarrow \mathbb{R}^{n+l+1} \times [0,1]$ sending (x,t) to $(h(x,t), t)$ and its normal bundle. It yields a bordism between the diagram (20.12) and a diagram of the shape

$$\begin{array}{ccc} E_{\nu(\iota_{n+l} \circ i')} = E_{\nu(i') \oplus \mathbb{R}} & \xrightarrow{\overline{f_1}} & E_{\mathrm{pr}_{l+1}^*\,\xi_{l+1}} \\ {\scriptstyle p_{\nu(i') \oplus \mathbb{R}}}\downarrow & & \downarrow{\scriptstyle p_{\mathrm{pr}_{l+1}^*\,\xi_{l+1}}} \\ M & \xrightarrow{f} & B_{l+1}. \end{array} \qquad (20.14)$$

Note that the lower horizontal arrow is unchanged and that the diagram (20.14) is another representative of y.

The map $i_{l,l+1}\colon \mathrm{BO}(l) \to \mathrm{BO}(l+1)$ is l-connected. Since $b\colon B \to \mathrm{BO}$ is a fibration, we conclude from the pullback diagram (20.5) that also the map $b_{l+1}\colon B_{l+1} \to \mathrm{BO}(l+1)$ is a fibration. Because of the pullback (20.6) the map $j_{l,l+1}\colon B_l \to B_{l+1}$ is l-connected. Since $l \geq (n+1)$ holds by assumption and M is n-dimensional, the map given by composition with $j_{l,l+1}$

$$(j_{l,l+1})_*\colon [M, \mathrm{BO}(l)] \to [M, \mathrm{BO}(l+1)]$$

is bijective. Hence we can find a map $f'\colon M \to B_l$ satisfying $f \simeq j_{l,l+1} \circ f'$. This implies that we can find a diagram

$$\begin{array}{ccc} E_{\nu(i')} \oplus \mathbb{R} & \xrightarrow{\overline{j_{l,l+1} \circ f'}} & E_{\mathrm{pr}_{l+1}^*\,\xi_{l+1}} \\ {\scriptstyle p_{\nu(i') \oplus \mathbb{R}}}\downarrow & & \downarrow{\scriptstyle p_{\mathrm{pr}_{l+1}^*\,\xi_{l+1}}} \\ M & \xrightarrow{j_{l,l+1} \circ f'} & B_{l+1}, \end{array} \qquad (20.15)$$

20.3 Bordism Groups Associated to Spaces over BO

which is homotopic to the diagram (20.14) and in particular also represents y in $\Omega_n(\mathrm{pr}_{l+1}^* \xi_{l+1})$. From (20.15) we get a diagram

$$
\begin{array}{ccc}
E_{\nu(i') \oplus \underline{\mathbb{R}}} & \xrightarrow{\overline{f_1}} & E_{\mathrm{pr}_l^* \xi_l \oplus \underline{\mathbb{R}}} \\
{\scriptstyle p_{\nu(i') \oplus \underline{\mathbb{R}}}} \downarrow & & \downarrow {\scriptstyle p_{\mathrm{pr}_l^* \xi_l \oplus \underline{\mathbb{R}}}} \\
M & \xrightarrow{f'} & B_l
\end{array}
\qquad (20.16)
$$

whose composite with the diagram (20.7) is (20.15). Since $l \geq (n+1)$ we can find a diagram (20.13) such that the diagram

$$
\begin{array}{ccc}
E_{\nu(i') \oplus \underline{\mathbb{R}}} & \xrightarrow{\overline{f' \oplus \mathbb{R}}} & E_{\mathrm{pr}_l^* \xi_l \oplus \underline{\mathbb{R}}} \\
{\scriptstyle p_{\nu(i') \oplus \underline{\mathbb{R}}}} \downarrow & & \downarrow {\scriptstyle p_{\mathrm{pr}_l^* \xi_l \oplus \underline{\mathbb{R}}}} \\
M & \xrightarrow{f'} & B_l
\end{array}
\qquad (20.17)
$$

is homotopic relative f' to the diagram (20.16). This finishes the proof of assertion (i).

(ii) The map $\Omega_n(\overline{j_{l,l+1}})$ is surjective by assertion (i). Injectivity is proved by an analogous argument applied to a bordism relative its boundary.

(iii) This is a direct consequence of assertion (ii). This finishes the proof of Lemma 20.11. □

Next we deal with functoriality and homotopy invariance of $\Omega_n(b)$. If we have a map of fibrations

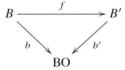

we get an induced map of abelian groups

$$\Omega_n(f) \colon \Omega_n(b) \to \Omega_n(b'), \qquad (20.18)$$

which comes from the various maps $\Omega(f_l, \overline{f_l}) \colon \Omega_n((b_l)^* \xi_l) \to \Omega_n((b'_l)^* \xi_l)$ for the obvious bundle maps

$$
\begin{array}{ccc}
E_{(b_l)^* \xi_l} & \xrightarrow{\overline{f_l}} & E_{(b'_l)^* \xi_l} \\
{\scriptstyle p_{(b_l)^* \xi_l}} \downarrow & & \downarrow {\scriptstyle p_{(b'_l)^* \xi_l}} \\
B_l & \xrightarrow{f_l} & B'_l.
\end{array}
$$

This is functorial in f. Moreover, if $f_0\colon b \to b'$ and $f_1\colon b \to b'$ are strongly fibre homotopy equivalent, then $\Omega_n(f_0) = \Omega_n(f_1)$.

Finally we want to get rid of the assumption that $b\colon B \to \mathrm{BO}$ is a fibration. Consider a map $f\colon X \to Y$. Then there exists a functorial construction of a commutative diagram, see [178, page 407] or [431, Theorem 7.30 on page 42]

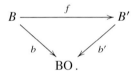
(20.19)

such that p_f is a fibration and u_f is a homotopy equivalence. So using (20.10) we define for a map $b\colon B \to \mathrm{BO}$, which is not necessarily a fibration, the abelian group

$$\Omega_n(b\colon B \to \mathrm{BO}) := \Omega_n(p_b\colon E_b \to \mathrm{BO}). \qquad (20.20)$$

Consider the commutative diagram

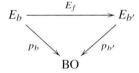

It induces a commutative diagram

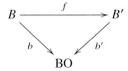

where p_b and $p_{b'}$ are fibrations. Now define using (20.18)

$$\Omega_n(f)\colon \Omega_n(b) := \Omega_n(p_b) \xrightarrow{\Omega_n(E_f)} \Omega_n(b') := \Omega_n(p_{b'}). \qquad (20.21)$$

Thus we obtain for $n \geq 0$ a covariant functor Ω_n from the category of spaces over BO to the category of abelian groups.

Exercise 20.22 Consider a commutative diagram

such that f is a homotopy equivalence. Show that $\Omega_n(f)\colon \Omega_n(b) \to \Omega_n(b')$ is an isomorphism.

Exercise 20.23 Let $b\colon B \to \mathrm{BO}$ be a fibration. Show

$$\Omega_n(b\colon B \to \mathrm{BO}) \cong \operatorname{colim}_{l\to\infty} \pi_{n+k}(\mathrm{Th}(\mathrm{pr}_k^* \xi_k)).$$

20.4 Modified Surgery

We first explain that we can find highly connected representatives for classes in $\Omega_n(b)$.

Lemma 20.24 *Let n be a natural number satisfying $n \geq 4$. Let k be the natural number, for which $n = 2k$ or $n = 2k + 1$. Consider a map $b\colon B \to \mathrm{BO}$ such that B has the homotopy type of a CW-complex with finite k-skeleton. Let x be an element of $\Omega_n(b)$.*

Then we can find for every natural number $l \geq (n + 1)$ a diagram

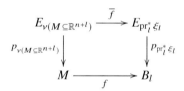

such that M is an n-dimensional closed manifold, f is k-connected and the canonical map $\Omega_n(\mathrm{pr}_l^* \xi_l) \to \Omega_n(b)$ sends the class represented by the diagram above to x.

Proof. Obviously we can assume without loss of generality that $b\colon B \to \mathrm{BO}$ is a fibration. The claim follows from Theorem 4.59 and Lemma 20.11 after we have proved that B_l is homotopy equivalent to a CW-complex with finite k-skeleton. Recall that B_l is defined by the pullback (20.5). As $i_l\colon \mathrm{BO}(l) \to \mathrm{BO}$ is l-connected, the map $j_{l,l+1}\colon B_l \to B$ is l-connected. Hence there is a CW-complex X with finite k-skeleton together with an l-connected map $f\colon B_l \to X$. Let X_k be the k-skeleton of X. Since $k \leq l$, we can find a map $u\colon X_k \to B_l$ such that the composite $f \circ u\colon X_k \to X$ is homotopic to the inclusion $X_k \to X$. Since this inclusion is k-connected and $k + 1 \leq l$, the map $u\colon X_k \to B_l$ is k-connected. By attaching cells of dimension $\geq k + 1$ to X_k we can extend u to a weak homotopy equivalence $v\colon Y \to B_l$. Hence it remains to show that B_l has the homotopy type of a CW-complex since then v is a homotopy equivalence and Y is CW-complex with finite k-skeleton.

Let $F \to E \to B$ be a fibration. If F and B have the homotopy type of a CW-complex, then E has the homotopy type of a CW-complex, see for instance [251, Lemma 7.2]. Note that there is a fibration $\Omega B \to F' \to E$ such that F' and F are homotopy equivalent, and that ΩB has the homotopy type of a CW-complex if B does, see [296]. We conclude that F has the homotopy type of a CW-complex if

E and B have. Since B, $BO(k)$, and BO have the homotopy type of a CW-complex and b and b_k are fibrations with the same fibre, B_k has the homotopy type of a CW-complex. □

Theorem 20.25 (Existence of stable diffeomorphisms) *Let l and k be natural numbers satisfying $k \geq 2$ and $l \geq (2k + 1)$. Let $b \colon B \to \mathrm{BO}$ be a map such that B has the homotopy type of a CW-complex with finite k-skeleton. Consider for $i = 0, 1$ a commutative diagram*

$$\begin{array}{ccc} E_{\nu(M_i \subseteq \mathbb{R}^{2k+l})} & \xrightarrow{\bar{f}_i} & E_{\mathrm{pr}_l^* \xi_l} \\ p_{\nu(M_i \subseteq \mathbb{R}^{2k+l})} \downarrow & & \downarrow p_{\mathrm{pr}_l^* \xi_l} \\ M_i & \xrightarrow{f_i} & B_l \end{array}$$

such that M_i is a $2k$-dimensional closed manifold, and f_i is k-connected for $i = 0, 1$. Suppose that the two diagrams define the same class in $\Omega_{2k}(b)$.

Then M_0 and M_1 are stably diffeomorphic.

Proof. This is proved in [225, Corollary 3]. There, an alternative proof is also outlined, which is due to Peter Teichner and which we describe next. We suppress the bundle data in the sequel. Since the two diagrams above represent the same class in $\Omega_n(b)$, we conclude from Lemma 20.24 that the classes represented by the two diagrams above in $\Omega_n(\mathrm{pr}_l^* \xi_l)$ already agree. Hence we can find a bordism $F \colon W \to B_l$ between $f_0 \colon M_0 \to B_l$ and $f_1 \colon M_1 \to B_l$. The obvious relative version of Lemma 20.11 applied to W shows that we can change W by surgery on its interior such that the map $F \colon W \to B_k$ is k-connected. Note that the inclusions of M_0 and M_1 into W induce isomorphisms on the homotopy groups in degree $i \leq (k - 1)$. By inspecting the proof of the s-Cobordism Theorem 2.1, in particular of Lemma 2.37, we see that we can assume that W is obtained from M_0 by attaching handles of index k and $k + 1$. Since W is $(2k + 1)$-dimensional, we can pass to the dual handle decomposition. In other words, we can attach k-handles to M_0 and k-handles to M_1 so that the resulting two closed manifolds M_0' and M_1' are related by a trivial h-cobordism and hence are diffeomorphic. The attaching maps of the k-handles to M_i become nullhomotopic in W. Since $\pi_{k-1}(M_i) \to \pi_{k-1}(W)$ is bijective for $i = 0, 1$, they become nullhomotopic in M_i and the bundle data become trivial. This implies that M_i' is obtained from M_i by taking the connected sums with finitely many copies of $S^k \times S^k$, see Lemma 2.16.

A similar argument was actually used in [156]. □

Remark 20.26 In Theorem 20.25 the k-connected maps f_i occur and some fibration $b \colon B \to \mathrm{BO}$. It is not clear what the right model for b is and whether the reference maps f_i exist. This problem is systematically addressed by introducing the notions of the normal $(k - 1)$-type and the notion of a $(k - 1)$-smoothing in Section 20.5, which leads to the improved version Theorem 20.44 of Theorem 20.25.

20.5 Normal k-Type

In the sequel we choose for convenience the standard model for $BO(l)$ given by Grassmann manifolds and the standard tautological bundle ξ_l over $BO(l)$. There are canonical cofibrations $BO(l) \to BO(l + 1)$, and we define $BO = \colim_{l \to \infty} BO(l)$. The advantage of this model is the following. Let $i \colon M \hookrightarrow \mathbb{R}^{n+l}$ be an embedding of an n-dimensional closed manifold M into $\mathbb{R}^{n+l}$. Then the Gauss map yields a unique map of vector bundles

$$\begin{array}{ccc} E_{\nu(i)} & \xrightarrow{\overline{f(i)}} & E_{\xi_l} \\ p_{\nu(i)} \downarrow & & \downarrow p_{\xi_l} \\ M & \xrightarrow{f(i)} & BO(l). \end{array} \qquad (20.27)$$

20.5.1 Embedding Spaces and Gauss Maps up to Contractible Choice

Note that the composite $j \circ f(i) \colon M \to BO$ is obviously unique up to homotopy. However, for the following we need more, namely that it is unique up to contractible choice, which we explain next. Let $\mathrm{Emb}(M, \mathbb{R}^{n+l})$ be the space of embeddings $M \hookrightarrow \mathbb{R}^{n+l}$ for $l \geq (n+1)$. It is non-empty, $(l-n-1)$-connected and has the homotopy type of a CW-complex, see [68, Appendix]. Let $\mathrm{Emb}(M, \mathbb{R}^{n+l}) \to \mathrm{Emb}(M, \mathbb{R}^{n+l+1})$ be the map given by composition with the standard embedding $\mathbb{R}^{n+l} \hookrightarrow \mathbb{R}^{n+l+1}$. It is a cofibration. Define

$$\mathrm{Emb}(M, \mathbb{R}^\infty) := \colim_{l \to \infty} \mathrm{Emb}(M, \mathbb{R}^{n+l}). \qquad (20.28)$$

This is a weakly contractible space of the homotopy type of a CW-complex and hence a contractible space. Now one can construct for each $l \geq (n + 1)$ a map

$$\nu_{M,l} \colon \mathrm{Emb}(M, \mathbb{R}^{n+l}) \times M \to BO(l) \qquad (20.29)$$

such that for every element $i \in \mathrm{Emb}(M, \mathbb{R}^{n+l})$ the restriction of $\nu_{M,l}$ to $M = \{i\} \times M$ is exactly $j_l \circ f(i)$. These maps $\nu_{M,l}$ fit together such that we can define a map

$$\nu_M \colon \mathrm{Emb}(M, \mathbb{R}^\infty) \times M \to BO \qquad (20.30)$$

by the zigzag

$$\mathrm{Emb}(M, \mathbb{R}^\infty) \times M = \left(\mathrm{colim}_{l \to \infty} \mathrm{Emb}(M, \mathbb{R}^{n+l})\right) \times M$$
$$\xleftarrow{\cong} \mathrm{colim}_{l \to \infty} \left(\mathrm{Emb}(M, \mathbb{R}^{n+l}) \times M\right)$$
$$\xrightarrow{\mathrm{colim}_{l \to \infty} \nu_{M,l}} \mathrm{colim}_{l \to \infty} \mathrm{BO}(l) = \mathrm{BO}. \quad (20.31)$$

20.5.2 The Moore–Postnikov Factorisation

Recall that a path connected space X is called k-*connected* for $k \geq 0$ if for one (and hence all) base point $x \in X$ and $1 \leq i \leq k$ we have $\pi_i(X, x) = \{1\}$. A map $f \colon X \to Y$ of path connected spaces is called k-*connected* for $k \geq 0$ if for one (and hence all) base point x in X the map $\pi_i(f, x) \colon \pi_i(X, x) \to \pi_i(Y, f(x))$ is bijective for $1 \leq i \leq k - 1$ and surjective for $i = k$. Note that f is k-connected for $k \geq 1$ if and only if the homotopy fibre of F_y over one (and hence all) $y \in Y$ is a $(k-1)$-connected space.

A path connected space X is called k-*coconnected* for $k \geq 0$ if for one (and hence all) base point $x \in X$ and $k \leq i$ we have $\pi_i(X, x) = \{1\}$. A map $f \colon X \to Y$ of path connected spaces is called k-*coconnected* for $k \geq 0$ if for one (and hence all) base point x in X the map $\pi_i(f, x) \colon \pi_i(X, x) \to \pi_i(Y, f(x))$ is bijective for $i \geq (k + 1)$ and injective for $i = k$. Note that f is k-coconnected for $k \geq 1$ if and only if the homotopy fibre of F_y over one (and hence all) $y \in Y$ is a k-coconnected space.

Let $f \colon X \to Y$ be a map of CW-complexes. The p-*stage in the Moore–Postnikov factorisation* of f consists of a factorisation $f \colon X \xrightarrow{u_f} B_f^p \xrightarrow{v_f} Y$ such that u_f is p-connected and v_f is p-coconnected. It can be constructed by attaching cells. If we have a second such factorisation $f \colon X \xrightarrow{u'_f} (B_f^p)' \xrightarrow{v'_f} Y$, then there exists a homotopy equivalence $g \colon B_f^p \to (B_f^p)'$, which is up to homotopy uniquely determined by the property that the following diagram commutes up to homotopy

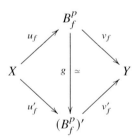

The map g is constructed by elementary obstruction theory. Since it is a weak homotopy equivalence of CW-complexes, it is a homotopy equivalence. If one has two such maps from B_f^p to $(B_f^p)'$, one can construct the homotopy between them again by elementary obstruction theory.

20.5 Normal k-Type

If $h\colon X\times[0,1]\to Y$ is a homotopy between $f_0, f_1\colon X\to Y$, then we get a diagram, which commutes up to homotopy

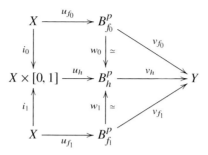

where i_k sends x to (x,k) for $k=0,1$ and whose middle vertical arrows are homotopy equivalences unique up to homotopy. Thus we get a homotopy equivalence $w_h\colon B^p_{f_0}\to B^p_{f_1}$ whose homotopy class depends only on h. If h' is another homotopy from f_0 and f_1 and h and h' are homotopic as homotopies from f_0 to f_1, then w_h and $w_{h'}$ are homotopic.

20.5.3 Normal k-Types

Definition 20.32 (Normal structures) Let M be an n-dimensional closed manifold. Let $b\colon B\to \mathrm{BO}$ be a fibration with path connected total space.

(i) We call a lift $\overline{\nu_M}$ of ν_M along b a *b-structure* on M, i.e., $\overline{\nu_M}$ is a map making the following diagram commutative

$$\begin{array}{ccc} & & B \\ & \overline{\nu_M}\nearrow & \downarrow b \\ \mathrm{Emb}(M,\mathbb{R}^\infty)\times M & \xrightarrow{\nu_M} & \mathrm{BO} \end{array} \qquad (20.33)$$

(ii) A b-structure $\overline{\nu_M}$ is called a *k-smoothing* if ν_M is $(k+1)$-connected;
(iii) We say that b is *k-universal* if $b\colon B\to \mathrm{BO}$ is $(k+1)$-coconnected.

Exercise 20.34 Let M be an n-dimensional connected closed smooth manifold. Let $b\colon \mathrm{BSO}\to\mathrm{BO}$ the obvious fibration. Show

(i) The following statements are equivalent:

(a) There exists a b-structure on M;
(b) For one $l\geq 0$ and one embedding $i\colon M \hookrightarrow \mathbb{R}^{n+l}$ the normal bundle $\nu(i)$ is orientable;
(c) For every $l\geq 0$ and every embedding $i\colon M \hookrightarrow \mathbb{R}^{n+l}$ the normal bundle $\nu(i)$ is orientable;
(d) The tangent bundle TM is orientable;

(e) $H_n(M;\mathbb{Z})$ is infinite cyclic;
(f) The first Stiefel–Whitney class $w_1(M) \in H^1(M;\mathbb{Z}/2)$ is trivial;

(ii) Suppose that there exists a b-structure on M. Then the following choices determine one another:

(a) A choice of vertical homotopy class of a b-structure;
(b) A choice of an orientation of $v(i)$ for some $l \geq 0$ and some embedding $i \colon M \hookrightarrow \mathbb{R}^{n+l}$
(c) A choice of an orientation of TM;
(d) A choice $[M]$ of a generator of the infinite cyclic group $H_n(M;\mathbb{Z})$.

By turning the map v_f appearing in the p-stage of a Moore–Postnikov decomposition into a fibration, we obtain the following result. This is also discussed in [225, Section 2], for part (iii) see also [4].

Lemma 20.35 *Let M be an n-dimensional connected closed manifold and let k be a natural number.*

(i) *There exists a k-universal fibration $b \colon B \to \mathrm{BO}$ together with a k-smoothing $\overline{v_M} \colon \mathrm{Emb}(M,\mathbb{R}^\infty) \times M \to B$;*
(ii) *Consider two k-universal fibrations $b_i \colon B_i \to \mathrm{BO}$, for which there exist k-smoothings $\overline{v_{M\,i}} \colon \mathrm{Emb}(M,\mathbb{R}^\infty) \times M \to B_i$ for $i = 0, 1$.*
Then there exists a strong fibre homotopy equivalence $u \colon B_0 \to B_1$ covering the identity on BO such that $u \circ \overline{v_{M\,0}} \simeq \overline{v_{M\,1}}$ holds;
(iii) *Let $b \colon B \to \mathrm{BO}$ be a k-universal fibration, for which there exists a k-smoothing $\overline{v_M} \colon \mathrm{Emb}(M,\mathbb{R}^\infty) \times M \to B$. Let $\mathrm{ho\text{-}aut}(b)$ be the group of strong fibre homotopy classes of strong fibre self-homotopy equivalences of b.*
Then $\mathrm{ho\text{-}aut}(b)$ acts freely and transitively on the set of vertical homotopy classes of k-smoothings of M.
Let F be the homotopy fibre of b and $G(F)$ be the topological monoid of self-homotopy equivalences of F. Suppose that F has the homotopy type of a CW-complex. Then $G(F)$ is an H-space and $[\mathrm{BO}, G(F)]$ is a group, which is isomorphic to $\mathrm{ho\text{-}aut}(b)$.

Because of Lemma 20.35 the following definition makes sense.

Definition 20.36 (Normal k-type) Let M be an n-dimensional connected closed manifold. Its *normal k-type* B_M^k is the strong fibre homotopy class of any k-universal b-structure $b \colon B \to \mathrm{BO}$, for which there exists a k-smoothing $\overline{v_M} \colon \mathrm{Emb}(M,\mathbb{R}^\infty) \times M \to B$.

Note that B_M^k is a diffeomorphism invariant of M.

Exercise 20.37 Consider natural numbers $n \geq 4$ and $k \geq 1$. Let M and N be two n-dimensional manifolds. Show:

(i) If M and N have the same normal $(k+1)$-type, then they have the same normal k-type;

(ii) Suppose that $k \geq (n + 1)$. Then M and N have the same normal k-type if and only if we can find embeddings $i_M \colon M \hookrightarrow \mathbb{R}^{n+k}$ and $i_N \colon N \hookrightarrow \mathbb{R}^{n+k}$ and a bundle map

$$\begin{array}{ccc} \nu(i_M) & \xrightarrow{\bar{f}} & \nu(i_N) \\ p_{\nu(i_M)} \downarrow & & \downarrow p_{\nu(i_N)} \\ M & \xrightarrow[f]{\simeq} & N \end{array}$$

covering a homotopy equivalence $f \colon M \xrightarrow{\simeq} N$.

Example 20.38 (The normal 1-type) Let M be an n-dimensional connected closed manifold. Let π be its fundamental group, and let $c_M \colon M \to B\pi$ be a classifying map, i.e., a map inducing the identity on fundamental groups. Note that c_M is unique up to homotopy. The Leray–Serre spectral sequence applied to the fibration $\widetilde{M} \xrightarrow{p_M} M \xrightarrow{c_M} B\pi$ yields a short exact sequence

$$\{1\} \to H^2(B\pi; \mathbb{Z}/2) \xrightarrow{c_M^*} H^2(M; \mathbb{Z}/2) \xrightarrow{p_M^*} H^2(\widetilde{M}; \mathbb{Z}/2). \tag{20.39}$$

Note that p_M^* sends $w_2(M)$ to $w_2(\widetilde{M})$.

Then precisely one of the following cases occurs:

(i) The first Stiefel–Whitney class $w_1(M) \in H^1(M; \mathbb{Z}/2)$ is non-trivial and $w_2(M)$ is not in the image of $c_M^* \colon H^2(B\pi; \mathbb{Z}/2) \to H^2(M; \mathbb{Z}/2)$;
(ii) The first Stiefel–Whitney class $w_1(M) \in H^1(M; \mathbb{Z}/2)$ is non-trivial and $w_2(M)$ is in the image of $c_M^* \colon H^2(B\pi; \mathbb{Z}/2) \to H^2(M; \mathbb{Z}/2)$;
(iii) We have $w_1(M) = 0$ and $w_2(M)$ is not in the image of $c_M^* \colon H^2(B\pi; \mathbb{Z}/2) \to H^2(M; \mathbb{Z}/2)$;
(iv) We have $w_1(M) = 0$, $w_2(M) \neq 0$, and $w_2(M)$ is contained in the image of $c_M^* \colon H^2(B\pi; \mathbb{Z}/2) \to H^2(M; \mathbb{Z}/2)$;
(v) We have $w_1(M) = 0$ and $w_2(M) = 0$.

Then the following holds depending on the case we are in.

(i) We conclude from (20.39) that $w_2(\widetilde{M}) \neq 0$ and hence $\widetilde{M}$ has no Spin structure. Since c_M is 2-connected, it induces an isomorphism $c_M^* \colon H^1(B\pi; \mathbb{Z}/2) \xrightarrow{\cong} H^1(M; \mathbb{Z}/2)$. Let $u_1 \colon B\pi \to K(\mathbb{Z}/2, 1)$ be a map whose associated class in $H^1(B\pi; \mathbb{Z}/2)$ is the preimage of $w_1(M)$ under the isomorphism c_M^*. We can turn u_1 into a fibration $p_{u_1} \colon E_{u_1} \to K(\mathbb{Z}/2, 1)$, see (20.19). Let $w_1 \colon BO \to K(\mathbb{Z}/2, 1)$ be the map given by the generator of $H^1(BO; \mathbb{Z}/2) \cong \mathbb{Z}/2$. Then a representative b for B_M^1 is given by the pullback

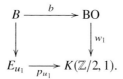

We have $\pi_2(B) \cong \mathbb{Z}/2$. Moreover, ν_M and c_M together with a choice of a homotopy $h\colon u_1 \circ c_M \simeq w_1 \circ \nu_M$ yield a map $\widehat{\nu_M}\colon \mathrm{Emb}(M,\mathbb{R}^\infty) \times M \to B$, which is a 1-smoothing.

(ii) We conclude from (20.39) that $w_2(\widetilde{M}) = 0$ and hence $\widetilde{M}$ has a Spin structure. Moreover, there is precisely one element $u_2 \in H^2(B\pi,\mathbb{Z}/2)$ with $w_2(M) = c_M^*(u_2)$. It can be realised as a map $u_2\colon B\pi \to K(\mathbb{Z}/2,2)$. We can turn the map $u_1 \times u_2\colon B\pi \to K(\mathbb{Z}/2,1) \times K(\mathbb{Z}/2,2)$ into a fibration $p_{u_1,u_2}\colon E_{u_1,u_2} \to K(\mathbb{Z}/2,1) \times K(\mathbb{Z}/2,2)$, see (20.19). Let $w_2\colon \mathrm{BO} \to K(\mathbb{Z}/2,2)$ be the map given by the generator of $H^2(\mathrm{BO};\mathbb{Z}/2) \cong \mathbb{Z}/2$. Then a representative b for B_M^1 is given by the pullback

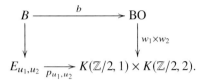

We have $\pi_2(B) \cong \{0\}$. The map $\widehat{\nu_M}$ and the composite $\mathrm{Emb}(M,\mathbb{R}^\infty) \times M \xrightarrow{\mathrm{pr}_M} M \xrightarrow{c_M} B\pi$ together with a choice of a homotopy $h\colon (u_1 \times u_2) \circ c_M \simeq (w_2 \times w_1) \circ \widehat{\nu_M}$ yield a 1-smoothing $\mathrm{Emb}(M,\mathbb{R}^\infty) \times M \to B$.

(iii) We conclude from (20.39) that $w_2(\widetilde{M}) \neq 0$ and hence $\widetilde{M}$ has no Spin structure. Moreover, a model for B_M^1 is given by $b\colon \mathrm{BSO} \times B\pi \to \mathrm{BSO} \to \mathrm{BO}$. We have $\pi_2(\mathrm{BSO} \times B\pi) \cong \mathbb{Z}/2$. There is an identification of bordism groups

$$\Omega_n(b) = \Omega_n(B\pi).$$

Since $w_1(M) = 0$, we can choose an orientation on M and thus we get a lift $\widehat{\nu_M}\colon \mathrm{Emb}(M,\mathbb{R}^\infty) \times M \to \mathrm{BSO}$ of $\nu_M\colon \mathrm{Emb}(M,\mathbb{R}^\infty) \times M \to \mathrm{BO}$. Then a 1-smoothing is given by $\widehat{\nu_M} \times c_M\colon \mathrm{Emb}(M,\mathbb{R}^\infty) \times M \to \mathrm{BSO} \times B\pi$.

Two 1-smoothings $\widehat{\nu_M}_i \times c_i\colon \mathrm{Emb}(M,\mathbb{R}^\infty) \times M \to \mathrm{BSO} \times B\pi$ for $i = 0,1$ are vertical homotopy equivalent if and only if $\widehat{\nu_M}_0$ and $\widehat{\nu_M}_1$ induce the same orientation on M and c_1 and c_2 are homotopic.

(iv) Since $w_2(M) \neq 0$, there is no Spin structure on M. We conclude from the exact sequence (20.39) that $w_2(\widetilde{M}) = 0$ and hence $\widetilde{M}$ has a Spin structure. Moreover, there is precisely one element $u_2 \in H^2(B\pi,\mathbb{Z}/2)$ with $w_2(M) = c_M^*(u_2)$. It can be realised as a map $u_2\colon B\pi \to K(\mathbb{Z}/2,2)$. We turn u_2 into a fibration $p_{u_2}\colon E_{u_2} \to K(\mathbb{Z}/2,2)$, see (20.19). Let $w_2\colon \mathrm{BSO} \to K(\mathbb{Z}/2,2)$ be the map given by the generator of $H^2(\mathrm{BSO};\mathbb{Z}/2) \cong \mathbb{Z}/2$. Then a representative b for B_M^1 is given by the pullback

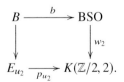

We have $\pi_2(B) \cong \{0\}$. The map $\widehat{\nu_M}$ and the composite $\mathrm{Emb}(M, \mathbb{R}^\infty) \times M \xrightarrow{\mathrm{pr}_M} M \xrightarrow{c_M} B\pi$ together with a choice of a homotopy $h \colon u_2 \circ c_M \circ \mathrm{pr}_M \simeq w_2 \circ \widehat{\nu_M}$ yield a 1-smoothing $\mathrm{Emb}(M, \mathbb{R}^\infty) \times M \to B$.

(v) Since $w_1(M) = 0$ and $w_2(M) = 0$, we can choose a Spin structure on M. A representative for B_M^1 is given by

$$\mathrm{BSpin} \times B\pi \to \mathrm{BSpin} \to \mathrm{BO}.$$

We have $\pi_2(\mathrm{BSpin} \times B\pi) \cong \{0\}$. There is an identification of bordism groups

$$\Omega_n(b) = \Omega^{\mathrm{Spin}}(B\pi).$$

The Spin structure on M yields a lift of $\widehat{\nu_M} \colon \mathrm{Emb}(M, \mathbb{R}^\infty) \times M \to \mathrm{BSpin}$ of $\nu_M \colon \mathrm{Emb}(M, \mathbb{R}^\infty) \times M \to \mathrm{BO}$. Then a 1-smoothing is given by the map $\widehat{\nu_M} \times c_M \colon \mathrm{Emb}(M, \mathbb{R}^\infty) \times M \to \mathrm{BSpin} \times B\pi$.

Two 1-smoothings $\widehat{\nu_M}_i \times c_i \colon \mathrm{Emb}(M, \mathbb{R}^\infty) \times M \to \mathrm{BSpin} \times B\pi$ for $i = 0, 1$ are vertical homotopy equivalent if and only if $\widehat{\nu_M}_0$ and $\widehat{\nu_M}_1$ induce the same Spin structure on M and c_1 and c_2 are homotopic.

20.6 Classification up to Stable Diffeomorphisms

Let M be an n-dimensional connected closed manifold. Fix a fibration $b \colon B \to \mathrm{BO}$. Let $\overline{\nu_M}$ be a b-structure, i.e., a lift of the map ν_M along b. Denote by $[\overline{\nu_M}]$ the vertical homotopy class of a b-structure $\overline{\nu_M}$ where we call two lifts $\overline{\nu_M}$ and $\overline{\nu_M}'$ *vertical homotopic* if there is a homotopy

$$h \colon \mathrm{Emb}(M, \mathbb{R}^\infty) \times M \times [0, 1] \to B$$

such that $h_0 = \overline{\nu_M}$, $h_1 = \overline{\nu_M}'$, and $b \circ h_t = \nu_M$ holds for $t \in [0, 1]$. Next we define an element

$$\eta(M, [\overline{\nu_M}]) \in \Omega_n(b). \tag{20.40}$$

Choose a representative $\overline{\nu_M}$ of $[\overline{\nu_M}]$ and an element $i \in \mathrm{Emb}(M, \mathbb{R}^\infty)$. Then we can find $l \geq (n+1)$ such that i is an embedding $i \colon M \hookrightarrow \mathbb{R}^{n+l}$. We get from the bundle map (20.27) associated to i and the diagram

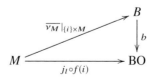

and from the definitions of B_l and $\mathrm{pr}_l \, \xi_l$ in terms of pullbacks a bundle map

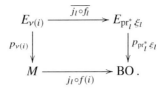

It defines an element in $\Omega_n(\mathrm{pr}_l^* \xi_l)$ and thus an element in $\Omega_n(b)$, which we define to be $\eta(M, [\overline{\nu_M}])$. It remains to check that it is well defined, namely, that it is independent of the choice of $i \in \mathrm{Emb}(M; \mathbb{R}^\infty)$ and $\overline{\nu_M} \in [\overline{\nu_M}]$. This follows by constructing appropriate bordisms, which come from homotopies and the fact that $\mathrm{Emb}(M, \mathbb{R}^\infty)$ is contractible. So we get a well-defined map

$$\eta \colon \mathrm{vholifts}(\nu_M) \to \Omega_n(b) \tag{20.41}$$

where $\mathrm{vholifts}(\nu_M)$ is the set of vertical homotopy classes of lifts $\overline{\nu_M}$ of ν_M along b. Using contractibility of $\mathrm{Emb}(M, \mathbb{R}^\infty)$ one easily checks the next lemma.

Lemma 20.42 *Consider any element $i \in \mathrm{Emb}(M; \mathbb{R}^\infty)$. Then the evaluation map*

$$\mathrm{vholifts}(\nu_M \colon \mathrm{Emb}(M, \mathbb{R}^\infty) \times M \to \mathrm{BO}) \xrightarrow{\cong} \mathrm{vholifts}(\nu_M|_{\{i\} \times M} \colon M \to \mathrm{BO})$$

is bijective.

Remark 20.43 (Discarding $\mathrm{Emb}(M; \mathbb{R}^\infty)$) Because of Lemma 20.42 we can ignore the contractible space $\mathrm{Emb}(M; \mathbb{R}^\infty)$ in the sequel and think of ν_M and $\overline{\nu_M}$ as maps $\nu_M \colon M \to \mathrm{BO}$ and $\overline{\nu_M} \colon M \to B$.

Theorem 20.44 (Stable diffeomorphism classification) *Consider the n-dimensional closed manifolds M_0 and M_1 for the even integer $n = 2k \geq 4$. Then M_0 and M_1 are stably diffeomorphic if and only if the following two conditions are satisfied:*

(i) *Their normal $(k-1)$-types $B_{M_0}^{k-1}$ and $B_{M_1}^{k-1}$ agree;*
(ii) *There exists a representative $b \colon B \to \mathrm{BO}$ of the normal $(k-1)$-type $B_{M_0}^{k-1} = B_{M_1}^{k-1}$ and $(k-1)$-smoothings $\overline{\nu_{M_0}}$ on M_0 and $\overline{\nu_{M_1}}$ on M_1 such that we get in $\Omega_n(b)$ the equality*

$$\eta(M_0, [\overline{\nu_{M_0}}]) = \eta(M_1, [\overline{\nu_{M_1}}]).$$

Proof. This follows from Theorem 20.25. □

Remark 20.45 (Compatibility with the $(k-1)$-smoothings) Consider the situation of Theorem 20.44. Suppose that the conditions appearing in (ii) are satisfied. Fix the choice of $b \colon B \to \mathrm{BO}$ and $(k-1)$-smoothings $\overline{\nu_{M_0}}$ on M_0 and $\overline{\nu_{M_1}}$ on M_1 such that $\eta(M_0, [\overline{\nu_{M_0}}]) = \eta(M_1, [\overline{\nu_{M_1}}])$ holds.

According to Theorem 20.44, we can find natural numbers g_0 and g_1 and a diffeomorphism

$$f \colon M_0 \sharp W_{g_0} \xrightarrow{\cong} M_1 \sharp W_{g_1}.$$

Then we get for $i = 0, 1$ from the $(k-1)$-smoothing $\overline{\nu_{M_i}}$ a $(k-1)$-smoothing $\overline{\nu_{M_i} \sharp W_{g_i}}$ whose vertical homotopy type is uniquely determined by the vertical homotopy type

20.6 Classification up to Stable Diffeomorphisms

of the $(k-1)$-smoothing $\overline{\nu_{M_i}}$, see Section 20.9. Moreover, the diffeomorphism f can be arranged to respect the vertical homotopy classes of the $(k-1)$-smoothings $\overline{\nu_{M_0 \sharp W_{g_0}}}$ and $\overline{\nu_{M_1 \sharp W_{g_1}}}$. This follows from inspecting the proof of Theorem 20.25 as explained, for instance, in [225].

The analogue in classical surgery is that the simple structure set $\mathcal{S}^s(M)$ decides whether a simple homotopy equivalence $N \to M$ with a closed manifold as source is homotopic to a homeomorphism or diffeomorphism where no stabilisation is allowed.

Example 20.46 Let M and N be two oriented connected closed 4-manifolds with fundamental classes $[M] \in H_4(M;\mathbb{Z})$ and $[N] \in H_4(N;\mathbb{Z})$ such that the second Stiefel–Whitney classes of their universal coverings $w_2(\widetilde{M})$ and $w_2(\widetilde{N})$ are both non-trivial. Let $c_M \colon M \to B\pi_1(M)$ and $c_N \colon N \to B\pi_1(N)$ be classifying maps for the universal coverings. Note that these maps are unique up to homotopy.

Suppose that M and N have the same signature and that there is an isomorphism $u \colon \pi_1(M) \xrightarrow{\cong} \pi_1(N)$ such that $(u \circ c_M)_*[M] = (c_N)_*[N]$ in $H_4(B\pi_1(N);\mathbb{Z})$ and $u^*(w_2(N)) = w_2(M)$ in $H^2(M;\mathbb{Z}/2)$ hold.

Then we can find natural numbers g_0 and g_1 and an orientation preserving diffeomorphism
$$f \colon M_0 \sharp W_{g_0} \xrightarrow{\cong} M_1 \sharp W_{g_1}$$
such that $\pi_1(f)$ and u agree up to inner automorphism under the obvious identifications $\pi_1(M_i \sharp W_{g_i}) = \pi_1(M_i)$ for $i = 0, 1$.

This follows from Theorem 20.44 and Remark 20.45 and the following facts. The normal 1-type of both M and N is given by $b \colon \mathrm{BSO} \times B\pi \to \mathrm{BSO} \to \mathrm{BO}$ and the bordism group $\Omega_4(b)$ agrees with the singular oriented bordism group $\Omega_4(B\pi_1(N))$. Moreover, we can find 1-smoothings for M_0 and M_1 whose composites with the projection $\mathrm{BSO} \times B\pi \to B\pi$ represent c_{M_0} and c_{M_1}. All this is explained in Example 20.38. The Atiyah–Hirzebruch spectral sequence yields an isomorphism
$$\Omega_4(B\pi_1(N)) \xrightarrow{\cong} \Omega_4(\{\bullet\}) \times H_4(B\pi_1(N);\mathbb{Z})$$
which sends the bordism class $[f \colon X \to B\pi_1(N)]$ of a map $f \colon X \to B\pi_1(N)$ with an oriented closed 4-manifold as source to $([X], f_*([X]))$. The signature induces an isomorphism $\Omega_4(\{\bullet\}) \xrightarrow{\cong} \mathbb{Z}$.

Exercise 20.47 Let M and N be connected closed oriented 4-manifolds whose fundamental groups are isomorphic and cyclic of odd order. Suppose that both admit no Spin structure and that their signatures agree.

Show that for any isomorphism $u \colon \pi_1(M) \xrightarrow{\cong} \pi_1(N)$ there is a stable diffeomorphism from M to N inducing u on the fundamental groups.

Remark 20.48 (Stabilisation with $\mathbb{CP}^2$) Two connected closed 4-manifolds M and N are called $\mathbb{CP}^2$-*stably diffeomorphic* if they become diffeomorphic after taking a connected sum with finitely many copies of $\mathbb{CP}^2$. This is a weaker condition than being stably diffeomorphic.

For simplicity we assume that M and N are orientable. Then M and N are stably $\mathbb{CP}^2$-diffeomorphic if and only if there is exists an isomorphism $u\colon \pi_1(M) \xrightarrow{\cong} \pi_1(N)$ and $\epsilon \in \{\pm 1\}$ such that $(u \circ c_M)_*([M]) = \epsilon \cdot (c_N)_*[N]$ holds in $H_4(B\pi_1(N); \mathbb{Z})$. This is proved in [208, Theorem 1.1] following [225].

Remark 20.49 Let M be an $2k$-dimensional closed manifold for the natural number $k \geq 2$. Let $b\colon B \to \mathrm{BO}$ be a representative of its normal $(k-1)$-type B_M^{k-1}. Recall that ho-aut(b) is the group of strong fibre homotopy classes of strong fibre self-homotopy equivalences of b. Then there is a bijective correspondence between the set of stable diffeomorphism classes of $2k$-dimensional manifolds with the same normal $(k-1)$-type as M and the set $\Omega_{2k}(b)/\text{ho-aut}(b)$. This follows essentially from Lemma 20.35 and Theorem 20.44. See also [396], where more information about ho-aut(b) and its action on $\Omega_{2k}(b)$, in particular for 4-dimensional manifolds, is given. Note that here we are only interested in the stable diffeomorphism class and not in the question whether any $(k-1)$-smoothings are respected, compare with Remark 20.45.

The analogue in classical surgery is that the quotient $\mathcal{S}^s(M)/\text{ho-aut}^s(M)$ for $\mathcal{S}^s(M)$ the simple structure set and ho-aut$^s(M)$ the group of homotopy classes of simple self-homotopy equivalences of M is in bijective correspondence to the (unstable) diffeomorphism or homeomorphism classes of n-dimensional closed manifolds which are homotopy equivalent to M, provided that $n \geq 5$.

20.7 Tangential Approach

One can also change the whole setup by replacing the normal bundle $\nu(i\colon M \hookrightarrow \mathbb{R}^{n+l})$ by the tangent bundle TM. This is very similar to our considerations for normal maps, see Remark 7.21. The advantage is obvious, the tangent bundle is intrinsically defined. A disadvantage is that the tangent bundle, in contrast to the normal bundle of a given embedding, has no preferred choice of a classifying map. But such a classifying map is unique up to contractible choice by the following argument, which is easier than the argument with $\mathrm{Emb}(M; \mathbb{R}^{n+l})$ and works for any l-dimensional vector bundle ξ over a CW-complex X.

20.7.1 Passing to the Tangent Bundle

Consider an l-dimensional vector bundle ξ over a CW-complex with projection $p_\xi \colon E_\xi \to X$. One can replace ξ by the principal $\mathrm{GL}(l, \mathbb{R})$ bundle $F_l(\xi)$ over X with projection $p_{F_l(\xi)} \colon E_{F_l(\xi)} \to X$ whose fibre over a point $x \in X$ is the space of linear $\mathbb{R}$-isomorphisms from $\mathbb{R}^l$ to the fibre $p_\xi^{-1}(x)$ over x. Since X is a CW-complex, the total space $E_{F_l(\xi)}$ of $F_l(\xi)$ is a free $\mathrm{GL}(l, \mathbb{R})$-$CW$-complex. Let $p_{\mathrm{GL}(l,\mathbb{R})} \colon \mathrm{EGL}(l, \mathbb{R}) \to \mathrm{BGL}(l, \mathbb{R})$ be the universal principal $\mathrm{GL}(l, \mathbb{R})$ bundle. Its total space is a free $\mathrm{GL}(l, \mathbb{R})$-$CW$-complex, which is contractible after forgetting the group action, and $p_{\mathrm{GL}(l,\mathbb{R})}$ is given by dividing out the $\mathrm{GL}(l, \mathbb{R})$-action. The space $\mathrm{map}_{\mathrm{GL}(l,\mathbb{R})}(E_{F_l(\xi)}, \mathrm{EGL}(l, \mathbb{R}))$ of

GL$(l, \mathbb{R})$-maps from $E_{F_l(\xi)}$ to EGL$(l, \mathbb{R})$ is contractible and in particular non-empty. We obtain a map

$$\mathrm{map}_{\mathrm{GL}(l,\mathbb{R})}(E_{F_l(\xi)}, \mathrm{EGL}(l, \mathbb{R})) \times E_{F_l(\xi)} \to \mathrm{EGL}(l, \mathbb{R})$$

by evaluation. Note that a principal GL$(l, \mathbb{R})$ bundle $p \colon E \to X$ over X determines an l-dimensional vector bundle $\widehat{p} \colon \widehat{E} := E \times_{\mathrm{GL}(l,\mathbb{R})} \mathbb{R}^n \to X$ and up to natural equivalence this passage is the inverse of the passage from ξ to $F_l(\xi)$. This is a functorial construction. Moreover, this construction applied to $p_{\mathrm{GL}(l,\mathbb{R})} \colon \mathrm{EGL}(l, \mathbb{R}) \to \mathrm{BGL}(l, \mathbb{R})$ yields the universal l-dimensional bundle ξ_l over BGL$(l, \mathbb{R})$ with projection $p_{\xi_l} \colon E_{\xi_l} \to B\,\mathrm{GL}(l, \mathbb{R})$. Hence we get a map

$$\mathrm{map}_{\mathrm{GL}(l,\mathbb{R})}(E_{F_l(\xi)}, \mathrm{EGL}(l, \mathbb{R})) \times E_\xi \to E_{\xi_l}$$

whose evaluation at an element $u \in \mathrm{map}_{\mathrm{GL}(l,\mathbb{R})}(E_{F_l(\xi)}, \mathrm{EGL}(l, \mathbb{R}))$ is a bundle map

$$\begin{CD}
E_\xi @>{\overline{f_u}}>> E_{\xi_l} \\
@V{p_\xi}VV @VV{p_{\xi_l}}V \\
X @>>{f_u}> \mathrm{BGL}(l, \mathbb{R}).
\end{CD}$$

Note that up to homotopy it does not matter whether we work with BGL$(l, \mathbb{R})$ or BSO(l).

If in the setup with normal bundle one does not want to use the Gauss map or the model for BSO(l), one can replace the contractible space Emb$(M; \mathbb{R}^{n+l})$ by an other contractible space $P(i)$, which is the total space of locally trivial fibre bundle $P(i) \to$ Emb$(M; \mathbb{R}^{n+l})$ whose fibre over $i \in \mathrm{Emb}(M; \mathbb{R}^{n+l})$ is $\mathrm{map}_{\mathrm{GL}(l,\mathbb{R})}(E_{\nu(i)}, \mathrm{EGL}(l, \mathbb{R}))$. It is not hard to check that the approach using the tangent bundle and the approach using the normal bundle are equivalent.

20.7.2 Connected Sum

For the reader's convenience we recall what happens when we want to take the connected sum of two closed n-dimensional manifolds with a preferred vertical homotopy type of a k-smoothing. In principle this has already been discussed in the surgery step, see Theorem 4.39, but it is worthwhile to discuss briefly what we get in the setting of this chapter.

Consider a fibration $b \colon B \to \mathrm{BO}$ and a natural number $k \geq 1$ such that b is k-universal. Then $\pi_1(b)$ is either surjective or trivial because of $\pi_1(BO) \cong \mathbb{Z}/2$.

Let us begin with the case where $\pi_1(b)$ is surjective. Then a closed manifold M with a k-smoothing is non-orientable, i.e., has non-vanishing first Stiefel–Whitney class, and there is precisely one k-smoothing on D^n up to vertical homotopy. If M and

N are two n-dimensional closed manifolds with preferred k-smoothing, then choose any two embeddings $i_M \colon D^n \hookrightarrow M$ and $i_N \colon D^n \hookrightarrow N$ and perform the connected sum $M \natural_{i_M, i_N} N$ with respect to these embeddings. Note that any two embeddings $i_0, i_1 \colon D^n \hookrightarrow M$ are isotopic and such an isotopy can be embedded into a diffeotopy on M, which is constant outside a tubular neighbourhood of an embedded arc joining $i_0(0)$ and $i_1(0)$. This implies that $M \natural_{i_M, i_N} N$ inherits a k-smoothing unique up to vertical homotopy and that for two different choices of embeddings i_M and i_N there is a diffeomorphism between the two models for the connected sum, which is compatible with the preferred vertical homotopy class of k-smoothings.

Suppose that $\pi_1(b)$ is trivial. Then a manifold M with a preferred k-smoothing is orientable and comes with a preferred orientation, and the group $\mathbb{Z}/2$ acts freely and transitively on vholifts(ν_{D^n}). Choose any two embeddings $i_M \colon D^n \hookrightarrow M$ and $i_N \colon D^n \hookrightarrow N$ with are both orientation preserving. Note that two embeddings $D^N \to M$ are isotopic if and only if they are both orientation preserving or both orientation reversing, and such an isotopy can be embedded into a diffeotopy as described above. Now one can proceed as above to define the connected sum $M \sharp N$ and a preferred vertical homotopy type of a k-smoothing on it.

20.8 Applications

As an illustration we describe a few of the numerous applications of modified surgery. More information can be found, for instance, in the survey article by Kreck [225].

20.8.1 Complete Intersections

A *complete intersection* of complex dimension n is a complex submanifold of $\mathbb{CP}^{n+r}$ defined by the transversal intersection of r non-singular hypersurfaces, which are given by homogeneous polynomials of degree $d_1, d_2, \ldots, d_r$ with linearly independent gradients. Thom discovered that their diffeomorphism class depends only on the set $\underline{d} = \{d_1, d_2, \ldots, d_r\}$. Therefore we denote the associated complete intersection by $X_n(\underline{d})$. The total degree $\deg(\underline{d})$ is defined to be $\prod_{i_1}^{r} d_i$.

The following was proved by Claudia Traving in her Diplom-Arbeit in Mainz from 1985 under the supervision of Matthias Kreck and Stephan Stolz using modified surgery and exploiting the fact that the inclusion $X_n(\underline{d}) \to \mathbb{CP}^{n+r}$ is n-connected, see also [225, Theorem A].

20.8 Applications

Theorem 20.50 (Complete Intersections) *Consider two complete intersections $X_n(\underline{d})$ and $X_n(\underline{d}')$ of dimension $n \geq 2$ such that for all primes p with $p(p-1) \leq n-1$ the total degrees $\deg(\underline{d})$ and $\deg(\underline{d})$ are divisible by p^k, where k is the smallest natural number satisfying $k \geq \frac{2n+1}{2(p-1)} + 1$.*

Then $X_n(\underline{d})$ and $X_n(\underline{d}')$ are diffeomorphic if and only if the total degrees, the Pontrjagin classes and the Euler characteristics agree.

Note that the k-th Pontrjagin class is a multiple of x^{2k} where x generates the second cohomology of the complete intersection. Thus we can compare this invariant for different complete intersections. Note that the Pontrjagin classes and the Euler characteristics can be computed in terms of n and $\underline{d}$, see [191, Subsection 22.1].

More information about the smooth classification of complete intersections, in particular in dimension 4, and further references can be found in [112].

20.8.2 Homogeneous Spaces

The following result is taken from [230, Theorem A], see also [225, Theorem G].

Theorem 20.51 (Homogeneous manifolds of dimension 7) *Two 7-dimensional simply connected homogeneous manifolds M and N with $SU(3) \times SU(2) \times U(1)$-symmetry, which are both Spin or both non-Spin, are diffeomorphic if and only if*

(i) $|H^4(M;\mathbb{Z})| = |H^4(N;\mathbb{Z})|$;
(ii) $s_i(M) = s_i(N) \in \mathbb{Q}/\mathbb{Z}$ for $i = 1, 2, 3$.

The homeomorphism type is determined by the same invariants except that $s_1(M)$ is replaced by $\widetilde{s}_i(M) := 28 \cdot s_i(M) \in \mathbb{Q}/\mathbb{Z}$.

The invariants $s_i(M)$ are linear combinations of η-invariants in the sense of [14] measuring the spectral asymmetry of certain differential operators and a secondary characteristic number.

Kreck–Stolz [231] classify certain homogenous spaces called *Wallach spaces* up to diffeomorphism and homeomorphism. Thus they construct the first examples of homeomorphic but not diffeomorphic manifolds admitting metrics with positive sectional curvature, as well as the first example of such manifolds which are homogeneous spaces of the form G/H with G simple and simply connected.

These investigation play a key role in the construction of manifolds whose space of metrics with positive sectional curvature is not connected, by Kreck-Stolz [232].

20.8.3 4-Manifolds

We have already discussed two applications to 4-manifolds in Example 20.46 and Remark 20.48.

A typical example where modified surgery works very well and classical surgery does not solve the problem is the proof in dimension 4 of the stable version of the prime decomposition and of the Kneser Conjecture, which are well-known and basic results in dimension 3, see [183, Chapter 3 and 7]. They roughly say that a connected closed oriented 4-manifold has a prime decomposition unique up to taking connected sum with certain simply connected 4-manifolds and that any decomposition of the fundamental group into free products is implemented by such a prime decomposition, see [229, Theorem 0.1 and Theorem 0.2].

We at least state the precise formulation in a special case. We call a connected compact orientable smooth four-manifold M (possibly with non-empty boundary) *stably prime* if M_i stably oriented diffeomorphic to $M_i' \natural M_i''$ implies that M_i' or M_i'' is simply connected and closed.

Theorem 20.52 (Stable prime decomposition for 4-manifolds) *Let M be a connected compact oriented smooth 4-manifold. Then:*

(i) *There are stably prime oriented 4-manifolds $M_1, M_2, \ldots, M_n$ and a stable oriented diffeomorphism*
$$f: M \xrightarrow{\cong} \natural_{i=1}^n M_i;$$

(ii) *Let $f': M \xrightarrow{\cong} \natural_{i=1}^{n'} M_i'$ be another stable oriented diffeomorphism for stably prime oriented 4-manifolds $M_1', M_2', \ldots, M_{n'}'$. Suppose that none of the M_i's or M_i''s is simply connected and closed. Then $n = n'$ and $M_i \natural S_i$ and $M_{\sigma(i)}' \natural S_i'$ are stably oriented diffeomorphic for $i \in \{1, 2, \ldots, n\}$, appropriate simply connected closed oriented 4-manifolds S_i and S_i', and a permutation σ.*

A group π is *indecomposable* if π is non-trivial and $\pi \cong \Gamma_1 * \Gamma_2$ implies that Γ_1 or Γ_2 is trivial.

Theorem 20.53 (Stable Kneser Decomposition for 4-manifolds) *If M is a closed connected smooth oriented 4-manifold with non-trivial fundamental group, then there are oriented smooth four-manifolds $M_1, M_2, \ldots, M_n$ with indecomposable $\pi_1(M_i)$ for $i \in \{1, 2, \ldots, n\}$, such that M and $\natural_{i=1}^n M_i$ are stably oriented diffeomorphic.*

If we have two splittings $\natural_{i=1}^n M_i$ and $\natural_{i=1}^{n'} M_i'$ of M as above, then $n = n'$ and $M_i \natural S_i$ and $M_{\sigma(i)}' \natural S_i'$ are oriented diffeomorphic for $i \in \{1, 2, \ldots, n\}$, appropriate simply connected closed smooth manifolds S_i, S_i' and a permutation $\sigma \in \Sigma_n$.

The following two theorems demonstrate the necessity of working stably, see [228, Theorem 0.1 and Theorem 0.2].

Theorem 20.54 (Counterexamples up to homotopy to prime decomposition in dimension 4) *For distinct prime numbers p_0 and p_1 there exists a connected closed smooth orientable 4-manifold M such that $\pi_1(M)$ is isomorphic to $(\mathbb{Z}/p_0 \times \mathbb{Z}/p_0) * (\mathbb{Z}/p_1 \times \mathbb{Z}/p_1)$ and if M is homotopy equivalent to a connected sum $M_0 \sharp M_1$, then M_0 or M_1 is homeomorphic to S^4.*

Theorem 20.55 (Four-manifolds splitting as a connected sum topologically but not smoothly) *There is a specific orientable closed 4-manifold with the following property. It is homeomorphic to $M_0 \sharp M_1$ for two closed topological 4-manifolds M_0 and M_1 with $\pi_1(M_i) = \mathbb{Z}/2$, but $M \sharp k(S^2 \times S^2)$ is diffeomorphic to $M_0 \sharp M_1$ for non-simply connected closed smooth 4-manifolds M_0 and M_1 if and only if $k \geq 8$.*

Here is an instance where one gets a classification up to homeomorphism without stabilisation, see [225, Theorem F] using modified surgery.

Theorem 20.56 (Classification of closed topological 4-dimensional spin manifolds with infinite cyclic fundamental group) *Two closed topological 4-dimensional spin manifolds with infinite cyclic fundamental group are homeomorphic if and only if their quadratic intersection forms on the second homotopy group with values in the group ring of their fundamental groups agree.*

Theorem 20.56 including an existence statement and without the assumption of a Spin structure but taking the Kirby–Siebenmann invariant into account was proved independently by Freedman–Quinn [159, Theorem 10.7 A] using classical surgery theory and Freedman's result about good fundamental groups, see Remark 8.30.

Further results about the stable classifications of 4-manifolds based on modified surgery theory can be found for instance in [88, 115, 173, 174, 206, 207, 209, 380, 396]. They are all based on Remark 20.49. The basic invariants are the fundamental group, the second homotopy group including the equivariant intersection form on it, and the Kirby–Siebenmann invariant.

20.9 Notes

If one wants to get an analogue of the surgery obstructions in L-groups in the setting of modified surgery, one has to consider certain l-monoids and replace the condition that the obstruction vanishes by the condition that it is elementary. We refer to [225, Theorem 3 and Theorem 4] for the details, see also [107, 113]. Unfortunately, these l-monoids are very hard to compute. The main tools for the computation of the L-groups such as Dress induction, localisation sequences, Rothenberg sequences, and the Farrell–Jones Conjecture are not available for the l-monoids.

For an example of an application to the problem whether a closed smooth manifold carries a metric of positive scalar curvature, we refer for instance to [225, Corollary 2].

Chapter 21
Solutions of the Exercises

Chapter 2

2.2. This follows directly from the s-Cobordism Theorem 2.1 since for any finitely presented group G and $m \geq 5$ there exists a closed connected m-dimensional manifold with G as fundamental group.

2.3. Let $(W; M_0, f_0, M_1, f_1)$ be an h-cobordism such that $\dim(W) = 2$ and W is connected. Since M_0, M_1, and W are homotopy equivalent, M_0 and M_1 are connected closed 1-dimensional manifolds and hence both diffeomorphic to S^1, and W is homotopy equivalent to $S^1 \times [0, 1]$. The classification of compact 2-manifolds implies that W is trivial, i.e., diffeomorphic to $S^1 \times [0, 1]$.

2.8. By passing to the induced $\mathbb{Z}$-equivariant diffeomorphisms $\overline{f} \colon \mathbb{R} \to \mathbb{R}$ on the universal covering of S^1 obtained by lifting f, one can find a diffeotopy of f to one of the diffeomorphisms $\mathrm{id} \colon S^1 \to S^1$ or $S^1 \to S^1, z \mapsto z^{-1}$. Both obviously extend to D^2.

2.31. By Poincaré duality the Euler characteristic of M is zero. The Euler characteristic can be computed from the cellular chain complex by the alternating sum of the dimensions of the chain modules. This is the alternating sum of the number of cells of a given dimension and hence the alternating sum of handles of a given index.

2.32. Since M is homotopy equivalent to a CW-complex with no 1-cells, it is simply connected.

2.33. The boundary is empty. So we have to start by attaching a zero-handle to the empty set. This gives D^{2n}. Since $S^n \times S^n$ has $\mathbb{Z}^2$ as homology group in dimension n, there must be at least be two n-handles. Attach an n-handle to D^{2n} by a standard embedding $S^{n-1} \times D^n \hookrightarrow S^{2n-1} = \partial D^{2n}$. The result is $D^n \times S^n$. Attach another n-handle with an appropriate embedding $S^{n-1} \times D^n \hookrightarrow S^{n-1} \times S^n = \partial(D^n \times S^n)$ so that the resulting manifold with boundary has S^{2n-1} as boundary. Finally attach a

2n-handle by id: $S^{2n-1} \to S^{2n-1}$. This final attachment of at least one 2n-handle is necessary since otherwise W would have non-empty boundary.

2.40. The ring $\mathbb{Z}$ has a Euclidean algorithm. Now apply standard results from linear algebra, see for instance [361, Proof of Theorem 2.3.2 on page 74].

2.41. Since $(1 - t - t^{-1}) \cdot (1 - t^2 - t^3) = 1$, the element $1 - t - t^{-1}$ is a unit in $\mathbb{Z}[\mathbb{Z}/5]$. Since $\mathbb{Z}[\mathbb{Z}/5]$ is commutative, it has a determinant $\det_{\mathbb{Z}[\mathbb{Z}/5]}$. If $B \in \mathrm{GL}(n, \mathbb{Z}[\mathbb{Z}/5])$ is a matrix representing zero in $\mathrm{Wh}(\mathbb{Z}/5)$, then standard properties of determinants over commutative rings imply that $\det_{\mathbb{Z}[\mathbb{Z}/5]}(B)$ is of the shape $\pm t^i$ for some $i \in \{0, 1, 2, 3, 4\}$. Hence the $(1, 1)$-matrix $A = (1 - t - t^{-1})$ does not represent zero in $\mathrm{Wh}(\mathbb{Z}/5)$.

Chapter 3

3.2. If $\dim(W) = 1$, already W is trivial. Hence we can assume $\dim(W) \geq 2$. Then $\dim(W \times S^n)$ is greater than or equal to six. By Theorem 2.1 (i) it suffices to show that the Whitehead torsion of $f_0 \times \mathrm{id}_{S^n}$ is trivial. This follows from the product formula appearing in Theorem 3.1 (iv) since the Euler characteristic of S^n vanishes for odd n.

3.10. The map ϕ is induced by the composite

$$K_1(\mathbb{Z}[\mathbb{Z}/5]) \xrightarrow{f_*} K_1(\mathbb{C}) \xrightarrow{\det} \mathbb{C}^\times \xrightarrow{||} \mathbb{R}^{>0}.$$

Since $(1 - t - t^{-1}) \cdot (1 - t^2 - t^3) = 1$, the element $1 - t - t^{-1}$ is a unit in $\mathbb{Z}[\mathbb{Z}/5]$ and defines an element in $\mathrm{Wh}(\mathbb{Z}/5)$. Its image under ϕ is $(1 - 2 \cdot \cos(2\pi/5))$ and hence different from 1.

3.15. Let $\gamma_* \colon \mathrm{cone}_*(f_*) \to \mathrm{cone}_*(f_*)$ be a chain contraction given by

$$\begin{pmatrix} h_{n-1} & g_n \\ l_{n-1} & k_n \end{pmatrix} \colon C_{n-1} \oplus D_n \to C_n \oplus D_{n+1}.$$

Then we obtain a chain map $g_* \colon D_* \to C_*$, and $h_* \colon g_* \circ f_* \simeq \mathrm{id}$ and $k_* \colon \mathrm{id} \simeq f_* \circ g_*$ are chain homotopies.

Conversely, given, g_*, h_* and k_*, define γ_* as above with $l_* = 0$. Then γ is a chain homotopy between a chain isomorphism $u_* \colon \mathrm{cone}_*(f_*) \to \mathrm{cone}_*(f_*)$ and the zero map. Hence $u_*^{-1} \circ \gamma_*$ is a chain contraction of $\mathrm{cone}_*(f_*)$.

3.23. Since all CW-complexes are finite, the image of the injective map $\pi_1(p_Y) \colon \pi_1(\overline{Y}) \to \pi_1(Y)$ has finite index in $\pi_1(Y)$. Hence restriction with $\pi_1(p_Y)$ defines a homomorphism $\mathrm{Wh}(\Pi(Y)) \to \mathrm{Wh}(\Pi(\overline{Y}))$. One easily checks that it sends $\tau(f)$ to $\tau(\overline{f})$. Hence $\tau(\overline{f}) = 0$ if $\tau(f) = 0$.

Solutions of the Exercises

The converse is not true in general. Suppose that $\pi_1(Y)$ is finite and $\mathrm{Wh}(\Pi(Y)) \neq \{1\}$. Then we can arrange that $\tau(f) \neq 0$. Let p_X and p_Y be the universal coverings. Then $\mathrm{Wh}(\Pi(\overline{Y})) \cong \mathrm{Wh}(\{1\}) = \{0\}$ and hence $\tau(\overline{f}) = 0$.

3.29. Let the homotopy equivalence $j_k \colon M_k \to W$ be the composite of the diffeomorphism $f_i \colon M_i \to \partial_i W$ and the inclusion $\partial W_i \to W$ for $k = 0, 1$. Let j_i^{-1} be a homotopy inverse. We conclude from Lemma 3.1 (ii) and (iii) and Lemma 3.28 (ii) that $\tau(j_1^{-1} \circ j_0) = 0$. Since $\tau \neq 0$, the h-cobordism W is non-trivial.

3.32. This is done by the following sequence of expansions. We describe the simplicial complexes obtained after each step:

(i) The standard 2-simplex spanned by v_0, v_1, v_2;
(ii) Three vertices v_0, v_1, v_2 and two edges $\{v_0, v_1\}$ and $\{v_0, v_2\}$;
(iii) The standard 1-simplex spanned by v_0, v_1;
(iv) The standard 0-simplex given by v_0.

3.63. By definition $\mathbb{RP}^3$ is the lens space $L(V)$ for the cyclic group $\mathbb{Z}/2$ where V has as underlying unitary vector space $\mathbb{C}^2$ and the generator s of $\mathbb{Z}/2$ acts on V by $-\mathrm{id}$. The cellular $\mathbb{Z}[\mathbb{Z}/2]$-chain complex $C_*(SV)$ is concentrated in dimensions $0, 1, 2, 3$ and is given by

$$\cdots \to 0 \to \mathbb{Z}[\mathbb{Z}/2] \xrightarrow{s-1} \mathbb{Z}[\mathbb{Z}/2] \xrightarrow{s+1} \mathbb{Z}[\mathbb{Z}/2] \xrightarrow{s-1} \mathbb{Z}[\mathbb{Z}/2] \to 0 \to \cdots.$$

Hence $\mathbb{R}^- \otimes_{\mathbb{Z}[\mathbb{Z}/2]} C_*(SV)$ is the $\mathbb{R}$-chain complex

$$\cdots \to 0 \to \mathbb{R} \xrightarrow{2 \cdot \mathrm{id}} \mathbb{R} \xrightarrow{0} \mathbb{R} \xrightarrow{2 \cdot \mathrm{id}} \mathbb{R} \to 0 \to \cdots.$$

It is contractible since a chain contraction γ_* is given by $\gamma_0 = \gamma_2 = 1/2 \cdot \mathrm{id}$ and $\gamma_n = 0$ for $n \neq 0, 2$. Hence $(c + \gamma)_{\mathrm{odd}} \colon \mathbb{R}^- \otimes_{\mathbb{Z}[\mathbb{Z}/2]} C_{\mathrm{odd}}(SV) \to \mathbb{R}^- \otimes_{\mathbb{Z}[\mathbb{Z}/2]} C_{\mathrm{ev}}(SV)$ is given by

$$\begin{pmatrix} 2 & 1/2 \\ 0 & 2 \end{pmatrix} \colon \mathbb{R}^2 \to \mathbb{R}^2.$$

This implies $\rho(\mathbb{RP}^3; V) = 4^2 = 16$.

3.73. (i) Theorem 3.49 (ii) implies that there is a map $L(7; k_1, k_2) \to L(7; k_1, k_2)$ of degree d if and only if $d \cdot l(L(7; k_1, k_2)) = u^2 \cdot l(L(7; k_1, k_2))$ holds for some $u \in \mathbb{Z}/7$ in $H_n(B\mathbb{Z}/7)$. Since $l(L(7; k_1, k_2))$ is a generator of $H_n(B\mathbb{Z}/7)$, we get $\overline{d} = u^2$ in $\mathbb{Z}/7$ for some $u \in \mathbb{Z}/7$. Hence $\overline{d} \in \{\overline{0}, \overline{1}, \overline{2}, \overline{4}\}$. Therefore $d = -1$ is not possible.

(ii) We have already explained that the homotopy groups and homology groups of a lens space depend only on its dimension. This applies in particular to $L(5; 1, 1)$ and $L(5; 2, 1)$. In order to check that they are not homotopy equivalent, it suffices by Theorem 3.49 (v) to show that there is no element $e \in \mathbb{Z}/5^*$ satisfying $\overline{1} = \pm e^2 \cdot \overline{2}$ in $\mathbb{Z}/5$. This follows from the fact that $\pm e^2 \in \{\overline{1}, \overline{4}\}$ holds for $e \in \mathbb{Z}/5^*$.

(iii) To show that $L(7; 1, 1)$ and $L(7; 2, 1)$ are oriented homotopy equivalent, it suffices by Theorem 3.49 (v) to find $e \in \mathbb{Z}/7^*$ with $\overline{1} = e^2 \cdot \overline{2}$. Take $e = \overline{2}$. Since we cannot find $e \in \mathbb{Z}/7^*$ such that $\overline{1} = \pm e \cdot \overline{1}$ and $\overline{1} = \pm e \cdot \overline{2}$ hold in $\mathbb{Z}/7$, we conclude from Corollary 3.71 that $L(7; 1, 1)$ and $L(7; 2, 1)$ are not diffeomorphic.

Chapter 4

4.5. If I is finite, the map f can be made 2-connected by attaching finitely many 2-cells because of Lemma 4.4 (ii). Suppose that f can be made 2-connected by attaching finitely many cells. The fundamental group $\pi_1(\bigvee_{i \in I} S^1)$ is a free group generated by the set I. Attaching finitely many 0- and 1-cells has the effect that the resulting space X_1 has as fundamental group the free group generated by the set $I \amalg I_0$ for some finite set I_0. We can kill the fundamental group of X_1 by attaching finitely many 2-cells by assumption. Hence $\pi_1(X_1)$ can be made trivial by adding finitely many relations $r_1, r_2, \ldots, r_s$. Choose a finite set $I' \subseteq I$ such that each relation r_i is a word in letters belonging to I'. Then $I = I'$ and hence I is finite.

4.7. Suppose that $f' \colon M' \to X$ is obtained from f by one surgery step. A direct computation shows if $M_0 = M - \mathrm{int}(S^k \times D^{n-k})$

$$\chi(M) - \chi(M')$$
$$= \chi\big(M_0 \cup_{S^k \times S^{n-k-1}} S^k \times D^{n-k}\big) - \chi\big(M_0 \cup_{S^k \times S^{n-k-1}} D^{k+1} \times S^{n-k-1}\big)$$
$$= \chi(M_0) - \chi(S^k \times S^{n-k-1}) + \chi(S^k \times D^{n-k}) -$$
$$\big(\chi(M_0) - \chi(S^k \times S^{n-k-1}) + \chi(D^{k+1} \times S^{n-k-1})\big)$$
$$= \chi(S^k \times D^{n-k}) - \chi(D^{k+1} \times S^{n-k-1})$$
$$= \chi(S^k) - \chi(S^{n-k-1}).$$

Since $\chi(S^k) \equiv 0 \bmod (2)$ holds, the claim follows.

4.10. If suffices to show that $f' \colon S^2 \to S^2$ has degree ± 1. Note that the map $f \colon T^2 \to S^2$ is transverse to $x_0 \in S^2$ and that $f^{-1}(x_0) = 0 \in \mathrm{int}(D^2) \subset T^2$. The surgery step does not alter f near 0, nor does it alter $f^{-1}(x_0)$. It follows that $f' \colon S^2 \to S^2$ is transverse to $x_0 \in S^2$ and that $|f^{-1}(x_0)| = 1$. So f' has degree ± 1, as required.

4.16. Remark 4.13 implies that M and M' are bordant. Hence M and M' have the same Stiefel–Whitney numbers. Since a homotopy equivalence of closed manifolds preserves the Stiefel–Whitney classes, see [308, § 11, Lemma 11.13 & Theorem 11.14], M' and N have the same Stiefel–Whitney numbers. Hence M and N have the same Stiefel–Whitney numbers.

4.24. We have $\pi_2(\mathbb{CP}^2) \cong \mathbb{Z}$ and $\pi_3(S^4) = \pi_2(S^4) = 0$. Hence the boundary map $\pi_3(f) = \pi_3(S^4, \mathbb{CP}^2) \to \pi_2(\mathbb{CP}^2)$ is an isomorphism. A generator of $\pi_2(\mathbb{CP}^2)$ is represented by the standard embedding $S^2 = \mathbb{CP}^1 \hookrightarrow \mathbb{CP}^2$, and hence we can find $(Q, q): (D^3, S^2) \to (S^4, \mathbb{CP}^2)$ representing a generator $\omega \in \pi_3(f)$ with q an embedding.

Let $q': S^2 \hookrightarrow \mathbb{CP}^2$ be any embedding representing a generator of $\pi_2(\mathbb{CP}^2)$, let $\nu(q')$ be the normal bundle of $q'(S^2) \subset \mathbb{CP}^2$ and $T\mathbb{CP}^2$, and let TS^2 be the tangent bundles of $\mathbb{CP}^2$ and S^2 respectively. There is a bundle isomorphism

$$(q')^*T\mathbb{CP}^2 \cong (q')^*\nu(q') \oplus TS^2.$$

Since all these bundles are orientable, their first Stiefel–Whitney classes vanish. The Cartan formula [308, § 4, Axiom 3] gives

$$(q')^*(w_2(\mathbb{CP}^2)) = (q')^*(w_2(\nu(q'))) + w_2(S^2),$$

where $(q')^*: H^2(\mathbb{CP}^2; \mathbb{Z}/2) \to H^2(S^2; \mathbb{Z}/2)$ is an isomorphism. But we have $w_2(S^2) = 0$ since S^2 is stably parallelisable whereas $(q')^*(w_2(\mathbb{CP}^2)) \neq 0$ [308, Theorem 14.10 and Problem 14B]. Hence $w_2(\nu(q')) \neq 0$ and so $\nu(q')$ is non-trivial.

4.27. Let $h: M \times [0, 1] \to N$ be a regular homotopy from f_0 to f_1. Let $\exp: D_\epsilon N \to N$ be an exponential map for appropriate $\epsilon > 0$. It has the property that $\exp \circ s = \mathrm{id}_N$ and $T(\exp \circ s) = \mathrm{id}_{TN}$ holds for the zero section $s: N \to TN$. Fix $t_0 \in [0, 1]$. Then we can find $\delta(t_0) > 0$ such that the image of $\exp_{h_{t_0}(x)}: D_\epsilon T_{h_{t_0}(x)} N \to N$ contains $h_t(x)$ for all $t \in [0, 1]$ with $|t - t_0| \leq \delta(t_0)$ and $x \in M$. Define for $t \in [0, 1]$ with $t - t_0 \leq \delta(t_0)$ a smooth map

$$H[t]: M \times [0, 1] \to N, \quad (x, r) \mapsto \exp_{h_{t_0}(x)}(r \cdot \exp^{-1}_{h_{t_0}(x)}(h_t(x))).$$

By construction $H[t]_0 = h_{t_0}$ and $H[t]_1 = h_t$. If $A \in M(m, n; \mathbb{R})$ is a matrix for which the induced map $r_A: \mathbb{R}^m \to \mathbb{R}^n$ is injective, then there exists an open neighbourhood U of A in $M(m, n; \mathbb{R})$ such that for any $B \in U$ and $r \in [0, 1]$ the map $r_{rA+(1-r)B}: \mathbb{R}^m \to \mathbb{R}^n$ is injective. This implies that we can by possibly shrinking $\delta(t_0)$ achieve that $H[t]_r: M \to N$ is an immersion for every $r \in [0, 1]$ since $T_{h_t(x)}$ depends continuously on (t, x) by the definition of a regular homotopy. Since $[0, 1]$ is compact, we can find a sequence $t_0 = 0 < t_1 < t_2 < \cdots < t_{l-1} < t_l = 1$ of elements in $[0, 1]$ such that $t_i - t_{i-1} \leq \delta(t_i)$ holds for $i = 1, 2, \ldots l$. Hence we get for $i = 1, 2, \ldots, l$ smooth regular homotopies $h[i]: M \times [0, 1] \to N$ with $h[i]_0 = h_{t_{i-1}}$ and $h[i]_1 = h_{t_i}$. Choose a smooth monotone map $\alpha: [0, 1] \to [0, 1]$ such that $\alpha(r) = 0$ for $0 \leq r \leq 1/3$ and $\alpha(r) = 1$ for $2/3 \leq r \leq 1$. Define a smooth map $H: M \times [0, 1] \to N$ by putting $H_r = h[i]_{\alpha((r-t_{i-1})/(t_i-t_{i-1}))}$ for $r \in [t_{i-1}, t_i]$. This is a smooth regular homotopy with $H_i = f_i$.

4.28. One easily checks that h is continuous since $||h_t(x)|| \leq t$ holds for $|x| \leq t$. By the chain rule we get for (x, t) with $|x| \leq t$ that the differential of h_t satisfies $T_x h_t = T_{t^{-1}x} f$ if $|x| \leq t$ and $t > 0$ hold, and $T_x h_t = (1, 0)$ otherwise. Hence each h_t is an immersion. Suppose that h is a regular homotopy. Recall that the function sending $t \in [0, 1]$ to $Th_t \in \text{map}(\mathbb{R}, M(1, 2; \mathbb{R}))$ is continuous by the definition of a regular homotopy. This is equivalent to the statement that the function $\mathbb{R} \times [0, 1] \to M(1, 2; \mathbb{R})$ sending (x, t) to $T_x h_t$ is continuous. Fix $y \in [-1, 1]$. Then $\lim_{t \to 0} T_{ty} h_t = T_0 h_0$. Since $T_{ty} h_t = T_y f$ and $T_0 h_0 = (1, 0)$ holds, we get $T_y f = (1, 0)$ for all $y \in [-1, 1]$. This implies that $f(x) = (x, 0)$ holds for $x \in [-1, 1]$ and hence for $x \in \mathbb{R}$.

4.37. This follows directly from Example 4.36.

4.45. Apply Theorem 4.44 in the special case when X is a point.

Chapter 5

5.17. Without loss of generality we can assume that X is connected, otherwise treat any component separately. Because of (5.12) this follows from

$$H_n(X; \underline{A}) \cong H_n(C_*(\widetilde{X}) \otimes_{\mathbb{Z}\pi} A) \cong H_n(C_*(X) \otimes_{\mathbb{Z}} A) = H_n(X; A),$$

where we equip A with the trivial π-action.

5.18. Because of (5.12) this follows from

$$H_n(X; \widetilde{O}) \cong H_n(C_*(\widetilde{X}) \otimes_{\mathbb{Z}\pi} \mathbb{Z}\pi) \cong H_n(C_*(\widetilde{X})) = H_n(\widetilde{X}; \mathbb{Z}).$$

5.24. Without loss of generality we can assume that X is connected. Then the element $w_1(\xi)$ in $H^1(X; \mathbb{Z}/2)$ is the same as a group homomorphism $w_1(\xi): \pi_1(X) \to \mathbb{Z}/2$. Obviously $w_1(\xi)$ sends the class $[u] \in \pi_1(X)$ of a map $u: S^1 \to X$ to $w_1(u^*\xi)$. Hence it suffices to consider the easy case $X = S^1$.

5.26. It follows immediately from the definition of the orientation covering that for any loop w in M at base point x the automorphism $O_{\widetilde{PM}}([w])$ of the infinite cyclic group $O_{\widetilde{PM}}(x)$ is $w_1(M)(x)$ if $w_1(M): \pi_1(X, x) \to \{\pm 1\}$ is the first Stiefel–Whitney class. Hence $w_1(O_{\widetilde{PM}}) = w_1(O_M)$.

5.33. (i) For every point $x \in \partial M$, there exists an open neighbourhood U of x in M and a diffeomorphism $\phi: \mathbb{R}^{d-1} \times \mathbb{R}_{\geq 0} \xrightarrow{\cong} U$ such that $\phi(\mathbb{R}^{d-1} \times \{0\}) = U \cap \partial M$. We obtain an outward normal vector field v_U on U by sending $x \in \partial U = U \cap \partial M$ to $(T_{\phi^{-1}(x)} \phi)((0, -1))$ for $0 \in T_{\phi^{-1}(x)} \mathbb{R}^{d-1}$ and $-1 \in \mathbb{R} = T_{\phi^{-1}(x)} \mathbb{R}$. Now use partitions of unity to construct an outward normal vector field v on M;

(ii) This is obvious by counting dimensions;

(iii) Take the isotopy given by $(t \cdot n_v + (1-t) \cdot n_{v'})$;
(iv) Use the isotopy appearing in the proof of assertion (iii);
(v) This is obvious.

5.35. If we vary the outward vector field, we vary the isomorphism $T\partial M \oplus \mathbb{R} \xrightarrow{\cong} i^*TM$ by an isotopy through bundle isomorphisms over ∂M. Hence the induced map $O_{\partial M} \to i^*O_M$ is independent of the choice of the outward vector field.

5.48. Since $H_n(X;\mathbb{Z}^w)$ is non-trivial, the dimension of X as a CW-complex has to be at least n. The mapping cone is simple homotopy equivalent to a point and hence a finite simple Poincaré complex of formal dimension 0.

5.50. Recall $\pi_1(\mathbb{R}P^n) = \mathbb{Z}/2$. Hence a homomorphism $w: \pi_1(\mathbb{R}P^n) \to \mathbb{Z}/2$ is either trivial or the identity. Let $w: \pi_1(\mathbb{R}P^n) \to \mathbb{Z}/2$ be any homomorphism. Compute by inspecting the cellular chain complex of the universal covering of $\mathbb{R}P^n$ that $H_n(\mathbb{R}P^n;\mathbb{Z}^w) \cong \mathbb{Z}$ holds if w is trivial and n is odd or if w is the identity and n is even, and $H_n(\mathbb{R}P^n,\mathbb{Z}^w) \cong 0$ holds if w is trivial and n is even or if w is the identity and n is odd. Hence $w_1(\mathbb{R}P^n)$ is trivial if n is odd and is the identity if n is even.

5.71. If the dimension of X is not $4k$, then there is nothing to prove. If the dimension of X is $4k$, then the intersection pairing of X^- is $-I^{2k}$. By definition, $\text{sign}(X^-) = \text{sign}(-I^{2k}) = -\text{sign}(I^{2k}) = -\text{sign}(X)$.

5.72. If we take $\mathbb{F}_2$ with trivial $\pi_1(X)$-action as coefficients, we conclude from Poincaré duality that $H^{d-i}(X;\mathbb{F}_2) \cong H_i(X;\mathbb{F}_2)$ holds for all $i \geq 0$ and $d = \dim(X)$. Since $\mathbb{F}_2$ is a field, we get $b_i(X;\mathbb{F}_2) := \dim_{\mathbb{F}_2}(H_i(X;\mathbb{F}_2))$ for $i \geq 0$. We conclude $b_i(X;\mathbb{F}_2) = b_{d-i}(X;\mathbb{F}_2)$ for $i \geq 0$. We can write $d = 2k+1$. Now we compute

$$\chi(X) = \sum_{i=0}^{2k+1}(-1)^i \cdot b_i(X;\mathbb{F}_2)$$

$$= \sum_{i=0}^{k}(-1)^i \cdot b_i(X;\mathbb{F}_2) + (-1)^{2k+1-i} \cdot b_{2k+1-i}(X;\mathbb{F}_2)$$

$$= \sum_{i=0}^{k}(-1)^i \cdot \big(b_i(X;\mathbb{F}_2) - b_i(X;\mathbb{F}_2)\big)$$

$$= 0.$$

5.74. Let $\overline{A} \to A$ and $\overline{C} \to C$ be the restrictions of the universal covering $\widetilde{X} \to X$ to A and C for $C \in \pi_0(A)$. If $\widetilde{C} \to C$ is the universal covering, then we get a π-homeomorphism $\pi \times_{\pi_1(C)} \widetilde{C} \xrightarrow{\cong} \overline{C}$. We obtain an identification

$$H_{n-1}(C; \mathbb{Z}^{w_1(C)}) = H_{n-1}(C_*(\widetilde{C}) \otimes_{\mathbb{Z}\pi_1(C)} \mathbb{Z}^{w_1(C)}) = H_{n-1}(C_*(\overline{C}) \otimes_{\mathbb{Z}\pi} \mathbb{Z}^w).$$

Applying $\mathbb{Z}\pi \otimes_{\mathbb{Z}\pi_1(C)} -$ to the $\mathbb{Z}\pi_1(C)$-chain map $- \cap [C] \colon C^{n-1-*}(\widetilde{C}) \to C_*(\widetilde{C})$ yields a $\mathbb{Z}\pi$ chain map $- \cap [C] \colon C^{n-1-*}(\overline{C}) \to C_*(\overline{C})$. The first one is a $\mathbb{Z}\pi_1(C)$-chain homotopy equivalence if and only if the second one is a $\mathbb{Z}\pi$-chain homotopy equivalence since $\pi_1(C) \to \pi$ is by assumption injective.

The following diagram of $\mathbb{Z}\pi$-chain complexes commutes

$$\begin{array}{ccc} C^{n-*}(\widetilde{X}, \overline{A}) & \xrightarrow{C^{n-*}(i)} & C^{n-*}(\widetilde{X}) \\ {\scriptstyle -\cap[X,A]}\downarrow & & \downarrow{\scriptstyle -\cap[X,A]} \\ C_*(\widetilde{X}) & \xrightarrow[C_*(i)]{} & C_*(\widetilde{X}, \overline{A}) \end{array}$$

where $i \colon \widetilde{X} \to (\widetilde{X}, \overline{A})$ is the inclusion and the vertical maps are $\mathbb{Z}\pi$-chain homotopy equivalences. It induces a $\mathbb{Z}\pi$-chain homotopy equivalence

$$\alpha \colon \operatorname{cone}(C^{n-*}(i)) \to \operatorname{cone}(C_*(i)).$$

Since we have the short exact sequences of $\mathbb{Z}\pi$-chain complexes

$$0 \to C^{n-*}(\widetilde{X}, \overline{A}) \to C^{n-*}(\widetilde{X}) \to C^{n-*}(\overline{A}) \to 0$$

and

$$0 \to C_*(\overline{A}) \to C_*(\widetilde{X}) \to C_*(\widetilde{X}, \overline{A}) \to 0,$$

we can construct a commutative diagram of $\mathbb{Z}\pi$-chain complexes with $\mathbb{Z}\pi$-chain homotopy equivalences:

$$\begin{array}{ccccc} \bigoplus_{C \in \pi_0(A)} C^{n-1-*}(\overline{C}) & \xrightarrow{\cong} & C^{n-1-*}(\overline{A}) & \xleftarrow{\simeq} & \Sigma^{-1}\operatorname{cone}(C^{n-1-*}(i)) \\ {\scriptstyle \bigoplus_{C \in \pi_0(A)} -\cap[C]}\downarrow & & & & \downarrow{\scriptstyle \Sigma^{-1}\alpha} \\ \bigoplus_{C \in \pi_0(A)} C_*(\overline{C}) & \xrightarrow[\cong]{} & C_*(\overline{A}) & \xleftarrow[\simeq]{} & \Sigma^{-1}\operatorname{cone}(C_*(i)). \end{array}$$

This implies for every $C \in \pi_0(A)$ that the $\mathbb{Z}\pi$-chain map $-\cap[C] \colon C^{n-1-*}(\overline{C}) \to C_*(\overline{C})$ is a $\mathbb{Z}\pi$-chain homotopy equivalence. Hence for every $C \in \pi_0(A)$ the $\mathbb{Z}\pi_1(C)$-chain map $- \cap [C] \colon C^{n-1-*}(\widetilde{C}) \to C_*(\widetilde{C})$ is a $\mathbb{Z}\pi$-chain homotopy equivalence.

The proof in the simple setting is analogous, one just has to ensure that all $\mathbb{Z}\pi$-chain homotopy equivalences involved in the argument above are simple.

5.76. Suppose that $(\operatorname{cone}(M), M)$ is a finite Poincaré pair of formal dimension n for some infinite cyclic local coefficient system $\mathcal{O}_{\operatorname{cone}(X)}$. Then M is a finite Poincaré complex of formal dimension $(n-1)$ for $\mathcal{O}_{\operatorname{cone}(M)}|_M$. Since $\operatorname{cone}(M)$ is simply connected, $w_1(\mathcal{O}_{\operatorname{cone}(M)})$ and hence also $w_1(\mathcal{O}_{\operatorname{cone}(M)}|_M) = w_1(M)$ are trivial. Hence M is an orientable smooth manifold of dimension $(n-1)$. Poincaré duality implies

$H_k(\mathrm{cone}(M), M; \mathbb{Z}) \cong H^{n-k}(\mathrm{cone}(M); \mathbb{Z})$. Since $H_k(\mathrm{cone}(M))$ and $H^k(\mathrm{cone}(M); \mathbb{Z})$ vanish for $k \geq 1$, we conclude from the long exact sequence of the pair $(\mathrm{cone}(M), M)$ that $H_k(M; \mathbb{Z}) \cong H_k(S^{n-1}; \mathbb{Z})$ holds for $k \in \mathbb{Z}$.

Suppose that M is $(n-1)$-dimensional, connected, and orientable and satisfies $H_k(M; \mathbb{Z}) \cong H_k(S^{n-1}; \mathbb{Z})$ for $k \in \mathbb{Z}$. Then one easily checks that with respect to the constant local coefficient system $\mathcal{O}_{\mathrm{cone}(M)}$ with value $\mathbb{Z}$ the pair $(\mathrm{cone}(M), M)$ is a finite Poincaré complex of formal dimension $(n-1)$ for any choice $[[\mathrm{cone}(M), M]]$ of generator of the infinite cyclic group $H_n(\mathrm{cone}(M), M; \mathbb{Z})$.

5.77. Because of Exercise 5.76 and the proof of the Poincaré Conjecture, these are all closed orientable smooth manifolds that are homeomorphic to S^{n-1}.

5.83. Let $b_i(M) := \dim_{\mathbb{R}}(H_i(M; \mathbb{R}))$ be the i-th Betti number. Poincaré duality implies $b_i(M) = b_{4k-i}(M)$ for all $i \geq 0$. We conclude directly from the definition of the signature that $\mathrm{sign}(M) \equiv b_{2k}(M) \bmod 2$. We get modulo 2

$$\chi(M) \equiv \sum_{i=0}^{4k} (-1)^i \cdot b_i(M)$$

$$\equiv \sum_{i=0}^{2k-1} (-1)^i \cdot b_i(M) + b_{2k}(M) + \sum_{i=2k+1}^{4k} (-1)^i \cdot b_i(M)$$

$$\equiv \sum_{i=0}^{2k-1} (-1)^i \cdot b_i(M) + b_{2k}(M) + \sum_{i=2k+1}^{4k} (-1)^i \cdot b_{4k-i}(M)$$

$$\equiv \sum_{i=0}^{2k-1} (-1)^i \cdot b_i(M) + b_{2k}(M) + \sum_{i=0}^{2k-1} (-1)^i \cdot b_i(M)$$

$$\equiv b_{2k}(M)$$

$$\equiv \mathrm{sign}(M).$$

5.89. Fix an orientation of M. Let M^- be obtained by reversing the orientation. Then $\mathrm{id} \colon \partial M \to \partial M^-$ is an orientation reversing diffeomorphism. Now apply Lemma 5.85 (i).

5.90.

(i) If n is odd, then $\dim(\mathbb{CP}^n)$ is not divisible by 4 and hence $\mathrm{sign}(\mathbb{CP}^n) = 0$. If n is even, then the intersection pairing of $\mathbb{CP}^n$ looks like $\mathbb{Z} \times \mathbb{Z} \to \mathbb{Z}$, $(a, b) \mapsto ab$ and hence $\mathrm{sign}(\mathbb{CP}^n) = 1$.
(ii) Since $S(TM \oplus \underline{\mathbb{R}})$ is the boundary of the total space of $D(TM \oplus \underline{\mathbb{R}})$, Theorem 5.79 together with Definition 5.73 implies $\mathrm{sign}(STM \oplus \underline{\mathbb{R}}) = 0$.
(iii) If f is an orientation preserving diffeomorphisms, then $H^*(f)$ is compatible with the cup product and $H_n(f)$ sends $[M]$ to $-[M]$. Hence the intersection

pairing I is isomorphic to $-I$. This implies $\text{sign}(M) = -\text{sign}(M)$ and hence $\text{sign}(M) = 0$.

5.91. If $n \neq 4k$, then there is nothing to prove. Assume that $n = 4k$ and put $M^\circ = M \setminus \text{int}(D^n)$. Then there is a homeomorphism $M \sharp N \to M^\circ \cup N^\circ$. It follows that $H^{2k}(M \sharp N; \mathbb{R}) \cong H^{2k}(M; \mathbb{R}) \oplus H^{2k}(N; \mathbb{R})$ and, since connected sum is compatible with orientations, we get $I(M \sharp N) = I(M) \oplus I(N)$. Hence

$$\text{sign}(M \sharp N) = \text{sign}(I(M) \oplus I(N)) = \text{sign}(I(M)) + \text{sign}(I(N))$$
$$= \text{sign}(M) + \text{sign}(N).$$

By Exercise 5.90 $\text{sign}(\mathbb{CP}^2) = 1$ and so we have $\text{sign}(\mathbb{CP}^2 \sharp \mathbb{CP}^2) = 2$ whereas $\text{sign}((\mathbb{CP}^2)^- \sharp \mathbb{CP}^2) = 0 \neq \pm 2$. Since the signature is a homotopy invariant up to sign, $\mathbb{CP}^2 \sharp \mathbb{CP}^2$ and $(\mathbb{CP}^2)^- \sharp \mathbb{CP}^2$ are not homotopy invariant.

Chapter 6

6.4. Check that the subspace of the total space $E_0 *_e E_1$ of the exterior fibrewise join given by $\{(e_0, e_1, t) \mid t \in [0, 1/2]\}$ is homeomorphic to $DE_0 \times E_1$, the subspace given by $\{(e_0, e_1, t) \mid t \in [1/2, 1]\}$ is homeomorphic to $E_0 \times DE_1$, and the subspace given by $\{(e_0, e_1, t) \mid t = 1/2\}$ is homeomorphic to $E_0 \times E_1$.

6.27. For a natural number k define X_k to be the quotient of S^1 by the subspace $\{\exp(2\pi i t) \mid t \in [0, 1 - (k+1)^{-1}]\} \subset S^1$. Let $f_k \colon X_k \to X_{k+1}$ be the projection. Since each f_n is a homotopy equivalence, $\text{hocolim}_{k \to \infty} X_k$ is homotopy equivalent to $X_0 = S^1$. Hence $\text{colim}_{k \to \infty} X_k$ is $\{\bullet\}$.

6.35. We conclude $\pi_1(BG) \cong \text{colim}_{k \to \infty} \pi_0(G(k))$ from Lemma 6.13 and from (6.26). The degree defines an isomorphism $\pi_0(G(k)) \cong \mathbb{Z}/2$ for all k, which is compatible with the map $G(k) \to G(k+1)$. Hence $\pi_1(BG) \cong \mathbb{Z}/2$. The degree also induces isomorphisms $\pi_k(S^k) \cong \mathbb{Z}$, which are compatible with the suspension homomorphisms $\pi_k(S^k) \to \pi_{k+1}(S^{k+1})$. This implies $\pi_0^s \cong \mathbb{Z}$.

6.46. This follows directly from Lemma 6.42 (i) since $H^0(f; \mathbb{Z}) \colon H^0(X'; \mathbb{Z}) \to H^0(X, \mathbb{Z})$ sends $1_{X'}$ to 1_X.

6.59. If the degree of f is 1, then w is trivial, f is homotopic to the identity, and T_f is homotopy equivalent to $S^1 \times S^k$. Hence one can just apply the Künneth formula. Suppose that the degree of f is -1. Then w is the non-trivial homomorphism. The pullback of the universal covering of S^1 to T_f by the projection $p \colon T_f \to S^1$ has a total space, which is homotopy equivalent to S^k and which is a model for the homotopy fibre. Hence the fibre of q is S^k. Obviously w is the first Stiefel–Whitney class of q. Since $k \geq 2$ we can identify $\pi = \pi_1(S^1) = \pi_1(T_F) = \pi_1(E) = \mathbb{Z}$.

We get from the Thom isomorphism of Theorem (6.54) isomorphisms

$$H_p(DE, E; \mathbb{Z}) \cong H^\pi_{p-k-1}(S^1; \mathbb{Z}^w);$$
$$H^\pi_p(DE, E; \mathbb{Z}^w) \cong H_{p-k-1}(S^1; \mathbb{Z});$$
$$H^p_\pi(DE, SE; \mathbb{Z}^w) \cong H^{p-k-1}(X; \mathbb{Z});$$
$$H^p(DE, SE; \mathbb{Z}) \cong H^{p-k-1}_\pi(X; \mathbb{Z}^w).$$

A direct computation shows

$$H^\pi_q(S^1; \mathbb{Z}^w) = \begin{cases} \mathbb{Z}/2 & \text{if } q = 0; \\ 0 & \text{otherwise;} \end{cases}$$

$$H_q(S^1; \mathbb{Z}) = \begin{cases} \mathbb{Z} & \text{if } q = 0, 1; \\ 0 & \text{otherwise;} \end{cases}$$

$$H^q_\pi(S^1; \mathbb{Z}^w) = \begin{cases} \mathbb{Z}/2 & \text{if } q = 1; \\ 0 & \text{otherwise;} \end{cases}$$

$$H^q(S^1; \mathbb{Z}) = \begin{cases} \mathbb{Z} & \text{if } q = 0, 1; \\ 0 & \text{otherwise.} \end{cases}$$

Since DE is homotopic to S^1, we get from the long exact sequences of the pair (DE, E) isomorphisms

$$H_p(E; \mathbb{Z}) \cong H_{p+1}(DE, E; \mathbb{Z}) \oplus H_p(S^1; \mathbb{Z});$$
$$H^\pi_p(E; \mathbb{Z}^w) \cong H^\pi_{p+1}(DE, E; \mathbb{Z}^w) \oplus H^\pi_p(S^1; \mathbb{Z}^w);$$
$$H^p(E; \mathbb{Z}) \cong H^{p+1}(DE, E; \mathbb{Z}) \oplus H^p(S^1; \mathbb{Z});$$
$$H^p_\pi(E; \mathbb{Z}^w) \cong H^{p+1}_\pi(DE, E; \mathbb{Z}^w) \oplus H^p_\pi(S^1; \mathbb{Z}^w).$$

Recall that T_f and E are homotopy equivalent. Now one just has to put these computations together.

6.81. This follows from Theorem 6.68.

6.94. The standard identification $S^n \wedge S^{m-n} \xrightarrow{\cong} S^m$ is an m-duality map. Hence the m-dual of the sphere S^n is S^{m-n}.

Chapter 7

7.7. Take g to be a strong fibre homotopy inverse of f. We have constructed in (6.106) an explicit strong fibre homotopy equivalence $\overline{R}$, which is strongly fibre homotopic to the identity. Now we get

$$f * g \simeq_{\text{sfh}} (\text{id}_{S^{k-1}} * g) \circ (f * \text{id}_{S^{k-1}})$$
$$\simeq_{\text{sfh}} (\text{id}_{S^{k-1}} * g) \circ \overline{R} \circ (f * \text{id}_{S^{k-1}})$$
$$\simeq_{\text{sfh}} \overline{R} \circ (\text{id}_{S^{k-1}} * g) \circ (\text{id}_{S^{k-1}} * f)$$
$$\simeq_{\text{sfh}} \text{id}_{S^{k-1} * S^{k-1}} \circ (\text{id}_{S^{k-1}} * (g \circ f))$$
$$\simeq_{\text{sfh}} \text{id}_{S^{k-1} * S^{k-1}}.$$

7.9. Because of Lemma 7.4 there is precisely one automorphism o' in $\text{aut}(O_X)$, for which $H_n(\text{id}_X, o')([[X]]) = [[X]]'$ holds. Now define the desired bijection on representatives by sending $[\xi, c, o]$ to $[\xi, c, o' \circ o]$.

7.18. The bundle map $\overline{f}$ induces a map of Thom spaces $\text{Th}(\overline{f}): \text{Th}(\nu(M)) \to \text{Th}(\xi)$. We obtain using Exercise 6.46 a commutative diagram

$$\begin{array}{ccc}
\pi_{n+k}(\text{Th}(\nu(M))) & \xrightarrow{\pi_{n+k}(\text{Th}(\overline{f}))} & \pi_{n+k}(\text{Th}(\xi)) \\
{\scriptstyle h_M} \downarrow & & \downarrow {\scriptstyle h_\xi} \\
H_{n+k}(\text{Th}(\nu(M))) & \xrightarrow{H_{n+k}(\text{Th}(\overline{f}))} & H_{n+k}(\text{Th}(\xi)) \\
{\scriptstyle -\cap \overline{U}_{S\nu(M)}} \downarrow \cong & & \cong \downarrow {\scriptstyle -\cap \overline{U}_{S\xi}} \\
H_n(M; O_M) & \xrightarrow{H_n(f, \alpha)} & H_n(X; O_X)
\end{array}$$

where h_M and h_ξ are the Hurewicz homomorphisms, $\overline{U}_{S\nu(M)}$ and $\overline{U}_{S\xi}$ are the intrinsic fundamental classes in the sense of Definition 6.45, and the isomorphism $\alpha: O_M \xrightarrow{\cong} O_X$ has been defined in Definition 7.13 (ii).

From the collapse map $c: S^{n+k} \to \text{Th}(\nu(M))$ we get an element in $\pi_{n+k}(\text{Th}(\nu(M)))$ whose image under the left vertical arrow is the intrinsic fundamental class $[[M]]$, see Lemma 6.67 (ii). The degree one condition says that the image of $[[M]]$ under $H_n(f, \alpha)$ is $[[X]]$. Let d be the composite of c with $\text{Th}(\overline{f})$. Then $[d] \in \pi_{n+k}(\text{Th}(\xi))$ satisfies $h_\xi([d]) \cap \overline{U}_{S\xi} = [[X]]$ and therefore is a componentwise generator. Hence $h_\xi([d])$ is a componentwise generator. The existence of o implies $w_1(\xi) = w_1(X)$.

7.20. Consider two normal k-maps of degree one $(M_m, i_m, f_m, \xi_m, \overline{f}_m, o_m)$ for $m = 0, 1$. Since in the bordism relation, see Definition 7.16, we require a vector bundle Ξ over $X \times [0, 1]$ whose restriction to $X \times \{m\}$ is isomorphic to ξ_m and the projection $X \times [0, 1] \to X$ is a homotopy equivalence, ξ_0 and ξ_1 are isomorphic. Since $\mathcal{N}_n(X)$ is defined to be $\text{colim}_{k \to \infty} \mathcal{N}_n(X, k)$, the map β is well defined.

A necessary condition for a class $[\xi] \in \text{VB}(X)$ to be in the image of β comes from Exercise 7.18, namely, for one and hence all representatives ξ of $[\xi]$ there exists a stable fibre homotopy equivalence $S\xi \to \nu_X$, where ν_X is the Spivak normal

Solutions of the Exercises

k-fibration $S\xi \to v_X$ for some natural number. It is also sufficient because of Theorem 7.19.

7.30. We indicate how the desired normal bordism with respect to the tangent bundle of degree one is constructed.

(i) We take $W = M \times [0, 1]$, $\partial_i W = M \times \{i\}$, $u_i : M \to \partial_i W$ sends x to (x, i), $F = f \times \mathrm{id}_{[0,1]}$, $\Xi = \mathrm{pr}^* \xi \oplus \underline{\mathbb{R}}$ for the projection $\mathrm{pr} \colon X \times [0, 1] \to X$,

$$(\underline{F}, F) \colon TW = TM \times [0,1] \xrightarrow{v} \mathrm{pr}^* TM \oplus \underline{\mathbb{R}} \xrightarrow{\mathrm{pr}^* \overline{f} \oplus \mathrm{id}_{\underline{\mathbb{R}}}} \mathrm{pr}^* \xi \oplus \underline{\mathbb{R}} = \Xi$$ for the canonical bundle isomorphism v, $\overline{l}_0 \colon \xi \to l_0^* \Xi = \xi$ is id_ξ, $\overline{l}_1 \colon \xi \to l_1^* \Xi = \xi$ is u, and define

$$O \colon O_\Xi = O_{\mathrm{pr}^* \xi \oplus \underline{\mathbb{R}}} \xrightarrow{\cong} O_{\mathrm{pr}^* \xi} = \mathrm{pr}^* O_X \xrightarrow{\mathrm{pr}^* O_o} \mathrm{pr}^* O_X = O_{X \times [0,1]}$$ using Lemma 5.23 (iv);

(ii) Define $W = M \times [0, 1]$ and use $(\overline{h}, h)$ to define the desired pair $(\overline{F}, F)$. The other data are obvious;

(iii) Take $W = M \times [0, 1]$ and now use g as the diffeomorphism u_0 in Definition 7.24. The other data are obvious;

(iv) Take $W = M \times [0, 1]$, $F = f \times [0, 1]$, $\Xi = \mathrm{pr}^* \xi \oplus \underline{\mathbb{R}}$ for the projection $\mathrm{pr} \colon X \times [0, 1] \to X$, and

$$\overline{F} \colon TW \oplus \underline{\mathbb{R}}^a = T(M \times [0, 1]) \oplus \underline{\mathbb{R}}^a \xrightarrow{\cong} \mathrm{pr}^* TM \oplus \underline{\mathbb{R}} \oplus \underline{\mathbb{R}}^a$$

$$\xrightarrow{\mathrm{id}_{\mathrm{pr}^* TM} \oplus \mathrm{flip}} \mathrm{pr}^* TM \oplus \underline{\mathbb{R}}^a \oplus \underline{\mathbb{R}} = \mathrm{pr}^*(TM \oplus \underline{\mathbb{R}}^a) \oplus \underline{\mathbb{R}}$$

$$\xrightarrow{\mathrm{pr}^* \overline{f} \oplus \mathrm{id}_{\underline{\mathbb{R}}}} \mathrm{pr}^* \xi \oplus \underline{\mathbb{R}} = \Xi.$$

The other data are now obvious.

7.35. Because of Theorem 7.34 it suffices to show that $[S, G/O]$ consists of two elements. The canonical map $G/O \to K(\mathbb{Z}/2, 2)$ is 3-connected and S is two-dimensional. Hence we obtain bijections

$$[S, G/O] \cong [S, K(\mathbb{Z}/2, 2)] \cong H^2(S; \mathbb{Z}/2) \cong H_0(S; \mathbb{Z}/2) \cong \mathbb{Z}/2.$$

Chapter 8

8.2. We have to show that $H_n^\Gamma(X; \mathbb{Z}^w)$ is infinite cyclic. This follows from the isomorphism (5.13) and the fact that $H_n(X; O_X)$ is isomorphic to $H^0(X)$ by Poincaré duality.

8.17. This follows from unravelling the definitions of the isomorphism $\beta_{m,o_m} \colon H_n(M; \mathbb{Z}^w) \xrightarrow{\cong} H_n(M; \mathcal{O}_M)$ of (5.13) and of the intrinsic fundamental class $[[M]] \in H_n(M; \mathcal{O}_M)$, see Definition 5.29.

8.37. Since $K_i(\widehat{M}) = 0$ for $i \leq j$, we have $H_i(\text{cone}(C_*(\widehat{f}))) = 0$ for $i \leq j$. This implies that $\text{cone}(C_*(\widehat{f}))$ is $\mathbb{Z}\Gamma$-chain homotopy equivalent to a finite projective $\mathbb{Z}\Gamma$-chain complex P_* with $P_j = 0$ for $i \leq j$. Since $\text{cone}(C^*(\widehat{f}))$ is $\text{cone}(C_*(\widehat{f}))^*$ and hence $\mathbb{Z}\Gamma$-chain homotopy equivalent to P^* and $P^i = 0$ for $i \leq j$, we get $K^i(\widehat{M}) := H^i(\text{cone}(C^*(\widehat{f}))) = 0$ for $i \leq j$.

8.74. This follows directly from (8.73).

8.75. There is a commutative diagram of $\Lambda[\mathbb{Z}/2]$-modules with $\Lambda[\mathbb{Z}/2]$-isomorphisms as vertical arrows

$$\begin{array}{ccc}
\bigoplus_{g \in S} \Lambda \oplus \bigoplus_{g \in T} \Lambda[\mathbb{Z}/2] & \xrightarrow{\cong} & \Lambda G \\
{\scriptstyle \bigoplus_{g \in S} (1-w_1(g)\cdot\epsilon)\,\mathrm{id}_\Lambda \oplus \bigoplus_{g \in T} L_{(1-w_1(g)\cdot\epsilon t)}} \Big\downarrow & & \Big\downarrow {\scriptstyle 1-\epsilon T} \\
\bigoplus_{g \in S} \Lambda \oplus \bigoplus_{g \in T} \Lambda[\mathbb{Z}/2] & \xrightarrow{\cong} & \Lambda G
\end{array}$$

where $\mathbb{Z}/2$ acts trivially on Λ and $L_{(1-w_1(g)\cdot\epsilon t)}$ is multiplication by the element $(1 - w_1(g) \cdot \epsilon t)$. Both the kernel and the cokernel of $L_{(1-w_1(g)\cdot\epsilon t)}$ are Λ-isomorphic to Λ.

8.85. This follows from the explicit description of $Q_{+1}(\Lambda G)$ and $Q^{+1}(\Lambda G)$ from Exercise 8.75 since $(1 + T)\colon Q_{+1}(\mathbb{Z}G) \to Q^{+1}(\mathbb{Z}G)$ is on the summand belonging to $g \in T$ given by $2 \cdot \mathrm{id}_{\mathbb{Z}}\colon \mathbb{Z} \to \mathbb{Z}$.

8.94. The only if part follows directly from the definitions. For the if part recall first from Remark 8.78 that the descriptions of quadratic forms as pairs (P, ψ) and as triples (P, λ, μ) are equivalent. If (P, ψ) is a form that corresponds to (P, λ, μ), then the fact that $L \subset P$ satisfies the four conditions in the exercise implies that $\lambda|_{L \times L} \equiv 0$ and $\mu|_L \equiv 0$. Hence L is a lagrangian of (P, ψ).

8.97. Define $L' = \{x \in P \mid n \cdot x \in L \text{ for some } n \in \mathbb{Z}, n \neq 0\}$. Then P/L' is a torsionfree finitely generated $\mathbb{Z}$-module. Hence $L' \subseteq P$ is a direct summand. Obviously we have $\lambda(x, y) = 0$ for $x, y \in L$. Hence L' is a lagrangian.

8.151. Suppose that $\widehat{F}$ exists. Let I be the groupoid with two objects 0 and 1 and precisely one morphisms between any two objects. Consider $g \in G$. Then $c_g \colon G \to G$ and $\mathrm{id}\colon G \to G$ considered as functors of groupoids are naturally equivalent. Hence there is an equivalence of groupoids $T \colon \mathcal{G} \times I \to \mathcal{G}$ such that the composite of T with the two inclusions $\mathcal{G} \to \mathcal{G} \times I$ is c_g and id_G. Since $\widehat{F}(T)$ is an isomorphism, we have $F(c_g) = \widehat{F}(c_g) = \widehat{F}(\mathrm{id}_G) = F(\mathrm{id}_G)$.

Solutions of the Exercises

Suppose that F is conjugation invariant. Define $\widehat{F}(\mathcal{G})$ for a groupoid $\mathcal{G}$ by the colimit of the functor $\mathcal{G} \to \mathbb{Z}\text{-MOD}$ that sends an object x to $F(\mathrm{aut}(x))$ and a morphism $u\colon x \to y$ in $\mathcal{G}$ to $F(c_u)\colon F(\mathrm{aut}(x)) \to F(\mathrm{aut}(y))$ where the group isomorphism $c_u\colon \mathrm{aut}(x) \to \mathrm{aut}(y)$ is given by conjugation with u.

8.161. This is obvious.

8.166. For a ring with involution R and a non-singular $(-1)^k$-quadratic form (K, λ, μ) over R with finitely generated free K, we get in $L_{2k}(R)$ the equality

$$[(K, -\lambda, -\mu)] = -[(K, \lambda, \mu)].$$

Changing in the choice (GBP), see (8.8), the orientation of $T_m M$ changes the resulting geometric intersection pairing and self-intersection pairing by putting minus in front.

8.170. For $m = 0, 1$ the maps $L_{2k}(l_m \circ v_m, O_{l_m \circ v_m})$ appearing in Theorem 8.168 are bijective by Lemma 8.157 (iv).

8.174. We have for a closed oriented manifold M of dimension $4k$ that $\mathrm{sign}(M \times \mathbb{CP}^2) = \mathrm{sign}(M)$ and $\mathrm{sign}(M \times \mathbb{CP}^1) = 0$. Now apply Theorem 8.173.

8.177. We have the isomorphism

$$a\colon \bigoplus_{C \in \pi_0(W)} L_{2k}(\mathbb{Z}\Pi(C), O_C) \xrightarrow{\cong} L_{2k}(\mathbb{Z}\Pi(W), O_W)$$

of Lemma 8.157 (iii). Since $w_1(W) = 0$ and hence $w_1(C) = 0$ for each component C, we can choose an isomorphism $b_C\colon L_{2k}(\mathbb{Z}\Pi_0(C), O_C) \xrightarrow{\cong} L_{2k}(\mathbb{Z})$ for each component C of W by Lemma 8.149 (i). Let C_m be the component of W that meets the image of X_m under the map $X_m \to \partial_m W \subseteq W$. Consider the composite

$$L_{2k}(\mathbb{Z}\Pi(X_m), O_{X_m}) \xrightarrow{L_{2k}(l_m \circ v_m, O_{l_m \circ v_m})} L_{2k}(\mathbb{Z}\Pi(W), O_W)$$

$$\xrightarrow{a^{-1}} \bigoplus_{C \in \pi_0(W)} L_{2k}(\mathbb{Z}\Pi(C), O_C) \xrightarrow{\mathrm{pr}} L_{2k}(\mathbb{Z}\Pi(C_m), O_{C_m}) \xrightarrow{b_{C_m}} L_{2k}(\mathbb{Z}).$$

Consider the case $C_0 = C_1$. Then we conclude from Theorem 8.168 that

$$s_0\big(\sigma(M_0, f_0, a_0, \xi_0, \overline{f}_0, o_0)\big) = s_1\big(\sigma(M_1, f_1, a_1, \xi_1, \overline{f}_1, o_1)\big).$$

Suppose that $C_0 \neq C_1$. Then we conclude from Theorem 8.168 for $m = 0, 1$

$$s_m\big(\sigma(M_m, f_m, a_m, \xi_m, \overline{f}_m, o_0)\big) = 0.$$

8.178. This follows from Theorem 8.168, as the homomorphism $L_n(\pi_1(N)) \to L_n(\pi_1(N \times S^n))$ induced by the inclusion is injective and S^n bounds D^{n+1}.

8.186. Since TM is trivial and TS^{2k} is stably trivial, there exist normal maps $\overline{f} \colon TM \oplus \underline{\mathbb{R}^a} \to \underline{\mathbb{R}^{2k+a}}$ which cover f, but the choice of such normal maps is not unique, even up to homotopy. Specifically, given two normal maps $\overline{f}_0, \overline{f}_1$ covering f, there is a bundle automorphism $\theta \colon TM \oplus \underline{\mathbb{R}^a} \cong TM \oplus \underline{\mathbb{R}^a}$ such that $\overline{f}_1 = \overline{f}_0 \circ \theta$. Now the bundle automorphism θ defines a homotopy class of maps, also denoted $\theta \in [S^k \times S^k, \mathrm{SO}]$, where as usual SO denotes the stable special orthogonal group. Since SO is an infinite loop space and $S^k \times S^k$ is stably a wedge of spheres, we have the isomorphism

$$[S^k \times S^k, \mathrm{SO}] \cong \pi_k(\mathrm{SO}) \oplus \pi_k(\mathrm{SO}) \oplus \pi_{2k}(\mathrm{SO}).$$

Since $k = 3, 7$, Bott periodicity gives $\pi_k(\mathrm{SO}) \cong \mathbb{Z}$ and $\pi_k(\mathrm{SO}) = 0$. Taking the primary obstruction for a nullhomotopy of map $S^k \times S^k \to \mathrm{SO}$, we obtain the first of the following isomorphisms

$$[S^k \times S^k, \mathrm{SO}] \cong H^k(S^k \times S^k; \pi_k(\mathrm{SO})) \cong H^k(S^k \times S^k; \mathbb{Z}).$$

It follows that we can realise every linear perturbation of a quadratic form $\mu \colon H_k(S^k \times S^k; \mathbb{Z}) \to \mathbb{Z}/2$ by a change of framing. Consulting the list of $\mathbb{Z}/2$-valued quadratic forms on $(\mathbb{Z}/2)^2$ in (8.105), we see that we can realise every quadratic form on $H_k(S^k \times S^k; \mathbb{Z})$ by altering the framings covering the collapse map $f \colon S^k \times S^k \to S^{2k}$, and, again by invoking (8.105), we can realise $\overline{f}_i$ with $\sigma(f_0, \overline{f}_0) = 0$ and $\sigma(f_1, \overline{f}_1) = 1$.

8.191. We can assume without loss of generality that f is highly connected since doing surgery on the interior changes neither $\mathrm{sign}(M, \partial M)$ nor $\mathrm{sign}(X, \partial X)$. Since X is simply connected, we do not have to pass to coverings. Recall from Theorem 8.99 that we get an isomorphism $L_0(\mathbb{Z}) \xrightarrow{\cong} \mathbb{Z}$ by sending an element in $L_0(\mathbb{Z})$ represented by a non-degenerate quadratic $\mathbb{Z}$-form to $1/8$ times the signature of its underlying non-degenerate symmetric $\mathbb{Z}$-form. By Theorem 8.189 it suffices to show for the surgery kernel $K_k(M)$ with its symmetric non-degenerate intersection pairing $s \colon K_k(M) \times K_k(M) \to \mathbb{Z}$ that its signature is the difference $\mathrm{sign}(M, \partial M) - \mathrm{sign}(X, \partial X)$.

Recall from Lemma 8.116 (i) that we obtain an isomorphism

$$K_k(M) \oplus H_k(X) \xrightarrow{\cong} H_k(M)$$

coming from the inclusion $K_k(M) \to H_k(M)$ and the composite

$$H_k(X) \xrightarrow{(-\cap [X, \partial X])^{-1}} H^k(X, \partial X) \xrightarrow{H^k(f, \partial f)} H^k(M, \partial M) \xrightarrow{(-\cap [M, \partial M])} H_k(M).$$

Finally one checks that this isomorphism is compatible with the various non-degenerate symmetric pairings. This implies

$$\operatorname{sign}(K_k(M), s) + \operatorname{sign}(X, \partial X) = \operatorname{sign}(M, \partial M).$$

8.204. We first attach a copy of $\partial_0 X \times I$ to X along $\partial_0 X \subset \partial X$ and obtain the Poincaré pair

$$(X', \partial X') := \left(X \cup_{\partial_0 X} (\partial_0 X \times I), \partial_1 X \cup (\partial_{01} X \times I) \cup (\partial_0 X \times \{1\})\right).$$

There is a obvious homotopy equivalence of pairs $(X', \partial X') \simeq (X, \partial X)$. Let $\omega' \in L_{2k}(\mathbb{Z}\pi_1(\partial_0 X), w_1(\partial_0 X))$ be some preimage of $\omega \in L_{2k}(\mathbb{Z}\pi, w)$. We apply Theorem 8.195 to construct a degree one normal map of pairs relative boundary $(f_0, \overline{f}_0) \colon (M_0, \partial M_0) \to (\partial_0 X \times I, \partial(\partial_0 X \times I))$ with surgery obstruction ω'. In the construction of $(f_0, \overline{f}_0)$, there is an underlying manifold triad $(M_0, \partial_0 M_0, \partial_1 M_0)$ such that $\partial_0 M_0 = \partial_0 X$ and $\partial_0 f_0$ is the identity diffeomorphism and such that the corresponding bundle map is the derivative of $\partial_0 f_0$ stabilised by the identity. Hence we can set $M := X \cup M_0$, glue $(f_0, \overline{f}_0)$ to the identity degree one normal map $\operatorname{id} \colon X \to X$, and define the relative boundary degree one normal map of pairs

$$(f, \overline{f}) := (f_0, \overline{f}_0) \cup (\overline{\operatorname{id}}, \operatorname{id}) \colon (X \cup M_0, \partial(X \cup M_0)) \to (X', \partial X') \simeq (X, \partial X).$$

It follows easily from the definitions that $\sigma(f, \overline{f}) = \omega$.

Chapter 9

9.2. Obviously $RG^a \cong_{RG} RG^b \implies R^a \cong_R R^b$.

9.11. This is shown by a direct calculation as follows. To check that Γ_μ is a symmetric lagrangian, we need to show that the sequence

$$0 \longrightarrow \Gamma_\mu \xrightarrow{\begin{pmatrix} 1 \\ \mu - \epsilon\mu^* \end{pmatrix}} Q \oplus Q^* \xrightarrow{\left((\epsilon\mu^* - \mu)\ 1\right)} \Gamma_\mu^* \longrightarrow 0$$

is exact. That the required composition is zero is a simple calculation. To complete the proof of exactness in the middle boils down to solving the equation

$$(\epsilon\mu^* - \mu) \cdot x + y = 0.$$

To check that Γ_μ is a quadratic lagrangian, we observe that the restriction of the standard representative of the standard hyperbolic form on $Q \oplus Q^*$ to Γ_μ is precisely

$$\mu - \epsilon\mu^*,$$

which is zero in $Q_\epsilon(\Gamma_\mu)$. This is what was required.

9.16. The inverse map is given by stabilisation with a hyperbolic form, which allows us to pass from stably finitely generated free modules to finitely generated free modules.

9.22. The only if statement has been proved in Lemma 9.20 (ii). For the if statement consider $(H_\epsilon(F); F, G')$ for G' the image of $\begin{pmatrix} \gamma \\ \delta \end{pmatrix} : G \to F \oplus F^*$.

9.45. We first show that b_C and q_C are well defined. Note that, as $\delta \colon G \to F^*$ is injective, $z \in G$ is determined uniquely by $x \in F^*$ and $s \in \mathbb{Z}$. Let $x' = x + \delta(u)$ and $y' = y + \delta(v)$ for $u, v \in G$ such that $[x] = [x']$ and $[y] = [y']$. Then $z' := z + su$ satisfies $sx' = \delta(z')$. If λ denotes the ϵ-symmetric form on $H_\epsilon(F)$, then we also have $\delta(v)(\gamma(z)) = \lambda(v, z) = 0$ since $v, z \in G$ and G is a lagrangian for λ. Hence

$$y'(\gamma(z')) = (y + \delta(v))(\gamma(z + su))$$
$$= y(\gamma(z)) + \delta(v)(\gamma(z)) + s(y + v)(\gamma(u))$$
$$\equiv y(\gamma(z)) \bmod s \cdot \mathbb{Z}.$$

This proves that b_C is well defined. Moving to q_C, recall that $\theta \in Q_{-\epsilon}(Q)$ is a $(-\epsilon)$-quadratic refinement of the $(-\epsilon)$-symmetric form $\gamma^*\delta \colon G \to G^*$. With z' as above, we have

$$\theta(z')(z') = \theta(z + su)(z + su) = \theta(z)(z) + s\big(\theta(z)(u) + \theta(u)(z)\big) + s^2\theta(u)(u).$$

Now $\theta(z)(u) = (\gamma\delta)(z)(u) = \delta^*(z)(\gamma(u)) = sx(\gamma(u))$ is divisible by s and so is $\theta(u)(z) = \epsilon\theta(z)(u)$. It follows that $\theta(z')(z') \equiv \theta(z, z) \bmod s^2 \cdot \mathbb{Z}$, and so q_C is well defined.

It is clear from the definition that $b_C \colon T_C \times T_C \to \mathbb{Q}/\mathbb{Z}$ is both bilinear and $(-\epsilon)$-symmetric and that q_C is a $(-\epsilon)$-quadratic refinement of b_C. To see that b_C is non-singular, it suffices to show that $\widehat{b}_C \colon T_C \to \widehat{T}_C$ is surjective since then it is bijective as T_C is finite. Let $\{z_1, \ldots, z_k\}$ be a basis for G where $\gamma(z_i) = r_iw_i$, $\delta(z_i) = s_ix_i$, $\{w_1, \ldots, w_k\}$ is a basis for F and $\{x_1, \ldots, x_k\}$ is a basis for F^*. We note that the set $\{[x_1], \ldots, [x_k]\}$ generates T_C, that $[x_i]$ has order s_i, and that, since $(\gamma \oplus \delta)(G) \subset F \oplus F^*$ is a summand, the natural numbers r_i and s_i are coprime for each $i = 1, \ldots, k$. Let $\{w_1^*, \ldots, w_k^*\}$ be the basis of F^* dual to the basis $\{w_1, \ldots, w_k\}$ of F. Hence $w_j^*(\gamma(z_i)) = r_i\delta_{ij}$, where δ_{ij} is the Kronecker delta function and so $b_C([x_i], [w_j^*]) = \frac{r_i}{s_i}\delta_{ij}$. Since r_i and s_i are coprime, the set $\{\widehat{b}_C([w_1^*]), \ldots, \widehat{b}_C([w_k^*])\}$ generates $\widehat{T}_C$. Therefore $\widehat{b}_C$ is onto, as required.

9.76. The result of the surgery is a degree one map $S^k \times S^{k+1} \to S^{2k+1}$.

9.77. The result of the surgery is the identity map $S^{2k+1} \to S^{2k+1}$.

9.83. The proof is by a straightforward diagram chase. Here are some hints.

To see injectivity of $\alpha \oplus \gamma$, proceed as follows. Let $a \in A$ be such that $(\alpha \oplus \gamma)(a) = 0$. Then $\alpha(a) = 0$, and hence there exists an $x' \in X'$ such that $i'(x') = a$. Since $\gamma(a) = \gamma(i'(x'))$ and $\gamma \circ i'$ is injective, we obtain $x' = 0$ and hence $a = 0$.

To see surjectivity of $\delta - \beta$, consider some $d \in D$. Define $y = j(d) \in Y$. Since $j \circ \delta$ is surjective, there exists a $c \in C$ such that $j(\delta(c)) = y$. Consider $\delta(c) - d \in D$. Since $j(\delta(c) - d) = 0$, there exists a $b \in B$ such that $\beta(b) = \delta(c) - d$. Hence $d = \delta(c) - \beta(b)$.

That $\text{im}(\alpha \oplus \gamma) \subseteq \ker(\delta - \beta)$ follows from the commutativity of the square which involves these maps. To see the other inclusion, consider $(b, c) \in B \oplus C$ such that $\beta(b) = \delta(c)$. Then there exist $a, a' \in A$ satisfying $\alpha(a) = b$ and $\gamma(a') = c$. Moreover, there exists an $x \in X$ such that $\alpha(i(x)) = \alpha(a - a')$ and $x' \in X'$ such that $i'(x') = i(x) - (a - a')$. Then $a'' = i'(x') + a = i(x) + a'$ has the property $(\alpha \oplus \gamma)(a'') = (b, c)$.

9.111. Via a diagram chase around Diagram (9.81) with $j = k$, one can show that the suggested surgery produces M' with trivial surgery kernels in all dimensions. In more detail, if $a \geq 0$ is the rank of $K_k(M)$, then the relevant part of the resulting diagram becomes:

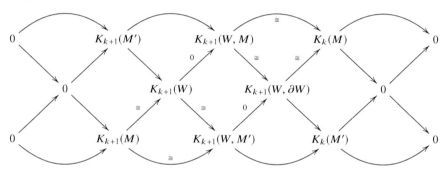

with $K_{k+1}(M) \cong K_k(M) \cong \mathbb{Z}^a$ and $K_{k+1}(M') \cong 0$ and $K_k(M') \cong 0$.

9.112. Represent a generator $x \in K_k(M) \cong \mathbb{Z}/s$ by $q \colon S^k \times D^{k+1} \hookrightarrow M$, perform surgery to obtain M', and consider the Heegaard decompositions $M = M_0 \cup U$ and $M' = M_0 \cup U'$. Our assumptions give exact sequences

$$0 \to K_{k+1}(U, \partial U) \cong \mathbb{Z} \xrightarrow{\partial} K_k(M_0) \xrightarrow{i_*} K_k(M) \cong \mathbb{Z}/s \to 0;$$

$$K_{k+1}(U', \partial U) \cong \mathbb{Z} \xrightarrow{\partial'} K_k(M_0) \xrightarrow{i'_*} K_k(M') \to 0.$$

Choose generators $\lambda \in K_{k+1}(U, \partial U) \cong \mathbb{Z}$ and $\lambda' \in K_{k+1}(U', \partial U) \cong \mathbb{Z}$ so that $x = i_*(\partial'(\lambda'))$. The first sequence yields $s \cdot \partial'(\lambda') = t \cdot \partial(\lambda)$ for some $t \in \mathbb{Z}$ and Lemma IV.3.8 in [55] implies $t = 0$. Hence $\ker(i'_*) = \text{im}(\partial') \subset \text{tors}(K_k(M_0))$. Therefore i'_* is injective on $\text{im}(\partial(\mathbb{Z}))$. We conclude that $y = i'_*(\partial(\lambda)) \in K_k(M')$ has infinite order.

Chapter 10

10.2. This follows from the fact that the vector bundle TS^2 is not trivial whereas the vector bundle $TS^2 \oplus \underline{\mathbb{R}}$ is trivial.

10.7. Choose natural numbers a and b, an R-basis B for $V \oplus R^a$ that represents the given U-equivalence class of stable bases on V, and an isomorphism $v \colon V \oplus R^a \xrightarrow{\cong} R^b$ sending the R-basis B to the standard R-basis of S^b of R^b. For a natural number n let $u(n) \colon R^n \xrightarrow{\cong} (R^n)^*$ be the R-isomorphism sending the standard basis S^n to its dual basis. We obtain an R-isomorphism

$$R^b \xrightarrow{u(b)} (R^b)^* \xrightarrow{v^*} (V \oplus R^a)^* \xrightarrow{w} V^* \oplus (R^a)^* \xrightarrow{\mathrm{id}_{V^*} \oplus u(a)^{-1}} V^* \oplus R^a$$

where w is the canonical isomorphism. Equip V^* with the U-equivalence class of bases represented by the image of the standard basis on $\mathbb{R}^b$ under the isomorphism above.

We leave it to the reader to check that this definition is independent of the choice of the representative of B for the given U-equivalence class of bases on V and that $*(\tau^U(f)) = \tau^U(f^*)$ holds.

10.16. This follows from the vanishing of $\mathrm{Wh}(\{1\})$.

10.24. The map $\det \colon K_1(\mathbb{Q}) \to \mathbb{Q}^\times$ is an isomorphism since every invertible square matrix over a field can be reduced by elementary row and column operation to a diagonal matrix. Since the determinant of an invertible matrix over $\mathbb{Z}$ is ± 1, the $\mathbb{Q}$-basis on $\mathbb{Q} \otimes C_i$ and $H_i(\mathbb{Q} \otimes C_*)$ are unique up to $\{\pm 1\}$-equivalence and

$$K_1(\mathbb{Q})/U \xrightarrow{\cong} \mathbb{Q}^{>0}, \quad [A] \mapsto |\det(A)|$$

is an isomorphism. Fix $i \geq 0$. Choose natural numbers r_i, $t_{i,1}$, $t_{i,2}$ and t_{i,j_i} such that the abelian group $H_i(C_*)$ is isomorphic to $\mathbb{Z}^{r_i} \oplus \mathbb{Z}/t_{i,1} \oplus \cdots \oplus \mathbb{Z}/t_{i,j_i}$. Let $E(i)_*$ be the $\mathbb{Z}$-chain complex concentrated in dimension 0 satisfying $E(i)_0 = \mathbb{Z}^{r_i}$. Let $T(i, k)_*$ be the $\mathbb{Z}$-chain complex concentrated in dimensions 0 and 1 with first differential $t_{i,k} \cdot \mathrm{id} \colon \mathbb{Z} \to \mathbb{Z}$ for $k \in \{1, 2, \ldots, j_i\}$. Now one easily constructs a $\mathbb{Z}$-chain map

$$f_* \colon \bigoplus_{i \geq 0} \left(\Sigma^i E(i)_* \oplus \bigoplus_{k=1}^{j_i} \Sigma^i T(i, k)_* \right) \to C_*$$

that induces an isomorphism on homology. It is a $\mathbb{Z}$-chain equivalence. Hence its Whitehead torsion lies in $K_1(\mathbb{Z})$. We conclude from Lemma 10.22 and Lemma 10.23

Solutions of the Exercises

$$\rho^U(\mathbb{Q} \otimes_{\mathbb{Z}} C_*) = \rho^U \left(\bigoplus_{i \geq 0} \left(\Sigma^i \mathbb{Q} \otimes_{\mathbb{Z}} E(i)_* \oplus \bigoplus_{k=1}^{j_i} \Sigma^i \mathbb{Q} \otimes_{\mathbb{Z}} T(i,k)_* \right) \right)$$

$$= \sum_{i \geq 0} (-1)^i \left(\rho^U(\mathbb{Q} \otimes_{\mathbb{Z}} E(i)_*) + \sum_{k=1}^{j_i} \rho^U(\mathbb{Q} \otimes_{\mathbb{Z}} T(i,k)_*) \right).$$

Obviously

$$\prod_{i \geq 0} |\operatorname{tors}(H_i(C_*))|^{(-1)^i} = \prod_{i \geq 0} \left(\prod_{k=1}^{j_i} |\operatorname{tors}(H_i(T(i,k)_*))| \right)^{(-1)^i}.$$

Hence it suffices to prove the claim

$$D\bigl(\rho^U(\mathbb{Q} \otimes_{\mathbb{Z}} C_*)\bigr) = \prod_{i \geq 0} |\operatorname{tors}(H_i(C_*))|^{(-1)^i}$$

in the special case when $C_* = E(i)_*$ and $C_* = T(i,k)_*$. These cases follow by a direct calculation.

Chapter 11

11.4. The natural map ho-aut$^s(X)\backslash \mathcal{S}^s(X) \to \mathcal{M}^s(X)$ is obviously surjective. To see that this map is injective, it is enough to show that any two simple structures $[f_0 \colon M \to X]$ and $[f_1 \colon M \to X]$ lie in the same orbit of the action of ho-aut$^s(X)$. Let $f_1^{-1} \colon X \to M$ be a homotopy inverse for f_1 and define the self-homotopy equivalence $g := f_0 \circ f_1^{-1} \colon X \to X$. Then $f_0 \simeq (f_0 \circ f_1^{-1}) \circ f_1 = g \circ f_1$, and so $[f_0] = [g][f_1]$, as required.

An action is transitive if and only if the set of orbits has one element.

11.13. The space $(D^p \times S^q)/(S^{p-1} \times S^q)$ is the Thom space of the trivial rank p vector bundled over S^q and so is homotopy equivalent to the wedge $S^p \vee S^{p+q}$. Now $[S^p \vee S^{p+q}, G/O] \cong \pi_p(G/O) \oplus \pi_{p+q}(G/O)$ and the first claim follows from (11.11).

For the second claim, it suffices to compute $[S^p \times S^q, G/O]$ because of (11.11). There is a space B(G/O) with a homotopy equivalence $G/O \to \Omega B(G/O)$ since G/O is a loop space. Since the suspension of $S^p \times S^q$ splits as a wedge of spheres $\Sigma(S^p \times S^q) \simeq S^{p+1} \vee S^{q+1} \vee S^{p+q+1}$ up to homotopy, we have

$$[S^p \times S^q, G/O] \cong [S^p \times S^q, \Omega B(G/O)] \cong [\Sigma(S^p \times S^q), B(G/O)]$$
$$\cong \pi_{p+1}(B(G/O)) \oplus \pi_{q+1}(B(G/O)) \oplus \pi_{p+q+1}(B(G/O))$$
$$\cong \pi_p(G/O) \oplus \pi_q(G/O) \oplus \pi_{p+q}(G/O).$$

11.33. This follows from the fact that any self-homotopy equivalence of S^n is homotopic to a homeomorphism.

11.37. The topological version of the s-Cobordism Theorem 2.1, the vanishing of Wh($\{1\}$), and the Alexander trick imply that $\mathcal{S}^{\text{TOP},s}(D^{5+i}, \partial D^{5+i})$ consists of one element for $i \geq 0$. From the topological surgery exact sequence applied to $(X, \partial X) = (D^5, \partial D^5)$, see Theorem 11.25, and Remark 11.36, we conclude that we get for $i \geq 1$ isomorphisms of abelian groups

$$\pi_{5+i}(\text{G/TOP}) \cong [D^{5+i}/\partial D^{5+i}, \text{G/TOP}] \cong \mathcal{N}^{\text{TOP}}(D^{5+i}, \partial D^{5+i}) \cong L_{5+i}(\mathbb{Z})$$

since G/TOP is simply connected and $D^{5+i}/\partial D^{5+i} \cong S^{5+i}$.

11.45. The surgery exact sequence in the topological category looks like $L_6(\mathbb{Z}) \to \mathcal{S}^{\text{TOP},h}(M) \to \mathcal{N}(M) \to L_5(\mathbb{Z})$. Since $L_6(\mathbb{Z}) \cong \mathbb{Z}/2$ by Theorem 8.111 and $L_5(\mathbb{Z}) = 0$ by Theorem 9.18, it suffices to show that $\mathcal{N}(M)$ is finite. Since M is simply connected, we get $H^4(M; \mathbb{Z}) \cong H_1(M; \mathbb{Z}) \cong \{0\}$. This implies that $H^{4j}(M; \mathbb{Z}_{(2)})$ vanishes for $j \geq 1$. Since M is homotopy equivalent to a finite CW-complex, $\widetilde{KO}^0(X)$ is a finitely generated abelian group and $H^{4j-2}(M; \mathbb{Z}/2)$ is finite for $j \geq 1$. We conclude from Theorem 11.24 and Section 11.9 that $[X, \text{G/TOP}]$ and hence $\mathcal{N}(M)$ are finite.

11.46. The surgery exact sequence in the topological category looks like $L_7^s(\mathbb{Z}) \to \mathcal{S}^{\text{TOP},s}(M) \to \mathcal{N}(M) \to L_6^s(\mathbb{Z})$. Since Wh($\{1\}$) vanishes, we obtain $L_6^s(\mathbb{Z}) \cong L_6(\mathbb{Z}) \cong \mathbb{Z}/2$ by Theorem 8.111 and $L_7^s(\mathbb{Z}) = L_7(\mathbb{Z}) \cong 0$ by Theorem 9.18. Hence suffices to show that $\mathcal{N}(M)$ is finite if and only if $\pi_2(M)$ is finite. Since M is simply connected, we get from the Hurewicz isomorphism and Poincaré duality $H^4(M; \mathbb{Z}) \cong H_2(M; \mathbb{Z}) \cong \pi_2(M)$. This implies that $H^{4j}(M; \mathbb{Z}_{(2)})$ is finite for all $j \geq 1$ if and only if $\pi_2(M)$ is finite. Since M is homotopy equivalent to a finite CW-complex, $\widetilde{KO}^0(X)$ is a finitely generated abelian group and $H^{4j-2}(M; \mathbb{Z}/2)$ is finite for $j \geq 1$. We conclude from Theorem 11.24 and Section 11.9 that $[X, \text{G/TOP}]$ and hence $\mathcal{N}(M)$ are finite.

Chapter 12

12.4. Let Σ be a homotopy n-sphere for $n \geq 6$. Let $D_0^n \to \Sigma$ and $D_1^n \to \Sigma$ be two disjoint embedded disks. Equip D_i^n with orientation coming from the one on Σ. Then $W = \Sigma - (\text{int}(D_0^n) \coprod \text{int}(D_1^n))$ is a simply connected h-cobordism. By the h-Cobordism Theorem 2.4 there is an orientation preserving diffeomorphism $(F, \text{id}, f) \colon (\partial D_0^n \times [0, 1], \partial D_0^n \times \{0\}, \partial D_0^n \times \{1\}) \to (W, \partial D_0^n, \partial D_1^n)$. Hence Σ is oriented diffeomorphic to $D_0^n \cup_f (D_1^n)^-$, and f is orientation preserving.

12.34. The claim is true for $n \in \{1, 3, 5, 7, 9, 11, 13, 15\}$ by (12.2). Hence we can assume $n \geq 17$ in the sequel. Theorem 12.1 (v) implies that Θ_n vanishes if and only if bP_{n+1} and $\operatorname{coker}(J_n)$ vanish.

Next we deal with the case $n = 4k-1$ for $k \geq 5$. Then bP_{4k} is non-trivial by Theorem 12.1 (i) and the claim follows.

Suppose that $n = 4k+1$ for $k \geq 4$. If bP_{4k+2} vanishes, we conclude from Theorem 12.1 (ii) that $4k+2 \in \{30, 62, 126\}$. If $4k+2 \in \{30, 62\}$, then $bP_{4k+2} = 0$ by Theorem 12.1 (ii). Since $\pi_{8l+5}(SO) \cong \{0\}$ for $l \geq 1$, the group $\operatorname{coker}(J_{8l+5})$ vanishes if and only if $\pi^s_{8l+5} = 0$. Now the claim follows.

12.35. The claim is true for $n \in \{2, 4, 6, 8, 10, 12, 14\}$ by (12.2). Hence we can assume $n \geq 16$ in the sequel.

Next we consider the case $n = 8k$ for $k \geq 2$. Then Θ_k vanishes if and only if $\pi^s_{4k} \cong \mathbb{Z}/2$ since $\pi_{8l}(SO) \cong \mathbb{Z}/2$ and we have Theorem 12.1 (iv) and Theorem 12.33 (i).

Suppose $n = 8k + 4$ for $k \geq 2$. Then Θ_k vanishes if and only if $\pi^s_{4k} = 0$ since $\pi_{8l}(SO) \cong 0$ and we have Theorem 12.1 (iv) and Theorem 12.33 (i).

Next we consider the case $n = 4k+2$ for $k \geq 4$. We conclude from Theorem 12.1 (iii) that Θ_{4k+2} vanishes if and only if either $(\operatorname{coker}(J_{4k+2}) \cong \mathbb{Z}/2$ and $bP_{4k+2} = 0)$ or $(\operatorname{coker}(J_{4k+2}) = 0$ and $bP_{4k+2} \neq 0)$. If we have $4k+2 \notin \{30, 62, 126\}$, then bP_{4k+2} is non-trivial by Theorem 12.1 (ii). If we have $4k+2 \in \{30, 62\}$, then bP_{4k+2} is trivial by Theorem 12.1 (ii). Since $\pi_{4k+2}(SO) \cong \{0\}$ for $k \geq 1$, the group $\operatorname{coker}(J_{4k+2})$ is isomorphic to $\pi^s_{4k+2} = 0$. Now the claim follows.

12.62. Consider the relevant piece of the Kervaire–Milnor Braid:

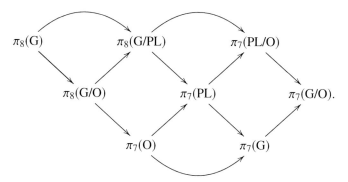

Based on the information $\pi_7(O) \to \pi_7(G)$ being onto and other information in Chapter 12, the braid above is isomorphic to the following braid:

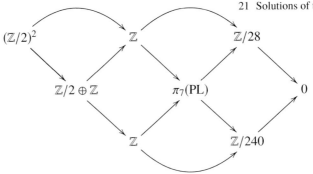

Elementary algebra now shows that $\pi_7(\mathrm{PL}) \cong \mathbb{Z} \oplus \mathbb{Z}/4$.

Chapter 13

13.3. We have the following commutative diagram

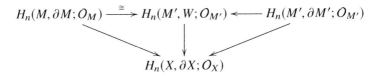

where the two horizontal arrows come from the inclusions, the left one is an isomorphism by excision, the image of $[[M, \partial M]]$ under the left horizontal arrow agrees with the image of $[[M', \partial M']]$ under the right horizontal arrow, and the other arrows are induced by $(f, \partial f)$, (f', F) and $(f', \partial f')$. Since $(f, \partial f)$ has degree one, $(f', \partial f')$ has degree one.

Since $V = M' \times [0, 1]$ and $E = f' \times \mathrm{id}_{[0,1]}$, one easily checks that $(E, \partial E)$ has degree one since $(f', \partial f')$ has degree one.

13.5. Choose a normal map of degree one $h\colon L \to S^{4k}$ such that $\mathrm{sign}(L) \neq 0$. This can be done by Lemma 12.12 and Lemma 12.46. Now define M to be the connected sum $N\#L$, taken on the interior of M, and define $f := \mathrm{id}_N \#h\colon N\#L \to N\#S^{4k} \xrightarrow{u} N$ and $\partial f = \mathrm{id}_{\partial N}$ for some orientation preserving diffeomorphism $u\colon N\#S^{4k} \xrightarrow{\cong} N$ that induces the identity on the boundary.

Obviously assertion (i), i.e., ∂f is a diffeomorphism, holds.

Suppose that assertion (ii) is not true, i.e., that the normal map $(f, \partial f)$ is normally bordant relative boundary to a normal map $(f', \partial f)\colon (M', \partial M) \to (N, \partial N)$ with f' a homotopy equivalence. Then M and M' are bordant relative boundary. This implies by Theorem 5.79 and Lemma 5.85

$$\begin{aligned}\operatorname{sign}(L) &= \operatorname{sign}(M \cup_{\partial f} N^-) - \operatorname{sign}(N \cup_{\partial N} N^-)\\ &= \operatorname{sign}(M' \cup_{\partial f} N^-) - \operatorname{sign}(N \cup_{\partial N} N^-)\\ &= \operatorname{sign}(M', \partial M') - \operatorname{sign}(N, \partial N)\\ &= 0.\end{aligned}$$

This contradicts $\operatorname{sign}(L) \neq 0$ and hence assertion (ii) is true.

Assertion (iii) follows directly from the π-π-Theorem 13.4.

13.36. This follows from the facts that for a compact subset $C \subseteq K$ there exists a compact CW-complex $L \subseteq K$ with $C \subseteq L$ and the image of a map from a finite CW-complex to K is compact.

13.43. The group homomorphism given by the determinant $\det\colon \operatorname{GL}(n,\mathbb{R}) \to \operatorname{GL}(1,\mathbb{R})$ is 1-connected. Hence the map $\operatorname{Bdet}\colon \operatorname{BGL}(n,\mathbb{R}) \to \operatorname{BGL}(1,\mathbb{R})$ is 2-connected. It corresponds to sending a bundle ξ to the associated determinant $\det(\xi)$, which is a 1-dimensional vector bundle and satisfies $\mathcal{O}_\xi = \mathcal{O}_{\det(\xi)}$. A 1-dimensional vector bundle μ carries the same information as the associated infinite cyclic coefficient system $\mathcal{O}_\mu$.

13.49. Consider the element $[u]$ in $\widehat{H}^n(\mathbb{Z}/2;A)$ represented by $u \in A$ satisfying $u - (-1)^n \cdot *(u) = 0$. Then $u + (-1)^n \cdot *(u) = u + u = 2u$ and hence $2 \cdot [u] = [2 \cdot u] = 0$ in $\widehat{H}^n(\mathbb{Z}/2;A)$.

13.53. We treat only the decoration h, since the modifications for the decoration s are obvious. We will give the explanation only in terms of the underlying maps, ignore all bundle data and take the degree one conditions for granted. By passing to the mapping cylinder we can arrange that K is a subspace of L and $v \colon K \to L$ is the inclusion.

An element in $\mathcal{L}^\epsilon_{n+1}(K \xrightarrow{(v,\overline{v})} L)$ is represented by a map of triads

$$(f; \partial_0 f, \partial_1 f) \colon (M; \partial_0 M, \partial_1 M) \to (X; X_0, X_1)$$

with a reference map $u \colon X \to L$ such that $\partial_0 f$ and $\partial \partial_0 f = \partial \partial_1 f$ are homotopy equivalences and $u(\partial_1 X) \subseteq K$ holds. We leave it to the reader to figure out the bordism relation.

The map j_{n+1} sends the class of $(f, \partial f) \colon (M, \partial M) \to (X, \partial X)$ with ∂f a homotopy equivalence to the class of the map of triads where we put $\partial_1 M = \partial_1 X = \emptyset$ and $\partial_0 M = \partial M$ and $\partial_0 X = \partial X$. The map ∂_{n+1} is given by sending the class of $(f, \partial_0 f, \partial_1 f) \colon (M; \partial_0 M, \partial_1 M) \to (X; X_0, X_1)$ with a reference map $u \colon X \to L$ to the class of $(\partial_1 f, \partial \partial_1 f) \colon (\partial_1 M, \partial \partial_1 M) \to (\partial_1 X, \partial \partial_1 X)$ with the reference map $u|_{\partial_1 X} \colon \partial_1 X \to K$.

We leave it to the reader to check that the sequence is exact and natural.

13.56. We conclude from Theorem 13.46 and Theorem 13.52 that the canonical map

$$L_n^s(\mathbb{Z}[\mathbb{Z}^m]) \xrightarrow{\cong} L_n^h(\mathbb{Z}[\mathbb{Z}^m])$$

is bijective for all n, m. We obtain by induction over n from Theorem 13.46 and Theorem 13.55 an isomorphism

$$L_n^h(\mathbb{Z}[\mathbb{Z}^m]) \cong \bigoplus_{i=0}^{n} \binom{n}{i} \cdot L_{n-i}^h(\mathbb{Z}).$$

Now use the computations of $L_{n-i}(\mathbb{Z})$ appearing in Theorem 8.99, Theorem 8.111 and Theorem 9.18.

Chapter 14

14.7. Observe that for any R-modules P and Q we have an isomorphism $(P \oplus Q)^{*t} \cong P^{*t} \oplus Q^{*t}$ compatible with the evaluation maps (14.6) in the sense that $e(P \oplus Q) = e(P) \oplus e(Q)$. For $M = R$ we obviously have that $e(R) \colon R \to (R^{*t})^{*t}$ is an R-isomorphism. By iterating the first observation, we obtain that $e(R^n)$ is an R-isomorphism for any $n \geq 0$. Next recall that a finitely generated R-module P is projective if and only if it is a direct summand of a finitely generated free R-module. Another application of the first observation now gives that $e(P)$ is an R-isomorphism as well.

The necessity of the projective assumption is seen by considering the example when $R = \mathbb{Z}$ and $M = \mathbb{Z}/2$. The necessity of the finitely generated assumption is seen by considering the example where R is a field and M is an infinite-dimensional vector space over R.

14.10. We know already that the slant map s is bijective. It remains to prove the compatibility with the involutions. Consider $m \in M$ and $n \in N$. We have to show that the elements $s(m \otimes n)^*$ and $\text{ev} \circ s(n \otimes m)$ in $\hom_R(M^*, (M^*)^*)$ agree. Fix $f \in M^*$. Then $\overline{s(m \otimes n)^*(f)} \colon M^* \to R$ sends ϕ to $f(s(m \otimes n)(\phi))$ and $\text{ev} \circ s(n \otimes m)(f)$ sends ϕ to $\overline{\phi(s(n,m)(f))}$. We compute in R

$$f(s(m \otimes n)(\phi)) = f(\overline{\phi(m)} \cdot n) = \overline{\phi(m)} \cdot f(n)$$

$$= \overline{f(n)\phi(m)} = \overline{\phi(\overline{f(n)} \cdot m)} = \overline{\phi(s(n,m)(f))}.$$

14.38. Take $C = 0[R] \oplus 0[R]$, $D = E = I$, and $f = g$ to be the obvious inclusion. Then E is the cellular R-chain complex of S^1 with respect to the corresponding CW-complex structure and is explicitly given by the 1-dimensional R-chain complex

$R \oplus R \xrightarrow{\begin{pmatrix} -1 & -1 \\ +1 & +1 \end{pmatrix}} R \oplus R$. In particular, E is not R-chain homotopy equivalent to $0[R]$. Now take $C' = C = 0[R] \oplus 0[R]$, $D = E = 0(R)$ and $f = g = (\mathrm{id}, \mathrm{id})$. Then $E' = 0[R]$. Take $u = \mathrm{id}_C$ and v and w to be the obvious projections. Then u, v and w are R-chain homotopy equivalences. But the induced map $z \colon F \to F'$ cannot be an R-chain homotopy equivalence.

14.51. Let $k \colon z(v, h') \simeq z(v, h'')$ be an R-chain homotopy. We can write $k_n = (a_{n-1}, b_n) \colon C_{n-1} \oplus D_n \to E_{n+1}$ in degree n. Since we have

$$e_{n+1} \circ (a_{n-1}, b_n) + (a_{n-2}, b_{n-1}) \circ \begin{pmatrix} -c_{n-1} & 0 \\ f_{n-1} & d_n \end{pmatrix} = (h''_{n-1}, v_n) - (h'_{n-1}, v_n),$$

we conclude

$$e_{n+1} \circ a_{n-1} - a_{n-2} \circ c_{n-1} + b_{n-1} \circ f_{n-1} = h''_{n-1} - h'_{n-1};$$
$$e_{n+1} \circ b_n + b_{n-1} \circ d_n = 0.$$

The second equation is equivalent to the assertion that b is an R-chain homotopy $b \colon 0 \simeq 0$ of R-chain maps $D \to E$ and the first equation is equivalent to the assertion that a is an R-chain homotopy $a \colon h' + b \circ f \simeq h''$ of R-chain homotopies of R-chain maps $C \to E$. This proves one of the implications. The other implication is obtained by performing the above calculation backwards.

14.53. The 2-dimensional $\mathbb{Z}$-chain complex C given by $\mathbb{Z} \xrightarrow{2\,\mathrm{id}_\mathbb{Z}} \mathbb{Z} \xrightarrow{\mathrm{pr}} \mathbb{Z}/2$ for pr the projection is acyclic, but not projective. It cannot be contractible since $C \otimes_\mathbb{Z} \mathbb{Z}/2$ has non-trivial homology. The standard resolution $\widehat{W}$ of Definition 15.9 is an acyclic projective $\mathbb{Z}[\mathbb{Z}/2]$-chain complex. It cannot be contractible since $\widehat{W} \otimes_{\mathbb{Z}[\mathbb{Z}/2]} \mathbb{Z}$ has non-trivial homology.

14.56. By the universal property of the mapping cone from Remark 14.45, the canonical R-chain map $z = z(e', -g \circ e' + h') \colon \mathrm{cone}(f) \to \mathrm{cone}(f')$ is given by the formula

$$z(x, y) = e'(y) + (h' - g \circ e')(x).$$

It is an isomorphism of R-chain complexes as its inverse is $z' = z(e, g \circ e + h)$, where $h \colon C \to \mathrm{cone}(f)$ is the canonical chain homotopy $h \colon 0 \simeq e \circ f$, as an easy calculation shows. The identity $z \circ e = e'$ follows immediately from the above formula for z.

14.57. (i) We have the exact sequence $0 \to C \xrightarrow{f \oplus g} D \oplus E \xrightarrow{\overline{g} - \overline{f}} F \to 0$. Since f is a cofibration, i.e., it is degreewise split injective, we can choose for each $n \in \mathbb{Z}$ an R-submodule D'_n of D_n such that $\mathrm{im}(f_n) \oplus D'_n = D_n$. Hence we obtain an R-isomorphism $(\overline{g}_n \circ i) \oplus \overline{f}_n \colon D'_n \oplus E_n \xrightarrow{\cong} F_n$, where i denotes the inclusion. We conclude that $\overline{f}_n$ is split injective and hence $\overline{f}$ is a cofibration.

(ii) Since f and $\overline{f}$ are cofibrations, we get a commutative diagram of R-chain complexes with degreewise split exact rows where the third vertical arrow is the R-chain map induced by g and $\overline{g}$ and is bijective since we started with a pushout

$$\begin{array}{ccccccccc} 0 & \longrightarrow & C & \xrightarrow{f} & D & \longrightarrow & \mathrm{coker}(f) & \longrightarrow & 0 \\ & & \downarrow g & & \downarrow \overline{g} & & \downarrow \cong & & \\ 0 & \longrightarrow & E & \xrightarrow{\overline{f}} & F & \longrightarrow & \mathrm{coker}(\overline{f}) & \longrightarrow & 0. \end{array}$$

Now apply Proposition 14.55.

14.58. Assume first that f is a cofibration. Consider the following pushout

$$\begin{array}{ccc} C & \xrightarrow{f} & D \\ \downarrow i & & \downarrow \overline{i} \\ I \otimes C & \xrightarrow{\overline{f}} & I \otimes C \cup_f D \end{array}$$

where i is the split injection coming from the obvious inclusion $0[\mathbb{Z}] \to I$. Since i is an R-chain homotopy equivalence, $\overline{i}$ is an R-chain homotopy equivalence by Exercise 14.57. Hence the obvious inclusion $j\colon I \otimes C \cup_f D \to I \otimes D$ is an R-chain homotopy equivalence. We have the exact sequence of R-chain complexes $0 \to I \otimes C \cup_f D \xrightarrow{j} I \otimes D \xrightarrow{p} \mathrm{coker}(j) \to 0$. We conclude from Proposition 14.55 that $\mathrm{coker}(j)$ is contractible. Hence there is an R-chain map $s\colon \mathrm{coker}(j) \to I \otimes D$ such that $j \oplus s\colon I \otimes C \cup_f D \oplus \mathrm{coker}(j) \xrightarrow{\cong} I \otimes D$ is an R-chain isomorphism, see Lemma 14.54. The triple (u, v, h) yields an R-chain map $w\colon I \otimes C \cup_f D \to E$. Let $\overline{w}\colon I \otimes D \to E$ be the composite of the inverse of the R-chain isomorphism $j \oplus s$ with $w \oplus 0\colon I \otimes C \cup_f D \oplus \mathrm{coker}(j) \to E$. Then $w = \overline{w} \circ j$. One easily checks that $\overline{w}$ yields the desired R-chain map u' and the desired R-chain homotopy h'.

To see the converse, recall that an R-chain homotopy of maps $C \to D$ can be viewed as an R-chain map of the shape $I \otimes C \to D$, and let us argue as follows. Consider the following homotopy extension problem with $E = \mathrm{cyl}(f)$:

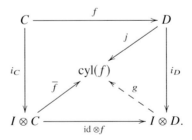

The chain maps $\overline{f}$, $\mathrm{id} \otimes f$, g are given by the following 3×3 matrices in each degree:

$$\overline{f}_n = \mathrm{id}_{C_{n-1}} \oplus \mathrm{id}_{C_n} \oplus f_n, \quad (\mathrm{id} \otimes f)_n = f_{n-1} \oplus f_n \oplus f_n,$$

and

$$g_n = \begin{pmatrix} \alpha_n & \beta_n & 0 \\ \gamma_n & \delta_n & 0 \\ \varepsilon_n & \varphi_n & \mathrm{id}_{D_n} \end{pmatrix} : D_{n-1} \oplus D_n \oplus D_n \to C_{n-1} \oplus C_n \oplus D_n.$$

The entries in the last column of g_n are forced by the commutativity of the right-hand triangle. The identity $\overline{f}_n = g_n \circ (\mathrm{id} \otimes f)_n$ then forces in particular that $\mathrm{id}_{C_n} = \delta_n \circ f_n$, giving the degreewise splitting of f.

14.62. We get from (14.43) the following commutative diagram of R-chain complexes

$$\begin{array}{ccccccccc}
0 & \longrightarrow & C & \xrightarrow{k} & \mathrm{cyl}(f) & \xrightarrow{q} & \mathrm{cone}(f) & \longrightarrow & 0 \\
& & \downarrow \mathrm{id} & & \downarrow p & & \downarrow z & & \\
0 & \longrightarrow & C & \xrightarrow{f} & D & \xrightarrow{g} & E & \longrightarrow & 0
\end{array}$$

whose rows are degreewise split exact and where z is the map given by the universal property of the mapping cone. Since the first two vertical arrows are R-chain homotopy equivalences, z is an R-chain homotopy equivalence by Proposition 14.55.

14.63. Fix an R-chain complex C, which is not contractible. Consider the sequence $C \to 0 \to \Sigma C$ and an R-chain map $h\colon \Sigma C \to \Sigma C$. We can think of h as an R-chain homotopy $0 \simeq 0$ of R-chain maps $C \to \Sigma C$. The R-chain map $z\colon \mathrm{cone}(0) \to \Sigma C$ coming from the universal property of $\mathrm{cone}(0)$ can be identified with $h\colon \Sigma C \to \Sigma$. Hence the sequence is not a homotopy cofibration sequence for $h = 0$ whereas it is a homotopy cofibration sequence for $h = \mathrm{id}$.

14.65. Let $C \xrightarrow{f} D \xrightarrow{g} E$ be a homotopy cofibration with respect to the R-chain homotopy $h\colon 0 \simeq g \circ f$. We get from (14.43) the following diagram of R-chain complexes

$$\begin{array}{ccccccccc}
0 & \longrightarrow & C & \xrightarrow{k} & \mathrm{cyl}(f) & \xrightarrow{q} & \mathrm{cone}(f) & \longrightarrow & 0 \\
& & \downarrow \mathrm{id} & & \downarrow p & & \downarrow z & & \\
0 & \longrightarrow & C & \xrightarrow{f} & D & \xrightarrow{g} & E & \longrightarrow & 0
\end{array}$$

where z is the map given by the universal property of the mapping cone. The first square commutes strictly. The second square commutes up to R-chain homotopy, namely, an R-chain homotopy $k\colon z \circ q \simeq g \circ p$ is given in degree n by $k_n = (0, h_n, 0)\colon C_{n-1} \oplus C_n \oplus D_n \to E_{n+1}$. All vertical arrows are R-chain homotopy equivalences and the upper row is a cofibration sequence. This finishes the proof of the implication (i) $\implies$ (ii).

The implication (ii) $\implies$ (i) follows from Lemma 14.64 (i).

14.69. Fix an R-chain complex C that is not contractible. Consider an R-chain map $h\colon \Sigma C \to \Sigma C$ and the square

$$\begin{array}{ccc} C & \longrightarrow & 0 \\ \downarrow & & \downarrow \\ 0 & \longrightarrow & \Sigma C. \end{array}$$

Think of h as an R-chain homotopy $0 \circ \mathrm{id} \simeq \mathrm{id} \circ 0$ of R-chain maps $C \to \Sigma C$. The R-chain map $z\colon \mathrm{cone}(0) \to \mathrm{cone}(0)$ agrees with $h\colon \Sigma C \to \Sigma$. Hence the square is not homotopy cocartesian for $h = 0$ whereas it is homotopy cocartesian for $h = \mathrm{id}$.

14.70. We use Lemma 14.68, which contains various equivalent conditions for a homotopy cocartesian square. The square

$$\begin{array}{ccc} C & \xrightarrow{f} & D \\ g\downarrow & & \downarrow \overline{g} \\ E & \xrightarrow{\overline{f}} & F \end{array}$$

is homotopy cocartesian if and only if the sequence $C \to D \oplus E \to F$ is a homotopy cofibration sequence. Hence we have an R-chain homotopy equivalence $z\colon \mathrm{cone}(f \oplus g) \to F$. The square

$$\begin{array}{ccc} C & \xrightarrow{f \oplus g} & D \oplus E \\ \overline{g} \circ f \downarrow & & \downarrow \overline{f} \oplus \overline{g} \\ F & \xrightarrow{\Delta} & F \oplus F \end{array}$$

is homotopy cocartesian if and only if the canonical map $a\colon \mathrm{cone}(f \oplus g) \to \mathrm{cone}(\Delta)$ is an R-chain equivalence. It is easy to see that there is an R-chain homotopy equivalence $z'\colon \mathrm{cone}(\Delta) \to F$. Some diagram chasing shows that $z = z' \circ a$. Hence z is an R-chain homotopy equivalence if and only if a is.

14.76. We need to show that under the assumptions the diagram

$$\begin{array}{ccc} A & \xrightarrow{z(\alpha,\beta)} & \mathrm{hofib}(f) \\ \mu\downarrow & & \downarrow \mathrm{hofib}(\kappa,\lambda) \\ A' & \xrightarrow{z(\alpha',\beta')} & \mathrm{hofib}(f') \end{array}$$

commutes up to homotopy.

Recall that $\mathrm{hofib}(f)_n = C_n \oplus D_{n+1}$ and that for $a \in A_n$ we have $z(\alpha, \beta)(a) = (\alpha(a), \beta(a))$, and the analogous formula holds for $z(\alpha', \beta')$. Also $\mathrm{hofib}(\kappa, \lambda)(x, y) = (\kappa(x), \lambda(y))$.

If $\lambda \circ \beta = \beta' \circ \mu$, then we get

$$\mathrm{hofib}(\kappa, \lambda) \circ z(\alpha, \beta)(a) = (\kappa(\alpha(a)), \lambda(\beta(a))) =$$
$$= (\alpha'(\mu(a)), \beta'(\mu(a))) = z(\alpha', \beta') \circ \mu(a).$$

If $\overline{\lambda}\colon \lambda \circ \beta \simeq \beta' \circ \mu$, then we have for each $a \in A_n$ the equation

$$d \circ \overline{\lambda}(a) + \overline{\lambda} \circ d(a) = \beta' \circ \mu(a) - \lambda \circ \beta(a).$$

Define

$$\overline{\overline{\lambda}}\colon A \to \mathrm{hofib}(f') \quad \text{by} \quad \overline{\overline{\lambda}}(a) = (0, \overline{\lambda}(a)).$$

Then

$$\overline{\overline{\lambda}}\colon z(\alpha', \beta') \circ \mu \simeq \mathrm{hofib}(\kappa, \lambda) \circ z(\alpha, \beta).$$

14.77. Direct verification shows that defining $z(\mu, \eta)\colon \mathrm{hofib}(f) \to \mathrm{hofib}(f')$ by

$$z(\mu, \eta)(x, y) = (\mu(x), \eta(y))$$

produces the desired chain homotopy $z(\kappa_0, \lambda_0) \simeq z(\kappa_1, \lambda_1)$.

14.107. The required map $\overline{u\{2\}}\colon \mathrm{cyl}(f\{1\}) \to \mathrm{cyl}(g\{1\})$ is given by the universal property of the mapping cylinder. Explicitly the formula is

$$\begin{pmatrix} u\{1\} & 0 & h\{1\} \\ 0 & u\{1\} & 0 \\ 0 & 0 & u\{2\} \end{pmatrix}\colon C\{1\}_{n-1} \oplus C\{1\}_n \oplus C\{2\}_n \to D\{1\}_{n-1} \oplus D\{1\}_n \oplus D\{2\}_n.$$

14.125. This follows from the long exact sequence in homology associated to the R-chain map f.

Chapter 15

15.39. By definition

$$d_{W^\%(C)}(\widetilde{\varphi})_s = d_{C \otimes C}(\widetilde{\varphi}_s) - (-1)^{n+1}(\widetilde{\varphi}_{s-1} + (-1)^s T(\widetilde{\varphi}_{s-1})) \in (C \otimes_R C)_{n+s}.$$

Now calculate that for s even

$$d_{C \otimes C}(\widetilde{\varphi}_s) - (-1)^{n+1}(\widetilde{\varphi}_{s-1} + (-1)^s T(\widetilde{\varphi}_{s-1})) = d_{C \otimes C}(\widetilde{\varphi}_s) + 0$$
$$= (-1)^n d_{C \otimes C}(\varphi_{s+1}) = (\varphi_s + (-1)^{s+1} T(\varphi_s)) = (\varphi_s - T(\varphi_s)),$$

and for s odd

$$d_{C \otimes C}(\widetilde{\varphi}_s) - (-1)^{n+1}(\widetilde{\varphi}_{s-1} + (-1)^s T(\widetilde{\varphi}_{s-1}))$$
$$= 0 - (-1)^{n+1}(\widetilde{\varphi}_{s-1} + (-1)^s T(\widetilde{\varphi}_{s-1}))$$
$$= -(-1)^{n+1}(-1)^n(\varphi_s + (-1)^s T(\varphi_s)) = (\varphi_s - T(\varphi_s)).$$

15.86. For the particular choice of ω' where $\omega'_s = T(\omega_s)$ for $s = 0, 1$ with ω as in Remark 15.81, we obtain the formulas:

$$h^{\%}(\varphi)_s = (f_0 \otimes h)(\varphi_s) + (h \otimes f_1)(\varphi_s) + (-1)^{n+s+1}(h \otimes h)(T(\varphi_{s-1}));$$
$$h_{\%}(\psi)_s = (f_0 \otimes h)(\psi_s) + (h \otimes f_1)(\psi_s) + (-1)^{n+s+1}(h \otimes h)(T(\psi_{s+1}));$$
$$\widehat{h}^{\%}(\theta)_s = (f_0 \otimes h)(\theta_s) + (h \otimes f_1)(\theta_s) + (-1)^{n+s+1}(h \otimes h)(T(\theta_{s-1})).$$

15.98. The map μ_1 can be viewed as a chain map of degree 2

$$\mu_1 \colon C \otimes_R C \to \Sigma C \otimes_R \Sigma C,$$

which in terms of the involutions satisfies $\mu_1(T(\lambda)) = -T(\mu_1(\lambda))$. Recall that $\varphi \in W^{\%}(C)_n$ is a polarised symmetric structure if it satisfies (15.21). For the chain $S(\varphi) \in W^{\%}(\Sigma C)_{n+1}$ to be a polarised symmetric structure means to satisfy the corresponding formula, which is:

$$d_{\Sigma C \otimes_R \Sigma C}(S(\varphi)_s) = (-1)^{n+1}(S(\varphi)_{s-1} + (-1)^s T(S(\varphi)_{s-1})) \in (\Sigma C \otimes_R \Sigma C)_{n+s}.$$

The left-hand side becomes

$$d_{\Sigma C \otimes_R \Sigma C}(S(\varphi)_s) = (-1)^{n+s} \mu_1 d_{C \otimes_R C}(T(\varphi_{s-1})).$$

The right-hand side becomes

$$(-1)^{n+1}(S(\varphi)_{s-1} + (-1)^s T(S(\varphi)_{s-1}))$$
$$= (-1)^{n+1}[(-1)^{n+s-1}\mu_1(T(\varphi_{s-2})) + (-1)^s T((-1)^{n+s-1}\mu_1(T(\varphi_{s-2})))]$$
$$= (-1)^s \mu_1(T(\varphi_{s-2})) + T(\mu_1(T(\varphi_{s-2})))$$
$$= \mu_1((-1)^s(T(\varphi_{s-2})) - \varphi_{s-2}).$$

Applying T to both sides of (15.21) yields the formula whose sides are equal the two sides calculated above.

Solutions of the Exercises

15.101. This is proved by induction using the formula $(\Delta \otimes \mathrm{id}_W) \circ \Delta = (\mathrm{id}_W \otimes \Delta) \circ \Delta$, which holds for the diagonal map $\Delta \colon W \to W \otimes W$ from (15.69).

15.102. Apply the formulas from Remark 15.97 twice.

15.103. Direct verification using the definition of $h^{\%}$ from (15.78) and S from (15.88).

15.112. For a homotopy fibration sequence $C \to D \to E$, we get a commutative diagram

$$\begin{array}{ccccc} C & \longrightarrow & \mathrm{cyl}(f) & \longrightarrow & \mathrm{cone}(f) \\ \mathrm{id} \downarrow \simeq & & p \downarrow \simeq & & z \downarrow \simeq \\ C & \longrightarrow & D & \longrightarrow & E \end{array}$$

whose upper row is a cofibration sequence and whose vertical arrows are chain homotopy equivalences. Now apply homotopy invariance, see Subsection 15.3.4, Lemma 14.64 (i), and Lemma 15.110.

15.197. Follow [344, Proposition 2.2].

15.204. Direct elementary calculation.

15.229. To show that $\mathrm{cone}(f^{\%}) = \hom_{\mathbb{Z}[\mathbb{Z}/2]}(W, \mathrm{cone}(f \otimes f))$, note that an $(n + 1)$-chain of the left-hand side consists of a pair $(\varphi, \delta\varphi)$ where $\varphi \in W^{\%}(C)_n$ and $\delta\varphi \in W^{\%}(D)_{n+1}$. The components of these are $\varphi_s \in (C \otimes C)_{n+s}$ and $\delta\varphi_s \in (D \otimes D)_{n+1+s}$. An $(n+1)$-chain of the right-hand side consist of an element $\overline{\varphi} \in \hom_{\mathbb{Z}[\mathbb{Z}/2]}(W, \mathrm{cone}(f \otimes f))_n$ whose components are $\overline{\varphi}_s$ in $(C \otimes C)_{n+s} \oplus (D \otimes D)_{n+1+s}$. The required identification is given by the formula $\overline{\varphi}_s = (\varphi_s, \delta\varphi_s)$. The boundary operators agree under this.

15.257. The proposed chain homotopy

$$(\mathrm{ev}_m)_r = (-1)^r h \circ \varphi_{0,r} \circ h^* \colon \mathrm{cone}(f)^{n+1-r} \to \mathrm{cone}(f)_{r+1}$$

is given by the matrix

$$(\mathrm{ev}_m)_r = \begin{pmatrix} (-1)^r \varphi_{0,r} & 0 \\ 0 & 0 \end{pmatrix} \colon C^{n-r} \oplus D^{n+1-r} \to C_r \oplus D_{r+1}.$$

The differentials have matrices

$$(d_{\mathrm{cone}(f)})_r = \begin{pmatrix} -(d_C)_{r-1} & 0 \\ f & (d_D)_r \end{pmatrix}$$

and

$$(d_{\text{cone}(f)^{n+1-*}})_r = \begin{pmatrix} (-1)^r d_C^* & (-1)^{r+1} f^* \\ 0 & (-1)^{r+1} d_D^* \end{pmatrix}.$$

Now we get

$$(d_{\text{cone}(f)})_{r+1} \circ (\text{ev}_m)_r + (\text{ev}_m)_{r-1} \circ (d_{\text{cone}(f)^{n+1-*}})_r$$

$$= \begin{pmatrix} -d_C & 0 \\ f & d_D \end{pmatrix} \begin{pmatrix} (-1)^r \varphi_{0,r} & 0 \\ 0 & 0 \end{pmatrix} + \begin{pmatrix} (-1)^{r-1} \varphi_{0,r-1} & 0 \\ 0 & 0 \end{pmatrix} \begin{pmatrix} (-1)^r d_C^* & (-1)^{r+1} f^* \\ 0 & (-1)^{r+1} d_D^* \end{pmatrix}$$

$$= \begin{pmatrix} 0 & \varphi_{0,r-1} \circ f^* \\ (-1)^r f \circ \varphi_{0,r} & 0 \end{pmatrix}$$

$$= \text{ev}_r(\varphi, \delta\varphi) \circ e^{n+1-*} - e \circ \text{ev}_l(\varphi, \delta\varphi).$$

15.258. Given R-chain maps $f \colon C \to D$ and $g \colon C' \to D'$, we have

$$T \circ (f \otimes g) = (-1)^{|f| \cdot |g|} (g \otimes f) \circ T.$$

The pair $(\varphi_0, \delta\varphi_0)$ is an $(n+1)$-cycle in $\text{cone}(f \otimes f)$ since

$$d_{\text{cone}(f \otimes f)}(\varphi_0, \delta\varphi_0) = \begin{pmatrix} -d_{C \otimes C} & 0 \\ f \otimes f & d_{D \otimes D} \end{pmatrix} \begin{pmatrix} \varphi_0 \\ \delta\varphi_0 \end{pmatrix} = (0, 0).$$

Using the definition $\text{ev}_l(\varphi_0, \delta\varphi_0) = (e \otimes \text{id})(\delta\varphi_0) + (h \otimes f)(\varphi_0)$ and the definition $\text{ev}_r(\varphi_0, \delta\varphi_0) = (\text{id} \otimes e)(\delta\varphi_0) + (f \otimes h)(\varphi_0)$, we see

$$T(\text{ev}_l(\varphi, \delta\varphi)) = (\text{id} \otimes e)(T(\delta\varphi_0)) + (f \otimes h)(T(\varphi_0)) = \text{ev}_r(T(\varphi), T(\delta\varphi)).$$

Next observe $(-\widetilde{\varphi}_0, \widetilde{\delta\varphi}_0) = (-1)^{n+1}(\varphi_1, \delta\varphi_1)$ and hence

$$d_{\text{cone}(f \otimes f)}((-1)^{n+1}(\varphi_1, \delta\varphi_1)) = (-1)^{n+1} \begin{pmatrix} -d_{C \otimes C} & 0 \\ f \otimes f & d_{D \otimes D} \end{pmatrix} \begin{pmatrix} \varphi_1 \\ \delta\varphi_1 \end{pmatrix}$$

$$= (\varphi_0, \delta\varphi_0) - (T(\varphi)_0, T(\delta\varphi)_0).$$

Thus applying the chain map ev_r to the $(n+2)$ chain $(-\widetilde{\varphi}_0, \widetilde{\delta\varphi}_0)$ yields the desired homology.

15.275. Exercise 15.258 implies that, if ev_r is an R-chain homotopy equivalence, then also $T(\text{ev}_l)$ is an R-chain homotopy equivalence and hence also ev_l is an R-chain homotopy equivalence. Finally, by Lemma 14.60 and Diagram (15.254) the map $\text{ev}(\varphi)$ is also an R-chain homotopy equivalence.

15.276. Consider the chain map

$$\text{id} \oplus \text{id} \colon C \oplus C \to C,$$

and choose an $(n+1)$-dimensional chain $\theta \in W^{\%}(C)_{n+1}$ such that $d_{W^{\%}(C)}(\theta) = \varphi' - \varphi$, which exists by the assumption. Then the pair $(-\varphi \oplus \varphi', \theta)$ is a polarisation of an $(n+1)$-dimensional symmetric structure since the above equation together with the fact that φ, φ' are n-dimensional cycles shows that it is an $(n+1)$-dimensional cycle in cone$((\mathrm{id} \oplus \mathrm{id})^{\%})$.

Suppose now that $(C, [\varphi])$ is Poincaré. Then the ladder of evaluation maps associated to the pair just obtained has the shape as follows

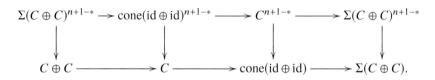

The outer two vertical arrows are R-chain homotopy equivalences by the assumption. Note that from the universal property of the mapping cone we obtain an R-chain homotopy equivalence cone$(\mathrm{id} \oplus \mathrm{id}) \to \Sigma C$. Using this to replace cone$(\mathrm{id} \oplus \mathrm{id})$ by ΣC in the above diagram and modifying the maps accordingly identifies the remaining arrows with the evaluation map of $(C, [\varphi])$, which is an R-chain homotopy equivalence by the assumption.

15.293. Consider the following two homotopy commutative diagrams whose rows and columns are homotopy cofibration sequences:

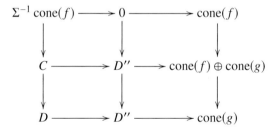

and

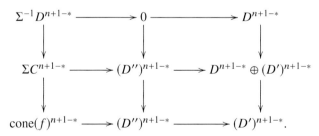

There are evaluation maps from the second to the first diagram and everything commutes either strictly or up to homotopy. The ladder obtained from the middle rows corresponds to the ladder obtained in the construction. The ladders corresponding to the top row and the middle and right columns are obvious. The ladder corresponding to the left column is that of the pair $(f: C \to D, (\varphi, \delta\varphi))$. The ladder corresponding to the bottom row is the desired ladder. Note that in this row the homotopy cofibration sequences are actually short exact sequences by direct observation.

15.304. We offer two arguments.

The first is by a calculation as follows. If $1/2 \in R$, then given an n-dimensional symmetric structure $\varphi \in W^{\%}(C)_n$ we can define $\psi \in W_{\%}(C)_n$ by $\psi_0 := (1/2) \cdot \varphi_0$ and $\psi_s = 0$ for $s \geq 1$. Then ψ is a quadratic structure on C whose symmetrisation is the symmetric structure $\varphi' \in W^{\%}(C)_n$ given by

$$\varphi'_0 := \psi_0 + T(\psi_0) = (1/2) \cdot (\varphi_0 + T(\varphi_0))$$

and $\varphi'_s = 0$ for $s \geq 1$. The symmetric complexes (C, φ) and (C, φ') are isomorphic since $d_{W^{\%}(C)}(\kappa) = \varphi - \varphi'$ for the chain $\kappa \in W^{\%}(C)_{n+1}$ given by $\kappa_s = (1/2) \cdot (-1)^n \varphi_{s+1}$. For each s the desired equation is

$$d_{C \otimes C}(\kappa_s) - (-1)^{n+1}(\kappa_{s-1} + (-1)^s T(\kappa_{s-1})) = \varphi_s - \varphi'_s.$$

That it holds follows from the equation (15.21).

The second argument is by general properties of Q-groups as follows. If $1/2 \in R$, then $\widehat{Q}^n(R) = 0$ for all $n \in \mathbb{Z}$ since multiplication by 2 is an isomorphism on R. This implies that $\widehat{Q}^n(C) = 0$ for any $C \in bR\text{-}CH_{\text{fgp}}$ by a spectral sequence argument as in [58, Chapter VII.7]. Therefore for a given chain complex $C \in bR\text{-}CH_{\text{fgp}}$ the symmetrisation map between Q-groups $(1+T): Q_n(C) \to Q^n(C)$ is an isomorphism for all $n \in \mathbb{Z}$. The same holds in the relative case, and this implies the symmetrisation map (15.303) is an isomorphism in this case.

15.334. This is

$$(\delta\varphi/\varphi)_{0,r} = \begin{pmatrix} 0 & 0 \\ (-1)^r f \circ \varphi_{0,r} & \delta\varphi_{0,r} \end{pmatrix} : C^{n-r} \oplus D^{n-r+1} \to C_{r-1} \oplus D_r$$

and

$$(\delta\varphi/\varphi)_{s,r} : C^{n+s-r} \oplus D^{n+1+s-r} \to C_{r-1} \oplus D_r$$

$$(\delta\varphi/\varphi)_{s,r} = \begin{pmatrix} (-1)^{(r+1)(n+s)+r} e(C_{r-1})^{-1} \circ \varphi^*_{s-1,n+s-r} & 0 \\ (-1)^r f \circ \varphi_{s,r} & \delta\varphi_{s,r} \end{pmatrix}.$$

15.335. To see the desired statement, we use Diagram (14.32). It enables us to convert the definition of $(\delta\varphi/\varphi)_s$, which is in terms of applying $(\alpha \otimes \beta)$ for suitable α, β to components $\delta\varphi_s$ and φ_s for suitable s, to an expression that is in terms of applying $\hom(I(\alpha), \beta)$ to components $\delta\varphi_{s,r}$ and $\varphi_{s,r}$ for suitable s, r. The respective calculations are as follows:

$$\hom(I(e), e)(\delta\varphi_{s,r}) = e \circ \delta\varphi_{s,r} \circ e^*;$$
$$\hom(I(ef), h)(\varphi_{s,r-1}) = h \circ \varphi_{s,r-1} \circ f^* \circ e^*;$$
$$\hom(I(h), h)(T_{\hom}(\varphi)_{s-1,r-1}) = h \circ (-1)^{r(n+s)+r} e(C_{r-1})^{-1} \circ \varphi^*_{s-1,n+s-r} \circ h^*,$$

where the last line uses that

$$T_{\hom}(\varphi)_{s-1,r-1} = (-1)^{r(n+s)-1} e(C_{r-1})^{-1} \circ \varphi^*_{s-1,n+s-r},$$

which is in turn obtained via the formula for $T_{\hom}$ from Remark 15.8.

15.336. Observe that we have

$$\widetilde{(\delta\varphi/\varphi)}_0 = (-1)^{n+1} \widetilde{(\delta\varphi/\varphi)}_1$$
$$= (e \otimes e)(\widetilde{\delta\varphi_0}) + ((e \circ f) \otimes h)(-\widetilde{\varphi_0}) + (h \otimes h)(T(\varphi_0)).$$

Moreover, we have:

$$\widetilde{\mathrm{ev}}_{(l,r)} \circ e = (e \otimes e)(\widetilde{\delta\varphi_0}) + ((e \circ f) \otimes h)(-\widetilde{\varphi_0});$$
$$T(\mathrm{ev}_m) = -(h \otimes h)(T(\varphi_0)).$$

15.352. This is

$$(\delta\psi/\psi)_{0,r} : C^{n-r} \oplus D^{n-r+1} \to C_{r-1} \oplus D_r;$$
$$(\delta\psi/\psi)_{0,r} = \begin{pmatrix} (-1)^{(r+1)n+r} e(C_{r-1})^{-1} \circ \psi^*_{1,n-r} & 0 \\ (-1)^r f \circ \psi_{0,r} & \delta\psi_{0,r} \end{pmatrix},$$

and

$$(\delta\psi/\psi)_{s,r} : C^{n+s-r} \oplus D^{n+1+s-r} \to C_{r-1} \oplus D_r;$$
$$(\delta\psi/\psi)_{s,r} = \begin{pmatrix} (-1)^{(r+1)(n+s)+r} e(C_{r-1})^{-1} \circ \varphi^*_{s+1,n-s-r} & 0 \\ (-1)^r f \circ \psi_{s,r} & \delta\psi_{s,r} \end{pmatrix}.$$

15.359. The desired isomorphism is the composite of the following isomorphisms of degree 0 where in the right column we make a calculation on homomorphisms in degree a of the complexes in the left column:

$$
\begin{array}{ccc}
\Sigma^m \hom_R(E,D) & & f_r: E_{r-(a-m)} \to D_r \\
\downarrow {\scriptstyle (15.25)} & & \downarrow \\
\hom_R(\Sigma^{-m}E, D) & & (-1)^{m(a-m)} f_r: E_{r-(a-m)} \to D_r \\
\downarrow {\scriptstyle \hom(I(e^{-1}),\mathrm{id})} & & \downarrow \\
\hom_R((\Sigma^{-m}E)^{-*-*}, D) & & (-1)^y f_r \circ e^{-1}: E^{**}_{r-(a-m)} \to D_r \\
\downarrow {\scriptstyle (15.25)} & & \downarrow \\
\hom_R((\Sigma^m E^{-*})^{-*}, D) & & (-1)^x f_r \circ e^{-1}: E^{**}_{r-(a-m)} \to D_r
\end{array}
$$

for $y = m(a-m) + (r-1)$ and $x = y + m(m+r-a) = r(m+1) + a$ and in the last isomorphism (15.25) is applied as $\Sigma^m E^{-*} \cong (\Sigma^{-m} E)^{-*}$.

Plugging in $D = E$, $a = m$ and $f_r = (\mathrm{id}_E)_r$ we calculate the sign

$$(-1)^{r(m+1)+m} = (-1)^{(r+1)(m+1)+1}.$$

15.368. The chain map $\hom(\mathrm{id}_D, f)\colon \hom_R(D,C) \to \hom_R(D,D)$ is a chain homotopy equivalence if f is an R-chain homotopy equivalence, and hence its homotopy fibre $\mathrm{hofib}(\hom(\mathrm{id}, f))$ is contractible. The statement about the pairs (g, h) follows immediately from the definition of 0-cycles and 1-chains in respective complexes.

15.372. Recall that φ is an $(n+1)$-dimensional cycle in the chain complex $W^{q_0}(E)$. This means that the collection of R-module homomorphisms $\varphi_{s,r}\colon E^{n+1+s-r} \to E_r$, satisfies

$$(d_E)_{r+1} \circ \varphi_{s,r+1} - (-1)^r \varphi_{s,r} \circ (d_E)^*_{n+1+s-r}$$
$$= (-1)^{n+1} \varphi_{s-1,r} + (-1)^{t(n+1,s,r)} \varepsilon e(E_r)^{-1} \circ \varphi^*_{s-1,n+1+s-r} \colon E^{n+s-r} \to E_r$$

where $t(n+1, r, s) = r \cdot (n+1+s) + 1$ and $e(C_r)\colon C_r \to (C_r^*)^*$ is the map appearing in the r-th component of the isomorphism (14.26).

It needs to be shown that $\partial\varphi$ is an n-dimensional cycle in $W^{q_0}(\partial E)$. This means that the collection of R-module homomorphisms

$$(d_{\partial E})_{r+1} \circ \partial\varphi_{s,r+1} - (-1)^r \partial\varphi_{s,r} \circ (d_{\partial E})^*_{n+s-r}$$
$$= (-1)^n \partial\varphi_{s-1,r} + (-1)^{t(n,s,r)} \varepsilon e(\partial E_r)^{-1} \circ \partial\varphi^*_{s-1,n+s-r} \colon \partial E^{n-1+s-r} \to \partial E_r$$

where $t(n, r, s) = r \cdot (n+s) + 1$ and $e(\partial E_r)\colon \partial E_r \to (\partial E_r^*)^*$ is the map appearing in the r-th component of the isomorphism (14.26).

The verification of these identities is obtained by plugging into it the formulas from Remark 15.370 and using the above stated identities for φ.

15.468. Since S^m is for $m \geq 2$ the boundary of D^{m+1} and both S^m and D^{m+1} are simply connected, $\text{ssign}_\pi(N) \in L^m(\mathbb{Z})$ is zero. We conclude from Theorem 15.467 that $\text{qsign}_{\pi \times \pi'}((f, \bar{f}) \times \text{id}_{S^m}) \in L_m(\mathbb{Z}[\pi_1(X \times S^m)])$ vanishes. Now apply Theorem 8.112, Theorem 9.60, and Theorem 15.4.

15.473. This follows from (15.470) and Theorem 15.472. We need the existence of a vector bundle reduction of the Spivak normal fibration of X/G to find an normal map of degree one with X/G as target.

15.476. Because of Remark 10.14 and Theorem 15.474, the map $L_n^{\mathbb{Z}}(\mathbb{Z}G) \to L_n^s(\mathbb{Z}G)$ is bijective for all $n \in \mathbb{Z}$ if and only if $\widehat{H}^n(\mathbb{Z}/2, G_{\text{ab}})$ is trivial for all $n \in \mathbb{Z}$. Obviously the generator of $\mathbb{Z}/2$ acts on G_{ab} by $-\text{id}$. Hence $\widehat{H}^n(\mathbb{Z}/2, G_{\text{ab}})$ vanishes for all $n \in \mathbb{Z}$ if and only if $\cdots \xrightarrow{0} G_{\text{ab}} \xrightarrow{2\,\text{id}} G_{\text{ab}} \xrightarrow{0} \cdots$ is exact.

Chapter 16

16.7. The canonical maps are given by the homomorphisms $L_n^s(\mathbb{Z}G, w) \to L_n^h(\mathbb{Z}G, w)$, $L_n^{\langle j+1 \rangle}(\mathbb{Z}G, w) \to L_n^{\langle j \rangle}(\mathbb{Z}G, w)$, $L_n^{\langle j \rangle}(\mathbb{Z}G, w) \to L_n^{\langle -\infty \rangle}(\mathbb{Z}G, w)$ for $j \leq 1$, and the fact that by definition $L_n(\mathbb{Z}G, w) = L_n^h(\mathbb{Z}G, w) = L_n^{\langle 1 \rangle}(\mathbb{Z}G, w)$ holds. Now apply the Rothenberg sequences and the fact that elements in the Tate cohomology of $\mathbb{Z}/2$ for any $\mathbb{Z}[\mathbb{Z}/2]$-module are of order two.

16.13. Define a $\mathbb{C}G$-isomorphism $V' \to \hom_{\mathbb{C}G}(V, \mathbb{C}G)$ by assigning to $f \in \hom_\mathbb{C}(V, \mathbb{C})$ the element in $\hom_{\mathbb{C}G}(V, \mathbb{C}G)$ that sends $v \in V$ to $\sum_{g \in G} f(g^{-1}v) \cdot g$. The inverse sends the map $h \colon V \to \mathbb{C}G$ to the map $V \to \mathbb{C}$ that assigns to v the coefficient of $e \in G$ in $h(v) \in \mathbb{C}G$.

16.17. The main point is to show that there is an isomorphism $V' \xrightarrow{\cong} V^*$ of $\mathbb{C}G$-modules if V is a finitely generated projective $\mathbb{C}G$-module. This has been explained in the solution to Exercise 16.13.

16.18. We get from Theorem 16.11 (vi) and (vii)

$$\text{tors}(L_n^s(\mathbb{Z}G)) \cong \begin{cases} \{1\} & \text{if } n \equiv 0, 1 \bmod 4; \\ \mathbb{Z}/2 & \text{if } n \equiv 2 \bmod 4; \\ \{1\} & \text{if } |G| \text{ is odd and } n \equiv 3 \bmod 4; \\ \mathbb{Z}/2 & \text{if } |G| \text{ is even and } n \equiv 3 \bmod 4. \end{cases}$$

Theorem 16.11 (iv) implies

$$\mathrm{rk}(L_n^s(\mathbb{Z}G)) = \begin{cases} \frac{|G|+1}{2} & \text{if } |G| \text{ is odd and } n \equiv 0 \bmod 4; \\ \frac{|G|-1}{2} & \text{if } |G| \text{ is odd and } n \equiv 2 \bmod 4; \\ \frac{|G|}{2} + 1 & \text{if } |G| \text{ is even and } n \equiv 0 \bmod 4; \\ \frac{|G|}{2} - 1 & \text{if } |G| \text{ is even and } n \equiv 2 \bmod 4; \\ 0 & \text{if } n \equiv 1, 3 \bmod 4. \end{cases}$$

16.21. We conclude from Conjecture 16.19 that the decoration does not matter. If $g = 0$, $\pi_1(F_g)$ is trivial and hence $L_n^{\langle -\infty \rangle}(\mathbb{Z}[\pi_1(F_g)]) = L_n^{\langle -\infty \rangle}(\mathbb{Z})$. Suppose $g \geq 1$. Then F_g itself is a model for $B\pi_1(F_g)$. Conjecture 16.19 implies

$$H_n(F_g; \mathbf{L}(\mathbb{Z})) \cong L_n^{\langle -\infty \rangle}(\mathbb{Z}[\pi_1(F_g)]).$$

We use the Atiyah–Hirzebruch spectral sequence to compute $H_n(F_g; \mathbf{L}(\mathbb{Z}))$. This is rather easy since F_g is 2-dimensional, the edge homomorphism, which describes $H_n(\{\bullet\}; \mathbf{L}^{\langle -\infty \rangle}) \to H_n(F_g; \mathbf{L}^{\langle -\infty \rangle})$, is split injective, and $L_n^{\langle -\infty \rangle}(\mathbb{Z})$ is $\mathbb{Z}$ if $n \equiv 0 \bmod 4$, $\mathbb{Z}/2$ if $n \equiv 0 \bmod 4$ and $\{0\}$ otherwise. The result is

$$L_n^{\langle -\infty \rangle}(\mathbb{Z}[\pi_1(F_g)]) \cong \begin{cases} \mathbb{Z} \oplus \mathbb{Z}/2 & n \equiv 0 \bmod 4; \\ \mathbb{Z}^{2g} & n \equiv 1 \bmod 4; \\ \mathbb{Z} \oplus \mathbb{Z}/2 & n \equiv 2 \bmod 4; \\ (\mathbb{Z}/2)^{2g} & n \equiv 3 \bmod 4. \end{cases}$$

16.23. Choose $m = p$ in Example 16.22, and let M be the aspherical closed orientable 3-manifold constructed there, for which $L_1^\epsilon(\mathbb{Z}[\pi_1(M)])$ contains $\mathbb{Z}/p$ as a subgroup. Now take $M = N \times T^k \times T^{d-k}$ for $k \in \{0, 1, 2, 3\}$ such that $k + 1 \equiv n \bmod 4$. Then M is an aspherical closed orientable manifold of dimension d. Moreover, the Shaneson splitting and the Rothenberg sequences, see Section 16.2, imply that $L_n^\epsilon(\mathbb{Z}[\pi_1(N)])$ contains $\mathbb{Z}/p$ as a subgroup.

16.24. Since G is a Farrell–Jones group, G satisfies Conjecture 16.19. Now use Remark 16.20 and the facts that $\mathbb{Q}$ is flat over $\mathbb{Z}$ and for any homology theory with values in rational vector spaces all the differentials in the Atiyah–Hirzebruch spectral sequence are trivial.

16.30. Each model for ED and hence each model for BD must be infinite-dimensional since there is no natural number N such that $H_n(BD)$ vanishes for $n \geq N$. A 1-dimensional model for $E_{\mathcal{FIN}}(D)$ is given by $\mathbb{R}$ with the standard D-action.

16.45. By Theorem 16.44 the group $\mathbb{Z}$ is a Farrell–Jones group. Hence $\mathbb{Z}$ satisfies Conjecture 16.19 by Corollary 16.41. Note that a model for $E\mathbb{Z}$ is $\mathbb{R}$ with the standard $\mathbb{Z}$-action. Hence we get for any decoration ϵ and integer n

$$H_n^{\mathbb{Z}}(\mathbb{R}; L(\mathbb{Z})^w) \cong L_n^{\epsilon}(\mathbb{Z}[\mathbb{Z}], w).$$

From Remark 16.20 we obtain a long exact sequence

$$\cdots \to L_n(\mathbb{Z}) \xrightarrow{2 \cdot \mathrm{id}} L_n(\mathbb{Z}) \to H_n^{\mathbb{Z}}(\mathbb{R}; L(\mathbb{Z})^w) \to L_{n-1}(\mathbb{Z}) \xrightarrow{2 \cdot \mathrm{id}} L_{n-1}(\mathbb{Z}) \to \cdots.$$

This implies using Theorem 16.11 (i)

$$L_n^{\epsilon}(\mathbb{Z}[\mathbb{Z}], w) \cong \begin{cases} 0 & n \equiv 1 \bmod 4; \\ \mathbb{Z}/2 & \text{otherwise.} \end{cases}$$

16.46. We conclude from Theorem 16.44 (ie) that $\pi_1(S)$ and $\pi_1(M)$ are Farrell–Jones groups. Consider any infinite cyclic subgroup $C \subseteq \pi_1(M)$. Then we get an extension $1 \to \pi_1(S) \to p^{-1} \to C \to 1$. Since any automorphism of $\pi_1(S)$ is induced by a surface homeomorphism $h \colon S \to S$, the group $p^{-1}(C)$ is isomorphic to the fundamental group of the mapping torus of h, which is a 3-manifold. Again by Theorem 16.44 (ie) the group $p^{-1}(C)$ is a Farrell–Jones group. Now G is a Farrell–Jones group by Theorem 16.44 (iic).

Chapter 17

17.5. Since $S^1 = K(\mathbb{Z}, 1)$, we only have to treat the case $n \geq 2$. There is a map $f \colon S^n \to K(\mathbb{Z}, n)$ inducing the identity on the n-th homology. It induces an isomorphism $\pi_m(S^n)_{\mathbb{Q}} \to \pi_m(K(\mathbb{Z}, n))_{\mathbb{Q}}$ for $m \geq 1$ if n is odd.

17.12. This follows from Theorem 17.10 as the conditions on M imply $H^4(M; \mathbb{Z}/2) = H^3(M; \mathbb{Z}/2) = 0$.

17.13. We have explained $\pi_k(\mathrm{TOP}/O) \cong \Theta_k$ in Section 17.5 and stated $|\Theta_{13}| = 3$ in (12.2).

17.18. This follows from Theorem 17.17 since we have $H_0(\mathbb{CP}^3 \times S^2; \mathbb{Q}) = \mathbb{Q}$ and $H_4(\mathbb{CP}^3 \times S^2; \mathbb{Q}) \cong \mathbb{Q} \oplus \mathbb{Q}$.

17.27. Take $a = r \times \mathrm{id}_{S^q}$ with r a self-diffeomorphism of S^p of degree -1.

17.32. Write $X = X_0 \cup D^n$ so that the collapse map can be viewed as sending X_0 to a point. For a compact manifold Y the correspondence $[Y/\partial Y, G/\mathrm{TOP}] \cong \mathcal{N}^{\mathrm{TOP}}(Y, \partial Y)$ sends the constant map to the identity degree one normal map on Y.

Chapter 18

18.5. The number is 15. This follows from Example 18.4 since Θ_7 is isomorphic to $\mathbb{Z}/28$ by (12.2) and Theorem 12.49.

18.8. This follows from Theorem 18.7 as $[g] = 0$ holds.

18.9. This follows from the definition of the structure set.

18.12. Observe that $(S^p \times D^q)/(S^p \times S^{q-1}) \simeq S^q \wedge S^{p+q}$. This allows us to use analogous arguments as in the proof of Theorem 18.11 to show that $S^{\text{TOP}}(S^p \times D^q, S^p \times S^{q-1}) \cong L_q(\mathbb{Z})$.

18.16. The map $\mathbb{CP}^d \to \mathbb{CP}^\infty$ is $(2d+1)$-connected since $\mathbb{CP}^\infty$ is obtained from $\mathbb{CP}^d$ by attaching cells of dimension $\geq 2d+2$. As $\mathbb{CP}^d$ is a $2d$-dimensional CW-complex, the inclusion $\mathbb{CP}^d \to \mathbb{CP}^\infty$ induces a bijection $[\mathbb{CP}^d, \mathbb{CP}^d] \xrightarrow{\cong} [\mathbb{CP}^d, \mathbb{CP}^\infty]$. Since $\mathbb{CP}^\infty$ is an Eilenberg–MacLane space of type $(\mathbb{Z}, 2)$, we obtain a bijection $[\mathbb{CP}^d, \mathbb{CP}^\infty] \xrightarrow{\cong} H^2(\mathbb{CP}^d)$, $[f] \mapsto H^2(f)(\iota)$ for fixed generator ι of the infinite cyclic group $H^2(\mathbb{CP}^\infty)$, which is isomorphic to $\hom_{\mathbb{Z}}(H_2(\mathbb{CP}^d), H_2(\mathbb{CP}^d))$. Hence we obtain a bijection

$$[\mathbb{CP}^d, \mathbb{CP}^d] \xrightarrow{\cong} \hom_{\mathbb{Z}}(H_2(\mathbb{CP}^d), H_2(\mathbb{CP}^d)), \quad [f] \mapsto H^2(f).$$

18.50. This follows from Lemma 3.57 and Theorem 3.61 (i).

Chapter 19

19.13. This follows directly from Example 19.12.

19.16. In view of Theorem 19.15 we have to show $H_5(M, \{x\}; \mathbf{L}(\mathbb{Z})\langle 1 \rangle) \cong H_3(M; \mathbb{Z}/2)$. We want to apply the Atiyah–Hirzebruch spectral sequence, which converges to $H_*(M, \{x\}; \mathbf{L}(\mathbb{Z})\langle 1 \rangle)$. By Theorem 16.11 (i) its E^2-term is

$$E^2_{p,q} = H_p(M, \{x\}, \pi_q(\mathbf{L}(\mathbb{Z})\langle 1 \rangle)) \cong \begin{cases} H_p(M; \mathbb{Z}/2) & p \geq 2, q \geq 1, q \equiv 2 \mod 4; \\ H_p(M; \mathbb{Z}) & p \geq 2, q \geq 1, q \equiv 0 \mod 4; \\ \{0\} & \text{otherwise.} \end{cases}$$

Hence on the line $p+q = 5$ there is only one no-trivial entry, namely at $(p,q) = (3,2)$, and all differentials starting or ending at this point are trivial.

19.18. The positive statements follow from the long exact homotopy sequence of a fibration. The desired example is for instance given by the universal principal S^1-bundle $S^1 \to ES^1 \to BS^1$ as S^1 and ES^1 are aspherical and $\pi_2(BS^1) \cong \pi_1(S^1) \cong \mathbb{Z}$ holds.

19.30. We proceed by contradiction. Suppose that M is aspherical. The obvious map $f\colon M_1 \sharp M_2 \to M_1 \vee M_2$ given by collapsing S^{n-1} to a point is $(n-1)$-connected, where n is the dimension of M_1 and M_2. Let $p\colon \widetilde{M_1 \vee M_2} \to M_1 \vee M_2$ be the universal covering. By the Seifert–van Kampen Theorem the fundamental group of $\pi_1(M_1 \vee M_2)$ is $\pi_1(M_1) * \pi_1(M_2)$ and the inclusion of $M_k \to M_1 \vee M_2$ induces injections on the fundamental groups for $k = 1,2$. We conclude that $p^{-1}(M_k) = \pi_1(M_1 \vee M_2) \times_{\pi_1(M_k)} \widetilde{M_k}$ for $k = 1,2$. Since $n \geq 3$, the map f induces an isomorphism on the fundamental groups and an $(n-1)$-connected map $\widetilde{f}\colon \widetilde{M_1 \sharp M_2} \to \widetilde{M_1 \vee M_2}$. Since $\widetilde{M_1 \sharp M_2}$ is contractible, $H_m(\widetilde{M_1 \vee M_2}) = 0$ for $1 \leq m \leq n-1$. Since $p^{-1}(M_1) \cup p^{-1}(M_2) = \widetilde{M_1 \vee M_2}$ and $p^{-1}(M_1) \cap p^{-1}(M_2) = p^{-1}(\{\bullet\}) = \pi_1(M_1 \vee M_2)$, we conclude $H_m(p^{-1}(M_k)) = 0$ for $1 \leq m \leq n-1$ from the Mayer–Vietoris sequence. This implies $H_m(\widetilde{M_k}) = 0$ for $1 \leq m \leq n-1$ since $p^{-1}(M_k)$ is a disjoint union of copies of $\widetilde{M_k}$.

Suppose that $\pi_1(M_k)$ is finite. Since $\pi_1(M_1 \sharp M_2)$ is torsionfree because of Lemma 19.29, $\pi_1(M_k)$ must be trivial and $M_k = \widetilde{M_k}$. Since M_k is simply connected and $H_m(M_k) = 0$ for $1 \leq m \leq n-1$, M_k is homotopy equivalent to S^n. Since we assume that M_k is not homotopy equivalent to a sphere, $\pi_1(M_k)$ is infinite. This implies that the manifold $\widetilde{M_k}$ is non-compact and hence $H_n(\widetilde{M_k}) = 0$. Since $\widetilde{M_k}$ is n-dimensional, we conclude $H_m(\widetilde{M_k}) = 0$ for $m \geq 1$. Since $\widetilde{M_k}$ is simply connected, all homotopy groups of $\widetilde{M_k}$ vanish by the Hurewicz Theorem [178, Theorem 4.32 on page 366] or [431, Corollary IV.7.8 on page 180]. Hence M_1 and M_2 are aspherical. Using the Mayer–Vietoris argument above, one shows analogously that $M_1 \vee M_2$ is aspherical. Since M is by assumption aspherical, $M_1 \sharp M_2$ and $M_1 \vee M_2$ are homotopy equivalent. Since they have different Euler characteristics, namely $\chi(M_1 \sharp M_2) = \chi(M_1) + \chi(M_2) - (1 + (-1)^n)$ and $\chi(M_1 \vee M_2) = \chi(M_1) + \chi(M_2) - 1$, we get a contradiction.

19.38. The long exact homotopy sequence of a fibration shows that E is aspherical. By Theorem 19.32 it suffices to show that $\pi_1(E)$ is a Farrell–Jones group. This follows from Theorem 16.44 (ie) and (iic) since any automorphism $\phi\colon \pi_1(F) \xrightarrow{\cong} \pi_1(F)$ can be realised by $\phi = \pi_1(f)$ for a self-homeomorphism $f\colon F \to F$ and the mapping torus of f is a closed manifold of dimension ≤ 3.

Chapter 20

20.3. The Euler characteristic is a homotopy invariant. We conclude $\chi(M\sharp N) = \chi(M) + \chi(N) - 1 - (-1)^n$ if M and N are n-dimensional closed manifolds. We have $\chi(S^k \times S^k) = \chi(S^k)^2 = ((1 + (-1)^k))^2 = 2 \cdot (1 + (-1)^k)$. This implies

$$\chi(M_0) + g_0 \cdot 2 \cdot (-1)^k = \chi(M_0 \sharp W_{g_0}) = \chi(M_1 \sharp W_{g_1}) = \chi(M_1) + g_1 \cdot 2 \cdot (-1)^k.$$

20.22. From the construction it is clear that we can assume without loss of generality that b and b' are fibrations. It suffices to show that the map $\Omega(f_l, \overline{f_l})\colon \Omega_n(b_l^*\xi_l) \to \Omega_n((b_l')^*\xi_l)$ is bijective. Since f is a homotopy equivalence, b and b' are fibrations and the map $f_l\colon B_l \to B_l'$ is obtained from f by a pullback construction, the map f_l is a homotopy equivalence. This implies that f_l is even a fibre homotopy equivalence. Let $g_l\colon B_l' \to B_l$ be a fibre homotopy inverse. Then we can find a bundle map $(g_l, \overline{g_l})\colon (\text{pr}_l')^*\xi_l \to \text{pr}_l\,\xi_l$ covering g_l such that both composites $(g_l, \overline{g_l}) \circ (f_l, \overline{f_l})$ and $(f_l, \overline{f_l}) \circ (g_l, \overline{g_l})$ are homotopic as bundle maps to the identity. Hence $\Omega_n(f_l, \overline{f_l})$ is bijective.

20.23. This follows from Theorem 4.55 and the definition of $\Omega_n(b)$, see (20.10).

20.34. (i) We only show the equivalence (a) $\iff$ (b). The other implications follows from Lemma 5.23 or standard algebraic topology. If we have a b-structure, then we obtain from (20.27) a vector bundle isomorphism

$$\begin{array}{ccc} E_{\nu(i)} & \xrightarrow{\overline{f}} & E_{\text{pr}_{l+1}^*\,\xi_{l+1}} \\ {\scriptstyle p_{\nu(i)}}\downarrow & & \downarrow{\scriptstyle p_{\text{pr}_l^*\xi_l}} \\ M & \xrightarrow{f} & \text{BSO}(l) \end{array}$$

Since $p_{\text{pr}_l^*\xi_l}$ is orientable, $\nu(i)$ is orientable.

Suppose that $\nu(i)$ is orientable. Since $p_{\text{pr}_l^*\xi_l}$ is the universal orientable l-dimensional vector bundle, we obtain a diagram as above and thus a b-structure.

(ii) We only show the equivalence (a) $\iff$ (b). The other implications follows from Lemma 5.23 or standard algebraic topology. For the implication (a) $\iff$ (b) one argues as above taking into account that for a fixed orientation the diagram above is unique up to homotopy and two diagrams as above induce the same orientation on $\nu(i)$ if they are homotopic.

20.37. (i) One has to show by elementary obstruction theory that for a map $f\colon X \to Y$ the $p+1$-stage and the p-stage of the Moore–Postnikov decomposition, see Subsection 20.5.2, are related by an up to homotopy unique map g for which the following

diagram commutes up to homotopy

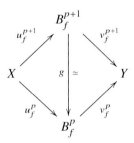

(ii) Suppose that M and N have the same normal k-type. Since $k \geq (n+1)$, we can find embeddings $i_M \colon M \to \mathbb{R}^{n+k}$ and $i_N \colon N \to \mathbb{R}^{n+k}$. Let $b \colon B \to BO$ be a representative of the common normal k-type. Since M and N have the same normal k-type, we can find k-smoothings $\overline{\nu_M} \colon M \to B$ and $\overline{\nu_N} \colon N \to B$ which are both $(k+1)$-connected. Since $n \leq (k+1)$ and M is n-dimensional, we can find a map $f \colon M \to N$ which induces isomorphisms on the homotopy groups in degrees $i \leq n$ such that $\overline{\nu_M} \circ f \simeq \overline{\nu_N}$. This implies that f is a homotopy equivalence. One easily checks that f is covered by the desired bundle map.

The other implication is now obvious.

20.47. Since π is cyclic of odd order, $H^2(\pi; \mathbb{Z}/2)$ vanishes. Now (20.39) implies that $H^2(M; \mathbb{Z}/2) \xrightarrow{\cong} H^2(\widetilde{M}; \mathbb{Z}/2)$ is injective. Hence $w_2(\widetilde{M})$ is non-trivial as $w_2(M)$ is non-trivial by assumption. The same argument shows that $w_2(\widetilde{N})$ is non-trivial. Since π is cyclic of odd order, $H_4(B\pi; \mathbb{Z})$ vanishes. Now the claim follows from Example 20.46.

References

1. J. F. Adams. On the groups $J(X)$. IV. *Topology*, 5:21–71, 1966.
2. J. F. Adams. *Stable Homotopy and Generalised Homology*. University of Chicago Press, Chicago, Ill., 1974. Chicago Lectures in Mathematics.
3. J. P. Alexander, G. C. Hamrick, and J. W. Vick. Linking forms and maps of odd prime order. *Trans. Amer. Math. Soc.*, 221(1):169–185, 1976.
4. G. Allaud. On the classification of fiber spaces. *Math. Z.*, 92:110–125, 1966.
5. R. C. Alperin, R. K. Dennis, R. Oliver, and M. R. Stein. SK_1 of finite abelian groups. II. *Invent. Math.*, 87(2):253–302, 1987.
6. K. K. S. Andersen, T. Bauer, J. Grodal, and E. K. Pedersen. A finite loop space not rationally equivalent to a compact Lie group. *Invent. Math.*, 157:1–10, 2004.
7. D. R. Anderson. The Whitehead torsion of the total space of a fiber bundle. *Topology*, 11:179–194, 1972.
8. C. Arf. Untersuchungen über quadratische Formen in Körpern der Charakteristik 2. I. *J. Reine Angew. Math.*, 183:148–167, 1941.
9. M. A. Armstrong and E. C. Zeeman. Piecewise linear transversality. *Bull. Amer. Math. Soc.*, 73:184–188, 1967.
10. M. Aschenbrenner, S. Friedl, and H. Wilton. *3-Manifold Groups*. Zürich: European Mathematical Society (EMS), 2015.
11. M. F. Atiyah. Thom complexes. *Proc. London Math. Soc. (3)*, 11:291–310, 1961.
12. M. F. Atiyah. Elliptic operators, discrete groups and von Neumann algebras. *Astérisque*, 32-33:43–72, 1976.
13. M. F. Atiyah and I. G. Macdonald. *Introduction to Commutative Algebra*. Addison-Wesley Publishing Co., Reading, Mass.-London-Don Mills, Ont., 1969.
14. M. F. Atiyah, V. K. Patodi, and I. M. Singer. Spectral asymmetry and Riemannian geometry. I. *Math. Proc. Cambridge Philos. Soc.*, 77:43–69, 1975.
15. M. F. Atiyah and I. M. Singer. The index of elliptic operators. III. *Ann. of Math. (2)*, 87:546–604, 1968.
16. A. Bak. Odd dimension surgery groups of odd torsion groups vanish. *Topology*, 14(4):367–374, 1975.
17. L. Balko, T. Macko, M. Niepel, and T. Rusin. Higher simple structure sets of lens spaces with the fundamental group of arbitrary order. *Arch. Math., Brno*, 2019.
18. L. Balko, T. Macko, M. Niepel, and T. Rusin. Higher simple structure sets of Lens spaces with the fundamental group of order 2^K. *Topology Appl.*, 263:299–320, 2019.
19. M. Banagl and A. Ranicki. Generalized Arf invariants in algebraic L-theory. *Adv. Math.*, 199(2):542–668, 2006.
20. D. Barden. The Structure of Manifolds. Ph. D. thesis, Cambridge, 1963.
21. M. G. Barratt, J. D. S. Jones, and M. E. Mahowald. Relations amongst Toda brackets and the Kervaire invariant in dimension 62. *J. London Math. Soc. (2)*, 30(3):533–550, 1984.

22. A. Bartels and M. Bestvina. The Farrell–Jones conjecture for mapping class groups. *Invent. Math.*, 215(2):651–712, 2019.
23. A. Bartels, S. Echterhoff, and W. Lück. Inheritance of isomorphism conjectures under colimits. In Cortinaz, Cuntz, Karoubi, Nest, and Weibel, editors, *K-Theory and Noncommutative Geometry*, EMS-Series of Congress Reports, pages 41–70. European Mathematical Society, 2008.
24. A. Bartels, F. T. Farrell, and W. Lück. The Farrell–Jones Conjecture for cocompact lattices in virtually connected Lie groups. *J. Amer. Math. Soc.*, 27(2):339–388, 2014.
25. A. Bartels and W. Lück. Induction theorems and isomorphism conjectures for K- and L-theory. *Forum Math.*, 19:379–406, 2007.
26. A. Bartels and W. Lück. On crossed product rings with twisted involutions, their module categories and L-theory. In *Cohomology of Groups and Algebraic K-Theory*, volume 12 of *Adv. Lect. Math. (ALM)*, pages 1–54. Int. Press, Somerville, MA, 2010.
27. A. Bartels and W. Lück. The Borel conjecture for hyperbolic and CAT(0)-groups. *Ann. of Math. (2)*, 175:631–689, 2012.
28. A. Bartels, W. Lück, and H. Reich. The K-theoretic Farrell–Jones conjecture for hyperbolic groups. *Invent. Math.*, 172(1):29–70, 2008.
29. A. Bartels, W. Lück, and H. Reich. On the Farrell–Jones Conjecture and its applications. *Journal of Topology*, 1:57–86, 2008.
30. A. Bartels, W. Lück, H. Reich, and H. Rüping. K- and L-theory of group rings over $GL_n(\mathbf{Z})$. *Publ. Math., Inst. Hautes Étud. Sci.*, 119:97–125, 2014.
31. H. Bass. *Algebraic K-Theory*. W. A. Benjamin, Inc., New York-Amsterdam, 1968.
32. S. Basu and R. Kasilingam. Inertia groups of high-dimensional complex projective spaces. *Algebr. Geom. Topol.*, 18(1):387–408, 2018.
33. S. Basu and R. Kasilingam. Inertia groups and smooth structures on quaternionic projective spaces. *Forum Math.*, 34(2):369–383, 2022.
34. T. Bauer, N. Kitchloo, D. Notbohm, and E. K. Pedersen. Finite loop spaces are manifolds. *Acta Math.*, 192(1):5–31, 2004.
35. H. J. Baues. Quadratic functors and metastable homotopy. *J. Pure Appl. Algebra*, 91(1-3):49–107, 1994.
36. M. Behrens, M. Hill, M. J. Hopkins, and M. Mahowald. Detecting exotic spheres in low dimensions using coker J. *J. Lond. Math. Soc. (2)*, 101(3):1173–1218, 2020.
37. S. Behrens, B. Kalmár, M. H. Kim, M. Powell, and A. Ray, editors. *The Disc Embedding Theorem*. Oxford University Press, Oxford, 2021.
38. I. Belegradek. Aspherical manifolds, relative hyperbolicity, simplicial volume and assembly maps. *Algebr. Geom. Topol.*, 6:1341–1354 (electronic), 2006.
39. M. Bestvina, K. Fujiwara, and D. Wigglesworth. The Farrell–Jones conjecture for hyperbolic-by-cyclic groups. *Int. Math. Res. Not. IMRN*, 7:5887–5904, 2023.
40. J. M. Boardman and R. M. Vogt. *Homotopy Invariant Algebraic Structures on Topological Spaces*. Lecture Notes in Mathematics, Vol. 347. Springer-Verlag, Berlin-New York, 1973.
41. M. Bökstedt and A. Neeman. Homotopy limits in triangulated categories. *Compos. Math.*, 86(2):209–234, 1993.
42. R. Bott. The stable homotopy of the classical groups. *Ann. of Math. (2)*, 70:313–337, 1959.
43. A. K. Bousfield and D. M. Kan. *Homotopy Limits, Completions and Localizations*. Springer-Verlag, Berlin, 1972. Lecture Notes in Mathematics, Vol. 304.
44. G. E. Bredon. *Sheaf Theory*. Springer-Verlag, New York, second edition, 1997.
45. G. E. Bredon. *Topology and Geometry*. Springer-Verlag, New York, 1997. Corrected third printing of the 1993 original.
46. S. Brendle and R. Schoen. Manifolds with 1/4-pinched curvature are space forms. *J. Amer. Math. Soc.*, 22(1):287–307, 2009.
47. E. Brieskorn. Beispiele zur Differentialtopologie von Singularitäten. *Invent. Math.*, 2:1–14, 1966.
48. T. Bröcker and K. Jänich. *Einführung in die Differentialtopologie*. Springer-Verlag, Berlin, 1973. Heidelberger Taschenbücher, Band 143.

49. T. Bröcker and K. Jänich. *Introduction to Differential Topology*. Cambridge University Press, Cambridge, 1982. Translated from the German by C. B. Thomas and M. J. Thomas.
50. T. Bröcker and T. tom Dieck. *Kobordismentheorie*. Springer-Verlag, Berlin, 1970.
51. J. Brookman, J. F. Davis, and Q. Khan. Manifolds homotopy equivalent to $P^n \# P^n$. *Math. Ann.*, 338(4):947–962, 2007.
52. W. Browder. Homotopy type of differentiable manifolds. Colloq. algebr. Topology, Aarhus 1962, 42-46 (1962)., 1962.
53. W. Browder. On the action of $\Theta^n (\partial \pi)$. In *Differential and Combinatorial Topology (A Symposium in Honor of Marston Morse)*, pages 23–36. Princeton Univ. Press, Princeton, N.J., 1965.
54. W. Browder. The Kervaire invariant of framed manifolds and its generalization. *Ann. of Math. (2)*, 90:157–186, 1969.
55. W. Browder. *Surgery on Simply-Connected Manifolds*. Springer-Verlag, New York, 1972. Ergebnisse der Mathematik und ihrer Grenzgebiete, Band 65.
56. W. Browder and M. W. Hirsch. Surgery on piecewise linear manifolds and applications. *Bull. Am. Math. Soc.*, 72:959–964, 1966.
57. W. Browder, T. Petrie, and C. T. C. Wall. The classification of free actions of cyclic groups of odd order on homotopy spheres. *Bull. Am. Math. Soc.*, 77:455–459, 1971.
58. K. S. Brown. *Cohomology of Groups*, volume 87 of *Graduate Texts in Mathematics*. Springer-Verlag, New York, 1982.
59. K. S. Brown. *Cohomology of Groups*. Springer-Verlag, New York, 1994. Corrected reprint of the 1982 original.
60. R. Brown. *Elements of Modern Topology*. McGraw-Hill Book Co., New York, 1968.
61. B. Brück, D. Kielak, and X. Wu. The Farrell–Jones conjecture for normally poly-free groups. *Proc. Amer. Math. Soc.*, 149(6):2349–2356, 2021.
62. G. Brumfiel. Differentiable S^1- actions on homotopy spheres. mimeographed. University of California, Berkeley, 1968.
63. G. Brumfiel. On the homotopy groups of BPL and PL/O. *Annals of Mathematics*, 22:73–79, 1968.
64. G. Brumfiel. On the homotopy groups of BPL and PL/O, II. *Topology*, 8:305–311, 1969.
65. G. Brumfiel. On the homotopy groups of BPL and PL/O, III. *Michigan Math. J.*, 17:217–224, 1969.
66. G. W. Brumfiel. Homotopy equivalences of almost smooth manifolds. In *Algebraic Topology (Proc. Sympos. Pure Math., Vol. XXII, Univ. Wisconsin, Madison, Wis., 1970)*, pages 73–79. Amer. Math. Soc., Providence R.I., 1971.
67. J. Bryant, S. Ferry, W. Mio, and S. Weinberger. Topology of homology manifolds. *Ann. of Math. (2)*, 143(3):435–467, 1996.
68. D. Burghelea, R. Lashof, and M. Rothenberg. *Groups of Automorphisms of Manifolds*. Lecture Notes in Mathematics, Vol. 473. Springer-Verlag, Berlin, 1975. With an appendix ("The topological category") by E. Pedersen.
69. B. Calmés, E. Dotto, Y. Harpaz, F. Hebestreit, M. Land, K. Moi, D. Nardin, T. Nikolaus, and W. Steimle. Hermitian K-theory for stable ∞-categories II: Cobordism categories and additivity. Preprint,arXiv:2009.07224 [math.KT], 2020.
70. B. Calmés, E. Dotto, Y. Harpaz, F. Hebestreit, M. Land, K. Moi, D. Nardin, T. Nikolaus, and W. Steimle. Hermitian K-theory for stable ∞-categories III: Grothendieck-Witt groups of rings. Preprint, arXiv:2009.07225 [math.KT], 2020.
71. B. Calmès, E. Dotto, Y. Harpaz, F. Hebestreit, M. Land, K. Moi, D. Nardin, T. Nikolaus, and W. Steimle. Hermitian K-theory for stable ∞-categories I: Foundations. *Selecta Math. (N.S.)*, 29, 2023.
72. S. Cappell, A. Ranicki, and J. Rosenberg, editors. *Surveys on Surgery Theory. Vol. 1*. Princeton University Press, Princeton, NJ, 2000. Papers dedicated to C. T. C. Wall.
73. S. Cappell, A. Ranicki, and J. Rosenberg, editors. *Surveys on Surgery Theory. Vol. 2*. Princeton University Press, Princeton, NJ, 2001. Papers dedicated to C. T. C. Wall.
74. S. Cappell and S. Weinberger. Homology propagation of group actions. *Comm. Pure Appl. Math.*, 40(6):723–744, 1987.

75. S. Cappell and S. Weinberger. Which H-spaces are manifolds? I. *Topology*, 27(4):377–386, 1988.
76. S. Cappell and S. Weinberger. Surgery theoretic methods in group actions. In *Surveys on Surgery Theory, Vol. 2*, volume 149 of *Ann. of Math. Stud.*, pages 285–317. Princeton Univ. Press, Princeton, NJ, 2001.
77. S. E. Cappell. Manifolds with fundamental group a generalized free product. I. *Bull. Amer. Math. Soc.*, 80:1193–1198, 1974.
78. S. E. Cappell. Unitary nilpotent groups and Hermitian K-theory. I. *Bull. Amer. Math. Soc.*, 80:1117–1122, 1974.
79. S. E. Cappell. A splitting theorem for manifolds. *Invent. Math.*, 33(2):69–170, 1976.
80. S. E. Cappell and J. L. Shaneson. The codimension two placement problem and homology equivalent manifolds. *Ann. of Math. (2)*, 99:277–348, 1974.
81. S. E. Cappell and J. L. Shaneson. Nonlinear similarity. *Ann. of Math. (2)*, 113(2):315–355, 1981.
82. S. E. Cappell and J. L. Shaneson. On 4-dimensional s-cobordisms. *J. Differential Geom.*, 22(1):97–115, 1985.
83. S. E. Cappell, J. L. Shaneson, M. Steinberger, S. Weinberger, and J. E. West. The classification of nonlinear similarities over Z_{2^r}. *Bull. Amer. Math. Soc. (N.S.)*, 22(1):51–57, 1990.
84. M. Cárdenas and E. K. Pedersen. On the Karoubi filtration of a category. *K-Theory*, 12(2):165–191, 1997.
85. H. Cartan and S. Eilenberg. *Homological Algebra*. Princeton University Press, Princeton, N. J., 1956.
86. D. W. Carter. Lower K-theory of finite groups. *Comm. Algebra*, 8(20):1927–1937, 1980.
87. A. J. Casson. Generalisations and applications of block bundles. In *The Hauptvermutung Book. A Collection of Papers on the Topology of Manifolds*, pages 33–67. Dordrecht: Kluwer Academic Publishers, 1996.
88. A. Cavicchioli, F. Hegenbarth, and D. Repovš. On the stable classification of certain 4-manifolds. *Bull. Austral. Math. Soc.*, 52(3):385–398, 1995.
89. A. Cavicchioli, Y. V. Muranov, and F. Spaggiari. Mixed structures on a manifold with boundary. *Glasg. Math. J.*, 48(1):125–143, 2006.
90. J. Cerf. Topologie de certains espaces de plongements. *Bull. Soc. Math. France*, 89:227–380, 1961.
91. J. Cerf. *Sur les difféomorphismes de la sphère de dimension trois* ($\Gamma_4 = 0$), volume No. 53 of *Lecture Notes in Mathematics*. Springer-Verlag, Berlin-New York, 1968.
92. S. Chang and S. Weinberger. On invariants of Hirzebruch and Cheeger-Gromov. *Geom. Topol.*, 7:311–319 (electronic), 2003.
93. S. Chang and S. Weinberger. *A Course on Surgery Theory*, volume 211 of *Annals of Mathematics Studies*. Princeton University Press, Princeton, NJ, 2021.
94. T. A. Chapman. Compact Hilbert cube manifolds and the invariance of Whitehead torsion. *Bull. Amer. Math. Soc.*, 79:52–56, 1973.
95. T. A. Chapman. Topological invariance of Whitehead torsion. *Amer. J. Math.*, 96:488–497, 1974.
96. R. M. Charney and M. W. Davis. Strict hyperbolization. *Topology*, 34(2):329–350, 1995.
97. M. Clapp. Duality and transfer for parametrized spectra. *Arch. Math. (Basel)*, 37(5):462–472, 1981.
98. M. M. Cohen. *A Course in Simple-Homotopy Theory*. Springer-Verlag, New York, 1973. Graduate Texts in Mathematics, Vol. 10.
99. P. E. Conner and E. E. Floyd. *Differentiable Periodic Maps*. Academic Press Inc., Publishers, New York, 1964.
100. P. E. Conner and E. E. Floyd. *The Relation of Cobordism to K-Theories*, volume 28 of *Lect. Notes Math.* Springer, Cham, 1966.
101. F. Connolly, J. F. Davis, and Q. Khan. Topological rigidity and H_1-negative involutions on tori. *Geom. Topol.*, 18(3):1719–1768, 2014.
102. F. Connolly, J. F. Davis, and Q. Khan. Topological rigidity and actions on contractible manifolds with discrete singular set. *Trans. Amer. Math. Soc. Ser. B*, 2:113–133, 2015.

103. F. Connolly and T. Koźniewski. Examples of lack of rigidity in crystallographic groups. In *Algebraic Topology Poznań 1989*, volume 1474 of *Lecture Notes in Math.*, pages 139–145. Springer, Berlin, 1991.
104. F. Connolly and A. Ranicki. On the calculation of UNil. *Adv. Math.*, 195(1):205–258, 2005.
105. F. X. Connolly. Linking numbers and surgery. *Topology*, 12:389–409, 1973.
106. F. X. Connolly and T. Koźniewski. Rigidity and crystallographic groups. I. *Invent. Math.*, 99(1):25–48, 1990.
107. A. Conway, D. Crowley, M. Powell, and J. Sixt. Simply connected manifolds with large homotopy stable classes. *J. Aust. Math. Soc.*, 115(2):172–203, 2023.
108. D. Crowley. The smooth structure set of $S^p \times S^q$. *Geom. Dedicata*, 148:15–33, 2010.
109. D. Crowley and C. M. Escher. A classification of S^3-bundles over S^4. *Differential Geom. Appl.*, 18(3):363–380, 2003.
110. D. Crowley and T. Macko. The additivity of the ρ-invariant and periodicity in topological surgery. *Algebr. Geom. Topol.*, 11(4):1915–1959, 2011.
111. D. Crowley and T. Macko. On the cardinality of the manifold set. *Geom. Dedicata*, 200:265–285, 2019.
112. D. Crowley and C. Nagy. The smooth classification of 4-dimensional complete intersections. Preprint, arXiv:2003.09216 [math.GT], 2020.
113. D. Crowley and J. Sixt. Stably diffeomorphic manifolds and $l_{2q+1}(\mathbb{Z}[\pi])$. *Forum Math.*, 23(3):483–538, 2011.
114. D. Crowley and D. J. Wraith. Positive Ricci curvature on highly connected manifolds. *J. Differential Geom.*, 106(2):187–243, 2017.
115. J. F. Davis. The Borel/Novikov conjectures and stable diffeomorphisms of 4-manifolds. In *Geometry and Topology of Manifolds*, volume 47 of *Fields Inst. Commun.*, pages 63–76. Amer. Math. Soc., Providence, RI, 2005.
116. J. F. Davis and W. Lück. Spaces over a category and assembly maps in isomorphism conjectures in K- and L-theory. *K-Theory*, 15(3):201–252, 1998.
117. J. F. Davis and W. Lück. The p-chain spectral sequence. *K-Theory*, 30(1):71–104, 2003. Special issue in honor of Hyman Bass on his seventieth birthday. Part I.
118. J. F. Davis and W. Lück. Manifolds homotopy equivalent to certain torus bundles over Lens spaces. *Commun. Pure Appl. Math.*, 74(11):2348–2397, 2021.
119. J. F. Davis and W. Lück. On manifold models for the classifying space for proper actions. preprint, arXiv:2303.15765 [math.GT], 2023.
120. J. F. Davis and R. Milgram. A survey on the space form problem. *Math. reports*, 2:223–283, 1985.
121. J. F. Davis, F. Quinn, and H. Reich. Algebraic K-theory over the infinite dihedral group: a controlled topology approach. *J. Topol.*, 4(3):505–528, 2011.
122. M. Davis. Exotic aspherical manifolds. In T. Farrell, L. Göttsche, and W. Lück, editors, *High Dimensional Manifold Theory*, number 9 in ICTP Lecture Notes, pages 371–404. Abdus Salam International Centre for Theoretical Physics, Trieste, 2002. Proceedings of the summer school "High dimensional manifold theory" in Trieste May/June 2001, Number 2. http://www.ictp.trieste.it/~pub_off/lectures/vol9.html.
123. M. W. Davis. Groups generated by reflections and aspherical manifolds not covered by Euclidean space. *Ann. of Math. (2)*, 117(2):293–324, 1983.
124. M. W. Davis. *The Geometry and Topology of Coxeter Groups*, volume 32 of *London Mathematical Society Monographs Series*. Princeton University Press, Princeton, NJ, 2008.
125. M. W. Davis, J. Fowler, and J.-F. Lafont. Aspherical manifolds that cannot be triangulated. *Algebr. Geom. Topol.*, 14(2):795–803, 2014.
126. M. W. Davis and T. Januszkiewicz. Hyperbolization of polyhedra. *J. Differential Geom.*, 34(2):347–388, 1991.
127. G. de Rham. Reidemeister's torsion invariant and rotations of S^n. In *Differential Analysis, Bombay Colloq.*, pages 27–36. Oxford Univ. Press, London, 1964.
128. G. de Rham, S. Maumary, and M. A. Kervaire. *Torsion et type simple d'homotopie*. Springer-Verlag, Berlin, 1967.

129. A. Dold. Über fasernweise Homotopieäquivalenz von Faserräumen. *Math. Z.*, 62:111–136, 1955.
130. A. Dold. Zur Homotopietheorie der Kettenkomplexe. *Mathematische Annalen*, 140:278–298, 1960.
131. A. Dold. Relations between ordinary and extraordinary homology. Colloq. alg. topology, Aarhus 1962, 2-9, 1962.
132. A. Dold. Partitions of unity in the theory of fibrations. *Ann. of Math. (2)*, 78:223–255, 1963.
133. A. Dold. *Halbexakte Homotopiefunktoren*, volume 12. Springer, Cham, 1966.
134. A. Dold. *Lectures on Algebraic Topology*. Springer-Verlag, Berlin, second edition, 1980.
135. S. K. Donaldson. Irrationality and the h-cobordism conjecture. *J. Differential Geom.*, 26(1):141–168, 1987.
136. K. H. Dovermann and T. Petrie. G surgery. II. *Mem. Amer. Math. Soc.*, 37(260):xxiii+118, 1982.
137. K. H. Dovermann and M. Rothenberg. An equivariant surgery sequence and equivariant diffeomorphism and homeomorphism classification (a survey). In *Topology Symposium, Siegen 1979 (Proc. Sympos., Univ. Siegen, Siegen, 1979)*, volume 788 of *Lecture Notes in Math.*, pages 257–280. Springer, Berlin, 1980.
138. K. H. Dovermann and M. Rothenberg. Equivariant surgery and classifications of finite group actions on manifolds. *Mem. Amer. Math. Soc.*, 71(379):viii+117, 1988.
139. K. H. Dovermann and R. Schultz. *Equivariant Surgery Theories and their Periodicity Properties*. Springer-Verlag, Berlin, 1990.
140. B. Eckmann and P. J. Hilton. Homotopy groups of maps and exact sequences. *Comment. Math. Helv.*, 34:271–304, 1960.
141. J. Eells, Jr. and N. H. Kuiper. An invariant for certain smooth manifolds. *Ann. Mat. Pura Appl. (4)*, 60:93–110, 1962.
142. Y. Eliashberg and N. Mishachev. *Introduction to the h-Principle*, volume 48 of *Graduate Studies in Mathematics*. American Mathematical Society, Providence, RI, 2002.
143. N.-E. Enkelmann, W. Lück, M. Pieper, M. Ullmann, and C. Winges. On the Farrell–Jones conjecture for Waldhausen's A–theory. *Geom. Topol.*, 22(6):3321–3394, 2018.
144. F. T. Farrell. *Lectures on Surgical Methods in Rigidity*, volume 86 of *Lect. Math. Phys., Math., Tata Inst. Fundam. Res.* Berlin: Springer; Bombay: Tata Institute of Fundamental Research, 1996.
145. F. T. Farrell, L. Göttsche, and W. Lück, editors. *High Dimensional Manifold Theory*. Number 9 in ICTP Lecture Notes. Abdus Salam International Centre for Theoretical Physics, Trieste, 2002. Proceedings of the summer school "High dimensional manifold theory" in Trieste May/June 2001, Number 1. http://www.ictp.trieste.it/˜pub_off/lectures/vol9.html.
146. F. T. Farrell, L. Göttsche, and W. Lück, editors. *High Dimensional Manifold Theory*. Number 9 in ICTP Lecture Notes. Abdus Salam International Centre for Theoretical Physics, Trieste, 2002. Proceedings of the summer school "High dimensional manifold theory" in Trieste May/June 2001, Number 2. http://www.ictp.trieste.it/˜pub_off/lectures/vol9.html.
147. F. T. Farrell and W. C. Hsiang. Rational L-groups of Bieberbach groups. *Comment. Math. Helv.*, 52(1):89–109, 1977.
148. F. T. Farrell and L. E. Jones. Examples of expanding endomorphisms on exotic tori. *Invent. Math.*, 45(2):175–179, 1978.
149. F. T. Farrell and L. E. Jones. Negatively curved manifolds with exotic smooth structures. *J. Amer. Math. Soc.*, 2(4):899–908, 1989.
150. F. T. Farrell and L. E. Jones. Isomorphism conjectures in algebraic K-theory. *J. Amer. Math. Soc.*, 6(2):249–297, 1993.
151. F. T. Farrell, L. E. Jones, and W. Lück. A caveat on the isomorphism conjecture in L-theory. *Forum Math.*, 14(3):413–418, 2002.
152. F. T. Farrell, W. Lück, and W. Steimle. Obstructions to fibering a manifold. *Geom. Dedicata*, 148:35–69, 2010.
153. S. C. Ferry, A. A. Ranicki, and J. Rosenberg, editors. *Novikov Conjectures, Index Theorems and Rigidity. Vol. 1*. Cambridge University Press, Cambridge, 1995. Including papers from the conference held at the Mathematisches Forschungsinstitut Oberwolfach, Oberwolfach, September 6–10, 1993.

154. S. C. Ferry, A. A. Ranicki, and J. Rosenberg, editors. *Novikov Conjectures, Index Theorems and Rigidity. Vol. 2*. Cambridge University Press, Cambridge, 1995. Including papers from the conference held at the Mathematisches Forschungsinstitut Oberwolfach, Oberwolfach, September 6–10, 1993.
155. D. Frank. The signature defect and the homotopy of BPL and PL/O. *Comment. Math. Helv.*, 48:525–530, 1973.
156. M. Freedman. Uniqueness theorems for taut submanifolds. *Pacific J. Math.*, 62(2):379–387, 1976.
157. M. H. Freedman. The topology of four-dimensional manifolds. *J. Differential Geom.*, 17(3):357–453, 1982.
158. M. H. Freedman. The disk theorem for four-dimensional manifolds. In *Proceedings of the International Congress of Mathematicians, Vol. 1, 2 (Warsaw, 1983)*, pages 647–663, Warsaw, 1984. PWN.
159. M. H. Freedman and F. Quinn. *Topology of 4-Manifolds*. Princeton University Press, Princeton, NJ, 1990.
160. M. H. Freedman and P. Teichner. 4-manifold topology. I. Subexponential groups. *Invent. Math.*, 122(3):509–529, 1995.
161. W. Fulton and J. Harris. *Representation Theory*. Springer-Verlag, New York, 1991. A first course, Readings in Mathematics.
162. S. Galatius and O. Randal-Williams. Moduli spaces of manifolds: a user's guide. In *Handbook of Homotopy Theory*, CRC Press/Chapman Hall Handb. Math. Ser., pages 443–485. CRC Press, Boca Raton, FL, [2020] ©2020.
163. S. Galatius, U. Tillmann, I. Madsen, and M. Weiss. The homotopy type of the cobordism category. *Acta Math.*, 202(2):195–239, 2009.
164. S. Gallot, D. Hulin, and J. Lafontaine. *Riemannian Geometry*. Springer-Verlag, Berlin, 1987.
165. S. Gitler and J. D. Stasheff. The first exotic class of BF. *Topology*, 4:257–266, 1965.
166. D. L. Gonçalves and M. Spreafico. Quaternionic line bundles over quaternionic projective spaces. *Math. J. Okayama Univ.*, 48:87–101, 2006.
167. D. H. Gottlieb. Poincaré duality and fibrations. *Proc. Amer. Math. Soc.*, 76(1):148–150, 1979.
168. M. Gromov. Volume and bounded cohomology. *Inst. Hautes Études Sci. Publ. Math.*, 56:5–99 (1983), 1982.
169. M. Gromov. Hyperbolic groups. In *Essays in Group Theory*, pages 75–263. Springer-Verlag, New York, 1987.
170. A. Haefliger. Differential embeddings of S^n in S^{n+q} for $q > 2$. *Ann. of Math. (2)*, 83:402–436, 1966.
171. I. Hambleton. Surgery obstructions on closed manifolds and the inertia subgroup. *Forum Math.*, 24(5):911–929, 2012.
172. I. Hambleton. Topological spherical space forms. In *Handbook of Group Actions. Vol. II*, volume 32 of *Adv. Lect. Math. (ALM)*, pages 151–172. Int. Press, Somerville, MA, 2015.
173. I. Hambleton and M. Kreck. On the classification of topological 4-manifolds with finite fundamental group. *Math. Ann.*, 280(1):85–104, 1988.
174. I. Hambleton, M. Kreck, and P. Teichner. Topological 4-manifolds with geometrically two-dimensional fundamental groups. *J. Topol. Anal.*, 1(2):123–151, 2009.
175. I. Hambleton and I. Madsen. Actions of finite groups on $\mathbf{r}^{n+k}$ with fixed set $\mathbf{r}^k$. *Canad. J. Math.*, 38(4):781–860, 1986.
176. I. Hambleton and E. K. Pedersen. Non-linear similarity revisited. In *Prospects in Topology (Princeton, NJ, 1994)*, pages 157–174. Princeton Univ. Press, Princeton, NJ, 1995.
177. I. Hambleton and L. R. Taylor. A guide to the calculation of the surgery obstruction groups for finite groups. In *Surveys on Surgery Theory, Vol. 1*, pages 225–274. Princeton Univ. Press, Princeton, NJ, 2000.
178. A. Hatcher. *Algebraic Topology*. Cambridge University Press, Cambridge, 2002.
179. J.-C. Hausmann and P. Vogel. *Geometry on Poincaré Spaces*, volume 41 of *Mathematical Notes*. Princeton University Press, Princeton, NJ, 1993.

180. F. Hebestreit, M. Land, W. Lück, and O. Randal-Williams. A vanishing theorem for tautological classes of aspherical manifolds. *Geom. Topol.*, 25(1):47–110, 2021.
181. S. Helgason. *Differential Geometry, Lie Groups, and Symmetric Spaces*. American Mathematical Society, Providence, RI, 2001. Corrected reprint of the 1978 original.
182. J. Hempel. *3-Manifolds*. Princeton University Press, Princeton, N. J., 1976. Ann. of Math. Studies, No. 86.
183. J. Hempel. *3-Manifolds*. AMS Chelsea Publishing, Providence, RI, 2004. Reprint of the 1976 original.
184. N. Higson and J. Roe. Mapping surgery to analysis. I. Analytic signatures. *K-Theory*, 33(4):277–299, 2005.
185. N. Higson and J. Roe. Mapping surgery to analysis. II. Geometric signatures. *K-Theory*, 33(4):301–324, 2005.
186. N. Higson and J. Roe. Mapping surgery to analysis. III. Exact sequences. *K-Theory*, 33(4):325–346, 2005.
187. M. A. Hill, M. J. Hopkins, and D. C. Ravenel. On the nonexistence of elements of Kervaire invariant one. *Ann. of Math. (2)*, 184(1):1–262, 2016.
188. M. W. Hirsch. Immersions of manifolds. *Trans. Amer. Math. Soc.*, 93:242–276, 1959.
189. M. W. Hirsch. *Differential Topology*. Springer-Verlag, New York, 1976. Graduate Texts in Mathematics, No. 33.
190. M. W. Hirsch and B. Mazur. *Smoothings of Piecewise Linear Manifolds*. Princeton University Press, Princeton, N. J.; University of Tokyo Press, Tokyo, 1974. Annals of Mathematics Studies, No. 80.
191. F. Hirzebruch. *Neue topologische Methoden in der algebraischen Geometrie*. Zweite ergänzte Auflage. Ergebnisse der Mathematik und ihrer Grenzgebiete, N.F., Heft 9. Springer-Verlag, Berlin, 1962.
192. F. Hirzebruch. *Topological Methods in Algebraic Geometry*, volume 131 of *Grundlehren*. Springer, 1978.
193. F. Hirzebruch. Singularities and exotic spheres. In *Séminaire Bourbaki, Vol. 10*, pages Exp. No. 314, 13–32. Soc. Math. France, Paris, 1995.
194. J. Hollender and R. M. Vogt. Modules of topological spaces, applications to homotopy limits and E_∞ structures. *Arch. Math. (Basel)*, 59(2):115–129, 1992.
195. W. C. Hsiang and R. W. Sharpe. Parametrized surgery and isotopy. *Pacific J. Math.*, 67(2):401–459, 1976.
196. R. Hu. Characteristic classes of L-theory. Preprint, arXiv:2309.05170 [math.AT], 2023.
197. J. F. P. Hudson. Piecewise linear topology. Mathematics Lecture Note Series. New York-Amsterdam: W.A. Benjamin, Inc. IX, 282 p. (1969)., 1969.
198. D. Husemoller. *Fibre Bundles*. Springer-Verlag, New York, second edition, 1975. Graduate Texts in Mathematics, No. 20.
199. D. C. Isaksen, G. Wang, and Z. Xu. Stable homotopy groups of spheres: from dimension 0 to 90. *Publ. Math. Inst. Hautes Études Sci.*, 137:107–243, 2023.
200. B. Jahren and S. Kwasik. Free involutions on $S^1 \times S^n$. *Math. Ann.*, 351(2):281–303, 2011.
201. I. M. James. On the iterated suspension. *Quart. J. Math., Oxford Ser. (2)*, 5:1–10, 1954.
202. I. M. James and J. H. C. Whitehead. The homotopy theory of sphere bundles over spheres. I. *Proc. Lond. Math. Soc. (3)*, 4:196–218, 1954.
203. P. J. Kahn. Self-equivalences of $(n-1)$-connected $2n$-manifolds. *Math. Ann.*, 180:26–47, 1969.
204. H. Kammeyer, W. Lück, and H. Rüping. The Farrell–Jones conjecture for arbitrary lattices in virtually connected Lie groups. *Geom. Topol.*, 20(3):1275–1287, 2016.
205. R. Kasilingam. Smooth structures on a fake real projective space. *Topology Appl.*, 206:1–7, 2016.
206. D. Kasprowski and M. Land. Topological 4-manifolds with 4-dimensional fundamental group. *Glasg. Math. J.*, 64(2):454–461, 2022.
207. D. Kasprowski, M. Land, M. Powell, and P. Teichner. Stable classification of 4-manifolds with 3-manifold fundamental groups. *J. Topol.*, 10(3):827–881, 2017.

208. D. Kasprowski, M. Powell, and P. Teichner. Four-manifolds up to connected sum with complex projective planes. *Amer. J. Math.*, 144(1):75–118, 2022.
209. D. Kasprowski and P. Teichner. $\mathbb{CP}^2$-stable classification of 4-manifolds with finite fundamental group. *Pacific J. Math.*, 310(2):355–373, 2021.
210. D. Kasprowski, M. Ullmann, C. Wegner, and C. Winges. The A-theoretic Farrell–Jones conjecture for virtually solvable groups. *Bull. Lond. Math. Soc.*, 50(2):219–228, 2018.
211. K. Kawakubo. Inertia groups of low dimensional complex projective spaces and some free differentiable actions on spheres. I. *Proc. Japan Acad.*, 44:873–875, 1968.
212. K. Kawakubo. On the inertia groups of homology tori. *J. Math. Soc. Japan*, 21:37–47, 1969.
213. A. Kawauchi and S. Kojima. Algebraic classification of linking pairings on 3-manifolds. *Math. Ann.*, 253:29–42, 1980.
214. M. A. Kervaire. Some nonstable homotopy groups of Lie groups. *Illinois J. Math.*, 4:161–169, 1960.
215. M. A. Kervaire. Le théorème de Barden-Mazur-Stallings. *Comment. Math. Helv.*, 40:31–42, 1965.
216. M. A. Kervaire and J. Milnor. Groups of homotopy spheres. I. *Ann. of Math. (2)*, 77:504–537, 1963.
217. Q. Khan. Free transformations of $S^1 \times S^n$ of square-free odd period. *Indiana Univ. Math. J.*, 66(5):1453–1482, 2017.
218. R. C. Kirby. *The Topology of 4-Manifolds*. Springer-Verlag, Berlin, 1989.
219. R. C. Kirby and L. C. Siebenmann. *Foundational Essays on Topological Manifolds, Smoothings, and Triangulations*. Princeton University Press, Princeton, N.J., 1977. With notes by J. Milnor and M. F. Atiyah, Annals of Mathematics Studies, No. 88.
220. E. Kjær Pedersen and A. Ranicki. Report 36/2006: Mini-workshop: The Hauptvermutung for high-dimensional manifolds (August 13th – August 19th, 2006). *Oberwolfach Rep.*, 3(3):2195–2225, 2006.
221. J. R. Klein. Poincaré duality spaces. In *Surveys on Surgery Theory. Vol. 1: Papers Dedicated to C. T. C. Wall on the Occasion of his 60th Birthday*, pages 135–165. Princeton, NJ: Princeton University Press, 2000.
222. B. Kleiner and J. Lott. Notes on Perelman's papers. *Geom. Topol.*, 12(5):2587–2855, 2008.
223. A. Kosiński. On the inertia group of π-manifolds. *Amer. J. Math.*, 89:227–248, 1967.
224. A. A. Kosinski. *Differential Manifolds*, volume 138 of *Pure Appl. Math., Academic Press*. Boston, MA: Academic Press, 1993.
225. M. Kreck. Surgery and duality. *Ann. of Math. (2)*, 149(3):707–754, 1999.
226. M. Kreck and W. Lück. *The Novikov Conjecture: Geometry and Algebra*, volume 33 of *Oberwolfach Seminars*. Birkhäuser Verlag, Basel, 2005.
227. M. Kreck and W. Lück. Topological rigidity for non-aspherical manifolds. *Pure and Applied Mathematics Quarterly*, 5 (3):873–914, 2009. special issue in honor of Friedrich Hirzebruch.
228. M. Kreck, W. Lück, and P. Teichner. Counterexamples to the Kneser conjecture in dimension four. *Comment. Math. Helv.*, 70(3):423–433, 1995.
229. M. Kreck, W. Lück, and P. Teichner. Stable prime decompositions of four-manifolds. In *Prospects in Topology (Princeton, NJ, 1994)*, pages 251–269. Princeton Univ. Press, Princeton, NJ, 1995.
230. M. Kreck and S. Stolz. A diffeomorphism classification of 7-dimensional homogeneous Einstein manifolds with $SU(3) \times SU(2) \times U(1)$-symmetry. *Ann. of Math. (2)*, 127(2):373–388, 1988.
231. M. Kreck and S. Stolz. Some nondiffeomorphic homeomorphic homogeneous 7-manifolds with positive sectional curvature. *J. Differential Geom.*, 33(2):465–486, 1991.
232. M. Kreck and S. Stolz. Nonconnected moduli spaces of positive sectional curvature metrics. *J. Amer. Math. Soc.*, 6(4):825–850, 1993.
233. V. S. Krushkal and F. Quinn. Subexponential groups in 4-manifold topology. *Geom. Topol.*, 4:407–430, 2000.
234. P. Kühl, T. Macko, and A. Mole. The total surgery obstruction revisited. *Münster J. Math.*, 6:181–269, 2013.

235. A. Kupers. The first two k-invariants of Top/O. Preprint, arXiv:2303.12854 [math.GT], 2023.
236. S. a. Kwasik and R. Schultz. Vanishing of Whitehead torsion in dimension four. *Topology*, 31(4):735–756, 1992.
237. J. Lafontaine. *An Introduction to Differential Manifolds*. Springer, Cham, second edition, 2015.
238. T. Y. Lam. *A First Course in Noncommutative Rings*. Springer-Verlag, New York, 1991.
239. T. Lance. Differentiable structures on manifolds. In *Surveys on Surgery Theory, Vol. 1*, pages 73–104. Princeton Univ. Press, Princeton, NJ, 2000.
240. T. Lawson. In which I try to get the signs right for once, or: a public exercise in futility. unpublished, 2013.
241. J. M. Lee. *Introduction to Topological Manifolds*, volume 202 of *Graduate Texts in Mathematics*. Springer, New York, second edition, 2011.
242. B. Leeb. 3-manifolds with(out) metrics of nonpositive curvature. *Invent. Math.*, 122(2):277–289, 1995.
243. S. Lefschetz. *Introduction to Topology*. Princeton University Press, Princeton, N. J., 1949.
244. J. Levine. Inertia groups of manifolds and diffeomorphisms of spheres. *Am. J. Math.*, 92:243–258, 1970.
245. J. Levine and K. E. Orr. A survey of applications of surgery to knot and link theory. In *Surveys on Surgery Theory, Vol. 1*, volume 145 of *Ann. of Math. Stud.*, pages 345–364. Princeton Univ. Press, Princeton, NJ, 2000.
246. J. P. Levine. Lectures on groups of homotopy spheres. In *Algebraic and Geometric Topology (New Brunswick, N.J., 1983)*, pages 62–95. Springer, Berlin, 1985.
247. S. López de Medrano. *Involutions on Manifolds*. Springer-Verlag, New York, 1971. Ergebnisse der Mathematik und ihrer Grenzgebiete, Band 59.
248. W. Lück. The transfer maps induced in the algebraic K_0-and K_1-groups by a fibration. I. *Math. Scand.*, 59(1):93–121, 1986.
249. W. Lück. The transfer maps induced in the algebraic K_0- and K_1-groups by a fibration. II. *J. Pure Appl. Algebra*, 45(2):143–169, 1987.
250. W. Lück. *Transformation Groups and Algebraic K-Theory*, volume 1408 of *Lecture Notes in Mathematics*. Springer-Verlag, Berlin, 1989.
251. W. Lück. Hilbert modules and modules over finite von Neumann algebras and applications to L^2-invariants. *Math. Ann.*, 309(2):247–285, 1997.
252. W. Lück. A basic introduction to surgery theory. In F. T. Farrell, L. Göttsche, and W. Lück, editors, *High Dimensional Manifold Theory*, number 9 in ICTP Lecture Notes, pages 1–224. Abdus Salam International Centre for Theoretical Physics, Trieste, 2002. Proceedings of the summer school "High dimensional manifold theory" in Trieste May/June 2001, Number 1. http://www.ictp.trieste.it/˜pub_off/lectures/vol9.html.
253. W. Lück. Chern characters for proper equivariant homology theories and applications to K- and L-theory. *J. Reine Angew. Math.*, 543:193–234, 2002.
254. W. Lück. L^2-*Invariants: Theory and Applications to Geometry and K-Theory*, volume 44 of *Ergebnisse der Mathematik und ihrer Grenzgebiete. 3. Folge. A Series of Modern Surveys in Mathematics [Results in Mathematics and Related Areas. 3rd Series. A Series of Modern Surveys in Mathematics]*. Springer-Verlag, Berlin, 2002.
255. W. Lück. K- and L-theory of the semi-direct product of the discrete 3-dimensional Heisenberg group by $\mathbb{Z}/4$. *Geom. Topol.*, 9:1639–1676 (electronic), 2005.
256. W. Lück. Survey on classifying spaces for families of subgroups. In *Infinite Groups: Geometric, Combinatorial and Dynamical Aspects*, volume 248 of *Progr. Math.*, pages 269–322. Birkhäuser, Basel, 2005.
257. W. Lück. K- and L-theory of group rings. In *Proceedings of the International Congress of Mathematicians. Volume II*, pages 1071–1098, New Delhi, 2010. Hindustan Book Agency.
258. W. Lück. Aspherical manifolds. *Bulletin of the Manifold Atlas 2012*, pages 1–17, 2012.
259. W. Lück. Assembly maps. In H. Miller, editor, *Handbook of Homotopy Theory*, Handbook in Mathematics Series, pages 853–892. CRC Press/Chapman and Hall, Boca Raton, FL, 2019.

260. W. Lück. On Brown's problem, Poincaré models for the classifying spaces for proper action and Nielsen realization. preprint, arXiv:2201.10807 [math.AT], 2022.
261. W. Lück. Isomorphism Conjectures in K- and L-theory. In preparation, see http://www.him.uni-bonn.de/lueck/data/ic.pdf, 2024.
262. W. Lück and I. Madsen. Equivariant L-theory. I. *Math. Z.*, 203(3):503–526, 1990.
263. W. Lück and I. Madsen. Equivariant L-theory. II. *Math. Z.*, 204(2):253–268, 1990.
264. W. Lück and A. A. Ranicki. Surgery transfer. In *Algebraic Topology and Transformation Groups (Göttingen, 1987)*, pages 167–246. Springer-Verlag, Berlin, 1988.
265. W. Lück and A. A. Ranicki. Surgery obstructions of fibre bundles. *J. Pure Appl. Algebra*, 81(2):139–189, 1992.
266. W. Lück and H. Reich. The Baum-Connes and the Farrell–Jones conjectures in K- and L-theory. In *Handbook of K-Theory. Vol. 1, 2*, pages 703–842. Springer, Berlin, 2005.
267. W. Lück and D. Rosenthal. On the K- and L-theory of hyperbolic and virtually finitely generated abelian groups. *Forum Math.*, 26(5):1565–1609, 2014.
268. W. Lück and R. Stamm. Computations of K- and L-theory of cocompact planar groups. *K-Theory*, 21(3):249–292, 2000.
269. W. Lück and W. Steimle. Non-connective K- and Nil-spectra of additive categories. In *An Alpine Expedition Through Algebraic Topology*, volume 617 of *Contemp. Math.*, pages 205–236. Amer. Math. Soc., Providence, RI, 2014.
270. W. Lück and W. Steimle. A twisted Bass–Heller–Swan decomposition for the algebraic K-theory of additive categories. *Forum Math.*, 28(1):129–174, 2016.
271. J. Lurie. Algebraic L-Theory and Surgery. Lecture notes URL: https://www.math.ias.edu/~lurie/287x.html, 2011.
272. S. Mac Lane. *Homology*. Academic Press Inc., Publishers, New York, 1963.
273. T. Macko. Fake lens spaces. *Bulletin of the Manifold Atlas*, 2013.
274. T. Macko and C. Wegner. On fake lens spaces with fundamental group of order a power of 2. *Algebr. Geom. Topol.*, 9(3):1837–1883, 2009.
275. T. Macko and C. Wegner. On the classification of fake lens spaces. *Forum Math.*, 23(5):1053–1091, 2011.
276. I. Madsen. Spherical space forms. In *Proceedings of the International Congress of Mathematicians (Helsinki, 1978)*, pages 475–479. Acad. Sci. Fennica, Helsinki, 1980.
277. I. Madsen. Spherical space forms: a survey. In *18th Scandinavian Congress of Mathematicians (Aarhus, 1980)*, volume 11 of *Progr. Math.*, pages 82–103. Birkhäuser Boston, Mass., 1981.
278. I. Madsen and R. J. Milgram. On spherical fiber bundles and their PL reductions. In *New Developments in Topology. The Edited and Revised Proceedings of the Symposium on Algebraic Topology, Oxford, June 1972*. Cambridge: At the University Press, 1974.
279. I. Madsen and R. J. Milgram. *The Classifying Spaces for Surgery and Cobordism of Manifolds*. Princeton University Press, Princeton, N.J., 1979.
280. I. Madsen and M. Rothenberg. On the classification of G-spheres. II. PL automorphism groups. *Math. Scand.*, 64(2):161–218, 1989.
281. I. Madsen, L. R. Taylor, and B. Williams. Tangential homotopy equivalences. *Comment. Math. Helv.*, 55:445–484, 1980.
282. I. Madsen, C. B. Thomas, and C. T. C. Wall. The topological spherical space form problem. II. Existence of free actions. *Topology*, 15(4):375–382, 1976.
283. I. Madsen, C. B. Thomas, and C. T. C. Wall. Topological spherical space form problem. III. Dimensional bounds and smoothing. *Pacific J. Math.*, 106(1):135–143, 1983.
284. M. Mahowald. The order of the image of the J-homomorphisms. *Bull. Amer. Math. Soc.*, 76:1310–1313, 1970.
285. M. E. Mahowald and M. Tangora. Some differentials in the Adams spectral sequence. *Topology*, 6:349–369, 1967.
286. C. Manolescu. Triangulations of manifolds. *ICCM Not.*, 2(2):21–23, 2014.
287. C. Manolescu. Pin(2)-equivariant Seiberg-Witten Floer homology and the triangulation conjecture. *J. Amer. Math. Soc.*, 29(1):147–176, 2016.
288. T. Matumoto and L. Siebenmann. The topological s-cobordism theorem fails in dimension 4 or 5. *Math. Proc. Cambridge Philos. Soc.*, 84(1):85–87, 1978.

289. J. P. May. Classifying spaces and fibrations. *Mem. Amer. Math. Soc.*, 1(1, 155):xiii+98, 1975.
290. J. P. May and K. Ponto. *More Concise Algebraic Topology*. Chicago Lectures in Mathematics. University of Chicago Press, Chicago, IL, 2012.
291. B. Mazur. Relative neighborhoods and the theorems of Smale. *Ann. of Math. (2)*, 77:232–249, 1963.
292. J. McCleary. *A User's Guide to Spectral Sequences*, volume 58 of *Cambridge Studies in Advanced Mathematics*. Cambridge University Press, Cambridge, second edition, 2001.
293. G. Mess. Examples of Poincaré duality groups. *Proc. Amer. Math. Soc.*, 110(4):1145–1146, 1990.
294. J. Milnor. On manifolds homeomorphic to the 7-sphere. *Ann. of Math. (2)*, 64:399–405, 1956.
295. J. Milnor. On simply connected 4-manifolds. In *Symposium Internacional de Topología Algebraica International Symposium on Algebraic Topology*, pages 122–128. Universidad Nacional Autónoma de México and UNESCO, Mexico City, 1958.
296. J. Milnor. On spaces having the homotopy type of cw-complex. *Trans. Amer. Math. Soc.*, 90:272–280, 1959.
297. J. Milnor. A procedure for killing homotopy groups of differentiable manifolds. In *Proc. Sympos. Pure Math., Vol. III*, pages 39–55. American Mathematical Society, Providence, R.I, 1961.
298. J. Milnor. *Morse Theory*. Princeton University Press, Princeton, N.J., 1963.
299. J. Milnor. Spin structures on manifolds. *Enseign. Math. (2)*, 9:198–203, 1963.
300. J. Milnor. Microbundles. I. *Topology*, 3(suppl. 1):53–80, 1964.
301. J. Milnor. *Lectures on the h-Cobordism Theorem*. Princeton University Press, Princeton, N.J., 1965.
302. J. Milnor. Whitehead torsion. *Bull. Amer. Math. Soc.*, 72:358–426, 1966.
303. J. Milnor. On characteristic classes for spherical fibre spaces. *Comment. Math. Helv.*, 43:51–77, 1968.
304. J. Milnor. *Singular Points of Complex Hypersurfaces*. Princeton University Press, Princeton, N.J., 1968.
305. J. Milnor. Classification of $(n-1)$-connected $2n$-dimensional manifolds and the discovery of exotic spheres. In *Surveys on Surgery Theory, Vol. 1*, pages 25–30. Princeton Univ. Press, Princeton, NJ, 2000.
306. J. Milnor and D. Husemoller. *Symmetric Bilinear Forms*. Springer-Verlag, New York, 1973. Ergebnisse der Mathematik und ihrer Grenzgebiete, Band 73.
307. J. Milnor and E. Spanier. Two remarks on fiber homotopy type. *Pacific J. Math.*, 10:585–590, 1960.
308. J. Milnor and J. D. Stasheff. *Characteristic Classes*. Princeton University Press, Princeton, N. J., 1974. Annals of Mathematics Studies, No. 76.
309. E. E. Moise. *Geometric Topology in Dimensions 2 and 3*. Springer-Verlag, New York, 1977. Graduate Texts in Mathematics, Vol. 47.
310. J. Morgan and G. Tian. *Ricci Flow and the Poincaré Conjecture*, volume 3 of *Clay Mathematics Monographs*. American Mathematical Society, Providence, RI, 2007.
311. J. Morgan and G. Tian. *The Geometrization Conjecture*, volume 5 of *Clay Mathematics Monographs*. American Mathematical Society, Providence, RI; Clay Mathematics Institute, Cambridge, MA, 2014.
312. J. W. Morgan and D. P. Sullivan. The transversality characteristic class and linking cycles in surgery theory. *Ann. Math. (2)*, 99:463–544, 1974.
313. H. K. Mukerjee. Classification of homotopy Dold manifolds. *New York J. Math.*, 9:271–293, 2003.
314. H. K. Mukerjee. Classification of homotopy real Milnor manifolds. *Topology Appl.*, 139(1-3):151–184, 2004.
315. H. K. Mukerjee. Classification of homotopy Wall's manifolds. *Topology Appl.*, 153(18):3467–3495, 2006.
316. A. Mukherjee. *Differential Topology.*, volume 72. New Delhi: Hindustan Book Agency, 2015.

317. J. R. Munkres. *Elementary Differential Topology*, volume 1961 of *Lectures given at Massachusetts Institute of Technology, Fall*. Princeton University Press, Princeton, N.J., 1966.
318. I. Namioka. Maps of pairs in homotopy theory. *Proc. London Math. Soc. (3)*, 12:725–738, 1962.
319. M. H. A. Newman. The engulfing theorem for topological manifolds. *Ann. of Math. (2)*, 84:555–571, 1966.
320. A. J. Nicas. Induction theorems for groups of homotopy manifold structures. *Mem. Amer. Math. Soc.*, 39(267):vi+108, 1982.
321. S. P. Novikov. A diffeomorphism of simply connected manifolds. *Dokl. Akad. Nauk SSSR*, 143:1046–1049, 1962.
322. S. P. Novikov. Homotopically equivalent smooth manifolds. I. *Izv. Akad. Nauk SSSR Ser. Mat.*, 28:365–474, 1964.
323. S. P. Novikov. Algebraic construction and properties of Hermitian analogs of K-theory over rings with involution from the viewpoint of Hamiltonian formalism. Applications to differential topology and the theory of characteristic classes. I. II. *Izv. Akad. Nauk SSSR Ser. Mat.*, 34:253–288; ibid. **34** (1970), 475–500, 1970.
324. R. Oliver. *Whitehead Groups of Finite Groups*. Cambridge University Press, Cambridge, 1988.
325. P. Orlik. Smooth homotopy lens spaces. *Mich. Math. J.*, 16:245–255, 1969.
326. R. S. Palais. Extending diffeomorphisms. *Proc. Amer. Math. Soc.*, 11:274–277, 1960.
327. D. S. Passman. *Infinite Group Rings*. Marcel Dekker Inc., New York, 1971. Pure and Applied Mathematics, 6.
328. E. K. Pedersen. Controlled methods in equivariant topology, a survey. In *Current Trends in Transformation Groups*, volume 7 of *K-Monogr. Math.*, pages 231–245. Kluwer Acad. Publ., Dordrecht, 2002.
329. E. K. Pedersen and C. A. Weibel. A non-connective delooping of algebraic K-theory. In *Algebraic and Geometric Topology; Proc. Conf. Rutgers Uni., New Brunswick 1983*, volume 1126 of *Lecture Notes in Mathematics*, pages 166–181. Springer, 1985.
330. E. K. r. Pedersen. Continuously controlled surgery theory. In *Surveys on Surgery Theory, Vol. 1*, volume 145 of *Ann. of Math. Stud.*, pages 307–321. Princeton Univ. Press, Princeton, NJ, 2000.
331. G. Perelman. The entropy formula for the Ricci flow and its geometric applications. Preprint, arXiv:math.DG/0211159, 2002.
332. G. Perelman. Finite extinction time for the solutions to the Ricci flow on certain three-manifolds. Preprint, arXiv:math.DG/0303109, 2003.
333. G. Perelman. Ricci flow with surgery on three-manifolds. Preprint, arxiv:math.DG/0303109, 2003.
334. T. Petrie. The Atiyah-Singer invariant, the Wall groups $L_n(\pi, 1)$, and the function $(te^x + 1)/(te^x - 1)$. *Ann. of Math. (2)*, 92:174–187, 1970.
335. T. Petrie and J. D. Randall. *Transformation Groups on Manifolds*. Marcel Dekker Inc., New York, 1984.
336. D. Quillen. The Adams conjecture. *Topology*, 10:67–80, 1971.
337. F. Quinn. A geometric formulation of surgery. In *Topology of Manifolds (Proc. Inst., Univ. of Georgia, Athens, Ga., 1969)*, pages 500–511. Markham, Chicago, Ill., 1970.
338. F. Quinn. Resolutions of homology manifolds and the topological characterization of manifolds. *Inventiones Mathematicae*, 72:267–284, 1983.
339. F. Quinn. An obstruction to the resolution of homology manifolds. *Michigan Math. J.*, 34(2):285–291, 1987.
340. A. Ranicki. A composition formula for manifold structures. *Pure Appl. Math. Q.*, 5(2, Special Issue: In honor of Friedrich Hirzebruch. Part 1):701–727, 2009.
341. A. A. Ranicki. Algebraic L-theory. I. Foundations. *Proc. London Math. Soc. (3)*, 27:101–125, 1973.
342. A. A. Ranicki. Algebraic L-theory. II. Laurent extensions. *Proc. London Math. Soc. (3)*, 27:126–158, 1973.

343. A. A. Ranicki. The total surgery obstruction. In *Algebraic Topology, Aarhus 1978 (Proc. Sympos., Univ. Aarhus, Aarhus, 1978)*, volume 763 of *Lecture Notes in Math.*, pages 275–316. Springer, Berlin, 1979.
344. A. A. Ranicki. The algebraic theory of surgery. I. Foundations. *Proc. London Math. Soc. (3)*, 40(1):87–192, 1980.
345. A. A. Ranicki. The algebraic theory of surgery. II. Applications to topology. *Proc. London Math. Soc. (3)*, 40(2):193–283, 1980.
346. A. A. Ranicki. *Exact Sequences in the Algebraic Theory of Surgery*. Princeton University Press, Princeton, N.J., 1981.
347. A. A. Ranicki. Additive L-theory. *K-Theory*, 3(2):163–195, 1989.
348. A. A. Ranicki. *Algebraic L-Theory and Topological Manifolds*. Cambridge University Press, Cambridge, 1992.
349. A. A. Ranicki. *Lower K- and L-Theory*. Cambridge University Press, Cambridge, 1992.
350. A. A. Ranicki. On the Hauptvermutung. In *The Hauptvermutung Book*, pages 3–31. Kluwer Acad. Publ., Dordrecht, 1996.
351. A. A. Ranicki. *High-dimensional knot theory*. Springer-Verlag, New York, 1998. Algebraic Surgery in Codimension 2, With an appendix by Elmar Winkelnkemper.
352. A. A. Ranicki. *Algebraic and Geometric Surgery*. Oxford Mathematical Monographs. The Clarendon Press Oxford University Press, Oxford, 2002. Oxford Science Publications.
353. A. A. Ranicki. The structure set of an arbitrary space,the algebraic surgery exact sequence and the total surgery obstruction. In T. Farrell, L. Göttsche, and W. Lück, editors, *High Dimensional Manifold Theory*, number 9 in ICTP Lecture Notes, pages 491–514. Abdus Salam International Centre for Theoretical Physics, Trieste, 2002. Proceedings of the summer school "High dimensional manifold theory" in Trieste May/June 2001, Number 2. http://www.ictp.trieste.it/~pub_off/lectures/vol9.html.
354. A. A. Ranicki, A. J. Casson, D. P. Sullivan, M. A. Armstrong, C. P. Rourke, and G. E. Cooke. *The Hauptvermutung Book*, volume 1 of *K-Monographs in Mathematics*. Kluwer Academic Publishers, Dordrecht, 1996. A collection of papers of the topology of manifolds.
355. A. A. Ranicki, A. J. Casson, D. P. Sullivan, M. A. Armstrong, C. P. Rourke, and G. E. Cooke, editors. *The Hauptvermutung Book. A Collection of Papers on the Topology of Manifolds*, volume 1 of *K-Monogr. Math.* Dordrecht: Kluwer Academic Publishers, 1996.
356. D. C. Ravenel. *Complex Cobordism and Stable Homotopy Groups of Spheres*. Academic Press Inc., Orlando, FL, 1986.
357. N. Ray and E. K. Pedersen. A fibration for Diff Σ^n. In *Topology Symposium, Siegen 1979 (Proc. Sympos., Univ. Siegen, Siegen, 1979)*, volume 788 of *Lecture Notes in Math.*, pages 165–171. Springer, Berlin, 1980.
358. E. Rees. Problems concerning embeddings of manifolds. *Adv. in Math. (China)*, 19(1):72–79, 1990.
359. V. A. Rohlin. New results in the theory of four-dimensional manifolds. *Doklady Akad. Nauk SSSR (N.S.)*, 84:221–224, 1952.
360. J. Rosenberg. Higher G-signatures for Lipschitz manifolds. *K-theory*, 7:101–132, 1993.
361. J. Rosenberg. *Algebraic K-Theory and its Applications*. Springer-Verlag, New York, 1994.
362. J. Rosenberg and S. Stolz. Metrics of positive scalar curvature and connections with surgery. In *Surveys on Surgery Theory, Vol. 2*, volume 149 of *Ann. of Math. Stud.*, pages 353–386. Princeton Univ. Press, Princeton, NJ, 2001.
363. C. P. Rourke and B. J. Sanderson. Block bundles. *Bull. Amer. Math. Soc.*, 72:1036–1039, 1966.
364. C. P. Rourke and B. J. Sanderson. Block bundles. I. *Ann. of Math. (2)*, 87:1–28, 1968.
365. C. P. Rourke and B. J. Sanderson. Block bundles. II. Transversality. *Ann. of Math. (2)*, 87:256–278, 1968.
366. C. P. Rourke and B. J. Sanderson. Block bundles. III. Homotopy theory. *Ann. of Math. (2)*, 87:431–483, 1968.
367. C. P. Rourke and B. J. Sanderson. *Introduction to Piecewise-Linear Topology*. Springer-Verlag, Berlin, 1982. Reprint.

368. C. P. Rourke and D. P. Sullivan. On the Kervaire obstruction. *Ann. Math. (2)*, 94:397–413, 1971.
369. Y. B. Rudyak. Piecewise linear structures on topological manifolds. Preprint, MPI Bonn Preprint Series 2001 (17), 2001.
370. H. Rüping. The Farrell–Jones conjecture for S-arithmetic groups. *J. Topol.*, 9(1):51–90, 2016.
371. M. Sapir. A Higman embedding preserving asphericity. *J. Amer. Math. Soc.*, 27(1):1–42, 2014.
372. J. A. Schafer. Topological Pontrjagin classes. *Comment. Math. Helv.*, 45:315–332, 1970.
373. A. Scorpan. *The Wild World of 4-Manifolds*. American Mathematical Society, Providence, RI, 2005.
374. P. Scott. The geometries of 3-manifolds. *Bull. London Math. Soc.*, 15(5):401–487, 1983.
375. H. Seifert and W. Threlfall. Topologie. In *Naturforschung und Medizin in Deutschland 1939–1946, Band 2*, pages 239–252. Dieterich'sche Verlagsbuchhandlung, Wiesbaden, 1948.
376. J. L. Shaneson. Wall's surgery obstruction groups for $G \times Z$. *Ann. of Math. (2)*, 90:296–334, 1969.
377. S. Smale. The classification of immersions of spheres in Euclidean spaces. *Ann. of Math. (2)*, 69:327–344, 1959.
378. S. Smale. Generalized Poincaré's conjecture in dimensions greater than four. *Ann. of Math. (2)*, 74:391–406, 1961.
379. S. Smale. On the structure of manifolds. *Amer. J. Math.*, 84:387–399, 1962.
380. F. Spaggiari. On the stable classification of Spin four-manifolds. *Osaka J. Math.*, 40(4):835–843, 2003.
381. E. H. Spanier. Function spaces and duality. *Ann. Math. (2)*, 70:338–378, 1959.
382. E. H. Spanier. *Algebraic Topology*. McGraw-Hill Book Co., New York, 1966.
383. M. Spivak. Spaces satisfying Poincaré duality. *Topology*, 6:77–101, 1967.
384. V. Srinivas. *Algebraic K-Theory*. Birkhäuser Boston Inc., Boston, MA, 1991.
385. J. Stallings. Whitehead torsion of free products. *Ann. of Math. (2)*, 82:354–363, 1965.
386. J. Stasheff. A classification theorem for fibre spaces. *Topology*, 2:239–246, 1963.
387. N. Steenrod. *The Topology of Fibre Bundles*. Princeton University Press, Princeton, N. J., 1951.
388. R. E. Stong. *Notes on Cobordism Theory*. Princeton University Press, Princeton, N.J., 1968.
389. D. Sullivan. *Geometric topology. Part I*. Massachusetts Institute of Technology, Cambridge, Mass., 1971. Localization, Periodicity, and Galois Symmetry, Revised version.
390. D. Sullivan. Genetics of homotopy theory and the Adams conjecture. *Ann. of Math. (2)*, 100:1–79, 1974.
391. D. P. Sullivan. *Triangulating Homotopy Equivalences*. ProQuest LLC, Ann Arbor, MI, 1966. Thesis (Ph.D.)–Princeton University.
392. D. P. Sullivan. Triangulating and smoothing homotopy equivalences and homeomorphisms. Geometric Topology Seminar Notes. In *The Hauptvermutung Book*, volume 1 of *K-Monogr. Math.*, pages 69–103. Kluwer Acad. Publ., Dordrecht, 1996.
393. R. M. Switzer. *Algebraic Topology—Homotopy and Homology*. Classics in Mathematics. Springer-Verlag, Berlin, 2002. Reprint of the 1975 original [Springer, New York; MR0385836 (52 #6695)].
394. L. Taylor and B. Williams. Surgery spaces: formulae and structure. In *Algebraic Topology, Waterloo, 1978 (Proc. Conf., Univ. Waterloo, Waterloo, Ont., 1978)*, volume 741 of *Lecture Notes in Math.*, pages 170–195. Springer, Berlin, 1979.
395. L. R. Taylor. Surgery groups and inner automorphisms. In *Algebraic K-theory, III: Hermitian K-Theory and Geometric Applications (Proc. Conf. Battelle Memorial Inst., Seattle, Wash., 1972)*, pages 471–477. Lecture Notes in Math., Vol. 343. Springer, Berlin, 1973.
396. P. Teichner. *Topological Four-Manifolds with Finite Fundamental Groups*. PhD thesis, Mainz, 1992.
397. H. Toda. *Composition Methods in Homotopy Groups of Spheres*. Annals of Mathematics Studies, No. 49. Princeton University Press, Princeton, N.J., 1962.

398. T. tom Dieck. *Transformation Groups*. Walter de Gruyter & Co., Berlin, 1987.
399. T. tom Dieck. *Algebraic Topology*. EMS Textbooks in Mathematics. European Mathematical Society (EMS), Zürich, 2008.
400. V. Turaev. Homeomorphisms of geometric three-dimensional manifolds. *Mat. Zametki*, 43(4):533–542, 575, 1988. translation in Math. Notes 43 (1988), no. 3-4, 307–312.
401. R. M. Vogt. A note on homotopy equivalences. *Proc. Amer. Math. Soc.*, 32:627–629, 1972.
402. C. T. C. Wall. The action of Γ_{2n} on $(n-1)$-connected $2n$-manifolds. *Proc. Amer. Math. Soc.*, 13:943–944, 1962.
403. C. T. C. Wall. Killing the middle homotopy groups of odd dimensional manifolds. *Trans. Amer. Math. Soc.*, 103:421–433, 1962.
404. C. T. C. Wall. On simply-connected 4-manifolds. *J. London Math. Soc.*, 39:141–149, 1964.
405. C. T. C. Wall. Quadratic forms on finite groups, and related topics. *Topology*, 2:281–298, 1964.
406. C. T. C. Wall. Surgery of non-simply-connected manifolds. *Ann. of Math. (2)*, 84:217–276, 1966.
407. C. T. C. Wall. Classification problems in differential topology. VI. Classification of $(s-1)$-connected $(2s+1)$-manifolds. *Topology*, 6:273–296, 1967.
408. C. T. C. Wall. Poincaré complexes. I. *Ann. of Math. (2)*, 86:213–245, 1967.
409. C. T. C. Wall. Non-additivity of the signature. *Invent. Math.*, 7:269–274, 1969.
410. C. T. C. Wall. On the axiomatic foundations of the theory of Hermitian forms. *Proc. Cambridge Philos. Soc.*, 67:243–250, 1970.
411. C. T. C. Wall. Norms of units in group rings. *Proc. London Math. Soc. (3)*, 29:593–632, 1974.
412. C. T. C. Wall. Formulae for surgery obstructions. *Topology*, 15(3):189–210, 1976.
413. C. T. C. Wall. Corrigendum: "Formulae for surgery obstructions" (Topology **15** (1976), no. 3, 189–210). *Topology*, 16(4):495–496, 1977.
414. C. T. C. Wall. *Surgery on Compact Manifolds*, volume 69 of *Mathematical Surveys and Monographs*. American Mathematical Society, Providence, RI, second edition, 1999. Edited and with a foreword by A. A. Ranicki.
415. C. T. C. Wall. *Differential Topology.*, volume 156. Cambridge: Cambridge University Press, 2016.
416. A. H. Wallace. Modifications and cobounding manifolds. *Canad. J. Math.*, 12:503–528, 1960.
417. G. Wang and Z. Xu. The triviality of the 61-stem in the stable homotopy groups of spheres. *Ann. of Math. (2)*, 186(2):501–580, 2017.
418. O. Wang. Torus bundles over lens spaces. Preprint, arXiv:2203.02566 [math.GT], 2022.
419. C. Wegner. The K-theoretic Farrell–Jones conjecture for CAT(0)-groups. *Proc. Amer. Math. Soc.*, 140(3):779–793, 2012.
420. C. Wegner. The Farrell–Jones conjecture for virtually solvable groups. *J. Topol.*, 8(4):975–1016, 2015.
421. C. A. Weibel. *An Introduction to Homological Algebra*. Cambridge University Press, Cambridge, 1994.
422. C. A. Weibel. *The K-Book*, volume 145 of *Graduate Studies in Mathematics*. American Mathematical Society, Providence, RI, 2013. An introduction to algebraic K-theory.
423. S. Weinberger. On smooth surgery. *Comm. Pure Appl. Math.*, 43(5):695–696, 1990.
424. S. Weinberger. *The Topological Classification of Stratified Spaces*. University of Chicago Press, Chicago, IL, 1994.
425. S. Weinberger. *Variations on a Theme of Borel: An Essay on the Role of the Fundamental Group in Rigidity*, volume 213 of *Cambridge Tracts in Mathematics*. Cambridge University Press, Cambridge, 2023.
426. S. Weinberger, Z. Xie, and G. Yu. Additivity of higher rho invariants and nonrigidity of topological manifolds. *Commun. Pure Appl. Math.*, 74(1):3–113, 2021.
427. M. Weiss. Surgery and the generalized Kervaire invariant. I. *Proc. London Math. Soc. (3)*, 51(1):146–192, 1985.

428. M. Weiss. Surgery and the generalized Kervaire invariant. II. *Proc. London Math. Soc. (3)*, 51(2):193–230, 1985.
429. M. Weiss. Visible L-theory. *Forum Math.*, 4(5):465–498, 1992.
430. M. Weiss and B. Williams. Automorphisms of manifolds. In *Surveys on Surgery Theory, Vol. 2*, volume 149 of *Ann. of Math. Stud.*, pages 165–220. Princeton Univ. Press, Princeton, NJ, 2001.
431. G. W. Whitehead. *Elements of Homotopy Theory*, volume 61 of *Graduate Texts in Mathematics*. Springer-Verlag, New York, 1978.
432. J. H. C. Whitehead. On C^1-complexes. *Ann. of Math. (2)*, 41:809–824, 1940.
433. H. Whitney. The general type of singularity of a set of $2n-1$ smooth functions of n variables. *Duke Math. J.*, 10:161–172, 1943.
434. H. Whitney. The self-intersections of a smooth n-manifold in $2n$-space. *Ann. of Math. (2)*, 45:220–246, 1944.
435. H. Whitney. The singularities of a smooth n-manifold in $(2n-1)$-space. *Ann. of Math. (2)*, 45:247–293, 1944.
436. R. E. Williamson, Jr. Cobordism of combinatorial manifolds. *Ann. of Math. (2)*, 83:1–33, 1966.
437. J. A. Wolf. *Spaces of Constant Curvature*. AMS Chelsea Publishing, Providence, RI, sixth edition, 2011.
438. Z.-i. Yosimura. Universal coefficient sequences for cohomology theories of CW-spectra. *Osaka J. Math.*, 12(2):305–323, 1975.
439. E. C. Zeeman. The Poincaré conjecture for $n \geq 5$. In *Topology of 3-Manifolds and Related Topics (Proc. The Univ. of Georgia Institute, 1961)*, pages 198–204. Prentice-Hall, Englewood Cliffs, N.J., 1962.
440. V. F. Zenobi. Mapping the surgery exact sequence for topological manifolds to analysis. *J. Topol. Anal.*, 9(2):329–361, 2017.

Notation

$A(R)$, 266
$\mathrm{Arf}(P, \lambda, \mu)$, 269
$\mathrm{Arf}(P, \lambda, \mu, B)$, 268
$\mathrm{Arf}'(Q, \lambda, \mu)$, 267
A^*, 179
$\mathrm{aut}(p)$, 182
$\mathrm{aut}_\infty(p)$, 183
B_n, 444
$bR\text{-CH}$, 507
$bR\text{-CH}_{\mathrm{fgf}}$, 507
$bR\text{-CH}_{\mathrm{fgp}}$, 507
BG, 150
$\mathrm{BG}(k)$, 145
B_M^k, 862
$\mathrm{BG/O}$, 169
BO, 148, 851
BPL, 415
BTOP, 415
bP^{n+1}, 434
$C_*(\widetilde{X}, \widetilde{A})$, 25
$C_*(\widetilde{W}, \widetilde{\partial_0 W})$, 27
$C_*^{\mathbb{Z}\Pi(X)}(X)$, 107
C_{ev}, 45
C_{odd}, 45
C^{n-*}, 121
$C_*^!(\widehat{f})$, 243
$C_!^{n-*}(\widehat{f})$, 243
(C, φ), 570
$(C, [\varphi])$, 570
(C, ψ), 570
$(C, [\psi])$, 570
(C, θ), 570
$(C, [\theta])$, 570
$C^s(X)$, 577
$c(L(\alpha))$, 59
$\widetilde{C}(X)$, 577
$\widetilde{C}^s(X)$, 577
$(c_* + \gamma_*)_{\mathrm{ev}}$, 45
$(c_* + \gamma_*)_{\mathrm{odd}}$, 45
$R\text{-CH}$, 507
$R\text{-CH}_{\mathrm{fgf}}$, 507
$R\text{-CH}_{\mathrm{fgp}}$, 507
$\mathrm{cone}(f)$, 55
$\mathrm{cone}_*(f_*)$, 44
$\mathrm{cone}^*(f^*)$, 239
$\mathrm{cone}^{n-*}(C_*^!(\widehat{f}))$, 243
$\mathbb{CP}^n$, 137
$\mathrm{cyl}(f)$, 54
$\mathrm{cyl}_*(f_*)$, 44
$\deg(f)$, 60
$\deg(f, \partial f; \overline{f})$, 138
$\det(\xi)$, 111
DV, 57
$D(G)$, 64
$\widetilde{D}(G)$, 825
$D_W(G)$, 824
$\mathbf{E}\langle 1 \rangle$, 834
$e(f)$, 59
$\mathrm{Emb}(M, \mathbb{R}^{n+l})$, 859
$\mathrm{Emb}(M, \mathbb{R}^\infty)$, 859

EG, 59
$E_n(i,j)$, 42
$e(P)$, 245
$E(R)$, 42
$F(k)$, 152
$f(\alpha)$, 59
$f^{c/o}$, 569
$f_{c/o}$, 569
$\widehat{f}^{c/o}$, 569
$f^{Q,n}$, 575
$f_{Q,n}$, 575
$\widehat{f}^{Q,n}$, 575
$(f : C \to D, [\varphi, \delta\varphi])$, 650
$(f : C \to D, (\psi, \delta\psi))$, 650
$(f, \overline{f})$, 87, 270
$(F, \overline{F})$, 279
$F_*(X, O_X)$, 291
$\widetilde{f_!}$, 241
$\widetilde{f}^!$, 242
$\widehat{G}$, 817
G/O, 169
G/PL, 415
G/TOP, 415
$G(k)$, 145
GROUPS, 286
GROUPS^w, 290
G-sign, 770
G-sign(B), 769
G, 152
$H^k(X)_F$, 246
I_n, 42
J, 169
J_n, 443
$\overline{J}$, 442
$GL(n, R)$, 41
$GL(R)$, 41
ho-aut$^s(X)$, 408
ho-aut(b), 862
hocolim$_{k \to \infty} X_k$, 150
hofib(f), 528
$H_\epsilon(P)$, 254
$H^\epsilon(P)$, 245
$H_n(X; O_X)$, 107
$H_n(X; \mathbb{Z}^w)$, 122
$H^0(X)^\times$, 197

$H^G_*(-, \mathbf{K}_{\mathcal{A}})$, 774
$\widetilde{H}_n(X)$, 178
$\widehat{H}^n(\mathbb{Z}/2; A)$, 127
hR-CH$_{\text{fgf}}$, 507
hR-CH$_{\text{fgp}}$, 507
$I(G)$, 774
$I(M)$, 831
$I_k(M)$, 228
$K_1(R)$, 41
$K_k(\widetilde{M})$, 240
$K^k(\widetilde{M})$, 240
$K_j(\widetilde{M})$, 272
$K^j(\widetilde{M})$, 272
$K_j(\widetilde{M}, \widetilde{\partial M})$, 272
$K^j(\widetilde{M}, \widetilde{\partial M})$, 272
$\text{KI}(f, \overline{f})$, 439
$\widetilde{K}_1(R)$, 41
$K(A, l)$, 423
$\mathbf{K}_{\mathcal{A}}$, 774
$\mathbf{K}_{\text{EM}}(G)$, 786
$L_{2k}(R)$, 259
$L_{2k+1}(R)$, 333
$L_n^p(R)$, 558
$L_{2k}^U(R)$, 395
$L_{2k}^{\widetilde{U}}(R)$, 396
$L_{2k}(\mathbb{Z}\Pi(X), O_X)$, 292
$L_n^h(\mathbb{Z}\Pi(X), O_X)$, 397
$L_n^s(\mathbb{Z}\Pi(X), O_X)$, 397
$L_n^h(\mathbb{Z}\pi, w)$, 397
$L_n^p(\mathbb{Z}\pi, w)$, 766
$L_n^s(\mathbb{Z}\pi, w)$, 397
$L_n^{\langle j \rangle}(\mathbb{Z}G, w)$, 766
$L_n^{\langle -\infty \rangle}(\mathbb{Z}G, w)$, 766
$L_n^\epsilon(\mathbb{Z}G, w)$, 766
$L^n(R)_{\text{chain}}$, 676
$L_n(R)_{\text{chain}}$, 676
$L_h^n(R)_{\text{chain}}$, 676
$L_n^h(R)_{\text{chain}}$, 676
$L_p^n(R)_{\text{chain}}$, 675
$L_n^p(R)_{\text{chain}}$, 676
$L_U^n(R)_{\text{chain}}$, 751
$L_n^U(R)_{\text{chain}}$, 752
$L_s^n(\mathbb{Z}\pi, w)_{\text{chain}}$, 752
$L_n^s(\mathbb{Z}\pi, w)_{\text{chain}}$, 752

Notation

$L_n^p(\mathbb{Z}G, w)$, 766
$\overline{L}_n^s(\mathbb{Z}G)$, 767
$\mathbf{L}(R)$, 759
$L(V)$, 57
$L(t; k_1, \ldots, k_b)$, 57
$L(\alpha)$, 58
$L^\perp$, 260
$Q_\epsilon(\mathbb{Z})$, 347
M^t, 502
M^-, 84
$(M, f, \overline{f})$, 97
$(M, f, a, \xi, \overline{f}, o)$, 205
$(M, i, f, \xi, \overline{f}, o)$, 201
$(M, [M], f, \widehat{f}, a, \xi, \overline{f})$, 223
$M + (\phi^q)$, 19
$(M; \partial_0 M, \partial_1 M)$, 311
MSO, 786
MO, 786
$n[M]$, 507
O, 148
$\mathrm{Or}(G)$, 774
PL, 415
(P, ϕ), 245
$(P, \psi; F, G)$, 330
(P, ψ), 253
$-(P, \psi)$, 338
P^*, 50
$p_{DE}: DE \to X$, 98
$p_{SE}: SE \to X$, 98
$p_j(\xi)$, 301
$Q_{(-1)^k}(\mathbb{Z}\pi, w)$, 233
$Q^\epsilon(P)$, 252
$Q_\epsilon(P)$, 252
$Q_\epsilon(\mathbb{Q}/\mathbb{Z})$, 346
$Q^n(C)$, 570
$Q_n(C)$, 570
$\widehat{Q}^n(C)$, 570
$Q^{n+1}(f)$, 649
$Q_{n+1}(f)$, 649
$\mathrm{qsign}((f, \overline{f}))$, 680
$\underline{\mathbb{R}}^a$, 75
R-MOD, 504
R-MOD$_{\mathrm{fgf}}$, 504
R-MOD$_{\mathrm{fgp}}$, 504
$R_\mathbb{C}(G)$, 769

$\overline{R_\mathbb{C}(G)}$, 813
$\overline{\mathbb{Q}R_\mathbb{C}(G)}$, 813
$R_\mathbb{C}^\pm(G)$, 769
$R_\mathbb{R}^\pm(G)$, 770
$\mathbb{Q}R_\mathbb{C}(G)^{\pm 1}$, 814
$\overline{\mathbb{Q}R_\mathbb{C}(G)}^{\pm 1}$, 814
$\mathbb{R}P^n$, 125
$\mathrm{RU}(R)$, 340
$\mathrm{RU}(r, R)$, 340
S^p, 586
$\mathrm{SF}_k(X)$, 144
S^n, 431
S_+^n, 21
$\mathrm{SF}(X)$, 148
$\mathrm{SF}_k(X)$, 144
$\mathrm{sign}(X)$, 132
$\mathrm{sign}(X, \partial X)$, 135
$\mathrm{ssign}(X)$, 679
$\mathrm{ssign}(X, [[X]])$, 679
$\mathrm{SU}(R)$, 339
$\mathrm{SU}(r, R)$, 339
SV, 57
s_X, 170
$\mathbf{s}_Y^N$, 795
$\mathbf{s}_Y$, 797
S^{k-1}, 142
T, 252
$\widehat{T}$, 346
$\mathrm{Th}(p)$, 142
$\mathrm{Th}(\xi)$, 98
$[t_M]$, 169
$[t_{M,n}]$, 169
TOP, 415
TOP/PL, 423
$\mathrm{TU}(R)$, 340
$\mathrm{TU}(r, R)$, 340
U_p, 158
$\overline{U_p}$, 157
$\mathrm{UU}(R)$, 340
$\mathrm{UU}(r, R)$, 340
V', 769
$\underline{V}$, 111
$\mathrm{vholifts}(\nu_M)$, 866
$\mathrm{VB}(X)$, 147
$\mathrm{VB}_k(X)$, 144

W, 568
$\widehat{W}$, 568
$W_{\mathcal{C}_0}(C)$, 569
$W^{\mathcal{C}_0}(C)$, 569
$\widehat{W}^{\mathcal{C}_0}(C)$, 569
$W_{\mathcal{C}_0}^{[0,p]}(C)$, 594
$\mathrm{Wh}(G)$, 35, 41
$\mathrm{Wh}(\Pi(Y))$, 49
$\mathrm{Wh}^{\mathrm{geo}}(X)$, 56
$(W; M_0, f_0, M_1, f_1)$, 15
$w_1(X)$, 125
$w_1(p)$, 144
W_q, 23
$(W; \partial_0 W, \partial_1 W)$, 15
$W(\mathbb{Z})$, 263
X^-, 124
$X * Y$, 60
$X \wedge Y$, 99
$\widetilde{X}$, 25, 107
$W^{2n-1}(d)$, 456
$(X, O_X, [[X]])$, 123
$(X; \partial_0 X, \partial_1 X)$, 311
$|x|$, 505
$\mathbb{Z}^w$, 108

$\Delta^U(C_*)$, 45
$\Delta^U(C_*)$, 395, 398
$\Delta(C_*, D_*, E_*)$, 398
$\Delta(LHS(C_*, D_*, E_*))$, 399
$\Delta(L(\alpha))$, 65
$\Delta(X)$, 65
$\Delta(X; U)$, 66
$\widetilde{\Delta}(L(\alpha))$, 825
λ, 229
$\mu(\alpha)$, 233
$\widehat{\mu}$, 111
$\nu(M)$, 98
$\nu(q)$, 88
$[\nu_M]$, 169
$[\nu_X]$, 169
$\overline{\psi}_{\{F\}}$, 611
$\overline{\psi}_{\{F, \partial F\}}$, 662
$\overline{\psi}_{\{F\}_\pi, w}$, 617
$\overline{\psi}_{\{F, \partial F\}_\Gamma, w}$, 663

$\pi_0(\mathrm{Imm}(M, N))$, 88
$\pi_0(\mathrm{Imm}_k(M))$, 90
$\pi_0(\mathrm{Imm}_k^{\mathrm{fr}}(M))$, 90
$\pi_n(\mathbf{E})$, 102
$\pi_n^s(X)$, 443
π_i^s, 153
$\pi_s^j(X)$, 188
$\Pi(X)$, 107
$\Pi(X, O_X)$, 290
$\sigma(\overline{f}, f)$, 270, 294, 312, 384
$\sigma^s(\overline{f}, f)$, 402
$\sigma_{\mathrm{geo}}^h(\overline{f}, f)$, 489
$\sigma_{\mathrm{geo}}^s(\overline{f}, f)$, 489
ΣC_*, 45, 515
Σf, 515
$\Sigma(f, \overline{f})$, 379
$\Sigma^k Y$, 99
Σ_X^p, 588
$\Sigma_{\widetilde{X}}^p$, 588
$\Sigma_{Y,\Gamma}^p$, 588
$\tau(f)$, 49
$\tau(f_*)$, 46
$\tau^U(f)$, 394
$\tau^U(f_*)$, 395
$\tau(W, M_0)$, 50
$\widehat{\tau}(X)$, 128
Θ^n, 433
φ_X, 578
$\varphi_{(X,A)}$, 661
$\widetilde{\varphi}_{(X,A)}$, 661
$\varphi_{Y,\Gamma}$, 579
$\overline{\varphi}_{(Y,B),\Gamma,w}$, 661
$\widetilde{\varphi}_{(Y,B),\Gamma,w}$, 661
$\varphi_{Y,\Gamma,w}$, 579
$\widetilde{\varphi}_{Y,\Gamma}$, 579
$\widetilde{\varphi}_{Y,\Gamma,w}$, 580
$\xi_0 \oplus \xi_1$, 143
(ξ, c, o), 198
$\chi_{\mathbb{R}}(D_*)$, 68
$\chi(X)$, 131
$\Omega_n(b \colon B \to \mathrm{BO})$, 856
$\Omega_n(\xi)$, 100
$\Omega_n(X)$, 101
Ω_n^{fr}, 436

Notation

Ω_n^{alm}, 436
Ω_n^{unor}, 786

$\mathcal{F}$, 774
$\mathcal{FIN}$, 774
$\mathcal{FJ}$, 779
$\mathcal{G}/O(X)$, 198
$\mathcal{H}^p(M)$, 131
$\mathcal{L}(\xi)$, 301
$\mathcal{L}_j(\xi)$, 301
$\mathcal{L}_n^h$, 481
$\mathcal{L}_n^s$, 481
$\mathcal{L}_n^{s,h}$, 491
$\mathcal{M}^s(X)$, 408
$\mathcal{N}(X)$, 210
$\mathcal{N}(X, \partial X)$, 410
$\mathcal{N}^{\text{PL}}(X, \partial X)$, 415
$\mathcal{N}_n^{\text{TOP}}(X, \partial X)$, 415
$\mathcal{NM}_n(X)$, 204
$\mathcal{NI}_n(X, k)$, 198
$\mathcal{NM}_n(X)$, 204
$\mathcal{NT}_n(X)$, 209
$\mathcal{O}_M$, 114
$\mathcal{O}_p$, 153
$\mathcal{O}_\xi$, 111
$\mathcal{R}$, 290
$\mathcal{S}^h(X)$, 409
$\mathcal{S}^s(X)$, 408
$\mathcal{S}^h(X, \partial X)$, 408
$\mathcal{S}^s(X, \partial X)$, 407
$\mathcal{S}^{\text{PL},h}(X, \partial X)$, 415
$\mathcal{S}^{\text{PL},s}(X)$, 415
$\mathcal{S}^{\text{TOP},h}(X, \partial X)$, 415
$\mathcal{S}^{\text{TOP},s}(X, \partial X)$, 415
$\mathcal{NT}_n(X)$, 209 $\mathcal{TR}$, 774
$\mathcal{VCY}$, 775
$\mathcal{VCY}_I$, 777
$\mathcal{VCY}_{II}$, 777

$\partial(Q, \mu)$, 331
$\partial_{0,1} M$, 311
$\partial_1 W_q$, 23
$\partial_1^\circ W_q$, 23
$\partial_{0,1} X$, 311

$\cap$, 122
$[M]_\mathbb{R}$, 131
$[M, \partial M]$, 119
$[[M, \partial M]]$, 117
$[X, A]$, 132
$[[X, A]]$, 132
$[X]$, 122
$[[X]]$, 122
$\{A, B\}$, 179
$\{X, Y\}_\Gamma$, 620
$\langle -, - \rangle$, 134, 246
$\simeq_{\text{sfh}}$, 142

Index

acyclic
   U-acyclic, 66
Alexander trick, 16
algebraic boundary
   of a quadratic complex, 706
   of a symmetric complex, 696
algebraic surgery exact sequence
   for closed topological manifolds, 760
   for simplicial complexes, 759
   intrinsic, 761
algebraic surgery quadratic case
   data, 715
   effect, 715
algebraic surgery symmetric case
   data, 709
   effect, 709
algebraic surgery transfer map, 754
algebraic Thom Construction
   quadratic version, 690
   symmetric version, 683
annihilator, 260
Arf invariant, 269
aspherical, 838
attaching a handle, 19
attaching maps, 25

b-structure, 861
   k-universal, 861
based exact sequence, 46
basis
   equivalent basis, 26
   stable basis, 394
   stably U-equivalent, 394
   symplectic, 265
   U-equivalent, 393
Bernoulli numbers, 444
bilinear form
   definite symmetric, 263
   indefinite symmetric, 263
bordism
   formation, 372
   normal ξ-bordism, 100
   normal bordism, 202
   normal bordism of normal Γ-maps of degree one, 225
   normal bordism of normal Γ-maps with cylindrical target of degree one, 227
   normal bordism of normal maps of degree one with respect to the tangent bundle relative boundary, 309
   normal bordism of normal maps with respect to the tangent bundle, 206
   normal bordism of normal maps with cylindrical target with respect to the tangent bundle (of degree one), 207
   normal bordism with cylindrical target (of degree one), 204
   normal bordism with respect to the tangent bundle of degree one, 207
   normal bordisms of rank k normal maps of degree one, 203
   tangential ξ-bordism, 97
boundary
   algebraic boundary of a quadratic complex, 706
   algebraic boundary of a symmetric complex, 696
   formation, 331
   of a quadratic complex, 706
   of a symmetric complex, 696
   of an $(n+1)$-dimensional quadratic pair, 650
   of an $(n+1)$-dimensional symmetric pair, 650
   of an ϵ-quadratic form, 331

bounded chain complex, 507
Brieskorn variety, 456

cell
  closed, 25
  open, 25
cellular
  basis, 26
  cellular $\mathbb{Z}\pi$-chain complex, 25
  map, 39
  pushout, 39
chain complex
  d-dimensional, 552
  k-connected, 552
  bounded, 507
  bounded below, 552
  cellular $\mathbb{Z}\pi$-chain complex, 25
  contractible, 45
  dual, 507
  elementary, 400
  finite, 45
  finite based free, 45
  free, 45
  handlebody $\mathbb{Z}\pi$-chain complex, 27
  positive, 552
  projective, 45
  reduced cellular, 577
  reduced singular, 577
  singular, 577
  stably U-based finite, 395
chain contraction, 45
chain homotopy
  of chain maps of degree n, 505
chain homotopy equivalence, 505
chain level Umkehr map, 243
  for pairs, 275
chain map, 505
  k-connected, 554
  composition chain map, 506
  hom, 506
  of degree n, 505
  switch, 506
  tensoring, 506
characteristic map, 25
Choice
  (ABP), 230
  (BP), 108
  (GBP), 228
  (NBP), 224
  (SBP), 158
classifying map of a principal G-bundle, 59
classifying space
  of a group G, 59
  for a family of subgroups, 775

clutching function, 446
cobordant
  cobordant symmetric Poincaré complexes, 666
  cobordant quadratic algebraic Poincaré complexes, 667
cobordism, 15
  symmetric between Poincaré chain complexes, 666
  diffeomorphic relative M_0, 15
  h-cobordism, 15
  over M_0, 15
  quadratic between Poincaré chain complexes, 666
  s-cobordism, 52
  trivial, 15
coconnected
  k-coconnected map, 860
  k-coconnected space, 860
cofibration
  of chain complexes, 520
cofibration sequence
  of chain complexes, 522
cohomology with compact support, 128
collapse map
  Hopf, 80
  Thom, 165
compatible local orientation of the tangent space, 231
complementary lagrangian, 331
composition chain map, 506
Conjecture
  Borel Conjecture, 843
  Full Farrell–Jones Conjecture, 779
  Inertia Conjecture for closed aspherical manifolds, 847
  K-theoretic Farrell–Jones Conjecture with coefficients in additive G-categories, 775
  L-theoretic Farrell–Jones Conjecture with coefficients in additive G-categories with involution, 776
  Manifold Hauptvermutung, 790
  Poincaré Conjecture, 16
  Rigidity for closed aspherical manifolds, 847
conjugation invariant functor, 286
connected
  k-connected map, 860
  k-connected space, 860
connected sum, 84
Construction
  Algebraic surgery – quadratic case, 715
  Algebraic surgery – symmetric case, 709
  Pontrjagin–Thom Construction, 101

Index   947

The algebraic boundary of a quadratic
   complex, 704
The algebraic boundary of a symmetric
   complex, 693
The algebraic Thom Construction –
   quadratic case, 690
The algebraic Thom Construction –
   symmetric case, 683
The quadratic construction, 596
The symmetric construction, 578
The symmetric construction – equivariant
   version, 579
The symmetric construction – pointed
   version, 579
The symmetric construction – pointed
   equivariant version, 579
contragredient representation, 769

degree
   of a map between w-oriented finite Poincaré
      complexes, 138
   of a map of lens spaces, 60
democratic invariant, 267
determinant line bundle, 111
diffeomorphism
   orientation preserving, 798
   orientation reversing, 798
   stable, 850
direct sum
   of structured chain complexes, 623
disk bundle, 98
dual
   chain complex, 51
   m-dual, 179
   R-module, 50
   S-dual, 179
dual chain complex, 507
duality
   m-duality, 179
duality map
   m-duality map, 179

effect of the algebraic surgery
   quadratic case, 716
   symmetric case, 712
Eilenberg–MacLane space, 423
elementary
   chain complex, 400
   collapse, 53
   expansion, 52
embedding
   standard, 85
   trivial, 23
Euler characteristic, 131

evaluation chain map, 508
even
   even ϵ-symmetric form, 255
   even non-singular symmetric form over $\mathbb{Z}$,
      255
exact sequence
   abelian, 420
   additive, 420
exotic N, 805
exotic sphere, 17
extended homotopy cofibration sequence, 535

fake space
   (simple smooth fake N, 804
   (simple) PL fake N, 804
   fake N, 804
   simple fake N, 804
family of subgroups, 774
Farrell–Jones group, 779
fibration
   coreducible, 182
   normal, 164
   of chain complexes, 529
   orientable, 144
fibration sequence
   of chain complexes, 529
fibre homotopy, 142
   strong, 142
fibre homotopy equivalence
   strong, 142
Figure
   Bordism W, 374
   Dual cell complex, 126
   Elementary expansion, 53
   Handle cancellation, 22
   Handle subtraction, 283
   Handlebody, 19
   Handlebody decomposition, 20
   Intersection pairing, 230
   Manifold W with boundary
      $\partial W = \partial_0 W \cup \partial_1 W \cup \partial_2 W$,
      310
   Mapping cylinder and mapping cone, 55
   Outward normal vector field, 118
   Poincaré Conjecture, 17
   Pontrjagin–Thom construction, 99
   Quadratic construction, 598
   Schematic drawing of a Heegaard splitting,
      362
   Self-intersections, 233
   Source of a surgery step for $M = T^2$, 81
   Surgeon's suitcase, 83
   Surgery along $S^0 \times D^1 \hookrightarrow S^1 \sqcup S^1$, 82
   Surgery Program, 18

The E_8-graph, 264
Wall realisation, 320
Whitney trick, 236
first Stiefel–Whitney class
    of a finite Poincaré complex, 125
    of a spherical fibration, 144
flat manifold, 839
form
    $\mathbb{Q}$-non-singular ϵ-quadratic, 354
    $\mathbb{Q}$-non-singular ϵ-symmetric, 354
    ϵ-quadratic, 253
    ϵ-symmetric, 245
    linear ϵ-quadratic form, 254
    non-singular ϵ-quadratic, 253
    non-singular ϵ-symmetric form, 245
    standard hyperbolic ϵ-quadratic form, 254
    standard hyperbolic ϵ-symmetric form, 245
    U-based non-singular ϵ-quadratic form, 395
formation
    $\mathbb{Q}$-trivial ϵ-quadratic formation, 344
    bordism formation, 372
    boundary formation, 331
    ϵ-quadratic formation, 330
    ε-symmetric formation, 637
    kernel formation, 365
    split ε-quadratic, 643
    stably U-based ϵ-quadratic, 396
    stably isomorphic ϵ-quadratic formations, 331
    stably isomorphic ε-symmetric formations, 638
    trivial ϵ-quadratic formation, 330
    trivial ε-symmetric formation, 638
    trivial ε-symmetric formation, 638
    trivial ϵ-quadratic formation, 330
split ε-quadratic boundary, 738
framing
    almost stable, 436
    stable, 435
free
    free unitary representation, 57
functor
    half-conjugation invariant functor, 290
fundamental class
    intrinsic, 117
    of a Poincaré pair, 132
fundamental groupoid, 107

G-signature, 769
G-signature homomorphism, 770
generalised lens space, 58
generator
    componentwise, 197
geometric

surgery exact sequence, 413
surgery obstruction, 489
surgery obstruction group, 482
group
    Farrell–Jones group, 779
    of homotopy spheres, 433

half-conjugation invariant functor, 290
handle, 19
    boundary of the core, 19
    cocore, 19
    core, 19
    dual, 34
    transverse sphere, 19
handlebody $\mathbb{Z}\pi$-chain complex, 27
handlebody decomposition, 20
    dual, 33
Heegaard splitting, 362
hessian, 336
hom chain map, 506
homological aspherical, 846
homology
    reduced, 178
    stably U-based, 398
    with coefficients in $\mathbb{Z}^w$, 122
    with coefficients in a coefficient system, 107
    with coefficients in a module, 123
homotopy
    regular, 88
homotopy N, 805
homotopy n-dimensional symmetric complex, 633
homotopy n-dimensional quadratic complex, 633
homotopy cartesian square of chain complexes, 539
homotopy cocartesian square of chain complexes, 526
homotopy cofibration sequence
    extended, 535
    of chain complexes, 522
homotopy cofibre
    of a chain map, 522
homotopy colimit, 150
    of a sequence of chain maps, 545
homotopy equivalence
    normal, 256
    simple, 53
homotopy fibration sequence
    of chain complexes, 529
homotopy fibre
    of a chain map, 528
homotopy n-sphere, 431
homotopy pullback

Index

of chain complexes, 538
homotopy pushout
  of chain complexes, 524
Hopf collapse map, 80
Hopf construction, 443
hyperbolic manifold, 839
hyperbolisation, 840
hyperquadratic
  complex, 570
  hyperquadratic structure, 570

immersion
  group of regular homotopy classes, 228
  pointed, 228
inertia subgroup, 831
intersection pairing
  for kernels, 249
  for manifolds with boundary, 135
  geometric, 229
  on singular cohomology, 130
  on singular cohomology with coefficients in R, 129
  on singular homology, 130
  on singular homology with coefficients in R, 130
intrinsic algebraic surgery exact sequence, 761
intrinsic surgery obstruction
  in even dimension, 293
  in odd dimension, 386
involution
  of rings, 120
  w-twisted involution of a group ring, 50
irreducible 3-manifold, 839

J-homomorphism, 443
join, 60
  fibrewise join of spherical fibrations, 143
  of (generalised) lens spaces and (fake) complex projective spaces, 811
join construction, 811

k-connected map, 76
k-smoothing, 861
k-universal b-structure, 861
kernel formation, 365
kernel groups
  relative boundary, 272
Kervaire invariant, 439
Kervaire invariant problem, 449
Kronecker pairing
  for chain complexes, 246
  for spaces, 134

L-class, 301

L-group
  (free) quadratic L-groups in terms of chain complexes, 676
  (free) symmetric L-groups in terms of chain complexes, 676
  projective quadratic L-group, 766
  decorated quadratic L-groups in even dimensions, 395
  decorated quadratic L-groups in odd dimensions, 396
  decorated quadratic L-groups in terms of chain complexes, 751
  projective quadratic L-groups in even dimensions, 558
  projective quadratic L-groups in terms of chain complexes, 675
  projective symmetric L-groups in terms of chain complexes, 675
  quadratic L-groups in even dimensions, 259
  quadratic L-groups in odd dimensions, 333
  reduced simple L-group, 767
  simple quadratic L-groups, 397
  simple quadratic L-groups in terms of chain complexes, 752
  simple symmetric L-groups in terms of chain complexes, 752
lagrangian, 260
  complementary lagrangians, 331
  sublagrangian of a quadratic form, 260
  sublagrangian of a symmetric form, 260
Lemma
  Elimination Lemma, 23
  Franz' Independence Lemma, 71
  Normal Form Lemma, 33
lens space, 57
  generalised, 58
linking form
  ϵ-quadratic linking form, 346
  ϵ-symmetric linking form, 346
  decomposable ϵ-quadratic linking form, 355
  indecomposable ϵ-quadratic linking form, 355
  non-singular ϵ-quadratic linking form, 346
  non-singular ϵ-symmetric linking form, 346
local coefficient system, 107
  infinite cyclic, 108
local orientation coefficient system, 114

m-dual, 179
manifold
  flat, 839
  hyperbolic, 839
  Kervaire manifold, 801
  Milnor manifold, 801

Manifold Hauptvermutung, 790
manifold set
  simple, 408
manifold structure, 408
  simple, 407
manifold triad, 311
mapping cone
  of a chain map, 45, 514
  of a cochain map, 239
  of a map of spaces, 55
mapping cylinder
  of a chain map, 45
  of a map of spaces, 54
mapping telescope, 544
metabolic
  quadratic form, 262
  symmetric form, 260
module
  stably finitely generated free, 246
  stably U-based, 394
  U-based, 393
Mostow rigidity, 844

non-degenerate symmetric bilinear form, 130
non-singular, 245
normal ξ-bordism, 100
normal ξ-map, 100
normal k-type, 862
normal bordism, 202
  $(k+1)$-normal bordism, 368
  $(k+1)$-trace normal bordism, 368
normal bordism of normal maps with respect to the tangent bundle, 206
normal bordism of normal maps with cylindrical target with respect to the tangent bundle (of degree one), 207
normal bordism with cylindrical target (of degree one), 204
normal bordism with respect to the tangent bundle of degree one, 207
normal bordisms of rank k normal maps of degree one, 203
normal bundle, 98
  of an immersion, 88
normal fibration, 164
normal homotopy equivalence, 256
normal invariant
  rank k normal invariant, 198
  set of normal invariants, 198
  set of rank k normal invariants, 198
normal map
  normal Γ-map of degree one, 223
  of degree one with reference, 481
  of degree one with respect to the tangent bundle, 205
  of degree one with respect to the tangent bundle relative boundary, 309
  rank k normal map, 201
  rank k normal map of degree one, 202
  with respect to the tangent bundle, 205
normal splitting along submanifold, 795
null space of a symmetric $\mathbb{R}$-bilinear pairing, 131
nullhomology, 554

orbit category, 774
orientable, 124
orientation
  conventions, 119
orientation homomorphism
  of a finite Poincaré complex, 125
  of a spherical fibration, 144
orientation preserving diffeomorphism, 798
orientation reversing diffeomorphism, 798
outward normal vector field, 118

pair
  quadratic, 650
  symmetric, 650
PL rigid, 847
Poincaré torsion, 127
Poincaré triad, 311
Poincaré complex
  n-dimensional polarised quadratic Poincaré complex, 618
  n-dimensional polarised symmetric complex Poincaré complex, 618
  n-dimensional quadratic complex Poincaré complex, 618
  n-dimensional symmetric complex Poincaré complex, 618
  w-oriented finite n-dimensional, 124
  oriented connected finite n-dimensional, 123
  connected finite n-dimensional, 122
  connected simple, 123
  finite n-dimensional, 123
  finite of formal dimension n, 125
  oriented, 124
  simple, 124
  w-oriented connected finite n-dimensional, 123
  with reversed orientation, 124
Poincaré pair
  finite n-dimensional, 132
  simple finite n-dimensional, 133
  special, 483
Poincaré $\mathbb{Z}\pi$-chain homotopy equivalence, 122

Index 951

pointed
  free Γ-space, 579
  immersion, 228
  regular homotopy, 228
polarisation, 570
Pontrjagin classes, 301
Pontrjagin dual, 817
pq-condition, 72
prime 3-manifold, 839
product
  slant, 178
pullback, 143
  of chain complexes, 527
pushout
  of chain complexes, 512

quadratic
  n-dimensional polarised quadratic Poincaré complex, 618
  n-dimensional quadratic complex Poincaré complex, 618
  cobordism, 666
  connected complex, 619
  evaluation map, 653
  pair, 650
  Poincaré pair, 663
  quadratic complex, 570
  quadratic structure, 570
  signature of a degree one normal map, 680
quadratic construction, 611
  equivariant, 617
  relative, 662
  relative equivariant, 662

regular homotopy, 88
  pointed, 228
Reidemeister stable U-equivalence class of stable $\mathbb{Z}\pi$-bases, 466
Reidemeister torsion
  of a contractible finite based free chain complex, 45
  of a finite CW-complex with respect to an acyclic representation, 66
  of a generalised lens space, 65
  Reidemeister U-torsion of a contractible finite U-stably based chain complex, 395
  Reidemeister U-torsion of a stably U-based finite chain complex with stably U-based homology, 398
representation
  contragredient, 769
representation of the surgery kernel, 361
Rho-invariant
  for finite groups, 815

rigid
  PL, 847
  smoothly, 419
  topologically, 419
ring
  with involution, 50
  with property (IBN), 329
rotation trick, 535

S-dual, 179
S-Γ-duality, 620
self-intersection element, 233
self-intersection function, 234
set of equivariant stable homotopy classes, 620
set of normal maps
  to a compact manifold, 410
  to a Poincaré complex, 204
set of stable homotopy classes, 179
signature
  of a finite oriented $4k$-dimensional Poincaré complex, 132
  of a finite Poincaré pair, 135
  of a symmetric $\mathbb{R}$-bilinear pairing, 132
  quadratic signature of a degree one normal map, 680
  symmetric signature of a geometric Poincaré complex, 679
simple
  geometric surgery exact sequence, 413
  geometric surgery obstruction, 489
  geometric surgery obstruction group, 482
  homotopy equivalence, 53
  manifold set, 408
  manifold structure, 407
singular chain complex, 577
slant maps, 509
slant product, 178
smash product, 99
smoothing
  k-smoothing, 861
smoothing theory, 457
smoothly rigid, 419
space
  pointed free, 579
Space Form Problem, 72
special Poincaré pair, 483
spectrum, 102
  1-connective cover, 834
sphere bundle, 98
spherical fibration, 142
Spherical Space Form Problem, 73
Spivak
  normal fibration, 167

normal structure for a finite Poincaré
    complex, 167
normal structure of closed manifold, 167
split ε-quadratic formation, 643
splitting invariant along submanifold, 797
stable
    basis, 394
    cohomotopy group, 188
    diffeomorphism, 850
    homotopy groups, 443
    homotopy groups of a spectrum, 102
    stem, 153
    Umkehr map of a normal map of degree one, 621
stable isomorphism
    of ϵ-quadratic formations, 331
    of ε-symmetric formations, 638
    of split ε-quadratic formations, 644
stably
    U-based ϵ-quadratic formation $(P, \psi; F, G)$, 396
    diffeomorphic, 851
    equivalent spherical fibrations, 148
    finitely generated free module, 246
    parallelisable, 434
    prime 4-manifold, 872
    U-based finite chain complex, 395
    stably U-based homology, 398
    U-based module, 394
    U-equivalent basis, 394
standard embedding, 85
strictly
    n-dimensional quadratic complex, 633
    n-dimensional symmetric complex, 633
strong
    fibre homotopy, 142
    fibre homotopy equivalence, 142
structure homomorphisms of a formation, 334
structure set, 408
    simple, 407
sublagrangian
    of a quadratic form, 260
    of a symmetric form, 260
surgery
    along an embedding, 80
    on the boundary, 461
    step, 97
    step on the boundary, 461
    The effect of algebraic surgery
        — quadratic case, 716
    The effect of algebraic surgery
        — symmetric case, 712
    trivial surgery, 86
    trivial surgery with trivial framing, 86

surgery exact sequence
    algebraic for closed topological manifolds, 760
    algebraic for simplicial complexes, 759
    geometric, 413
    geometric for the PL category and the topological category, 416
    intrinsic algebraic, 761
    simple geometric, 413
surgery kernel, 240
    representation of the surgery kernel, 361
surgery obstruction
    for manifolds with boundary and simple homotopy equivalences, 402
    geometric, 489
    intrinsic surgery obstruction in even dimension, 293
    intrinsic surgery obstruction in odd dimension, 386
    simple geometric, 489
    total, 427, 763
surgery obstruction group
    geometric, 482
    simple geometric, 482
surgery on the boundary, 461
surgery problem, 214
Surgery Program, 17
Surgery Program recognising manifolds, 217
suspension
    for (generalised) lens spaces, 812
    of a chain complex, 45, 515
    of a pointed space, 99
    of a spherical fibration, 148
suspension map
    for hyperquadratic structures, 585
    for quadratic structures, 585
    for symmetric structures, 585
switch chain map, 506
symmetric
    n-dimensional polarised symmetric complex Poincaré complex, 618
    n-dimensional symmetric complex Poincaré complex, 618
    Poincaré pair, 663
    cobordism, 666
    connected complex, 619
    definite symmetric bilinear form, 263
    evaluation map, 653
    indefinite symmetric bilinear form, 263
    pair, 650
    signature of a geometric Poincaré complex, 679
    structure, 570
    symmetric complex, 570

Index 953

symmetric construction, 578
  pointed relative, 661
  equivariant symmetric construction, 579
  pointed, 579
  pointed equivariant symmetric construction, 580
  pointed relative equivariant, 661
  relative, 661
symmetrisation map, 576
symplectic basis, 265

tangential
  ξ-map, 97
  ξ-bordism, 97
Tate cohomology, 127
tensor product
  over a category, 107
tensoring chain map, 506
Theorem
  G-signature homomorphism, 770
  $L_{4k+2}(\mathbb{Z})$, 269
  $L_{4k}(\mathbb{Z})$, 263
  $\mathbb{Q}$-trivial formations versus boundary formations, 345
  TOP/PL is an Eilenberg–MacLane, 423
  Algebraic L-theory of $\mathbb{Z}G$ for finite groups, 768
  Algebraic Rothenberg sequence, 755, 757
  Algebraic Shaneson splitting, 757
  Aspherical closed manifolds with exotic fundamental groups, 842
  Aspherical closed non-PL manifolds, 841
  Basic properties of Whitehead torsion, 39
  Bordism invariance of the degree, 211
  Bordism invariance of the intrinsic surgery obstruction in even dimensions, 299
  Bordism invariance of the intrinsic surgery obstruction in odd dimensions, 387
  Bordism invariance of the signature, 134
  Browder's Theorem about the Arf invariant, 449
  Browder–Novikov Theorem, 417
  Characterisation of Poincaré duality in terms of the normal fibration, 175
  Classification of closed topological 4-dimensional spin manifolds with infinite cyclic fundamental group, 873
  Classification of Homotopy Spheres, 432
  Classification of spherical fibrations, 145
  Classification of stable homotopy classes of spherical fibrations over finite CW complexes, 151
  Classification of stable isomorphisms classes of vector bundles over finite CW-complexes, 148
  Classification of vector bundles, 144
  Closed manifolds are Poincaré complexes, 125
  Compact manifolds are Poincaré pairs, 133
  Comparing geometric and algebraic Whitehead torsion, 56
  Complete Intersections, 871
  Composition formula in the topological structure set, 806
  Computation of Θ_n/bP_{n+1}, 450
  Computation of bP_{4k+2} in terms of the Arf-invariant-one-problem, 449
  Computation of bP_{4k}, 448
  Computations of the J-homomorphism, 444
  Counterexamples up to homotopy to prime decomposition in dimension 4, 873
  Diffeomorphism classification of lens spaces, 69
  Differentiable Sphere Theorem, 457
  Existence and uniqueness of Spivak normal fibrations, 168
  Existence of stable diffeomorphisms, 858
  Exotic aspherical closed manifolds in dimension 4, 841
  Exotic aspherical closed manifolds with hyperbolic fundamental group, 842
  Exotic universal covering, 842
  Extended realisation surgery obstructions, 323
  Four-manifolds splitting as a connected sum topologically but not smoothly, 873
  Franz' Independence Lemma, 71
  Generalised Atiyah duality, 180
  Geometric Rothenberg sequence, 496
  Geometric Rothenberg sequence for structure sets, 496
  Geometric Shaneson splitting, 497
  Geometric Shaneson splitting for structure sets, 499
  Geometric surgery exact sequence, 413
  Geometric Surgery Obstruction Theorem, 489
  h-Cobordism Theorem, 16
  Hirzebruch Signature Theorem, 301
  Homogeneous manifolds of dimension 7, 871
  Homology isomorphisms and topologically rigidity, 846
  Homotopy cartesian and homotopy cocartesian agree for chain complexes, 539

Homotopy classification of (generalised) lens spaces, 60
Homotopy cofibrations and homotopy fibrations agree for chain complexes, 532
Homotopy type of G/TOP, 785
Homotopy type of G/TOP as infinite loop space, 792
Homotopy-theoretic interpretation of the Kervaire–Milnor braid, 452
Identification of the form/formation and the chain complex version of quadratic L-groups, 558
Identification of the form/formation and the chain complex version of quadratic L-groups in the simple case, 752
Identification of the geometric and algebraic surgery exact sequence, 760
Identification of the simple surgery obstruction with the simple quadratic signature, 753
Identification of the surgery obstruction with the quadratic signature, 558
Identifying the geometric surgery obstruction groups with the algebraic L-groups, 490
Immersions and bundle monomorphisms, 89
Indecomposable linking forms over $\mathbb{Z}$, 355
Intrinsic surgery obstruction in even dimension, 298
Intrinsic surgery obstruction in even dimension relative boundary, 311
Intrinsic surgery obstruction in odd dimension, 387
Intrinsic surgery obstruction in odd dimension relative boundary, 389
Inverse of the Thom Isomorphism Theorem, 161
Kervaire–Milnor braid, 451
Killing the absolute kernels of a normal map of pairs, 275
Killing the surgery kernels of a normal map, 243
Kirby–Siebenmann invariant, 790
Localisation of simply connected spaces, 785
Making a normal ξ-bordism highly connected, 102
Making a normal map highly connected], 214
Making a tangential ξ-map highly connected, 98
Maslov identity, 341
Non-triangulable aspherical closed manifolds, 841

Normal invariants and G/O, 200
Normal invariant and surgery problems, 204
Normal invariants in the topological category after rationalisation, 793
Normal Maps and G/O, 210
Normal maps, G/PL and G/TOP, 416
Normal splitting along a submanifold, 795
Odd local part of oriented bordism groups, 788
Odd-dimensional L-groups via automorphisms, 340
π-π-Theorem, 462
π-π-Theorem for triads, 463
PL and topological structure sets in the simply connected case, 417
Poincaré Conjecture, 16
Poincaré duality and S-duality, 180
Pontrjagin Thom construction and oriented bordism, 102
Pontrjagin–Thom Construction, 101
Product formula for surgery obstructions, 753
Products in L-groups, 753
Puppe sequences for chain complexes, 533
Realisation of even-dimensional surgery obstructions, 317
Realisation of odd-dimensional surgery obstructions, 390
Realisation of simple surgery obstructions, 405
Recognising the S-dual of a map, 180
Reducing the family, 778
Reflection trick, 842
s-Cobordism Theorem, 15
Simple homotopy classification of generalised lens spaces, 65
Simple surgery obstruction for manifolds with boundary, 404
Simply connected surgery obstruction maps are surjective, 800
Solution to the simply connected surgery problem in odd dimensions, 388
Solution to the surgery problem in even dimensions, 302
Space Form Problem, 72
Spherical Space Form Problem, 73
Stable diffeomorphism classification, 866
Stable Kneser Decomposition for 4-manifolds, 872
Stable parallelisability of homotopy spheres, 440
Stable prime decomposition for 4-manifolds, 872
Status of the Borel Conjecture, 843

Sullivan splitting of G/O$_{(p)}$, 424
Surgery and hyperbolic kernels, 256
Surgery exact sequence for PL and TOP, 416
Surgery obstruction in even dimensions, 270
Surgery obstruction in odd dimensions, 360
Surgery obstruction maps for lens spaces, 826
Surgery step, 94
Surjectivity of the surgery obstruction for PL and TOP, 417
The 2-local homotopy type of bordism spectra, 787
The long exact sequence for homotopy spheres, 440
The simple topological structure set of a generalised lens space with fundamental group of odd order, 822
The structure set for simply connected closed manifolds, 838
The structure set for simply connected manifold times aspherical manifold, 837
The surgery exact sequence for homotopy spheres, 439
The topological classification of fake lens spaces with fundamental groups of odd order, 824
Thom chain homotopy equivalence, 159
Thom Isomorphism Theorem, 160
Topological Browder–Novikov Theorem, 417
Topological rigidity and connected sums, 846
Topological rigidity and homology spheres, 846
Topological rigidity and products of two spheres, 845
Topological structure sets of certain torus bundles over lens spaces, 830
Topological structure sets of closed aspherical manifolds, 828
Topological structure sets of complex projective spaces, 809
Topological structure sets of products of spheres, 807
Topological structure sets of quaternionic projective spaces, 811
Topological structure sets of real projective spaces, 820
Topologically rigid 3-manifolds, 845
Turning an immersion into an embedding, 236
up-down formula for the surgery transfer, 754
Vanishing of bP_{2k+1}, 449
Vanishing of $L_{2k+1}(\mathbb{Z})$, 334
Vanishing of the surgery transfer, 755
Whitney's Approximation Theorem, 89
Zero criterion for formations, 337
thickening
  of an immersion, 91
Thom class
  intrinsic, 157
Thom collapse map, 99, 165
Thom isomorphism, 161
Thom space
  of a spherical fibration, 142
  of a vector bundle, 98
topological rigid, 419
torsion dual of a finite abelian group, 346
total surgery obstruction, 427, 763
trace of surgery
  along the embedding, 81
  on a normal map, 97
transport groupoid, 774
triad
  of manifolds, 311
  Poincaré, 311
trivial surgery, 86
  with trivial framing, 86

U-based module, 393
U-equivalence class
  Reidemeister stable U-equivalence class of stable $\mathbb{Z}\pi$-bases, 466
stable U-equivalence class
  of stable bases, 394
U-equivalent
  basis, 393
U-simple, 751
U-torsion
  of a stable isomorphism of stably U-based modules, 394
Umkehr map, 241
  chain level Umkehr map, 243
  chain level Umkehr map for pairs, 275
  for pairs, 273
  stable Umkehr map of a normal map of degree one, 621
unit disk, 57
unit sphere, 57
universal covering functor, 107
universal principal G-bundle, 59

vector bundle reduction of the Spivak normal fibration, 168
vector field
  outward normal vector field, 118

Whitehead group, 41
  of a space, 49
Whitehead torsion
  of a chain homotopy equivalence, 46
  of a homotopy equivalence, 49
  Whitehead U-torsion of a chain homotopy equivalence of stably U-based finite R-chain complexes, 395
Witt group, 262

wreath product, 779
Wu class, 625
Wu formula, 167

ξ-bordism
  normal, 100
  tangential, 97
ξ-map
  normal, 100
  tangential, 97

GPSR Compliance

The European Union's (EU) General Product Safety Regulation (GPSR) is a set of rules that requires consumer products to be safe and our obligations to ensure this.

If you have any concerns about our products, you can contact us on ProductSafety@springernature.com

In case Publisher is established outside the EU, the EU authorized representative is:

Springer Nature Customer Service Center GmbH
Europaplatz 3
69115 Heidelberg, Germany

Batch number: 08086107

Printed by Printforce, the Netherlands